May 1985

MOLECULAR CELL GENETICS

MOLECULAR CELL GENETICS

Edited by

Michael M. Gottesman
National Institutes of Health

A WILEY-INTERSCIENCE PUBLICATION
JOHN WILEY & SONS
New York • Chichester • Brisbane • Toronto • Singapore

Library of Congress Cataloging in Publication Data:

Main entry under title:

Molecular cell genetics.
"A Wiley-Interscience Publication."
Includes index.
1. Molecular genetics. 2. Cytogenetics. I. Gottesman, Michael M.
QH430.M63 1985 574.87′322 84-21007
ISBN 0-471-87925-8

Printed in the United States of America

10 9 8 7 6 5 4 3 2 1

CONTRIBUTORS

IRENE ABRAHAM
Laboratory of Molecular Biology
National Cancer Institute
National Institutes of Health
Bethesda, Maryland

GERALD M. ADAIR
Science Park—Research Division
University of Texas System Cancer Center
Smithville, Texas

IRENE L. ANDRULIS
Department of Medical Genetics
The Hospital for Sick Children
Toronto, Ontario, Canada

FERNANDO CABRAL
Division of Endocrinology
University of Texas Medical School
Houston, Texas

LAWRENCE CHASIN
Department of Biological Sciences
Columbia University
New York, New York

JUNG CHOI
Department of Genetics
University of California, Berkeley
Berkeley, California

ERNEST H.Y. CHU
Department of Human Genetics
University of Michigan Medical School
Ann Arbor, Michigan

L. SCOTT CRAM
Los Alamos National Laboratory
Los Alamos, New Mexico

PAUL J. DOHERTY
Laboratory of Molecular Biology
National Cancer Institute
National Institutes of Health
Bethesda, Maryland

RAYMOND G. FENWICK
Department of Biochemistry
Dalhousie University
Halifax, Nova Scotia Canada

MICHAEL M. GOTTESMAN
Laboratory of Molecular Biology

National Cancer Institute
National Institutes of Health
Bethesda, Maryland

STEPHEN G. GRANT
Genetics Department and Research Institute
The Hospital for Sick Children
Toronto, Ontario, Canada

BRUCE H. HOWARD
Laboratory of Molecular Biology
National Cancer Institute
National Institutes of Health
Bethesda, Maryland

C. JAMES INGLES
Best Institute
University of Toronto
Toronto, Ontario, Canada

RAMAN M. KOTHARI
Sarabhai Research Center
Baroda, India

VICTOR LING
Ontario Cancer Institute
Toronto, Ontario, Canada

MENASHE MARCUS
Department of Genetics
Hebrew University
Jerusalem, Israel

MARY MCCORMICK
Laboratory of Molecular Biology
National Cancer Institute
National Institutes of Health
Bethesda, Maryland

DAVID PATTERSON
Department of Biophysics and Genetics
Eleanor Roosevelt Institute for Cancer Research
University of Colorado Medical Center
Denver, Colorado

THEODORE PUCK
Department of Biophysics and Genetics
Director, Eleanor Roosevelt Institute for Cancer Research
University of Colorado Medical Center
Denver, Colorado

CHARLES W. ROTH
Unité d'Immunoparasitologie
Institut Pasteur
Paris, France

RUTH SAGER
Dana Farber Cancer Institute
Harvard Medical School
Boston, Massachusetts

IMMO E. SCHEFFLER
Department of Biology
University of California, San Diego
LaJolla, California

MATTHEW J. SCHIBLER
Division of Endocrinology
University of Texas Medical School
Houston, Texas

JERRY W. SHAY
Department of Cell Biology
University of Texas Health Sciences Center at Dallas
Dallas, Texas

MICHAEL J. SICILIANO
Department of Genetics
M.D. Anderson Hospital
University of Texas
Texas Medical Center
Houston, Texas

LOUIS SIMINOVITCH
Department of Medical Genetics
University of Toronto
Hospital for Sick Children
Toronto, Ontario, Canada

Anne E. Simon
Department of Biology
Indiana University
Bloomington, Indiana

Raymond L. Stallings
Genetics Group LS-3 MS 886
Los Alamos National Laboratories
Los Alamos, New Mexico

Pamela Stanley
Department of Cell Biology
Albert Einstein College of Medicine
Bronx, New York

Carolyn Steglich
Department of Microbiology
East Carolina School of Medicine
Greenville, North Carolina

Milton W. Taylor
Department of Biology
Indiana University
Bloomington, Indiana

Lawrence H. Thompson
Biomedical and Environmental Sciences Division
Lawrence Livermore National Laboratory
Livermore, California

John J. Wasmuth
Department of Biological Chemistry
University of California
Irvine, California

Carolyn D. Whitfield
Department of Biochemistry
Howard University College of Medicine
Washington, D.C.

Ronald G. Worton
Department of Medical Genetics
University of Toronto
Hospital for Sick Children
Toronto, Ontario, Canada

George Yerganian
Cytogen/Research and Development, Inc.
West Roxbury, Massachusetts

To
my father,
my wife,
and my children

PREFACE

Molecular cell genetics is the synthesis of somatic cell genetics with molecular biology. This volume focuses on the major role that Chinese hamster cell lines have had in the development of this new biological discipline. For more than two decades, Chinese hamster cells have provided model systems for the study of genetic alterations in cultured mammalian cells. For the somatic cell geneticist, Chinese hamster cells serve the same function that *Escherichia coli* serve for the molecular biologist. Genetic studies using Chinese hamster cells have provided important insights and a logical, more rigorous structure to the study of the function of various metabolic pathways in all mammalian cells. More recently, dramatic developments in recombinant DNA technology have added molecular detail to our understanding of the mechanisms of somatic cell genetic alterations.

This volume will examine a wide variety of genetic systems developed in Chinese hamster cells. Each of the authors addresses the rationale for development of these systems, the details of mutant isolation and analysis, the major conclusions derived from these genetic studies, and, where possible, the molecular basis for the expression of mutant phenotypes. Comparison is made with genetic analysis developed in non-Chinese hamster cell systems, so that the reader has an overview of the state of knowledge for each genetic locus.

The first section of the book contains several personal perspectives on the historical development and characterization of cultured Chinese hamster cells by the scientists who developed these cell lines. The emphasis here is on the Chinese hamster cell lines (CHO, V79, and CHEF) that have been the mainstays of somatic cell genetics. The second section focuses on more technical aspects of genetic manipulation involving somatic cells including somatic cell hybridization techniques, chromosome isolation, cDNA cloning and expression, and DNA and vector-mediated gene transfer. The final section explores a wide variety of genetic systems in detail and demonstrates how Chinese hamster cell genetics can be used to study practically any problem in cell biology. The genetic loci discussed in detail here are those that have been most thoroughly analyzed as of this date. Other systems, still in the development stage, are either mentioned in the text or appear in Appendix II, which contains a list of major classes of Chinese hamster cell mutants isolated to date. Appendix I describes the lineage of currently available Chinese hamster cell lines that are referred to throughout the book.

This book is intended as a resource for the practicing somatic cell geneticist and for the cell biologist with an interest in the manner in which genetics can be used to elucidate biological problems. The content is presented in a style that should be easily accessible to an advanced undergraduate, graduate student, or post-doctoral fellow. It is anticipated that *Molecular Cell Genetics* will also be useful as a text of somatic cell genetics for use in the classroom.

MICHAEL M. GOTTESMAN

Bethesda, Maryland
January 1985

CONTENTS

SECTION 2: GENETIC MANIPULATION OF CHINESE HAMSTER CELLS

SECTION 3: GENETIC SYSTEMS DEVELOPED IN CHINESE HAMSTER CELL LINES

MOLECULAR CELL GENETICS

SECTION 1

THE DEVELOPMENT AND CHARACTERIZATION OF CHINESE HAMSTER CELL LINES

CHAPTER 1

THE BIOLOGY AND GENETICS OF THE CHINESE HAMSTER

George Yerganian
Cytogen Research and Development, Inc.
West Roxbury, Massachusetts
and Harvard School of Public Health
Department of Population Sciences
Boston, Massachusetts

I. INTRODUCTION

The Chinese hamster (CH), *Cricetulus griseus*, is undoubtedly the least common rodent in the laboratory today. Its limited numbers contrast with the extensive use of cell lines derived from the CH for somatic cell genetics, mutational and gene transfer studies, and short-term genetic toxicity (gen-tox) assays. In anticipation of this trend, our laboratory initiated whole animal studies with the intent of establishing additional cultured cell types to accommodate the needs of various *in vitro* applications. In the future, "second-generation" cell lines derived from current inbred strains will be used to extend genetic analysis of cultured cells to include differentiated functions of tissues.

Some 65 years have passed since the CH was first used as a laboratory specimen. Just three decades ago, reviews of the sparse literature centered around the problem of domestication. Shortly thereafter, the early descriptions of chromosomal features and the success in establishing cell lines led to the rapid technical adaptations encompassed by this volume. In this introductory chapter, coverage of the biology and genetics of the CH will be limited to representative examples of observations relevant to the establishment of new cell lines.

II. BRIEF HISTORY

A. 1919–1948

Introductions of hamster species to the laboratory date back to 1919 when Hsieh first reported the substitution of wild specimens of CH in place of mice for typing pneumococci and applying serum for complement fixation. During the 1920s, independent activities of British and American investigators led to implicating the CH as the rodent vector for the transmission of kala azar or leishmaniasis. In turn, the epidemiological aspects of leishmaniasis spurred interest in the Near East, thereby leading to trapping of the endemic Syrian and Armenian hamsters (Yerganian, 1972).

While earlier field workers in China had to compensate for civil strife and, later, the war with Japan, the rise of the People's Republic of China (PRC) was accompanied by the CH ascending to the political scene. The shipment of the ancestors of today's CH to the United States in December, 1948, was followed by the imprisonment of Dr. C.H. Hu of the Department of Pathology at the Peking Union Medical College. Also, Dr. Robert Briggs Watson was sentenced to be hanged *in absentia* for events that took place during his heroic dash from The Rockefeller Foundation Laboratory in Nanking to Shanghai to meet the last PanAm flight to the United States before the fall of the "Bamboo Curtain." The PRC's Germ Warfare Commission's condemnation of Drs. Hu and Watson was based on the assumption that the animals were to be bred in the United States, infected with cholera and plague, and parachuted into Manchuria. These accusations were the forerunners of misjudgements that plagued the CH until recently.

B. 1949–1983

The events surrounding earlier efforts at domestication are detailed elsewhere (Yerganian, 1972). Observations leading to proposing the CH for karyological studies are provided here for the first time.

During the fall–winter of 1949, the writer was screening European journals which had been delayed enroute because of World War II. In general, most volumes were thin, and combined several years of publications. Among the titles that appeared in the *Proceedings of The Royal Society of Edinburgh (B)*, the one by George Pontecorvo (1943), entitled "Meiosis in the Striped Hamster [CH] and the Problem of Heterochromatin in Mammalian Sex-Chromosomes," attracted my attention. Pontecorvo indicated the diploid number of the CH to be 14, whereas the correct diploid number of 22 was reported independently by Matthey (1952), Sachs (1952), and Yerganian (1952).

During World War II, George Pontecorvo and his physicist brother, Bruno, were alien residents in Great Britain, and were encouraged to conduct their respective studies. The testicular material used by Pontecorvo had been collected by E. Hindle during the kala azar expedition (1926–1927) sponsored by the Royal Society. The animal carcasses and tissues had been stored in fixatives or paraffin blocks for some 16 years prior to the time Pontecorvo examined the testicular material. Undoubtedly, the long-term storage led to artifacts that influenced his erroneous chromosome counts. Nonetheless, his treatise on the heterochromatin of mammalian sex chromosomes was superb and is meaningful to this day.

Subsequent to completing my thesis requirements (1950), and relating to Pontecorvo's paper while with the Botany Department at the University of Minnesota (1950–1951), the advantages of using an experimental animal with a low chromosome number in radiation cytogenetics and tissue culture soon became obvious. As an AEC postdoctoral fellow at the Brookhaven National Laboratory (1951–1952), I finally acquired the CH. The pugnacity and solitary nature of the females were controlled sufficiently to foster the first of 17 litters (Fig. 1.1A). All existing CH colonies stem from these particular animals (Yerganian, 1958, 1972).

My request to apply the second postdoctorate year at the Brookhaven National Laboratory to evaluating the CH for radiation mutagenesis was, paradoxically, refused on the ground that I was "wasting my time," and "how did I expect to find a job while training as a botanist and involved with mammals, etc." I decided to continue this effort elsewhere. This decision proved to be favorable, since Schwentker discontinued his stocks soon thereafter without revealing his breeding method (Yerganian, 1972). Efforts to transfer the fellowship to one of the several institutions pursuing mammalian studies failed to materialize. Arrangements were finally made to move the breeding stocks to the Biology Department at Boston University, where fundamental studies on microcirculation in the

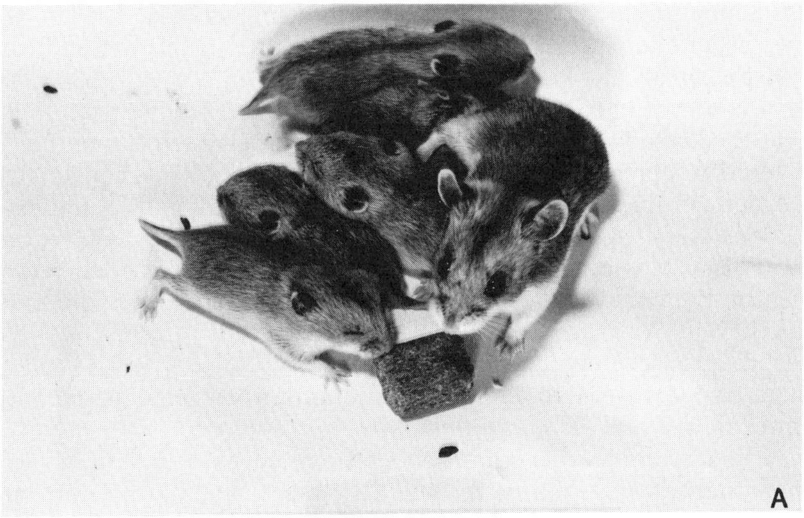

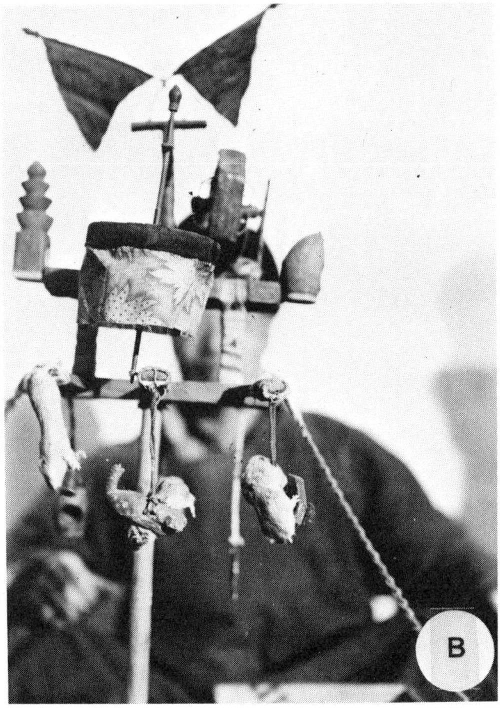

Figure 1.1. (A) First of 17 litters born and, later, "deported" from the Brookhaven National Laboratory to Boston University. (Born February 27, 1952; 14 days old.) Descendants ot existing CH stem from matings limited at the time to two sets of males and females (photo No. 3-108-2, Brookhaven National Laboratory). (B) "Hamster Circus," circa 1930, Beijing. Wild specimens (fearful of height) trained to perform trapeze "acts" (among the sights that attracted Hsieh to purchase and substitute CH for mice in 1919).

cheek pouch of the Syrian hamster are still being pursued. The fact that there was another species of hamster with a cheek pouch was sufficient to foster a constructive and lasting relationship.

A brief hiatus in funding after some 27 years of a productive relationship at the Dana-Farber Cancer Institute necessitated relocating the colony. Interim support was provided by the Rippel Foundation and, later, by the American Cancer Society to house reduced numbers of inbreds at three local sites. A renewal of funding from the NIH fostered administering the reassembled colony through Northeastern University. Presently, the colony is housed at the Foster Animal Facility, Brandeis University in Waltham, Massachusetts. The recently formed Cytogen Research and Development, Inc. ensures the availability of CH and Armenian Hamsters (AH) to the scientific community without undue interruptions.

It is only fitting on this occasion to acknowledge the individuals who, unknowingly at times during the past 30 years, shared in my efforts and, indirectly, contributed toward the productivity of many investigators. They are Robert Briggs Watson (deceased, Fig. 1.2); C.H. Hu; Victor Schwentker; Brenton R. Lutz (deceased); George P. Fulton (Emeritus) and Donald I. Patt (Emeritus) of Boston University; Sidney Farber (deceased); Walter S. Jones and Karl Weiss of Northeastern University; and K. C.

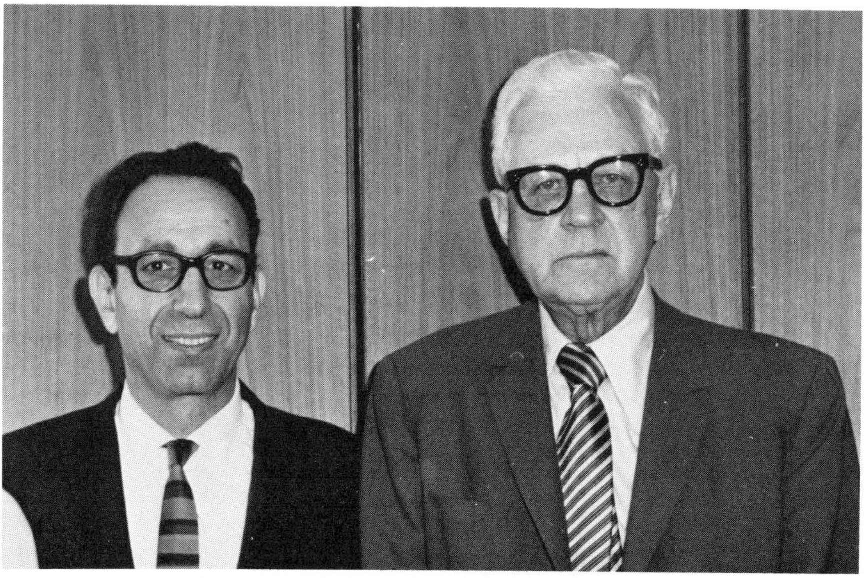

Figure 1.2. "... Someday, when I am in Boston, I may see some of the descendants of 'my' hamsters." Excerpt from letter written by Dr. Robert Briggs Watson (right) to the author (left), dated February 15, 1962. Details of the search for Dr. Watson, covering a decade of correspondence, are provided elsewhere (Yerganian, 1972).

Hayes of Brandeis University. With reference to Pontecorvo's paper (1943) and the above acknowledgments, another segment of the events surrounding the CH is recorded.

References to earlier experimental infections (diphtheria, rabies, influenza, monilia, encephalitis, and trichinella), breeding procedures, growth and reproduction, sperm and testicular development, fertilization and early embryonal development, cheek pouch and microcirculation, and behavioral aspects and breeding in simulated environments are provided elsewhere (Yerganian, 1979). Earlier references describing juvenile and adult-onset diabetes mellitus (pathology, genetic aspects, renal lesions, periodontia and dental caries, pancreatic adenocarcinomas, serum proteins, stress and glycosuria, retinopathy, and fatty acid metabolism) are also provided in the above reference. Gerritsen (1982) has provided an in-depth assessment of the diabetic strains presently available. References detailing the meiotic and mitotic cycles, and some cell lines, are also listed by Yerganian (1979).

The historic relationships and detailed personal accounts leading to the introductions of the Chinese, Syrian, and Armenian hamsters as laboratory specimens are documented elsewhere (Yerganian, 1972). Reference to the pugnacious disposition of the early laboratory progeny (Yerganian, 1958, 1967) deserves revision to reflect the three decades of inbreeding and the potential for domestication suggested in Figure 1.1B. Inbred and randombred stocks maintained by the author qualify as the most docile strains of laboratory rodents.

III. CYTOTAXONOMY OF THE GENUS *CRICETULUS*

The Chinese hamster, *Cricetulus barabensis griseus*, represents one of four subgroups of the genus *Cricetulus* or Old World dwarf hamsters (Allen, 1932; Argyropulo, 1933; Ellerman, 1941; and Ellerman and Morrison-Scott, 1951). Comparative chromosome banding studies (trypsin–Giemsa and quinacrine mustard) have disclosed several unique relationships among members of the genus *Cricetulus* (dwarf hamsters) and the giant European hamster, *Cricetus cricetus* (Fig. 1.3). The latter species is presently classified as the sole member of its genus. Yet, its chromosomes and banding patterns are identical to those of the dwarf hamsters. Independent observations on the CH by the author and others (Gamperl et al.; 1976; Vistorin; et al., 1977); its immediate subspecies, *Cricetulus barabensis barabensis* (Radjabli and Kriukova, 1973); and (Lavappa, 1977: Lavappa and Yerganian, 1970; Yerganian and Papoyan, 1965); the present prototype of the *migratorius* group, the Armenian hamster are compared in Table 1.1 with the findings on the giant European hamster, *Cricetus cricetus* (Gamperl, et al.; 1976; Vistorin; et al.; 1976, 1977).

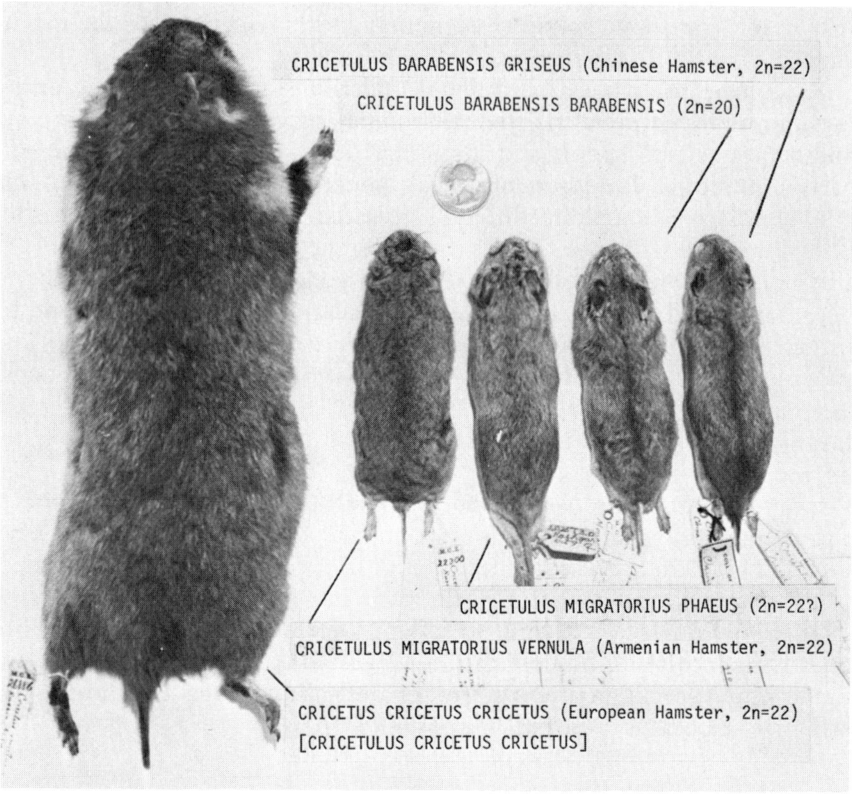

Figure 1.3. Representative collection of skin specimens of Old World hamsters. (Courtesy of the Museum of Comparative Zoology, Harvard University.) Giant European Hamster (left), presently the sole member of the genus *Cricetus*, is flanked by four examples of dwarf species of the subgroups *barabensis* and *migratorius*. See text for details and cytotaxonomic considerations in support for the reclassification of the European hamster as *Cricetulus cricetus cricetus*. Relative sizes indicated by U.S. quarter coin (upper center).

The karyotype of the CH serves as the ancestral prototype to construct plausible evolutionary relationships. The chromosome types and numbers listed under each species in Table 1.1 are those assigned by the various investigators. The recorded homologies in chromosome banding patterns are arranged horizontally to demonstrate the karyotypic stability during evolution and the suggested migration patterns. In this instance, a westward migration is supported, that is, from eastern China (CH) to Outer Mongolia and the Gobi Desert (BH), Siberia and Eastern Europe (AH), and onto Western Europe when the European hamster (EH) is transposed from its present taxonomic isolation to the proposed subgrouping within the genus *Cricetulus*.

A major Robertsonian centric fusion of chromosomes 6 and 7 of the

TABLE 1.1
Comparative Karyology of Some Hamsters

Cricetulus Griseus Chinese (CH) (2n=22)	*Cricetulus Barabensis* Barabensis (BH) (2n=20)	*Cricetulus Migratorius* Armenian (AH) (2n=22)	*Cricetus Cricetus* European (EH) (2n=22)
SA1	SA1	6	5
LA1	LA1	7	6
2	2	1	1
3	3	3	3
4	4	4	4
5	5	5[a]	7
6	6:7	2	2
7			
8	8	8[a]	8[a]
9	9	9[a]	9[a]
10	10	10[a]	10[a]
X	X	X[a]	X[a]
Y	Y	Y[a]	Y[a]

[a] Pericentric inversions.

CH led to a reduction in the chromosome number of the BH (2n = 20) and the formation of a key metacentric marker (Table 1.1; 6:7 under BH), which is noted as chromosome 2 in the AH (Fig. 1.4; C6:A2:C7) and the EH. The single marker which separates the CH from its subspecies BH is also present in the AH, and the taxonomically isolated EH—the sole member of *Cricetus*.

A closer relationship between the AH and EH is evidenced by the two identical large subtelocentrics derived from the short (SA) and long (LA) arms of the metacentric chromosome 1 of the CH/BH (Table 1.1), as depicted in Figure 1.4 (A6:C1:A7). Thus, the AH and EH stem from a common ancestral intermediate derived from the BH. Their separation during the westward migrations took place when the telocentric chromosome 5 (CH/BH) underwent a pericentric inversion in association with similar events involving sex chromosomes, to give rise to the *migratorius* group (Fig. 1.4; A5:C5). Since chromosome 5 of the EH is "intact," that is identical to the CH, the EH shares a greater degree of autosomal banding homology with the CH/BH than the AH. Yet, in taxonomic terms only, the EH is presently separated as distantly from the CH as the rat is from the mouse.

Vistorin et al. (1976, 1977) and Gamperl et al. (1976) have discovered that the EH karyotype consists of extensive segments of tandem duplications of centromeric heterchromatin, in addition to the striking ho-

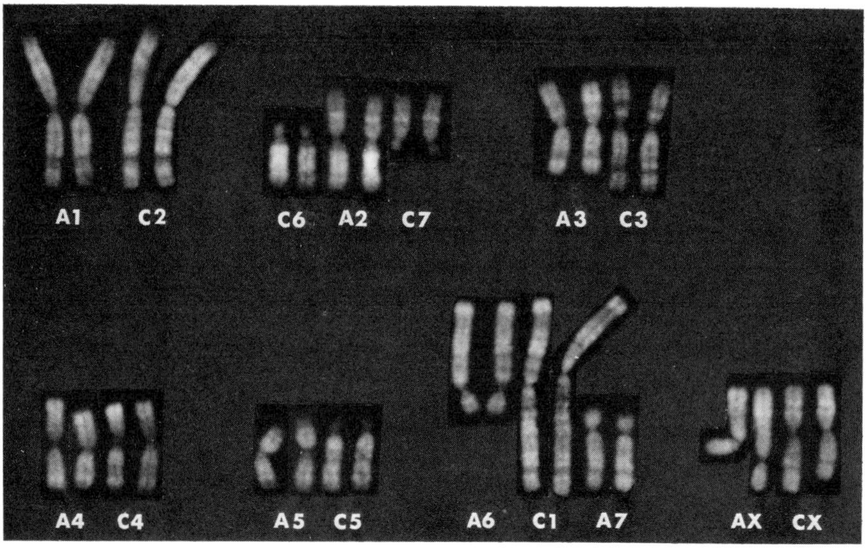

Figure 1.4. Quinacrine mustard chromosome banding patterns of major chromosome types of the Chinese (C) and Armenian (A) Hamsters. See text for descriptions relative to chromosome homologies assembled in Table 1.1.

mology of the chromosome banding pattern with genus *Cricetulus*. Setting aside the ill-defined role of centromeric heterochromatin in the evolutionary process, the prospects for reclassifying the EH along cytotaxonomic terms are in order. Its karyological features are those of the genus *Cricetulus*. Understandably, its relative size, pelt colorations, hibernating physiology, and geographic distribution differ strikingly from the CH, BH, and AH. Nevertheless, basic morphoanatomical features were recognized by the earlier taxonomists as being distinctly *Cricetulus*. These considerations support placing the EH as a new subgroup, *cricetus*, of the genus *Cricetulus* and renaming it *Cricetulus cricetus cricetus*. More importantly, chromosome banding patterns have remained undisturbed during the evolution of these hamster species. The role of centromeric heterochromatin and the mechanism for tandem duplication may be subjects of innovative investigations along the lines demonstrated by Biedler and Spengler (1976) for facultative/obligatory heterochromatin associated with the X chromosomes.

IV. GENETIC ASPECTS

A. Domestication and Inbreeding

Chinese and Armenian hamsters were inbred by means of full-sib matings immediately following their domestication (Yerganian, 1958, 1967;

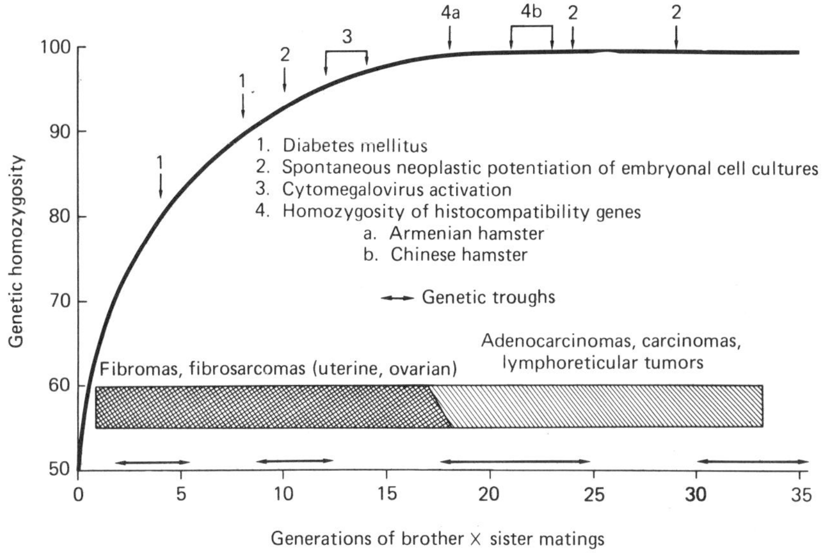

Figure 1.5. Onset of multifactorially inherited features in dwarf hamsters. Sequence of events (described in text) that accompanied full-sib matings of the Chinese and Armenian hamsters. Tumor spectrum shown is that of the CH. Autoimmune-prone AH fails to develop spontaneous tumors during a life-span of 800–1000 days.

Yerganian and Papoyan, 1964). During the course of four to eight generations of inbreeding (Fig. 1.5), all sublines of CH developed hereditary (spontaneous) insulin-dependent diabetes mellitus (Meier and Yerganian, 1959, 1961). Genetic studies disclosed a complex inheritance pattern involving recessive loci and ill-defined modifiers (Yerganian, 1964; Gerritsen, 1982). The veritable "fall-out" of the juvenile form of diabetes, involving 50–90% of the offspring uniformly among the five initial sublines, strongly suggested that genetic factors were present with the wild-type specimens introduced into the laboratory. Onset of the diabetic syndrome within the several generations of inbreeding also led to the revelation that wild populations of dwarf hamsters may serve as natural reservoirs of clinically important syndromes (Yerganian, 1972).

Excessive female infertility occurred at various stages of inbreeding, owing to suspected physioanatomical manifestations associated with progressive genic homozygosity (Fig. 1.5; genetic troughs). The most critical phase took place during the 22nd–23rd generations. Only two strains among many initiated during 1952–1955 bridged the critical 22nd–23rd full-sib matings. Each of the two successful strains (A/GY and B/GY) were bred on by one female among numerous sublines that had progressed to the 22nd generation. Our compilations indicated that this success reflected the raising and breeding of over 20,000 females for each strain in the course of 22 generations of full-sib matings. Strains involving lesser numbers of animals failed to progress beyond this criti-

TABLE 1.2
Distribution of Spontaneous Neoplasms in Aging (250–1000 days old) Chinese Hamsters (1970–1975)

Types of Neoplasms	Number	Percentage
Reticular cell neoplasia and myeloproliferative disorders	9	24
Associated with hepatocarcinomas	1	3
Hepatocarcinomas only	7	18
Uterine-cervical (adeno)carcinomas[a,b]	13	34
Ovarian neoplasms[b]	4	10
Leiomyosarcomas	2	5
Others		
Pancreas	1	3
Osteogenic sarcoma	1	3
TOTAL[c]	38	

[a] For details and relationships with human endomyometrical neoplasms, see Brownstein and Brooks (1980).

[b] Twenty-five percent with associated lymphomatoid and/or sarcomatoid granulomatoses of salivary gland, pancreas, and/or lung.

[c] See Table 1.3 for representative cell lines.

cal period. Fortunately, the Armenian hamster did not undergo as drastic a test. Nevertheless, advanced stocks of both species require attention during breeding, in contrast to the readily managed randombreds.

B. Spontaneous Neoplasms

The incidence of spontaneous neoplasms among randombred and inbred males is virtually nil. In contrast, inbreeding of females was accompanied by a shift from an otherwise monotonous series of uterine fibrosarcomas to the frequent development of (adeno)carcinomas upon achieving the inbred status (Fig. 1.5). Similarly, hepatic cirrhosis gave way to hepatocarcinomas when inbreeding was attained. Heterozygosis resulting from matings of unrelated strains correlated with a reversion to uterine fibrosarcomas among $F_{1\text{-}2}$, followed by a gradual drift toward the adenocarcinomatous and mixed-cell patterns as inbreeding was advanced among new sublines. Spontaneous tumors develop when animals are well into the second half of their life-span (2.5–3.0 years). Baker et al. (1974); Resznik et al. (1976); Benjamin and Brooks (1977), and Brownstein and Brooks (1980) have also recorded a low background incidence of neoplasia among control animals in long-term aging and carcinogenesis studies. The incidences and histopathological features of spontaneous uterine endomyometrial neoplasms and similarities with clinical counterparts are detailed by Brownstein and Brooks (1980). The occasional ductal and exocrine (adeno)carcinomas of the pancreas noted

TABLE 1.3
Cell Lines of the Chinese Hamster

Origin	Nonneoplastic	Neoplastic
Embryonal fibroblast (CHEF cells) (see Table 1.4)	F4224[b,d] F4228[d]	F4225[b,d] (3 MCA transformant) F4226[b,d] (Spontaneous F4231[b,d] transformant)
Embryonal epithelioid	DP13[d]	SV40 transformant[d]
Undifferentiated mesenchymal derivative (adult smooth muscle cell with chondrogenic potential)	T233[a,b,c,d]	
Adult smooth muscle cell (with latent cytomegalovirus)	T219[a,b,c,d]	
Histiocytic	T337[d]	
Macrophage/monocytic	XT57[a,b,c,d]	
Hepatocarcinomas		T246[d] T264[d] T268[d] T280[d] T316[d]
Uterine carcinoma (mesodermal component)	T310[a,b,c,d]	
Ovarian carcinoma		T251[d]
Cervical carcinoma		T257[d] T304[d]
Testicular rhabdomyosarcoma		T250[c,d]

[a] Tested by continuous animal passaging in syngeneic recipients.
[b] Classic diploid: others pseudodiploid or low aneuploids (2s = 23,24).
[c] Suspension-type culture.
[d] Monolayer-type culture.

among aging, nonsymptomatic members of diabetogenic sublines (Poel and Yerganian, 1961) may have reflected partial homozygosity of the multifactorial hereditary mechanism for diabetes mellitus.

The distribution of spontaneous neoplasms and the derived cell lines noted during 1970–1975 (30–35 generations of inbreeding) are listed in Tables 1.2 and 1.3, respectively. On two occasions (Fig. 1.5), the cytomegalovirus of the Chinese hamster (CHCMV), described as inclusion-body disease by Kuttner and Wang (1934), was isolated from benign granulomatous explants (T89 and T219) that appeared in 11th and 14th generation animals (Yerganian, 1967, 1972).

C. Pancreatic Lesions

The pancreas of the CH is highly responsive to the carcinogenic and adjuvant-promoter effects of different hydrocarbons. Pancreatic neoplasms were initiated in the course of applying 3,4-benzo-(*a*)-pyrene to the skin of nonsymptomatic segregants of diabetic-prone strains (Poel and Yerganian, 1961). More recent studies disclosed the pancreas also to have a predisposition for progressive pancreatitis and insulitis (Fig. 1.6) in response to mineral oil or pristane administered intraperitoneally (Yerganian et al.; 1979). The premise for the study was based on the characteristic islet-associated lymphocytic infiltrates that appear in the exocrine pancreas and liver with aging. The precise nature of the inflammation remains obscure. It may reflect either a mild pancreatitis or an autoimmune response. Since this is a feature of otherwise normal-appearing animals and is virtually absent in other laboratory species, we assumed the predisposition for pancreatic involvement to reflect a persisting segment of the multifactorial juvenile diabetes genotype.

Subcutaneous administration of mineral oil and pristane resulted in early manifestations of sarcomatoid and atherosclerotic lesions. While the intraperitoneal route led to otherwise rare instances of pleomorphic pancreatic (soft tissue) carcinomas (Yerganian et al.; 1979), the subcutaneous route proved to be an excellent source for permanent classic diploid phagocytic cell lines derived from reactive noncaseating granulomas (Table 1.3).

V. CELL LINE CHARACTERIZATION

A. Influence of Culture Media Components on Chromosome Stability

Early studies suggested the presentation of abnormal marker chromosomes in normal CH cell lines followed instances of culture neglect or the relaxation of technical steps. In addition, follow-up communications with laboratories provided earlier with representative sublines of classic diploid cultures disclosed cultures were either lost, experienced aneuploidization, or continued to retain the classic diploid chromosome complement for several years of additional passaging. We suspected that the failure to maintain viable cultures or to lose diploidy was associated with the use of media formulated to maintain aneuploid mouse and human cell lines. It was reasoned that any modification of a medium that optimized the proliferation of aneuploid/tetraploid cell lines may also accelerate the selection for aneuploid variants among essentially classic diploid cell lines.

Subsequently (Yerganian and Lavappa, 1971), three particular components of cell culture media formulations were found to be "toxic" for classic diploid cell lines: (a) $MgSO_4$, at levels employed in Earle's balanced salt solution; (b) serum, when employed in amounts exceeding 10%; and (c) sodium bicarbonate, whenever equaling or exceeding the

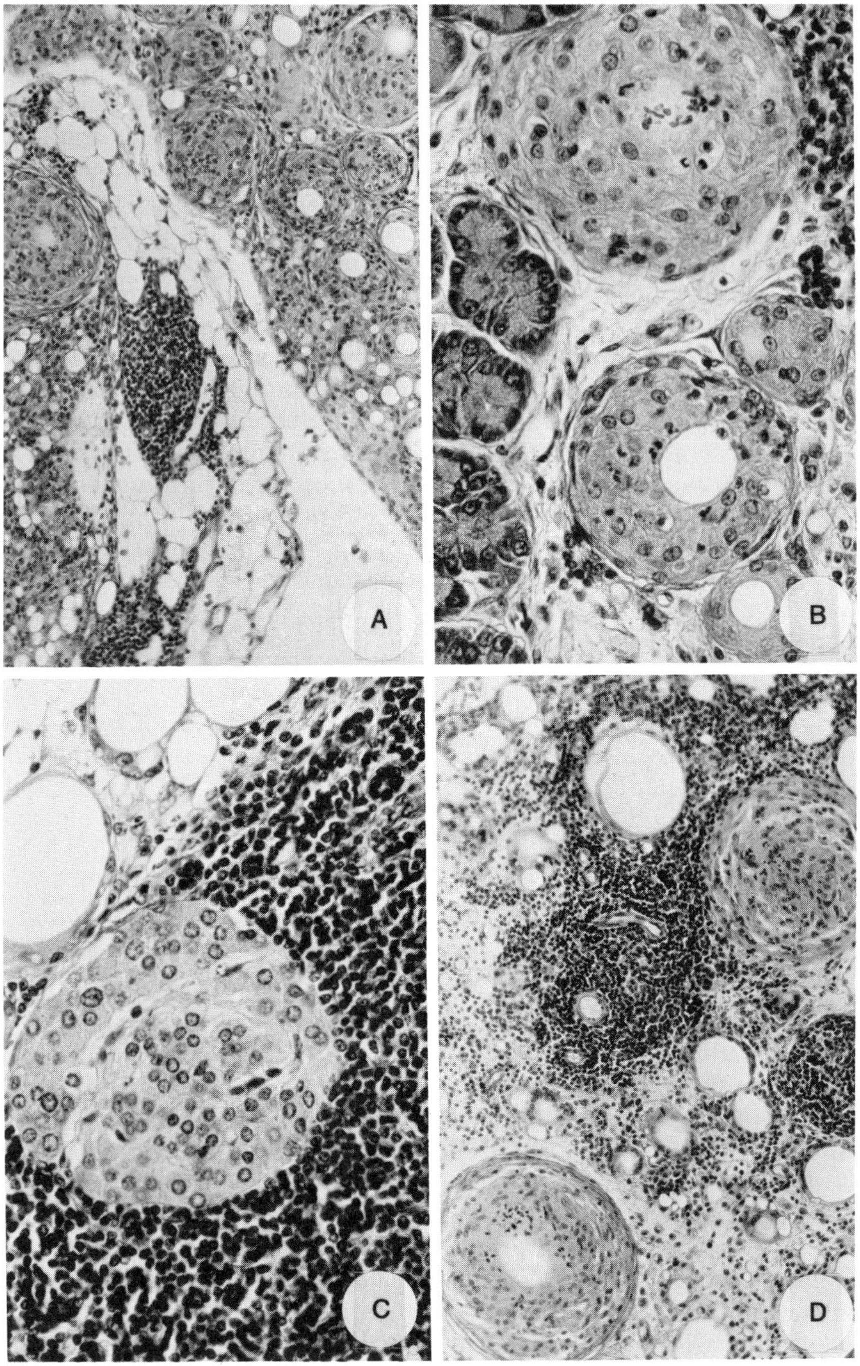

Figure 1.6. Mineral oil- and pristane-induced sarcomatoid and lymphomatoid pancreatic granulomatoses of the Chinese hamster. (A) Multiples of intravascular occlusions of mesenteric tissue adjacent to pancreas, with early lymphocytic infiltration (2 weeks posttreatment with 0.5 mL oil, intraperitoneally). (B) Nodular, noncaseatinglike granulatomous response at higher magnification. Increasing lymphoplasmacytoid infiltration and proliferative granulation of the exocrine (2–4 months following above treatment). (C) Peripheral islet surrounded by extensive lymphocytic infiltration. (D) Persistence of vascular lesions, increasing lymphoplasmacytoid proliferation, and complete granulation of the exocrine. Ductal features remain normal (4–6 months following above treatment).

amount employed for Eagle's MEM prepared with Earle's balanced salt solution. The salt components are present in excess amounts in such formulations as modified Dulbecco's and McCoy's 5A, which were employed in laboratories that experienced a loss of cultures or aneuploidization. When Earle's and Hank's salt solutions were employed proportionally, the adverse effects on classic diploid cell lines were all but eliminated.

More importantly, classic diploid cell lines are sensitive to the "toxic" effects of serum. This ill-defined tendency became apparent when single cell platings were employed to test fresh lots of fetal bovine serum. A consistent elevation in plating efficiency is readily manifested with each lot of serum. This encourages one to order or reserve large volumes of a particular lot to ensure its continued availability. However, as long as the cells remain diploid, the culture will begin to reflect a slow build up of reactivity to the new lot of serum within some 30 days of continuous exposure. This is manifested by a reduction in plating efficiency and growth rates. As a consequence of repeated exposure to the same lot of serum, classic diploid cells are challenged to stagnate, mutate, or succumb, depending on the levels of serum employed and the buffering capacity of the medium. *It must be emphasized that the detrimental aspects noted above are not exhibited by aneuploid cell lines. Also, cell membrane alterations of classic diploid cells following transformation by viral or chemical agents alleviate the observed toxicity of the various ingredients mentioned.*

The latent toxicity of serum on diploid cell lines is overcome by combining different batches from different animal sources (Yerganian and Lavappa, 1971). Aside from ensuring optimal growth, classic diploidy is fostered also by adjusting the pH of the medium to remain at 7.2–7.3 after some 24 hours of incubation. A simple way to adjust the pH when employing Eagle's MEM, for example, is to prepare media separately using Hank's and Earle's balanced salt formulations. The two are mixed proportionally (1:1, 1:2, etc.) to provide for adjustments favorable to either cloning or maintaining monolayer cultures. The lower buffering capacity of Hank's salt solution encourages cloning, while the elevated amounts of $MgSO_4$ and sodium bicarbonate in Earle's balanced salt solution are detrimental to classic diploid cells unless diluted, pregassed, or used as conditioned medium. Moreover, the use of proportioned growth media encourages optimal performance of cells maintained in different types of culture vessels and incubated in a nonregulated type of CO_2 incubator.

The elevated stability of the karyotype noted for the above adjustments has been confirmed recently. The use of mixed sera over single sources narrowed down the range of aneuploidy following chemical transformation of Chinese hamster cells *in vitro* (Kirkland, 1976; Kirkland and Venitt, 1976). Similarly, Papadopoulo et al. (1977) reported that mixed sera would prolong normal embryo cell cultures of the Syrian hamster. They were able to passage cultures for 2½ years while retaining classic diploidy. This represents a 10-fold increase in longevity over the average survival time attained with other media formulations.

B. Spontaneous versus Chemical Transformation, *In vitro*

Invariably, embryo cultures are initiated with the mixed sexes of a given pregnancy. Earlier studies indicated that the potential for spontaneous neoplastic transformation, as expressed in syngeneic recipients, resides in cells derived from only one of several embryos employed to prepare cultures. This is evidenced when mixtures of cells displaying the XX and XY sex chromosomes are implanted into host animals and the resulting tumors are entirely and consistently composed of cells of the same sex. While adult-derived cultures may also express spontaneous neoplastic potentials (Yerganian and Leonard, 1961), the majority of the embryonal cell lines remain normal until transformed by an oncogenic agent (Yerganian, Freeman and Gagnon, 1968).

The latter study involved two male embryonal cell lines (I and II) derived from the same pregnancy but maintained separately (Table 1.3, embryonal fibroblasts). Several heretofore unrecognized aspects of *in vitro* transformation were revealed by this simple measure to prepare individual cultures from each embryo. Namely, spontaneous transformation can, indeed, be expressed very early on rare occasions, fixation of the neoplastic potential is delayed until transformed cultured cells are presented to the selective pressures of immunologically tolerant host animals, and spontaneously transformed cells share nonhistocompatibility membrane antigens with normal cells. These features, when related to studies employing mixed embryo cultures for *in vitro* transformation, could explain in part why some cultures failed to respond to chemical carcinogens (in the absence of potential spontaneous transformants to serve as sensitive target cells), why the use of newborn and immunosuppressed animals as hosts fails to give rise to routinely transplantable sarcomas, and why present methods fail to distinguish between spontaneous versus chemically induced transformations occurring in the same plate.

Sublines of Embryo I (Table 1.4) consistently failed to elicit tumors in hosts during 25 transfer generations of intermittent implantations. Sublines of Embryo I became transformed morphologically following exposure to 3MCA (1.0 μg/mL for 3–7 days) at the second passage. At the same time, subtle focal proliferations appeared in both the control and 3MCA-treated sublines of Embryo II during the 30-day posttreatment period. Subsequent subcultures and animal inoculations revealed: (a) all

TABLE 1.4
Neoplastic Expression of "F" Cell Lines (Derivation of Sager's CHEF Cells)

Cell Source	Treatment	Passage Number	*In vitro* Morphology	Tumor Incidence[a] 3rd Passage	Tumor Incidence[a] 4th–6th Passages	Passage Numbers of Additional Tests	Days for Tumor Growth to 8×15 mm size[b]
Embryo I							
(F3940 Series)							
F4224A,B	None	1–25	Normal	0/7	—	8–10, 19, 25	—
F4228A,B							
F4225A,B	3MCA[a]	2	Transformant	4/5	6/6	8–18	31–74
Embryo II							
(F3941 Series)							
F4226A,B	None	2	Spontaneous transformants	5/5	5/5	8, 18, 23	30–60
F4231A,B							
F4227A,B	3MCA	2	Spontaneous transformants	2/2	3/4	17	33–50

[a] Starting some 35–40 days after exposure to 3MCA (subcultures twice weekly at 1:3 or 1:4 splits).
[b] Cell inoculations ranged from 5×10^6 to 10^7 cells subcutaneously.
[c] 1.0μg/mL 3MC for 3–7 days.

sublines featured the classic diploid karyotype (90%); (b) only 3MCA transformed Embryo I cells expressed neoplastic properties, whereas (c) both the control and 3MCA-treated Embryo II sublines were equally neoplastic. Thus, separate propagation of embryonal derivatives is conducive toward the early onset of rare instances of spontaneous neoplastic transformation masked, most likely, by the presence of numerous normal cells when employing mixed embryo cultures.

Additional revelations were disclosed in the course of inoculating immunologically tolerant hosts. Upon reculturing primary tumors, each tumor featured a distinctively different aneuploid karyotype or stem cell. In essence, the primary tumor mass stemmed from the select proliferation of a variant single cell regardless of the size of cell inoculum. This further suggested that fixation of the neoplastic process takes place *in situ* and only one of the many potentially neoplastic cells gives rise to the primary tumor mass. The role of aneuploidy or structural change, which failed to challenge the classic diploid status *in vitro*, was reversed the moment immunological parameters were introduced following cell inoculations.

During the course of inoculating untreated and 3MCA-treated cell lines into syngeneic recipients, animals failing to develop tumors were reserved and challenged once again, as depicted in Figure 1.7. As expected, previously negative hosts, when inoculated a second time with normal cells, remained negative. However, previously negative hosts inoculated with spontaneous transformants also failed to develop tumors, in contrast to the rapid formation of transplantable fibrosarcomas when employing normal hosts. 3MCA transformants of both derivatives formed tumors in syngeneic hosts, irrespective of the number of times animals had been negative to either normal cells or spontaneous transformants. Thus, tumor formation by spontaneous transformants was restricted to normal hosts. "Preimmunization" of hosts with normal cells was reflected by selective suppression of the neoplastic potential of spontaneous transformants. The latter undoubtedly reflected the capacity of normal cells to provoke antibodies against cell membrane antigens (acquired or exposed *in vitro*) that are shared with spontaneous transformants. Hence, immunologically tolerant syngeneic adult hosts can distinguish spontaneous from chemical transformants when following the steps depicted in Fig. 1.7.

The more enlightening aspect of this study is the need to initiate embryonal cultures from individual embryos to guard against rare instances of an early acquisition of neoplastic potentiation by otherwise normal-appearing classic diploid cells. The incidence of such an event is estimated at one for every 25–40 explanted embryos. Rare instances of early neoplastic potentiation of embryonal male cells does not reflect an acceleration of the neoplastic process, since males rarely develop tumors during a life-span of 900–1000 days.

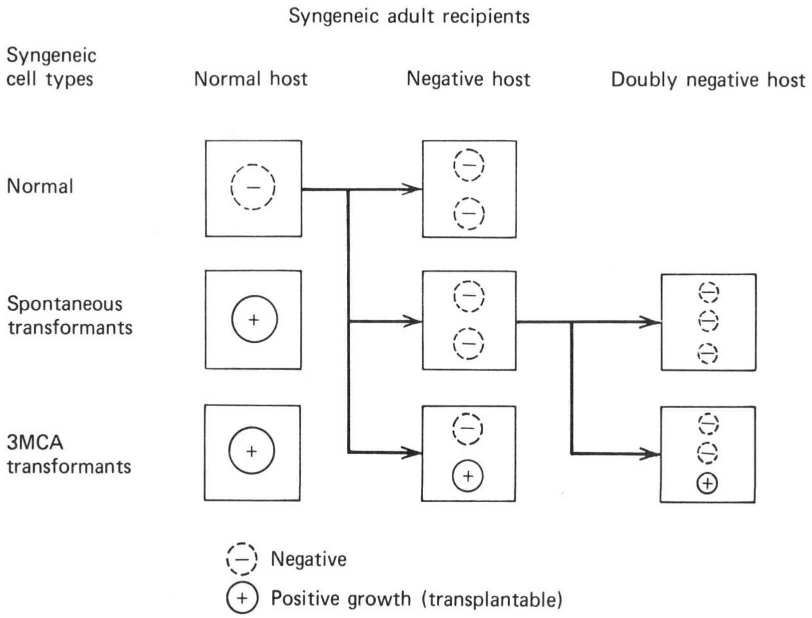

Figure 1.7. Suppression of primary tumor induction by spontaneous transformants following immunization of hosts with syngeneic normal cells. Scheme by which spontaneous transformants can be detected as bearing normallike antigens. This is the procedure most applicable for detecting "false positives" during the course of applying CH embryonal cultures for *in vitro* transformation assays.

C. *In vitro* Activation of the Chondrogenetic DNA in Smooth Muscle Cells

1. Background

Earlier, in Section I, reference was made to "second-generation" cell lines that lend themselves to studies relating to gene activation and extracellular matrices. These promising cell lines reflect the overall experiences gained in the course of routine culturing of normal, benign, and neoplastic tissues in parallel with inbreeding and fixation of the diabetes mellitus, ovarian and uterine (adeno)carcinomas, hepatic cirrhosis and hepatocarcinomas, and mineral oil/pristane-induced granulomatoses. The reason for this dramatic transition from the monotonous series of tough, resilient fibromatous uterine masses and fibrosarcomas to the soft, easily minced (adeno)carcinosarcomas, after reaching the 19th–22nd generation of full-sib matings, cannot be readily deciphered (Fig. 1.5). In particular, the pathology of the mixed Mullarian mesodermal tumor explants which are composed of carcinomatous elements is confusing. The "fibroblastoid" or stromal components are undifferentiated smooth muscle cells (SMC), and not derivatives of the resilient fibrosarcomatous elements that prevailed prior to attaining the higher levels of genic homozygosity. Consequently, cell culture procedures had to be altered to satisfy the propagation of normal, benign, and neoplastic SMC.

We were able to establish two benign granulomatous proliferations, T219 and T233, consisting of SMC, as continous diploid cell lines with high plating efficiencies (65–95%). Unlike the typical "fibroblasts" of lung or embryonal origin, SMC cell lines of granulomatous proliferations are highly stable and clone readily on bacterial (plastic) plates. Moreover, SMC requires less calcium for optimal proliferation. In fact, the T219 cell line proliferates optimally in Eagle's MEM formulated for suspension cultures and supplemented with 5% fetal calf serum. Mouse chondrocytes, for example, tend to float and/or display limited growth and differentiation (Lowe et al., 1978), whereas CH SMC "ball-up" and fail to attach to the substrate, unless Ca^{2+} is reduced. Since many of the growth media were formulated for existing permanent cell lines, it was reasonable to assume that CH SMC and chondroblasts may fare better in medium with reduced concentrations of Ca^{2+}. Currently, Ca^{2+} concentrations of 0–30 mg/L are employed to initiate primary cultures, followed by interim increases to 50–70 mg/L and, later, 120 mg/L after cultures are established.

More recently, the SMC line, T233, derived from a glomuslike granulomatous extension of a cavernous artery of the liver, was shown to respond to the phorbol ester 12-*O*-tetradecanoyl-phorbol-13-acetate (TPA) by converting irreversibly to a chondroblast/cyte. Unlike the temporal differentiation of "fibroblastic" cultures to the adipocyte, chondrocyte, and myotubular forms in response to promoters, growth factors, or purine analogs (Jones and Taylor, 1980; Sager and Kovac, 1982), the chondroblast/cyte phenotype of the T233 cell is irreversible and perpetuated by cell recruitment (data to be presented elsewhere). The incidence of chondroblast/cyte transformation of T233 cells in response to 24 hours of exposure to 1×10^{-8} *M* TPA was estimated to be about 1×10^{-6} or less. Control cell cultures were entirely negative. Fortunately, the high plating efficiency of T233 cells facilitated cloning. The differentiated clones contrasted with the parental forms that exhibit the "sheaves of wheat" morphology characteristic of classic diploid "fibroblasts." While parental clones display multilayering centripetally when grown in media with fetal calf serum, monolayer cultures exhibit contact inhibition and cells fail to survive when in suspension beyond several hours. In contrast, chondroblast transformants, as clonal proliferations or monolayers, remain sessile and multilayered, lack contact inhibition, survive as "floaters," and exhibit a progressive assembly of extracellular (metachromatic) matrix. Moreover, the mitotic apparatus of the chondroblast phenotype is rotated 90° so that the plane of cytokinesis is along the short axis of the cell. This orientation reflects the "stacks of coins" assembly that chondrocytes assume in epiphyseal cartilage.

The low incidence of chondroblast/cyte transformation is increased by recruitment of the entire population of parent SMC, in the course of a single passage. The phenomenon can be readily followed by seeding parent SMC cultures with 10–25 "inducer"chondroblasts. Contact between the two cell forms results in the retraction of cytoplasmic processes of the parental SMC. As transformed cells divide, the reoriented spindle apparatus gives rise to mirror-imaged daughter cells. In turn, newly transformed chondroblasts participate in the recruitment process until each culture is transformed within 5–7 days postseeding.

Primary embryonal and uterine SMC cultures respond even more rapidly to the recruitment phenomenon when seeding sparsely with inducer cells. Aggregates of embryonal chondrocytes or limb buds in minced fragments virtually explode into mitotic activity. Sheets of trypsin-digested embryonal derivatives featuring tight junctions, transform to bipolar perichondriallike cells, one-by-one, in the absence of mitosis. Here, too, entire cultures are transformed to the chondroblast/cyte phenotype within 10 days. Cloning and chromosome sexing ensure the isolation of new sublines. Female "inducer" cells are employed to transform male derivatives, and vice versa. The recruitment process can be perpetuated indefinitely by seeding fresh explants as well as near-terminal cultures. Upon assuming the chondrogenic phenotype, the cells are immortalized, and animal inoculations have yet to produce a tumor.

Biochemical characterizations of the associated extracellular matrices of parent "fibroblasts" and derived chondroblasts are scheduled shortly. In the event the type I and III collagens, suspected in association with the fibroblast phenotype, are replaced by type II collagen and associated cartilage-associated proteoglycans, the observed phenotypic manifestations will have been substantiated.

2. *Plausible Relationships and Reflections*

a. Chondrogenesis and Bone Repair. Mesenchymal (perivascular) cells are recruited at wound sites to undergo the osteogenetic (reparative) pathway of postnatal development. Cell recruitment is initiated in response to bone morphogenetic protein (BMP) binding to the membrane receptors of competent mesenchymal cells to activate a gene-regulator molecule and the chondroosteogenetic DNA (Ham, 1974; Urist et al., 1983). *In vitro* studies have generally relied on the competency of embryonal myoblasts of the mouse and rat to transform into cartilage in response to the BMP of demineralized bone matrix (Nogami and Urist, 1974; Nathanson and Hay, 1980a,b). With BMP defined as a trypsin-sensitive, noncollagenous protein that persists in demineralized bone matrix, the genesis of this elusive growth determinant *in vitro* from nonosteogenetic cells has yet to be demonstrated. Our limited experiences with SMC cultures have disclosed that the equivalent of the *in situ* response to BMP (Urist et al., 1983) is readily and predictably duplicated *in vitro*. In all instances to date, cell recruitment and gene activation have been induced by the sparse use of "inducer" cells. In addition, the developmental signal appears to reside permanently as a cell-transmissible membrane factor of chondroblasts derived from T233 cells, embryonal derivatives,, and uterine SMC. The latter affinity may explain, in part, why the chondrogenetic phenotype is irreversible. In addition, activation of the chondroosteogenetic DNA can be initiated and sustained *in vitro* without the use of bone matrix. Thus, this new cultured CH cell system promises to be a useful model for studying the process of mesenchymal cell differentiation.

b. Neoplastic Transformation. The phenotypic manifestations of chondrogenic transformation parallel those assigned to cells transformed by oncogenic agents (see Chapters 26 and 27). Loss of contact inhibition, multilayering, and the substrate independence of neoplastic cells are shared by nonneoplastic chondroblasts/cytes derived from SMC and embryonal "fibroblasts." The major difference between malignantly transformed cells and chondrogenic transformation is the absence of "inducer" properties among agent-transformed cells.

Presently, some five to six genes are known to code for the various collagen types (Eyre, 1980), and cartilage-forming cells feature a 10-fold increase in protein synthesis over typical fibroblasts (Von der Mark and Conrad, 1979). The likelihood that some of the genes activated in the course of neoplastic progression may be related to growth-promoting factors has gained support recently. Doolittle et al. (1983) have presented amino acid sequencing data which suggest that the Simian sarcoma virus *onc* gene, v-sis, is derived from the gene(s) that encode for platelet-derived growth factor (PDGF). Similarly, the viral oncogene v-*erb*-B is closely related to the epidermal growth factor (EGF) receptor (Downward et al., 1984). While growth stimulation is featured by *in vitro* oncogenic transformants, it is nevertheless accompanied by a loss of contact inhibition, substrate independence, and various cell membrane alterations also associated with exposure to tumor promoters and cell differentiation per se (Hynes, 1976; Colburn, 1980; and Horowitz & Weinstein, 1983). The immortality of chondrogenetic transformants is best illustrated by the manner in which near-terminal fibroblast cultures can be "salvaged" following the addition of a few "inducer" cells and, later, cloning new (sexed) transformants. Thus, "fibroblastoid" cell types can attain "instant" immortality as chondrogenetic transformants. The extent to which nonmesodermal derivatives may respond in this manner remains to be tested.

As a consequence of the above findings, we now have a better understanding of the events that took place when uterine tumors suddenly became soft and adenomatous during inbreeding. More than likely, the involvement of SMC as stromal elements was accompanied by a striking increase in the amounts of type III collagen over the type I collagen suspected of the earlier and more resilient primary fibromas and fibrosarcomas. The histopathology of mixed mesodermal Mullarian uterine tumors have included instances where the suspected predominance of type III collagen formed extensive myxoid patterns and unmineralized osteoid (Brownstein and Brooks, 1980).

In summary, these studies demonstrate the latent competency of particular undifferentiated mesenchymal derivatives to express activated segments of the genome at elevated levels. More than likely, particular collagen genes are not subject to regulation or repression when activated *in vitro* and the stimulus is removed. The absence of revertants to the fibroblast/SMC phenotype further suggests that the *in vitro* environment (5% CO_2 in air) is highly supportive of chondroblast proliferation and hypertrophy of chondrocytes. Accumulation of embryonal-like hyaline cartilage is disrupted during trypsin passaging and routine media changing. However, progression to osteogenic differentiation is expected to be fostered by single cell clonings; exposure of colonies to supplements, such as ascorbic acid and alpha-ketoglutarate; elevations in glycine and lysine; and providing atmospheric conditions with 30–50% O_2.

D. Cytomegalovirus-Associated Benign Granulomatous Derivatives

Primary explants of benign granulomatous proliferations of SMC origin adapt readily as permanent or immortalized mesenchymallike cell lines. Classic diploidy is exceptionally stable, and single cell plating efficiencies range from 65 to 95%. Primary cultures of two granulomatous proliferations, T89 (Yerganian, Nell, Cho, Hayford, and Ho, 1968) and T219, yielded rare isolates of the cytomegalovirus of the Chinese hamster (CHCMV) by the third week of explantation (Fig. 1.8). The frequency of *in vitro* activation of CHCMV is estimated to be one for every 300 normal, benign, and neoplastic cell explants. Despite the progressive cytopathic effects (CPE), trypsin passaging of primary explants resulted in sparse subcultures and a brief remission of CPE. Cyclic cell proliferation and CPE characterized the resulting continuous cell lines. This pattern facilitated the isolation of T219 carrier-state (cyclic) and nonproducer (reinfectable) clonal sublines. Activation of cytomegalovirus by means of allogeneic stimulations, described earlier for murine bone marrow lymphocyte cocultures (Olding et al., 1975), was applied to assess the inactive status of the CHCMV genome in nonproducer T219 cells. Cocultures of T219 and T233 SMC failed to result in CPE. In con-

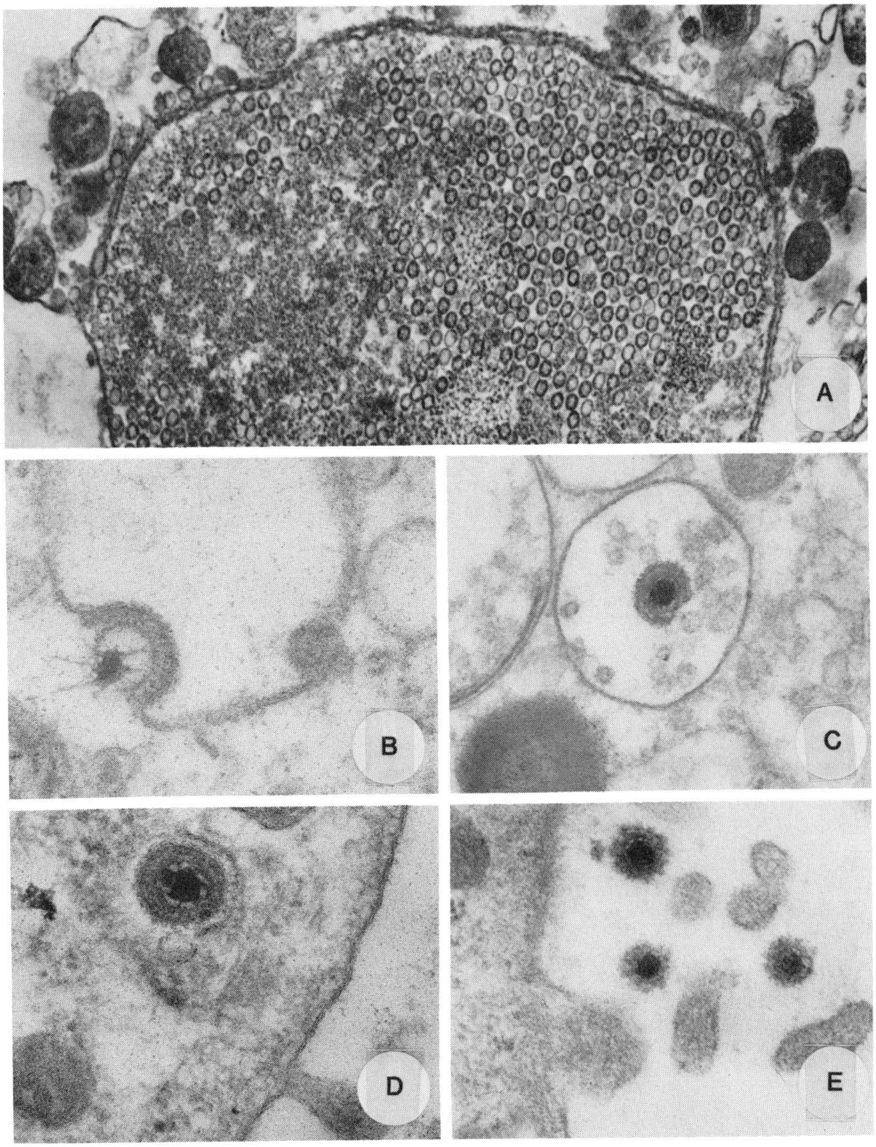

Figure 1.8. Ultrastructural features of the Chinese hamster cytomegalovirus (CHCMV). (A) Accumulation of incomplete virion within nucleus, and relatively few particles in the cytoplasm adjacent to nuclear (doubled) membrane. (B), (C), and (D). Assembly of complete virus in association with cytoplasmic vesicles. (E) Infectious virus release from the cell surface membrane.

trast, cocultures of T219 cells and T233 chondroblast transformants resulted in an intense and sustained CPE limited to T219 cells, leaving T233 chondroblasts to flourish as nonlytic producers of infectious CHCMV.

Independent ultrastructural studies conducted by Drs. A. Krishan and B. Hampar disclosed cell-associated virions and capsids in the nucleus and cytoplasm. Intranuclear particles are double membraned, and have an average diameter of 840 Å (Fig. 1.8A), while cytoplasmic elements may have electron-dense nucleoids and an average diameter of 1400 Å, due to additional membranes acquired when associating with vesicles (Fig. 1.8B,C,D), and as cell-free infectious particles (Fig. 1.8E).

Infectious virus has a strong tendency to remain cell-associated in the parent cell lines and infected embryo cultures. CPE is limited to Chinese, Armenian and Syrian hamster and monkey (BSC-1) cells (in decreasing order of sensitivity). Mouse and human cells are totally resistant to CHCMV CPE. Karyotypes of chronically infected CH cell lines that undergo cyclic CPE remained classic diploid during several years of continuous propagation.

VI. NUCLEOLAR ORGANIZING REGIONS

The ammoniacal-silver-staining procedure, employed to reveal reticulin fibers in connective tissue, also stains the chromosomal localization of ribosomal DNA, synaptonemal complexes, and the meiotic X and Y axes (Goodpasture and Bloom 1975; Dresser and Moses, 1979). In spermatocytes and cultured cells, the silver-stained bodies correspond to the nucleolar organizing regions (NORs) positioned terminally on the long arms of chromosomes 3,4,5,7,8, and 9. Similar bodies are revealed with the use of Coomasie Blue (Wang and Jurrlink, 1979). Earlier studies (Yerganian Griffen, Ho, and Nell, 1968) revealed the NORs of the CH participated in sharing in the formation of common nucleoli. The latter is evidenced by active sites remaining aligned, end-to-end, to form telomeric associations (TAs) that persist in colchicine-arrested metaphases. Different patterns of TAs characterize classic diploid cell lines of differing experimental backgrounds. Prior to the advent of chromosome banding techniques (Fig. 1.9), localization of the suspected NORs was depicted as seen in Figure 1.10, with chromosome types 8 and 9 pooled as one locus for purposes of constructing the pictorial scheme of Figure 1.11 to illustrate cell line differences. Normal embryonal classic diploid cell lines exhibit persisting TAs that involve the NORs preferentially in the 3:7, 7:7 and 1:8–9 pairings (Fig. 1.11B). Primary–secondary embryonal cultures transformed by the small-plaque polyoma virus featured an altered TA pattern before, as well as after, establishing clonal sublines (Fig. 1.11C,D). The 7:7 type of TA featured in normal embryonal cells

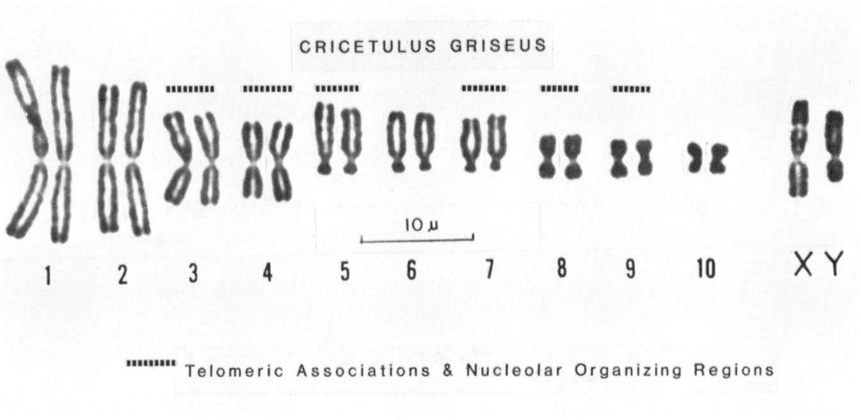

Figure 1.9. Idiogram of classic diploid male karyotype (quinacrine mustard—UV fluorescence).

was replaced by the acquisition (activation?) of the 3:9 pairing. Adult-derived normal fibroblastic cell lines, such as DON and DEDE, differ in both the numbers and distribution of NORs (Goodpasture and Bloom, 1975). Fibrosarcomas of adult origin featuring the classic karyotype also display different TA patterns. For example, T171 (Fig. 1.11E) featured a prominent 3:4 TA, and chromosome 7 sharing nucleolar formations with chromosomes 3, 4, and 9. The presence of 3:4 TAs suggests activation of the latter NOR, if not an amplification of the ribosomal RNA production at this site. Tumor T173 was secondarily exposed to SV40 virus and featured an exaggerated incidence of the 7:7 TA pattern that is noted at a lesser degree for embryonal cells. The distinctive TA patterns are suspected of reflecting cell types of origin. In the case of normal epithelial cell lines, such as DP13, the TA pattern is a mirror image of that

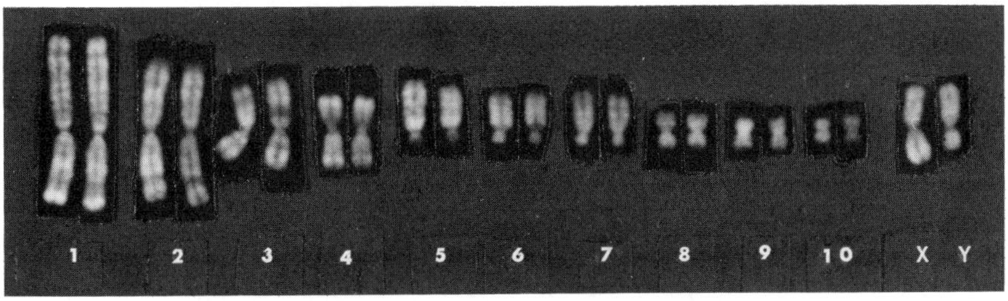

Figure 1.10. Idiogram of classic diploid male karyotype. Telomeric associations (TAs) and nucleolar organizing regions (NORs) at ends of long arms indicated (acetocarmine, phase contrast).

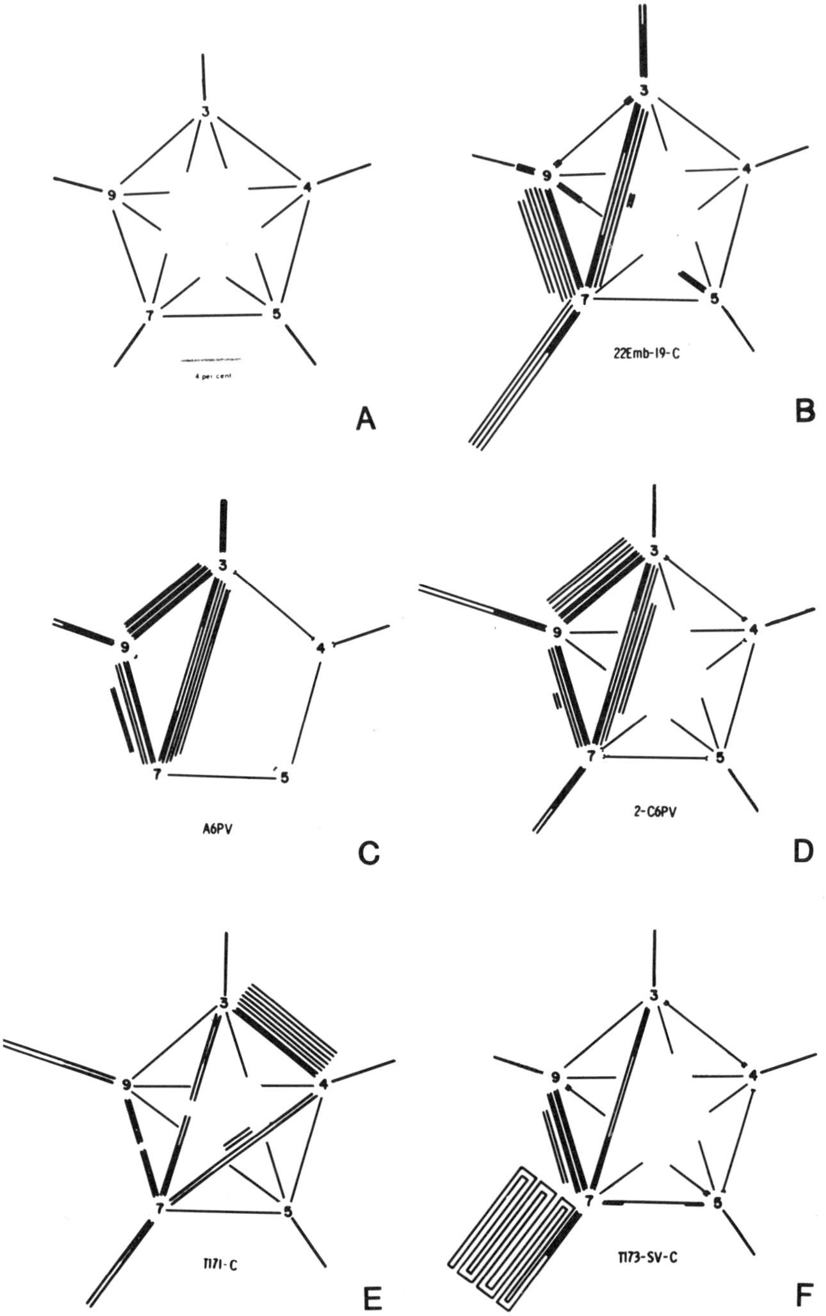

Figure 1.11. Patterns of persisting telomeric associations (TAs) among classic diploid cell lines of different backgrounds. Graphic representation of the distribution of intrachromosomal (3:3, 7:7, etc.) and interchromosomal (3:7, 7:9, etc.) TAs. (A) Configuration representing NORs of chromosomes 3,4,5,7, and 8–9. Relative distributions of TAs among classic diploid cell lines: (B) normal embryonal fibroblasts; (C) polyoma virus-transformed clonal subline, A6PV; (D) same as (C) following two rounds of single cell cloning; (E) spontaneous uterine fibrosarcoma, T171, and (F) spontaneous uterine fibrosarcoma, T173, transformed secondarily with SV40 virus.

noted for embryonal fibroblasts, namely, the 3:3, 3:4, 3:5,4:5, and 5:5 pairings dominate (now shown).

The stabilizing nonrandom patterns TAs suggested the possibility that different nucleolar genes were repressed, activated, of amplified, with excess ribosomal RNAs reflected as persisting TAs. An accurate assessment of the latter necessitates uniform or governed proliferation rates for all cell lines. Slowly proliferating cells or an extended G_1 maximizes TAs; whereas rapid proliferation is generally accompanied by fewer TAs. Nevertheless, persisting TAs may reflect an amplification of the associated NORs. The above discussion has not taken into account the NOR at the secondary constriction of the long arm of the X chromosome, nor the prospects of activating NOR in colchicine-induced micronuclei (Phillips and Phillips, 1969).

VII. TRENDS RELATIVE TO MEIOSIS, EARLY EMBRYONAL DEVELOPMENT, AND FUTURE GENERATION RISK ASSESSMENTS

Technical aspects for preparing meiotic divisions have been advanced considerably. Preparations of spermatogonial and spermatocyte metaphases by procedures outlined by Hsu et al. (1979) maximize the numbers of metaphases isolated from testicular material. Sugawara and Mikamo (1980) have applied Kamiguchi's oocyte preparation technique most effectively for scoring colchicine-induced meiotic nondisjunction in primary oocytes of normally cycling female CH. However, an average yield of only 7.5 eggs per virgin female employed when 5–9 months of age earmarks the Japanese testing procedure as a costly one.

Earlier attempts to superovulate immature and mature female CH and AH, (Pickworth et al., 1968) were inconsistent. The time intervals for administering pregnant mare's serum (PMS) and human chorionic gonadotropin (hCG) were the same (48–56 hours) as those employed successfully by others for mice, rats, and Syrians hamsters.

Recently, Yerganian et al. (unpublished) have successfully applied superovulation of immature and mature female CH to assess tubulin-interacting properties of test compounds (colchicine, vincristine, sodium nitrite, Benomyl, MBC, *p*-fluorophenylalanine, cadmium chloride, and DES). Superovulation was achieved routinely, and predictably, by administering 20 IU PMS and 20 IU hCG in concert with the corresponding endogenous surges of follicle-stimulating (FSH) and luteinizing (LH) hormones of the four-day estrus cycle. This entails a 72–76-hour PMS–hCG time interval, which can be scheduled conveniently to meet personal and experimental needs. PMS is administered to cycling females (multiparous and nonparous) at metestrus [morning-after females are usually set with males for mating (Yerganian, 1958)], followed 72–76 hours later by administering hCG (late afternoon of proestrus, prior to normally setting females with males to mate overnight). Sacrifices to isolate and process oocytes in metaphase II are conducted some 20–22 hours after administering hCG. Immature (2-month-old) or noncyclic females are processed in the same manner. Immature and mature females yield 20–70 superovulated ova. The use of immature females in conjunction with the Kamiguchi oocyte technique results in a very efficient, cost effective assay for detecting dyad nondisjunction resulting from tubulin-interacting properties of non-mutagens and noncarcinogens.

The CH, with its readily identified 11 chromosome types, can now be applied most effectively in conjunction with superovulation for routine use in genetic toxicological assays relating to future generation risks. Current and projected applications include (1) nondisjunction in MII oocytes; (2) embryo culture (Basler and Rohrborn, 1977); (3) activation of human sperm following penetration of zona-free ova for karyotyping (Rudak et al., 1978; Martin et al., 1982); and (4) gene transfer assays that include embryo reimplantations for phenotypic manifestations. Presently, comparative tests using prepubertal Syrian hamsters are underway to determine if the follicular maturation associated with the 72–76-hour PMS–hCG time interval may be more favorable than 56-hour-derived superovulated ova for crypreservation (zona-intact) and penetration of thawed (zona-free) ova by heterologous (CH and human) sperm.

VIII. DISCUSSION

Undifferentiated cells persist as cell lines, usually without tissue-specific markers, whereas cell lines from tumors may retain specific differentiated characters, both functional and morphological. Cell lines derived from different normal organs are similar in morphology. Endothelial cells and pericytes of small blood vessels are the two predominant cell types of all mouse cultures (Franks and Cooper, 1972). Similarly, these authors noted that the ultrastructural features of cell strains from human embryonal lung were vasoformative in origin. The same may be said for cell lines derived from the CH. Nevertheless, the pericyte is a relatively undifferentiated mesenchymal cell and is capable of differentiating to SMC.

Recently, Michttinen et al. (1983) demonstrated the SMC component of glomus tumor cells to differ from other types of SMC; it lacks desmin and instead features vimentin, the fibroblast-type of intermediate filament associated with vascular SMC. In addition, normal vascular and glomus SMC feature a continuous basal lamina (laminin): the major glycoprotein constituent of muscle cells, adipocytes, and the basal aspects of endothelial cells. Presently, absence of the muscle-type intermediate filament, composed of desmin, serves to distinguish vascular SMC from similar elements comprising the stroma of tissues and organs.

We suspect this cytoskeletal difference among SMC derivations is implicated in the failure of T219 (adult stromal SMC) to assume the irreversible chondroblast/cyte phenotype. The ease with which embryonal "fibroblasts" and myoblasts transform irreversibly to the chondroblast/cyte phenotype following contact with "inducer" T233 cells suggests that a "vimentin" cytoskeletal makeup may predominate. Current reproducible examples of competency are (a) the chondroosteogenic pathway of perivascular cells in response to BMP *in situ* and (b) the readily activated expression of chondrogenetic DNA following cell–cell interactions with T233 "inducer" cells *in vitro*. A similar consideration may yet be extended to relating stability and efficiency of host cells in expressing DNA after chromosome transfection/transformation. The extent to which well-recognized karyological entities, such as NORs and TAs, may serve as meaningful markers for cell differentiation remains to be determined.

In retrospect, earlier cell lines, such as the undifferentiated V79 and CHO cells, which are featured in many of the other chapters in this volume, were derived from noninbred animals. Inbred-derived cell and tumor lines have, for the most part, stemmed from activities of this laboratory. The advent of SMC involvement among uterine tumors necessitated a series of minor adjustments in cell culture media formulations. Fibroblasts and SMC derivatives, notably from benign tumors and granulomatous reactions, proliferate readily, retain the classic diploid status, and display elevated plating efficiencies. In light of recent observations, the rare isolations of CHCMV from primary explants of granulomatous proliferations suggest mesenchymal elements can harbor, activate, and disseminate CHCMV systemically without cell lysis or exhibiting nuclear inclusion bodies *in situ*. In addition, one cannot rule out the possibility that the expression of early CMV antigens in pathognomic SMC is an expression of the competency of such cells to readily activate cellular DNA in response to a spectrum of endogenous and exogenous signals.

The historic relationship of inbreeding the CH and the presentation of type I juvenile diabetes; its clinically-oriented metabolical, pathological, and vascular manifestations; and the mixed cellularity of uterine Mullarian tumors tend to focus on the uniqueness of vascular SMC. In recognition of the rapid loss of differentiated properties during the early passages of mesenchymal derivatives, the SMC derivatives of benign tumors and granulomatous proliferations are excellent sources of permanent cell lines that retain the competency of undifferentiated mesenchyme.

ACKNOWLEDGMENTS

The author expresses his sincerest gratitude to Professor Hilton A. Salhanick for the use of his laboratory. Supported in part by Grant 1R43 HD 18529-01 from the National Institute of Child Health and Human Development, Grant CA 25772 from the National Cancer Institute, EPA Contracts 68-01-6321 and 68-01-3732, and the Newton-Wellesley In Vitro Fertilization Laboratories, Inc.

REFERENCES

Allen, J. A. (1932). In *Natural History of Central Asia. Part II. The Mammals of China and Mongolia*, American Museum of Natural History, New York.

Argyropulo, A. (1933). *Z. Saugetierk* **8**, 129–149.

Baker, J. R., Mason, M. M., Yerganian, G., Weisburger, E. K., and Weisburger, J. H. (1974). *Proc. Soc. Exptl. Biol. Med.* **146**, 291–293.

Basler, A. and Rohrborn, G. (1977). *Mutation Res.* **42**, 373–378.

Benjamin, S. A. and Brooks, A. L. (1977). *Vet. Pathol.* **14**, 449–462.

Biedler, J. L. and Spengler, B. A. (1976). *J. Nat. Cancer Inst.* **57**, 683–695.

Brownstein, D. G. and Brooks, A. L. (1980). *J. Nat. Cancer Inst.* **64**, 1209–1214.

Colburn, N. H. (1980). In *Carcinogenesis* (T. J. Slaga, ed.), Raven Press, New York, pp. 33–56.

Doolittle, R. R., Hunkapiller, M. W., Devare, S. G., Robbins, K. C., Aronson, S. A., and Antoniades, H. N. (1983). *Science* **221**, 275–277.

Downward, J., Yarden, Y., Mayes, E., Scrace G., Totty, N., Stockwell, P., Ullrich, A., Schlessinger, J., and Waterfield, M. D. (1984). *Nature* **307**, 521–527.

Dresser, M. E. and Moses, M. J. (1980). *Chromosoma* **76**, 1–22.

Ellerman, J. R. (1941). In *The Families and Genera of Living Rodents*, British Museum of Natural History, London.

Ellerman, J. R. and Morrison-Scott, T. C. S. (1951). In *Check List of Paleoartic and Indian Mammals*; British Museum of Natural History, London.

Eyre, D. R. (1980). *Science* **207**, 1315–1322.

Franks, L. M. and Cooper, T. W. (1972). *Inst. J. Cancer* **9**, 19–29.

Gamperl, R., Vistorin, G., and Rosenkranz, W. (1976). *Chromosoma* **55**, 259–265.

Gerritsen, G. C. (1982). *Diabetes* **31**, 14–23.

Goodpasture, C. and Bloom, C. E. (1975). *Chromosoma* **53**, 37–50.

Ham, A. W. (1974). *Histology*, 7th ed., Lippincott, Philadelphia.

Horowitz, A. D. and Weinstein, I. B. (1983). In *Growth and Maturation Factors* (G. Guroff, ed.), Wiley, New York, pp. 155–191.

Hsu, T. C., Elder, F., and Pathak, S. (1979). *Environmental Mutagenesis* **1**, 291–294.

Hynes, R. O. (1976). *Biochem. Biophys. Acta* **458**, 73–107.

Jones, P. A. and Taylor, S. M. (1980). *Cell* **20**, 85–94.

Kirkland, D. J. (1976). *Br. J. Cancer* **34**, 134–144.

Kirkland, D. J. and Venitt, S. (1976). *Br. J. Cancer* **34**, 145–152.

Kuttner, A. G. and Wang, S. H. (1934). *J. Exptl. Med.* **60**, 773–792.

Lavappa, K. S. (1974). *Lab. Anim. Sci.* **24**, 817–819.

Lavappa, K. S. (1977). *Cytologia* **42**, 65–72.

Lavappa, K. S. and Yerganian, G. (1970). *Exptl. Cell Res.* **61**, 159–172.

Lowe, M. E., Pacifici, M., and Holtzer, H. (1978). *Cancer Res.* **38**, 2350–2356.

Martin, R. H., Lin, C. C., Balkan, W., and Burns, K. (1982). *Am. J. Human Genetics* **34**, 459–468.

Matthey, R. (1952). *Chromosoma* **5**, 113–138.

Meier, H. and Yerganian, G. (1959). *Proc. Soc. Exptl. Biol. Med.* **100**, 810–815.

Meier, H. and Yerganian, G. (1961). *Diabetes* **10**, 12–18.

Michttinen, M., Lehto, V.-P., and Virtanen, I. (1983). *Virchow Archiv B* **43**, 139–149.

Nathanson, M. A. and Hay, E. D. (1980a). *Develop. Biol.* **78**, 301–331.

Nathanson, M. A. and Hay, E. D. (1980b). *Develop. Biol.* **78**, 332–351.

Nogami, H. and Urist, M. R. (1974). *J. Cell Biol.* **62**, 510–519.

Olding, L. B., Jensen, F. C., and Oldstone, M. B. A. (1975). *J. Exp. Med.* **141**, 561–572.

Papadopoulo, D., et al. (1977). *Br. J. Cancer* **36**, 65–71.

Phillips, S. G. and Phillips, D. M. (1969). *J. Cell Biol.* **40**, 248–268.

Pickworth, S., Yerganian, G., and Chang, M. C. (1968). *Anat. Rec.* **162**, 197–207.

Poel, W. E. and Yerganian, G. (1961). *Am. J. Med.* **31**, 861–863.

Pontecorvo, G. (1943). *Proc. Roy. Soc. Edin. B* **62**, 32–42.

Radjabli, S. I. and Kriukova, E. P. (1973). *Tsitologiya* **15**, 1527–1531.

Resznik, G., Mohr, U., and Kmoch, N. (1976). *Br. J. Cancer* **33**, 411–418.

Rudak, E., Jacobs, P. A., and Yanagimachi, R. (1978). *Nature* **274**, 911–913.

Sachs, L. (1952). *Heredity* **6**, 357–363.

Sager, R. and Kovac, P. (1982). *Proc. Natl. Acad. Sci.* **79**, 480–484.

Sugawara, S. and Mikamo, K. (1980). *Jpn. J. Human Genet.* **25**, 235–240.

Urist, M. R., DeLange, R. J., and Finerman, G. A. M. (1983). *Science* **220**, 680–686.

Vistorin, G., Gamperl, R., and Rosenkranz, W. (1976). *Z. Saugertierk* **41**, 342–348.

Vistorin, G., Gamperl, R., and Rosenkranz, W. (1977). *Cytogenet. Cell Genet.* **18**, 24–32.

Von Der Mark, K. and Conrad, G. (1979). *Clin. Orthopaedics* **139**, 185–205.

Wang, H. C. and Jurrlink, B. H. J. (1979). *Chromosoma* **75**, 327–332.

Yerganian, G. (1952). *Genetics* **37**, 638.

Yerganian, G. (1958). *J. Nat. Cancer Inst.* **20**, 705–727.

Yerganian, G. (1964). In *Aetiology of Diabetes Mellitus and Its Complications* (M. P. Cameron and M. O'Connor, eds.), J. and A. Churchill, London, pp. 25–41.

Yerganian, G. (1967). In *UFAW Handbook on the Care and Management of Laboratory Animals*, 3rd ed., (W. Lane-Petter et al., eds.) E. and S. Livingstone, Ltd., Edinburgh, pp. 340–352.

Yerganian, G. (1972). In *Pathology of Hamsters*, Prog. Exptl. Tumor Res. Res. (F. Homburger, ed.), S. Kargel, Basel, pp. 2–41.

Yerganian, G. (1979). In *Inbred and Genetically Defined Strains of Laboratory Animals, Part 2* (P. L. Altman and D. D. Katz, eds.), FASEB, Bethesda, Maryland, pp. 500–504.

Yerganian, G., Freeman, A., and Gagnon, H. J. (1968). *J. Cell Biol.* **39**, 146a.

Yerganian, G., Griffin, R., Ho, T., and Nell, M. (1968). *Genetics* **60**, 240–241.

Yerganian, G. and Lavappa, K. S. (1971). In *Chemical Mutagens* (A. Hollaender, ed.), Plenum, New York, pp. 387–409.

Yerganian, G. and Leonard, M. J. (1961). *Science* **133**, 1600–1601.

Yerganian, G., Nell, M. A., Cho, S. S., Hayford, A. H., and Ho, T. (1968). *Nat. Cancer Inst. Monogr.* **29**, 241–268.

Yerganian, G., Paika, I., Gagnon, H. J., and Battaglino, A. (1979). *Progr. Exp. Tumor Res.* **24**, 424–434.

Yerganian, G. and Papoyan, S. (1964). *Hereditas* **52**, 307–319.

CHAPTER 2

DEVELOPMENT OF THE CHINESE HAMSTER OVARY (CHO) CELL FOR USE IN SOMATIC CELL GENETICS

Theodore T. Puck

The Lita Annenberg Hazen laboratory for the Study
of Human Development
The Florence R. Sabin Laboratories for Genetic
and Developmental Medicine
The Eleanor Roosevelt Institute for Cancer Research
and the Department of Biochemistry, Biophysics,
and Genetics
and the Department of Medicine of the University
of Colorado Health Sciences Center
Denver, Colorado

I. INTRODUCTION

Somatic cell genetics was conceived as a new genetic discipline designed to overcome the limitations posed by application of classical genetic procedures to humans. These limitations involve the large generation time of humans (approximately 25 years as opposed to several weeks in the mouse, and 20 minutes in *Escherichia coli*), and the fact that it is impossible to carry out human matings designed to answer definitively specific questions in human genetics. The procedure involved taking minute cell samples, preferably from the skin from any human subject; reliable growth of such biopsies into large populations by the methods of tissue culture; and devising a technique whereby a monodisperse suspension of such cells could be quantitatively and conveniently grown into isolated clonal colonies in a petri dish so that mutant clones could readily be recognized, picked, and established into new genetic stocks. Such mutants could then be used as markers to illuminate genetic mechanisms (Puck and Marcus, 1955; Puck et al., 1956; Puck and Kao, 1982). These procedures permitted the concepts of microbial genetics to be applied to mammalian somatic cells.

A cell most suited for somatic cell genetic studies with particular references to human genetics should have the following characteristics: easy and reliable preparation of stable cell cultures from biopsies; a small and reasonably stable set of chromosomes; ability of single cells to grow rapidly and quantitatively into discrete, recognizable colonies, both on solid surfaces and in nutrient supension; ability to grow in reasonably well-defined media containing a minimal amount of serum components; ability to hybridize readily with human cells; and an immunological structure readily differentiable from that of human cells. Therefore, after having established the practicality of somatic cell genetics as a discipline, we prepared cell cultures from a variety of animals (Tjio and Puck, 1958) and studied their usefulness for genetic studies.

In 1957, Dr. George Yerganian of the Boston Children's Cancer Research Foundation provided us with a female Chinese hamster, which was one of a colony of outbred animals that he was breeding (see Chapter 1, Section II B). We removed an ovary, trypsinized 0. lg of tissue, and established a cell culture. The cells which grew out were of a fibroblast-like morphology. Unlike cells taken from biopsies of various human and opposum tissues, those from the Chinese hamster failed to show the strictly euploid chromosomal constitution. The chromosomal variations observed after the cell line had become established were small but definite. Thus, approximately 1% of the cell population displayed chromosome numbers of 21 or 23, instead of the normal number of 22. This small degree of chromosomal variation has persisted (see Chapter 5) and may have contributed by selection to the ultimate change of the culture to the transformed state. Fibroblastic cellular and colonial morphologies were maintained for more than 10 months. At some unknown time thereafter, the change in morphology and other properties became the dominant characteristic of the culture as a whole. Subsequent recloning yielded the culture which is used today (see Appendix I for pedigrees of Chinese hamster ovary cell lines in current use).

The Chinese hamster ovary (CHO) cells derived from this original explant have been adopted for use in a large number of other laboratories, as evidenced by many of the chapters in this volume. This widespread use has made it possible to pool the information about growth, mutation, hybridization, and other properties obtained in various laboratories and so expedite progress in somatic cell genetics (Puck and Kao, 1982).

II. GROWTH CHARACTERISTICS, NUTRITIONAL REQUIREMENTS, AND CHROMOSOMAL CONSTITUTION

The CHO cell is a hardy cell which readily lends itself without trauma to the various procedures required for its biopsy and cultivation *in vitro* and dispersal into single cell suspension for large-scale clonal growth. It can be grown with a generation time of 10–11 hours in standard media, and readily yields a plating efficiency in the neighborhood of 100%. While a variety of different media can be used for its cultivation as single cells, the one used routinely by us consists of the F12 medium (Ham, 1965) (Table 2.1) supplemented with fetal calf serum in a final concentration of 4–8% to increase its growth rate. Where a somewhat more defined molecular environment is necessary, the fetal calf serum can be replaced by its dialyzed macromolecular fraction. Single cells can still be grown with high efficiency into colonies in plastic dishes when the concentration of the macromolecular fraction of fetal calf serum has been reduced to as little as 0.5–1%. Newborn calf serum, horse serum, human serum, and mixtures thereof can also be used as serum supplements to obtain reliable growth of the cell. The CHO cell was demonstrated to be a proline auxotroph (Ham, 1963; Kao and Puck, 1967; Baich, 1977).

The use of serum-free media for growth of CHO cells has been actively pursued in this and other laboratories. The CHO cell lends itself

TABLE 2.1
Composition of Medium F12

Stock	Compound	Concentration (*M*/L) in F12
12-1 (100×)[a]	L-Arginine HCl	1.0×10^{-3}
	Choline chloride	1.0×10^{-4}
	L-Histidine HCl	1.0×10^{-4}
	L-Isoleucine	3.0×10^{-5}
	L-Leucine	1.0×10^{-4}
	L-Lysine HCl	2.0×10^{-4}
	L-Methionine	3.0×10^{-5}
	L-Phenylalanine	3.0×10^{-5}
	L-Serine	1.0×10^{-4}
	L-Threonine	1.0×10^{-4}
	L-Tryptophane	1.0×10^{-5}
	L-Tyrosine	3.0×10^{-5}
	L-Valine	1.0×10^{-4}
12-2 (100×)	Biotin	3.0×10^{-8}
	Calcium pantothenate	1.0×10^{-6}
	Niacinamide	3.0×10^{-7}
	Pyridoxine HCl	3.0×10^{-7}

Table 2.1 (*Continued*)

Stock	Compound	Concentration(M/L) in F12
	Thiamin HC1	1.0×10^{-6}
	KC1	3.0×10^{-3}
12-3 (100 ×)	Folic acid	3.0×10^{-6}
	$Na_2HP0_4 \cdot 7H_20$	1.0×10^{-3}
12-4 (100 ×)[b]	$FeS0_4 \cdot 7H_20$	3.0×10^{-6}
	$MgC1_2 \cdot 6H_20$	6.0×10^{-4}
	$CaC1_2 \cdot 2H_20$	3.0×10^{-4}
12-5 (1000 ×)	Phenol red	3.3×10^{-6}
	Glucose	1.0×10^{-2}
	L-Glutamine	1.0×10^{-3}
	Riboflavin	1.0×10^{-7}
	Sodium pyruvate	1.0×10^{-3}
12-7 (100 ×)[c]	L-Cysteine HC1	1.0×10^{-4}
12-8 (100 ×)	L-Proline	3.0×10^{-4}
	Putrescine dihydrochloride[h]	1.0×10^{-6}
	Vitamin B_{12}	1.0×10^{-6}
12-9 (100 ×)	L-Alanine	1.0×10^{-4}
	L-Aspartic acid	1.0×10^{-4}
	L-Glutamic acid	1.0×10^{-4}
	Glycine	1.0×10^{-4}
12-10 (100 ×)	Hypoxanthine	3.0×10^{-5}
	myo-Inositol	1.0×10^{-4}
	Lipoic acid	1.0×10^{-6}
	Thymidine	3.0×10^{-6}
	$CuS0_4 \cdot 5H_20$	1.0×10^{-8}
12-11 (1000 ×)	$ZnS0_4 \cdot 7H_20$	3.0×10^{-6}[g]
No stock[d]	NaC1	1.3×10^{-1}
No stock[e]	$NaHC0_3$	1.4×10^{-2}
12-12 (1000 ×)[f]	Linoleic acid	3.0×10^{-7}

[a] Freshly thawed solution 12-1 frequently contains a precipitate that will redissolve on gentle warming.

[b] One drop concentrated HC1 is added to each 100 mL of stock 12-4 to prevent precipitation.

[c] Stock 12-7 must be discarded if it contains a precipitate.

[d] Solid sodium chloride is carefully weighed and added directly to the medium.

[e] The medium is neutralized to pH 7.2–7.4 (orange by phenol red indicator) with 1.0 *N* NaOH before the sodium bicarbonate is added.

[f] Stock 12-12 is prepared in absolute ethanol and is normally not added until immediately before the medium is to be used. Stock 12-12 is stored at −20°C. Stock 12-12 should be prepared fresh at least once every month.

[g] The concentration of zinc sulfate is reduced to 5×10^{-7} *M* in medium F12M.

[h] Spermidine HC1 is equivalent to putrescine in serum-free systems (Chasin et al., 1974), and was used in most of the experiments reported in this chapter. However, the toxicity of spermidine in the presence of serum proteins (Hamiton and Ham, 1977) has just been confirmed in this laboratory. Therefore, the use of putrescine rather than spermidine in F12 is recommended.

fairly well for cultivation in media with greatly reduced serum or even no serum (Hamilton and Ham, 1977). More detailed information about growth of CHO cells under a variety of culture conditions is given in Chapter 6.

The chromosomal constitution of the transformed CHO cell was demonstrated to contain a reasonably constant karyotype in which, however, were included chromosomes of abnormal constitution (see Chapter 5). Some of these are readily relatable to the normal Chinese hamster chromosomes. Others which had undergone extensive rearrangement were arbitrarily designated by us as the Z group (Tjio and Puck, 1958). Recently, detailed characterization of these chromosomes has been carried out (Stallings and Siciliano, 1980) and is reported in Chapter 5. Dr. Kao selected for special study the subclone CHO-K1 in the spontaneously transformed CHO culture because it had a modal chromosome number of 21. Many of the mutant cells whose isolation and characterization are reported in this volume are direct descendants of this cell line. Other CHO lines are derived from our original cultures, which were sent to Dr. Tobey at Los Alamos National Laboratory (see Appendix I).

The role of the cytoskeleton in affecting chromosomal stability of the CHO cell was first demonstrated by Cox and Puck (1969) who found that nondisjunctional effects occur at high frequency in daughter cells at mitosis if the microtubular structure is only slightly damaged. Colcemid added to CHO cells in concentrations of approximately $^{1}/_{100}$th–$^{1}/_{50}$th that usually used to block cells in metaphase, almost completely destroyed the orderliness of distribution of the mitotic chromosomes between the two daughter cells resulting in an exceedingly large range of chromosome numbers in the resulting progeny cells. Recent studies by Cabral and Gottesman, and by Ling, reported in Chapter 22, indicate that colcemid-resistant mutants can be isolated, and these mutants carry alterations in one of the subunits of the microtubule, β-tubulin.

III. AUXOTROPHIC AND REGULATORY MUTANTS OF CHO

In the history of microbial genetics one of the most useful families of mutants has been the auxotrophs, that is, mutants in which an extra nutritional requirement has been acquired by the cell. Usually such mutants have suffered loss of a particular step of a biosynthetic pathway and can be used to delineate the pathway. An effective and convenient method for securing auxotrophic mutants of the CHO cell was developed (Puck and Kao, 1967; Kao and Puck, 1968) by adapting to mammalian cells the principle underlying the penicillin method of mutant selection in microbial genetics. This procedure, the BrdU–visible-light methodology, utilizes the principle that incorporation of BrdU into DNA does not impair its genetic functions, but shifts its 260-nm absorption band toward the region of visible light. Therefore, cells that have multiplied in the presence of BrdU incorporate that compound into their DNA and are selectively photolyzed by means of visible light. Thus, to prepare mutants requiring any given nutrilite for growth, a mutagenized population is grown in the presence of the minimal nutritional medium to which BrdU is also added. The cell population is then exposed to visible light in an intensity sufficient to destroy only those cells that have reproduced, incorporating BrdU in their DNA, leaving behind as potentially viable cells auxotrophs for which the previous medium was insufficient to support growth. The desired metabolite is then added to the nutrient medium, and only the specifically desired auxotrophic cells will grow into discrete colonies. These can be picked and established as stable, mutant stocks (Fig. 2.1). This method has made pos-

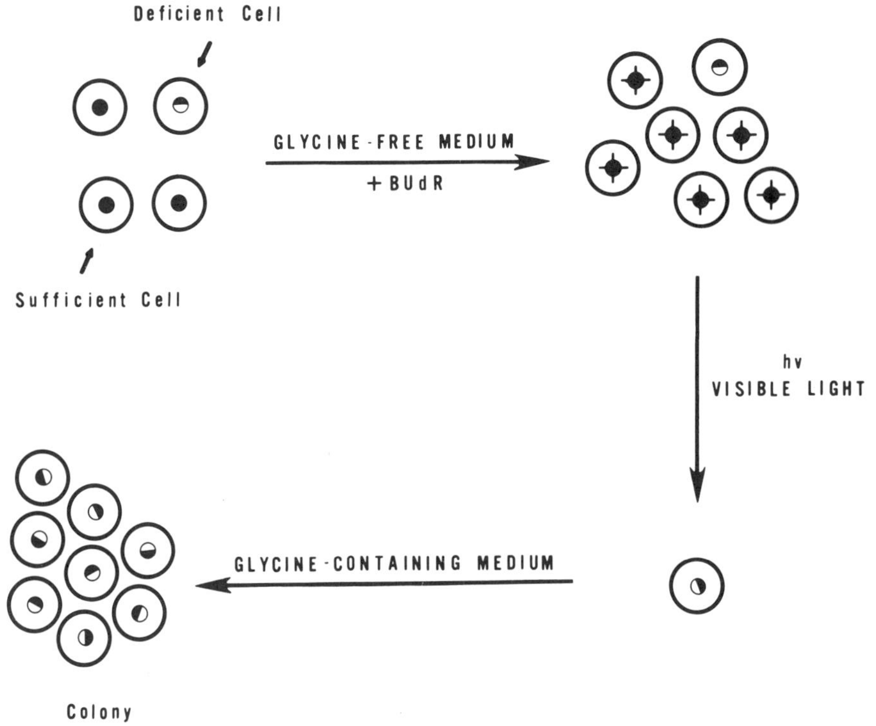

Figure 2.1. Schematic representation of the BrdU–visible-light technique for isolation of nutritionally deficient mutant clones. The mutated cell population is exposed to BrdU in a deficient medium in which only the normal, wild-type cells can grow. These alone incorporate BrdU into their DNA and are killed on subsequent exposure to a standard fluorescent lamp. The medium is then changed to a composition lacking BrdU but enriched with the specific nutrient which would permit the desired auxotrophic cells to grow into colonies.

sible preparation of scores of different auxotrophic mutants of the CHO cell that have been used as markers for elucidation of biochemical pathways and other genetic experiments (Table 2.2) (Patterson et al., 1974; Patterson and Carnright, 1977; Hankinson, 1976; Patterson, this volume, Chapter 11). Many of these auxotrophic mutants are described in greater detail elsewhere in this volume. It has been brought to the author's attention that a BrdU–visible-light method was described previously and independently for use in bacteria (Bonhoeffer and Schaller, 1965).

A typical example of an auxotrophic CHO cell ($gly^{-}A$) produced by this method is shown in Fig. 2.2. Other mutants and mutational isolation systems in the CHO cell have been described (Table 2.2; Appendix II). A particularly interesting set of mutants with respect to transport velocity has been described by Moffett et al. (1983).

Early in the development of somatic cell genetics, criteria were described by which one could decide with reasonable confidence whether a given CHO auxotroph was or was not a single gene mutation (Kao and Puck, 1975). The development of single gene complementation analysis in mammalian cells made possible confirmation of the single gene nature of particular mutations as well as genetic–biochemical analysis of a variety of pathways (Kao et al., 1969). These complementation techniques are described in detail by Patterson in Chapter 11.

Another powerful experimental tool developed in these laboratories is the use of replica plating for isolation of CHO mutants (Stamato and Hohman, 1975; Stamato and Jones, 1977). This procedure is necessary for mutants in which demonstration of the possession of an altered genetic constitution requires killing of the cells. This development, which again followed concepts familiar from microbial genetics, has made possible isolation of a number of important mutants among which are UV-resistant mutants that have contributed in important ways to elucidation of mammalian DNA repair mechanisms (Stamato et al., 1981). Recently, Raetz et al. (1982) have utilized a polyester fiber for replica plating and achieved close to 100% faithful reproduction of CHO clones in as many as four replicas formed at the same time.

A particularly interesting principle for the production of regulatory mutants in the CHO cell was developed by Dr. Sinensky (1978) of our laboratories. Consider that it is desired to isolate a mutant that has lost regulation of the biosynthesis of a particular metabolite, for example, cholesterol. Such a mutant will continue to synthesize cholesterol even when adequate amounts of this substance are provided in the nutrient medium. Sinensky's procedure involves screening of a series of chemical analogs of cholesterol to obtain one which cannot be used by the cell for membrane synthesis, but which is sufficiently similar to cholesterol so that its presence induces the regulatory system to shut off biosynthesis of this metabolite. 25-hydroxy-cholesterol functions in this

Table 2.2
Auxotrophic Mutants in CHO and Chinese Hamster Cells[a]

Parent	Mutant	Nutritional Requirement	Enzyme Defect	References
CHO	pro^-	Proline or Δ^1-pyrroline-5-carboxylic acid	Defective in converting glutamic acid to glutamic γ-semialdehyde	Ham (1963); Jones (1975); Kao and Puck (1967)
CHO-K1/pro^-	gly^-A	Glycine	Serine hydroxymethyltransferase	Chasin et al. (1974); Jones et al. (1972); Kao et al. (1969)
	gly^-B	Glycine or folinic acid		Kao et al. (1969)
	gly^-C	Glycine		Kao et al. (1969)
	gly^-D	Glycine		Kao et al. (1969)
	ade^-A	Adenine, hypoxanthine, their ribonucleosides, or ribonucleotides, or 5-aminoimidazole-4-carboxamide	Amidophosphoribosyltransferase	Kao and Puck (1972b); Patterson (1975); Patterson et al. (1974)
	ade^-B	Like ade^-A	Phosphoribosylformylglycinamidine synthetase	Kao (1980); Kao and Puck (1972a, 1972b); Patterson (1975)
	ade^-C	Like ade^-A	Phosphoribosylglycinamide synthetase	Moore et al. (1977); Patterson (1975); Patterson et al. (1974)
	ade^-D	Like ade^-A	Phosphoribosylaminoimidazole carboxylase	Patterson (1975); Patterson et al. (1974)
	ade^-E	Like ade^-A	Phosphoribosylglycinamide formyltransferase	Jones et al. (1981); Patterson (1975); Patterson et al. (1974)
	ade^-F	Like ade^-A except unable to utilize 5-aminoimidazole-4-carboxamide	Phosphoribosylaminoimidazole carboxamide formyltransferase	Patterson (1975); Patterson et al. (1974)

TABLE 2.2 (*Continued*)

Parent	Mutant	Nutritional Requirement	Enzyme Defect	References
CHO-K1/pro^-	ade^-G	Like ade^-A	Phosphoribosylaminoimidazole synthetase	Patterson et al. (1981)
	ade^-H	Adenine	Adenylosuccinate synthetase	Patterson (1976); Waldren (1981)
	ade^-I	Adenine	Adenylosuccinate lyase	Patterson (1976)
	ade^-P_{AB}	Like ade^-A	Amidophosphoribosyltransferase and phosphoribosylformylglycinamidine synthetase, not complementing either ade^-A or ade^-B	Oates et al. (1980)
	Urd^-A	Uridine, orotic acid or dihydroorotic acid	Carbamyl phosphate synthetase, aspartate transcarbamylase, and dihydroorotase	Davidson and Patterson (1979); Davidson et al. (1979); Patterson and Carnright (1977)
	Urd^-B	Slower growth, not requiring uridine	Dihydroorotate dehydrogenase	Taylor et al. (1971); Stamato and Patterson (1979)
	Urd^-C	Uridine	Orotate phosphoribosyltransferase and OMP decarboxylase	Patterson (1980)
	$trans^-$	Unable to synthetize valine, leucine, or isoleucine from its respective α-keto acid	Branched-chain amino acid transaminase	Jones and Moore (1976)
	ala^r	Resistant to inhibition of growth in alanine	Increased transport of proline	Moffet et al. (1983)
	ala^-	Alanine	Alanyl-tRNA synthetase	Hankinson (1976)
	glu^-	Glutamate		Hankinson (1976)
	ino^-	Inositol		Kao and Puck (1970, 1974b, 1975)

TABLE 2.2 (*Continued*)

Parent	Mutant	Nutritional Requirement	Enzyme Defect	References
CHO-K1/pro$^-$	GAT$^-$	Glycine–adenine–thymidine	Folylpolyglutamate synthetase	Jones et al. (1980); Kao and Puck (1969a, 1972b)
	pur$^-$	Like ade$^-$A	Phosphoribosylformylglycinamide synthetase	Feldman and Taylor (1974); Taylor and Hanna (1977)
	tsGAT$^-$	Glycine–adenine–thymidine; auxotrophic at 39.5°C, prototrophic at 34°C		Schroder and Hsie (1975)
	Mutant #49	Unsaturated fatty acid	Microsomal stearoly-CoA desaturase	Chang and Vagelos (1976)
	Mutant #215	Cholesterol	Defective in lanosterol demethylation	Chang and Vagelos (1976)
CHO-K1/ pro$^-$gly$^-$A	ser$^-$	Serine		Jones and Puck (1973)
CHO/pro$^-$	AUXB1/ GAT$^-$	Glycine–adenosine–thymidine	Folypolyglutamate synthetase	McBurney and Whitmore (1974a); Thompson et al. (1973)
	AUXB3/	Glycine–adenosine	Defective in folate metabolism; not complementing AUXB1/GAT$^-$	McBurney and Whitmore (1974a)
CHO/pro$^-$ AUXB1/GAT$^-$	tsAUXB1/ GAT$^-$	Glycine–adenosine–t ymidine; auxotrophic at 38.5°C, prototrophic at 34°C	Defective in folate metabolism at nonpermissive temperature	

(*Continued on next page*)

TABLE 2.2 (*Continued*)

Parent	Mutant	Nutritional Requirement	Enzyme Defect	References
CHO/pro^-	$TSH1/leu^-$	Auxotrophic for leucine at 39.5°C, prototrophic at 34°C	Defective in leucyl-tRNA synthetase at 39.5°C	Farber and Deutschu (1976); Molnar and Rauth (1975); Thompson et al. (1975)
	PSV3/asn	Auxotrophic for asparagine at 39.5°C, prototrophic at 34°C	Defective in aspara-gly-tRNA synthetase at 39.5°C	Thompson et al. (1975)
	arg^-	Inability to utilize citrulline	Argininosuccinate synthetase or argininosuccinase or both	Naylor et al. (1976)
	cys	Inability to utilize cystathionline in place of cystine	Cystathionase	Naylor et al. (1976)
	$Polyamine^-$	Polyamine		Pohjanpelto et al. (1981)
	$Inosine^-$	Unable to grow on inosine as the only carbon source	Purine nucleoside phosphorylase	Hoffee (1979)
	TdR^-	Thymidine deoxycytidine or deoxyuridine	Inability to reduce UDP to dUDP	Meuth et al. (1979)

[a] For methods in isolating auxotrophic mutant cells see Kao and Puck (1974b) and Taylor and Hanna (1977). No distinction has been made between mutant and variant. All the mutants listed have been characterized at least in part as to their specific nutritional requirements, enzymes affected, reverse mutation frequencies, genetic complementation, dominance and recessiveness, gene mapping, etc.

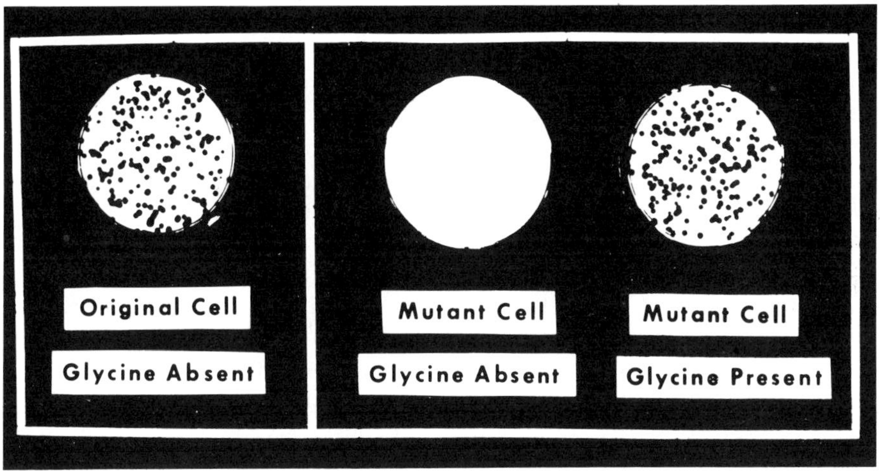

Figure 2.2. Demonstration of the typical all-or-none response of a glycine mutant in the presence and absence of glycine. Such mutants are readily prepared by the BrdU–visible-light procedure.

manner. Therefore, a mutagenized cell population was placed in medium deprived of cholesterol, but containing 25-hydroxy-cholesterol. The normal cells, sensing the presence of the 25-hydroxy derivative, activate the regulatory system to shut off cholesterol synthesis. As a result these cells cannot form colonies. Mutants lacking this regulatory system, however, cannot shut off cholesterol biosynthesis, and so these cells alone form colonies. This system is contributing fundamental understanding of the cholesterol metabolic regulation in mammalian cells (Sinensky, 1977, 1980). Another important instance of genetic regulation in auxotrophic mutants lies in the finding of Harris (1982 and unpublished; see also Chapter 11, Section IV-E and Fig. 11.5) that the proline auxotrophy of the CHO cell can be reversed by 8-azacytidine, demonstrating the regulatory nature of this mutant. This observation also suggests that some auxotrophic mutants in CHO cells may result from methylation of structural genes or regulatory sequences. The experiments of Cox and Puck (1969) with colcemid and Harris (1982) with azacytidine are extremely important in demonstrating that mutants or mutation-like behavior can be produced (or reversed) by agents that do not directly affect the DNA sequence. Implicit in these phenomena may lie the explanation of important differentiation phenomena. Intensive molecular study of these and similar phenomena are needed in order to clarify the full powers of the methylation and other reactions in the expression in mammalian cells of their genetic potential.

IV. HYBRIDIZATION INVOLVING THE CHO CELL

There are two important kinds of hybridizations involving the CHO cell. When mutants of the CHO cell are fused with each other, the resultant hybrid contains the sum of the chromosomes of each parental cell and so constitutes a polyploid cell which is reasonably stable. This kind of hybridization is important in complementation analysis of CHO and other cells (Kao et al., 1969). General priniciples concerning formation of somatic cell hybrids are presented in Chapter 7.

When cells from two different species are hybridized, chromosomes are lost fairly rapidly and often the chromosomes of one species are lost preferentially. We have found that in fusion of a CHO cell with a human cell, the human chromosomes are lost particularly rapidily. Eventually, chromosome stablization occurs after which further chromosome loss may slow down considerably (Puck and Kao, 1970; Puck, 1972).

These features have made it possible to use the auxotrophic mutants to prepare stable hybrids containing single human chromosomes as well as particular combinations of human chromosomes. For example, if one hybridizes the CHO gly A mutant lacking the gene for serine hydroxymethylase, with the human lymphocyte, and places the resulting hybrid population in a medium lacking glycine, the only colonies appearing will be those in which the glycine-deficient CHO cell has been complemented by the human chromosome containing the serine hydroxymethylase gene. After a period of some weeks, cytogenetic analysis of such cultures reveals that many of them contain only the single human chromosome which can be identified as number 12. Thus, one can conclude that the human gene for serine hydroxymethylase resides on human chromosome 12. By this means, a variety of single human chromosome hybrids containing chromosomes 3, 8, 21, as well as combinations like 8 and 14 have been prepared. We presume that it would be possible by selecting the appropriate mutants to obtain hybrids with any desired single human chromosome or particular combinations of chromosomes. Reasonably stable hybrids containing unselected groups of human chromosomes have also been prepared. By means of batteries of such hybrids it has been possible to identify the human chromosome on which particular human genes are contained (Puck and Kao, 1982; Ruddle, 1981; Kao, 1983).

Various human–CHO hybrids have been used in a variety of new biochemical–genetic and immunogenetic approaches, some of which will be illustrated here. The presence of a single human chromosome in a CHO hybrid means that any human gene that it expresses or that can be identified by molecular hybridization with an appropriately prepared DNA or RNA probe automatically makes possible assignment of that gene to its particular human chromosome.

By identifying a series of such markers that are contained on a given human chromosome, one can proceed to mutagenize the hybrid and select clones which have lost particular markers. Cytogenetic analysis of deletion mutant clones makes possible regional mapping of the given marker to a particular segment of the appropriate arm (Kao et al., 1977; Kao et al., 1978; Meisler et al., 1981). Fine-structure mapping is also possible as described in Section VI.

V. IMMUNOGENETIC ASPECTS OF CHO CELLS AND HYBRIDS

A. The Antigenic Specificity of Human and CHO Cell Surface Antigens

In a critical experiment carried out by Oda and Puck (1961), it was shown that immunological reactivity of cell membranes of Chinese hamster and human cells exhibit a high degree of species specificity. Human cells were injected into one set of rabbits and Chinese hamster ovary cells were injected into another set and the separate antisera were collected. It was found that human cells were killed by extremely small concentrations of antihuman cell serum in the presence of complement and Chinese hamster cells were similarly killed at very small concentrations of the antiserum produced in response to themselves. However, virtually no killing of either cell occurred when they were exposed to quite high concentrations of the heterologous antiserum. Such killing was shown to result from cell cytolysis arising from rupture of the membrane. Therefore, neither cell membrane contains an appreciable number of antigens sufficiently similar to those of cells from the other species to cause lysis by the heterologous antiserum. These cell surface antigens have proved to be remarkably useful markers for many kinds of experiments.

We shall discuss here experiments in which these immunological features have been combined with CHO hybridization to provide a number of illuminating experiments that appear applicable to a variety of mammalian systems. We shall only consider antigens which are expressed without specific inducing agents when the appropriate genetic determinants are present.

B. Detection and Identification of Tissue-Specific Human Cell Surface Antigens

One of the most direct experimental approaches used in this system has permitted identification and genetic analysis of human cell surface antigens associated with cells in particular states of differentiation. The hybrid containing the single human chromosome 11 has been adopted as a model system for a variety of studies. When human fibroblasts are used as an antigen for injection into rabbits, an antiserum results which can be tested for killing of a variety of human–CHO hybrids. For example, when this antiserum is added to the hybrid containing only human chromosome 12 in the presence of complement, no killing whatever is observed, even in fairly high concentrations of the antisera. We conclude that the normal human fibroblast contains no cell surface antigen in common with those which human chromosome 12 lends to this particular hybrid in the standard growth medium (Jones et al., 1975). When we add the same antiserum to the hybrid containing chromosome 11 as its sole human chromosomal component, this hybrid is killed to an extent of more than 99% by extremely small concentrations of the antiserum. The further conclusion may then be drawn that human fibroblasts do express cell surface antigens with immunological specificity resembling that which chromosome 11 contributes to its CHO hybrid (Kao et al., 1976). Obviously, this approach can be extended to many tissues. For example, we have found that human chromosome 12 is responsible for contributing to its monochromosome hybrid antigens that cross react with those of the human brain and human kidney (Puck and Nielson, unpublished).

An extension of this approach makes possible resolution into individual antigens of immunogenetic activity associated with a given chromosome. The preceding paragraph described how antigenic activity shared by the membrane of the human fibroblasts could also be demonstrated on the membranes of hybrids containing chromosome 11. To resolve such activities into individual component antigens, human tissues were tested to find cells that contain some but not all of the cell surface antigens resulting from the presence of human chromosome 11. Such tissue cells can readily be identified: When injected into a rabbit, they should produce antisera lethal to the hybrid with human chromosome 11. However, when antiserum against the human fibroblasts (or the 11 monochromosome hybrid, itself) is exhaustively adsorbed with the given human tissue cells, it should not be possible to remove all of the killing activity of that antiserum for the number 11 hybrid. This kind of analysis proved to be simple and effective. For example, antiserum prepared in the rabbit against human red cells is highly lethel to the hybrid with human 11. However, exhaustive adsorption of antiserum against human fibroblasts with red cells leaves behind considerable lethal activity (Wuthier et al., 1973). We have demonstrated that the human red cell membrane contains two antigens, named a_1 and a_3, with genetic loci on human chromosome 11, whereas other cell surface antigens whose genetic loci reside on this chromosome are contained on other human tissues (Jones and Puck, 1977).

A separate antigenic activity, named a_2, was shown to result from a locus on human chromosome 11, but not to be present on human red cells (Jones et al., 1975, 1979).Obviously, this technique offers a method of some generality for identifying tissue-specific cell surface antigens. The a_2 antigen has subsequently been resolved into three different individual antigens (Jones, unpublished). Another approach which separates different antigenic activities due to loci on the same chromosome involves the use of monoclonal antibodies (Epstein and Clevenger, 1985; Jones et al., 1983).

The simplicity, specificity, and selectively of the cell surface antigenic activities makes them ideal as genetic markers so that one can readily prepare mutants that have lost any desired antigen or combination of antigens. Thus, treating a population of human 11 hybrid cells with a mutagenic agent and plating the cells in the presence of a specific antiserum plus complement will yield colonies only from the hybrids which have lost the antigen in question. Cytogenetic examination of such hybrids reveals that many of such mutants have suffered terminal deletions so that it became possible to establish that the loci for antigens 1 and 3 are located on the short arm of chromosome 11, while the locus for the number 2 complex is on the long arm (Jones et al., 1980; Meisler et al., 1981).

It is possible to carry out complementation analysis on such mutants exactly as was previously described for CHO cells in this chapter. Such studies revealed that more than one genetic locus is often required for development of a particular cell membrane activity. Of particular interest was the demonstration that full development of such antigenic activity occasionally requires participation by Chinese hamster genes as well as the genes contained on the human chromosome present in the hybrid cell (Jones et al., 1979, 1980).

VI. REGIONAL AND FINE-STRUCTURE MAPPING OF HUMAN GENES USING RECOMBINANT DNA APPROACHES

By the use of mutagenic agents like X-rays, which produce large numbers of terminal deletions in the human chromosomal constituents of CHO hybrids, and application of cytogenetic analysis to mutants selected for loss of particular antigens, regional mapping of these antigens on the included human chromosome has been carried out (Fig. 2.3) (Jones and Kao, 1978). Such mutants can also be examined for loss of any other human markers that are available for the chromosome in question, and regional mapping of these markers as well as of the immunogenetic ones can be achieved (Gusella at al., 1979). However, regional mapping has a resolving power limited to that achievable by microscopic examination and is far too gross for the kinds of genetic analysis ultimately required for the human genome. Moreover, such

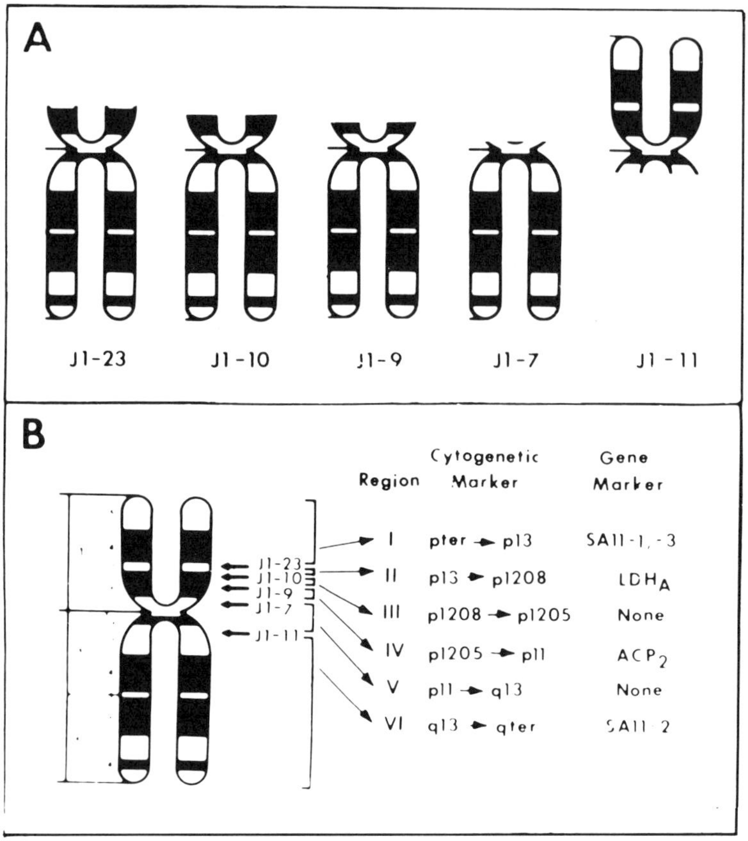

Figure 2.3. (A) Diagram showing various terminal deletions of human chromosome 11 in five clones. (B) Scheme of human chromosome 11; arrows indicate the breakpoints at which terminal deletions occurred in five clones. These five breakpoints divide chromosome 11 into six regions, each of which is characterized by cytogenetic and gene markers as indicated. (SA11-1_1, -3 refers to cell surface antigens a_1 and a_3, while SA11-2 refers to a_2.)

regional mapping yields only geometrical relaionships referred to the mitotic chromosome whereas true genetic mapping must be based on interphase chromosomes whose linkage relationships may depart appreciably from the apparent linearity of the condensed mitotic chromosome.

A new procedure, which has been developed over a period of several years in a collaborative study with Drs. James Gusella and David Housman (Gusella et al., 1979, 1980, 1982), permits fine-structure mapping in any single human chromosome incorporated in the CHO cell as a monochromosome hybrid. An important feature of this procedure involves the demonstation that the repetitive sequences of the CHO cell and the human cell are sufficiently different from each other so as to resist cross-species molecular hybridization under standard experimental condition (Gusella et al., 1982). This makes possible elegant mapping manipulations. The method consists of the following steps:

1. Two markers are identified, one on each arm of the human chromosome in question and preferably as near to the telomere as possible.
2. The hybrid cell population is subjected to mutagenesis under conditions that maximize the yield of terminal deletions, and mutants are selected that have lost one or the other, or both markers.
3. DNA is prepared from the original hybrid and from each of the mutants and each such preparation is subjected to restriction by a convenient nuclease.
4. The resulting preparations from the original hybrid and each of its deletion mutants is subjected to gel electrophoresis so that DNA fragments are sorted according to size.
5. Any convenient human repetitive DNA probe is then applied to the gels by the Southern blot technique. Since these repetitive sequences are highly specific, the moderately repetitive human DNA sequence hybridizes only with human DNA ignoring that of the CHO.
6. The bands that result from the original unmutagenized hybrids are then numbered in order of decreasing molecular weight and a table is constructed in which the presence or absence of each of the bands in the parental hybrids indicated as shown in Table 2.3. The groups I, II, and III indicate mutants which have lost a_1, a_2, or both markers as shown in Table 2.4.
7. The principle of minimal marker loss is then applied so that each mutant is ordered with respect to the length of the deletion it has suffered (Table 2.4). Thus, the various bands and other markers can be arranged with respect to their true chromosomal mapping order.
8. One can achieve any desired degree of resolving power simply by increasing the number of mutant cells selected. It is possible to focus on one particular area of the chromosome by demanding loss of a particular distal marker and retention of a specific proximal marker.

This principle of human gene mapping does not require isolation of messenger RNA; it appears applicable widely and perhaps anywhere throughout the human genome; it is capable at least theoretically of extremely high resolving power; and it is virtually independent of cytogenetics. In addition, it promises to furnish many new cloned DNA sequences.

TABLE 2.3
Presence of Fragments Hybridizing to the Standard Human Probe in Hybrid Cell Lines

Marker or Fragment	J1	J1-1	J1-4B	J1-7	J1-8	J1-10	J1-10[a]	J1-11	J1-21	J1-23	J1-24	J1-27	J1-37	J1-39
1	+	−	−	+	+	+	−	+	+	+	+	+	+	+
2	+	−	−	+	+	+	−	−	+	+	+	+	+	+
3	+	−	−	+	+	+	−	+	+	+	+	+	+	+
4	+	−	−	+	+	+	−	−	+	+	+	+	+	+
5	+	−	−	−	−	−	−	+	+	+	+	−	−	−
6	+	−	−	+	+	+	+	+	+	+	+	+	+	+
7	+	+	+	−	+	+	+	+	+	+	+	+	+	+
8	+	−	−	+	+	+	−	−	+	+	+	+	+	+
9	+	−	+	+	+	+	+	+	+	+	+	+	+	+
10	+	−	−	−	−	−	−	+	−	−	−	−	−	−
11	+	−	−	+	+	+	−	−	+	+	+	+	+	+
12	+	−	−	−	−	−	−	+	+	−	−	−	−	−
14	+	−	−	−	+	−	−	+	+	+	+	−	−	−
15	+	−	−	+	+	+	−	+	+	+	+	+	+	+
16	+	−	+	−	+	−	−	+	+	+	+	−	+	−
20	+	−	−	−	−	−	−	+	−	−	−	−	−	−
22	+	−	−	+	+	+	−	−	+	+	+	+	+	+
23	+	−	−	−	+	−	−	+	+	+	+	−	+	−
24	+	−	−	+	+	+	−	−	+	+	+	+	+	+
LDHA[b]	+	−	−	−	−	−	−	+	+	+	+	−	−	−
β-Globin	+	+	+	−	+	+	+	+	+	+	+	+	+	+
a_1	+	−	−	−	−	−	−	+	−	−	−	−	−	−
a_2	+	−	−	+	+	+	−	−	+	+	+	+	+	+
a_3	+	−	−	−	−	−	−	+	−	−	−	−	−	−
Missing markers, number	0	22	20	12	7	10	20	7	4	5	5	10	8	10

[a] Gusella et al. (1982). [b] Lactate dehydrogenase A.

TABLE 2.4
Rearranged Data of Table 2.2 to Demonstrate Linkage

Marker or Fragment	Group I Mutants										Group II Mutant	Group III Mutants		
	J1	J1-21	J1-24	J1- 3	J1-8	J1-37	J1-39	J1-27	J1-10	J1-7	J1-11	J1-4B	J1-10[a]	J1-1
a_1	+	−	−	−	−	−	−	−	−	−	+	−	−	−
a_2	+	−	−	−	−	−	−	−	−	−	+	−	−	−
10	+	−	−	−	−	−	−	−	−	−	+	−	−	−
20	+	−	−	−	−	−	−	−	−	−	+	−	−	−
12	+	+	−	−	−	−	−	−	−	−	+	−	−	−
5	+	+	+	+	−	−	−	−	−	−	+	−	−	−
LDHA[b]	+	+	+	+	−	−	−	−	−	−	+	−	−	−
14	+	+	+	+	+	−	−	−	−	−	+	−	−	−
23	+	+	+	+	+	+	−	−	−	−	+	−	−	−
15	+	+	+	+	+	+	−	−	−	−	+	+	−	−
7	+	+	+	+	+	+	+	+	+	−	+	+	+	+
β-Globin	+	+	+	+	+	+	+	+	+	−	+	+	+	+
9	+	+	+	+	+	+	+	+	+	+	+	+	+	−
6	+	+	+	+	+	+	+	+	+	+	+	−	+	−
1	+	+	+	+	+	+	+	+	+	+	+	−	−	−
3	+	+	+	+	+	+	+	+	+	+	+	−	−	−
13	+	+	+	+	+	+	+	+	+	+	+	−	−	−
a_2	+	+	+	+	+	+	+	+	+	+	−	−	−	−
2	+	+	+	+	+	+	+	+	+	+	−	−	−	−
4	+	+	+	+	+	+	+	+	+	+	−	−	−	−
5	+	+	+	+	+	+	+	+	+	+	−	−	−	−
11	+	+	+	+	+	+	+	+	+	−	−	−	−	−
22	+	+	+	+	+	+	+	+	+	−	−	−	−	−
34	+	+	+	+	+	+	+	+	+	+	−	−	−	−
Missing markers, number	0	4	5	5	7	8	10	10	10	12	7	20	20	22

[a] Gusella et al. (1982). [b] Lactate dehydrogenase A.

VII. THE REVERSE TRANSFORMATION REACTION IN THE CHO CELL

The CHO cell arose from a fibroblast which became spontaneously transformed in the course of its cultivation in our laboratory. It exhibits all of the stigmata of transformation including the characteristic changes in cell membrane properties, the cellular and colonial morphologies typical of malignant cells, loss of growth regulation as evidenced by its ability to grow with 100% plating efficiency in suspension in low serum concentrations, and the ability to kill nude mice or cheek-pouch-inoculated hamsters by a malignant process. Reverse transformation is the name which we coined to characterize the regaining of normal cell characteristics by a tumor cell treated with specific chemical agents (Hsie and Puck, 1971; Hsie and Waldren, 1972; Puck, 1979). Chapter 23 of this volume is devoted to the manner in which cyclic AMP may bring about this complex phenotypic change. A few features of this reaction which we have characterized will be discussed here.

Most of our studies were carried out with the CHO cell. Some of the major conclusions and theoretical postulates in this devolpment are as follows: (1) Cyclic AMP, presumably through its action in promoting specific phosphorylations (Gottesman et al., 1980), is required for organization of the cell cytoskeleton; (2) the organized cell cytoskeleton makes specific connections with receptor sites and other macromolecules in the cell membrane and also is directly or indirectly linked to specific nuclear structures in a way which makes cyoskeletal integrity necessary for regulation of growth and specific biochemical actions in interphase cells (Rumsby and Puck, 1982); (3) at least part of the regulatory action on cell metabolism and reproduction achieved by cyclic AMP through its mediation of cytoskeletal structure occurs through change in exposure of specific gene regions in interphase cells (Schonberg et al., 1983); (4) distortion of cytoskeletal organization is the underlying damage which causes many transformed cells to display a set of common characteristics including compact morphology, loss of specific cell–cell interactions that produce the typical tightly bound, organized colonial morphology, the presence of rapidly oscillating knob-like structures in the membrane, and a variety of other biochemical changes in the membrane of the transformed cell; (5) the fact that the cytoskeleton in mitosis is normally responsible for precise distribution of the chromosomes to the daughter cells suggests that the karyotype instability of transformed cells may be due to a defect in the cytoskeleton which in turn is responsible for the rapid development of resistance to chemotherapeutic agents by transformed cells and may contribute to their acquisition of invasive properties (Puck, 1979); (6) in addition to producing reverse transformation in a spontaneously transformed cell like the CHO, dibutyryl/cyclic AMP can reverse the malignant characteristics of vole cells transformed by introduction of the *Src* oncogene. Presumably, then, the phosphorylating action of cyclic AMP can compete with the abnormal phosphorylation due to the *Src* oncogene and overcome the cytoskeletal damage induced by the latter (Puck et al., 1981). Paradoxically, CHO cells transformed by RSV are relatively resistant to the growth-inhibitory effects of 8-Br–cyclic AMP (Roth et al., 1982). This is thought to be due to the direct stimulatory effect of cyclic-AMP-dependent phosphorylation of the protein product of the *Src* oncogene (Roth et al., 1983; Chapter 26). (7) CHO mutants resistant to the reverse transformation action of cyclic AMP have been produced. Their analysis promises to unlock the specific steps of the pathway leading to regulation of cell reproduction (Hsie and Puck, 1972; Gabrielson et al., 1982); and (8) these considerations lead to the prediction that cancer cells whose growth regulation is blocked in a step other than that affected by cyclic AMP may yield reverse transforming actions through the influence of other agents. Such actions have been found (Moore, unpublished).

VIII. DETECTION AND QUANTITATION OF MUTAGENESIS USING CHO-K1 AND ITS HYBRIDS

From the very beginning, the CHO cell has lent itself to the detection of mutagenic agents. The first mammalian system for the demonstration of single gene mutagenesis by a variety of agents utilized the CHO cell and afforded both qualitative and crudely quantitative methods for measuring mutation in specific genes (Kao and Puck, 1968, 1969).

The need to be able quantitatively to measure mutagenic effects of environmental agents has become critical in modern industrial society. Mutagenesis is a necesary step in production of genetic disease, cancer, and a variety of teratogenic processes. The pioneering work of Ames (Ames et al., 1975) in developing a microbial test for measuring mutagenic effects of externally applied agents constituted a major milestone in this field. A number of different mutagenesis tests utilizing the Ames principle have been developed with mammalian cells and have demonstrated the ability to detect the presence of externally applied mutagens (Ames et al., 1975; Ames, 1979).

Careful consideration of these various procedures, however, revealed many if not most of them to be unable effectively to detect deletions or chromosomal nondisjunctional events. Such tests involve a selected marker gene which can exist in alternative allelic forms. Cells are exposed to suspected mutagens and the relative number of colonies that have lost the marker is considered to be a measure of the amount of mutagenesis which has resulted from the given exposure. However, in most such tests the marker gene is contained on a chromosome which also contains other genes necessary for cell reproduction. Therefore, while point mutations are readily detected, deletions involving any of these other essential genes cause death of the cell in question so that it is prevented from making its contribution to the mutational score. But deletion and chromosomal loss (which accompanies a corresponding gain when nondisjunction occurs) are exceedingly important causes of disease in humans so that failure to score these particular processes constitutes a serious limitation in any system designed to assess health threats of environmental agents.

Hybrids of the CHO cell containing single human chromosomes have been used to remedy this deficiency, and a new system has been developed for monitoring of environmental mutagenicity, which appears to offer a more comprehensive approach (Waldren et al., 1979; Waldren et al., 1983). The principle adopted utilizes marker genes carried on a chromosome unnecessary for cell reproduction. Obviously, such a chromosome is furnished by the human monochromosome hybrids of the CHO cell. In early experiments, markers identified on long and short arms of human chromosome 11 were used for these experiments. These immunogenetic markers make it possible to identify point mutations, small deletions, large deletions, and loss of all or most of the chromosome. This system also shows a more sensitive response to mutagenic agents than any of the other mammalian cell systems previously proposed. In its original form, the system had a drawback due to the fact that the hybrid culture employed contained an appreciable background level of cells that had already lost the human chromosome. It has now been possible to reduce this background level to negligible proportions by the use of flow fluorometry to reject cells which have spontaneously lost the marker chromosome. This system appears able to score quantitatively mutagenic effects of X-irradiation in the range of 10 rad and may be capable of still further increases in sensitivity.

IX. SUMMARY

The CHO cell has made possible simple and rapid somatic cell genetic studies in many fields including gene identification, synteny determination, analysis of biochemical pathways, radiation biology, immunogenetics, gene mapping, cancer, and measurement of mutagenesis. It has established somatic cell genetics on a firm basis making possible rapid and reliable studies of a wide range of genetic phenomena in somatic cells *in vitro*. The results of such varied studies have established the validity of a thesis that could not be accepted without demonstrable proof: For large areas of genetics the results of classical and somatic cell genetics are mutually consistent. In the period that lies ahead, this cell promises to aid in unlocking major features of the biochemical genetics of differentiation. These advances in somatic cell genetics combined with those of recombinant DNA technology and new developments in protein chemistry and molecular immunology promise great new achievements in solving human health problems. The sharing of these achievements by all mankind requires the establishment of peace throughout the world.

ACKNOWLEDGMENTS

This is contribution number 478 of the Eleanor Roosevelt Institute for Cancer Research. This work was supported by grants from NIH HD02080, Reynolds Industries, and Monsanto Company. Dr. Theodore T. Puck is a Research Professor of the American Cancer Society.

REFERENCES

Ames, B. N. (1979). *Science* **204**, 587–593.

Ames, B. N., McCann, J. E., and Yamaski, E. (1975). *Mut. Res.* **31**, 347–364.

Baich, A. (1977). *Somat. Cell Gen.* **3**, 529–538.

Bonhoeffer, F., Schaller, H. (1965). *Biochem. Biophys. Res. Comm.* **20**, 93–96.

Chang, T. Y., Telakowski, C., Heuel, M. V., Alberts, A. W., and Vagelos, P. R. (1977). *Proc. Natl. Acad. Sci. USA* **74**, 832–836.

Chang, T. Y. and Vagelos, P. R. (1976). *Proc. Natl. Acad. Sci. U.S.A.* **73**, 24–28.

Chasin, L. A., Feldman, A., Konstam, M., and Urlaub, G. (1974). *Proc. Natl. Acad. Sci. USA* **71**, 718–722.

Cox, D. M. and Puck, T. T. (1969). *Cytogen.* **8**, 158–169.

Davidson, J. N., Carnright, D. V., and Patterson, D. (1979). *Somat. Cell Gen.* **5**, 175–191.

Davidson, J. N. and Patterson, D. (1979). *Proc. Natl. Acad. Sci. USA* **76**, 1731–1735.

Epstein, A. L. and Cleavenger, C. (1985). in *Adv. Non-Hist. Pro. Res.* (I. Bekhor, ed.) CRC Press, in press.

Farber, R. A. and Deutschu, M. P. (1976). *Somat. Cell Gen.* **2**, 509–520.

Feldman, R. I. and Taylor, M. W. (1974). *Bioc. Gen.* **12**, 393–405.

Gabrielson, E. G., Scoggin, S. H., and Puck, T. T. (1982). *Exp. Cell. Res.* **142**, 63–68.

Gottesman, M. M., LeCam, A., Bukowski, M., and Pastan, I (1980). *Somat. Cell Gen.* **6**, 45–61.

Gusella, J. F., Jones, C., Kao, F. T., Housman, D., and Puck, T. T. (1982).*Proc. Natl. Acad. Sci USA* **79**, 7804–7808.

Gusella, J. F., Keys, C., Varsanyi-Breiner, A., Kao, F. T., Jones, C., Puck, T. T., and Housman, D. (1980). *Proc. Natl. Acad. Sci. USA* **77**, 2829–2833.

Gusella, J. F., Varsanyi-Breiner, A., Kao, F. T., Jones, C., Puck, T. T., Keys, C., Orkin, S., and Housman, D. (1979). *Proc. Natl. Acad. Sci. USA* **76**, 5239–5243.

Ham, R. G. (1963). *Exp. Cell. Res.* **29**, 515–526.

Ham, R. G. (1965). *Proc. Natl. Acad. Sci. USA* **53**, 288–293.

Hamilton, W. G. and Ham, R. G. (1977). *In Vitro* **13**, 537–547.

Hankinson, O. (1976). *Somat. Cell Gen.* **2**, 497–507.

Harris, M. (1982). *Cell* **29**, 483–492.

Hoffee, P. (1979). *Somat. Cell Gen.* **5**, 319–328.

Hsie, A. W. and Puck, T. T. (1971). *Proc. Natl. Acad. Sci. USA* **68**, 358–361.

Hsie, A. W. and Puck, T. T. (1972). *J. Cell Biol.* **55**, 118a.

Hsie, A. W. and Waldren, C. A. (1972). *J. Cell Biol.* **47**, 922.

Jones, C. (1975). *Somat. Cell Gen.* **1**, 345–354.

Jones, C. and Kao, F. T. (1978). *Hum. Gen.* **45**, 1–10.

Jones, C., Kao, F. T., and Taylor, R. T. (1980). *Cytogen. Cell. Gen.* **28**, 181–194.

Jones, C., Kimmel, K. A., Carey, T. E., Miller, Y. E., Lehman, D. W., and Mackenzie, D. (1983). *Somat. Cell Gen.* **9**, 489–496.

Jones, C. and Moore, E. E. (1976). *Somat. Cell Gen.* **2**, 235–243.

Jones, C., Moore, E. E., and Lehman, D. W. (1979). *Proc. Natl. Acad. Sci. USA* **76**, 6491–6495.

Jones, C., Moore, E. E., and Lehman, D. W. (1980). *Adv. Path.* **7**, 309–317.

Jones, C., Patterson, D., and Kao, F. T. (1981). *Somat. Cell Gen.* **7**, 399–409.

Jones, C. and Puck, T. T. (1973). *J. Cell. Phys.* **81**, 299–304.

Jones, C. and Puck, T. T. (1977). *Somat. Cell Gen.* **3**, 407–420.

Jones, C., Wuthier, P., Kao, F. T., and Puck, T. T. (1972). *J. Cell. Phys.* **8**, 291–298.

Jones, C., Wuthier, P., and Puck, T. T. (1975). *Somat. Cell Gen.* **1**, 235–246.

Kao, F. T. (1980). *J. Cell. Biol.* **87**, 291a.

Kao, F. T. (1983). *Intl. Rev. Cytology* **85**, 169–146.

Kao, F. T., Chasin, L., and Puck, T. T. (1969). *Proc. Natl. Acad. Sci. USA* **64**, 1284–1288.

Kao, F. T., Jones, C., Law, M. L., and Puck, T. T. (1978). *Cytogen Cell. Gen.* **22**, 474–477.

Kao, F. T., Jones, C., and Puck, T. T. (1976). *Proc. Natl. Acad. Sci. U.S.A.* **73**, 193–197.

Kao, F. T., Jones, C., and Puck, T. T. (1977). *Somat. Cell Gen.* **3**, 421–429.

Kao, F. T. and Puck, T. T. (1967). *Genetics* **55**, 513–524.

Kao, F. T. and Puck, T. T. (1968). *Proc. Natl. Acad. Sci. USA* **60**, 1275–1281.

Kao, F. T. and Puck, T. T. (1969). *J. Cell. Phys.* **74**, 245–258.

Kao, F. T. and Puck, T. T. (1970). *Nature* **228**, 329–332.

Kao, F. T. and Puck, T. T. (1972a). *Proc. Natl. Acad. Sci. USA* **69**, 3273–3277.

Kao, F. T. and Puck, T. T. (1972b). *J. Cell. Phys.* **80**, 41–50.

Kao, F. T. and Puck, T. T. (1974a). *Bioc. Gen.* **12**, 393–405.

Kao, F. T. and Puck, T. T. (1974b). In *Methods in Cell Biology* (D.M. Prescott, ed.), Academic Press, New York, Vol. 8, pp. 23–29.

Kao, F. T. and Puck, T. T. (1975). *Genetics* **79**, 343–352.

McBurney, M. W. and Whitmore, G. F. (1974a). *Cell* **2**, 173–182.

McBurney, M. W. and Whitmore, G. F. (1974b). *Cell* **2**, 183–188.

Meisler, M. H., Wanner, L., Kao, F. T., and Jones, C. (1981). *Cytogen. Cell. Gen.* **31**, 124–128.

Meuth, M. Trudel, M., and Siminovitch, L. (1979). *Somat. Cell Gen.* **5**, 303–518.

Moffet, J., Curriden, S., Ertsey, R., Mendiaz, E., and Englesberg, E. (1983). *Somat. Cell. Gen.* **9**, 189–213.

Molnar, S. J. and Rauth, A. M. (1975). *J. Cell. Phys.* **85**, 173–178.

Moore, E. E., Jones, C., Kao, F. T., and Oates, D. C. (1977). *Am. J. Hum. Gen.* **29**, 389–396.

Naylor, S. L., Busby, L. L., and Klebe, R. J. (1976). *Somat. Cell Gen.* **2**, 93–111.

Oates, D. C., Vannais, D., and Patterson, D. (1980). *Cell* **20**, 797–805.

Oda, M. and Puck, T. T. (1961). *J. Exp. Med.* **113**, 599–610.

Patterson, D. (1975). *Somat. Cell Gen.* **1**, 91–110.

Patterson, D. (1976). *Somat. Cell Gen.* **2**, 189–203.

Patterson, D. (1980). *Somat. Cell Gen.* **6**, 101–114.

Patterson, D. and Carnright, D. V. (1977). *Somat. Cell Gen.* **3**, 483–495.

Patterson, D., Graw, S., and Jones, C. (1981). *Proc. Natl. Acad. Sci. USA* **78**, 405–409.

Patterson, D., Kao, F. T., and Puck, T. T. (1974). *Proc. Natl. Acad. Sci. USA* **71**, 2057–2061.

Pohjanpelto, P., Virtanen, I., and Holtta, E. (1981). *Nature* **293**, 475–478.

Puck, T. T. (1972). The Mammalian Cell as a Microorganism: Genetic and Biochemical Studies *in vitro*, Holden-Day, San Francisco.

Puck, T. T. (1979). *Somat. Cell Gen.* **5**, 973–990.

Puck, T. T., Erikson, R. L., Meek, W. D., and Nielson, S. E. (1981). *J. Cell. Phys.* **107**, 399–412.

Puck, T. T. and Kao, F. T. (1967). *Proc. Natl. Acad. Sci. USA* **58**, 1227–1234.

Puck, T. T. and Kao, F. T. (1982). *Ann. Rev. Gen.* **16**, 225–271.

Puck, T. T. and Marcus, P. I. (1955). *Proc. Natl. Acad. Sci. USA* **41**, 432–437.

Puck, T. T., Marcus, P. I., and Cieciura, S. J. (1956). *J. Exp. Med.* **103**, 273–283.

Raetz, C., Wermuth, M. M., McIntyre, T. M., Esko, J., and Wing, D. (1982). *Proc. Natl. Acad. Sci. USA* **79**, 3223–3227.

Roth, C., Pastan, I., and Gottesman, M. M. (1982). *J. Cell. Physiol.* **111**, 42–48.

Roth, C., Richert, N., Pastan, I., and Gottesman, M. M. (1983). *J. Biol. Chem.* **258**, 10768–10773.

Ruddle, F. H. (1981). *Nature* **294**, 115–120.

Rumsby, G. and Puck, T. T. (1982). *J. Cell. Physiol.* **111**, 133–138.

Schonberg, S., Patterson, D., and Puck, T. T. (1983). *Exp. Cell Res.* **145**, 57–62.

Schroder, C. H. and Hsie, A. W. (1975). *Exp. Cell. Res.* **9**, 170–174.

Sinensky, M. (1977). *Biochem. Biophys. Res. Comm.* **78**, 863–867.

Sinensky, M. (1978). *Proc. Natl. Acad. Sci. USA* **75**, 1247–1249.

Sinensky, M. (1980). *J. Cell Biol.* **85**, 166–169.

Stallings, R. L. and Siciliano, M. J. (1983). In *Isozymes: Current Topics in Biological and Medical Research*, Alan R. Liss, New York, Vol. 10, pp. 313–321.

Stamato, T. D., Hinkle, L., Collins, A. R. S., and Waldren, C. A. (1981). *Somat. Cell Gen.* **7**, 307–320.

Stamato, T. D. and Hohmann, K. (1975). *Cytogen. Cell Gen.* **15**, 372–379.

Stamato, T. D. and Jones, C. (1977). *Somat. Cell Gen.* **3**, 639–647.

Stamato, T. D. and Patterson, D. (1979). *J. Cell. Phys.* **98**, 459–468.

Taylor, R. T. and Hanna, M. L. (1977). *Arch. Bioc. Biop.* **181**, 331–344.

Taylor, M. W., Souhrada, M., and McCall, J. (1971). *Science* **172**, 162–163.

Thompson, L. H. and Baker, R. M. (1973). In *Methods in Cell Biology*, (D. M. Prescott, ed) **6**, 299–315, Academic Press, New York.

Thompson, L. H., Harkins, J. L., and Stanners, C. P. (1973). *Proc. Natl. Acad. Sci. USA* **70**, 3094–3098.

Thompson, L. H., Stanners, C. P., and Siminovitch, L. (1975). *Somat. Cell Gen.* **1**, 187–208.

Tjio, J. H. and Puck, T. T. (1958). *J. Exp. Med.* **108**, 259.

Tu, A. and Patterson, D. (1977). *Bioc. Gen.* **15**, 195–210.

Waldren, C. A., Jones, C., and Puck, T. T. (1979). *Proc. Natl. Acad. Sci. USA* **76**, 1358–1362.

Waldren, C. A., Puck, T. T., and Cram, S. (1983). *Env. Mut.* **5**, Cd-6.

Wuthier, P. Jones, C., and Puck, T. T. (1973). *J. Exp. Med.* **138**, 229–244.

CHAPTER 3

THE ESTABLISHMENT OF THE V79 CHINESE HAMSTER LUNG CELL LINE

Ernest H.Y. Chu
Department of Human Genetics
University of Michigan Medical School
Ann Arbor, Michigan

I. THE ADVANTAGES OF CELL CULTURES FOR GENETIC STUDIES

Experimental studies with higher eukaryotes have been aided greatly by the development of cell culture techniques. Under the controlled *in vitro* conditions, each individual somatic cell of a multicellular organism becomes an independent entity of life, whose reproductive history, morphology, structure, and function in response to its immediate milieu can be analyzed in detail, in much the same as with microoganisms (Puck, 1972). In contrast to the intact organism, such as man, somatic cells in culture are particularly favorable for genetic studies because of a number of biological features including (1) a distinctive chromosome morphology characteristic of the species of origin; (2) the short generation time; (3) attainability of large population sizes; (4) development of pure lines through single cell cloning; (5) ability to enhance the genetic variability by experimental mutagenesis; (6) genetic complementation, recombination, and segregation in cell hybrids and their progeny; and (7) the feasibility of other experimental manipulations.

The earliest applications of mammalian cell cultures were largely limited to studies on cell physiology, nutrition, and response to viral infections. In fact, in addition to primary cell cultures, only a few long-term cell lines were established. Among the best known of these were the L cell, isolated by Wilton Earle from a cell of the mouse strain C3H (Earle, 1943), and the HeLa cell, isolated by George Gey from a biopsy of a human carcinoma of the uterus (Gey et al., 1952).

In the mid-1950s, two technical developments, namely, the spreading of mammalian mitotic chromosomes by hypotonic pretreatment (Hsu and Pomerat, 1953) and the cloning of single mammalian cells by the use of the plating technique (Puck et al., 1956), mark the beginning of a surge of research activities in the following decades in the areas of mammalian cytogenetics, radiation biology, and somatic cell genetics. In 1958, Joshua Lederberg made a summary comment of a symposium on "genetic approaches to somatic cell variation." He wondered why biologists showed such a strong antisexual bias in the consideration of somatic cells. He projected that some transductive phenomena exclusive of mating may occur in mammalian somatic cells through fusion. He also predicted the possibility of somatic segregation as well as mitotic crossing over in mammalian cell hybrids, akin to the "parasexual" genetic recombition in the genetic system of filamentous fungi (Pontecorvo, 1954, 1958). It is remarkable that most of these putative phenomena have been observed in subsequent years.

In this chapter I will describe the early history of the V79 Chinese hamster lung cell line and why it was chosen for radiobiological and genetic experiments. It should be pointed out that there are other cell lines established from the lung tissue of the Chinese hamster, including notably the Don cells (Hsu and Zenzes, 1964) and the CCL 39 or Dede cells (Robert de Saint Vincent and Buttin, 1973) that have also been instrumental to many fruitful studies. However, these and other Chinese hamster cell lines will not be described here.

II. THE ORIGIN AND EVOLUTION OF THE V79 LUNG CELL LINE

Normal tissue explanted *in vitro* should, at least initially, give a true representation of the normal somatic chromosomes of the animal or plant species in question. Attempts to maintain the normal karyotypes of human cell lines, however , had met with varying degrees of success (Chu, 1962). Although the availability of diploid cell strains of various species, including human fibroblasts (Hayflick and Moorhead, 1961), was a definite improvement over the use of mixed, undefined populations of heteroploid cell lines, karotypic abnormalities do exist even in diploid cell strains (Chu, 1962). In that same article, I wrote: "Certain practical measures, such as prevention of cross-contamination, rigid control of culture conditions, and repeated clonal isolation should help to preserve the genetic purity of the cell line. Additionally, a repository of frozen stock cultures can serve not only as controls but also as sources of supply of uniform material adequate for many types of biological experiments." These remarks remain true today. It may also be interesting to note that V79 cells were first used for establishing the fast-cooling–fast-thawing technique for cold storage of mammalian cell stocks retaining a high viability (Mazur et al., 1969).

During the period of 1954–1959, I was with Norman Giles in the Department of Botany, Yale University. I am indebted to Dr. Giles who inspired and encouraged me to enter the field of mammalian somatic cell genetics. We were at that time searching for appropriate mammalian euploid cell material for a study on the spontaneous and radiation-induced chromosome aberrations. Our chromosome analysis of some 35 mammalian "normal" cell strains or lines revealed that all were heteroploid and thus unsuitable for our purposes. We were convinced that we must start from primary cell cultures. These early studies have led to a description of the karyotype of man (Chu and Giles, 1959; Chu, 1960) and other primates (Chu and Giles, 1957; Chu and Bender, 1961; Chu and Swomley, 1961), and to a report on the types and frequencies of X-ray-induced chromosome aberrations in diploid human fibroblasts (Chu et al., 1961). Similar approaches to the genetic analysis of cultured mammalian cells were undertaken at that time in the laboratories of Bentley Glass, Michael Bender, and William Young at the John Hopkins University, T. T. Puck and his associates at the University of Colorado, Robert De Mars and Robert Krooth at the National Institutes of Health, Louis Siminovitch and Klaus Rothfels at the University of Toronto, among others.

The problem of chromosome stability in cell cultures was a problem of immediate concern. For instance, Hayflick and Moorhead (1961) reported that they were able to maintain a number of human diploid cell strains of both embryonic and adult origins for as long as 50 subculture passages over a period of 1 year. Ruddle (1960) found that the pig kidney strain he studied remained predominantly diploid for almost five years. Tjio and Puck (1958) observed no appearance of deviation from the original chromosomal pattern in tissue culture cells from the American opposum, but found chromosomal irregularities in Chinese hamster cell cultures even during early *in vitro* life. Yerganian, Ford, and their coworkers (1958, 1959, 1961) noted extensive karyotypic variation in cultures of Chinese hamster cells, but were able to isolate by cloning a stable, actively proliferating diploid line (Yerganian and Leonard, 1961).

Although I have stressed the importance of using euploid cells for genetic studies, aneuploid cell lines with a low chromosome number may be better suited for certain specific purposes. The diploid chromosome number of the Chinese hamster (*Cricetulus griseus*) is 22. The average lengths of chromosomes range from 2.4 ± 0.4 to 13.7 ± 2.6 μm (Hsu and Zenzes, 1964). Shortly after cultivation *in vitro*, the cells of this species become near-diploid, with a small percentage of polyploid cells in the population (Yu, 1963). The Don cell line, however, remained essentially diploid with no detectable karyotypic modifications (Hsu and Zenzes, 1964). Most of the established Chinese hamster cell lines exhibit a near-diploid chromosome number, a well-defined chromosome morphology, active mitotic activity, and a short generation time. These features appear to be particularly suitable for such studies as mitotic synchronization and chromosome aberration analysis. Working at the Oak Ridge National Laboratory, I obtained from George Yerganian aneuploid cell lines of the Chinese hamster for several investigations, including the wavelength dependence of ultraviolet-light- (UV) induced chromosome aberrations (Chu, 1965a), differential UV and X-ray sensitivity of chromosomes to breakage in synchronized cell populations (Chu, 1965b), and the first demonstration of the lack of excision repair of UV-induced pyrimidine dimers in rodent cells (Trosko et al., 1965). Having previously measured the mitotic chromosomes and scored for aberrations in thousands of human diploid fibroblasts, it was refreshing to see the abundant number of elegant metaphase spreads prepared from the cultured Chinese hamster cells.

My search for suitable cell material for genetic analysis continued. It soon became apparent to me that permanent cell lines established from animals or humans would be useful if one is simply interested in a particular well-characterized genetic variation which arises *in vivo* and expresses at the cellular level *in vitro*. Furthermore, for isolation of new mutants immortal cell lines with abnormal karyotypes may actually facilitate the recognition and recovery of recessive mutations, in addition to the advantage of having an indefinite life span. Pontecorvo (personal communication, 1961) attempted to isolate autosomal recessive mutants through the use of mitotic poisons to induce abnormal chromosome segregation. De Mars and Hooper (1960) pioneered the indirect selection of auxotrophic mutants of HeLa cells. In George Klein's laboratory at Karolinska Institute in Stockholm where I spent my sabbatical year in 1965, I started using mouse tumor cells for a study of H-2 variation. Upon returning to this country the following year, I paid a visit to Warren Sinclair and C. K. Yu at the Argonne National Laboratory. I was impressed by the high plating efficiency (nearly 100%), excellent colony morphology (perfect circles of monolayer cells), and short generation time (approximately 12 hr) of a line of Chinese hamster cells they were studying. I immediately requested the cells and was granted a subculture. Spontaneous variants either auxotrophic for L-glutamine or resistant to 8-azaguanine were initially selected from this cell line (Chu et al., 1969).

This cell line (V79-122D1) was originally cultured from the lung tissue of a male Chinese hamster and designated V strain by Ford and Yerganian (1958); it was subcultured and named V79 by Mortimer Elkind of the National Cancer Institute in December, 1958. It was with this cell line that the discovery was made of X-ray-induced damage and recovery in mammalian cells in culture (Elkind and Sutton, 1959). A subculture was given to Sinclair and has been maintained at Argonne since May

1961. The results of X-ray-induced heritable damage (Sinclair and Morton, 1963) and the karyotype of the V79 cells (Yu, 1963; Yu and Sinclair, 1964) have appeared.

III. BIOLOGICAL CHARACTERISTICS OF V79 CELLS

A. Nutritional Requirements

The cell line was cultured in the laboratories of Elkind and Sinclair in HUT-15 medium (cf. Elkind and Sutton, 1959), but could be maintained in most types of culture media containing serum.

In our laboratory, we have used the Dulbecco modified Eagle's minimum essential medium (MEM) supplemented with 5% v/v fetal calf serum (FCS). The modified MEM contains Earle's salt solution, a 50% increase of essential amino acids, a 100% increase of nonessential amino acids, 2 *mM* L-glutamine and 0.1 *mM* sodium pyruvate. No antibiotics are included for routine stock maintenance. In MEM supplemented with 2% dialyzed FCS, the cells suffer no loss in plating effeciency but grow more slowly than in medium with 5% undialyzed serum (unpublished data). The V79 cells show normal plating efficiency in medium in which glucose is replaced by a equimolar concentration of galactose, but show only about 5% plating efficiency if glucose is substituted by mannose. Surviving clones exhibit a similarly low plating efficiency in mannose medium. In contrast, Chinese hamster ovary (CHO) cells show an opposite growth response, that is, normal plating efficiency in media containing mannose but not galactose.

B. Growth Pattern *in vitro* and *in vivo*

When growing in the same medium, V79 cells appear to attach more firmly than CHO cells to the glass or plastic surface, thus giving rise to a less serious problem of satellite colony formation. On the other hand, V79 cells can be, but are more difficult to be, adapted to grow in suspension cultures (G. Milman, personal communication).

The average generation time of V79 cells is about 12 hr, consisting of approximately 2 hr in G_1, 6–7 hr in S, 2 hr in G_2, and 1 hr in mitosis. The cell population doubling time is about 16 hr (Kimball et al., 1971). Various types of chemical and physical methods have been successfully applied to synchronize V79 cells mitotically (Prescott, 1976). Clones have been isolated in which the G_1 phase is completely absent (Robbins and Scharff, 1967).

A combination of (1) microphotometric determinations of the DNA and protein content and dry mass of single cells and (2) autoradiography was used to follow the changes in cell cycle parameters and cell growth in cultures of V79 cells as they passed from the initial stage of growth, through the exponential phase, into decline (Kimball et al., 1971). A description of the events were given in terms of the rates of three processes: protein synthesis, initiation of DNA synthesis, and initiation of mitosis. There are indications that the rate of initiation of mitosis is less closely associated with the rate of protein synthesis than is the rate of initiation of DNA synthesis.

The plating efficiency of the V79 cell line and various derivative clones approaches unity. The cells can also be plated without appreciable loss of plating efficiency in semisolid medium or on the surface of solidified agar submerged under a layer of liquid medium (unpublished data). Isolation of clonal colonies originated from single cells can be achieved either by the use of the ring method or by direct picking.

V79 cells have little capacity to metabolize xenobiotics (Huberman, 1978). In order to extend the usefulness of this *in vitro* cell culture system for testing the mutagenicity of chemical carcinogens, Heinrich Malling and I in the late 1960s developed (1) a liver microsome hydroxylation system and (2) a mammalian host-mediated assay with V79 cells (Chu, 1972). Our preliminary studies with V79 cells in culture have shown that a combined treatment with dimethylnitrosamine and mouse liver microsomes gives a significant increase in the frequency of azaguanine-resistant mutants over treatments with the components alone. The second modification we developed was to enclose a hamster cell suspension in a dialysis bag and implant it surgically into the cavities of rats. The host animals were treated or untreated with test compounds. After various periods of time the bags were removed, and the recovered V79 cells were plated *in vitro* to determine cell survival and the frequency of mutation. Similarily, Huberman (1978) cocultivated V79 with irradiated primary rodent fibroblasts to assay for the frequency of mutations in V79 cells induced by chemical carcinogens. To expand the efficiency of the cell-mediated mutagenesis, Huberman and Jones (1980) substituted primary hepatocytes for fibroblasts in carcinogen-activation experiments.

V79 cells can be injected subcutaneously into and propagated in athymic *nude* mice (Shin et al., 1975). Solid cell mass reached 15 mm or larger in diameter in 6–8 weeks after injection. It has been shown (Freedman et al., 1976) that mammalian cells, including V79 cells, and cell hybrids propagated in *nude* mice retained and expressed the respective genetic markers and that the solid tumor can reach a size larger than the host (e.g., 20 g tumor in a 15-g mouse). These studies clearly indicate that (1) V79 cells have been neoplastically transformed and (2) the *in vivo* propagation of V79 cells and the retention of cellular phenotypes may be useful in situations when large amounts of cells and cell products are needed.

C. Karyological Characteristics

After some 5 years of transfer and subculture from time to time in different laboratories, Yu (1963) reported that the V79 cell line still contained the same percentage of near-diploid cells. The majority (>95%) of the cells contained 23 chromosomes, and most of those possess the same chromosome pattern, which is therefore taken as representative of the line. However, 34 different types of chromosome constitution were found in the cells studied, and cells having the same number of chromosomes may be of different pattterns. More recently, cells of the V79-4 clone isolated in my laboratory were found to have a consistent set of 20 chromosomes (Thacker, 1981). G- and C-band analysis showed that, compared to the chromosome set of freshly isolated Chinese hamster somatic cells, the V79-4 chromosomes had various characteristics deletions and rearrrangements. However, it is probable that at least one copy of each autosome was still present, and three pairs of autosomes appeared unchanged form those in freshly isolated cells. The establishment of this karyotype for V79-4 allows mutant sublines to be screened for chromosomal changes associated with the altered phenotype.

D. Genetic and Epigenetic Variation

Spontaneous variation of somatic cells *in vitro* has been observed and well summarized (Harris, 1964; Littlefield, 1976). After the introduction of the quantitative techniques for the mass cloning of single cells (Puck et al., 1956), it was immediately apparent that considerable variation exists in the ability of individual cells to form colonies and in the morphology of individual colonies. The heteroploidy of these mammalian cell lines was revealed shortly thereafter when cytogenetic studies demonstrated extensive variation in chromosome complement (Chu and Giles, 1958; Hsu, 1961). In addition, the surprisingly high frequency of biochemical variants in such cultures was recognized (cf. Littlefield, 1976).

The independent demonstration of experimental mutagenesis in two laboratories (Chu and Malling, 1968; Kao and Puck, 1968), using V79 and CHO cells, respectively, has greatly increased the spectrum of genetic variability in culture mammalian cells. Genetic markers are a prerequisite to progress in somatic cell genetics, as evidenced by numerous studies and exemplified by other contributions in this volume. The genetic and epigenetic nature of variants isolated from V79 and other mammalian cell lines have been discussed elsewhere (Chu, 1974; Chu et al., 1975; Siminovitch, 1976). Mutagenesis studies with cultured mammalian cells, especially V79 cells have been reviewed recently (Bradley et al., 1981; Chu et al., in press). It suffices to say that the V79 Chinese hamster cells and their derivatives have contributed to the beginning of mammalian cell genetics and undoubtedly will continue to be useful for cell and molecular genetic studies because of their characteristic biological features and the wealth of information and the collection of genetic mutants that have been accumulated.

REFERENCES

Bradley, M. O., Bhuyan, B., Francis, M. C., Langebach, R., Peterson, A., and Huberman, E. (1981). *Mutation Res.* **87**, 81–142.

Chu, E. H. Y. (1960). *Am. J. Human Genet.* **12**, 97–103.

Chu, E. H. Y. (1962). In *Analytic Cell Culture,* National Cancer Institute Monograph No. 7, pp. 55–71.

Chu, E. H. Y. (1965a). *Mutation Res.* **2**, 75–94.

Chu, E. H. Y. (1965b). *Genetics* **52**, 1279–1294.

Chu, E. H. Y. (1972). In *Environment and Cancer,* Willians and Wilkins, Baltimore, Maryland, pp. 198–213.

Chu, E. H. Y. (1974). *Genetics* **78**, 115–132.

Chu, E. H. Y. and Bender, M. A. (1961). *Science* **133**, 1399–1405.

Chu, E. H. Y., Brimer, P., Jacobson, K. B., and Merriam, E. V. (1969). *Genetics* **62**, 359–377.

Chu, E. H. Y. and Giles, N. H. (1957). *Am. Naturalist* **91**, 273–282.

Chu, E. H. Y. and Giles, N. H. (1958). *J. Natl. Cancer Inst.* **20**, 383–401.

Chu, E. H. Y. and Giles, N. H. (1959). *Am. J. Human Genet.* **11**, 63–79.

Chu, E. H. Y., Giles, N. H., and Passano, K. W. (1961). *Proc. Natl. Acad. Sci. USA* **47**, 830–839.

Chu, E. H. Y., Li, I.-C., and Fu, J. (in press). In *Mutation, Cancer and Malformation* (E. H. Y. Chu and W. M. Generoso, eds.), Plenum, New York.

Chu, E. H. Y. and Malling, H. V. (1968). *Proc. Natl. Acad. Sci. USA* **61**, 1306–1312.

Chu, E. H. Y. and Swomley, B. A. (1961). *Science* **133**, 1925–1926.

Chu, E. H. Y., Sun, N. C., and Chang, C. C. (1975). In *Mammalian Cells: Problems and Probes* (C. R. Richmond, D. P. Peterson, P. F. Mullaney, and E. C. Anderson, eds.), National Technical Information Service, U.S. Department of Commerce, Springfield, Virginia, pp. 228–238.

De Mars, R. and Hooper, J. L. (1960). *J. Exp. Med.* **111**, 559–572.

Earle, W. R. (1943). *J. Nat. Cancer Inst.* **3**, 555–558.

Elkind, M. M. and Sutton, H. (1959). *Nature* **184**, 1293–1295.

Ford, D. K., Boguszewski, C., and Auersberg, N. (1961). *J. Natl. Cancer Inst.* **26**, 691–706.

Ford, D. K., Wakonig, R., and Yerganian, G. (1959). *J. Natl. Cancer Inst.* **22**, 765–799.

Ford, D. K. and Yerganian, G. (1958). *J. Natl. Cancer Inst.* **21**, 393–425.

Freedman, V. H., Brown, A. L., Klinger, H. P., and Shin, S.-I. (1976). *Exp. Cell Res.* **98**, 143–151.

Gey, G. O., Coffman, W. D., and Kubicek, M. T. (1952). *Cancer Res.* **12**, 264–265.

Harris, M. (1964). *Cell Culture and Somatic Variation,* Holt, Rinehart, and Winston, New York.

Hayflick, L. and Moorhead, P. S. (1961). *Exper. Cell Res.* **25**, 585–621.

Hsu, T. C. (1961). *Intern. Rev. Cytol.* **12**, 69–161.

Hsu, T. C. and Pomerat, C. M. (1953). *J. Hered.* **44**, 23–29.

Hsu, T. C. and Zenzes, M. T. (1964). *J. Natl. Canc. Inst.* **32**, 857–869.

Huberman, E. (1978). *J. Environ. Pathol. Toxicol.* **2**, 29–42.

Huberman, E. and Jones, C. (1980). *Ann. New York Acad. Sci.* **349**, 264–272.

Kao, F.-T. and Puck, T. T. (1968). *Proc. Natl. Acad. Sci. USA* **60**, 1275–1281.

Kimball, R. F., Perdue, S. W., Chu, E. H. Y., and Ortiz, J. R. (1971). *Exper. Cell Res.* **66**, 17–32.

Lederberg, J. (1958). *J. Cell. Comp. Physiol.* **52** (Supplement 1), 383–401.

Littlefield, J. W. (1976). *Variation, Senescence and Neoplasia in Cultured Somatic Cells.* Harvard University Press, Cambridge, Massachusetts.

Mazur, P., Farrant, J., Leibo, S. P., and Chu, E. H. Y. (1969). *Cryobiology* **6**, 1–9.

Pontecorvo, G. (1954). *Caryologia. Suppl.* **6**, 192–200.

Pontecorvo, G. (1958). *Trends in Genetic Analysis*, Columbia University Press, New York.

Prescott, D. M. (1976). *Reproduction of Enkaryotic Cells*, Academic Press, New York.

Puck, T. T. (1972). *The Mammalian Cell as a Microorganism*, Holden-Day, San Francisco.

Puck, T. T., Marcus, P. I., and Cieciura, S. J. (1956). *J. Exp. Med.* **103**, 273–284.

Robbins, E. and Scharff, M. D. (1967). *J. Cell Biol.* **34**, 684–688.

Robert de Saint Vincent, B. and Buttin, G. (1973). *Eur. J. Biochem.* **37**, 481–488.

Ruddle, F. H. (1960). Ph.D. dissertation, University of California.

Shin, S.-I., Baum, S. G., Fleischer, N., and Rosen, O. M. (1975). *J. Cell. Sci.* **18**, 199–206.

Siminovitch, L. (1976). *Cell* **7**, 1–11.

Thacker, J. (1981). *Cytogenet. Cell Genet.* **29**, 16–25.

Tjio, J. H. and Puck, T. T. (1958). *J. Exp. Med.* **108**, 259–268.

Yerganian, G. and Leonard, M. J. (1961). *Science* **133**, 1600–1601.

Yu, C. K. (1963). *Canad. J. Genet. Cytol.* **5**, 307–317.

Yu, C. K. and Sinclair, W. K. (1964). *Canad. J. Cytol.* **6**, 109–116.

CHAPTER 4

THE DEVELOPMENT AND USE OF THE CHINESE HAMSTER EMBRYO FIBROBLAST (CHEF) CELL LINE

Ruth Sager
Dana-Farber Cancer Institute
Harvard Medical School
Boston, Massachusetts

I. INTRODUCTION

The fundamental challenge of mammalian cell genetics is to apply the concepts and methodologies of microbial genetics in a manner as rigorous as has been possible with microbial systems. This approach was laid out with clarity and optimism by Puck (1972) in his path-finding monograph *The Mammalian Cell as a Microorganism*. Impelled by Puck's vision, I set out in 1975 to look for a suitable mammalian cell system in which to apply microbial genetics to the analysis of tumorigenesis.

Having initially been trained as a maize cytogeneticist, I was not attracted by aneuploid or pseudodiploid cell lines. In pursuit of a diploid cell line, I went for advice to Dr. George Yerganian in February, 1976. To my great good fortune, he gave me a frozen ampule of his F4224A cells, and they became the starting material from which the CHEF cell lines were developed, as discussed below. Through the wizardry of Yerganian's touch, this cell line was diploid, nontumorigenic, and immortal, and has remained so in subsequent years of passage and investigation. To my knowledge, this cell line is unique in its chromosome stability and nontumorigenicity.

In this chapter, the properties of CHEF cells will be summarized, and the use of CHEF cells in the analysis of selected problems will be discussed. CHEF cells were initially developed for the genetic analysis of tumorigenicity and results pertaining to that subject will be reviewed in Chapter 26.

II. ORIGIN AND GROWTH OF CHEF CELLS

The F4224A cell line was established by Dr. George Yerganian from a male Chinese hamster embryo, and it was obtained by this laboratory as a frozen ampule. Our records consider the time of receipt as passage 1. All studies in this labotatory have been carried out using alpha-MEM plus 10% fetal bovine serum, with added 2 *mM* glutamine, streptomycin (100 μg/mL), and penicillin (100U/mL); cells are incubated in a humidified incubator with 6.5% CO_2 at 37°C.

Cells were grown from the original frozen ampule, then frozen down for storage, and one of the new ampules was used for cloning. Cells were plated at cloning densities, and colonies with two clearly different morphologies were identified. Ten colonies of each type, chosen for futher study, maintained their identity in subsequent plating, and one colony of each morphology was chosen to initiate new stocks. The colony morphologies, shown in Figure 4.1, distinguish CHEF/18-1 and CHEF/16-2, the doubly cloned lines that are the source of cells for all further studies (Sager and Kovac, 1978). Although colony morphologies are distinctive, cells of the two types are quite similar in spindle-shape; the 16-2's are somewhat plumper than the 18-1's and do not attach quite as firmly to plastic.

Typical growth curves are shown in Figure 4.2. CHEF/18's grow slightly slower than do CHEF/16's. Average growth rates based on numerous experiments are given in Table 4.1, and together with a summary of the properties of the two cell lines (Sager and Kovac, 1978). Included in the table are comparable data for two mutant cell lines used in many experiments to be discussed below: CHEF/205-30, a thioguanine-resistant (HPRT-) mutant of CHEF/18 and CHEF/204-Bu50, a BrdU-resistant (TK-) mutant derived from CHEF/16.

With respect to the serum requirement for growth, CHEF/18's plate with 30–40% efficiency in 10% fetal calf serum, but not at all in 3% serum, whereas CHEF/16's plate with high efficiency in 10% serum and more poorly in 3% serum. As discussed below, using a defined serum-free medium, it was shown that CHEF/16's have lost the EGF requirement, but retain other growth factor requirements of CHEF/18 cells.

The assay for anchorage-independence is based on colony formation

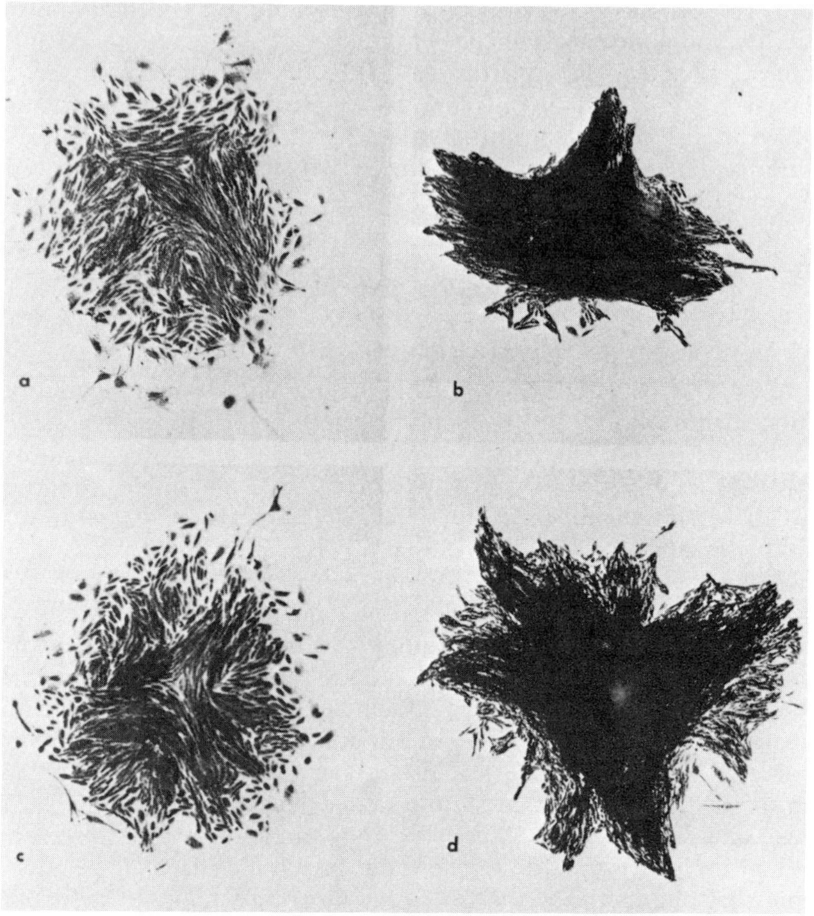

Figure 4.1. Seven-day colonies from (*a*) CHEF/18; (*b*) CHEF/16; (*c*) 205-30; (*d*) 204-Bu50 [from Sager and Kovac (1978)].

by cells seeded into 1.3% methylcellulose in 60 mm dishes over a 0.6% agar base. Fresh medium is added weekly, and macroscopically visible colonies are counted after 4 weeks. Microcolonies are rare with CHEF cells, thus permitting an unambiguous quantitation.

The properties listed in Table 4.1 distinquish the nontumorigenic CHEF/18's from the CHEF/16's, which are anchorage independent and tumorigenic. Neither CHEF/18 nor CHEF/16 cells have prominant actin cables or fibronectin, assessed by indirect immunofluorescence (Chen and Sager, unpublished), nor do they produce much plasminogen activator (Rivkin and Sager, unpublished).

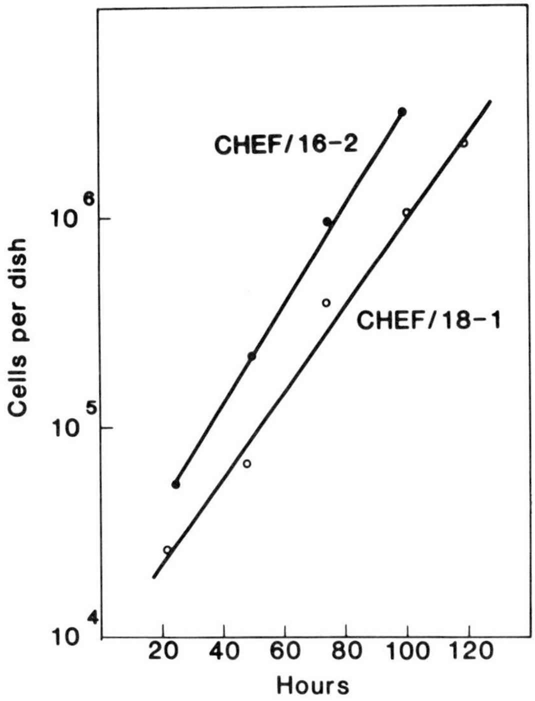

Figure 4.2. Growth rates of CHEF/18-1 and CHEF/16-2.

III. THE NUDE MOUSE ASSAY OF TUMOR-FORMING ABILITY

The quantitative assay of tumor-forming ability is critically important in the genetic analysis of tumorigenicity. One wishes to assay the tumor-forming ability of single cells, but the methods are inadequate, first because the assay is at the population level rather than the single cell level and, second, because the cells must proliferate, thus undergoing selection during tumor formation. Low numbers of cells cannot be tested in the nude mouse assay, because the animals possess a vestigial immune response, probably the result of natural killer (NK) cell activity. As a consequence, most tumorigenic cells must be introduced at high titer, at least 10^5 cells per site, to obtain a single tumor from cells injected subcutaneously.

The coinjection procedure, in which 10^7 X-irradiated or mitomycin-treated normal cells are coinjected with the test cells, provides a means to assay a dilution series of test cells, and thereby to determine the lowest number of test cells that give rise to a tumor. This method, developed initially by Stiles and Kawahara (1978), has been used extensively in CHEF cell studies. For example, it has been shown that

TABLE 4.1
Comparison of Hamster Cell Lines[a]

	18-1 (nontumorigenic)	16-2 (tumorigenic)	205-30(18-1 TG^R)	204-Bu50(16-2 $BrdU^R$)
Monolayer morphology	Loosely packed	Densely packed	Loosely packed	Densely packed
Colony morphology	Loosely packed Swirling edges	Tightly packed Compact edges	Loosely packed Swirling edges	Tightly packed Compact edges
Population doubling time (hr)	14–16	10–12	14–15	11–12
Saturation density (cells/cm^2)	$(1–2) \times 10^5$	$(6–8) \times 10^5$	$(1–2) \times 10^5$	$(2–3) \times 10^5$
Plating efficiency				
in 10% serum	30–40%	60–70%	30–40%	45–55%
in 3% serum	0%	3%	0%	2%
in methylcellulose	$<10^{-5}$	60%	$<10^{-5}$	25%
Tumors in nude mice				
1×10^6 cells/site		3/3	0/2	6/6
4×10^6	0/3		0/8	
1×10^7	0/30			
Coinjection of 16-2 with 10^7 X-irradiated 18-1: 10, 10^2, 10^3, 10^4, 10^5, 10^6 cells/site		12/12		
Chromosomes				
mode	22	22	22	22
range	21–24	21–23	21–24	20–22
tetraploid (%)	2	3	8	8

[a] From Sager and Kovac (1978).

CHEF/16-2 cells are tumorigenic to the level of 10 cells, when they are coinjected with 10^7 X-irradiated CHEF/18 cells (Sager and Kovac, 1978). Thus, by introducing high titers of cells, the coinjection procedure effectively circumvents the low level of NK-based immune rejection in nude mice, at least in the Balb/c nudes, which are bred and utilized in our laboratory. It then becomes possible to distinquish heterogeneous cell populations containing both tumorigenic and nontumorigenic cells from populations that are homogeneous for either phenotype.

The question of NK cell rejection was also addressed by comparing the tumor-forming ability of three anchorage-independent mutants that did not form tumors in the standard nude mouse assay (10^7 cells injected subcutaneously) with nude mice that had received 450 rad whole-body X-irradiation 24–72 hr prior to injection. No tumors appeared in any of the injected X-irradiated mice, supporting the validity of the nude mouse assay (Smith and Sager, 1982).

The time course of tumor formation in the nude mouse assay is influenced by the malignancy of the injected cells. CHEF/16 cells and their mutant derivatives injected subcutaneously at 4×10^6 cells/site make tumors of 0.5 cm or larger in 3–4 weeks, and when CHEF/16 cells are coinjected in a dilution series with 10^7 X-irradiated CHEF/18 cells, tumors arise just as rapidly except at the lowest dilution, 10 and 10^2, which take an additional week. As discussed in Chapter 27, these tumors are of clonal origin, as shown by chromosome analysis. Highly malignant tumorigenic cells recovered in some experiments make tumors in 1–2 weeks, whereas some cell populations require 2–3 months. Tumors arising in 4 months or longer, when excised, grown in culture, and retested, give rise to rapidly forming tumors. This result is evidence that selection occurred during growth of the initial tumor and thus the cell population that produced the tumor is different from the inoculum. Further and more direct evidence on this point comes from comparison of chromosomes of the injected cells with those in cells derived from the resulting tumors. In slow growing tumors as discussed in Chapter 27, chromosome changes occur, reflecting selection of new evolving genotypes and phenotypes during tumor growth.

Tumors derived from CHEF cells have been excised and characterized as fibrosarcomas, consistent with other evidence that the cells are fibroblastic.

IV. CHROMOSOMES

An important advantage of Chinese hamsters for genetic analysis over other rodents is their low chromosome content: 10 pairs of autosomes plus two sex chromosomes, XX or XY. The chromosomes are distinctive in morphology (Fig. 4.3) and after Giemsa banding each can be readily

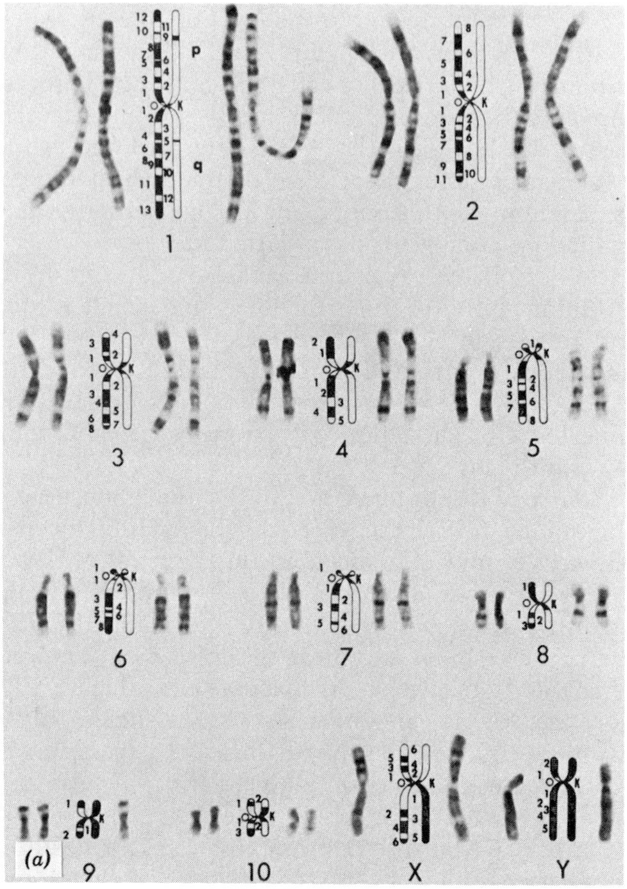

Figure 4.3(*a*). Giemsa banding patterns of CHEF/18 and CHEF/205-30. In each group the chromosome pair on the left (from CHEF/18) and the chromosome pair on the right (from CHEF/205-30) are compared to a schematic representation of the banding patterns of normal Chinese hamster cells [from Stubblefield (1974)] in which the left-hand chromatid shows the G-banding pattern.

identified in the light microscope (Sager and Kovac, 1978; Kitchin and Sager, 1980). Because of their distinctive sizes and banding patterns, the chromosomes are also readily separable in the FACS (fluorescence-activated cell sorter) as shown in Figure 4.4 (Cram et al., 1983; and Chapter 7). The sorting capability is making possible rapid mapping by DNA hybridization of any genes that have been cloned. Sorting is also an invaluable tool for cloning of individual chromosomes into vectors such as lambda or cosmids.

In our initial studies of karyotypes of the CHEF cell lines (Sager and Kovac, 1978), it was ascertained that the chromosomes are intrinsically very stable, as shown by the frequency distributions of chromosome

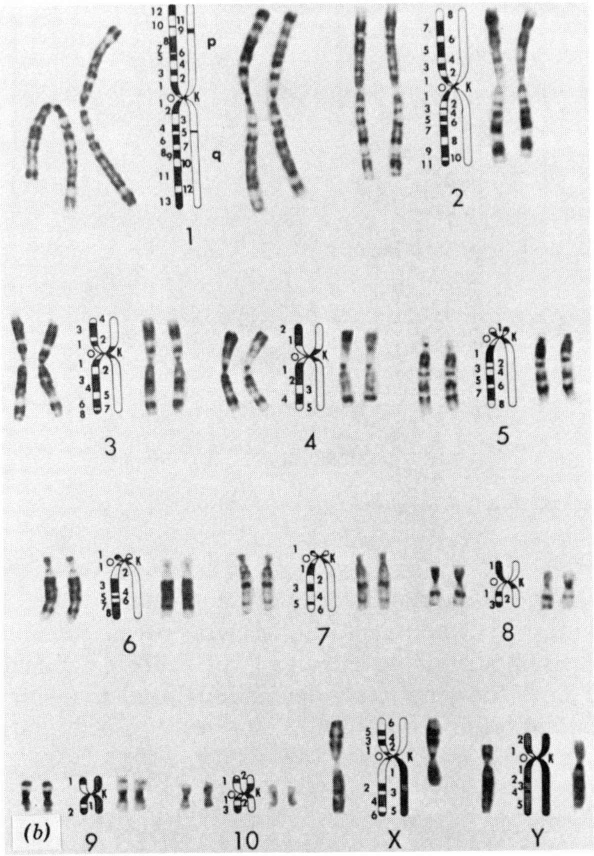

Figure 4.3(*b*). Giemsa banding patterns of CHEF/16 and CHEF/204-Bu50. In each group the chromosome pair on the left (from CHEF/16) and the chromosome pair on the right (from CHEF/204-Bu50) are compared to a schematic representation of the banding patterns of normal Chinese hamster cells [from Stubblefield (1974)] in which the left-hand chromatid shows the G-banding pattern.

numbers per cell in 18-1's, 16-2's, and mutants 205-30, a TG-resistant derivative of 18-1, and 204-Bu50, a BrdU-resistant of 16-2 (Fig. 4.5).

Subsequently, chromosomes from these four cell lines were examined in detail by Giemsa banding (Kitchin and Sager, 1980). It was shown that CHEF/18 and CHEF/16 have normal Giemsa banding patterns and constitutive heterochromatin distributions characteristic of normal diploid Chinese hamster cells and that they exhibit relatively little chromosomal variation within growing populations of cells. The two mutant lines, later used for cell fusion studies (Chapter 27) showed somewhat more variability than did the parental lines from which they were selected. In particular, the CHEF/204-Bu50 cells have a deletion of either part or all of the heterochromatic long arm of the X chromosome.

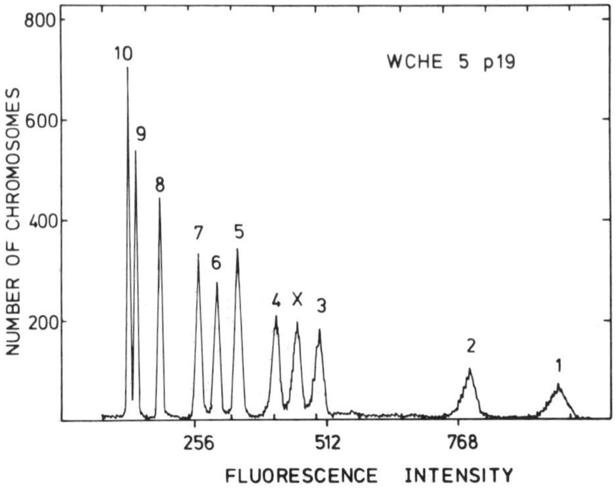

Figure 4.4. FACS distribution of Chinese hamster chromosomes [from Cram et al. (1983)].

Because of their karyotypic normality and relative homogeneity, reflecting chromosome stability during growth in cell culture, the CHEF cell lines offer important advantages over other established Chinese hamster cell lines for genetic studies and, in particular, for the experimental investigation of chromosome changes occurring during tumorigenesis (Chapter 27).

V. GROWTH IN SERUM-FREE DEFINED MEDIUM

Cherington et al. (1979) established the growth factor requirements of CHEF/18 and CHEF/16 cells in serum-free medium. The method used was the following. Cell lines to be tested were plated into salts consisting of alpha-MEM and F12 (1:1) with 10% serum at 3×10^3 cells/cm^2 in 35 mm or 60 mm dishes and incubated for 1 day. Each dish was then washed with alpha-F12 and then fed with this medium plus the appropriate growth factor additions to be tested. At appropriate intervals, usually once per day, sets of dishes were trypsinized and cells counted. Under these conditions, confluence was reached in 3–4 days, and samples were counted for a total of 5 days.

The only essential factors for growth of CHEF/18 cells under these conditions are EGF (10 ng/mL), insulin (10 μg/mL), transferrin (5 μg/mL), and a supplement of $FeSO_4$ (2.5 μM). A major difference was found in the requirement of CHEF/16 cells, namely, the loss of the EGF requirement, and a similar loss was found with a tumor-derived CHEF/18 cell line called T30-4. The growth rate of CHEF/18 cells in the three-factor medium was about one-half that in 10% fetal calf serum.

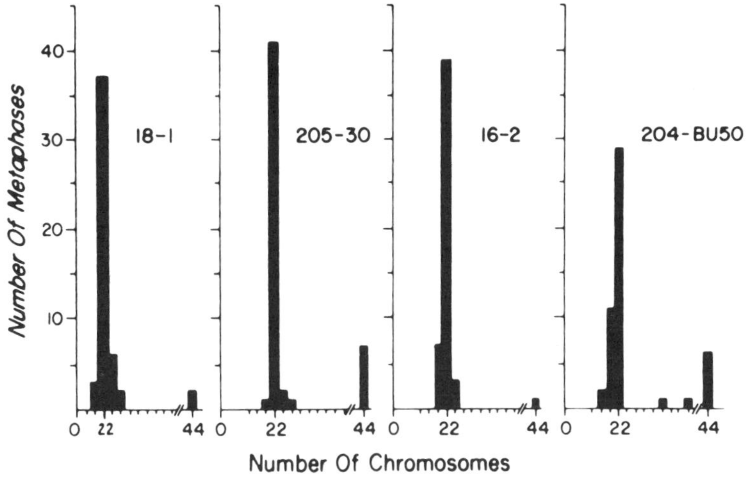

Figure 4.5. Frequency distribution of chromosome numbers per cell in 50 metaphase samples of CHEF cell lines.

In further studies (Cherington and Pardee, 1980) it was shown that addition of thrombin (10–100 ng/mL; 0.3–3 n*M*) permitted growth in the defined medium at the same rate as in serum. In addition, thrombin was shown to compensate partially for the absence of EGF, suggesting that the two factors act through a related mechanism.

On the basis of these studies, a survey was carried out (Sager et al., 1982) to examine the growth factor requirements of a series of "low-serum" and anchorage-independent mutants (Chapter 27) selected after mutagenesis (Smith and Sager, 1982). In these studies, each cell line was tested for growth in 10% serum, in complete serum-free medium (with all additions), and in serum-free media from which individual growth factors had been omitted. Since transferrin and $FeSO_4$ were required by all cells, the studies were limited to EGF, insulin, and thrombin.

The 12 anchorage-independent mutants examined, each derived from CHEF/18, had not lost the growth requirement for either EGF or thrombin, or in three mutants for both factors. These findings clearly dissociate the EGF requirement from the anchorage requirement in CHEF/18 cells. This result is in line with the recent evidence that a *different* growth factor produced by certain sarcoma cells regulates anchorage independence from that which appears to substitute for EGF (Roberts et al., 1983).

Among the 12 low serum mutants of CHEF/18 that were examined, three had a reduced insulin requirement and three resembled the anchorage mutants in growing with either EGF or thrombin. The other six were surprising. Three would not grow serum-free, and thus had acquired a new requirement despite their ability to grow on low serum, and the others grew in complete serum-free medium, but required all the added growth factors.

Although most of these findings have not yet been followed up, the results to date point up the usefulness of a serum-free-defined medium for the study of specific growth factor requirements. The fact that mutants with changes in individual growth factor requirements have been recovered, shows that these growth requirements are under multiple gene control.

One would have hoped that the development of a serum-free medium would liberate the investigator from the expense and unpredictability of serum. Unfortunately, this expectation has not been realized with CHEF cells for a curious reason. During the serum-free studies, it was noted that cells grown continuously in serum-free medium developed large foci of rounded cells, many of which detached from the plastic (Sager et al., 1982). In subsequent studies (Sager and Kovac, 1982) we found that CHEF/18 cells grown in serum with added insulin undergo differentiation to form mature adipocytes, and the strange foci previously seen were actually clusters of differentiating adipocytes. This observation has led to the use of CHEF cells in the study of differentiation (see Section VII-B) but has made it impossible to use serum-free medium for routine growth. Because the insulin requirement is so high (10 μg/mL), it is very likely that the true growth requirement is for somatomedin C or MSA (Leof et al., 1982) rather than insulin, but neither of these factors are yet available as media supplements for routine growth.

VI. MUTAGENESIS AND MUTANTS

A. Mutagenesis Procedures

In this laboratory mutagenesis experiments have been carried out with three drugs: EMS (ethylmethansulfonate), MNNG (*N*-methyl-*N*′nitro-*N*-nitrosoguanidine, and 4-NQO 4(nitroquinolinoxide). Mutagen treatments are designed to achieve 33% survival, which is optimal for single-hit mutagenesis. Typically, recloned subconfluent cells are seeded at 10^6 per 75 cm^2 flask, and after 24–48 hr incubation, mutagen is added. The treatment period is 16 hr for EMS (200–400 μg/mL), 1 hr for MNNG ($10^{-5}M$), and 30 min for 4-NQO (0.4–0.75 μg/mL). Cells are then washed with PBS, trypsinized, subcultured in flasks for intermediate cultivation, and plated at $(2-5) \times 10^2$ per 100 mm dish to measure survival. After appropriate time for expression (depending on the mutation), selection is initiated.

B. CHEF Mutants

Mutants resistant to thioguanine, to BrdU, and to ouabain were chosen as selectable markers for cell fusion experiments. Other drug-resistant and electrophoretic mutants were selected for other genetic studies. Characterized mutants are listed in Table 4.2.

Mutant frequencies were determined with recently recloned populations of CHEF/18 cells. Typical frequencies, after EMS mutagenesis (20–50% survival) and intermediate cultivation, were (i) thioguanine resistance, 5×10^{-5} (5×10^{-7} spontaneous); (ii) ouabain resistance, $(1-2) \times 10^{-5}$ (4×10^{-8} spontaneous).

Recovery of BrdU resistant cells required stepwise selection after EMS mutagenesis, presumably because the phenotype is recessive and the TK gene is autosomal. EMS-induced mutants resistant to diptheria toxin (DT) and to α-amanitin were recovered after one-step selection. Electrophoretic mutants were identified in a collaboration with M. Siciliano. CHEF/18 cells were mutagenized with EMS (14% survival) and survivors plated for single colonies. One hundred and twenty clones were isolated, electrophoretic mobilities were determined for forty enzymes, and six mutants were identified (listed in Table 4.2). This yield of mutants is significantly lower than that obtained by Siciliano with CHO cells (UV treated), and supports the idea that CHO cells may be functionally hemizygous for many loci as a result of chromosome rearrangements (Siminovitch, 1976), whereas CHEF cells are functionally diploid.

VII. USE OF CHEF CELLS IN SELECTED RESEARCH PROJECTS

A. Use of Cell Fusion for Genetic Analysis

With the development of methods for recovery of somatic cell hybrids (reviewed in Shay, 1982; see also Chapter 7) it became feasible to apply cell hybridization to genetic analysis with somatic cells. A highly successful application has been to gene mapping. In addition, cell fusion can be used for studies of complementation between independently isolated mutants with similar phenotypes to determine how many complementation groups (presumably different genes) regulate that phenotype. This method has been used with notable success in distinguishing different complementation groups in the human genetic disease *Xeroderma pigmentosum*. With CHEF cells it has been used to study complementation of anchorage and low-serum mutants (Marshall and Sager, 1981; Smith and Sager, 1985).

A further application, not well recognized in the literature, but potentially very useful, is in determining whether a particular mutant phenotype segregates as a single gene difference in the progeny of mutant ×

TABLE 4.2
Mutant Cell Lines

Cell Line	Origin	Mutations	Selection	Tumorigenicity
205-30	CHEF/18	$HPRT^-$	One step	No
18Bu-1	CHEF/18	TK^-	Multistep	No
18oua-r	CHEF/18	Ouabain-r	One step	ND[a]a
18DM	CHEF/18	TK^-; oua-r	Multistep (TK); one step (sp)	No
294-7	CHEF/18	$HPRT^-$; CAP-r	One step ($HPRT^-$); Multistep (CAP)	No
18-DT-10	CHEF/18	DT-r	One step	ND
18ama-4	CHEF/18	α-Amanitin-r	One step	ND
18mtx-8	CHEF/18	mtx-r	Multistep	No
18c5	CHEF/18	Adenylate kinase 2	One step	ND
18c8	CHEF/18	Triosephosphate isomerase	One step	ND
18c21	CHEF/18	Isocitrate dehydrogenase	One step	ND
18c65	CHEF/18	Peptidase-2	One step	ND
18c88	CHEF/18	α-Glucosidase	One step	ND
18c91	CHEF/18	Peptidase-3	One step	ND
204-Bu50	CHEF/16	TK^-	Multistep	Yes
16DM	CHEF/16	$HPRT^-$;oua-r	One step	Yes
204mtx-20	CHEF/16	mtx-r	Multistep	Yes
16oua-4	CHEF/16	oua-r	One step	ND

[a] ND = not determined.

normal cell hybrids. This method has been useful with CHEF mutants in demonstrating the segregation of mutations to drug resistance in hybrids that are selected with other markers. If the mutation results from a change on one chromosome, then sensitivity versus resistance will segregate unselected in reduced hybrids. This method was applied successfully with thioguanine, ouabain, α-amanitin, and diphtheria-toxin-resistant mutants, as well as with anchorage independence and LS mutants (unpublished). The segregation may be skewed from 1:1 depending on other genes linked on the same chromosome (see below for discussion of TK segregation).

Interspecies fusions have been used for gene-mapping purposes ever since the discovery of chromosome elimination in clones of mouse × rat cell hybrids (Weiss and Green, 1967). Using intraspecies cell hybrids derived from fusions between suitably marked CHEF/18 mutants, we examined the feasibility of developing a mapping procedure based on chromosome elimination.

Cell fusion procedures used in this laboratory follow the general protocol outlined in Chapter 7. Briefly, 10^6 cells of one parent are seeded in a 25 cm^2 flask and incubated overnight. The following day 10^6 cells of the second parent are added to the flask. Four hours later fusion is induced by adding polyethylene glycol (PEG 1000) mixed 1:1 w/v with phosphate-buffered saline, Ca^{2+} and Mg^{2+} free (PBS-CMF). Each flask receives 2 mL of this mixture for 35–45 s, and is then quickly washed three times with PBS-CMF, then alpha-MEM + 10% FCS is added, and cells are incubated overnight. Control flasks are treated identically, omitting the exposure to PEG.

The following day control and PEG-treated cells are trypsinized, counted, and replated at appropriate cell densities to determine hybrid recovery frequencies in selective medium on plastic. In addition to assay for anchorage independence, the fusion mixtures can be plated directly into selective medium containing methylcellulose (Marshall and Sager, 1981).

Initially, hybrids were selected from fusions of 205-30 (CHEF/18, HPRT −) × 204-Bu50 (CHEF/16, TK −) by HAT selection (Sager and Kovac, 1978). (Results of this cross bearing on transformation and tumorigenicity are discussed in Chapter 27.) Extensive chromosome reduction was observed in a fraction of cells from each of 20 hybrid clones, following back selection with thioguanine or BrdU. Approximately half of the hybrid subclones examined were in the diploid range (22–28 chromosomes per cell) and are referred to as reduced hybrids.

Segregation of $HPRT^+/HPRT^-$ was about 1:1 in reduced hybrids selected in BrdU, indicating that selection for TK^- did not affect the segregation of the X chromosome carrying the *HPRT* gene. The coordinate loss of the X chromosome was verified in the thioguanine-resistant subclones.

All reduced hybrids regardless of how selected, that is, with thioguanine or BrdU, were TK$^-$, indicating the loss of two homologous chromosomes from TK$^+$. This unexpected result may reflect the presence of other genes with selective advantage on the TK chromosome from the CHEF/16 parent, since in the reciprocal cross (CHEF/18 TK$^-$ × CHEF/16 TK$^+$ oua-r HPRT$^-$) the hybrid clones were mainly TK$^+$. Hybrid subclones were used to assign the TK gene to chromosome 7 (unpublished). In subsequent studies using microcell transfer for mapping, this assignment was confirmed (unpublished).

A different approach to mapping with cell hybrids was taken in localizing to chromosome 1 a gene involved in the anchorage requirement for growth. In this study (Marshall et al., 1982), a set of hybrids from the fusion 205-30 × 204-Bu50 that were previously shown to be anchorage dependent were plated into HAT-methylcellulose medium to select subclones that had regained anchorage independence of the 204-Bu50 parent. A set of 26 subclones derived from five different hybrids were examined by chromosome banding. The analysis revealed that only chromosome 1 showed a consistent reduction in copy number coincident with the appearance of anchorage independence, a trait sensitive to gene dosage in its expression. Thus, in this study, the correlation between loss of a chromosome and reappearance of a previous suppressed phenotype provided the evidence to localize a gene.

B. Differentiation

In a series of important papers, Jones and Taylor (Taylor and Jones, 1979; Jones and Taylor, 1980, 1981, 1982) demonstrated that a short treatment of 3T3 or 10T ½ cells with 5-azacytidine (azaC) could induce these embryo fibroblastic cells to differentiate into various mesenchymal histotypes including myoblasts/myotubules, adipocytes, and chondrocytes. Since the principal mode of action of azaC is to interfere with methylation of newly replicated DNA (Santi et al., 1983), these results have led to renewed interest in the role of methylation changes in the regulation of differentiation.

CHEF cells respond to azaC in much the same manner as do the mouse cell lines studied by Taylor and Jones (Taylor and Jones, 1979; Jones and Taylor, 1980). We found that different concentrations of azaC preferentially induced particular mesenchymal histotypes: 2 μM had no effect, 3 μM gave primarily adipocytes, 10 μM gave primarily myoblasts, and chondrocytes were seen following treatment with 30 μM azaC. The toxicity increased with the dose, being minimal at 2–3 μM. Mature fibroblasts, unresponsive to azaC, were also found. Presence of azaC during one round of the cell cycle was sufficient, consistent with the detailed studies of Jones and Taylor (1981, 1982).

CHEF cells are especially suitable for studies of differentiation because they are diploid; chromosome rearrangements if present are submicroscopic. Thus, one can investigate whether particular chromosome changes may be involved in commitment or in terminal differentiation. Diploidy also ensures a homogeneous cell population for studies of biochemical or genomic molecular changes occurring during the differentiation process.

CHEF cells have been used for studies of adipocyte differentiation (Sager and Kovac, 1982) but not as yet for other histotypes. With CHEF cells, we have found that the azaC induction of adipocytes, but not of other cell types, can be carried out with insulin in the absence of azaC. Insulin is effective when added as a supplement to serum; effects of concentrations as low as 0.01 μg/mL can be detected on a serum background, although for maximal effects one needs 1–10 μg/mL (unpublished work). Furthermore, as noted above, when CHEF cells are grown in serum-free medium, for which 10 μg/mL insulin is required, adipocyte formation occurs rapidly in confluent cultures. When CHEF/18 cells are grown at cloning density in serum-free medium, adipocytes arise during colony formation.

By growing CHEF cells in the presence of insulin, but removing insulin before confluence, preadipocyte cultures can be established. These cultures differ from the CHEF stem cell population in several respects: (1) They form adipocytes rapidly at confluence following addition of insulin, whereas CHEF stem cells require at least 2 weeks at confluence with added insulin before detectable adipocyte differentiation occurs. (2) They do not respond to azaC by differentiating into other mesenchymal histotypes. (3) They contain a low but measurable amount of the heat-stable isozyme of GPDH (α-glycerophosphate dehydrogenase) not present in CHEF/18 stem cells (unpublished work).

Thus, with these cells the commitment step of differentiating to the preadipocyte stage can be established experimentally and distinguished from further stages of adipocyte development. One may thus investigate the molecular basis of commitment per se, independent of later stages in differentiation.

C. Susceptibility to Natural Killer (NK) Cell Lysis

Natural killer (NK) cells, a subpopulation of lymphocytes, have potent lytic activity against tumor cells (Herberman et al., 1978). Several reports suggest that a target antigen (NK-TA) recognized by NK cells is present on cells that are sensitive to NK-mediated lysis. Some identified antigens with target activity appear to be conserved in evolution, since a variety of T-cell lymphoma tumor cells from several mammalian species have been recognized and killed by xenogeneic NK cells (Hansson et al., 1978). Recent studies have suggested that NK target antigens are differentiation-type antigens present on embryonic cells and reappearing during tumorigenesis (Ahrlund-Richter et al., 1980; Stern et al., 1980; Gidlund et al., 1981).

We have found recently that NK cells of human origin can mediate lysis of tumor-derived CHEF cells (Dubey et al., 1983). This activity has been used to distinguish the tumorigenic potential of a series of transformed cell lines derived by mutation from CHEF/18 cells. In a study of anchorage-independent and low-serum mutants we found that these transformation traits did not correlate well with tumor-forming ability: most transformed lines were nontumorigenic as assayed in the nude mouse. Using human NK-mediated lysis, we then found that only the strongly tumorigenic and tumor-derived CHEF cells were susceptible to lysis, whereas the transformed but nontumorigenic cells were resistant (Dubey et al., 1983). Certain tumorigenic CHEF lines were resistant to lysis, indicating that a positive result may be more significant with this assay than a negative result.

Our initial interest in this system was (i) to show that the presence of NK susceptibility in CHEF cells did not interfere with the nude assay: all NK-susceptible cell lines were also tumorigenic in the nude mouse; and (ii) to develop a cell culture method to identify tumorigenic cells. Since some tumorigenic cell lines were resistant to NK-mediated lysis, the screen may be too stringent for detection of all tumorigenic cells, but further work may produce a reliable alternative to tumor testing in the animal. The nature of the conserved antigen and the regulation of its expression remain problems of great interest.

D. DNA Transfer

CHEF/18 cells have proven to be excellent recipients for DNA transfer by transfection (Smith et al., 1982) using the $CaPO_4$–DNA coprecipitation procedure (Graham and van der Eb, 1973). Subsequent studies with CHEF cells have been carried out principally with cloned recombinant DNA vectors (Sager et al., 1983) although genomic DNA has been successfully transferred at low frequencies, comparable to those reported with NIH/3T3 cells (Smith et al., 1982).

The procedures used do not vary significantly from those of other investigators. Cells are plated at 10^6 per 100 mm dish, incubated overnight, refed with fresh medium 4 hr prior to addition of DNA, which was prepared according to Wigler et al. (1979). The following day (16–18 hr after DNA addition), 3.2 mL of 40% DMSO in cold medium is added dropwise to each dish. After incubation for 30 min, dishes are washed once with medium without serum, then complete medium is added and cells are returned to incubator. Two days later, cells are refed with selective medium. If cells are to be split (e.g., 1:4 or 1:5), it is done at this time.

Splitting of cultures to permit further doublings was found to be important in assays for focus formation with genomic DNA. However, splitting is unnecessary with drug selection, and marginal when plasmid DNA is used as source of transforming genes in the focus assay.

Typical yields of foci following transfection with linear pEJ (Shih and Weinberg, 1982) or pSV2gpt-EJ (Sager et al., 1983) are in the range of 10^3 foci per picomole transforming DNA. Yields are consistently higher with linear than with circular plasmids.

VIII. SUMMARY

In summary, CHEF/18 cells are diploid and nontumorigenic, derived from embryonic secondary mesenchymal stem cells. In culture they appear fibroblastic and give rise to fibrosarcomas in nude mice, following tumorigenic transformation. Tumor-derived CHEF/18 cells contain surface target antigens recognized by human NK cells that are absent from nontumorigenic CHEF cell surfaces.

The principal work with CHEF cells has been devoted to the genetic analysis of tumorigenesis. Studies of transformed mutants derived in one-step mutagenesis, studies of suppression of transformation and of tumor-forming ability in hybrids and cybrids, chromosome studies of transformed and tumor-derived cells, and DNA transfection experiments are discussed in Chapter 27.

In this chapter, the origin of CHEF cells, and their utilization in mutagenesis, cell fusion, gene mapping, growth factor studies, chromosome analysis, and DNA transfection experiments has been presented. In addition, the conversion of the CHEF/18 line from stem cells to preadipocytes by a short treatment with either 5-azacytidine or insulin has been discussed. This finding has provided novel material for investigation of the commitment process in development.

REFERENCES

Ahrlund-Richter, L., Masucci, G., and Klein, G. (1980). *Somatic Cell Genet.* **6**, 89–99.

Cherington, P. V. and Pardee, A. B. (1980). *J. Cell Physiol.* **105**, 25–32.

Cherington, P. V., Smith, B. L., and Pardee, A. B. (1979). *Proc. Natl. Acad. Sci. USA* **76**, 3937–3941.

Cram, L. S., Bartholdi, M. F., Ray, F. A., Travis, G. I., and Kraemer, P. M. (1983). *Cancer Res.* **43**, 4828–4837.

Dubey, D. P., Staunton, D. E., Smith, B. L., Yunis, E. J., and Sager, R. (1983). *Proc. Natl. Acad. Sci. USA* **80**, 7303–7307.

Gidlund, M., Orn, A., Pattenagale, P. K., Hansson, M., Wigzell, H., and Nilsson, K. (1981). *Nature* **292**, 848–849.

Graham, F. L. and van der Eb, A. J. (1973). *Virology* **52**, 456–467.

Hansson, M., Karre, K., Kiessling, R., and Klein, G. (1978). *J. Immunol.* **121**, 6–12.

Herberman, R. B., Nunn, M. E., and Holden, H. T. (1978). *J. Immunol.* **121**, 304–309.

Jones, P. A. and Taylor, S. M. (1980). *Cell* **20**, 85–93.

Jones, P. A. and Taylor, S. M. (1981). *Nucleic Acids Res.* **9**, 2933–2947.

Jones, P. A. and Taylor, S. M. (1982). *J. Mol. Biol.* **162**, 679–692.

Kitchin, R. M. and Sager, T. (1980). *Somatic Cell Genet.* **6**, 75–87.

Leof, E. B., Wharton, W., VanWyk, J. J., and Pledger, W. J. (1982). *Exp. Cell Res.* **141**, 107–115.

Marshall, C. J. and Sager, R. (1981). *Somatic Cell Genet.* **7**, 713–723.

Marshall, C. J., Kitchin, R. M., and Sager, R. (1982). *Somatic Cell Genet.* **8**, 709–722.

Puck, T. T. (1972). *The Mammalian Cell As a Microorganism*, Holden-Day, San Francisco.

Roberts, A. B., Frolik, C. A., Anzano, M. A., and Sporn, M. B. (1983). *Fed. Proc.* **42**, 2621–2626.

Sager, R. and Kovac, P. (1978). *Somatic Cell Genet.* **4**, 375–392.

Sager, R. and Kovac, P. (1982). *Proc. Natl. Acad. Sci. USA* **79**, 480–484.

Sager, R., Bennett, F., and Smith, B. L. (1982). In *Growth of Cells in Hormonally Defined Media* (G., Sato, A. B. Pardee, and D. Sirbasku, eds.), Cold Spring Harbor Laboratory, New York, pp. 231–241.

Sager, R., Tanaka, K., Lau, C. C.,, Ebina, Y., and Anisowicz, A. (1983). *Proc. Natl. Acad. Sci. USA* **80**, 7601–7605.

Santi, D. V., Garrett, C. E., and Barr, P. J. (1983). *Cell* **33**, 9–10.

Shay, J. W. (1982). *Techniques in Somatic Cell Genetics*, Plenum Press, New York.

Shih, C. and Weinberg, R. A. (1982). *Cell* **29**, 161–169.

Siminovitch, L. (1976). *Cell* **7**, 1–11.

Sinclair, W. K. and Morton, R. A. (1963). *Nature* **199**, 1158–1160.

Smith, B. L. and Sager, R. (1982). *Cancer Res.* **42**, 389–396.

Smith, B. L. and Sager, R. (1985). *Somatic Cell Mol. Genet.* (in press).

Smith, B. L., Anisowicz, A., Chodosh, L. A., and Sager, R. (1982). *Proc. Natl. Acad. Sci. USA* **79**, 1964–1968.

Stern, P., Gidlund, M., Orn, A., and Wigzell, H. (1980). *Nature* **285**, 341–342.

Stiles, C. D. and Kawahara, A. A. (1978). In *The Nude Mouse in Experimental and Clinical Research* (J. Fogh, and B. C. Giovanella, eds.), Academic Press, New York.

Stubblefield, E. (1974). In *The Cell Nucleus* (H. Busch, ed.), Academic Press, New York, Vol. 2, pp. 149–162.

Taylor, S. M. and Jones, P. A. (1979). *Cell* **17**, 771–779.

Trosko, J. E., Chu, E. H. Y., and Carrier, W. L. (1965). *Radiation Res.* **24**, 667–672.

Weiss, M. C. and Green, H. (1967). *Proc. Natl. Acad. Sci. USA* **58**, 1104–1111.

Wigler, M., Pellicer, A., Silverstein, S., Axel, R., and Urlaub, G. (1979). *Proc. Natl. Acad. Sci. USA* **76**, 1373–1376.

CHAPTER 5

THE GENETIC MAP OF THE CHINESE HAMSTER AND THE GENETIC CONSEQUENCES OF CHROMOSOMAL REARRANGEMENTS IN CHO CELLS

Michael J. Siciliano
Department of Genetics
University of Texas System Cancer Center
M. D. Anderson Hospital & Tumor Institute
Houston, Texas

Raymond L. Stallings
Genetics Group LS-3 MS 886
Los Alamos National Laboratories
Los Alamos, New Mexico

Gerald M. Adair
Science Park—Research Division
University of Texas System Cancer Center
Smithville, Texas

I. GENE MAPPING METHODOLOGIES IN CHINESE HAMSTER CELLS

A. Problems

Since the birth of somatic cell genetics with the first introduction of sublines of HeLa cells genetically variant for growth under different serum conditions (Puck and Fisher, 1956), the field has grown and approached a wide variety of genetic phenomena including mutation, gene mapping, and gene expression. Particularly, cell lines derived from the Chinese hamster *(Cricetulus griseus)* have been widely used in these studies, because of their excellent adaptation to tissue culture conditions, high plating efficiency, and stable karyotype with a low number of readily identifiable chromosomes. Yet, it is only since 1980 that significant numbers of genetic loci have been mapped onto Chinese hamster chromosomes. Why?

In any classical genetic-mapping study (breeding or somatic cell hybridization) two characteristics must be present with respect to the genetic loci being studied. They must have different identifiable alleles (polymorphism), and alleles must segregate. Limited polymorphism and/or lack of segregation has inhibited both breeding studies and somatic cell hybridization studies aimed at identifying the chromosomal location of genetic loci in the Chinese hamster. These problems are detailed below.

1. Electrophoretic Polymorphisms of Enzyme Loci in Chinese Hamsters

Products of enzyme loci detectable by histochemical staining on nondenaturing gels following electrophoresis of crude extracts from cells or tissues have been used successfully for years as genetic markers. In 1957, Hunter and Markert first applied histochemical staining procedures to starch gels after electrophoresis for the *in situ* demonstration of enzyme activities. The position of the enzyme is marked by a band or zone of stain directly in the gel. Slight differences in migration between two specimens can be easily detected by simple inspection of the gel [see reviews by Shaw (1965); Markert and Whitt (1968); Siciliano et al., (1974)]. These differences have been shown to be due to amino acid substitutions resulting from variant alleles coding for the enzymes. Enzyme loci with a significant frequency of variant alleles (>0.01) within a species are polymorphic and may be studied for linkage and chromosome assignment by standard breeding, cytogenetic, and linkage analyses. This approach has proven to be effective in mapping several enzyme loci in mice [e.g., Womack and Sharp (1976)] where biochemical polymorphisms have been shown to exist in feral mice and between inbred strains (Roderick et al., 1971).

However, since the Chinese hamsters used for research in the West have been established and inbred from a limited number of animals (Chapter 1), extensive electrophoretic polymorphism at enzyme loci for linkage breeding studies would not seem to be available. We studied that possibility (Stallings and Siciliano, 1981*a*) by surveying the electrophoretic mobilities of 43 enzyme gene products in 11 cell lines derived from 9 different animals and in 11 different tissues from 26 additional hamsters. Polymorphisms were detected at only two loci—*ADA* and *AK2*. Electrophoretic banding patterns from tissues, sublines, and subclones of the same genetic source and Hardy–Weinberg distribution of phenotypes from different sources confirmed that

Abbreviations for enzyme loci follow the recommendations of the Committee for Human Gene Nomenclature (Shows et al., 1979). For the enzyme loci referred to in this paper, they are: *ACP1, -2,* acid phosphatase 1, -2; *ADA,* adenosine deaminase; *ADK,* adenosine kinase; *AK1, -2,* adenylate kinase 1, -2; *APRT,* adenine phosphoribosyl transferase; *DHFR,* dihydrofolate reductase; *ENO1,* enolase 1; *ESD,* esterase D; *GAA,* α-glucosidase; *GALK,* galactokinase; *GALT,* galactose-l-phosphate uridyl transferase; *GAPD,* glyceraldehyde-3-phosphate dehydrogenase; *GLO1,* glyoxalase 1; *G6PD,* glucose-6-phosphate dehydrogenase; *GPI,* glucose phosphate isomerase; *GSR,* glutathione reductase; *HPRT,* hypoxanthine phosphoribosyl transferase; *IDH2,* isocitrate dehydrogenase 2; *ITPA,* inosine triosephosphatase; *LARS,* leucyl-tRNA synthetase; *LDHA, -B,* lactate dehydrogenase A, -B; *ME1,* malic enzyme 1; *MPI,* mannose phosphate isomerase; *NP,* nucleoside phosphorylase; *PEPA, -B, -C, -D, -S,* peptidases -A, -B, -C, -D, -S; *PGD,* 6 phosphogluconate dehydrogenase; *PGK,* phosphoglycerate kinase; *PGM1, -2, -3,* phosphoglucomutase 1, -2, -3; *PKM2,* pyruvate kinase M2; *SOD1,* superoxide dismutase 1; *TK,* thymidine kinase; *TPI,* triose phosphate isomerase.

the polymorphisms were based on the inheritance of codominant, autosomal *ADA* and *AK2* alleles that specified electrophoretically variant enzymes. While these results were sufficient to discourage breeding studies to map enzyme loci in the Chinese hamster material available to us, the finding of polymorphic loci in these inbred stocks implied that wild populations of Chinese hamsters may be highly polymorphic and might be of interest to population geneticists and breeders.

It was also of interest to learn that the product of the most common Chinese hamster *ADA* allele comigrated with mouse ADA and was present in 10 of the 11 cell lines studied. The Chinese hamster ovary (CHO) cell line was the only Chinese hamster line homozygous for the less common allele (see Fig. 5.1). We have recently taken advantage of this informative allele to map *ADA* onto chromosome 2 of the mouse in a series of mouse × CHO somatic cell hybrids which segregated mouse chromosomes (Siciliano et al., 1984).

2. Interspecific Electrophoretic Differences and Chromosome Segregation from Somatic Cell Hybrids

The problem of lack of polymorphism is readily solved by looking for electrophoretic differences between homologous loci of different species. Between different mammalian species, there has been ample time for genetic drift and fixation of alleles which specify proteins with different electrophoretic mobilities. Consequently, it has not been surprising to find that out of approximately 40 enzyme loci, the products of which are readily detectable following electrophoresis, all but a few Chinese hamster enzymes are resolvable from human and mouse gene products. Some notable exceptions are the most common hamster ADA and the mouse form (noted in Fig. 5.1), LDHB which is the same in all three species, APRT which cannot be distinguished between human and Chinese hamster, and G6PD which cannot be readily distinguished between mouse and Chinese hamster.

This rich source of variation would theoretically be applicable to the linkage and mapping analysis of Chinese hamster enzyme loci in interspecific somatic cell hybridization experiments. The approach would be analogous to procedures that have been very successful in mapping such loci onto human chromosomes (Ruddle, 1972; Shows, 1974). In those experiments, the procedure involves fusion of human cells with rodent cells, producing proliferating hybrid somatic cells. Such cells would then more or less randomly lose human chromosomes (Weiss and Green, 1967). One then assigns the gene to the specific human chromosome, or syntenic group, by correlating the presence of the human gene product with the retention of a specific human chromosome and/or by correlating the absence of the human gene product with the loss of the specific human chromosome (concordant segregation). Enzyme loci and

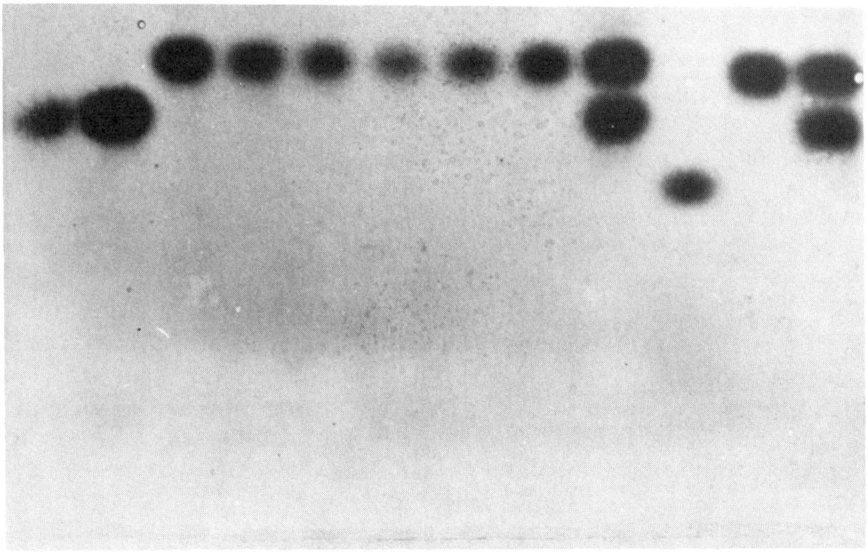

Figure 5.1. *ADA* zymogram. Anodal end is toward the top and the origin is at the cathodal end at the bottom. Tissue samples are lung from a Chinese hamster (Lu1), and lung and liver from a different Chinese hamster (Lu2, Li2). Chinese hamster cell lines CHO and V79 are represented as well as five subclones of the CHEF line (A, B, C, D, E). Also represented are the mouse cell line, Cl1D, and the human cell line, HeLa. Two different electrophoretic forms of *ADA* are seen in Chinese hamster material (fast and slow). Samples have one, the other, or both forms. Samples from the same genetic source have the same form(s). The fast Chinese hamster form is not resolved from the mouse form, and all rodent forms are resolved from human *ADA* [from Stallings and Siciliano (1981*a*)].

chromosomes that demonstrate this concordant segregation are therefore recognized as being members of the same syntenic group. As one can see, the essence of the success of this methodology depends not only on the polymorphisms exhibited between the gene products but also on the fact that the chromosomes of the species of interest (in the preceding case, human) segregate. Therefore, in order to use this method to map genes on Chinese hamster chromosomes, one needs to develop a hybridization scheme in which Chinese hamster chromosomes segregate.

Throughout the 1970s, in a large number of studies where Chinese hamster cells were used in such interspecific fusions, between either human cells or mouse cells, the Chinese hamster chromosomes were retained and the human or the mouse chromosomes segregated. While this phenomenon has been extremely useful in mapping genes onto human and mouse chromosomes, it has frustrated those interested in the assignment of these loci onto Chinese hamster chromosomes.

B. Successful Somatic Cell Procedures in Mapping Chinese Hamster Chromosomes

1. Fusion Between Culture-Established Mouse Cell Lines and Primary Chinese Hamster Cells

In the past, somatic cell hybrids made between mouse and Chinese hamster cells were almost invariably made between tissue-culture-established hamster cell lines and primary mouse cells. As indicated above, this fusion scheme resulted in the segregation of mouse chromosomes and therefore has been useful in the assignment of mouse genes [e.g., Franke et al. (1977); Lalley et al., (1978*a–d*)]. It was noted by Roberts and Ruddle (1980) that the direction of segregation could be reversed if the mouse parent was an established cell line and the hamster parent was derived from primary cells. It was further recognized by these workers, as it was by us (Stallings and Siciliano, 1981*b*), that after picking such hybrids, enzymes for both species were present with hamster enzymes in lesser amounts. We found the expected number and location of intermediately migrating heteropolymeric bands for multimeric enzymes (see Fig. 5.2). C-band chromosome analysis revealed 11–16 Chinese hamster chromosomes/cell/clone. These data indicated that the clones were true hybrids slowly segregating Chinese hamster genes, with each clone extremely heterogeneous with respect to the hamster genes segregated.

In our experiments, we fused Chinese hamster spleen cells with mouse TK-deficient Cl1D cells (Dubbs and Kitt, 1964), by modification of the procedure of Gefter et al. (1977). The modifications consisted of using RPMI medium rather than DME as a diluent of polyethylene glycol (PEG)—the fusing agent—and using spleen cells (rather than tissue culture cells) as the parental Chinese hamster cell stock. Spleen cells were obtained by mincing, in a Petri dish, a spleen which was removed aseptically from a female Chinese hamster sacrificed by cervical dislocation. Following fusion with PEG, hybrids were selected in HAT medium (10^{-4} *M* hypoxanthine, 10^{-5} *M* aminopterin, and 10^{-5} *M* thymidine) (Littlefield, 1964). This scheme allowed the growth of only hybrid cells since the primary hamster spleen cells are incapable of unlimited growth under these conditions and the parental mouse Cl1D cells, since they lack TK, are killed by HAT medium which blocks the *de novo* pathway for purine and pyrimidine synthesis. Hybrid cells are capable of growing because they receive the *TK* gene from the hamster cells and apparently retain the trait for unlimited proliferation from the mouse Cl1D cells.

While it was apparent in this fusion scheme that Chinese hamster chromosomes were segregating, it was impossible to map Chinese ham-

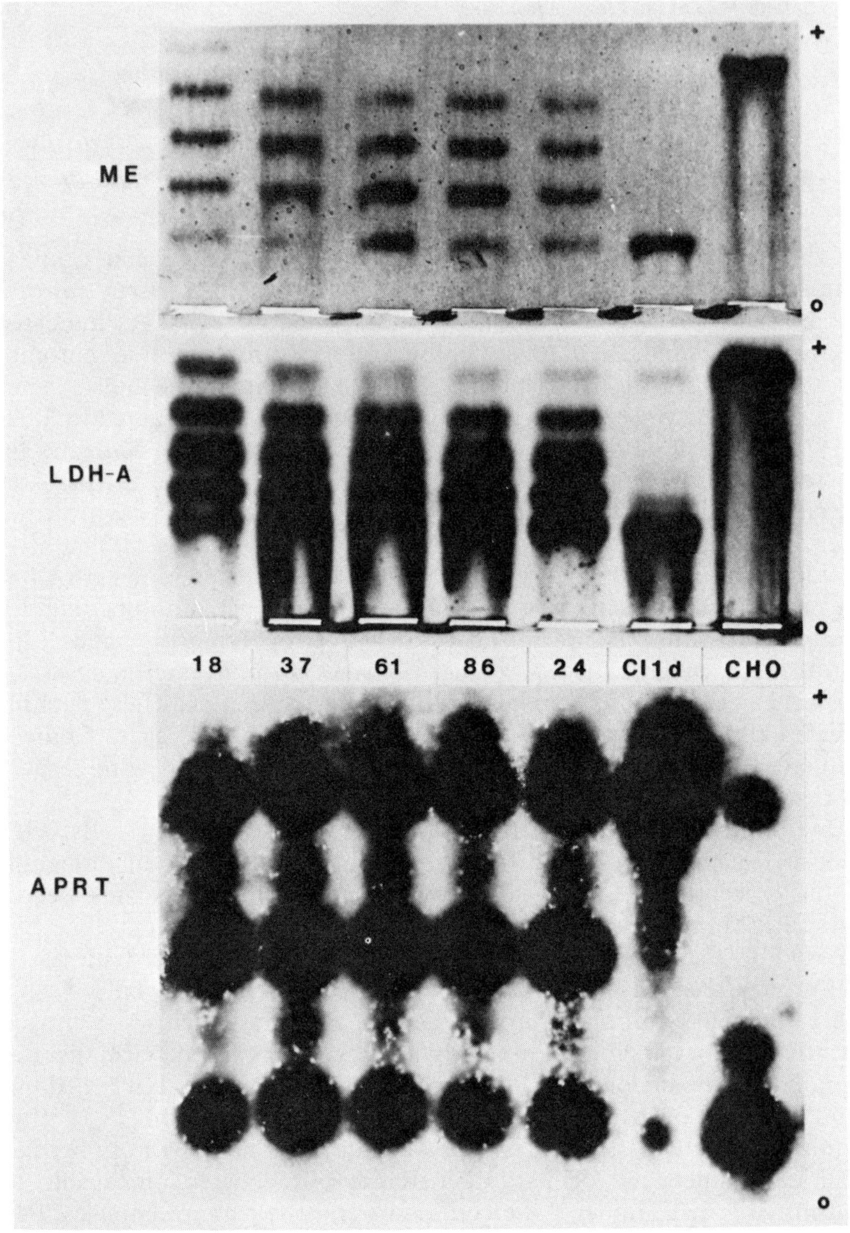

Figure 5.2. Zymograms of *ME* (top), *LDHA* (middle), and *APRT* (bottom) demonstrating the location of the mouse forms (Cl1D), Chinese hamster forms (CHO) and the patterns in five primary, interspecific, somatic cell hybrids sampled at passage 4 (18, 37, 61, 86, and 24). Anodal ends (+) and origins (O) are indicated at the right. Since *ME* and *LDH* are tetramers, three heteropolymeric bands are expected, migrating between the mouse and hamster homopolymers, in true hybrids in which both mouse and hamster subunits are being produced in the same cell. Since *APRT* is dimeric, only one intermediately migrating heteropolymer is expected. These expectations being met, the true nature of the hybrids is validated. Lesser or absent expression of the hamster homopolymer in many hybrids suggests the segregation of hamster chromosomes. Presence of heteropolymers in all hybrids indicates that segregation is incomplete.

ster genes using those initial primary hybrid clones because segregation was slow and no clone was composed of cells all of which had segregated the same Chinese hamster chromosome. Clearly, further steps were going to need to be taken for Chinese hamster gene assignments to be made.

2. *Subcloning and Clone Panel Formation*

In order to allow segregants for Chinese hamster chromosomes to accumulate, each hybrid clone was allowed to grow for 30 passages and was then subcloned. Subcloning was accomplished by the limiting dilution technique. In it, a suspension of cells was made such that one hybrid cell was present in 0.2 ml of medium. Then 0.2ml of medium was deposited into each well of a 96-well Falcon microwell test plate. After a week, the plates were scanned and each well containing a single colony of growing cells was identified. Each of those colonies or subclones was expanded for both isozyme and chromosome analysis. From the independent hybrid clones and subclones, a clone panel was established to assign Chinese hamster enzyme loci to chromosomes. Included in the panel were at least one representative subclone from each of the independent hybrids and as many additional subclones as had different complements of Chinese hamster chromosomes. An example of the first clone panel we produced (Stallings and Siciliano, 1981b) is presented in Table 5.1.

As can be seen from Table 5.1, we were able to make assignments of isozyme loci that were located on chromosomes that segregated in a significant number of the members of the hybrid clone panel. Therefore, by this technique we were able to assign *ADK, ESD, GLO, GSR, NP, PEPB, PEPS,* and *PGM2* to chromosome 1; to confirm the assignments of *ENO1, PGD,* and *PGM1* to chromosome 2; and to provisionally assign *GALT* to chromosome 2. Although there was an indication that *TPI* might be on Chinese hamster chromosome 8, the presence of only two segregants precluded a firm assignment at that time.

These data indicated some further problems lie in mapping Chinese hamster genes. Clearly, chromosomal segregation was not random. For chromosomes which segregated readily, for example the 1 and 2, assignments were not too difficult to make. However, for enzymes on chromosomes which tended not to segregate, many more hybrids and rounds of subcloning would be necessary in order for the assignments to be made. By those procedures, we were eventually able to make the assignments to *TPI* and other isozyme loci—these and other assignments are described in the next section. However, the necessity to devise methods either to speed up segregation or to at least select for segregants which may be in low frequency in the heterogeneous hybrid population of cells, had been recognized.

TABLE 5.1
Chinese Hamster Chromosomes Segregated and Isozymes Present in Interspecific Hybrid Clone Panel Members[a]

Clone Panel Member	Chromosomes Segregated	Presence (+) or absence (−) of Informative[b] Chinese Hamster Loci															
		ADK	*ENO1*	*ESD*	*GAA*	*GALT*	*GLO1*	*GSR*	*NP*	*PEPA*	*PEPB*	*PEPS*	*PGD*	*PGM1*	*PGM2*	*SOD1*	*TPI*
C1/4	X	+	+	+	+	+	+	+	+	+	+	+	+	+	+	+	+
C2/10	None	+	+	+	+	+	+	+	+	+	+	+	+	+	+	+	+
C2/A1	X	+	+	+	+	+	+	+	+	+	+	+	+	+	+	+	+
C2/E1	1, 2, X	−	−	−	+	−	−	−	−	+	−	−	−	−	−	+	+
C3/E1	None	+	+	+	+	+	+	+	+	+	+	+	+	+	+	+	+
C4/3/E1	None	+	+	+	+	+	+	+	+	+	+	+	+	+	+	+	+
C5/4	X	+	+	+	+	+	+	+	+	+	+	+	+	+	+	+	+
C6/1	2	+	−	+	+	−	+	+	+	+	+	+	−	−	+	+	+
C8/1	X	+	+	+	+	+	+	+	+	+	+	+	+	+	+	+	+
C8/E2	None	+	+	+	+	+	+	+	+	+	+	+	+	+	+	+	+
C9/2	X	+	+	+	+	+	+	+	+	+	+	+	+	+	+	+	+
C9/E1	None	+	+	+	+	+	+	+	+	+	+	+	+	+	+	+	+
C10/4	None	+	+	+	+	+	+	+	+	+	+	+	+	+	+	+	+
C11/3	1, X	−	+	−	+	+	−	−	−	+	−	−	+	−	−	+	+
C11/3B5	1, 2, 5, 8, X	−	−	−	+	−	−	−	−	−	−	−	−	−	−	−	−
C11/3E6	1, 8, X	−	+	−	+	+	−	−	−	+	−	−	+	−	−	+	−
C12/9	1, 2, 3, X	−	−	−	−	−	−	−	−	+	−	−	−	−	−	+	+
C14/2/A1	1, 2, X	−	−	−	+	−	−	−	−	+	−	−	−	−	−	+	+
C14/2/E1	None	+	+	+	+	+	+	+	+	+	+	+	+	+	+	+	+
C16/1	None	+	+	+	+	+	+	+	+	+	+	+	+	+	+	+	+
C17/A1	2, X	+	−	+	+	−	+	+	+	+	+	+	−	−	+	+	+
C17/E1	None	+	+	+	+	+	+	+	+	+	+	+	+	+	+	+	+

[a] From Stallings and Siciliano (1981b).

[b] Noninformative isozyme loci did not segregate from any clone panel members. They were *ACP2, APRT, AK1, GPI, GAA, IDH2, ITPA, LDHA, ME1, MPI, PGM3, PKM2, PEPD,* and *PEPC*.

3. Selective Systems for Informative Segregants

Roberts and Ruddle (1980) constructed a series of somatic cell hybrids between primary Chinese hamster embryo fibroblasts and mouse A9 cells deficient in HPRT. They were the first to note that in such a fusion scheme each clone became a heterogeneous population of cells with every cell containing an intact mouse genome but which lost several Chinese hamster chromosomes. Rather than going through a whole series of subclonings and relying on random segregation of a particular Chinese hamster chromosome, they devised a method to select for hybrid subclones that had lost a particular Chinese hamster chromosome. They took advantage of earlier observations (Creagan et al., 1975; Draper et al., 1979) that mouse cells are approximately 10^4-fold more resistant to diphtheria toxin (DT) than Chinese hamster cells. Chinese hamster cells may therefore be viewed as being DT sensitive (DTS) due to genetic factors at the *DTS* locus. Consequently, by subjecting hybrids to levels of DT that normally kill Chinese hamster cells but which were nontoxic to mouse cells, they were effectively able to select for hybrids that segregated the Chinese hamster chromosome bearing the *DTS* locus. By this procedure, they determined that *DTS* segregated concordantly with Chinese hamster chromosome 2 and that these same hybrids also segregated Chinese hamster *PGM1, PGD,* and *ENO1.* They were therefore successfully able to provisionally assign *DTS, PGM1, PGD,* and *ENO1* to Chinese hamster chromosome 2—assignments that we were later able to confirm (Stallings and Siciliano, 1981b; Stallings et al., 1982).

We applied the concept of selection for hybrid clones that may have randomly segregated certain Chinese hamster chromosomes from hybrids to our hybrid clone panel described above. We took advantage of the fact that the mouse cells used in the fusion were TK deficient, and that the *TK* gene present in the hybrids that allowed them to survive in HAT medium was on a Chinese hamster chromosome. Since bromodeoxyuridine (BrdU) will select against cells that have TK (Marin, 1969; Green et al., 1971), we selected a matched set of subclones in either HAT or BrdU for survivors. Of the 24 BrdU-resistant colonies picked, 23 segregated Chinese hamster chromosome 7 whereas all 13 HAT-selected colonies retained a 7. Also segregating concordantly with BrdU resistance were Chinese hamster isozymes for GALK and ACP1. Therefore, by this technique, we were able to rapidly assign *TK, GALK,* and *ACP1* to Chinese hamster chromosome 7. The one BrdU-resistant clone which had the chromosome 7 also had Chinese hamster forms of GALK and ACP1. Since we could observe no gross morphological alterations in chromosome 7 of that clone, its BrdU resistance was due to some mechanism other than segregation of a visible chromosomal segment bearing the *TK* gene.

Cases of discordance, where a chromosomal alteration is seen, are often useful in making a regional assignment of a gene to a particular segment of a chromosome. For instance, in this same study (Stallings and Siciliano, 1981b) 22 subclones of one particular independent clone which retained a chromosome 2 had, as expected, the Chinese hamster forms of *ENO1, PGD,* and *GALT.* However, they were deficient in Chinese hamster *PGM1.* As opposed to other clone panel members retaining Chinese hamster chromosome 2, all subclones of this particular clone had an interstitial deletion in the long arm of the chromosome (see Fig. 5.3) which enabled the regional assignment of *PGM1* to that particular segment.

We used a variation of this procedure to obtain segregants for *APRT* and loci that might be linked to it, since we had obtained no Chinese hamster segregants for that important enzyme (see Fig. 5.2). Since azaadenine (AA) will select against cells with APRT, mutagenesis and AA selection were used (Adair et al., 1980) to isolate an APRT-deficient derivative of mouse Cl1D cells (LTAO). LTAO cells were fused with primary Chinese hamster cells to generate a series of hybrids, in which random segregation was allowed followed by selection in AA for hamster *APRT* segregants (Adair et al., 1983a). Segregants were obtained (Fig. 5.4) concordantly with the loss of Chinese hamster chromosome 3 and the Chinese hamster forms of LDHA, IDH2 and GAA. Exceptional subclones, in which coordinate segregation of these syntenic markers was disrupted by chromosome breakage or deletions, allowed further localization of these genes to specific regions of the 3 chromosome. These will be described further in Section II in the discussion of Chinese hamster chromosome 3 loci.

4. *Fusion of CHO Cells with Culture-Adapted Mouse Cells*

A major method of regionally assigning genes or blocks of genes to segments of chromosomes is to map the genes in cells in which rearrangements and translocations have taken place. CHO cells are a potentially rich source for this type of variation. Detailed C-band and G-band analysis of the karyotype of this cell line revealed only eight of the chromosomes to be normal when compared with euploid Chinese hamster chromosomes. In the remaining chromosomes (Z chromosomes), evidence was found of translocations, deletions, and pericentric inversions (Deaven and Petersen, 1973). Furthermore, it was found that these variations were quite stable in that the same basic karyotype was found in a majority of uncloned cells as well as in most cells of several independent clones (Worton et al., 1977). Clearly, if the isozyme loci which were successfully mapped onto euploid Chinese hamster chromosomes from normal diploid cells could be assigned onto the identified translocations in CHO cells, the regions involved in those translocations

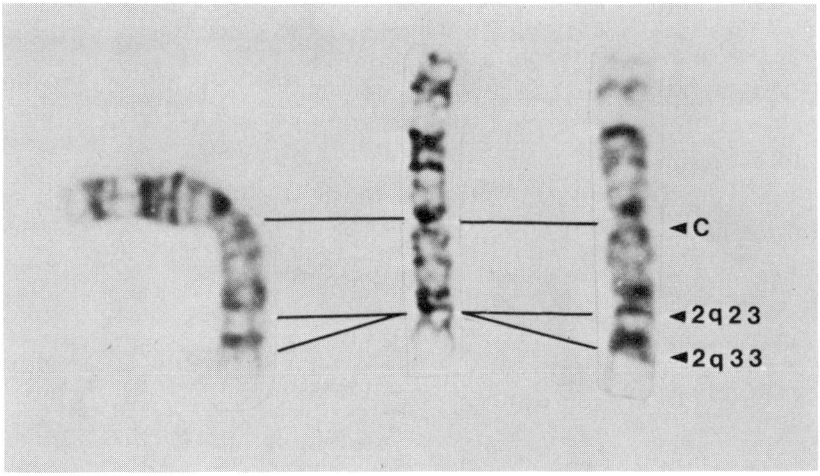

Figure 5.3. Chromosomes on the left and right are normal-appearing G-band chromosome 2s derived from hybrid clones expressing all four genetic loci (*PGM1, PGD, ENO1,* and *GALT*) asssigned to chromosome 2. The center chromosome is derived from a hybrid subclone which expresses all of the chromosome 2 loci except *PCG1*. The centromere (C) and the boundaries of the deleted region (2q23 to 2q33) are indicated. This deletion permits regional assignment of the *PGM1* locus to the deleted region [from Stallings and Siciliano (1981*b*)].

would indicate the region of Chinese hamster genome in which these loci were located.

In addition, knowledge of the location of enzyme loci on CHO chromosomes would determine the impact of the rearrangements on the expression and mutability of those loci, and provide biochemical markers in this cell line, which has been and likely will continue to be used for a wide variety of somatic cell genetic experiments. Unfortunately (for this purpose), CHO cells are a classical, culture-adapted cell line, the chromosomes of which had been shown to be invariably retained in interspecific hybrids. It was therefore much to our surprise and delight to find that hybrids made with our HPRT-deficient CHO subline (TGA102a) and mouse TK^- Cl1D cells segregated CHO chromosomes (Stallings et al., 1982; Siciliano et al., 1983a).

Subsequent experiments (Stallings et al., 1983; Adair et al., 1983b; Adair et al., 1984; Stallings et al., 1984a) using at least five different CHO cell lines derived from various laboratories in North America revealed that the extraordinary segregation characteristics of CHO × Cl1D hybrids was not an aspect of the particular CHO cell line that we initially used in our experiment because, irrespective of the particular subline of CHO cells used, CHO chromosomes always segregated. We therefore assume that the direction of segregation was dic-

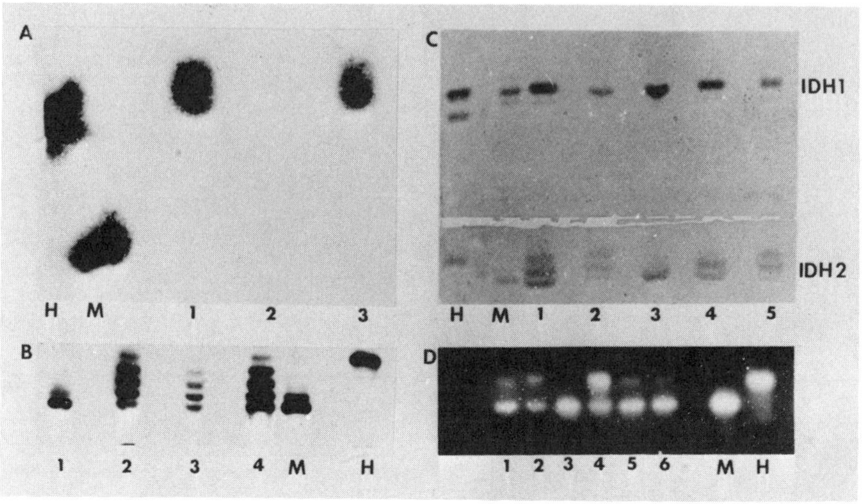

Figure 5.4. Zymograms showing APRT (A), LDHA (B), IDH (C), and GAA (D) activity. Channels containing hamster and mouse controls are labeled H and M, respectively, while representative members of a hybrid clone panel are labeled with Arabic numerals. All enzymes, except IDH2, migrate anodally. (A) Mouse control is the $APRT^{+}$ Cl1D cell line. No mouse APRT is observed in hybrids (channels 1–3) since an $APRT^{-}$ derivative of Cl1D (LTAO) was use for hybridization. Hybrids retaining hamster APRT are in channels 1 and 3, while channel 2 contains an AA^{r} hybrid having segregated hamster APRT. (B) Since LDHA is a tetrameric enzyme, hybrids retaining both hamster and mouse genes (channels 2, 3, and 4) possess heteropolymeric bands. A segregant for LDHA is in channel 1. (C) Positions of both the IDH1 and IDH2 enzymes are labeled on the right. Since IDH2 is dimeric, hybrids with the Chinese hamster gene product express at least the intermediately migrating heteropolymer (channels 1, 2, 4, and 5). A segregant for hamster IDH2 is in channel 3. (D) GAA, being monomeric, possess a two-banded pattern in hybrids retaining hamster GAA (channels 1, 2, 4, 5, and 6). Channel 3 contains a GAA segregant [from Adair et al. (1983*a*)].

tated by the Cl1D genome and suggest that mouse cell line would probably be an appropriate fusion partner for mapping Chinese hamster chromosomes present in other long-term culture-adapted Chinese hamster cell lines. While we might like to claim otherwise, we must admit that our initial choice of Cl1D as a fusion partner was purely fortuitous.

As in the hybrids made with Chinese hamster primary cells, the segregation of CHO chromosomes from the hybrids was slow, requiring many passages and subcloning cycles to isolate hybrids that completely segregated particular CHO chromosomes. Segregation was also nonrandom. It is curious to note, however, that CHO chromosomes derived from euploid Chinese hamster chromosomes which segregated quickly, also segregated quickly and the derivatives of euploid Chinese hamster chromosomes which tended to segregate slowly, also segregated slowly. This information implies that the rate of segregation is intrinsic to the genetic material on the chromosomes themselves rather than to any extrinsic factors. Nonrandom segregation has been shown to be generally widespread phenomena in somatic cell genetic experiments; it has been observed for mouse chromosomes segregating from primary mouse cells fused with Chinese hamster V79 cells (Franke et al., 1977) and for human chromosomes segregating from rodent × human hybrids (Croce et al., 1973; Norum and Migeon, 1974). Franke et al. (1977) observed, however, that which individual chromosomes are preferentially segregated is very often a function of the segregating set's fusion partner. Clearly, the whole question of direction of chromosomal segregation as well as the question of which chromosomes of a set are preferentially segregated remains one of the mysteries of somatic cell genetics.

As with hybrids made with euploid Chinese hamster cells, drug selection can once again be used to isolate hybrids rapidly that segregate CHO chromosomes responsible for drug sensitivity and to identify the chromosomes bearing those and linked loci. We therefore repeated the protocols described above for DT, BrdU, and AA selection to identify the CHO chromosomes carrying the appropriate loci. Results of the experiments will be described in Section II.

5. *Introduction of Selectable Markers into Culture-Adapted Cells*

After using the *DTS, TK,* and *APRT* loci to select for segregants that had lost particular CHO chromosomes, we ran out of useful loci for that purpose. However, the advantage of having culture-adapted CHO cells, which segregate chromosomes after fusion with mouse cells, is that selectable markers for further genetic analysis can be introduced into the lines. There are two general ways to do this: by mutagenesis followed by selection for a particular marker or by the introduction of a selectable marker by DNA transfection and subsequent random incorporation of the marker onto CHO chromosomes. We have taken advantage of this latter procedure by using CHO TK^- cells that were transformed by Abraham et al. (1982) with the cloned gene for herpes simplex virus TK. Stable TK-transformed CHO cells were fused with TK^- mouse Cl1D cells, and hybrids were allowed to segregate randomly followed by the usual protocol of BrdU or HAT selection. BrdU-selected hybrid subclones tended to lose the CHO Z5 chromosome implying that that was the site at which the herpes TK gene had been incorporated and aided in the identification of hamster enzyme loci located on that chromosome (Adair et al., 1984). Details of that assignment are described subsequently in the section on gene mapping.

6. *Electrophoretic Shift Mutations*

Electrophoretic shift mutations induced at isozyme loci in CHO cells have been produced at 24 different loci (Siciliano et al., 1978, 1983*b*; Stallings et al., 1982, 1983; Adair et al., 1984; Stallings et al., 1984*a*). In

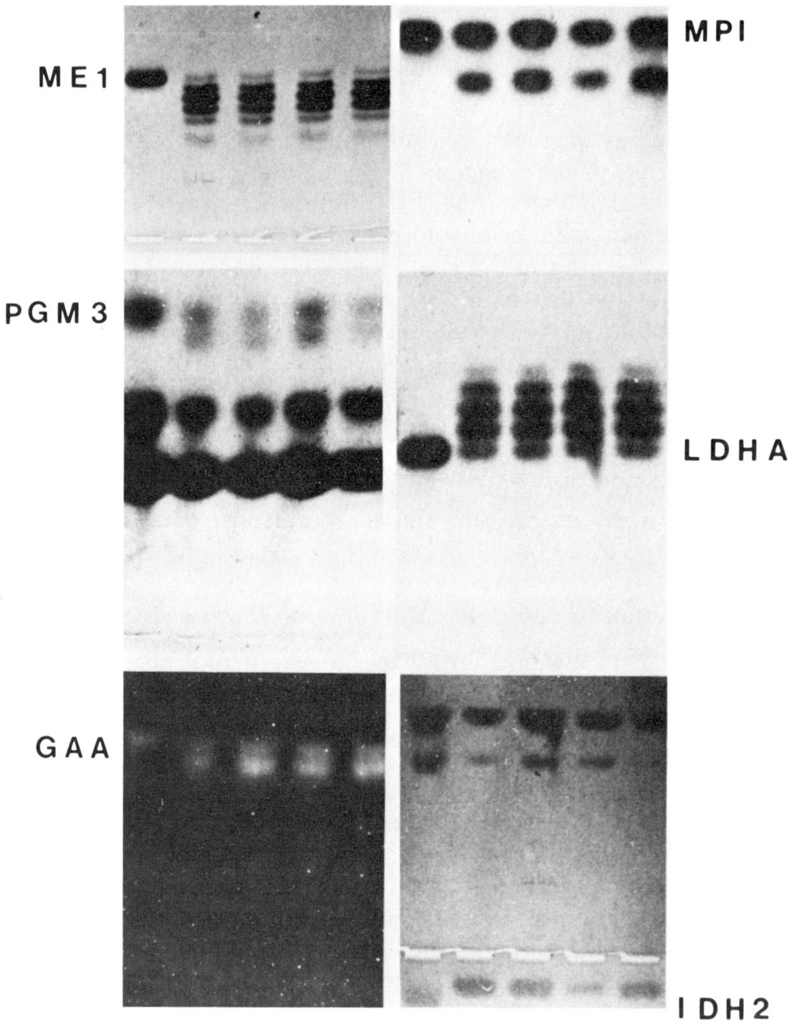

Figure 5.5. Zymograms of LDHA, GAA, IDH2, ME1, PGM3, and MPI, showing the characteristic multibanded or single-band shift isozyme patterns observed for electrophoretic mobility shift mutants for these enzyme loci. In each panel, the leftmost channel is the wild-type CHO cell control; the other samples represent subclones of a representative mutant isolate. Multiple band shift mutants, if they are truly indicative of induced heterozygosity, have patterns consistent with the subunit structures of the enzyme involved. For tetrameric enzymes (e.g., ME1 and LDHA) the patterns are five banded, for dimeric enzymes (not shown here but exemplified in Fig. 5.8) the patterns are three banded, and for monomeric enzymes (e.g., MPI, PGM3, and GAA) the patterns are two banded. IDH2 is an example of a single-band shift mutation faithfully replicated in all four subclones. Additional zones of activity for PGM, ME, and IDH represent products of other isozyme loci (from Adair et al., 1984).

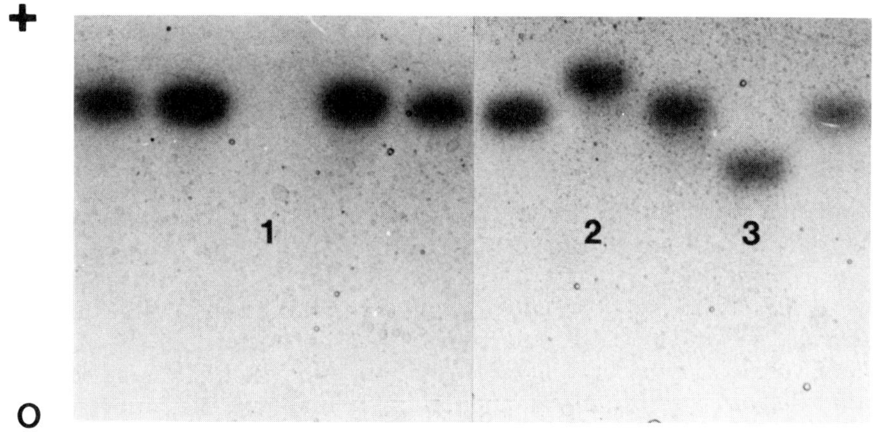

Figure 5.6. Zymogram showing three (1, 2, and 3) *PEPD* mutants indicative of the presence of only a single functional allele for this locus in CHO cells. Mutant 1 is a null (no activity), while 2 and 3 indicate two different single-band shift mutations.

these studies, CHO cells have been exposed to a mutagen—either ultraviolet light (UV) or ethylmethane sulfonate (EMS). Following mutagenic exposure, cells were allowed to divide twice and then were cloned by limiting dilution as described previously. Individual clones were expanded to approximately 60 million cells and subjected to electrophoretic analysis for variation at approximately 40 isozyme loci. Most variants fell into one of three categories: a multiple-banded electrophoretic shift consistent with subunit structure of the enzyme indicating induced heterozygosity, a single band shift indicative of hemizygosity, or null (gene product is no longer produced). Variants were proven to be mutants by analysis of subclones (see Fig. 5.5 for examples). In a combination of mapping and electrophoretic shift analysis of three loci in CHO cells *(TPI, PEPD,* and *GPI),* multiple-banded shift patterns were shown to be indicative of functional dizygosity of the enzyme locus in CHO cells whereas single-band shift and null mutants (Fig. 5.6) were shown to be due to the presence of only a single allele in those cells (Siciliano et al., 1983a). Such mutants therefore help verify the chromosomal assignments of enzyme loci mapped in CHO cells.

7. Intraspecific Hybridization

Some of the initial chromosomal assignments of Chinese hamster loci were obtained as a result of intraspecific hybridization of Chinese hamster cells. Farrell and Worton (1977) fused CHO cells with CHO cells, or CHO cells with CHW cells [a male culture-established line with a marker X chromosome (Lin et al., 1971)]. In these fusions, one line was $HPRT^+$ and the other parent was $HPRT^-$ producing a hybrid heterozygous for the marker and able to grow in HAT, indicating dominance of the $HPRT^+$ phenotype. Segregants resistant to 6-thioguanine (6TG) were selected (selects for $HPRT^-$ cells), and karyotype analysis of survivors indicated the correlation of the segregation of the $HPRT^+$ phenotype with the X chromosome. In previous intraspecific hybridization experiments (Rosenstrauss and Chasin, 1975), CHO cells deficient for G6PD as well as HPRT were fused with wild-type CHO cells, and chromosomes were allowed to segregate. Hybrids were then selected for 6TG resistance ($HPRT^-$), and *G6PD* was shown to segregate concordantly with *HPRT.* Therefore, in these intraspecific fusions, *HPRT* and *G6PD* were shown to be X-linked in the Chinese hamster.

Using similar approaches of intraspecific hybridization allowing segregation and then selection for drug-sensitive markers, Wasmuth and Chu (1980) and Campbell and Worton (1980) took advantage of the deletion (2q1–q24) in the long arm of one of the chromosome 2s (the Z2) in CHO cells to locate the genes for *LARS, emetine resistance (EMT),* and *chromate resistance (CHR)* to that region of chromosome 2 in the Chinese hamster.

Adair and Siciliano (manuscripts in preparation) have recently used the combination of intraspecific hybridization, electrophoretic shift mutants, and drug-resistance markers to approach the assignment of a large number of loci responsible for various forms of drug resistance in CHO cells. The primary focus of this work is the mapping of hemizygous drug-resistance markers and associated enzyme loci. In a recent experiment (which exemplifies others that are in progress), an *IDH2* electrophoretic mobility shift mutant was hybridized with a *GPI* electrophoretic mobility shift mutant carrying a methyl-glyoxalbisguanylhydrozone- (MBG-) resistance marker. As previously indicated, both mapping and electrophoretic mutation analysis indicated *GPI* to be a hemizygous locus in CHO cells so that the electrophoretic marker in one cell type was a single-band shift for GPI. As will be seen in the next section (and in Fig. 5.5), *IDH2* is also hemizygous in CHO cells, located on a different chromosomal segment than *GPI,* with the cells bearing a mutation also resulting in a single-band shift. Hybrids between the two cell types produced three-banded patterns for each enzyme, consistent with their dimeric structures, and the presence of wild-type and mutant alleles for each locus in the hybrid. Since MBG resistance is selected with high frequency in CHO cells (Gupta and Singh, 1982), we suspected that it may also be hemizygous and may segregate with either *IDH2* or *GPI.* Hybrids were cultured for several passages to allow segregation, and then independent populations were plated into MBG-containing medium to select for hybrid cells that had lost the chromosome carrying the wild-type allele for MBG. MBG-resistant clones were subsequently analyzed electrophoretically to determine whether the wild-type allele for *GPI* or the variant allele for *IDH2* had cosegregated with *MBG.* Concordant segregation of the variant allele for *IDH2* was observed in all 16 MBG-selected hybrid segregants, indicating the probable linkage of *MBG* with *IDH2* onto CHO chromosome Z3. This result not only provided an assignment for *MBG* in the Chinese hamster genome, but also indicated the chromosomal basis for that locus's hemizygosity and high mutation frequency.

8. *Microcell-Mediated Gene Transfer*

A useful mechanism to get around problems of segregation would be to presegregate chromosomes into microcells before fusion. The principles of this technique were first described by Fournier and Ruddle (1977), who used the method to transfer murine chromosomes into mouse, Chinese hamster, and human somatic cells. By this procedure, cells were cultured for 24–48 hr in the presence of a mitotic arrestor (colchicine). During this prolonged treatment, small groups of chromosomes become enveloped by nuclear membrane, as it reforms, resulting in a cell with multiple small nuclei each containing a subset of the chromosome complement (Stubblefield, 1964). These microcells, as they are now called, are centrifuged out of the cell in the presence of cytochalasin B and collected as a pellet. The microcells may then be used for intra- or interspecific fusion. Worton et al. (1981) used the procedure to transfer methotrexate resistance (*DHFR* locus) from micronucleated methotrexate-resistant CHO cells into sensitive CHO cells. The transfer of the marker correlated with the transfer of Chinese hamster chromosome 2. While this procedure might also be considered a shortcut to obtaining segregants in interspecific hybridizations, to date we know of no other case where that has been done to map Chinese hamster chromosomes.

9. *In situ Hybridization*

The ultimate method, and as yet the only method, to avoid all problems associated with polymorphisms or segregation in mapping studies would be the ability to view directly the site of a gene of interest on a metaphase spread of chromosomes. This procedure has been used to identify the site of *ribosomal RNA (rRNA)* genes in the Chinese hamster genome (Hsu et al., 1975). The basis of all *in situ* hybridization procedures involves the use of a radioactive nucleic acid probe for the gene or genes under study. The hot probe is added directly to slides of cells in metaphase upon which the DNA of the chromosomes has been denatured. The probe then selectively hybridizes to the regions of the genome that code for it, and those regions are visualized by autoradiography. While this has proven successful in mapping the *rRNA* genes in Chinese hamster cells, that success was probably in no small measure owing to the fact that these genes are in multiple copies in the regions where they are located so that hybridization is relatively specific.

The method therefore could also be used to identify the location of amplified sequences. This has been done by Nunberg et al. (1978) using a probe for a *DHFR* gene on metaphase spreads of Chinese hamster cells in which the gene was amplified by selection for methotrexate resistance producing a homogeneously staining region (HSR). The probe localized on chromosome 2, consistent with its location determined by other methods (Worton et al., 1981).

Theoretically, the procedure could be used to identify the location of any gene for which a probe was available. However, for single copy genes the hybridization would be much less specific, and one would need a statistical approach in scoring the location of silver grains among the different chromosomes in as many as 100 or more different metaphases (Harper et al., 1981). To date, we do not know of any study where this latter approach has been used to map Chinese hamster loci.

II. THE CHINESE HAMSTER GENE MAP

In this section, we will describe the location of all genes mapped to date on euploid Chinese hamster chromosomes and on the normal and rearranged Z-group chromosomes in CHO cells. We shall therefore also be considering the genetic basis for the origin of the Z-group chromosomes and be able to compare those conclusions with the cytogenetic observations of Deaven and Peterson (1973) and Worton et al. (1977). First, however, it is necessary to discuss some inconsistencies in the literature with respect to the nomenclature of Chinese hamster chromosomes.

A. Karyotypes and Nomenclature

Ray and Mohandas (1976) proposed a banding nomenclature for Chinese hamster chromosomes which was consistent with the procedure adopted for human chromosomes at the Paris conference (1972). There are two aspects of this new nomenclature that represent departures from what was formerly in the literature and what is occasionally seen after 1976.

By the former nomenclature, chromosomes were arranged by decreasing size and numbered 1, 2, X, 4, . . . , 11. In the new, approved system, the sex chromosomes were moved to the end and autosomes 4–11 were renumbered 3–10.

The second change involved chromosome 2. Besides the G-band patterns, there are two major landmarks which help distinguish the p (short) and q (long) arms of this slightly submetacentric chromosome. On one arm is the *DHFR* gene, which when amplified produces an HSR

(Biedler and Spangler, 1976; Nunberg et al., 1978). On the other arm is the large region which is deleted (forming the Z2 chromosome) in CHO cells (Deaven and Peterson, 1973) containing *EMT* as well as other genes (Campbell and Worton, 1980). Formerly, the arm with *EMT* was called p, and the arm with *DHFR* was called q. After G-band analysis and measuring both arms, the p and q arms were reversed by Ray and Mohandas (1976) so that now the arm bearing *EMT* (and the deletion on the Z2 in CHO) is the q, and the arm to which *DHFR* has been mapped (and produces an *HSR* after methotrexate selection) is the p. We measured 29 Chinese hamster chromosome 2s from Chinese hamster cell lines (DEDE, DON, and CHO) and determined p/q ratios using the revised nomenclature. For nine chromosomes, the ratio was < 0.95, for 18 chromosomes the ratio was 0.95–1.05, and for two chromosomes the ratio was > 1.05 (the range was 0.85–1.09). These data are consistent with the new nomenclature, which is now being widely accepted [e.g., compare Worton et al. (1977) with Campbell and Worton (1980)]. G-banded karyotypes of normal Chinese hamster and CHO cells are shown in Fig. 5.7.

B. Mapping and Activity of Genes on Normal and Derived Z-Group Chromosomes

1. Chromosome 1

Clone panel analysis of hybrids made between primary Chinese hamster spleen cells × mouse Cl1D cells (Stallings and Siciliano, 1981*b*) revealed that the hamster chromosome 1 segregated in 6 of 22 panel members. Segregating concordantly and therefore assigned to chromosomes 1 were *GLOl, GSR, PEPB, PEPS, PGM2, ADK, NP,* and *ESD* (Table 5.1).

In hybrids made with the CHO-LA line × Cl1D cells, all markers segregated concordantly with the normal chromosome 1 of CHO cells, confirming their assignment (Stallings et al., 1984*a*). However, only seven markers, all but *ESD,* segregated concordantly with the Z1. Deaven and Peterson (1973) noted by G-band analysis of CHO-LA cells that the Z1 is an intact 1 except for a missing dark band at the end of the short (p) arm (compare 1 and Z1 on Fig. 5.7). Together, these data allowed the regional assignment of *ESD* to the dark band at the end of the normal Chinese hamster 1p.

Electrophoretic shift mutations for *ESD* in CHO-LA cells, however, indicated that the *ESD* allele from Z1 was not simply deleted but was translocated since the patterns for the mutants indicated diploidy in the cells (Fig. 5.8). Since chromosome Z6 in CHO-LA cells is a chromosome 5 with an extra band on the q arm (Deaven and Peterson, 1973), and since the Toronto strain of CHO cells has two normal 1s and two normal 5s (Worton et al., 1977), we suggested that the Z1 and Z6 in

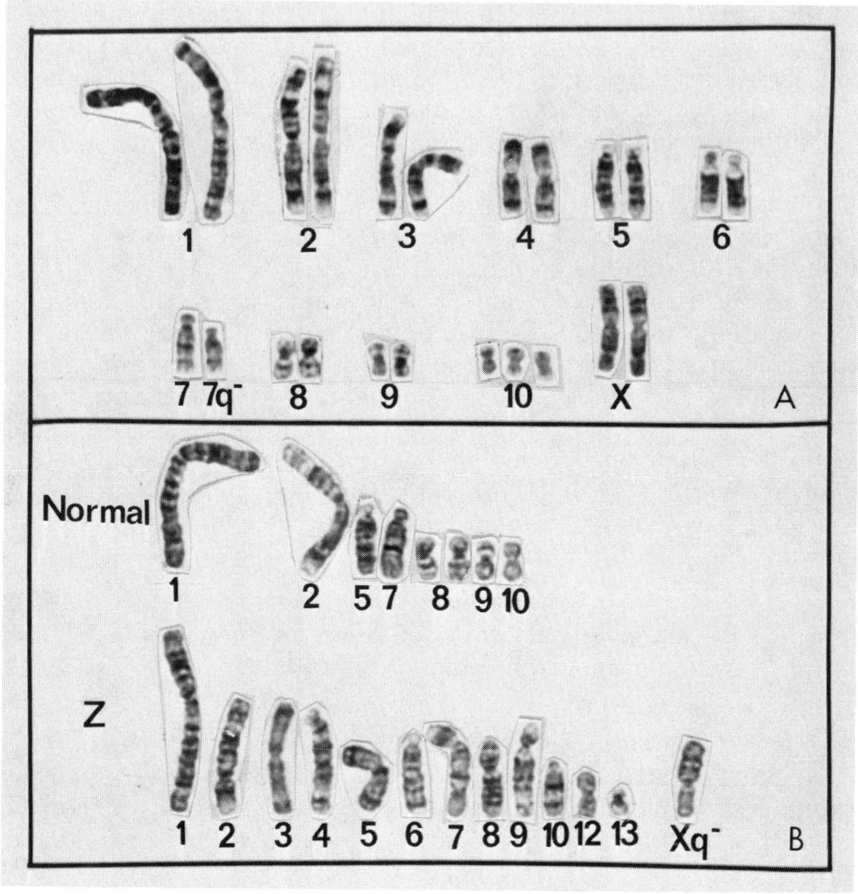

Figure 5.7. G-banded karyotypes of the near-euploid Chinese hamster cell line DEDE (A) and of the CHO subline TGA102a (B). The DEDE karyotype has a normal set of Chinese hamster chromosomes except for a deletion on 7q and trisomy for chromosome 10. The TGA102a karyotype has eight normal chromosomes (upper row) and 13 rearranged chromosomes (lower row). The rearranged or Z-group chromosomes are similar to those published by Deaven and Petersen (1973) except for an extra band on Z9q and for a portion of Xq translocated to Z7q.

CHO-LA arose by reciprocal translocation between the ends of normal 1p and 5q. Two lines of evidence demonstrated this to be the case (Stallings et al., 1984a). The Z6q lacked the Ag-NOR stained region (indicative of *rRNA* genes; Goodpasture and Bloom, 1975) normally present on Chinese hamster chromosome 5, and the Z1p tip was Ag-NOR positive. Finally, among the CHO-LA × Cl1D hybrid clone panel members, which segregated the normal chromosome 1, *ESD* segregated concordantly with the Z6.

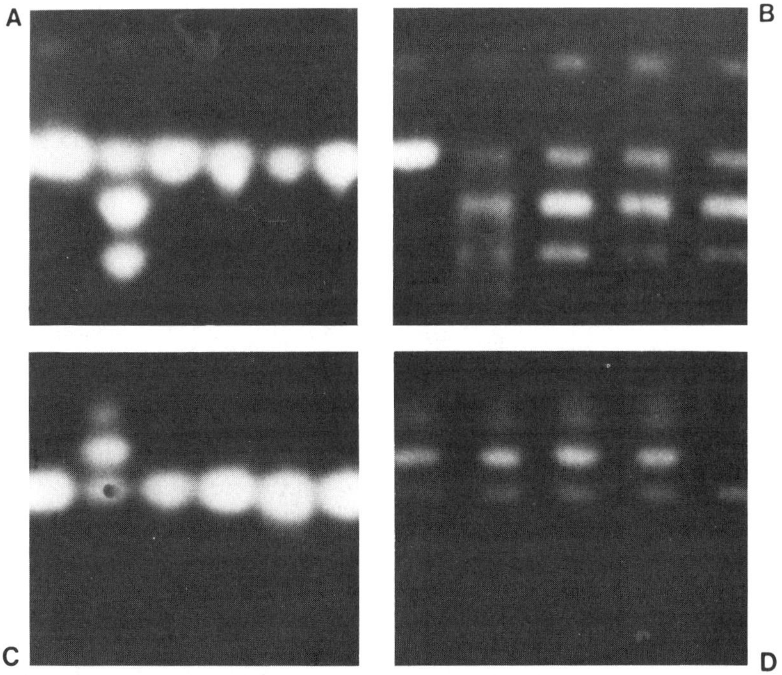

Figure 5.8. Zymograms demonstrating two *ESD* electrophoretic shift mutations in CHO cells exposed to mutagen and in subclones of clones in which variants were detected. Anodal ends of gels are toward the top and origins toward the bottom of each panel. Of the two zones of activity, the slower-migrating, more intensely stained zone is the product of the *ESD* locus. The faster migrating, less intensely stained zone seen in many samples we refer to as ES2. ES2 is informative on these gels only as an internal standard. In panel A is a clone with a three-banded pattern for *ESD*indicating heterozygosity at the locus for this dimeric enzyme. The other samples on the gel display single-band patterns characteristic of CHO cells. Panel B shows four subclones of the mutant with a normal control verifying the variant as a true mutation. Panels C and D present the same information for a seccond mutant. In this case, the variant form of the enzyme is anodal of control (panel C). All four subclones repeat the mutant pattern (panel D) [from Stallings et al. (1984*a*)].

2. *Chromosome 2*

This is one of the more interesting Chinese hamster chromosomes because several selectable markers have been assigned to it. As indicated previously (Section I B 7), taking advantage of the q1-q24 deletion on chromosome Z2 of CHO cells and using intraspecific hybridization, drug selection, and chromosome analysis Campbell and Worton (1980) located *EMT* and *CHR* to that region of chromosome 2. Using similar procedures, Wasmuth and Chu (1980) confirmed the *EMT* and *CHR* assignments and placed *LARS* in the same region. By microcell-mediated gene transfer (Section I B 8), and *in situ* hybridization of a cloned probe to amplified sequences (Section I B 9), Worton et al. (1981) and Nunberg et al. (1978) regionally assigned *DHFR* to the p arm of chromosome 2.

Using the DT selection system on primary Chinese hamster × mouse A9 hybrids, Roberts and Ruddle (1980) assigned *DTS, PGM1, PGD,* and *ENO1* to the 2. We confirmed the *PGM1, PGD,* and *ENO1* assignments in hamster spleen cell × mouse Cl1D hybrids and also assigned *GALT* to that chromosome (Section I B 2 above and Table 5.1). An interstitial deletion (q23-q33—see Fig. 5.3) in one hybrid clone enabled the assignment of *PGM1* to that region [Section I B 3., and Stallings and Siciliano (1981*b*)]. Roberts et al. (1983*a*) confirmed the location of *PGM1* on the q arm and also placed *PGD* and *ENO1* on the same arm (proximal to *PGM1* according to our data).

Since *PGD, PGM1,* and *ENO1* are also syntenic in humans and mice along with *AK2* (Lalley et al., 1978a), we tested for *AK2* linkage to Chinese hamster chromosome 2 by fusion of hamster × mouse Cl1D cells and selection for segregants of the hybrids obtained with DT, as above. The relatively euploid female Chinese hamster line DEDE (Stich et al., 1964) was used since it has an AK2 phenotype (AK2*1) that is electrophoretically separable from mouse AK2. [CHO would not do because its phenotype, AK2*2, comigrates with mouse (Stallings and Siciliano, 1981*a*)]. Hamster *AK2* segregated concordantly with chromosome 2 and all its markers allowing provisional assignment to that chromosome (Stallings and Siciliano, 1982).

By applying DT selection to CHO × mouse Cl1D hybrids, we found that *DTS* and *GALT* could also be regionally assigned to 2q1–24 and considered structurally hemizygous in CHO—11 of 13 hybrid subclones selected in DT that segregated the normal 2 but retained the Z2 were DT resistant and missing the hamster form of *GALT*. Since those same 11 hybrid subclones retained hamster *PGM1, PGD,* and *ENO1,* they were assigned outside the q1–24 region and are structurally dizygous in CHO cells. The assignment of these loci outside the region of the Z2 deletion allowed the previous regional assignment of *PGM1* (2q23–q33), partially overlapping the Z2 deletion, to be refined to 2q25–q33 and placed *PGD* and *ENO1* proximal to *PGM1* in the same region. Functional dizygosity of *ENO1, PGD,* and *PGM1* in CHO cells, as determined by the isolation of diploid heterozygous electrophoretic shift mutants following UV and EMS exposure, confirmed their location outside the Z2 deletion (Stallings et al., 1982).

3. *Chromosomes 3 and 4*

a. Chromosome 3. As indicated previously (Section I B 3), fusion of primary hamster spleen cells with the $APRT^{-}$ variant of mouse Cl1D

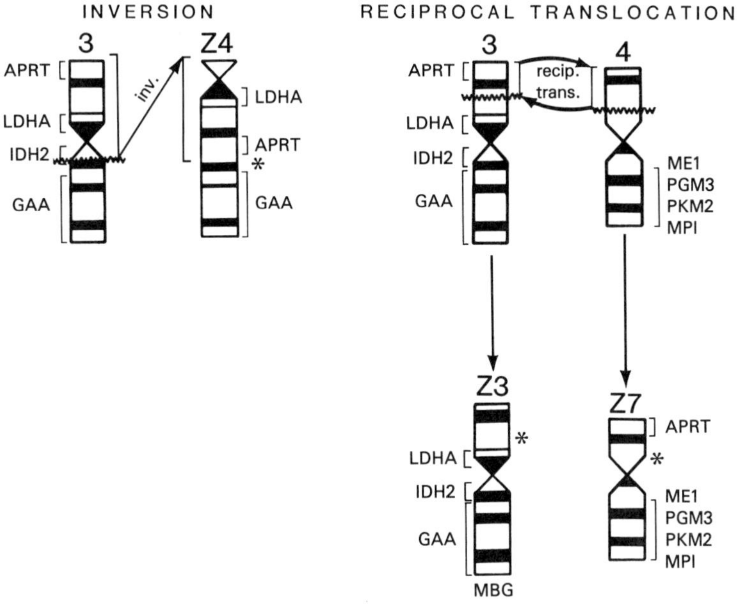

Figure 5.9. Ideogram demonstrating the position of enzyme loci on hamster chromosomes 3 and 4, as well as the inversion and translocations involving these chromosomes and the redistribution of enzyme loci onto chromosomes Z3, Z4, and Z7 in CHO cells. The asterisk indicates the position of the inversion and translocation breakpoints on the Z chromosomes.

cells (LTAO), and subsequent selection of hybrid segregants with AA resulted in the assignments of *LDHA, IDH2, GAA,* and *APRT* to chromosome 3 (Adair et al., 1983a). While loss of the entire number 3 chromosome was the most common event associated with segregation of these four loci, discordant segregation of *APRT* and one or more of the other three loci was observed in each of eight clone panel members that had segregated only part of chromosome 3. In each case, discordancies could be attributed to linkage disruption by chromosome 3 breakage or deletion. Analysis of overlapping deletions and isozyme expression of the four loci allowed their regional assignment as indicated on the ideogram (Fig. 5.9). By *in situ* hybridization and silver-staining methods, *rRNA* genes have been mapped to the end of chromosome 3q (Hsu et al., 1975; Goodpasture and Bloom, 1975).

b. Chromosome 4. Chromosome 4 has no markers for selecting segregant hybrid cells and is an extremely slow random segregator. Therefore, it was necessary to map enzyme loci on it largely by exclusion. In a Chinese hamster primary cell × mouse Cl1D hybrid clone panel consisting of 49 members, only a single clone segregated hamster chromosome 4. It was also the only clone panel member to segregate the hamster isozymes for *PGM3, MPI, PKM2,* and *ME1.* Furthermore, since segregation of hamster chromosomes other than the four excluded linkage of *MPI, PKM2, PGM3,* and *ME1* to chromosomes 1, 2, 3, 5, 6, 7, 8, 9, 10, and X, we provisionally assigned these markers to chromosome 4 (Stallings et al., 1984*b*).

By *in situ* hybridization and silver-staining methods, *rRNA* genes have been mapped to the end of chromosome 4q (Hsu et al., 1975; Goodpasture and Bloom, 1975).

c. CHO-Z Group Chromosomes Derived from Chromosomes 3 and 4. Deaven and Peterson (1973) suggested that in CHO cells a pericentric inversion involving the entire p arm and the proximal portion of the q arm of the submetacentric chromosome 3 gave rise to the subtelocentric Z4. The other chromosome 3, they suggested, participated in a reciprocal translocation between the distal half of its p arm and the distal half of chromosome 4p, giving rise to the CHO Z3 (containing the q arm, centromere, and portion of the p arm of the 3 as well as a portion of the p arm of a normal 4), and the CHO Z7 (containing normal chromosome 4 material except for most of the p arm which was from the 3p) (See Fig. 5.9.). The other chromosome 4 was seen to have undergone a small deletion near the centromere, giving rise to the Z5.

These rearrangements and our gene assignments to chromosomes 3 and 4 could both be validated by concordant segregation analysis of isozymes and chromosomes from CHO × mouse Cl1D somatic cell hybrids. This work has recently been concluded (Adair et al., 1984). Since spontaneous segregation of the candidate CHO chromosomes (Z3, Z4, Z5, and Z7) tended to be slow, in addition to our standard fusion protocol (CHO-TGA102a × Cl1D), we fused CHO × mouse LTAO and AA selected hybrids for segregation of *APRT.* We also fused *TK*$^-$ CHO cells transfected with herpes *TK* gene (Abraham et al., 1982) and selected hybrids in BrdU that segregated the chromosome into which the gene has become incorporated. This latter method proved valuable in obtaining Z5 segregants.

Concordant segregation analysis of CHO chromosomes and isozymes nicely met the expectations of what one would predict based on our gene assignments to normal Chinese hamster chromosomes 3 and 4 and the rearrangements of these chromosomes originally suggested by Deaven and Peterson (1973) and later supported by Worton et al. (1977). *APRT* segregated concordantly with Z4 and Z7, verifying its position at the distal end of the p arm of the normal chromosome 3 and verifying chromosome 3 input in CHO Z4 and Z7. *LDHA* and *GAA* segregated concordantly with Z3 and Z4, placing these loci proximal to the breakpoint on the chromosome 3 responsible for the reciprocal translocation with the 4 and verifying chromosome 3 involvement in the Z3. *MPI, PKM2, PGM3,* and *ME1* segregated concordantly with the Z5 and Z7, placing all four loci proximal to the breakpoint on the chromosome 4 involved in the reciprocal translocation with the 3, and verifying the facts that the Z5 is a slightly modified 4 and chromosome 4 material participates in the Z7.

There were some surprises. *IDH2* segregated concordantly only with the Z3, indicating that the locus is functionally hemizygous in CHO cells and is linked to *LDHA* and *GAA*. *IDH2* appears to be located proximal to the inversion breakpoint on chromosome 3 that was responsible for the generation of the Z4 chromosome. This rearrangement may have caused either the loss of the *IDH2* allele or modified its expression. Concordance of *ME1* with Z5 and Z7 was not maintained among members of the hybrid clone panel derived from fusions made with one CHO subline (TGA102a). TGA102a hybrids segregated *ME1* concordantly only with the Z7. This suggested that *ME1* was functionally hemizygous in TGA102a, with only the allele on the Z7 being expressed. Since the line was obtained after UV mutagenesis and 6TG selection, the Z5 *ME1* allele may have been mutated.

Electrophoretic shift mutations induced in CHO-LA cells at six of these loci are consistent with these conclusions. *LDHA, GAA, MPI, ME1,* and *PGM3* have all produced multiple-band shifts indicating dizygosity, whereas *IDH2* mutants were single-band shifts and nulls indicating hemizygosity (see Fig. 5.5). In 73 CHO-TGA102a clones picked and analyzed after EMS treatment, the only *ME1* mutant was a null, consistent with hemizygosity of that locus in that cell line as indicated above. We have not obtained *PKM2* or *APRT* mutants; however, diploid shift mutants for *APRT* in CHO cells were isolated by Simon et al. (1982) and functional dizygosity for *APRT* in CHO was further supported by the isolation of restriction-site polymorphisms in the *APRT* genomic sequence (Nalbantoglu et al., 1983; Simon and Taylor, 1983).

As with the hybrids made with euploid hamster cells, deletions in these chromosomes in certain exceptional members of the hybrid clone panel have resulted in further refinements of the regional location of these genes. These assignments are summarized in Fig. 5.9.

As indicated previously (Section I B 7), intraspecific fusion of a CHO, *IDH2* shift mutant clone with a MBG^R CHO line, and subsequent selection for hybrid segregants in MBG resulted in concordant segregation of *IDH2* with MBG. *MBG* is therefore also on the CHO Z3. Since the Z3 has elements of the normal hamster 3 and 4 chromosomes, we cannot assign MBG to the euploid genome as yet.

4. Chromosome 5

In all of our hybrids, we have been able to locate no genes on this chromosome even though segregants have been obtained. As indicated in the chromosome 1 discussion, *rRNA* genes have been located at the tip of the q arm, and in CHO cells a reciprocal translocation between that region and the tip of the 1p landed *ESD* on the modified 5 (Z6). This is consistent with the observation of Deaven and Peterson (1973) that CHO-LA cells have a normal 5 and a Z6 (which is a 5 plus an extra dark and light band attached to the long arm).

5. Chromosome 6

Nine members of a hybrid clone panel constructed from hybrids resulting from fusion of primary Chinese hamster spleen cells or fibroblasts and mouse Cl1D cells segregated Chinese hamster chromosome 6, as well as hamster *AK1, ITPA,* and *ADA* (Stallings and Siciliano, 1982). *AK1* and *ITPA* were therefore assigned to the hamster 6. However, the hamster spleen cells and fibroblasts used in the fusion were heterozygous for *ADA* *(ADA**1/ADA*2).* Since the product of *ADA*1* comigrates with mouse *ADA* (Stallings and Siciliano, 1981a) (Fig. 5.1), the absence of electrophoretically resolvable ADA from the nine panel members could have been due to *ADA*'s presence on another chromosome (other than 6) which was not present in more than one copy. If that one copy of this other chromosome always carried the *ADA*1* allele, these nine panel members would not have electrophoretically resolvable hamster *ADA.* Therefore, those data were consistent with, but did not prove, *ADA*–chromosome 6 synteny. ADA synteny to the Chinese hamster chromosome 6 was determined definitively by recognizing that all clone panel members which retained both homologs of the 6 expressed ADA*2, half the clone panel members with one chromosome 6 expressed ADA*2, and clone panel members lacking both chromosome 6s did not express ADA*2. *In situ* hybridization and silver staining has also placed *rRNA* genes at the tip of the q arm of chromosome 6 (Hsu et al., 1975; Goodpasture and Bloom, 1975).

In CHO cells, since no normal number 6 chromosomes are present, one would expect these loci to be located on Z-group chromosomes. To map *ADA* and *ITPA* in CHO cells [*AK1* is not informative in CHO cells because it is not expressed (Stallings and Siciliano, 1981a)], we analyzed a panel of 72 CHO × mouse Cl1D hybrids (Stallings et al., 1983). Both enzymes segregated concordantly with CHO chromosomes Z8 and Z9 confirming the observations of Deaven and Peterson (1973) that the Z8 is a normal 6 with a light and dark band added to the end of the short arm (origin unknown, but see discussion of chromosome 7 and Fig. 5.10), and that Z9 is also a normal 6 with a light band added to the short arm (origin unknown). Electrophoretic shift mutants for *ADA* and *ITPA* confirmed the presence of two functional alleles for each locus in CHO cells.

6. Chromosome 7

As indicated previously (Section I B 3), we (Stallings and Siciliano, 1981b) applied BrdU selection to primary hamster spleen cells × mouse Cl1D

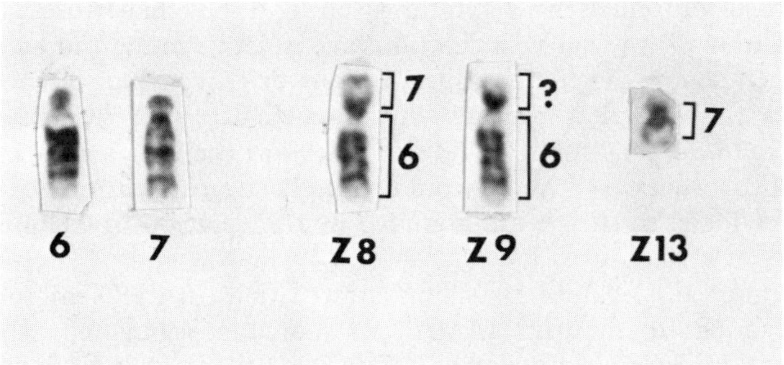

Figure 5.10. G-banded normal Chinese hamster chromosomes 6 and 7 and derivative CHO chromosomes Z8, Z9, and Z13. Based on G-banding patterns, the long arms of chromosomes Z8 and Z9 originated from chromosome 6. This was confirmed genetically by the assignment of *ADA* and *ITPA* to each of these chromosomes. The chromosome 7 *ACP1* locus, also assigned to chromosome Z8, is probably located on the short arm that contains unidentified material. The other chromosome 7 loci, *TK* and *GALK*, were assigned to chromosome Z13 [from Stallings et al. (1983)].

hybrids to isolate hybrids that segregated *TK*. *TK* segregated with chromosome 7 along with *GALK* and *ACP1* allowing assignment of those three loci to the 7. Roberts et al. (1983b) confirmed these assignments in microcell-mediated transfers from primary hamster cells in mouse LTK$^-$, APRT$^-$ cells.

For CHO assignments, BrdU- and HAT-selected subclones of CHO × mouse Cl1D hybrids indicated that all three Chinese hamster chromosome 7 markers were present on the normal chromosome 7 present in CHO cells. *TK* and *GALK* were also shown to be together on the Z13 while *ACP1* segregated with the Z8 (Stallings et al., 1983). While Deaven and Peterson (1973) identified one normal 7 in CHO cells, they suggested that the other 7 had a portion of the short arm deleted to give rise to the Z10. Our analyses confirmed the normal 7 but suggested there was a translocation of a small piece from the 7, carrying *ACP1* to the 6 forming the CHO Z8. Another portion of the 7 appears to have gone into the formation of the Z13. We find no evidence for chromosome 7 material in the CHO Z10. See Figure 5.10 for summary of these chromosomal rearrangements.

Fusing the CHO line EDGAL1-1, which is TK$^{+/-}$, GALK$^{+/-}$ and has an interstitial deletion near the end of the long arm of chromosome 7, with mouse Cl1D cells contributed to the regional assignment of these loci. Since, in the hybrid clone panel, the functional *TK* allele segregated with the Z13 chromosome, it appears that the genetic basis for functional hemizygosity at the *TK* locus in this cell line may be a cytologically detectable interstitial deletion on the distal portion of the long arm of chromosome 7 (Stallings et al., 1983). Experiments are in progress to confirm that this deletion has, in fact, resulted in the loss of the *TK* gene using the cloned Chinese hamster *TK* coding sequence. Nevertheless, these results, together with previous data supporting the provisional regional assignment of *TK* and *GALK* to the long arm of chromosome 7 (Roberts et al., 1983b), would strongly suggest further localization of the *TK* locus to the region bounded by the deletion in CHO EDGAL 1-1.

Roberts et al. (1983b) have demonstrated that, in CHO microcell-derived hybrids, α-amanitin (AMA) resistance cosegregated with *TK, GALK, ACP1,* and chromosome 7. The *AMA* locus, which codes for a subunit for RNA polymerase II, has been shown to be likely functionally hemizygous in CHO cells (Gupta et al., 1978). Since Roberts et al. have mapped the functional *AMA* allele to the single unrearranged chromosome 7 in CHO cells, the other allele appears to have been either lost or inactivated as a consequence of the rearrangements of the other chromosome 7 homolog that resulted in the distribution of chromosome 7 material to the Z13 and Z8 chromosomes. As soon as a cloned probe for the RNA polymerase II coding sequence becomes available, it would be most enlightening to determine whether this gene is present, but inactive, in interspecific somatic cell hybrids that have segregated chromosome 7 but retain either the Z13 or Z8 chromosome. We have previously used this type of approach to determine that one of the *APRT* alleles had been lost in a CHO $APRT^{+/-}$ heterozygote (Adair et al., 1983b).

7. Chromosome 8

A clone panel constructed from primary Chinese hamster spleen cells or fibroblasts fused with mouse Cl1D cells revealed the concordant segregation of *TPI* with hamster chromosome 8 (Siciliano et al., 1983a). This was subsequently confirmed by a single *TPI* and chromosome 8 segregant identified in a panel of Chinese hamster microcell hybrids fused with mouse cells (Roberts et al., 1983b). Chromosome 8 was also shown to contain *rRNA* genes at the tip of the p arm (Hsu et al., 1975; Goodpasture and Bloom, 1975).

Worton et al. (1977) identified two normal 8 chromosomes in CHO cells, and we verified concordant segregation between *TPI* and chromosome 8 in hybrids made between CHO cells and mouse Cl1D cells (Siciliano et al., 1983a). In this same work, electrophoretic shift mutations for *TPI* in CHO cells were consistent with the presence of two functional alleles present in the cell line. In a second panel of CHO × Cl1D hybrids, *GAPD* was also shown to segregate concordantly with chromosome 8 and was therefore also assigned to that chromosome in CHO cells and in the Chinese hamster (Greenspan and Siciliano, in preparation).

8. Chromosome 9

In hybrids made with euploid Chinese hamster cells and mouse Cl1D cells, Siciliano et al. (1983a) showed concordant segregation of hamster *GPI, PEPD,* and chromosome 9. Deaven and Peterson (1973) could identify only one normal chromosome 9 in CHO cells. There were suggestions of chromosome 9 material possibly involved in the Z13. We have found (Siciliano et al., 1983a) no remnant of the missing 9 translocated to any other CHO chromosome by all modern banding techniques. Furthermore, in this same study we presented two lines of evidence that indicated that the chromosome 9 linkage group was hemizygous in CHO cells. Among 28 CHO × mouse Cl1D clone panel members, six segregated chromosome 9 along with *GPI* and *PEPD,* while every other CHO genetic element was represented at least twice. All other 22 clone panel members had hamster chromosome 9, *GPI,* and *PEPD.* We recovered single-band shift and null mutants for *GPI* and *PEPD* in CHO clones exposed to EMS (Fig. 5.6). The single-band shift and null *GPI* and *PEPD* mutants were consistent with the presence of only one functional allele for each of these two enzymes in CHO cells. The functional haploidy for the *GPI-PEPD* linkage group correlated with monosomy for chromosome 9 in CHO cells.

9. Chromosome 10

No gene assignments have been made for this chromosome, although it had segregated in several of our clone panels. One normal 10 has been observed in CHO cells with no identifiable remnant of the other chromosome (Deaven and Peterson, 1973).

10. X Chromosome

Westerveld et al. (1972) suggested linkage of *PGK, HPRT,* and *G6PD* with the hamster X chromosome by studying the relationship between gene multiplicity and enzyme activity in cells. As indicated in Section I B 7, elegant intraspecific fusion experiments with genetically marked cells and subsequent selection, enzyme, and chromosome analyses of segregants enabled the definitive assignments of *HPRT* and *G6PD* in the Chinese hamster (Farrell and Worton, 1977; Rosenstrauss and Chasin, 1975). Deaven and Peterson showed the presence of only one X chromosome in CHO cells (obviously active from the mapping experiments) with little evidence for a portion of the other X translocated to any of the Z chromosomes.

11. Summary of Section

To conclude this section, it can be stated that great strides have been made recently in mapping Chinese hamster chromosomes. Genetic markers have now been assigned to all chromosomes with the exception of the 5 and 10. Clone panels have been established to assign additional genes as methodologies for their visualization become available. Multiply marked chromosomes have been identified so that questions of somatic recombination can be approached using cells of this organism. The mapping of the same genes onto the rearranged Z group chromosomes not only gives us an awesome appreciation of the accuracy of the original cytogenetic work suggesting their origin (Deaven and Peterson, 1973) but helps sort out and focus attention onto regions where hemizygosity may be due to deletion or gene inactivation events.

III. EVOLUTION OF MAMMALIAN LINKAGE GROUPS

A. Conservation of Synteny of Chromosomally Assigned Loci

It surely has not escaped the notice of even the most casual observer that many of the loci that are syntenic in the Chinese hamster are also linked in other mammalian species. The relationships of the syntenic groups demonstrated in the hamster with homologous loci assigned to mouse and human chromosomes are indicated in Table 5.2.

Extensive conservation of autosomal synteny is also seen between humans and other major orders of mammals, such as felines (O'Brien and Nash, 1982). It is not known whether this linkage conservation is physiologically significant or simply reflects closely linked loci not yet dispersed onto separate chromosomes by random processes. The latter possibility is supported by the many observations that genes involved in common metabolic pathways are usually not linked in mammals. The extent of linkage homology between species appears to reflect phylogenetic similarities—for instance, the *ADK-ESD-NP* group shared in hamsters and mice is dispersed onto three separate human chromosomes. Since the Chinese hamster has approximately one-half the number of chromosomes as other mammalian species in which genes have been assigned, it is not surprising that loci syntenic on Chinese hamster chromosomes are located on separate chromosomes in primates and other rodents.

B. Application to Future Studies

Knowing the relationships provides a perspective which can be extremely useful in further genetic studies to study their bases and the application of Chinese hamster cell lines in approaching human biomedical problems.

1. Mapping and Role of Protooncogenes in Linkage-Group Conservation

Having mapped *AK1, ADA,* and *ITPA* to Chinese hamster chromosome 6 (Stallings and Siciliano, 1982), and noting both the linkage of *AK1* to mouse chromosome 2 (Franke et al., 1977), as well as the similarity of G-band pattern between mouse chromosome 2 and the long arm of Chinese hamster chromosome 6, we suggested that *ADA* and *ITPA* might also be located on mouse chromosome 2. To assign the genes coding for these mouse isozymes, we made a series of mouse spleen × CHO somatic cell hybrids that segregate mouse chromosomes and utilized a series of hybrids made from mouse microcells and CHO cells (Siciliano et al., 1984). As predicted, *ADA* and *ITPA* segregated concordantly with the mouse chromosome 2. In addition to extending the concept of mammalian autosomal linkage group conservation, since *ADA* and *ITPA* are also syntenic on the human chromosome 20 (Hopkinson et al., 1976; Meera Khan et al., 1976), these assignments are significant in light of the locations of the cellular homologs of the viral oncogenes c-*abl* and c-*src* to human and mouse chromosomes. Human chromosome 9 is the site of c-*abl* (Heisterkamp et al., 1982) and is also the location of *AK1* (Westerveld et al., 1976). Since both c-*abl* and *AK1* are on chromosome 2 of the mouse (Goff et al., 1982), it can be seen that the linkage between a protooncogene had been maintained in mouse and man. A similar relationship has also been observed between the human chromosome 11 group containing c-Ha-*ras* 1 (deMartinville et al., 1983), *insulin* (Harper et al., 1981), β-*globin* (Gusella et al., 1979), and *LDHA* (Boone et al., 1972), and mouse chromosome 7 containing homologs of the same genes (Chirgwin et al., 1983; Human Gene Mapping VI, 1982; O'Brien et al., 1978; Sakaguchi et al., 1984).

The assignment of *ADA* and *ITPA* to chromosome 2 of the mouse defined another such relationship, since c-*src,* on human chromosome 20 (Sakaguchi et al., 1983) along with *ADA* and *ITPA,* has recently been located on mouse chromosome 2 (Sakaguchi et al., 1984).

The maintenance of the genetic relationships of oncogenes with multiple loci having "housekeeping" functions indicates that the evolutionary conservation of synteny of the latter loci extends to and includes protooncogenes. An alternative hypothesis might suggest that protooncogenes could behave evolutionarily like transposable elements (Fink *et al.,* 1980), occupying different positions in the genomes of even closely related forms. That idea does not appear to be consistent with the existing facts. Indeed, considering the detrimental survival effects of protooncogene translocations associated with oncogenesis (Crews et al., 1982; Dalla-Favera et al., 1982), protooncogenes might actually act as stabilizing factors in the genome and be positive forces for maintaining the evolutionary conservation of mammalian autosomal linkage groups. Consid-

TABLE 5.2
Comparative Synteny of Loci that Have Been Mapped in Chinese Hamster, Mouse, and Human

Loci	Chromosome Assignment in: Chinese Hamster	Mouse	Human	References[a]: Mouse	Human
PEPS ⎤		⎤ 5	⎤ 4	Lalley et al. (1977)	Brown et al. (1977)
PGM2		⎦	⎦	Hutton and Roderick (1970)	Wijnen et al. (1977)
ADK		⎤	10	Leinwand et al. (1978)	Klobutcher et al. (1975)
ESD	1	14	13	Womack et al. (1977)	Van Heyningen et al. (1975)
NP		⎦	14	Womack et al. (1977)	Ricciuti and Ruddle (1973)
PEPB		10	12	Franke et al. (1977)	Chen et al. (1973)
GSR		8	8	Nichols and Ruddle (1975)	de la Chapell et al. (1976)
GLO ⎦		17	6	Minna et al. (1978)	Bender and Grzeschik (1976) Giblett and Lewis (1976)
PGM1 ⎤		⎤	⎤	Lalley et al. (1978*a*)	Van Cong et al. (1971)
ENO1		4	1	Lalley et al. (1978*a*)	Giblett et al. (1973)
PGD				Lalley et al. (1978*a*)	Weitkamp et al. (1971)
AK2		⎦	⎦	Lalley et al. (1978*a*)	Van Cong et al. (1972)
DTS	2	?	⎤	—	Creagan et al. (1975)
LARS		?	5	—	Giles et al. (1980)
EMT		?		—	Dana and Wasmuth (1982)
CHR		?	⎦	—	Dana and Wasmuth (1982)
DHFR		?	?	—	—
GALT ⎦		?	9	—	Benn et al. (1979)
LDHA ⎤		⎤ 7	11	O'Brien et al. (1978)	Boone et al. (1972)
IDH2	3	⎦	15	Lalley et al. (1978*b*)	Grzeschik (1976)
GAA		?	17	—	Solomon et al. (1979)
APRT ⎦		8	16	Kozak et al. (1975)	Tischfield and Ruddle (1974)

Table 5.2 (*Continued*)

Loci	Chromosome Assignment in: Chinese Hamster	Mouse	Human	References[a]: Mouse	Human
ME1			6	Henderson (1968)	Chen et al. (1973)
PGM3				Nadeau et al. (1981)	Van Someran et al. (1974)
	4	9		Johnson et al. (1981)	
MPI			15	Lalley et al. (1978*c*)	Shows (1974)
PKM2				Lalley et al. (1978*c*)	Shows (1974)
ADA			20	Siciliano et al. (1984)	Tischfield et al. (1974)
ITPA	6	2		Siciliano et al. (1984)	Hopkinson et al. (1976)
					Meera Khan et al. (1976)
AK1			9	Franke et al. (1977)	Westerveld et al. (1976)
TK		11	17	Kozak and Ruddle (1977)	Miller et al. (1971)
GALK	7			Kozak and Ruddle (1977)	Elsevier et al. (1974)
ACP1		12	2	Franke et al. (1977)	Hamerton et al. (1975)
AMA		?	?	—	—
GAPD	8	6	12	Bruns et al. (1979)	Bruns and Gerald (1976)
TPI				Minna et al. (1978)	Jongsma et al. (1973)
GPI				Hutton and Roderick (1970)	Hamerton et al. (1973)
	9	7	19		McMorris et al. (1973)
PEPD				Lalley et al. (1978*d*)	McAlpine et al. (1976)
HPRT				Chapman and Shows (1976)	Nyhan et al. (1967)
PGK	X	X	X	Chapman and Shows (1976)	Chen et al. (1971)
G6PD				Chapman and Shows (1976)	Childs et al. (1958)

[a] Only mouse and human assignments are referenced. Chinese hamster assignments are documented in Section II B.

ering these concepts, the loci now shown to be on mouse chromosome 2 and hamster 6, and the similar banding patterns between those two chromosomes, we would next predict that c-*abl* and c-*src* will map to Chinese hamster chromosome 6. Experiments to determine that are now in progress.

2. Monosomic Regions in CHO Cells

For many years auxotrophic and drug-sensitive, recessive mutations in CHO cells have been useful in identifying the location of homologous loci in the human genome. The protocol has been to isolate the mutant CHO line after exposure to mutagen and selection in appropriate medium conditions, fuse with human cells and select for complementation of the CHO defect. This demands the retention of the human chromosome (which is readily identifiable by cytogenetic, biochemical, and/or molecular techniques) carrying the wild-type genes. Since two mutational events are needed at diploid autosomal loci, one might expect such selectable mutations to occur preferentially at loci in the CHO genome which are functionally haploid (either because of gene inactivation or deletion events associated with the chromosomal rearrangements in CHO cells—see Section II). As Dana and Wasmuth (1982) have shown for the *LARS, EMT,* and *CHR* loci, loci that are syntenic on a human chromosome may be readily mapped if their syntenic relationship has been conserved in the Chinese hamster and if CHO cells are hemizygous for the region of the genome in which they are located.

A recent example of this application has been our demonstration (in collaboration with Larry Thompson and Anthony Carrano of the Lawrence Livermore Laboraratories—manuscripts in preparation) that two of three CHO DNA repair-deficient mutant lines (from different complementation groups—see Chapter 21) studied in fusion with normal human cells have their phenotypes complemented by genes on human chromosome 19. As can be seen in Table 5.2, there is homology between the human 19 and the Chinese hamster chromosome 9—*GPI* and *PEPD* are on both chromosomes. It appears, therefore, that the hamster chromosome 9 has at least two DNA repair gene loci. Since chromosome 9 is monosomic in CHO cells (see Section II), there was an advantage for isolation of recessive mutations for repair genes located on that chromosome, and subsequent analysis of those genes and their human homologs.

This raises certain questions and suggests definite approaches. The first question is: How many DNA repair genes are there? Approaches to this and other problems would be to identify CHO sublines that might be hemizygous for different regions of the CHO genome. Such lines are identifiable by the electrophoretic shift mutants recovered in them. For instance, *ME1,* generally a diploid locus, was shown to be hemizygous in CHO-TGA102a (see Section II B 3 b). Since *ME1* is part of a relatively well-conserved linkage group (Table 5.2), mutants isolated in it may be useful in identifying genes associated with human chromosome 6. Further analyses of our shift mutants and subsequent isolation of selectable mutants in them is underway for this purpose. One can see that the combination of gene mapping, electrophoretic shift, and selectable mutation studies, coupled with our understanding of conserved linkage groups, may provide a better understanding of human biomedical problems. That CHO cells are now, more than ever, a uniquely appropriate cell line to approach these areas should be an obvious conclusion from the work summarized in this chapter.

ACKNOWLEDGMENTS

This work was supported in part by NIH research grants CA-28711, CA-04484, CA-34936, and a gift from the Exxon Corporation. We thank Ms. Katherine Arnsworth for her help in preparing the manuscript.

REFERENCES

Abraham, I., Tyagi, J. S., and Gottesman, M.M. (1982). *Somat. Cell Genet.* **8,** 23–40.

Adair, G. M., Carver, J. H., and Wandres, D. L. (1980). *Mutation Res.* **72,** 187–205.

Adair, G. M., Stallings, R. L., Friend, K. K., and Siciliano, M. J. (1983a). *Somat. Cell Genet.* **9,** 477–487.

Adair, G. M., Stallings, R. L., Nairn, R. S., and Siciliano, M. J. (1983b). *Proc. Natl. Acad. Sci. USA* **80,** 5961–5964.

Adair, G. M., Stallings, R. L., and Siciliano, M. J. (1984). *Somat. Cell and Molec. Genet.* **10,** 283–295.

Bender, K., and Grzeschik, K. H. (1976). *Cytogenet. Cell Genet.* **16**, 93–96.

Benn, P. A., D'Ancona, G. G., Croce, C. M., Shows, T. B., and Mellmann, W. J. (1979). *Cytogenet. Cell Genet.* **24**, 37–41.

Biedler, J. L. and Spangler, B. A. (1976). *Science* **191**, 185–187.

Boone, C. M., Chen, T. R., and Ruddle, F. H. (1972). *Proc. Natl. Acad. Sci. USA* **69**, 510–514.

Brown, S., Lalley, P. A., and Minna, J. D. (1977). *Hum. Gene Mapping* **4**, 167–171.

Bruns, G. A. P. and Gerald, P. S. (1976). *Science* **192**, 54–56.

Bruns, G., Gerald, P. S., Lalley, P., Francke, U., and Minna, J. (1979). *Cytogenet. Cell Genet.* **25**, 139.

Campbell, C. E. and Worton, R. G. (1980). *Somatic Cell Genet.* **6**, 215–224.

Chapman, V. M., and Shows, T. B. (1976). *Nature* **259**, 665–667.

Chen, S. H., Malcom, L. A., Yosida, A., and Giblett, E. R. (1971). *Am. J. Hum. Genet.* **23**, 87–91.

Chen, T. R., McMorris, F. A., Creagan, R., Ricciuti, R., Tischfield, J., and Ruddle, F. H. (1973). *Am. J. Hum. Genet.* **25**, 200–207.

Childs, B., Zinkham, W., Brown, E. A., Kimbro, E. L., and Torbet, J. V. (1958). *Bull. Johns Hopkins Hosp.* **102**, 21.

Chirgwin, J. M., Ancheone, T. L., Rosenbaum, A. L., Diaz, J. A., and Lalley, P. A. (1983). *Diabetes* **32**, 46A.

Creagan, R. P., Chen, S., and Ruddle, F. H. (1975). *Proc. Natl. Acad. Sci. USA* **72**, 2237–2241.

Crews, S., Barth, R., Hood, L., Prehn, J., and Calame, K. (1982). *Science* **218**, 1319–1321.

Croce, D. M., Knowles, B. B., and Koprowski, H. (1973). *Exp. Cell Res.* **82**, 457–461.

Dalla-Favera, R., Bregni, M., Erikson, J., Patterson, D., Gallo, R. C., and Croce, C. M. (1982). *Proc. Natl. Acad. Sci. USA* **79**, 7824–7827.

Dana, S. and Wasmuth, J. J. (1982). *Somat. Cell Genet.* **8**, 245–264.

Deaven, L. L. and Peterson, D. F. (1973). *Chromosoma* **41**, 129–144.

de la Chapelle, A., Icen, A., Aula, P., Leisti, J., Turleau, C., and De Grauchy, J. (1976). *Ann. Genet.* **19**, 253–256.

deMartinville, B., Giacalone, J., Shih, C., Weinberg, R. A., and Franke, U. (1983). *Science* **219**, 498–501.

Draper, R., Chin, D., Euvey-Owens, D., Scheffler, I., and Simon, M. (1979). *J. Cell Biol.* **83**, 116–125.

Dubbs, D. R. and Kit, S. (1964). *Exp. Cell Res.* **33**, 19–27.

Elsevier, S. M., Kucherlapati, R. S., Nichols, E. A., Creagen, R. P., Giles, E., Ruddle, F. H., Willecke, K., and McDougall, J. K. (1974). *Nature* **251**, 633–635.

Farrell, S. A. and Worton, R. G. (1977). *Somatic Cell Genet.* **3**, 539–551.

Fink, G., Farabaugh, P., Roeder, G., and Chaleff, D. (1980). *Cold Spring Harbor Symp. Quant. Biol.* **45**, 575–580.

Fournier, R. E. K. and Ruddle, F. H. (1977). *Proc. Natl. Acad. Sci. USA* **74**, 319–323.

Francke, U., Lalley, P. A., Moss, W., Ivey, J., and Minna, J. D. (1977). *Cytogenet. Cell Genet.* **19**, 57–84.

Gefter, M. L., Margulies, D. H., and Scharff, M. D. (1977). *Somat. Cell Genet.* **3**, 231–236.

Giblett, E. R., and Lewis, M. (1976). *Cytogenet. Cell Genet.* **16**, 313.

Giblett, E. R., Chen, S. H., Anderson, J. E., and Lewis, M. (1973). *Hum. Gene Mapping* **1**, 91.

Giles, R. E., Shimizu, N., and Ruddle, F. (1980). *Somat. Cell Genet.* **5**, 667–686.

Goff, S. P., D'Eustachio, P., Ruddle, F. H., and Baltimore, D. (1982). *Science* **218**, 1317–1319.

Goodpasture, C. and Bloom, S. E. (1975). *Chromosoma* **53**, 37–50.

Grzeschik, K.-H. (1976). *Hum. Genet.* **34**, 23–28.

Green, H., Wang, R., Kehinde, O., and Meuth, M. (1971). *Nature (London), New Biol.* **234**, 138–140.

Gupta, R. S. and Singh, B. (1982). *Mutation Res.* **94**, 449–466.

Gupta, R. S., Chan, D. Y. H., and Siminovitch, L. (1978). *J. Cell Physiol.* **97**, 461–468.

Gusella, J., Varsanyi-Breiner, A., Kao, F. T., Jones, C., Puck, T., Keys, C., Orkin, S., and Housman, D. (1979). *Proc. Natl. Acad. Sci. USA* **76**, 5239–5243.

Hamerton, J. L., Douglas, G. R., Gee, P. A., and Richardson, B. J. (1973). *Cytogenet. Cell Genet.* **12**, 128–135.

Hamerton, J. L., Mohandas, T., McAlpine, P. J., and Douglas, G. R. (1975). *Am. J. Hum. Genet.* **27**, 595–608.

Harper, M. E., Ulrich, A., and Saunders, G. F. (1981). *Proc. Natl. Acad. Sci. USA* **78**, 4458–4460.

Heisterkamp, N., Groffen, J., Stephenson, J. R., Spurr, N. K., Goddfellow, P. M., Solomon, E., Carritt, B., and Bodmer, W. F. (1982). *Nature (London)* **299**, 747–749.

Henderson, N. S. (1968). *Ann. N. Y. Acad. Sci.* **151**, 429–440.

Hopkinson, D. A., Povey, S., Solomon, E., Bobrow, M. and Gormley, I. P. (1976). *Cytogenet. Cell Genet.* **16**, 159–160.

Hsu, T. C., Spirito, S. E., and Pardue, M. L. (1975). *Chromosoma* **53**, 25–36.

Human Gene Mapping VI (1982). *Cytogenet. Cell Genet.* **32**, 111–245.

Hunter, R. L. and Markert, C. L. (1957). *Science* **125**, 1294–1295.

Hutton, J. J. and Roderick, T. H. (1970). *Biochem. Genet.* **4**, 339–350.

Johnson, F. M., Hendren, R. W., Chasalow, F., Barnett, L. B., and Lewis, S. E. (1981). *Biochem. Genet.* **19**, 599–615.

Jongsma, A. P. M., Los. W. R. T., and Hagermeijer, J. (1973). *Hum. Gene Mapping* **1**, 106–107.

Klobutcher, L. A., Nichols, E. A., Kucherlapati, R. S., and Ruddle, F. H. (1975). *Hum. Gene Mapping* **3**, 171–174.

Kozak, C. A, and Ruddle, F. H. (1977) *Somat. Cell Genet.* **3**, 121–134.

Kozak, C., Nichols, E., and Ruddle, F. H. (1975). *Somat. Cell Genet.* **1**, 371–382.

Lalley, P. A, Franke, U., and Minna, J. D. (1977). *Cytogenet. Cell Genet.* **22**, 573–576.

Lalley, P. A., Franke, U., and Minna, J. D. (1978a). *Proc. Natl. Acad. Sci. USA* **75**, 2382–2386.

Lalley, P. A., Franke, U., and Minna, J. D. (1978b). *Cytogenet. Cell Genet.* **22**, 577–580.

Lalley, P. A., Franke, U., and Minna, J. D. (1978c). *Cytogenet. Cell Genet.* **27**, 281–284.

Lalley, P. A., Minna, J. D., and Franke, U. (1978d). *Nature* **274**, 160–163.

Leinwand, L., Fournier, R. E. K., Nichols, E. A., and Ruddle, F. H. (1978). *Cytogenet. Cell Genet.* **21**, 77–85.

Lin, C. C., Chang, T., and Nieweczas-Late, V. (1971). *Can. J. Genet. Cytol.* **13**, 9–13.

Littlefield, J. W. (1964). *Science* **145**, 709–710.

Marin, G. (1969). *Exp. Cell Res.* **59**, 29–36.

Markert, C. L. and Whitt, G. S. (1968). *Experientia* **24**, 977–1088.

McAlpine, P. J., Mohandas, T., Ray, M., Wang, H., and Hamerton, J. L. (1976). *Cytogenet. Cell Genet.* **16**, 204–205.

McMorris, F. A., Chen, T. R., Ricciuti, R., Tischfield, J., Creagan, R., and Ruddle, F. H. (1973). *Science* **179**, 1129–1131.

Meera Khan, P. and Robson, E. B. (1978) *Cytogenet. Cell Genet.* **22**, 106–110.

Meera Khan, P., Pearson, P. L., Wijnen, L. L. L., Doppert, B. A., Westerveld, A., and Bootsma, D. (1976). *Cytogenet. Cell Genet.* **16**, 420–421.

Miller, O. J., Allderdice, P. W., Miller, D. A., Breg, W. R., and Migeon, B. R. (1971). *Science* **173**, 244.

Minna, J. D., Bruns, G. A. P., Krinsky, A. H., Lalley, P. A., Franke, U., and Gerald, P. S. (1978). *Somat. Cell Genet.* **4**, 241–252.

Nadeau, J. H., Kompf, J., Siebert, G., and Taylor, B. A. (1981). *Biochem. Genet.* **19**, 465–474.

Nalbantoglu, J., Goncalves, O., and Meuth, M. (1983). *J. Mol. Biol.* **167**, 575–594.

Nichols, E. A. and Ruddle, F. H. (1975). *Biochem. Genet.* **13**, 323–329.

Norum, R. A. and Migeon, B. R. (1974). *Nature* **251**, 42–74.

Nunberg, J. H., Kaufman, R. J., Schimke, R. T., Urlaub, G., and Chasin, L. A. (1978). *Proc. Natl. Acad. Sci. USA* **75**, 5553–5556.

Nyhan, W. L., Pesek, J., Sweetman, L., Carpenter, P. G., and Carter, C. H. (1967). *Pediat. Res.* **1**, 5.

O'Brien, S. J. and Nash, W. G. (1982). *Science* **216**, 257–265.

O'Brien, D., Linnenbach, A., and Croce, C. M. (1978). *Cytogenet. Cell Genet.* **21**, 72–76.

Paris Conference (1972). *Birth Defects: Original Article Series,* The National Foundation, New York, Vol. 8, No. 7.

Puck, T. T. and Fisher, H. W. (1956). *J. Exp. Med.* **104**, 427–434.

Ray, M. and Mohandas, T. (1976). *Cytogenet. Cell Genet.* **16**, 83–91.

Ricciuti, F. and Ruddle, F. H. (1973). *Nature* **241**, 180–182.

Roberts, M. and Ruddle, F. H. (1980). *Exp. Cell Res.* **127**, 47–54.

Roberts, M., Melera, P. W., Davide, J. P., Hart, J. T., and Ruddle, F. H. (1983a) *Cytogenet. Cell Genet.* **36**, 599–604.

Roberts, M., Scangos, G. A., Hart, J. T., and Ruddle, F. H. (1983b). *Somat. Cell Genet.* **9**, 235–248.

Roderick, T. H., Ruddle, F. H., Chapman, V. M., and Shows, T. B. (1971). *Biochem. Genet.* **5**, 457–466.

Rosenstrauss, M. and Chasin, L. A. (1975). *Proc. Natl. Acad. Sci. USA* **72**, 493–497.

Ruddle, F. H. (1972). *Advan. Hum. Genet.* **3**, 173–235.

Sakaguchi, A. Y., Naylor, S. L., and Shows, T. B. (1983). *Prog. Nuc. Acid Res. Mol. Biol.* **29**, 279–283.

Sakaguchi, A. Y., Lalley, P. A. Zabel, B. U., Ellis, R. W., Scolnick, E. M., and Naylor, S. L. (1984). *Proc. Natl. Acad. Sci. USA* **81**, 525–529.

Shaw, C. R. (1965). *Science* **149**, 936–943.

Shows, T. B. (1974). In *Somatic Cell Hybridization* (R. L. Davidson and F. de La Cruz, eds.), Raven Press, New York, pp. 15–25.

Shows, T. B., et al. (1979). *Cytogenet. Cell Genet.* **25**, 96–116.

Siciliano, M. J., Wright, D. A., and Shaw, C. R. (1974). In *Progress in Medical Genetics X* (A. C. Steinberg, and A. Bearn, eds.), Grune and Stratton, New York, pp. 17–53.

Siciliano, M. J., Siciliano, J., and Humphrey, R. M. (1978). *Proc. Natl. Acad. Sci. USA* **75**, 1919–1923.

Siciliano, M. J., Stallings, R. L., Adair, G. M., Humphrey, R. M., and Siciliano, J. (1983a) *Cytogenet. Cell Genet.* **35**, 15–20.

Siciliano, M. J., White, B. F., and Humphrey, R. M. (1983*b*). *Mutation Res.* **107**, 167–176.

Siciliano, M. J., Fournier, R. E. K., and Stallings, R. L. (1984). *J. Hered.* **75**, 175–180.

Simon, A. E. and Taylor, M. W. (1983). *Proc. Natl. Acad. Sci. USA* **80**, 810–814.

Simon, A. E., Taylor, M. W., Bradley, W. E. C., and Thompson, L. H. (1982). *Mol. Cell Biol.* **2**, 1126–1133.

Solomon, E., Swallow, D., Burgess, S., and Evans, L. (1979). *Ann. Hum. Genet.* **42**, 273–281.

Stallings, R. L. and Siciliano, M. J. (1981a). *Somat. Cell Genet.* **7**, 295–306.

Stallings, R. L. and Siciliano, M. J. (1981b). *Somat. Cell Genet.* **7**, 683–698.

Stallings, R. L. and Siciliano, M. J. (1982). *J. Hered.* **73**, 399–404.

Stallings, R. L., Siciliano, M. J., Adair, G. M., and Humphrey, R. M. (1982). *Somat. Cell Genet.* **8**, 413–422.

Stallings, R. L., Adair, G. M., Siciliano, J., Greenspan, J., and Siciliano, M. J. (1983). *Molec. Cell Biol.* **3**, 1963–1974.

Stallings, R. L., Adair, G. M., Lin, J., and Siciliano, M. J. (1984*a*). *Cytogenet. Cell Genet.* **38**, 132–137.

Stallings, R. L., Adair, G. M., and Siciliano, M. J. (1984*b*). *Somat. Cell Molec. Genet.* **10**, 109–111.

Stich, H. F., Van Hoosier, G. L., and Trenton, J. J. (1964). *Exp. Cell Res.* **34**, 400–403.

Stubblefield, E. (1964). In *Cytogenetics of Cells in Culture* (R. J. C. Harris, ed.), Academic Press, New York, pp. 223–253.

Tischfield, J. A., and Ruddle, F. A. (1974). *Proc. Natl. Acad. Sci. USA* **71**, 45–49.

Tischfield, J. A., Creagan, R. P., Nichols, E. A., and Ruddle, F. H. (1974). *Human Heredity* **24**, 1–11.

Van Cong, N. Billardon, C., Picard, J. Y., Feingold, J., and Frezal, J. (1971). *C. R. Acad. Sci.* **272**, 485–487.

Van Cong, N., Billardon, C., Rebourcet, R., Le Borgne de Kaouel, C., Picard, J. Y., Weil, D, and Frezal, J. (1972). *Ann. Genet.* **15**, 213–218.

Van Heyningen, V., Borrow, M., Bodmer, W. F., Gardiner, S. E., Povey, S., and Hopkinson, D. A. (1975). *Ann. Hum. Genet. Lond.* **38**, 295–303.

Van Someren, H., van Henegouwen, H. B., Los, W., Wurzer-Figurelli, E., Doppert, B., Vervloet, M., and Meera Khan, P. (1974). *Humangenetik* **25**, 189–201.

Wasmuth, J. J., and Chu, L.-Y. (1980). *J. Cell Biol.* **87**, 697–702.

Weiss, M. C. and Green, H. (1967). *Proc. Natl. Acad. Sci. USA* **58**, 1104–1111.

Weitkamp, L.-R., Guttormsen, S. A., and Gutendyke, R. M. (1971). *Am. J. Hum. Genet.* **23**, 462–470.

Westerveld, A., Visser, R. P. L. S., Freeke, M. A., and Bootsma, D. (1972). *Biochem. Genet.* **7**, 33–40.

Westerveld, A., Jongsma, A. P. M., Meera Khan, P., Van Someren, H., and Bootsma, D. (1976). *Proc. Natl. Acad. Sci. USA* **73**, 895–899.

Wijnen, L. M. M., Grzeschik, K.-H., Pearson, P. L. and Meera Khan, P. (1977). *Hum. Genet.* **37**, 271–278.

Womack, J. E. and Sharp, M. (1976). *Genetics* **82**, 665–675.

Womack, J. E., Davisson, M. T., Eicher, E. M., and Kendall, D. A. (1977). *Biochem. Genet.* **15**, 347–355.

Worton, R. G., Ho, C. C., and Duff, C. (1977). *Somat. Cell Genet.* **3**, 27–45.

Worton, R., Duff, C., and Flintoff, W. (1981). *Mol. Cell Biol.* **1**, 330–335.

SECTION 2

GENETIC MANIPULATION OF CHINESE HAMSTER CELLS

CHAPTER 6

GROWTH PROPERTIES OF CHINESE HAMSTER OVARY (CHO) CELLS

Michael M. Gottesman
Laboratory of Molecular Biology
National Cancer Institute
National Institutes of Health
Bethesda, Maryland

I. INTRODUCTION

The purpose of this chapter is to give the reader some practical hints about the growth of one subline of Chinese hamster ovary (CHO) cells. Although information contained in this chapter is specifically derived only from the study of CHO cells, the principles which are touched upon concerning growth, storage, and use in mutant isolation are relevant to most Chinese hamster cells and, indeed, to most cultured mammalian cells.

The reader is also referred to Chapter 2 by Dr. Theodore Puck, who discusses the origin of the CHO line and its many advantages for genetic analysis. Chapter 2 also contains information about the nutritional requirements of CHO cells. Appendix I reviews the pedigree of the cell line used for these studies (subclone 10001 of CHO Pro^-5, Toronto substrain). The majority of the chapters in this volume make specific reference to the use of CHO cells for mutant isolation. For general principles of growth of cells in tissue culture see Jakoby and Pastan (1979), Paul (1975), and Tooze (1973).

II. NUTRITIONAL REQUIREMENTS

A. Medium

The CHO line originally established by Puck et al. (1958) was grown in a semidefined medium known as Ham's F12 (see Chapter 2, Table 2.1) supplemented with fetal bovine serum. This medium is a rich medium which contains proline, required by all CHO cells (but which is not a nutritional requirement of most other established mammalian cell lines) as well as a variety of other nutrilites (i.e., nutrients needed in small amounts). The reason for establishing these cells in a rich medium was to provide them with a maximum number of nutrilites so that it would be possible to develop cell lines with a large number of specific nutritional auxotrophies. To a great extent, this hope has been realized through the work of Puck and in many other laboratories (see Chapter 2, Table 2.2; Chapter 11; and Appendix II). We routinely grow our CHO cells in another rich medium known as alpha-modified minimal essential medium without ribonucleosides or deoxyribonucleosides (α-MEM) (Stanners et al., 1971). The contents of this medium are listed in Table 6.1. As shown, it contains a variety of nutrients not found in MEM. We supplement the medium with 2 m*M* glutamine before use and add penicillin (50 units/mL) and streptomycin (50 μg/mL).

CHO cells will grow in MEM supplemented only with proline, but their growth rate is 25% slower in such minimal media, and for most purposes the rapid doubling time ($t_{1/2}$ of 12 hr) found in α-MEM is an advantage that offsets the small additional expense of the richer medium. There are two formulations of α-MEM which are commercially available. We use α-MEM from Flow Laboratories, which contains 2 g $NaHCO_3$/L. Medium obtained from most other suppliers contains 2.6 g $NaHCO_3$/L and the small difference in pH at the same CO_2 concentration may be enough to affect the growth of the cells (see Section III B).

B. Serum

Under routine conditions of monolayer culture, CHO cells are not too fastidious with respect to their requirement for serum. They will grow in 10% fetal bovine or 10% calf serum with a doubling time of approximately 12 hr at 37°C. However, the cells appear to be fairly sensitive to toxic effects of some lots of serum, different sera have different growth-promoting capacity, and growth in suspension without Spinner medium is possible only with carefully selected lots of fetal bovine serum. We buy fetal bovine serum in large lots after carefully checking individual lots for the following properties:

1. CHO cells should be able to form colonies with an efficiency of 80–90% and grow with a doubling time of 12 hr at 37°C when the se-

TABLE 6.1
Formulation of alpha-Modified MEM and Dulbecco's MEM[a]

Component	alpha-MEM (mg/L)	MEM (mg/L)
Salts		
$CaCl_2 \cdot 2H_2O$	264.9	264.9
$Fe(NO_3)_3 \cdot 9H_2O$	—	0.1
KC1	400.0	400.0
$MgSO_4 \cdot 7H_2O$	200.0	200.0
NaCl	6800.0	6400.0
$NaHCO_3$	2000.0	3700.0
$NaH_2^-PO_4 \cdot H_2O$	140.0	125.0
Amino Acids		
L-Alanine	25.0	—
L-Arginine	126.4	84.0
L-Asparagine	50.0	—
L-Aspartic acid	30.0	—
L-Cysteine·HCl·H_2O	100.0	—
L-Cystine	24.02	48.0
L-Glutamic acid	75.0	—
Glycine	50.0	30.0
L-Histidine·HCl·H_2O	41.9	42.0
L-Isoleucine	52.5	104.8
L-Leucine	52.5	104.8
L-Lysine·HCl	73.06	146.2
L-Methionine	14.9	30.0
L-Phenylalanine	33.02	66.0
L-Proline	40.0	—
L-Serine	25.0	42.0
L-Threonine	47.64	95.2
L-Tryptophan	10.2	16.0
L-Tyrosine	32.22	72.0
L-Valine	46.9	93.6
Vitamins		
Ascorbic acid	50.0	—
Biotin	0.1	—
D-Ca pantothenate	1.0	—
Choline chloride	1.0	4.0
Folic acid	1.0	4.0
i-Inositol	2.0	7.0
Nicotinamide	1.0	4.0
Pyridoxal·HCl	1.0	4.0
Riboflavin	0.1	0.4
Thiamin·HCl	1.0	4.0
Vitamin B_{12}	1.36	—

TABLE 6.1 (*Continued*)

Component	alpha-MEM (mg/L)	MEM (mg/L)
Other Components		
Dextrose	1000.0	4500.0
Lipoic acid	0.2	—
Phenol red, Na	10.0	15.0
Sodium pyruvate	110.0	110.0

[a] Adapted from the Flow Laboratories product catalog.

rum is present at a concentration of 10% in α-MEM. Toxic sera will give lower cloning efficiencies and longer doubling times (manifest also as small clones after 7 days). Cloning efficiency is measured by plating 200 cells (diluted into complete medium) in a 100-mm tissue culture dish, incubating for 5–7 days, and then removing the medium and staining the colonies with 0.5% methylene blue in 50% ethanol. In this type of cloning experiment, care should be taken not to move the dishes during the incubation time since CHO cells (especially mitotic cells) readily detach from the dishes and will form secondary colonies. Doubling times are determined by plating 2×10^4 cells in each well of a 24-well tissue culture dish (surface area, approximately 2 cm^2). Duplicate wells are treated with 0.25% trypsin in a buffer containing 2 m*M* EDTA for at least 15 min at 37°C, and the entire contents of each well is counted in a Coulter Counter (Gottesman et al., 1980).

2. Colony formation should be efficient at concentrations of serum as low as 0.5%. This ability of CHO cells to grow in low serum concentrations is characteristic of malignantly transformed cells. Below this serum concentration, CHO cells will not clone efficiently, although it is possible to get growth of mass populations for several weeks in α-MEM medium supplemented with insulin and transferrin (D. Evain, personal communication). Fibroblast growth factor (FGF) and epidermal growth factor (EGF) seem to inhibit CHO growth under these conditions. Hamilton and Ham (1977) have published a completely defined medium, lacking serum, with which clonal growth of CHO cells is possible. However, for unknown reasons possibly related to contaminants in the water supply, our laboratory has not succeeded in growing our strain of CHO cells in this medium, which also reflects the experiences of most other laboratories currently using CHO cells. Commercial preparations of so-called "completely defined serum-substitute" usually contain 0.5% serum supplemented with growth factors, transferrin, and insulin. Although these preparations may be more uniform than different lots of serum, they do not seem to stimulate growth of CHO cells more than the 0.5% content of serum would predict. Carver et al. (1983) have recently analyzed the growth

characteristics of CHO cells in low serum concentrations and have confirmed the requirement for at least 0.5% serum for efficient cell cloning. The development of a completely defined medium in which CHO cells would clone with high efficiency and reliability would be a great boon to geneticists working with this cell line.

3. If growth in suspension is intended, fetal bovine sera should be prechecked for ability to support such growth (see Section III C). It is our experience that only one of three randomly chosen lots of serum will allow the growth of CHO cells in suspension in regular medium without clumping of the cells. The ability to support suspension growth without clumping seems to be due to the presence of an additional serum component, since, in mixing experiments with inadequate sera, the ability to prevent clumping may be found when the adequate serum makes up only one-third of the total serum. For experiments in which specific nutrients are deleted from medium (such as in the isolation of drug-resistant mutants in which the drugs are analogs of nutrients or in metabolic-labeling experiments), it is essential that serum be dialyzed prior to use. To accomplish this, we place fetal bovine serum in sterile dialysis tubes and dialyze at 4°C against phosphate buffered saline. Three changes of buffer with a 20-fold excess of buffer in each dialysis is usually sufficient. Serum should be filtered through a 0.45-μm Millipore filter after this procedure to remove precipitated protein and adventitious bacterial contamination.

4. Sera should be checked for special requirements of the investigator. For example, if cells are to be used as recipients in DNA-mediated transformation experiments, it is essential to check that the serum supports such transformation with high efficiency.

III. GROWTH CONDITIONS

A. Temperature

CHO cells will grow over a fairly wide range of temperatures. This ability allows the isolation of temperature-sensitive and cold-sensitive mutants, some of which are described in this book. Cell are grown routinely at 37°C, but they will grow fairly well from 34 to 40°C (cloning efficiency, $> 50\%$). It is a good idea when growing cells at the extremes of this temperature range to monitor the temperature of the incubators with a thermistor and continuous recorder, since minor variations in temperature (> 40.5°C or < 32°C) interfere dramatically with cell growth and cloning efficiency.

We have measured the doubling time of CHO cells as a function of temperature (Fig. 6.1). Growth curves in complete medium containing 10% fetal bovine serum were obtained as described above (Section II B). Between 34 and 39°C growth of cells is logarithmic, and it is a rela-

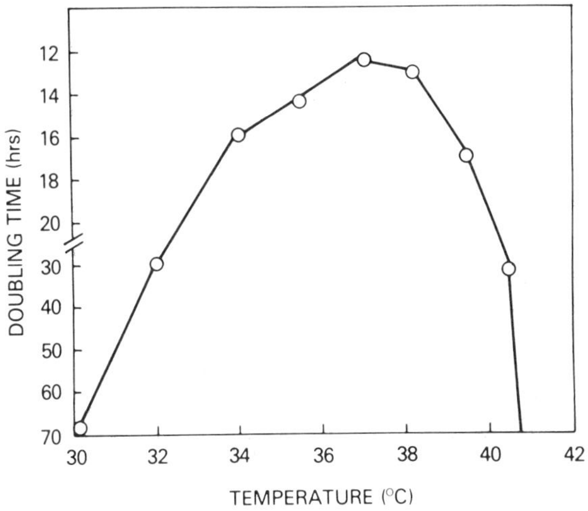

Figure 6.1. Effect of temperature on the growth of CHO cells. Doubling times of cells were determined as previously described (Gottesman et al., 1980).

tively simple matter to obtain doubling times. There is a fairly broad temperature optimum, with doubling times of approximately 12–14 hr obtained between 36 and 38.5°C and an increase of doubling time at either end of this range. Below 34 or above 39°C, cell growth is not logarithmic, and doubling time increases with time at the indicated temperature. The numbers given for doubling times at the extreme ranges of temperature are approximate only and represent an average doubling time between 24 and 72 hr after shift from 37°C. Cells will clone with an efficiency of 20% at 40.5°C, but cease growing abruptly above 40.5°C. They clone with high efficiency and will continue to grow slowly even at 28–30°C. This ability of CHO cells to grow and survive at temperatures not much above room temperature means that mass populations of cells can be stored in a 30°C incubator for 1–2 weeks without passaging (see Section IV).

B. pH

CHO cells are relatively sensitive to variations in pH of the medium. The cells do not like extremely acid medium (see Fig. 6.2), but will grow fairly well in relatively alkaline medium. To measure the effect of pH on growth of CHO cells, we grow them in complete medium with 10% fetal bovine serum and vary the concentration of CO_2 in the incubator. A sample of medium in the incubator is tested for pH throughout the course of each experiment. As can be seen in Figure 6.2, CHO cells will grow from pH 7.0 to pH 8.2, but will not grow well outside of this range. A frequent cause of poor growth of CHO cells is incorrect CO_2 concentration in incubators. If CHO cells are cultivated in α-MEM medium in the same incubator with other cells growing in MEM, the CHO medium will either be too acid or the MEM will be too basic to allow optimal cell growth, since MEM contains almost twice as much $NaHCO_3$ as α-MEM (Table 6.1).

C. Monolayer and Suspension Growth

The CHO cell line we used for all of our studies was obtained from Louis Siminovitch at the University of Toronto. This Toronto subclone is a descendant of strain CHO-S which Larry Thompson had adapted for growth in suspension. CHO-K1 lines in current use have not generally been adapted to suspension growth, and hence the comments in this section pertain only to the Toronto subline of CHO cells.

Monolayer culture of CHO cells is no different from culture of any other transformed cell type. We use complete α-MEM medium supplemented with 10% fetal bovine serum (see Section II), but cells growing in monolayers will do quite well in calf serum which has been prescreened to eliminate toxic sera. The major caution is that cells should be passaged frequently enough to be sure that they do not overgrow and die, presumably from accumulation of toxic metabolites and low pH (see Section III B). Cells should not be allowed to grow to densities beyond 1×10^5 cells/cm^2. If cells are growing exponentially when harvested, there will be no delay in their growth after plating in fresh medium, but there will be a lag in growth if they were not growing well at the time of replating. We remove cells from tissue culture dishes using 0.25% trypsin and 0.2 m*M* EDTA in Tris–dextrose buffer (NaC1, 8 g/L; KCl, 0.38 g/L; Na_2HPO_4, 0.1 g/L; Tris–HCl, 3 g/L; and dextrose, 1.0 g/L adjusted to pH 7.4 with HCl). This buffer preserves high viability of cells even for 1 hr at 37°C. CHO cells can be grown in monolayer flasks or plates, or in roller bottles at 1–2 rpm. CHO cells attach easily to a variety of commercially available tissue culture beads and can be grown to high yield. For roller bottles, we inoculate approximately 2×10^6 cells per 100 mL of medium per 750 cm^2 roller bottle and can harvest 5×10^7 cells per roller bottle after 3 days of growth.

When large numbers of different subclones of CHO cells are being carried in tissue culture, it is convenient to grow each strain in a separate well of a 24-well tissue culture dish. Cells carried this way can be passaged weekly by scraping the cells off the surface of the dish and suspending them in the medium in the well. A small sample of the resulting cell suspension (approximately 0.05 mL) can be used to inoculate a new well. It is possible to carry several dozen independent cell lines in this way, as might happen when mutants are being screened for a specific phenotype. Since genetic drift of mutant cell lines is possible, it is always desirable to freeze mutant isolates as soon as possible after isolation (see Section IV).

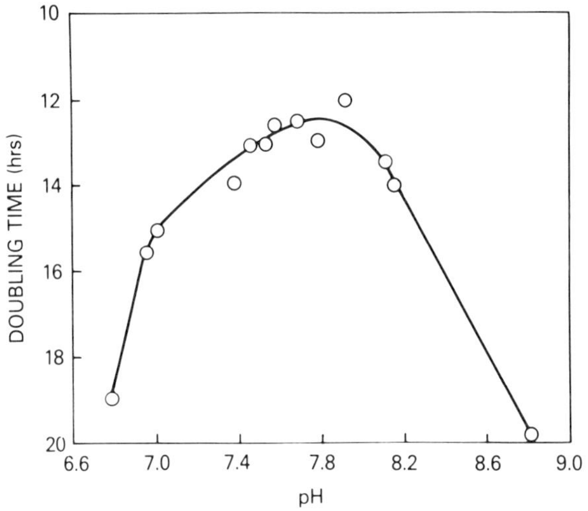

Figure 6.2. Effect of medium pH on the growth of CHO cells. The pH of the medium was altered by changing the concentration of CO_2 in the incubator. All growth curves were obtained at 37°C.

It is possible to obtain large quantities of CHO cells in monolayer culture using roller bottles or carrier beads in suspension (R. Padmanabhan, personal communication). Cytodex beads (Pharmacia) should be washed with phosphate-buffered saline and sterilized by autoclaving. Beads are resuspended in α-MEM containing 10% fetal calf serum and inoculated with CHO cells (approximately 15 cells/bead). The beads can be kept in suspension by stirring at 20–30 rpm in a siliconized spinner flask. Approximately 10^8 cells can be grow on 10^6 beads in 100 mL of medium.

CHO cells can be grown in suspension in semisolid medium such as agar or agarose (methylcellulose can also be used, but the universal presence of visible particles in methylcellulose suspensions interferes with identification of CHO clones). Procedures for growth in semisolid medium have been recently published (Gottesman, 1983) and involve preparing a suspension of cells in 0.35% nontoxic agar (Difco) or agarose and pipetting this suspension on top of a bottom layer of prehardened 0.5% agar. CHO cells will grow to sizable colonies in agar and can be easily picked from the soft agar layer with a pipet. This is a good way to clone cells and can also be used for certain types of selections which require growth in suspension (see Chapter 26 for use to isolate transformed cells and Chapter 22 for use to select cAMP resistant cells). Agar lots should be screened for toxicity, and if none can be found, highly purified agarose such as Indubiose (Fisher) can be used.

Cells may be grown in suspension without semisolid medium in bottles in a gyrorotatory shaker bath or in Spinner medium in Spinner bottles. For growth in a gyrorotatory shaker bath, cells are inoculated in 100-mL bottles containing 20 mL of medium ($> 2 \times 10^4$ cells/mL), gassed with CO_2, and shaken at 160 rpm. Use of reciprocal shakers or other agitation speeds results in cell lysis and poor growth.

CHO cells will grow in Spinner culture using Eagle's modified essential spinner medium (MEM without Ca^{2+}) supplemented with proline (1 mg/mL) and 10% fetal calf serum (a mixture of fetal calf serum, 8%, and horse serum, 2%, is also recommended). Using these conditions with standard spinner bottles at a medium speed setting (20–30 rpm), CHO cells will grow with a doubling time of approximately 12 hr (R. Padmanabhan, personal communication).

D. Replica Plating

In recent years, many reports have appeared describing procedures for replica plating of CHO cells using materials such as nylon (Stamato and Waldren, 1977), filter paper (Esko and Raetz, 1978) and polyester (Raetz et al., 1982). We have found that the most faithful transfer of CHO cells occurs with the procedure recently published by Raetz et al. using polyester sheets. Approximately 200–500 cells per 100-mm tissue culture dish are plated and allowed to grow into microcolonies for 24 hr. At the end of this time, up to four pieces of sterilized polyester are placed on top of these colonies and weighted down with small glass beads. After 1 week, the polyester sheets can be peeled off. Each contains a precise replica of the original colonies and can be stained to identify colonies or further incubated at a different temperature (to test for cold- or heat-sensitive mutants) or used to detect the presence of enzymatic activity in the colonies or some other phenotype of interest. The master plate can be used to store the colonies at 30°C until they are picked for testing. This technique has immense potential for the isolation of conditional lethal mutants or for screening large numbers of colonies (up to 10,000) for specific phenotypes for which selections do not exist.

E. Tumor Formation

CHO cells will form tumors in Swiss nude (nu/nu) female mice (Croy and Osoba, 1974; Gottesman, et al., 1983). When inoculated subcutaneously, 10^4 CHO cells will form a visible tumor in 20% of the mice; 3×10^5 cells will form tumors with an efficiency of 50%; 10^6 cells will form tumors 80% of the time; and 10^7 cells will form tumors with 100% efficiency. This degree of tumorigenicity is comparable to that found for many other transformed cell lines.

The tumors that appear after inoculation of CHO cells in nude mice are encapsulated and do not generally metastasize. They are composed of nests of basophilic, spindle-shaped cells similar in appearance to the

CHO cells in culture. The tumors are highly vascular, and much of their mass is due to large dilated vessels (Gottesman et al., 1984; Gottesman and Vlahakis, unpublished data).

Since CHO cells are tumorigenic, it is possible to use them to study mechanisms of tumorigenicity. Chapters 23, 26, and 27 discuss genetic alterations which affect the ability of cells to form tumors in nude mice.

IV. STORAGE

To be useful for genetic analysis a cell line must grow quickly, clone with high efficiency, have a relatively stable karyotype, and store easily. Most mutants which are isolated are stored for future analysis, and it is indeed fortunate that CHO cells survive a variety of storage procedures easily.

For short-term storage of CHO cells (i.e., while analyzing for specific mutant phenotypes), several options are available. As noted in Section III C, it is possible to store CHO sublines in multiwell dishes and passage them weekly by scraping the cells off the dishes. For longer storage, cells can be incubated at 30°C, where they will grow slowly (doubling time, approximately 70hr) and can be passaged every 2 weeks or so. Cells can also be stored at 4°C if the cultures are gassed with CO_2 so as to maintain a medium pH of 7.4–7.8. Under these conditions, approximately one-half of the cells die every 24 hours. It is possible, starting with 10^6 cells, to rescue viable cells from the refrigerator after 2 weeks. This procedure is not recommended, however, as it may select for variant CHO cells.

For long-term storage of CHO cells, freezing is necessary. We have successfully used 10% glycerol (dilute cells in complete medium supplemented with 10% fetal bovine serum with one-fifth volume of 50% filter sterilized glycerol) and 7% DMSO (add appropriate volume of DMSO directly to cell suspension in complete medium plus serum). If DMSO is used, it is essential to cool the cells immediately to 4°C so as to prevent cell killing by DMSO. Cells are frozen slowly by wrapping them in an insulating material (such as polystyrene) and placing them at − 20°C (for glycerol) or − 70°C (for DMSO). Cells can be stored at − 70°C for up to 1 year without loss of viability, but for truly long-term storage should be kept in a liquid-nitrogen tank. Our frozen cultures have been fully viable for at least 8 years when stored at liquid-nitrogen temperature.

Cells are defrosted rapidly by warming to 37°C and plating immediately in a tissue culture dish with complete medium. For maximum viability, we remove the medium after 1 hr and replace it with fresh medium. Greater than 50% of the cells should survive the freezing procedure. Although viability seems to be somewhat better using 7% DMSO, there is one advantage to the use of glycerol. If a freezer defrosts unexpectedly, cultures frozen in 10% glycerol will maintain some viability for up to 24 hr, whereas cells frozen in 7% DMSO will die rapidly at room temperature.

Very early after establishment of a new cell line it is critical to freeze many vials of these cells for future use. There is little question that cultured mammalian cells alter their phenotypic properties after long-term cultivation owing to factors such as incubator temperature, alteration in content of media, different lots of serum, the schedule of cell passage, and viral or mycoplasma infections. For experiments in which single-step mutants are compared with parental cell lines, both the mutants and parental lines must be in their pristine state for these comparisons to be meaningful. We defrost new vials of cells after approximately 100 doublings (50 days for CHO cells) of growth to be sure that genetic drift does not confuse interpretation of comparative data.

V. ISOLATION OF MUTANTS

It is beyond the scope of this chapter to review all of the techniques available for the isolation of mutants of Chinese hamster cells. Many of these techniques are well described in this book, or have been reviewed elsewhere (Thompson and Baker, 1973; Thompson, 1979). This section will briefly consider general principles involved in the mutagenesis of CHO cells and the subsequent isolation of mutants from mutagenized populations.

A. Mutagenesis

Many altered phenotypes of CHO cells cannot be demonstrated in unmutagenized cell populations because the rate of spontaneous appearance of these mutations is quite low and the frequency of mutants may be less than 1 in 10,000,000 cells. In these cases, it is necessary to mutagenize cells in order to increase the rate of appearance of the mutations of interest. The usual mutagens used for CHO cells are ethylmethane sulfonate, *N*-methyl-*N*-nitro-*N*-nitrosoguanidine and ultraviolet light, although other mutagens can be used. The first two named compounds are alkylating agents, and, on the basis of theoretical considerations and evidence derived from prokaryotic systems, would be expected to induce single-base-change mutations. To date, however, there is insufficient sequence data from mammalian cell mutants to be certain that these agents do act in the predicted manner. Ultraviolet light would be expected to produce deletions, but again, there are no data to support or refute this prediction.

In order to reduce the likelihood of multiple mutations in mutagenized cells, it is desirable to use the minimum amount of mutagen which will produce a measurable frequency of mutants. In practice, this amount is usually the quantity of mutagen which reduces survival of the cells to 20% of their usual cloning efficiency. Since the contents of the medium affects the efficiency of mutagenesis [e.g., media containing thymidine makes cells more susceptible to mutagenesis (Peterson et al., 1978)], it is a good idea to determine a killing curve for the mutagen which is to be used to find the dose of mutagen that will allow clonal growth of 20% of the treated cells. To isolate independent mutants, it is necessary to mutagenize separate flasks of cells.

After mutagenesis, cells are allowed to grow in nonselective medium for several days. This expression period may vary from 2 to 10 days depending on the selection; optimum expression time must be determined in each case. It is a good idea to monitor effectiveness of mutagenesis by determining the frequency of appearance of mutants of a well-defined phenotype. We use resistance to 2 m*M* ouabain for this purpose (Baker et al., 1974). In general, mutagenesis should result in at least a 10-fold increase in ouabain resistance in the mutagenized population.

B. Selection of Mutants

Efficient selection of mutants requires efficient killing or removal of nonmutant cells. In designing a mutant selection, it is essential to choose conditions in which wild-type cells will not survive the selection. To determine these conditions, we generate survival curves under increasingly stringent selection conditions. It is important to be certain that cell density, ratio of medium volume to cells, and preparation of selective medium are identical to that which will be used for the actual selection. The majority of selections employed to date for CHO cells are very sensitive to the density of cells used for selection; thus, whereas a specific drug concentration will kill 5×10^5 cells on a 100-mm tissue culture dish, the same drug concentration will have no effect when $(1-2) \times 10^6$ cells are plated on the same tissue culture dish. In practice, we always select cells at 5×10^5 cells per 100-mm dish with 15 mL of selective medium. To determine the minimum drug concentration needed for quantiative killing of cells, we set up several such dishes with doses of drug increasing by small increments, and choose the minimum drug concentration in which there is no obvious mass growth or survival of the cells.

In order to get independent mutants, it is necessary to use cell populations that have been independently mutagenized (i.e., different flasks). Once a mutant has been obtained, it is useful to try to isolate spontaneous mutants of the same type in order to be certain that the complete phenotype of the mutant in question is due to a single mutation and not due to multiple mutations induced by the mutagen. Where this is feasible, independent spontaneous mutants can only be derived from truly independent cell populations. These independent populations can be obtained by reducing the number of cells in an inoculum to 1 or a very low number, such as 100, where the likelihood of finding a mutant cell line in the inoculum is essentially zero.

This approach is also used to determine the intrinsic *rate* of appearance of mutants (as opposed to the frequency, which is simply the percentage of mutant cells in any given population). The principles of this approach were first worked out by Salvador Luria and Max Delbruck looking at the spontaneous appearance of resistance to bacteriophage by bacteria (Luria and Delbruck, 1943). If mutants arise in a spontaneous fashion, as opposed to arising only in response to selective pressure, then they should appear randomly in populations of cells grown from small inoculums such that some of these populations will contain large numbers of mutants (i.e., those populations in which mutants appeared early during the growth of the population), whereas other populations should contain few mutants, or no mutants at all. Luria and Delbruck developed a statistical approach using this *fluctuation analysis* to prove the spontaneous origin of mutants and to calculate the mutation rate using the distribution of mutants in multiple populations and the number of populations with no mutants at all. This approach has been taken in somatic cell genetics to demonstrate the spontaneous appearance of a large number of mammalian mutants and has been used effectively for CHO cells to calculate mutation frequencies (Baker et al., 1974; Meuth et al., 1979; Rabin and Gottesman, 1979).

A number of criteria have been established to prove that a phenotypic change in a cell population is truly the result of a mutation, as opposed to an epigenetic change. This issue is discussed extensively by Siminovitch (1976). In general, if an alteration in a cell breeds true, if this alteration appears spontaneously as indicated by a Luria–Delbruck fluctuation analysis, and if the rate of appearance of this mutation can be increased by mutagens, it is considered a likely candidate for a true mutation. Additional data which prove that a phenotypic change is due to a mutation include the demonstration of an altered protein product, such as an isoelectric shift variant, or an enzyme whose activity is temperature sensitive in a manner consistent with the temperature-sensitive phenotype of the cell. In these cases, it is important to prove the linkage of the biochemical phenotype with the genetic change; this proof can be obtained by reversion analysis (selection for revertant phenotype results in reversion of the biochemical defect) or, as has been possible only very recently, by cotransfer of the biochemical phenotype with the DNA encoding the mutant protein (see Chapter 8). Final, conclusive demonstration that a mutation has occurred depends on isolation and sequencing of the mutant gene, or restriction enzyme analysis demonstrating an altered gene segment on a Southern blot using a probe for the gene in question (see Chapter 12). Reversible changes in DNA, such as methylation, qualify by this criterion as mutations since they involve demonstrable changes in DNA on Southern blots, but most workers consider such changes to be epigenetic.

Occasionally, mutants are isolated and found to have higher rates of mutation for other loci not clearly related to the locus involved in the selection for the original mutation. Such mutants may express *mutator* genes, either as a result of the original alteration or as the reason why the mutant was isolated in the first place. A report has appeared indicating that a mutation in CHO cells that affects ribonucleotide–diphosphate reductase and renders cells resistant to arabinosyl cytosine and auxotrophic for thymidine may have mutator effects (Meuth et al., 1979). Mutator effects have also been described in an aphidicolin-resistant Chinese hamster V79 line with an altered DNA polymerase (Liu et al., 1983). A recent paper by Drobetsky and Meuth (1983) suggests that sequential selection of CHO cells for gradually increasing drug resistance (see, for example, Chapter 25) results in the isolation of cell lines with increased mutation rates at a variety of unrelated loci. These data raise the possibility that any selection in tissue culture may predispose to the isolation of cell lines carrying mutator genes, and, indeed, suggest that all cultured cell lines, which probably needed to mutate in order to adapt to tissue culture, may carry mutator genes. This is a caution that must be kept in mind when considering the rates and types of mutation found among somatic cells in culture (see Chapter 29) and attempting to extrapolate these data to cells in their normal environment in animals.

REFERENCES

Baker, R.M., Brunette, D.M., Mankovitz, R., Thompson, L.H., Whitmore, G.F., Siminovitch, L., and Till, J.E. (1974). *Cell* **1**, 9–21.

Carver, J.H., Salazar, E.P., and Knize, M.G. (1983). *In Vitro* **19**, 699–706.

Croy, B.A. and Osoba, D. (1974). *J. Immun.* **113**, 1626–1634.

Drobetsky, E. and Meuth, M. (1983). *Mol. Cell. Biol.* **3**, 1882–1885.

Esko, J.D. and Raetz, C.R.H. (1978). *Proc. Natl. Acad. Sci. USA* **75**, 1190–1193.

Gottesman, M.M., Roth, C. Leitschuh, M., Richert, N., and Pastan, I. (1983). In *Tumor Viruses and Differentiation,* (E. Scolnick and A. Levine, eds.), Alan R. Liss, New York, pp. 365–380.

Gottesman, M.M., Vlahakis, G., and Roth, C. (1984). In *Peptide Hormones, Biomembranes and Cell Growth* (R. Verna, ed.) Plenum Publishing, New York.

Gottesman, M.M., LeCam, A., Bukowski, M., and Pastan, I. (1980). *Somat. Cell Genet.* **6**, 45–61.

Gottesman, M.M. (1983). In *Methods in Enzymology,* (J.D. Corbin, and J.G. Hardman, eds.), Academic Press, New York, Vol. 99, pp. 197–206.

Hamilton, W.G. and Ham, R.G. (1977). *In Vitro* **13**, 537–547.

Jakoby, W.B. and Pastan, I.H. (eds.) (1979). *Methods in Enzymology* (S. P. Colowick and N. O. Kaplan, series eds.), Academic Press, New York, Vol. 58.

Liu, P. K., Chang, C.-C., Trosko, J. E., Dube, D. K., Martin, G. M., and Loeb, L. (1983). *Proc. Natl. Acad. Sci. USA* **80**, 797–801.

Luria, S. E. and Delbruck, M. (1943). *Genetics* **28**, 491–511.

Meuth, M., L'Heureux-Huard, N., and Trudel, M. (1979). *Proc. Natl. Acad. Sci. USA* **76**, 6506–6509.

Paul, J. (1975). *Cell and Tissue Culture,* Churchill Livingstone, Edinburgh.

Peterson, A. R., Landolph, J. R., Peterson, H., and Heidelberger, C. (1978). *Nature* **276**, 508–510.

Puck, T. T., Ciecuira, S. J., and Robinson, A. (1958). *J. Exp. Med.* **108**, 945–955.

Rabin, M. S. and Gottesman, M. M. (1979). *Somat. Cell Genet.* **5**, 571–583.

Raetz, C. R. H., Wermuth, M. M., McIntyre, T. M., Esko, J. D., and Wing, D. C. (1982). *Proc. Natl. Acad. Sci. USA* **79**, 3223–3227.

Siminovitch, L. (1976). *Cell* **7**, 1–11.

Stamato, T. D. and Waldren, C. A. (1977). *Somat. Cell Genet.* **3**, 431–440.

Stanners, C. P., Eliceiri, G. L., and Green, H. (1971). *Nature New Biology* **230**, 52–54.

Thompson, L. H. and Baker, R. M. (1973). In *Methods in Cell Biology,* (D. Prescott, ed.), Academic Press, New York, Vol. 6, pp. 209–281.

Thompson, L. H. (1979). In *Methods in Enzymology,* (W. B. Jakoby and I. Pastan, eds.), Academic Press, New York, Vol. 58, pp. 308–322.

Tooze, J. (ed.) (1973). In *The Molecular Biology of Tumour Viruses,* Cold Spring Harbor Laboratory, Cold Spring Harbor, New York, pp. 74–172.

CHAPTER 7

CELL FUSION AND CHROMOSOME SORTING

Jerry W. Shay
University of Texas Health Science Center at Dallas
Department of Cell Biology
Dallas, Texas

L. Scott Cram
Los Alamos National Laboratory
Los Alamos, New Mexico

I. INTRODUCTION

Adaptation to environmental change is believed to occur by either mutation or amplification of DNA or by the parasexual acquisition of new genetic material. Even though much is known about the role of mutation rates as the basis of variation in somatic cells (see Worton and Grant, Chapter 28, and Siminovitch, Chapter 29), only recently have a variety of methodologies been developed for parasexually introducing genetic material into eukaryotic cells. The first part of this chapter will describe some of the methods used for introducing new genetic information into animal cells, while the second part will describe the use of the fluorescence-activated cell sorter for isolating specific genetic material for use in gene transfer experiments. The topics of DNA-mediated and vector-mediated gene transfer, will be covered extensively in other chapters (see Abraham, Chapter 8, and Howard and McCormick, Chapter 9) so these subjects will not be covered here. Since Chinese hamster cells in general have a high plating efficiency, a short generation time, and a small and stable karyotype, they have proven to be a convenient model for investigations into such diverse topics as mutagenesis, gene transfer, genetic complementation, differentiation, and regulation of gene expression. Thus, many of the methodologies described in this chapter were often initially "worked out" using Chinese hamster cells.

II. CELL FUSION TECHNIQUES AND SELECTIVE SYSTEMS

Somatic cell hybridization was one of the first systems used to analyze interactions between somatic cells. Cell hybridization effectively results in the initial formation of a product containing the entire nuclear genome and cytoplasmic substance of two different cells (Okada, 1958; Barski et al., 1960; Sorieul and Ephrussi, 1961; Harris and Watkins, 1965). By treating cell populations with membrane fusing agents such as inactivated Sendai virus (Harris and Watkins, 1965) or polyethylene glycol (Pontecorvo, 1975), the efficiency of cell fusion and the selection of viable cell hybrids can be increased greatly. There have been many technical improvements in basic cell fusion procedures but these are beyond the scope of this chapter, and the reader is encouraged to refer to a compilation of reviews on this subject (Shay, 1982). Since cell fusion is less than 100% efficient, it is necessary to isolate the hybrid cells of interest from the parental cells that either do not participate in cell fusion or that fuse to a like cell (e.g., homokaryon). Several methods have been developed to select only true hybrid cells, most of which depend on the existence of complementing recessive mutations in each of the parental cell lines (Littlefield, 1964). In addition, there have been techniques using dominant markers (Siminovitch, 1976), and as with the recessive markers these methods rely on the ability of the fusion products to grow and divide in a special medium in which the parental cells cannot grow (see Fenwick, Chapter 13). More recently, methods have been described for using the fluorescence-activated cell sorter or specific biochemical inhibitors to isolate fusion products (Jongkind et al., 1979; Jongkind and Verkerk, 1982; Wright, 1978, 1981, 1982). These techniques do not require cell division of the fusion product, therefore allowing studies on the immediate products of cell fusion (heterokaryons).

A. Cell Hybridization

Shortly after the initial observation that cells could be fused to each other, it became clear that interspecific hybrid cells are not mere curiosities, since they have at least two properties that make them suitable for genetic analysis. First, both sets of chromosomes are functional and the hybrids therefore exhibit the hereditary characteristics of both parents. Second, as the hybrids multiply they lose some of their chromosomes spontaneously, resulting in many different combinations of parental genes. Even though the underlying mechanism of chromosome segregation remains poorly defined (Chapter 28), this process has provided a rapid and efficient method for gene mapping. Gene mapping is accomplished by correlating the expressed phenotypes (gene products) with the retention of whole or parts of chromosomes in the hybrids (Ruddle and Creagan, 1975; McKusick and Ruddle, 1977) (Chapter 5). Recently, mapping of unexpressed (silent) phenotypes has become possible using restriction enzyme analysis and cloned recombinant DNA probes (Ruddle, 1981). With relatively straightforward techniques we are now able to determine (1) which chromosome a gene is on, (2) which genes are on the same chromosome, and (3) where on the chromosome the gene is located. Using these methods, we are developing a more precise knowledge of how genes function individually and as coordinated sets.

Heterokaryons, which are the immediate product of the fusion of two different cells, have proven useful in understanding certain aspects of differentiation. Since selective systems are used in most cell hybridization studies, the ability of the fusion product to divide is required. Differentiated cells are usually postmitotic or only slowly dividing. By limiting analysis to only those fusion products that are capable of rapid division, in many cases, one may inadvertently bias the results against the expression of differentiated function. Since it is known that less than 1% of the heterokaryons formed ultimately give rise to dividing hybrid clones, it is clear that such hybrids are a highly selected subset, and making generalizations about control of expression of differentiation in such hybrids may be misleading. In order to study heterokaryons techniques have been developed using irreversible biochemical inhibitors and the fluorescence-activated cell sorter to isolate the fusion products in pure form (Wright, 1978; Jongkind et al., 1979). Heterokaryon analysis has proven to be useful for several reasons, including obtaining information on the regulatory changes that occur during the transition from precursor cells to terminally differentiated cells. In addition, postmitotic differentiated cells can be used for analysis since cell division is not a prerequisite for isolating fusion products. Heterokaryon analysis avoids the problem of chromosome segregation, and in many cases allows normal diploid cells to be used without selectable genetic markers (Wright, 1982). Therefore, cell hybrid and heterokaryon analysis are complementary in the types of information they can provide. In addition to the mapping of genes, these techniques have provided much information concerning the expression, coexpression, reexpression, activation, cross activation, and extinction of genes in eukaryotic cells.

B. Cell Enucleation

Even though there is little doubt that cell development is ultimately controlled by nuclear genes, it is known that some aspects of cellular inheritance cannot be satisfactorily explained solely in terms of the nuclear genome. Cell enucleation and fusion of mammalian cells in culture have provided a convenient and direct technology to determine in a wide variety of systems if the mammalian cell cytoplasm can modify nuclear gene activity.

Using cytochalasin B in combination with mild centrifugation, it has become possible to separate populations of mammalian cells into nuclear and cytoplasmic parts (Prescott et al., 1972; Wright and Hayflick, 1972). The nucleated part, called the karyoplast (Shay et al., 1973, 1974) or minicell (Ege et al., 1974), consists of a nucleus enclosed in a thin shell of cytoplasm (2–10%, depending on the technique) and an intact plasma membrane. The cytoplasmic part, called the cytoplast (Shay et al., 1973, 1974), contains the bulk of the cytoplasm and includes all the types of organelles normally found in the cytoplasm, e.g. centrioles, Golgi, endoplasmic reticulum, ribosomes, mitochondria, lysosomes, and cytoskeleton (Wise and Prescott, 1973; Shay et al., 1974). It has been demonstrated that cytoplasts can at least temporarily attach, spread, elongate, respond to hormonal stimulation, round-up, agglutinate, communicate, and translocate (Carter, 1967; Poste and Reeve, 1972; Goldman et al., 1973; Schroder and Hsie, 1973; Pollack et al., 1974; Miller and Ruddle, 1974; Shay et al., 1975, 1977; Cox et al., 1976; Wise and Larsen, 1976; Clark and Shay, 1981). Such studies with enucleated cells suggest that the information necessary for many aspects of cell behavior is present in the cytoplasm. Karyoplasts, which do not contain centrioles or cytoskeletal components, are incapable of spreading or locomotion, which is added support for the hypothesis that form-determining and motility mechanisms operate in the cytoplast without nuclear participation during the short period of cytoplast viability. In addition, with one exception (Zorn et al., 1980), investigators have observed that although the karyoplasts are endowed with everything usually considered essential for viability, they do not survive or regenerate their lost parts. Karyoplasts have the entire nuclear genetic material, an intact nuclear envelope, some cytoplasmic organelles, and a continuous plasma membrane. Why then do the karyoplasts fail to regenerate? One is led to conclude that the missing cytoplasmic elements or the organization of cytoplasmic elements is essential for cell viability.

C. Cybrids and Reconstituted Cells

Using cytoplasts and karyoplasts in combination with cell fusion techniques, it has become possible to produce cytoplasmic hybrids (cybrids) and reconstituted cells in order to better understand the nature of cytoplasmic substances that might act to control their own synthesis as well as to regulate the expression of other genes (Fig. 7.1). Cybrids are defined as the viable proliferating fusion product of a cytoplast with a whole cell (Bunn et al., 1974), while reconstituted cells are the viable proliferating fusion product of a cytoplast with a karyoplast (Veomett et al., 1974). Even though both of these products permit investigation into nuclear–cytoplasmic interactions, the main difference between reconstituted cells and cybrids is that the former does not involve as much dilution of the donor cytoplasm by the recipient cytoplasm (Bunn, 1982).

There are several methods for selecting cybrids and reconstituted cells from a mixed population of parental cells and fusion products, but again these are beyond the scope of this chapter, and the reader is encouraged to refer to the original articles or a recent compilation of techniques for additional details (Shay, 1982). Clearly, however, the development of selectable cytoplasmic genetic markers has proven to be a powerful tool enabling studies into cytoplasmic modification of nuclear gene expression (see Whitfield, Chapter 19).

Analysis of cybrid and reconstituted cells has provided much information on the regulation of gene expression in normal and transformed cells. Such techniques avoid chromosome segregation and gene dosage effects that may complicate cell hybrid analysis. In addition, these techniques have provided a valid model system to discover the existence of stable factors present in the cytoplasm which may modify nuclear gene expression. Although many examples of cytoplasmic modification of nuclear gene expression have been reported to date (Howell and Sager, 1978; Gopalakrishnan and Anderson, 1979; Lipsich et al., 1979; Lindler, 1980; Kahn et al., 1981; Shay et al., 1981; Clark and Shay, 1982b), definitive information on the nature of these regulatory molecules is still very limited.

D. Nuclear Hybrids

Nuclear hybrids are the viable proliferating fusion products of nonregenerating karyoplasts to other whole cells (Fig. 7.1). These techniques allow the characterization of nuclear–nuclear interactions in the presence of predominantly one cytoplasm (Weide et al., 1982a, 1982b). Although these techniques have only had limited experimental use to date, they should permit the manipulation of mitochondrial populations and may enhance the study of mitochondrial genetics in the future.

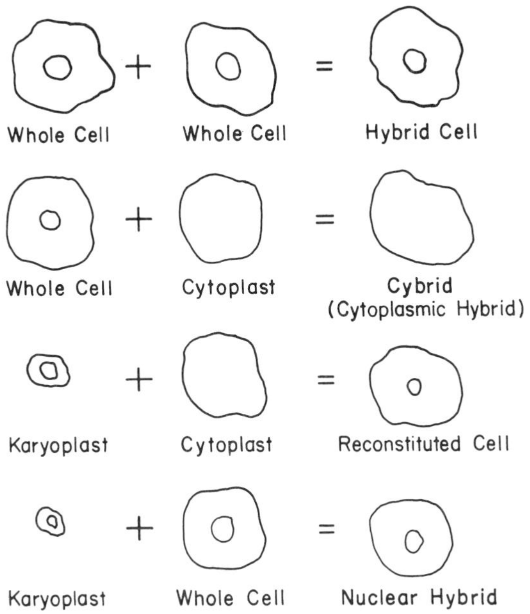

Figure 7.1. Diagramatic representation of techniques to produce hybrids, cybrids, reconstituted cells, and nuclear hybrids.

E. Microcells

Microcell-mediated gene transfer permits the introduction of only one or a few donor chromosomes into the recipient cells. Microcells (Fournier and Ruddle, 1977; Fournier, 1982) or microkaryoplasts (Shay and Clark, 1977) are produced by treating the donor cells with a mitotic inhibitor (usually colcemid) for a prolonged period of time (36–48 hr). Since colcemid prevents the assembly of microtubules, donor cells accumulate in metaphase (Fig. 7.2). For reasons that are still unclear, individual or small clusters of chromosomes eventually become surrounded by a nuclear envelope (micronuclei) as if the cells were attempting to reenter interphase even though chromosome separation did not occur. This process, however, serves to partition the donor cell's chromosomes into discrete subnuclear packets, which can then be physically removed (Fournier, 1982). The removal of the packages of chromosomes (micronuclei) is accomplished by centrifugation in the presence of cytochalasin B, similar to the procedure for isolating intact nuclei (karyoplasts). The micronuclei are pulled from the cells on a long stalk of cytoplasm, which eventually breaks resulting in the production of microcells or microkaryoplasts. Microcells consist of a single or a few chromosomes surrounded by a nuclear

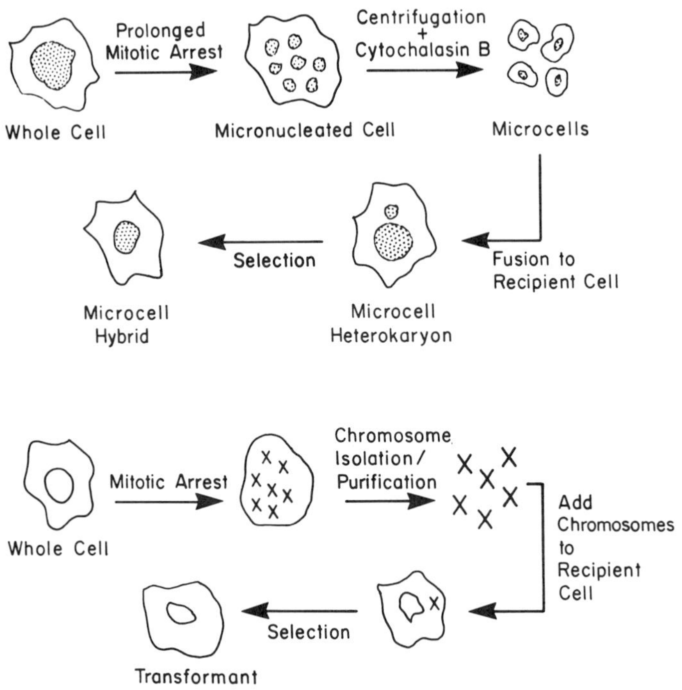

Figure 7.2. Diagramatic representation of microcell-mediated and chromosome-mediated gene transfer techniques.

envelope, a small rim of cytoplasm, and an intact plasma membrane. The microcells can be immediately used or further purified to specific sizes by various techniques such as using Ficoll gradients or the cell sorter. The microcells can then be fused to the recipient cells using standard techniques. Even though the microcells are only viable for a few hours, they may be rescued by fusion to intact cells. By use of appropriate selective growth conditions, specific genes can be expressed in microcell hybrids. This technique, although similar to whole cell fusion, essentially results in the determination of chromosome segregation prior to fusion. There are several other advantages of this procedure. For example, microcell hybrids can be generated with constellations of chromosomes not easily obtained using standard hybridization procedures (Fournier, 1982). In addition, by introducing only a small amount of cytoplasmic components one can avoid introducing epigenetic or mitochondrial components into the recipient cell. In this regard microcell hybrids and nuclear hybrids are similar. Microcell hybrids have also been useful in normal and regional gene mapping (Worton et al., 1981; Fournier, 1982) and have even been used to analyze the chromosomal integration site of foreign DNA (Smiley et al., 1978; Fournier et al., 1979). An alternative to producing microcells is the production of chromosome-containing minisegregant cells, which involves arresting cells in mitosis by colcemid or hyperbaric nitrous oxide treatment and then subjecting the cells to prolonged cold shock (Schor et al., 1975; Johnson and Mullinger, 1982). When the cells are returned to 37°C, a severely altered cytokinesis occurs resulting in the formation of minisegregants containing subnuclear packets of chromosomes.

III. TECHNIQUES FOR TRANSFERRING SUBCELLULAR FRACTIONS

A. Chromosome-Mediated Gene Transfer

This technique utilizes isolated metaphase chromosomes as vectors for the transfer of genetic information (McBride and Ozer, 1973; McBride, 1982). The techniques for isolating metaphase chromosomes and their purification will be discussed subsequently. As illustrated in Figure 7.2, isolated metaphase chromosomes are added to recipient cells and are taken up by phagocytosis. Most of the ingested metaphase chromosomes are degraded into small inactive fragments. However, depending on the recipient cell type and the sensitivity of the selectable marker, chromosome-mediated gene transfer occurs with a frequency between 10^{-7} and 10^{-5}. Donor chromosomal fragments (transgenome) can be observed in the recipient cells for several generations after uptake if selection is maintained (McBride, 1982). In the absence of selection it has been estimated that between 1% and 10% of the cells lose the ability to express the selectable gene in each generation (Klobutcher and Ruddle, 1979). The stable retention of transferred genes appears with a much lower frequency, and the retained genes appear to integregate into nonhomologous sites in the recipient cell genome. Closely linked genes are cotransferred by this method which permits the regional mapping of linked genes (Miller and Ruddle, 1978; Klobutcher and Ruddle, 1979). In addition, this procedure may be useful in bridging the gap between recombinant DNA and somatic cell hybridization gene-mapping techniques.

B. Mitoplasts, Microcytospheres, and Microplasts

Various types of cytoplasmic fragments can be obtained by exposing cells to cytochalasin B. Mitoplasts are obtained from mitotic cells by cytochalasin B removal of chromosomes (enucleation). The cytoplasm of mitotic cells (mitoplasts) may be useful in studying the nature of mitotic factors involved in chromosome condensation (Sunkara et al., 1977; Rao et al., 1982). Microcytospheres are obtained by treating cells in suspension with high concentrations of cytochalasin B (25 μg/mL) at 37°C for 4–5 min

(Clark and Shay, 1982b; Maul and Weibel, 1982). This procedure results in many blebs over the surface of the cells, which can be removed by mixing (vortexing) for 4 mins. The microcytospheres can be separated from the whole cells by a low-speed spin (400g for 5 min) and further purified by overlaying the enriched microcytosphere preparation on a 10% sucrose solution and centrifuging 800g for 10 min. The microcytospheres do not penetrate the sucrose and they can be further analyzed on a fluorescence-activated cell sorter. By staining with R123 (a mitochondrial-specific dye), microcytospheres can be obtained with or without mitochondria (Clark and Shay, 1982). Microcytospheres can be used like cytoplasts for fusion experiments thereby enabling investigations into cytoplasmic factors which may modify nuclear gene expression.

Using a similar technique, isolation of microplasts has also been described by other investigators (Albrecht-Buehler, 1980). Exposing cells growing in monolayer culture to cytochalasin B until they arborize and then washing away the cells leaves only tiny attached fragments (microplasts) behind. These small cytoplasmic fragments can move filopodia, produce blebs, and ruffle lamellipodia, and may offer new insights into certain aspects of cell motility (Albrecht-Buehler, 1980).

C. Mitochondrial-Mediated Gene Transfer

Chloramphenicol (CAP) resistance was the first cytoplasmic drug-resistance marker to be described for mammalian cells (see Whitfield, Chapter 19). CAP is a potent inhibitor of mitochondrial protein synthesis capable of killing sensitive (wild-type) mammalian cells. Mammalian cell mutants can be isolated that are resistant to CAP, and the following experiments show that CAP resistance is inherited cytoplasmically. In cytoplasmic transfer experiments, the nucleus of the CAP^r cell is physically removed by cytochalasin-B-induced enucleation, and the membrane-bound cytoplasmic fragment (cytoplast) containing the mitochondria and other organelles are fused either to a CAP^s whole cell (cybrid) or to a karyoplast obtained from a CAP^s whole cell (reconstituted cell). In both types of experiments CAP^r colonies are obtained. Contaminating non-enucleated parental cells are not responsible for these observations, since the recipient cells or karyoplasts have specific nuclear markers, allowing positive identification of cell types. Since these experiments only suggest that mitochondria are responsible for CAP resistance, additional experiments were required to confirm that this conclusion was correct.

It was demonstrated that rhodamine 6G (R6G), a toxic mitochondria-specific fluorescent compound, greatly reduces the transmission of CAP resistance when cells carrying the CAP^r marker are treated with R6G prior to hybrid or cybrid formation (Ziegler and Davidson, 1981; Ziegler, 1982). In addition, it has recently been demonstrated that mixing purified mitochondria obtained from CAP^r cells with CAP^s cells results in transfer of CAP^r (Clark and Shay, 1982; Clark and Shay, 1982a). Finally, it has now been established that the mitochondrial DNA of CAP^r mouse and human cells contains a single base change in the region encoding the 3′ end of the large ribosomal RNA (Blanc et al., 1981; Kearsey and Craig, 1981; Wallace, 1982) and this may be responsible for resistance to the drug.

D. Comparison of Microinjection, Liposomes, and Red-Blood-Cell Ghosts

Biochemical purification of DNA, RNA, and various proteins has become more sophisticated so that a need has developed for additional techniques to analyze the biological activity of these purified molecules. Purified nucleic acids can be introduced into certain cells as calcium phosphate precipitates (Abraham, Chapter 8), but direct introduction of genetic material via microinjection with small microcapillary needles is more efficient. The major disadvantage of microneedle injection is the relatively small number of cells that can be injected. However, microinjection is a versatile method in that almost any molecule can be directed to specific areas within the cell (Graessmann and Graessmann, 1982). Liposome and red-blood-cell-mediated transfer of molecules are alternatives to microneedle injection in which large numbers of cells can be treated (Rechsteiner, 1982; Straubinger and Papahadjopoulous, 1982). In addition, specialized equipment and skills that are required for microneedle injection are avoided using these techniques. Only small volumes can be introduced via liposomes, while red blood cells permit substantial amounts of certain molecules to be introduced. Even though the term "microinjection" using red-blood-cell ghosts is still employed, it is perhaps more appropriate to refer to this procedure as red-blood-cell-ghost-mediated transfer since cell fusion and not a microneedle is used to transfer the molecules. Additional information about these techniques can be found in Chapter 8.

IV. FLOW CYTOMETRY AND SORTING

Flow cytometry is a technique for the rapid and quantitative analysis of single cells and chromosomes. Sorting, as the name implies, is the separation of cells or chromosomes based on individual properties. Flow cytometry and sorting combine to provide the biologist with an effective method for separating unique cell populations in a manner similar to those used by the biochemist to separate specific macromolecules.

The following sections describe flow cytometry and sorting techniques used to analyze and sort Chinese hamster cells and Chinese hamster

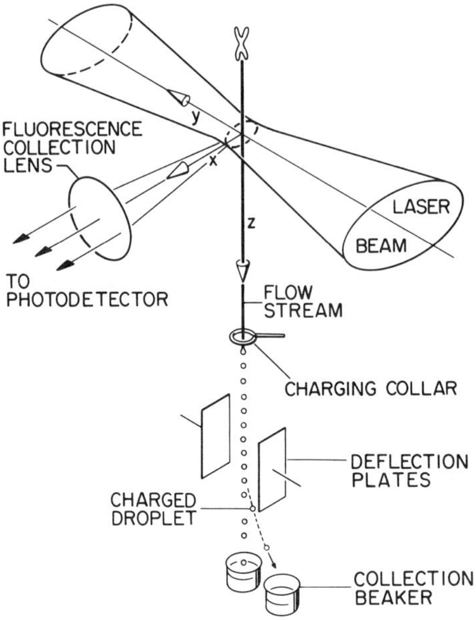

Figure 7.3. Diagramatic representation of flow karyotype analysis and chromosome sorting. Chromosomes flow along the *z* axis, the laser beam points in the *y* direction, and the fluorescence detector is located in the *x* direction. The flow stream is broken into uniform droplets, some of which contain chromosomes. Droplets containing chromosomes to be sorted are charged and electrostatically deflected as shown.

chromosomes. Advanced developments will be outlined at the end of this section. Several excellent review articles are available which describe the general principles of flow cytometry and sorting (Herzenberg and Sweet, 1976; Arndt-Jovin and Jovin, 1978; Melamed et al., 1979).

The basic principle for the analysis of isolated chromosomes is illustrated in Figure 7.3. Single chromosomes tagged with a fluorochrome specific for DNA, for example, flow single file, one at a time, through a focused laser beam. The fluorescence signal is optically collected and analyzed. The liquid stream, which moves the chromosomes along like beads on a string, can be enclosed in a flow chamber or can exist as a stream in air. The stream jets into air, breaking into uniform droplets, some of which will contain a chromosome. The properties of the droplet are measured as it passes through the laser beam and, if a chromosome of interest is present, the droplet will be electrostatically charged and deflected. Droplets containing unwanted chromosomes are allowed to fall in a straight trajectory.

The pulse of fluorescent light emitted by a cell or chromosome as it passes through the laser beam is measured and recorded. Depending on the specificity of the fluorochrome, properties such as DNA, RNA, membrane antigenic sites, enzyme activity, and membrane potential can be rapidly measured on a per-cell basis. Typical analysis rates are 1000–2000 cells/sec. Sorting rates can be as high as 1000 cells/sec depending on the frequency with which the subpopulation being sorted occurs.

Multivariate analyses have been added to the basic concept illustrated in Figure 7.3. These additional variables greatly expand the capability of flow cytometers to resolve subpopulations of cells or chromosomes. One example is dual-beam systems, in which the object being analyzed sequentially intersects two laser beams tuned to different wavelengths, which excites two fluorochromes bound to the same object. A recent review by Shapiro (1983) describes most of the common fluorochromes used for flow cytometric analysis. Another example, slit scanning [see review by Cram et al. (1985)], measures gross morphological features and is particularly useful for chromosomes. Slit scanning requires that the laser beam be focused to a slit of illumination so that asymmetric objects such as chromosomes that align in flow perpendicular to the slit are scanned as they flow through the ribbon of illumination.

Chinese hamster cells, such as CHO, and chromosomes isolated from the Chinese hamster cell line M3-1 have been analyzed extensively using flow cytometry. These cell lines have been used as the equivalent of a reference standard for evaluating instrument performance because of their ready availability, short doubling times, and relatively low number of chromosomes.

A. Flow Karyotype Analysis

Each of the Chinese hamster euploid autosomes and the X chromosome can be distinguished based on propidium iodide fluorescence. Figure 7.4 illustrates the flow karyotype of chromosomes isolated from a near diploid Chinese hamster cell line. The Y chromosome, when present, is located in the same peak as the number 5 chromosome. Identification of the chromosome type constituting each peak in the flow karyotype has been done based on the results of sorting (Gray et al., 1975), Q banding (Carrano et al., 1979), comparing male and female flow karyotypes, and following sequential changes with flow karyotype analysis and G banding. Several types of information can be obtained from a flow karyotype. For example, the area of each peak is proportional to the number of chromosomes of that type, the peak mean is proportional to DNA content, and the background continuum is a reflection of culture heteroploidy and debris resulting from sample preparation (Bartholdi et al., 1984).

1. Data Analysis

Analysis of a univariate (propidium iodide fluorescence) flow karyotype such as shown in Figure 7.4 consists of fitting a Gaussian function to each

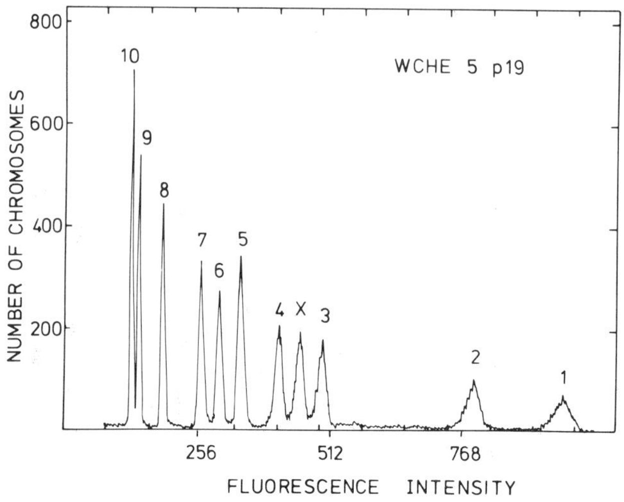

Figure 7.4. Fluorescence distribution of chromosomes isolated from WCHE/5 cells (female) at passage 19 and stained using the hypotonic propidium iodide technique (Aten et al., 1980). Analysis was performed on a high illumination Los Alamos flow cytometer (Bartholdi et al., 1983). The distribution is a histogram of the number of chromosomes per channel as a function of channel number, which is proportional to fluorescence intensity. Each Chinese hamster chromosome type is indicated by its identification number just above the peaks. Each chromosome type is present in equal numbers (equal peak area) except number 5. Approximately 50,000 chromosomes were analyzed. Excitation was at 488 nm.

peak and a polynomial to the background continuum (Jett, personal communication). Excluding chromosome 5, which is present in three copies in most cells of the type analyzed here, all the peak areas agree to within 10%. Typically, standard deviations are 10% or less if several flow karyotypes are analyzed. Selective chromosome aggregation, breakage, and differential large chromosome settling contribute to errors in peak area. Peak means are normalized to a percentage of autosomal total for comparing flow karyotypes. Peak means have been found to be remarkably reproducible. The number of chromosome fragments or debris counted as background in Figure 7.4 is between 8% and 10% of the total number of events. In other cloned Chinese hamster cell lines the background is as low as 2% of the total number of events.

The reproducibility with which peak means can be measured was demonstrated by Ray et al. (1984). Using chromosomes isolated from several Chinese hamster cell strains over a period of several months, the autosomal normalization varied by about 1.0% (Table 7.1). This level of reproducibility makes the technique particularly sensitive for detecting polymorphic differences.

Flow karyotype resolution can be noticeably influenced by spontaneous changes occurring in the cells from which the chromosomes are iso-

TABLE 7.1

Autosomal Normalization of Flow Karyotypes from 14 Individual Chinese Hamster Early Passage Cell Cultures

Cell Lineage	Chromosome Number												
	1	2	3	4	5	6	7	8	9			10	X	Y
									High	Medium	Low			
WCHE/5	12.2	10.1	6.4	5.4	4.4	3.9	3.5	2.5	—	—	1.86	1.7	5.9	—
CCHE/27	12.2	10.1	6.4	5.4	4.5	3.9	3.4	2.5	2.22	2.02	—	1.7	6.0	—
FECH/2	12.0	10.0	6.5	5.4	4.5	3.9	3.5	2.5	—	—	1.81	1.7	6.1	4.5
FECH/3	12.0	10.1	6.4	5.4	4.6	3.9	3.5	2.5	2.17	—	1.87	1.7	5.9	4.6
FECH/4	12.0	10.0	6.5	5.4	4.6	3.9	3.5	2.5	—	2.07	1.85	1.7	6.1	4.6
FECH/5	12.2	10.1	6.5	5.4	4.5	3.9	3.5	2.5	—	2.03	1.85	1.7	6.0	—
FECH/6	12.1	10.0	6.4	5.4	4.5	3.9	3.5	2.5	—	—	1.86	1.7	6.0	4.5
FECH/7	12.1	10.1	6.5	5.3	4.5	3.9	3.5	2.5	—	—	1.86	1.7	6.0	—
FECH/9	12.0	10.0	6.4	5.4	4.6	4.0	3.5	2.5	2.24	—	1.89	1.7	6.0	4.6
FECH/10	11.8	9.9	6.5	5.5	4.6	4.0	3.5	2.5	2.26	—	1.92	1.7	6.1	—
FECH/11	12.1	10.1	6.5	5.4	4.5	3.9	3.5	2.5	—	—	1.86	1.7	6.0	4.5
FECH/12	12.2	10.0	6.5	5.4	4.5	3.9	3.5	2.5	2.22	—	—	1.7	6.0	—
FECH/13	11.9	10.0	6.5	5.4	4.6	4.0	3.5	2.5	—	2.06	1.88	1.7	6.0	4.6
FECH/14	12.3	10.2	6.4	5.4	4.5	3.9	3.4	2.4	—	1.98	—	1.6	6.1	4.5
Mean	12.1	10.1	6.5	5.4	4.5	3.9	3.5	2.5	2.22	2.03	1.87	1.7	6.0	4.6
S.D.[a]	±0.14	±0.08	±0.05	±0.04	±0.06	±0.04	±0.04	±0.03	±0.03	±0.04	±0.03	±0.03	±0.07	±0.05

[a] Standard deviation.

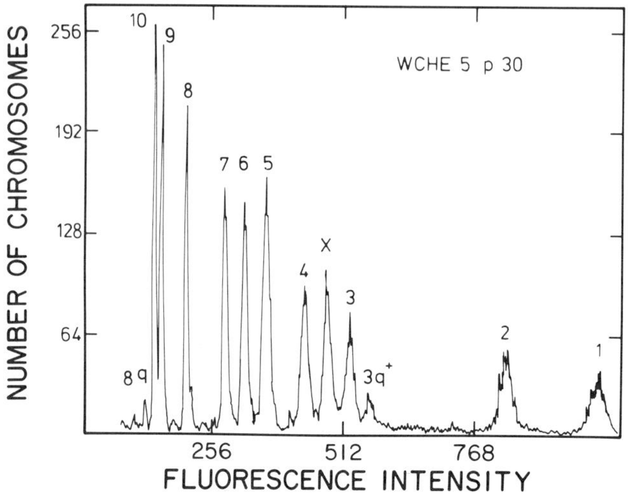

Figure 7.5. Flow karyotype of WCHE/5 passage 30 cells. Chromosomes were isolated and analyzed as described in Figure 7.3. Aberrant chromosome types in this population of cells were identified by combining flow karyotype information with G banding.

lated. Bartholdi et al. (1984) have found periods during the spontaneous progression of Chinese hamster cells in culture from which high-resolution flow karyotypes can be reproducibly obtained.

2. Aberration Detection

Univariate Chinese hamster flow karyotypes combined with classical banding procedures have been used to detect chromosome aberrations such as insertion elements, breaks, nonreciprocal translocations, and polymorphisms. The appearance of an insertion element in the long arm of the number 3 chromosome isolated from a whole Chinese hamster embryo cell line (WCHE/5) was first detected in a flow karyotype (Fig. 7.5) as a new peak. Identification of the specific chromosome involved and the reason for the shift in peak mean required G banding. The area of the $3q^+$ peak (relative to the area of the normal number 3 peak) is proportional to the number of chromosomes containing the insertion element. G-banded chromosomes from this particular cell line appeared to have an insertion element that varied in size from those that did not have the extra band to those that had a very distinctive band. The value of the flow karyotype in this instance was (1) to demonstrate the presence of cells in the culture with the aberration and (2) to verify that the insertion element was occurring as a tightly regulated entity (Cram et al., 1983). Figures 7.4 and 7.5 contain examples of trisomy (the number 5 chromosome) and broken chromosome arms (8q). Both of these flow karyotypes were obtained from aneuploid Chinese hamster cell lines. In contrast, a hetero-

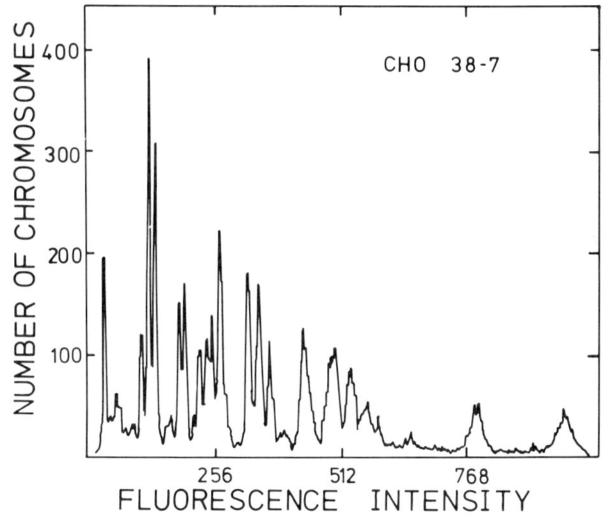

Figure 7.6. Flow karyotype of chromosomes isolated from the heteroploid cell line CHO. Approximately 25,000 chromosomes were analyzed. Data courtesy of Deaven *et al.* (1983).

ploid line such as CHO presents a complex flow karyotype, as illustrated in Figure 7.6. The emergence of a new stem line with an altered karyotype might be difficult to detect in such a distribution. Banding analysis suffers from the same problem.

A chromosome polymorphism in the number 9 chromosome of the Chinese hamster has been found by Ray et al. (1984) using flow karyotype analysis. The autosomal normalization of 14 flow karyotypes from 14 cell lines (Table 7.1) indicated that one or the other, neither, or both homologs contained either one or two extra segments of DNA that are closely regulated in size. G banding confirmed the number 9 homolog difference only in the instance where two extra segments were present in one homolog and none in the other. Human polymorphisms are being quantified in a similar fashion using bivariate flow karyotypes (Langlois et al., 1982).

3. *Bivariate Analysis*

Bivariate flow karyotypes require staining chromosomes with two fluorochromes of differing binding specificities which are independently excited as the chromosome flows through two laser beams. Differences in base ratios between human chromosome types can be measured using Hoechst 33258 [AT binding preference (Comings and Drets, 1976; Latt and Wohlleb, 1975)] and chromomycin A3 [GC binding preference (Behr et al., 1969)]. In contrast to human chromosomes, Chinese hamster chromosomes have a relatively constant AT/GC ratio (Fig. 7.7). The peaks in

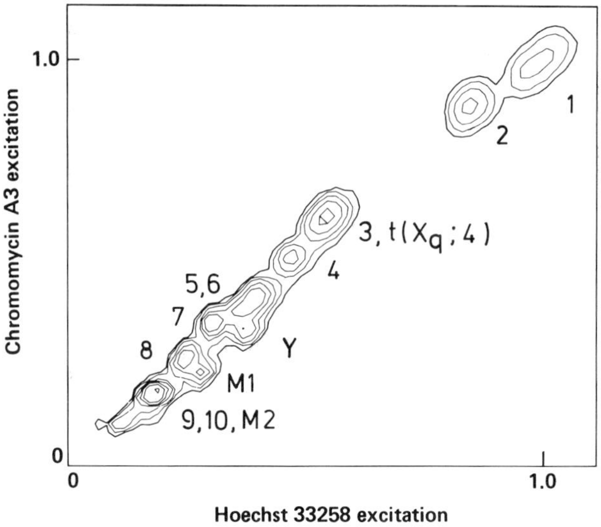

Figure 7.7. Bivariate flow karyotype (histogram) of double-stained Chinese hamster chromosomes. Chromosomes were stained with Hoechst 33258 (3.7 μ*M*) and chromomycin A_3 (70 μ*M*). Dual-parameter flow cytometry was performed using excitation wavelengths as noted on each axes (Langlois et al., 1980). Peaks are represented by contours. Data courtesy of R. Langlois, Lawrence Livermore National Laboratory.

the contour map remain clustered along the 45° axis, which is interpreted to mean that all chromosome types have similar dye-binding specificities (Langlois et al., 1980). Cremer and Gray (1983) have used a similar staining procedure with 5-bromo-2′-deoxyuridine-treated cells to measure the replication kinetics of Chinese hamster chromosomes.

4. Sample Preparation

Flow karyotype preparation protocols are summarized in Table 7.2. All the steps of sample preparation (cell culture, mitotic blocking, hypotonic cell swelling, chromosome release, and staining) may require modification for the cell line being used. A variety of cell types have been used for chromosome isolation including lymphocytes, fibroblasts, lymphoblasts, and bone marrow. Table 7.2 outlines some of the known advantages and disadvantages of the most common procedures. A few guidelines are: (1) cells should be maintained in exponential growth for four to six population doublings prior to blocking; (2) mitotic-arresting agents should be used at low concentration and as briefly as possible; (3) remove, as uniformly as possibly, all serum prior to mitotic swelling; (4) microscopically monitor swelling; and (5) monitor the number of mitotic cells used and the number of chromosomes recovered. Uniform swelling of the mitotic cells is the most critical step for all the procedures. The initial

TABLE 7.2
Chromosome Preparation and Staining Procedures Commonly Used for Flow Karyotype Analysis and Sorting

Common Name	Chromosome Features	Advantages	Limitations[a]	References
Hexylene glycol	Normal length maintained	Sorted chromosomes can be Q banded	Molecular weight is unknown	Wray and Stubblefield (1970) Carrano et al. (1978)
		Preparations store well at $-70°C$		
Hypotonic propidium iodide	Extended	Fast	Banding is uncertain	Aten et al. (1980)
		High-resolution flow karyotypes	Limited staining options	Bijman (1983)
Magnesium sulfate	Contracted	Applicable to very small samples	Chromosomes too contracted for banding	van den Engh et al. (1984)
		High-resolution flow karyotypes	Mg available for nuclease activity	
Polyamines	Highly contracted	Chromosomes are well stabilized	Chromosomes too contracted for banding	Sillar and Young (1981) Blumenthal et al. (1979)
FUS (fixation and ultrasound)	Normal	Low debris, stable over months at 4°C	Use of acetic acid fixation	Stoehr et al. (1982)

[a] Limitations refer to current procedures only.

spreading of chromosomes within the cell occurs as the mitotic cell swells. If the plasma membrane ruptures before swelling is complete, the chromosomes remain in close proximity and will not completely separate, even after shearing. A convenient method for monitoring cell swelling during the development or modification of a procedure is to add the fluorochromes Hoechst 33342 (10 μM final concentration) and propidium iodide (50 μg/mL final concentration) to the sample. Hoechst will be taken up by all the cells, while cells with intact membranes will exclude propidium iodide. Ideally, mitotic cells will swell uniformly then slowly permeabilize. As the cell membrane breaks, the chromosome fluorescence changes from blue to red.

B. Chromosome Sorting

Flow cytometry is the preferred method for sorting individual chromosome types for several reasons: (1) chromosomes are rapidly sorted on an individual basis, (2) the method of sorting is independent of the property measured, and (3) the purity of the sorted fraction is continually reassessed as the separation process takes place. Bulk isolation techniques do not provide these advantages, but should be useful as an adjunct for preenrichment prior to flow sorting.

Any chromosome type clearly resolved in a flow karyotype can be sorted in large amounts (10^6–10^7 copies) and with high purity (greater than 95%). Assuming an analysis rate of 2000 chromosomes/sec and 11 Chinese hamster chromosome types present (female), a single chromosome type would be sorted at a rate of 180 chromosomes/sec or 6.5×10^5 chromosomes/hr.

The major uses for sorted chromosomes are (1) to identify flow karyotype peaks and (2) to obtain purified chromosomes for further analysis. Identification procedures are extremely important for flow karyotype analysis because logical peak assignments based on arm length measurements are not always accurate, particularly for human flow karyotypes where polymorphisms have been shown to be responsible for the movement of some peaks relative to others from individual to individual (Langlois et al., 1982). Purified chromosomes are being used for (1) chromosome-mediated gene transfer, (2) dot blotting, (3) construction of chromosome-specific DNA sequence libraries, and (4) *in situ* hybridization experiments. Categories 2 and 4 require relatively small numbers of chromosomes. Construction of chromosome-specific DNA sequence libraries can, depending on the cloning vector selected, require tens of micrograms of purified chromosomes.

The construction of chromosome-specific DNA sequence libraries will facilitate (1) linkage and pedigree analyses by the study of DNA restriction fragment length polymorphisms and (2) gene mapping and cloning by combined somatic cell and molecular genetic techniques. The study of

DNA sequence organization among individual chromosomes within a given species and studies of interspecies genetic comparisons will also benefit. DNA sequence libraries constructed from flow-sorted human chromosomes will first become widely available in late 1984. Lawrence Livermore National Laboratory and Los Alamos National Laboratory are collaborating to sort all the human chromosome types possible, to construct the DNA sequence libraries, and to make the libraries available to the biomedical research community.

To date, 12 human-chromosome-specific libraries have been constructed, four of which were derived from flow-sorted chromosomes: the human X chromosome (Kunkel et al., 1982), human chromosomes 21 and 22 (Krumlauf et al., 1982), and an abnormal human chromosome number 1 (Kanda et al., 1983). Some of the practical and theoretical considerations for construction of DNA sequence libraries from flow-sorted Chinese hamster chromosomes have been described by Griffith et al. (1984).

Chinese-hamster–human hybrid cells allow certain human chromosomes to be resolved when superimposed on a background of Chinese hamster chromosomes. This approach has been used to sort the human chromosome 13 from a Chinese-hamster–human hybrid cell containing two or more normal human 13 chromosomes. This method is useful when the human chromosome of interest is located in a region of the Chinese hamster karyotype where it can be resolved. However, such hybrids frequently end up with a more heteroploid Chinese hamster karyotype background then did the original hamster cell used to form the hybrid.

Chromosomes can be sorted either directly into centrifuge tubes containing the isolation buffer or onto a thermoelectric cooled block that freezes the sorted droplets as they hit. Chromosome recovery is good (up to 90%), and chromosome morphology depends on the isolation buffer. We and others have consistently found that the sheath fluid has to be matched with the sample buffer to maintain morphology. Chromosomes isolated by the Aten technique (Aten et al., 1980) and stained with propidium iodide have a molecular weight of greater than 50,000 kb. Major damage to the chromosomes has apparently not occurred.

The procedures listed in Table 7.2 have certain advantages and disadvantages, some of which are of importance for chromosome sorting. To confirm sorting purity cytogenetically one would probably choose the hexylene glycol procedure. However, the fact that high-quality G banding has not been achieved using chromosomes isolated by an alternative technique means only that a procedure has yet to be developed.

C. Future Developments

The amount of sorter time required for many chromosome-sorting experiments remains high. Preenrichment as described by Collard et al. (1980) may reduce the time required for sorting by as much as a factor of 5 (Collard and Gray, 1984). High-speed chromosome sorting is being developed jointly by the Livermore and Los Alamos groups. This technology should decrease sorting time by a factor of 7. If prefractionation and high-speed sorting can be made mutually compatible, effective sorting rates could be increased by a factor of 35.

The reproducibility of chromosome preparations is one of the limiting factors in all the applications discussed above. Progress in this area continues to be made while additional developments add new variables for resolving chromosome types. For example, preferential staining of heterochromatic regions in human chromosome 1, 9, and Y by DAPI (4′-6-diamidino-2-phenylindole), chromomycin A_3, and the counter stain netropsin, resolves these chromosomes for sorting (Meyne et al., 1984). *In situ* hybridization techniques offer the potential for uniquely identifying single chromosomes. These techniques as well as others, such as slit scanning of chromosomes (Lucas et al., 1981), and biophysical measurements, such as emission anisotropy (Cram et al., 1979), will expand the tools available for the identification, analysis, and sorting of normal and aberrant chromosome types.

ACKNOWLEDGMENTS

We are indebted to numerous colleagues for supplying manuscripts, including Andy Ray, Marty Bartholdi, Paul Kraemer, and Julie Meyne.

This work was supported by grants from NSF PCM 8317788 to J. W. Shay and through the Los Alamos National Flow Cytometry Resource sponsored by the Division of Research Resources of The National Institutes of Health (Grant RR01315) and the Department of Energy to L. S. Cram.

REFERENCES

Albrecht-Buehler, G. (1980). *Proc. Natl. Acad. Sci. USA* **77**, 6639–6644.

Arndt-Jovin, D. J. and Jovin, T. M. (1978). *Ann. Rev. Biophys. Bioeng.* **7**, 527–558.

Aten, J. A., Kipp, J. B. A., and Barendsen, G. W. (1980). In *Flow Cytometry IV*, (O. D. Laerum, T. Lindmo, and E. Thorud, eds.), Universitetsforlaget, Oslo, pp. 287–292.

Barski, G., Sorieul, S., and Cornefert, F. (1960). *C. R. Acad. Sci (Paris)* **251**, 1825–1827.

Bartholdi, M. F., Ray, F. A., Jett, J. H., Cram, L. S., and Kraemer, P. M. (1984). *Cytometry* **5**, 534–538.

Bartholdi, M. F., Sinclair, D. C., and Cram, L. S. (1983). *Cytometry* **3**, 395–401.

Behr, W., Honikel, K., and Hartman, G. (1969). *Europ. J. Biochem.* **9**, 82–92.

Bijman, J. T. (1983). *Cytometry* **3**, 354–358.

Blanch, H., Adams, C. A., and Wallace, D. C. (1981). *Nucleic Acids Res.* **9**, 5785–5795.

Blumenthal, A. B., Dieden, J. D., Kapp, L. N., and Sedat, J. W. (1979). *J. Cell Biol.* **80**, 255.

Bunn, C. L. (1982). In *Techniques in Somatic Cell Genetics* (J. W. Shay, ed.), Plenum Press, New York, pp. 189–199.

Bunn, C. L., Wallace, D. C., and Eisenstadt, J. M. (1974). *Proc. Natl. Acad. Sci. USA* **71**, 1681–1685.

Carrano, A. V., Gray, J. W., Langlois, R. G., Burkhart-Schultz, K. J., and Van Dilla, M. A. (1979). *Proc. Natl. Acad. Sci. USA* **76**, 1382–1384.

Carrano, A. V., Gray, J. W., and Van Dilla, M. A. (1978). In *Mutagen-Induced Chromosome Damage in Man* (H. J. Evans and D. C. Lloyd, eds.), Edinburgh University Press, Edinburgh, pp. 326–338.

Carter, S. E. (1967). *Nature* **213**, 261–266.

Clark, M. A. and Shay, J. W. (1981). *Endocrinology* **109**, 2261–2263.

Clark, M. A. and Shay, J. W. (1982a). *Nature* **295**, 605–607.

Clark, M. A. and Shay, J. W. (1982b). *Proc. Natl. Acad. Sci. USA* **79**, 1144–1148.

Collard, J. G. and Gray, J. W. (1984). *Cytometry* **5**, 9–10.

Collard, J. G., Tulp, A., Stegman, J., Bauer, F. W., Jongkind, J. F., and Verkerk, A. (1980). *Exp. Cell Res.* **130**, 217–228.

Comings, D. E. and Drets, M. E. (1976). *Chromosoma* **56**, 199–211.

Cox, R. P., Krauss, M. R., Bolis, M. E., and Dancis, J. (1976). *J. Cell Biology* **71**, 693–703.

Cram, L. S., Arndt-Jovin, D. T., Grimwade, B. G., and Jovin, T. M. (1979). *J. Histochem. Cytochem.* **27**, 445–453.

Cram, L. S., Bartholdi, M. F., Wheeless, L. L., and Gray, J. W. (1985). In *Methods of Flow Cytometry* (J. Ploem, J. Van Dilla, and P. Dean, eds.), Academic Press, New York.

Cremer, C. and Gray, J. W. (1983). *Cytometry* **3**, 282–286.

Deaven, L. L., Campbell, E., and Bartholdi, M. F. (1983). In *Cancer: Etiology and Prevention* (R. G. Crispen, ed.), The University of Chicago Press, Chicago.

Ege, T., Hamberg, H., Krondhal, U., Ericsson, J., and Ringertz, N. R. (1974). *Exp. Cell Res.* **87**, 365–377.

Fournier, R. E. K. (1982). In *Techniques in Somatic Cell Genetics* (J. W. Shay, ed.), Plenum Press, New York, pp. 309–327.

Fournier, R. E. K., Juricek, D. K., and Ruddle, F. H. (1979). *Somat. Cell Genet.* **5**, 1061–1077.

Fournier, R. E. K., and Ruddle, F. H. (1977). *Proc. Natl. Acad. Sci. USA* **74**, 319–323.

Goldman, R. D., Pollack, R., and Hopkins, N. H. (1973). *Proc. Natl. Acad. Sci. USA* **70**, 750–754.

Gopalakrishnan, T. V. and Anderson, W. F. (1979). *Proc. Natl. Acad. Sci. USA* **76**, 3932–3936.

Graessmann, A. and Graessmann, M. (1982). In *Techniques in Somatic Cell Genetics* (J. W. Shay, ed.), Plenum Press, New York, pp. 463–470.

Gray, J. W., Carrano, A. V., Steinmetz, L. L., Van Dilla, M. A., Moore, D. H., Mayall, B. H., and Mendelsohn, M. L. (1975). *Proc. Natl. Acad. Sci. USA* **72**, 1231–1234.

Griffith, J. K., Cram, L. S., Crawford, B. D., Jackson, P. J., Schilling, J., Schmike, R. T., Walters, R. A., Wilder, M. E., and Jett, J. H. (1984). *Nucleic Acids Res.* **12**, 4019–4034.

Harris, H. and Watkins, J. F. (1965). *Nature* **205**, 640–646.

Herzenberg, L. A. and Sweet, R. G. (1976). *Sci. Am.* **234**, 108–117.

Howell, A. N., and Sager, R. (1978). *Proc. Natl. Acad. Sci. USA* **75**, 2358–2362.

Johnson, R. T. and Mullinger, A. M. (1982). In *Techniques in Somatic Cell Genetics* (J. W. Shay, ed.), Plenum Press, New York, pp. 329–347.

Jongkind, J. F. and Verkerk, A. (1982). In *Techniques in Somatic Cell Genetics* (J. W. Shay, ed.), Plenum Press, New York, pp. 81–100.

Jongkind, J. F., Verkerk, A., and Tanke, H. (1979). *Exp. Cell Res.* **120**, 444–448.

Kahn, C. R., Bertolotti, R., Ninio, M., and Weiss, M. C. (1981). *Nature* **290**, 717–720.

Kanda, N., Schreck, R., Alt, F., Burns, G., Baltimore, D., and Latt, S. (1983). *Proc. Natl. Acad. Sci. USA* **80**, 4069–4073.

Kearsey, S. E. and Craig, I. W. (1981). *Nature* **290**, 607–608.

Klobutcher, L. A. and Ruddle, F. H. (1979). *Nature* **280**, 657–660.

Krumlauf, R., Jeanpierre, M., and Young, B. D. (1982). *Proc. Natl. Acad. Sci. USA* **79**, 2971–2975.

Kunkel, L. M., Tantravaki, V., Eisenhard, M., and Latt, S. (1982). *Nucleic Acids Res.* **10**, 1557–1578.

Langlois, R. G., Carrano, A. V., Gray, J. W., and Van Dilla, M. A. (1980). *Chromosoma* **77**, 229–252.

Langlois, R. G., Yu, L. C., Gray, J. W., and Carrano, A. V. (1982). *Proc. Natl. Acad. Sci. USA* **79**, 7876–7880.

Latt, S. A. and Wohlleb, J. C. (1975). *Chromosoma* **52**, 297–316.

Linder, S. (1980). *Exp. Cell Res.* **130**, 159–167.

Lipsich, L. A., Kates, J. R., and Lucas, J. L. (1979). *Nature* **281**, 74–76.

Littlefield, J. W. (1964). *Science* **145**, 709–710.

Lucas, J. N., Peters, D., Van Dilla, M. A., and Gray, J. W. (1981). *Cytometry* **2**, 113.

Maul, G. G. and Weibel, J. (1982). In *Techniques in Somatic Cell Genetics* (J. W. Shay, ed.), Plenum Press, New York. pp. 237–243.

McBride, O. W. (1982). In *Techniques in Somatic Cell Genetics* (J. W. Shay, ed.), Plenum Press, New York, pp. 375–383.

McBride, O. W. and Ozer, H. L. (1973). In *Possible Episomes in Eukaryotes* (L. G. Silvestri, ed.), North-Holland, Amsterdam, pp. 255–267.

McKusick, V. A. and Ruddle, F. H. (1977). *Science* **196**, 390–405.

Melamed, M. R., Mullaney, P. F., and Mendelsohn, M. L. (eds.) (1979). *Flow Cytometry and Sorting*, Wiley, New York.

Meyne, J., Bartholdi, M., Travis, G., and Cram, L. S. (1984). *Cytometry* **5** (in press).

Miller, R. A. and Ruddle, F. H. (1974). *J. Cell Biol.* **63**, 295–299.

Miller, C. L. and Ruddle, F. H. (1978). *Proc. Natl. Acad. Sci. USA* **75**, 3346–3350.

Okada, Y. (1958). *Biken's J.* **1**, 103–110.

Pollack, R., Goldman, R. D., Conlon, S., and Chang, C. (1974). *Cell* **3**, 51–54.

Pontecorvo, G. (1975). *Somat. Cell Genet.* **4**, 397–400.

Poste, G. and Reeve, P. (1972). *Exp. Cell Res.* **72**, 556–560.

Prescott, D. M., Meyerson, D., and Wallace, J. (1972). *Exp. Cell Res.* **71**, 480–485.

Rao, P. N., Sunkara, P. S., and Al-Bader, A. A. (1982). In *Techniques in Somatic Cell Genetics* (J. W. Shay, ed.), Plenum Press, New York, pp. 245–254.

Ray, F. A., Bartholdi, M. F., Kraemer, P. M., and Cram, L. S. (1984). *Cytogenet. Cell Genet.* (in press).

Rechsteiner, M. C. (1982). In *Techniques in Somatic Cell Genetics* (J. W. Shay, ed.), Plenum Press, New York, pp. 385–398.

Ruddle, F. H. (1981). *Nature* **294**, 115–120.

Ruddle, F. H. and Creagan, R. P. (1975). *Ann. Rev. Genet.* **9**, 407–486.

Schor, S. L., Johnson, R. T., and Mullinger, A. M. (1975). *J. Cell Sci.* **19**, 281–303.

Schroder, C. H. and Hsie, A. W. (1973). *Nature* **246**, 58–60.

Shapiro, H. M. (1983). *Cytometry* **3**, 227–243.

Shay, J. W. (ed.) (1982). *Techniques in Somatic Cell Genetics*, Plenum Press, New York.

Shay, J. W. and Clark, M. A. (1977). *J. Ultrast. Res.* **58**, 155–159.

Shay, J. W., Gershenbaum, M. R., and Porter, K. R. (1975). *Exp. Cell Res.* **94**, 47–55.

Shay, J. W., Lorkowski, G., and Clark, M. A. (1981). *J. Supramolec. Struct.* **16**, 75–82.

Shay, J. W., Porter, K. R., and Prescott, D. M. (1973). *J. Cell Biol.* **59**, 311a.

Shay, J. W., Porter, K. R., and Prescott, D. M. (1974). *Proc. Natl. Acad. Sci. USA* **71**, 3059–3063.

Shay, J. W., Porter, K. R., and Krueger, T. C. (1977). *Exp. Cell Res.* **105**, 1–8.

Sillar, R. and Young, B. D. (1981). *J. Histochem. Cytochem.* **29**, 74–78.

Siminovitch, L. (1976). *Cell* **7**, 1–11.

Smiley, J. R., Steege, D. A., Juricek, K. D., Summers, W., and Ruddle, F. H. (1978). *Cell* **15**, 455–468.

Sorieul, S. and Ephrussi, B. (1961). *Nature* **190**, 653–654.

Stoehr, M., Hutter, K. J., Frank, M., and Goerttler, K. (1982). *Histochem.* **74**, 57–61.

Straubinger, R. M. and Papahadjopoulos, D. (1982). In *Techniques in Somatic Cell Genetics* (J. W. Shay, ed.), Plenum Press, New York, pp. 399–414.

Sunkara, P. S., Al-Bader, A. A., and Rao, P. N. (1977). *Exp. Cell Res.* **107**, 444–448.

van den Engh, G., Trask, B., Cram, L. S., and Bartholdi, M. F. (1984). *Cytometry* **5**, 108–117.

Veomett, G., Prescott, D. M., Shay, J., and Porter, K. R. (1974). *Proc. Natl. Acad. Sci USA* **71**, 1999–2002.

Wallace, D. C. (1982). In *Techniques in Somatic Cell Genetics* (J. W. Shay, ed.), Plenum Press, New York, pp. 159–188.

Weide, L. G., Clark, M. A., Rupert, C. S., and Shay, J. W. (1982b). *Somat. Cell Genet.* **8**, 15–21.

Weide, L. G., Clark, M. A., and Shay, J. W. (1982a). In *Techniques in Somatic Cell Genetics* (J. W. Shay, ed.), Plenum Press, New York, pp. 281–289.

Wise, G. E. and Larsen, R. (1976). *Exp. Cell Res.* **97**, 141–150.

Wise, G. E. and Prescott, D. M. (1973). *Exp. Cell Res.* **81**, 63–72.

Worton, R., Duff, C., and Flintoff, W. (1981). *Mol. Cell Biol.* **1**, 330–335.

Wray, W. and Stubblefield, E. (1970). *Exp. Cell Res.* **59**, 469–478.

Wright, W. E. (1978). *Exp. Cell Res.* **112**, 395–407.

Wright, W. E. (1981). *Somat. Cell Genet.* **7**, 769–775.

Wright, W. E. (1982). In *Techniques in Somatic Cell Genetics* (J. W. Shay, ed.), Plenum Press, New York, pp. 47–65.

Wright, W. E. and Hayflick, L. (1972). *Exp. Cell Res.* **74**, 187–194.

Ziegler, M. L. (1982). In *Techniques in Somatic Cell Genetics* (J. W. Shay, ed.), Plenum Press, New York, pp. 211–220.

Ziegler, M. L. and Davidson, R. L. (1981). *Somat. Cell Genet.* **7**, 73–88.

Zorn, G. A., Lucas, J. J., and Kates, J. R. (1980). *Cell* **18**, 659–672.

CHAPTER 8

DNA-MEDIATED GENE TRANSFER

Irene Abraham
Laboratory of Molecular Biology
National Cancer Institute
National Institutes of Health
Bethesda, Maryland

I. INTRODUCTION

The usefulness and promise of the gene transfer technique in studying mammalian cells is made abundantly clear by surveying the large number of studies that use this method. Gene transfer can be divided into two main types: transfer of cloned genes and transfer of uncloned genes from whole genomic DNA. In this chapter we will be concerned principally with the transfer of uncloned DNA's, while the transfer of cloned genes and vectors will be covered in Chapter 9. Another related technique, which is beyond the scope of this chapter, is the transfer of RNA molecules (Liu et al., 1979; Lin et al., 1982; Straubinger and Padahadjopoulos, 1982). Introduced RNA can be translated in mammalian cells, and this property can be used as a basis for identifying a particular species of RNA (Liu et al., 1979; Chapter 20). An expressing cDNA library can be made from the selected RNA, and this potentially can be used to select and clone the gene in question (see Chapter 20). Some of the considerations discussed in this chapter may also apply to this interesting approach.

By uncloned whole genomic DNA we will mean DNA that is isolated from mammalian cells without being previously identified and purified by cloning in a bacterial plasmid or viral vector. There can be several goals in transferring uncloned whole genomic DNA. DNA-mediated gene transfer (DMGT) can be used to gain information similar to that obtained by the hybridization of somatic cells; that is, to gain information about dominance, or to study the genetic complementation of related phenotypes. Because in DMGT we are dealing with only one molecule, DNA, purified away from the other molecules and organelles of the cell, this system is significantly cleaner than traditional somatic cell hybridization. It adds conclusive proof that a trait is being transmitted through the DNA and not through other cellular molecules.

DNA transformation has also been used to study the mechanism of X-chromosome inactivation in females. One of the two X chromosomes in cells from females is inactivated. Investigators have asked the question of whether this inactivation will remain constant after the transfer of DNA to another cell. This can be done, for example, by following the activity of an identifiable X-linked HGRPT (hypoxanthine guanine phosphoribosyl transferase) gene that is known to be inactive in the DNA donor. In most cases the property of inactivation is passed on after DNA transformation (Liskay and Evans, 1980; Chapman et al., 1982; deJonge et al., 1982; Lester et al., 1982; Venolia and Gartler, 1983)

Another exciting area that can be approached through the use of the gene transfer methodologies is that of gene recombination. The advent of reliable techniques for DNA transfer have made these studies possible. It was apparent early in the use of this technique that the DNA was integrating in the host chromosomes through some recombinational mechanism (Perucho et al., 1980*a*; Robins et al., 1981). Many recent studies have looked at the recombination of viral or cloned DNA's introduced into the cell and followed the occurrence of nonhomologous and homologous recombination (Botchan et al., 1980; Jackson, 1980; Perucho et al., 1980a; Sompayrac and Danna, 1981; Anderson et al., 1982; de Saint Vincent and Wahl, 1982; Folger et al., 1982; Wilson et al., 1982; Liskay and Stachelek, 1983; Miller and Temin, 1983; Pomerantz, 1983; Ruley and

Fried, 1983; Small and Scangos, 1983; Lin, Sperle, and Sternberg, 1984, Lin and Steinberg, unpublished). Liskay and Stachelek (1983) have followed the recombination between two complementary and defective tk genes located in direct repeat on a plasmid after its stable integration into an L cell. They found apparent homologous recombination between these two genes occurring at the rate of 10^{-5}–10^{-4}. From their analysis, they suggest that half of these recombinational events involve gene conversion. Smith and Berg (unpublished) have reached similar conclusions using a different system to detect recombination in chromosomally integrated DNA segments. Other studies (Calos et al., 1983; Razzaque et al., 1983) have shown that some plasmids are subjected to very high frequencies of mutation, about 1% per plasmid. The significance of this mutagenesis and its presence in other modes of DNA transformation is not clear, but poses provocative questions. Further studies of recombination and mutagenesis should enlarge our general knowledge of recombination and repair events in mammalian cells as well as elucidating the events occurring during DNA transfer.

Perhaps the most widely used application of DMGT is as a method to identify, clone, and purify genes. Many genes produce their products in very small amounts per cell. Because small amounts of mRNA make cloning the gene via traditional methods that depend on large mRNA abundance very difficult, if not impossible, the genes must be cloned by other means. If the genes produce a product that can be selected for (or against), then transformants carrying the gene of interest can be selected. Selection of transformed cells can be based on the biochemical characteristics of the gene transferred, as in drug selections, or can be based on affinity of the transferred gene product for antibody, as in methods utilizing the fluorescence-activated cell sorter (FACS). Various schemes, some of which will be discussed in this chapter, can be utilized to purify and clone the genes in question.

DMGT can also be used to transfer genes, either cloned or uncloned to new backgrounds. For example, genes can be transferred to new cell types or to cells of different species. This could add a powerful technique for studying the regulation of specific genes in different host backgrounds. Approaches of this type have already been used for many cloned genes (Corces et al., 1981; Green et al., 1983).

A potential use of DMGT of great interest is gene therapy. That is, it is possible that cells deficient in a particular function can be complemented by the addition of new genes via DMGT. This last possibility would of course have tremendous clinical implications.

Another very important field that has appeared with the advent of this powerful technique is the study into the biological mechanism of transformation itself. The questions of how the DNA enters the cell and the nucleus, how it becomes stabilized and integrated into the host chromosomes, and how the expression of the introduced DNA is controlled are fascinating ones. Amplification, methylation, and control of transferred DNA will be considered briefly in Section V. The use of DMGT opens up new ways of asking these questions, which are fundamental ones in genetics and cell biology. The answers should have wide application to understanding these basic problems in both cultured cells and the whole organism.

The purpose of this chapter will be to introduce the technique of DMGT by acquainting the reader with the methodologies involved, the biological properties of transformation, the techniques for cloning genes via DMGT, and the future prospects of DMGT, with special emphasis on the Chinese hamster ovary (CHO) cell.

II. GENES TRANSFERRED BY DMGT

Genomic transfers have successfully been executed in several rodent and primate cell lines such as the mouse L, mouse 3T3, African green monkey CV-1, and CHO cells. In a very early report, DNA transformation of mammalian cells by genomic DNA was reported (Szybalska and Szybalski, 1962). D98S human cells negative for HGPRT were transformed with DNA from cells positive for HGPRT, after treating the cells with spermine. Since this initial study, there has been no repetition of DNA transformation using this method. The method most widely used currently for DNA transformation is based on a procedure devised for transforming cells with isolated adenovirus and SV40 DNA, using a $CaPO_4$–DNA precipitate (Graham and van der Eb, 1973; Graham et al., 1974). This approach was subsequently used to transfer the purified herpes simplex virus *(HSV)* thymidine kinase *(tk)* gene to mouse L tk^- cells using isolated viral DNA (Bacchetti and Graham, 1977; Maitland and McDougal, 1977) or the isolated *HSV-tk* gene fragment (Wigler et al., 1977). Transformed cells were selected for *tk* activity on HAT (hypoxanthine, aminopterin, thymidine) media (Wigler et al., 1977). The next step was to transform tk^- cells with cellular DNA containing the transferred viral *tk* gene or the native eukaryotic gene (Wigler et al., 1978). Since then, rapid progress has been made and many other defined genes such as alpha-amanitin resistance (Ingles and Shales, 1982; Chapter 15), adenine phosphoribosyl transferase (Wigler et al., 1979b; Chapter 12), hypoxanthine phosphoribosyl transferase (Peterson and McBride, 1980; Lester et al., 1980; Jolly et al., 1982; Chapter 13), and dihydrofolate reductase (Lewis et al., 1980; Wigler et al., 1980; Chapter 16) have been transferred. Several undefined genes, that is, genes whose mode of action is not known, such as onc genes [for reviews see Weinberg (1983) and Cooper (1982); see also, Shih et al., (1979); Cooper et al., (1980); Cooper and Neiman (1980); Hopkins et al., (1981); Krontiris and Cooper (1981); Lane et al. (1981); Murray et al. (1981); Perucho et al. (1981); Shih et al. (1981);

Shilo and Weinberg (1981); Goldfarb et al. (1982); Goubin et al. (1983); Lane et al. (1982); Pulciani et al. (1982); Smith et al. (1982); Souyri and Fleissner (1983); Shimizu et al. (1983)] have been transferred. Other undefined genes that have been isolated are gene(s) conferring multiple drug resistance (Debenham et al., 1982; Chapter 25), genes providing capacity for excision repair (Takano et al., 1982; Rubin et al., 1983; MacInnes et al., 1984a, b; Waldren et al., 1984; Chapter 21), genes conferring sensitivity to tumor promoters (Colburn et al., 1983), and DNA altering reversion frequency of spontaneously transformed NIH/3T3 cells (Baker, 1983). Fried et al. (1983) have transferred DNA sequences that appear to be involved in controlling expression of portions of viral polyoma DNA. Beta-2 microglobulin 1 and T-cell differentiation antigens Leu-1 and Leu-2 have been transferred using gene transfer followed by selection of transformed cells by FACS (Kavathas and Herzenberg, 1983a). *HLA* genes have also been transferred and selected using FACS (Kamarck et al., 1984; Kuhn et al., 1984). A more complete list of the native uncloned mammalian genes that have been transmitted by DMGT to date can be found in Table 8.1.

Initially it was thought that CHO cells might be refractory to DNA transformation. Happily, this has proved not to be the case. Chinese hamster cells can be readily transformed with plasmid DNA (Abraham et al., 1982; Nairn et al., 1982; Smith et al., 1982; Tindall and Hsie, 1984) as well as with genomic DNA (Peterson and McBride, 1980; Srinivasan and Lewis, 1980; Andrulis and Siminovitch, 1981; Abraham et al., 1982; Smith et al., 1982; MacInnes et al., 1984; Rubin et al., 1983, 1984; Waldren et al., 1984).

III. METHODS FOR THE INTRODUCTION OF DNA

Of the various methods used for the transfer of DNA into mammalian cells, only that using $CaPO_4$ precipitation has received wide application for the transfer of genomic DNA. Chromosome transfer techniques are of interest, but are of less use for cloning genes. Some of the other methods used for transferring DNAs such as protoplast fusion, microinjection and liposomes, and DEAE dextran have mainly been used to transfer cloned or viral DNAs. They will also be briefly considered here since they may have the potential to be used for genomic transfers.

A. $CaPO_4$

The most common procedure used for introducing DNA into foreign cells is by adding DNA directly to cells in the form of a DNA–$CaPO_4$ precipitate (Graham and van der Eb, 1973; Graham et al., 1974). In this method, high-molecular-weight DNA is mixed with $CaCl_2$ and Na_2HPO_4 in Hepes buffered saline and allowed to precipitate for 15–30 min. The usual precipitate added to cells has a final DNA concentration of 2 μg/mL. The precipitate is left on the cells, usually for between 4 and 24 hr. After removal of the precipitate, the cells are allowed to recover for a variable expression time, and then transformants are selected with appropriate selection medium. Generally, this procedure has been performed on cells in monolayer culture, but it has also been used on cells in suspension (Chu and Sharp, 1981; Shen et al., 1982). There have been many published modifications of this method to increase the frequency of transformation. After adding the $CaPO_4$–DNA precipitate to cells, investigators have treated the cells with polyethylene glycol (Shen et al., 1982) or used glycerol shock (Frost and Williams, 1978; Gorman et al., 1983); dimethyl sulfoxide (DMSO) (Stow and Wilkie, 1976); butyrate (Gorman and Howard, 1983); verapamil, a calcium antagonist (Akiyama et al., 1983); or chloroquine treatment (Luthman and Magnusson, 1983). Generally, these posttransformation procedures have been tested with cloned or viral DNAs. Inhibition of transformation of host cells after treatment with interferon has been reported (Dubois et al., 1983). The $CaPO_4$ method has also been used to transfer lambda phage particles containing the *HSV tk* gene (Ishiura et al., 1982). This particular application may be important for transferring lambda libraries of eukaryotic DNA.

B. DEAE Dextran

In an early study, McCutchan and Pagano (1968) have shown that treatment with diethylaminoethyl-dextran (DEAE-D) enhances infectivity of simian virus 40 (SV40) DNA. This technique, in modified form, has been used in later studies of DNA transformation (Milman & Herzberg, 1981; Sompayrac and Danna, 1981) to transfer cloned DNAs. It has not been used to date for transferring genomic DNA.

C. Cell Hybridization and Chromosome Fragmentation

Fusing two cells to form a new hybrid is one of the oldest techniques used for genetic analysis in somatic cell genetics (see Chapter 7). This classical technique is now being combined with modern molecular biology to present an alternative means for gene transfer and source for gene isolation. Cells from the same or different species can be fused, and the resulting hybrid is at least, initially, a combination of the two separate cytoplasms and genomes. In many interspecific crosses after formation of a hybrid nucleus, one of the parent's chromosomes may be lost preferentially. This is often the case in human-rodent hybrids. Goss and Harris (1975) have used this technique to study the linkage of thymidine kinase and galactokinase. They hybridized X-irradiated human cells to rodent cells and were able to select hybrids that had only a small amount of human DNA. A similar technique has been used recently with multi-

TABLE 8.1
Native Chromosomal Genes Transferred by DNA-Mediated Gene Transfer[a]

Genes Transferred	DNA Donor Cell Type	Host Cell	Reference
Thymidine kinase	Mouse L	Mouse L	Wigler et al. (1978)
	Mouse liver		Wigler et al. (1978)
	Calf thymus		Wigler et al. (1978)
	Chicken RBC		Wigler et al. (1978)
	CHO		Wigler et al. (1978) Lewis et al. (1980); Corsaro and Pearson (1981) Srinvasan and Lewis (1980)
	Human HeLa		Wigler et al. (1978); Peterson and McBride (1980)
	CH-V79		Wigler et al. (1978); Peterson and McBride (1980)
	CHO	CHO	Abraham et al. (1982)
	CHO	CH-V79	Srinivasan and Lewis (1980)
Adenine phosphoribosyl transferase	CHO	Mouse L	Wigler et al. (1979b)
	HeLa	Mouse 613	Lester et al. (1980)
Hypoxanthine phosphoribosyl transferase	Human	Human	Szybalska and Szybakski (1962)
	CH-V79	Mouse L (A9)	Peterson and McBride (1980)
	HeLa		Peterson and McBride (1980)
	CHO		Graf et al. (1979)
	Mouse BW5147-V1		Willecke et al. (1979)
	CH-V79	Mouse TG8	Willecke et al. (1979)
	HeLa		Willecke et al. (1979)
	Mouse L929	CHTG49	Willecke et al. (1979)
	HeLa		Willecke et al. (1979)
	CHO	Mouse 3T6	Graf et al. (1979)
		Human KB	Graf et al. (1979)
	Human fibroblast	Mouse 613	Lester et al. (1980)
	Human placental	Mouse L(A9)	Jolly et al. (1982)
	CHEF	CHEF/18	Smith et al. (1982)
Alpha-amanitin resistance RNA polymerase II	Syrian Hamster BHK-21	BHK-21	Ingles and Shales (1982)
HLA genes	Human JM	Mouse	Kavathas and Herzenberg (1983a)
	Human MOLT-4		Kuhn et al. (1984)
4F2 surface antigen	Human MOLT-4	Mouse	Kuhn et al. (1984)

TABLE 8.1 (*Continued*)

Genes Transferred	DNA Donor Cell Type	Host Cell	Reference
T-cell differentiation antigens (Leu-1, Leu-2)	Human JM	Mouse	Kavathas and Hertzenberg (1983a)
Beta$_2$-microglobulin	Human JM	Mouse	Kavathas and Hertzenberg (1983a)
Aspartylhydroxamate resistance	CHO	CHO	Andrulis and Siminovitch (1981)
cAMP resistance	CHO	CHO	Abraham et al. (unpublished)
Methotrexate resistance (altered DHFR)	CHO	Mouse L	Lewis et al. (1980); Wigler et al. (1980)
		Mouse NIH 3T3	Lewis et al. (1981); Wigler et al. (1980)
		CH-V79	Srinivasan and Lewis (1980)
		CHO	Srinivasan and Lewis (1980)
Ouabain resistance	CHO	Mouse L	Corsaro and Pearson (1981)
onc genes	Chemically transformed rat, mouse fibroblasts	Mouse 3T3	Shih et al. (1979) Shilo and Weinberg (1981)
	Chicken LLV induced tumor		Cooper (1980)
	Human bladder carcinoma		Krontiris and Cooper (1981); Murray et al. (1981); Perucho et al. (1981); Shih et al. (1981); Goldfarb et al. (1982);
	Human colon carcinoma		Murray et al. (1981); Perucho et al. (1981)
	Human lung carcinoma		Perucho et al. (1981)
	Mouse lung carcinoma		Shih et al. (1981)
	Mouse, human mammary carcinoma		Lane et al. (1982)
	Human neuroblastoma		Shimizu et al. (1983); Perucho et al. (1981)
	Rat neuroblastoma		Shih et al. (1981)
	Mouse glioma		Shih et al. (1981)
	Human sarcoma		Pulciani et al. (1982)
	Mouse sarcoma		Hopkins et al. (1981)
	Human promyelocytic leukemia		Murray et al. (1981)
	Human T-cell leukemia		Souyri and Fleissner (1983)

TABLE 8.1 (*Continued*)

Genes Transferred	DNA Donor Cell Type	Host Cell	Reference
	Human pre-B cell lymphocyte neoplasm		Lane et al. (1982)
	Human, mouse cell lymphoma		Lane et al. (1982)
	Chicken B-cell lymphoma		Goubin et al. (1983)
	Human,mouse plasmacytoma/myeloma		Lane et al. (1982)
	Human, mouse T-cell lymphoma		Lane et al. (1982)
	Human, mouse mature T-helper cell neoplasm		Lane et al. (1982)
	Human bladder carcinoma	CHEF	Smith et al. (1982)
	Tumorigenic CHEF		Smith et al. (1982)
	Mouse Ha-8	Nonmalignant human fibroblasts "Bloom's syndrome"	Doniger et al. (1983)
Gene expression control sequences	Mouse DNA	Rat-1	Fried et al. (1983)
Reversion frequency control	Mouse, human	Spontaneously transformed 3T3	Baker (1983)
Sensitivity to tumor promoters	Sensitive mouse JB6	Mouse JB6	Colburn et al. (1983)
Multiple drug resistance	CHO	Mouse	Debenham et al. (1982)
UV–excision–repair	CHO AA8	CHO UV-135	MacInnes et al. (1984)
	Human HeLa	CHO UV-20	Rubin et al. (1983, 1984)
	Human A204 or MRC5	Human XP20S(SV)	Takano et al. (1982)
β-tubulin	CHO	CHO	Whitfield et al. (unpublished)

[a] CHO = Chinese hamster ovary cell line; CH-V79 = Chinese hamster lung cell line; CHTG49 = Chinese hamster fibroblast line; CHEF = Chinese hamster embryo fibroblast cell line.

ple cycles of irradiation, fusion, and selection to create cell lines that have a very small amount of human DNA, with the idea of using these lines to clone human genes (Cirullo et al., 1983). This technique is described in detail in Chapter 14. Hybrids were made between normal human leukocytes and temperature-sensitive CHO mutants with defects in *leucyl-tRNA* or *asparaginyl-tRNA* genes. Selected hybrids that contained only one or a few human chromosomes were put through several rounds of gamma-irradiation and refusion to CHO cells and were reselected. This had the result of fragmenting the chromosomes and limiting the amount of human DNA retained in the subsequent hybrids to small fragments presumably linked to the *leucyl-tRNA* or *asparaginyl-tRNA* genes. Since less than 0.1% of the human genome is retained in the hybrids, the authors feel that they can now make a genomic lambda library from this DNA and will be able to pick out the human sequences by using probes of total genomic DNA. If this approach is successful, it represents an alternative to the now standard method of transferring relatively small stretches [about 50 kb (Ruddle, 1981)] of purified DNA by $CaPO_4$–DNA transfer. This and similar techniques may also have the advantage of preserving longer pieces of the chromosomal region and thus enabling studies of gene linkage (Goss and Howard, 1975), and the transfer of very large genes.

D. Chromosome Transfer

In this method, intact chromosomes are isolated from donor cells and are used to transfer genes to host cells (McBride and Ozer, 1973). This method is considered more fully in Chapter 7. One distinct advantage of this method is that it can give up to 100-fold higher frequencies of gene transfer (Lewis et al., 1980) than using purified high-molecular-weight DNA for DMGT. However, the very fact that accounts for the success of this method, that is, the maintenance of the chromosomal structure, makes it less useful for applications such as gene cloning. Since large chromosomal-size pieces of DNA (Ruddle, 1981) are carried to the host cell in this method, it would be difficult to isolate the particular gene of interest since this gene will be embedded in a large body of donor sequences. A way of possibly circumventing this problem, by chromosomal fragmentation, was suggested in the preceding section. In DMGT, on the other hand, gene cloning can be readily accomplished by using molecular markers such as bacterial gene sequences or, in the case of heterologous transfers, repetitive DNA sequences from the donor that differ from those of the host (see Section VI).

E. Protoplast Fusion

Protoplasts are bacterial cells that have had their cell walls removed. This method involves fusing bacterial cells that contain cloned eukaryotic bacterial or viral DNA with mammalian cells (Schaffner, 1980). Schaffner introduced cloned SV40 genes carried on the bacterial plasmid pBR322 (Bolivar et al., 1977) in *Escherichia coli* cells directly into CV-1 monkey cells. Cells are fused with bacteria with polyethylene glycol (PEG) treatment or the bacterial cells are layered on the mammalian cells in the form of a $CaPO_4$ precipitate. PEG fusion yields the highest frequency of transformation, one SV40 infection per 15 monkey cells per 10^4 protoplasts (Schaffner, 1980), and this is the method most commonly used by other workers. After fusion of the bacterial and mammalian cells, some of the foreign DNA enters the eukaryotic cell nucleus and is incorporated. Transformants can then be selected by the appropriate biochemical selections or through use of FACS (see Section IV and Chapter 7). This protoplast fusion method has been used successfully with CHO cells (de Saint Vincent et al., 1981. Milbrandt et al., 1983; Lewis et al., 1984) and cells such as monkey CV-1 and rat FR3T3G (Sandri-Goldin et al., 1981; Rassoulzadegan et al., 1982). This method has also been used for transferring lambda phage vectors to eukaryotic cells (Lewis et al., 1984) to introduce the viral *thymidine kinase* and bacterial *xanthine–guanine phosphoribosyl transferase* genes into CHO and Chinese hamster lung cells. To date this method of protoplast fusion has only been used for cloned DNAs. Presumably this method could be used for directly transferring and selecting DNA from genomic libraries that can act as shuttle vectors. Potentially, then, selection for uncloned genes could be performed after the fusion of bacterial cells containing libraries of mixed, cloned cellular DNAs. One potential difficulty with this approach is that bacterial chromosomal DNA, which is taken up after protoplast fusion along with the plasmid DNA, has been reported to be toxic to mammalian cells (Shen et al., 1982)

F. Microcell-Mediated Gene Transfer

Microcells are small portions of plasma membrane that surround a small number of chromosomes that are created after prolonged treatment of cells with mitotic-blocking agents (Fournier and Ruddle, 1977). These microcells can be fused with host cells, and the transfer of a limited number of chromosomes will result (Fournier and Ruddle, 1977). This approach is basically a modification of standard cell hybridization techniques with the advantage that only a few chromosomes at a time are transferred. A more detailed description of this approach is given in Chapter 7.

G. Microinjection

Microinjection of molecules into the cell nucleus or cytoplasm is a very precise and specific method for introducing foreign molecules into the cell (Diacumakos, 1973). Capecchi (1980) and Anderson et al. (1980) have shown that single cells individually injected with the cloned *HSV-tk* gene DNA will subsequently express the DNA at high frequency. The transformation frequency was greatly enhanced (Capecchi, 1980) by the presence on the plasmid of viral SV40 sequences, including the enhancer region. Other studies have confirmed the utility of the microinjection technique (Yamaizumi et al., 1983). This method of transformation is highly efficient per cell, although a limited number of cells can be injected at one time. Because of these characteristics of this system, it has been the method of choice for introducing DNA into mouse embryos in order to form transgenic mice (Gordon et al., 1980; Wagner et al., 1981).

H. Liposomes

Liposomes are vesicles made of lipids that can be formed containing a variety of molecules. They can then be fused with the plasma membranes of cells, delivering to them by very gentle means the liposome content. This method has been utilized for the delivery of DNA to cells (Fraley et al., 1980; Wong et al., 1980; Schaefer-Ridder et al., 1981; Straubinger and Papahadajopoulos, 1982). One of the current limitations of this technique is the difficulty of packaging very high-molecular-weight DNAs into liposomes. To date this method has only been used for transferring cloned DNAs.

I. Other Methods

Several other new methods have been suggested for the transfer of DNA to cultured cells. These have generally involved the transfer of cloned genes or relatively short viral DNAs. Straus and Raskas (1980) reported the transfer of adenovirus type-2 DNA using erythrocyte ghosts as the carrier. Wiberg et al. (1983) have used a similar method to transfer polyoma virus DNA and cloned portions of the virus. Wong and Neumann (1982) have demonstrated gene transfer after short pulses in an electric field. Stable transformants were obtained that express viral *HSV-tk* gene. Lo (1983) has demonstrated stable gene transfer after a combined microinjection–electric (iontophoretic microinjection) impulse technique.

J. Summary with Specific Reference to CHO Cells

As is apparent from this brief overview of techniques for gene transfer, there is no lack of techniques to try. However, DNA-mediated transformation of CHO cells has generally been achieved through the $CaPO_4$–DNA precipitate method. Transformation of CHO cells with cloned DNAs reaches levels comparable to those with mouse L cells, possibly the best host cell yet identified. Transformations with genomic DNA have a frequency about 10-fold lower than that in L cells (Abraham et al., 1982; unpublished). In our laboratory we have relied on the $CaPO_4$ method, since this has been shown to be useful in transferring purified genomic DNA and has been the most widely used method. The DEAE-dextran method has not produced stable transformation frequencies as high as with the $CaPO_4$ method in CHO cells. Efforts to enhance the $CaPO_4$ method with agents such as glycerol, DMSO, or lysosomotropic agents such as chloroquine have not proved successful in our experience with genomic DNA transfers in CHO cells. Studies with butyrate post-treatment have shown a threefold to fourfold increase in transformation with cloned DNAs but no alteration or even a decrease in the frequency of transformation with genomic DNA (Abraham and Gottesman, unpublished). It is certainly possible that the increases in frequencies investigators see with these various treatments are cell specific, are only effective on transient transformation, or are only active with the specific cloned genes and promoters that have been used (Gorman and Howard, 1983). Significant increases in transformation frequency of CHO cells with genomic and cloned DNAs were, however, achieved by increasing incubation time of the cells with the $CaPO_4$-DNA precipitate to between 16 and 24 hr (Corsaro and Pearson, 1981; Abraham et al., 1982).

Many of the other methods such as microinjection and liposome transfer have not been used for genomic transfers. The method of protoplast fusion may have great potential for efficient transferring of libraries of cloned cDNAs or genomic DNAs directly from the bacterium. This would have the distinct advantage of eliminating the DNA purification step that is now necessary for DNA transfers. However, as mentioned above (Shen et al., 1982) there is some evidence that bacterial DNA is toxic to mammalian cells. Transfer of phage libraries (Lewis et al., 1984) may avoid this problem.

As can be seen from Table 8.1, many investigators have been successful in transferring genomic DNA to CHO cells, and this tally is sure to grow.

IV. STRATEGIES FOR TRANSFORMANT SELECTION

Possibly as important as the method for transformation is the approach for selection of the transformant. In many cases the selection method is dictated by the gene of interest, but other factors such as selection timing and use of cotransformation may be of critical importance in obtaining a successful transfer. Strategies for selecting transformants are generally the same as those for selecting mutants (Thompson, 1979; Ray and Siminovitch, 1982). This section will include a brief review of the most popular schemes that have been used for selecting DNA transformants, as well as some more novel approaches that may be successful.

A. Drug Resistance

Drug resistance is probably the most widely used characteristic with which to select transformants. Selections of this type are based on transferring DNA to a host cell that is sensitive to some selective medium. The DNA that is transferred is from cells that are insensitive to this medium. The DNA can be from uncloned genomic DNA or can be cloned DNA. The most widely used system of this type is the HAT (hypoxanthine, aminopterin, thymidine) system. Thymidine kinase negative cells can be transformed with thymidine kinase genes of viral [herpes simplex virus (HSV)] or eukaryotic origin. Tk^- cells cannot survive on exogenous thymidine if their *de novo* pyrimidine pathway is blocked with aminopterin. Similar selections for APRT (adenine phosphoribosyl transferase) (Chapter 12) and HGPRT (Chapter 13) and the bacterial enzyme Eco-gpt (xanthine–guanine–phosphoribosyltransferase; XGPRT) (Mulligan and Berg, 1980; Chapter 9) are also based on blocking *de novo* purine or pyrimidine pathways and forcing the cells to use these enzymes for salvage pathways. When the appropriate cells are used (i.e., tk^-, $aprt^-$, $hgprt^-$, or $xgprt^-$), these enzymes are lacking in these cells and they must be supplied by the appropriate genes taken up during DNA transfer in order to survive. The *Eco-gpt* gene has the advantage of being able to be transferred to and selected for in virtually any mammalian cell; that is, it is a dominant selectable marker. This is because this enzyme is a novel one to mammalian cells, all mammalian cells being necessarily negative for it. Therefore, a new negative cell line does not have to be selected, as is the case with TK, APRT, and HGPRT.

Any gene that confers drug resistance can theoretically be transferred via DNA transfer, if expression of this gene is a dominant characteristic. Resistance to the neomycin analog G-418 (geneticin) is a good example of such a dominant phenotype, in this case encoded by an aminoglycoside phosphotransferase gene cloned from *E. coli* (Southern and Berg, 1982). This marker can also be transferred into virtually any mammalian cell regardless of the cell's phenotype. Strategies for creating and using vectors for DMGT carrying dominant selectable markers such as G-418 resistance and Eco-gpt are discussed further in Chapter 9.

B. Morphological Selection

This type of selection has been used extensively for isolating cells that have undergone malignant transformation after being transformed by DMGT. Flat NIH3T3 cells are transformed with DNA from tumor lines or primary tumors and transformants are selected on the basis of a change in colony morphology (Shih et al., 1979; Cooper et al., 1980; Cooper, 1982). This approach has been used with dramatic success to isolate a variety of oncogenes associated with human tumor cell lines.

C. Replica Plating

Another method of potential interest might be that of replica plating (Raetz et al., 1982). In replica plating, cells are transferred to a filter as a direct replica of the pattern of cells or clones of cells grown as a monolayer on the initial plate. Either the cells on the filter or those transferred to a new plate can be tested for various characteristics such as antibody binding, or conditional lethals such as heat or cold sensitivity, or drug sensitivity. This method has proved very useful for selecting traits that cannot be selected for by traditional methods. Replica plating might also be used as an alternative to FACS for the second step of isolating transformants after primary selection for cloned genes (Section IV E). The method of replica plating, while classical in bacterial genetics, has only recently begun to be used frequently in somatic cell genetic studies, and has not yet been used for DNA transfer experiments.

D. FACS

Transformant cells can be selected (Kavathas and Herzenberg, 1983a; Kamarck et al., 1984) on the basis of the expression of surface antigens through analysis by FACS. After transformation, cells are selected for expression of surface antigens by complexing them with fluorescently labeled antibody to the desired antigen, and processing them through the FACS. The FACS will be able to detect transformants that are now expressing a new antigen and have the specific antibody bound to them. This is a very exciting and promising technique that should be useful for any number of proteins that are expressed at the cell surface and are unable to be selected by other routes. These investigators (Kavathas and Herzenberg, 1983a, b; Kamarck et al., 1984) initially obtained their transformants through a novel process of cotransformation and sequential selection that will be described in the next section.

E. Cotransformation and Sequential Selection

One very successful strategy for identifying transformants carrying characteristics that are not easily selected has been devised. This strategy can presumably be utilized with all the other selection techniques previously mentioned. The approach (Kavathas and Hertzenberg, 1983; Kamarck et al., 1984; Kuhn et al., 1984a, b) involves cotransforming L cells with a cloned selectable marker, as well as uncloned native genomic DNA, and sequentially selecting for the cloned marker and then the genomic gene of interest. After about 2 weeks transformants expressing the cloned DNA are initially selected. These cells are pooled and then selected for the particular genomic gene of interest (illustrated in Fig. 8.1). The first step enriches for cells that have taken up DNA. This is feasible since cotransformation is a very common event, and cells that do take up DNA generally take up a large amount of DNA, up to 2000 kb (Perucho et al., 1980*a*). Since the genome size of a mammalian cell is about 3×10^6 kb a pool of transformed cells containing at least 1.5×10^3 cells

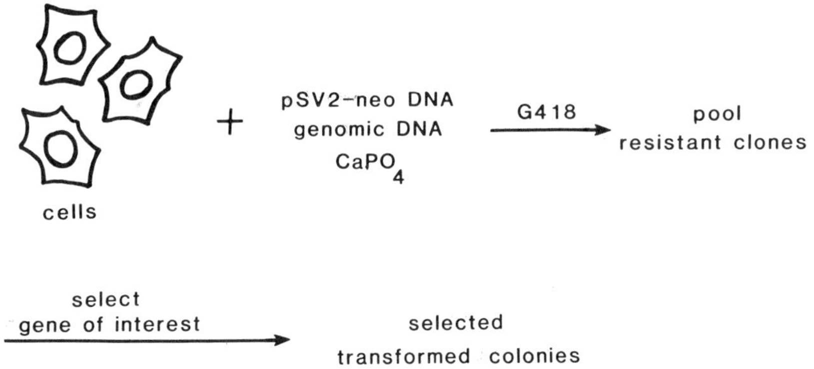

Figure 8.1. A scheme for selecting cells transformed with genomic DNA by using selectable cloned plasmid DNA and eukaryotic genomic DNA.

should contain a representation of the entire transformed genome. Therefore, in a pool of transformed clones somewhat larger than this number, one can expect to find the desired transformed gene. This indeed appears to be the case (Kavathas and Herzenberg, 1983a). In the initial experiments using this strategy, the cells were then further selected after pooling for expression of surface antigens by complexing them with fluorescently labeled antibody to the desired antigen, and processing them through a FACS, as described in the previous section. This two-step process serves the purpose of enriching in the first step for cells that have taken up DNA. In the second step, the number of cells that must be tested in order to find a positive transformant is relatively small, since the frequency is approximately 10^{-4}–10^{-3} (Kavathas and Herzenberg, 1983a; Abraham et al., unpublished). This frequency is much higher than the rate of spontaneous mutation, which also presents a great advantage. In conventional one-step selections the frequency of transformant recovery is often uncomfortably close to that of the spontaneous mutation rate for a particular trait.

This two-step strategy to DNA transformation can also be used with the other selection approaches previously mentioned if the genomic gene of interest is suitable. This in fact has been done for the selection of transformants for cAMP resistance (Abraham et al. unpublished) and for UV resistance (MacInnes et al., 1984). In the case of the transfer of cAMP resistance, DNA was isolated from a mutant CHO cell (10215) carrying dominant resistance to cAMP owing to an altered catalytic subunit in cAMP-dependent protein kinase (Gottesman et al., 1980; Chapter 23). Cells were cotransformed with this genomic DNA and with plasmid DNA carrying the gene for G-418 resistance *(pSV2-neo)*. Clones were selected for G-418 resistance for 10 days and a total of 5000 G-418–resistant clones were pooled. Representatives of this pool were selected for resistance to 1 m*M* 8Br-cAMP in agar. Seven colonies were recovered that were resistant to the growth-inhibitory effects of 8Br-cAMP. One of these clones has been studied in detail and was shown to be a bona fide transformant on the basis of resistance to cAMP growth inhibition, phosphorylation characteristics, and cAMP-dependent protein kinase activities. A similar cotransformation was performed with another dominant mutant, 10248, which is resistant to 8 Br-cAMP owing to an alteration in the RI regulatory subunit of the cAMP-dependent protein kinase with similar results. This strategy has also been used to transfer mutant tubulin genes in CHO cells (Whitfield et al., unpublished).

V. BIOLOGICAL PROPERTIES OF TRANSFORMATION AND TRANSFORMANTS

A. Entry of DNA into the Nucleus

Although many workers have used the methodology of DNA transformation, there has been relatively little attention paid to studying the events that occur in the cell after DNA transformation. Capecchi (1980) has shown that no transformation could be detected when DNA was microinjected into L cell cytoplasm (1000 cells) as opposed to transient transformation in 50–100% of the cells after direct microinjection into the nucleus. Apparently, the DNA has to be in the nucleus to be expressed, and very few molecules are able to enter the nucleus intact from the cytoplasm. Very little is known about the pathway the DNA follows on its way from the cytoplasm to the nucleus. Loyter et al. (1982*a*) have shown by fluorescent probes of the DNA–$CaPO_4$ precipitate (DAPI, which becomes fluorescent after binding to double-stranded DNA, and chlorotetracycline, which fluorescently stains calcium complexes) that practically every cell in the treated monolayer will take up the complex in the cytoplasm. A much smaller number (1–5%) show the precipitate in the nucleus. Therefore, the bottleneck in arriving at a transformed cell via the $CaPO_4$–DNA method is in the transfer of the precipitate from the cell cytoplasm to the nucleus. The basis of this loss of DNA during the traverse from the cytoplasm to the nucleus is not known, but presumably is due either to the action of enzymes in the cytoplasm or to the barrier imposed by the nuclear membrane. The well-known differences in transformation frequency between different cells, up to 100- to 1000-fold, are probably due to differences in the cytoplasm, since frequencies of transformation of different cell types microinjected into the nucleus are remarkably similar (Shen et al., 1982).

The $CaPO_4$–DNA precipitate itself has several characteristics that explain its usefulness as a vehicle for DNA (Loyter et al., 1982*a*). First, the complex is very nuclease resistant. DNA once in the complex can only be completely digested by a combination of DNase, micrococcal nuclease, and EDTA. Second, the precipitate is avidly taken up by the cells, with or without DNA. The ease with which the precipitate enters the cell varies with the DNA concentration within the precipitate. Increased fluorescence in the cells reached a maximum at about 20 μg DNA/60 mm dish. The pH at which the precipitate was formed was also found to be of significant importance in uptake, being optimal between 7.1 and 7.5. This is similar to results obtained after looking at the effect of pH on transformation efficiency (Graham and van der Eb, 1973). Loyter et al. (1982*b*) present evidence that some of the preciptate is taken up by a receptor-mediated endocytic process that is obliterated by depleting ATP stores, or preincubation of cells in high levels of colcemid. ATP depletion interfered with adsorption as well as uptake of the DNA. Uptake was also inhibited by treatment of the cells with colcemid. Electron micrographs of cells treated with [^{3}H]DNA–$CaPO_4$ precipitates show considerable radioactivity in vesicles in the cytoplasm. The fate or identity of these vesicles is unclear, but presumably they may be involved in the pathway by which the DNA enters the nucleus.

B. Fate of DNA in the Nucleus

What happens to DNA in DMGT after it enters the nucleus? There is much evidence from DNA transformation experiments that indicates that after the DNA enters the nucleus it is initially present in an unstable state (Wigler et al., 1977; Graf et al., 1979; Wigler et al., 1979*a*; Willecke et al., 1979; Capecchi, 1980). The DNA that has entered the nucleus appears to be transiently active at a relatively high level for a short time. This transient activity decays rapidly and the level of stably expressing cells is much lower (Linsley and Siminovitch, 1982). Using autoradiography to detect enzyme activity, or direct biochemical assays, the transient expression of transferred genes for thymidine kinase or Eco-gpt can be rapidly measured (Linsley and Siminovitch, 1982). These authors found that transient expression measured after 48 hr of transferred HSV-tk or Eco-gpt DNAs was from 10- to 100-fold higher than stable expression in CHO, mouse L, and mouse 3T3 cells. Transient expression is detectable after 12 hr and reaches a peak in CV-1 cells after 48 hr and then declines (Gorman and Howard, 1983; Howard and Gorman, personal communication). After this point, the DNA appears to be integrated, but the integrated DNA is still subject to a high degree of instability. This instability will be discussed below.

Unstable expression of genes transferred from genomic DNA has also been detected (Graf et al., 1979; Wigler et al., 1979a; Willecke et al., 1979; Lewis et al., 1980; Lester et al., 1980; Pellicer et al., 1980; Peterson and McBride, 1980; Scangos & Ruddle, 1981; Chang et al., 1982). Early experiments (Wigler et al., 1977) demonstrated that while the transferred

HSV-tk gene was relatively stable under selective conditions, revertants to the *tk*$^-$ genotype could be readily selected with a frequency of about 1% when the cells were plated in BrdU medium, which allows only tk$^-$ cells to survive. This initial observation was confirmed by many other workers in the field.

What are the molecular events involved after the DNA enters the nucleus? An early event is probably the ligation of the incoming DNA into long concatameric structures by the host. When L tk$^-$ cells are transformed with two different cloned DNAs, such as *HSV-tk* and φ-*X174*, and cells are selected for resistance to HAT, it is found that a high percentage of these cells have also taken up the unselected DNA as well (15 out of 16). This has also been found to be true in CHO cells (Abraham et al., 1982; and unpublished). Wigler et al. (1979a) have found that when the L cell is cotransformed with two different DNAs, such as *HSV-tk* gene and φ-*X174*, revertants that have lost the *tk* gene may also lose, in many cases, the φ-*X174* genes. This is taken as evidence that the cotransformed DNAs are indeed linked together and inserted near each other in the chromosome. Thus, a revertant caused by a deletion of tk sequences would also result in a loss of the cotransformed sequences if the deletion were of large enough size. More direct evidence of ligation of transformed sequences comes from the experiments of Perucho et al. (1980*a*) showing that the DNA flanking transferred exogeneous plasmid sequences consists of carrier DNA rather than host DNA. Scangos et al. (1981) found the transferred *tk* gene linked to high-molecular-weight DNA in both stable and unstable lines. Corroborating evidence for the linkage of the transformed DNAs comes from studies of the chromosomal localization of transferred DNA in buffalo rat liver (BRL) cells (Robins et al., 1981). BRL cells were cotransformed with *HSV-tk* and human growth hormone *(HGH)* genes. Transformants were selected for *tk* expression on HAT medium, and were examined for the presence of the HGH DNA by *in situ* hybridization. In each of four transformant lines, the HGH DNA was seen at a single (and different) chromosomal site. Since multiple copies of the *HGH* gene were present in these transformants, this suggests that in a particular transformant all copies are inserted at one single site in a chromosome. In two of the lines, the DNA was found inserted at the site of gross chromosomal rearrangements. All the sites of insertion were different, but because of the small sample size it is impossible to say that the DNA insertion sites are truly random.

Many important events that occur in cultured mammalian cells have been discovered or been studied intensively in cells after DNA transformation. For example, ligation of entering DNA, recombination of homologous and nonhomologous DNAs, integration of transformed DNAs into the host chromosomes, amplification of genes, and methylation of genes. These have all been observed after DMGT and the mechanisms of these processes are under investigation now. These events are closely involved in the process that allows transformation and also may tell us eventually something about what occurs in normal mammalian cells. Some examples of such studies are cited below.

Along with cotransformation, another phenomenon common to DMGT that was noted early was that of gene amplification. Transformants that have been transformed with the genomic DNA carrying an altered dihydrofolate reductase (Perucho et al., 1980a; Wigler et al, 1980) can be subsequently selected in medium containing methotrexate. Sublines of these transformants that have amplified this gene, and in some cases carrier or unselected cotransformed DNAs, can be selected after exposure of the cells to increasing levels of methotrexate. Other workers (de Saint Vincent et al, 1981; Milbrandt et al., 1983; Wahl et al., 1984) have shown the amplification of genes in transformants from the cloned Chinese hamster dihydrofolate reductase gene and the cloned Syrian hamster *CAD* gene (coding for the multifunctional protein containing the enzyme activities of carbamyl phosphate synthetase, aspartate transcarbamylase, dihydroorotase). Interestingly, the frequency of amplification of the *CAD* genes varies with the chromosomal location of the inserted genes (Wahl et al., 1984). Kavathas and Herzenberg (1983*b*) have shown that they can select for a subline of DNA-transformed cells with amplified exogenous DNA by sorting with the FACS. In an interesting study of gene amplification, Roberts and Axel (1982) transformed tk^- $aprt^-$ cells with a plasmid containing the hamster *aprt* gene and an incomplete *HSV-tk* gene. They selected cells that were aprt positive, and then selected cells that were also tk positive. The tk positive cells arose through an amplification event that occurred at a frequency of 1.8×10^{-7} to 1.6×10^{-5} and allowed for expression of large quantities of the defective tk enzyme. The amplification unit was between 40 and 200 kb. They found that they could back select for $aprt^-$ tk^+ lines at a frequency of 10^{-4}. There were two classes of such lines. The first class had lost all the *aprt* gene copies, while retaining the *tk* gene. The second, less frequent, class retained multiple copies of the *tk* gene, and also multiple copies of the *aprt* gene. The *aprt* genes, however, appeared to all have the same mutation, which rendered the gene inactive. Because the same defect appeared in all the *aprt* copies, the simplest explanation would be that a mutation occurred in one of the repeated *aprt* sequences and then spread to the other copies via some sort of correction mechanism. In a genomic transfer to CHO cells, Whitfield (Whitfield et al, unpublished) has found a several-fold amplification of a mutant tubulin gene.

Methylation of DNA has also been studied by the use of DMGT. Methylation at HpaII sites is generally heritable. Prior methylation of the cloned chicken *tk* gene is correlated with a decrease in the transformation efficiency when compared to transformation with the unmethylated gene (Wigler et al., 1981). Methylation has also been found to be correlated with altered expression of genes already transformed into cells. In cells transformed with the gene *HSV-tk*, cells that are phenotypically thymidine kinase negative can be selected after growth on BrdU. One class of such revertants that appear have retained their *HSV-tk* genes, but these genes are now extensively methylated. This methylation correlates with the loss of gene activity (Ostrander et al., 1982). Another cause for loss of gene activity after transformation appears to be changes in chromatin structure (Davies et al., 1982). These workers found a change in DNAase I senstivity of chromatin at the site of the transferred *tk* gene after rereversion of cells from a tk^- phenotype to a tk^+ phenotype. They found a correlation of lack of thymidine kinase activity with a decrease of DNase sensitivity.

VI. RECOVERY OF TRANSFERRED DNA

For many investigators, transforming the recipient cells with foreign DNA is merely the first step toward cloning the gene. Recovering a gene that has previously been cloned or one that has been cloned in a closely related organism can be approached by making a genomic plasmid or phage library, and recovering the proper clones by hybridization with the reference clone. This relatively straightforward method will not be considered here. Instead, we will consider approaches to recovering a previously uncloned gene that has been transferred via genomic DMGT. Since the transformed gene of interest is usually one that is represented poorly by mRNA, making cDNA libraries of the host or recipient cells is generally not adequate. There will be too few cDNA copies of the desired gene in the library. The cloning must proceed through direct cloning of the DNA; this can be accomplished through several means.

A. Recovery through Linked DNA Sequences

One method of gene rescue that has frequently been used is to ligate the DNA to be transferred with marker sequences such as the bacterial plasmid pBR322. After the ligation, the cells are transformed with the DNA as usual and selected for the desired phenotype. This will result in cells carrying the gene of interest, along with many kilobases of extraneous, cotransformed DNA, and many copies of the marker DNA. Presumably, some marker sequences will have been ligated near the gene of interest that has been transferred. To eliminate the extraneous DNA, DNA is isolated from the selected transformants and new untransformed host cells are transformed with the DNA for a second and sometimes third round of transformation. This repeated transformation will eliminate most of the extraneous genomic and marker DNA simply by dilution.

At this point, several different strategies can be employed. If pBR322 is used as the marker DNA, the DNA from the selected transformant, known to have a minimal number of copies of the marker DNA as determined by Southern analysis, can be cut with an appropriate restriction enzyme, ligated and circularized and used to transform *E. coli* (Perucho et al., 1980*b*). Bacteria that express ampicillin resistance that is coded for by pBR322 can be selected. This method has been used for isolation of the chicken *tk* gene (Perucho et al., 1980*b*). The gene was identified by isolating ampicillin-resistant plasmids and testing for the ability to transform L tk^- cells to tk^+. A similar method has been used to clone the hamster *aprt* gene after transfer into mouse $aprt^-$ cells (Lowy et al., 1980). In this case, however, after secondary transformants were produced, the DNA from these was partially cleaved with EcoRI and cloned in lambda phage charon 4A. The resulting recombinant library was then hybridized with pBR322 DNA to pick up lambda clones that contained pBR322 and the hamster *aprt* gene. The advantages of this second method are that no functional sequences from the original pBR322 marker DNA need to be maintained. Neither the origin of replication nor the drug-resistant genes need be present, since there is no selection step in *E. coli.* Secondly, since one can clone larger pieces of DNA into lambda (up to about 40 kb) than in pBR322, one has a greater chance of recovering the entire sequence of the desired eukaryotic genomic DNA. A modification of this last technique has been used (Goldfarb et al., 1982) to clone a transforming gene from human T24 bladder carcinoma cells and to clone the Chinese hamster *tk* gene (Lewis et al., 1983). In this technique, an *E. coli* tRNA amber suppressor gene was used as the initial marker DNA that was ligated to the eukaryotic genomic DNA to be transferred, instead of pBR322. When a recombinant lambda library is made with DNA from secondary transformants, in a lambda strain that contains an amber mutation, only those recombinant phages that now contain the suppressor gene will be able to survive on a suppressor minus host.

In intraspecific gene transfers it is often unnecessary to ligate a marker to the transforming DNA, since DNA from the donor may contain endogenous repetitive sequences that can themselves serve as markers. For example, in the transfer of human DNAs the near ubiquitous presence of Alu sequences (Houck et al., 1979) near structural genes in human DNA (Jolly et al., 1982; Pulciani et al, 1982; Shih & Weinberg, 1982; Lin et al., 1983) can be utilized. When human DNA is transferred to mouse cells, many Alu sequences are also transferred. These can be identified in transformants by the use of appropriate probes, such as cloned Alu sequences. After a second round of transformation, the number of Alu sequences in the mouse decreases drastically, leaving presumably a small number closely associated with the genomic gene that has been selected. Thus the linked Alu sequence has the same function as ligating a bacterial or phage marker to the DNA, except that it is an endogenous marker that can be detected because of the interspecific differences between the human and the mouse. From the secondary transformants a lambda library can be made and the eukaryotic gene of interest, linked to the Alu sequence, can again be identified by hybridization to an Alu-specific probe. The selected lambda clones can then be tested for biological activity by transforming them into the host cells.

B. Shuttle Vectors

Attempts have been made to construct vectors that can be used as the initial cloning vectors and also as shuttle vectors that can be easily recovered from mammalian cells and propagated in bacterial cells. These vectors could be used to clone directly whole genomic DNAs and these libraries can then be used to transform host cells. The transformants selected after several rounds presumably would contain only the gene of interest and these could be directly rescued out. These techniques would eliminate the need to make a library after the transformants were selected, and also would eliminate the need to ligate the DNA initially to marker DNA. The SV40-based vector, pSV2-gpt (Mulligan and Berg, 1980), can be rescued out from monkey CV-1 cells after transformation by fusing them with COS-1 cells (Breitman et al., 1982). COS-1 cells are simian cells that constitutively produce T antigen (Gluzman, 1981). This allows the pSV2-gpt vector to replicate autonomously as an episome and be recovered as low-molecular-weight DNA. The limits on size of inserted DNA that would still be able to be rescued after attachment to an SV40 origin of replication is unknown (Breitman et al., 1982). Shuttle vectors are described in greater detail in Chapter 9.

One vector system that is designed to have a large capacity for inserted DNA is called the cosmid system (Collins and Hohn, 1978). Cosmids are pBR322-based plasmids that can be packaged into lambda particles since they contain the cohesive end site (cos) of lambda. It has been shown that mammalian cells transformed with cosmid vectors containing either *HSV-tk* or lambda vectors containing chicken *tk* genes (Lindenmaier et al., 1982) can effectively be rescued from the cells by *in vitro* packaging of the integrated cosmid or lambda DNA. This approach presents several advantages over other gene-rescue techniques. First, cosmids can be packaged that accommodate up to 45 kb of inserted DNA. Thus, considerably more DNA can fit in these vectors than in the standard pBR322 plasmid. The cosmids are still efficient for replication in *E. coli*. Since transformation is initially performed with the cosmid library, the mammalian DNA does not have to be previously cut with a restriction enzyme to link it to marker DNA. This is useful for the majority of cases where it is initially unknown which restriction enzymes can be used to cut the DNA without abolishing function of the gene of interest. Lau and Kan (1983) have constructed similar cosmid vectors that contain the SV40 origin sequences and the selectable markers pSV2-gpt, pSV2-DHFR (mouse dihydrofolate reductase), or pSV2-neo. These markers, as discussed earlier, are dominant selectable markers in mammalian cells. The SV40 sequences also allow for autonomous replication in COS-1 cells that constitutively produce T antigen. They constructed libraries of these cosmids with mammalian DNA and selected for cosmids that carried the alpha-globin gene cluster by hybridization with previously cloned alpha-globin probes. They report that they were able to effectively transform mammalian cells with these vectors. They could also efficiently recover the cosmids containing the alpha-globin complex by *in vitro* packaging. These studies suggest that it would be possible directly to recover genes after transformation of cells with a cosmid library constructed with the mammalian DNA of interest. The transformants containing the desired genes could be selected by hybridization with previously cloned probes, as demonstrated (Lau and Kan, 1983), or by selection on appropriate media for selectable traits such as thymidine kinase (Lau and Kan, 1984), or on the basis of antibody binding (Kavathas and Hertzenberg, 1983a). After the appropriate transformant is identified, the desired gene could be efficiently cloned out using *in vitro* packaging. Other vectors that might be useful for making libraries with which to transform cells, and recover genes, might be those based on the BPV-virus-, Bk-virus, or cDNA-based expression libraries. These will be considered further in the Chapters 9 and 10.

VII. CONCLUSION AND PROSPECTS FOR CHO CELLS

DMGT, although a relatively young technique, has been shown to be of great value in attempts to study the genetics of cells in culture. Its greatest triumph has been in the discovery of the numerous oncogenes, which is still continuing. It has been also very useful in facilitating the cloning of other genes that would have been considered impossible just a few years ago. Gene transfer can be used to answer questions about gene regulation and can enable the mammalian cell to act as an assay system for *in vitro* altered genes. The process of gene transfer itself has drawn much attention and its study should answer many questions about recombination and DNA repair as well as questions about the uptake of macromolecules and their fate in the cell.

The experience of our laboratory and that of others shows that the CHO cell is an excellent cell in which to pursue studies of DMGT. The cells can be transformed with both cloned and genomic DNAs at frequencies approaching those of mouse L cells. The DMGT technique should see great use in the future for the study of cellular processes, of gene regulation, and especially in the isolation of genes, both mutant and wild-type, in the CHO cell.

REFERENCES

Abraham, I., Tyagi, J. S., and Gottesman, M. M. (1982). *Som. Cell Gen.* **8**, 23–39.

Akiyama, S., Ono, M., and Kuwano, M. (1983). *Bioch. Biophys. Res. Comm.* **110**, 783–788.

Anderson, R. A., Krakauer, T., and Camerini-Otero, R. D. (1982). *Proc. Natl. Acad. Sci. USA* **79**, 2748–2752.

Anderson, W. F., Killos, L., Sanders-Haigh, G., Kretschmer, P. J., Diacumakos, E. G. (1980). *Proc. Natl. Acad. USA* **77**, 5399–5403.

Andrulis, I. L. and Siminovitch, L. (1981). *Proc. Natl. Acad. Sci. USA* **78**, 5724–5728.

Bacchetti, S. and Graham, F. L. (1977). *Proc. Natl. Acad. Sci. USA* **74**, 1590–1594.

Baker, R. F. (1983). *Proc. Natl. Acad. Sci. USA* **80**, 1174–1178.

Bolivar, F., Rodriguez, R. L., Greene, P. J., Betlach, M. C., Heyneker, H. L., Boyer, H. W., Crosa, J. H., and Falkow, S. (1977). *Gene* **2**, 95–113.

Botchan, M., Stringer, J., Mitchison, T., and Sambrook, J. (1980). *Cell* **20**, 143–152.

Breitman, M. L., Tsui, L.-C., Buchwald, M., and Siminovitch, L. (1982). *Mol. Cell. Biol.* **2**, 966–976.

Calos, M. P., Lebkowski, J. S., and Botchan, M. R. (1983). *Proc. Natl. Acad. Sci. USA* **80**, 3015–3019.

Capecchi, M. R. (1980). *Cell* **22**, 479–488.

Chang, L. J.-A., Gamble, C., Izaguirre, C. A., Minden, M. D., Mak, T., and McCulloch, E. A. (1982). *Proc. Nat. Acad. Sci. USA* **79**, 146–150.

Chapman, V. M., Kratzer, P. G., Siracusa, L. D., Quarantillo, B. A., Evans, R., and Liskay, R. M. (1982). *Proc. Natl. Acad. Sci. USA* **79**, 5357–5361.

Chu, G. and Sharp, P. A. (1981). *Gene* **13**, 197–202.

Cirullo, R. E., Dana, S., and Wasmuth, J. J. (1983). *Mol. Cell. Biol.* **3**, 892–902.

Colburn, N. H., Talmadge, C. B., and Gindhart, T. D. (1983). *Mol. Cell. Biol.* **3**, 1182–1186.

Collins, J. and Hohn, B. (1978). *Proc. Natl. Acad. Sci. USA* **75**, 4242–4246.

Cooper, G. M. (1982). *Science* **218**, 801–806.

Cooper, G. M. and Neiman, P. E. (1980). *Nature* **287**, 656–659.

Cooper, G. M., Okenquist, S., and Silverman, L. (1980). *Nature* **284**, 418–421.

Corces, V., Pellicer, A., Axel, R., and Meselson, M. (1981). *Proc. Natl. Acad. Sci. USA* **11**, 7038–7042.

Corsaro, C. M. and Pearson, M. L. (1981). *Som. Cell Genet.* **7**, 617–630.

Davies, R. L., Fuhrer-Krusi, S., and Kucherlapati, R. S. (1982). *Cell* **31**, 521–529.

Debenham, P. G., Kartner, N., Siminovitch, L., Riordan, J. R., and Ling, V. (1982). *Mol. Cell. Biol.* **2**, 881–889.

deJonge, A. J. R., Abrahams, P. J., Westerveld, A., and Bootsma, D. (1982). *Nature* **295**, 624–626.

de Saint Vincent, B. R. and Wahl, G. M. (1983). *Proc. Natl. Acad. USA* **80**, 2002–2006.

de Saint Vincent, B. R., Delbruck, S., Eckhart, W., Meinkoth, J., Vitto, L., and Wahl, G. (1981). *Cell* **27**, 267–277.

Diacumakos, E. G. (1973). *Methods in Cell Biology* (D. M. Prescott, ed.), Academic Press, New York, vol. 7, pp. 287–311.

Doniger, J., Di Paolo, J. A., and Popescu, N. C. (1983). *Science* **222**, 1144–1146.

Dubois, M.-F., Vignal, M., Le Cunff, M., and Chany, C. (1983). *Nature* **303**, 433–435.

Folger, K. R., Wong, E. A., Wahl, G., and Cappechi, M. R. (1982). *Mol. Cell. Biol.* **2**, 1372–1387.

Fournier, R. E. K. and Ruddle, F. (1977). *Proc. Natl. Acad. Sci. USA* **74**, 319–323.

Fraley, R., Subramani, S, Berg, P., and Papahadjopoulos, D. (1980). *J. Biol. Chem.* **255**, 10431.

Fried, M., Griffiths, M., Davies, B., Bjursell, G., La Mantia, G., and Lania, L. (1983). *Proc. Natl. Acad. Sci. USA* **80**, 2117–2121.

Frost, E. and Williams, J. (1978). *Virol.* **91**, 39–50.

Gluzman, Y. (1981). *Cell* **23**, 175.

Goldfarb, M., Shimizu, K., Perucho, M., and Wigler, M. (1982). *Nature* **296**, 404–409.

Gordon, J. W., Scangos, G. A., Plotkin, D. J., Barbosa, J. A., and Ruddle, F. H. (1980). *Proc. Natl. Acad. Sci. USA.* **77**, 7380–7384.

Gorman, C. M. and Howard, B. H. (1983). *Nuc. Acids Res.* **11**, 7631–7648.

Gorman, C. M., Padmanabhan, R., and Howard, B. H. (1983). *Science* **221**, 551–553.

Goss, S. J. and Harris, H. (1975). *J. Cell Sci.* **25**, 17–37.

Gottesman, M. M., LeCam, A., Bukowski, M., and Pastan, I. (1980). *Somat. Cell Genet.* **6**, 45–61.

Goubin, G., Goldman, D. S., Luce, J., Neiman, P. E., and Cooper, G. M. (1983). *Nature* **302**, 114–119.

Graf, L. H., Urlaub, G., and Chasin, L. A. (1979). *Somat. Cell Genet.* **5**, 1031–1044.

Graham, F. L. and van der Eb., A. J. (1973). *Virology* **52**, 456–467.

Graham, F. L., van der Eb, A. J., and Heijneker, H. L. (1974). *Nature* **251**, 687–691.

Green, M. R., Treisman, R., and Maniatis, T. (1983). *Cell* **35**, 137–148.

Hopkins, N., Besmer, P., De Leo, A. B., and Law, L. W. (1981). *Proc. Nat. Acad. Sci. USA* **78**, 7555–7559.

Houck, C. M., Rinehart, F. P., and Schmid, C. W. (1979). *J. Mol. Biol.* **132**, 289–306.

Ingles, C. J. and Shales, M. (1982). *Mol. Cell. Biol.* **2**, 666–673.

Ishiura, M., Hirose, S., Uchida, T., Hamada, Y., Suzuki, Y., and Okada, Y. (1982). *Mol. Cell. Biol.* **2**, 607–616.

Jackson, D. A. (1980). In *Introduction of Macromolecules into Viable Mammalian Cells* (R. Baserga, C. Croce, and G. Rovera, eds.), Alan R. Liss, New York.

Jolly, D. J., Esty, A. C., Bernard, H. U., and Friedmann, T. (1982). *Proc. Natl. Acad. Sci. USA* **79**, 5038–5041.

Kamarck, M. E., Barbosa, J. A., Kuhn, L., Peters, P. G. M., Shulman, L., and Ruddle, F. H. (1984). *Somatic Cell Genetics and Flow Cytometry.*

Kavathas, P. and Herzenberg, L. A. (1983*a*). *Proc. Nat. Acad. Sci. USA* **80**, 524–528.

Kavathas, P. and Herzenberg, L. A. (1983*b*). *Nature* **306**, 385–387.

Krontiris, T. G. and Cooper, G. M. (1981). *Proc. Natl. Acad. Sci. USA* **78**, 1181–1184.

Kuhn, L. C., Barbosa, J. A., Kamarck, M. E., and Ruddle, F. H. (1984a). *Molec. Biol. Med.* **1**, 335–352.

Kuhn, L. C., McClelland, A., Ruddle, F. H. (1984b). *Cell* **37**, 95–103.

Lane, M.-A., Sainten, A., and Cooper, G. M. (1981) *Proc. Natl. Acad. Sci. USA* **78**, 5185–5189.

Lane, M.-A., Sainten, A., and Cooper, G. M. (1982). *Cell* **28**, 873–880.

Lau, Y.-F. and Kan, Y. W. (1983). *Proc. Natl. Acad. Sci. USA* **80**, 5225–5229.

Lau, Y.-F. and Kan, Y. W. (1984). *Proc. Natl. Acad. Sci. USA* **81**, 414–418.

Lester, S. C., Korn, N. J., and DeMars, R. (1982). *Som. Cell Gen.* **8**, 265–284.

Lester, S. C., LeVan, S. K., Steglich, C., and DeMars, R. (1980). *Somat. Cell Genet.* **6**, 241–259.

Lewis, W. H., Bevilacqua, P. J., and Bradley, W. E. C. (1984), *Proceedings of the 13th International Cancer Congress*, Alan R. Liss, New York.

Lewis, J. A., Shimizu, K., and Zipser, D. (1983). *Mol. Cell. Biol.* **3**, 1815–1823.

Lewis, W. H., Srinivasan, P. R., Stokoe, N., and Siminovitch, L. (1980). *Som. Cell Genet.* **6**, 333–347.

Lin, F.-L., Sperle, K., Sternberg, N. (1984) *Mol. Cell. Biol.* **4**, 1020–1034.

Lin, P.-F., Yamaizumi, M., Murphy, P. D., Egg, A., and Ruddle, F., (1982). *Proc. Natl. Acad. Sci. USA* **79**, 4290–4294.

Lin, P.-F., Zhao, S.-Y., and Ruddle, F. H. (1983). *Proc. Natl. Acad. Sci. USA* **80**, 6528–6532.

Lindenmaier, W., Hauser, J., de Wilke, I. G., and Schutz, G. (1982). *Nuc. Acids. Res.* **10**, 1243–1256.

Linsley, P. S. and Siminovitch, L. (1982). *Mol. Cell. Biol.* **2**, 593–597.

Liskay, R. M. and Evans, R. J. (1980). *Proc. Natl. Acad. Sci. USA* **77**, 4895–4898.

Liskay, R. M. and Stachelek, J. L. (1983). *Cell* **35**, 157–165.

Liu, C. P., Slate, D. L., and Ruddle, F. H. (1979). *Proc. Natl. Acad. Sci. USA* **76**, 4503–4506.

Lo, C. W. (1983). *Mol. Cell. Biol.* **3**, 1803–1814.

Lowy, I., Pellicer, A., Jackson, J. F., Sim, G.-K., Silverstein, S., and Axel, R. (1980). *Cell* **22**, 817–823.

Loyter, A., Scangos, G., Juricek, D., Keene, D., and Ruddle, F. H. (1982*b*). *Exp. Cell. Res.* **139**, 223–234.

Loyter, A., Scangos, G. A., and Ruddle, F. H. (1982a). *Proc. Natl. Acad. Sci. USA* **79**, 422–426.

Luthman, H. and Magnusson, G. (1983). *Nucl. Acids. Res.* **11**, 1295–1308.

MacInnes, M. A., Bingham, J. M., Strniste, G. F., and Thompson, L. H. (1984a). In *UCLA Symposium on Molecular and Cellular Biology, New Series, Volume 11, Cellular Responses to DNA Damage* (E. C. Friedberg and B. A. Bridges, eds.), Alan R. Liss, New York.

MacInnes, M. A., Bingham, J. M., Thompson, L. H., Strniste, G. F. (1984b). *Mol. Cell. Biol.* **4**, 1152–1158.

Maitland, N. J. and McDougall, J. K. (1977). *Cell* **11**, 233–241.

McBride, O. W. and Ozer, H. L. (1973). *Proc. Natl. Acad. Sci. USA* **70**, 1258–1262.

McCutchan, J. H. and Pagano, J. S. (1968). *J. Natl. Can. Inst.* **41**, 351–357.

Milbrandt, J. D., Azizkhan, J. C., Greisen, K. S., and Hamlin, J. L. (1983). *Mol. Cell. Biol.* **3**, 1266–1273.

Milbrandt, J. D., Azizkhan, J. C., and Hamlin, J. L. (1983). *Mol. Cell. Biol.* **3**, 1274–1282.

Miller, C. K. and Temin, H. M. (1983). *Science* **220**, 606–609.

Milman, G. and Herzberg, M. (1981). *Somat. Cell Genet.* **7**, 161–170.

Mulligan, R. C. and Berg, P. (1980). *Science* **209**, 1422–1427.

Murray, M., Shilo, B., Shih, C., Cowing, D., Hsu, H. W., and Weinberg, R. A. (1981). *Cell* **24**, 355–361.

Nairn, R. S., Adair, G. M., and Humphrey, R. M. (1982). *Mol. Gen. Genet.* **187**, 384–390.

Ostrander, M., Vogel, S., and Silverstein, S. (1982). *Mol. Cell. Biol.* **2**, 708–714.

Pellicer, A., Robins, D., Wold, B., Sweet, R., Jackson, J., Lowy, I., Roberts, J. M., Sim, G.-K., Silverstein, S., and Axel, R. (1980). *Science* **209**, 1414–1422.

Perucho, M., Hanahan, D., and Wigler, M. (1980*a*). *Cell* **22**, 309–317.

Perucho, M., Hanahan, D., Lipsich, L., and Wigler, M. (1980*b*). *Nature* **285**, 207–210.

Perucho, M., Goldfarb, M. P., Shimizu, K., Lama, C., Fogh, J., and Wigler, M. H. (1981). *Cell* **27**, 467–476.

Peterson, J. L. and McBride, O. W. (1980). *Proc. Natl. Acad. Sci. USA* **77**, 1583–1587.

Pomerantz, B. J. (1983). *Mol. Cell. Biol.* **3**, 1680–1685.

Pulciani, S., Santos, E., Lauver, A. V., Long, L. K., Robbins, K. C., and Barbacid, M. (1982). *Proc. Natl. Acad. Sci. USA* **79**, 2845–2849.

Raetz, C. R. H., Wermuth, M. M., McIntyre, T. M., Esko, J. D., and Wing, D. C. (1982). *Proc. Natl. Acad. Sci. USA* **77**, 5192–5196.

Rassoulzadegan, M., Binetruy, B., and Cuzin, F. (1982). *Nature* **295**, 257–260.

Ray, P. N. and Siminovitch, L. (1982). In *Somatic Cell Genetics* (C. T. Caskey and D. C. Robbins, eds.), Plenum Press, New York.

Razzaque, A., Mizusawa, H., and Seidman, M. M. (1983). *Proc. Natl. Acad. Sci. USA* **80**, 3010–3014.

Roberts, J. M. and Axel, R. (1982). *Cell* **29**, 109–119.

Robins, D. M., Ripley, S., Henderson, A. S., and Axel, R. (1981). *Cell* **23**, 29–39.

Rubin, J. S., Joyner, A. L., Bernstein, A., and Whitmore, G. F. (1984). In *Cellular Responses to DNA Damage, UCLA Symposium on Molecular and Cellular Bilogy., New Series* (E. C. Friedberg and B. A. Bridges, eds.), Alan Liss, New York, Vol. 11.

Rubin, J. S., Joyner, A. L., Bernstein, A., and Whitmore, G. F. (1983) *Nature* **306**, 206–208.

Ruddle, F. H. (1981). *Nature* **294**, 115–120.

Ruley, H. E. and Fried, M. (1983). *Nature* **304**, 181–184.

Sandri-Goldwin, R., Goldin, A., Levine, M., and Glorioso, J. (1981). *Mol. Cell. Biol* **1**, 743–752.

Scangos, G., Huttner, K. M., Juricek, D. K., and Ruddle, F. H. (1981). *Mol. Cell Biol* **1**, 111–120.

Scangos, G. and Ruddle, F. H. (1981). *Gene* **14**, 1–10.

Schaefer-Ridder, M., Wang, Y., and Hofschneider, P. H. (1981). *Science* **215**, 166–168.

Schaffner, W. (1980). *Proc. Natl. Acad. Sci. USA* **77**, 2163–2167.

Shen, Y., Hirschhorn, R. R., Mercer, W. E., Surmacz, E., Tsutsui, Y., Soprano, K. J., and Baserga, R. (1982). *Mol. Cell. Biol.* **2**, 1145–1154.

Shih, C., Padhy, L. C., Murray, M., and Weinberg, R. A. (1981). *Nature* **290**, 261–264.

Shih, C., Shilo, B.-Z., Goldfarb, M. P., Dannenberg, A., and Weinberg, R. A. (1979). *Proc. Natl. Acad. Sci. USA* **76**, 5714–5718.

Shih, C., and Weinberg, R. A. (1982). *Cell* **29**, 161–169.

Shilo, B.-Z. and Weinberg, R. A. (1981). *Nature* **289**, 607–609.

Shimizu, K., Goldfarb, M., Perucho, M., and Wigler, M. (1983). *Proc. Natl. Acad. Sci. USA* **80**, 383–387.

Small, J. and Scangos, G. (1983). *Science* **219**, 174–176.

Smith, B. L., Anisowicz, A., Chodosh, L. A., and Sager, R. (1982). *Proc. Natl. Acad. USA* **79**, 1964–1968.

Sompayrac, L. M. and Danna, K. J. (1981). *Proc. Natl. Acad. Sci. USA* **78**, 7575–7578.

Southern, P. J. and Berg, P. (1982). *J. Mol. App. Genet.* **1**, 327–341.

Souyri, M. and Fleissner, E. (1983). *Proc. Natl. Acad. Sci. USA* **80**, 6676–6679.

Srinivasan, P. R. and Lewis, W. H. (1980). In *Introduction of Macromolecules into Viable Mammalian Cells* (R. Baserga, C. Croce, and G. Rovera, eds.) Alan R. Liss, New York.

Stow, N. D. and Wilkie, N. M. (1976). *J. Gen. Virol.* **33**, 447–458.

Straubinger, R. M. and Padahadjopoulos, D. (1982). In *Techniques in Somatic Cell Genetics* (J. W. Shay, ed.), Plenum Press, New York.

Straus, S. E. and Raskas, H. J. (1980). *J. Gen. Virol.* **48**, 241–245.

Szybalska, E. H. and Szybalski, W. (1962). *Proc. Natl. Acad. Sci. USA* **48**, 2026–2034.

Takano, T., Noda, M., and Tamura, T. (1982) *Nature* **296**, 269–271.

Thompson, L. H. (1979). In *Methods in Enzymology* (W. H. Jakoby and I. H. Pastan, eds.), Academic Press, New York, Volume LVIII. Cell Culture, pp. 308–321.

Tindall, K. R. and Hsie, A. W. (1984). In *UCLA Symposium on Molecular and Cellular Biology, New Series, Volume 11, Cellular Responses to DNA Damage* (E. C. Friedberg and B. A. Bridges, eds.), Alan R. Liss, New York.

Venolia, L. and Gartler, S. M. (1983). *Nature* **302**, 82–83.

Wagner, E. F., Stewart, T. A., and Mintz, B. (1981). *Proc. Natl. Acad. Sci. USA* **78**, 5016–5020.

Wahl, G. M., de Saint Vincent, B. R., and DeRose, M. L. (1984) *Nature* **307**, 516–520.

Waldren, C., Snead, D., Stamato, T. (1984). In *UCLA Symposium on Molecular and Cellular Biology, New Series, Volume 11, Cellular Responses to DNA Damage* (E. C. Friedberg and B. A. Bridges, eds.), Alan R. Liss, New York.

Warrick, H., Hsiung, N., Shows, T. B., and Kucherlapati, R. (1980) *J. Cell Biol.* **86**, 341–346.

Weinberg, R. A. (1983). *J. Cell. Biol.* **97**, 1661–1662.

Wiberg, F. C., Sunnerhagen, P., Kaltoft, K., Zeuthen, J., and Bjursell, G. (1983). *Nuc. Acids Res.* **11**, 7287–7302.

Wigler, M., Levy, D., and Perucho, M. (1981). *Cell* **24**, 33–40.

Wigler, M., Pellicer, A., Silverstein, S., and Axel, R. (1978). *Cell* **14**, 725–731.

Wigler, M., Silverstein, S., Lee, L.-S., Pellicer, A., Cheng, Y., and Axel, R. (1977). *Cell* **11**, 223–232.

Wigler, M., Sweet, R., Sim, G. K., Wold, B., Pellicer, A., Lacy, E., Maniatis, T., Silverstein, S., and Axel, R. (1979*a*). *Cell* **16**, 777–785.

Wigler, M., Pellicer, A., Silverstein, S., Axel, R., Urlaub, G., and Chasin, L. (1979*b*). *Proc. Natl. Acad. Sci. USA* **76**, 1373–1376.

Wigler, M., Perucho, M., Kurtz, D., Dana, S., Pellicer, A., Axel, R., and Silverstein, S. (1980). *Proc. Natl. Acad. Sci. USA* **77**, 3567–3570.

Willecke, K., Klomfass, M., Mierau, R., and Dohmer, J. (1979). *Mol. Gen. Genet.* **170**, 179–185.

Wilson, J. H., Berget, P. B., and Pipas, J. M. (1982). *Mol. Cell. Biol.* **2**, 1258–1269.

Wong, T.-K., and Neumann, E. (1982). *Bioc. Bioph. Res. Comm.* **107**, 584–587.

Wong, T.-K., Nicolau, C., and Hofschneider, P. H. (1980). *Gene* **10**, 87–94.

Yamaizumi, M., Horwich, A. L., and Ruddle, F. H. (1983). *Mol. Cell. Biol.* **3**, 511–522.

CHAPTER 9

VECTOR-MEDIATED GENE TRANSFER

Bruce H. Howard
Mary McCormick
Laboratory of Molecular Biology
National Cancer Institute
National Institutes of Health
Bethesda, Maryland

I. INTRODUCTION

Chinese hamster cells have played an important role in the development of vector-mediated gene transfer. They have a stable, near diploid karyotype and are readily transformed by standard techniques such as microinjection of DNA, protoplast fusion, and calcium-phosphate-mediated DNA transfection. Moreover, numerous mutant Chinese hamster cell lines are available, and these have formed the basis of selectable marker systems. For such systems, conditions are devised such that introduction of a specific cellular or recombinant gene results in complementation of the host cell mutation and, consequently, cell growth. This type of selection has at least three applications. First, the mutant gene may represent the gene of interest, in which case there is a direct selection for that gene. Second, a plasmid vector containing the selectable marker may be used as a vehicle to introduce a gene for which there is no selection. This may be accomplished by cloning the gene of interest into the vector. Third, the vector containing the selectable marker can be cotransformed (not covalently linked, but mixed during the gene transfer procedure) with DNA containing the gene of interest. Selection for the marker yields a high proportion of cells that also carry the nonselectable marker. This third possibility, cotransformation of a nonselectable gene of interest with a selectable marker, greatly extends the applicability of the selectable marker systems. This approach is discussed extensively in Chapter 8.

The advantages of vector-mediated gene transfer can be contrasted with current limitations of genomic DNA gene transfer in Chinese hamster cells. With the use of plasmid vectors, genes of interest can be engineered for efficient gene expression by introducing heterologous promoters, transcription enhancers, splice sites, and polyadenylation sites. These regulatory signals may be adjoined to a cDNA form of the gene that contains the entire coding region, but none of the possibly extensive intervening sequences found in many eukaryotic genes. In contrast, when total genomic DNA is used as the source material, the gene of interest may contain regulatory elements that result in low level expression. The gene, including all of its untranslated regions, may be quite large and difficult to isolate intact. Finally, if the gene of interest is present as single or low copy in the genome, its representation in a total genomic DNA preparation is very low, especially when compared to the number of copies of a cloned gene present in an equivalent microgram amount of a plasmid DNA preparation. Since the effects of low copy number and poor gene expression can drastically reduce the efficiency of transfection, the ability to remedy these problems through gene cloning makes vector-mediated gene transfer the method of choice over genomic DNA transfer in many cases.

II. CURRENTLY USED GENETIC MARKERS AND SELECTION SYSTEMS

A. Dihydrofolate Reductase

Dihydrofolate reductase (DHFR) is a quite versatile genetic marker. As shown in Figure 9.1, DHFR catalyzes the reduction of folate to tetrahydrofolate (FH_4). FH_4 in turn reacts with serine to produce glycine and methylene-FH_4; methylene-FH_4 reacts with deoxyuridine monophosphate (dUMP) to form thymidine monophosphate (dTMP). FH_4 intermediates are also required for purine biosynthesis.

With these biosynthetic pathways in mind, Urlaub and Chasin (1980) isolated Chinese hamster ovary (CHO) cell mutants deficient in DHFR activity (see Chapter 16). DHFR-deficient mutants are triple auxotrophs requiring glycine, hypoxanthine, and thymidine (Fig. 9.1). To obtain such *dhfr* mutants, CHO cells were mutagenized with ethyl methanesulfon-

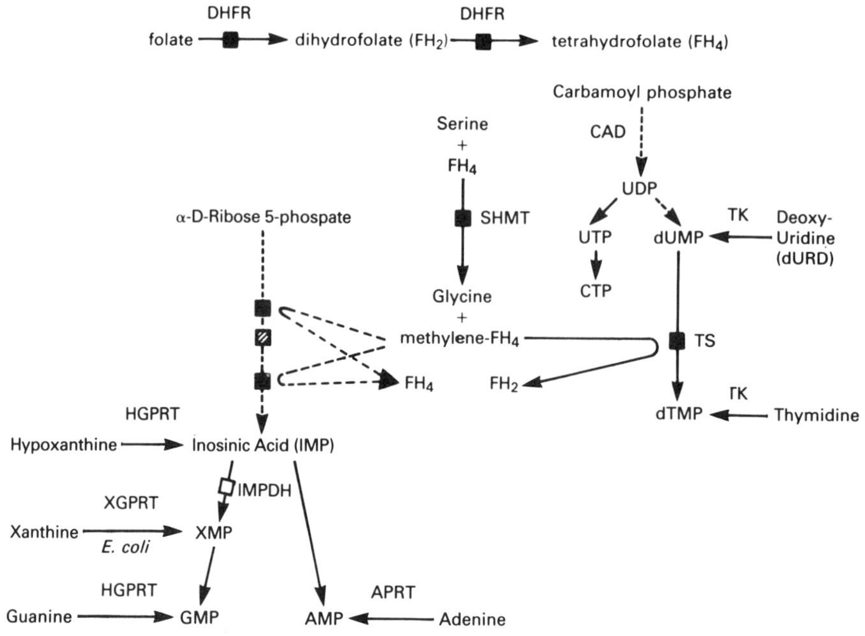

Figure 9.1. Simplified representation of biosynthetic pathways involving available selectable markers for CHO cells. Abbreviations for enzymes: DHFR, dihydrofolate reductase; CAD, carbamoyl-P synthetase, aspartate transcarbamylase, and dihydroorotase (only the latter two activities are indicated in this figure); TK, thymidine kinase; TS, thymidylate synthetase; SHMT, serine hydroxymethyl transferase; HGPRT, hypoxanthine–guanine phosphoribosyl transferase; XGPRT, xanthine–guanine phosphoribosyl transferase; APRT, adenine phosphoribosyl transferase; and IMPDH, inosine monophosphate dehydrogenase. Solid arrows (→) indicate single reactions. Dashed arrows indicate multiple reactions. The solid square (■) indicates reactions that will be inhibited by folate analogs such as aminopterin and methotrexate. The hatched square (▨) indicates the principal reaction inhibited by azaserine. The open square (□) indicates the inhibition of IMPDH by mycophenolic acid.

ate, and mutants deficient in DHFR were selected, in a two-step process, for their ability to grow in the presence of high specific activity [^{3}H]dUrd. [^{3}H]dUrd is toxic to wild-type cells if it is converted to [^{3}H]dTMP via [^{3}H]dUMP, and ultimately incorporated into DNA, owing to the deleterious effects of radioactive decay. Only cells with functional DHFR convert dUMP to dTMP (Fig. 9.1). A single-step selection for *dhfr*$^-$ cells was unsuccessful, presumably because *dhfr* is present in CHO cells in diploid amounts and the frequency of mutagenizing both alleles simultaneously is very low. Therefore, heterozygotes with one defective *dhfr* gene were isolated initially, followed by selection for cells completely deficient in *dhfr*. [^{3}H]dUrd selection for the heterozygote included intermediate amounts of methotrexate (a folate analog), allowing killing of cells with two active *dhfr* genes, but providing enough of the DHFR inhibitor for the survival of cells with reduced DHFR activity. CHO cells that were heterozygous for *dhfr* were then grown on [^{3}H]dUrd without methotrexate, and survivors that grew only when supplied with glycine, hypoxanthine, and thymidine were found to completely lack functional DHFR, as measured by the inability to bind [^{3}H]methotrexate, and loss of catalytic activity.

Using the *dhfr*-deficient CHO cell line described above, Kaufman and Sharp (1982) described the construction of a modular *dhfr* mouse cDNA gene. The cDNA copy of this gene was made by reverse transcribing partially purified DHFR mRNA (Chang et al., 1978). Therefore, it does not contain any of the intervening, nontranslated regions (introns) commonly present in eukaryotic genes. This cDNA was cloned into a portion of the bacterial plasmid pBR322 containing the origin of replication and the gene for tetracycline resistance (Chang et al., 1978). Kaufman and Sharp (1982) examined the promoter, RNA splicing, and polyadenylation signals required to render this cloned *dhfr* cDNA capable of efficient gene expression. As shown in Figure 9.2, various constructs were made that were tested for their ability to express *dhfr*. The regulatory and recognition signals that were used for these constructs were the following: (1) the adenovirus major late promoter, (2) a hybird intron formed by the 5′-donor splicc sitc of thc adcnovirus major latc lcadcr and a 3′-acceptor splice site from a variable-region immunoglobulin gene, (3) the entire SV40 genome, (4) the early SV40 polyadenylation sequence, and (5) the 72-bp repeat known to be a transcriptional enhancer. Figure 9.2 shows that *dhfr* expression is the highest when the SV40 enhancer sequence is inserted upstream from the adenovirus major late promoter, and the hybird splice site and the early SV40 polyadenlyation sequence are included in the construct (see construct 7). The expression is assayed by the frequency with which DHFR$^-$ CHO cells are converted to the DHFR$^+$ phenotype after transfer of the cloned gene into the cells by calcium-phosphate-mediated DNA transfection according to the method of Graham and van der Eb (1973). In the above optimal case, the ratio of DHFR$^+$ transformants to the original number of DHFR$^-$ CHO cells plated was 2×10^{-3} as opposed to $<1 \times 10^{-7}$ for the simple *dhfr* cDNA insertion into pBR322 (see Fig. 9.2, constructs 1 and 7). Analysis of the mRNA produced from constructs 3 and 4 in Figure 9.2 showed that the SV40 polyadenylation site and the hybrid splice site were functional in RNA processing, and that transcription was initiated at the adenovirus major late leader sequence.

Lee et al. (1981) constructed a series of vectors containing the mouse *dhfr* cDNA (Chang et al., 1978) under the control of the mouse mammary tumor virus long terminal repeat (MMTV-LTR). It was known that MMTV transcription can be greatly increased by treating infected cells with a glucocorticoid such as dexamethasone. These authors demonstrated that the MMTV-LTR contains the sequences necessary for both

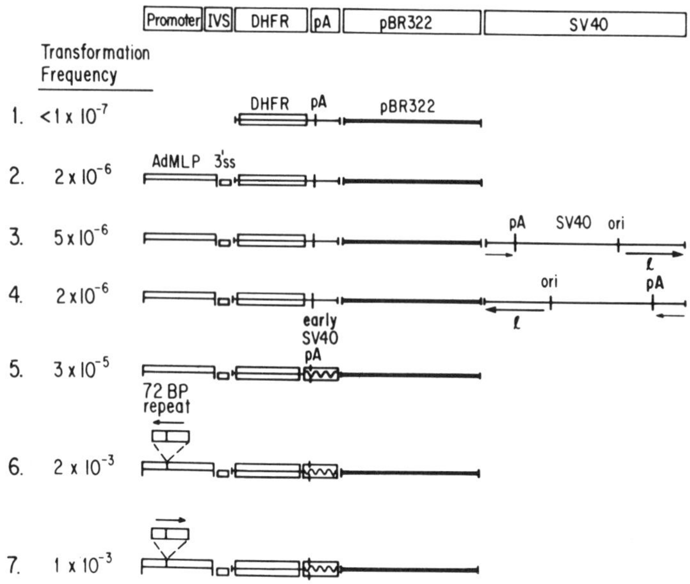

Figure 9.2. $DHFR^-$ to $DHFR^+$ transformation frequency from various cDNA genes. Cells were transfected with plasmid DNA (1 μg DNA/10^6 cells) in the absence of carrier DNA. Transformation frequencies were determined from the number of $DHFR^+$ transformants arising per number of cells plated into selective media. To ensure internal consistency, the results reported here are from a single experiment in which 3×10^6 cells were plated into selective media. All results have been duplicated in other experiments. Recombinants indicated are (1) pDHFR26; (2) pAdD26-1 constructed from the DHFR cDNA clone pDHFR26, the Ad2 major late promoter, and a 3′ splice site isolated from a variable-region immunoglobulin gene; (3) pASD11; (4) pASD12; (5) pAdD26SV(A) No. 1 or 3, which contains an SV40 early polyadenylation (pA) site; (6) pCVSVE and (7) pCVSVL, which contain the 72-bp repeat of SV40. The arrows on the 72-bp repeat segment indicate the direction of late SV40 transcription, and those on the SV40 DNA segment indicate the late transcription unit (from Kaufman and Sharp, 1982; copyright held by American Society for Microbiology).

expression of *dhfr* (promoter) and glucocorticoid regulation when transfected into $DHFR^-$ CHO cells (Urlaub and Chasin, 1980). Thus, they can stimulate DHFR production three- to five-fold with dexamethasone. However, the MMTV promotor is weak, so that the five-fold stimulation represents a significant drop in DHFR production when compared to that of their pSV2dhfr clone which uses the SV40 promotor (see Fig. 9.4).

Overlapping partial DNA fragments of the *dhfr* gene were isolated from CHO cells containing 150 copies of *dhfr*. This gene has been found to be minimally 26 kb and contains six exons and at least five introns (Carothers et al., 1983). The *dhfr* gene from a murine cell line is at least 31 kb (Nunberg et al., 1980; and Crouse et al., 1982). Amplification of the *dhfr* gene to hundreds of copies per genome has been found to occur in both murine and Chinese hamster cells, primarily when they are grown on the folate analog, methotrexate (Alt et al., 1978; Schimke et al., 1978; Milbrant et al., 1981; Flintoff et al., 1982; Johnston et al., 1983). This amplification is associated with chromosomal anomalies such as appearance of homogeneously staining regions (HSR) and double minute chromosomes, and increase in chromosome length (Biedler and Spengler, 1976; Nunberg et al., 1978; Dolnick et al., 1979; Kaufman et al., 1979; Bostock and Clark, 1980; Milbrandt et al., 1981; Flintoff et al., 1982). It is estimated that a 135-kb region, including the *dhfr* gene, is amplified in a methotrexate-resistant CHO cell line; and there is evidence that this region contains an origin of DNA replication that may be involved in the amplification process (Milbrandt et al., 1981; Heintz et al., 1983).

Using one of the above-mentioned MMTV-LTR-controlled *dhfr* clones, pMDSG (Lee et al., 1981), Ringold et al. (1981) showed that cDNA clones of *dhfr* can also be amplified with methotrexate. Similarly, Kaufman and Sharp (1982) have shown that they can amplify their cloned cDNA *dhfr* from several copies to several hundred copies by growing their CHO DHFR$^+$ transformants on methotrexate. In addition to *dhfr*, pMDSG also contained the gene for the *Escherichia coli* xanthine–guanine phosphoribosyl transferase (see that XGPRT section of this chapter) under the control of the SV40 promoter. Upon methotrexate amplification of *dhfr* cDNA, the XGPRT gene is also increased in copy number approximately 50 times. Wigler et al. (1980) showed that amplification of pBR322 sequences could be achieved by ligating pBR322 to a mixture of CHO genomic DNA fragments containing a mutant *dhfr* gene, followed by transfection into mouse cells and growth on methotrexate. Thus, it appears to be possible to increase the expression of other genes of interest by linking them to *dhfr*, and growing transformants on methotrexate.

In addition to yielding amplified wild-type *dhfr* genes, methotrexate selection can also result in mutant *dhfr* with a reduced affinity for methotrexate. Simonsen & Levinson (1983) have constructed a cDNA clone of a mouse DHFR that binds methotrexate with only $^1/_{270}$th the affinity of the wild-type enzyme. DNA sequence analysis revealed a single base-pair change in the coding region of the mutant protein. Expression vectors based on the SV40 early promoter were constructed with both the wild-type and mutant *dhfr* cDNA so that the single base-pair change was the only difference between them. CHO cells deficient for DHFR (Urlaub and Chasin, 1980) were transfected with each of these constructs, and DHFR$^+$ transformants were grown in various levels of methotrexate. The results demonstrated that the mutant DHFR preferentially conferred resistance to methotrexate, which lead the investigators to propose that the mutant DHFR could be used as a dominant selectable marker (i.e., a marker that requires no particular host cell mutation, such as auxotrophy, for selection). Thus, wild-type CHO cells were transfected with the mutant *dhfr* cDNA, followed by selection in 250–500 n*M* methotrexate. Colonies arose at a frequency of $9–27 \times 10^{-5}$ cells plated, with no background colonies from cells transfected with the wild-type *dhfr* clone. The authors state that this represents at least a 100-fold increase over the rate of appearance of methotrexate-resistant colonies due to gene amplification.

B. CAD

CAD is a multifunctional protein that catalyzes the first three steps of UMP biosynthesis (Figure 9.1 and Chapter 11). This three-enzyme complex includes carbamoyl-P synthetase, aspartate transcarbamylase, and dihydroorotase. PALA (*N*-phosphonacetyl-L-aspartate) is an analog inhibitor of the aspartate transcarbamylase and has been used to select for Chinese hamster cells that overproduce CAD (Swyryd et al., 1974; Kempe et al., 1976; Coleman et al., 1977; Padgett et al., 1979; and Wahl et al., 1979). Wahl et al. (1979) showed that a major cause of the overproduction of CAD protein in PALA-resistant Syrian hamster cells is CAD gene amplification. These results were obtained by determining the reassociation kinetics of CAD-specific DNA from $PALA^r$ and $PALA^s$ hamster cells. By comparing the rates of hybridization to a ^{32}P-labeled CAD partial cDNA probe, they determined that the $PALA^r$ cells contained up to 191 times the number of CAD genes present in wild-type hamster cells. The increased number of genes in each case roughly corresponded to increased CAD enzyme levels as measured by ATCase (aspartate transcarbamylase) activity.

de Saint Vincent et al. (1981) cloned the CAD gene from the $PALA^r$ Syrian hamster cells containing approximately 200 CAD genes described above (Wahl et al., 1979). A λ-derived cosmid vector (Collins and Bruning, 1978; Collins and Hohn, 1978), which can be packaged *in vitro* with inserts of 40–45 kb, replicates as plasmid after infection, and confers resistance to tetracycline, was used to construct the library from the $PALA^r$ cells. *E. coli* harboring the full-length CAD gene were identified by a series of hybridizations using ^{32}P-labeled CAD DNA probes. CHO cells auxotrophic for uridine ($UrdA^-$, CAD deficient) were fused with bacterial protoplasts containing the CAD gene, and uridine prototrophs were recovered at an average frequency of 2.8×10^{-3} transformants/cells plated. In this study, ATCase levels were measured, comparing the $UrdA^-$ cell line with that of the CAD transformants which no longer require uridine. The transformants had 2–15 times the ATCase activity; and resistance to PALA was significantly increased in these cases. The transformants also have both Syrian and Chinese hamster CAD gene sequences as assayed by Southern blot hybridization, and the copy number of the cloned Syrian hamster gene is greater than 10 copies per CHO $UrdA^-$ cell. (Note that the Syrian hamster and CHO CAD genes are different enough to be differentiated by restriction analysis, and that the UrdA$^-$ CHO cells contain at least some CAD sequences.)

Since the cloned CAD gene was amplified in the transformants, yielding cells that were PALA^r, these investigators attempted to use CAD as a dominant selectable marker, that is, to introduce this gene into wild-type CHO cells and select for resistance to high levels of PALA. In such experiments it was found that after 20–30 generations in 250 μM PALA the cloned Syrian hamster CAD gene was amplified, indicating that this increased copy number was responsible for PALA resistance. The copy number of the endogenous CAD gene in these transformants was similar to that of wild-type CHO cells. However, it was not clear from this report that the number of PALA^r colonies was significantly above the nontransformed (negative) control background, so that the applicability of this wild-type CAD gene as a dominant selectable marker is in question.

Structural information on the CAD gene derives from the work of Padgett et al. (1982), who have cloned two overlapping fragments comprising the entire Syrian hamster CAD gene from a PALA^r cell line. The entire transcriptional unit for this gene is 25 kb, and it contains approximately 37 intervening sequences, with 7.9 kb of coding sequence (mRNA) and 17 kb of intervening sequence. This was determined by heteroduplex analysis between CAD mRNA and CAD genomic DNA clones, and by probing gel fractionated CAD mRNA with small subclones of the CAD gene. In addition, from the work of de Saint Vincent et al. (1981) it is known that the entire functional gene is no larger than 40 kb.

By *in situ* autoradiography on spread chromosomes, Wahl et al. (1982) found CAD-amplified genes within expanded chromosomes and showed that these regions had a banded pattern after staining with trypsin-Giemsa, in contrast to the homogeneously staining regions associated with amplified *dhfr*. As with *dhfr*, CAD amplification is associated with chromosome elongation, double minute chromosomes, and other chromosomal abnormalities. However, Wahl et al. (1979) state that CAD gene amplification is stable in the absence of PALA selection, whereas this is not generally true of *dhfr* amplification in the absence of methotrexate. From chromosome spreads, they estimate that approximately 500 kb copy of CAD gene is amplified. The work of Wahl et al. (1984) demonstrates that the chromosomal location of CAD genes introduced into CHO cells by $CaPO_4$ coprecipitation can have a significant effect on the propensity of these genes to amplify in the presence of PALA.

There are many analogies between the DHFR and CAD systems. Accordingly, it is presumed that the advantages of using *dhfr* as a genetic marker, as cited in the previous section, will also apply to the CAD system. For example, cotransformation of CAD and nonselectable genes, and amplification of genes linked to CAD appear to be feasible approaches. It also seems likely that an appropriately engineered PALA^r mutant CAD gene could be used as a dominant selectable marker, as demonstrated by Simonsen and Levinson (1983) for *dhfr*.

C. Adenine Phosphoribosyl Transferase

Adenine phosphoribosyl transferase (APRT) catalyses the conversion of adenine to adenosine monophosphate (AMP), as illustrated in Figure 9.1. $APRT^-$ cells will not grow in the presence of azaserine, which blocks *de novo* purine biosynthesis, since they are deficient in the *aprt* purine salvage pathway. $APRT^+$ cells will grow in the presence of azaserine if they are provided with the adenine substrate enabling them to use the *aprt* purine salvage pathway. The details of this selection system can be found in Chapter 12.

Wigler et al. (1979a) used this scheme to select for mouse Ltk^- $aprt^-$ cells that had been converted to the $APRT^+$ phenotype by DNA-mediated transfer of the *aprt* locus from wild-type CHO cells. Using the calcium phosphate coprecipitation transfection procedure (Grahm and van der Eb, 1973), they transformed the $aprt^-$ cells with total CHO high-molecular-weight DNA, and selected for $APRT^+$ cells with azaserine and adenine. The frequency of this transformation was approximately 6×10^{-6} colonies/cells plated, with a background reversion frequency to $APRT^+$ of 3×10^{-8} for the Ltk^- $aprt^-$ cells.

Lowy et al. (1980) isolated the hamster *aprt* gene from CHO total genomic DNA through a series of cloning and transformation steps, depicted in Figure 9.3. First, they digested CHO DNA to completion with Hind III, and ligated these fragments into the unique *Hind*III restriction site in pBR322. This ligation mix was used to transfect mouse *Ltk*– *aprt*– cells as described above (Wigler et al., 1979a). Genomic DNA from these primary transformants was used to again transform Ltk^- $aprt^-$ cells so that the amount of nonessential CHO::pBR322 DNA, present in the cells as a result of the first transfection, would be significantly reduced. Genomic DNA from these secondary transformants was then partially digested with *Eco*RI and cloned into a λ phage vector, Charon 4A. This vector accepts approximately 20-kb inserts, and yields phage plaques upon infection in *E. coli* (Maniatis et al., 1978; de Wet et al., 1980). The recombinant phage were screened with a radioactive pBR322 probe, and a single hybridizing clone, λHaprt-1, was identified from a total of 6×10^5 plaques.

To prove that λHaprt-1 contained functional *aprt*, DNA was isolated and used to transfect $aprt^-$ L cells. One microgram of this DNA gave a transfection frequency of 4.5×10^{-4} colonies/cells plated; and 2×10^9 λHaprt-1 phage particles, equivalent to 0.1 μg of DNA, yielded a frequency of 1.8×10^{-4}. In these cases, Ltk^- $aprt^-$ genomic carrier DNA was added to make a total of 20 μg DNA. Twenty

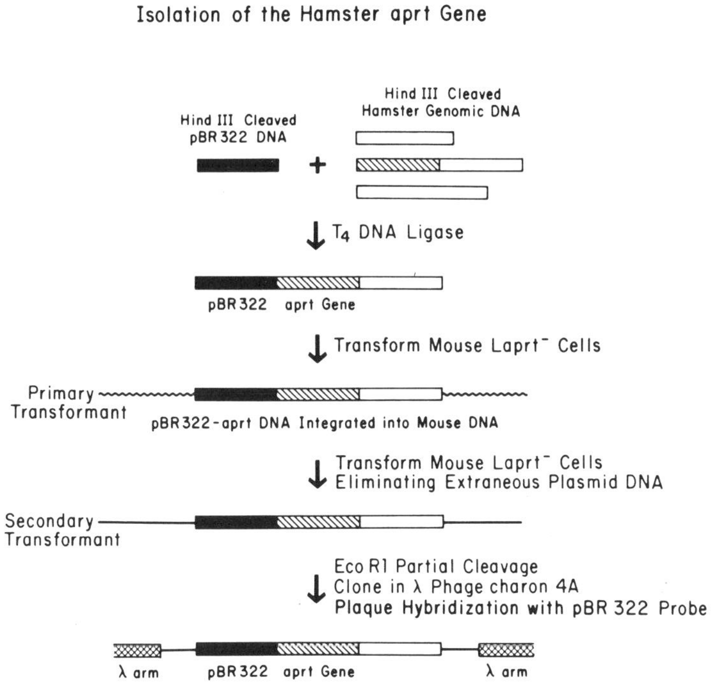

Figure 9.3. Experimental scheme for the isolation of the hamster *aprt* gene. See text for explanation [from Lowy et al. (1980); copyright held by MIT].

micrograms of wild-type CHO genomic DNA resulted in a transformation frequency of 4.1×10^{-6}, demonstrating that under the conditions used, cloning improved the transfection frequency by approximately two orders of magnitude. APRT$^+$ transformants were shown to express the CHO-specific *aprt* gene by protein isoelectric focusing, followed by enzyme assay *in situ*.

Further subcloning of the CHO *aprt* gene into pBR322 and restriction endonuclease gene inactivation experiments located functional *aprt* to a 4.3-kb *Hind*III/*Bgl*II DNA fragment, and this fragment was found in all APRT$^+$ transformants by Southern analysis. Threfore, the maximum size for the complete CHO *aprt* gene is 4.3 kb.

Sikela et al. (1983) used a fragment from the above CHO *aprt* clone (Lowy et al., 1980) as a radioactive probe to screen for *aprt* in a mouse genomic DNA library constructed in the λ phage vector, Charon 4A. Hybridizing clones were used to transform *aprt*$^-$ CHO cells. Using 50 ng of transforming DNA plus 20 μg of carrier DNA, a transformation frequency of approximately 1×10^{-5} colonies/cells plated was observed. After starch gel electrophoresis, an *in situ* assay for APRT enzyme activity showed that the transformants were producing mouse APRT.

From restriction endonuclease gene inactivation studies, it was found that the mouse *aprt* sequence required for transformation was at most 3.1 kb. Comparison of an *aprt* cDNA from a mouse liver cDNA library with the 3.1-kb genomic *aprt* clone demonstrated that the intact gene contains at least three intervening sequences.

D. Thymidine Kinase

Thymidine kinase (TK) performs two enzymatic functions, as shown in Figure 9.1. It converts deoxyuridine (dUrd) to deoxyuridine monophosphate (dUMP), and thymidine (dThy) to thymidine monophosophate (dTMP). Cells that are deficient for TK cannot grow in media containing hypoxanthine, aminopterin, and thymidine [HAT media, Wigler et al. (1970)]. The folate analog, aminopterin, blocks the thymidylate-synthetase-catalyzed synthesis of dTMP from dUMP, and the tk^- cells cannot utilize thymidine (see Fig. 9.1). Aminopterin also inhibits *de novo* purine biosynthesis, but addition of hypoxanthine allows AMP and GMP production through the HGPRT purine salvage pathway. In the presence of functional TK, cells can grow in HAT media because dTMP is produced from thymidine.

Much work has been done on the transfer of the *tk* gene from herpes simplex virus 1 (HSV-1) into mouse Ltk^- cells (Wigler et al., 1977; Pellicer et al., 1978). Using Ltk^- cells, Perucho et al. (1980) isolated the chicken *tk* gene by a plasmid rescue technique, as discussed in a later section of this chapter. The gene for the HSV-1 thymidine kinase has been cloned in pBR322 (Enquist et al., 1979). This plasmid, pX1, has been used to transform tk^- CHO cells to the TK^+ phenotype using the DNA–calcium-phosphate coprecipitation method (Graham and van der Eb, 1973; Abraham et al., 1982; Linsley and Siminovitch, 1982; Nairn et al., 1982). In such gene transfer experiments, Abraham et al. (1982) and Linsley and Siminovitch (1982) report that with 10 μg of pX1, TK^+ colonies occurred at a frequency of about 5×10^{-5} colonies/cells plated, upon selection in HAT medium. Linsley and Siminovitch (1982) also report the frequency of transformation of tk^- CHO cells, as measured by incorporation of [^{3}H]thymidine, to be 60-fold higher than with HAT selection. However, this figure reflects transient, not stable, gene expression. Nairn et al. (1982) showed that there is clonal variation in the transformation frequency of different sublines of tk^- CHO cells, using pX1. For one particular subline they report a frequency of approximately 3×10^{-4} colonies/cells plated, while other sublines were 10 times lower in frequency. In general, the stability of the *tk* marker, upon removal from HAT selection, is high for the TK^+ CHO transformants (Abraham et al., 1982; Nairn et al., 1982).

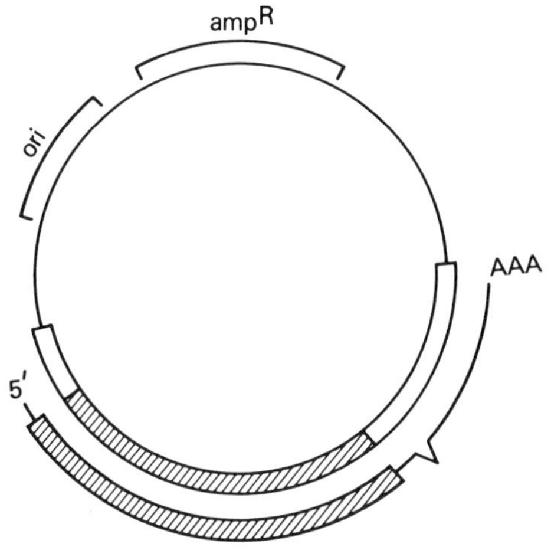

Figure 9.4. Schematic representation of pSV2. The regions marked ori and amp^R are from pBR322. The open segments indicate the position of the SV40 early promoter, small t intron, and poly(A) addition site. The cross-hatched segment indicates the position of inserts coding for one of the following: *E. coli* xanthine–guanine phosphoribosyl transferase (pSV2gpt), Tn*5* amino glycoside phosphotransferase (pSV2neo), *E. coli* galactokinase (pSVK), Tn*9* chloramphenicol acetyltransferase (pSV2cat), and *E. coli* β-galactosidase. The mRNA transcript for these genes is indicated outside the plasmid circle and shows the small t intron splice and poly(A) tail.

E. Xanthine and Hypoxanthine–Guanine Phosphoribosyl Transferases

Functioning in an analogous manner to APRT, hypoxanthine–guanine phosphoribosyl transferase (HGPRT) is a constituent of the purine salvage pathway. As shown in Figure 9.1, in eukaryotes, HGPRT converts hypoxanthine to inosinic acid (IMP), and guanine to guanosine monophosphate (GMP). The *E. coli* enzyme, xanthine–guanine phosphoribosyl transferase (XGPRT), converts xanthine to xanthine monophosphate (XMP), a reaction that is poorly carried out by eukaryotic HGPRT.

Mulligan and Berg (1980, 1981) cloned the *E. coli xgprt* using an SV40/pBR322 recombinant vector, pSV2. As shown in Figure 9.4, this vector contains the SV40 early promoter and origin of replication, small t intron, and polyadenylation site, and pBR322 sequences including those encoding the β-lactamase (Ampicillin) and the origin of replication. The *xgprt* gene is under the control of the SV40 promoter. Mulligan and Berg (1980, 1981) demonstrated that pSV2gpt could complement $hgprt^-$ human cells, and could be used as a dominant selectable marker in animal cells containing wild-type HGPRT, under the appropriate selective conditions. Modifications of the HAT media (hypoxanthine, aminopterin, and thymidine) used for thymidine kinase (*tk*) selection (see *tk* section) are required. With Figure 9.1 as a guide, conditions permitting the use of *xgprt* as a dominant selectable marker can be understood. Aminopterin, a folate analog, blocks all reactions requiring tetrahydrofolate, including *de novo* purine biosynthesis, and glycine and thymidine production. Therefore, glycine and thymidine must be provided in addition to the substrates for the purine salvage pathway, when aminopterin is used. In the presence of active HGPRT, hypoxanthine is eventually converted to GMP and AMP. However, in order to use *xgprt* as a dominant selectable marker, the conversion of IMP to XMP by the host cell IMP dehydrogenase must be blocked by mycophenolic acid, so the only GMP produced is via xanthine conversion to XMP. In some instances, serum that has been dialyzed must be used in the growth media to remove low-molecular-weight precursors such as guanine, which would interfere with the selection. In summary, the components required for *xgprt* as a dominant selectable marker are hypoxanthine, glycine, thymidine, xanthine, aminopterin, and mycophenolic acid (XMHAT media). Abraham et al. (1982) used pSV2gpt to transform wild-type and UV-sensitive CHO cells and obtained frequencies of approximately 1×10^{-4} colonies/cells plated, using 1 μg of plasmid and 30 μg of carrier DNA.

Breitman et al. (1982) used *hgprt*$^-$ CHO cells to develop a shuttle vector system whereby they could introduce and then recover pSV2gpt from CHO cells. Transformation was by the method of Graham and van der Eb (1973), and selection was in the XMHAT medium described above. By Southern analysis it was determined that pSV2gpt DNA was integrated and intact in all transformants, and that no free plasmid existed in the CHO cells. These CHO GPT$^+$ transformants were then fused with COS 1 (Gluzman, 1981) cells by treatment with polyethylene glycol (PEG). COS 1 is a line of monkey cells that constitutively produces the SV40 T antigen, so it was presumed that cell fusion would induce excision and allow for free replication of pSV2gpt. In such experiments, 0.1 – 2 ng of pSV2gpt plasmid DNA was rescued per 10 cm culture dish, and plasmid was reestablished in *E. coli* by transformation and selection with Ampicillin. By restriction analysis it was found that most plasmids did not undergo rearrangement, and retransformation of these plasmids into a purE$^-$ (*de novo* purine biosynthesis) gpt$^-$ strain of *E. coli* yielded Gpt$^+$ colonies in all cases.

F. Galactokinase

Schumperli et al. (1982) cloned the *E. coli* galactokinase (*galK*) gene under the control of the SV40 early promoter in the pSV2 vector described in the XGPRT section. This construct, pSVK, was transfected into both wild-type and *galK*$^-$ CHO cells. Transient *galK* expression was measured by enzyme assay *in situ*, after starch gel electrophoretic separation of the CHO and *E. coli* galactokinases. Results showed that both wild-type and *galK*$^-$ CHO cells were capable of transient *E. coli galK* expression. In addition, it was shown that pSVK transfected *galK*$^-$ CHO cells could utilize galactose. In this case, the cells were selected in glucose-free galactose-containing medium, and colonies appeared at a frequency of 5×10^{-4} colonies/cells plated (10 μg pSVK). It is important to note that cotransformation of NIH3T3 cells with pSVK and pSV2gpt (see section on XGPRT), followed by selection for the *gpt* marker, results in stable expression of *E. coli galK*. Such enhancement of stable transformation of an unselected marker by cotransfection with a dominant selectable marker can be equally applied to CHO cells.

G. Tn5 Aminoglycoside Phosphotransferase

Tn5 aminoglycoside phosphotransferase (*neo*) confers resistance to neomycin, kanamycin, and related compounds, in bacterial cells. A compound similar to neomycin, G418, blocks protein synthesis in eukaryotic cells, and has been used to select for eukaryotic cells expressing the bacterial aminoglycoside phosphotransferase gene (Jimenex and Davies, 1980; Colbere-Garapin et al., 1981; Southern and Berg, 1982). Southern and Berg (1982) cloned the *neo* gene from Tn5 into the pSV2 vector, under the control of the SV40 early promoter (see Fig. 9.4). They have also constructed pSV2neo-SVgpt, a vector containing both dominant selectable markers. Gorman et al. (1982a, 1982b, 1983) have transfected wild-type CHO cells with pSV2neo and obtained a stable transformation frequency of 1×10^{-3} colonies/cells plated.

III. OPTIMIZATION OF DNA-MEDIATED GENE TRANSFER

Although the previous discussion documents that eukaryotic vectors and previously cloned genes are readily introduced into Chinese hamster cells, there remain many interesting gene transfer experiments that are inaccessible using present techniques. Currently attainable transfection efficiencies still limit direct detection of single copy genes, especially where no strong selection is available. Similarly, detection of specific cDNA clones in eukaryotic expression vector libraries remains very difficult. Given this situation, there is a continuing need to evaluate and optimize new methods for DNA-mediated gene transfer.

One approach to evaluate gene transfer methods is to measure transient expression of specialized markers carried on eukaryotic vectors. Vectors carrying the *E. coli* chloramphenicol acetyltransferase (CAT) gene, for example, provide a very easily assayed marker function (Gorman et al., 1982a,b; Laimins et al., 1982; Rosenthal et al., 1983; Walker et al., 1983). CAT activity has not been detected in mammalian cells; therefore, expression of exogenously introduced vector DNA may be measured unambiguously. The absence of interference by endogenous enzyme activities greatly increases the convenience and sensitivity of the CAT assay. Other genes for determining transient vector expression have been described and may be used in place of, or in addition to, CAT vectors (e.g., as internal controls). This group includes vectors that code for SV40 T antigen (Benoist and Chambon, 1981), *E. coli* guanine xanthine phosphoribosyltransferase (Mulligan and Berg, 1980, 1981), *E. coli* galactokinase (Schumperli et al., 1982), and *E. coli* beta-galactosidase (Hall et al., 1983).

Information on transient vector expression obtained using the CAT or alternative systems has, in some instances, proved useful in predicting stable transfection frequencies. In CHO cells two- to three-fold higher CAT expression levels were observed when the SV40 early-promoter-directed transcription of the CAT gene than when the same gene was placed under control of the 3′ long terminal repeat of Rous sarcoma virus (Gorman et al., 1982b). This difference correlated with several-fold higher *neo* or *gpt* stable transformation efficiencies when these selectable markers were under transcriptional control of the SV40 early promoter rather than the Rous long terminal repeat. Conversely, in a number of primate cell types both CAT expression levels and neo stable transformation efficiencies were 5- to 10-fold higher under control of the Rous long terminal repeat.

Gene transfer efficiencies depend not only on methods for introduction of DNA, but also on transfection characteristics of recipient cell types (Abraham et al., 1982; Nairn et al., 1982; Gorman et al., 1983). Thus, in principle, it should be possible to select variant cell lines exhibiting increased competence for DNA uptake and/or expression. Experiments using mouse L tk^- cells and the herpes thymidine kinase gene as a selectable marker indeed suggested that DNA-mediated gene transfer could be used to enrich for such variants (Corsaro and Pearson, 1982a,b). In those studies, however, it was necessary to isolate tk^- revertants prior to rechecking transformation frequencies with the herpes tk^- marker. Perhaps owing to this requirement the phenotype of increased transfection competence appeared to be relatively unstable.

Finally, it is important that reports concerning optimum DNA transfection parameters (Wigler et al., 1977, 1979b; Pellicer, et al., 1978; Parker and Stark, 1979; Lewis et al., 1980; Abraham et al., 1982; Corsaro and Pearson, 1982a,b; Gorman et al., 1982a,b; Shen et al., 1982) be kept in proper perspective. Although these reports are valuable, they should be interpreted cautiously. Generalization of data obtained, for example, using the herpes tk marker with mouse L tk^- cells to transfection of Chinese hamster cells with other vectors is not always justified. For this reason, in particular, eukaryotic vectors will continue to be valuable for investigators attempting to improve DNA-mediated gene transfer methods or set up new applications.

IV. ANIMAL VIRUSES AS GENE TRANSFER VECTORS

To date, vectors for introduction of DNA into Chinese hamster cells have consisted primarily of *E. coli* cloning vehicles carrying mammalian selectable markers. Emphasis here and in the previous gene transfer chapters reflect this relatively simple approach. In other mammalian cell systems, however, more complete mammalian vectors, possessing origins of replication or facilitated integration mechanisms, have been used. Several examples of such vectors will be briefly described, with special reference to potential application for cloning in Chinese hamster cells.

Papovaviruses, especially simian virus 40 (SV40), have been widely used as vectors for introduction and propagation of recombinant genomes in mammalian cells (Gluzman, 1982). In permissive cell types, for example monkey kidney CV-1 cells in the case of SV40 recombinants, these vectors permit isolation of high titer viral lysates. Such lysates may be used productively to infect virtually all cells in a recipient culture. Where replication in the infected cells ensues, the amplified template copy numbers further contribute to high levels of vector-encoded mRNA and polypeptides.

Although high expression levels make SV40 viral vector systems attractive for some studies, multiple disadvantages of this approach have severely limited its use. First, the range of appropriately permissive cell types is narrow; in particular, Chinese hamster cells are only semipermissive for SV40 and for polyoma virus, another potential lytic papovavirus vector (Topp et al., 1980). Second, lysis of the host cell occurs after about 3 days, limiting studies on gene regulation to transient (i.e., 24–72 hr) expression. Third, requirements for viral packaging and obligatory complementation with a helper virus restrict the size of insert DNA fragments to no more than 2.5-kb pairs (Goff and Berg, 1976). Last, and least widely appreciated, deletions and rearrangements of insert DNA frequently occur as recombinant viral stocks are propagated.

An interesting nonlytic papovavirus vector is the bovine papilloma virus (BPV) replicon. This genome does not generate infectious viral stocks in tissue culture, but has been demonstrated to persist as a stable plasmid (copy number 20-200) in mouse NIH/3T3 and C127 cells (Law et al., 1981; Sarver et al., 1981). Stably transfected cells are detected as a result of BPV-induced formation of transformed foci on confluent monolayers of these cell types. Unfortunately, CHO and numerous other widely used Chinese hamster lines do not form flat monolayers suitable for the focus forming assay. A number of laboratories are thus interested in incorporating *neo* or another dominant selectable marker into BPV vectors (Law et al., 1983). The incorporation of such markers should enable determination of whether BPV recombinants can persist as stable, nontransforming plasmid structures in a wider range of cell lines. Insertion of foreign sequences into BPV vectors occasionally predisposes to structural rearrangement or insertion into host high-molecular-weight DNA (DiMaio et al., 1982; Law et al., 1982).

At present, retrovirus vectors seem to offer the most promising animal virus system for very high efficiency gene transfer in Chinese hamster cells. Several investigators have demonstrated that selectable markers may be inserted into cloned, modified retrovirus genomes, and that, following DNA-mediated cotransfection with a cloned helper virus into permissive cells, high titer stocks of recombinant retroviruses can be obtained (Shimotohno and Temin, 1981; Wei et al., 1981; Tabin et al., 1982; Joyner and Bernstein, 1983a,b; Mann et al., 1983; Miller et al., 1983). The host cell range of retroviruses is determined largely by the composition of the viral envelope protein. Hybrid or "pseudotype" retroviruses that are composed of viral genomic RNA packaged in envelope glycoproteins from a second coinfecting helper virus generally exhibit a host range characteristic of the latter virus. Accordingly, murine sarcoma virus (MSV) can be rendered competent to infect normally nonpermissive hamster cells by coinfection of mouse cells with an endogenous hamster retrovirus (Teich, 1982). Retrovirus vectors pseudotyped by coinfection with broad host range "amphotropic" retrovirus helpers should similarly exhibit infectivity for Chinese hamster cells.

In common with other mammalian viral systems, retrovirus vectors manifest a number of limitations imposed by viral replication and packaging mechanisms. The occurrence of polyadenylation signals in a DNA fragment can dramatically impair propagation of the retrovirus vector into which it has been inserted (Shimotohno and Temin, 1981). It is also likely that the occurrence of strong promoters or RNA splicing signals in an insert fragment will interfere with retrovirus function. The maximum size fragment that can be inserted into retrovirus vectors is probably about 7-kb pairs, but will vary depending on the specific vector system utilized. Inserts larger than several kilobase pairs may be prone to deletion and rearrangement if viral stocks are passaged.

V. VECTORS FOR GENE SHUTTLING

Whereas use of vectors for DNA-mediated transfection has become relatively commonplace, there are relatively few reports describing techniques for efficient "gene shuttling" between *E. coli* and mammalian cells. Two alternative "eukaryotic vector-independent" gene transfer methods for isolating single copy sequences are more frequently employed: (1) primary and secondary transfection with total genomic DNA, followed by screening bacteriophage lambda libraries for highly repetitive sequences carried on donor DNA; (2) sib selection (viz., purification of specific gene sequences from a genomic library by successive transfection cycles with pooled library subfractions). Gene shuttling, nonetheless, has the potential to increase in importance as more investigators attempt to isolate genes for which no cDNA probe exists. With this potential in mind, a discussion of current and potential approaches seems appropriate.

The first report of molecular cloning by gene shuttling was that of Perucho et al. (1980), who used "plasmid rescue" to isolate the chicken thymidine kinase gene. Chicken genomic DNA restriction fragments ligated to linearized pBR322 molecules were introduced into mouse L tk^- cells by $CaPO_4$–DNA coprecipitation. Primary and secondary clones carrying pBR322 and chick tk sequences colinearly integrated into high-molecular-weight cellular DNA were then selected in HAT medium. To recover the chick *tk* gene, total genomic DNA from a tk^+ transformant was cleaved with a restriction endonuclease known from previous experiments to have a single cutting site in pBR322 but no sites in the chicken *tk* gene; the resulting fragments were recircularized with T4 ligase and introduced by $CaCl_2$ transfection into *E. coli.* The two pBR322-chick *tk* recombinants isolated demonstrated the feasibility of this approach. Nevertheless, the frequency of recovering such recombinants was much lower than predicted from "reconstruction" experiments (presumably carried out by mixing appropriate amounts of purified plasmid DNA with high-molecular-weight mammalian genomic DNA).

The plasmid rescue method has the advantage that it is applicable with virtually all mammalian cell types, including, of course, Chinese hamster cells. It also seems possible, in retrospect, that the low efficiency observed by Perucho et al. was due in part to the presence of "poison" sequences present in the pBR322 vector that they employed. These poison sequences have been shown to reduce the transfection competence of a recombinant genome propagated in mammalian cells by up to 100-fold (Lusky and Botchan, 1981). On the other hand, the plasmid rescue approach is likely to remain limited in the future by the requirement for using a restriction enzyme that cleaves neither the insert nor the vector. For most larger DNA segments, an appropriate restriction enzyme will be difficult to identify or not available.

An approach which attempts to circumvent the requirement for excision by a specific restriction endonuclease is based on *in situ* replication of recombinants followed by homologous or illegitimate recombination. In this approach replication of an integrated SV40- or polyoma-containing recombinant is generally achieved by fusion of nonpermissive or semipermissive transformants, for example, mouse fibroblasts or Chinese hamster cells, with permissive cells (Breitman et al., 1982). Where the recombinant vector moiety carries a polyoma or SV40 origin of replication, replication may be achieved by fusion with permissive helper cells that carry origin-defective viral early transcription units coding for the appropriate large T antigen (Gluzman, 1981; Binetruy et al., 1982). It is postulated that, following *in situ* replication, recombination between flanking sequences occurs to yield circular molecules suitable for transfection directly into *E. coli.* Tandem integration of recombinant molecules or integration by a retrovirus mechanism (generating long terminal repeats) may favor homologous recombination between sequences flanking an insert DNA segment of interest. It remains uncertain whether integrated recombinants carrying relatively large inserts, for example, 30–50-kb pairs, can be recovered intact by this method.

At present the only published approach for shuttling recombinants carrying larger genomic DNA inserts is based on *in vitro* bacteriophage lambda packaging. Cosmid cloning vehicles contain a cos cleavage site for packaging and carry exogenous DNA fragments of up to 45-kb pairs (Collins and Hohn, 1978). Following $CaPO_4$ transfection of mammalian cells, cosmid recombinants may be integrated into host high-molecular-weight chromosomal DNA as tandem multimers; this tandem repeat arrangement places cos sites at an appropriate distance (35–50-kb pairs), which in turn permits *in vitro* packaging of an intact cosmid recombinant from total genomic DNA. Cosmid shuttling is generally applicable and several groups have used it successfully (Lindemaier et al., 1982; Lau and Kan, 1983).

ACKNOWLEDGMENTS

We thank Ray Steinberg for the photography, and Raji Padmanabhan for many helpful discussions. We are also grateful to Dr. Randal J. Kaufman and American Society for Microbiology Press, and Dr. Israel Lowy and M.I.T. Press for the use of previously published figures. We especially thank Dr. Michael Gottesman for his patience, enthusiasm, and well-timed threats.

REFERENCES

Abraham, I., Tyagi, J. S., and Gottesman, M. M. (1982). *Somat. Cell Genet.* **8**, 23–29.

Alt, F. W., Kellems, R. E., Bertino, J. R., and Schimke, R. T. (1978). *J. Biol. Chem.* **253**, 1357–1370.

Benoist, C. and Chambon, P. (1981). *Nature* **290**, 304–310.

Biedler, J. L. and Spengler, B. A. (1976). *Science* **191**, 185–187.

Binetruy, B., Rautmann, G., Meneguzzi, G., Breathnach, R., and Cuzin, F. (1982). In *Eukaryotic Viral Vectors* (Y. Gluzman, ed.), Cold Spring Harbor Laboratory, New York, pp. 87–92.

Bostock, C. J. and Clark, E. M. (1980). *Cell* **19**, 709–715.

Breitman, M. L., Tsui, L. C., Buchwald, M., and Siminovitch, L. (1982). *Mol. Cell. Biol.* **2**, 966–976.

Carothers, A. M., Urlaub, G., Ellis, N., and Chasin, L. A. (1983). *Nucl. Acids Res.* **11**, 1997–2012.

Chang, A. C. Y., Nunberg, J. H., Kaufman, R. J., Erlich, H. A., Schimke, R. T. and Cohen, S. N. (1978). *Nature* **275**, 617–624.

Coleman, P. F., Suttle, D. P., and Stark, G. R. (1977). *J. Biol. Chem.* **252**, 6379–6385.

Colbere-Garapin, F., Horodniceanu, F., Kourilsky, P., and Garapin, A. C. (1981). *J. Mol. Biol.* **150**, 1–14.

Collins, J. and Bruning, H. H. (1978). *Gene* **4**, 85–107.

Collins, J. and Hohn, B. (1978). *Proc. Natl. Acad. Sci. USA* **75**, 4242–4246.

Corsaro, C. M. and Pearson, M. L. (1982*a*). *Somat. Cell Genet.* **7**, 603–616.

Corsaro, C. M. and Pearson, M. L. (1982*b*). *Somat. Cell Genet.* **7**, 617–630.

Crouse, G. F., Simonsen, C. C., McEwan, R. N., and Schimke, R. T. (1982). *J. Biol. Chem.* **257**, 7887–7897.

de Saint Vincent, B. R., Delbruck, S., Eckhart, W., Meinkoth, J., Vitto, L. and Wahl, G. (1981). *Cell* **27**, 267–277.

de Wet, J. R., Daniels, D. L., Schroeder, J. L., Williams, B. G., Thompson, K. D., Moore, D. D., and Blattner, F. R. (1980). *J. Virol.* **33**, 401–410.

DiMaio, D., Treisman, R., and Maniatis, T. (1982). *Proc. Natl. Acad. Sci. USA* **79**, 4030–4034.

Dolnick, B. J., Berenson, R. J., Bertino, J. R., Kaufman, R.J., Nunberg, J. H., and Schimke, R. T. (1979). *J. Cell Biol.* **83**, 394–402.

Enquist, L. W., Vande Woude, G. F., Wagner, M., Smiley, J. R., and Summers, W. C. (1979). *Gene* **7**, 335–342.

Flintoff, W. F., Weber, M. K., Nagainis, C. R., Essani, A. K., Robertson, D., and Salser, W. (1982). *Mol. Cell. Biol.* **2**, 275–285.

Gluzman, Y. (1981). *Cell* **23**, 175–182.

Gluzman, Y. (1982). *Eukaryotic Viral Vectors*, Cold Spring Harbor Laboratory, New York, pp. 1–5.

Goff, S. and Berg, P. (1976). *Cell* **9**, 695–705.

Gorman, C. M., Moffat, L. F., and Howard, B. H. (1982a). *Mol. Cell. Biol. 2,*, 1044–1051.

Gorman, C. M., Merlino, G. T., Willingham, M. C., Pastan, I., and Howard, B. H. (1982b). *Proc. Natl. Acad. Sci. USA* **79**, 6777–6781.

Gorman, C., Padmanabhan, R., and Howard, B. H. (1983). *Science* **221**, 551–553.

Graham, F. L. and van der Eb, A. J. (1973). *Virology* **52**, 456–457.

Hall, C., Jacob, E., Ringold, G., and Lee, F. (1983). *J. Mol. Appl. Gen.* **2**, 101–109.

Heintz, N. H., Milbrandt, J. D., Greisen, K. S., and Hamlin, J. L. (1983). *Nature* **302**, 439–441.

Jimenez, A. and Davies, J. (1980). *Nature* **287**, 869–871.

Johnston, R. N., Beverly, S. M., and Schimke, R. T. (1983). *Proc. Natl. Acad. Sci. USA* **80**, 3711–3715.

Joyner, A. L. and Bernstein, A. (1983a). *Mol. Cell. Biol.* **3**, 2180–2190.

Joyner, A. L. and Bernstein, A. (1983b). *Mol. Cell. Biol.* **3**, 2191–2202.

Kaufman, R. J., Brown, P. C., and Schimke, R. J. (1979). *Proc. Natl. Acad. Sci. USA* **76**, 5669–5673.

Kaufman, R. J. and Sharp, P. A. (1982). *Mol. Cell Biol.* **2**, 1304–1319.

Kempe, T. D., Swyryd, E. A., Bruist, M., and Stark, G. R. (1976). *Cell* **9**, 541–550.

Laimins, L. A., Khoury, G., Gorman, C., Howard, B., and Gruss, P. (1982). *Proc. Natl. Acad. Sci. USA* **79**, 6453–6457.

Lau, Y. F. and Kan, Y. W. (1983). *Proc. Natl. Acad. Sci. USA* **80**, 5225–5229.

Law, M. F., Byrne, J. C., and Howley, P. M. (1983). *Mol. Cell. Biol.* **3**, 2110–2115.

Law, M. F., Lowry, D. R., Dvoretzky, K. R., and Howley, P. M. (1981). *Proc. Natl. Acad. Sci. USA* **78**, 2727–2731.

Law, M. F., Howard, B., Sarver, N., and Howley, P. M. (1982). In *Eukaryotic Viral Vectors* (Y. Gluzman, ed.), Cold Spring Harbor Laboratory, New York, pp. 79–85.

Lee, F., Mulligan, R., Berg, P., and Ringold, G. (1981). *Nature* **294**, 228–232.

Lewis, W. H., Srinivasan, P. R., Stokoe, N., and Siminovitch, L. (1980). *Somat. Cell Genet.* **6**, 333–347.

Lindemaier, W., Hauser, H., deWilke, I.G., and Schutz, G. (1982). *Nucl. Acid Res.* **10**, 1243–1256.

Linsley, P. S. and Siminovitch, L. (1982). *Mol. Cell. Biol.* **2**, 593–597.

Lowy, I., Pellicer, A., Jackson, J. R., Sim, G. K., Silverstein, S., and Axel, R. (1980). *Cell* **22**, 817–823.

Lusky, M. and Botchan, M. (1981). *Nature* **293**, 79–81.

Maniatis, T., Hardison, R., Lacy, E., Lauer, J., O'Connell, C., Quon, D., Sim, G. K., and Efstratiadis, A. (1978). *Cell* **15**, 687–701.

Mann, R., Mulligan, R. C., and Baltimore, D. (1983). *Cell* **33**, 153–159.

Milbrandt, J. D., Heintz, N. H., White, W. C., Rothman, S. M., and Hamlin, J. L. (1981). *Proc. Natl. Acad. Sci. USA* **78**, 6043–6047.

Miller, A. D., Jolly, D. J., Friedmann, T., and Verma, I. M. (1983). *Proc. Natl. Acad. Sci. USA* **80**, 4709–4713.

Mulligan, R. C. and Berg, P. (1980). *Science* **209**, 1422–1427.

Mulligan, R. C. and Berg, P. (1981). *Proc. Natl. Acad. Sci. USA* **78**, 2072–2076.

Nairn, R. S., Adair, G. M., and Humphrey, R. M. (1982). *Mol. Gen. Genet.* **187**, 384–390.

Nunberg, J. H., Kaufman, R. J., Schimke, R. T., Urlaub, G., and Chasin, L. A. (1978). *Proc. Natl. Acad. Sci. USA* **75** 5553–5556.

Nunberg, J. H., Kaufman, R. J., Chang, A. C. Y., Cohen, S. N., and Schimke, R. T. (1980). *Cell* **19**, 355–364.

Padgett, R. A., Wahl, G. M., Coleman, P. F., and Stark, G. R. (1979). *J. Biol. Chem.* **254**, 974–980.

Padgett, R. A., Wahl, G. M., and Stark, G. R. (1982). *Mol. Cell. Biol.* **2**, 293–301.

Parker, B. and Stark, G. (1979). *J. Virol.* **31**, 360–369.

Pellicer, A., Wigler, M., Axel, R., and Silverstein, S. (1978). *Cell* **14**, 133–141.

Perucho, M., Hanahan, D., Lipsich, L., and Wigler, M. (1980). *Nature* **285**, 207–210.

Ringold, G., Dieckmann, B. and Lee, F. (1981). *J. Mol. Appl. Genet.* **1**, 165–175.

Rosenthal, N., Kress, M., Gruss, P., and Khoury, G. (1983). *Science* **222**, 749–755.

Sarver, N., Gruss, P., Law, M. F., Khoury, G., and Howley, P. (1981). *Mol. Cell. Biol.* **1**, 486–496.

Schimke, R. T., Kaufman, R. J., Alt., F. W., and Kellems, R. F. (1978). *Science* **202**, 1051–1055.

Schumperli, D., Howard, B. H., and Rosenberg, M. (1982). *Proc. Natl. Acad. Sci. USA* **79**, 257–261.

Shen, Y., Hirschhorn, R. R., Mercer, W. E., Surmacz, E., Tsutsui, Y., Soprano, K. J., and Baserga, R. (1982). *Mol. Cell. Biol.* **2**, 1145–1154.

Shimotohno, K. and Temin, H. M. (1981). *Cell* **26**, 67–77.

Sikela, J. M., Khan, S. A., Feliciano, E., Trill, J., Tischfield, J. A., and Stambrook, P. J. (1983). *Gene* **22**, 219–228.

Simonsen, C. C. and Levinson, A. D. (1983). *Proc. Natl. Acad. Sci. USA* **80**, 2495–2499.

Southern, P. G. and Berg, P. (1982). *J. Mol. Appl. Gen.* **1**, 327–341.

Swyryd, E. A., Seaver, S., and Stark, G. R. (1974). *J. Biol. Chem.* **249**, 6945–6950.

Tabin, C. J., Hoffmann, J. W., Goff, S. P., and Weinberg, R. A. (1982). *Mol. Cell. Biol.* **2**, 426–436.

Teich, N. (1982). In *RNA Tumor Viruses* (R. Weiss, N. Teich, H. Varmus, and J. Coffin, eds.), Cold Spring Harbor Laboratory, New York, p. 105.

Topp, W. C., Lane, D., and Pollack, R. (1980). In *DNA Tumor Viruses*, 2nd ed. (J. Tooze, ed.), Cold Spring Harbor Laboratory, New York, pp. 205–296.

Urlaub, G. and Chasin, L. A. (1980). *Proc. Natl. Acad. Sci. USA* **77**, 4216–4220.

Wahl, G. M., de Saint Vincent, B. R., and De Rose, M. L. (1984). *Nature* **307**, 516–520.

Wahl, G. M., Padgett, R. A., and Stark, G. R. (1979). *J. Biol. Chem.* **254**, 8679–8689.

Wahl, G. M., Vitto, L., Padgett, R. A., and Stark, G. R. (1982). *Mol. Cell. Biol.* **2**, 308–319.

Walker, M. D., Edlund, T., Boulet, A. M., and Rutter, W. J. (1983). *Nature* **306**, 557–561.

Wei, C., Gibson, M., Spear, P. G., and Scolnick, E. M. (1981). *J. Virol.* **39**, 935–944.

Wigler, M., Pellicer, A., Silverstein, S., Axel, R., Urlaub, G., and Chasin, L. (1979*a*). *Proc. Natl. Acad. Sci. USA* **76**, 1373–1376.

Wigler, M., Silverstein, S., Lee, L. S., Pellicer, A., Cheng, Y., and Axel, R. (1977). *Cell* **11**, 223–232.

Wigler, M., Perucho, M., Kurtz, D., Dana, S., Pellicer, A., Axel, R. and Silverstein, S. (1980). *Proc. Natl. Acad. Sci. USA* **77**, 3567–3570.

Wigler, M., Sweet, R., Sim, G. K., Wold, B., Pellicer, A., Lacy, E., Maniatis, T., Silverstein, S., and Axel, R. (1979*b*). *Cell* **16**, 777–785.

CHAPTER 10

CLONING AND EXPRESSION OF cDNAs

Paul J. Doherty
Laboratory of Molecular Biology
National Cancer Institute
National Institutes of Health
Bethesda, Maryland

The cloning of DNAs (cDNAs) complementary to messenger RNA (mRNAs) representing specific gene products is discussed in this chapter. The emphasis here is on the synthesis and subsequent isolation of a specific cDNA. Manipulation of cDNAs into cloning vectors for purposes of sequencing or for expression in bacterial or eukaryotic cells (Chapter 9) is discussed here only insofar as these methods are used for positive identification of a particular clone. These latter approaches combine somatic cell genetic analysis with molecular cloning technology.

The basic strategy for any protocol of cloning cDNA requires the ability to synthesize a complementary DNA copy from polyadenylated mRNA; the choice of a vector and host cell; the introduction of the cDNA into the vector; the transformation of the host with the recombinant vector; and, finally, the identification of the host clones carrying the recombinant vector of interest.

I. CHOICE OF BACTERIAL HOST CELL

A number of *Escherichia coli* strains are in common usage for cDNA cloning. Generally *recA*$^-$ hosts are preferred (e.g., HB101, DH1) because of the occasional rearrangement of cDNA inserts by *recA*$^+$ hosts (e.g., C600, N38, X1776). The improved transformation efficiency observed with *recA*$^+$ hosts compared to *recA*$^-$ hosts using $CaCl_2$ treatment (Cohen et al., 1972; Dagert and Erhlich, 1979) is largely overcome by the high-efficiency transformation method devised by Hanahan (Hanahan, 1983).

Some bacterial hosts have been engineered to provide specialized functions that can be used to advantage in cDNA cloning. For example, cloning vectors pUC8 and pUC9 (Viera and Messing, 1982) carry a *lac*Z fragment which will complement a defective β-galactosidase under the control of the F episome in hosts JM101 and JM103 and present on the chromosome of host JM83. On induction of β-galactosidase by IPTG in the presence of the indicator Xgal, transformation by the pUC8 or pUC9 plasmid will result in the appearance of deep blue colonies. Insertion of DNA into the *lac*Z fragment will inactivate this complementation giving rise to white colonies. Consequently, these specialized hosts provide an assay for the successful insertion of DNA into the plasmid (Messing, 1984). Some exceptions to this scenario have been noted (Close et al., 1983).

II: CHOICE OF PLASMID

To be useful for cDNA cloning, the plasmid that acts as a vector in carrying the cDNA into the host cell and allows for the amplification of the inserted sequence should have a number of properties. The plasmid should have unique restriction sites for the introduction of the cDNA into single sites. This allows for the orderly insertion of cDNA sequences and reclosing of the vector. In addition, the plasmid requires selectable markers so that only host cells carrying the plasmid are able to grow after transformation. A number of plasmid vehicles are commonly used which provide any of a variety of selections. PBR322, the most commonly used vector to date, provides the host cell with resistance to the antibiotics ampicillin and tetracycline (Bolivar et al., 1977; Sutcliffe, 1978,1979). pMK16 (Kahn et al., 1979) renders the cell resistant to kanamycin, tetracyline, and the colicin E1. pACYC 184 has selectable markers for tetracyline and chloramphenicol resistance (Chang and Cohen, 1978). For more detailed discussion of the array of phenotypes produced by the large number of vectors available, refer to Kahn et al. (1979) and Maniatis et al. (1982). It is most useful if one of the unique restriction sites for insertion of cDNA inactivates one of the selectable markers. For example, insertion into the Pstl site of pBR322 inactivates the β-lactamase gene responsible for ampicillin resistance. Consequently, resistance to tetracycline and sensitivity to ampicillin acts as a screen for pBR322 plasmids carrying cDNA inserts. This gives an indication of the background of plasmids without inserted cDNA. In general, the mode of insertion of cDNA into the vector is one factor that dictates which plasmid to choose. If cDNA is to be introduced using EcoRI linkers (refer to discussion of linkers), then it may be useful to use pACYC 184. Insertion into the EcoRI site inactivates chloramphenicol resistance leaving the host cell resistant to tetracycline. Direct positive selections for cDNA insertions into the Tet site of pBR322 have also been developed (Bochner et al., 1980; Maloy and Nunn, 1981). Bochner et al. (1980) argue that tetracycline requires neutralization by cations to cross cell membranes. Tet^r cells have a putative TET protein which inserts itself into the membrane and effectively lowers the cation concentration. Consequently, the cells are Tet resistant. The authors argue that the lowering of cation concentrations in the membrane makes the cells hypersensitive to lipophilic chelators like fusaric acid owing to the lowering of cation concentration below a critical level.

A third factor in the choice of a cloning vehicle relates to the type of replicon in the plasmid. Plasmids carrying replicons of the Col E1 type have two advantages. The replication of the plasmids is not under stringent control. The relaxed mode of replication allows the plasmid to multiply to high copy number independent of the replication of the chromosomal DNA. In contrast, plasmids with replicons of p15A type (i.e., pACYC 184) are under stringent control and are present at 2–3 copies/cell (Chang and Cohen, 1978). Second, plasmids of Col E1 type can be further amplified with chloramphenicol giving very high copy number (Clewell, 1972). Consequently, the Col E1 plasmids provide an advantage in terms of signal strength in the initial identification of the clone and in terms of quantity of plasmid recoverable from cultures of a particular clone.

Finally, the choice of a cloning vector is dictated by the means available to identify the correct clone. If a DNA probe is available either in the form of a synthetic oligonucleotide or a cloned cDNA, then any vector permitting the insertion and subsequent amplification of the cDNA in a host cell is sufficient. (Refer to Fig. 10.1 for strategies of clone identification.) If no hybridizable probe is available, use of antibody to the protein product of the gene by the strategy listed in Fig. 10.1, strategy II, can be quite tedious. Initial identification of a cDNA clone may be less cumbersome with the successful utilization of bacterial expression vectors. In addition to the previously discussed factors for choice of a cloning vehicle, expression vectors come equipped with any of a wide array of transcriptional and translational machinery. This will be discussed in greater detail (Section VI C).

III. cDNA SYNTHESIS

Two general strategies have evolved to accomplish insertion of cDNA into plasmids. The traditional approach has involved the ligation or an-

Strategy	Starting Material	Initial Screening	Confirmation
I	Synthetic Oligonucleotide or cDNA Probe	Colony Hybridization (i)	If antibody available a) Hybrid Selection and Translation (ii) b) Bacterial Expression Vector Without antibody c) Sequencing (iii,iv,v,vi)
II	Antibody	Use cDNA probe prepared from an enriched RNA fraction to prepare sublibrary Screen sublibrary by Hybrid Selection and Translation	a) Sequencing b) Bacterial Expression Vector
III	Antibody	Bacterial Expression Vector	a) Hybrid Selection and Translation b) Sequencing
IV	Antibody	Eucaryotic Expression Vector	a) Hybrid Selection and Translation b) Sequencing
V	No probe of any kind	Eucaryotic Expression Vector if a phenotype can be predicted	a) Sequencing

Figure 10.1 Strategies for screening cDNA libraries: i. Grunstein and Hogness (1975); ii. Parnes et al. (1981); iii. Maxam and Gilbert (1977); iv. Maxam and Gilbert (1980); v. Sanger and Coulson (1975); vi. Sanger et al. (1977).

nealing of presynthesized double-stranded cDNA into a suitably modified vector (Fig. 10.2). With this strategy the first-strand cDNA is primed from oligo dT (12–18 base pairs) or a specific nucleotide fragment complementary to a unique mRNA. Oligo dT has the advantage of priming from the 3′ end of the mRNA. The alternative approach (Okayama and Berg, 1982; Claudio et al., 1983; Gunning et al., 1983) employs a dT-tailed linearized vector to anneal polyadenylated mRNA (Fig. 10.3). Subsequently, the cDNA is constructed within the plasmid which is ultimately recircularized by ligation. This approach, referred to as vector-mediated cDNA synthesis, has several advantages which will be discussed below. Both methods require polyadenylated mRNA which is either endogenous to the message or is enzymatically added. The discussion below on the synthesis of first-strand cDNA is relevant to both approaches.

A. Isolation and Assessment of mRNA Quality

The mRNA used for cDNA synthesis is a limiting factor with regard to the quality of the cDNA library produced. In order to limit ribonuclease activity a number of protocols have evolved to isolate RNA from tissues

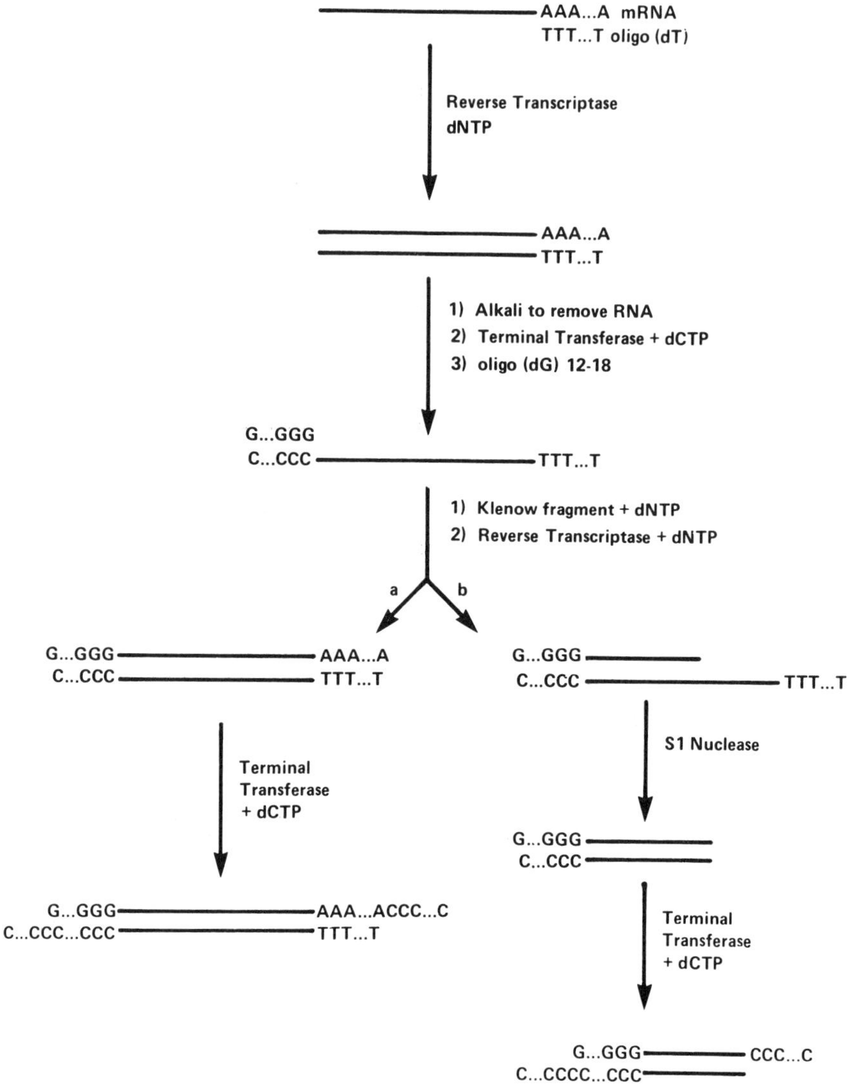

Figure 10.2. cDNA synthesis using polyadenylated RNA as substrate and oligo(dT) as primer: Option *a* theoretically will select for full or nearly full length second-strand cDNA as partial copies of first-strand cDNA will produce a hybrid that is a poor substrate for terminal transferase. Option *b* will produce a considerably larger library but may compromise the length of the cDNA clone.

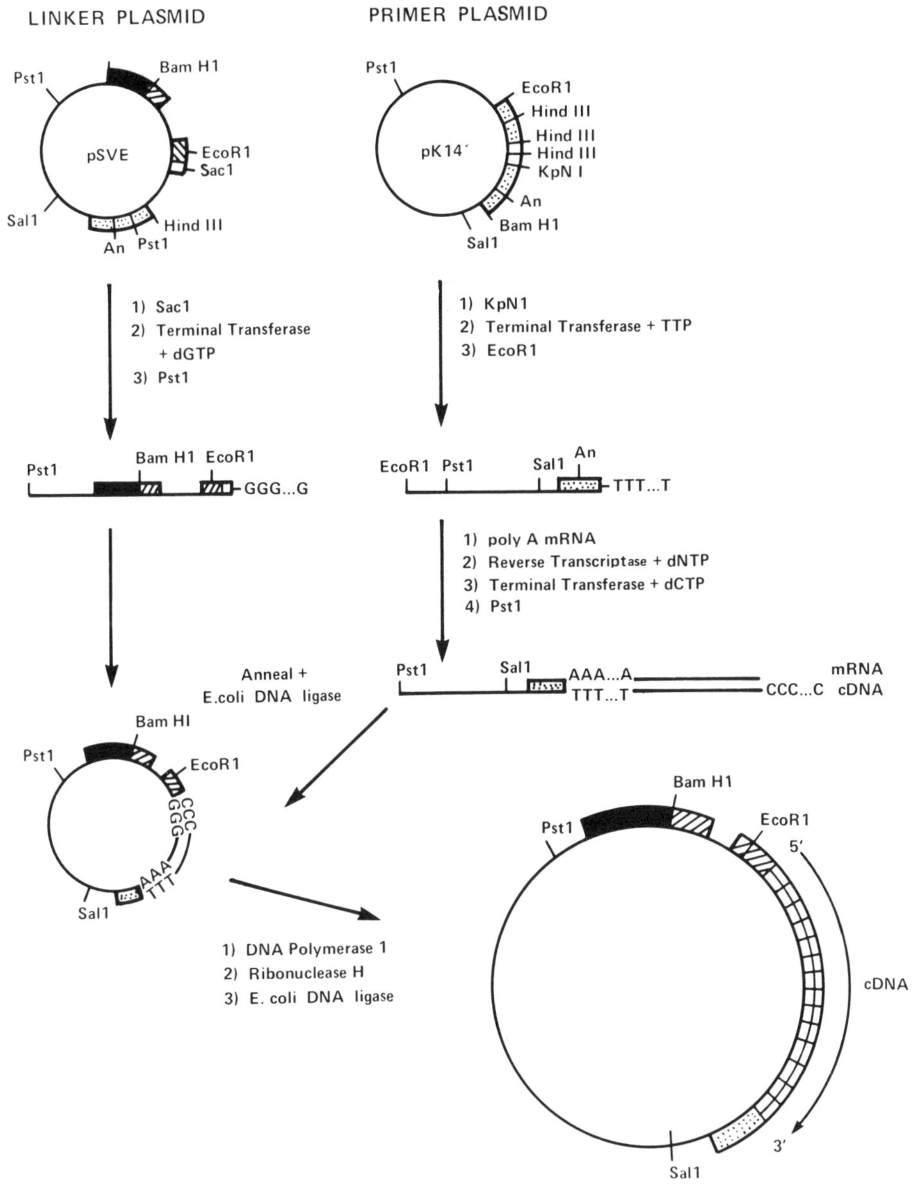

Figure 10.3. Vector-mediated cDNA synthesis (Breathnach and Harris, 1983): The primer plasmid contains the polyadenylation site (An) in the final construct. The linker provides both the promoter and splice sites. The promoter sequence is represented by the solid box, the splice sites by the diagonal lines, and the polyadenylation site by the dotted box. The linker plasmid, pSVE, can be used to express full length cDNA prepared by classical techniques (such as shown in Fig. 10.2) when inserted into the vector and introduced into eucaryotic cells.

or cells in culture using highly chaotropic agents such as guanidine isothiocyanate or guanidine hydrochloride (Glisin et al., 1974; Ullrich et al., 1977; Chirgwin et al., 1979; Feramisco et al., 1982; Maniatis et al., 1982). Whatever the method employed, the eventual outcome can be enhanced by thoughtful assessment of the polyA mRNA produced by two passages through oligo dT cellulose (Edmonds et al., 1971; Aviv and Leder, 1972; Maniatis et al., 1982). The laboratories which consistently produce high-quality cDNA libraries (Okayama and Berg, 1982, 1983; Gunning et al., 1983; Jolly et al., 1983) recommend the following approach for evaluation.

To determine whether the RNA is of good quality:

1. **a.** Reverse transcribe the mRNA.
   b. Resolve the single-stranded DNA products (ssDNA) by denaturing agarose gel electrophoresis.
   c. Transfer the ssDNA by the Southern blot technique (Southern, 1975).
   d. Hybridize with a cloned radioactively labeled probe.

This should produce a discrete band comigrating with molecular-weight markers consistent with the size of the full length message. Smearing is indicative of mRNA which is not appropriate for cDNA synthesis.

2. Reverse transcribe the mRNA under conditions of DNA primer excess. Under these conditions all of the mRNA which can be transcribed is evaluated. Assuming an average length of mRNA of approximately 1200 base pairs, then the fraction of mRNA that is able to be transcribed can be calculated from incorporation of labeled deoxynucleotides into the high-molecular-weight DNA fraction. Approximately 20% of the mRNA should be transcribed.

3. Reverse transcribe the mRNA under conditions of RNA excess. Under these conditions it is assumed that the molar amount of RNA involved in cDNA synthesis is equivalent to the limiting amount of primer DNA added. Consequently, the average length of cDNA synthesized can be calculated from considerations of the amount of incorporated deoxynucleotide; 1200 base pairs is the suggested target figure.

RNA preparations which satisfy these criteria are suitable for cDNA synthesis.

B. Synthesis of First-Strand cDNA

In recent years a number of laboratories have attempted to optimize the conditions for reverse transcription of mRNA into single-stranded cDNA. A very detailed study of reverse transcriptase and its associated ribonuclease H activity has been presented by Berger et al. (1983). This group was able to demonstrate that a major limitation to yield of cDNA per unit mRNA was the deadenylation of polyadenylated mRNA by the inherent ribonuclease H activity of reverse transcriptase.

In this same study Berger et al., (1983) were also able to show that the limiting factor in terms of synthesis of full length cDNA was the initiation of polymerization. Once initiated, the reverse transcription of full length globin mRNA was quite rapid (1–2 min). The authors were thus able to provide an explanation for the stochiometric quantities (as opposed to catalytic quantities) of reverse transcriptase commonly recommended for cDNA synthesis (Buell et al., 1978; Wickens et al., 1978; Yoo et al., 1982; Okayama and Berg, 1982; Berger et al., 1983). To promote initiation and polymerization, high levels of deoxynucleotides are also suggested. Berger et al. (1983) suggest that some of the fragmented cDNAs commonly observed may be the consequence of transcription from random primers produced by ribonuclease H action binding to mRNA. Buell et al. (1978) have pointed out that the quality of reverse transcriptase is critical. Commercial sources not free of ribonuclease activity need to be purified further. Use of a ribonuclease inhibitor such as RNAsin may also alleviate the problem (Christophe et al., 1980; Michelson et al., 1983).

Studying the parameters that affected synthesis of full length cDNA rather than total incorporation into cDNA, Buell et al. (1978) note that with highly purified reverse transcriptase parameters including time of incubation, temperature, salt, or even nucleotide concentration could be varied widely without altering the proportion of full length cDNAs. Some mRNAs may be particularily difficult to transcribe. Lysozyme, ovomucoid, and ovalbumin mRNAs (630–1900 base pairs) are transcribed to full length, whereas conalbumin (2500 base paris) could not be transcribed to full length in this study. Some authors have pointed out critical requirements for full length cDNA which may be peculiar to the RNA studied. Retzel et al. (1980) in transcribing avian myeloblastosis virus RNA (7500 base pairs) noted a narrow window of permissible salt concentrations close to that of physiological saline. On the other hand, Vassart and Brocas (1980) have managed to synthesize an 8000-base-pair cDNA for thyroglobulin using 52 m*M* KCl and the aid of a ribonuclease inhibitor. This suggests that conditions may have to be optimized for individual mRNAs to achieve full length cDNA synthesis.

C Synthesis of Double-Stranded cDNA (dscDNA)

A number of strategies exist for the synthesis of the second strand. Figure 10.2 provides a schematic summary of one such strategy. RNA template for the first strand can be removed by alkali or heat denaturation. The cDNA strand remaining will generally and fortuitously form a hairpin loop at the 5′ end of the cDNA. This presents a 3′-OH able to prime the second-strand synthesis using the first strand as template. At the completion of the second-strand synthesis, which may be accomplished with either reverse transcriptase or DNA polymerase I (or both in combination), the cDNA is covalently joined at the 5′ end by the hairpin loop. To introduce the cDNA into a plasmid via linkers or by homopolymer tails (see subsequent discussion), the single-stranded hairpin loop has to be removed. This is accomplished enzymatically with carefully titrated levels of S1 nuclease. This has the disadvantage of removing portions of the cDNA representing the 5′ end of the mRNA. To circumvent this problem the cDNA can be tailed with terminal transferase creating a homopolymer at the 5′ end. This is subsequently annealed to a complementary homopolymer which acts as a primer for the second-strand synthesis. In this way the two strands are not covalently joined and the S1 nuclease is avoided. Land et al. (1981), after removing the template mRNA with alkali, tailed the cDNA with dCTP using terminal transferase and primed the second strand with oligo(dG)12–18 (Fig. 10.2). They report high efficiency cloning of 5′ terminal sequences with this strategy. Yoo et al. (1982) similarly removed the mRNA template with alkali and tailed the cDNA with dATP, generating a homopolymer that could be annealed with olig(dT)$_{15}$ which primed the second-strand synthesis. With the approach of vector-mediated cDNA synthesis previously alluded to (Okayama and Berg, 1982), the RNA–DNA hybrid which presents after the first-strand synthesis is immediately tailed without prior removal of the RNA (Fig. 10.3). The authors argue that the terminal transferase reaction prefers blunt or nearly blunt double-stranded substrates. Since the addition of the oligo(dC) is ultimately required to recircularize the vector, the terminal transferase reaction should accordingly act as a selection for full length cDNA. It is difficult to speculate on how powerful a selection it is for full or nearly full length cDNA. Terminal transferase is thought to add a single nucleotide to the growing chain before dissociating and binding to an alternate 3′-OH terminus. Substrates with 3′-OH extensions should most rapidly form a stable complex with terminal transferase and will more frequently be subject to nucleotide addition. If terminal transferase acts in a strictly nonprocessive manner and prefers 3′-OH terminal extensions even in the presence of Co^{2+} (Michelson and Orkin, 1982), then on theoretical grounds those cDNAs which most rapidly form 3′-OH extensions in the presence of terminal transferase (full or nearly full length cDNAs) would become preferred substrates. On the same grounds it can be argued that in the presence of Mg^{2+}, which exaggerates the preference for 3′-OH extensions, the selection would be considerably more stringent (Roychoudhury et al., 1976).

Various enzyme activities have been used to synthesize the second strand from whatever primer is chosen. Initially DNA polymerase I, an *Escherichia coli* enzyme, was employed (Efstradiatis and Villa-Komaroff, 1979). This enzyme has 5′→3′ polymerase activity in addition to 5′→3′ and 3′→5′ exonuclease activities. Subsequently, use of the Klenow fragment has superceded use of DNA polymerase I. This polypeptide is a DNA polymerase I fragment generated by subtilisin digestion (Jacobsen et al., 1974). This fragment lacks the 5′→3′ exonuclease activity. This latter activity may be important in generating suitable primers for second-strand synthesis when the primer results from hairpin looping as the exonuclease can remove unmatched base pairs at the 3′ side of the hairpin loop. Wickens et al. (1978) were able to show that large amounts of partially double-stranded DNA are present throughout the reaction suggesting that initiation is not rate limiting, but that elongation is rate limiting with DNA polymersae I. This is in contrast to first-strand synthesis by reverse transcriptase where little in the way of intermediates is seen and initiation is rate limiting (Berger et al., 1983). Reverse transcriptase has also been used for second-strand cDNA synthesis. Studying second-strand cDNA synthesis of globin, Kay et al. (1980) reported no intermediates between full length first-strand and full length second-strand cDNA synthesis suggesting that initiation may be rate limiting for second-strand synthesis by reverse transcriptase as well. With the different behavior of the two enzymes, it has been suggested that both be used sequentially (Maniatis et al., 1982).

With the approach of vector-mediated cDNA synthesis (Okayama and Berg., 1982) (shown schematically in Fig. 10.3), *E. coli* ribonuclease H activity is used to nick the RNA strand of the RNA/DNA hybrid. Subsequently, the 5′→3′ exonuclease activity of DNA polymerase I clears the path for the growing second strand catalyzed by the 5′→3′ polymerase activity and primed by a linker fragment annealed to the oligo(dC)-tailed cDNA. The nicks in the double-stranded vector remaining after this process are removed by *E. coli* DNA ligase which will not ligate RNA to DNA in these constructs. However, T4 DNA ligase, which does have this capacity, has also been used successfully with the vector-mediated approach (Gunning et al., 1983).

An interesting approach to the synthesis of second-strand cDNA has recently been reported by Gubler and Hoffman (1983). As an alternative to removing mRNA by alkali treatment before proceeding with the synthesis of the second strand, the authors borrow from the vector-mediated strategy of Okayama and Berg (1982) and use a combination of ribonuclease H and DNA polymerase I to nick translate the second strand. This considerably simplifies the process of cDNA synthesis since S1 nuclease treatment is unnecessary prior to tailing the cDNA.

IV. INTRODUCTION OF cDNA INTO PLASMID

When the approach of Okayama and Berg (1982) is not used, a method for insertion of the cDNA into the plasmid must be found. It is generally advisable to preselect full length cDNAs derived from a particular mRNA by preparative agarose gel electrophoresis. Once this is accomplished, a variety of techniques are available to introduce cDNA into vectors.

Two approaches are used. One involves the construction of complementary homopolymer tails on the cDNA and the linearized plasmid. The alternate approach involves the addition of linkers to the ends of the cDNA which can be ligated with T4 DNA ligase to an appropriate site in the vector.

With the former alternative the most commonly used method uses terminal transferase to add oligo(dC) to the ends of the dscDNA. Plasmids like pBR322, pUC 8, or pUC 9, if linearized at the PstI site and tailed with dGTP, will anneal to the dC-tailed cDNA and, without the requirement of ligation, form plasmids competent to transfect the *E. coli* host. The advantage of this particular approach is that the PstI site is reconstituted at either end of the insert allowing for easy excision of the insert.

When linkers are used, a high molar ratio of linkers are ligated to blunt-ended cDNA. If the linkers are prepared with 5′-OH termini, they are initially treated with polynucleotide kinase and ATP to generate 5′-phosphorylated termini to allow efficient ligation. Conversely, the vector is treated with phosphatase to remove the 5′-phosphate to prevent reclosing of the vector. After the blunt-end ligation of linkers to cDNA, the linkers need to be restricted to generate cohesive ends to enable insertion into a complementary restriction site. This can have the unfortunate result of causing endonucleolytic fragmentation of the cDNA if the cDNAs contain restriction sites for the enzymes being used. If a restriction map of the dscDNA is known, then judicious choice of linkers circumvents the problem. Alternatively, methylating the cDNA or using linkers containing rare restriction sites are possible solutions if full length cDNAs are desired.

A further problem with the use of linkers is the ligation between cDNAs containing the cohesive ends. This can be in part overcome by the sequential addition of two different linkers to the cDNAs. The first is added before the S1 nuclease digestion. Consequently, only one end is accessible to ligation. After the S1 nuclease digestion, a second different linker is added and the cDNA ligated into a plasmid with heterogeneous ends complementary to the heterogeneous ends of the cDNA. This permits directional insertion of cDNA into the plasmid (Kurtz and Nicodemus, 1981;Helfman et al., 1983). Since the orientation of the cDNA can be controlled, this approach offers advantages if bacterial expression vectors are used as a cloning strategy (Fig. 10.1).

V. TRANSFORMATION PROCEDURES

Once the plasmids carrying the cDNA are constructed, they are propagated and amplified in *E. coli*. Plasmids are initially introduced into the host by one of two commonly used procedures. The more common of the two procedures prepares bacterial cells competent to take up DNA by resuspending the cells in $CaCl_2$ (Mandel and Higa, 1970; Clewell, 1972; Cohen et al., 1972; Dagart and Ehrlich, 1979). This procedure should yield 10^6–10^7 transformants per microgram of covalently closed plasmid. Plasmids consisting of two annealed DNA fragments will transfect at a rate 1 to 2 orders of magnitude less. A second method recently described by Hanahan (1983) involving growth of cells in medium with elevated levels of Mg^{2+} and treatment of the cells at 0°C with a solution of Mn^{2+}, Ca^{2+}, Rb^+ or K^+, dimethyl sulfoxide, dithiothreitol, and hexamine cobalt (III), is more complex but under optimal conditions gives 1 to 2 orders of magnitude higher yields of transformants.

VI. IDENTIFICATION OF POSITIVE CLONES

The strategy for screening a cDNA library to isolate individual clones depends on the nucleic acid or antibody probes available. If a hybridizable cDNA probe (Yoo et al., 1982) or synthetic oligonucleotide probe (Suggs et al., 1981; Wallace et al., 1981; Michelson et al., 1983; Schulze et al., 1983) is available, then a number of approaches are possible. As mentioned above, cDNAs enriched for a particular sequence can be isolated by resolving the total cDNA fraction on an agarose gel and eluting the sequence of interest. The migration of the relevant sequences is demonstrated by Southern analysis (Southern, 1975). Alternatively, once the library is constructed, a plasmid preparation from the pooled library can be linearized by an appropriate restriction enzyme and resolved by electrophoresis through agarose. The plasmid fraction containing the full-sized clones of interest is again identified by Southern analysis (Southern, 1975). If a restriction map of the desired clone is not known, then it is necessary to attempt a number of restriction cuts to find a site which cuts only the plasmid. Plasmids carrying polylinkers, that is, pUC 8 or pUC 9 (Vieira and Messing, 1982), or rare restriction sites, that is, Sal 1 in pcDV1 (Okayama and Berg, 1983) are most useful for this approach.

A. Strategy I: Colony Hybridization

A large number of individual bacterial clones can be analyzed with specific nucleic acid probes using the procedure of colony hybridization (Figs. 10.4 and 10.5) (Grunstein and Hogness, 1975; Hanahan and Meselson, 1980). Individual clones are grown on, or transferred to, nitrocellulose sheets. The bacterial cells are lysed and plasmids are denatured with alkali treatment. After neutralizing the filters, the DNA is baked irreversibly onto the nitrocellulose filters. Labeled probe is added under condi-

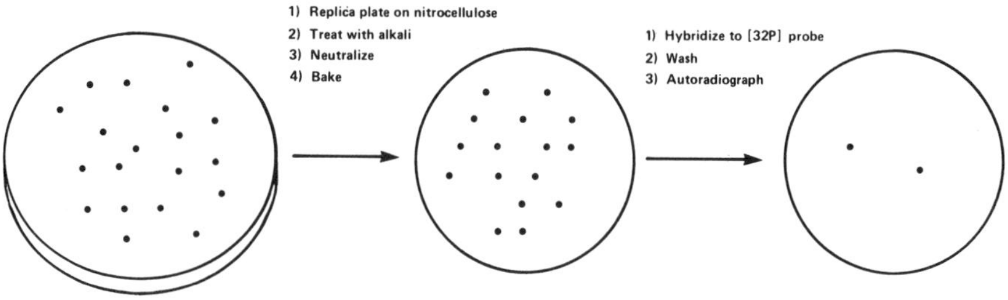

Figure 10.4. Colony hybridization: Colonies are grown on nitrocellulose sheets or on agar plates and transferred to nitrocellulose by pressing a nitrocellulose sheet to the agar plate supporting the growth of bacterial colonies. Alkali is introduced from beneath the filter by layering the nitrocellulose on a puddle of the solution. Similarly the filter is treated subsequently with a neutralizing solution. The DNA is irreversibly baked onto the filter. ^{32}P-labeled probe is added to the filters under conditions conducive to specific hybridization of probe to plasmid DNA containing sequences complementary to the probe.

tions that are sufficiently stringent to permit only specific hybridizations to remain after washing the filters. Autoradiography identifies the clones that recognize the probe.

Corroborative evidence requires the availability of a specific antibody or information regarding the amino acid sequence of the protein encoded by the desired clone. With antibody available, the technique of hybrid selection and translation provides a straightforward verification (Figs. 10.6 and 10.7). This technique has been described and modified by a variety of authors (Prives et al., 1974; Paterson et al., 1977; Ricciardi et al., 1979; Cleveland et al., 1980). Refer to Williams (1981) for a review. In general, mRNA is bound to recombinant plasmid previously denatured and immobilized on nitrocellulose or activated cellulose filter papers. Alternatively, RNA is hybridized to plasmid DNA in solution and the partially duplex DNA/RNA hybrids formed is immobilized on nitrocellulose (Nagata et al., 1980). In either case the RNA specifically bound to cDNA is eluted and translated using any of a variety of commercial or laboratory preparations of reticulocyte lysates or wheat germ extracts. The eluted mRNA should be translatable into a protein recognized by a specific antibody to the protein encoded by the cDNA clone. Alternatively, the hybrid selected mRNA can be injected into *Xenopus* oocytes and the translated product identified with antibody (Long et al., 1982).

If amino acid sequence data are available, then sequencing the clone by one of the commonly used techniques provides the best evidence to verify a particular clone (Sanger and Coulson, 1975; Maxam and Gilbert, 1977,1980; Sanger et al., 1977). As an alternate confirmatory technique the cloned gene can be subcloned into a bacterial or eukaryotic expression vector (discussed below) and the host cell induced to produce a polypeptide detectable by antibody.

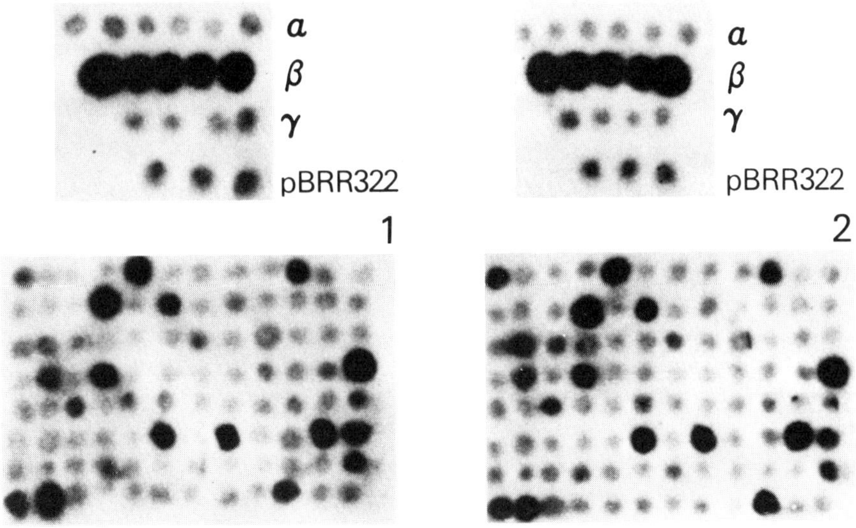

Figure 10.5. Colony hybridization: The autoradiograph demonstrates use of this technique to identify positive clones for β casein. The top two panels show the control data giving hybridizations with α, β, γ casein and pBR322. The bottom two panels represent duplicate filters containing clones from a cDNA library prepared from lactating mouse mammary gland using vector-mediated cDNA synthesis (P. Doherty, unpublished data).

B. Strategy II: Hybrid Selection and Translation

If no probe is available, then antibody alone can be used to identify positive clones. One commonly used but quite tedious approach uses hybrid selection and translation as a screening procedure (see Fig. 10.6). As this is cumbersome with a large number of clones, the library is generally prescreened by a cDNA probe enriched for a particular sequence. One application of this is to resolve an RNA preparation on a denaturing agarose gel using glyoxal (McMaster and Carmichael, 1977), formaldehyde (Lehrach et al., 1977; Goldberg, 1980), or methlymercuric hydroxide (Bailey and Davidson, 1976) and to isolate RNA fractions from individual gel slices. The eluted proteins are translated and the protein products are immunoprecipitated with specific antibody. An example of this approach is shown in Figures 10.7 and 10.8. A labeled cDNA is prepared from the fraction most enriched in transcripts coding for the protein of interest. The cDNA probe then selects for a subset of clones. Enrichment of particular mRNA species through immunopurification of polysomes has also been used successfully (Korman et al., 1982). A demonstration of this approach is illustrated in Figures 10.9 and 10.10.

An alternate approach can be taken if two similar cell lines or tissues at different stages of differentiation exist (Garfinkel et al., 1982; Hastings

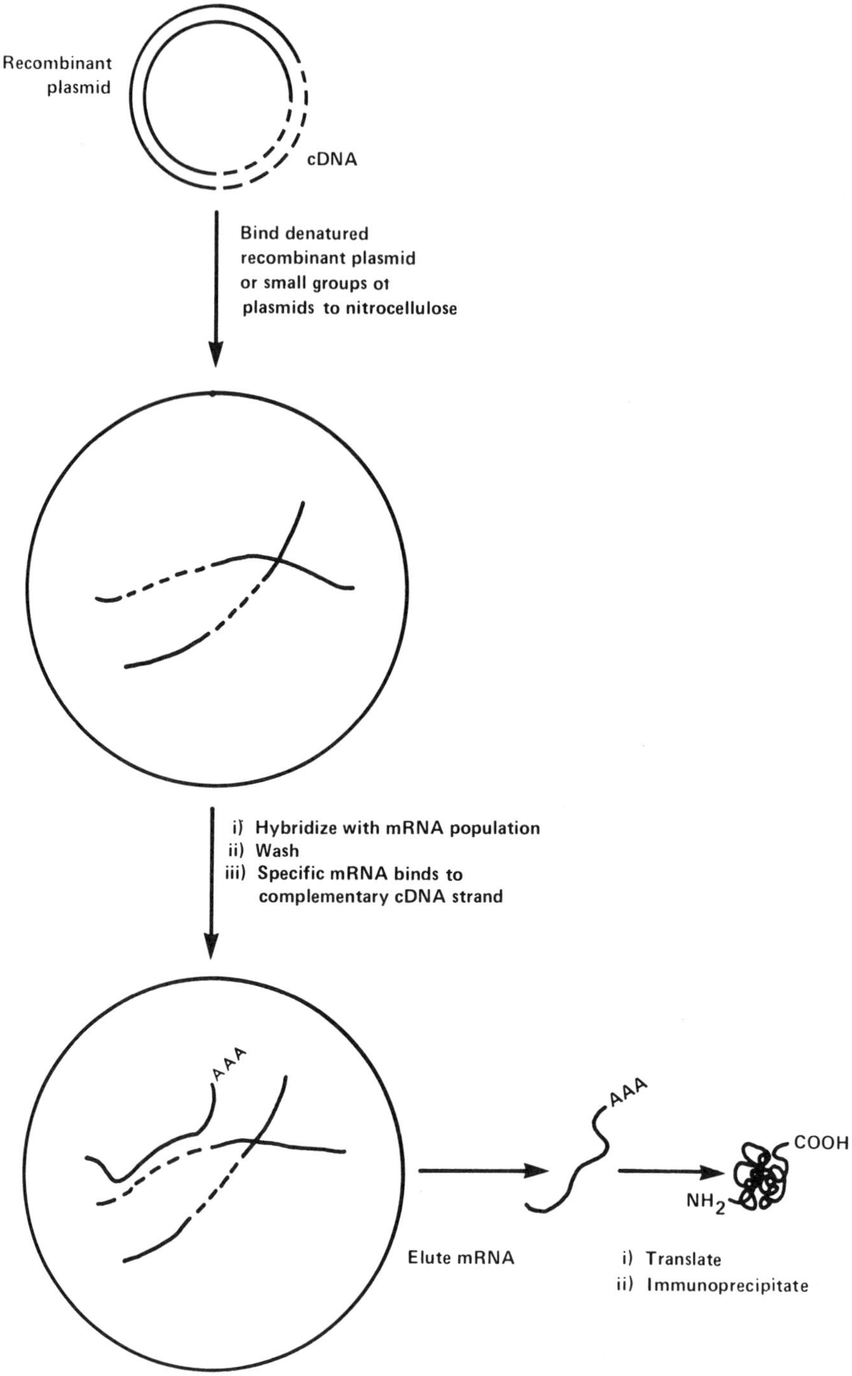

Figure 10.6. Hybrid selection and translation: Recombinant plasmid DNA is denatured and baked onto nitrocellulose. The filter is incubated with a hybridization solution containing a general mRNA population. Nonspecifically bound mRNA is washed from the filters prior to the elution of specifically bound mRNA. The RNA is identified by immunoprecipitation of the translated product.

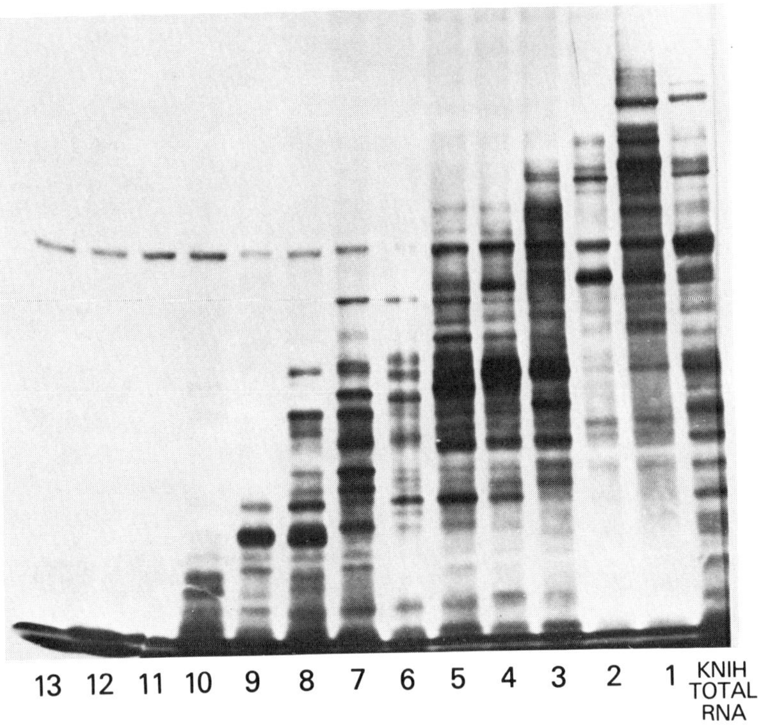

Figure 10.7. Preparation of cDNA probe by enriching for a particular mRNA. Polyadenylated mRNA was resolved on a low-melting agar methylmercuric hydroxide gel (Bailey and Davidson, 1976). mRNA was isolated from 13 individual gel slices and translated with rabbit reticulocye lysate using [^{35}S]methionine as label. The protein products were resolved on a 7.5% polyacrylamide gel and the dried gel flurographed. Lane 1 represents the top most slice of the denaturing gel (P. Doherty, unpublished data).

and Emerson, 1982) only one of which contains mRNA (usually assumed by the presence or absence of the protein product) for the gene to be cloned. cDNA probes are prepared from polyadenylated mRNA from both the positive and negative cell lines or tissues and are used to screen the cDNA library. Only clones giving a positive hybridization exclusively with the cDNA prepared from the positive cell line or tissue are selected for further analysis by hybrid selection and translation. Differential colony hybridization is also employed if a particular RNA species is inducible under a given set of conditions (St. John and Davis, 1979; Cochran et al., 1983).

Another approach to producing enriched cDNA probes employs the method of cascade hybridization (Timberlake, 1980; Zimmerman et al., 1980). cDNA from unfractionated polyadenylated mRNA is sequentially

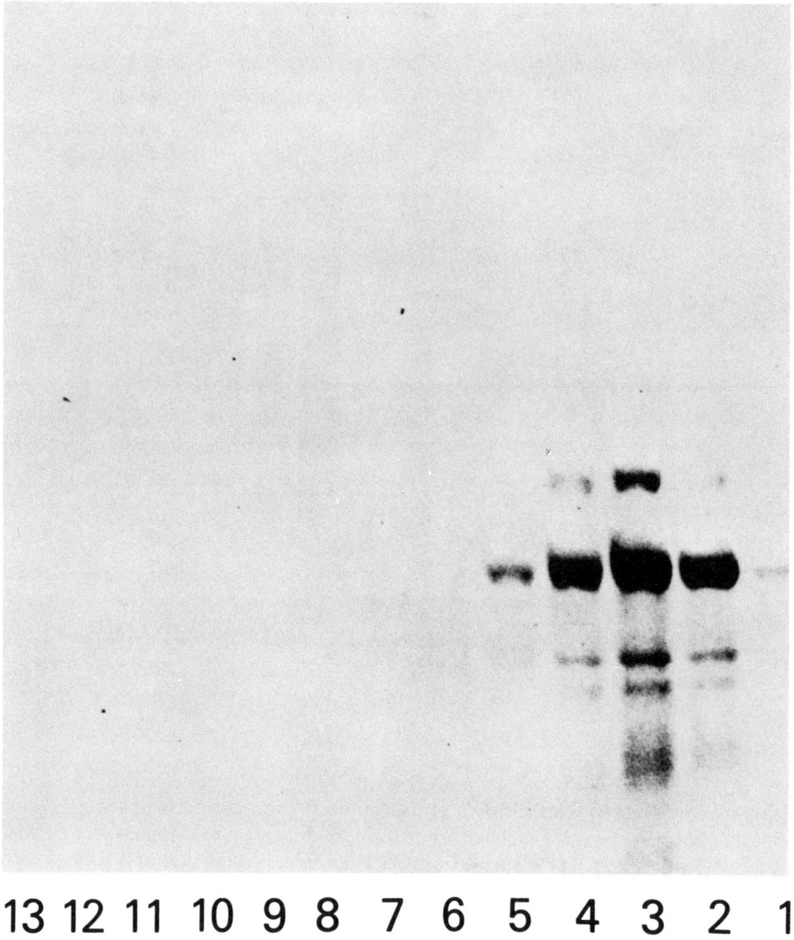

Figure 10.8. The translated products discussed in Figure 10.7 were immunoprecipitated with an antibody to a major excreted protein (MEP) (Gottesman, 1978; Gottesman and Sobel, 1980; Gottesman and Cabral, 1981) of Kirsten virus transformed NIH-3T3 cells (KNIH 3T3 cells) (P. Doherty, unpublished data).

hybridized to increasing mass excess of polyadenylated mRNA from a similar cell line not containing the sequence to be cloned. Hybrids are removed by chromatography on hydroxyapatite. Unreacted cDNA is hybridized to mRNA from cells which do contain the sequence and the hybrids fractionated by hydroxyapatite.

C. Strategy III: Bacterial Expression Vectors

Where antibody is the only probe available, direct identification of positive clones through the use of bacterial expression vectors as cloning ve-

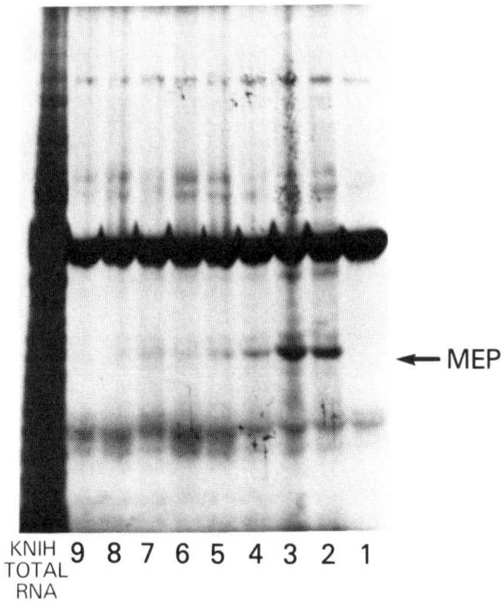

Figure 10.9. Enrichment of a particular mRNA species using polysome immunopurification (Korman et al. 1982). Polysomes isolated from KNIH-3T3 cells were reacted with antibody directed against MEP (see Fig. 10.8). The immune complex was isolated by protein A sepharose chromatography and the eluted mRNA passed over oligo dT cellulose. Individual fractions were collected and translated (see Fig. 10.7). Bands common to all fractions are endogenous to the lysate. Lane 3 has the highest concentration of highly enriched MEP mRNA (P. Doherty, unpublished data).

hicles for the construction of cDNA libraries is becoming increasingly popular. A description of a few of these will illustrate the general principal.

Young and Davis (1983a) have used the vector λgtll for this purpose. This approach departs somewhat from that in the previous discussion in that a λ bacteriophage is used instead of a plasmid. This vector has a unique EcoRI restriction site located 53 base pairs upstream from the β-galactosidase termination codon (*lacZ*). Insertion into this site causes inactivation of β-galactosidase. λgtll creates a blue patch on the bacterial lawn if X-gal indicator plates are used. The recombinant phage lack this ability. λgtll has further properties that may be useful. First it produces a temperature-sensitive repressor (*c*I857), which is inactive at 42°C. The temperature shift results in an induction of the lysogen and a rapid increase in copy number and transcription of the foreign DNA. Secondly, the λgtll contains an amber mutation (S100) that prevents the phage from lysing the host. Consequently, large numbers of phage accumulate because of the absence of lysis. To further enhance the signal, protein degradation is minimized using lon$^-$ host cells defective in one protein degradation pathway (Gottesman et al., 1981; Mizusawa and Gottesman, 1983). With the temperature shift the lysogens are induced, the hybrid β-galactosidase/cDNA transcript is accumulated, and large quantities of hybrid protein are produced in host cells defective in protein degradation. To screen with antibody, infected cells are grown on nitrocellulose sheets at 32°C, replica plated, induced at 42°C, lysed with a chloroform atmosphere, treated with DNAase, reacted with specific antibody, and

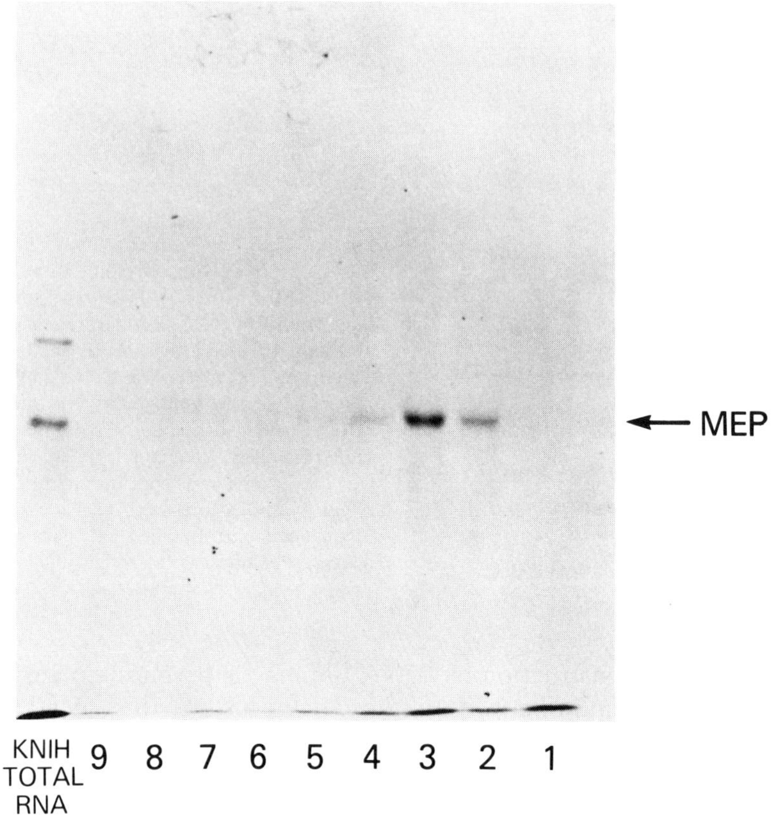

Figure 10.10. Immunoprecipitation of the translated products discussed in Fig. 10.9 (P. Doherty, unpublished data) using antibody against MEP.

then iodinated protein A. Positive phage are identified by autoradiography. Young and Davis (1983a) have used this protocol to isolate clones of α-amylase and ovalbumin.

Young and Davis (1983*b*) have recently published a second protocol which they found necessary for the isolation of yeast RNA polymerase II genes. In this case the host cell chosen did not cause lysogeny but allowed plaque formation. Because of the amber mutation (S100) discussed above in the λgtll, the host cell carries a *supF* gene in order that the S100 mutation does not interfere with lysis. The host cell also carries a plasmid, pMC9, which has *lac*I repressor in order to suppress the expression of cDNAs coding for lethal proteins during the initial growth phase of the phage. IPTG is used to induce the activity at the opportune point in the procedure.

Helfman et al. (1983) have used pUC8 and pUC9 as bacterial expression vectors. This plasmid carries an amino-terminal fragment of the *lac*Z gene which will complement a defective β-galactosidase gene under the control of the F episome in hosts JM101 and JM103 and present on the chromosome of host JM83. As for λgtll, insertion into the polylinker at the N-terminus of the *lac*Z coding region generally prevents cells carrying these plasmids from converting X-gal, the chromogenic substrate, to a blue color. Usually, the inserted fragment will be out of phase with the N-terminal *lac*Z fragment. This problem is overcome by the sequential addition of linkers (Helfman et al., 1983). In any case, the cDNAs will be in phase in only one-third of the clones whenever a hybrid gene is constructed. In addition, cDNAs randomly inserted into expression vectors will be in proper orientation to be read from the *lac*Z promoter only 50% of the time, so that only one-sixth of these recombinants can be expected to express the inserted gene.

Other bacterial expression vectors reported vary with regard to the promoter (Rosenberg and Court, 1979; Hawley and McClure, 1983), the ribosome binding site [Shine and Delgarno (1975); for review see Gold et al. (1981)], and the amount of transcribed plasmid sequences translated with those of the cDNA.

A considerable number of bacterial expression vectors have been produced primarily to express large amounts of a particular protein by inserting previously cloned cDNA upstream from a promoter and ribosomal binding site functional in *E. coli*. Some of these may have application to the construction of cDNA libraries or the confirmation of clones selected by an alternate technique.

Use of the P_L promoter (lambda) inducible by inactivation of the mutant repressor, *c*I857, at 41°C produces high levels of the upstream gene products (Shimatake and Rosenberg, 1981). Backman and Ptashne (1978) achieved high levels of gene expression using the *lac* promoter and a ribosome binding site which was a hybrid between *lac* and lambda sequences. Variations in the location and sequence of the ribosome binding site proved crucial to gene expression. High levels of expression of the human leukocyte interferon gene has been observed using the *E. coli trp* promoter and *trp* leader ribosome binding sequences (Goeddel et al., 1980). Similar constructs have been used to express hepatitis B surface and core antigens (Edman et al., 1981), human plasminogen activator (Pennica et al., 1983), and human immunoglobulin E ε chain cDNA (Kurokawa et al., 1983). Expression of antigenic determinants of the hemagglutinin gene of human influenza virus was achieved using both the *trp* promoter and *trp* leader ribosome binding sequences and the *lac* promoter and ribosome binding sequences. Stable fusion proteins result in either case with the *trp*E or β-galactosidase gene products (Davis et al., 1981). Using the tryptophan operator/promoter system the capsid viral protein of foot and mouth disease, VP3, has been expressed as a fusion protein with the *trp*E gene product inducible by growth in tryptophan-depleted medium (Kleid et al., 1981). Human interleukin 2 cDNA has been expressed in *E. coli* using the *trp* system and the λ P_L promoter and a ribosome binding site from phage mu (Devos et al., 1983).

D. Strategies IV and V: Eukaryotic Expression Vectors

A final approach to cloning cDNA involves the expression of the cDNA in eukaryotic cells. This approach does not require a nucleic acid probe or antibody if the expression of the cDNA gives rise to a predictable and selectable phenotype (i.e., drug resistance). At the time of this writing there have been no reports of successful initial cDNA clone identification directly by transfer of cDNA clones into eukaryotic cells.

The most promising approach would appear to be that of vector-mediated cDNA synthesis (Fig. 10.3) (Breathnach and Harris, 1983;Jolly et al., 1983;Okayama and Berg, 1983). The plasmid primer and linker DNA-containing plasmids originally described by Okayama and Berg (1982) become pK14 and pSVE in the hands of Breathnach and Harris (1983) (Fig. 10.3). The linker fragment from pSVE (or pSVT) in the final construction provides the SV40 early (or late) promoter and β-globin splice sites upstream from the cDNA sequences. The pK14 primer fragment provides a downstream SV40 polyadenylation site. The splice sites are not positioned downstream from the cDNA in the final construction to avoid removing the splice site as part of a polyadenylation event in the event that a full length cDNA has its own polyadenylation site. Although the requirment and positioning of splice and polyadenylation sites are largely conjecture at this point, the empirical result is that the constructs are functional. Breathnach and Harris (1983) report that when ovalbumin cDNA-containing vectors represent 2% of the clone population, they were able by indirect immunofluorescence to detect the gene transferred into eukaryotic cells by protoplast fusion. The authors suggest that a sibling selection transferring groups of 50 cloned plasmids should result in successful clone identification by indirect immunofluorescence.

Breathnach and Harris (1983) have constructed the pSVE plasmid in such a way that cDNA made by classical methods can be annealed or ligated into the vector and still be expressed provided the orientation of the cDNA relative to the promoter is correct.

Okayama and Berg (1983) have modified their own plasmids used for vector-mediated cDNA synthesis and produced a primer plasmid, pCDV1, which provides the SV40 early region promoter and SV40 splice sites, and a linker plasmid, pL1, which provides an SV40 polyadenylation signal. The orientation of the cDNA relative to the promoter, splice, and polyadenylation sites are the same as those discussed above for the Breathnach and Harris vectors. The pL1 and pCDV1 vectors have successfully been used to express hypoxanthine–guanine phosphoribosyl transferase allowing the survival of $HGPRT^-$ host cells.

As with bacterial expression vectors, many variations in eukaryotic expression vectors are possible. For a detailed analysis, see Chapter 9. Undoubtedly each combination of cDNA clone and recipient cell will determine the optimal vector construction. For example, Hamer and Leder (1979) have reported the requirement of introns for the expression of the β^{maj} globin gene from an SV40-derived vector. Laub and Rutter (1983) report that the absence of introns did not affect the expression of full length insulin cDNA from an SV40 early region promoter using COS cells. Schumperli et al. (1982) have inserted the *E. coli* galactokinase gene (*gal*K) into a eukaryotic expression vector using the SV40 early promoter and found no requirement for SV40 splice signals for the expression in monkey, mouse, or hamster cell lines.

With regard to the placement of the splice signals 5′ or 3′ from the cDNA in the final construction, it is not clear that a downstream splice would necessarily be removed by processing of a polyadenylation site endogenous to a full length cDNA. Montell et al. (1983) have pointed out that the AAUAAA hexanucleotide cannot be the only signal, since these sequences are observed in coding regions of mRNA. They speculate that other secondary or tertiary structures may be required for processing. Consequently, with cDNA which is transcribed from processed mRNA, these secondary signals may be absent.

Brennand et al. (1983) have shown that hypoxanthine–guanine phosphoribosyl transferase (HPRT) cDNA can be expressed in Chinese hamster and human fibroblasts which are $HPRT^-$ and allow growth of the transformants using the retroviral long terminal repeat (LTR) promoter in the absence of both splice and polyadenylation signals. The authors conclude that in this case neither is an absolute requirement.

The Rous sarcoma virus LTR (Gorman et al., 1982) and the C3H murine mammary tumor virus LTR (Prakash et al., 1983) act as promoters in eukaryotic cells and bacterial cells and, fortuitously, also provide a ribosome binding site (Mermer et al., 1983). Consequently, plasmids constructed as expression vectors using retroviral LTRs should be effective in shuttling genes through both bacterial and eukaryotic cells.

A further subtlety involves enhancers [refer to Khoury and Gruss (1983) for review]. Berg et al. (1983) have shown that the relative enhancer activities of the SV40 72 base pair repeat and the Harvey sarcoma 73 base pair repeat on the mouse β^{maj} globin gene product was species specific. Consequently, considerably more information is required on the variety of vector signals in order to optimize the chances of isolating some of the more intractable cDNA clones by what is potentially a powerful cloning procedure.

The availability of dozens of mutant somatic cells carrying dominant selectable markers suggests that once full length cDNA cloning becomes routine, this approach may be the approach of choice for isolating cDNAs encoding biologically important proteins present in small quantities in cells. Such proteins include regulatory proteins, such as activators and repressors, and enzymes, whose alteration and function may be readily detected by genetic analysis.

REFERENCES

Aviv, H. J. and Leder, P. (1972). *Proc. Natl.Acad. Sci. USA* **69**, 1408–1412.

Backman, K. and Ptashne, M. (1978). *Cell* **13**, 65–71.

Bailey, J. M. and Davidson, N. (1976). *Anal. Biochem.* **70**, 75–78.

Berg, P. E., Yu, J. K., Popovic, Z., Schumperli, D., Johansen, H., Rosenberg, M., and Anderson, W.F. (1983). *Mol. Cell. Biol.* **3**, 1246–1254.

Berger, S. L., Wallace, D. M., Puskas, R. S., and Eschenfeldt, W. H. (1983). *Biochemistry* **22**, 2365–2372.

Bochner, B. R., Huang, H.C., Schieven, G. L., and Ames, B. N. (1980). *J. Bacteriol.* **143**, 926–933.

Bolivar, F., Rodriguez, R. L., Greene, P. J. Betlach, M. C., Heynecker, H. L., Boyer, H. W., Crosa, J. H., and Falkow, S. (1977). *Gene* **2**, 95–113.

Breathnach, R. and Harris, B. A. (1983). *Nucl. Acids Res.* **11**, 7119–7136.

Brennand, J., Konecki, D. S., and Caskey, C. T. (1983). *J. Biol. Chem.* **258**, 9593–9596.

Buell, G. N., Wickens, M. P., Payvar, F., and Schimke, R. T. (1978). *J. Biol. Chem.* **253**, 2471–2482.

Chang, A. C. Y. and Cohen, S. N. (1978). *J. Bacteriol.* **134**, 1141–1156.

Chirgwin, J. M., Przybyla, A. E., Macdonald, R. J., and Rutter, W. J. (1979). *Biochemistry* **18**, 5294–5299.

Christophe, D., Brocas, H., Gannon, F., Martynoff, G. D., Pays, E., and Vassart, G. (1980). *Eur. J. Biochem.* **111**, 419–423.

Claudio, T., Ballivet, M., Patrick, J., and Heinimann, S. (1983). *Proc. Natl. Acad. Sci. USA* **80**, 1111–1115.

Cleveland, D. W., Lopata, M. A, MacDonald, R. J., Cowan, N. J., Rutter, W. J., and Kirschner, N. W. (1980). *Cell* **20**, 95–105.

Clewell, D. B. (1972). *J. Bacteriol.* **110**, 667–676.

Close, T. J., Christmann, J. L., and Rodriguez, R. L. (1983). *Gene* **23**, 131–136.

Cochran, B. H., Reffel, A. C., and Stiles, C. D. (1983). *Cell* **33**, 939–947.

Cohen, S. N. and Chang, A. C. Y. (1977) *J. Bacteriol.* **132**, 734–737.

Cohen, S. N., Chang, A. C. Y., and Hsu, L. (1972). *Proc. Natl. Acad. Sci. USA* **69**, 2110–2114.

Dagert, M., and Ehrlich, S. D. (1979). *Gene* **6**, 23–28.

Davis, A. R., Nayak, D. P., Ueda, M., Hiti, A. L., Dowbenko, D., and Kleid, D. G. (1981). *Proc. Natl. Acad. Sci.USA* **78**,5376–5380.

Devos, R., Pletinck, G. Cheroutre, H. Simons, G., Degrave, W. Tavernier, J., Remaut, E., and Fiers, W. (1983). *Nucl. Acids Res.* **11**, 4307–4323.

Edman, J. C., Hallewell, R. A., Valenzuela, P., Goodman, H. M., and Rutter, W. J. (1981). *Nature* **291**, 503–506.

Edmonds, M., Vaughn Jr., M. H., and Nakazato, H. (1971). *Proc. Natl. Acad. Sci. USA* **68**, 1336–1340.

Efstradiatis, A. and Villa-Komaroff, L. (1979). In *Genetic Engineering* (J. K. Setlow and A. Hollaender, eds.), Plenum Press, New York, Vol. 1, p.15.

Feramisco, J. R., Smart, J. E., Burridge, K., Helfman, D. M., and Thomas, G. P. (1982). *J. Biol. Chem.* **257**, 11024–11031.

Garfinkel, L. I., Periasamay, M., and Nadel-Ginard, B. (1982). *J. Biol. Chem.* **257**, 11078–11086.

Glisin, V., Crkvenajakov, R., and Byus, C. (1974). *Biochemistry* **13**, 2633–2637.

Goeddel, D. V., Yelverton, E., Ullrich, A., Heynecker, H. L., Mioazari, G., Holmes, W., Seeburg, P. H., Dull, T. May, L. Stebbing, N., Crea, R. Maeda, S., McCandliss, R., Sloma, A., Tabor, J. M., Gross, M., Familleti, P. C., and Pestka, S. (1980). *Nature* **287**, 411–416.

Gold, L., Pribnow, D., Schneider, T., Shinedling, S., Singer, B. S., and Stromo, G. (1981). *Annu. Rev. Microbiol.* **35**, 365–403.

Goldberg, D. A. (1980). *Proc. Natl. Acad. Sci. USA* **77**, 5794–5798.

Goldberg, M. L., Lefton, R. P., Stark, G. R., and Williams, J. G. (1979). *Methods Enzymol.* **68**, 206–220.

Gorman, C. M., Merlino, G. T., Willingham, M. C., Pastan, I., and Howard, B. H. (1982). *Proc. Natl. Acad. Sci. USA* **79**, 6777–6781.

Gottesman, M. M. (1978). *Proc. Natl. Acad. Sci. USA* **75**, 2767–2771.

Gottesman, M. M. and Cabral, F. (1981). *Biochemistry* **20**, 1659–1665.

Gottesman, S., Gottesman M., Shaw, J. E., and Pearson, M. L. (1981). *Cell* **24**, 225–233.

Gottesman, M. M. and Sobel, M. E. (1980). *Cell* **19**, 449–455.

Grunstein, M. and Hogness, D. S. (1975). *Proc. Natl. Acad. Sci. USA* **72**, 3961–3965.

Gubler, U. and Hoffman, B. J. (1983). *Gene* **25**, 263–269.

Gunning, P., Ponte, P., Okayama, H., Engel, J., Blau, H., and Kedes, L. (1983). *Mol. Cell. Biol.* **3**, 787–795.

Hamer, D. H. and Leder, P. (1979). *Cell* **17**, 737–747.

Hanahan, D. (1983). *J. Mol. Biol.* **166**, 557–580.

Hanahan, D. and Meselson, M. (1980). *Gene* **10**, 63–67.

Hastings, K. E. M. and Emerson, C. P. (1982). *Proc. Natl. Acad. Sci. USA* **79**, 1553–1557.

Hawley, D. K. and McClure, W. R. (1983). *Nucl. Acids Res.* **11**, 2237–2255.

Helfman, D. M., Feramisco, J. R., Fiddes, J. C., Thomas, G. P., and Hughes, S. H. (1983). *Proc. Natl. Acad. Sci. USA* **80**, 31–35.

Jacobsen, H., Klenow, H., and Overgaard-Hansen, K. (1974). *Eur. J. Biochem.* **45**, 623–627.

Jolly, D. J., Okayama, H., Berg, P., Esty, A. S., Filpula, D., Bohlen, P., Johnson, G. G., Shively, J. E., Hunkapillar, T., and Friedman, T. (1983). *Proc. Natl. Acad. Sci. USA* **80**, 477–481.

Kahn, M., Kolter, R., Thomas, C., Figursky, D., Meyer, R., Remaut, D., and Helenski, D. R. (1979). *Methods Enzymol.* **68**, 268–280.

Kay, R. M., Harris, R., Patient, R. K., and Williams, J. G. (1980). *Nucl. Acids Res.* **8**, 2691–2707.

Khoury, G. and Gruss, P. (1983). *Cell* **33**, 313–314.

Kleid, D. G., Yansura, D., Small, B., Dowenko, D., Moore, D. M., Grubman, M.J., McKercher, P. D., Morgan, D. O., Robertson, B. H., and Bachrach, H. L. (1981). *Science* **214**, 1125–1129.

Korman, A. J., Knudsen, P. J., Kaufman, J. R., and Strominger, J. L. (1982). *Proc. Natl. Acad Sci.USA* **79**, 1844–1849.

Kurokawa, T., Seno, M., Sasada, R., Ono, Y., Onda, H., Igarishi, K., Kikuchi, M., Sugino, Y., and Honjo, T. (1983). *Nucl. Acids Res.* **11**, 3077–3085.

Kurtz, D. T. and Nicodemus, C. F. (1981). *Gene* **13**, 145–152.

Land, H., Grez, M., Hauser, H., Lindenmaier, W., and Schutz, G. (1981). *Nucleic Acids Res* **9**, 2251–2266.

Laub, O. and Rutter, W. J. (1983). *J. Biol. Chem* **258**, 6043–6050.

Lehrach, H., Diamond, D., Wozney, J. M., and Boedtker, H. (1977). *Biochemistry* **16**, 4743–4751.

Long, E. O., Wake, C. T., Strubin, M., Gross, N., Accolla, R. S., Carrel, S., and Mach, B. (1982). *Proc. Natl. Acad. Sci. USA* **79**, 7465–7469.

Maloy, S. R. and Nunn, W. D. (1981). *J. Bacteriol.* **145**, 1110–1111.

Mandel, M. and Higa, A. (1970). *J. Mol. Biol.* **53**, 154–162.

Maniatis, T., Fretsch, E. F., and Sambrook, J. (1982). *Molecular Cloning (A Laboratory Manual).* Cold Spring Harbor Press, Cold Spring Harbor

Maxam, A. M. and Gilbert, W. (1977). *Proc. Natl. Acad. Sci. USA* **74**, 560–564.

Maxam, A. M. and Gilbert, W. (1980). *Methods Enzymol.* **65**, 499–560.

McMaster, G. K. and Carmichael, G. G. (1977). *Proc. Natl. Acad. Sci, USA* **74**, 4835–4838.

Mermer, B., Malamy, M., and Coffin, J. M. (1983). *Mol. Cell. Biol.* **3**, 1746–1758.

Messing, J. (1984). *Methods in Enzymology*, Academic Press, New York.

Michelson, A. M., Markham, A. F., and Orkin, S. H. (1983). *Proc. Natl. Acad Sci. USA* **80**, 472–476.

Michelson, A. M. and Orkin, S. H. (1982). *J. Biol. Chem.* **257**, 14773–14782.

Mizusawa, S. and Gottesman, S. (1983). *Proc. Natl. Acad. Sci. USA* **80**, 358–362.

Montell, C., Fischer, E. F., Caruthers, M. H., and Berk, A. J. (1983). *Nature* **305**, 600–605.

Nagata, S., Taira, H., Hall, A., Johnsrud, L., Streuli, M., Escodi, J., Boll, W., Cantell, K., and Weissman, C. (1980). *Nature* **284**, 316–320.

Okayama, H., and Berg, P. (1982). *Mol. Cell. Biol.* **2**, 161–170.

Okayama, H., and Berg, P. (1983). *Mol. Cell. Biol.* **3**, 280–289.

Parnes, J. R., Velan, B., Felsenfeld, A., Ramanathan, L., Ferrini, U., Appella, E., and Sidman, J. G. (1981). *Proc. Natl. Acad. Sci. USA* **78**, 2253–2257.

Paterson, B. M., Roberts, B. E., and Kuff, E. L. (1977). *Proc. Natl. Acad. Sci. USA* **74**, 4370–4374.

Pennica, D., Holmes, W. E., Kohr, W. J., Harkins, R. N., Vehar, G. A., Ward, C. A., Bennet, W. F., Yalverton, E., Seeburg, P. H., Heyneker, H. L., Goeddel, D. V., and Collen, D. (1983). *Nature* **301**, 214–221.

Prakash, O., Guntaka, R. V., and Sarkar, N. H. (1983). *Gene* **23**, 117–130.

Prives, C. L., Aviv, H., Paterson, B. M., Roberts, B. E., Shmuel, R., Rozenblatt, S., Revel, M., and Winocour, E. (1974). *Proc. Natl. Acad. Sci. USA* **71**, 302–306.

Retzel, E. F., Collet, M. S., and Faras, A. J. (1980). *Biochemistry* **19**, 513–518.

Ricciardi, R. P., Miller, J. S., and Roberts, B. E. (1979). *Proc. Natl Acad. Sci. USA* **76**, 4927–4931.

Rosenberg, M. and Court, D. (1979). *Annu. Rev. Genet.* **13**, 319–353.

Roychoudhury, R., Jay, E., and Wu, R. (1976). *Nucleic Acids Res.* **3**, 863–877.

Sanger, R. and Coulson, A. R. (1975). *J. Mol. Biol.* **94**, 441–448.

Sanger, F., Nicklin, S. and Coulson, A. R. (1977). *Proc. Natl. Acad. Sci. USA* **74**, 5463–5467.

Schumperli, D., Howard, B. H., and Rosenberg, M. (1982). *Proc. Natl. Acad. Sci. USA* **79**, 257–261.

Shimatake, H. and Rosenberg, M. (1981). *Nature* **292**, 128–132.

Shine, J. and Delgarno, L. (1975). *Nature* **254**, 34–38.

Schulze, D. H., Pease, L. R., Obata, Y., Nathenson, S. G., Reyes, A. A., Ikuta, S., and Wallace, R. H. (1983). *Mol. Cell Biol.* **3**, 750–755.

Southern, E. (1975). *J. Mol. Biol.* **98**, 503.

St. John, T. P. and Davis, R. W. (1979). *Cell* **16**, 443–452.

Suggs, S. V., Wallace, R. B., Hirose, T., Kawashima, E. H., and Itakura, K. (1981). *Proc. Natl. Acad. Sci. USA* **78**, 6613–6617.

Sutcliffe, J. G. (1978). *Nucleic Acids Res.* **5**, 2721–2728.

Sutcliffe, J. G. (1979). *Cold Spring Harbor Symp. Quant. Biol.* **43**, 77–90.

Timberlake, W. E. (1980). *Dev. Biol.* **78**, 497–510.

Ullrich, A., Shine, J., Chirgwin, J., Pictet, R., Tischer, E., Rutter, W. J., and Goodman, H. M. (1977). *Science* **196**, 1313–1319.

Vassart, G. and Brocas, H. (1980). *Bichim. Biophys. Acta* **610**, 189–194.

Vicira, J. and Mcssing, J. (1982). *Gene* **19**, 259–268.

Wallace, R. B., Johnson, M. J., Hirose, T., Miyake, T., Kawashima, E. H., and Itakura, K. (1981). *Nucl. Acids Res.* **9**, 879–894.

Wickens, M. P., Buell, G. N., and Schimke, R. T. (1978). *J. Biol. Chem.* **253**, 2483–2495.

Williams, J. G. (1981). In *Genetic Engineering* (R. Williamson, ed.), Academic Press, New York, vol. 1, pp. 1–59.

Yoo, O. J., Powell, T., and Agarwal, K. L. (1982). *Proc. Natl. Acad. Sci. USA* **79**, 1049–1053.

Young, R. A. and Davis, R. W. (1983a). *Proc. Natl. Acad. Sci. USA* **80**, 1194–1198.

Young, R. A. and Davis, R. W. (1983b). *Science* **222**, 778–782.

Zimmerman, C. R., Orr, W. C., Leclerc, R. F., Barnard, E. C., and Timberlake, W. E. (1980). *Cell* **21**, 709–715.

SECTION 3

GENETIC SYSTEMS DEVELOPED IN CHINESE HAMSTER CELL LINES

A. Intermediary Metabolism

CHAPTER 11

De novo PURINE AND PYRIMIDINE BIOSYNTHESIS

David Patterson
Eleanor Roosevelt Institute for Cancer Research
Florence R. Sabin Laboratories for Genetic
and Developmental Medicine
Department of Biochemistry, Biophysics and Genetics
and Department of Medicine
University of Colorado Health Sciences Center
Denver, Colorado

I. BACKGROUND

A. Biochemistry

1. Identification and Characterization of the Steps of the Pathways

The critical importance of regulation of purine and pyrimidine nucleotide synthesis has long been recognized. The end products of these pathways are (1) the precursors of DNA and RNA synthesis; (2) critical for energy metabolism in cells; (3) important as cofactors in many enzyme reactions; (4) critical as nucleoside diphosphate sugar intermediates in the synthesis of membrane components including glycolipids and glycoproteins; and (5) used by most living cells as regulatory molecules in the form of cyclic nucleotides (see Chapter 23). The identification of the individual steps of purine and pyrimidine nucleotide synthesis was carried out in the 1950s and 1960s, both in bacterial systems and in avian liver systems in extremely elegant fashion and represents a major achievement of biochemical research. These studies have been extensively reviewed (Buchanan, 1959, 1960; Buchanan and Hartman 1959; Hartman and Buchanan, 1959; Hartman 1970; Henderson and Paterson, 1973; Makoff and Radford 1978; Shambaugh, 1979; Jones, 1980). Therefore, no attempt will be made to describe them in detail here. In most cases, however, complete purification of the relevant enzymatic steps and characterization of the relevant enzymes has not yet been accomplished. In fact, in some cases, surprising results continue to be uncovered regarding such fundamental aspects of these reactions as cofactor requirements. In the case of the purine biosynthetic pathway depicted in simplified form in Figure 11.1, this is largely due to the difficulty of assaying many of the enzymes of the pathway. Indeed, in most cases, the intermediates of the pathway are not commercially available, are difficult to synthesize in quantity, and, in many cases, are remarkably unstable. In fact, it is unclear whether the first intermediate unique to the purine biosynthetic

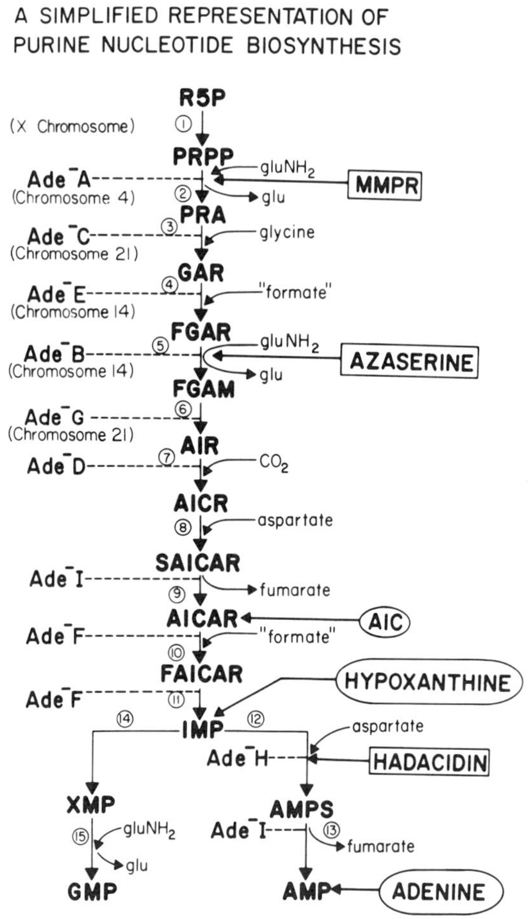

Figure 11.1. The pathway of purine nucleotide biosynthesis in mammalian cells.

pathway, phosphoribosylamine (PRA), has ever been isolated due to its great instability.

The pathway of *de novo* pyrimidine biosynthesis through the completion of synthesis of UMP requires only six steps, depicted in simplified form in Figure 11.2, and has been somewhat more tractable to biochemical analysis. In this case, complete unraveling of the biochemistry of the pathway awaited the realization that the first committed step in the pathway, carbamylphosphate synthetase, exists in two forms in mammalian cells (Jones, 1972). One is a mitochondrial form involved in the urea cycle and not involved in pyrimidine biosynthesis, and one is carbamylphosphate synthetase-2, a cytoplasmic enzyme which is involved in pyrimidine nucleotide synthesis (Hager and Jones, 1965, 1967; Ito and Tatibana, 1966; Tatibana and Ito, 1967, 1969).

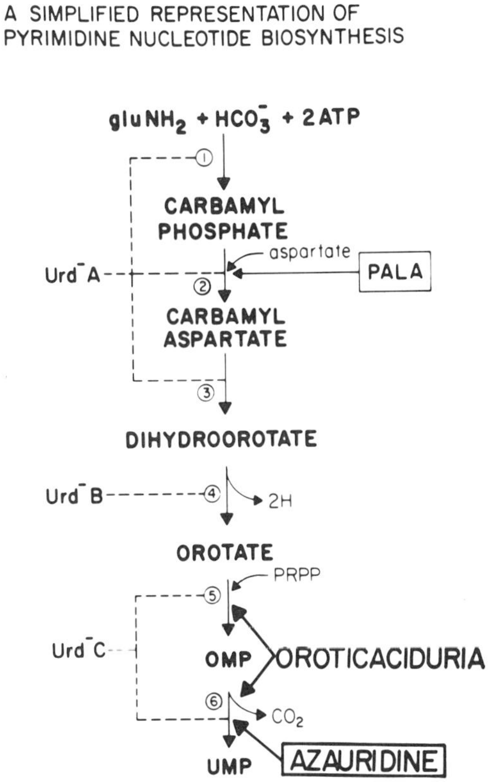

Figure 11.2. The pathway of pyrimidine nucleotide biosynthesis in mammalian cells.

2. *Copurification of Enzymes, Multienzyme Complexes, and Multifunctional Proteins*

One of the most intriguing aspects of the biochemical analysis of the purine and pyrimidine biosynthetic pathways in many animal systems has been the repeated observation that certain of the enzymes of these pathways copurify. It has been hypothesized, and enzyme purification data have been presented to support the hypothesis, that all of the enzymes required for IMP synthesis exist in a large multienzyme complex (Rowe et al., 1978). Recently, evidence has appeared suggesting that three enzymes of tetrahydrofolate cofactor metabolism and two enzymes of *de novo* purine metabolism, phosphoribosylglycineamide (GAR) formyltransferase and phosphoribosylaminoimidazole carboxamide (AICAR) formyltransferase, are located in a multienzyme complex (Capperelli et al., 1980; Smith et al., 1980; Mueller and Benkovic, 1981; Smith et al., 1981). There is good evidence that AICAR formyltransferase and the next enzyme in the pathway, inosinicase, also copurify (Flaks and Lukens, 1963) and that the enzyme adenylosuccinase carries out two steps of purine biosynthesis, steps 9 and 13 of Figure 11.1 (Giles et al., 1957; Dorfman, 1969; Westby and Gots, 1969; Armitt and Woods, 1970; Patterson, 1976; Tu and Patterson, 1977). There is somewhat weaker evidence that steps 6 and 7 (Fig. 11.1) are also carried out by a multifunctional enzyme complex (Patey and Shaw, 1973).

In general, the structure of the enzymes existing in these complexes remains undefined. It may well be that certain of the biochemically observed multienzyme complexes are artifactual and might not reflect *in vivo* associations but simply pecularities of the purification procedure employed. It is clear that some of the enzyme activities, which in some instances are observed to be associated, have been purified away from each other. However, it may be that it was the attempts to separate the various activities which were artifactual. One can imagine reasons for having the enzymes of a pathway associated in complexes, for example, efficient use of unstable intermediates. In some systems, experimental evidence exists supporting these hypotheses.

Since the late 1960s and early 1970s, there have been repeated reports of the biochemical association of the first three enzymatic steps of pyrimidine biosynthesis (depicted in Fig. 11.2) copurifying and being associated as a multienzyme complex (Jones, 1972; Shoaf and Jones, 1973; Coleman et al., 1977). It now seems clear through the biochemical analysis of many laboratories, and through biochemical–genetic analysis by Stark and his collaborators and in this laboratory, that these three enzymatic reactions are carried out on a single polypeptide in mammalian and other animal systems (Kent et al., 1975; Mori et al., 1975; Mori and Tatibana, 1975, 1978; Jarry, 1976; Kempe et al., 1976; Coleman et al., 1977; Patterson and Carnright, 1977; Davidson and Patterson, 1979; Davidson et al., 1979; Padgett et al., 1979; Wahl et al., 1979; deSaint Vincent et al., 1981). The fourth step of this pathway, dihydroorotate dehydrogenase, is not a cytoplasmic enzyme but appears to be associated with the outer surface of the inner mitochondrial membrane and may well be complexed with the electron transport system (Forman and Kennedy, 1975, 1976; Chen and Jones, 1976).

There have been numerous reports of biochemical association of the last two steps of this pathway, orotate phosphoribosyltransferase (OPRT) and OMP decarboxylase, and the best biochemical evidence now strongly indicates that, again, these two enzyme activities are carried out by a multifunctional protein (Suttle and Stark, 1979; McClard et al., 1980). This conclusion is strongly supported by somatic cell genetic data as well (Krooth et al., 1979; Levinson et al., 1979; Suttle and Stark, 1979; Patterson, 1980; Patterson et al., 1983).

3. Regulation of the Pathways in Animals

Given the critical nature of these two pathways, it would seem reasonable to expect that they would be under relatively careful biochemical genetic regulation. Opportunities for regulation of *de novo* purine biosynthesis by various biochemical means exist at virtually every step of the pathway. Regulatory mechanisms might include genetic repression and derepression, end product feedback inhibition, regulation by substrate or cofactor concentration, and perhaps other mechanisms (Henderson, 1972). There have been indications of genetic repression and derepression in microorganisms (Bach et al., 1979; Losson and Lacroute, 1983), but only a few studies have been carried out in any mammalian cell or animal system to assess whether such regulation occurs in animals. One report by Martin and Owen (1972) demonstrated loss of ability of cells to synthesize one of the early intermediates of the pathway, phosphoribosylformylglycineamide (FGAR), upon addition of exogenous purine. Restoration of this ability could be inhibited by presence of actinomycin when purines were removed from the medium. It was suggested that this inhibition of restoration of enzyme activity by actinomycin D might be evidence for transcriptional control of purine synthesis. Unfortunately, a number of other possibilities exist and were not ruled out. For example, it might be that actinomycin D itself inhibits FGAR accumulation.

At least two steps of the purine biosynthetic pathway are carried out by enzymes well known to have regulatory properties, namely, PRPP synthetase, the first enzyme depicted in Figure 11.1, and amidophosphoribosyltransferase, the second enzyme depicted in Figure 11.1. Both these enzymes are subject to end product inhibition. PRPP synthetase is extremely sensitive to inorganic phosphate concentration, and amidophosphoribosyltransferase has been shown to undergo allosteric and structural changes in response to substrate and end product concentrations (Wyngaarden and Ashton, 1959; Henderson, 1962; Holmes et al., 1973; Wood and Seegmiller, 1973; Green and Martin, 1974; Itoh et al., 1976; Planet and Fox, 1976; Itakura and Holmes, 1979; Prajda et al., 1979; Tsuda et al., 1979; Danks and Scholar, 1982). The relevance of any of these observations to regulation of purine synthesis *in vivo* is as yet unclear (with one exception to be discussed later).

Another line of evidence suggesting that purine synthesis may be regulated at the transcriptional or translational level involves a series of observations on various enzymes of this pathway, most particularly amidophosphoribosyltransferase, IMP dehydrogenase, and adenylosuccinate synthetase, in which the levels of enzymatic activity and, in some cases of actual enzyme molecules, were measured in cells from normal tissues and cells from malignancies with different growth rates (Reem and Friend, 1967; Katunuma and Weber, 1974; Jackson and Weber, 1975; Prajda et al., 1979; Tsuda et al., 1979). In general, the observations are that the higher the growth rate of a particular cell type, the higher the level of these enzymes. This has been interpreted to mean that transcriptional or translational control of enzyme levels in the purine pathway exists and that such regulation is somehow relevant to malignancy.

Regulation of the pyrimidine biosynthetic pathway is also complex. In animals it is clear that the first enzyme of the pathway, carbamylphosphate synthetase (Fig. 11.2), is allosterically regulated, its activity being decreased by UTP and increased by ATP and by phosphoribosylpyrophosphate (Jones, 1972; Tatibana and Shigesada, 1972; Tatibana and Mori, 1975). This situation is in contrast to pyrimidine synthesis in bacteria such as *Escherichia coli* in which aspartate transcarbamylase is allosterically regulated and is presumed to be the rate-limiting and regulated step of pyrimidine synthesis (Jones, 1972; Gerhart and Pardee, 1972). In animal cells, aspartate transcarbamylase is, as far as we can tell, biochemically unregulated. There is also some evidence that OPRT, the fifth enzyme of UMP synthesis, has regulatory properties (Shoaf and Jones, 1973; Hoogenraad and Lee, 1974).

Five of the six enzymes of UMP synthesis occur in two multifunctional protein molecules. It has been argued that such multifunctional proteins may serve the purpose of allowing for channeling of unstable intermediates or intermediates present in low concentration through a biochemical pathway (Jones, 1972; Traut and Jones, 1977; Traut, 1982). Such an organization of enzyme activities might also present animal cells with an alternative to an operon type of regulatory mechanism since regulation of a single protein would be sufficient to coordinately regulate several enzymatic functions. Evidence for such coordinate regulation of enzymes of pyrimidine synthesis has been obtained from studies on differentiating and malignant tissues (Reyes and Intress, 1970; Weber et al., 1978; Weber et al., 1981). The fourth enzyme of this pathway, DHO dehydrogenase, is unlikely to be in a tight complex with the other enzymes of the pathway since it is not a cytoplasmic enzyme (Chen and Jones, 1976). The significance of this finding for mechanisms of regulation is unclear.

B. Genetics

1. Prokaryotes and Lower Eukaryotes

There is a rich genetics of purine and pyrimidine biosynthetic pathways in prokaryotes and in certain lower eukaryotes. A thorough analysis of this literature is beyond the scope of this chapter, but a few points of particular relevance will be stressed. First, in *Salmonella* and in *E. coli*, all of the genes coding for the enzymes of purine synthesis are not linked in an operon configuration. There are, however, subgroups of genes which do appear to be in operonlike groups. The genes coding for GARS, AICAR formyltransferase, and inosinicase appear to form an operon in these organisms (Westby and Gots, 1969; Bachman, 1983). Adenylosuccinase, which carries out two enzymatic steps of the pathway, appears to be coded for by a single gene (Westby and Gots, 1969; Bachman, 1983). Similarly, the genes coding for the enzymes of pyrimidine biosynthesis ap-

TABLE 11.1
Listing of Chromosomal Location (Human) and Map Position (*E. coli*) of Genes for Enzymes of Purine Synthesis

Human		E. coli	
Name	Chromosome	Name	Locus (min)
Ade – A	4	purF	50
Ade – C	21	purD	90
Ade – E	14	—	—
Ade – B	14	purG(L)	55
Ade – G	21	purI(M)	54
Ade – D	?	purE	12
—		purC	53
Ade – I	?	purB	25
Ade – F	?	purH	90
Ade – F	?	purJ	90
Ade – H	?	purA	95

pear not to be linked in prokaryotic organisms (Jones, 1972; Trotta et al., 1971; Mergeay et al., 1974) and no evidence for structural association of the enzymes has been found. In *Neurospora*, the DNA regions coding for the first two enzyme activities are linked, while the third activity is not linked to these two. In this organism, apparently the first two activities are carried on a multifunctional protein, while the third activity is on a separate molecule (Denis-Duphil and Lacroute, 1971; Jones, 1972; Palmer and Dove, 1974; Makoff et al., 1978). Table 11.1 summarizes the known CHO and *E. coli* purine-requiring mutant groups and compares their chromosomal location, on the one hand, and their map position on the *E. coli* chromosome, on the other hand.

2. *Drosophila*

Of particular interest is the genetic and biochemical analysis of UMP synthesis in *Drosophila.* The first three enzymatic activities all are encoded by regions of the rudimentary locus in this organism (Norby, 1970). This was one of the earliest mutations observed in *Drosophila* and results in a characteristic phenotype in which the flies have short stubby or absent wings and which is associated with female sterility (Morgan, 1915). It is an X-linked trait. Nutritional and biochemical analysis of rudimentary locus mutants has demonstrated that these flies are lacking one or more of the first three enzymes of UMP synthesis (Norby, 1973; Falk and Nash, 1974; Jarry and Falk, 1974; Okada et al., 1974; Falk, 1976; Fausto-Sterling, 1977; Jarry, 1978; Rawls and Porter, 1979; Falk and DeBoer, 1980). More recent genetic and biochemical analysis indicates that the rudimentarylike locus, characterized by a phenotype very similar to that found in rudimentary, codes for the last two enzymes of UMP biosynthesis (Rawls, 1979, 1981; Conner and Rawls, 1982). A third locus, the DHOD null locus, codes for the fourth enzyme in the pathway, DHO dehydrogenase, and also results in similar phenotypic abnormalities (Rawls, et al., 1981). Thus, in *Drosophila* it appears that, again, there are multifunctional proteins carrying out the first three and the last two steps of UMP biosynthesis. Deficiencies in any of these steps yield characteristic morphological defects in the development of the fly.

3. Animals and Humans

Defects in the purine and pyrimidine biosynthetic pathways which lead to inherited metabolic disorders in humans have been observed. Thus, at least two different types of alterations in PRPP synthetase appear to lead to overproduction of purines and gout (Sperling et al., 1972; Becker et al., 1979). In one case, biochemical analysis of the altered enzyme suggests that the alteration results in decreased response of the enzyme to end product inhibition (Sperling et al., 1972). The second alteration, observed in an unrelated family, seems to result in an increased specific activity of the enzyme (Becker et al., 1979). These inherited metabolic disorders argue strongly that PRPP synthetase is one regulatory site of purine biosynthesis which is relevant in humans. There is also some evidence from what appear to be developmentally significant alterations in levels of enzymes of purine and pyrimidine synthesis that these pathways can be genetically regulated. For example, it appears that at least the last two enzyme activities of UMP synthesis are not found in rat intestinal mucosa (Raisonnier et al., 1981).

Humans defective in the last two enzymes of UMP synthesis are affected with the rare recessive inborn error of metabolism, hereditary orotic aciduria type I. If untreated, the disease is quite serious with symptoms including infantile hyperchromic anemia with megaloblastic marrow, leukopenia, retarded growth and development, and excessive excretion of orotic acid (Kelley, 1983). Treatment with large doses of uridine in the diet reverses these symptoms. This disease represents perhaps the only known human nutritional auxotrophic mutation. Very recently, evidence has been presented that a similar inborn error of metabolism may be carried in certain strains of cattle (Robinson et al., 1983).

II. RATIONALE FOR A GENETIC APPROACH USING CHO CELLS

In spite of the great deal of work carried out on analysis of these pathways, many questions still remain. In particular, the structures of the enzymes, the structures of the genes, the organization of the enzymes into larger complexes, the effects of end products, methods of alteration of gene expression, the chromosomal and regional location of most of the genes, and the relationship of gene and protein structure to expression of enzyme activity in the broad sense, including enzyme synthesis and degradation, are all still not understood in detail, especially in mammalian cells. It was felt by investigators in a number of laboratories several years ago that to alleviate this situation would require a combination of genetic and biochemical analysis. The rationale for this is that genetic analysis has a unique ability to reveal gene linkages and gene map positions and also to uncover genetic regulatory phenomena as demonstrated in the analysis of many metabolic pathways in microorganisms. Generation of mutant organisms simplifies analysis of the existence and role of multienzyme complexes or multifunctional proteins in regulation of metabolic pathways. In addition, it is possible to isolate and study mutants defective in regulatory as well as structural components of a pathway and therefore to uncover unexpected regulatory phenomena.

A system in which genetic biochemical regulation can best be studied must have the following properties: (1) it must be possible to isolate mutants, variants, and revertants with altered levels of the relevant activities; (2) the system should be known to involve regulation of an important metabolic pathway; (3) there should be some indication that relevant enzyme activities can be purified and that the relevant genes can be purified; (4) it should be possible to assign the relevant genes to particular chromosomes and to regionally map the genes at a finer level of resolution; and (5) the biochemical steps in the pathways should be known so that analysis of the pathways at the biochemical level can be undertaken. All of these conditions are met for analysis of purine and pyrimidine nucleotide synthesis in Chinese hamster cells.

III. SELECTION SCHEMES

A. The "BrdU Visible-Light" Selection

There have been four general schemes for obtaining mutants of Chinese hamster and other mammalian cells with aberrant purine or pyrimidine synthesis. The most useful of these has been the so-called "BrdU visible-light" selection procedure described in Chapter 2. Briefly, this selection procedure, devised by Drs. Puck and Kao (1967, 1968), depends on the inability of nongrowing cells to incorporate bromodeoxyuridine (BrdU) into their DNA. Thus, cells unable to grow without a particular nutrient will, when exposed to BrdU, be resistant to incorporation of this thymidine analog into their DNA. Wild-type cells not requiring the particular nutrilite will incorporate BrdU. After incorporation of BrdU, the culture may

TABLE 11.2
Distribution by Complementation Group of Purine-Requiring and Pyrimidine-Requiring Auxotrophs of CHO-K1

Complementation Group	Number of Independent Clones
Purine Requiring	
Ade – A	17
Ade – B	46
Ade – C	5
Ade – D	3
Ade – E	6
Ade – F	6
Ade – G	7
Ade – H	8
Ade – I	1
Complex	6
Total	105
Pyrimidine Requiring	
Urd – A	2
Urd – B	1
Urd – C	8
Total	11
Grand total	116

then be exposed to visible light, which will result in photolysis of BrdU-containing DNA and death of wild-type cells. Growth in complete medium will result then in the appearance of mutant colonies. In our laboratory, this procedure has been exceedingly effective and has allowed the isolation of over 100 independent clones of purine-requiring auxotrophic mutants and several clones of independently isolated pyrimidine-requiring mutants as shown in Table 11.2. Similarly, Taylor et al. (1970) reported the use of the BrdU visible-light procedure to isolate purine-requiring auxotrophs of CHO cells. A modification of this procedure in which black light was used instead of white fluorescent light and in which medium was altered was used by Chu et al. (1972) to isolate both purine and pyrimidine auxotrophs from Chinese hamster lung V79 cells. Uridine-requiring auxotrophs of V79 cells have also been isolated by Kusano et al. (1976), using a procedure in which diaminopurine lethality was substituted for BrdU visible-light lethality as the selection against wild-type cells.

B. Selection Schemes Based on Drug Resistance

1. Analogues

Several laboratories have isolated mutants and variants with altered purine or pyrimidine metabolism using drug-resistance selection protocols. Primarily, these protocols have resulted in alterations in *de novo* pyrimidine biosynthesis. Three laboratories have reported the isolation of mammalian cell mutants with defective pyrimidine synthesis and altered levels of both OPRT and OMP decarboxylase based on the resistance of cells to pyrimidine-base analogs, 5-fluorouracil or 5-fluoroorotic acid (Krooth et al., 1979; Levinson et al., 1979; Patterson, 1980). This selection appears to be due to the requirement in Chinese hamster and mouse cells for OPRT activity to activate these base analogs to their toxic form. Eight independently isolated clones of CHO cells isolated in this laboratory demonstrate a complete requirement for uridine for growth after this selection protocol (Patterson, 1980).

2. Inhibitors

Other drug-resistant mutants with aberrant purine or pyrimidine metabolism have also been selected in Chinese hamster and other mammalian cells in culture. The most well characterized are Syrian hamster mutants resistant to the transition state analog, *N*-(phosphonacetyl)-L-aspartate, or PALA. These mutants, first reported by Stark and colleagues, have been extensively characterized and arise by gene amplification (Kempe et al., 1976; Wahl, et al., 1979; deSaint Vincent et al., 1981). Thus, the gene for the first three enzymes of pyrimidine synthesis, the activity of the second of which is inhibited by PALA, is amplified extensively in PALA-resistant mutants. Such mutants can be isolated from a variety of mammalian cells in culture including Chinese hamster cells. In addition, drug-resistant gene amplification mutants have been isolated in response to treatment with increasing levels of azauridine together with or sequentially with pyrazofurin (Suttle and Stark, 1979). These drugs inhibit the last enzyme of UMP biosynthesis, OMP decarboxylase. The most highly characterized mutants of this type have been isolated from Syrian hamster or rat hepatoma cells (Suttle and Stark, 1979), but mutants have also been isolated in this laboratory from Chinese hamster. At present, PALA-resistant and azauridine-resistant gene amplification mutants represent the most well-understood mutants of mammalian cells in these pathways.

Additionally, mutants have been isolated from Chinese hamster cells in the laboratory of DeMars (Lester et al., 1980), and also in this laboratory, which are resistant to the glutamine analog azaserine that inhibits at least two steps of the *de novo* purine biosynthetic pathway steps, 2 and

5 (Fig. 11.1). Preliminary characterization seems to indicate that these mutants also are gene amplification mutants.

C. Replica Plating

A third method of isolation of Chinese hamster ovary mutants with defective pyrimidine synthesis is a replica plating screening method (Stamato and Hohmann, 1975; Esko and Raetz, 1978). Thus if one can devise an assay which distinguishes between mutant and wild-type cell colonies, even if this assay is lethal when applied, it is possible simply to make a replica plate, to carry out the assay on one copy of the replica, and return to the master replica to pick the colony with the mutant phenotype. This method has been used to isolate only a single type of mutant in pyrimidine biosynthesis in CHO cells (Stamato and Patterson, 1979) largely because it has not proven readily possible to devise assays which can be used on the single colony level to discern mutant and wild-type cells. It would seem likely that such methods could be developed for many of the enzymes of purine and pyrimidine synthesis using isotopic procedures modeled after those employed by Raetz's laboratory to isolate mutants of lipid metabolism (Esko and Raetz, 1980; Polokoff et al., 1981).

D. Culture of Cells from Patients with Inherited Metabolic Disorders

The final method for obtaining cells with aberrant purine or pyrimidine metabolism is culture of cells from patients with inherited metabolic disorders. This procedure is obviously limited by the number of such patients available and until recently by the fact that such cultures, if started from fibroblasts, would be of limited life-span. Nevertheless, cultures of cells from patients with orotic aciduria, deficiency of the last two steps of UMP synthesis, have been isolated and are available for study (Krooth, 1964). These cultures, of course, only exist for humans, and at the moment no such cultures have been obtained from Chinese hamsters.

IV. RECENT STUDIES DEFINING THE MOLECULAR BASIS OF MUTATIONS

A. General Strategies for Defining a Defect

By far the largest number of mutants defective in purine or pyrimidine metabolism have been isolated as auxotrophic mutants of Chinese hamster cells requiring purines for growth. Well over 100 independent isolates of such mutants have been obtained using the BrdU visible-light procedure by investigators in several different laboratories. The first such mutants were isolated by Dr. Fa-Ten Kao and Dr. Theodore T. Puck and were reported in 1969 (Kao and Puck, 1969). The original mutants were designated hyp$^-$, a designation which was later changed to Ade$^-$. While almost all mutants reported to date have been derived from CHO cells, at least one mutant has been isolated in Chinese hamster V79 cells by the laboratory of Ernest Chu and was reported in 1972 (Chu et al., 1972). (See Chapter 3.)

The isolation of a large number of such mammalian cell mutants, all of which possess a similar phenotype, namely, the requirement of purines for growth, requires a general strategy for determination of the defect in each mutant. As developed in this laboratory, the strategy contains four steps: (1) nutritional analysis; (2) analysis of intermediate accumulation; (3) complementation analysis; and (4) analysis of enzyme activity in cell-free extracts.

1. Growth and Nutritional Analysis

First, each mutant is subjected to a detailed analysis of its growth requirements with all available intermediates in the pathway. For the purine biosynthetic pathway, this is fairly limited, and generally four compounds are employed: aminoimidazole carboxamide, which detects mutants defective in enzymatic steps prior to step 10 in Figure 11.1; hypoxanthine, which detects mutants defective in enzymatic steps prior to step 12 in Figure 11.1; adenine, which detects all mutants except those deficient in the guanine-specific branch of the pathway; and guanine, which should detect those mutants deficient in the guanine-specific branch of the pathway. For the pyrimidine biosynthetic pathway, all intermediates are available, and analysis can, therefore, be more informative.

Analysis by growth in the presence of various intermediates is limited by a number of possible pitfalls. First, all the intermediates may not be available. Second, especially for the purine biosynthetic pathway, all of the intermediates are phosphorylated. This means that mammalian cells may not be particularly permeable to these intermediates. Also, some of the intermediates of both pathways appear to be unstable, especially in tissue culture medium. For example, phosphoribosylamine, the first unique intermediate in the purine pathway, and carbamylphosphate, the first intermediate of pyrimidine biosynthesis, are both noted for an extremely short half-life. Moreover, there are phosphatases and phosphorylases in fetal calf serum commonly used to grow CHO cells in culture which may dephosphorylate some of these intermediates. The use of nonphosphorylated precursors of the intermediates of the purine pathway has been tried. For example, attempts have been made to grow purine-requiring mutants on glycineamide. These experiments have met with no success even in situations where the defect is known to be prior to the appearance of the glycineamide-containing intermediate phosphorybosylglycineamide (GAR). It may be that mammalian cells do not have appropriate enzymes to convert what one might expect to be precursors of intermediates to the intermediates themselves.

2. *Intermediate Accumulation*

In both pathways, it has been possible to use accumulation of radio-labeled intermediates by mutants as a method of analysis. Most useful in the purine biosynthetic pathway has been the accumulation of intermediates labeled with [^{14}C]formate. Formate enters the purine biosynthetic pathway at step 4 in Figure 11.1. All intermediates after this are labeled when CHO cells are fed [^{14}C]formate. This approach has been used extensively in our laboratory and also by Taylor to analyze purine-requiring auxotrophs (Feldman and Taylor, 1974; Patterson et al., 1974; Patterson, 1975). Of course, such an approach requires a biochemical method to separate each of the labeled intermediates in the pathway and also requires some method to identify each intermediate accumulated by a specific mutant. Fortunately, all of the intermediates of the pathway are separable by combinations of various thin-layer chromatographic (TLC) systems. Moreover, the intermediates of the purine biosynthetic pathway, although not commerically available, have been well characterized by a number of investigators. Descriptions of these intermediates and the methods for their preparation published by Flaks and Lukens (1963) and Lukens and Flaks (1963) in *Methods in Enzymology* have been most valuable.

Of particular use in characterization of purine biosynthetic mutants has been the colorimetric reaction first characterized by Bratton and Marshall (1939). This reaction is extraordinarily simple to carry out, is extremely sensitive, and identifies unambiguously four of the intermediates of purine biosynthesis: phosphoribosylaminoimidazole (AIR), phosphoribosylaminoimidazolecarboxylic acid (AICA), phosphoribosylsuccinylaminoimidazolecarboxamide (SAICAR), and phosphoribosylaminoimidazolecarboxamide (AICAR). All these compounds give a slightly different color with the Bratton–Marshall reaction, which is clearly distinguishable to the eye or to the spectrophotometer. It is possible, in the case of the Ade$^-$D, Ade$^-$I, and Ade$^-$F complementation groups, to detect and distinguish easily the Bratton–Marshall product using a single 35 mm dish of CHO cells. Thus far this reaction has given completely unambiguous results as long as appropriate precautions are taken to maintain the correct pH of the reaction and to eliminate chemicals which may potentially interfere with the development of color. The reaction has been used by us to quantitate the appearance of these intermediates (Waldren and Patterson, 1979).

Experiments analogous to formate-labeling experiments can be carried out using radio-labeled glycine, which enters the purine biosynthetic pathway at step 3 catalyzed by phosphoribosylglycineamide synthetase (GARS). In general, any observations made with [^{14}C]formate will be confirmed if [^{14}C]glycine is used as a label. Often, however, [^{14}C]glycine contains radio-labeled contaminants which must be removed before its use, and the glycine itself interferes with ready interpretation of thin-layer chromatograms in some chromatographic systems. On the other hand, labeling with glycine does allow the detection of the appearance of GAR, an intermediate which would be missed by labeling with [^{14}C]formate. Labeling with [^{14}C]glycine has been used in the analysis of the Ade$^-$E mutant of CHO cells for the detection of accumulation of GAR by these mutants (Patterson, 1975).

The original labeling experiments using formate or glycine were carried out with CHO cells which had been removed from the Petri dishes on which the cells had been grown. More recently we have found that modifications of this procedure increase its sensitivity by approximately 20-fold. Thus, if one adds excess glycine to the tissue culture medium in which cells are grown, removes all sources of purines during the labeling reaction, carries out the labeling reaction in the plates in which the cells have been grown, and prewarms all medium used both for washes and incubations, accumulation of intermediates is markedly elevated. This modification, first reported by Oates, Vannais, and Patterson (Oates et al., 1980) seems only to affect the quantitation of intermediate accumulation, since accumulation of all intermediates by all of the mutants thus far examined has been qualitatively similar using this modification and using the originally reported procedure of Patterson (1975). It should be kept in mind, however, that detailed comparison of these two methods for assessment of intermediate accumulation has not been carried out for all complementation groups. It is not possible to state with certainty that any single experimental procedure will be best in all cases.

A most important part of analysis by intermediate accumulation is the use of specific inhibitors of purine biosynthesis. In particular, we have made use of the glutamine analog azaserine and also of the guanosine analog methylmercaptopurine riboside (MMPR). Azaserine inhibits very strongly the fifth step of the pathway, FGAR amidotransferase, and to a lesser extent the second step of the pathway, amidophosphoribosyltransferase (Patterson et al., 1974; Patterson, 1975; Oates et al., 1980). In CHO cells, it is possible to inhibit step five virtually completely with essentially no inhibition of step two using appropriate concentrations of azaserine, as best shown in Oates et al. (1980). Treatment of wild-type CHO-K1 cells with appropriate concentrations of azaserine causes the rapid and quantitative accumulation of FGAR, an intermediate readily detectable on several TLC systems after labeling with radio-labeled formate. If mutants are so treated, one finds that they fall into two classes. Either the mutant is able to synthesize FGAR or it is not able to synthesize FGAR. By this strategy, the purine pathway is conveniently broken down into two parts, one containing the first four steps, the other containing the rest. The use of azaserine also allows one to deduce whether a particular intermediate accumulated by a mutant and detected by TLC is in fact an intermediate of purine biosynthesis. If the accumulation of this intermediate is inhibited by azaserine, then this is good evidence that this intermediate is a purine biosynthesis intermediate.

The use of MMPR allows an independent means of gathering the same type of information. MMPR is phosphorylated by CHO cells and, as in other mammalian cells, is, in its nucleotide form, a potent inhibitor of amidophosphoribosyl transferase (Hill and Bennett 1969; Shantz et al., 1972; Patterson, 1975; Oates, 1976). If a mutant accumulates a new metabolically labeled compound and if the accumulation of this intermediate is eliminated by treatment of the mutant with MMPR, this is strong evidence that this intermediate appears in the purine biosynthetic pathway. This is especially the case if an intermediate is eliminated by treatment with either azaserine or MMPR or if the mobility of the intermediate is changed by treatment with azaserine such that a compound with the mobility of FGAR now accumulates. Perhaps the best example of the use of intermediate accumulation and inhibition in definition of mutants remains the initial report of Patterson (1975).

3. *Complementation Analysis*

Especially in analysis of the initial mutants, genetic complementation analysis was an invaluable aid to the characterization of the mutants. The general outline of a complementation analysis is described in Figure 11.3. Briefly, if one has two mutants, Ade^-1 and Ade^-2, requiring the same compound, in this case, for example, adenine, and one wishes to know if these mutants are defective in the same enzymatic step or in a different enzymatic step in the biosynthetic pathway, complementation analysis offers a simple and rapid way to determine this question. One simply fuses the two mutants in question and selects for growth of hybrids in the absence of added adenine. If hybrids grow, then one can conclude that the two mutants used as parents must be defective in different enzymatic steps in the biochemical pathway and the mutants are said to complement one another. That is, in the hybrid cell one mutant supplies the enzyme missing in the other and vice versa. Such studies were first reported in 1969 for glycine requiring auxotrophs (Kao et al., 1969). The first complementation groups for Ade^- mutants were defined this way in 1972 (Kao and Puck, 1972).

One also can ask whether a particular mutant is dominant or recessive. Generally, if two mutants complement, this is a good indication that both mutants are recessive. However, in cases where complementation is not observed or in cases where one has reason to doubt a straightforward interpretation of a complementation analysis, one may wish to fuse a par-

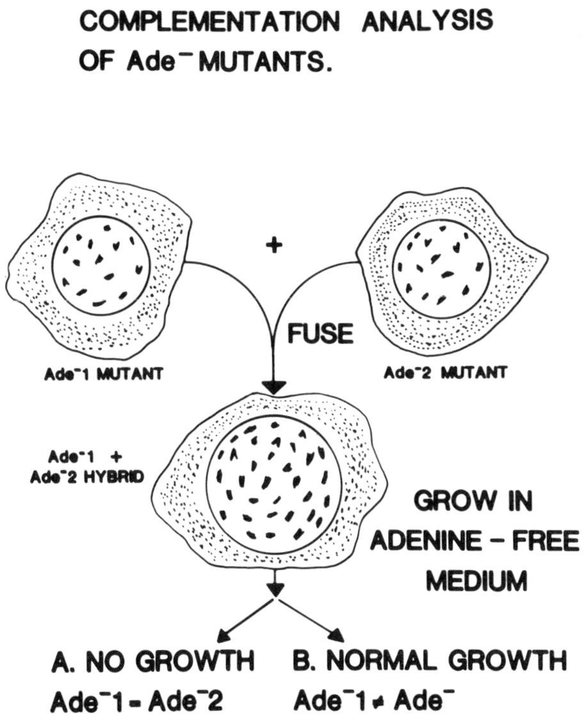

Figure 11.3. Schematic representation of the complementation analysis of adenine auxotrophic CHO mutants.

ticular purine-requiring mutant, for example, with a cell wild type for the purine biosynthetic pathway using markers unrelated to the purine biosynthesis pathway and to select hybrids on the basis of these other outside markers. Since all of the CHO-K1 cell mutants isolated are descendents of the original CHO-K1 cell which has an additional auxotrophy, namely, a requirement for proline for growth, this experiment is relatively straightforward. One requires a derivative of CHO-K1 cells that is proline$^+$ which may be obtained simply by reversion of the proline locus as first reported by Kao and Puck (1969) and which contains a mutation conferring auxotrophy for another marker, for example, glycine. It is then possible to select hybrids in glycine-free, proline-free medium between this pro$^+$ gly$^-$ revertant and the pro$^-$ Ade$^-$mutant. The medium will contain adenine. After hybrids are selected, they can then be tested for whether or not they possess a requirement for purine. If the requirement is present, the mutant is said to be dominant. If the requirement is absent, the mutant is said to be recessive. This protocol avoids problems due to potential reversion at the purine locus or due to potential second site restoration of activities in one or another of the hybrid cells. Thus far, all purine-requiring mutants analyzed this way have been found to be recessive. Certain pyrimidine-requiring mutants may, however, show a dominant phenotype (Patterson et al., 1983b).

Complementation analysis is particularly useful in situations where intermediate accumulation or growth experiments yield little information. For example, it is possible to discriminate between mutants at step 2 and step 3 of the purine pathway using complementation analysis, although neither of these mutants accumulates any intermediates when labeled with either [^{14}C]glycine or [^{14}C]formate (Patterson, 1975). Complementation analysis also allows a more rapid initial analysis of a group of newly isolated mutants. In our laboratory, we now have a standard battery of one of each of the nine known complementation groups of purine-requiring auxotrophs. When new mutants are isolated, they are fused to this battery of cells. It is thus possible to assign a mutant preliminarily to a particular complementation group without carrying out nutritional or intermediate accumulation analysis. This procedure should be recognized simply as a rapid screening since unusual complementation behavior, such as that shown by the Ade$^-$P$_{AB}$ and Ade$^-$P$_{CG}$ mutants, to be described later, would be missed. Its advantage is that it does allow us to focus on more informative new types of mutants more rapidly.

4. *Enzyme Assay*

Once analysis of a mutant by growth, complementation, and intermediate accumulation has been carried out, it is of course necessary then to assay at least representative members of each complementation group by direct enzyme assay in cell-free extracts for the step or steps likely to be defective as revealed by these other forms of analysis. Direct enzyme assay of all complementation groups except Ade$^-$D and Ade$^-$E has confirmed their defects as predicted by other methods of analysis. Direct enzyme assay was the only procedure that allowed the placement of the Ade$^-$A and Ade$^-$C complementation groups in steps 2 and 3 of Figure 11.1, respectively, since no intermediates were accumulated by these mutants and since their growth characteristics were identical (Oates and Patterson, 1977). The enzymatic defect in each of the Ade$^-$ complementation groups is shown in Table 11.3.

B. Unique Aspects of the General Strategy Applied to the Pyrimidine Pathway

The logic of the analysis of mutants of pyrimidine biosynthesis is much the same as that used for analysis of purine biosynthesis. However, in practice there are many substantial differences. First, the intermediates for this pathway are all commercially available. This allows for more simple analysis of growth requirements and also for more simple direct

TABLE 11.3
Summary of Defects in CHO-K1 Cell Purine-Requiring Mutants

Complementation Group	Presumed or Established Site of Defect	Criteria
Ade – A	Amidophosphoribosyltransferase (EC 2.4.2.14)	Enzyme assay
Ade – B	FGAR amidotransferase (ED 6.3.5.3)	Enzyme assay
Ade – C	GAR synthetase (EC 6.3.4.13)	Enzyme assay
Ade – D	AIR carboxylase (EC 4.1.1.21)	Intermediate accumulation
Ade – E	GAR formyltransferase (EC 2.1.2.2)	Intermediate accumulation
Ade – F	AICAR formyltransferase (EC 2.1.2.3)	Enzyme assay
	Inosinicase (EC 3.5.4.10)	Enzyme assay
Ade – G	AIR synthetase (EC 6.3.3.1)	Enzyme assay
Ade – H	Adenylosuccinate synthetase (EC 6.3.4.4)	Enzyme assay
Ade – I	Adenylosuccinate lyase (EC 4.3.2.2)	Enzyme assay
Ade – P_{CG}	GAR synthetase (EC 6.3.4.13)	Enzyme assay
	AIR synthetase (EC 6.3.3.1)	Enzyme assay
Ade – P_{AB}	Amidophosphoribosyltransferase (EC 2.4.2.14)	Enzyme assay
	FGAR amidotransferase (EC 6.3.5.3)	Enzyme assay

enzyme assay. The fact that one is dealing with six enzymatic steps instead of 15 means that direct enzyme assay is a more tractable approach. In addition, isolation of pyrimidine-requiring mutants has been less successful than isolation of purine-requiring mutants and only approximately 15 such mutants have been isolated. Finally, the nature of the mutants in the pyrimidine pathway by and large has been easier to deduce since in most cases the procedure by which the mutant was isolated limits the likely steps at which the lesion occurs. For example, the single mutant in step 4 of this pathway, DHO dehydrogenase (Figure 11.2), was isolated by a replica plating procedure that screened for mutants which possessed decreased levels of this particular enzyme activity (Stamato and Patterson, 1979). Mutants isolated in the last two steps of the pathway were isolated by virtue of their resistance to 5-fluorouracil, and this was carried out because it was hypothesized that OPRT was required for the lethal action of this particular agent (Patterson, 1980). While complementation analysis, growth requirement analysis, and intermediate accumulation were carried out for these mutants, it was possible to carry out direct enzymological assays and to determine their defects at an early stage of analysis quite rapidly. Of course, it is often informative and intellectually satisfying to carry out assays of all the steps of the relevant pathway for each mutant, a task of considerable magnitude considering the number of mutants and enzymatic steps involved.

C. Summary of Results of Studies to Define Enzymatic Defects

Using the previously described approaches, mutants falling into nine different complementation groups have been isolated from CHO-K1 cells and assigned to 11 of the steps of *de novo* purine biosynthesis. The four steps that are missing include the first step catalyzed by PRPP synthetase. Mutants in this step do exist in the human population as described earlier (Sperling et al., 1972; Becker et al., 1979) and have been isolated from rat hepatoma cells in culture (Green and Martin, 1973). As of yet, no mutants defective in step 8, phosphoribosylsuccinylaminoimidazolecarboxamide synthetase, have been reported. However, it has been reported that this step and the step catalyzed by reaction 7 are carried out on a multifunctional enzyme (Patey and Shaw, 1973). Thus it may be that at least certain members of the Ade$^-$D complementation group, assigned to step 7 by virtue of their accumulation of AIR, may in fact be defective in both steps 7 and 8. No Chinese hamster mutant cells have been described which are defective in the GMP-specific branch of purine biosynthesis, steps 14 and 15 of Figure 11.1.

Mutants falling into three complementation groups have been established as defective in UMP biosynthesis. The Urd$^-$A complementation group is defective in the first three enzymatic reactions (Patterson and Carnright, 1977), the Urd$^-$B group is defective in the fourth step (Stamato and Patterson, 1979), and the Urd$^-$C group is defective in steps 5 and 6 (Patterson, 1980). The definition of the Urd$^-$B complementation group is somewhat unusual in that no mutants in this step which require uridine for growth have yet been found, although such a requirement can be induced by treatments which do not affect wild-type cells (Stamato and Patterson, 1979). Therefore, standard complementation analysis has not been possible, although hybrid cells between members of this complementation group and others using outside markers do possess the enzyme activity. The lesion in these mutants was identified by enzyme assay, intermediate accumulation, and nutritional analysis. Its designation as a separate complementation group is, therefore, primarily to maintain a consistent terminology.

D. Analysis of Specific Mutants

It is beyond the scope of this chapter to discuss in detail the assignment of each complementation group to a particular enzymatic step. However, four systems will be discussed in some detail for illustrative purposes, because these systems reveal unique and unexpected complexities regarding purine or pyrimidine biosynthesis and because they offer exceptional opportunities for future study.

1. *Pyrimidine Pathway Mutants*

Somatic cell biochemical genetic analysis of pyrimidine synthesis in CHO cells has proceeded extremely rapidly to the molecular level due largely to the availability of informative mutants and appropriate nucleic acid probes.

a. The Urd⁻A Complementation Group. By appropriate analysis of the two mutants of the Urd⁻A complementation group and revertants of one of these, it has been possible to demonstrate that the first three enzyme activities of UMP biosynthesis are carried on a single multifunctional protein with a molecular weight of 220,000 (Patterson and Carnright, 1977; Davidson and Patterson, 1979; Davidson et al., 1979; Davidson et al., 1981). In both mutants, residual enzyme activity of approximately 10% or less for all three enzymes has been detected. It has been possible to demonstrate in one mutant that the structure of the protein is altered such that a molecule of approximately 190,000 daltons remains which contains the first and third activities while the second activity has been cleaved away from this molecule (Davidson and Patterson, 1979; Davidson et al., 1979; Davidson, et al. 1981). Revertants of this mutant can be isolated in which the amounts of the first and third activities are elevated (Davidson and Patterson, 1979; Davidson et al., 1979). In other cases, revertants in which a protein with apparently normal molecular weight and possessing all three activities coordinately restored have been isolated (Davidson et al., 1979). Elegant studies by Davidson's group have demonstrated that it is possible to reproduce the lesions seen in the mutants qualitatively by treatment of wild-type enzymes with proteases (Davidson et al., 1981; Rumsby et al, 1984). Using this approach, Davidson has been able to deduce the domain structure of this protein (Rumsby et al, 1984). In very recent studies, this group has been able to correct the defect in *E. coli* mutants lacking the second of these enzyme activities using a partial cDNA derived from the Syrian hamster (Davidson and Niswander, 1983). These studies confirm and extend the studies on the protein itself and offer a new approach to the study of this and other multifunctional proteins.

Earlier work by Stark and his collaborators has clearly demonstrated that a similar protein which they named the CAD protein exists in Syrian hamster cells (Kempe et al., 1976; Coleman et al., 1977; Wahl et al., 1979; deSaint Vincent et al., 1981). This work takes advantage of the fact that it is possible to amplify the gene for CAD protein by treatment of cells with progressively increasing concentrations of the inhibitor PALA. Dr. Stark's group and also the group of Dr. Geoffrey Wahl have isolated partial cDNA clones and partial and complete functional genomic clones containing DNA sequences of the Syrian hamster CAD gene (deSaint Vincent et al., 1981). These and related studies have shed considerable

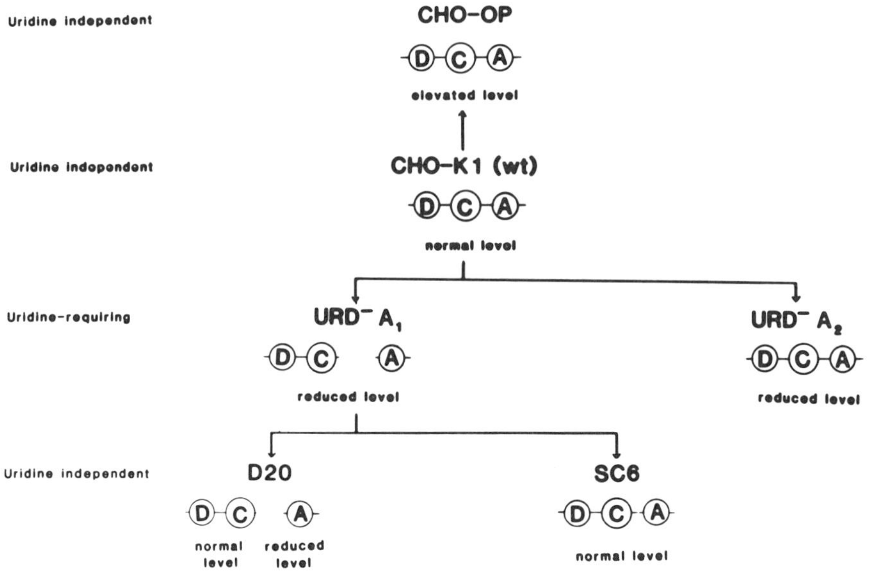

Figure 11.4. Genetic alterations affecting CAD in CHO cells.

light on the structure of the Syrian hamster gene. The types of mutants and revertants which exist at the CAD locus in CHO cells are shown in Figure 11.4.

Drs. Stark and Wahl have generously made available to us certain recombinant DNA molecules carrying fragments of the CAD gene or fragments of the CAD cDNA. These have been of great use in analysis of the gene in CHO cells and have allowed for the first time a combination of somatic cell genetic and recombinant approaches to the study of auxotrophic mutants in this biochemical pathway in Chinese hamster cells. Eventually this approach should allow the definition of the exact nature of the lesion in the CAD gene, the effect of this lesion on transcription of the gene, and a comparison of the lesion found in the DNA with the lesion found in the CAD protein.

b. The Urd$^-$C Complementation Group. Analysis of mutants defective in the last two steps of UMP synthesis, OPRT and OMP decarboxylase, has also proceeded rapidly. The nature of the defect in these mutants was determined using approaches of mutant analysis described earlier. As first pointed out by Krooth (Krooth et al., 1979), this system presents certain advantages over other systems for study of purine or pyrimidine synthesis. Of primary importance, it is relatively simple to isolate mutants defective in these enzymatic steps. OPRT appears to be required to activate either 5-fluorouracil or azaorotate to their lethal forms.

Therefore, simply by selection of the mutants resistant to the killing action of these agents, mutants deficient in OPRT can be selected (Krooth et al., 1979; Levinson et al., 1979; Patterson, 1980). In every case, mutants isolated using this protocol have shown a coordinate loss of OPRT and also OMP decarboxylase, even though 5-fluoro UMP, for example, would not be a substrate for the latter enzyme since it is not carboxylated. Such CHO mutants have an absolute requirement for uridine for growth and have been designated Urd^-C. Revertants of these mutants are easy to isolate simply by selecting for growth in the absence of uridine. Thus it is not necessary to go through a complicated and difficult negative selection scheme such as the BrdU visible-light selection to isolate large numbers of mutants in this enzymatic system. The coordinate loss of both activities and restoration of activities in revertants demonstrates coordinate regulation of these two enzyme activities. It has been possible to isolate overproducers of these enzymes by selection for resistance to progressively increasing concentrations of azauridine or pyrazofurin or both (Suttle and Stark, 1979). In all cases reported to date, coordinate overproduction of both enzyme activities has been observed. These findings coupled with biochemical analyses, notably purification of the enzyme activities to apparent homogeneity, make it virtually certain that OPRT and OMP decarboxylase are carried on a single multifunctional polypeptide. Mutants lacking those activities are analogous to cells obtained from patients with orotic aciduria type 1 (Krooth, 1964; Kelley, 1983).

An electrophoretic mobility difference between human and hamster enzyme activities can be demonstrated (Patterson et al., 1983a). Moreover, it has been possible to create CHO-human cell hybrids between Urd^-C mutants and normal cells. Certain of these hybrids have retained only human chromosome 3, or a fraction of chromosome 3, as their human genetic material. These hybrids express the human electrophoretic form of OPRT/OMP decarboxylase. Thus we can conclude that the structural gene for this multifunctional enzyme resides on the long arm of human chromosome 3 between the centromere and chromosome band q21 (Jones et al., 1984; Patterson et al., 1983a).

The ability to discriminate between human and hamster activity has allowed us to attempt DNA-mediated gene transfer of the human gene for OPRT/OMP decarboxylase into Urd^-C mutants. (See Chapter 8 for a more complete discussion of this approach to genetic analysis.) This has been accomplished, and we now have Urd^+C transferrents expressing human OPRT/OMP decarboxylase activity. It should be possible to isolate from these transferrents a functional copy of the human OPRT/OMP decarboxylase gene and to undertake an in-depth analysis of mutations at this locus in both human and Chinese hamster cells. Such probes have already been isolated from rat hepatoma cells (Suttle, 1983). It seems highly likely that probes from other species will now become available rapidly.

2. *Purine Pathway Mutants*

The study of purine synthesis has several advantages to offer and is significant in some ways beyond that of the study of pyrimidine synthesis. The pathway of purine biosynthesis is considerably more complex than the pathway of pyrimidine biosynthesis, and the organization of the various enzymes in this pathway not only reflects this complexity but apparently varies from enzyme to enzyme and from organism to organism. There are at least three proposed mechanisms by which synthesis of phosphoribosylamine, the second step depicted in Figure 11.1 and the first committed step of the pathway, can occur and until recently there was little evidence regarding which of these was significant in mammalian cells. Work on two different systems in the purine biosynthetic pathway is of particular relevance to these questions and will be discussed below.

a. The Ade^-P_{CG} System. The most straightforward and internally consistent complex system in the purine pathway involves the Ade^-C and Ade^-G complementation groups which catalyze steps 3 and 6 of purine biosynthesis, respectively (Oates and Patterson, 1977; Irwin et al., 1979). While most mutants of the Ade^-C complementation group are missing the activity of GAR synthetase (GARS) but possess the activity of AIR synthetase (AIRS), and most mutants of the Ade^-G group lack AIRS activity but possess GARS activity, one mutant, Ade^-P_{CG}, has been isolated which is defective in both activities (Patterson et al., 1981). It is possible to restore by reversion both activities in this mutant. Thus, it is possible to demonstrate coordinate regulation of the genes for GARS and AIRS. It would be of great importance to evaluate whether or not these enzyme activities are indeed on multifunctional peptides, are associated in a noncovalent enzyme complex, or are not functionally associated. In *Schizosaccaromyces pombe*, evidence similar to that obtained in CHO cells has been interpreted to mean that GARS and AIRS activities are present together on a multifunctional polypeptide (Fluri et al., 1976). There is no definitive evidence as yet on this point in mammalian cells, although certain of the data obtained, namely, a reduced GARS activity in at least one Ade^-G mutant, the coordinate loss of both activities in Ade^-P_{CG}, and coordinate restoration of activities of GARS and AIRS in the revertant are consistent with such a hypothesis. It should be noted that while one Ade^-G mutant has reduced GARS activity, the other three isolates have now been checked and have normal GARS activity (Kato and Patterson, unpublished observations).

Genetic analysis of the Ade^-C and Ade^-G mutants has allowed the assignment of genes correcting both these defects to human chromosome 21 (Moore et al., 1977; Patterson et al., 1981). In recent unpublished experiments, we have been able to obtain evidence that a gene on human chromosome 21 corrects the defect in the Ade^-P_{CG} mutant as well (Jones, personal communication). The assignment of these genes to the same chromosome is consistent with the possibility that a single gene might be involved in all mutants. Recently, we have been able to demonstrate that human GARS appears to have a markedly increased heat sensitivity when compared with the CHO enzyme (Patterson and Schandle, 1983). The enzyme in human/CHO hybrids containing human chromosome 21 as their only cytogenetically identifiable human material has heat sensitivity identical to that of human enzyme. Therefore, it appears that the gene on human chromosome 21 which corrects the defect in the Ade^-C mutant must code for the GARS enzyme. It appears likely that the most direct approach to future progress in this area will involve isolation of the relevant enzymes and DNA segments, their characterization, and fine-structure mapping.

b. The Ade^-P_{AB} System. The first complex system of biosynthesis to be observed, and still the most complicated and least well-understood collection of mutants in either biochemical pathway, involves the purine complementation groups Ade^-A and Ade^-B. These were the first two complementation groups of purine-requiring mutants to be isolated and were first reported by Kao and Puck in 1969 and 1971. These mutants were shown to belong to two separate complementation groups in 1972 (Kao and Puck). The Ade^-A and Ade^-B complementation groups apparently represent the only adenine-requiring complementation groups that have been reported in the literature by laboratories other than our own. Taylor et al. (1970) reported the isolation of purine-requiring mutants of CHO cells in 1971 and later, in 1974, Feldman and Taylor demonstrated that at least some of these mutants possessed a defect in FGAR amidotransferase. This defect corresponds to our Ade^-B complementation group (Patterson et al., 1974; Patterson, 1975; Oates et al., 1980). In 1972, Chu et al. reported the isolation of an adenine-requiring mutant of V79 cells, which was shown by Feldman and Taylor in 1975 to be defective in amidophosphoribosyltransferase activity. This mutant was subjected to more extensive biochemical analysis in 1976 by Holmes et al.

As discussed previously, the enzymatic reaction catalyzed by amidophosphoribosyltransferase is of particular interest with respect to purine biosynthesis because it is thought to be the rate-limiting step of the *de novo* purine biosynthetic pathway and catalyzes the first unique reaction in the pathway (Wyngaarden and Ashton, 1959). Elegant biochemical studies by a number of laboratories have demonstrated that this enzyme has many characteristics of a feedback-inhibition-regulated enzyme including allosteric structural changes and inhibition by end products (Wyngaarden and Ashton, 1959; Henderson, 1962; Holmes et al., 1973; Wood and Seegmiller, 1973; Tsuda et al., 1979). In addition, this enzyme has an unusual sensitivity to molecular oxygen which may have regulatory significance (Itakura and Holmes, 1979). This latter characteristic makes the enzyme difficult to study.

One particularly intriguing aspect of this step of purine biosynthesis, which results in the synthesis of phosphoribosylamine (PRA), is that there are at least three mechanisms by which PRA may be synthesized in cell-free extracts of mammalian cells. These include (1) the utilization of glutamine and phosphoribosylpyrophosphate (PRPP) as substrates; (2) the use of ammonia and PRPP as substrates; and (3) the use of ammonia and ribose-5-phosphate (R5P) as substrates. Until mutants defective in PRA biosynthesis became available in mammalian cells, it was exceedingly difficult to determine which of these reactions actually plays a role in purine biosynthesis in intact cells and in animals including humans. Isotopic experiments indicate that glutamine is the primary amino donor at the amidophosphoribosyltransferase step but that a substantial fraction of nitrogen found in PRA may come from ammonia (Sperling et al., 1973). Holmes et al. (1976) assessed amidophosphoribosyltransferase, aminophosphoribosyltransferase, and ammonia-R5P aminotransferase activities in wild-type Chinese hamster cells and in Ade^-A mutants. They demonstrated that ammonia-R5P aminotransferase activity was normal but that glutamine-dependent amidophosphoribosyltransferase or ammonia-dependent aminophosphoribosyltransferase were both deficient in the mutant (Holmes et al., 1976). They concluded that the simultaneous disappearance of both PRPP-dependent activities suggested that these two enzyme activities were closely related structurally or genetically in Chinese hamster cells. A similar conclusion was reached by this laboratory in analysis of the Ade^-A complementation group of CHO cells (Oates and Patterson, 1977). Later, it was demonstrated that the Ade^-B complementation group is deficient in FGAR amidotransferase (Oates, 1976; Oates, et al., 1980).

These mutants took on added interest and importance with the description in 1980 of a new type of mutant of CHO cells, the Ade^-P_{AB} mutant (Oates et al., 1980). This mutant shows unusual complementation behavior in that it fails to complement all members of the Ade^-A or Ade^-B complementation groups thus far tested but does complement all other hypoaxanthine-requiring Ade^- complementation groups. Biochemical analysis of this mutant demonstrates that it has indeed lost FGAR amidotransferase activity as have members of the Ade^-B complementation group and that it has lost amidophosphoribosyltransferase activity as have members of the Ade^-A complementation group. Unlike Ade^-A mutants, however, this mutant has retained essentially normal levels of aminophosphoribosyltransferase activity. Thus, both glutamine-dependent activities, steps 2 and 5 of purine biosynthesis, have been lost in this mutant. The ammonia-dependent activity which can carry out PRA synthesis remains. This considerably complicates any genetic or regulatory interpretations regarding synthesis of PRA in mammalian cells. One conclusion which is clear from studies of this mutant is that ammonia can indeed be used as a source of nitrogen in the second step of *de novo* biosynthesis, namely, synthesis of PRA. The level of FGAR synthesis in $Ade^{-}P_{AB}$ is very similar to the observed levels of ammonia-dependent FGAR synthesis in either CHO-K1 or $Ade^{-}B$ cells. Therefore, it is probable that $Ade^{-}P_{AB}$ uses the ammonia-dependent reaction for FGAR synthesis even in the presence of glutamine. This is a strong indication that ammonia-dependent synthesis of PRA can be sufficient for mammalian purine synthesis and cell growth.

Considerable further analysis of this type of mutant has been undertaken. Of particular significance is the analysis of purine-independent revertants of $Ade^{-}P_{AB}$. A large number of these have been isolated and a few have been analyzed in some detail. They are able to grow in purine-free medium, although with reduced growth rate. Growth rate can be restored to normal by addition of hypoxanthine to the medium. Thus, purine synthesis is growth limiting in these cells. They appear to have complete or nearly complete restoration of FGAR amidotransferase activity. Early attempts to detect glutamine-dependent PRA synthesis in these revertants were unsuccessful. More detailed experiments on the enzyme activities in one of these revertants carried out in the laboratory of Dr. Ed Holmes now suggests that glutamine-dependent PRA synthesis has been partially restored. Moreover, it appears that at extremely high glutamine concentrations activity may be present at low levels in the $Ade^{-}P_{AB}$ mutant, although there is a clear difference between the mutant and the revertant. The K_m's for glutamine for the enzyme from wild type are approximately 1 m*M* and for the revertant greater than 10 m*M*. The K_m of the residual activity in the mutant is not possible to determine because of solubility limits of glutamine. The K_m of 1 m*M* observed in wild-type CHO-K1 cells is similar to that observed for amidophosphoribosyltransferase from a wide variety of sources. The K_m for ammonia for PRA synthesis appears to be unaltered in wild-type, mutant, or revertant cells, as expected.

These data allow the conclusion that both the forward mutation and the reversion event have affected both amidophosphoribosyltransferase activity and FGAR amidotransferase activity in the $Ade^{-}P_{AB}$ mutant. This information is of great significance, since otherwise one could argue that two independent mutations in the same cell, or a deletion of two closely linked genes, could result in the phenotype observed in the $Ade^{-}P_{AB}$ mutant. There are also several other lines of evidence against this interpretation. First, we have obtained three independent isolates of purine auxotrophs of CHO-K1, all isolated after treatment with ethylmethane sulfonate, which have complementation behavior like that of $Ade^{-}P_{AB}$. This is approximately the frequency at which most of our purine auxotrophs appear. If $Ade^{-}P_{AB}$ were a double mutant, it would be expected to be extremely rare. Second, the existence of partial revertants of $Ade^{-}P_{AB}$ demonstrates that deletion of all the relevant genes is not involved in the original mutation. Third, the complementation pattern observed with Ade^-P_{AB}, Ade^-A, and Ade^-B is not easily explained by assuming that Ade^-P_{AB} is the result of two independent mutations.

It would of course be extremely useful to know the location of the gene or genes involved in the Ade^-A, Ade^-B, and Ade^-P_{AB} complementation groups. At the moment, no information is available on this point in CHO cells. However, two genes have been assigned to human chromosomes. A gene correcting the defect in Ade^-B cells has been assigned to human chromosome 14 (Kao, 1980), and Stanley and Chu (1978) assigned the Ade^-A gene to human chromosome 4. Therefore, in humans at least, two of the genes involved in this system are not even located on the same chromosome.

In very recent preliminary experiments carried out in this laboratory, further evidence for a relationship between the Ade^-A and the Ade^-B genetic loci has been obtained. It was decided to attempt to isolate mutants in which one or both of these genes are amplified by selection for resistance to increasing concentrations of the glutamine analog, azaserine, which as mentioned previously inhibits both glutamine-dependent enzymatic steps. Such mutants have now been obtained and show resistance to several-hundred-fold the normal lethal levels of azaserine. In addition, these mutants show resistance to several-hundred-fold the normal lethal concentrations of MMPR, a very potent inhibitor of PRA synthesis, as described above. However, the mode of action of these two drugs is likely to be unrelated. Since azaserine is at least 100-fold more effective at inhibition of the FGAR amidotransferase, it seems likely that both FGAR amidotransferase and amidophosphoribosyltransferase have been affected in these mutants. These characteristics are most easily explained by hypothesizing amplification of both FGAR amidotransferase and amidophosphoribosyltransferase activities. Again, it is clear that progress in unraveling this complex biochemical–genetic system awaits purification of the relevant enzymes and DNA sequences.

E. Genomic (DNA) Alterations in Mutants

To date there is no direct published evidence that any of the mutants so far isolated in purine or pyrimidine synthesis result from DNA changes. However, there is considerable circumstantial evidence that at least some of these mutants actually result from changes in DNA sequence. For example, Patterson et al. (1974) reported that certain members of the Ade^-A complementation group that were generated with ICR191, a mutagen known to cause frame shift mutations, could be reverted by treatment with ICR191 but not with mutagens thought to cause primarily point mutations or deletions. In two cases it has been possible to correct the nutritional requirements of these mutants by DNA-mediated transfer of a gene apparently coding for enzyme activity. In the case of the Urd$^-$A mutant, this transfer has been carried out with a cloned functional gene (deSaint Vincent et al., 1981). In the case of the Urd$^-$C mutants, DNA-mediated gene transfer using whole human cell DNA as a donor and Urd$^-$C cells as recipients has resulted in growth of uridine-independent colonies expressing enzymes of human electrophoretic mobility and carrying human DNA as assessed by Southern analysis using human middle repetitive DNA as a probe (Gusella et al., 1982; Davis, Bleskan, and Patterson, unpublished observations).

Finally, in very recent experiments, we have been able to identify in the DNA of Urd$^-$A cells a specific restriction site lesion in the Urd$^-$A gene using cloned cDNA probes supplied by Drs. Stark and Wahl. This lesion appears to cause the loss of a *PstI* restriction site in the DNA of the Urd$^-$A cell. This restriction site has been localized to within an approximately 500-base-pair region of the gene, and upon preliminary analysis, appears to be near that part of the gene that codes for aspartate transcarbamylase activity. A lesion in this region of the gene would be consistent with our hypotheses regarding the nature of this mutation as deduced from the alterations observed in the CAD protein in this mutant. More definitive experiments along these lines will require sequencing of the relevant region of the affected gene, cloning of the mutant gene, and transfer of this gene into some other host followed by demonstration that the recipient cell which has taken up the mutated gene now synthesizes an appropriately altered protein.

The question of whether alterations in methylation of DNA play any role in isolation of purine and pyrimidine-requiring mutants of CHO cells has begun to be addressed. Harris has elegantly shown that in fact it is possible to revert the proline auxotrophy of CHO-K1 cells as well as alterations at the other loci by treatment of these cells with nonlethal concentrations of azacytidine (Harris, 1982; Harris, 1984). The hypothesized mechanism of this reversion involves decreased methylation of a gene coding for an enzyme required for growth in the absence of proline, although this has not yet been directly demonstrated.

In our laboratory, we have asked whether the Urd$^-$A mutant can be reverted by treatment with 5-azacytidine. In this experiment, Urd$^-$A CHO cells were treated with 5-azacytidine and after treatment the culture was split into two aliquots. One was examined for reversion at the proline locus and the other was examined for reversion at the uridine locus. In agreement with Harris's results, we found very high reversion at the proline locus. It was impossible to detect any increased reversion at the Urd$^-$A locus after treatment with 5-azacytidine. Our preliminary conclusion from this experiment (shown in Fig. 11.5) is that alteration in methylation which can be reverted by azacytidine treatment is not the primary cause of inactivation of the Urd$^-$A locus in Urd$^-$A mutants. There is, however, good evidence that at

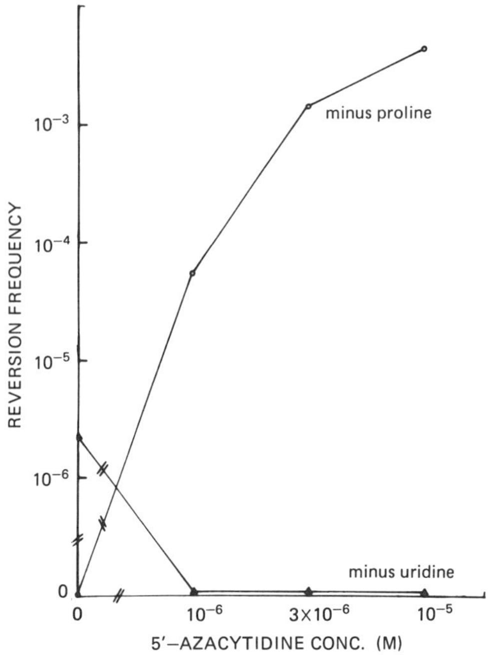

Figure 11.5. Reversion of proline auxotrophy and failure of reversion of uridine auxotrophy by 5′-azacytidine. See text for details. Urd^-A (pro^-) CHO cells were treated with 5′-azacytidine and analyzed for reversion of auxotrophy in medium minus uridine (Δ) or proline (○).

least some methylations which have regulatory significance are not revertable by azacytidine treatment (Stein et al., 1982). Therefore, further experimentation is essential to clarify this critical point.

F. Assignment of Genes to Specific Chromosomes

A question of fundamental importance in analyzing any biochemical pathway by somatic cell genetic approaches is the locations of the genes involved. To date there is no information available regarding the location of the genes for purine or pyrimidine biosynthesis in CHO cells. Mutants isolated in CHO cells have been used extensively for assignment of the genes coding for these enzymes to particular human chromosomes by the procedure of fusion of Chinese hamster mutants with wild-type human cells, usually either lymphocytes or normal human fibroblasts. In this way, the genes correcting the enzymatic defects observed in the mutants localized to five of the first six enzymatic steps of the *de novo* purine biosynthetic pathway have been assigned to specific chromosomes. The gene coding for PRPP synthetase, the first enzyme of the pathway, has been assigned to the X chromosome by family studies (Becker et al., 1979). Stanley and Chu (1978) first reported assignment of the Ade^-A locus to human chromosome 4. Two chromosomes appear to carry more than one gene for enzymes of purine biosynthesis. The Ade^-E and Ade^-B loci appear to be on chromosome 14 in human cells (Kao, 1980; Jones et al., 1981) the Ade^-C and Ade^-G loci appear to be on chromosome 21 (Moore et al., 1977; Patterson et al., 1981). There is now evidence that the gene correcting the Ade^-C defect is in fact the structural gene for the GARS enzyme, since somatic cell hybrids containing chromosome 21 express GARS with human heat lability rather than hamster heat lability (Patterson and Schandle, 1983).

Assignment of genes to particular chromosomes and regional mapping of genes is of great importance in formulation of hypotheses regarding regulation of metabolic pathways. For example, any hypotheses regarding the nature of the Ade^-P_{AB} mutant must take into account the observation that it is likely that genes coding for components of these two enzymes will be on separate chromosomes, at least in some species. On the other hand, the assignment of the Ade^-C, Ade^-G, and Ade^-P_{CG} genes all to chromosome 21 is certainly consistent with the possibility that these two enzymes are carried by a multifunctional protein or at least that the genetic regions defining their activities are contiguous on the DNA. It has also been possible to carry out mapping of genes for the purine pathway in other species using a cell hybridization approach. Broad et al. (1984) were able to create CHO/sheep cell hybrids with Ade^-C and Ade^-G mutants and to demonstrate that these genes are syntenic to the sheep equivalent of SOD1. It appears, therefore, that a conserved linkage group has been observed between sheep and humans.

Assignment of the three genes involved in synthesis of enzymes of pyrimidine biosynthesis is also underway. Thus far it has been possible to regionally map the genes for OPRT/OMP decarboxylase to a small region of human chromosome 3 using cell hybrids between the Urd^- mutants and various human cells (Patterson et al., 1983a; Jones et al., 1984).

V. USE OF CHO MUTANTS FOR STUDIES OF NUCLEOTIDE METABOLISM

The Chinese hamster ovary cell mutants thus far isolated and characterized with defects in purine or pyrimidine metabolism have been very useful in studies of various aspects of nucleotide metabolism in several experimental systems. For example, the mutants are very useful in analysis of the effects of various drugs on nucleotide metabolism. The observation that mutants lacking OPRT are resistant to 5-fluorouracil shows that this enzyme must play a critical role at least in some mammalian cells in the toxicity and metabolism of this widely used cancer chemotherapeutic agent. Both purine- and pyrimidine-requiring mutants have been used for analysis of the mechanism of action of caffeine. In particular, it has been possible to use the mutants to localize the sites of action of caffeine. It was possible to demonstrate that caffeine interferes with the accumulation of FGAR in Ade$^-$B cells (Waldren and Patterson, 1979). Since this experiment can be done by adding caffeine directly to such cultures without adding any other inhibitor, it is likely that the effect of caffeine is a fairly direct one and that the effect must be prior to step 5 of the *de novo* purine biosynthetic pathway. Similar experiments allowed the localization of the effect of caffeine on pyrimidine synthesis to the step of this pathway involving PRPP and OPRT. No effect is observed on the early steps of the pyrimidine biosynthetic pathway (Rumsby et al., 1982). In addition, Chinese hamster cells have been used to assess the metabolism of caffeine directly (Waldren and Patterson, 1979; Simons et al., 1980; Tu et al., 1982). These experiments have direct relevance to the effect of caffeine on repair of DNA lesions induced by such agents as ultraviolet light (Waldren et al., 1983).

Other aspects of nucleotide metabolism may also be amenable to analysis by somatic cell genetic approaches. For example, it has recently been possible using a modification of the BrdU visible-light procedure to isolate a uridine-requiring mutant of CHO cells which accumulates large amounts of UDP glucuronic acid, this compound making up as much as 60% of acid soluble nucleotides in these cells (Patterson et al., 1983b). It may well be possible by extension of this approach and by analysis of mutants such as this to analyze nucleotide sugar metabolism in Chinese hamster cells.

In unpublished experiments, we have been able to assess aspects of the interconversion of guanine and adenine nucleotides using Ade$^-$H and Ade$^-$I mutants of CHO cells. This is of importance because of the hypothesis that regulation of the AMP- and GMP-specific branches of the pathway are coordinated in order to maintain an appropriate balance of adenine and guanine nucleotides (Crabtree and Henderson, 1971). Moreover, the possible interconversion of guanine and adenine nucleotides is of both theoretical and practical importance, for example, in the design of selection schemes to isolate specific types of mutants. It appears that at least in CHO cells, conversion of guanine nucleotides to adenine nucleotides is essentially nonexistent. It is possible to demonstrate that addition of guanine to cultures of Ade$^-$I cells has a sparing effect on the adenine requirement, but guanine cannot replace adenine. In fact, guanine cannot supply the growth requirements of any of our purine-requiring mutants. Guanine nucleotide levels do have a direct effect on AMP synthesis, however, which apparently is mediated through modulation of AMPS synthetase by guanine nucleotides. This can be deduced from the characterization of mutants of CHO cells which belong to the Ade$^-$H complementation group, but whose auxotrophic phenotype is contingent upon the presence of guanine in the culture medium (Tu and Patterson, 1978). The characteristics of these mutants lend direct support to the hypothesis that the GMP and AMP branches of this pathway are cross coordinately regulated.

Similarly, it has been possible to demonstrate that cytidine, cytosine, or uracil cannot support growth of uridine-requiring CHO cells (Patterson and Carnright, 1977). This means that CHO cells do not have a functional cytidine deaminase activity, nor do they possess uracil phosphoribosyltransferase activity sufficient to support growth. Such information is of fundamental importance for understanding regulation of these important pathways and demonstrates that a genetic approach can simplify such biochemical studies.

It is also possible to use mutants deficient in one step of a metabolic pathway to isolate mutants not isolatable from wild-type cells, an approach used by us to isolate HPRT negative mutants on a nutritional basis (Patterson and Jones, 1976) and by others to isolate a variety of important mutants of nucleotide metabolism in other mammalian cells (deSaint Vincent et al., 1980; Ullman et al., 1982).

In very recent experiments, the laboratory of Dr. Holmes in collaboration with our laboratory has used several of the Chinese hamster ovary cell mutants to assess the metabolic fate of aminoimidazolecarboxamide (AICA) riboside. It appears that AICA riboside may be useful in this regard because it is a compound which may be used to restore ATP levels which are depleted in ischemic episodes (Swain et al., 1982a, b; Sabina et al., 1983).

Both purine-requiring and pyrimidine-requiring mutants of CHO cells have been used to study nucleotide synthesis and salvage in the intracellular parasite *Toxoplasma gondii.* These studies have allowed the conclusion that these parasites must be incapable of synthesizing purines and must therefore depend on their host cells for purines (Schwarzman and Pfefferkorn, 1982). On the other hand, *Toxoplasma gondii* is capable of *de novo* pyrimidine biosynthesis (Schwartzman and Pfefferkorn, 1981). Such studies might potentially allow for design of therapeutic protocols for treatment of parasitic diseases.

VI. MAJOR CONCLUSIONS AND PROSPECTS FOR THE FUTURE

It seems clear that these extensive and varied uses of CHO mutants in purine and pyrimidine metabolism demonstrate the validity of a somatic cell genetic approach to analysis of metabolic pathways in mammalian cells. Moreover, work thus far demonstrates the critical need for a combined molecular, biochemical, and genetic approach. Although great progress has been made, it is only through such approaches that further progress on some of the most interesting questions of molecular genetics will be possible. Of particular importance for the immediate future will be the isolation of the functional genes coding for the various enzymes of these pathways, as has already been done for the CAD gene (deSaint Vincent et al., 1981). In addition, it would be important to isolate cDNA's in various vectors for analysis of gene structure and for assessment of gene regulation. Such studies will likely be the most straightforward way for rigorous confirmation and analysis of the molecular nature of the lesions in the mutants.

It will be necessary, of course, to continue to combine these approaches with classical somatic cell genetic approaches. For example, it was by complementation analysis that the unusual mutants Ade^-P_{AB} and Ade^-P_{CG} were uncovered, both of which demonstrate coordinate regulation of various enzymatic steps of the purine pathway. The mechanisms of coordinate regulation, especially in the case of Ade^-P_{AB}, remain a mystery. At present, only approximately 13% of the possible complementations between our purine-requiring auxotrophs have been carried out. It would be particularly informative, for example, to carry out all possible complementation analyses between Ade^-E and Ade^-B since these two loci have been mapped to the same human chromosome. The Ade^-E activity is one of those activities found in a multifunctional complex by Caparelli et al. (1980), Smith et al. (1980), Mueller and Benkovic (1981), Smith et al. (1981). It would not be surprising, then, to find unusual complementation between members of the Ade^-E and Ade^-B complementation groups which would indicate the need for a more penetrating analysis of the genetic and biochemical mechanisms by which expression of these genes might be regulated. There is now good evidence that the regulation of this locus is indeed complex, at least in *Drosophila*, based on analysis of primary transcription products using cloned DNA probes (Henikoff et al., 1983).

Isolation of the Urd^-A and Urd^-C mutants has given us firm genetic evidence for the multifunctional nature of the proteins carrying out the activities deficient in these mutants. It seems likely that similar observations may be found in the purine biosynthetic process and that isolation of additional mutants will aid greatly in the genetic and biochemical analysis and assessment of these possibilities.

Molecular genomic and cDNA probes of the various genes of the pathways should be useful for a number of studies including the assessment of the role of methylation on control of transcription of these genes and the significance of and mechanisms behind observations of a number of investigators that alterations in levels of many of the enzymes of *de novo* purine and pyrimidine pathways occur in malignant cells and that these alterations seem quantitatively related in some way to the growth rate of the particular malignant cells under study.

Aberrant regulation of these metabolic pathways is likely to be relevant not only to malignancy but to normal human development and to a number of developmental and genetic disease situations. The relationship of the $Urd^{-}C$ complementation group which is defective in OPRT and OMP decarboxylase activity to orotic aciduria in humans is of course obvious. It also seems highly likely that aberrant regulation of purine biosynthesis in humans may be responsible for at least a substantial fraction of patients with primary gouty arthritis of unknown etiology. Gouty arthritis represents one of the most common forms of arthritis affecting the population of the United States (Wyngaarden and Kelley, 1972). Mutants of CHO cells likely will be useful in understanding these regulatory phenomena and in isolating molecular probes that can be used for study of purine regulation in cells from a wide variety of individuals and animals.

The Chinese hamster cell mutants in the $Ade^{-}C$ and $Ade^{-}G$ complementation groups have already been instrumental in undertaking a molecular analysis of Down syndrome, the most common genetic chromosomal cause of mental retardation among the population of the United States (Adams et al., 1981). The localization of at least two genes involved in purine biosynthesis on chromosome 21 may be of particular significance to Down syndrome, since elevated serum purines, in particular uric acid, are often seen in patients with Down syndrome (Fuller et al., 1962; Goodman et al., 1966). It may well be that mutants in this complementation group can be used for the isolation of aberrant chromosome 21's from patients with Down syndrome owing to translocation rather than simple trisomy. By utilization of recently developed recombinant DNA technology, it has been possible to use hybrids between these CHO mutants and human cells to isolate DNA segments derived from chromosome 21 which may be directly relevant to the pathology seen in Down syndrome. Such probes and cell types should be invaluable in answering many as yet unanswered questions regarding Down syndrome in particular and chromosomal and other diseases in general (Patterson et al., 1982; Scoggin and Patterson, 1982; Patterson and Schandle, 1983).

Many of these future prospects are directly relevant to the phenomenon of aging in humans and other animals. On theoretical grounds, it would not be surprising if abnormalities in nucleotide biosynthesis played a role in the developmental changes seen during aging. On a more practical level, it is clear that several of the developmental and disease processes that are being studied using CHO mutants defective in purine biosynthesis are directly related to problems seen with increasing frequency in an aging population. Gouty arthritis prevalence increases greatly with aging and especially increases in women after menopause. Ischemia and other forms of heart disease are of particular concern to the aging population. Down syndrome, which may be classified a segmental progeriod syndrome (Martin, 1978), seems to have a direct relationship to presenile dementia of the Alzheimer's type, since Down syndrome patients who have reached approximately age 40 have psychological and pathological characteristics indistinguishable from those seen in patients with presenile dementia of the Alzheimer's type (Olson and Shaw, 1969; Rosner and Lee, 1972; Neibuhr, 1974). Malignancy of many types appears to be more common in older populations. Thus, while a great deal of fundamental importance has been accomplished in the study of the purine and pyrimidine biosynthetic pathways in Chinese hamster ovary cells, it seems clear that this only represents a beginning and the establishment of the experimental system. The most exciting work, both from a fundamental molecular genetic point of view and from the point of view of using this system to investigate related issues of fundamental biomedical importance, remains for the future.

ACKNOWLEDGMENTS

This is publication #486 of the ERICR and FSL and was supported by grants from NIH (AG00029, HD02080, HD13423) and the National Foundation, March of Dimes. The excellent editorial assistance of Ms. D. Hess is gratefully acknowledged.

REFERENCES

Adams, M. M., Erickson, J. D., Layde, P. M., and Oakley, G. P. (1981). *J. Am. Med. Assoc.* **246**, 758–760.

Armitt, S. and Woods, R. A. (1970). *Genet. Res. Camb.* **15**, 7–17.

Bach, M. L., Lacroute, F., and Botstein, D. (1979). *Proc. Natl. Acad. Sci. USA,* **76**, 386–390.

Bachmann, B. J. (1983). *Microbiol. Rev.* **47**, 180–230.

Becker, M. A., Yen, R. C. K., and Itkin, P. (1979). *Science* **203**, 1016–1019.

Bratton, A. C. and Marshall, E. K. (1939). *J. Biol. Chem.* **128**, 537–550.

Broad, T. E., Jones, C., and Patterson D. (1984). *Cytogenetics and Cell Genetics* **37**, 427.

Buchanan, J. M. (1959). *Harvey Lect.* **54**, 104–130.

Buchanan, J. M. (1960). In *Nucleic Acids* (E. Chargaff and J. N. Davidson, eds.), Academic Press, New York, Vol. 3, p. 303.

Buchanan, J. M. and Hartman, S. C. (1959). *Advan. Enzymol. Relat. Areas Mol. Biol.* **21**, 199–261.

Capperelli, C., Benkovic, P., Chettur, G., and Benkovic, S. (1980). *J. Biol. Chem.* **255**, 1885–1890.

Chen, J. J. and Jones, M. E. (1976). *Arch. Biochem. Biophys.* **176**, 82–90.

Chu, E. H. Y., Sun, N. C., and Chang, C. C. (1972). *Proc. Natl. Acad. Sci. USA* **69**, 3459–3463.

Coleman, P. F., Suttle, D. P., and Stark, G. R. (1977). *J. Biol. Chem.* **252**, 6379–6385.

Conner, T. W. and Rawls, Jr., J. M. (1982). *Biochem. Genet.* **20**, 607–619.

Crabtree, G. W. and Henderson, J. F. (1971). *Can. Res.* **313**, 985–991.

Danks, M. K. and Scholar, E. M. (1982). *Biochem. Pharm. (Great Britain)* **31**, 1687–1691.

Davidson, J. N. and Niswander, L. (1983). *Proc. Natl. Acad. Sci. USA* **80**, 6897–6901.

Davidson J. and Patterson, F. (1979). *Proc. Natl. Acad. Sci. USA* **76**, 1731–1735.

Davidson, J. N., Carnright, D. V., and Patterson, D. (1979). *Som. Cell Genet.* **5**, 175–191.

Davidson, J. N., Rumsby, P. C., and Tamaren, J. (1981). *J. Biol. Chem.* **256**, 5220–5225.

Denis-Duphil, M. and Lacroute, F. (1971). *Mol. Gen. Genet.* **112**, 354–364.

deSaint Vincent B. R., Dechamps, M., and Buttin, G. (1980). *J. Biol. Chem.* **255**, 162–167.

deSaint Vincent, B. R., Delbruck, S. Eckhart, W., Meintoth, J., Vitto, L., and Wahl, G. (1981). *Cell* **27**, 257–277.

Dorfman, B. Z. (1969). *Genet.* **67**, 377–389.

Esko, J. D. and Raetz, C. R. H. (1978). *Proc. Natl. Acad. Sci. USA* **75**, 1190–1193.

Esko, J. D. and Raetz, C. R. H. (1980). *Proc. Natl. Acad. Sci. USA* **77**, 5192–5196.

Falk, D. R. (1976). *Molec. Gen. Genet.* **148**, 1–8.

Falk, D. R. and DeBoer III, E. A. (1980). *Molec. Gen. Genet.* **180**, 419–424.

Falk, D. R. and Nash, D. (1974). *Genet.* **76**, 755–766.

Fausto-Sterling, A. (1977). *Biochem. Genet.* **15**, 803–813.

Feldman, R. I. and Taylor, M. W. (1974). *Biochem. Genet.* **12**, 393–405.

Feldman, R. I. and Taylor, M. W. (1975). *Biochem. Genet.* **13**, 227–234.

Flaks, J. and Lukens, L. (1963). In *Methods in Enzymology* (S. P. Colowick and N. O. Kaplan, eds.), Academic Press, New York, Vol. 6, pp. 52–95.

Fluri, R., Coddington, A., and Flury, U. (1976). *Mol. Gen. Genet.* **147**, 272–282.

Forman, H. J. and Kennedy J. (1975). *J. Biol. Chem.* **250**, 4322–4326.

Forman, H. J. and Kennedy, J. (1976). *Arch. Biochem. Biophys.* **173**, 219–224.

Fuller, R. W., Luce, M. M., and Mertz, E. T. (1962). *Science* **137**, 868–869.

Gerhart, J. C. and Pardee, A. B. (1972). *J. Biol. Chem.* **237**, 891–896.

Giles, N. H., Partridge, C. W. H., and Nelson, N. J. (1957). *Proc. Natl. Acad. Sci. USA* **43**, 305–317.

Goodman, H. O., Lafland, H. B., and Thomas, J. J. (1966). *Am. J. Mental. Defic.* **71**, 427–432.

Green, C. D. and Martin, Jr., D. W. (1973). *Proc. Natl. Acad. Sci USA* **70**, 3698–3702.

Green, C. D. and Martin, Jr., D. W. (1974). *Cell* **2**, 241–245.

Gusella, J. F., Jones, C., Kao, F. T., Housman, D., and Puck, T. T. (1982). *Proc. Natl. Acad. Sci. USA* **79**, 3418–3422.

Hager, S. E. and Jones, M. E. (1965). *J. Biol. Chem.* **240**, 4556–4563.

Hager, S. E. and Jones, M. E. (1967). *J. Biol. Chem.* **242**, 5667–5674.

Harris, M. (1982). *Cell* **29**, 483–492.

Harris, M. (1984). *Somat. Cell and Molec. Genetics* (in press).

Hartman, S. C. (1970). In *Metabolic Pathways*, 3rd ed. (G. M. Greenberg, ed.), Academic Press, New York, Vol. 4, p. 1–68.

Hartman, S. C. and Buchanan, J. M. (1959). *Ergeb. Physiol. Biol. Chem. Exp. Pharmakol.* **50**, 75–121.

Henderson, J. F. (1962). *J. Biol. Chem.* **237**, 2631–2635.

Henderson, J. F. (1972). *ACS Monograph 170*, American Chemical Society, Washington, D.C.

Henderson, J. F. and Paterson, A. R. P. (1973). *Nucleotide Metabolism*, Academic Press, New York.

Henikoff, S., Sloan, J. S., and Kelly, J. D. (1983). *Cell* **34**, 405–414.

Hill, D. L. and Bennett, L. L. (1969). *Biochem.* **8**, 122–130.

Holmes, E. W., McDonald, J. A., McCord, J. M., Wyngaarden, J. B., and Kelley, W. N. (1973). *J. Biol. Chem.* **248**, 144–150.

Holmes, E. W., King, G. L., Leyva, A., and Singer, S. C. (1976). *Proc. Natl. Acad. Sci. USA* **73**, 2458–2461.

Hoogenraad, N. J. and Lee, D. C. (1974). *J. Biol. Chem.* **249**, 2763–2768.

Itakura, M. and Holmes, E. W. (1979). *J. Biol. Chem.* **254**, 333–338.

Ito, K. and Tatibana, M. (1966). *Biochem. Biophys. Res. Commun.* **23**, 672–678.

Itoh, R., Holmes, E. W., and Wyngaarden, J. B. (1976). *J. Biol. Chem.* **251**, 2234–2240.

Irwin, M., Oates, D. C., and Patterson, D. (1979). *Som. Cell Genet.* **5**, 203–216.

Jackson, R. C. and Weber G. (1975). *Nature* **256**, 331–333.

Jarry, B. (1976). *FEBS Lett.* **70**, 71–75.

Jarry, B. (1978). *Eur. J. Biochem.* **87**, 533–540.

Jarry, B. and Falk, D. (1974). *Molec. Gen. Genet.* **135**, 113–122.

Jones, M. E. (1972). *Curr. Top. Cell Regul.* **6**, 227–264.

Jones, M. E. (1980). *Ann. Rev. Biochem.* **49**, 253–279.

Jones, C., Miller, Y. E., Palmer, D., Morse, H., Kirby, M., and Patterson, D. (1984). *Cytogenetics and Cell Genetics* **37**, 500.

Jones, C., Patterson, D., and Kao, F. T. (1981). *Som. Cell Genet.* **7**, 399–409.

Kao, F. T. (1980). *J. Cell Biol.* **87**, 291a.

Kao, F. T., Chasin, L., and Puck, T. T. (1969). *Proc. Natl. Acad. Sci. USA* **64**, 1284–1291.

Kao, F. T. and Puck, T. T. (1967). *Genet.* **55**, 513–524.

Kao, F. T. and Puck, T. T. (1968). *Proc. Natl. Acad. Sci. USA* **60**, 1275–1281.

Kao, F. T. and Puck, T. T. (1969). *J. Cell Physiol.* **74**, 245–257.

Kao, F. T. and Puck, T. T. (1971). *J. Cell Physiol.* **78**, 139–143.

Kao, F. T. and Puck, T. T. (1972). *J. Cell Physiol.* **80**, 41–49.

Katunuma, N. and Weber, G. (1974). *FEBS Lett. (Amsterdam)* **49**, 53–56.

Kelley, W. N. (1983). In *The Metabolic Basis of Inherited Disease,* 5th ed. (J. B. Stanbury, J. B. Wyngaarden, D. S. Fredrickson, J. L. Goldstein, and M. S. Brown, eds.) McGraw-Hill, New York, pp. 1202–1226.

Kempe, T. D., Swyryd, E. A., Bruist, M., and Stark, G. R. (1976). *Cell* **9**, 541–550.

Kent, R. J., Lin, R. L., Sallach, H. J., and Cohen, P. P. (1975). *Proc. Natl. Acad. Sci. USA* **72**, 1712–1716.

Krooth, R. S. (1964). *CSHSGB* **29**, 189–212.

Krooth, R. S., Hsiao, W. L., and Potvin, B. W. (1979). *Som. Cell Genet.* **5**, 551–569.

Kusano, T., Kato, M., and Yamane, I. (1976). *Cell Struc. Func.* **1**, 393–396.

Lester, S. C., LeVan, S. K., Steglich, C., and DeMars, R. (1980). *Somat. Cell Genet.* **6**, 241–260.

Levinson, B. B., Ullman, B., and Martin, Jr., D. W. (1979). *J. Biol. Chem.* **254**, 4396–4401.

Losson, R. and Lacroute, F. (1983). *Cell* **32**, 371–377.

Lukens, L. and Flaks, J. (1963). In *Methods in Enzymology* (S. P. Colowick, and N. O. Kaplan, eds.), Academic Press, New York, Vol. 6, pp. 671–713.

Makoff, A. J., Buxton, F. P., and Radford A. (1978). *Mol. Gen. Genet.* **161**, 297–304.

Makoff, A. J. and Radford, A. (1978). *Microbio. Rev.* **42**, 307–328.

Martin, D. W. and Owen, N. T. (1972). *J. Biol. Chem.* **247**, 5477–5485.

Martin, G. M. (1978). In *Genetic Effects on Aging* (D. Bergsma and D. E. Housman, eds.), Alan R. Liss, New York, pp. 5–39.

McClard, R. W., Black, M. J., Livingstone, L. R., and Jones, M. E. (1980). *Biochem.* **19**, 4699–4706.

Mergeay, M., Gigot, D., Beckmann, J., Glansdorff, N., and Pierard, A. (1974). *Mol. Gen. Genet.* **133**, 299–316.

Moore, E. E., Jones, C., Kao, F. T., and Oates, D. C. (1977). *Am. J. Hum. Genet.* **29**, 389–396.

Morgan, T. H. (1915). *Amer. Natur.* **49**, 240–250.

Mori, M. and Tatibana, M. (1975). *Biochem.* **78**, 239–242.

Mori, M. and Tatibana, M. (1978). *M. Eur. J. Biochem.* **86**, 381–388.

Mori, M. Ishida, H., and Tatibana, M. (1975). *Biochem.* **14**, 2622–2630.

Mueller, W. T. and Benkovic, S. J. (1981). *Biochem.* **20**, 337–344.

Niebuhr, E. (1974). *Hum. Genet.* **21**, 99–101.

Norby, S. (1970). *Hereditas* **66**, 205–214.

Norby, S. (1973). *Hereditas* **73**, 11–16.

Oates, D. C. (1976). Ph.D. Thesis, University of Colorado.

Oates, D. C. and Patterson, D. (1977). *Som. Cell Genet.* **3**, 561–577.

Oates, D. C., Vannais, D., and Patterson, D. (1980). *Cell* **20**, 797–805.

Okada, M., Kleinman, I. A., and Schneiderman, H. A. (1974). *Devel. Biol.* **37**, 55–62.

Olson, M. I. and Shaw, C. W. (1969). *Brain* **92**, 147–156.

Padgett, R., Wahl, G., Coleman, P., and Stark, G. (1979). *J. Biol. Chem.* **254**, 974–980.

Palmer, L. M. and Dove, D. J. (1974). *Mol. Gen. Genet.* **138**, 243–255.

Patey, C. A. H. and Shaw, G. (1973). *Biochem. J.* **135**, 543–545.

Patterson, D. (1975). *Som. Cell Genet.* **1**, 91–110.

Patterson, D. (1976). *Som. Cell Genet.* **2**, 189–203.

Patterson, D. (1980). *Som. Cell Genet.* **6**, 101–114.

Patterson, D. and Carnright, D. V. (1977). *Som. Cell Genet.* **3**, 483–495.

Patterson, D., Graw, S., and Jones, C. (1981). *Proc. Natl. Acad. Sci. USA* **78**, 405–409.

Patterson, D. and Jones, C. (1976). *Som. Cell. Genet.* **2**, 429–239.

Patterson, D., Jones, C., Scoggin, C., Miller, Y. E., and Graw, S. (1982). *Annals NYAS* **372**, 69–81.

Patterson, D., Jones, C., Morse, H. Rumsby, P., Miller, Y., and Davis R. (1983*a*). *Somat. Cell Genet.* **9**, 359–374.

Patterson, D., Kao, F. T., and Puck, T. T. (1974). *Proc. Natl. Acad. Sci. USA* **71**, 2057–2061.

Patterson, D. and Schandle, V. B. (1983). *Banbury Report 14*, Cold Spring Laboratory, New York, 215–233.

Patterson, D., Vannais, D. B., and Laas W. (1983*b*) *JCP* **116**, 4066.

Planet, G. and Fox, I. H. (1976) *J. Biol. Chem.* **251**, 5839–5844.

Polokoff, M. A., Wing, D. C., and Raetz, C. R. H. (1981). *J. Biol. Chem.* **56**, 7687–7690.

Prajda, N., Morris, H. P., and Weber, G. (1979). *Cancer Res.* **39**, 3909–3914.

Puck, T. T. and Kao, F. T. (1967). *Proc. Natl. Acad. Sci. USA* **58**, 1227–1234.

Raisonnier, A., Bouma, M. E., Salvat, C., and Infante, R. (1981). *Eur. J. Biochem.* **118**, 565–569.

Rawls, Jr., J. M. (1979). *Comp. Biochem. Physiol.* **62B**, 207–216.

Rawls, Jr., J. M. (1981). *Molec. Gen. Genet.* **184**, 174–179.

Rawls, Jr., J. M. and Porter, L. A. (1979). *Genet.* **93**, 143–161.

Rawls, Jr., J. M., Chambers, C. L., and Cohen, W. S. (1981). *Biochem. Genet.* **19**, 115–127.

Reem, G. H. and Friend, C. (1967). *Science* **157**, 1203–1204.

Reyes, P. and Intress, C. (1970). *Life Sc.* **22**, 577–582.

Robinson, J. L., Drabik, M. R., Dombrowski, D. B., and Clark, J. H. (1983). *Proc. Natl. Acad. Sci. USA* **80**, 321–323.

Rosner, F. and Lee, S. L. (1972). *Am. J. Med.* **53**, 203–218.

Rowe, P., McCairns, E., Madsen, G., Sauer, D., and Elliott, H. (1978). *J. Biol. Chem.* **253**, 7711–7721.

Rumsby, P. C., Hirohisa, K., Waldren, C. A., and Patterson, D. (1982). *J. Biol. Chem.* **257**, 11364–11367.

Rumsby, P. C., Campbell, P. C., Niswander, L., and Davidson, J. N. (1984). *Biochem. J.* (in press).

Sabina, R. L., Patterson, D., and Holmes, E. W. (1983). *Fed. Proc.* **42**, 2209 (Abstract).

Schwarzman, J. D. and Pfefferkorn, E. R. (1981). *J. Parasitol.* **67**, 150–158.

Schwarzman, J. D. and Pfefferkorn, E. R. (1982). *Exper. Parasit.* **53**, 77–86.

Scoggin, C. H. and Patterson, D. (1982). *Arch. Intern. Med.* **142**, 462–464.

Shambaugh, G. E. (1979). *Am. J. Clin. Nutr.* **32**, 1290–1297.

Shantz, G. C., Fontennelle, L. J., and Henderson, J. F. (1972). *Biochem. Pharmacol.* **21**, 1203–1206.

Shoaf, T. and Jones, M. E. (1973). *Biochem.* **12**, 4039–4051.

Simons, I., Tu, A. S., Robertson, R. S., Callahan, M. M., and Sivak, A. (1980). *J. Cell Biol.* **87**, 317a.

Smith, G. K., Mueller, W. T., Wasserman, G. F., Taylor, W. D., and Benkovic, S. J. (1980). *Biochem.* **19**, 4313–4321.

Smith, G. K., Benkovic, P. A., and Benkovic, S. J. (1981). *Biochem.* **20**, 4034–4036.

Sperling, O., et al. (1972). *Rev. Eur. Etud. Clin. Biol.* **17**, 703–706.

Sperling, O., Wyngaarden, J. B., and Starmer, C. F. (1973). *J. Clin. Invest.* **52**, 2468–2485.

Stamato, T. D. and Hohmann, L. K. (1975). *Cytogenet. Cell Genet.* **15**, 372–379.

Stamato, T. D. and Patterson, D. (1979). *J. Cell Physiol.* **98**, 459–468.

Stanley, W. and Chu, E. H. Y. (1978). *Cytogenet. Cell Genet.* **22**, 228–231.

Stein, R., Razin, A., and Cedar, H. (1982). *Proc. Natl. Acad. Sci. USA* **79**, 3418–3422.

Suttle, D. P. (1983). *J. Biol. Chem.* **258**, 7707–7713.

Suttle, D. P. and Stark, G. R. (1979). *J. Biol. Chem.* **254**, 4602–4607.

Swain, J. L., Hines, J. J., Sabina, R. L., and Holmes, E. W. (1982a). *Circ. Res.* **51**, 102–105.

Swain, J. L., Sabina, R. L., Peyton, R. B., Jones, R. N., Wechsler, A. S., and Holmes, E. W. (1982b) *Proc. Natl. Acad. Sci. USA* **79**, 655–659.

Tatibana, M. and Ito, K. (1967). *Biochem. Biophys. Res. Commun.* **26**, 221–227.

Tatibana, M. and Ito, K. J. (1969). *Biol. Chem.* **244**, 5403–5413.

Tatibana, M. and Mori, M. (1975). *J. Biochem.* **78**, 239–242.

Tatibana, M. and Shigesada, K. (1972). *J. Biochem.* **72**, 549–560.

Taylor, M. W., Souhrada, M., and McCall, J. (1970). *Science* **172**, 162–163.

Traut, T. W. (1982). *Trends Biochem. Sciences* **7**, 255–257.

Traut, T. W. and Jones, M. E. (1977). *J. Biol. Chem.* **252**, 8374–8381.

Trotta, P. P., Burt, M. E., Haschemeyer, R. H., and Meister, A. (1971). *Proc. Natl. Acad. Sci. USA* **68**, 2599–2603.

Tsuda, M., Katunuma, N., Morris, H. P., and Weber, G. (1979). (EC 2.4.2.14). *Cancer Res.* **39**, 305–311.

Tu, A. S. and Patterson, D. (1977). *Biochem. Genet.* **15**, 195–210.

Tu, A. S. and Patterson, D. (1978). *J. Cell Physiol.* **96**, 123–132.

Tu, A. S., Robertson, R. S., Callahan, M. M., and Thayer, P. S. (1982). *Mutat. Res.* (in press).

Ullman, B., Wormstred, M. A., Cohen, M. B., and Martin, Jr., D. W., (1982). *Proc. Natl. Acad. Sci. USA* **79**, 5127–5131.

Wahl, G. M., Padgett, R. A., and Stark, G. R. (1979). *J. Biol. Chem.* **254**, 8679–8689.

Waldren, C. A. and Patterson D. (1979). *Cancer Res.* **39**, 4975–4892.

Waldren, C. A., Patterson, D., Kato, H., and Rumsby, P. (1983). In *Radioprotectors and Anticarcinogens* (O. F. Nygaard and M. G. Simic, eds.), Academic Press, New York, pp. 357–361.

Weber, G., Kizaki, H., Tzeng, D., Shiotano, T., and Olah, E. (1978). *Life Sc.* **23**, 729–736.

Weber, G., Hager, J. C., Lui, M. S., Prajda, N., Tzeng, D. Y., Jackson, R. C., Takeda, E., and Eble, J. N. (1981). *Cancer Res.* **41**, 854–859.

Westby, C. A. and Gots, J. S. (1969). *J. Biol. Chem.* **244**, 2095–2102.

Wood, A. W. and Seegmiller, J. E. (1973). *J. Biol. Chem.* **248**, 138–143.

Wyngaarden, J. B. and Ashton, D. M. (1959). *J. Biol. Chem.* **234**, 1492–1496.

Wyngaarden, J. B. and Kelley, W. N. (1972). In *The Metabolic Basis of Inherited Disease* (J. B. Stanbury, J. B. Wyngaarden, and D. C. Frederickson, eds.) McGraw-Hill, New York, pp. 889–968.

CHAPTER 12

THE APRT SYSTEM

Milton W. Taylor
Anne E. Simon
Raman M. Kothari*
Department of Biology
Indiana University
Bloomington, Indiana

*Current address: Sarabhai Research Center, Baroda, India.

I. INTRODUCTION

There has been much debate over the last few years whether "variants" isolated as drug-resistant cells in culture are the result of true mutational events or are due to some unexplained epigenetic effect. This controversy has arisen because of the high frequency with which cell strains carrying recessive markers can be selected and the lack of direct correlation between ploidy and mutation rates in some cell lines (Harris, 1971; Morrow et al., 1978). It has been suggested that such variants may arise by mechanisms involving continuous chromosomal segregation, duplication, and mitotic recombination as well as mutation. Support for such mechanisms for loci other than APRT is discussed elsewhere in this volume (see Chapters 28 and 29).

Most of the research on mutation frequencies in CHO cells has been done with the X-linked HGPRT locus, where it is known that there is only one functional allele (see Chapter 13). However, surprisingly high mutation frequencies have been found at a number of autosomal loci selected on the basis of drug resistance. Indeed, the frequency of mutation to 2,6-diaminopurine resistance [a mutation at the autosomal adenine phosphoribosyl transferase locus (APRT)] is as high as that at the single active sex-linked hypoxanthine guanine phosphoribosyl transferase (HGPRT) locus.

As pointed out by Siminovitch (1976), this high mutation frequency could occur if substantial regions of the CHO genome were physically or functionally hemizygous. However, using "wild-type" CHO cells, a number of groups (Chasin, 1974; Jones and Sargent, 1974) have reported the isolation of presumptive heterozygotes as single-step mutants at the APRT locus, indicating that two functional copies of the APRT locus are present in the wild-type cells. Yet single-step mutants with no APRT activity can be isolated from wild-type cells (Taylor et al., 1977). Table 12.1 summarizes the frequency of single-step (selection at low concentrations of analogues) and two-step mutations (selection at high concentrations of analogue) at the APRT locus as reported by a number of laboratories. Theoretically, one should obtain autosomal recessive mutants only following the intermediate formation of a heterozygote. As will be discussed subsequently, this mechanism does occur, but other mechanisms of generating autosomal recessive mutations do occur in CHO and other cell lines at this locus. As we shall see, current methods of molecular analysis have allowed us to unravel the paradoxes associated with mutation at the APRT locus in CHO cells.

The APRT system has many advantages over other systems for studying the molecular mechanism of mutation. These are: (1) Simple methods for selecting heterozygotes, homozygotes, and revertants. (2) The enzyme subunit is small enough (~ 20,000 daltons) so that it should be easy to sequence the amino acids. (3) The CHO gene has been cloned and is sufficiently small to facilitate sequencing of the DNA of the complete gene and associated regions. Moreover, it has been reported that the CHO genomic probe hybridizes to human DNA, and this may be useful in selecting the homologous gene from other organisms. (4) The *apt* gene has been used in gene transformation experiments in mouse *apt*$^-$ L cells and CHO *apt*$^-$ cells and has been shown to transform cells from *apt*$^-$ to *apt*$^+$ at high frequencies.

II. BIOCHEMISTRY OF THE APRT SYSTEM

A. The Enzymes

APRT catalyzes the conversion of adenine to the nucleotide, adenosine 5′-monophosphate, utilizing 5-phosphoribosyl-1-pyrophosphate and

TABLE 12.1
Frequency (Rate) of Mutations at the APRT Locus in CHO Cells[a]

Wild-Type Cell Line	Partial Resistance: Concentration of Analogue	Partial Resistance: Spontaneous Mutation Frequency	Complete Resistance: Concentration of Analogue	Complete Resistance: Spontaneous Mutation Frequency
1. CHO	7–11 μg/mL AA	10^{-5} (r)[b]	5 μg/mL	2×10^{-7} (r)
2. SC1	8 μg/mL AA	(1–2) × 10^{-5} (f)	——	——
3. CHO-ClOA	4–5 μg/mL AA	10^{-4} (f)	20–80 μg/mL	3×10^{-7} (r)
4. CHO-pro$^-$	6–7 μg/mL DAP	10^{-5} (f)	30 μg/mL	10^{-9}–10^{-4} (r)
5. CHOK1			20 μg/mL	3×10^{-5}
6. CHOK1			20 μg/mL	2×10^{-7} (r)

[a] References: (1) Jones and Sargent (1974); (2) Thompson et al. (1980); (3) Adair et al. (1980); (4) Bradley and Letovance (1982); (5) Taylor et al. (1977); (6) Chasin (1974).
[b](r) = rate; (f) = frequency; AA = 8-azadenine; DAP = 2,6-diaminopurine.

adenine as substrates. Since all organisms have a *de novo* pathway for the formation of AMP, APRT is classified as a "salvage enzyme." In CHO cells, it provides the only known mechanism whereby free adenine can be converted to a utilizable nucleotide. The relationship of APRT to other purine salvage enzymes and to the *de novo* purine biosynthetic pathway is illustrated in Figure 12.1.

APRT has been purified to near homogeneity by taking advantage of an enzyme site exhibiting affinity for both the product of the reaction and for the substrate (Hershey and Taylor, 1978). By passing a crude preparation of cell extract over an AMP-agarose column and eluting with substrate PRPP, a 3000-fold purification is achieved. Purification from a crude supernatant can be improved by a prior 75% ammonium sulfate precipitation. After dialysis, the APRT is bound to the AMP-agarose, the column is washed with high salts (0.5 *M* KC1) to remove the bulk of the proteins, and the enzyme is specifically eluted with 0.5 m*M* PRPP. When AMP-agarose-purified enzymes from CHO cells are subjected to SDS-polyacrylamide gel electrophoresis, the major protein bands at ~ 20,000 daltons. When run under nondenaturing conditions or on a G100 gel filtration column, the active protein has a molecular weight of 40,000 daltons and is thus a dimer of identical subunits (Taylor et al. 1979). The K_m for PRPP and adenine is 10^{-6} *M* and 3 x 10^{-6} *M*, respectively, for the CHO APRT. This is similar to that found for other mammalian sources of the enzyme.

APRT is routinely assayed by a radiochemical method based on the precipitation of labeled AMP as a lanthanum salt (Bakay et al., 1969).

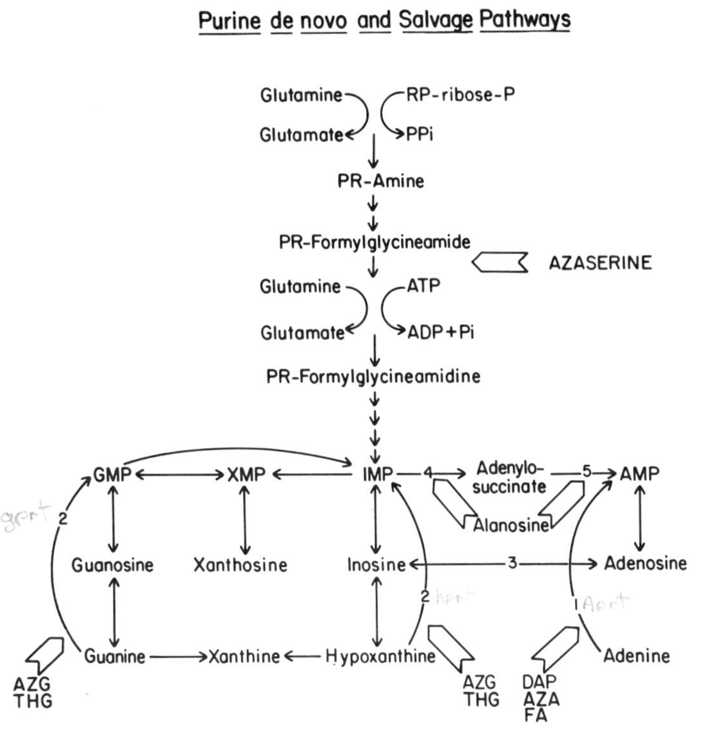

Figure 12.1. Purine *de novo* and salvage enzyme pathways. 1. Adenine phosphoribosyl transferase (APRT). 2. Hypoxanthine–guanine phosphoribosyl transferase (HGPRT) 3. Adenosine deaminase. 4. Adenylosuccinate synthetase. 5. Adenylosuccinate lyase.

B. Electrophoretic Variants

Two methods have been used to detect electrophoretic variants of APRT from CHO cells and to distinguish CHO APRT from other species of the enzyme. Tischfield et al. (1973) developed a method for identifying APRT in polyacrylamide gels and cellulose acetate by allowing the reaction to proceed after electrophoresis in the gel. Radioactive adenine in the presence of the reaction mixture is added to the gel, and the reaction is terminated by placing the gel in a lanthanum chloride solution to precipitate the radioactive reaction products. The gel is then washed free of unreacted reaction mixture, dried, and the AMP that has been formed is identified by autoradiography. Electrophoretic variants can be detected since they migrate differently into the gel.

Electrophoretic variants have also been detected by two-dimensional gel electrophoresis of APRT immunoprecipitates (Simon et al., 1982). In this method, APRT is precipitated from crude cell extracts by the addition of antibody for 2 hr, followed by incubation for 1 hr with prewashed protein-A Sepharose to enhance the precipitation. The protein-A antigen–antibody complex is pelleted by centrifugation in a microfuge, washed twice in cold phosphate-buffered saline, followed by three washes in 0.25 *M* LiC1, and a final wash in PBS. The pellet is extracted three times with 0.5 mL 1 *M* acetic acid, and the supernatants are combined, frozen, and lyophilized.

Two-dimensional gel electrophoresis is performed according to O'Farrell (1975), and the protein is stained with silver nitrate according to Switzer et al. (1979) as modified by Oakley et al. (1980). An example of wild type and an electrophoretic variant of APRT is reproduced in Figure 12.2. The pI of CHO *apt* is 5.8.

III. GENETIC ANALYSIS OF THE APRT SYSTEM

A. Selection Procedures

The purine analogues 2,6-diaminopurine, 8-azaadenine, and 8-fluoroadenine have all been used in the selection of cells defective or lacking APRT activity. Resistance to these drugs arises from an inability of the cell to metabolize cytotoxic nucleotides from the base analogs. Cells resistant to any one analogue show cross resistance to the other. Table 12.1 summarizes the concentrations used to select partial and complete drug-resistant mutants. CHO heterozygotes with approximately 50% wild-type APRT activity do survive in low concentrations (4–8 μg/mL) of 2,6-diaminopurine or 8-azaadenine. This may reflect competition between adenine in the medium and the small amount of base analog. The affinity of APRT for the base analogues is lower than the affinity for adenine (Thomas et al., 1973). One can thus select mutants at this locus by stepwise selection (Jones and Sargent, 1974; Chasin, 1974). Mutants resistant to low concentrations of the drug are either hemizygotes or true heterozygotes (Simon et al., 1983). Mutants obtainable at the second step are true homozygotes (or hemizygotes containing only a mutant copy of the enzyme). Single-step mutants are selected by bypassing selection for heterozygotes (i.e., at high concentrations of 2,6-diaminopurine, heterozygotes are killed, and only homozygotes are selected). The molecular mechanism of this selection is discussed subsequently.

The standard selection procedure used in this laboratory is to mutagenize cells with the desired mutagen, to allow the cells to recover for 3–7 days, and to plate the cells at a density of 5×10^5 cells/100 cm^2 Petri dish in selective medium (20 μg/mL 2,6-diaminopurine), changing the medium at 3-day intervals. The purine analog is made up as a 100X solution in H_2O and diluted into the medium, normally F12 + 10% calf serum (or MEM + 10% calf serum) before use. Batches of calf se-

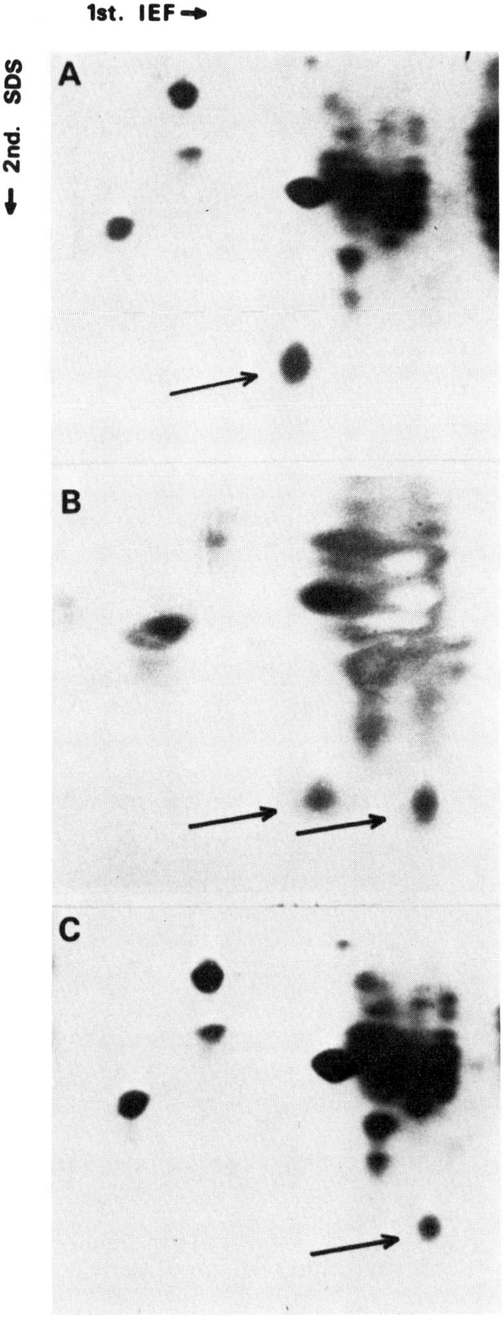

Figure 12.2. Two-dimensional gel electrophoresis of immunoprecipitates: (A) wild-type CHO APRT, (B) 416 APRT, (C) 416d^{r}c25 APRT.

rum are pretested for the presence of adenine by testing the growth of a pur$^-$ hpt$^-$ *Escherichia coli* strain with aliquots of serum. If necessary, serum is dialyzed against three consecutive 40-volume aliquots of phosphate-buffered saline.

After 14–21 days of incubation, resistant colonies are picked for subcloning. In order to avoid multiple mutational events, the cells to be mutagenized are grown in adenine, azaserine medium for a week before mutagenesis. This eliminates any APRT$^-$ cells present in the culture before mutagenesis, since such cells die in azaserine (Fig. 12.1). Only one mutant is picked from each individual plate, so that mutants are independent and cannot be sister cells.

A systematic analysis of the expression time of APRT$^-$ in mutant cells following mutagenesis indicates that a lag of 2–4 days is sufficient for gene expression (Adair et al., 1980). Extending the recovery period to longer periods does not increase the number of mutants recovered.

The density of the cell population appears to be crucial for the recovery of mutants. Cell killing in 2,6-diaminopurine is not efficient at cell populations above 5×10^5 cells/plate (100 mm^2). Jones and Sargent (1974), in reconstruction experiments in which a known number of mutant cells (heterozygotes) were plated with wild-type CHO, showed that recovery of mutants decreased significantly with increasing numbers of cells in the presence of 25 μg/mL azaadenine. Although the recovery of homozygotes was not so drastically affected, their recovery from a mixed population decreased above a population of 2×10^6/dish. Carver et al. (1980) showed by reconstruction experiments using 1×10^5–1×10^6 cells/100 cm^2 plate that optimum recovery of azaadenine-resistant mutants occurred between $(1–4) \times 10^5$ cells/plate, using 40–80 μg/mL AA. There was a sharp decline in the observed frequency as the number of cells increased. These results probably reflect cross feeding of the cytotoxic nucleosides and nucleotides at high cell concentrations. In calculating mutation rates, one has to consider the recovery rate from the total population.

Induced mutants in all strains of CHO can be isolated following treatment with mutagens such as EMS, or ICR-170 at frequencies of 1×10^{-6}–1×10^{-5}. Such mutants have been found to contain from 35% to < 1% APRT activity (Taylor et al., 1977; Bradley and Letovanec, 1982; Meuth and Arrand, 1982). These latter (< 1%) drug-resistant isolates are thus not heterozygotes and probably arise as the result of the loss of one allele and mutation of the other (see below). Spontaneous mutants, resistant to high concentrations of 2,6-diaminopurine, occur very rarely in Chinese hamster cell lines, except in one clonal line derived from CHO-K1, called CHO-P1. The reason for the high frequency of spontaneous mutants with this strain is unknown (Taylor et al., 1977).

B. Reversion of APRT Mutants

If most mutants at the APRT locus arise by point mutation, reversion (i.e., growth in medium containing an inhibitor of purine biosynthesis plus adenine) should occur at a relatively high frequency. Although revertants have been found in a few cases, the number of mutant clones giving rise to revertants has been much less than predicted. This may be due in part to the fact that cells carrying point mutations are likely to contain low amounts of APRT activity which are sufficient to allow growth of some mutant cells in selective medium containing adenine, particularly in the presence of analogues blocking *de novo* purine biosynthesis. These cells constitute a background which obscures the lower frequency of true revertants. From our experiences, mutant strains with greater than 1% residual APRT activity grow slowly in the selective media normally used, and are thus not suitable for reversion studies.

Three inhibitors of *de novo* purine biosynthesis are normally used in reverse selection: (1) azaserine, which blocks the synthesis of formylglycineamide ribotide; (2) aminopterin, which blocks both *de novo* purine and pyrimidine biosynthesis by blocking tetrahydrofolate reductase and CH_3 transfer; and (3) alanosine, which blocks the interconversion of IMP to AMP (Graff and Plagemann, 1976) (Fig. 12.1). All three selective media—"AAT" medium (adenine, aminoptein, thymidine), "AAA" medium (alanosine, azaserine, adenine), or AA (azaserine, adenine)—have been used for the selection of revertants. Utilizing fully resistant mutants, Jones and Sargent (1974) selected revertants in AAT medium at frequencies of 1.6×10^{-4} for one cell line and 1.3×10^{-5} for a second cell line. These revertants had about 35% wild-type activity and were presumably the result of reversion of a single point mutation.

C. Role of APRT in Cellular Metabolism

The only known and confirmed role for APRT in CHO cells is to salvage preformed adenine and recycle adenine arising from nucleic acid catabolism. Although the detailed mechanisms by which purine salvage enzymes in general, and APRT in particular, function are not clearly understood, it undoubtedly maintains cellular homeostasis. This must be particularly true in these tissues, such as human leukocytes and blood platelets in which no *de novo* purine pathway is functioning (Murray, 1971).

Human red cells contain transport systems for both purine bases and nucleosides [reviewed in Murray (1971)]. Evidence has been provided that purines enter and leave rabbit red cells as free base rather than nucleosides. However, the mechanism of transport, whether by active or passive diffusion, is still unclear. In studies with membrane vesicles prepared from *E. coli*, Hochstadt-Ozer and Stadtman (1971) showed that transport was dependent on transferase activity and that purine nucleotides accumulated in the vesicle as the transport product. Quinlan and Hochstadt (1974) have reported that hypoxanthine uptake was dependent on phosphoribosyl pyrophosphate, a substrate of the transferase reaction. However, Zylka and Plagemann (1975), using HGPRT$^-$ Novikoff hepatoma cells, and Alford and Barnes (1976), using HGPRT$^-$ Chinese hamster fibroblasts proposed that facilitated diffusion preceded phosphorylation during purine accumulation.

Adenine uptake was measured in a series of CHO APRT$^-$ mutants (DAP-2 with 5% residual APRT activity and DAP-12 with no detectable APRT activity) (Witney and Taylor, 1978). In APRT$^-$ CHO cells, adenine uptake was found (a) to follow biphasic saturation kinetics, (b) to strongly correlate with residual APRT activity after normal and rapid initial uptake, and (c) to be independent of phosphorylation by APRT in mutants with reduced activity. These studies suggested that adenine was transported as a free base by facilitated diffusion and subsequently phosphorylated by APRT. In support of this, Zylka and Plagemann (1975) have shown that adenine phosphoribosylation is preceded by facilitated diffusion and, at low concentrations of adenine, transport is a rate-limiting step. There is no evidence from CHO cells that APRT is located on the cell membrane. Indeed, human APRT activity is localized exclusively in the cytoplasm (Arnold and Kelly, 1978).

D. Gene Isolation and Characterization

The CHO *apt* genomic DNA was cloned by Lowy et al. (1980). This was achieved by cleavage of the total CHO DNA with Hind III and ligation of Hind III cut CHO DNA to Hind III cut *E. coli* plasmid pBR322. APRT$^-$ mouse L cells were transformed with the hybrid molecules and APRT$^+$ clones selected in AA medium. DNA from such primary transformants was isolated and used in a second round of transformation to enrich for CHO DNA-APRT sequences and to eliminate extraneous plasmid DNA.

DNA from such transformants was partially cleaved with EcoRI and cloned into λ phage Charon 4A. This phage library was then screened with a pBR322 probe and a single hybridizing clone was identified from 6×10^5 plaques. This clone was plaque purified and designated λ Haprt-1. It could be shown by transformation activity, by APRT activity in transformed cells, and by the detection of λ Haprt-1 sequences in transformed cells that this recombinant clone contained a functioning CHO *apt* gene. By restriction enzyme mapping, the *apt* gene was shown to reside on a 7.8 kb Hind III fragment, which was subsequently subcloned into the Hind III site of pBR322 and the resultant plasmid

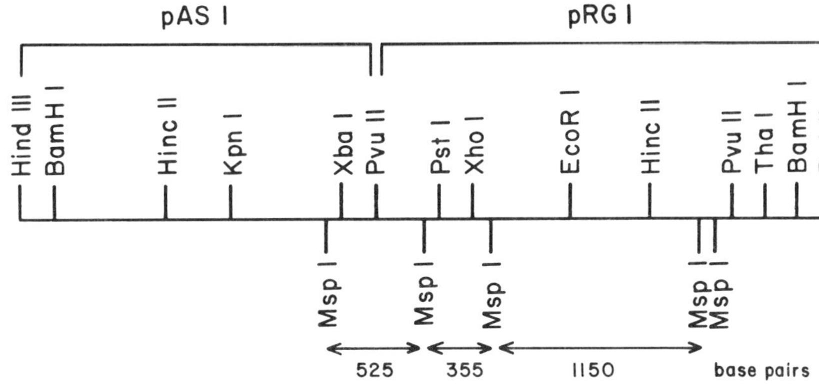

Figure 12.3. Partial restriction map of CHO-*apt* genomic DNA.

designated pH *apt*-1. The *apt* gene was localized to a 4.3 kb Hind III-Bgl II fragment thought to be the minimum size of the gene by transformation experiments (Fig. 12.3). More recently (Tang and Taylor, unpublished; Meuth and Arrand, 1982), the *apt* transforming activity was shown to reside in the 3.8 kb Bam HI fragment. CHO APRT mRNA has approximately 950 base pairs (Lowy et al., 1980) and is present at between 10–20 copies per cell.

E. Heterozygosity and Homozygosity

Although fully drug-resistant mutations at the *apt* locus can be isolated in a single step in CHO cells and in other Chinese hamster cell lines (Taylor et al., 1977; Meuth and Arrand 1982; Tischfield et al., 1982), presumptive CHO *apt* heterozygotes resistant to only low concentrations of either azaadenine or 2,6-diaminopurine have been isolated by several groups (Adair et al., 1980; Bradley and Letovanec, 1982; Chasin,

1974; Thompson et al., 1980). Heterozygosity (+/−), the presence of two alleles, can be recognized by the subsequent isolation of fully drug-resistant strains, by intermediary levels of APRT activity, by the presence of a normal enzyme, and by the presence of 50% immunoprecipitable material (Simon et al., 1982). Bradley and Letovanec (1982) first reported the occurrence of two types of heterozygotes in CHO cells selected as resistant to 6–7 μg/mL DAP in α-medium. Two of the heterozygotes (D416 and G1441) were selected by sib selection (Rosenstraus and Chasin, 1975) following EMS treatment. These lines were phenotypically stable, had 30–50% wild-type APRT-specific activity, and gave rise to fully DAP-resistant mutants at relatively high frequencies (7×10^{-5}). However, some subclones of these heterozygotes (D416 and G1441) gave rise to "jackpots"—spontaneous mutations at frequencies as high as 10^{-3}. The mutation frequency in heterozygotes other than D416 and G1441 was 10–100-fold lower. In this class of heterozygotes mutation frequency could be enhanced by mutagenesis. On the basis of this analysis, Bradley and Letovanec (1982) proposed that two "genetic" pathways are available to wild-type CHO cells undergoing mutation to the heterozygous state, namely, a high-frequency and a low-frequency event. The high-frequency event occurs only once, so that the $APRT^{-}/^{-}$ state can only be achieved at a high frequency if mutation to heterozygosity has occurred by the low-frequency mechanism. Conversely, if mutation to heterozygosity had occurred by the high-frequency mechanism, subsequent conversion to the fully resistant state would require a low-frequency event. Bradley and Letovanec (1982) suggested that a physical difference in the genomic environment of the two alleles may be such that only one allele can undergo the high-frequency event. Evidence supporting this hypothesis is discussed later.

Thus, there appear to be two classes of heterozygotes: Heterozygotes of Class I (HET-I) are usually isolated from nonmutagenized cultures, have 50% wild-type APRT activity, and give rise at a low spontaneous rate (3×10^{-7} per cell per generation) to fully DAP^{R} $APRT^{-}$ cells. This second step can be enhanced by treatment with a mutagen. Class II heterozygotes (HET-II) are usually isolated following mutagenesis and give rise to fully DAP^{R} $APRT^{-}$ cells at a high spontaneous rate of 10^{-6}–10^{-3} per cell per generation, a rate not strongly influenced by mutagenesis.

Simon et al. (1982) quantitated and analyzed using two-dimensional gel electrophoresis and thermal inactivation studies the APRT protein produced by a large group of homozygotes generated from both classes of heterozygotes. It could be shown that HET D416 (a Het-II line) contained two proteins precipitable by anti-APRT antiserum. One of these proteins was wild-type, the other an electrophoretic variant (EV-1) (Fig. 12.2). All fully DAP^{R} mutants of D416, isolated independently, con-

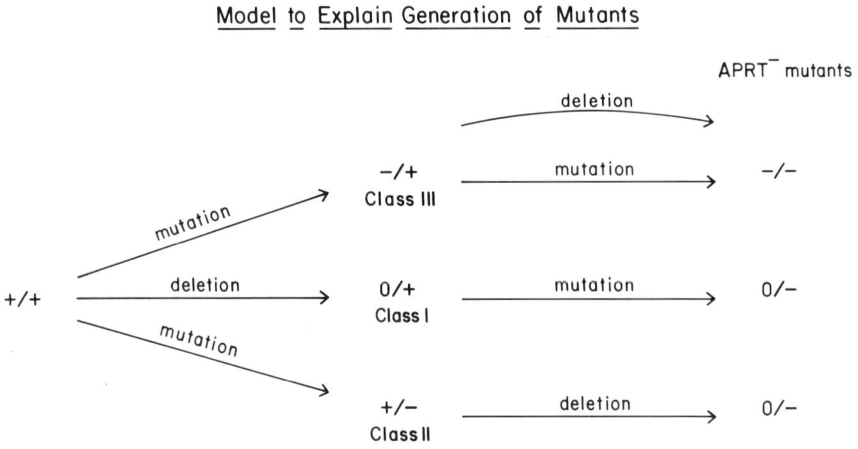

Figure 12.4. Model for expression of recessive phenotype at the *apt* locus in CHO cells. Symbols: +, wild = type allele; -, mutated allele; 0, deletion of *apt* allele.

tained only the EV-1 protein. Similar results were found with derivatives of a second Class II heterozygote (GM1441). These results indicated that the high-frequency event, leading to homozygotes of Class II HETS, was either a deletion or inactivation of one of the two *apt* alleles, with the retention of a mutant allele.

Class I HETS, on the other hand, gave rise to a variety of mutant proteins at a low frequency. The Class I HETS themselves all contained 50% APRT wild-type activity. Thus the initial event in the formation of the Class I heterozygotes was the loss (deletion) or inactivation (or possibly mutation) of one allele. A model illustrating the generation of high-frequency mutation at the *apt* locus is presented in Figure 12.4. This model has been confirmed at the DNA level utilizing a plasmid containing the whole *apt* gene, pH APRT-1, and two probes derived from pH APRT-1 (pAS-1 and pRG-1) (Fig. 12.3). Utilizing restriction-enzyme-cut DNA extracted from the CHO *apt* mutants and probes pAS-1 and pRG-1, differences in Msp I- and Hpa II-cut DNA can readily be distinguished between wild type, HET D416, and its derivatives (Fig. 12.5). Similar results were obtained by Meuth and Arrand (1982) with other single step mutants (AA^r16 and AA^r33). AA^r16 had lost a Taq/XhoI site (these enzymes share a central tetranucleotide sequence, TCGA). AA^r33 has lost the same MspI site as 416 and its derivatives, suggesting that this site is a "hot spot" for mutation. Both the results from this laboratory and those of Meuth and Arrand (1982) suggest that these mutants arose by single base pair alterations rather than large DNA rearrangements, since the new fragments are of the size expected from a change of a single restriction enzyme site. Most CHO $APRT^-$ mutants do

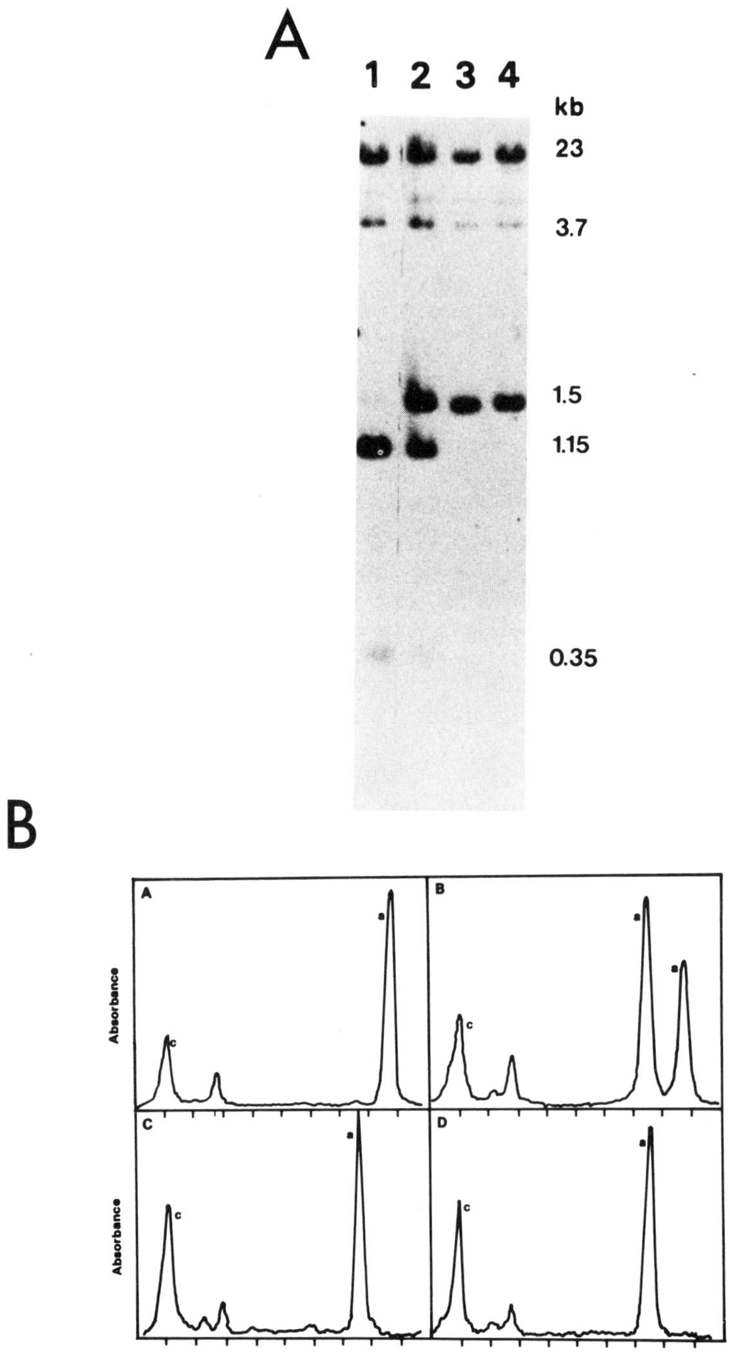

Figure 12.5. (A) Southern blot of MspI digested DNA. (1) CHO, (2) 416, (3) 416d^{r}c25, (4) 416^{r}c26: Probe was ^{32}P-labeled pRG-1. (B) Densitometer tracing illustrating relationship of c (*cad* gene) to a (*apt* gene).

not show alterations in restriction patterns, indicating that they probably arise by base pair changes that do not involve a restriction enzyme site.

The data presented in Figure 12.5 are interpreted as showing that the Class II heterozygote D416 arose from a mutational event at the Msp I site in one allele (lane 2) and that the fully resistant derivatives of D416 arose from a *deletion* of genetic material on the other allele (lane 3, 4). This interpretation was confirmed by measuring the number of copies of the *apt* gene present in the cell lines by probing with the CHO *cad* gene (23 kb in Fig. 12.5) as an internal control. The fully resistant derivatives contained only one copy of the *apt* gene.

Similar analysis was done with Class I HETS. In fully resistant D422d^{r}G·1, an electrophoretic variant (EV-1) identical to that formed in D416 was detected. Since the heterozygote D422 had a single-copy wild-type band and the fully resistant derivative D422d^{r}G·1 had an altered Msp I site, D422d^{r}G·1 could only have arisen by a mutational event on the single wild-type allele present in the Het I D422 strain. Thus, a group of Class I heterozygotes arose by the deletion of one allele, and fully resistant cells arose from this heterozygote by mutation in the remaining allele. Thus, these partially resistant strains are hemizygous rather than heterozygous at the *apt* locus. Other HETS which like Class I Hets have a low spontaneous mutation rate to fully resistant cell lines have been shown to retain two copies of the allele, one of which must be mutant. Such HETS form Class III in this model (Simon et al., 1983) (Fig. 12.4).

Thus, at a single gene locus it is possible to obtain both true heterozygotes (+/−) and hemizygotes (+/0), both of which can give rise to fully DAP^{r} mutants by different mechanisms. These results may not be unique to the *aprt* locus since similar high frequencies of mutation have been described (Rabin and Gottesman, 1979; Gupta and Siminovitch, 1978) for tubercidin and toyocamycin resistance. Resistance to these drugs is due to a defect in the purine salvage enzyme, adenosine kinase, and the high-frequency mutation may result from deletion of one copy of the *adk* gene.

F. Gene Mapping

The *apt* locus has recently been mapped by cell hybridization to LTAO, an APRT^{-}, ouabain-resistant subline of LMTK^{-} C1.1D (clone 1D). In Chinese hamsters the *apt* locus has been located to the short arm of chromosome 3, in the region 3p2→pter (Adair et al., 1980) (see Chapter 5 for chromosome nomenclature and techniques for mapping). Isozyme analysis of hybrid clones indicated that APRT, LDHA, IDH2, and GAA are syntenic in Chinese hamsters. In CHO cells, both homologs of chromosome 3 have undergone rearrangement. The Z3 and Z7 chromosomes in CHO appear to be the products of a reciprocal translocation between chromosomes 3 and 4, while the Z4 appears to have arisen by pericentric inversion following a break in the proximal region of the short arm of the other chromosome 3 homolog (Deaven and Peterson, 1973; Worton et al., 1977). Thus the CHO *apt* genes should map to the distal portion of the short arm of chromosome Z7, and to the mid region of the long arm of chromosome Z4. This has been confirmed by analysis of interspecific hybrids between CHO and LTAO cells hybrids. Chinese hamster APRT was present in all hybrid clones that retained either a Z4 or Z7 chromosome. Clones that segregated APRT had lost both these chromosomes.

Hybrids have been constructed between LMTK $APRT^-$ and CHO-AT3-2, an APRT heterozygote (Adair et al., 1980; Carver et al., 1980). A total of 22 independent hybrid clones were isolated. Azaadenine-resistant segregants were obtained, and cytogenetic analysis indicated that each was missing the CHO-AT3-2 Z7 chromosome. Thus, APRT expression in the hybrids was concordant with the retention of Z7. Thus the Z4 *apt* locus is either deleted or inactive in CHO-AT3-2.

By using Southern blot technique and plasmids pAS-1 and pRG, it could be shown that the *apt* gene was missing in hybrids that retained Z4. Thus the above heterozygote arose by a deletion within the Z4 chromosome.

These data, combined with the observations of Simon et al. (1982, 1983), suggest that deletion occurs at only one of the *apt* loci. None of the Class I heterozygotes have given rise to $APRT^-$ mutants at high spontaneous frequencies like Class II heterozygotes. This would suggest that deletion occurs at a high frequency in the Z4 chromosome, perhaps due to physical differences in the genomic environment. Based on mapping studies, the Z4 *apt* gene in CHO appears to be located very near the site of inversion addition of 3q material while on Z7; the *apt* gene is located far from any translocation breakpoints.

Thus, one might interpret the model in Figure 12.4, in terms of chromosome Z4 instability. One can expect this model to be tested in the near future, but the size of the deletion is still unknown, and may vary among mutants.

G. DNA-Mediated Gene Transfer

Although DNA transformation was first discovered in bacterial systems in the 1940s, it has not been possible to transform mammalian cells with DNA until recently. In many cases the *apt* gene has been used in DNA-mediated gene transfer experiments because of the availability of apt^- competent recipient cells and strong selective systems. Using the cloned CHO *apt* gene (Bam fragment), a high frequency of transformation is obtainable in mouse *tk*⁻ *apt*⁻ L cells (Tang and Taylor, unpublished).

Transformation is usually carried out using the calcium phosphate precipitation technique (Graham and van der Eb, 1973) as modified by Wigler et al. (1979). One day prior to transformation, cells are seeded at 8×10^5 cells/100 mm Petri dish in Dulbecco's MEM. Medium is changed 4 hr prior to the addition of DNA. Plasmid DNA (linearized) is mixed with 20 μg/mL sheared calf thymus DNA. Sterile H_2O and 50 μL/mL of 2.5 *M* $CaCl_2$ are added to a final volume of 0.5 mL (250 m*M* $CaCl_2$). The DNA/$CaCl_2$ mixture is added dropwise through a siliconized Pasteur pipette to an equal volume of 2X HBS [50 m*M* Hepes, 280 m*M* $NaCl_2$, 1.5 m*M* sodium phosphate (equal amounts of mono and dibasic), pH 7.1]. The DNA is precipitated by bubbling with another siliconized Pasteur pipette during the addition of the Hepes buffer. The calcium phosphate DNA precipitate is allowed to form without agitation for 30–45 min at room temperature. One milliliter of the precipitate is added per dish of semiconfluent cells directly in the growth medium. After approximately 18 hr the transformation medium is replaced by fresh medium containing antibiotics, and incubation continued for 24 hr. At this time, cells are trypsinized, counted, and plated at 5×10^5 cells/100 mm dish for the selection of transformant clones. Selective medium (AA) contains 0.05 m*M* azaserine, 0.1 m*M* adenine, and 10% dialyzed bovine serum and is changed every 3 days for 3 weeks while transformant clones develop. The number of clones can be scored by staining with crystal-violet. Complete details regarding optimization of gene transfer into CHO cells and properties of the transformants can be found in Chapter 8.

Utilizing the APRT system, the efficiency of DNA uptake (and transient expression) can be separated from the efficiency of integration and stable expression (as expressed by clonal growth) by initially assaying [^{3}H]adenine uptake by the *apt*⁻ cells. The ability of such cells to take up adenine is due to the "transient" expression of APRT (phenotypic expression).

Phenotypic expression is measured by seeding cells in Lab-Tek slide chambers for 24 hr in DMEM. At this time, the medium is replaced with 0.3 mL DMEM containing 10 μCi/mL [^{3}H]adenine per chamber. Following 24 hr of labeling, cells are washed with PBS and fixed by rinsing in absolute methanol, incubating for 1 min in absolute methanol, and then air dried. The fixed slides are dipped into Kodak NTB2 emulsion (1:1 dilution) which has been liquefied by standing for 1 hr at 42°C. All development is done in a dark room. The slides are drained, allowed to dry, and exposed for 24 hr at 4°C in a light-tight slide box with a dessicant. The exposed slides were developed for 2 min at room temperature in Kodak D-19 developer, rinsed for 10 s in water, fixed with Kodak fixer for 5 min, rinsed with water for 5 min, and air dried. The *apt*$^+$ transformants appear as dark black cells against the lightly colored background under the microscope.

As shown in Table 12.2, approximately 10% of the cells expressing APRT activity after 24 hr go on to form stable transformants (genotypic expression). Similar results have been reported by Linsley and Siminovitch (1982) using L cells and herpes simplex type 1 TK and *E. coli gpt* genes. CHO cells showed a phenotypic response that was considerably lower than mouse L cells. This reduction in phenotypic expression of CHO cells relative to L cells may reflect differences in the conditions needed to optimize DNA-mediated transformation. Indeed, Howard and his coworkers have recently found that transient expression of the clonal bacterial chloramphenicol transacetylase gene (CAT) in CHO cells occurs at very high frequency if conditions for transformation and cell growth are optimal (see Chapter 9).

Gene transfer and genotypic expression of CHO *apt* is very much more efficient than gene transfer with the herpes virus *tk* gene. When *apt* is placed in a plasmid proximal to murine sarcoma virus LTR sequences, there is no enhancement of DNA transformation (Tang, thesis, 1983) (Table 12.2). This may be due to the presence of an enhancer sequence in the DNA fragment containing the *apt* gene (Capecchi, personal communication).

IV. APRT$^-$ MUTANTS IN OTHER SYSTEMS

A. Prokaryotes

The *apt* locus of *E. coli* K12 maps close to the dnaZ locus (Kocharyon et al., 1975). *Apt*$^-$ mutants of *E. coli* have been selected as resistant to 2-fluoroadenine (10 μg/mL) following UV irradiation (Levine and Taylor, 1982*a*, 1982*b*). A very fast qualitative assay for APRT has been developed in which *E. coli* was rendered permeable by treatment with 50 μl 0.1% SDS and 100 μl $CHCl_3$/mL cells and the solubilized extract was used for the assay. The *E. coli apt* gene has been cloned into pBR322, and a preliminary restriction map has been published (Hershey et al., 1982).

Kalle and Gots (1963) reported the isolation of two DAP-resistant mutants of *Salmonella typhimurium*. Both mutants were unable to convert diaminopurine to its ribonucleotide. One mutant (dap-r-3) had no detectable APRT activity; the other (dap-r-6) was unable to utilize diaminopurine but it had normal activity with adenine. This mutant was shown to contain an altered form of the enzyme.

TABLE 12.2
Transformation Efficiencies of the Sal I-Cleaved Plasmids Containing the Hamster *aprt* Gene and the Moloney Sarcoma Virus LTR Element[a]

Amount of Donor Plasmid (ng)	pHaprt-2		pAPRTL-1		pHaprt-3		pAPRTL-2	
	A[b]	B[c]	A	B	A	B	A	B
5	1,000	60	720	52	640	30	NC[d]	NC
10	960	52	2,000	160	2,200	100	1,000	200
50	7,100	680	2,500	300	7,200	450	NC	NC
100	14,000	1,200	9,200	600	6,800	600	7,800	800

[a] Each number represents an independent experiment. Plasmids pAPRTL-1 and pAPRTL-2 contain Moloney sarcoma virus LTR sequences.
[b] Number of total $aprt^+$ transformants/10^6 cells labeled with [^{3}H]adenine.
[c] Number of stable $aprt^+$ clones/10^6 cells in AA media.
[d] Not counted.

B. Eukaryotes

1. *Human*

Complete APRT deficiency has been noted in a small number of families (Van Acker et al., 1977; Delbarre et al., 1974). Such individuals sometimes suffer from renal stones composed of 2,8-dihydroxyadenine and acute renal failure. In the families studied for complete deficiency of APRT, several cases of partial deficiency were found. All the heterozygous patients were clinically normal. The human *aprt* gene maps to chromosome 16 (Tischfield and Ruddle, 1974). The human gene has been cloned and the restriction map determined (Stambrook et al., 1984).

2. *Mouse*

A large number of studies have been done with DAP^R mouse cells. $APRT^-$ cells were first isolated in a mouse cell line of uncertain ancestry by Atkins and Gartler (1968), in 3T6 cells (Kusano et al., 1971), in mouse teratocarcinoma (Reuser and Mintz, 1979), Ehrlich ascites carcinoma (Hori and Henderson, 1966; Murray, 1967), and in mouse L cells (Tischfield et al., 1982). In the latter cell line the frequency of mutation is abnormally high ($\sim 10^{-3}$ per cell per generation). More recently, we have isolated both heterozygotes and homozygotes in the mouse lymphoma cell line, L5178Y (Paerantakul and Taylor, 1983). In all of these mouse cell studies, the mutation frequency is higher than predicted for an autosomal locus.

A functional mouse *apt* gene has been cloned from a mouse sperm genomic DNA library in λ Charon 4A (Sikela et al., 1983). Like the CHO gene, the functional expression gene is small (3.1 kb). Digestion of the plasmid DNA with EcoRI had no effect on transformation efficiency. Digestion with a series of restriction enzymes (Pst I, Bam HI, Ava I, Bal I, Pvu II, or Hind III) abolished transforming activity. The mouse *apt* gene contains at least three introns.

3. *Drosophila melanogaster*

Mutants of *Drosophila melanogaster* that survive the lethal effects of purine (7H-imidazo[4–5] pyrimidine) are correlated with a reduction in APRT activity. Mapping data placed *apt* approximately 0.1 cm to the right of *R* on chromosome three (map position 3:3, 03) (Johnson and Friedman, 1981, 1983). As in the case of other eukaryotic systems, the APRT protein is a dimer of two identical 23,000-dalton molecular weight subunits. The level of APRT activity shows gene dose dependence in APRT heterozygotes.

4. Other

APRT has also been characterized from monkey liver (Krenitsky et al., 1969), rat liver (Krenimer et al., 1975), and beef liver (Flaks et al., 1957). Two species of APRT have been reported to be present in the yeast *Saccharomycetes pombe*, differing by molecular weight and pH (Nagy and Ribet, 1977).

REFERENCES

Adair, G. M., Carver, J. H., and Wandres, D. L. (1980). *Mutat. Res.* **72**, 187–205.

Alford, B. L. and Barnes, Jr., E. M. (1976). *J. Biol. Chem.* **251**, 4823–4827.

Arnold, W. J. and Kelley, W. H. (1978). *Methods in Enzymology* (P. A. Hoffee and M. E. Jones, eds.), Academic Press, New York, Vol. 51, 568–574.

Atkins, J. H. and Gartler, S. M. (1968). *Genetics* **60**, 781–787.

Bakay, B., Telfer, M. A., and Nyhan, W. L. (1969). *Biochem. Med.* **3**, 230.

Bradley, W. E. C. and Letovanec, D. (1982). *Somat. Cell Genet.* **8**, 51–66.

Carver, J. H., Adair, G. M., and Wandres, D. L. (1980). *Mutat. Res.* **72**, 207–230.

Chasin, L. A. (1974). *Cell* **2**, 37–41.

Deaven, L. L. and Peterson, D. F. (1973). *Chromosoma* **41**, 129–144.

Delbarre, F., Auscher, C., Amor, B., DeGrey, A., Cartier, P., and Hamet, M. (1974). *Biomed.* **21**, 82–85.

Flaks, J. G., Erwin, M. J., and Buchanan, J. M. (1957). *J. Biol. Chem.* **228**, 201–213.

Graff, J. C. and Plagemann, P. G. W. (1976). *Cancer Res.* **36**, 1428–1440.

Graham, F. L. and van der Eb, A. J. (1973). *Virology* **52**, 456–467.

Gupta, R. S. and Simonovitch, L. (1978). *Somat. Cell Genet.* **4**, 715–735.

Harris, M. (1971). *J. Cell. Phys.* **78**, 177–186.

Hershey, H. V. and Taylor, M. W. (1978). *Prep. Biochem.* **8**, 453–462.

Hershey, H. V., Gutstein, R., and Taylor, M. W. (1982). *Gene* **19**, 89–92.

Hochstadt-Ozer, J. and Stadtman, E. R. (1971). *J. Biol. Chem.* **246**, 5304–5311.

Hochstadt-Ozer, J. (1972). *J. Biol. Chem.* **247**, 2419–2426.

Hori, M. and Henderson, J. F. (1966). *J. Biol. Chem.* **241**, 1406–1411.

Johnson, D. H. and Friedman, T. B. (1981). *Science* **212**, 1035–1036.

Johnson, D. H. and Friedman, T. B. (1983). *Proc. Natl. Acad. Sci. USA* **80**, 2990–2994.

Jones, G. E. and Sargent, P. A. (1974). *Cell* **2**, 43–54.

Kalle, G. P. and Gots, J. S. (1963). *Science* **142**, 680–681.

Kocharyon, S. M., Livshits, V. A., and Sukhodolets, V. V. (1975). *Sov. Genet.* **11**, 1417–1425.

Krenimer, J. G., Young, L. G., and Groth, D. P. (1975). *Biochim. Biophys. Acts* **384**, 37–101.

Krenitsky, T. A., Neil, S. M., Elion, G. B., and Hitchings, G. H. (1969). *J. Biol. Chem.* **244**, 4779–4784.

Kusano, T., Long, C., and Green, H. (1971). *Proc. Natl. Acad. Sci. USA* **68**, 82–86.

Levine, R. A. and Taylor, M. W. (1982*a*). *J. Bacteriol.* **149**, 923–930.

Levine, R. A. and Taylor, M. W. (1982*b*). *J. Bacteriol.* **149**, 1041–1049.

Lieberman, I. and Ove, P. (1960). *J. Biol. Chem.* **235**, 1765–1768.

Linsley, P. S. and Siminovitch, L. (1982). *Mol. Cell. Biol.* **2**, 593–597.

Lowy, I., Pellicer, A., Jackson, J. F., Sim, G. K., Silvestein, S., and Axel, R. (1980). *Cell* **22**, 817–823.

Meuth, M. and Arrand, J. E. (1982). *Mol. Cell Biol.* **2**, 1459–1462.

Morrow, J., Stocco, D., and Barron, E. (1978). *J. Cell. Phys.* **96**, 81–86.

Murray, A. W. (1967). *Biochem. J.* **103**, 271–279.

Murray, A. W. (1971). *Adv. Biochem.* **40**, 811–826.

Nagy, M. and Ribet, A. M. (1977). *Eur. J. Biochem.* **77**, 77–85.

O'Farrell, P. H. (1975). *J. Biol. Chem.* **250**, 4007–4021.

Oakley, B. R., Kirsch, D. R., and Morris, N. R. (1980). *Anal. Biochem.* **105**, 361–363.

Paeratakul, U. and Taylor, M. W. (1983). *Mutat. Res.* (in press).

Quinlan, D. C. and Hochstadt, J. (1974). *Fed. Proc.* **33**, 1359.

Rabin, M. S. and Gottesman, M. M. (1979). *Som. Cell Genet.* **5**, 571–583.

Reuser, A. J. J. and Mintz, B. (1979). *Som. Cell Genet.* **5**, 781–792.

Rosenstraus, M. J. and Chasin, L. A. (1975). *Proc. Natl. Acad. Sci. USA* **72**, 493–497.

Sikela, J. M., Khan, S. A., Feliciano, E., Trill, J., Tischfield, J. A., and Stanbrook, P. J. (1983). *Gene* **22**, 219–228.

Siminovitch, L. (1976). *Cell* **7**, 1–11.

Simon, A. E., Taylor, M. W., and Bradley, W. E. C. (1983). *Mol. Cell. Biol.* **3**, 1703–1710.

Simon, A. E., Taylor, M. W., Bradley, W. E. C., and Thompson, L. H. (1982). *Mol. Cell. Biol.* **2**, 1126–1133.

Stambrook, P. J., Dush, M. K., Trill, J. J., and Tischfield, J. A. (1984). *Somat. Cell and Molec. Genet.* **10**, 359–368.

Switzer, R. C., Merril, C. R., and Shifrin, S. (1979). *Anal. Biochem.* **98**, 231–237.

Taylor, M. W., Hershey, H. V., and Simon, A. E. (1979). *Banbury Report No. 2, Mammalian Cell Mutagenesis: The Maturation of Test Systems* (A. W. Hsie, J. P. O'Neill, and V. K. McElheny, eds.), Cold Spring Harbor Laboratory, pp. 211–223.

Taylor, M. W., Pipkorn, J. H., Tokito, M. K., and Pozzatti, Jr., R. O. (1977). *Somat. Cell Genet.* **3**, 195–206.

Thomas, C. B., Arnold, W. J., and Kelley, W. N. (1973). *J. Biol. Chem.* **248**, 2529–2535.

Thompson, L. U., Fong, S., and Brookman, K. (1980). *Mutat. Res.* **74**, 21–36.

Tischfield, J. A., Bernhard, H. P. and Ruddle, F. H. (1973). *Anal. Biochem.* **53**, 545–554.

Tischfield, J. A. and Ruddle, F. H. (1974). *Proc. Natl. Acad. Sci. USA* **71**, 45–49.

Tischfield, J. A., Trill, J. S., Lee, Y. I., Coy, K., and Taylor, M. W. (1982). *Mol. Cell Biol.* **2**, 250–257.

Van Acker, K. J., Simmonds, H. B., Potter, C., and Cameron, J. S. (1977). *New England J. Med.* **297**, 127–132.

Wigler, M., Pellicer, A., Silvestein, S., Axel, R., Urlaub, G., and Chasin, L. (1979). *Proc. Natl. Acad. Sci. USA* **76**, 1373–1376.

Witney, F. R. and Taylor, M. W. (1978). *Biochem. Genet.* **16**, 917–926.

Worton, R. G., Ho, C. C., and Duff, C. (1977). *Somat. Cell Genet.* **3**, 37–45.

Zylka, J. M. and Plagemann, P. G. W. (1975). *J. Biol. Chem.* **250**, 5756–5767.

CHAPTER 13

THE HGPRT SYSTEM

Raymond G. Fenwick
Department of Biochemistry
Dalhousie University
Halifax, Nova Scotia, Canada

Prepared with support from the National Cancer Institute of Canada.

I. INTRODUCTION

The reaction catalyzed by hypoxanthine-guanine phosphoribosyltransferase (HGPRT, EC 2.4.2.8) was first described three decades ago by Kornberg et al. (1955). Over the subsequent years many researchers have focused their attention on HGPRT and the gene that encodes the enzyme. Those interested in the molecular genetics of cultured mammalian cells were attracted by the early development of simple schemes which could be used to select for or against the presence of cellular HGPRT activity and thus to isolate cells with variant phenotypes. This along with associations made between HGPRT variation and heritable diseases of humans have prompted the development of experimental tools and methodologies for the analysis of the system. As is often the case, those tools have also been applicable to experiments dealing with other topics of basic biological interest including mutagenesis, X-chromosome inactivation, cell fusion, and gene transfer.

Chinese hamster cells, of course, have not been the only types of cells used for the study of the HGPRT system and in this chapter I will note important observations made in other genetic backgrounds. However, experiments using those cells have been central to the development of the system and in keeping with the theme of this volume we can illustrate most important aspects with examples involving Chinese hamster cells. For almost 10 years I had the pleasure of collaborating with C. Thomas Caskey and others in his group on the analysis of the HGPRT system in the V79 clone of Chinese hamster lung cells. Therefore, to illustrate specific points I will often draw upon data from the results of those efforts.

II. OVERVIEW OF THE HGPRT SYSTEM

A. Properties of the Enzyme

1. Reaction, Substrates, and Inhibitors

HGPRT catalyzes transfer of the phosphoribosyl moiety from 5-phosphoribosyl-1-pyrophosphate (PP-ribose-P) to the 9 position of hypoxanthine or guanine to yield IMP or GMP plus pyrophosphate. This provides an energy efficient alternative to the de novo synthesis of purine ribonucleotides. Using HGPRT purified from Chinese hamster brain or liver, Olsen and Milman (1974a) demonstrated that a single enzyme utilizes both hypoxanthine and guanine as substrates. This has also been found to be the case for HGPRT from other mammals and lower eukaryotes such as yeasts. However, two enzymes are produced by bacteria such as *Escherichia coli* and *Salmonella typhimurium*, the first having high affinities for guanine and xanthine but little activity with hypoxanthine and a second which has the opposite pattern (see Kelley and Wyngaarden, 1983). The dimagnesium salt of PP-ribose-P appears to be the actual phosphoribosyl donor in the reaction and optimum activity is obtained when the magnesium concentration is about 5 mM (Krenitsky et al., 1969). In addition, by varying the ratio of magnesium of PP-ribose-P it is possible to alter both the kinetics and mechanism of the reaction (see Kelley and Wyngaarden, 1983).

In kinetic studies of purified Chinese hamster HGPRT, Olsen and Milman (1974a) found K_m values for hypoxanthine, guanine, and PP-ribose-P of 0.52, 1.1 and 5.3 μM, respectively. They observed similar results using extracts of cultured Chinese hamster cells (V79) but Fenwick et al. (1977a) obtained somewhat higher values. Higher values have also been reported for HGPRT from other mammalian sources (Kelley and Wyngaarden, 1983). Olsen and Milman (1974a) found a pH optimum of about 10.5 for HGPRT activity but they and other authors have conducted kinetic analyses at pH 7.5 to 8. Kong and Parks (1974) demonstrated that the effect of pH on human HGPRT was a function of the pK_a of the purine being used as a substrate and from their results concluded that nonionized forms of the purines are the effective substrates for HGPRT.

In addition to hypoxanthine and guanine, HGPRT from eukaryotic cells will utilize a variety of purines with an oxo or thio but not an amino group at position 6 (Krenitsky et al., 1969; Miller and Bieber, 1969). Binding of the purine is enhanced by a 2-amino group but decreased by a 2-hydroxyl group. Thus 6-mercaptopurine, 6-thioguanine, and 8-azaguanine are substrates for eukaryotic HGPRT whereas adenine and xanthine are not. 6-thioguanine and 8-azaguanine are routinely used as agents for the selection of HGPRT-deficient cells because they are converted into toxic nucleotides by the enzyme (see Elion, 1967; Roy-Burman, 1970). However, under physiological conditions 8-azaguanine is a much less effective substrate or inhibitor of HGPRT than is 6-thioguanine (Miller and Bieber, 1969; Gillin et al., 1972; Sharp et al., 1973). This is consistent with the neutral species of purines serving as substrates since the pK_a of 8-azaguanine is 6.4 while that of 6-thioguanine is 8.2.

HGPRT is inhibited by its products IMP and GMP but not, as might be expected, by AMP (Olsen and Milman, 1974a). Although purine analogs such as 6-thioguanine are often described in the literature as competitive inhibitors, they are, as pointed out by Kong and Parks (1974), actually alternative substrates. Nonmetabolized inhibitors especially noncompetitive inhibitors have been sought because they might have therapeutic value (Piper et al., 1980) but to my knowledge none have been identified.

2. *Localization*

HGPRT is a ubiquitous enzyme found throughout mammalian tissues and cell types (Krenitsky 1969; Lo and Palmour, 1979). The most extensive studies of HGPRT activities in specific tissues have been conducted with the human system and have shown the specific activity to be highest in brain, especially in the basal ganglia (Kelley and Wyngaarden, 1983). High specific activity has also been noted in the Chinese hamster brain (Olsen and Milman, 1974a) and this has been shown to correlate with high levels of HGPRT mRNA (Melton et al., 1981). Thus, expression of the HGPRT gene appears to be regulated in a cell- or tissue-specific fashion. For cultured cells, however, I am not aware of any evidence that enzyme activity varies with the cell cycle, presence of substrates, and so on, so expression of the gene within particular types of cells seems to be constitutive.

HGPRT is usually isolated as and considered to be a soluble cytoplasmic protein. However, at least one group has argued for an association between the enzyme and the cell membrane (Hochstadt and Quinlan, 1976).

3. *Structure*

SDS-polyacrylamide gel electrophoresis of Chinese hamster HGPRT purified from brain (Olsen and Milman, 1974a) or immunoprecipitated from cultured V79 cells (Fenwick et al., 1977a) indicated that the molecular weight of the enzyme protein is 25,000–26,000. Similar values have been determined for the human (Olsen and Milman, 1974b) and mouse (Hughes et al., 1975) proteins. Studies of cloned cDNA copies of HGPRT mRNA revealed that the Chinese hamster and mouse proteins contain 218 amino acids (including the initiator methionine) and have molecular weights of 24,500 (Konecki et al., 1982). Wilson et al. (1982) determined the amino acid sequence of HGPRT purified from human erythrocytes and showed it to contain 217 residues including an acetylated alanine at its amino terminus. Comparison of the determined and inferred amino acid sequences demonstrated that the proteins from Chinese hamster, mouse, and human are identical in at least 206 positions and thus possess remarkably high sequence conservation (Konecki et al., 1982).

The functional form of mammalian HGPRT is an oligomer of the protein described above. By exclusion chromatography or polyacrylamide gel electrophoresis, the native molecular weight was originally estimated to be 75,000–85,000 which led to a conclusion that the enzyme is a trimer of the subunit protein (Olsen and Milman, 1974a). However, subsequent studies in which subunits of human HGPRT were chemically cross-linked indicated a tetrameric or dimer of dimers structure (Holden and Kelley, 1978). With electrophoretic and isoelectric analyses of the heteropolymeric enzyme formed in a human–mouse somatic cell hybrid, Johnson et al (1978) obtained clear evidence of a dimer. The last authors also demonstrated that the sedimentation rate of the human enzyme could be changed by altering the ionic strength of the centrifugation medium and concluded that the dimers associate to form tetramers at high ionic strength. Thus the earlier conclusions of a trimeric structure may have reflected an equilibrium between those two forms.

Chinese hamster HGPRT can be separated into multiple species or isozymes by electrophoresis or isoelectric focusing (Olsen and Milman, 1974a; Chasin and Urlaub, 1976). Some of the heterogeneity, which also has been observed for the enzymes from other mammalian sources, may be due to aging of the enzyme or alterations incurred during purification. For human erythrocytes, a more acidic form of the HGPRT subunit protein accumulates as the cells age (Johnson et al., 1982) and gives rise to homo- and heterodimers which can be separated on the basis of charge. Wilson et al. (1982) have shown that this alteration is due to the deamidation of the asparagine which is the 106th amino acid in the protein. The significance of this alteration for the structure and function of HGPRT is unknown.

B. Relationship to Human Disease

The relevance of HGPRT to genetic diseases of humans was identified when Seegmiller et al. (1967) demonstrated that virtually complete deficiency of the enzyme was the primary biochemical defect in patients suffering from Lesch–Nyhan syndrome. That disease is characterized by the excessive production of uric acid and certain characteristic neurological features such as choreoathetosis, spasticity, mental retardation, and self-mutilation. Death usually occurs in the second and third decade from infection or renal failure. How HGPRT deficiency causes the abnormalities of the central nervous system is not understood. A partial deficiency of the enzyme has been associated with a severe form of gout (Kelley et al., 1967). Excessive production of uric acid is also observed in such patients but they do not suffer the severe neurological abnormalities associated with the Lesch–Nyhan syndrome. A thorough discussion of these clinical syndromes has been presented by Kelley and Wyngaarden (1983).

C. Location of the HGPRT Gene

Once HGPRT deficiency was shown to be associated with the Lesch–Nyhan syndrome the HGPRT gene was thought to be located on the X chromosome because the syndrome was known to be inherited in an X-linked, recessive fashion (Seegmiller et al., 1967). That assignment was confirmed by following segregation of the HGPRT phenotype from human–mouse somatic cell hybrids (Ricciuti and Ruddle, 1973). The results demonstrated that HGPRT was syntenic with two other X-linked genes, glucose-6-phosphate dehydrogenase (GGPD) and phosphoglycerate kinase (PGK). Those loci have also been shown to be linked in Chinese hamster (Westerveld et al., 1972) and mouse (Chapman and Showes, 1976).

Farrell and Worton (1977) localized the HGPRT gene to the distal end of the short arm on the Chinese hamster X chromosome and the genes for G6PD and PKG have also been assigned to that arm (Fenwick, 1980). The order of those markers along the long arm of the human X is centromere-PGK-HGPRT-G6PD (Miller et al., 1978). HGPRT and G6PD are both located in the distal region of Xq but they segregate independently in informative families (Francke et al., 1974) indicating that they are still far apart in genetic terms. Additional studies will be needed to determine the order of those commonly used loci along the Chinese hamster X chromosome.

The knowledge that HGPRT is encoded by an X-linked gene served to focus attention on that locus as a target for genetic alterations in mammalian cells. Since male cells have a single X chromosome and one X is inactivated in female cells during embyronic development (Lyon, 1972), the HGPRT locus is either actually or functionally hemizygous in diploid or pseudodiploid cells. Thus, inactivation of the HGPRT gene by genetic or epigenetic mechanisms will dramatically alter the phenotype of a cell. Such would not be the case if the gene were autosomal and diploid. This greatly facilitates isolation of variants from a population of normal cells.

D. HGPRT and Cellular Utilization of Purines

Most cells other than mature erythrocytes (Fontenelle and Henderson, 1969) or terminally differentiated Friend erythroleukemia cells (Reem and Friend, 1975) are able to satisfy their requirements for purines through the de novo synthesis of purine ribonucleotides. Thus, under normal conditions, HGPRT is not required for cellular viability or growth. If, however, the de novo pathway is inhibited by mutations (Patterson et al., 1974) or drugs (Szybalski et al., 1962) cells become dependent on exogenous purines. When the only purines available are hypoxanthine or guanine, HGPRT becomes an essential function.

Although cultured cells derived from Lesch–Nyhan patients have been reported to have accelerated rates of de novo purine synthesis, Hershfield and Seegmiller (1977) have demonstrated that normal human lymphoblasts and HGPRT mutants synthesize purines at similar rates when they are grown in purine-free medium. Those authors did find, however, that the mutant cells secreted 7- to 10-fold more purines (mostly hypoxanthine) than normal cells, but the amount secreted was less than 10% of the purines synthesized by the de novo pathway. Still, this is consistent with the overproduction of uric acid, a catabolite of hypoxanthine, in Lesch–Nyhan patients. They also found that addition of hypoxanthine to the growth medium or the use of undialyzed serum caused a substantial inhibition of de novo purine synthesis in wild-type but not HGPRT-deficient cells. Thus, hypoxanthine is a normal component of serum, and although its concentration might vary in different types and batches, the action of HGPRT on that substrate results in the formation of purine nucleotides and inhibition of de novo purine synthesis in normal but not HGPRT-deficient cells. A full discussion of the possible mechanisms for that inhibition is beyond the scope of this chapter but has been the subject of other reviews (Kelley and Wyngaarden, 1983; Wyngaarden and Kelley, 1983). However, it should be noted that the first and rate-limiting step in the de novo pathway, phosphoribosylpyrophosphate amidotransferase, utilizes PP-ribose-P and is inhibited by purine nucleotides. Thus, HGPRT activity might inhibit de novo synthesis by competing for PP-ribose-P or increasing the levels of purine nucleotides. These alternatives have been often investigated but difficult to confirm or differentiate.

III. GENETICS AND THE HGPRT SYSTEM

As described in Section II, protocols for selection of HGPRT-deficient and HGPRT-proficient mammalian cells had been developed by 1962. At that time it was not clear whether phenotypic variation in mammalian cells was caused by genetic or epigenetic events. Thus, a major and historically important use of the HGPRT system has been to validate the field of somatic cell genetics by examining HGPRT variants for evidence of mutational alterations.

A natural extension of such studies has been to ask what types of mutations can be detected with the system. This has been done for two purposes. One is to gather information on the process of mutation in mammalian cells by determining the spectrum of DNA alterations which can cause loss or restoration of HGPRT activity. The second is to increase our understanding of the HGPRT system by identifying mutations that alter specific aspects of the structure and function of HGPRT or its gene.

Finally, genetic investigations of the HGPRT system are also of value in that they provide cell lines and methods that can be utilized in other genetic and biochemical studies. Examples include the use of the HGPRT locus as a test system for the analysis of mutagens and carcinogens; the use of clones carrying well characterized alleles for construction of somatic cell hybrids, analysis of linkage, and gene transfer experiments; and the application of mutant cell lines plus analytical tools for HGPRT to the study of biological phenomena such as X chromosome inactivation, regulation of purine utilization, and purine transport.

IV. SELECTIVE SCHEMES

Perhaps the most important lesson one learns from the evaluation of efforts to select mutant or variant mammalian cells is that you invariably get out of a selection exactly what you have asked for. The real test, however, is understanding or appreciating what types of phenotypic alterations are favored or demanded by a particular set of selective conditions. Even a causal review of the literature reveals that the results of both forward and reverse selective schemes for altered HGPRT phenotypes are markedly influenced by choices of selective agents or conditions. Several factors which affect selective schemes have been documented in early publications or reviews (Thompson and Baker 1973; Clements, 1973), but they are not universally appreciated because they are periodically "rediscovered." Thus I will take some time to outline what I believe to be important variables in the selective procedures.

A. Forward Selection of the HGPRT-Negative Phenotype

1. Choice of Selective Agents

A number of purine analogs act as substrates for HGPRT which converts them to toxic nucleotides (Elion and Hitchings, 1965; Elion, 1967). Those which will kill HGPRT-positive cells and thus act as selective agents for cells with HGPRT-deficient phenotypes include 8-azaguanine, 8-azahypoxanthine, 6-mercaptopurine, and 6-thioguanine, but 8-azaguanine and 6-thioguanine have been used most frequently. Early selections for resistance to 8-azaguanine employed D98 human cells (Szybalski, 1959; Szybalski and Smith, 1959) or mouse L cells (Lieberman and Ove, 1959). Chu and Malling (1968), using the V79-122D1 clone, isolated the first resistant Chinese hamster cells. Likewise the use of 6-thioguanine to select variants of Chinese hamster cells (Chasin, 1972) followed the work of Subak-Sharpe (1965) who had used the drug to select resistant Syrian hamster cells.

Clements (1975) compiled and tabulated the results of early experiments in which the various selective agents were used to isolate resistant clones from a variety of mammalian cell lines. In an important observtion he made note of the fact that many of the isolates resistant to 8-azaguanine retained significant levels of HGPRT activity while those selected for resistance to 6-thioguanine usually did not. Although it is not always appreciated, this stems from the fact mentioned above (see Section II.A.2) that 6-thioguanine is a much more effective substrate for HGPRT than 8-azaguanine. Although there is ample evidence for HGPRT mutations in 8-azaguanine-resistant clones retaining HGPRT activity (Fenwick et al., 1977a) this difference means that the two drugs select for quite different phenotypes and potentially different but overlapping sets of mutants.

A point of some importance which is passed on by word of mouth but virtually never mentioned in publications is the preparation, characterization, and storage of stock solutions of the selective agents. The final concentrations of 8-azaguanine and 6-thioguanine in our selective media are routinely 140 μM and 27 μM, respectively. We have been able to prepare 25-fold concentrates of 8-azaguanine and 100-fold concentrates of 6-thioguanine by dissolving the analogs in 0.01 N KOH. After filter sterilization the 8-azaguanine stock must be stored at room temperature since it precipitates at 4°C but we have detected little or no loss of activity over periods of several months. The 6-thioguanine stock degenerates quite rapidly at 4°C but is stable for months at −20°C. As can be seen in Figure 13.1, both analogs have distinctive absorption spectra which can be used to characterize the stock solutions. As 6-thioguanine degenerates, the absorption maximum at 320 nm decreases and we use only those stocks which have an $A_{320}/A_{260} \geq 2.5$.

2. *Important Parameters in Protocols*

A number of factors have been found to alter significantly the selection of HGPRT-deficient cell lines. However those of importance to a particular experiment are dependent on the goals of that study. If you wish to accurately determine the frequency of spontaneous or mutagen-induced mutants in a population you must be aware of factors that might cause you to over or underestimate their numbers. If, on the other hand, you want to determine the spectrum of mutations which arise spontaneously or after mutagen treatment you may be more concerned with the purity of your initial cell population, specificity of selective media, and selection of independent isolates.

Most laboratories continue to use serum as a source of growth factors and macromolecular components in their media but early work pointed out that fetal bovine serum antagonized inhibition of cell growth by selective agents such as 8-azaguanine (see Thompson and Baker, 1973). Studies by Hershfield and Seegmiller (1977) demonstrated that fetal calf

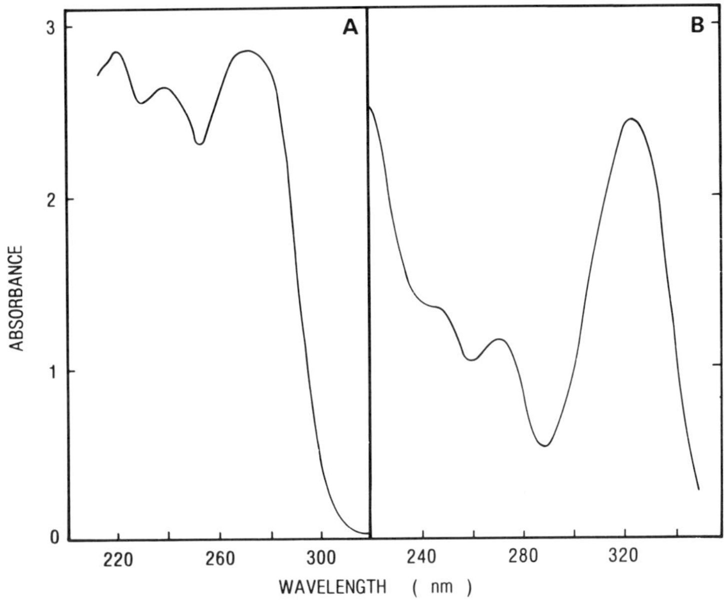

Figure 13.1. Absorption spectra of selective agents. Stock solutions of 140 μ*M* 8-azaguanine (A) and 27 μ*M* 6-thioguanine (B) prepared in 0.01 *N* KOH were diluted 10-fold with water prior to measurement of their absorption spectra.

serum contains enough hypoxanthine to cause inhibition of de novo purine synthesis in HGPRT-positive cells. Thus the antagonist is most likely hypoxanthine. Since 8-azaguanine has been shown to be a relatively weak competitive inhibitor of HGPRT, its effectiveness is reduced until the exogenous hypoxanthine is consumed. This allows cell numbers to increase before the analog begins to exert its effect and thus leads to overestimations of mutant frequencies. Thompson and Baker (1973) pointed out that this problem can be reduced or eliminated by dialysis of serums used in selective media. Although 6-thioguanine is an efficient competitive inhibitor of HGPRT (K_i = 1.8 μ*M*; Krenitsky et al., 1969) it is still affected by this phenomenon especially when cells are plated at low densities (Thacker et al., 1976). Since purines derived from sera interfere with analog inhibition of the cell growth, it should go without saying that the growth medium used in such experiments should be one of the formulations that does not contain purine bases or nucleotides.

A related phenomenon that can increase the apparent number of resistant clones, especially when cells are plated at high densities, is reduction of the effective drug dose through interference by metabolites released from dead or dying cells. This problem can be eliminated by refeeding of the cultures with fresh selective medium at 2 or 3 day intervals (Littlefield, 1963). Although I am not aware of direct comparisons, one would expect that selections involving 8-azaguanine would be affected to a greater extent than those using 6-thioguanine because of the affinity differences noted above. Frequent feeding of selective plates, however, can make enumeration of drug-resistant colonies difficult when using cells, such as Chinese hamster ovary (CHO), that are easily displaced from the substrate by physical agitation. Displaced cells can form multiple satellite colonies, which if initiated early in the selection may be hard to distinguish from original colonies. To solve this problem some authors have eliminated the need for feeding by conducting selections using relatively low cell densities (Thompson and Baker, 1973). Others have utilized periodic feedings and determined the frequency of drug-resistant colonies by using the Poisson equation to calculate the average number of drug-resistant colonies per culture or plate from the fraction of replicate cultures having no colonies (Gillin et al., 1972). That fraction can not be affected by colony splitting.

The experimental yield of drug-resistant colonies can be reduced due to metabolic cooperation between sensitive and resistant cells (Dancis et al., 1969; Subak-Sharpe et al., 1969) during which the analogs metabolically activated to toxic nucleotides in the former are transferred to and reduce the viability of the latter. This process occurs at tight junctions formed between cells and not all cell lines will form that type of junction (Wright et al., 1976). Thus the extent of experimental problems caused by this phenomenon is dependent on the type of cell being used and the density at which the cells are inoculated. Reconstruction experiments in which low numbers of HGPRT cells are subjected to 8-azaguanine or 6-thioguanine selection in the presence of large and varying numbers of wild-type cells have demonstrated that Chinese hamster cells are subject to this problem but have also shown that it can be minimized by controlling the size of the inoculum (Chasin, 1973).

When mutagens are used to induce HGPRT-deficient cells or suspect compounds are being tested for mutagenic activity several new components enter into the design and evaluation of experiments. Since most mutagens also reduce cell viability one must always be aware of the fact that the selective pressure may be either increased or decreased if the cytotoxic effects of mutagens alter cell density, metabolic cooperation, or the release of competitors from dead cells. In addition, newly induced mutations will not be fully expressed until preexisting levels of HGPRT are diminished by either turnover or cell division. It may take as many as 8 or 9 days after cells have been mutagenized before the frequency of 6-thioguanine-resistant or HGPRT-deficient cells reaches a stable maximum (Chasin, 1973). This usually means that the mutagenized population needs to be subcultured one or more times before the cells are subjected to selective conditions. For CHO cells this task can be made easier by maintaining the cells as suspension (Thompson et al., 1980) or unattached cultures (Li, 1981). As common sense would dictate, the optimum expression time will vary with the stringency of the selective conditions (Thilly et al., 1978; Bonatti et al., 1980; Thacker et al, 1982) and one must bear in mind that very different sets of mutant phenotypes may be selected when either the selective stringency or expression time are varied.

B. Counter Selection of the HGPRT-Positive Phenotype

1. Choice of Selective Agents

When the de novo pathway of purine biosynthesis is inhibited cell growth becomes dependent on the presence of HGPRT if hypoxanthine or guanine provides the sole source of purines. This permits selection of revertants from populations of HGPRT-deficient cells as well as the elimination of preexisting mutants from populations of normal cells. Szybalski et al. (1962) applied this strategy by supplementing medium with hypoxanthine, aminopterin, and thymidine (HAT). The folate antagonist aminopterin (or the related compound methotrexate) blocks synthesis of reduced folates which are required for the biosynthesis of purines, thymidylate, and glycine. Thus glycine must also be included (THAG) if it is not already present in the medium being used.

Because of the additional requirements imposed by folate analogs, situations arise in which HAT medium cannot be used to select $HGPRT^+$ cells. In such cases azaserine can be used to block de novo purine synthesis. As an analog of glutamine it inhibits 5′-phosphoribosylformyl-glycineamide L-glutamine amidoligase (EC 6.3.5.3), the fourth enzyme in the de novo pathway (Levenberg et al., 1957). In the presence of hypoxanthine this allows selection of cells with HGPRT-positive phenotypes (Siniscalco et al., 1969). As a glutamine analog, azaserine can cause toxic effects at other metabolic steps. Fortunately, this occurs at concentrations higher than that required to inhibit purine synthesis and azaserine plus hypoxanthine has proven to be an effective selective system for HGPRT-positive Chinese hamster cells (Chasin, 1973; Fuscoe et al., 1982).

2. Important Parameters in Protocols

Some of the factors that influence selection of HGPRT-deficient cells also have effects on selection of HGPRT-positive cells. Since serum can contain factors that block or reduce the toxic effects of aminopterin (Peterson et al., 1974) one must be concerned about the presence of purines in either media or sera, especially those that could satisfy the requirement for purine ribonucleotides imposed by aminopterin or azaserine without activation by HGPRT (e.g., adenine or adenosine). In practice, however, I have never encountered a lot of undialyzed serum which would allow growth of HGPRT-negative cells when de novo synthesis of purines was inhibited. This indicates that the purines contained in serum are or give rise to substrates of HGPRT (hypoxanthine or inosine) and can be ignored unless you wish to question whether a particular clone has survived selection in HAT medium because it has acquired resistance to aminopterin (see Section IV.B.3) in which case you must use dialyzed serum (Fenwick, 1980).

Metabolic cooperation can also occur between cells in HAT medium. In this instance it is possible that $HGPRT^-$ cells may be spared the lethality of HAT medium by growing in proximity to $HGPRT^+$ cells. Unlike the situation for the selection of $HGPRT^-$ cells this should not lead to an underestimation of the frequency of variants in a population since the growth of HAT-resistant cells should not be affected. However, it may be the basis underlying the high frequency of 6-thioguanine-resistant cells among HAT-resistant revertants of HGPRT mutants (Hodgkiss et al., 1980). This phenomenon will be most prevalent at high cell densities and thus might also prevent the elimination of preexisting mutants from populations of wild-type cells when they are simply passaged in HAT medium. When using HAT medium to isolate somatic cell hybrids, I have noted the density-dependent appearance of colonies in control experiments involving cells that have not been treated with agents to promote cell fusion (unpublished observations). Unlike the more frequent colonies that arise after treatment of the parents with Sendai virus or polyethylene glycol, these clones do not survive isolation by trypsinization plus subsequent propagation in HAT medium. This indicates that they are simply mixtures of parental cells which survived the initial selection by metabolic cooperation. Since cells can escape the selective consequences of HAT medium it is very important to reclone initial isolates in either selective or nonselective media before trying to define their phenotypes.

3. *Artifacts of HAT Selection*

A problem that has affected a number of laboratories has been an inability to obtain cell growth of somatic cell hybrids, revertants of HGPRT or thymidine kinase mutants, and even wild-type cells in HAT medium. Although few of these instances have reached the literature, one source of the difficulties is mycoplasma contamination. Stanbridge (1971) reviewed reports that cells of infected cultures are inhibited or killed in HAT medium because the mycoplasma degrade thymidine to thymine. Thymine can not be used to satisfy the TMP requirement imposed on cells by aminopterin. In addition, HGPRT from the mycoplasma can be detected in extracts prepared from infected cultures (Stanbridge et al., 1975). When HGPRT-deficient cells are infected, they can appear to have low levels of HGPRT activity but they retain resistance to 8-azaguanine and 6-thioguanine (van Diggelen et al., 1977). Since the microbial enzyme is quite unlike that from the wild-type host cells, it can be mistakenly identified as a mutationally altered mammalian enzyme (see Section V.A). Furthermore, because the infection prevents growth in HAT, infected HGPRT mutants will not give rise to phenotypic revertants in that selective system and may thus be classified in error as nonrevertable mutants.

There are at least three mechanisms by which cells can become resistant to the toxic effects of aminopterin and thus avoid the selective nature of HAT medium. These include mutation or over expression of dihydrofolate reductase and altered transport of the drug (see Schimke, 1984, and Chapter 16). These alterations can provide a significant source of HAT-resistant clones when revertants are being selected from relatively stable HGPRT mutants (Fenwick, 1980). This means that such isolates must be screened for HGPRT activity and/or aminopterin resistance before they can actually be classified as HGPRT-positive revertants.

V. STUDIES TO DEFINE THE MOLECULAR BASIS OF VARIANT PHENOTYPES

For studies involving variation of the HGPRT phenotype in cultured cells, the primary goal has often been to determine whether mutations are the cause of variation and more specifically whether such mutations can be localized to the HGPRT gene. Thompson and Baker (1973) outlined criteria which should be satisfied if a variant clone is to be defined as a mutant. Those included mutagen enhancement of variant frequencies, stability of variant phenotypes, presence of an altered gene product in variant cells, and association of the altered phenotype with a specific region of the variant's genome. By the time their review had been written, the first two criteria, which relate to the process of mutation in general, had been satisfied (also see Clements, 1975, for a thorough review of the early work dealing with HGPRT variation). However, it should be remembered that stable changes in phenotypes may occur by epigenetic mechanisms (e.g., see Harris, 1971, and see Section VI.E). Over the subsequent 10 years, a variety of methods have been developed and used to demonstrate the presence of altered forms of HGPRT in variant cells. The advent of recombinant DNA technologies has also made possible analysis of the primary product of the HGPRT gene, its mRNA, as well as the HGPRT gene itself. Since it has not been possible to map genetic alterations in mammalian cells by measuring recombination within the HGPRT gene or along the X chromosome, those new techniques have finally provided direct evidence for mutations affecting the HGPRT gene. In this section I will describe the assays which are

TABLE 13.1
Isolation and Analysis of Clones Resistant to 8-Azaguanine[a]

Mutagenesis[b]		Examples[c]	
Mutagen	Stimulation	Isolate	Growth in HAT
None	1	RJK78	S
		RJK88	S
MNNG	70	RJK3	R
		RJK10	S
EMS	70	RJK36	S
		RJK39	S
		RJK43	S
		RJK44	R
UV	20	RJK62	S
		RJK71	S
ICR191	8	RJK460	R
		RJK463	S

[a] A summary of the efforts of Gillin et al. (1972), Chiang (1977), and Fuscoe et al. (1983) to isolate spontaneous and mutagen-induced derivatives of V79 Chinese hamster cells.
[b] Mutagens used were *N*-methyl-*N'*-nitro-*N*-nitrosoguanidine (MNNG), ethyl methanesulfonate (EMS), ultraviolet light (UV), and ICR191. The frequency of spontaneous 8-azaguanine resistance was about 10^{-5} and fold stimulations by the mutagens are shown.
[c] The examples shown are from the collection of C.T. Caskey, Howard Hughes Medical Institute, Houston, Texas, and they are either sensitive (S) or resistant (R) to the inhibition of growth by HAT medium.

particularly useful for the analysis of variants and illustrate their application to cell lines we have studied.

A. Types of Assays

1. Isolation and Initial Characterization

As summarized in Table 13.1, Gillin et al. (1972), Chiang (1977), and Fuscoe et al. (1983) have presented evidence that a variety of mutagens increase the frequency of 8-azaguanine-resistant cells in cultures of RJKO. RJKO is a single colony isolate of the V79 line of Chinese hamster lung fibroblasts and V79 was derived from tissues of a male fetus (Ford and Yerganian, 1958). RJKO is pseudodiploid and the results of many investigations including those of Fenwick et al. (1977a,b) and Fuscoe et al. (1983) indicate that it continues to be haploid for the HGPRT locus.

Table 13.1 also lists examples of spontaneous and mutagen-induced variants isolated from RJKO. As noted, some of the variants will grow in counterselective HAT medium. Ideally, cells selected for HGPRT deficiency using a purine analog should simultaneously become sensitive to

TABLE 13.2
Growth Phenotypes of Chinese Hamster Clones Selected for Resistance to 8-Azaguanine[a]

		Response to Selective Media			
Group	Number	8AG	6TG	HAT	Examples
A	18	R	R	S	RJK10, 36, 39, 43
B	11	R	R	R	RJK3, 44
C	6	R	S	R	
	35				

[a] A summary of the sensitivity (S) or resistance (R) to growth inhibition by 8-azaguanine (8AG), 6-thioguanine (6TG), and HAT for 8-azaguanine-resistant clones isolated by Gillin et al. (1972) after MNNG or EMS mutagenesis as described in Table 13.1.

HAT medium and acquire unselected resistances to other analogs which are selective for the same phenotype (or vice versa when HAT-resistant revertants are selected from HGPRT-deficient cell lines). Although those types of events have been identified (Szybalski et al., 1962) the situation is often more complex. For instance, Table 13.2 illustrates that only half of the induced derivatives isolated from RJKO by Gillin et al. (1972) had the expected phenotype for HGPRT-negative cells. Similarly, when I selected HAT-resistant revertants from one of those clones, RJK39, I found that only 10 of 24 isolates had returned to the phenotype of wild-type cells (Table 13.3). Thus both forward and reverse selective protocols generate heterogeneous populations of isolates. Biochemical and molecular tests can be used to determine whether the heterogeneity

TABLE 13.3
Growth Phenotypes of Revertants Isolated from the HGPRT-Deficient Clone RJK39[a]

		Response to Selective Media			
Group	Number	HAT	8AG	6TG	Aminopterin
A	10	R	S	S	S
A	7	R	R	S	S
C	4	R	R	R	S
D	3	R	R	R	R
	24				

[a] After mutagenesis, HAT-resistant subclones of RJK39 (see Tables 13.1 and 13.2) were tested for growth inhibition by the designated selective agents (see Fenwick, 1980, for details).

is consistent with the acquisition of mutations affecting the structure or expression of the HGPRT gene.

2. Enzyme Activity

HGPRT activity can be measured in intact cells by following the uptake or incorporation of radiolabeled hypoxanthine or guanine. It can also be quantitated in cell lysates by following the conversion of those radiolabeled purines, in the presence of PP-ribose-P, to their corresponding ribonucleotides. Our protocols for each have been described (Fenwick, 1980) but similar procedures for one or both assays are included in many of the papers cited in this chapter. Early studies from a number of laboratories demonstrated that variants resistant to 8-azaguanine, 6-thioguanine, or related selective agents were often found to be HGPRT deficient by one or both of the assays. However, applications of the assays to large collections of variants revealed that many of the isolates retained measurable, although usually reduced levels of HGPRT activity, especially when 8-azaguanine had been used for selection of the variants (Gillin et al., 1972; Sharp et al., 1973).

The cellular assay has proven to be a particularly sensitive and informative method of detecting HGPRT activity. It can also be used to identify cell lines which have a unique HGPRT phenotype. Studies of HGPRT variants isolated from both forward (Gillin et al. 1972) and reverse (Chu et al., 1969) selections revealed that some clones will only incorporate the purine precursors of HGPRT when the de novo synthesis of purines is inhibited. For example, Table 13.4 illustrates that aminopterin stimulates hypoxanthine incorportion by several of the variant clones identified in Table 13.1. With the exception of RJK62, they are the clones that are resistant to HAT medium. However, aminopterin

TABLE 13.4
The Cellular Assay for HGPRT Activity[a]

	Incorporation of [^{14}C]hypoxanthine (cpm)	
Isolate	Control	Plus Aminopterin
RJKO	204,299	208,358
RJK3	3,127	110,406
RJK44	770	82,807
RJK62	−1,031	41,991
RJK88	312	127
RJK460	32,495	132,298

[a] Using the procedure described by Fenwick (1980), wild-type Chinese hamster cells (RJKO) and cells of 8-azaguanine-resistant clones described in Table 13.1 were plated in nonselective medium at 10^5 per 60 mm dish and 48 hr later incorporation of [^{14}C]hypoxanthine was measured during a 5 hr incubation in the presence or absence of 10 μM aminopterin.

TABLE 13.5
In Vitro Analysis of HGPRT[a]

Isolate	Kinetic Properties: V_{max}	Kinetic Properties: K_m PP-ribose-P	Kinetic Properties: Hill Coefficient	Electrophoretic Mobility
RJKO	1.44 $\frac{\text{nmol}}{\text{mg min}}$	30 μM	1.0	0.76
RJK3	0.32	50	1.2	0.61
RJK44	2.84	360	1.9	0.76

[a] As described by Fenwick et al. (1977a), HGPRTs in extracts of wild-type Chinese hamster cells (RJKO) and 8-azaguanine-resistant but HGPRT-positive subclones (see Table 13.4) were assayed for kinetic properties and relative electrophoretic mobilities in polyacrylamide gels.

neither increases the incorporation by wild-type cells, RJKO, nor stimulates incorporation by an enzyme-negative variant such as RJK88. By the cellular assay clones can thus be identified as having positive, negative, or conditional phenotypes for HGPRT activity.

Since substrate concentrations can be varied in the in vitro assay for HGPRT it can be used to quantitate both specific activities and affinities for substrates. A number of enzyme-positive variants have been shown to produce forms of HGPRT which have reduced affinities for PP-ribose-P and/or the purine substrates (Sharp et al., 1973; Chasin and Urlaub, 1976; Epstein et al., 1977; Fenwick et al., 1977a). As can be seen in Table 13.5, these include clones such as RJK3 and 44 which have increased K_m values and Hill coefficients for PP-ribose-P. The latter point indicates activation of the variant enzymes by that substrate and might be a reflection of defective interactions between subunits of the enzyme.

From the data in Tables 13.4 and 13.5, it is clear that kinetic abnormalities such as those expressed by RJK3 and 44 can virtually inactivate HGPRT in the normal cellular environment. This causes the conditional HGPRT phenotype in the cellular assay and can permit cell growth in both forward or reverse selective media. The utility of the cellular assay for defining molecular alterations in variant clones is thus obvious.

3. *Enzyme Protein*

A variety of immunological and physical techniques have been used to detect HGPRT protein in enzyme-negative cells, quantitate the concentration of HGPRT protein, and identify abnormal forms of HGPRT produced by variant cells. HGPRT protein was first identified in variants

TABLE 13.6
Quantitation of HGPRT Protein

Isolate[a]	Immunoprecipitation[b]	*In Vitro* Translation[c]
RJK0	0.57	1.35
RJK10	0.00	Not done
RJK36	0.00	0.00
RJK39	0.44	0.89

[a] The amounts of HGPRT protein produced by wild-type Chinese hamster cells (RJK0) and HGPRT-deficient subclones were determined using specific antisera.
[b] Data taken from Beaudet et al. (1973) who measured inhibition of HGPRT precipitation by cell extract proteins and expressed their results as units of HGPRT activity released per mg of extract added.
[c] Data taken from Melton et al. (1981) who measured *in vitro* translation of HGPRT protein directed by purified mRNA and expressed their results as percentage $\times$ 10^2 of total protein synthesis.

lacking HGPRT enzymatic activity by demonstrating that cellular proteins would block the immunoprecipitation of normal HGPRT (Beaudet et al., 1973; Ghangas and Milman, 1975; Wahl et al., 1975). In a more sensitive assay, highly specific antisera have been used to immunoprecipitate HGPRT which has been radiolabeled metabolically (Wahl et al., 1975; Fenwick et al., 1977a) or by in vitro translation of mRNA (Melton et al., 1981). The results from such assays of our wild-type cells, RJKO, and three enzyme-negative variants are listed in Table 13.6 and demonstrate that the protein can be detected in RJK39 but not RJK10 or 36.

Surveys of HGPRT-negative variants have demonstrated the presence of enzyme protein in up to 40% of the isolates (Wahl et al., 1975; Milman et al., 1976; Fenwick et al., 1982) but such results are probably underestimates. For instance, our analysis of revertants isolated from RJK10 indicates that the strain must make HGPRT protein (Fenwick et al., 1984) but we have been unable to detect the protein in RJK10 cells. One factor which limits the immunological analysis of HGPRT variation is that specific sera may not recognize individual subunits of the enzyme (Caskey et al., 1979). Thus protein alterations which prevent dimeric or tetrameric assembly of the enzyme might escape detection.

Alterations of the enzyme protein have also been documented by electrophoretic or isoelectric focusing assays. For variants that retain enzyme activity this has been accomplished by using the enzyme assay to locate HGPRT molecules which have been separated under nondenaturing conditions (Chasin and Urlaub, 1975; Fenwick et al., 1977a). When this technique was applied to the two kinetic variants described in Table 13.5, only the enzyme from RJK3 was found to have an altered

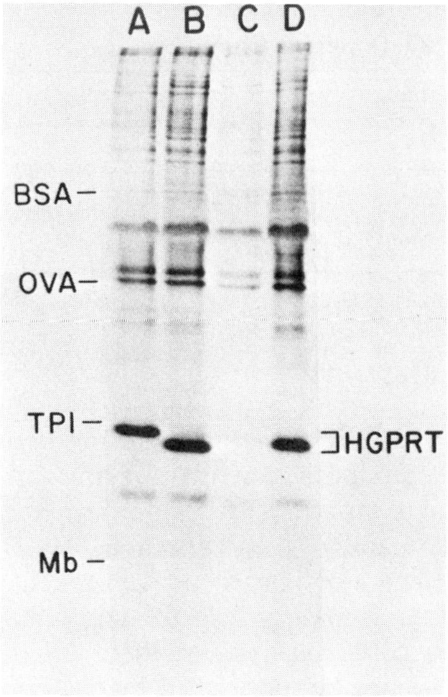

Figure 13.2. Detection of HGPRT protein produced by Chinese hamster cells. Using the procedure described by Fenwick et al. (1977a) extracts of cellular proteins labeled with [^{35}S]methionine were subjected to immunoprecipitation and the precipitates were analyzed by SDS-polyacrylamide electrophoresis and autoradiography. The cell lines examined were wild-type cells (RJKO, A) and the 8-azaguanine-resistant isolates RJK3 (B), RKJ36 (C), and RJK39 (D). The positions of protein standards are noted.

electrophoretic mobility. This plus quantitative aspects of the kinetic changes indicate that RJK3 and RJK44 express different alterations. To analyze the HGPRT subunit protein, enzymes immunopurified from cell extracts have been studied by SDS-polyacrylamide gel electrophoresis (Capecchi et al., 1974; Fenwick et al., 1977a). The results of such an experiment are shown in Figure 13.2. Consistent with the data in Table 13.6, the enzyme-negative variant RJK39 contains detectable enzyme protein but a similar variant, RJK36, does not. Furthermore, the proteins from RJK39 and the kinetic variant RJK3 migrated further than the wild-type protein from RJK0. Altered mobilities can be caused by either changes in molecular weight or amino acid substitutions (see Fenwick, 1980). The alterations in some variant proteins, including that from RJK39, have been assigned to specific tryptic peptides by using high pressure liquid chromatography to analyze digests of immunopurified HGPRT subunits (Capecchi et al., 1977; Milman et al., 1977; Kruh et al., 1981).

4. Reversion Analysis

By studying phenotypic reversion of HGPRT-deficient variants it is possible to gather supporting evidence for the mutational basis of HGPRT variation and to tentatively identify the alterations carried by particular cell lines. Table 13.7 shows that mutagens induced reversion of two

TABLE 13.7
Reversion Analysis of HGPRT-Deficient Clones

	Revertant Frequency[a]			
Clone	Spontaneous	MNNG	EMS	UV
RJK10	9×10^{-7}	4×10^{-5}	2×10^{-5}	4×10^{-5}
RJK39	$<2 \times 10^{-7}$	2×10^{-6}	8×10^{-7}	

[a] The spontaneous and mutagen-induced frequencies of HAT-resistant revertants were determined for RJK10 (Fenwick et al., 1977b) and RJK39 (Fenwick, 1980).

HGPRT-deficient clones mentioned above, RJK10 (negative for both enzyme activity and protein) and RJK39 (produces inactive enzyme protein). However, we have never been able to isolate revertants from some strains listed in Table 13.1 such as RJK36 and 43 (Fenwick et al., 1982). In an early study, Chu (1971) characterized a collection of 72 8-azaguanine-resistant clones and identified isolates which reverted spontaneously to HAT resistance, others which could be induced to revert by mutagens, and a final group which could not be induced to revert. Furthermore, some of his inducible clones responded specifically to mutagens known to promote base substitutions while others were only affected by frameshift mutagens. These results imply that different types of point mutations can cause HGPRT deficiency and that the negative phenotypes of nonrevertible variants may be caused by major genetic alterations, such as deletions, which cannot be repaired by reverse mutations.

The phenotypes of HGPRT-positive revertants can often be distinguished from that of wild-type cells. For revertants of RJK39, the heterogeneous patterns of growth in selective media were already noted in Table 13.3. An extensive analysis of those revertants (Fenwick, 1980) revealed that restoration of HGPRT activity was usually accompanied by elimination of the abnormal electrophoretic mobility which characterizes the subunit protein from RJK39 (see Fig. 13.2) but half of the revertants appear to produce kinetically altered HGPRT because they were found to respond to aminopterin during the cellular HGPRT assay described in Table 13.4. In addition, Kruh et al. (1981) showed that the tryptic peptide abnormality present in RJK39 HGPRT was either reversed or additionally modified in enzymes from the revertants. For the spontaneous and induced revertants of RJK10, Fenwick et al. (1977b) used a quantitative immunoprecipitation assay to demonstrate that one or more of the antigenic determinants present on Chinese hamster HGPRT are either absent or present in an altered form on HGPRT from the revertants. The identification of altered HGPRT in revertant cells indicates that the HGPRT-deficient clones from which they arose carry mutations affecting the HGPRT gene and that second-site or intragenic suppressor mutations can cause phenotypic reversion.

5. Nucleic Acids

Through the application of recombinant DNA technologies a variety of DNA molecules containing sequences from the HGPRT gene or its mRNA have been cloned and they are now being applied to the analysis of variant phenotypes. Melton (1981) isolated NBR4 as a phenotypic revertant of a HGPRT-deficient clone of mouse neuroblastoma cells and found that it overproduced a variant form of HGPRT. In vitro translation studies (Melton et al., 1981) indicated that NBR4 also overproduced HGPRT mRNA and the clone was used as an enriched source of HGPRT mRNA during the isolation of cDNA clones (Brennand et al., 1982). Using those cDNA sequences as probes for blot hybridization procedures Brennard et al. (1982) demonstrated that the HGPRT gene was amplified about 50-fold in NBR4 which explained its overproduction (at least 20-fold) of the mRNA. They also found that the murine cDNA probes could be used to detect HGPRT sequences in nucleic acids from other mammalian sources including human and Chinese hamster. Fuscoe et al. (1983) used the murine probe to identify variant clones of Chinese hamster cells in which the HGPRT gene had been altered. For example, Figure 13.3A shows a Southern blot (Southern, 1975) investigation of DNA from the HGPRT-negative clone RJK88 (see Tables 13.1 and 13.4). Since a majority of the restriction fragments identified by the probe in DNA from the wild-type cell line RJK0 are not present in RJK88 DNA, the variant can be classified as a deletion mutant. A limited restriction map of the region of RJK88 DNA recognized by the probe is shown in Figure 13.3B. Additional studies have revealed that it is not part of the functional HGPRT gene but it is present in DNA from Chinese hamster tissues (Fuscoe et al., 1983; Fenwick et al., 1984). Thus, it may be a HGPRT pseudogene.

As illustrated in Figure 13.4, the nucleic acid probes and the Northern blot procedure (Thomas, 1980) can be used to identify HGPRT mRNA. As would be expected, the figure shows that the mRNA is not produced by the deletion mutant RJK88. Since one can measure both the amount and size of HGPRT mRNA with this technique, it can be used to screen HGPRT-negative clones for evidence of alterations that affect the synthesis or processing of HGPRT mRNA.

Additional cDNA and genomic clones from the Chinese hamster, mouse, and human HGPRT systems have been isolated (Jolly et al., 1982; Konecki et al., 1982; Brennand et al., 1983; Jolly et al., 1983; Melton et al., 1984). The results of DNA sequencing efforts have led to the identification of mutations carried by some HGPRT-deficient cell lines and provided information that can be used for the analysis of other vari-

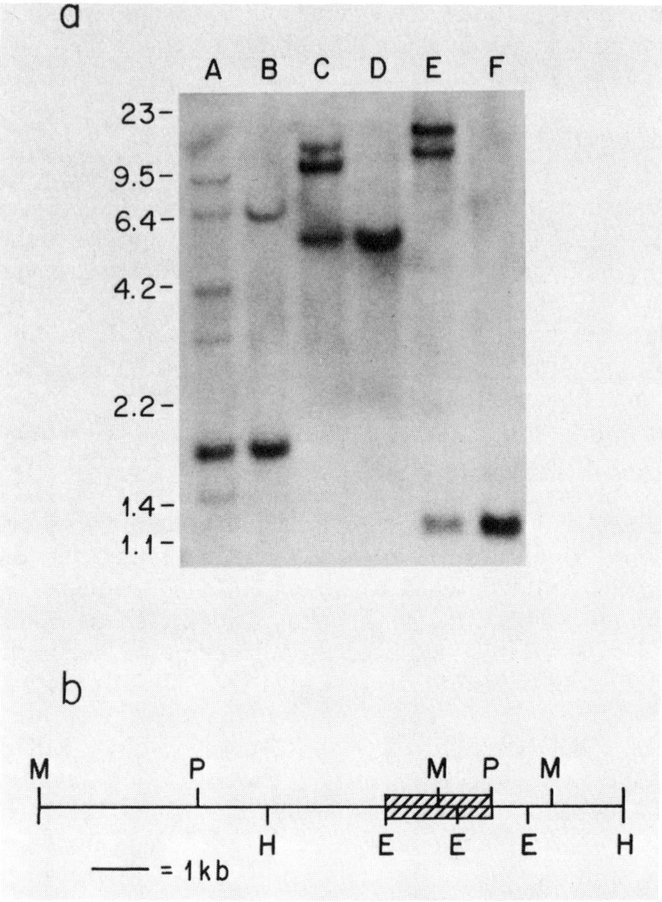

Figure 13.3. Blot hybridization analysis of the Chinese hamster HGPRT gene. (a) DNAs isolated from wild-type (RJKO) cells (lanes A, C, and E) or cells of the HGPRT-deficient isolate RJK88 (lanes B, D, and F) were digested with the restriction endonucleases *Msp* I (A and B), *Hind* III (C and D), or *Eco* RI (E and F) and the status of the HGPRT gene was determined using the cDNA probe and procedures described by Fuscoe et al. (1983). The positions of size markers (in kilobase pairs) are noted. (b) A limited restriction map of the region of Chinese hamster DNA which is recognized by the HGPRT cDNA probe and retained in the genome of RJK88. The hatched box indicates the portion to which the HGPRT probe hybridizes and E = *Eco* RI, H = *Hind* III, M = *Msp* I, and P = *Pst* I.

ants. For example, Melton et al. (1984) have determined that the mutation carried by the murine cell line NBR4 is a guanine to adenine transition which changes amino acid 201 in HGPRT from aspartic acid to asparagine. In addition, Wilson et al., (1982) knew from protein sequence data that the human allele $\text{HGPRT}_{\text{Toronto}}$ carried by a patient with gout caused a substitution of glycine for arginine at position 50 in

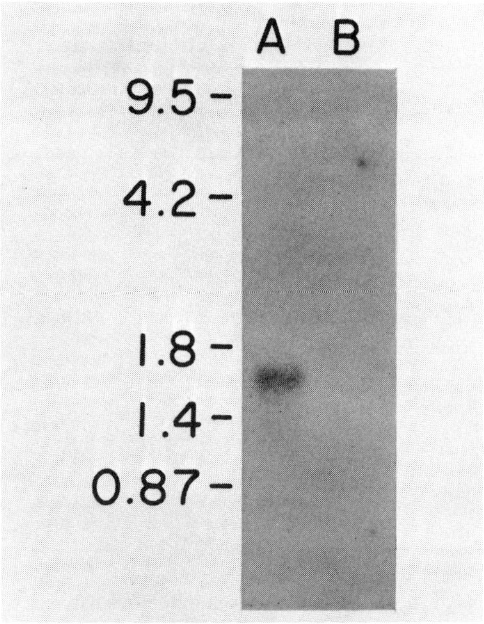

Figure 13.4. Blot hybridization analysis of Chinese hamster HGPRT mRNA. Poly(A) containing RNA was isolated from cells and HGPRT mRNA was measured using a cDNA probe and the procedures described by Fuscoe et al. (1983). In preparations from wild-type (RJKO) cells the probe detected a single transcript having a length of about 1600 nucleotides (A) but no mRNA was detected in preparations from the HGPRT-deficient isolate RJK88 (B). The positions of size markers (in kilobases) are noted.

the protein. From the available nucleotide sequence data they were able to predict that the mutation carried by the patient would abolish a genomic recognition site for *Taq* I restriction endonuclease and they verified the restriction site abnormality by blot hybridization analysis of the patient's DNA. Thus, given enough labor, the tools are now in hand for the precise definition of the mutational mechanisms underlying HGPRT variation in cultured cells as well as whole animal systems.

B. Analysis of Variants Selected for HGPRT Deficiency

To identify as efficiently as possible the molecular alterations causing HGPRT deficiency in particular variants one can use a sequential application of the techniques outlined in Section V.A. For the cell lines identified in Table 13.1, the results of this approach are illustrated in Table 13.8. Strains that produce altered forms of the enzyme are probable missense mutants. These can be identified by first using the growth (Table 13.2) and/or cellular HGPRT (Table 13.4) assays to determine which

TABLE 13.8
Analysis of HGPRT-Deficient Variants[a]

Isolate	HGPRT Activity	HGPRT Protein	HGPRT mRNA	HGPRT DNA	Reversion
RJK3	+				
RJK10	−	−	+		Yes
RJK36	−	−	+[b]	"Normal"	No
RJK39	−	+			Yes
RJK43	−	−	−	"Normal"	No
RJK44	+				
RJK62	+				
RJK71	−	−	−	Deletion	
RJK88	−	−	−	Deletion	
RJK460	+				
RJK463	−	+			

[a] A summary of the data obtained when the techniques described in Section V.A were applied to the analysis of the 8-azaguanine-resistant clones listed (see Table 13.1 for origins of the isolates). Interpretations of the results are discussed in Section V.B.
[b] Although HGPRT mRNA from RJK36 was detected with the technique illustrated in Figure 13.4, its electrophoretic mobility indicated that the molecule has a lower molecular weight than the normal mRNA (J. Fuscoe, personal communication).

clones retain enzyme activity (e.g., RJK3, 44, 62, and 460) and then the immunological assays (Fig. 13.2 and Table 13.6) to determine which enzyme-negative clones produce inactive forms of the protein (e.g., RJK39 and 463). As was illustrated for RJK3 and 44 (Table 13.5) assays such as those for kinetic or electrophoretic variation can be used to define further the enzyme alterations expressed by particular clones. Cell lines that have neither HGPRT activity nor protein can next be assayed for HGPRT mRNA by the Northern blot technique (Fig. 13.4). Positive clones such as RJK10 and 36 probably carry mutations that affect translation, assembly, or stability of the enzyme protein. The mRNA-negative phenotypes of RJK43, 71, and 88 could be caused by genetic or epigenetic alterations that reduce synthesis, processing, or stability of the transcript. Finally, the mRNA-deficient clones can be subclassified by using the Southern blot procedure (Fig. 13.3) to identify clones such as RJK71 and 88 in which major DNA alterations have disrupted gene expression.

As noted in Table 13.8, we have been unable to isolate revertants from either RJK36 and 43 and HGPRT mRNA from RJK36 appears to have a reduced molecular weight. This suggests that major DNA alterations might be the cause of their phenotypes. Our preliminary blot hybridization assays of the HGPRT genes in those strains have yet to reveal any abnormalities but the size and complexity of the HGPRT gene (see Section VI.C) make identification of other than large rearrangements very difficult. Thus, more sophisticated recombinant DNA techniques including the isolation of cDNA and genomic clones from interesting mutants will have to be employed to determine the location and types of mutations affecting specific cell lines.

C. Analysis of Phenotypic Revertants

Since selective protocols for the isolation of HGPRT positive cells usually involve inhibitors of de novo purine synthesis it is important to confirm that clones which arise have, in fact, regained HGPRT activity. As was illustrated in Table 13.3, 3 of 24 mutagen-induced "revertants" isolated from RJK39 were found by the growth assay to be resistant to aminopterin. Using the cellular HGPRT assay (Table 13.4) they were shown to be HGPRT deficient and were subsequently found to be defective in the uptake of aminopterin (Fenwick, 1980). The use of selective medium containing azaserine rather than aminopterin may reduce this problem (Fuscoe et al., 1982).

Efforts to understand the mechanisms by which revertants arise have been complicated by the recent observations that murine and Chinese hamster missense mutants can revert by gene amplification (Brennand et al., 1982; Fuscoe et al., 1983; Fenwick et al., 1984; Zownir et al., 1984). Some of these revertants are unstable because the amplified copies of the gene are not linked to the X chromosome and apparently do not segregate equally during cell division (Fenwick et al., 1984). Reversion by amplification also raises the possibility that missense mutants might revert through regulatory mutations that cause overproduction of HGPRT mRNA but such revertants have not been described. Blot hybridization procedures can be used to quantitate the number of HGPRT genes and the relative amount of the mRNA in a revertant (Brennand et al., 1982; Fuscoe et al., 1983) and, as described in Section V.A, a variety of assays can be used to determine whether revertants produce normal or altered forms of HGPRT. Conclusions about reversion mechanisms, the effects of mutagens, and so on can not be drawn until the status of the gene, mRNA, and enzyme have been determined.

A second complicating factor for reversion studies may be the presence of epigenetically inactivated but otherwise normal HGPRT alleles in some cell lines, especially female cells bearing an inactive X chromosome. Milman et al. (1976) isolated mutagen-induced revertants from a HGPRT-deficient mutant of HeLa cells which produces an inactive form of the enzyme protein. By two-dimensional electrophoresis they demonstrated that all five revertants isolated synthesized wild-type plus the mutant form of the enzyme. In addition, the HGPRT gene on an inactive X chromosome can be reactivated by treating the cell with agents such as 5-azacytidine (Mohandas et al., 1981). This should not be a problem with Chinese hamster cell lines such as CHO and V79 which appear to have a single X chromosome (Worton et al., 1977; Thacker, 1981) but it should always be borne in mind when evaluating revertants.

VI. MAJOR CONTRIBUTIONS OF THE HGPRT SYSTEM

Because the HGPRT system has been used to study biochemical, genetic, and clinical questions many important pieces of information have been gathered during the period of investigation. Any listing will be incomplete but I would like to note a number of points I consider to be pertinent to molecular cell genetics.

A. Validation of Somatic Cell Genetics

Although early work (Harris, 1971) raised some doubt as to whether genetic alterations were the events underlying the origin of HGPRT-deficient cells, there is now a wealth of information demonstrating that such is the case. Important milestones include mutagen stimulation of HGPRT deficiency (Chu and Malling, 1968), detection of altered HGPRT in variant cells (Beaudet et al., 1973), linkage of HGPRT deficiency to another X chromosome marker (Chasin and Urlaub, 1975), and the use of recombinant DNA techniques to identify deletions (Fuscoe et al., 1983) and point mutations (Melton et al., 1984) affecting the HGPRT gene. Evidence that genetic alterations are also involved in reversion of HGPRT deficiency includes specificity for mutagen induction of reversion (Chu, 1971), characterization of altered HGPRT produced by revertants (Sharp et al., 1973; Fenwick et al., 1977b), linkage of the revertant phenotype to changes on the X chromosome (Fenwick, 1980), and demonstrations that amplification of mutant HGPRT alleles can cause phenotypic reversion (Brennand et al., 1982; Fenwick et al., 1984).

It is important to stress that a wide range of genetic alterations can be detected with the HGPRT locus. When collections of HGPRT mutants have been isolated and characterized (Chu, 1971; Gillin et al., 1972; Sharp et al., 1973) the individual clones have been found to have heterogeneous phenotypes with respect to retention of HGPRT activity, reversion, and so on. Applications of the analytical procedures described in Section V to the analysis of the mutants have emphasized the degree of the observed phenotypic and genetic heterogeneity. In practice it is difficult to identify two mutants from such collections which can not be differentiated from one another (see Fenwick et al., 1982, for a summary of the diversity observed in a set of Chinese hamster HGPRT mutants). Mechanisms for the reversion of HGPRT mutants are also very heterogeneous. For example, I noted that over half of the revertants isolated from the missense mutant RJK39 (see Table 13.3 and 13.8) could be distinguishable from wild-type cells (Fenwick, 1980). This is indicative that the reversion events in those clones were actually second-site mutations. A second mutant RJK10 (Table 13.8) which we also know to be a missense mutant was found to revert either by second-site mutation or gene amplification (Fenwick et al., 1984). Given the experience of geneticists with bacterial and other well developed systems, it is not surprising that mutation and reversion of HGPRT can involve many sites and mechanisms. However, the potential for heterogeneity should always be borne in mind when isolating and characterizing variant clones.

B. The HGPRT Enzyme

As has been the case with many other proteins, analysis of mutant forms of HGPRT should help us to understand the relationship between the structure and function of the enzyme. Before cDNA or genomic clones of HGPRT sequences were isolated it was difficult to make use of the many mutant alleles that have been characterized in Chinese hamster and other cultured cells because the low concentration of the enzyme made fine structure analysis of the protein difficult. However, amino acid substitutions can now be deduced from the sequences of DNA clones. To date the most extensively studied example is the mutation carried by a mouse neuroblastoma line (NB^-; Melton et al., 1984). The guanine to adenine transition mutation carried by that line causes amino acid 217 to be asparagine rather than aspartic acid and the resulting enzyme has a reduced affinity for PP-ribose-P, an altered electrophoretic mobility, and an increased lability (Melton, 1981). Since a mutation in humans which changes amino acid 193 is known to reduce the affinity of the enzyme for both substrates (Wilson and Kelley, 1983), the region near the carboxyl terminus of the 218 amino acid protein may be involved in substrate binding (Melton et al., 1984). Initial studies involving Chinese hamster cell lines have indicated that the mutation which causes the negative HGPRT phenotype of RJK10 (see Table 13.8 and Section V.A.5) affects amino acid 152 or 155. This type of analysis is in its infancy but as it is continued we should be able to develop a functional map for HGPRT.

With both mouse and Chinese hamster cell lines it has been shown that somatic cell hybrids formed by fusion of HGPRT-deficient clones can exhibit complementation or restoration of HGPRT activity (Sekiguchi and Sekiguchi, 1973; Sekiguchi et al., 1974). This only occurred, however, when one of the parent clones had been selected with 8-azaguanine and the second selected with 6-thioguanine. Such results were initially taken as evidence that two or more genes or cistrons were required for expression of HGPRT but Chasin and Urlaub (1975) demonstrated that the phenomenon involved intracistronic complementation. Thus the defective subunit proteins from complementing mutants can associate to form functional hybrid enzyme molecules. There has also been one report of dominant HGPRT deficiency in mouse cells (Kadouri et al., 1978). Hybrids formed by fusing those clones to wild-type cells were found to be resistant to 8-azaguanine and sensitive to HAT. Again this was interpreted as evidence of multiple cistrons but the data would also be consistent with negative intracistronic complementation or inactivation of enzyme molecules containing a subunit protein encoded by the allele from the defective parent. Such studies point to functional interactions between the subunits of HGPRT molecules and perhaps more importantly stress that 8-azaguanine and 6-thioguanine can select quite different sets of mutants. The latter point has caused some confusion in the literature but the availability of complementing alleles means that it should be possible to identify regions of the protein that are involved in subunit interactions or selective resistance to the purine analogs.

C. The HGPRT Gene

The initial isolations of recombinant DNA molecules containing HGPRT sequences were dependent on the availability of genetically altered cell lines (Brennand et al., 1982) and procedures for DNA mediated gene transfer (Jolly et al., 1982). Those studies provided nucleic acid probes which have facilitated the analysis and isolation of the HGPRT gene. The wild-type HGPRT gene from mouse has been isolated from genomic libraries and has been found to be more than 33 kilobases long (Melton et al., 1984). Analysis of the gene has revealed that the 1307 base pairs sequence present in murine cDNA clones is distributed into nine exons one of which is only 18 base pairs long. Although less well characterized, the genes from Chinese hamster (Fuscoe et al., 1983) and human (Nussbaum et al., 1983) seem to be equally large and complex. In the genomes of all three species these studies have identified homologous DNA sequences which are not part of the functional gene. They have been mapped to at least two human autosomes (Nussbaum et al., 1983) and are not transcribed in Chinese hamster cells (Fuscoe et al., 1983). Thus HGPRT pseudogenes may be a common feature of mammalian genomes.

The 5′ region of the murine gene has been linked to human cDNA sequences to construct a functional minigene (Melton et al., 1984). This demonstrated the presence of a functional promoter of transcription but sequences normally associated with eukaryotic promoters are not present in the immediate 5′ flanking region of the gene. In addition, the region is very rich in guanine plus cytosine residues. Melton et al. (1984) have pointed out that the type of promoter required for

HGPRT, which is expressed at low levels in most cells, may not resemble those identified in genes which are highly expressed in differentiated cells.

Studies of cDNA clones indicate that a 3′ noncoding sequence of at least 600 nucleotides is present in mouse, hamster, and human mRNAs (Konecki et al., 1982; Jolly et al., 1983). In mouse this is encoded in a single exon which contains a signal sequence for polyadenylation (Melton et al., 1984). However, hamster cDNA clones indicate that there may be at least two alternative sites for polyadenylation (Konecki et al., 1982). This means that additional 3′ sequences may be part of the gene and that there may be more than one pattern for processing the HGPRT transcript.

D. Gene Transfer

The powerful selective systems for HGPRT phenotypes and the availability of well characterized cell lines have made the HGPRT locus an important part of many efforts to transfer genes from one cell to another (see Chapter 8). These include the fusions of entire genomes to form somatic cell hybrids (Littlefield, 1964), the transfer of one or a few chromosomes by microcell fusions (Fournier and Ruddle, 1977), transfer of portions of the X chromosome by chromosome mediated gene transfer (McBride and Ozer, 1973), and transfer of the HGPRT gene via DNA mediated gene transfer (Szybalska and Szybalski, 1962; Willecke et al., 1979). As discussed by Ruddle (1981) these techniques can be used to assign genes from donor cells to a particular chromosome, determine linkage relationships along a chromosome (X), or identify DNA sequences within a few thousand base pairs of a selected gene (HGPRT). Weissman and Stanbridge (1980) have noted that the isolation of somatic cell hybrids is facilitated by using a parental clone which is both ouabain-resistant and HGPRT-deficient because it allows selection of hybrids formed by fusion to unmarked, wild-type cells.

When the HGPRT gene is transferred via purified metaphase chromosomes or isolated DNA the HGPRT phenotype is often found to be unstable, apparently because the transferred DNA is not associated with a chromosome. Association, probably by integration, with recipient chromosomes can give rise to stable subclones (Fournier and Ruddle, 1977) but homologous recombination does not appear to be involved because association with the X chromosome has not been observed (Willecke et al., 1981). However, the resulting cell lines might be useful in that they contain new linkage groups which can be studied with the HGPRT selective protocols.

Amplification of mutant HGPRT alleles has been documented to occur in at least two types of gene transfer experiments. Chromosome-mediated transfer of a mutationally altered mouse allele (HGPRT-positive but 8-azaguanine-resistant) has resulted in overproduction of the altered HGPRT (Degnen et al., 1977) due to gene amplification (Linder et al., 1984) and amplification of the mutant mouse allele recipient NB^- cells occurred during an attempt to move the human X chromosome into that line by fusion with HeLa microcells (Melton, 1981). Since there have been other reports of fusion- or gene-transfer-induced HGPRT reversion (Watson et al., 1972; Shin et al., 1973; Bakay et al., 1975) the mixing of genomes or the addition of DNA to a cell may promote amplification and other mutational events. This might be related to the destabilization of chromosomes in somatic cell hybrids noted by Worton et al. (1977).

Cloned HGPRT sequences have also been utilized for DNA-mediated gene transfer. These include human cDNA sequences which were isolated in an expression vector (Jolly et al., 1983), human and Chinese hamster cDNA sequences which were recombined to put them under the transcription control of retroviral sequences (Brennand et al., 1983), and a minigene which was formed by combining the 5′ region of the mouse gene, the protein coding sequence from human cDNA, and the polyadenylation signal from Chinese hamster cDNA (Melton et al., 1984). These constructs will be useful for the study of HGPRT transcription, developing model protocols for genetic therapy, and efforts to probe HGPRT functions by site-directed mutagenesis.

E. The X Chromosome

Significant differences exist between the two X chromosomes in somatic cells of female placental mammals in that one of the chromosomes at random is inactivated during early development (see Martin, 1982, for a brief review). Thus only one of the two HGPRT alleles present in female cells is expressed (Rosenbloom et al., 1967; Chapman et al., 1983). The inactivation persists in cultured cells but rare spontaneous reactivation of genes along the X does occur (Kahan and Demars, 1975).

To probe the molecular mechanism of inactivation investigators have used DNA-mediated transfer of the HGPRT gene to determine whether DNA of the inactive X is altered. With one exception (de Jonge et al., 1984), the results indicate that transformation is markedly inhibited by X inactivation (Liskay and Evans, 1980; Chapman et al., 1982; Venolia and Gartler, 1983a). Treatment of cells with chemicals, especially 5-azacytidine which when incorporated into DNA leads to hypomethylation, can derepress genes on inactive X chromosomes including HGPRT (Mohandas et al., 1981; Graves, 1982; Jones et al., 1982) and reactivated HGPRT genes can once again be transferred via DNA to another cell (Lester et al., 1982; Venolia et al., 1982). These findings demonstrate that X inactivation involves DNA alterations and imply that methylation of the DNA is involved in that process.

Yen et al. (1984) and Wolf et al. (1984) have used cloned HGPRT-specific probes, restriction enzymes that are sensitive to methylation at their recognition sites, and DNA containing HGPRT genes in various states of activity to measure methylation around the HGPRT gene. The analysis has been difficult because there are many potential sites of methylation, especially in the 5′ region of the gene. However, the pattern that has emerged is one of general hypomethylation in the 5′ region of active genes. Although that region is more highly methylated in inactive genes, no specific pattern of methylated sites has been observed. In addition, reactivation of genes by 5-azacytidine was associated with general but neither complete nor specific demethylation of the sites investigated. Thus the overall pattern of methylation rather than methylation of specific sites may be involved in maintaining the active or inactive states of HGPRT.

To confirm that methylation plays an important role in somatic X inactivation additional loci along the chromosome will have to be investigated. If the hypothesis is supported, the next major goal will be to determine the mechanism by which this mode of inactivation is simultaneously applied to genes along a single X chromosome. However, there are other types of X inactivation which may involve different mechanisms. In extraembryonic tissues of mice the paternal X chromosome is inactivated but Kratzer et al. (1983) have found that the HGPRT gene from that inactivated chromosome can be transferred via DNA to recipient cells. Likewise, inactivation of the X chromosome during spermatogenesis does not affect DNA-mediated transformation of HGPRT (Venolia and Gartler, 1983b). Such results stress both the utility of the HGPRT system for studies of X inactivation and the complexity of that problem.

F. Analysis of Mutagens

The utility of the HGPRT locus for identifying the genotoxic effects of mutagens and carcinogens is based on the fact that HGPRT is not usually required for cell viability. This means that there are virtually no restrictions on the types of genetic alterations which might cause HGPRT deficiency and thus no limits on the types of mutational or premutational events that can be studied. For instance, HGPRT deficiency can be induced by mutagens that promote base substitutions, frameshift mutations, and major DNA rearrangements but the induction of a marker that requires missense mutations, ouabain resistance, is limited to mutagens that promote base substitutions (Arlett et al., 1975; Friedrich and Coffino, 1977a). As would be expected, however, mutations that alter but do not eliminate HGPRT activity are specifically induced by agents that promote base substitutions (Friedrich and Coffino, 1977b).

Mutagenesis studies have not been limited, of course, to Chinese hamster cells but the V79 and CHO clones have been used by many laboratories for that purpose. The list of agents examined is substantial and includes alkylating agents (Chu and Malling, 1968) UV- and X-radiation (Bridges and Huckle, 1970), reactive derivatives of hydrocarbons (Duncan and Brookes, 1973), viruses (Theile et al., 1976), fluorescent light (Bradley and Sharkey, 1977), various ionizing radiations (Cox et al., 1977), and pyrimidine deoxynucleotides (Peterson et al., 1978). Mutagens which require metabolic activation by pathways not present in Chinese hamster cells have been detected by using other cells or cellular homogenates to activate the agents (Huberman and Sachs, 1976; Huberman et al., 1976; Krahn and Heidelberger, 1977; Langenbach et al., 1978). Hsie et al. (1981) reviewed the use of CHO cells to determine mutagenicity of chemicals and made suggestions for the standardization of protocols and presentation of data. As noted by those authors, the time required to test a suspected mutagen in mammalian cells has limited the number of agents which have been examined and will continue to do so. In fact, most studies have simply confirmed the activities of known mutagens. Thus, the HGPRT system may be better suited for the analysis of mutagenic mechanisms unique to mammalian cells rather than surveys of potential mutagens which can be accomplished more rapidly in bacterial test systems (McCann and Ames, 1976).

Because the HGPRT system is so well characterized, it is often used to measure the influence of environmental or genetic conditions on mutational activity. For instance, it has been used to demonstrate altered mutation rates in Chinese hamster cells which are deficient in DNA repair (Stamato et al., 1981; Thompson et al., 1982) as well as in cells isolated from humans who are sensitive to agents that damage DNA (Maher et al., 1976; Warren et al., 1981). In addition, a direct method for testing mutagenicity in humans measures HGPRT-deficient lymphocytes which have arisen in vivo (Albertini et al., 1982). The locus has also been used to study questions such as which of the several DNA adducts formed by alkylating agents are involved in mutagenesis (Suter et al., 1980; Heflich et al., 1982) or whether HGPRT deficiency induced by radiation arises by mutation or X chromosome rearrangements (Cox and Masson, 1978).

With synchronous populations of cells, induction of HGPRT mutations by incorporation of bromodeoxyuridine (Aebersold and Burki, 1976) UV radiation (Riddle and Hsie, 1978), and alkylating agents (Tong et al., 1980; Jenssen, 1982) have been found to be most effective during the early part of S phase. These results indicated that the HGPRT gene is replicated during the early portion of S and that point has been confirmed by using recombinant DNA probes to measure the time of

HGPRT replication (Holmquist et al., 1982; Goldman et al., 1984). As discussed by Tong et al. (1980) enhanced mutagenesis during S phase may reflect the inability of a cell to repair premutagenic damage prior to DNA replication. Thus the effects of cell synchrony on induced mutagenesis can be employed to evaluate the capacity of cells to repair particular types of premutational damage (Jenssen, 1982).

VII. OUTSTANDING ISSUES

Investigators will obviously continue to exploit the HGPRT system for the study of fundamental biological, biochemical, and genetic questions. In addition, there are a number of specific issues relating to the HGPRT gene which have been raised over the years but never satisfactorily answered.

A purely technical point is the description at the molecular level of mutants carrying some of the classical types of genetic alterations. Frameshift mutants have been tentatively identified by their sensitivities to the induction of reversion by specific mutagens (Fuscoe et al., 1982). Once these have been confirmed and defined at the molecular level they can be used to study the specificities and molecular mechanisms of particular mutagens in mammalian cells. Nonsense mutants have been sought for years. The few tentative identifications (Capecchi et al., 1977; Celis et al., 1979) have proven to be false (Fenwick, 1980; Konecki et al., 1982). The initial goal of those searches, to have markers that might be used for the selection of cell lines expressing suppressor tRNA mutations, may have been obviated by the use of recombinant DNA technology to introduce suppressor tRNA genes into cells (Hudziak et al., 1982). However, the identification of nonsense mutants will not only be satisfying to those who have pursued them but will again provide cell lines that can be used for a variety of purposes.

Efforts to study somatic recombination are described elsewhere in this volume (see Chapter 28) but attempts to measure intragenic recombination between mutant HGPRT alleles (Tarrant and Holliday, 1977; Rosenstraus and Chasin, 1978) or integration of a transferred gene by homologous recombination (Willecke et al., 1981) have not met with success. However, we are now accumulating a collection of mutant alleles that have been defined at the molecular level and it should soon be possible to attempt such experiments using alleles known to carry different but single mutations at specific sites within the gene. Thus for the first time it will be possible to approach recombination experiments with the knowledge that recombination between the alleles under study might actually produce a wild-type gene.

A final question of considerable interest is regulation of HGPRT expression. The absence of characteristic promoter sequences near the

5′ end of the gene (Section VI.C), tissue specific differences in enzyme activity and mRNA content (Section II.A.2), a lack of any clear evidence as to whether expression of the gene in cultured cells is modulated by other genes (Section VI.B), and the identification of mRNA-deficient mutants (Section V.B) combine to provide an interesting puzzle. Attempts to order the pieces of the puzzle will help us to understand regulation of ubiquitously expressed genes such as HGPRT.

One point that is abundantly clear is that the HGPRT system has been either the focal point or an important experimental tool for a large number of investigators. Although some of its lessons have been difficult to master it is obvious that the system merits continued attention.

REFERENCES

Aebersold, P. M. and Burki, H. J. (1976). *Mutat. Res.* **40**, 63–66.

Albertini, R. J., Castle, K., and Borcherding, W. R. (1982). *Proc. Natl. Acad. Sci. USA* **79**, 6617–6621.

Arlett, C. F., Turnbull, D., Harcourt, S. A., Lehmann, A. R., and Colella, C. M. (1975). *Mutat. Res.* **33**, 261–278.

Bakay, B., Nyhan, W. L., Croce, C. M., and Koprowski, H. (1975). *J. Cell Sci.* **17**, 567–578.

Beaudet, A. L., Roufa, D. J., and Caskey, C. T. (1973). *Proc. Natl. Acad. Sci. USA* **70**, 320–324.

Bonatti, S., Abbondandolo, A., Mazzaccaro, A., and Fiorio, R. (1980). *Mutat. Res.* **72**, 475–482.

Bradley, M. O. and Sharkey, N. A. (1977). *Nature* **266**, 724–726.

Brennand, J., Chinault, A. C., Konecki, D. S., Melton, D. W., and Caskey, C. T. (1982). *Proc. Natl. Acad. Sci. USA* **79**, 1950–1954.

Brennand, J., Konecki, D. S., and Caskey, C. T. (1983). *J. Biol. Chem.* **258**, 9593–9596.

Bridges, B. A. and Huckle, J. (1970). *Mutat. Res.* **10**, 141–151.

Capecchi, M. R., Capecchi, N. E., Hughes, S. H., and Wahl, G. M. (1974). *Proc. Natl. Acad. Sci. USA* **71**, 4732–4736.

Capecchi, M. R., Vonder Haar, R. A., Capecchi, N. E., and Sveda, M. M. (1977). *Cell* **12**, 371–381.

Caskey, C. T., Fenwick, R. G., Jr., and Kruh, G. D. (1979). In *Mammalian Cell Mutagenesis: The Maturation of Test Systems* (A. W. Hsie, J. P. O'Neill, and V. K. McElheny, eds.), pp. 23–34, Cold Spring Harbor Laboratory, Cold Spring Harbor, N.Y.

Celis, J. E., Kaltoft, K., Celis, A., Fenwick, R., and Caskey, C. T. (1979). In *Nonsense Mutations and tRNA Suppressors* (J. E. Celis and J. D. Smith, eds.), pp. 255–276, Academic Press, London.

Chapman, V. M. and Showes, T. B. (1976). *Nature* **129**, 665–667.

Chapman, V. M., Kratzer, P. G., Siracusa, L. D., Quarantillo, B. A., Evans, R., and Liskay, R. M. (1982). *Proc. Natl. Acad. Sci. USA* **79**, 5357–5361.

Chapman, V. M., Kratzer, P. G., and Quarantilo, B. A. (1983). *Genetics* **103**, 785–795.

Chasin, L. A. (1972). *Nature* **240**, 50–52.

Chasin, L. A. (1973). *J. Cell Physiol.* **82**, 299–308.

Chasin, L. A. and Urlaub, G. (1975). *Science* **187**, 1091–1093.

Chasin, L. A. and Urlaub, G. (1976). *Somat. Cell Genet.* **2**, 453–467.

Chiang, C. S. (1977). Ph.D. thesis, Baylor College of Medicine, Houston, Texas.

Chu, E. H. Y. (1971). *Mutat. Res.* **11**, 23–34.

Chu, E. H. Y. and Malling, H. V. (1968). *Proc. Natl. Acad. Sci. USA* **61**, 1306–1312.

Chu, E. H. Y., Brimer, P., Jacobson, K. B., and Merriam, E. V. (1969). *Genetics* **62**, 359–377.

Clements, G. B. (1975). *Adv. Cancer Res.* **21**, 273–390.

Cox, R. and Masson, W. K. (1978). *Nature* **276**, 629–630.

Cox, R. Thacker, J., Goodhead, D. T., and Munson, R. J. (1977). *Nature* **267**, 425–427.

Dancis, J., Cox, R. P., Berman, P. H., Jansen, V., and Balis, M. E. (1969). *Biochem. Genet.* **3**, 609–615.

Degnen, G. E., Miller, I. L., Adelberg, E. A., and Eisenstadt, J. M. (1977). *Proc. Natl. Acad. Sci. USA* **74**, 3956–3959.

de Jonge, A. J. R., Abrahams, P. J., Westerveld, A., and Bootsma, D. (1984). *Nature* **295**, 624–626.

Duncan, M. E. and Brookes, P. (1973). *Mutat. Res.* **21**, 107–118.

Elion, G. B. (1967). *Fed. Proc.* **26**, 898–901.

Elion, G. B. and Hitchings, G. H. (1965). *Adv. Chemother.* **2**, 91–177.

Epstein, J., Leyva, N., Kelley, W. N., and Littlefield, J. W. (1977). *Somat. Cell Genet.* **3**, 135–148.

Farrell, S. A. and Worton, R. G. (1977). *Somat. Cell Genet.* **3**, 539–551.

Fenwick, R. G., Jr. (1980). *Somat. Cell Genet.* **6**, 477–494.

Fenwick, R. G., Jr., Sawyer, T. H., Kruh, G. D., Astrin, K. H., and Caskey, C. T. (1977a). *Cell* **12**, 383–391.

Fenwick, R. G., Jr., Wasmuth, J. J., and Caskey, C. T. (1977b). *Somat. Cell Genet.* **3**, 207–216.

Fenwick, R. G., Jr., Konecki, D. S., and Caskey, C. T. (1982). In *Somatic Cell Genetics* (C. T. Caskey and D. C. Robbins, eds.), pp. 19–41, Plenum Publishing, New York.

Fenwick, R. G., Jr., Fuscoe, J. C., and Caskey, C. T. (1984). *Somat. Cell Mol. Genet.* **10**, 71–84.

Fontenelle, L. J. and Henderson, J. F. (1969). *Biochim. Biophys. Acta* **177**, 175–176.

Ford, D. K. and Yerganian, G. (1958). *J. Natl. Cancer Inst.* **21**, 393–425.

Fournier, R. E. K. and Ruddle, F. H. (1977). *Proc. Natl. Acad. Sci. USA* **74**, 319–323.

Francke, U., Bakay, B., Conuor, J. D., Goldwell, J. G., and Nyhan, W. L. (1974). *Am. J. Hum. Genet.* **26**, 512–522.

Freidrich, U. and Coffino, P. (1977a). *Proc. Natl. Acad. Sci. USA* **74**, 679–683.

Friedrich, U. and Coffino, P. (1977b). *Biochim. Biophys. Acta* **483**, 70–78.

Fuscoe, J. C., O'Neill, J. P., Machanoff, R., and Hsie, A. W. (1982). *Mutat. Res.* **96**, 15–30.

Fuscoe, J. C., Fenwick, R. G., Jr., Ledbetter, D. H., and Caskey, C. T. (1983). *Mol. Cell. Biol.* **3**, 1086–1096.

Ghangas, G. S. and Milman, G. (1975). *Proc. Natl. Acad. Sci. USA* **72**, 4147–4150.

Gillin, F. D., Roufa, D. J., Beaudet, A. L., and Caskey, C. T. (1972). *Genetics* **72**, 239–252.

Goldman, M. A., Holmquist, G. P., Gray, M. C., Caston, L. A., and Nag, A. (1984). *Science* **224**, 686–692.

Graves, J. A. M. (1982). *Exptl. Cell Res.* **141**, 99–105.

Harris, M. (1971). *J. Cell Physiol.* **78**, 177–184.

Harris, M. (1982). *Cell* **29**, 483–492.

Heflich, R. H., Beranek, D. T., Kodell, R. L., and Morris, S. M. (1982). *Mutat. Res.* **106**, 147–161.

Hershfield, M. S. and Seegmiller, J. E. (1977). *J. Biol. Chem.* **252**, 6002–6010.

Hochstadt, J. and Quinlan, J. C. (1976). *J. Cell Physiol.* **89**, 839–852.

Hodgkiss, R. J., Brennand, J., and Fox, M. (1980). *Carcinogenesis* **1**, 175–187.

Holden, J. A. and Kelley, W. N. (1978). *J. Biol. Chem.* **253**, 4459–4463.

Holmquist, G., Gray, M., Porter, T., and Jordan, J. (1982). *Cell* **31**, 121–129.

Hsie, A. W., Casciano, D. A., Couch, D. B., Krahn, D. F., O'Neill, J. P., and Whitfield, B. L. (1981). *Mutat. Res.* **86**, 193–214.

Huberman, E. and Sachs, L. (1976). *Proc. Natl. Acad. Sci. USA* **73**, 188–192.

Huberman, E., Sachs, L., Yang, S. K., and Gelboin, H. V. (1976). *Proc. Natl. Acad. Sci. USA* **73**, 607–611.

Hudziak, R. M., Laski, F. A., RajBhandary, U. L., Sharp, P. A., and Capecchi, M. R. (1982). *Cell* **31**, 137–146.

Hughes, S. H., Wahl, G. M., and Capecchi, M. R. (1975). *J. Biol. Chem.* **250**, 120–126.

Jenssen, D. (1982). *Mutat. Res.* **106**, 291–296.

Johnson, G. G., Eisenberg, L. R., and Migeon, B. R. (1978). *Science* **203**, 174–176.

Johnson, G. G., Ramage, A. L., Littlefield, J. W., and Kazazian, H. H., Jr. (1982). *Biochemistry* **21**, 960–966.

Jones, P. A., Taylor, S. M., Mohandas, T., and Shapiro, L. J. (1982). *Proc. Natl. Acad. Sci. USA* **79**, 1215–1219.

Jolly, D. J., Esty, A. C., Bernard, H. U., and Friedman, T. (1982). *Proc. Natl. Acad. Sci. USA* **79**, 5038–5041.

Jolly, D. J., Okayama, H., Berg, P., Esty, A. C., Filpula, D., Bohlen, P., Johnson, G. G., Shively, J. E., Hunkapiller, T., and Friedman, T. (1983). *Proc. Natl. Acad. Sci. USA* **80**, 477–481.

Kadouri, A., Kunce, J. J., and Lark, K. G. (1978). *Nature* **274**, 256–259.

Kahan, B. and DeMars, R. (1975). *Proc. Natl. Acad. Sci. USA* **72**, 1510–1514.

Kelley, W. N. and Wyngaarden, J. B. (1983). In *The Metabolic Basis of Inherited Disease*, 5th ed. (J. B. Stanbury, J. B. Wyngaarden, D. S. Fredrickson, J. L. Goldstein, and M. S. Brown, eds.), pp. 1115–1143, McGraw-Hill, New York.

Kelley, W. N., Rosenbloom, F. M., Henderson, J. F., and Seegmiller, J. E. (1967). *Proc. Natl. Acad. Sci. USA* **57**, 1735–1739.

Konecki, D. S., Brennand, J., Fuscoe, J. C., Caskey, C. T., and Chinault, A. C. (1982). *Nucleic Acids Res.* **10**, 6763–6775.

Kong, M. and Parks, R. E., Jr. (1974). *Mol. Pharmacol.* **10**, 648–656.

Kornberg, A., Lieberman, I., and Simms, E. S. (1955). *J. Biol. Chem.* **215**, 417–427.

Krahn, D. F. and Heidelberger, C. (1977). *Mutat. Res.* **46**, 27–44.

Kratzer, P. G., Chapman, V. M., Lambert, H., Evans, R. E., and Liskay, R. M. (1983). *Cell* **33**, 37–42.

Krenitsky, T. A. (1969). *Biochim. Biophys. Acta* **179**, 506–509

Krenitsky, T. A., Papaioannou, R., and Elion, G. B. (1969). *J. Biol. Chem.* **244**, 1263–1270.

Kruh, G. D., Fenwick, R. G., Jr., and Caskey, C. T. (1981). *J. Biol. Chem.* **256**, 2878–2886.

Langenbach, R., Freed, H. J., and Huberman, E. (1978). *Proc. Natl. Acad. Sci. USA* **75**, 2864–2867.

Levenberg, B., Melnick, I., and Buchanan, J. M. (1957). *J. Biol. Chem.* **225**, 163–176.

Lester, S. C., Korn, N. J., and DeMars, R. (1982). *Somat. Cell Genet.* **8**, 265–284.

Li, A. P. (1981). *Mutat. Res.* **85**, 165–175.

Lieberman, I. and Ove, P. (1959). *Proc. Natl. Acad. Sci. USA* **45**, 867–872.

Linder, S., Coleman, A. W., and Eisenstadt, J. M. (1984). *Mol. Cell. Biol.* **4**, 618–624.

Littlefield, J. (1963). *Proc. Natl. Acad. Sci. USA* **50**, 568–576.

Littlefield, J. W. (1964). *Science* **145**, 709–710.

Liskay, R. M. and Evans, R. J. (1980). *Proc. Natl. Acad. Sci. USA* **77**, 4895–4898.

Lo, Y.-F. V. and Palmour, R. M. (1979). *Biochem. Genet.* **17**, 737–746.

Lyon, M. F. (1972). *Biol. Rev.* **47**, 1–35.

Maher, V. M., Ouellette, L. M., Curren, R. D., and McCormick, J. J. (1976). *Nature* **261**, 593–595.

Martin, G. R. (1982). *Cell* **29**, 721–724.

McBride, O. W. and Ozer, H. L. (1973). *Proc. Natl. Acad. Sci USA* **70**, 1258–1262.

McCann, J. and Ames, B. N. (1976). *Proc. Natl. Acad. Sci. USA* **73**, 950–954.

Melton, D. W. (1981). *Somat. Cell Genet.* **7**, 331–344.

Melton, D. W., Konecki, D. S., Ledbetter, D. H., Hejtmancik, J. F., and Caskey, C. T. (1981). *Proc. Natl. Acad. Sci. USA* **78**, 6977–6980.

Melton, D. W., Konecki, D. S., Brennand, J., and Caskey, C. T. (1984). *Proc. Natl. Acad. Sci. USA* **81**, 2147–2151.

Miller, O. J., Sanger, R., and Siniscalco, M. (1978). *Cytogenet. Cell Genet.* **22**, 124–128.

Miller, R. L. and Bieber, A. L. (1969). *Biochemistry* **8**, 603–608.

Milman, G., Lee, E., Ghangas, G. S., McLaughlin, J. R., and George, M., Jr. (1976). *Proc. Natl. Acad. Sci. USA* **73**, 4589–4593.

Milman, G., Krauss, S. W., and Olsen, A. S. (1977). *Proc. Natl. Acad. Sci. USA* **74**, 926–930.

Mohandas, T., Sparks, R. S., and Shapiro, L. J. (1981). *Science* **211**, 393–396.

Nussbaum, R. L., Crowder, W. E., Nyhan, W. L., and Caskey, C. T. (1983). *Proc. Natl. Acad. Sci. USA* **80**, 4035–4039.

Olsen, A. S., and Milman, G. (1974a). *J. Biol. Chem.* **249**, 4030–4037.

Olsen, A. S. and Milman, G. (1974b). *J. Biol. Chem.* **249**, 4038–4040.

Patterson, D., Kao, F.-T., and Puck, T. T. (1974). *Proc. Natl. Acad. Sci. USA* **71**, 2057–2061.

Peterson, A. R., Peterson, H., and Heidleberger, C. (1974). *Mutat. Res.* **24**, 25–33.

Peterson, A. R., Landolf, J. R., Peterson, H., and Heidelberger, C. (1978). *Nature* **276**, 508–510.

Piper, J. R., Laseter, A. G., and Montgomery, J. A. (1980). *J. Med. Chem.* **23**, 357–364.

Reem, G. H. and Friend, C. (1975). *Proc. Natl. Acad. Sci. USA* **72**, 1630–1634.

Ricciuti, F. C. and Ruddle, F. H. (1973). *Nature New Biol.* **241**, 180–182.

Riddle, J. C. and Hsie, A. W. (1978). *Mutat. Res.* **52**, 409–420.

Rosenbloom, F. M., Kelley, W. N., Henderson, J. F., and Seegmiller, J. E. (1967). *Lancet* **ii**, 305–306.

Rosenstraus, M. J. and Chasin, L. A. (1978). *Genetics* **90**, 735–760.

Roy-Burman, P. (1970). *Recent Results Cancer Res.* **25**, 1–111.

Ruddle, F. H. (1981). *Nature* **294**, 115–120.

Schimke, R. T. (1984). *Cell* **37**, 705–713.

Seegmiller, J. E., Rosenbloom, F. M., and Kelley, W. N. (1967). *Science* **155**, 1682–1684.

Sekiguchi, T. and Sekiguchi, F. (1973). *Exptl. Cell Res.* **77**, 391–403.

Sekiguchi, T., Sekiguchi, F., and Tomii, S. (1974). *Exptl. Cell Res.* **88**, 410–414.

Sharp, J. D., Capecchi, N. E., and Capecchi, M. R. (1973). *Proc. Natl. Acad. Sci. USA* **70**, 3145–3149.

Shin, S., Caneva, R., Schildkraut, C. L., Klinger, H. P., and Siniscalco, M. (1973). *Nature New Biol.* **241**, 194–196.

Siniscalco, M., Klinger, H. P., Eagle, H., Koprowski, H., Fujimoto, W. Y., and Seegmiller, J. E. (1969). *Proc. Natl. Acad. Sci. USA* **62**, 793–749.

Southern, E. M. (1975). *J. Mol. Biol.* **98**, 503–517.

Stamato, T. D., Hinkle, L., Collins, A. R. S., and Waldren, C. A. (1981). *Somat. Cell Genet.* **7**, 307–320.

Stanbridge, E. (1971). *Bacteriol. Rev.* **35**, 206–227.

Stanbridge, E. J., Tischfield, J. A., and Schneider, E. L. (1975). *Nature* **256**, 329–331.

Subak-Sharpe, J. H. (1965). *Exptl. Cell Res.* **38**, 106–119.

Subak-Sharpe, H., Burk, R. R., and Pitts, J. D. (1969). *J. Cell Sci.* **4**, 353–367.

Suter, W., Brennand, J., McMillan, S., and Fox, M. (1980). *Mutat. Res.* **73**, 171–181.

Szybalska, E. H. and Szybalski, W. (1962). *Proc. Natl. Acad. Sci. USA* **48**, 2026–2034.

Szybalski, W. (1959). *Exptl. Cell Res.* **18**, 588–591.

Szybalski, W. and Smith, M. J. (1959). *Proc. Soc. Exptl. Biol. Med.* **106**, 662–666.

Szybalski, W., Szybalska, E. H., and Ragni, G. (1962). *Natl. Cancer Inst. Monogr.* **7**, 75–78.

Tarrant, G. M. and Holliday, R. (1977). *Mol. Gen. Genet.* **156**, 273–279.

Thacker, J. (1981). *Cytogenet. Cell Genet.* **29**, 16–25.

Thacker, J., Stephens, M. A., and Stretch, A. (1976). *Mutat. Res.* **35**, 465–478.

Thacker, J., Stretch, A., and Brown, R. (1982). *Mutat. Res.* **103**, 371–378.

Theile, M., Scherneck, S., and Geissler, L. (1976). *Mutat. Res.* **37**, 111–124.

Thilly, W. G., Deluca, J. G., Hoppe IV, H., and Penman, B. W. (1978). *Mutat. Res.* **50**, 137–144.

Thomas, P. S. (1980). *Proc. Natl. Acad. Sci. USA* **77**, 5201–5205.

Thompson, L. H. and Baker, R. M. (1973). In *Methods in Cell Biology* (D. M. Prescott, ed.), Vol. VI, pp. 209–281, Academic Press, New York.

Thompson, L. H., Fong, S., and Brookman, K. (1980). *Mutat. Res.* **74**, 21–36.

Thompson, L. H., Brookman, K. W., Dillehay, L. E., Mooney, C. L., and Carrano, A. V. (1982). *Somat. Cell Genet.* **8**, 759–773.

Tong, C., Fazio, M., and Williams, G. M. (1980). *Proc. Natl. Acad. Sci. USA* **77**, 7377–7379.

van Diggelen, O. P., Phillips, D. M., and Shin, S.-I. (1977). *Exptl. Cell Res.* **106**, 191–203.

Venolia, L. and Gartler, S. M. (1983a). *Nature* **302**, 82–83.

Venolia, L. and Gartler, S. M. (1983b). *Somat. Cell Genet.* **9**, 616–627.

Vinolia, L., Gartler, S. M., Wassman, E. R., Yen, P., Mohandas, T., and Shapiro, L. J. (1982). *Proc. Natl. Acad. Sci. USA* **79**, 2352–2354.

Wahl, G. M., Hughes, S. H., and Capecchi, M. R. (1975). *J. Cell. Physiol.* **85**, 307–320.

Warren, S. T., Schultz, R. A., Chang, C.-C., Wade, M. H., and Trosko, J. E. (1981). *Proc. Natl. Acad. Sci. USA* **78**, 3133–3137.

Watson, B., Gormley, I. P., Gardiner, S. E., Evans, H. J., and Harris, H. (1972). *Exptl. Cell Res.* **75**, 401–409.

Weissman, B. and Stanbridge, E. J. (1980). *Cytogenet. Cell Genet.* **28**, 227–239.

Westerveld, A., Visser, R. P. L. S., Freeke, M. A., and Bootsma, D. (1972). *Biochem. Genet.* **7**, 33–40.

Willecke, K., Klomfass, M., Mierau, R., and Dohmer, J. (1979). *Mol. Gen. Genet.* **170**, 179–185.

Willecke, K., Klomfass, M., and Schafer, R. (1981). *Mol. Gen. Genet.* **182**, 70–76.

Wilson, J. M. and Kelley, W. N. (1983). *J. Clin. Invest.* **71**, 1331–1335.

Wilson, J. M., Tarr, G. E., Mahoney, W. C., and Kelley, W. N. (1982). *J. Biol. Chem.* **257**, 10978–10985.

Wolf, S. F., Jolly, D. J., Lunnen, K. D., Friedman, T., and Migeon, B. R. (1984). *Proc. Natl. Acad. Sci. USA* **81**, 2806–2810.

Worton, R. G., Ho, C. C., and Duff, C. (1977). *Somat. Cell Genet.* **3**, 27–45.

Wright, E. D., Stack, C., Goldfarb, P. S. G., and Subak-Sharpc, J. H. (1976). *Exptl. Cell Res.* **103**, 79–91.

Wyngaarden, J. B. and Kelley, W. N. (1983). In *The Metabolic Basis of Inherited Disease*, 5th ed. (J. B. Stanbury, J. B. Wyngaarden, D. S. Fredrickson, J. L. Goldstein, and M. S. Brown, eds.), pp. 1043–1114, McGraw-Hill, New York.

Yen, P. H., Patel, P., Chinault, A. C., Mohandas, T., and Shapiro, L. J. (1984). *Proc. Natl. Acad. Sci. USA* **81**, 1759–1763.

Zownir, O., Fuscoe, J. C., Fenwick, R., and Morrow, J. (1984). *J. Cell. Physiol.* **119**, 341–348.

CHAPTER 14

CHINESE HAMSTER CELL PROTEIN SYNTHESIS MUTANTS

John J. Wasmuth
Department of Biological Chemistry
California College of Medicine
University of California, Irvine
Irvine, California

I. INTRODUCTION

Protein synthesis is quite obviously a complex and intricate process, not only from biochemical and mechanistic standpoints, but also from the genetic or gene regulation point of view. More than 150 different gene products are directly involved in protein biosynthesis, including initiation, elongation, and termination factors; at least 20 different aminoacyl-tRNA synthetases; more than 60 species of tRNA; and ribosomes, which are composed of 3 different RNA species and over 70 different proteins. The regulatory mechanisms involved in coordinating the expression of these ~ 150 genes, so that a cell has the appropriate amount of each gene product required for protein synthesis, must be extraordinarily complicated. In such a complex system, genetics and mutant methodology can play an important role in understanding both the biochemical and gene regulation aspects of protein synthesis. Protein synthesis is such a tightly coupled system that in many cases it is likely that all the functions of an individual component, and its interactions with other components, can only be elucidated when it is altered via mutation. Depending on the nature of the alterations and the component that is affected, one then has an opportunity to examine how an alteration or defect in one component affects its interaction with other components and to determine whether other steps in protein synthesis are altered. In addition, once a gene encoding a protein synthesis component has been identified and "tagged" via mutation, it can be studied genetically to determine its physical location in the genome as well as its location relative to other genes, including those encoding other products involved in protein synthesis. Studies of this kind are essential to determine how this group of genes is distributed and organized within the genome, which is an absolute prerequisite to really understanding the mechanisms involved in regulating the entire process.

In this chapter I have concentrated on describing and discussing two particular classes of protein synthesis mutants, aminoacyl-tRNA synthetase mutants with conditionally lethal, temperature-sensitive phenotypes and mutants resistant to the protein synthesis inhibitor emetine. The selection, biochemical characterization, and genetic characterization of the relatively large number of different Chinese hamster cell mutants belonging to one of these two classes have provided interesting and informative insights not only into the organization of the complex protein synthetic machinery and the genes encoding its various components, but also into more basic genetic problems in mammalian cell genetics including the mechanisms involved in gene segregation and the basis of hemizygosity of certain autosomal loci (see Chapters 28 and 29 as well). In addition, the various temperature-sensitive aminoacyl-tRNA synthetase mutants have proven useful in at least two other regards: (1) as recipients in gene transfer experiments aimed at cloning the corresponding normal genes and (2) as a means to isolate interspecific Chinese hamster cell–human hybrids that retain, under selective pressure, single human chromosomes. These latter cell lines provide a means to construct a series of recombinant DNA libraries, each specific for a different, single human chromosome, which are extremely useful in fine-structure mapping of the human genome. All of these areas will be discussed in some detail in the sections that follow.

Among the several types of protein synthesis mutants not discussed in detail in this chapter, the most notable omission is the class selected as resistant to diphtheria toxin. Elegant genetic and biochemical analyses of diptheria-toxin-resistant mutants by several laboratories have demonstrated there are three distinct classes of such mutants: those with alterations in internalization of the toxin; those with alterations that affect elongation factor EF-2 (the target of action of the toxin) directly; and those that affect the posttranslational modification of EF-2 (Gupta and Siminovitch, 1978c; Draper et al., 1979; Moehring and Moehring, 1979; Moehring et al., 1979; Gupta and Siminovitch, 1980; Moehring et al., 1980).

II. SELECTIVE PROCEDURES FOR ISOLATING PROTEIN SYNTHESIS MUTANTS

A complete deficiency in any cellular component required for protein synthesis would obviously be lethal. Therefore, selections to isolate mutants with alterations in various of the components of the protein synthetic machinery are limited to those designed to recover protein synthesis inhibitor-resistant mutants or mutants with alterations that result in conditionally lethal phenotypes.

A. Selection of Conditionally Lethal, Temperature-Sensitive Protein Synthesis Mutants

Selective procedures designed specifically to recover temperature-sensitive mutants that are defective in protein synthesis are analogous to those procedures used to isolate mutants that are auxotrophic for various nutrients (Puck and Kao, 1967), except that the selective agents are different. The selective agents of choice in experiments to isolate temperature-sensitive protein synthesis mutants kill cells when, and only when, they are incorporated into protein. Thus, the selections are designed to kill normal cells and spare only mutant cells in a population that do not synthesize protein at some predetermined nonpermissive temperature, usually 39°C. The mutant cells are subsequently recovered by removing the selective agent and transferring cultures to some predetermined permissive temperature, usually 33–34°C.

Two different selective agents that have been utilized in selections designed to recover temperature-sensitive Chinese hamster cell mutants with defects in protein synthesis are high-specific-activity [^{3}H]amino acids (Thompson et al., 1975) and a synthetic analog of lysine, *S*-2-aminoethyl-L-cysteine (thialysine) (Wasmuth and Caskey, 1976). Cells that incorporate large amounts of [^{3}H]amino acids into protein are killed, during a subsequent low-temperature storage step, by radiation damage resulting from the decay of the isotope (Thompson et al., 1975). Thialysine is incorporated into protein, presumably in place of the natural amino acid lysine, apparently resulting in the synthesis of many nonfunctional, misfunctional, or rapidly degraded proteins, which results in cell death.

Since the rationales and basic protocols used in selections employing the two different selective agents are similar, only the procedure employing thialysine, which is utilized in this laboratory, will be described here. The procedure for utilizing [^{3}H]amino acids as the selective agent have been described by Thompson et al. (1975) and Adair et al. (1978). It should be noted that the latter procedure has been utilized almost exclusively with cells grown in suspension culture, while the former procedure has been utilized only for cells grown in monolayer. However, both procedures would probably work on cells grown in either manner.

The procedure described here, which is diagrammed in Figure 14.1, is a slight modification of the procedure described by Wasmuth and Caskey (1976) for isolating temperature-sensitive protein synthesis mutants of V-79 Chinese hamster lung (CHL) cells. A mutagenized or unmutagenized population of cells is grown for at least four generations at the permissive temperature (33°C) in medium supplemented with

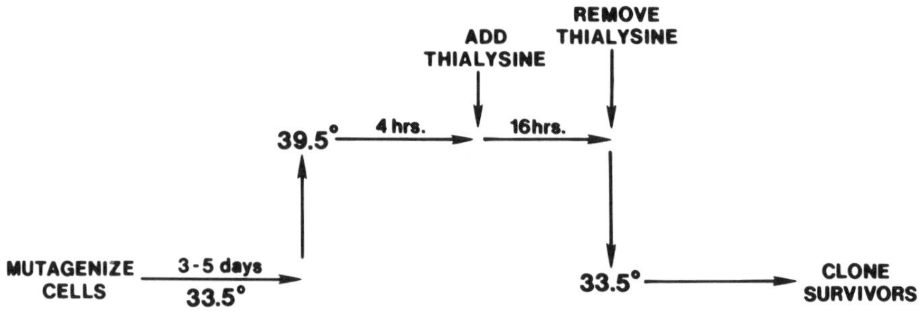

Figure 14.1. Selection procedure for isolating temperature-sensitive protein synthesis mutants using thialysine.

nonessential amino acids. The day before the selection is to begin, cells are dispersed, diluted, seeded into 100-mm culture dishes at a density of $(1-2) \times 10^6$cells/dish, and maintained at 33°C. The following day, the monolayers are washed with a salts solution, and 10 mL of a modified DME medium, which contains one-tenth the normal concentration of the essential amino acids, no nonessential amino acids, and completely lacks lysine, is added to each dish and the cultures are transferred to 39°C. Two to four hours later thialysine is added to cultures at a final concentration of 1.5 m*M*, and the cultures are maintained at 39°C in the presence of the analog for an additional 14–20 hr. At that time, the toxic medium is removed, the cultures are washed several times to remove thialysine, and the cells are returned to 33°C in complete medium. Colonies that arise 14–20 days later are isolated, and following expansion of the cultures, cells are analyzed to determine if they have a temperature-sensitive phenotype. Those cell lines that are temperature sensitive can be analyzed further to determine the nature of the defect, as described subsequently. Using this procedure with CHL cells or CHO cells, the frequency of cells that survive a single round of selection ranges from 1×10^{-8} to 1×10^{-5}. The exact frequency is very dependent on subtle alterations in the selective medium. In our experience, usually 5–95% of the colonies that arise following the selection are indeed temperature-sensitive mutants.

Several features of this selection, some theoretical and some practical, should be discussed further. (1) The cytotoxicity of thialysine is almost completely dependent on its being incorporated into protein. This statement is based on the observation that reversible and specific inhibitors of protein synthesis, such as emetine, protect ~ 50% of the cells in a culture from death in a selection similar to that just described. When the inhibitor is removed, the cells are perfectly viable and grow. (2) The omission of lysine from the selective medium at 39°C greatly enhances the cytotoxicity of thialysine, presumably because the natural amino acid competes effectively with its analog for uptake and incorporation into protein. In the absence of lysine, cells capable of synthesizing protein are, in essence, forced to utilize the analog to make protein, which is lethal. (3) Theoretically, a cell with an alteration in any gene that causes a great reduction in the rate of protein synthesis at 39°C should be recoverable from this type of selection provided that (a) the cell remains viable for the period of time it is exposed to the nonpermissive temperature and (b) the period of time the cell is at 39°C before the addition of the analog is sufficient for its defect in protein synthesis to be expressed. (4) Perhaps, most importantly, variations in the composition of the medium used before, during, and after the selection, especially the concentrations of the amino acids, can greatly affect the number and types of mutants recovered.

It should be noted that temperature-sensitive protein synthesis mutants have also been recovered from selections designed to isolate nondividing or noncycling cells (Thompson et al., 1973; Hankinson, 1976; Ashman, 1978). While these selective procedures can be used to isolate protein synthesis mutants, they are much less specific for this class of mutants than the [^{3}H]amino acid or thialysine "suicide" selections.

B. Selection of Protein Synthesis Inhibitor-Resistant Mutants

Any drug or toxin which is an inhibitor of a specific step in protein synthesis could be utilized as a selective agent in attempts to isolate mutants resistant to its inhibitory effects. While these selections are very straightforward, there are two points to keep in mind. First, because of the mechanism of action of some of the inhibitors of protein synthesis used in these selections, one should attempt to isolate mutants that are partially, not completely, resistant to those compounds. In the case of emetine in Chinese hamster cells, the optimal selective concentration of the drug, one which yields a reasonable number of mutants with levels of resistance clearly above that of wild-type cells, is in the range of 0.1–0.3 μM. For wild-type Chinese hamster cells, the LD_{10} for emetine is approximately 0.02 μM. As discussed later, attempts to isolate single-step mutants resistant to concentrations of emetine of 1 μM or above have been uniformly unsuccessful. Second, as is probably obvious to most readers, mutants resistant to protein synthesis inhibitors can sometimes arise by alterations, most notably those in uptake of the drug into the cell, that do not directly affect the target of the drug.

III. ISOLATION AND CHARACTERIZATION OF MUTANTS

A. Temperature-Sensitive Aminoacyl-tRNA Synthetase Mutants

In every temperature-sensitive mutant defective in protein synthesis in which the biochemical lesion has been determined, the affected component has been identified as one of the aminoacyl-tRNA synthetases (aaRS). This is likely because mutants with alterations in any one of these enzymes can easily be identified using a simple *in vivo* growth assay, which will be described subsequently. In addition, two types of aaRS mutants arise at extremely high frequencies in certain Chinese hamster cell lines, probably because the genes encoding these two enzymes are physically or functionally haploid (see Chapter 29). The high "background" of these common mutants can make it difficult to identify other types of mutants.

One very important point concerning aaRS mutants is that the expression of the conditionally lethal defect in protein synthesis can be dependent on growth temperature, the concentrations of specific amino acids in the growth medium, or both (Molnar and Rauth, 1975; Thompson et al., 1977; Adair et al., 1978). While some aaRS mutants have temperature-sensitive phenotypes that are independent of amino acid concentrations, for many mutants the temperature-sensitive phenotype can be either suppressed or enhanced by increasing or decreasing, respectively, the medium concentration of the amino acid correponding to the specific aaRS that is altered. In other aaRS mutants, the severity of inhibiton of protein synthesis is dependent only on the concentration of the cognate amino acid and is independent of the growth temperature. Mutants displaying a marked dependency on increased concentrations of an amino acid for normal growth at either a low or a high temperature have been termed hyperauxotrophic (Adair et al., 1978). This common characteristic of hyperauxotrophy can be utilized very effectively to determine easily and quickly if a particular temperature-sensitive mutant has a defect in any aaRS, and if so, which one, using the procedure described by Thompson et al. (1975). This growth assay simply determines whether 10 times the normal concentration of any of the 20 amino acids will enhance the growth or suppress the temperature-sensitive phenotype of mutant cells incubated at 33°C or 39°C, respectively. Since most aaRS mutants lyse and detach from culture dishes within 2 days after being transferred to nonpermissive conditions, and even a partial suppression of the phenotype by any one of the 20 amino acids is obvious upon microscopic examination of cultures, this test usually gives a very clear-cut result. For the vast majority of aaRS mutants, this test indicates protection by a single, specific amino acid. I am not aware of a single case where subsequent biochemical or genetic analysis of such a mutant identified a component other than the aaRS corresponding to the protecting amino acid as being the defective step in protein synthesis.

1. High Frequency of Leucyl- and Asparaginyl-tRNA Synthetase Mutants in CHO and V-79 CHL Cells

The first temperature-sensitive Chinese hamster cell line to have its biochemical defect defined was a CHO leucyl-tRNA synthetase (leuRS) mutant,TSH-1, which was isolated by Thompson et al. (1973) using a selection not specifically designed to recover protein synthesis mutants. Subsequently, Thompson et al. (1975) reported the isolation of a large number of temperature-sensitive protein synthesis mutants from mutagenized CHO cells using the more specific [^{3}H]amino acid selection. Surprisingly, all the mutants examined in this study had alterations that affected either leucyl- or asparaginyl-tRNA synthetase (asnRS) (see Chapter 17). In each mutant examined, the temperature-sensitive phenotype was suppressed by elevated concentrations of either leucine or asparagine and was recessive in interspecific cell hybrids. Complementation tests were performed by fusing the difficult mutants in pairwise combinations and determining whether the resultant hybrids were temperature resistant (complementation) or temperature sensitive (no complementation). By these tests, all the leucine-protected mutants belonged to one complementation group and all the asparagine-protected mutants belonged to a second complementation group. Further studies by Adair et al. (1979), using selective conditions designed to optimize the recovery of leuRS and asnRS mutants demonstrated that mutants with alterations in either leuRS or asnRS arise at frequencies of about 1×10^{-5} in mutagenized cultures of CHO cells. This is 100–1000 times higher than the frequency with which other aaRS mutants are recovered from CHO cells (Adair et al., 1978, 1979). The high frequency with which the recessive leuRS and asnRS mutants arose suggested that the affected genes might be X linked or otherwise physically or functionally haploid. Both loci were found to segregate independently of the X-linked *hprt* locus from cell hybrids, discounting the former possibility and indicating both genes were, in fact, autosomal but hemizygous (Adair et al., 1979).

Wasmuth and Caskey (1976) reported the isolation of temperature-sensitive protein synthesis mutants from V-79 CHL cells using thialysine as the selective agent. In view of the results discussed above, it was surprising to find that the vast majority of mutants recovered from these selections with CHL cells were asnRS mutants, which belonged to the same complementation group as the CHO asnRS mutants. Thus, in two different cell lines, mutants with alterations in this gene arise at a high frequency even though the gene is autosomal and the phenotype of the mutants is recessive.

To determine accurately the apparent frequency of asnRS mutants in CHL cells and to enhance the probability of isolating other classes of aaRS mutants, we have performed thialysine suicide selections using conditions designed to optimize for, or preclude the recovery of, asnRS mutants. The designs of the different selective conditions are based on the phenotypic modification of most aaRS mutants by the cognate amino acid. To optimize the recovery of induced asnRS mutants, cells are grown at 33°C in medium supplemented with 3 m*M* asparagine both before and after the selection to allow the recovery of asnRS mutants with an asparagine auxotrophic phenotype. (Hyperauxotrophic is not the proper term in this case since asparagine is a nonessential amino acid.) However, the selective medium at 39°C contains no asparagine which, if present, would suppress the temperature-sensitive phenotype of most asnRS mutants, causing them to synthesize protein and be killed by the incorporation of thialysine. Selections designed to preclude the recovery of most asnRS mutants and enhance the recovery of others utilize an analogous strategy. Cells are grown at 33°C in medium without asparagine but including elevated concentrations of all other amino acids to enhance recovery of other aaRS mutants which might have hyperauxotrophic phenotypes even at the low temperature. The selective medium at 39°C contains 3 m*M* asparagine, which will suppress the phenotype of most asnRS mutants resulting in their being killed as effectively as wild-type cells by thialysine, but contains only one-tenth the standard concentration of the 18 other amino acids. (Lysine is omitted altogether for the reasons discussed in Section II A.) The reduction in the concentrations of the 18 other amino acids enhances the possibility of recovering mutants with temperature-sensitive phenotypes that might be suppressed by the normal concentration of a particular amino acid.

Typical results of these two types of selections using CHL cells mutagenized with EMS are summarized in Table 14.1. In selections to optimize recovery of asnRS mutants, the frequency of survivors is high, approximately 2×10^{-5}. Thirty clones picked at random in this particular experiment were all found to be temperature sensitive and in 29 of thc 30, the temperature-sensitive phenotype was suppressed by asparagine. Thus, virtually all the survivors were probably asnRS mutants, and the frequency of this type of mutant is about 2×10^{-5}. However, for reasons discussed subsequently, this number has to be a significant underestimate of the real proportion of asnRS mutants in a mutagenized population of cells. In the second type of selection, the survival frequency was about 100 times below that observed in the first experiment, which points out two things. First, the selection effectively eliminates the recovery of most asnRS mutants. Second, the frequency with which other types of temperature-sensitive mutants are recovered is at least two orders of magnitude below the frequency of asnRS mutants.

We were once again surprised when we found that three of the six surviving clones in the second selection described above were leuRS mutants, the other common type of aaRS mutant in CHO cells, and that they belonged to the same complementation group as the CHO leuRS mutants. Although the recovery of leuRS mutants from selections with CHL cells is about 100-fold lower than in selections with CHO cells, we have some data which indicate that the true frequency of these mutants in mutagenized CHL cells may be considerably higher, but many mutants are not recovered because the thialysine selection is less efficient

TABLE 14.1
Recovery of Temperature-Sensitive Mutants from Thialysine Selections

Amino Acid Elevated in Selective Media at 39°C	Selective Pressure for Recovery of AsnRS^{ts} Mutants	Number of Cells Screened	Number of TS Clones Recovered	Frequency of TS Mutants
None	Positive	2.5×10^{7}	644	2.5×10^{-5}
asn	Negative	1.9×10^{8}	6	3.2×10^{-7}

in recovering them than the [^{3}H]amino acid selection. Regardless of absolute frequencies, we were struck by the fact that the same two types of mutants, those with alterations in asnRS or leuRS, were quite common in two independent Chinese hamster cell lines, suggesting that both loci might be hemizygous in both cell lines. Since we considered it unlikely that the two genes had been rendered hemizygous via two separate events in two different cell lines, it seemed possible that the two genes might be closely linked. If this were the case, a single event in each cell line, such as a small deletion, could have resulted in both loci becoming hemizygous simultaneously. This idea was tested directly after constructing a hybrid between wild-type CHL cells and a double mutant, which was isolated in two sequential selections and has mutations in the genes encoding both leuRS and asnRS. Using this hybrid, we found that the two affected genes segregate independently, demonstrating they are not closely linked and are most likely on different chromosomes. Further experiments, some of which are described below, provided good evidence that the gene encoding asnRS is hemizygous in both CHO and CHL, while the gene encoding leuRS is hemizygous in CHO but dizygous in CHL.

One last point concerns the true mutation frequency at the locus encoding asnRS. The frequency with which these mutants are recovered from selections in CHO and CHL cells is quite high, $(1\text{-}2) \times 10^{-5}$. However, if one takes into account that a reasonable number of mutants do not survive the selective procedure because they rapidly lose viability at 39°C (Wasmuth and Caskey, 1976), the true frequency of these mutants in a population of mutagenized cells must be closer to 10^{-4}. Furthermore, it is important to remember that in these selections one is asking for the recovery of only those mutants with specific types of alterations, those that render the enzyme nonfunctional at high temperature but leave it functional at a lower temperature. Intuitively, one would expect that the number of possible amino acid changes that would alter the enzyme in this manner, rather than making it totally nonfunctional, would be very limited. While this might suggest there is a single site (base) in the gene that is a mutational "hot spot," distinct phenotypic differences among many of the asnRS mutants indicate they have mutations at different sites within the gene (Adair et al., 1978). Thus, in mutagenized populations of CHO or CHL cells, the proportion of cells with a mutation somewhere in the gene encoding asnRS must be extraordinarily high when one considers that many of the mutations in this gene will result in the loss of cell viability. Similar arguments could be made for the gene encoding leuRS, at least in CHO cells. It would be of great interest to determine whether these genes are just extraordinarily susceptible to chemical mutagenesis, or some other, as yet undefined, genetic phenomenon is involved in generating these mutants. In any case, these two genes are among the most amenable in mammalian cells to genetic manipulation through mutant selection and, as discussed subsequently, the isolation of revertants. This makes them ideal systems for studying the relationship among gene structure, mutation, and gene function at the DNA level. For this reason, among others, this laboratory is currently attempting to isolate cloned genomic sequences encoding asnRS and leuRS using approaches that have been described (Cirullo et al., 1983b).

2. Isolation and Genetic Characterization of Other Aminoacyl-tRNA Synthetase Mutants

Although mutants with defects in aaRS's other than leuRS or asnRS have been isolated from CHO and CHL cells, the frequencies with which they are recovered suggests that the genes encoding them are probably not hemizygous in the parental cell lines (Adair et al., 1979). Using selective conditions designed to reduce the recovery of leuRS and asnRS mutants and enhance the recovery of other classes of temperature-sensitive aaRS mutants, Adair et al. (1978,1979) isolated and identified five new types of CHO mutants. The phenotype of most of these new mutants was modified by a specific amino acid, either methionine, arginine, histidine, glutamine, or lysine, which suggested that the mutants had defects in the corresponding aaRS. These results reemphasize the importance of being able to phenotypically modify most aaRS mutants with amino acids and design selections that, at the same time, avoid recovering a high "background" of common mutants and increase the chances of obtaining and identifying less common mutants.

Ashman (1978), using a selection designed to recover cells unable to divide and cycle at 39°C, recovered three classes of temperature-sensitive CHO mutants, which were phenotypically modified by histidine, leucine, or valine. In order to identify eight such mutants, over 600 surviving colonies had to be screened. The reason leuRS mutants were recovered from these experiments at a low frequency and asnRS mutants were not recovered at all is probably due in part to two things. First, in the selective protocol of Ashman (1978) cells were exposed to the nonpermissive temperature for much longer times than is required in the [^{3}H]amino acid or thialysine selections, which can greatly reduce the viability of most aaRS mutants. Second, the medium used in these selections contained normal concentrations of leucine and asparagine, which would greatly reduce recovery of leuRS and asnRS mutants for reasons described above. In addition to these mutants, Hankinson (1976) reported the isolation of a CHO mutant selected as an alanine auxotroph, in which alanyl-tRNA synthetase activity was not detectable *in vitro*. In this laboratory, we recently have identified a CHL mutant whose temperature-sensitive phenotype is suppressed by a high concentration of tryptophan, suggesting an alteration in tryptophanyl-tRNA synthetase (trpRS) (Chang and Wasmuth, 1984). In support of this notion, the activity of trpRS in cell-free extracts prepared from the mutant is greatly reduced.

In at least one mutant of each amino-acid-responsive class except the single lysine-responsive mutant, a defect in the *in vivo* or *in vitro* functioning of the corresponding aaRS has been confirmed. Thus, mutants with alterations affecting at least 9 and most likely 10 of the possible 20 different aaRS's have been isolated to date. More extensive biochemical characterizations of the various mutants will be discussed in Section III A 3.

a. Complementation Analysis of Mutants. The temperature-sensitive or hyperauxotrophic phenotype of every aaRS mutant examined thus far is recessive. Extensive complementation tests have also been performed on a large number of different aaRS mutants by various workers, especially Adair et al. (1978,1979). In all cases examined, mutants belonging to one class as defined by the amino acid to which they respond, also belong to a single and nonoverlapping group in complementation studies. For example, if any two mutants with a methionine-responsive phenotype are fused, the resultant hybrids are hyperauxotrophic for methionine and/or temperature sensitive (depending on the phenotypes of the parents). However, mutants in any one of the different amino-acid-responsive classes will complement any mutant from another class, and produce cell hybrids that are phenotypically temperature resistant and nonhyperauxotrophic. While not every single mutant isolated has been tested in this manner, there are no reported exceptions to these rules as yet.

These studies demonstrate that all mutants affecting a particular aaRS have alterations in the same gene and thus the same polypeptide component of the enzyme. Thus, there are no examples of two mutants with defects in the same aaRS, which have alterations in different, nonidentical subunits. However, Ashman (1978) reported evidence suggesting that intracistronic complementation occurred in hybrids between certain pairs of histidyl-tRNA synthetase mutants. Such a result is not surprising if the enzyme is a dimer or tetramer composed of identical subunits.

The availability of a battery of 10 mutants, each belonging to a different asRS-specific complementation group, can greatly aid anyone in trying to identify a temperature-sensitive mutant which may have a defective aaRS but which shows no response to an amino acid. Thus, Adair et al. (1978) were able to easily identify a non-amino-acid-responsive mutant as a glutaminyl-tRNA synthetase mutant based on this mutant's inability to complement a previously defined glutaminyl-tRNA synthetase mutant.

b. Isolation of Double Mutants with Alterations in Two Aminoacyl-tRNA Synthetases. For certain types of genetic and biochemical experiments, the availability of mutants with alterations in two different aaRS's is very useful. As mentioned earlier, one such double mutant, with alterations affecting both leuRS and asnRS, was very useful in the experiments that enabled us to determine that the genes encoding these enzymes were not linked. The construction of a cell line with independent mutations in genes encoding different aaRS's takes advantage, once again, of the amino-acid-suppressible temperature-sensitive phenotype of most aaRS mutants. For these constructions one simply starts with a temperature-sensitive amino-acid-suppresible aaRS mutant, and subjects this cell line to another [^{3}H]amino acid or thialysine selection using medium in which the concentration of the suppressing amino acid is elevated. Under these conditions, the mutant cell line will behave as wild type, grow at 39°C, and be killed by the selective agent except in those cells in the population that are temperature sensitive due to a second mutation. Depending on specific needs and which two aaRS's one wants altered in the same cell, it is certainly easiest to start with an uncommon mutant and isolate from it in the second selection, a more common type of mutant. In addition to the leuRS, asnRS double mutant mentioned earlier, Adair et al. (1979) have isolated asnRS, metRS, and lysRS, metRS double mutants, and a trpRS, asnRS double mutant has recently been constructed in this laboratory (Chang and Wasmuth, 1984).

3. *Biochemical Characterization of Aminoacyl-tRNA Synthetase Mutants*

As might be expected, the response of different aaRS mutants to changes in temperature and/or the concentration of the appropriate amino acids in the medium varies greatly, even among independent mutants belonging to the same complementation group. For many mutants, the rate of protein synthesis can be precisely manipulated over a very broad range by appropriate manipulations of growth temperature and amino acid supplementations. In mutants with very strong temperature-sensitive phenotypes, the rate of protein synthesis declines to almost undetectable levels within a few minutes after transfer of cells to very restrictive conditions. The decreased rate of protein synthesis in aaRS mutants under restrictive conditions is invariably accompanied by a decrease in the rate of DNA synthesis, although this latter response is somewhat delayed (Thompson et al., 1975; Wasmuth and Caskey, 1976; Ashman, 1978). This observation explains why aaRS mutants can be recovered from selections designed to recover cells that do not synthesize DNA under restrictive conditions.

For the vast majority of mutants examined, the activity of the suspect aaRS is reduced in cell-free extracts, relative to the activity observed in extracts from wild-type cells, even when extracts are prepared from mutant cells grown under fully permissive conditions (Hankinson, 1976;

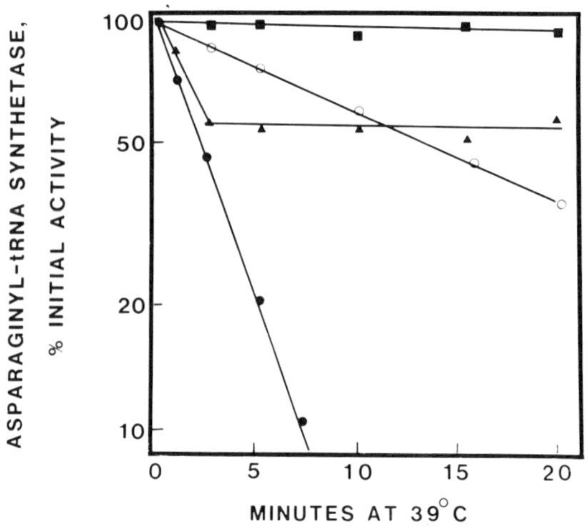

Figure 14.2. Thermal stability of asparaginyl-tRNA synthetase activity in cell-free extracts of wild-type, mutant, and heterozygous hybrid cell lines. Extracts were prepared from cells grown at 34°C and asparaginyl-tRNA synthetase activity was determined as described by Cirullo et al. (1983*a*). Cell-free extracts were incubated at 39°C for the indicated length of time, then samples were removed and assayed for asparaginyl-tRNA synthetase activity at 33°C. The amount of activity without preincubation at 39°C is defined as 100% for each extract: (■) wild-type CHL; (●) CHL asnRS mutant, UCW 132; (○) CHL asnRS mutant, UCW 132, 1 m*M* Mg-ATP added to extract during incubation at 39°C; (▲) W.T. CHL × UCW 132 hybrid.

Wasmuth and Caskey, 1976; Thompson et al., 1977; Ashman, 1978). In no case has a mutant been observed to have significantly lowered activity for more than a single aaRS. In some mutants, the activity of the altered aaRS is below detectable levels in cell-free extracts prepared from cells grown under permissive conditions, even though the enzyme in question must function to some extent *in vivo* or the cells would not be viable. For a large number of aaRS mutants, including those with alterations in asnRS, hisRS, glnRS, leuRS, metRS, and valRS, it has been possible to demonstrate that not only the activity but also a physical property (thermal stability) of the aaRS in question is altered (Wasmuth and Caskey, 1976; Ashman, 1978; Thompson et al., 1978). The results of *in vitro* heat inactivation experiments, like those summarized in Fig. 14.2 for wild-type CHL cells, a CHL asnRS mutant (UCW 132), and a hybrid between these two cell lines, illustrate several points concerning the nature of the altered enzyme. The asnRS activity from the mutant is much more sensitive to thermal inactivation *in vitro* than the enzyme from normal cells, the former enzyme having a half-life of 2 min at 39°C while the latter enzyme has a half-life of over 90 min. In addition, it can be seen that the enzyme from this particular mutant is partially protected from thermal inactivation by high concentrations of Mg-ATP.

Neither of the other substrates of the enzyme, asparagine or $tRNA^{asn}$, has any stabilizing effect on the mutant enzyme *in vitro*, even though aspargine completely suppresses the temperature sensitivity of the mutant *in vivo*. The results shown in Fig. 14.2 further demonstrate that in an extract from a UCW 132 × wild-type hybrid, which has been grown at 34°C, both the thermal stable and thermolabile forms of asnRS, characteristic of the wild-type and mutant parents, respectively, can be detected. Thus, the two forms of asnRS produced by the genomes of the different parents are expressed independently of one another in heterozygous cell hybrids. The simplest interpretation of these results is that the mutation rendering the asnRS mutant temperature sensitive is in the structural gene (*asnS*) encoding asnRS, or a polypeptide component of asnRS (the quaternary structure of the enzyme is unknown).

An unusual result was recently obtained in this laboratory in experiments examining the thermal stability of metRS activity from the CHO mutant Met-1. The enzyme from wild-type cells was found to be considerably more thermolabile *in vitro* (half-life of 13 min at 39°C) than the enzyme from the mutant, which has a half-life of approximately 100 min (Cirullo and Wasmuth, 1984a). However, the ability of metRS in the mutant to aminoacylate tRNA at 39°C *in vivo* is clearly impaired (Thompson et al., 1977) and Met-1 fails to complement another mutant, Met-2, which has a metRS activity that is more thermolabile than the wild-type enzyme (Thompson et al., 1978).

Studies on the CHO leuRS TSH-1 by Haars et al. (1976), Hampel et al. (1978), and Ritter et al. (1976) have provided two interesting results. First, leuRS activity from this cell line was found to be quite thermolabile *in vitro* if tRNA bound to the enzyme was not removed prior to initiating the heat-inactivation experiments. In the absence of bound tRNA, the enzyme was markedly less sensitive to heat inactivation in cell-free extracts. Second, in wild-type CHO cells the majority of the leuRS activity is associated with two high-molecular-weight complexes, which sediment as ~ 21S and 30S particles upon sucrose density gradient centrifugation of cell-free extracts and which contain several other aaRS activities (Hampel et al., 1978; Ritter et al., 1976). In TSH-1, however, leuRS activity is absent from the 21S and 30S complexes and exists primarily as a lower-molecular-weight form (Ritter et al., 1976; Hampel et al., 1978). Although the significance of this latter finding is unclear at present, further experiments of this type with other aaRS mutants may help to explain the functional significance of the large complexes containing many aaRS activities.

At least one leuRS and one asnRS mutant have been known to produce enzymes with altered catalytic and/or kinetic properties. Haars et al. (1976) found the leuRS from TSH-1 has a fourfold higher K_m for leucine than the enzyme from wild-type cells. Similarly, Andrulis et al. (1978) demonstrated that the asnRS from the CHO mutant Asn-5 has a threefold higher K_m for asparagine than the normal enzyme. In each of these mutants, the K_m of the defective enzyme for the cognate tRNA was unaltered.

Various biochemical studies on aaRSs extracted from mutant cells with alterations affecting leuRS, asnRS, glnRS, hisRS, metRS, and valRS have thus demonstrated that the enzymes examined have distinctly altered physical and/or kinetic properties. Although not absolutely conclusive, these data strongly suggest that the mutation in each of these cell lines is in the structural gene encoding the altered aaRS. After stating this assumption, I should mention that regardless of how strong or weak the data might be, calling the affected locus in these various mutants a structural gene for the affected aaRS may be a bit misleading since the quaternary structure of none of these enzymes is known. Thus, any one of the enzymes might be an oligomer and contain nonidentical subunits encoded by different structural genes, an alteration in either of which could potentially render the enzyme nonfunctional at 39°C *in vivo*. It must be remembered, however, that the genetic complementation tests have clearly demonstrated that this latter situation does not exist for any mutants studied thus far. Thus, there are no examples of mutants that complement one another yet have alterations affecting the same enzyme.

4. *Genetic Mapping of Genes Encoding Aminoacyl-tRNA Sythetases*

a. Gene Mapping in Chinese Hamsters. As discussed earlier, one goal of studying protein synthesis mutants is to use them to determine the organization of the genes encoding protein synthesis components in mammalian cells. In addition, the localization of genes that can be altered to produce cells with selectable phenotypes, such as those encoding certain of the aaRS's, can greatly increase our ability to perform sophisticated experiments to study more basic genetic problems, such as gene inactivation, hemizygosity, and gene segregation in mammalian cells. Toward these goals, this laboratory has spent considerable time over the past several years in attempts to determine the chromosomal location of genes encoding various protein synthesis components and to examine possible linkage relationships that might exist.

The only gene encoding an aaRS localized to a specific chromosome thus far in Chinese hamsters is *leuS* (also referred to as LARS) which encodes leuRS (Wasmuth and Chu, 1980). The assignment of the *leuS* gene to chromosome 2 in Chinese hamster cells is based on the observations of Wasmuth and Chu (1980) that this locus segregates concordantly from cell hybrids with two other genes, *emtB* and *chr*, which had previously been assigned to this chromosome (Worton et al., 1980) (Chapter 28). Studies on these three genes, their linkage relationship, and their locations have provided insights and experimental approaches to a number of different genetic problems in mammalian cells, some of which warrant further discussion here.

As already discussed, the *leuS* gene is defined by mutations that affect leuRS and renders cells temperature sensitive. Mutations in the *emtB* locus, which will be discussed in detail later, render cells resistant to the protein synthesis inhibitor, emetine, while mutations in the *chr* locus render cells resistant to normally cytotoxic concentrations of sodium chromate (Campbell et al., 1981). The phenotypes resulting from mutations in each of these genes are recessive in cell hybrids. Wasmuth and Chu (1980) constructed a cell line, starting with the CHO leuRS mutant, TSH-1, that has mutations in each of these three genes. Hybrids between this triple mutant, UCW 56, and wild-type CHO or CHL cells were isolated and, as expected, were phenotypically temperature resistant ($LeuS^+$), emetine sensitive (Emt^s), and chromate sensitive (Chr^s). In agreement with the results of Campbell and Worton (1980), we found that the wild-type *emtB* and *chr* genes segregate concordantly from these hybrids. That is, the vast majority of the segregants selected as resistant to emetine (loss of the wild-type *emtB* gene) are also resistant to chromate (loss of the wild-type *chr* gene) and the majority of the segregants selected as resistant to chromate simultaneously became resistant to emetine. Furthermore, we found that over 90% of the segregants selected as having lost either the wild-type *emtB* or *chr* gene simultaneously became temperature sensitive, demonstrating that the wild-type *leuS, emtB,* and *chr* genes are syntenic. This result is especially interesting since Worton et al. (1981) have demonstrated that the *emtB* and *chr* genes are located in a region on the long arm of chromosome 2 that is known to be physically haploid in the CHO cell line as the result of a large interstitial deletion of one chromosome 2 homolog. More recent experiments in this laboratory have provided evidence that strongly suggests the *leuS* gene is, in fact, also located in this region of chromosome 2 that is haploid in CHO cells. These observations clearly suggest that the *leuS*, *emtB*, and *chr* genes are all physically hemizygous in CHO cells, which would explain why recessive mutants with alterations in the *leuS* gene (and the *emtB* and *chr* genes) arise at such high frequencies in this cell line. Some of the studies to be discussed in later sections of this chapter have been greatly aided by, and in some cases totally dependent on, the knowledge that these three genes are linked and hemizygous.

The genes encoding other aaRS's have not yet been assigned to specific chromosomes in the Chinese hamster, but we have accumulated some negative data which should ultimately prove useful. The genes encoding trpRS and asnRS are not linked to one another, nor is either located on chromosome 2. In addition, neither of these two genes is linked to the *emtA* locus, a gene that will be discussed in Section III B 2.

b. Gene Mapping in Humans. Not surprisingly, the Chinese hamster cell aaRS mutants have proven very useful in determining the chromosomal locations of the corresponding genes in humans. Interspecific cell hybrids between any of the aaRS mutants and normal human cells can be isolated using conditions, such as growth at 39°C or in medium with a reduced concentration of the appropriate amino acid, which allow for the growth of only those cells that retain and express the human gene complementing the defective Chinese hamster gene. Subsequent karyological and biochemical analysis of such hybrids, which retain very few human chromosomes besides the one containing the selected marker, can enable one to localize the complementing human gene to a specific chromosome or region of a chromosome. Using such an approach, Giles et al. (1980) provisionally assigned the human LARS (*leuS* in our terminology) gene, which complements the mutation in a CHO leuRS mutant, to chromosome 5. This assignment was confirmed by Dana and Wasmuth (1982a,b), using temperature-resistant hybrids between human leukocytes and the CHO triple mutant previously described, UCW 56. In these experiments, it was found that in 10 of 10 heat-resistant (human LeuS^{+}) hybrids examined, the human *emtB* and *chr* genes were also present and expressed, which resulted in the cell lines becoming sensitive to both emetine and chromate (Dana and Wasmuth, 1982a). Furthermore, the three human genes, *leuS*, *emtB*, and *chr*, were observed to segregate concordantly with one another and with human chromosome 5. The linkage relationships between these three genes has thus been retained in at least two diverse mammalian species.

More recent studies in this laboratory on asnRS mutant × human cell hybrids (Cirullo et al., 1983a) and metRS mutant × human cell hybrids (Cirullo and Wasmuth, 1984a) have resulted in the provisional assignment of the human genes (*asnS* and *metS*) that complement the mutant genes in the two CHO mutants to chromosome 18 (Cirullo et al., 1983a) and 12 (Cirullo and Wasmuth, 1984a), respectively. Biochemical analysis of these hybrids, and those that express the human *leuS* gene, have demonstrated that the expression of the human *leuS*, *asnS*, or *metS* gene is associated with the appearance in the hybrids of a second form of the aaRS in question that is dramatically different from the mutant CHO form of the corresponding aaRS with respect to thermal stability *in vitro* (Dana and Wasmuth, 1982a; Cirullo et al., 1983a; Cirullo and Wasmuth, 1984). These results suggest, although do not conclusively prove, that the human *leuS*, *asnS*, and *metS* genes encode structural components of the respective aaRS's. Since all these hybrids grow vigorously under conditions that are nonpermissive for the CHO parents, it appears that the human forms of at least three aaRS's can function perfectly well in cells where the majority of the other protein synthesis components, probably including the cognate tRNAs, are products of the Chinese hamster genome. It should also be pointed out that the human gene encoding tryptophanyl-tRNA synthetase has been localized to chromosome 14, although these studies did not utilize Chinese hamster cell mutants (Denney and Craig, 1976). Thus, four genes encoding aaRS's have been mapped in humans so far and each is on a different chromosome.

As mentioned in Section I hybrids between human cells and Chinese hamster cell aaRS mutants, which are selected under conditions that require retention and expression of the human gene encoding the appropriate aaRS, allow for the isolation of cell lines that retain, under selective pressure, a single human chromosome. Genomic DNA libraries from cell lines can be screened using single and sensitive procedures (Gusella et al., 1979; Gusella et al., 1980) to identify recombinant phage-containing human DNA specifically, providing sources of single-human-chromosome-specific DNA probes which can be extremely useful for fine-structure mapping of the human genome. In view of the observation that the genes encoding the various aaRS's appear to be randomly distributed throughout the human genome and since 10 different types of Chinese hamster cell aaRS mutants have been isolated, a large number of single-human-chromosone-specific cell lines and libraries might be constructed using these mutants. In this laboratory, we have identified hybrids that express the human *leuS, asnS,* or *metS* genes and contain only the human chromosome on which the selected gene is located, number 5 (Dana and Wasmuth, 1982b), 18, or 12, respectively. A genomic DNA library which is complete and specific for human chromosome 5 has recently been prepared in his laboratory from the former hybrid and the preparation of a chromosome-18-specific library should be completed shortly.

5. Characterization of Temperature-Resistant Revertants

As alluded to earlier, aaRS mutants, as a class, represent excellent genetic systems in that one can not only isolate a relatively large number of mutants, but a simple counterselection, growth at an elevated temperature, also exists for the isolation of revertants. The ability to isolate revertants from aaRS mutants represents a very good means, at least potentially, to examine the regulation of many cellular processes using a genetic approach. Thus, it is important to recognize that phenotypic reversion of a temperature-sensitive or hyperauxotrophic aaRS mutant can conceivably result from any one of several different kinds of alterations, including (1) second site mutations within the altered structural gene, which partially restore normal properties to the defective enzyme (2) increased intracellular concentration of the suppressing amino acid, which could result from either increased transport or, in the case of a nonessential amino acids, an increased rate of biosynthesis; (3) increased cellular concentrations of the cognate tRNAs, which might protect certain defective enzymes from thermal inactivtion; (4) synthesis of greatly increased amounts of the partially defective enzyme. It is equally important to realize that the selective conditions used to isolate revertants can at least partially determine which of the potentially different classes of phenotypically suppressed revertants might be recovered. Selective conditions that are extremely nonpermissive for a mutant, for example, at 39°C in medium with greatly reduced amino acid concentrations, favor reocvery of any of those revertants with a more normal form of the altered enzyme. However, less stringent selective conditions might allow recovery of revertants in which the temperature-sensitive defect is suppressed, not corrected.

To date, only a relatively small number of temperature-resistant revertants have been characterized biochemically and the majority of these appear to have a secondary alteration in the defective enzyme. These various types of alterations including restoring an altered enzyme's K_m for the cognate amino acid to a more normal value or partially decreasing the thermolability of a defective enzyme Farber and Deutscher, 1976; Ashman, 1978; Thompson et al., 1978; Molnar et al., 1979). In addition, Molnar et al. (1979) described a revertant isolated from the CHO leuRS mutant TSH-1, in which the defect in leuRS remained but the cell line had twofold higher cell levels of the altered enzyme.

Very recently, in this laboratory, we characterized three temperature-resistant revertants isolated from the CHO asnRS mutant, Asn-5, which all appear to have arisen as a result of their overproducing the defective asnRS of the Asn-5 parent (Cirullo and Wasmuth, 1984b). The specific activity of asnRS in the three revertants is elevated 18-, 24-, and 30-fold, respectively, above that observed in the Asn-5 parent, which corresponds to 9-, 12-, and 15-fold higher activities than observed in wild-type CHO cells. Furthermore, the kinetic and thermolability properties of the enzymes from the three revertants and the Asn-5 parent are identical. SDS polyacrylamide gel electrophoresis of total cell protein has enabled us to identify a 56,000 molecular weight protein that is present in the revertants in much higher quantities than in the Asn-5 parent. However, whether or not this protein corresponds to a component of asnRS remains to be determined. Further analysis of these revertants is currently in progress but several lines of evidence suggest they have all arisen as a result of amplification of the *asnS* gene. The isolation of other revertants from different aaRS mutants should provide important information about protein synthesis in mammalian cells.

B. Emetine-Resistant Mutants

The ribosome is a very complex organelle; the interactions among its approximately 80 different components are also very complex. A complete understanding of these intricate interactions has been greatly aided by the use of genetics and mutant methodology, but the interpretation of genetic and biochemical studies with mutants that affect ribosomes is complex. It is important to keep in mind that any mutation altering the primary structure of any ribosomal component may well have profound and pleiotropic effects on the entire organelle. In addition, in a cell which is heterozygous for a mutant allele at a locus encoding a ribosomal component, two populations of ribosomes will exist, one class with the normal gene product and one class with the mutant gene product. In such a cell, each class of ribosomes can potentially affect the functioning of the other class since they cotranslate mRNA as polysomes. These kinds of possible interactions add yet another level of complexity to a genetic analysis of the ribosome.

The ipecac alkaloid, emetine, is a potent inhibitor of peptide chain elongation in mammalian cells, and the site of action of this drug has been localized to the ribosome (Vasquez, 1979). Emetine most likely affects the 40S ribosomal subunit specifically and prevents translocation of sensitive ribosomes along mRNA, which "freezes" polysomes (Vasquez, 1979). Besides being useful in genetically dissecting individual components of the ribosomes, the selection and characterization of Chinese hamster cell mutants resistant to the cytostatic action of emetine has emphasized two important general concepts in Chinese hamster cell genetics: (1) The ability to identify different classes of mutants among a group with phenotypes that are virtually identical is greatly facilitated by, and in some cases almost completely dependent on, careful genetic characterization of mutants, especially complementation analysis. (2) The use of several different established Chinese hamster cell lines for the selection of mutants markedly enhances the chances of isolating different types of mutants.

1. Isolation and Characterization of Emetine-Resistant Mutants from CHO Cells

Gupta and Siminovitch (1976) first reported the isolation of CHO mutants resistant to emetine, which arise at a very high frequency in mutagenized cultures of this cell line. These mutants are resistant to 10–30-fold higher concentrations of emetine than the wild-type parent. Emetine-resistant (Emt^r) CHO mutants have been characterized extensively by a number of laboratories. These studies have provided the following information.

1. In every CHO Emt^r studied, the phenotype of emetine resistance is recessive in intraspecific cell hybrids (Gupta and Siminovitch, 1976, 1977b; Campbell and Worton, 1979). The biochemical basis for the recessive nature of the Emt^r phenotype can be explained by the findings, discussed in detail subsequently, that the alteration in this class of mu-

tants affects a specific protein of the 40S ribosomal subunit. In cell lines with an alteration in this protein, ribosomes are less sensitive to emetine inhibition than in wild-type cells, whose ribosomes contain the normal form of this protein. Intraspecific $Emt^r \times Emt^s$ CHO cell hybrids synthesize (presumably) equal amounts of both forms of the protein. Since each ribosome contains a single copy of the protein, 50% of the ribosomes in such a hybrid will contain the altered protein, and will be partially resistant to the effects of emetine, while 50% of the ribosomes will contain the normal protein and be sensitive to the drug. When these cells are challenged with an appropriate concentration of emetine, translocation of emetine-sensitive ribosomes along mRNA is blocked, thereby preventing translation by resistant ribosomes that follow, even though they are resistant to direct inhibition by the drug.

2. Complementation analysis of a large number of independent CHO Emt^r mutants have shown all these cell lines to belong to a single complementation group (Gupta and Siminovitch, 1977b; Campbell and Worton, 1979). That is, hybrids between two different CHO Emt^r mutants are as resistant to the drug as the least resistant of the two parents. We have designated this complementation group of Emt^r mutants as EmtB, and the affected locus *emtB*, to distinguish them from other classes of Emt^r mutants which, as described below, have been isolated from other cell lines (Wasmuth et al., 1980, 1981).

3. The segregation frequency of the *emtB* locus from intraspecific CHO $Emt^r \times Emt^s$ hybrids is approximately the same as the segregation frequency for the X-linked *hprt* locus from analogous, heterologous hybrids (Gupta et al., 1978). These data, together with mutation frequency data (Campbell and Worton, 1979), suggested that the *emtB* locus was physically or functionally hemizygous in CHO cells. In contrast, the frequency of Emt^r segregants from similar hybrids in which the wild-type Emt^s parent was a Chinese hamster cell line other than CHO is two-three orders of magnitude below that observed in CHO × CHO hybrids. This result suggested that in these hybrids two segregation events were required for the reexpression of the mutant *emtB* allele derived from the CHO parent, indicating the wild-type parents in this latter class of hybrids were dizygous for the *emtB* locus (Gupta et al., 1978).

4. As mentioned in Section III A 4, the *emtB* locus is located in the region on the long arm of chromosome 2 that is physically haploid in CHO cells, and is linked to the *leuS* and *chr* genes (Campell and Worton, 1980; Wasmuth and Chu, 1980). The fact that these three genes are linked and segregate concordantly from cell hybrids has proven very valuable for several types of experiments to be discussed later. The locali-

zation of the *emtB* locus to a haploid region of the CHO genome fits nicely with the mutation and segregation frequency data just discussed, which indicated that CHO cells contain only a single functional copy of this gene.

2. Isolation and Characterization of Emetine-Resistant Mutants from V-79 CHL Cells and Chinese Hamster Peritoneal Cells: Identification of Different Complementation Groups

In view of the complex structure of the ribosome and the interactions among its various protein and RNA components, it seemed likely that alterations in any one of several of these components might decrease the binding of emetine to the ribosome. This idea assumes that emetine does not bind to a single, specific protein or RNA species on the ribosome but rather to a site that is defined in part by several different proteins or RNAs. Thus, it seemed possible that mutations in genes other than *emtB* might alter the affinity of the ribosome for emetine and result in cell lines becoming resistant to the drug. However, it was obvious that the ability to identify loci other than *emtB* that might be altered to give rise to the phenotype of emetine resistance would be difficult, at best, using CHO cells simply because of the high frequency of *emtB* mutants in this cell line. For example, if recessive mutations in either one of two genes could lead to emetine resistance and CHO cells are hemizygous for one locus (*emtB*) but dizygous for the other, the vast majority of mutants isolated from this cell line will obviously be those with alterations in *emtB*. In contrast, if another established Chinese hamster cell line was dizygous for both loci, the overall frequency of Emt^r mutants would be lower, but the chances of being able to isolate and identify mutants with alterations in the gene other than *emtB* would be greatly increased.

With this idea in mind, we isolated Emt^r mutants from both V-79 CHL lung cells and another established Chinese hamster fibroblast cell line (obtained from the American Type Culture Collection, cell line CCL 14.1) derived from peritoneal tissue (Wasmuth et al., 1980, 1981). The D_{10} for emetine for both the wild-type CHL and wild-type Chinese hamster peritoneal (CHP) cell lines is identical to the D_{10} for wild-type CHO cells, which is $(1-3) \times 10^{-8}$ *M*. Emt^r CHL and CHP mutants were isolated using a drug concentration of 3×10^{-7} *M*, which is in the range used to isolate Emt^r mutants from CHO cells. The mutagen-induced frequency of Emt^r mutants in both cell lines is $\sim 5 \times 10^{-7}$, 50–100-fold lower than the frequency observed for CHO cells, which indicated the *emtB* locus was probably not hemizygous in either the CHL or CHP cell lines. In addition, for every Emt^r CHL and CHP mutant examined, the phenotype of emetine resistance is recessive, as is the case for the CHO mutants (Wasmuth et al., 1980, 1981).

Preliminary biochemical analysis of CHL and CHP Emt^r mutants, which consisted of examining the effect of various concentrations of emetine on the rate of protein synthesis *in vivo*, demonstrated that these mutants were very similar to one another and to CHO Emt^r mutants, with respect to the degree of resistance to emetine (Wasmuth et al., 1980, 1981). For the different mutants, the ID_{50} for emetine (the concentration of emetine required to reduce the rate of protein synthesis *in vivo* to 50% the rate observed in the absence of the drug) ranges from 0.6 to 2.0 μM, while the ID_{50} for wild-type CHO, CHL, or CHP cells ranges from 0.09 to 0.12 μM (Wasmuth et al., 1980, 1981; Chang and Wasmuth, 1983a). It should be mentioned that similar analyses of hybrids between wild-type cells and any of the Emt^r mutants demonstrated that the ID_{50} for emetine in the various hybrids was very close to the ID_{50} for the wild-type parent, which again confirmed the recessive nature of the Emt^r phenotype of each mutant. The results of experiments of this type with Emt^r × Emt^s hybrids are the same, regardless of whether the Emt^s parent is derived from the CHO, CHL, or CHP cell line.

Although all of the Emt^r mutants we examined were phenotypically indistinguishable from one another, regardless of which cell line they were derived from, complementation tests performed with a large number of these Emt^r mutants provided interesting and somewhat surprising results. All 11 CHL Emt^r mutants we have analyzed fail to complement one another (hybrids are resistant to emetine), demonstrating that they all belong to a single complementation group (Wasmuth et al., 1980, 1981). Similarly, all seven CHP Emt^r mutants we examined failed to complement one another (Wasmuth et al., 1981). However, when complementation analyses were performed among mutants isolated from the different cell lines, strikingly different results were obtained. Several cloned, independent hybrids were isolated from the following three kinds of crosses: CHO Emt^r × CHL Emt^r; CHO Emt^r × CHP Emt^r; CHL Emt^r × CHP Emt^r. In each case, hybrids between at least two different, independent mutants from each of the three different cell lines were isolated and examined. The hybrids were selected using combinations of various genetic markers in the different cell lines which are unrelated to the Emt^r phenotype. As shown in Figure 14.3, in each type of hybrid, protein synthesis *in vivo* was found to be much more sensitive to emetine inhibition than in either of the two parental Emt^r mutants (Wasmuth et al., 1980, 1981). The ID_{50} for emetine in each hybrid is close to that for the wild-type Emt^s cell lines. This is in marked contrast to the results obtained with hybrids between any two CHO Emt^r mutants, any two CHL Emt^r mutants, and any two CHP Emt^r mutants, in which protein synthesis is as resistant to emetine as in the least resistant of the parents. These results clearly demonstrated, at least for those mutants examined thus far, that any CHO Emt^r mutant comple-

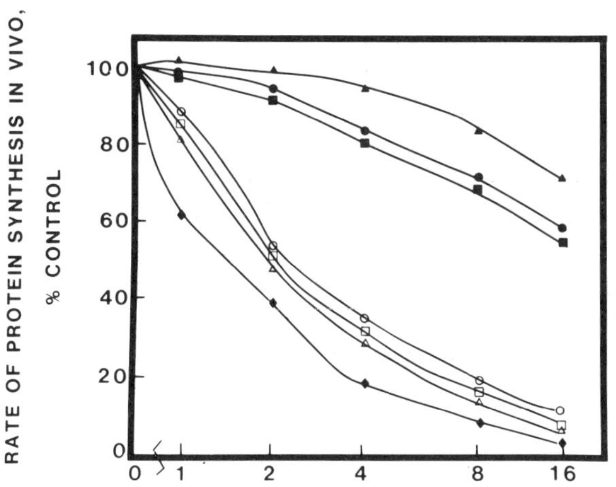

Figure 14.3. Inhibition by emetine of protein synthesis *in vivo* in emetine-resistant mutants and hybrids between mutants belonging to different complementation groups. The effect of the indicated concentration of emetine on the rate of incorporation of [^{3}H]amino acids into protein *in vivo* was determined as described by Wasmuth et al. (1980, 1981). The rate of protein synthesis in the absence of the drug is defined as 100% for each cell line: (♦) wild-type CHO, CHL, or CHP cell lines; (■) EmtA mutant, UCW 181; (▲) EmtB mutant, UCW 10; (●) EmtC mutant, UCW 282. Hybrids: (○) EmtA (UCW 181) × EmtB (UCW 10); (△) EmtA (UCW 181) × EmtC (UCW 282); (□) EmtB (UCW 10) × EmtC (UCW 282).

ments any CHL or CHP Emt^r mutant, any CHL Emt^r mutant complements any CHO or CHP Emt^r mutant, and any CHP Emt^r mutant complements any CHO or CHL Emt^r mutant. However, any two Emt^r mutants isolated from the same cell line fail to complement one another (Wasmuth et al., 1981). These results, therefore, defined three distinct complementation groups of Emt^r mutants, each showing a striking cell line specificity. We have designated the CHL-specific complementation group as EmtA, and the affected locus *emtA*; the CHP-specific complementation group at EmtC and the affected locus *emtC*; and as already mentioned, the CHO-specific complementation group as EmtB, and the affected locus, *emtB*. The term cell line specificity is not meant to imply that it is possible to isolate only one type of mutant from a given cell line but rather that each cell line gives rise to one particular type of mutant most often. Thus, while in this laboratory the cell line specificity of the different complementation groups has been absolute, Campbell and Worton (1979) reported the isolation of a V-79 CHL Emt^r mutant that failed to complement a CHO Emt^r mutant.

The striking cell line specificity of the different complementation groups of Emt^r mutants poses several interesting questions. As already discussed, the preponderance of *emtB* mutants can be explained by the finding that this locus is physically hemizygous in this cell line, which raised the possibility of analogous situations existing for the *emtA* locus in CHL cells and the *emtC* locus in CHP cells. When the various cell lines were established in culture, each apparently "evolved" a somewhat different karyotype that was compatible with rapid growth in culture. Although the karyotype of each is now relatively stable, they are all distinct from one another. It seems very likely, therefore, that different regions of the genome have by chance been rendered haploid in the different cell lines as a result of these karyotypic changes. However, two lines of evidence argue against the idea that the CHL and CHP cell lines are hemizygous for the *emtA* and *emtC* loci, respectively. As mentioned previously, the frequency with which Emt^r mutants arise in the CHL and CHP cell lines is much lower than the corresponding frequency in CHO cells, suggesting that more than a single event is required to generate an Emt^r mutant in the former two cell lines. These data could be misleading, however, if there are very stringent constraints on which sites within the *emtA* locus or the *emtC* locus can be altered via mutation to produce a cell line that is both viable and emetine resistant. Segregation experiments, which provide another means to approach the question of hemizygosity, are easier to interpret since one is asking for reexpression of a recessive mutation that already exists in a cell line. In these experiments, the frequency with which Emt^r segregants, for example, arise from heterozygous $EmtA^r \times Emt^s$ cell hybrids is dependent almost solely on how many wild-type *emtA* genes are donated to the hybrid by the Emt^s parent. We have recently completed experiments of this type using a set of nine different types of hybrids: $EmtA^r$ × wild-type CHO, CHL, or CHP; $EmtB^r$ × wild-type CHO, CHL, or CHP; $EmtC^r$ × wild type CHO, CHL, or CHP. As expected, the segregation frequency for the *emtB* locus in the $EmtB^r$ × wild-type CHO hybrid was high, 1.8×10^{-3}. Thus, only the single wild-type *emtB* gene donated by the wild-type CHO parent must be lost, or inactivated, in order for the $EmtB^r$ phenotype of the other parent to be expressed. In contrast, in each of the other eight types of hybrids, the frequency of Emt^r segregants was 2×10^{-5} or below, suggesting that in each case two independent segregation events were required to generate an Emt^r segregant. The simplest interpretation of these results is that the *emtA* and *emtC* loci are dizygous in the wild-type CHO, CHL, and CHP cell lines, making it seem very unlikely that hemizygosity is the basis for the preponderance of *emtA* mutants in CHL cells or the preponderance of *emtC* mutants in CHP cells. At present, there is no satisfactory explanation for the cell line specificity of the *emtA* and *emtC* mutants. One possibility, which is difficult to test experimentally, is that subtle differences in the genetic background of the two different cell lines is involved in the preferential expression of mutations at the different loci. For example, CHL but not CHP cells may have a mutant allele at some undefined locus which by itself produces no discernible phenotype but which is lethal in combination with most mutations at the *emtC* locus. This would obviously preclude the isolation of such mutants from the former cell line. Untestable hypotheses of this sort are certainly not very pleasing, but, unfortunately, are the best that can be put forward at present.

3. Biochemical Characterization of EmtA, EmtB, and EmtC Mutants

a. *In Vitro* Characterization of Mutants. Gupta and Siminovitch (1977a) demonstrated that protein synthesis in cell-free extracts prepared from CHO (*emtB*) mutants, was considerably (~ 10-fold) more resistant to inhibition by emetine than in extracts from wild-type cells. These authors also localized emetine resistance *in vitro* in *emtB* mutants to the 40S ribosomal subunit, a result confirmed by a direct analysis of ribosomal proteins from mutant cells, which will be described later. In addition, Wejksnora and Warner (1979) demonstrated that the 40S ribosomal subunit from EmtB mutants was unstable upon centrifugation through sucrose gradients containing 0.5 *M* KCl. The emetine-resistant phenotypes of both *emtA* and *emtC* mutants are also expressed *in vitro*, in cell-free protein synthesis assays, demonstrating that alterations in these loci also affect components of the protein synthetic machinery directly (Wasmuth et al., 1980, 1981) and not uptake of the drug. For technical reasons, it was not possible to localize the lesions in these two classes of mutants to specific subcellular components. However, an examination of the effect of emetine on protein synthesis in cell-free extracts derived from EmtA × Emt^s hybrids, EmtC × Emt^s hybrids, and hybrids between mutants belonging to the different complementation groups provided indirect evidence that the lesion in both EmtA and EmtC mutants affected the ribosome. In cell-free extracts from hybrids of these types, protein synthesis is considerably more resistant to inhibition by emetine than in extracts from wild-type Emt^s cell lines when mRNA (supplied to extracts in the form of polyuridylic acid) is present in two- to three-fold excess over the amount required to saturate the systems (Wasmuth et al., 1980, 1981). This is the expected result of such experiments if the mutation in the Emt^r parent(s) of the hybrids affects a component of the ribosome and the hybrid contains a mixture of emetine-resistant and emetine-sensitive ribosomes. Thus, in cell-free protein synthesizing extracts in which mRNA is in considerable excess the ratio of ribosomes to mRNA molecules is low, and many mRNA molecules will contain only a single or very few ribosomes. Under these conditions, in the presence of emetine concentrations that affect the normal but not the resistant ribosomes, most of the resistant ribosomes can translate mRNA unimpeded by emetine-blocked, sensitive ribosomes. However, when mRNA is rate limiting

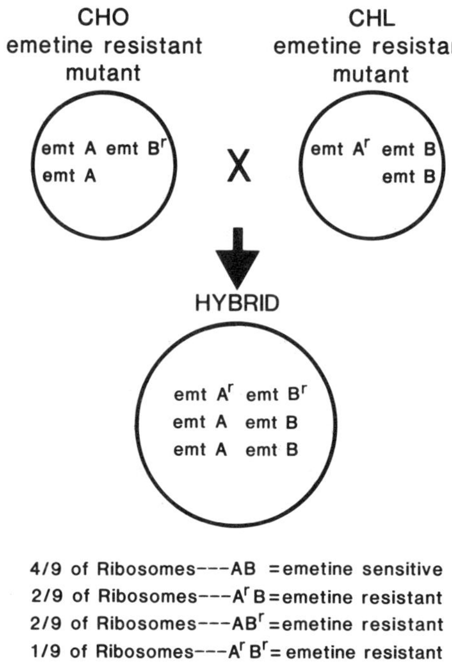

Figure 14.4. Model to account for the emetine-sensitive phenotype of EmtA^r × EmtB^r hybrids [from Wasmuth et al. (1980)].

and the ratio of ribosomes to mRNA molecules is high, as it is *in vivo*, most mRNA molecules will contain many ribosomes. As discussed earlier, under these conditions, the translation of mRNA by resistant ribosomes is prevented by the presence of emetine-blocked, sensitive ribosomes. Although certainly not conclusive, the response of cell-free extracts from hybrids to emetine is compatible with the hypothesis that each of these types of hybrids contains two populations of ribosomes; some resistant to emetine and some sensitive to the drug.

Additional lines of evidence, which will be discussed later, are consistent with the idea that EmtA mutants and EmtC mutants, like the EmtB mutants, have alterations that affect the ribosome directly. However, it must be kept in mind that the different classes of mutants complement one another, clearly demonstrating that a different gene product is altered in each of the three classes of mutants. Based on these facts and assumptions, one can formulate a straightforward and simple model to account for the emetine-sensitive phenotype of hybrids between two mutants which belong to different complementation groups. This model is diagrammed in Figure 14.4, using an EmtA^r × EmtB^r hybrid as an example. It is assumed that the CHO EmtB mutant is hemizygous for a resistant allele at the *emtB* locus but is dizygous and homozygous for the wild-type *emtA* allele. The CHL EmtA mutant is presumed to have become hemizygous for a resistant allele at the *emtA* locus but is dizygous and homozygous for the wild type *emtB* allele. Furthermore, it is assumed that each ribosome contains a single copy of the *emtA* gene product (protein A) and a single copy of the *emtB* gene product (protein B). A hybrid between these two cell lines would contain three *emtA* genes, one resistant allele and two sensitive alleles, and three *emtB* genes, one resistant allele and two sensitive alleles. Ribosomes containing either the mutant A protein, the mutant B protein, or both mutant proteins would be resistant to emetine. Ribosomes containing the wild-type A protein and the wild-type B protein would be sensitive to emetine. Since each ribosome must contain a single A protein and a single B protein, one would predict that four-ninths of the ribosomes in such a hybrid would have either a mutant A protein or a mutant B protein, one-ninth would have both of the mutant proteins, and four-ninths would have both of the wild-type proteins. The presence of the normal, emetine-sensitive ribosomes in such a hybrid would make it sensitive to emetine inhibition for the same reasons discussed above, for $Emt^r \times Emt^s$ hybrids. This same basic model would explain the emetine-sensitive phenotype of $EmtA^r \times EmtC^r$ and $EmtB^r \times EmtC^r$ hybrids, making the assumption that the *emtC* gene product is also a ribosomal protein which is present in one copy per ribosome. It is also important to note that if a cell-free extract from an $EmtA^r$ mutant and a cell-free extract from an $EmtB^r$ mutant are mixed *in vitro*, protein synthesis in the mixed extract is much more resistant to emetine inhibition than in an extract from an $EmtA^r \times EmtB^r$ hybrid (Wasmuth et al., 1980). Thus, each mutant appears to contain a single, emetine-resistant population of ribosomes which remain resistant to the drug when mixed *in vitro*. Only when the genomes of the two mutants are combined in a cell hybrid are emetine-sensitive ribosomes formed.

b. Cross Resistance of Emetine-Resistant Mutants to Protein Synthesis Inhibitors Structurally Related to Emetine. Gupta and Siminovitch (1977b) reported that CHO mutants (all of which are presumably EmtB mutants) are cross resistant to several other inhibitors of protein synthesis that have some structural similarities to emetine, and all of which appear to also affect the 40S ribosomal subunit. We examined the effect of one of the compounds tested by Gupta and Siminovitch, cryptopleurine, on several independent EmtA and EmtC mutants to determine if these classes of Emt^r mutants like the EmtB mutants were cross resistant to this drug. The three EmtB mutants and two EmtC mutants we examined were considerably (fivc- to seven-fold) more resistant to cryptopleurine than wild-type CHO or CHP cells (Chang and Wasmuth, 1983a). In contrast, all three independent EmtA mutants examined were as sensitive as wild-type cell lines to this compound, even though their resistance to emetine was comparable to that of the EmtB and EmtC mutants tested (Chang and Wasmuth, 1983a). Thus, lack of cross resistance to cryptopleurine appears to be relatively specific to the EmtA complementation group, the significance of which will be discussed later.

Although a mutant *emtA* locus by itself does not result in cross resistance to cryptopleurine, mutant alleles at this locus have a synergistic effect on the phenotypic expression of cryptopleurine resistance that results from mutations at the *emtB* or *emtC* loci. These observations will be discussed in detail in Section III B 7, which describes the construction and characterization of EmtA, EmtB double mutants and EmtA, EmtC double mutants.

4. Genetic Mapping of the emtA, emtB, and emtC Loci

The relationships among the *emtA*, *emtB*, and *emtC* loci has been determined by experiments examining linkages between various of these loci. Such analyses in this laboratory have been greatly facilitated by the findings, alluded to earlier, that the *emtB* locus is linked to two other genetic markers, *leuS* and *chr*, in the region on the long arm of chromosome 2 that is hemizygous in CHO cells. If either the *emtA* locus or the *emtC* locus were located in the region of chromosome 2 that is haploid in CHO cells, these genes should segregate concordantly with the *leuS* and *chr* genes as does the *emtB* gene. To examine these questions, two types of hybrids were utilized: (1) a hybrid between wild-type CHO cells and a CHP cell line with mutations in both the *emtC* and *chr* loci, and (2) a hybrid between wild-type CHO cells and a CHL cell line with mutations in the *emtA* and *leuS* genes (Chang and Wasmuth, 1983a). In the former type of hybrid, the wild-type *emtC* and *chr* loci from the CHO parent segregated independently (10 of 10 independent Chr^r segregants remained emetine sensitive and seven of seven independent $EmtC^r$ segregants remained chromate sensitive). In the latter type of hybrid, the wild-type *emtA* and *leuS* genes of the CHO parent segregated independently (eight of eight independent $EmtA^r$ segregants remained temperature resistant). These results demonstrated that neither the *emtA* nor the *emtC* locus is linked to hemizygous region of chromosome 2 in CHO cells (Chang and Wasmuth, 1983a). Thus, neither the *emtA* locus nor the *emtC* locus is closely linked to the *emtB* locus, clearly distinguishing the former two genetic loci from the latter one. Other, more indirect lines of evidence indicate that the three different loci are in fact located on three different chromosomes. Nielson-Smith et al. (1983) reached similar conclusions concerning the nonlinkage of *emtA* and *emtC* to *emtB* using a different experimental approach.

The CHO triple mutant, UCW 56, was also useful in determining the chromosomal location of the *emtB* locus in humans. Intraspecific cell hybrids between UCW 56 and normal human leukocytes were selected for retention and expression of the human *leuS* gene, which renders these hybrids temperature resistant (Dana and Wasmuth, 1982a). Ten out of ten hybrids also expressed the human *emtB* and *chr* genes, rendering them sensitive to emetine and chromate, respectively (Dana and Wasmuth, 1982a). In addition, the three human genes segregate concordantly, demonstrating that the linkage relationship among these genes has been retained during the evolution of the human karyotype. All three genes are located on the long arm of human chromosome 5, with the *emtB* locus being localized to the region between bands q23 and q34 (Dana and Wasmuth, 1982b). As discussed subsequently, the synthesis of the ribosomal protein gene product of the human *emtB* locus and its incorporation into functional ribosomes has been confirmed directly in one such hybrid which contains only human chromosome 5. Thus, this human ribosomal protein functions normally when incorporated into ribosomes in which most, if not all, the rest of the components are products of the CHO cell genome.

5. Characterization of High-Level, Two-Step-Selected Emetine-Resistant Mutants

Several laboratories have tried unsuccessfully to isolate, in single-step selections, Chinese hamster cell mutants resistant to emetine concentrations of 1 μM or higher, which is ~ 100-fold above the ID_{10} for wild-type cells. All of the Emt^r mutants described thus far were isolated at emetine concentrations ranging from 0.1 to 0.4 μM. Even in CHO cells, in which EmtB mutants resistant to these lower levels of emetine arise at a frequency of $\sim 1 \times 10^{-5}$, not a single high-level emetine-resistant mutant has ever been isolated from wild-type cells (frequency of $< 2 \times 10^{-9}$) (Gupta and Siminovitch, 1978a). We have made the same observation in this laboratory for both the CHL and CHP cell lines. However, Gupta and Siminovitch (1978a) demonstrated that CHO mutants resistant to 1 μM emetine could readily be isolated starting with low-level or first-step EmtB mutants. Thus, resistance to a high level of emetine appears to occur in two discrete steps. The low-level CHO Emt^r mutants will be referred to as $EmtB^{rI}$ mutants as distinguished from the two-step-selected, or Emt^{rII}, mutants. In every Emt^{rII} CHO mutant examined, resistance to the higher concentration of emetine is also expressed *in vitro*, demonstrating that the second as well as the first mutation in these cell lines affects the protein synthetic machinery (Gupta and Siminovitch, 1978*a*). In addition, the Emt^{rII} phenotype is recessive in either $Emt^{rII} \times Emt^s$ or $Emt^{rII} \times EmtB^{rI}$ hybrids (Gupta and Simonovitch, 1978*a*). The former type of hybrid is as sensitive to emetine as the wild-type parent, while the second type of hybrid is resistant to only the lower level of emetine characteristic of the EmtB^{rI} parent. Considering the complex structure of the ribosome and the interactions among its components, Gupta and Siminovitch reasoned that the second alteration in Emt^{rII} mutants could be either a second mutation in the *emtB* locus or a mutation in a second locus distinct from *emtB* which might have affected a second ribosomal component. This question is especially significant in view of the fact that at least two loci, *emtA* and *emtC*, besides *emtB* can be altered to give rise to low-level emetine-resistant mutants. Some evidence for the former possibility was obtained for three Emt^{rII} mutants examined by Gupta and Siminovitch (1978a) since resistance to low levels and high levels of emetine segregated concordantly 100% of the time from hybrids between each of these cell lines and wild-type cells. While the segregation experiments for these three mutants provided firm data that the two mutations in each were on the same chromosome, they provided no evidence as to whether or not the mutations were in the same gene. In a fourth Emt^{rII} mutant studied by Gupta and Siminovitch (1978a) evidence was obtained that the two mutations were in fact in different genes that are linked to the same chromosome. As discussed in Section III B 6, a direct examination of ribosomal proteins extracted from four independent Emt^{rII} mutants, using two-dimensional polyacrylamide gel electrophoresis, has shown that each of these cell lines has two mutations in the *emt^B* locus. Attempts in this laboratory to isolate Emt^{rII} mutants from three different first-step CHL EmtA mutants have been unsuccessful. However, analogous experiments with two different first-step EmtC mutants yielded many Emt^{rII} mutants. In one of five such mutants examined thus far, the first and second mutations segregate independently, indicating the second mutation affects a gene other than *emtC*. None of these mutants have yet been characterized further. A more direct and controlled approach to constructing cell lines with single mutations in two different genes that affect resistance to emetine is described in Section III B 7.

6. Analysis of Ribosomal Proteins Extracted from Mutants Using Two-Dimensional Polycrylamide Gel Electrophoresis

The genetic and biochemical analyses of Emt^r mutants described in preceding sections are consistent with the idea that all three complementation groups have alterations that affect the ribosome; the evidence being strongest for the EmtB mutants. This hypothesis has been confirmed for this class of mutants through a direct examination of ribosomal proteins extracted from EmtB mutants and wild-type cells using two-dimensional polyacrylamide gel electrophoresis. Boersma et al. (1979a) and, shortly thereafter, Reichenbecher and Caskey (1979) found a single ribosomal protein to be electrophoretically altered in different CHO Emt^r mutants. The altered protein in at least four independent EmtB^r mutants has since been identified as a component of the 40S subunit (Madjar et al., 1983; Chang and Wasmuth, 1983a) designated ribosomal protein S14 according to the standard nomenclature for mammalian ribosomal proteins proposed by McConkey et al. (1979). In addition, similar analysis of ribosomal proteins extracted from several second-step Emt^{rII} CHO mutants has provided firm evidence that the Emt^{rII} phenotype of these cell lines is the result of two discrete alterations in ribosomal protein S14 (Madjar et al., 1983). These points are illustrated in Figure 14.5, which shows two-dimensional gel electropherograms of ribosomal proteins extracted from wild-type CHO cells, an EmtB mutant, UCW 56, and UCW 600, an Emt^{rII} mutant which was isolated from UCW 56 in this laboratory. In this gel system, number II of Madjar et al. (1979), the S14 from wild-type CHO, CHL, or CHP cells, migrates to a position such that a reference line drawn between the spots corresponding to proteins S15 and L25 bisects the S14 spot. Ribosomal protein S14 extracted from ribosomes from UCW 56 is shifted to the left relative to the normal protein, such that it lies in the middle of the "V" generated by the reference lines between proteins L25 and S14, and proteins L25 and L11. This shift indicates the altered protein is less basic than the normal one. Ribosomal protein S14 extracted from UCW 600 is shifted even further to the left than the protein from UCW 56, such that the spot touches the reference line between proteins L25 and L11. The S14 form UCW 600 therefore appears to be even less basic than the corresponding protein from the first-step EmtB mutant, UCW 56. Thus, a mutation in the *emtB* locus in UCW 56 rendering it resistant to a low level of emetine is associated with a distinct electrophoretic alteration in protein S14 and the second mutation in UCW 600, rendering it resistant to a high level of emetine, is associated with a second electrophoretic alteration in the same protein. Analysis of other first-step EmtB mutants by Madjar et al. (1983) produced very surprising results in that all four independent mutants examined synthesize S14's with electrophoretic alterations apparently identical to that shown in Figure 14.5 for UCW 56. Furthermore, all four independent CHO Emt^{rII} examined by Madjar et al. (1983) produce S14's that are electrophoretically indistinguishable from the S14 shown in Figure 14.5 from UCW 600. While it is not surprising to find electrophoretic alterations in the S14 from some mutants, the striking consistency of the alterations among mutants of the two types seems remarkable. At first glance, these findings might suggest that the alterations in these cell lines affect not the structural gene encoding protein S14 but rather genes involved in posttranslational modification of the protein. However, an examination of ribosomal proteins extracted from EmtB × Emt^s or EmtB^{rII} × Emt^s hybrids dem-

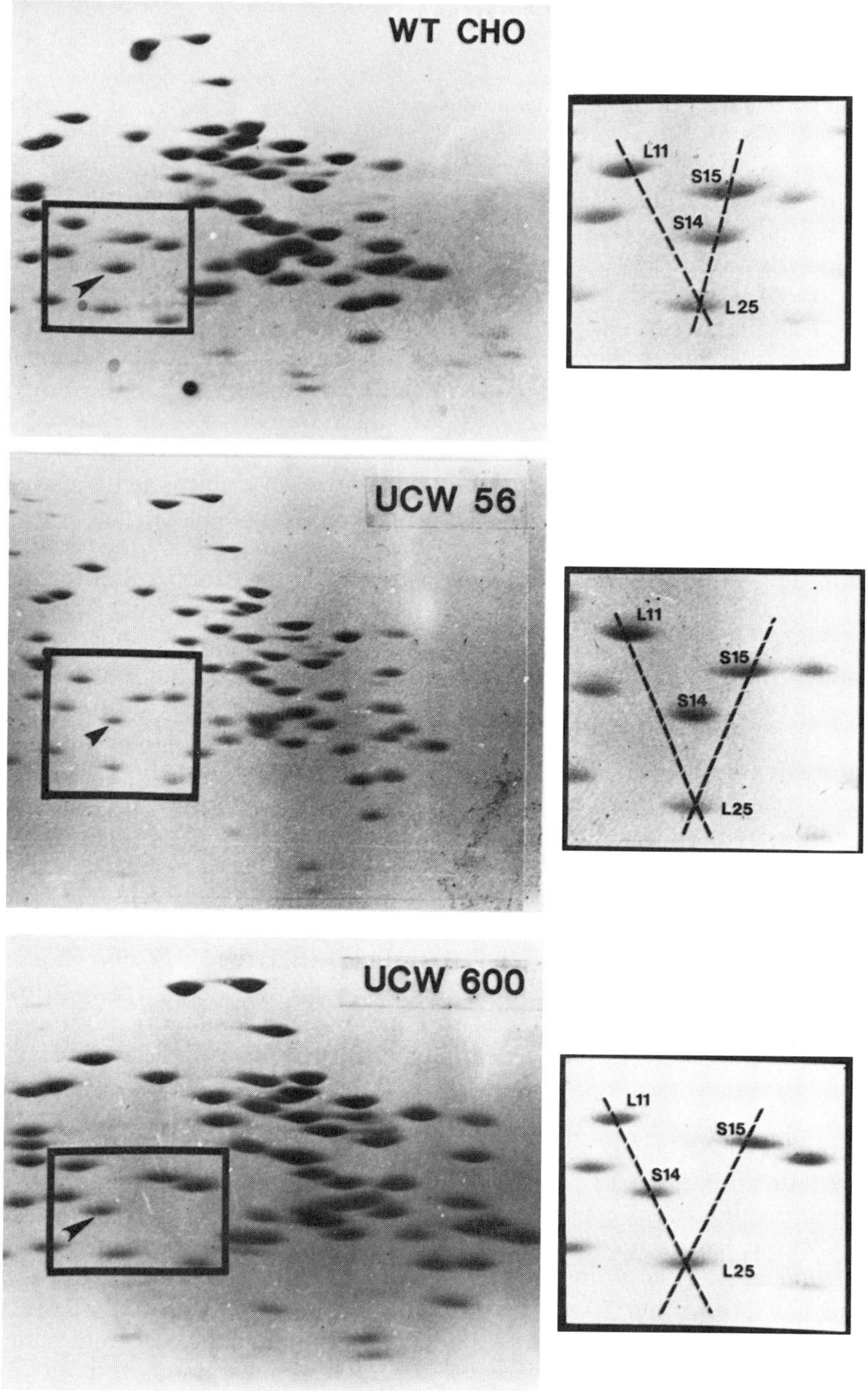

Figure 14.5. Two-dimensional gel electropherograms of ribosomal proteins extracted from wild-type CHO, an EmtB^{rI} mutant (UCW 56), and an EmtB^{rII} mutant (UCW 600). Extraction and electrophoresis of ribosomal proteins were performed as described by Chang and Wasmuth (1983*a*), using the two-dimensional gel system II of Madjar et al. (1979). The arrowhead on each gel indicates the position of the protein spot corresponding to S14. To the right of each electropherogram, the region of the gel enclosed within the box (which contains the spot corresponding to ribosomal protein S14) is enlarged to show more clearly the difference in the migration of the normal and the two mutant forms of S14.

onstrates that this is not the case. As shown in Figure 14.6, both the normal and altered forms of ribosomal protein S14 can be visualized in two-dimensional gel electropherograms of ribosomal proteins extracted from the cell line UCW 639, which is a hybrid between wild-type CHO and the Emt^{rII} mutant, UCW 600. Keeping in mind that the Emt^{rII} phenotype is recessive, if the lesions in UCW 600 had altered the posttranslational modification of S14, the normal gene functions supplied to the hybrid by the wild-type CHO parent of the UCW 639 hybrid would have carried out the proper posttranslational modification of all the ribosomal protein S14 synthesized in the hybrid. Therefore, only the normal electrophoretic form of this protein would have been present in these cells. This is clearly not the case. In addition, in an Emt^{rII} segregant, which was selected from UCW 639 as being resistant to a low level of emetine and subsequently confirmed to be resistant to the high level of emetine characteristic of UCW 600, only the doubly altered form of protein S14 is present. Thus, loss of the wild-type *emtB* locus from this hybrid is accompanied by loss of the wild-type form of protein S14. These experiments were performed on a hybrid involving an Emt^{rII} mutant only because the doubly altered S14 separates more clearly from the normal S14 than the singly altered form of S14 produced by EmtB^{rI} mutants. Boersma et al. (1979b) obtained analogous results in experiments with an $\mathrm{EmtB}^{rI} \times \mathrm{Emt}^{s}$ hybrid. Taken together, these results provide firm evidence that the CHO Emt^{rII} mutants examined in these studies have two mutations in the *emtB* locus and that this locus is the structural gene for ribosomal protein S14. Madjar et al. (1982) have also demonstrated that mRNA from an EmtB mutant, when translated *in vitro*, produces the same altered form of S14 synthesized *in vivo*.

The reasons why mutations in the *emtB* locus in so many different mutants produce S14s with apparently identical electrophoretic shifts remain unclear. However, as discussed by Madjar et al. (1983), there may exist a combination of stringent constraints on the structure of protein S14 such that only a very few specific alterations result in both cell viability and resistance to emetine.

We have recently analyzed ribosomal proteins from an interspecific hybrid between the EmtB^{rII} mutant UCW 600 and normal human leukocytes. This hybrid has retained only the human chromosome, number 5, which contains the human *emtB* gene and confers emetine sensitivity to this cell line. Both the mutant CHO form and normal human form of ribosomal protein S14 (which migrates coincidently with the normal Chinese hamster form of this protein) are clearly visible on gel electropherograms of ribosomal protein extracted from this hybrid. In addition, loss of the normal human form of S14 occurs concomitantly with the loss, or segregation, of human chromosome 5 from the hybrid. These results confirm that the human *emtB* locus, like its counterpart in Chinese hamster, is the structural gene for ribosomal protein S14.

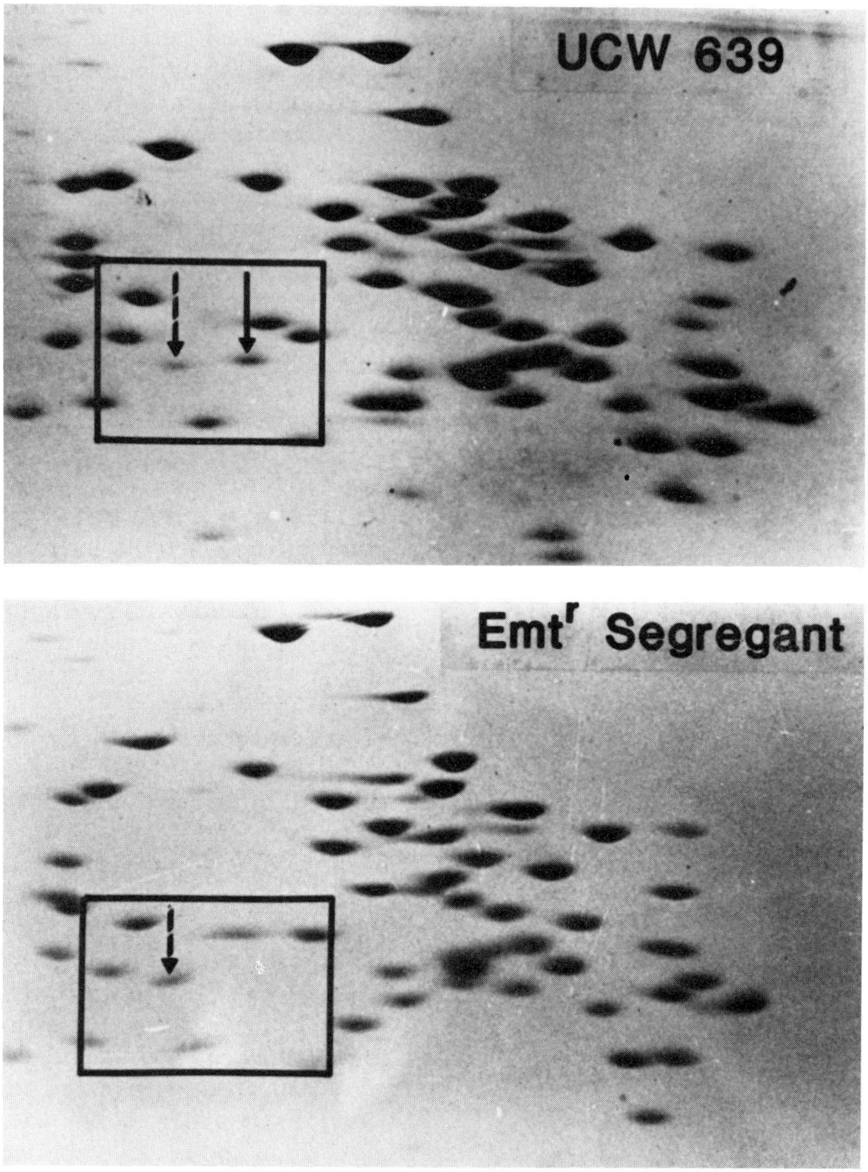

Figure 14.6. Two-dimensional gel electropherograms of ribosomal proteins extracted from an $EmtB^{rII}$ × Emt^s hybrid (UCW 639) and an $EmtB^{rII}$ segregant. The doubly altered S14 protein is indicated by a dashed arrow, while the wild-type S14 protein is indicated by a solid arrow.

Although mutations in the *emtB* locus are associated with alterations in a specific ribosomal protein, no such clear correlation can be made for *emtA* or *emtC* mutants (Chang and Wasmuth, 1983a). That definitive alterations in ribosomal proteins have not yet been observed in *emtA* and *emtC* mutants does not argue against the idea that these genes encode ribosomal proteins. If the mutations in these cell lines alter ribosomal proteins, they, like mutations in the *emtB* locus, must leave the ribosome functional as well as resistant to emetine. Thus, there could also be stringent constraints on the types of amino acid substitutions that can occur in the putative protein products of the *emtA* and *emtC* loci. If these alterations are such that they result in no charge changes in the proteins, they would not be easily detectable in gel electropherograms.

7. Construction and Characterization of EmtA, EmtB and EmtA, EmtC Double Mutants

In view of the negative results of gel analyses of ribosomal protein from *emtA* and *emtC* mutants, this laboratory took another approach in attempts to answer the question of whether or not the mutations in these cell lines affect ribosomal proteins. These experiments involve constructing and characterizing a series of Chinese hamster cell lines which express recessive mutations in two different loci, either *emtA* and *emtB*, *emtB* and *emtC*, or *emtA* and *emtC* (Chang and Wasmuth, 1983b). This type of genetic approach in a lower eukaryote has shown that combinations of various mutations affecting ribosomal components result in cold sensitivity, high levels of resistance to ribosomal inhibitors, and, in some cases, double mutant lethality (Crouzet and Beguerot, 1978; Picard-Bennoun, 1981). We were hopeful that the biochemical characterization of analogous Chinese hamster cell double mutants would provide information that might enable us to determine whether the altered products of mutant *emtA* or *emtC* genes, in combination with an altered ribosomal protein S14 produced by a mutant *emtB* allele, affected the function or integrity of the ribosome. Such a result would provide strong evidence that the products of the *emtA* and *emtC* genes interact directly with a known component of the ribosome, protein S14 (the product of the *emtB* gene), indicating that the former genes encode ribosomal components as well.

The way in which these double mutants are constructed enables one to predetermine exactly which mutant alleles at the two different loci will be combined and expressed in the same cell line. For example, five different EmtA mutants and five different EmtB mutants can each be extensively characterized individually with respect to different phenotypes such as degrees of resistance to emetine or related compounds and other growth properties. Cell lines can then be constructed which ex-

TABLE 14.2
ID_{50}'s for Emetine and Cryptopleurine in EmtA Mutant, EmtB Mutant, and EmtA, EmtB Double Mutant Cell Lines

Strain Number	Description	ID_{50} Emetine (μM)	ID_{50} Cryptopleurine (pM)
UCW 1 or UCW 100	Wild-type CHO or CHL	0.1	1.2
UCW 266	EmtA^r	1.8	1.6
UCW 932	EmtB^r	2.0	7.0
UCW 1075	UCW 266 × UCW 932 hybrid	0.2	1.4
UCW 1137	EmtA^r, EmtB^r double segregant from UCW 1075	15.5	12.5

press any one of the mutant *emtA* alleles together with any one of the mutant *emtB* alleles. The phenotypes and biochemical properties of the various double mutants can be compared to those of the appropriate EmtA mutant and the appropriate EmtB mutant. Similar constructions can be done for cell lines expressing various combinations of mutant *emtA* and *emtC* or *emtB* and *emtC* alleles.

The construction of double mutants involves a two-step segregation protocol, starting with hybrids between two selected mutants, each belonging to a different complementation group. The starting hybrids are sensitive to emetine and heterozygous at the two loci in question. The procedures used to derive EmtA, EmtB double mutants from one such hybrid have been described in detail (Chang and Wasmuth, 1983b) and will be discussed here only briefly. Unique selective pressures can be applied to an EmtA × EmtB hybrid, providing it has been constructed from cell lines with other, appropriate and selectable genetic markers, which enables one, in the first selective or segregation step, to isolate a segregant that reexpresses the mutant *emtA* allele specifically and in the second selective step (which does not require the use of emetine as a selective agent) to isolate a segregant that also reexpresses the mutant *emtB* allele (Chang and Wasmuth, 1983b). As shown in Table 14.2, in the one type of EmtA, EmtB double mutant characterized thus far, the ID_{50} for emetine is 1.5×10^{-5} M, which is ~ eight-fold higher than the ID_{50} of either parental mutant and 100-fold above the ID_{50} for wild-type cells. These double mutants, therefore, are resistant to approximately the same concentration of emetine as most of the second-step CHO EmtB^{rII} mutants described earlier. Thus resistance to high levels of emetine can result from two mutations in one gene or by single alterations in two different genes.

The EmtB parent of the double mutant is sixfold more resistant to cryptopleurine than wild-type cells, whereas the EmtA parent, like all EmtA mutants we have examined, is not cross resistant to this drug. As shown in Table 14.2, however, the double mutant is twice as resistant to cryptopleurine as the EmtB parent. This demonstrates that expression of the mutant *emtA* allele, which does not by itself result in increased resistance to cryptopleurine, enhances the resistance to cryptopleurine that results from the expression of the mutant *emtB* gene. This synergistic effect suggests that the gene products of the two loci interact directly with one another. Since the product of the *emtB* locus is ribosomal protein S14, it seems very probable that the product of the *emtA* locus is also a component of the ribosome, most likely a protein. It will be of interest to determine whether other EmtA, EmtB double mutants, with different mutant alleles at both loci, behave differently than the one characterized thus far. EmtB, EmtC double mutants have also been constructed recently using the same rationale and protocols used to construct the EmtA, EmtB double mutants but have not yet been characterized.

The construction of one set of EmtA, EmtC double mutants has also been completed recently, and initial characterization of these cell lines has provided very striking results. The construction of these double mutants also utilizes a two-step segregation protocol but one which is somewhat different from that used to construct EmtA, EmtB or EmtB, EmtC double mutants. In the first step, Emt^r segregants are selected from a hybrid between (a) an EmtA mutant, which is not cross resistant to cryptopleurine and (b) an EmtC mutant which is cross resistant to cryptopleurine. Segregants that are emetine resistant because they reexpress the mutant *emtC* allele are also resistant to cryptopleurine, while segregants that reexpress the mutant *emtA* allele remain sensitive to cryptopleurine, enabling one to easily determine which mutant allele is expressed in the first-step segregants. From first step, $EmtA^r$ segregants (which are sensitive to cryptopleurine) one can the select segregants that reexpress the mutant *emtC* allele by virtue of their becoming resistant to cryptopleurine. One EmtA, EmtC double mutant constructed in this manner has been partially characterized, with respect to its degree of resistance to emetine and cryptopleurine. These results are summarized in Table 14.3. In the double mutant, the ID_{50} for emetine is 3.5×10^{-4} *M*, 100–300 times higher than the ID_{50} for either of the single mutant parents and approximately 3500 times higher than the ID_{50} for wild-type cells. The ID_{50} for cryptopleurine in the double mutant is at least 80 times higher than the ID_{50} of the $EmtC^r$ parent and at least 300 times higher than in either the EmtA mutant or wild-type cells. This result is most intriguing in view of the fact that the mutation in the *emtA* locus, by itself, does not render the $EmtA^r$ cell line any more resistant to cryptopleurine than wild-type cells. Obviously, the two altered gene products (of the *emtA* and *emtC* loci) have a profound

TABLE 14.3
ID_{50}'s for Emetine and Cryptopleurine in EmtA Mutant, EmtC Mutant, and EmtA, EmtC Double Mutant Cell Lines

		ID_{50}	
Strain Number	Description	Emetine (μM)	Cryptopleurine (pM)
UCW 100 or UCW 104	Wild-type CHL or CHP	0.1	1.2
UCW 266	$EmtA^r$	1.8	1.6
UCW 282	$EmtC^r$	1.8	5.0
UCW 1093	UCW 266 × UCW 282 hybrid	0.2	1.5
UCW 1161	$EmtA^r$, $EmtC^r$ double segregant from UCW 1093	350	>400

synergistic effect on one another which again suggest they interact directly with one another in some fashion. In addition, we were surprised to find that the EmtA, EmtC double mutant has a pronounced cold-sensitive phenotype, a characteristic not observed for either of the single mutants. Thus, the plating efficiency of the EmtA, EmtC double mutant is reduced by three orders of magnitude at 33°C, relative to 37°C. Cold sensitivity is a phenotype long associated in *E. coli* with mutations that result in defects in the assembly of ribosomes (Jaskunas et al., 1974). Whether defects in the assembly or the function of ribosomes in the double EmtA, EmtC mutant might be involved in producing the cold-sensitive phenotype remains to be determined. However, this result together with the fact that this double mutant is resistant to extraordinarily high concentrations of two different compounds that affect the 40S ribosomal subunit, suggest very strongly that the *emtA* and *emtC* genes encode proteins that interact with one another as integral components of the 40S ribosomal subunit. As in the case of EmtA, EmtB double mutants, it will be of interest to determine if other EmtA, EmtC double mutants, which express various other mutant alleles at both loci, have phenotypes similar to or distinct from the double mutants described here.

8. Summary and Future Studies

Given the information presented in this section on emetine-resistant mutants, much of which admittedly provides suggestive rather than conclusive evidence, one can propose several hypotheses concerning the interrelationships among the gene products of the *emtA*, *emtB*, and *emtC* loci. If emetine binds not to a specific protein on the ribosome but rather to a site that is defined by several structural components, one could imagine protein products of the *emtA*, *emtB*, and *emtC* genes are all located at or near this site. An alteration in any one of these proteins could alter the structure of the emetine binding site slightly, rendering ribosomes partially resistant to the drug. The differences among the three complementation groups of Emt^r mutants with respect to cross resistance to cryptopleurine is also of interest in this regard. If some of the structural determinants that define the emetine and cryptopleurine binding sites on the ribosome are common, but each binding site also has some unique structural determinants, it is not surprising that different classes of mutants are affected by cryptopleurine to different degrees. Alterations in either the *emB* gene product or the *emtC* gene product might alter structural determinants common to both the emetine and the cryptopleurine binding sites, rendering cells resistant to both drugs. Alterations in the *emtA* gene product could, however, alter a determinant that was unique to the emetine binding site, leaving the cryptopleurine binding site relatively unaltered. When two of these proteins are altered in the same cell, which may well be the case in the double mutants just described, the architecture of the ribosome would be changed even further, which could have various effects on the function and structure of the ribosome, depending upon which proteins were altered and the nature of their alterations. The various, rather striking phenotypes exhibited by EmtA, EmtC double mutants in particular suggests that there are dramatic alterations in the structure and function of ribosomes having both an altered *emtA* gene product and an altered *emtC* gene product. Thus, although the cryptopleurine binding site is unaltered by the presence in ribosomes of only an altered *emtA* gene product, the more profound changes in the ribosomal architecture that result from the incorporation of both the mutant *emtA* and *emtC* gene products drastically alters the site to which cryptopleurine would normally bind.

It should be noted that two other types of Chinese hamster cell ribosomal mutants have been isolated in which the alterations affect the 60S ribosomal subunit. Haralson and Roufa (1975) described a temperature-sensitive CHL mutant in which the 60S ribosomal subunit was unstable at high temperatures. Gupta and Siminovitch (1978b) found that in CHO mutants selected as resistant to the protein synthesis inhibitor, trichodermin, the lesion could be localized to the 60S subunit of the ribosome. There are several reasons why it is important to continue attempts to expand the number of mutants available with alterations in different ribosomal proteins and to determine with certainty whether the *emtA* and *emtC* loci encode ribosomal proteins.

1. In order to even begin to understand how the expression of the more than 70 different genes encoding different ribosomal proteins is coordinately regulated in mammalian cells, it is essential that we learn

as much as possible about the organization of this group of genes within the genome. The ability to genetically "tag" various of these genes, through the selection of appropriate mutants, provides the best and most straightforward approach to this problem.

2. The ability to identify mutants with recessive phenotypes as having alterations in genes encoding ribosomal proteins has an important bearing on the possibility, suggested by the results of D'Eustachio et al. (1981) and Monk et al. (1981), that for each ribosomal protein there exists multiple (an average of 10), unlinked genes that encode them (at least for those they examined). The question of multiple versus single loci encoding each individual ribosomal protein is of obvious importance in considering and examining the possible control mechanisms that might be involved in regulating the coordinate synthesis of this group of functionally related proteins. If there are multiple functional and unlinked loci encoding a given ribosomal protein, it should virtually be impossible to isolate single-step mutants with alterations in their proteins in which the phenotype is recessive, unless one assumes in some cells only one or two of these genes is active. For the *emtB* locus, there is very strong evidence that it is a structural gene for ribosomal protein S14, only a single functional copy exists in CHO cells and most likely only two functional copies exist in CHL and CHP cells. In all three cell lines, this locus is located on chromosome 2. Therefore, without making unwarranted assumptions, the evidence for this ribosomal protein suggests it is encoded by a single genetic locus in Chinese hamsters, which is present in a single copy/haploid genome. However, determining if the *emtB* locus represents an exception or the rule will depend on whether or not the same strong evidence can be obtained for other ribosomal proteins and their genes, which points out the importance of obtaining further and unequivocal data concerning the *emtA* and *emtC* loci.

3. The isolation, from mutants, of cloned and expressible cDNAs, or genes, encoding electrophoretically altered ribosomal proteins that confer drug resistance to the ribosome would enable one to perform detailed molecular studies into the mechanisms that regulate the synthesis of ribosomal proteins and their incorporation into the ribosome in mammalian cells. In this regard, the cloning of the *EmtB* gene has recently been reported (Nakamichi et al., 1983). Thus, the identification and characterization of Chinese hamster cell lines with mutations in genes encoding ribosomal components represent not end points but rather a starting point that it is hoped will help to define the complex genetic and biochemical mechanisms that surely operate in mammalian cells to regulate the expression of these genes and the assembly of their individual gene products to form the ribosome.

ACKNOWLEDGMENTS

I am extremely grateful to the following colleagues and co-workers who contributed, both experimentally and intellectually, to the work performed in my laboratory: Linda Vock Hall, Janet Kolb, Douglas Skarecky, Dr. Sharon Dana, Dr. Ronald Cirullo, Dr. Stephen Chang, Dr. Leon Carlock, and Dr. Lee-Yun Chu. The excellent assistance of Darlene Wise in the preparation of the manuscript is also gratefully acknowledged. Various aspects of the work performed in the author's laboratory were supported by a Public Health Service grant from the National Institute of General Medical Sciences, a grant from the Genetic Biology Program of the National Science Foundation, a Basil O'Conner Starter Grant from the March of Dimes Birth Defects Foundation, and a Junior Faculty Research Award from the American Cancer Society.

REFERENCES

Adair, G. M., Thompson, L. H., and Lindl, P. A. (1978). *Somat. Cell Genet.* **4**, 27–44.

Adair, G. M., Thompson, L. H., and Font, S. (1979). *Somat. Cell Genet.* **5** 329–344.

Andrulis, I. L., Chiang, C. S., Arfin, S. M., Miner, T. A., and Hatfield, G. W. (1978). *J. Biol. Chem.* **253**, 58–62.

Ashman, C. R. (1978). *Somat. Cell Genet.* **4**, 294–312.

Boersma, D., McGill, S., Mollenkamp, J., and Roufa, D. J. (1979a). *J. Biol. Chem.* **254**, 559–567.

Boersma, D., McGill, S. M., Mollenkamp, J. W., and Roufa, J. (1979b). *Proc. Natl. Acad. Sci. USA* **76**, 415–419.

Campbell, C. E. and Worton, R. G. (1979). *Somat. Cell Genet.* **5**, 51–65.

Campbell, C. E. and Worton, R. G. (1980). *Somat. Cell Genet.* **6**, 215–224.

Campbell, C. E., Gravel, R. G., and Worton, R. G. (1981). *Somat. Cell Genet.* **7**, 535–546.

Chang, S. and Wasmuth, J. J. (1983a). *Molec. Cell. Biol.*. **3**, 429–438.

Chang, S. and Wasmuth, J. J. (1983b). *Molec. Cell Biol.* **3**, 761–772.

Chang, S. and Wasmuth, J. J. (1984). *Somat. Cell Molec. Genet.* **10**, 161–170.

Cirullo, R. E., Arrendondo-Vega, F. X., Smith, M., and Wasmuth, J. J. (1983a). *Somat. Cell Genet.* **9**, 215–233.

Cirullo, R. E., Dana, S., and Wasmuth, J. J. (1983b). *Molec. Cell. Biol.* **3**, 892–902.

Cirullo, R. E. and Wasmuth, J. J. (1984a). *Somat. Cell Molec. Genet.* **10**, 225–234.

Cirullo, R. E. and Wasmuth, J. J. (1984b). *Molec. Cell. Biol.* **4**, 1939–1941.

Crouzet, M. and Beguerot, J. (1978). *Mol. Gen. Genet.* **165**, 283–288.

Dana, S. and Wasmuth, J. J. (1982a). *Somat. Cell Genet.* **8**, 245–264.

Dana, S. and Wasmuth, J. J. (1982b). *Mol. Cell. Biol.* **2**, 1220–1228.

Denney, R. M. and Craig, I. W. (1976). *Biochem. Gent.* **14**, 99–117.

D'Eustachio, P., Meyuhas, O., Ruddle, F., and Perry, R. P. (1981). *Cell* **24**, 307–312.

Draper, R. K., Chin, D., Eurey-Owens, D., Scheffler, I., and Simon, M. I. (1979). *J. Cell Biol.* **83**, 116–125.

Farber, R. A. and Deutscher, M. (1976). *Somat. Cell Genet.* **2**, 509–520.

Giles, R. E., Shimizu, N., and Ruddle, F. H. (1980). *Somat. Cell Genet.* **6**, 667–686.

Gupta, R. S. and Siminovitch, L. (1976). *Cell* **9**, 213–219.

Gupta, R. S. and Siminovitch, L. (1977*a*). *Cell* **10**, 61–66.

Gupta, R. S. and Siminovitch, L. (1977*b*). *Biochem.* **16**, 3209–3214.

Gupta, R. S. and Siminovitch, L. (1978*a*). *Somat. Cell Genet.* **4**, 77–94.

Gupta, R. S. and Siminovitch, L. (1978*b*). *Somat. Cell Genet.* **4**, 355–374.

Gupta, R. S. and Siminovitch, L. (1978*c*). *Somat. Cell Genet.* **4**, 553–571.

Gupta, R. S. and Siminovitch, L. (1980). *Somat. Cell Genet.* **6**, 361–379.

Gupta, R. S., Chan, D. Y. H., and Siminovitch, L. (1978). *Cell* **14**, 1007–1013.

Gusella, J., Varsanyi-Breiner, A., Kao, F.-T., Jones, C., Puck, T., Keys, C. Orkin, S., and Housman, D. (1979). *Proc. Natl. Acad. Sci. USA* **76**, 5239–5243.

Gusella, J. F., Keys, C., Varanyi-Breiner, A., Kao, F.-T., Jones, C., Puck, T., and Housman, D. (1980). *Proc. Natl. Acad. Sci. USA* **77**, 2829–2833.

Haars, L., Hampel, A., and Thompson, L. (1976). *Biochim. Biophys. Acta* **454**, 493–503.

Hampel, A. E., Ritter, P. O., and Enger, M. D. (1978). *Nature* **276**, 844–845.

Hankinson, O. (1976). *Somat. Cell Genet.* **2**, 497–507.

Haralson, M. A. and Roufa, D. J. (1975). *J. Biol. Chem.* **250**, 8618–8623.

Jaskunas, S. R., Nomura, M., and Davies, J. (1974). In *Ribosomes* (M. Nomura, A. Tissieres, and P. Lengyel, eds.), Cold Spring Harbor Laboratory, New York, pp. 333–368.

Madjar, J.-J., Arpin, M., Buisson, M., and Reboud, J.-P. (1979). *Mol. Gen. Genet.* **171**, 121–134.

Madjar, J.-J., Neilsen-Smith, K., Frahm, M., and Roufa, D. J. (1982). *Proc. Natl. Acad. Sci. USA* **79**, 1003–1007.

Madjar, J.-J., Frahm, M., McGill, S., and Roufa, D. J. (1983). *Mol. Cell. Biol.* **3**, 190–197.

McConkey, E. H., Bielka, H., Gordon, J., Laskey, S. M., Lin, A., Oagata, K., Reboud, J.-P., Traugh, J. A., Traut, R. R., Warner, J. R., Welfe, H., and Wool, I. G. (1979). *Mol. Gen. Genet.* **169**, 1–6.

Moehring, J. M. and Moehring, T. J. (1979). *Somat. Cell Genet.* **5**, 453–468.

Moehring, T. J., Danley, D. E., and Moehring, J. M. (1979). *Somat. Cell Genet.* **5**, 469–480.

Moehring, J. M., Moehring, T. J., and Danley, D. E. (1980). *Proc. Natl. Acad. Sci. USA* **77**, 1010–1014.

Molnar, S. J. and Rauth, A. M. (1975). *J. Cell. Physiol.* **85**, 173–178.

Molnar, S. J., Thompson, L. H., Lofgren, D. J., and Rauth, A. M. (1979). *J. Cell. Physiol.* **98**, 327–340.

Monk, R. J., Meyuhas, O., and Perry, R. P. (1981). *Cell* **24**, 301–306.

Nakamichi, N., Rhoads, D. D., and Roufa, D. J. (1983). *J. Biol. Chem.* **258**, 13236–13242.

Nielsen-Smith, K., McGill, S., Frahm, M., and Roufa, D. J. (1983). *Mol. Cell. Biol.* **3**, 198–202.

Picard-Bennoun, M. (1981). *Mol. Gen. Genet.* **183**, 175–180.

Puck, T. T. and Kao, F. T. (1967). *Proc. Natl. Acad. Sci. USA* **58**, 1227–1234.

Reichenbecher, V. E. and Caskey, C. T. (1979). *J. Biol. Chem.* **254**, 6207–6210.

Ritter, P., Enger, M., and Hampel, A. (1976). In *Onco-Developmental Gene Expression* (W. H. Fishman and S. Sell, eds.), Academic Press, New York, pp. 47–56.

Thompson, L. H., Harkins, J. L., and Stanners, C. P. (1973). *Proc. Natl. Acad. Sci. USA* **70**, 3094–3098.

Thompson, L. H., Stanners, C. P., and Siminovitch, L. (1975). *Somat. Cell Genet.* **1**, 187–208.

Thompson, L. H., Lofgren, D. J., and Adair, G. M. (1977). *Cell* **11**, 156–168.

Thompson, L. H., Lofgren, D. J., and Adair, G. M. (1978). *Somat. Cell Genet.* **4**, 423–435.

Vasquez, D. (1979). *Inhibitions of Protein Biosynthesis*, Springer-Verlag, (New York, pp. 162–163.

Wasmuth, J. J. and Caskey, C. T. (1976). *Cell* **9**, 655–662.

Wasmuth, J. J. and Chu, L.-Y. (1980). *J. Cell. Biol.* **87**, 697–702.

Wasmuth, J. J., Hill, J. M., and Vock, L. S. (1980). *Somat. Cell Genet.* **6**, 495–516.

Wasmuth, J. J., Hill, J. M., and Vock, L. S. (1981). *Mol. Cell. Biol.* **1**, 58–65.

Wejksnora, P. J. and Warner, J. R. (1979). *Proc. Natl. Acad. Sci. USA* **76**, 5554–5558.

Worton, R. G., Duff, C., and Campbell, C. E. (1980). *Somat. Cell Genet.* **6**, 199–213.

Worton, R., Duff, C., and Flintoff, W. (1981). *Molec. Cell. Biol.* **1**, 330–335.

CHAPTER 15

RNA Polymerases

C. James Ingles
Banting and Best Department of Medical Research
University of Toronto
Toronto, Ontario, Canada

I. INTRODUCTION

There is abundant evidence that the regulation of gene expression at the level of RNA synthesis is of major importance in establishing altered cellular states of physiological adaptation or cytodifferentiation in eukaryotic cells. The mechanisms by which such alterations in transcription specificities and activities are brought about are, however, far from understood. A genetic approach to the study of RNA polymerase structure and function may make contributions to an elucidation of the molecular mechanisms regulating gene expression. As will be detailed in this chapter, Chinese hamster cell lines have proven to be particularly well suited for the initial forays into the molecular genetics of the transcription apparatus in mammalian cells.

II. BIOCHEMISTRY OF RNA POLYMERASES

A. RNA Polymerase Activities in Mammalian Cells

While virtually all cellular RNA synthesis is carried out by a single enzyme in bacterial cells, multiple forms of RNA polymerase (I, II, and III or A, B, and C) have been described for all eukaryotic species as divergent as plants, fungi, insects, and mammals. These enzymes differ in their chromatographic and catalytic properties, their subunit structures, their intracellular localization, and their transcriptive functions [for reviews see Chambon (1975); Roeder (1976); Lewis and Burgess (1982); and Sentenac and Hall (1982)]. RNA polymerase I (or A), the nucleolar enzyme, catalyzes the synthesis of the 40–45S precursor of 28S and 18S ribosomal RNA. RNA polymerase II (or B) synthesizes heterogeneous nuclear RNA (hnRNA), the precursor of cytoplasmic messenger RNAs, and RNA polymerase III (or C), synthesizes a variety of small cellular RNAs such as 4.5S pretransfer RNA and 5S ribosomal RNA.

Not only do all eukaryotes possess these same three RNA polymerase activities, but the analogous forms of the enzyme in different species have remarkably well-conserved subunit structures throughout the entire eukaryotic kingdom. Although the enzymes of Chinese hamster cells themselves have not been the subject of extensive protein purification and structural studies, the enzymes are so similar in different species that studies of the RNA polymerases of other mammalian cells and tissues can provide an adequate characterization of these Chinese hamster enzymes. Each of the eukaryotic RNA polymerases consists of a number (variously estimated to be 8–13) of subunit polypeptides, two of these for each enzyme being greater than 100,000 daltons in size. Thus RNA polymerase I has large subunits of about 190,000 and 125,000 daltons; RNA polymerase II, 214,000 and 140,000 daltons; and RNA polymerase III, 160,000 and 135,000 daltons. The array of smaller-molecular-weight subunits range in size between about 53,000 and 10,000 daltons; some of these are unique to each enzyme, while others appear to be shared among different forms of RNA polymerase. It should be noted that definitive proof that all of these polypeptides are in fact subunits of functional RNA polymerase complexes is lacking. Unlike the case of bacterial RNA polymerase, reconstitution of active enzyme from separated and purified subunits of eukaryotic polymerases has not been achieved. For some of the polypeptides there is, however, genetic evidence that clearly implicates them in catalytic roles.

B. α-Amanitin—A Specific Inhibitor of RNA Polymerase II

The different forms of RNA polymerases in mammalian cells were initially distinguished by their behavior during ion-exchange chromatography on DEAE-Sephadex. The three separate peaks of enzymatic activity eluted were designated polymerase I, II, and III (Roeder and Rutter, 1969). These enzymes can also be readily distinguished by their sensitivity to inhibition by α-amanitin, the bicyclic octapeptide toxin from the poisonous mushroom *Amanita phalloides* (Lindell et al., 1970). In *in vitro* assays of enzyme activity, mammalian RNA polymerase I is resistant to inhibition, polymerase II is inhibited by low (0.1 μg/ml) concentrations, and RNA polymerase III requires concentrations of α-amanitin greater than 100 μg/ml for similar inhibition.

α-Amanitin binds reversibly to the RNA polymerase II enzyme with a 1:1 stoichiometry (Cochet-Meilhac and Chambon, 1974), blocking the chain elongation phase of the RNA synthesis reaction as well as inhibiting the initiation of new chain synthesis. This selective inhibition of RNA polymerase II activity by low concentrations of α-amanitin provided the cornerstone for the genetic studies which followed.

III. MUTANT RNA POLYMERASE II

A. α-Amanitin-Resistant CHO Cell Lines

As suggested in Section I, a major reason for pursuing a genetic study of RNA polymerases in mammalian cells is that it provides an alternative approach which complements biochemical studies of these enzymes. The impetus, however, to isolate and characterize mutations affecting RNA polymerase activities received its initial strength from an additional source. At the time the first mutants were isolated (early 1970s) there was considerable skepticism regarding the existence of mutations in laboratory-cultured somatic cells (Harris, 1971; Metzger-Freed, 1972). Many compounds toxic to mammalian cells were being used as selective agents to derive cell lines with resistant phenotypes. Characterizing these altered cell lines with the aim of establishing correlations between the phenotypes and a biochemically demonstrated genotypic change was a priority. In the absence of traditional methods of genetic analysis such as recombination and segregation, it was hoped that these correlations could provide some much needed evidence that *bona fide* structural gene changes were involved in the events leading to altered cellular phenotypes.

Since the compound α-amanitin from poisonous mushrooms was both a cytotoxic substance and a specific inhibitor of RNA polymerase activities *in vitro*, it was a good candidate substance for use in deriving cell lines with drug-resistance phenotypes. It was hoped that α-amanitin-resistant cell lines might contain an easily demonstrated α-amanitin-resistant RNA polymerase II activity.

The first α-amanitin-resistant mammalian cell line was the Chinese hamster ovary cell line AR1/9-5B described by Chan et al. (1972). This mutant line has since been renamed Ama1. In this original study a number of resistant lines were selected from CHO cells treated with the mutagen ethylmethane sulfate (EMS) by plating and growth in the presence of α-amanitin in the medium. Unfractionated extracts derived from several of these isolates appeared to contain RNA polymerase activities with increased resistance to α-amanitin. When RNA polymerase activities in the cell line Ama1 in particular were fractionated by chromatography on DEAE-cellulose, the activity tentatively identified as RNA polymerase II was shown to be completely resistant to inhibition by 0.1 μg/ml α-amanitin, a concentration of α-amanitin that did inhibit the RNA polymerase II activity of wild-type CHO cells.

This pioneering study with CHO cell lines was followed in the next few years by other studies in which mutant lines with analogous α-amanitin-resistant (Ama^R) phenotypes were obtained. Mutants of the rat myoblast L6 (Somers et al., 1975a), Syrian hamster BHK-T6 (Amati et al., 1975), and mouse myeloma (Wulf and Bautz, 1976) lines and from human diploid fibroblast cell strains (Buchwald and Ingles, 1976) were in each case characterized by the presence of altered α-amanitin-resistant RNA polymerase II activities. Other indirect evidence also suggested that genotypic changes were involved. The frequency of occurrence of the α-amanitin-resistant phenotypes was generally in the range of 10^{-6}–10^{-5} after EMS mutagenesis. This mutagenesis usually increased the frequency of mutant isolation about 5- to 20-fold. In addition, the resistant phenotypes were stable in the absence of selection.

B. Altered RNA Polymerase II Activities in α-Amanitin-Resistant Lines

To characterize the RNA polymerase II activities in cell lines, a simple assay of RNA polymerase activities in whole cell lysates (Somers et al., 1975a) can be used. The assay favors detection of RNA polymerase II activity because of the presence of Mn^{2+} ions and high ammonium sulfate (0.4 *M*) concentrations. Under these conditions only about 20–25% of the incorporation of labeled precursors into RNA is due to polymerase I

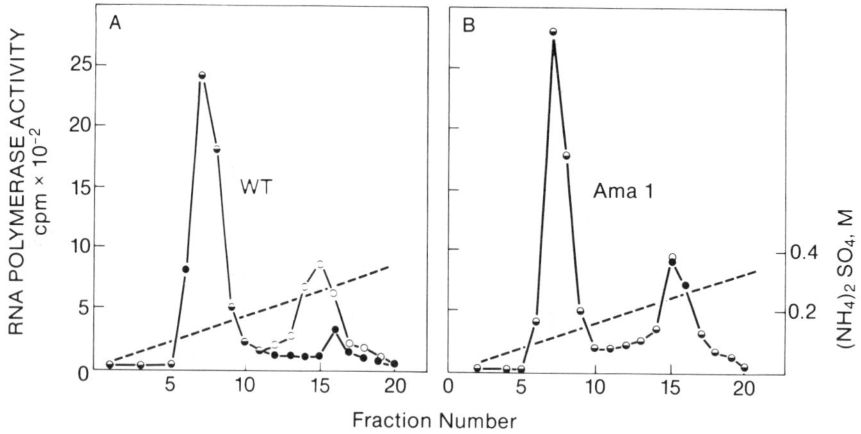

Figure 15.1. Resolution of CHO cell RNA polymerase activities by DEAE-Sephadex chromatography. Wild-type (panel A) and α-amanitin-resistant Ama1 (panel B) cell extracts were chromatographed on DEAE-Sephadex A-25 and eluted with a gradient of ammonium sulfate. Aliquots of each fraction were then assayed for RNA polymerase II activity in the absence (○—○) and the presence (●—●) of 0.10 μg/mL α-amanitin.

and III activities, the rest is polymerase II activity. The incorporation of ^{3}H-labeled ribonucleotides into RNA is measured by spotting reaction mixtures on Whatman DEAE-cellulose filter (DE81) disks and washing off unincorporated radioactive precursors. Under these reaction conditions the RNA synthesis largely represents elongation of nascent RNA chains. These assays as well as more definitive assays of RNA polymerase II activities in mutant cell lines after resolution of the multiple polymerase activities on ion-exchange chromatography clearly show that the resistance to α-amanitin is not due to quantitative changes in an otherwise unaltered enzyme in a manner similar to that which accounts for some other drug-resistant phenotypes (e.g., methotrexate resistance). An example of the chromatographic resolution of CHO cell RNA polymerase activities on DEAE-Sephadex is shown in Figure 15.1. Both wild-type CHO and mutant Ama1 cells have two major peaks of activity, RNA polymerase I and II. For wild-type cells (Fig. 15.1A), polymerase I is insensitive to α-amanitin and the second peak, polymerase II, is completely inhibited by 0.1 μg/ml α-amanitin. A minor, and usually much more unstable peak of activity, is due to RNA polymerase III. It is resistant to this concentration of α-amanitin but inhibited by much higher concentrations. This polymerase III activity can be seen to elute at the trailing edge of the polymerase II peak. The RNA polymerase activities of the CHO mutant Ama1 (Fig. 15.1B) are resolved in a similar fashion, but in contrast the α-amanitin sensitivity of the RNA polymerase II peak is drastically altered. The enzyme is entirely resistant to inactivation by low concentrations of α-amanitin.

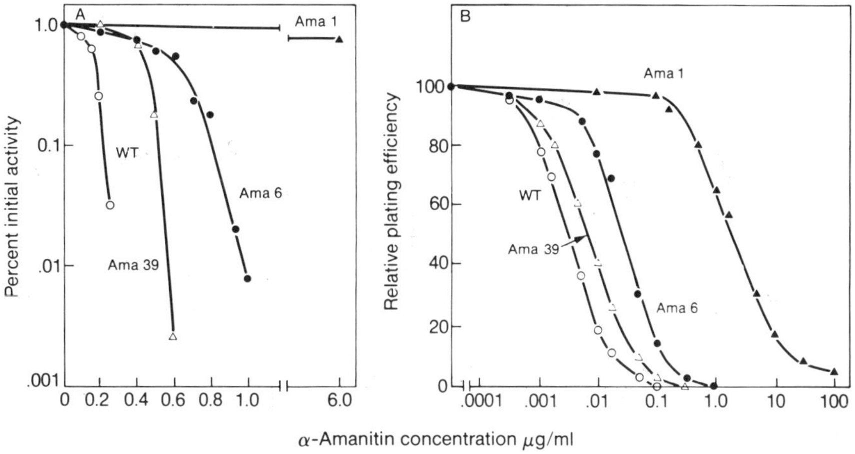

Figure 15.2. (A) Plating efficiencies of CHO wild-type and mutant cell lines in α-amanitin. CHO cell lines of wild-type sensitivity to α-amanitin (○—○) and three α-amanitin-resistant mutants—Ama 39 (Δ—Δ), Ama 6 (●—●), and Ama 1 (▲—▲) were seeded (10^3 cells/60 mm dish) with 5mL of medium containing increasing quantities of α-amanitin. Colonies were fixed, stained, and counted after 9 days growth at 34°C. (B) Sensitivity of CHO cell RNA polymerase II activities to inhibition by α-amanitin. DEAE-Sephadex fractioned RNA polymerase II activities from wild type and the same three mutant lines—Ama 39, Ama 6, and Ama 1—described in (A) were assayed with increasing concentrations of α-amanitin in the assay mixes.

Since the mutations to α-amanitin resistance could involve changes in a structural gene of RNA polymerase II, it was expected that different mutations in this gene might give rise to a spectrum of enzyme alterations. Cell lines might differ in their relative resistance to the cytotoxic effects of α-amanitin. If the resistance phenotypes are solely a function of altered RNA polymerase II activities and not due to changes in cell permeability to α-amanitin, then these differences in cytotoxicity to α-amanitin could be expected to correlate with differences in the sensitivity of mutant RNA polymerase II activities to inhibition by α-amanitin.

An examination of the survival curves for a series of cell lines selected in a single step from mutagen-treated populations of CHO cells (Fig. 15.2A) indicated that just such a range of mutants could be obtained in selection experiments. The three mutant CHO lines shown here—Ama39, Ama6 (Ingles et al., 1976), and Ama1—differ in their resistance to the cytotoxicity of α-amanitin. While parental cells were effectively killed at α-amanitin concentrations as low as 0.2 μg/ml α-amanitin, the mutants Ama39, Ama6, and Ama1 were increasingly more resistant. In fact, the original CHO mutant Ama1 (Chan et al., 1972) remains to this day the most resistant of any of the CHO mutants selected in these and other studies.

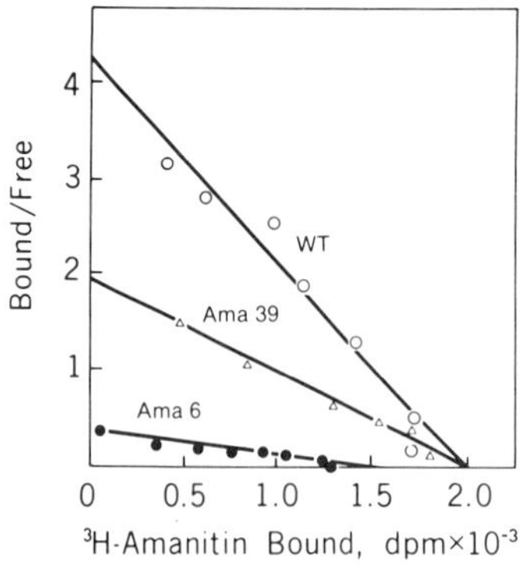

Figure 15.3. Binding of [^{3}H]amanitin to RNA polymerase II in cell lysates. Increasing concentrations of *0*-[^{3}H]methyl-demethy-γ-amanitin were incubated with cell lysates from wild-type (o—o), Ama 39 (Δ—Δ) and Ama 6 (●—●) cells and the [^{3}H]amanitin bound at equilibrium by RNA polymerase II determined. The equilibrium dissociation constants (K_D) of complexes between [^{3}H]amanitin and RNA polymerase II was determined from these Scatchard plots of the data. The slope of the line is $-1/K_D$.

These three mutant lines each have an RNA polymerase II activity with a correspondingly altered sensitivity to inhibition by α-amanitin (Fig. 15.2B). Thus Ama39, Ama6, and Ama1 cells required 2- to 3-fold, 8- to 10-fold, and 500–800-fold higher concentrations of α-amanitin than that required to inhibit the parental wild-type polymerase II activity. The increased resistance of RNA polymerase II activity to inhibition by α-amanitin correlated with the increase in resistance to the cytotoxicity of α-amanitin.

C. Altered [^{3}H]Amanitin Binding to Mutant RNA Polymerase II

α-Amanitin inhibition of RNA polymerase II activity results from the binding of a single molecule of α-amanitin to the enzyme (Cochet-Meilhac and Chambon, 1974). Since the RNA polymerase II in the mutants Ama39, Ama6, and Ama1 showed increased resistance to inhibition by α-amanitin (Fig. 15.2B), the altered resistance might be reflected in a reduced ability to bind α-amanitin. The binding of the amatoxin can be studied with the derivative *0*-[^{3}H]methyl-demethyl-γ-amanitin (Wieland and Fahrmeir, 1970). The binding of the ^{3}H-labeled amanitin to RNA polymerase II present in the cell lysates of various CHO cell lines as a function of increasing concentrations of [^{3}H]amanitin was determined. Ammonium sulfate precipitation procedures (Cochet-Meilhac and Chambon, 1974) were used to separate bound and free [^{3}H]amanitin. The equilibrium dissociation constants (K_D) of complexes between *0*-[^{3}H]methyl-demethyl-γ-amanitin and RNA polymerase II in parental (wild-type) and α-amanitin-resistant mutant lines was determined from Scatchard analysis. This analysis (Fig. 15.3) showed that a single component in each cell was responsible for the observed binding of [^{3}H]amanitin. Decreases in the affinity for binding amanitin by mutant RNA polymerases were readily apparent. From the slopes of the Scatchard plots shown in Fig. 15.3, the K_D for the binding of [^{3}H]amanitin by wild-type RNA polymerase II was determined to be 3.8×10^{-11} *M*. The mutant Ama39, whose RNA polymerase II activity had a two- to threefold increased resistance to inhibition by α-amanitin, had a K_D of 8.5×10^{-11} M. Ama6 RNA polymerase II, 8–10-fold more resistant than the wild-type enzyme had correspondingly altered binding properties, a K_D of 29×10^{-11} M. No binding to the 800-fold resistant Ama1 RNA polymerase II could be observed with the range of [^{3}H]amanitin concentrations used. Thus we had found a correlation between degree of resistance of the cells to the cytotoxicity of α-amanitin (Fig. 15.2A), the relative resistance of their RNA polymerase II activities (Fig. 15.2B), and the ability of the enzymes to bind amanitin (Fig. 15.3). This was a strong argument that alterations in RNA polymerase II do indeed account for the drug-resistance phenotypes of the cells. These data suggest that the different mutations to α-amanitin resistance involve different alterations in the amino acid sequence of the subunit polypeptide(s) of RNA polymerase II which bind α-amanitin. As will be described subsequently the molecular cloning of RNA polymerase II DNA has identified this polypeptide as the largest (210,000–220,000 dalton) subunit of the enzyme.

D. Ama^R Codominant Expression and Functional Hemizygosity in CHO Cells

The studies of α-amanitin-resistant RNA polymerase II in CHO cell lines was followed in the next few years by the selection of a variety of other mutant mammalian cell lines with analogous α-amanitin-resistant mutations. The mutants in other cell lines were in each case characterized by the presence of altered, α-amanitin-resistant, RNA polymerase II activities. But an important difference between CHO cell mutants and similar mutants in these other lines was soon noted. As can be seen in Figure 15.2B the inhibition of mutant CHO cell RNA polymerase II was in each case monophasic. From these data and the [^{3}H]amanitin binding data in Figure 15.3 one can conclude that *all* of the RNA polymerase II appears to be of the mutant type. In contrast, the inhibition of polymerase II activity in analogous mutants of the rat myoblast L6 cells (Somers et al., 1975a) and human cell strains (Buchwald and Ingles, 1976) were biphasic. Only a portion of the RNA polymerase II activity was altered in its α-amanitin sensitivity, while a portion had the same α-amanitin sensitivity as the wild-type parental enzyme. These differences between CHO cells and the other mammalian cell lines are due to the presence, in the mutants derived from other cell lines, of both sensitive and resistant forms of RNA polymerase II, and, therefore, of wild-type and mutant alleles coding for the α-amanitin binding subunit of RNA polymerase II. These observations clearly established two features of mutations to α-amanitin resistance. CHO cells appear to be functionally hemizygous for this gene; CHO cell mutants obtained in a single step possess only the mutant form of the enzyme. Second, α-amanitin resistance is a codominantly expressed mutation; mutant cells that possess both resistant and sensitive forms of RNA polymerase II have a drug-resistant phenotype. This behavior has been confirmed directly by constructing hybrids between each of the CHO α-amanitin-resistant cell lines Ama1 (Lobban and Siminovitch, 1975; Guialis et al., 1977), Ama6 and Ama39 (Ingles et al., 1976), and CHO cells of wild-type sensitivity. The hybrids all had RNA polymerase II activities that give rise to biphasic inhibition curves with α-amanitin. The hybrids had increased levels of resistance to the cytotoxicity of α-amanitin; hybrid cells made with the most resistant mutant Ama1 being almost as resistant as the resistant parent itself (Lobban and Siminovitch, 1975). It should be noted, however, that the apparent functional hemizygosity of this polymerase II gene is not a feature of all Chinese hamster cell lines. As Gupta et al. (1978) have shown, the Chinese hamster CHW and M3-1 cell lines appear to be functionally diploid at this locus. α-Amanitin-resistant mutants of these cell lines, unlike the CHO mutants, contained a mixture of Ama^R and Ama^S RNA polymerase II activities.

The codominant behavior of mutations conferring α-amanitin resistance means of course that α-amanitin-resistance mutations can provide a good genetic marker for selections in cell hybridization experiments. Ama^R mutations have also proven extremely valuable as a codominant transferable marker in the DNA-transformation experiments discussed subsequently. In addition, the unique property of this RNA polymerase II gene in CHO cells, namely, its apparent functional hemizygosity, has made the selection of recessive temperature-sensitive (TS) mutations in RNA polymerase II feasible.

E. Temperature-Sensitive RNA Polymerase II Mutations

A genetic analysis of RNA synthesis in mammalian cells will only be effective if the range of mutations available can be increased. Most mutations affecting an essential cellular function such as RNA synthesis will be lethal. We have therefore turned to the isolation of conditional lethal (temperature-sensitive) mutations in RNA polymerase II. The isolation of many mutations in specific functions in bacteria and yeast has been facilitated by the development of techniques for localized mutagenesis. Localized mutagenesis implies that mutations are not induced at random throughout the entire genome. The isolation of a mutant cell with an alteration in a specific function then does not solely depend on ingenuity in designing appropriate selection conditions. Rather, since the mutagenesis has been targeted to a specific region of the genome, it becomes feasible to rely on screening limited numbers of individual isolates to identify those possessing desired mutant phenotypes. Most of the classical techniques for localized mutagenesis involve isolation of the gene of interest or a small region of the chromosome containing it, mutagenesis of this gene, and reintroduction of this mutated gene back into the cell. An alternative method for localized mutagenesis has been described by Oeschger and Berlyn (1975). The mutagen nitrosoguanidine induces double mutations at a high frequency in prokaryotes, and these mutations are often closely linked on the bacterial chromosome. The induction of mutations with nitrosoguanidine coupled with positive selections for mutations in a specific locus can therefore be an effective method to enrich for mutations that map close to the selected function. Since α-amanitin provides a good selection for mutations in an RNA polymerase II locus in mammalian cells, it was hoped that among a spectrum of α-amanitin-resistant mutations induced by nitrosoguanidine there might be a subset that would contain additional mutations closely linked to the selected α-amanitin-resistance mutation. Some of these linked mutations if also in the RNA polymerase II gene might confer a TS phenotype to the cells. These mutants might contain TS RNA polymerase II activities. Since the studies discussed previously indicated that this RNA polymerase II locus in CHO cells was functionally hemizygous, recessive TS phenotypes due to mutations in this locus would be expressed. RNA polymerase II is an essential cellular function synthesizing mRNAs and the cells should be TS for growth.

This was the rationale for an attempt to isolate CHO cells lines with TS RNA polymerase II mutations (Ingles, 1978). Wild-type CHO cells were mutagenized with nitrosoguanidine, and a large number of α-amanitin-resistant isolates were selected at the permissive temperature of 34°C. About 200 α-amanitin-resistant isolates were then screened for conditional-lethal TS mutations by testing for growth at 39.5°C. About 5% of these mutant isolates were unable to grow at this nonpermissive temperature. As for the previous selected α-amanitin resistant mutants, the Ama^R and TS phenotypes were stable in the absence of selective conditions.

It was clear that the TS mutations could be either in RNA polymerase II or in any other unrelated essential cell function. These possibilities were distinguished by making use of the codominant behavior of α-amanitin resistance, and the recessive behavior of the TS mutations. As already indicated hybrid Ama^R/Ama^S cells are α-amanitin resistant and contain a mixture of Ama^R and Ama^S forms of polymerase II. The putative TS polymerase II mutants were expected to have an Ama^R and TS polymerase II activity. Hybrid cells formed by fusion of the Ama^R and TS isolates with α-amanitin-sensitive temperature-resistant (TS^+) cells, that is, cells with wild-type RNA polymerase II, should show a unique phenotype. If the TS mutation was in RNA polymerase II, then the hybrid $TSAma^R \times Ama^S$ cells would not grow at the elevated temperature in the presence of α-amanitin. If the TS mutation involved another gene and was recessive then the hybrid cells would grow at both 34 and 39.5°C in the presence of α-amanitin.

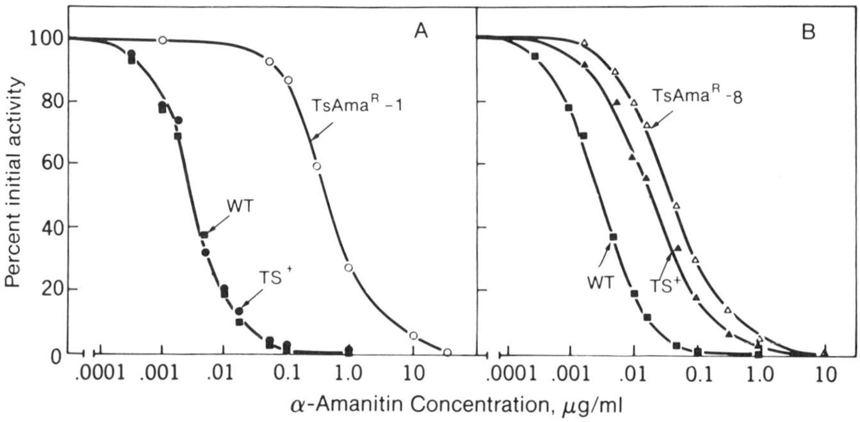

Figure 15.4. Inhibition of TS mutant and TS^+ revertant cell RNA polymerase II activities by α-amanitin. RNA polymerase II activities were assayed in the presence of increasing concentrations of α-amanitin: parental wild-type CHO cells (■—■); mutant $TsAma^R$-1 (○—○); TS^+ revertant of $TsAma^R$1 (●—●); mutant $TsAma^R$-8 (Δ—Δ); TS^+ revertant of $TsAma^R$-8 (▲—▲).

Of seven TS lines identified after selection of a large number of α-amanitin-resistant isolates at 34°C, four of these appeared to be TS RNA polymerase mutations (Ingles, 1978). Hybrid cells formed by fusion of these mutant clones with Ama^S cells were α-amanitin resistant at 34°C and extremely α-amanitin-sensitive at 39.5°C. In contrast, the other three TS mutants appeared to contain unlinked TS mutations; hybrids made with these isolates remained α-amanitin resistant at both 34 and 39.5°C. An examination of the RNA polymerase activities in these isolates and the respective hybrid lines confirmed these observations. The Ama^R and TS isolates all contained RNA polymerase II activities with altered sensitivity to inhibition by α-amanitin. The titration of RNA polymerase II activity in two of the mutant TS lines is shown in Figure 15.4. $TsAma^R$-1 cells contained an enzyme that was 100-fold more resistant than the wild-type Ama^S polymerase II and $TsAma^R$-8 polymerase II required 15–20-fold higher concentrations of α-amanitin than that required to inhibit Ama^S polymerase II. The hybrid cell lines formed by fusion of each of these TS lines with an Ama^S CHO line when grown at 34°C contained both the resistant polymerase II contributed by the parental Ama^R TS isolates and α-amanitin-sensitive activity contributed by the Ama^S cells. When grown at 39.5°C, however, the hybrid cells constructed with the four TS polymerase II isolates contained only the α-amanitin-sensitive component (Ingles, 1978).

In other experiments demonstrating the linkage of the TS and α-amanitin-resistance mutations, the effect of reversion of the TS phenotype was examined. TS^+ revertants able to grow at 39.5°C were selected after nitrosoguanidine or EMS mutagenesis of the TS lines. For one mutant line $TsAma^R$-1, the TS^+ revertant had an unexpected phenotype. TS^+ revertants of this line did not grow in α-amanitin either at 34 or 39.5°C. The RNA polymerase II in the TS^+ revertants had a sensitivity to inhibition exactly like that of wild-type Ama^S CHO cells (Fig. 15.4A). TS^+ revertants of another line, $TsAma^R$-8, also showed an altered α-amanitin sensitivity. The TS^+ revertants of $TsAma^R$-8 remained somewhat α-amanitin resistant, but the RNA polymerase II in these TS^+ revertants, although more resistant than wild-type Ama^S polymerase II, was not as resistant as that in the parental $TsAma^R$-8 cells (Fig. 15.4B). Reversion of the TS phenotype in both these cases was accompanied by a nonselected alteration in the α-amanitin sensitivity of polymerase II. This is convincing evidence that the mutations causing the TS phenotype in these lines involved changes in the enzyme RNA polymerase II.

Because the TS^+ reversion event restored wild-type sensitivity to inhibition by α-amanitin to the RNA polymerase II in revertants of $TsAma^R$-1, it is most likely that a single point mutation was responsible for both the increased drug resistance and the TS phenotype of $TsAma^R$-1 cells. The nitrosoguanidine comutagenic selection as a means of enriching for mutations in a limited region of the genome does not appear to be as effective in mammalian cells as it is in bacteria. Rather, the structural changes due to the missense mutations selected with drug resistance may on occasion also be responsible for a TS phenotype. The approach is actually then a screen for structural gene mutations to see if any confer TS phenotypes. It would appear that a useful proportion of the mutations conferring an α-amanitin-resistant phenotype do. Similar findings have been reported for TS tubulin mutants selected for drug resistance at a permissive temperature (see Chapter 22).

These TS mutations were identified as polymerase II mutations by a genetic analysis of the behavior of the TS phenotype in somatic cell hybrids and in TS^+ revertants (Ingles, 1978). These experiments did not address the question of the biochemical basis of this TS behavior. The TS mutations may affect catalytic activity directly, the enzyme may be denatured at elevated temperatures, or, alternatively, enzyme synthesis or enzyme assembly may be affected at the nonpermissive temperature. Preliminary studies were made on the mutants $TsAma^R$-1 and $TsAma^R$-8 (C. J. Ingles, unpublished). RNA synthesis did not stop abruptly upon shift of either mutant to the nonpermissive temperature. The TS defects did not appear to affect the catalytic activity of the enzyme directly. Nor were we able to demonstrate an increased thermal sensitivity of solubilized polymerase II from the mutant lines. With multisubunit enzymes such as RNA polymerase II, TS mutations frequently affect enzyme activity indirectly by interfering with important subunit–subunit interactions at elevated temperatures. For example, TS defects which affect enzyme assembly have already been documented for the *Escherichia coli* RNA polymerase (Kirschbaum et al., 1975). In such cases a TS defect may not be demonstrable *in vitro* with the solubilized enzyme.

F. Suppressors of TS RNA Polymerase Mutations

TS cell lines can each serve as parentals for the selection of additional mutations. Growth at the nonpermissive temperature provides a facile selection protocol for a second generation of mutant lines. The TS^+ revertants can be the result of true reversion events to the wild-type phenotype, or they may be the result of suppression by new mutations that correct, replace, or bypass the original defect [see Hartman and Roth (1973) for an excellent discussion of suppressor mutations]. As Jarvik and Botstein (1975) showed in an analysis of revertants of missense mutations, new TS or cold-sensitive phenotypes are often coacquired with the reversion event. These new phenotypes are often due to second site suppressing mutations in genetically unlinked functions. Most importantly, extragenic suppressors can often be in a gene whose products are known to interact physically with the original gene product. By this means mutations affecting other subunits of an enzyme or other proteins interacting with that enzyme may be isolated.

The CHO RNA polymerase II mutant $TsAma^R$-1 is uniquely suited for selections of second site TS suppressor mutations. As was shown previously, reversion of the TS mutation was accompanied by a coreversion of the α-amanitin resistance to a wild-type α-amanitin-sensitive phenotype (Ingles, 1978). Therefore, a simple selection scheme, selecting for growth at nonpermissive temperatures in the presence of α-amanitin ensures that the true wild-type TS^+Ama^S revertants do not survive. TS^+Ama^R revertants of $TsAmA^R$-1 were obtained at a frequency of 10^{-7}–10^{-6} only after EMS mutagenesis of the parental $TsAma^R$-1 cells (Wong and Ingles, unpublished). All the TS^+ isolate cells contained an α-amanitin-resistant RNA polymerase II with a sensitivity to inhibition by α-amanitin which was identical to that of the polymerase II in the parental $TsAma^R$-1 cell line. The original missense mutation which simultaneously gave a TS and Ama^R phenotype appeared to be still present; but the cells grew at 40°C, and the enzyme no longer had a TS defect. The nature of the suppressing mutations are not yet known. Suppression could be due to a second mutation in the same polypeptide (an intragenic suppressor) or in another subunit of RNA polymerase II or even in a protein factor that interacts with RNA polymerase II. It might also be a regulatory mutation that overcomes the TS defect by causing sufficient overproduction of the TS polymerase so that the cells now survive at 40°C. The availability of the cloned

RNA polymerase II DNA for this and other mutant RNA polymerase II alleles will permit an assessment of these alternatives.

IV. REGULATION OF RNA POLYMERASE II SYNTHESIS

The selection and characterization of mutations in somatic cells grown in laboratory culture laid a secure genetic foundation for more recent studies which employed DNA-mediated gene transfer to identify and molecularly clone human genes such as those conferring oncogenically transformed phenotypes. It is disappointing that a wider use has not yet been made of many of the mutants that somatic cell geneticists have generated. The RNA polymerase mutations in CHO cells have similarly been underused. The mutations to α-amanitin resistance in CHO and rat myoblast cells did however provide a novel opportunity to examine the regulation of RNA polymerase II subunit polypeptide synthesis. As discussed above, mutant cell lines such as the α-amanitin-resistant rat myoblast lines (Somers et al., 1975*a*), or certain hybrid CHO cell lines (Lobban and Siminovitch, 1975; Guialis et al., 1977), contained both α-amanitin-sensitive (Ama^S) and α-amanitin-resistant (Ama^R) forms of RNA polymerase II. The α-amanitin-sensitive enzyme is inactivated by the growth of such heterozygous cells in culture medium containing α-amanitin. Near normal growth of the rat myoblast L6 mutants (Somers et al., 1975*b*) and CHO hybrid lines (Guialis et al., 1977) was seen in α-amanitin, despite there being initially only 25–33% of RNA polymerase II activity resistant to α-amanitin. Either the enzyme is in excess or the level of resistant RNA polymerase II must increase when the mutant lines are grown in α-amanitin. Three different approaches to quantitating RNA polymerase II levels have revealed the existence of a regulatory mechanism which serves to keep the intracellular amount of active enzyme relatively constant. RNA polymerase II activities were measured both in cell lysate "run off" reactions or after solubilization, partial purification and assay on exogenous DNA templates. Cellular RNA polymerase II contents were estimated both immunologically and with a quantitative [^{3}H]amanitin binding assay. Finally, the rates of synthesis and degradation of RNA polymerase II subunit polypeptides were determined.

With the rat myoblast cell lines (Somers et al., 1975b) and with hybrid CHO cell lines (Guialis et al., 1977), we have shown that, when Ama^R/Ama^S cell lines were grown in the presence of α-amanitin, the α-amanitin-sensitive component was inactivated. At the same time there was a compensatory increase in the level of RNA polymerase II activity which was resistant to α-amanitin. The total level of RNA polymerase II activity remained constant.

Such an increase in Ama^R enzyme activity could be brought about either by the accumulation of more enzyme or by activation of preexisting molecules. Under the conditions where there was a two- to threefold increase in Ama^R polymerase II activity, a radioimmunoassay employing anti-RNA polymerase II antibodies showed no major change in the total mass of RNA polymerase II. The quantitation of [^{3}H]amanitin binding, a selective measure of only the Ama^S form of polymerase II indicated, on the other hand, that the loss of α-amanitin-sensitive RNA polymerase II activity was accompanied by a loss of [^{3}H]amanitin binding capacity in these cells. Together these results indicated that the increase in Ama^R polymerase II activity involves accumulation of more α-amanitin-resistant polymerase II enzyme molecules (Guialis et al., 1977).

This increase in RNA polymerase II molecules was subsequently shown to be the result of a coordinate increase in the rate of synthesis of RNA polymerase II subunit polypeptides (Guialis et al., 1979). The rates of synthesis and degradation of RNA polymerase II polypeptides were estimated by using [^{35}S]methionine labeling and immunoprecipitation followed by SDS gel electrophoresis and fluorographic quantitation. A rapid degradation of the RNA polymerase II polypeptides was associated with the inactivation of the Ama^S enzyme. At the same time there was an increase in the rate of synthesis of at least three different (214,000-, 25,000-, and 20,500-dalton) RNA polymerase II subunits.

The similarity of this regulation of RNA polymerase II in Ama^R/Ama^S heterozygous cell lines to the regulation of RNA polymerase levels in rifampicin^s/rifampicin^r merodiploid strains of *E. coli* (Hayward et al. 1973) and in *E. coli* strains carrying TS RNA polymerase II mutations (Kirschbaum et al. 1975) is striking. The increased synthesis of RNA polymerase in mammalian cells as in bacteria may be either the result of an increased rate of synthesis of subunit mRNAs or an increase in their rate of translation. Even in bacteria the molecular details of this mechanism remain uncertain. It has even been suggested that subunits of RNA polymerase themselves could play an autoregulatory role in this gene regulation (Goldberger, 1974). Our animal cell studies have not yet examined the transcriptional and translational regulation of RNA polymerase II mRNAs, although the availability of recombinant DNA clones encoding RNA polymerase II will soon make these experiments feasible.

V. RNA POLYMERASE II GENE TRANSFER

Gene isolation has in the majority of cases been accomplished to date using the cDNA cloning of mRNA species. Even genes represented only infrequently in mRNA populations can now often be isolated using sensitive methods to detect either oligonucleotide sequences in cDNAs or expression of cDNA encoded proteins. DNA transfer of genes expressing selectable phenotypes into appropriate recipient cells followed by recombinant DNA screening to identify and rescue transforming DNA has also proven to be a powerful method for the isolation of genes encoding enzymes or proteins that would otherwise be difficult to clone.

Since α-amanitin resistance is inherited codominantly, the gene encoding the subunit polypeptide which confers this resistant phenotype was a candidate for early gene transfer and rescue experiments. However, the DNA-mediated transfer of the Ama^R phenotype initially proved unsuccessful in this and other laboratories. In these early studies both mouse Ama^S tk^- and Ama^S CHO cell lines were used as recipient for transfer of DNA from the Ama^R CHO mutant Ama1. It is still not clear why these initial experiments were unsuccessful.

Because CHO cell lines had gained a certain notoriety as being difficult to transform with DNA (but see Chapter 8 for a more-reasoned counterview), we have used an alternative cell as recipient in RNA polymerase II gene transfer experiments. The Syrian hamster BHK21 cell line TsAF8 is a TS mutant isolated by Meiss and Basilico (1972). The TsAF8 cell line was isolated by virtue of its resistance to 5-fluoro-2′-deoxyuridine killing of dividing cells at nonpermissive temperatures (see Chapter 20). TsAF8 cells arrest at elevated temperatures in the middle of the G1 phase of the cell cycle (Burstin et al., 1974). A selective reduction in the activity of the enzyme RNA polymerase II measured in both isolated nuclei (Rossini and Baserga, 1978) and in solubilized cell extracts (Rossini et al., 1980) suggested that the TS defect might be in RNA polymerase II itself. The loss of RNA polymerase II activity at nonpermissive temperatures was shown to represent an actual loss of polymerase II molecules as quantitated by $[^3H]$amanitin binding; it did not reflect a generalized loss of major cellular proteins.

Definitive evidence that the TS defect in the mutant TsAF8 cells was an RNA polymerase II defect soon followed. We made hybrid cell lines by fusing the known CHO TS RNA polymerase II mutant $TsAma^R$-1 with TsAF8 cells. Both Ama^R polymerase II from the CHO allele and Ama^S polymerase II from the TsAF8 allele were shown to be present in the hybrid lines. The two TS mutations, however, failed to complement; the hybrid cells would not grow at 40°C (Shales et al., 1980). Furthermore, a characterization of RNA polymerase in several Ama^R derivatives of TsAF8 showed that just as in CHO cells this RNA polymerase II locus was present only in a single or functionally hemizygous state. The TS mutation, having a recessive phenotype was expected to lie in just such a functionally hemizygous locus. The introduction of an Ama^R mutation in TsAF8 in one case led to partial reversion of the TS phenotype. Taken together these data provided evidence that the TS mutation in TsAF8 cells is the result of a TS mutation in the same gene as the TS and Ama^R RNA polymerase II mutations in CHO cells (Shales et al., 1980).

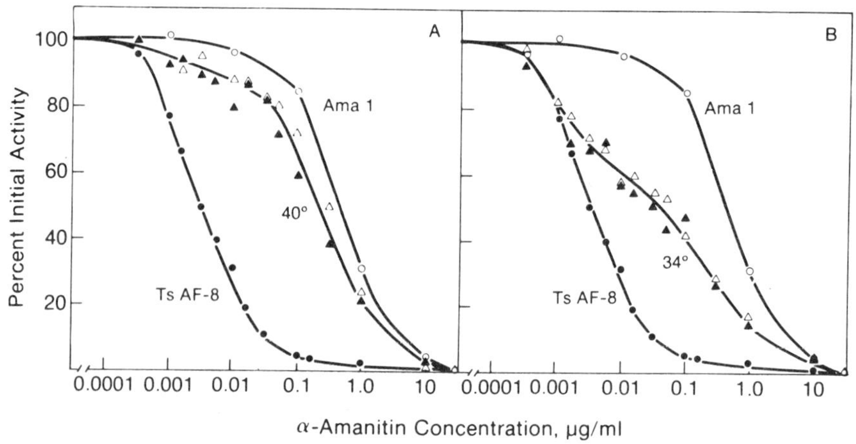

Figure 15.5. Inhibition of and TS^+ transformed cell RNA polymerase II activities by α-amanitin. RNA polymerase II activities present in cell lysates of recipient TsAF8 (●—●), donor CHO Ama 1 (○—○), and two independent TS^+ transformants (Δ—Δ, ▲—▲) of TsAF8 cells were assayed in the presence of increasing concentrates of α-amanitin. The TS^+ transformed cell lines were grown at 40°C (panel A) or 34°C (panel B) prior to enzyme assay.

TsAF8 cells can thus serve as recipients for gene transfer of the dominant TS^+ alleles of RNA polymerase II. These BHK21 cells are readily transformed by calcium-phosphate-coprecipitated DNA. DNA's from either TS^+ CHO, BHK21, or human (HeLa) cells were all effective at inducing the appearance of TS^+ transformants able to grow at the nonpermissive temperature of 40°C (Ingles and Shales, 1982).

To clearly distinguish TS^+ transformants from spontaneous TS^+ revertants of TsAF8, we used DNA from cell lines with biochemically well-characterized RNA polymerase II phenotypes, that is, those that were TS^+ (wild-type) but also, and more importantly, Ama^R. In most transformation experiments a mixture of transformants, TS^+ and also Ama^R at 40°C, and spontaneous revertants, TS^+ but Ama^S at 40°C, were found. An examination of the RNA polymerase II activities in these TS^+Ama^R transformed lines provided even more stringent evidence for gene transfer. The recipient cells contained an Ama^S polymerase II activity; the polymerase II activity in extracts of TsAF8 cells was inhibited by concentrations of α-amanitin that inhibited other wild-type mammalian polymerase II activities (Fig. 15.5). The transformants, on the other hand, when grown at 40°C contained an RNA polymerase II activity with a sensitivity to inhibition by α-amanitin characteristic of the particular DNA used to transform the TS cells. For example, when CHO Ama1 DNA was used as donor DNA, most of the activity in cells grown at 40°C is α-amanitin resistant just like that in Ama1 cells (Fig. 15.5A). When these same transformants were grown at 34°C, however, the RNA polymerase II titrations with α-amanitin were clearly biphasic (Fig. 15.5B); about one-half the activity had a sensitivity to α-amanitin inhibition like the Ama^S TsAF8 enzyme and one-half like the Ama^R polymerase II of Ama1 cells. The expression of donor DNA sequences had provided the new RNA polymerase II activity. These data of course provide even more convincing evidence that the TS defect in TsAF8 cells is, quite fortuitously, a defect in RNA polymerase II. Transfer of a gene for Syrian or Chinese hamster polymerase II and also human polymerase II was effective in complementing the TS defect in the Syrian hamster TsAF8 cells (Ingles and Shales, 1982). Since these DNA transfer experiments likely only involve DNA coding for a single subunit of RNA polymerase II, hybrid CHO-BHK or human-BHK enzymes must be functional. The hybrid enzyme must be capable of effective gene transcription. This implies that mammalian RNA polymerase II structure is extremely well conserved—a finding in accordance with both the antigenic homologies detected between RNA polymerases of different species (Ingles, 1973; Huet et al. 1982; Weeks et al. 1982) and the similarity in the subunit polypeptide sizes of these RNA polymerase II enzymes.

VI. MOLECULAR CLONING OF RNA POLYMERASE II DNA

A. Isolation of the RNA Polymerase II Gene by DNA Transfer

The successful DNA-mediated transfer of this Ama^R polymerase II gene was a crucial step in setting up a viable approach to obtaining recombinant clones of this gene. In order to isolate polymerase II DNA, it is necessary that we be able to identify transferred DNA sequences that confer this Ama^R phenotype on recipient TsAF8 cells. We have adopted an approach essentially similar to that used to clone a number of other mammalian genes such as those encoding thymidine kinase (Perucho et al., 1980) and several human oncogenes (Goldfarb et al., 1982; Weinberg, 1982). The scheme we are using is summarized in Figure 15.6. Briefly, DNA of the donor CHO cell line Ama1 has been partially cut with the restriction enzyme *Mbo*1 and ligated to *Bam*H1 cut marker bacterial plasmid DNA sequences prior to gene transfer. pGA276 is a derivative of pBR322 containing Amp^R, tet^R, SV40 *ori*, *Ecogpt*, and λ *cos* DNA sequences. Transformants were identified as TS^+Ama^R colonies and were shown to have acquired the new TS^+ and Ama^R polymerase II activity characteristic of Ama1 cells. In addition, these transformants were shown to have acquired donor DNA sequences by Southern blot experiments (Ingles et al., 1983). Genomic DNA from primary transformants and from a number of secondary TS^+Ama^R transformants of TsAF8 resulting from a second cycle of transformation with DNA of

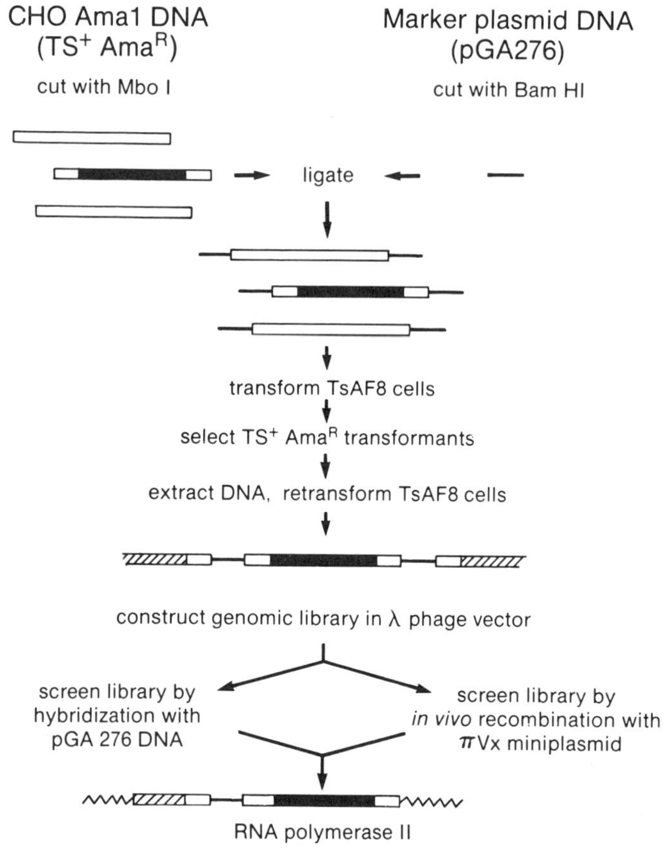

Figure 15.6. Scheme for the isolation of the RNA polymerase II DNA by DNA-mediated gene transfer and rescue of transforming DNA. Donor CHO DNA (▭) used in gene transfer; RNA polymerase II DNA (▬); recipient TsAF8 cell DNA (▨); marker bacterial plasmid DNA (—); λ phage vector DNA (〰).

a primary TS^+Ama^R transformant were isolated, cut with *Eco*RI, and probed for the presence of the market pGA276 plasmid sequences. The recipient cell line TsAF8 contained no crosshybridizing DNA (Fig. 15.7). The primary transformants contained an array of fragments hybridizing to pGA276 DNA. Three different secondary transformants contained only a limited number of fragments of pGA276-related DNA. The sizes of these fragments varied in different secondary TS^+Ama^R transformants as did the copy number of particular fragments as estimated by the intensity of the radioautographic signal. The second cycle of transfer had largely removed the cotransferrred DNA sequences.

The next step in cloning the Ama^R polymerase II DNA was to construct a genomic library in a λ phage vector. Partially cut *Mbo*I DNA

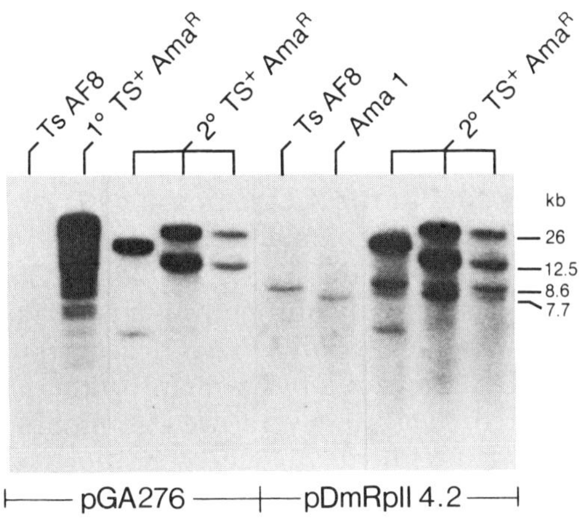

Figure 15.7. Marker bacterial and RNA polymerase II DNA in TS^+ Ama^R transformed cell lines. Twenty micrograms of genomic DNA from recipient TsAF8 cells, donor CHO Ama1 cells, a primary TS^+ Ama^R transformant and three independent secondary transformants were digested with *Eco*RI and probed with nick-translated DNAs for the presence of pGA276 DNA or *Drosophilia melanogaster RpII* p4.2 DNA.

from a secondary transformant shown in Figure 15.7 was size fractionated (15–20 kb), ligated to *Bam*H1 cut λ EMBL3B DNA (Murray, 1983), *in vitro* packaged, and amplified by plating on *E. coli* cells. This library has been screened both by standard Benton–Davis (1977) hybridization techniques using nick translated pGA276 DNA as probe, and also by the *in vivo* recombination technique developed by Seed (1983). Since the marker plasmid pGA276 contains the *col*E1 origin of replication and is identical to about 400 bp of DNA in the miniplasmid πVX, growth of our EMBL3B genomic library on πVX-containing *E. coli* strains permits *in vivo* plasmid/phage recombination to occur across these homologous DNAs. Identical phage in the genomic library have been isolated by both the hybridization approach and by the *in vivo* recombination technique. At present we are analyzing the DNA flanking these marker plasmid sequences. Some of it is derived from the original CHO Ama1 donor DNA used in the TS^+Ama^R transformation of TsAF8 cells. We have yet to clearly identify the Ama^R RNA polymerase II DNA in these clones.

B. Identification of CHO RNA Polymerase II DNA Sequences

In the preceding section an approach to recombinant DNA cloning of Ama^R RNA polymerase II DNA was described. It will be necessary to use this cloned DNA in gene transfer experiments to identify which portions encode the structural gene for an RNA polymerase II polypeptide. However, we have already identified DNA encoding this same CHO RNA polymerase II locus by quite another approach.

Mutations analogous to the CHO Ama^R and TS polymerase II have also been described in *D. melanogaster* (Greenleaf et al., 1979; Coulter and Greenleaf, 1982). They have been mapped to a single genetic locus *RpII* (or L5) which resides in the X chromosome (Greenleaf et al., 1980). This *RpII* locus DNA has been cloned (Searles et al., 1982). A number of $polyA^+$ RNAs originate from this region of cloned DNA (Ingles et al., 1983) and thus it was initially difficult to unambiguously establish which of these was the RNA polymerase II gene transcript. Two genomic subclones of *RpII* region DNA which together encode most of a ~ 7-kb polyadenylated transcript were used as probes in cross-species DNA hybridization experiments. Under conditions of reduced stringency only these DNAs detected related sequences in the DNA of several mammalian species (Ingles et al., 1983). As shown in genomic Southern blots (Fig. 15.7), the *D. melanogaster RpII* region clone p4.2 detected a single 7.7-kb fragment in *Eco*RI digested CHO DNA, and an 8.6-kb fragment in the Syrian hamster TsAF8 DNA. The different size of this related sequence in the cell lines of Chinese and Syrian hamster origin suggested immediately an obvious experiment. If the p4.2 and its related mammalian DNA encodes an RNA polymerase II polypeptide, then the RNA polymerase II gene transfer described above should show the transfer of the CHO-specific 7.7-kb fragment from Ama^R CHO cells to the recipient Syrian hamster TsAF8 cell line. DNAs of each of the cells, donor Ama1, recipient TsAF8, and the DNAs of several TS^+Ama^R transformants of TsAF8 were probed with nick translated *D. melanogaster RpII* region p4.2 DNA. The 7.7-kb CHO DNA sequence was present in the DNA of each of the secondary TS^+Ama^R transformants (Fig. 15.7). Since the nick translated p4.2 probe included vector pBR325 sequences, the homologous sequences in cotransferred marker pGA276 DNA were also detected. The copy number of these marker DNAs had increased in some lines and in one case there apparently was coamplification of the 7.7-kb Ama1 DNA fragment. Similar Southern blot experiments have also been performed with transformants derived with Ama^R human DNA as donor. A human DNA-specific 5.7-kb *Eco*RI fragment was detected in the donor human DNA and in the DNA of the resulting TS^+Ama^R transformants (Ingles et al., 1983).

Thus for all primary and secondary TS^+Ama^R transformants of TsAF8 cells, the acquisition of the TS^+ and Ama^R polymerase II phenotype and enzyme activity was accompanied by the appearance of a new *Eco*RI genomic fragment with sequence homology to the *D. melanogaster* p4.2 DNA sequence. The size of this new fragment in each case was identical to that detected in the DNA of the cell line and species used as donor for the gene transfer. These studies indicate that a portion of the RNA polymerase II gene in CHO cells is on a 7.7-kb EcoRI fragment and, furthermore, that within the cloned *RpII* region DNA of *D. melanogaster* the transcript encoded by p4.2 DNA (Searles et al., 1982; Ingles et al., 1983) is the *D. melanogaster* RNA polymerase II gene. Further studies have indicated that the p4.2 DNA encodes a 5′ domain of the *D. melanogaster RpII* gene (Greenleaf, 1983). A second genomic subclone p4.1 flanks p4.2 and encodes a 3′ domain of the *RpII* transcript. This p4.1 DNA also has a mammalian homolog. The p4.1 DNA detected in *Eco*RI digested CHO DNA a 12.0-kb fragment. This 12.0-kb fragment was also cotransferred with the p4.2-related DNA to $TS^{+}Ama^{R}$ transformants of TsAF8 cells (Ingles et al., 1983). Thus in CHO cells the structural gene encoding the polypeptide of RNA polymerase II that determines sensitivity to inhibition by α-amanitin is located on two *Eco*RI fragments 7.7 and 12.0 in size.

C. Identification of the RNA Polymerase II Polypeptide Conferring Ama^R Phenotypes

Although the *D. melanogaster* gene cloning and mammalian DNA transfer experiments identified DNAs encoding an RNA polymerase II polypeptide, the size of this subunit was not ascertained. The *RpII* region transcript encoded by p4.2 and p4.1 DNA *D. melanogaster* was about 7 kb in size, sufficient to encode the largest, 215,000-dalton subunit of polymerase II. However, amatoxins can be chemically cross-linked to the 140,000-dalton subunit of polymerase II (Brodner and Wieland, 1976), and it was long suspected that this subunit would therefore bear the Ama^R and TS polymerase II mutations. This issue has been clarified by two further lines of experiments.

Portions of p4.1 *D. melanogaster RpII* region DNA were subcloned into a bacterial hybrid protein expression vector. Several constructions were obtained that produced hybrid proteins that cross reacted with anti *D. melanogaster* RNA polymerase II antibodies. Use of subunit-specific antibodies indicated that these hybrid proteins display antigenic determinants unique to the largest polymerase II polypeptide, 215,000 daltons in size (Greenleaf, 1983).

When these *D. melanogaster* polymerase II DNAs were used as hybridization probes in cross-species hybridization experiments with *Saccharomyces cerevisiae* DNA, three different regions of homologous DNA were identified. Two of these loci have been molecularly cloned (Ingles et al., 1984). Each contains a sequence related not only to the *D. melanogaster* DNA fragment (p4.2) used as a probe in its isolation, but also to the immediately adjacent p4.1 *D. melanogaster* polymerase II DNA. One of these yeast loci, named *RP021*, encodes a 5.9-kb transcript, sufficient to encode a protein about 220,000 daltons in size. *In vitro* translation of mRNA hybrid-selected with *RP021* DNA indicated that this locus does indeed encode a polypeptide of this size, equal in size to the largest polypeptide subunit of yeast RNA polymerase II. This same DNA has also been shown to encode a polypeptide that bears antigenic determinants recognized by antibodies directed against the yeast 220,000-dalton RNA polymerase II polypeptide (Young and Davis, 1983). Thus this cloned *D. melanogaster* RNA polymerase II DNA, the site of Ama^R mutations, encodes the largest subunit of RNA polymerase II in *D. melanogaster*, in yeast, and, most likely, in mammalian cells. It is this largest subunit of the mammalian enzyme that is antigenically related to the *D. melanogaster* 215,000-dalton polypeptide. This largest subunit must therefore play the major role in determining the enzyme's affinity for α-amanitin. That amatoxins were cross-linked to the second largest (140,000-dalton) subunit of polymerase II (Brodner and Wieland, 1976) may only indicate that these two subunits are physically close in the active assembled enzyme. Alternatively, the mutations in the largest subunit may have pleiotropic effects on the binding of α-amanitin by a second subunit.

Immunological studies had indicated that there is a conservation of antigenic determinants among RNA polymerases I, II, and III (Ingles, 1973). These different RNA polymerases in different eukaryotic species are also sensitive to varying extents to inhibition by α-amanitin. They may share with RNA polymerase II a closely related α-amanitin-binding domain. Thus it is likely that there are conserved nucleotide sequences among the genes encoding some subunits of RNA polymerases I, II, and III. We suspect that the family of *S. cerevisiae* loci detected by *D. melanogaster* polymerase II DNA likely encodes the largest subunit of each of the enzymes, RNA polymerases I, II, and III. These DNAs may be members of a gene family which has evolved from a single primordial ancestor RNA polymerase.

VII. FUTURE PROSPECTS

The identification and molecular cloning of RNA polymerase genes will provide new tools to be used in gaining insights into other aspects of RNA polymerase function. The ability to use gene cloning to define genetically various components of the transcription complexes of eukaryotic cells and to analyze in detail mutations affecting RNA synthesis will be an asset in studies aimed at understanding the molecular details of events regulating the activity and selectivity of RNA polymerases. That some of the transcription machinery is so similar in eukaryotic species as divergent as yeast and CHO cells is fortuitous. It may well be that the genetic studies of transcription in yeast where gene cloning, mapping, and replacement with *in vitro* mutated DNAs can be done using a number of newly developed chromosomal, episomal, and integrating recombinant DNA vectors, will provide other cloned yeast DNAs that, like RNA polymerase II DNA, cross-hybridize with DNA of higher eukaryotes. DNA encoding CHO RNA polymerase polypeptides, identified and cloned initially by making use of the Ama^R CHO mutations, may be studied in the future making use of this cross-species homology. The molecular genetics of transcription in the Chinese hamster cell may get a strong assist from studies in other eukaryotic species.

ACKNOWLEDGMENT

Research in the author's laboratory has been generously supported by grants from the Medical Research Council (Canada) and the National Cancer Institute of Canada.

REFERENCES

Amati, P., Blasi, F., DiPorzio, U., Riccio, A., and Treboni, C. (1975). *Proc. Natl. Acad. Sci. USA* **72**, 753–757.

Benton, W. D. and Davis, R. W. (1977). *Science* **126**, 180–182.

Brodner, O. G. and Wieland, T. (1976). *Biochem.* **15**, 3480–3484.

Buchwald, M. and Ingles, C. J. (1976). *Somat. Cell Genet.* **2**, 225–233.

Burstin, S. J., Meiss, H. K., and Basilico, C. (1974). *J. Cell Physiol.* **84**, 397–408.

Chambon, P. (1975). *Ann. Rev. Biochem.* **44**, 613–638.

Chan, V. L., Whitmore, G. F., and Siminovitch, L. (1972). *Proc. Natl. Acad. Sci. USA* **69**, 3119–3123.

Cochet-Meilhac, M. and Chambon, P. (1974). *Biochim. Biophys. Acta* **353**, 160–184.

Coulter, D. E. and Greenleaf, A. L. (1982). *J. Biol. Chem.* **257**, 1945–1952.

Greenleaf, A. L. (1983). *J. Biol. Chem.* **258**, 13403–13406.

Greenleaf, A. L., Borsett, L. M., Jiamachello, P. F., and Coulter, D. E. (1979). *Cell* **18**, 613–622.

Greenleaf, A. L., Weeks, J. R., Voelker, R. A., Ohnishi, S., and Dickson, B. (1980). *Cell* **21**, 785–792.

Goldberger, R. F. (1974). *Science* **183**, 810–816.

Goldfarb, M. P., Shimizu, K., Perucho, M., and Wigler, M. H. (1982). *Nature* **296**, 404–409.

Guialis, A., Beatty, B. G., Ingles, C. J., and Crerar, M. M. (1977). *Cell* **10**, 53–60.

Guialis, A., Morrison, K. E. and Ingles, C. J. (1979). *J. Biol. Chem.* **254**, 4171–4176.

Gupta, R. S., Chan, D. H. Y., and Siminovitch, L. (1978). *J. Cell Physiol.* **97**, 461–468.

Harris, M. (1971). *J. Cell Physiol.* **78**, 177–184.

Hartman, P. E. and Roth, J. R. (1973). *Adv. Genet.* **17**, 1–105.

Hayward, R. S., Tittawella, I. P. B., and Scaife, J. G. (1973). *Nature New Biol.* **243**, 6–9.

Ingles, C. J. (1973). *Biochem. Biophys. Res. Commun.* **55**, 364–371.

Ingles, C. J. (1978). *Proc. Natl. Acad. Sci. USA* **75**, 405–409.

Ingles, C. J., Guialis, A., Lam, L., and Siminovitch, L. (1976). *J. Biol. Chem.* **251**, 2729–2734.

Ingles, C. J., Biggs, J., Wong, J. K-C., Weeks, J. R., and Greenleaf, A. L. (1983). *Proc. Natl. Acad. Sci. USA* **80**, 3396–3400.

Ingles, C. J., Himmelfarb, H. J., Shales, M., Greenleaf, A. L., and Friesen, J. D. (1984). *Proc. Natl. Acad. Sci. USA* **81**, 2157–2161.

Ingles, C. J. and Shales, M. (1982). *Molec. Cell Biol.* **2**, 666–673.

Jarvik, J. and Botstein (1975). *Proc. Natl. Acad. Sci. USA* **72**, 2738–2742.

Kirschbaum, J. B., Claeys, I. V., Nasi, S., Molholt, B. and Miller, J. H. (1975). *Proc. Natl. Acad. Sci. USA* **72**, 2375–2379.

Lewis, M. K. and Burgess, R. R. (1982). In *The Enzymes* (P. D. Boyer, ed.), Academic Press, New York, Vol. xv, pp. 109–152.

Lindell, T. J., Weinberg, F., Morris, P. W., Roeder, R. G., and Rutter, W. J. (1970). *Science* **170**, 447–449.

Lobban, P. E. and Siminovitch, L. (1975). *Cell* **4**, 167–172.

Meiss, H. K. and Basilico, C. (1972). *Nature New Biol.* **239**, 66–68.

Metzger-Freed, L. (1972). *Nature New Biol.* **235**, 245–246.

Murray, N. (1983). In *Lambda II* (R. W. Hendrix, J. W. Roberts, F. W. Stahl, and R. A. Weisberg, eds.), Cold Spring Harbor Laboratory, Cold Spring Harbor, New York, pp. 395–432.

Oeschger, M. P. and Berlyn, M. K. B. (1974). *Mol. Gen. Genet.* **134**, 77–83.

Perucho, M., Hanahan, D., Lipsich, L., and Wigler, M. (1980). *Nature* **285**, 207–210.

Roeder, R. G. (1976). In *RNA Polymerase* (R. Losick, and M. Chamberlin, eds.), Cold Spring Harbor Laboratory, Cold Spring Harbor, New York, pp. 285–329.

Roeder, R. G. and Rutter, W. J. (1969). *Nature* **224**, 234–237.

Rossini, M. and Baserga, R. (1978). *Biochem.* **17**, 858–863.

Rossini, M., Baserga, S., Huang, C. H., Ingles, C. J., and Baserga, R. (1980). *J. Cell Physiol.* **103**, 97–103.

Searles, L. L., Jokerst, R. S., Bingham, P. M., Voelker, R. A., and Greenleaf, A. L. (1982). *Cell* **31**, 585–592.

Seed, B. (1983). *Nucl. Acids Res.* **11**, 2427–2445.

Sentenac, A. and Hall, B. (1982). In *The Molecular Biology of the Yeast Saccharomyces, Metabolism and Expression* (J. N. Strathern, E. W. Jones, J. R. Broach, eds.), Cold Spring Harbor Laboratory, Cold Spring Harbor, New York, pp. 561–606.

Shales, M., Bergsagel, J., and Ingles, C. J. (1980). *J. Cell Physiol.* **105**, 527–532.

Somers, D. G., Pearson, M. L., and Ingles, C. J. (1975a). *J. Biol. Chem.* **250**, 4825–4831.

Somers, D. G., Pearson, M. L., and Ingles, C. J. (1975b). *Nature* **253**, 372–374.

Weinberg, R. W. (1982). *Adv. Cancer Res.* **36**, 149–163.

Wieland, Th. and Fahrmeir, A. (1970). *Leibigs Ann. Chem.* **736**, 95–99.

Wulf, E. and Bautz, H. (1976). *FEBS Lett.* **69**, 6–10.

Young, R. A. and Davis, R. W. (1983). *Science* **222**, 778–782.

CHAPTER 16

THE DIHYDROFOLATE REDUCTASE LOCUS

Lawrence Chasin
Department of Biological Sciences
Columbia University
New York, New York

I. INTRODUCTION

The enzyme dihydrofolate reductase (DHFR, E.C. 1.5.1.3) plays a central role in the intermediary metabolism of all cells, and for many years has been a focus of attention for studies of cellular metabolism, cell cycle regulation, mechanism of enzyme action, and cancer chemotherapy. More recently, the genetic locus specifying this enzyme has been the object of many somatic cell genetic studies. An attractive feature of the *dhfr* locus for genetic experiments has been the ability to isolate recombinant DNA clones containing *dhfr* gene sequences and the availability of powerful selection techniques for the isolation of several different types of mutants. Three basic mutant phenotypes have been studied: (1) enzyme deficiency, (2) enzyme structural alteration, and (3) enzyme overproduction. It is this third phenotype that has been most extensively investigated. The mechanism by which mutant cells overproduce DHFR has been shown to be gene amplification, that is, the accumulation of multiple copies of the *dhfr* gene. Indeed, the phenomenon of gene amplification as a mutational event in cultured mammalian cells was first demonstrated at the *dhfr* locus (Alt et al., 1978). The availability of amplified mutants containing hundreds of copies of the *dhfr* gene and hundreds of times more *dhfr* mRNA and enzyme protein than wild-type cells have facilitated the study of the other two mutant phenotypes (enzyme deficiency and enzyme alteration) as well as experiments on the regulation of *dhfr* gene expression and on gene transfer. In this chapter, the isolation and characterization of DHFR-deficient mutants of Chinese hamster ovary (CHO) cells will be described in some detail first. This work is mostly that of the author's own laboratory. This will be followed by a description of mutants with altered DHFR activity and a review of gene amplification at the *dhfr* locus. Finally, the results of gene transfer experiments involving the *dhfr* gene will be summarized. The mammalian cell lines that have been used for these genetic studies include human, mouse, Syrian hamster, and Chinese hamster. This chapter will deal principally with results from Chinese hamster cells, where all three phenotypes have been studied.

DHFR is a small (MW 21,000) monomeric protein that catalyzes the reduction of folic acid to dihydrofolic acid and thence to tetrahydrofolic acid. The substrate folic acid is provided as a vitamin in the growth medium; NADPH provides the reducing power for each step. The metabolic role of DHFR is to maintain adequate cellular levels of tetrahydrofolic acid. This cofactor is the carrier of one-carbon units for a variety of biosynthetic reactions, including the synthesis of glycine from serine, the formation of the purine ring in nucleotide biosynthesis, and the addition of the 5-methyl group of thymidylic acid. The synthesis of methionine and possibly tyrosine (Nichol et al., 1983) also requires tetrahydrofolate, but since most cell lines exhibit a requirement for these amino acids (Naylor et al., 1976), these roles are usually not considered.

In the absence of folate in the medium, cells exhibit a requirement for glycine, a purine, and thymidine. The purine requirement is usually satisfied with hypoxanthine or adenine. These same requirements can be induced by including an inhibitor of DHFR in the growth medium. An "antifolate" drug such as methotrexate (MTX, amethopterin) binds the enzyme with great affinity (Williams et al., 1979) and results in a depletion of cellular tetrahydrofolate. The resulting inhibition of TMP (and perhaps purine) synthesis leads to a cessation of DNA synthesis; it is this consequence that presumably underlies the effectiveness of MTX as a cancer chemotherapeutic agent (Bertino, 1979). As will be seen subsequently, the availability of a potent inhibitor of DHFR such as MTX has played an important role in the isolation of negative, altered, and amplified mutants at the *dhfr* locus.

II. REGULATION OF *dhfr* GENE EXPRESSION

The important role of DHFR in thymidylate synthesis is underscored by the fact that its activity increases when DNA synthesis is initiated during the cell cycle or after the stimulation of quiescent cells to divide. The cell cycle control of DHFR in amplified CHO cells has been studied by Mariani et al. (1981) using fluorescein-MTX and the fluorescence-activated cell sorter. They found that cellular DHFR content did not increase in a synchronized population until DNA synthesis had been initiated. This result was confirmed by the direct measurement of DHFR protein synthesis. It is not yet known at what level this effect is exerted.

The response of DHFR to changes in the cell growth phase (as opposed to the cell cycle *per se*) has been more extensively investigated. For some time it has been known that DHFR activity is substantially higher in exponentially growing cells than in stationary cultures (Hillcoat et al., 1967), and that this effect is caused by differences in the rate of enzyme synthesis (Alt et al., 1976). Working with amplified mouse mutants, Kellems and his colleagues have shown that this increase is due to corresponding changes in the level of *dhfr* mRNA. This increase can be induced in stationary phase (confluent) cultures by replating at a lower density or by infection with polyoma or adenovirus (Kellems et al., 1976, 1982; Yoder et al., 1983). The increase due to replating could be inhibited by cyclic AMP. The increased steady-state level of *dhfr* mRNA is accompanied by an increased rate of appearance of this RNA in the cytoplasm (Leys and Kellems, 1981). However, by performing experiments in which very short RNA pulse labeling times were used, it could be shown that transcription rates are the same in stimulated versus unstimulated cells. Thus under these conditions the synthesis of DHFR may be controlled at the level of intranuclear transcript stability.

Johnson and his co-workers have examined the regulation of DHFR synthesis in an amplified mutant of mouse 3T6 cells. The relative rate of DHFR synthesis is stimulated about fivefold upon adding serum back to cultures arrested by serum deprivation. The rate of DHFR synthesis and *dhfr* mRNA levels increase 10–20 hr after serum stimulation, at which time DNA synthesis begins (Johnson et al., 1978; Weidemann and Johnson, 1979; Wu and Johnson, 1982). Here, the evidence suggests that transcriptional stimulation is the primary event (Wu and Johnson, 1982; Collins et al., 1983). When these same cells are growth stimulated after arrest by amino acid deprivation, a fourfold increase in the rate of *dhfr* mRNA labeling is observed. In this case the rate of transcription is not affected, but rather there is an increased rate of processing of nuclear precursor RNA, with the accumulation of RNA sequences in the nucleus (Collins et al., 1983).

While it is not yet clear that all of these different mechanisms represent physiologically significant regulatory processes, the potential for control at a variety of levels obviously exists. Moreover, the regulation of DHFR activity within the cell may be spatially as well as temporally regulated, since a portion of the cellular DHFR activity is associated with a nuclear multienzyme complex involved in DNA synthesis (Prem veer Reddy and Pardee, 1980, 1982).

III. DHFR-DEFICIENT MUTANTS

For most genetic loci and for most genetic studies, the fundamental mutant phenotype is the inability to express the gene residing at that locus. The most obvious kind of lesion that can confer this phenotype on an enzyme-coding gene is one that alters the amino acid sequence of the protein through a missense, nonsense, frameshift, or deletion mutation. Point mutations that alter the activity of DHFR through single amino acid substitutions should be useful for confirming and extending what is known about the active site of this well-studied enzyme. One might also expect mutations that prevent the expression of the *dhfr* gene without impinging on protein-coding information. In theory, any of the steps in the flow of information from DNA to protein could be affected: transcription, termination, polyadenylation, splicing, transport of mRNA to the cytoplasm, translation, and mRNA stability. The definition of aspects of gene structure that are important for these processes may be provided by an analysis of mutants blocked at one or more of these steps. A third type of information that can be derived from an analysis of DHFR-deficient mutants is the elucidation at the DNA level of the types of genetic change that take place in the genome of somatic mammalian cells. These include alterations induced by the action of chemical and physical agents as well as spontaneous mutational events.

The most complete fine-structure genetic analysis of mutations occurring in mammals has been the definition of the nucleotide changes in the globin genes of individuals with thalassemia [see Maniatis et al. (1980) for a review]. This type of *in vivo* mutational analysis can now be extended to mutants generated in cell culture, as long as two requirements are met: There must be a selective system for the efficient isolation of negative mutants and cloned gene sequences must be available for use in localizing the mutations. As described subsequently, these conditions are satisfied by the *dhfr* system in CHO cells. In addition, several other genes coding for household enzymes are now amenable to this type of fine-structure mutational analysis. These include the genes for the adenine and hypoxanthine phophoribosyl transferases in Chinese hamster cells that are described in other chapters.

A. Selection Methods

1. *Radioactive Deoxyuridine as a Suicide Selection Agent*

MTX-treated cells serve as a convenient phenocopy for many of the properties expected of a DHFR-deficient mutant. Thus, just as MTX-induced growth inhibition can be reversed by provision of salvageable sources of the end products of one-carbon metabolism (Hakala and Taylor, 1959), mutants lacking DHFR activity should be viable as long as glycine, a purine, and thymidine are added to the medium. That is, the growth phenotype of DHFR-deficient mutants should be a triple auxotrophy relative to parental wild-type cells.

A method that selects against DHFR-positive CHO cells has been devised based on the role of the enzyme in the *de novo* synthesis of TMP (Urlaub and Chasin, 1980). A population of cells is exposed to tritated deoxyuridine labeled in the 6 position ([^{3}H]dUrd). Like tritiated thymidine, the [^{3}H]dUrd is toxic to wild-type cells by virtue of its incorporation into DNA and its subsequent radioactive decay. In order to be incorporated into DNA, [^{3}H]dUrd must first be converted to TMP. Four reactions are necessary for this conversion: the synthesis of tetrahydrofolate by DHFR, the attachment of a one-carbon unit to tetrahydrofolate by serine hydroxymethyltransferase, the phosphorylation of [^{3}H]dUrd by thymidine kinase, and the methylation of dUMP by thymidylate synthetase. A deficiency in any one of these four steps should lead to resistance to the toxic effects of [^{3}H]dUrd, and, in fact, mouse cell mutants lacking thymidylate synthetase activity have been isolated using this selective agent (Ayusawa et al., 1980). Mutants lacking serine hydroxymethyltransferase activity have been isolated using different methods. These mutants are not triple auxotrophs, but require only glycine (Kao et al., 1969); they contain residual enzyme levels due to the presence of at least two isozymes for this enzymatic activity (Chasin et al., 1974). Thymidine-kinase-deficient mutants would also be resistant, but have not been found by this method.

The exposure to [^{3}H]dUrd must take place in a medium that also contains glycine, a purine, and thymidine, since starvation for the last results in rapid cell death. Even low levels of thymidine (100 n*M*) compromise the effectiveness of [^{3}H]dUrd killing. Nevertheless, the use of high specific activity [^{3}H]dUrd (25 Ci/mmole) permits killing down to 0.01% after a 24-hr exposure (Urlaub and Chasin, 1980). With longer or higher exposures even mutant cells begin to die, perhaps because tritium begins to enter the general metabolic pool. Lower amounts of [^{3}H]dUrd can be used if the exposed cells are stored frozen for several weeks to allow the accumulation of radioactive disintegrations. As expected, the [^{3}H]dUrd-induced killing can be completely prevented by the inclusion of MTX in the medium to inhibit DHFR.

2. Selection of a *dhfr* Heterozygote from Pseudodiploid CHO Cells

The application of this selection technique to wild-type cells of the CHO-K1 line (Kao and Puck, 1968) does not yield any DHFR-deficient mutants, a result consistent with the presence of two copies of this gene per cell. The *dhfr* gene is known to be located on an autosome in Chinese hamster cells (Roberts et al., 1980; Worton et al., 1981), and these cells contain an approximately diploid amount of DNA. It is not feasible to screen a sufficient number of cells to isolate double mutants using the [^{3}H]dUrd technique. Therefore, a modified version of the [^{3}H]dUrd selection was used to isolate a heterozygous mutant containing one active and one mutant *dhfr* gene, and containing approximately one-half of the DHFR activity of wild-type cells. In this selection, a low concentration of MTX was included along with the [^{3}H]dUrd. The rationale was that at an appropriate concentration of MTX, all of the DHFR activity of a heterozygous mutant would be inhibited, rendering these cells resistant to [^{3}H]dUrd, while the wild-type cells would still contain sufficient active enzyme to be killed by the radioactive drug. Mutagenesis of CHO-K1 cells followed by several rounds of this selection regimen yielded one heterozygous mutant (Urlaub and Chasin, 1980). As predicted, this mutant (UKB25) contains one-half of the DHFR-specific activity of the wild-type parental cells. When the original selection method using [^{3}H]dUrd alone is applied to mutagenized populations of clone UKB25, resistant mutants are isolated at the high frequencies expected for a single gene target (e.g., 10^{-4} after ethyl methanesulfonate treatment). Most of these resistant mutants exhibit requirements for glycine, a purine, and thymidine, and all of these auxotrophs contain little or no DHFR activity. This strategy for selecting recessive mutations at diploid loci by first isolating a heterozygote using a marginal selection method has been previously used at the adenine phosphoribosyltransferase locus in CHO cells (Jones and Sargent, 1974; Adair et al., 1980; Bradley and Letovanec, 1982) and for the glucocorticoid receptor in mouse lymphoma cells (Bourgeois and Newby, 1977). It thus may be generally applicable as long as one has the means of selecting for a reduced level of the gene product.

3. *Selection of dhfr Hemizygotes*

The availability of the heterozygous mutant UKB25 allows the isolation of large numbers of DHFR-deficient mutants whose phenotype can be easily studied at the enzyme level. However, the analysis of the effects of the second mutation at the level of *dhfr* mRNA content and *dhfr* DNA sequence change is complicated by the fact that the first mutant allele (most likely containing a point mutation induced by ethyl methanesulfonate) is still present. The continued presence of the first mutant allele has been directly demonstrated by the analysis of restriction fragments in DHFR-deficient mutants induced by ionizing radiation. In two such mutants, new bands appear in Southern blots probed for *dhfr* sequences, indicating that a gross alteration has taken place at one allele. However, none of the wild-type fragments disappears in these mutants; the other allele is nonfunctional but still physically present (Graf and Chasin, 1982).

This problem has been solved by the isolation of truly hemizygous mutants that carry only one copy of the *dhfr* gene (Urlaub et al., 1983). In this case, the starting cell line was not CHO-K1, but a MTX-resistant CHO mutant, Mtx-RIII, isolated by Flintoff et al. (1976a). Mtx-RIII was originally selected in two steps. The first step yielded a mutant that produces wild-type levels of a structurally altered DHFR enzyme whose catalytic activity is resistant to low levels of MTX (Flintoff and Essani, 1980). A subsequent selection for resistance to higher concentrations of MTX resulted in an amplified mutant carrying about 20 copies of this

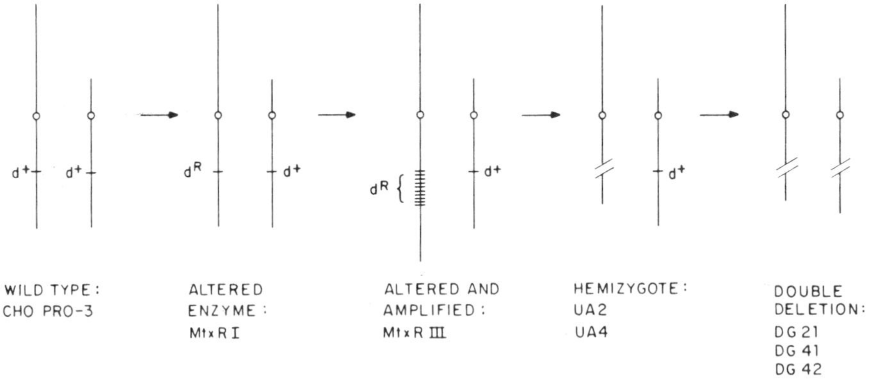

Figure 16.1. Interpretation of mutant cell lineages. The left chromosome in each pair represents Chinese hamster chromosome 2. The MTX-resistance marker of clone pro-3 Mtx-RIII has been mapped to one arm of this chromosome (Worton et al., 1981). The chromosome on the right represents Z2 of the CHO karyotype. Whereas this chromosome has undergone a deletion in one arm, the other arm retains a banding pattern identical to the homologous arm in chromosome 2 and is presumed to carry the second *dhfr* allele. The diagonal hatch marks indicate a deletion; R indicates a mutant allele coding for a MTX-resistant form of DHFR. From Urlaub et al. (1983).

structurally altered allele (Wigler et al., 1980; Flintoff et al., 1982). The hemizygote selection scheme was based on the assumption of diploidy at the *dhfr* locus in these cells; in addition to the amplified allele a second unamplified wild type allele would be present. By applying the combination of MTX plus [^{3}H]dUrd described previously, cells having even one copy of the gene coding for the altered enzyme should be resistant to MTX and be killed. In contrast, a mutant that had lost all copies of the resistant allele would be spared: The relatively low level of wild-type DHFR specified by the one remaining *dhfr* allele would be inhibited by the MTX in the medium. This selection scheme is depicted in Figure 16.1.

Two independent clones that survived this selection were isolated from populations treated with gamma rays to promote deletion mutations. These mutants behave like the expected hemizygotes according to several criteria (Urlaub et al., 1983). First, enzyme levels have dropped to somewhat less than the wild-type value, or one-tenth of that found in the parental amplified line. Second, the enzyme activity that remains is fully sensitive to MTX, indicating that it is being specified by a wild-type allele. Third, *dhfr* gene copy number, as estimated from Southern blots, has returned to somewhat less than wild-type levels. Finally, a second round of gamma-ray mutagenesis yielded DHFR-deficient mutants containing no detectable *dhfr* DNA sequences by Southern blot analysis (e.g., Fig. 16.3, lane 2). Thus all traces of the amplified *dhfr* allele have been deleted in the first mutagenesis, and the second mutagenesis deleted the remaining wild-type allele. This last result rules out the possibility that the amplified *dhfr* genes were simultaneously inactivated by gene conversion (Roberts and Axel, 1982).

These hemizygous clones have served as the starting point for the isolation of a number of mutants affected in the single remaining allele. The properties of these DHFR-deficient mutants are described subsequently.

4. Selection of MTX-Sensitive Cells from MTX-Resistant Populations

The same strategy that was used to isolate hemizygotes has also been used for segregation analysis of somatic cell hybrids. Intraspecific CHO hybrids have been formed where one parent was a MTX-resistant mutant carrying amplified copies of the *dhfr* gene and the other parent was either a wild-type cell or a DHFR-deficient mutant. Although these hybrids exhibit an intermediate level of resistance to MTX, their resistance is still much greater than that of the wild-type parent (Flintoff et al., 1976b). Segregants that have lost the amplified genes have been selected using a combination of MTX and [^{3}H]dUrd as described previously (Urlaub et al., 1981). When the nonresistant parent was a wild-type cell, segregants contained approximately wild-type levels of DHFR activity. When the nonresistant parent was a DHFR-deficient mutant, DHFR-specific activities in segregants were only about one-third that of wild-type cells. This amount of enzyme is close to that predicted on a gene dosage basis if these near-tetraploid segregants contain only one functional gene. The origin of this gene is presumably the unamplified wild-type allele of the MTX-resistant parent. Karyotypic examination showed that all of these MTX-sensitive segregants had lost the homogeneously staining chromosome region that harbors the amplified *dhfr* genes (see below). Curiously, this deletion almost always occurred by chromosome breakage rather than chromosome loss (Urlaub et al., 1981).

Flintoff and Weber (1980) used this selection technique to isolate wild-type cells from a population of MTX-resistant mutant cells containing an altered DHFR (pro-3 MtxRI). These revertants contain DHFR activity that exhibits wild-type sensitivity to MTX. They could have arisen by true reversion of the altered *dhfr* gene. Alternatively, if the pro-3 MtxRI mutant contains one wild-type allele in addition to the structurally altered gene (as proposed in the scheme of Fig. 16.1), then reversion to a MTX-sensitive phenotype could have arisen via a new mutation that destroyed the activity of the resistant enzyme. This latter interpretation is supported by the isolation of hemizygotes from a derivative of this mutant, as described previously.

TABLE 16.1
Characteristics of DHFR-Deficient Mutants

Parent	Mutagen	Number of Examples	DHFR[a]	mRNA[b]	DNA[c]
UKB25 (heterozygote)	Gamma rays	7	≤0.01(7)	NA[d]	WT
		2	≤0.01(2)	NA	WT
	EMS	1	≤0.01	NA	WT
		1	0.02	NA	WT
UA2 or UA4 (hemizygotes)	Gamma rays	7	≤0.01(1)	0.02	Deleted
		3	NA	0.02	Interrupted
		1	NA	0.02	5′ deleted
	UV	5	0.003(1)	WT	WT
		8	≤0.001(3)	0.03–0.20	WT
		2	≤0.001(1)	0.02	Deleted

[a]DHFR activity is expressed as the proportion of the parental value; the number in parentheses indicates the number of mutants analyzed.
[b]Messenger RNA levels are estimates from the intensity of autoradiographic bands of Northern blots, and are expressed as the proportion of the parental value; WT indicates a mRNA level indistinguishable from the parental value.
[c]The interpretation of DNA restriction fragment patterns is given.
[d]NA, not analyzed.

B. Biochemical Characterization of DHFR-Deficient Mutants

1. DHFR Enzyme Levels

DHFR activity in wild-type and deficient cells has been quantitated either by a standard catalytic assay (Frearson et al., 1966) or by a more convenient radioactive MTX binding assay (Johnson et al., 1978; Urlaub and Chasin, 1980). Not all clones that survive the [^{3}H]dUrd selection exhibit the auxotrophy characteristic of a DHFR deficiency; these presumably represent wild-type cells that have slipped through the selection. Upon testing, all clones that do exhibit nutritional requirements are found to lack DHFR activity (Table 16.1). In most cases the sensitivity of the assay is sufficient to detect 1% of wild-type DHFR levels. In only one case (DUK22) has a residual level of DHFR activity been detected in a DHFR-deficient mutant. This EMS-induced mutant exhibits 2% of wild-type DHFR-specific activity by either assay. The residual activity can be qualitatively distinguished from that of wild type on the basis of its increased heat lability at 43°C (Chasin et al., 1982). The appearance of this altered activity in one mutant indicates that at least some of these mutations are directly affecting the structural gene for *dhfr*; this idea is supported by direct examination of *dhfr* DNA sequences described subsequently.

The lack of significant MTX binding activity in DHFR-deficient mutants suggests that no other binding protein with a high affinity for tetrahydrofolate is present in CHO cells; such proteins have been reported to be present in some mammalian tissues [see Huennekens et al. (1976) for a review].

2. Growth Response

The growth requirements of DHFR-deficient mutants can be met by providing sources of the end products of one-carbon metabolism; in the presence of glycine, hypoxanthine, and thymidine growth rates and plating efficiencies are close to that of wild-type cells. It might also be expected that these growth requirements could be met by the addition of a source of tetrahydrofolic acid. Tetrahydrofolic acid itself cannot be provided for growth because of its instability. However, two tetrahydrofolate derivatives carrying one-carbon units are sufficiently stable to act as a source of the reduced cofactor: folinic acid (5-formyl-tetrahydrofolate) and 5-methyl-tetrahydrofolate. These compounds will substitute for the three end products of one-carbon metabolism for CHO cells that are grown in the presence of MTX or that are grown in folate-free medium. In fact, folinic acid is usually used to offset the toxic effects of MTX in cancer chemotherapy. It was therefore surprising to find that even high concentrations of these two compounds would not satisfy the hypoxanthine or thymidine requirements of DHFR-deficient mutants (Chasin et al., unpublished results). In contrast, the glycine requirement of the mutants can be satisfied by 100 n*M* folinic acid. This last result indicates that folinate enters these cells and is metabolized to yield the free tetrahydrofolate necessary for glycine synthesis. This inability of the mutant cells to utilize exogenous tetrahydrofolate may be related to the fact that a significant fraction of the DHFR activity in mammalian cells is part of a "replitase" complex that contains several enzymes involved in deoxynucleotide metabolism and DNA synthesis (Prem veer Reddy and Pardee, 1980, 1982). One characteristic of the replitase is that substrates are channeled into a sequence of reactions, all catalyzed by members of the complex. Thus it is possible that the synthesis of thymidylate from uridine takes place in several steps, all of which must be catalyzed by enzymes located within the replitase. The absence of active DHFR in the mutants may alter the replitase structure such that it can no longer function in thymidylate synthesis. In particular, it would be interesting to examine the state of the replitase in the DHFR double deletion mutant, since one of the structural components of the complex would be missing in this case.

C. Genetic Properties of DHFR-Deficient Mutants

The *dhfr* locus has been assigned to chromosome 2 in CHO cells on the basis of segregation studies using mouse–Chinese hamster hybrids (Roberts et al., 1980). The altered *dhfr* genes in the MTX-resistant mutant pro-3 MtxRIII have also been mapped to chromosome 2 (Worton et al., 1981). Chromosome 2 often contains amplified *dhfr* sequences in expanded homogeneously staining regions (HSRs) in MTX-resistant Chinese hamster lung cells (Biedler and Spengler, 1976; Lewis et al., 1982a) and CHO cells (Nunberg et al., 1978). In CHO cells, one chromosome 2 homologue has undergone a major deletion early in the evolution of this cell line. The affected chromosome, called Z-2 (Deaven and Peterson, 1973), retains apparently intact the arm thought to contain the *dhfr* locus. The nomenclature here can be confusing since this arm was originally designated as the long arm but more recently has been judged to be the short arm (Ray and Mohandas, 1976). Thus it is probable that the two *dhfr* alleles in wild-type CHO cells reside on chromosomes 2 and Z-2, but the latter location has not been directly demonstrated. Chapter 5 discusses nomenclature of Chinese hamster chromosomes in more detail.

Prototype DHFR-deficient mutants have been fused to wild-type cells and to each other to test for complementation in cell hybrids. The mutant growth phenotype (triple auxotrophy) is recessive to wild type, and no complementation has been observed (Urlaub and Chasin, 1980). These results are consistent with the idea that the DHFR-deficient mutants represent lesions in the structural gene for DHFR. However, since no systematic complementation analysis has been carried out, the existence of regulatory mutants cannot be ruled out. Cell hybrids have also been used to test for linkage between *dhfr* and the *glyB* locus. The latter represents a class of mutants affected in folic acid metabolism, since they respond to folate derivatives as well as glycine for growth (Kao et al., 1969). These two loci were found to segregate independently from hybrids (Urlaub et al., 1981), indicating a lack of linkage.

Mutation rates to DHFR deficiency in cells having one functional copy of the *dhfr* gene (heterozygous or hemizygous) are similar to those observed at other single gene loci in these CHO cells, for example, the X-linked hypoxanthine phosphoribosyltransferase locus (Chasin, 1973) and the adenine phosphoribosyltransferase locus in cells heterozygous for that gene (Chasin et al., 1974); Jones and Sargent, 1974; Adair et al., 1980). Spontaneous mutants appear at a frequency of $(2-5) \times 10^{-7}$, EMS and UV irradiation yield mutants at a frequency of $(1-2) \times 10^{-4}$, and gamma rays induce mutants at a frequency of about 2×10^{-5} (Urlaub and Chasin, 1980; Urlaub et al., 1983; and P. Mitchell, unpublished results).

DHFR-deficient mutants are very stable. Over a dozen mutants have been tested for spontaneous and induced reversion. Revertants can be selected in medium lacking any one or all three of the nutrients required by DHFR-deficient cells. When spontaneous revertants can be obtained, they occur at a frequency of about 10^{-7}. In the case of two EMS-induced mutants derived from the heterozygote (i.e., probably carrying two missense mutations), the revertant frequency can be increased 10-fold by another EMS treatment. A double deletion mutant lacking all traces of the *dhfr* gene has never been found to revert (less than 10^{-8}). However, selection of mutagenized (EMS) populations of the double deletion mutant for glycine independence alone can yield slow-growing colonies. These clones are able to grow without glycine even in folate-free or MTX-containing medium (unpublished results). They have apparently acquired another pathway for glycine biosynthesis, perhaps from threonine in the medium, since this amino acid can act as a substrate for serine hydroxymethyltransferase in the absence of tetrahydrofolate (Schirch, 1982).

D. Molecular Biological Characterization of DHFR-Deficient Mutants

1. Structure of the dhfr Gene in CHO Cells

Genomic sequences corresponding to the CHO *dhfr* gene have been cloned and a restriction map of the locus constructed. This work has been carried out independently in our own laboratory, where initial recombinants were cloned in lambda (Carothers et al., 1983), and in the laboratory of J. Hamlin, where a cosmid library was constructed (Milbrandt et al., 1983a). Both studies benefited from the availability of mouse *dhfr* cDNA and genomic clones (Chang et al., 1978; Crouse et al., 1982) which could be used as initial probes to detect Chinese hamster *dhfr* DNA sequences. Moreover, the use of amplified mutants as a source of DNA also facilitated the isolation of recombinant clones. Complementary DNA clones have also been isolated from amplified mutants of CHO (Carothers et al., 1983) and Chinese hamster lung (CHL) cells (Lewis et al., 1981; Melera et al., 1984). In the case of CHO, four clones spanning 1900 of the 2100 bases of the largest species of *dhfr* mRNA (see below) have been isolated; in the case of CHL, complete protein coding sequences have been cloned that are able to express active DHFR enzyme in *Escherichia coli.* These cDNA clones, along with mouse cDNA clones, have been used to identify exon sequences in the genomic restriction map.

A restriction map of the CHO *dhfr* gene is shown in Figure 16.2. The gene is relatively large, with 26 kb of DNA divided into six exons and five introns. The 5′ end of the gene has not yet been definitely established in either mouse or Chinese hamster. The most 5′ cDNA

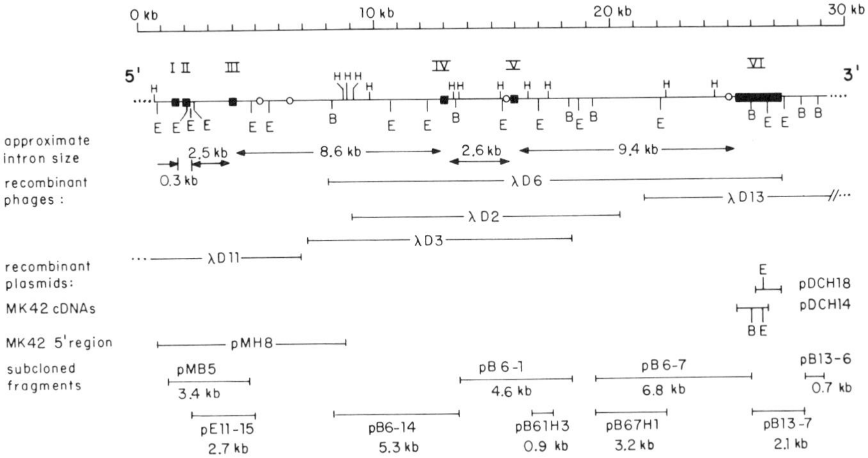

Figure 16.2. The *dhfr* locus of CHO cells. A restriction map of the gene is diagrammed showing the location of *Eco*RI (E), *Bam*HI (B), and *Hin*dIII (H) cleavage sites. The exons are represented as filled boxes. Repeated sequences are represented as open circles. Below the map, the approximate intron sizes are shown. The map was generated by restriction analysis of several *dhfr* recombinant lambda phages. The size and location of their inserts are drawn below the corresponding genomic regions. Also shown are the inserts of two cDNA sequences (pDCH14 and pDCH18) cloned in pBR322, and plasmid subclones derived from the recombinant phages. From Carothers et al. (1983).

cloned is from the mouse and extends 112 bp upstream of the ATG translation initiation codon (Simonsen and Levinson, 1983). The introns range in size from 0.3 to 9.4 kb; three introns contain sequences highly repeated in the genome. Repeated sequences are also found in the 5′ flank. The restriction fragments defined by this map have been shown to exist in the same arrangement in wild-type (unamplified) CHO cells and in the independently amplified CHO line pro-3 MtxRIII of Flintoff et al. (1976a). The map of the gene cloned by Milbrandt et al. (1983a) from another independently amplified line of CHO cells (CHOK-400) agrees well with the map shown in Figure 16.2. Moreover, these authors have cloned more than 135 kb of DNA from this region and have extended the map shown in Figure 16.2 considerably in both the 5′ and 3′ directions. The *dhfr* gene in CHL cells has been less extensively mapped, but it is already clear that there is at least one deviation from the CHO map. CHL cells exhibit a restriction fragment length polymorphism at the 3′ end of the gene: one allele yields a 20-kb *Hin*dIII fragment like CHO, while another allele contains an 8-kb fragment in this region (Lewis et al., 1982b). The overall structure of the CHO *dhfr* gene is very similar to that described for the mouse (Crouse et al., 1982), which also has six exons divided by five introns at apparently analogous locations. The sizes of the CHO introns are also similar to those in the mouse. The most notable difference between the two rodents lies in the size of the mRNA species. The two major forms of *dhfr* mRNA in Chinese hamster cells are 1100 and 2300 bases long, including any poly(A) tails (Lewis et al., 1982b; Carothers et al., 1983; see Fig. 16.4), while the two predominant forms in the mouse are 750 and 1600 bases (Setzer et al., 1980, 1982). Most of this difference is due to the more extensive 3′ untranslated region in the case of the Chinese hamster. The size of human *dhfr* mRNA is apparently even larger, approximately 3800 bases (Morandi et al., 1982). Nothing is known about the function, if any, of these 3′ untranslated sequences, all of which are specified by the sixth exon in both rodents. It is known that all three mRNAs detectable in Chinese hamster cells are associated with polysomes and are translatable in a cell-free system (Melera et al., 1982).

At least two pseudogenes for *dhfr* are present in the human genome (Chen et al., 1982). One of these has no introns and contains a 3′ poly(A) tract, indicating a mRNA origin. No evidence for *dhfr* pseudogenes has been reported for mouse or Chinese hamster cells.

2. Structural Changes at the dhfr Locus Induced by Ionizing Radiation

Twenty DHFR-deficient mutants that were isolated following gamma-ray treatment have been examined for gross changes in *dhfr* gene sequences by Southern blot analysis. Since the gamma-ray treatment increases mutant frequencies from less than 10^{-6} to more than 10^{-5}, it is likely that most of these mutations have been induced by the radiation. The most informative mutants have been those originating from the hemizygous clones UA2 and UA4, since the effects on the single remaining *dhfr* gene can be clearly seen. All 11 mutants that have been analyzed exhibit deletions or rearrangements of the *dhfr* gene sequence (G. Urlaub et al., unpublished results). Seven have lost the gene completely; these lack any unique DNA sequence from a 35-kb region that includes the *dhfr* gene (e.g., Fig. 16.3, lane 2). In one mutant a deletion has occurred that extends past the 5′ end to about the middle of the structural gene (Urlaub et al., 1983). Three mutants exhibit an interruption in the gene without the detectable loss of any gene sequences. Restriction digests of total DNA from these mutants have been analyzed by Southern blotting using probes from the *dhfr* regions neighboring the new joints. In no case could an insert of defined length be demonstrated; the minimum size of the interrupting DNA ranged from 8 to 14 kb in the three mutants. These mutations therefore represent either translocations or very large insertions.

In an earlier study, nine gamma-ray-induced mutants originating from a heterozygous, rather than a hemizygous, clone were similarly examined (Graf and Chasin, 1982). Two of these nine mutants exhibited new *dhfr* restriction fragments, indicating that deletions or translocations had occurred. The remaining seven mutants showed no difference from wild-type patterns. These seven mutants could represent complete deletions of a *dhfr* gene, since the continued presence of one nonfunctional *dhfr* allele in this set of mutants would give rise to "wild-type" fragments. These earlier results are therefore not inconsistent with the more recent data obtained with the hemizygous clone.

Considering only the mutations that have been induced in the hemizygous clones, 7 out of 11 have resulted in the loss of at least 35 kb of DNA. It therefore appears that the majority of lesions induced by this type of radiation are extensive deletions. It follows that there must be no genes required for cell viability that are closely linked to the *dhfr* locus in CHO cells, since these double deletion mutants are able to grow as long as the end-products of one-carbon metabolism are supplied. Finally, none of the eleven mutations analyzed in this series proved to be of a single base substitution type.

3. *Mutations Induced by Ultraviolet Light*

UV-treatment induced DHFR-deficient mutants in the hemizygous clone UA2 at a frequency of 10^{-4}, more than 100-fold over spontaneous levels (P. Mitchell et al., unpublished results). Fifteen of these mutants have been analyzed for changes at the *dhfr* locus. A Southern blot showing the presence of *dhfr* restriction fragments from a number of these mutants is shown in Figure 16.3. Thirteen mutants show no detectable difference when compared to the parental line (or to wild-type cells). In the remaining two, the entire *dhfr* gene has been deleted. UV therefore can produce at least two very different kinds of mutations in mammalian cells: large deletions and probably point mutations. Small deletions cannot be ruled out in this type of analysis: the size range of restriction fragments screened was from 300 to 6900 bp.

4. *dhfr mRNA Levels in DHFR-Deficient Mutants*

The level of *dhfr* mRNA has been measured in the gamma-ray- and UV-induced mutants described above. The poly(A)$^+$ fraction of total cellular DNA was examined by Northern blot analysis for size and for amount (as estimated from the intensity of bands) of *dhfr* specific sequences. These Northern blots are shown in Figure 16.4. The mutants fall into three classes (Mitchell et al., unpublished results). Class 1 mutants exhibit mRNA patterns indistinguisable from the parental cells in size, in amount, and in the ratio between the different molecular-weight forms of *dhfr* mRNA. This class is made up of five UV-induced mutants. Class 2 mutants contain decreased amounts of *dhfr* mRNA; all three size species appear coordinately reduced. This reduction can be quite drastic, down to less than 3% of parental levels, or less than one mRNA molecule per cell. The highest amount of *dhfr* mRNA in these

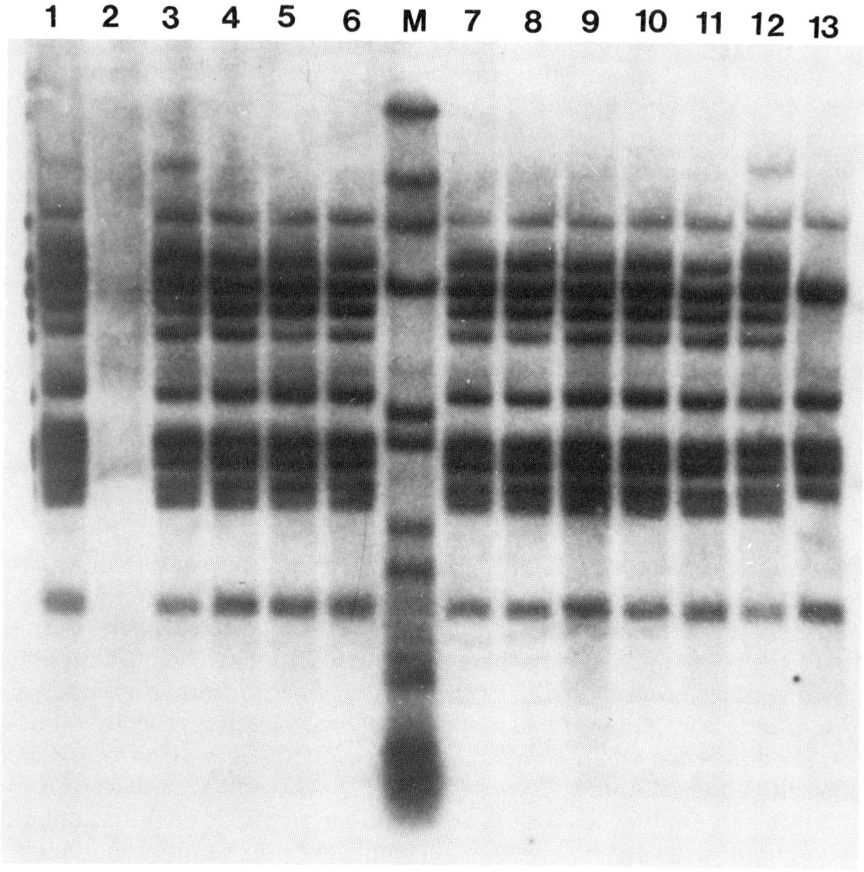

Figure 16.3. Example of a Southern blot of DNA from various DHFR-deficient mutants. Total cellular DNA (15 μg) was digested with *Kpn*I and *Bst*EII. The probe was a collection of cloned fragments comprising 13 kb of DNA from the *dhfr* locus. These fragments were devoid of repeated sequences. Lane 1: UA2, the parental hemizygote (DHFR$^+$); lane 2: DG21, a gamma-ray-induced complete deletion mutant; lanes 3 through 12: 10 UV-induced mutants showing no change in *dhfr* gene structure; lane 13: DG22, a gamma-ray-induced mutant lacking the 5′ half of the gene; lane M: size markers, end-labeled lambda *Hin*dIII and $\Phi \times 174$ *Hae*III fragments.

mutants is approximately 20% of the parental level. Eight of the UV-induced mutants fall into this category. The low levels of *dhfr* mRNA in these mutants could be due to a lower transcription rate, a lower maturation rate, or increased degradation. One argument in favor of the last possibility is the fact that the DHFR enzyme level is lower than the relative mRNA level in these mutants. Thus at least two phenotypes are produced by this type of mutation: a lower amount of a mRNA and a mRNA that cannot code for a functional enzyme. Both of these effects could be explained by a nonsense mutation that blocked translation, resulting in a mRNA molecule that was more vulnerable to degradation.

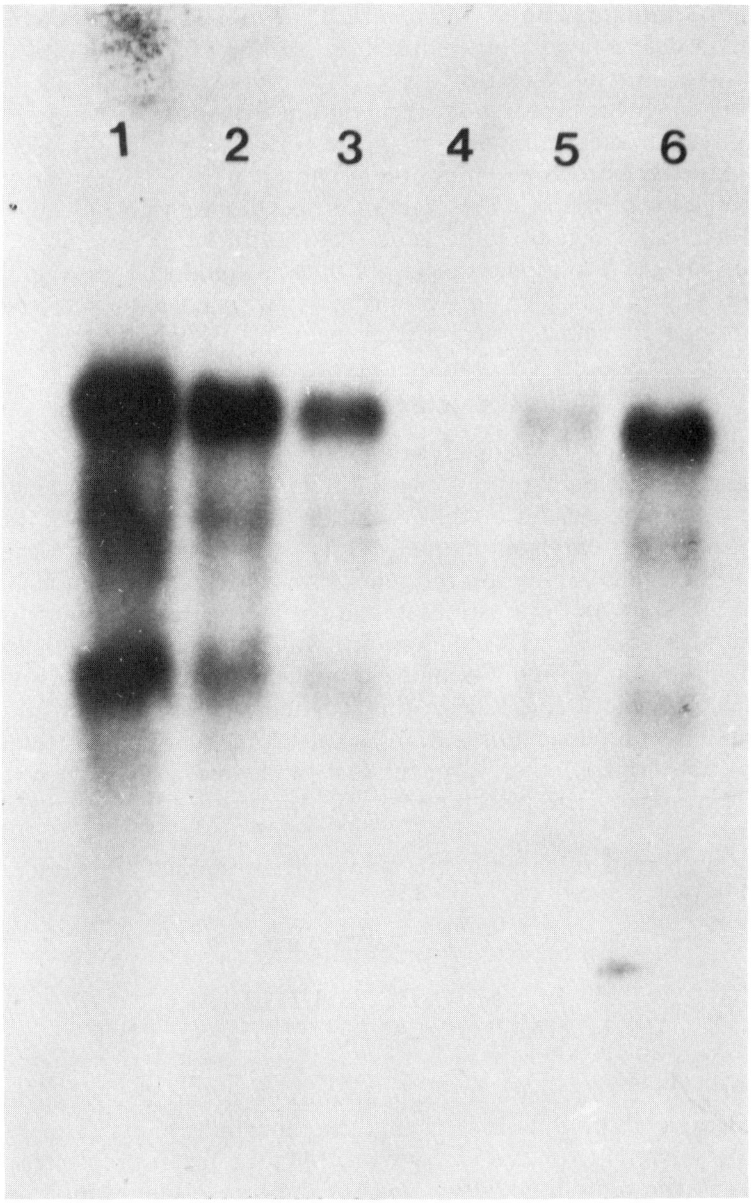

Figure 16.4. Example of a Northern blot of RNA from various mutants. Total poly(A)$^+$ RNA was isolated and 12 μg samples were electrophoresed in formaldehyde agarose gels. After blotting to nitrocellulose, hybridization was carried out using a mixture of cloned cDNA and genomic probes. Lane 1, amplified mutant MK42 (0.05 μg); lane 2, parental hemizygote clone UA2 (12 μg); lane 3, parental hemizygote UA2 (3 μg); lane 4, a UV-induced deletion mutant (12 μg); lane 5, a UV-induced low-level mRNA mutant (12 μg); lane 6, a UV-induced wild-type-level mRNA mutant (12 μg).

Class 3 mutants contain no detectable *dhfr* mRNA. Predictably, the complete deletion mutants induced by gamma rays or UV are in this class. The mutant lacking the 5′ half of the gene also lacks *dhfr* mRNA sequences. Apparently, the deletion that took place here has not connected the remaining portion of the gene to any other transcription unit. Finally, the three mutants in which a gene interruption has occurred also contain no detectable mRNA, despite the continued presence of the 5′ portion of the gene. Perhaps the *dhfr* promoter is still active, but the abnormal transcripts that are produced are rapidly degraded. Alternatively, new neighboring downstream sequences could be interfering with transcription of the gene.

5. *Methylation of the dhfr Gene*

The *dhfr* gene in CHO cells is extensively methylated at *Msp*I and *Hha*I sites throughout much of its length (Stein et al., 1983). However, there is a considerable region at the 5′ end of the gene that is not methylated. This nonmethylated domain starts at least 0.8 kb 5′ of the first exon and extends 2.0 kb into the gene (P. Mitchell et al., unpublished results). The state of methylation of many of these sites has been checked in a number of the DHFR-deficient mutants described above. In no case has an alteration in methylation pattern been found. This result is especially interesting in the case of the translocation or insertion mutants that no longer produce *dhfr* mRNA. If this absence of mRNA is due to a lack of transcription, then it means that methylation does not automatically accompany the placement of DNA into inactive regions of the genome.

The properties of the mutants previously described are summarized in Table 16.1.

IV. MUTATIONS AFFECTING THE ENZYMOLOGICAL PROPERTIES OF DHFR

In addition to DHFR-deficient cells, mutations in the *dhfr* gene can lead to structural alterations that change the enzyme's physical or kinetic properties. Mutants of this type are probably the result of missense mutations in the protein coding regions of the gene. One example has already been described in the discussion of DHFR-deficient clones: a mutant containing a low level of DHFR activity that exhibits an increased heat sensitivity (see Section II B 1). A more common alteration leads to an increased resistance of the enzyme to inhibition by antifolates such as MTX. Albrecht et al. (1972) isolated a MTX-resistant CHL cell mutant expressing high levels of a DHFR activity that is more resistant to inhibition by the drug. The resistant enzyme also displays altered chromatographic properties. Working with CHO cells, Flintoff et al. (1976a) isolated an EMS-induced mutant in a single step that contains normal levels of an enzyme that is 10- to 20-fold more resistant to MTX inhibition. The purified enzyme also displays an increased heat sensitivity (Gupta et al., 1977) and binds MTX with a lower affinity (Flintoff and Essani, 1980). As expected, these mutants are expressed codominantly in somatic cell hybrids (Flintoff et al., 1976b). Similar drug-resistant mutants have also been isolated in mouse (Goldie et al., 1981; Haber et al., 1981) and in human cells (Jackson et al., 1976).

V. GENE AMPLIFICATION AT THE *dhfr* LOCUS

A. Multistep Development of MTX Resistance Is Usually Due to Amplification of the *dhfr* Gene

By far the most widespread studies of genetic events at the *dhfr* locus have dealt with gene amplification. Cells in which the *dhfr* gene has been amplified exhibit a selectable phenotype, resistance to antifolate drugs such as MTX or methasquin. As described above, resistance to these drugs can also occur by mutations that alter the structure of the DHFR enzyme. Yet another mechanism is decreased permeation of MTX into the mutant cells (Fisher, 1962; Flintoff et al., 1976a). However, gene amplification appears to be the most common cause of resistance.

Early studies on the development of MTX-resistant clones in populations of mouse or hamster cells pointed out a characteristic feature of the origin of amplified mutants: resistance develops gradually. Populations are usually subjected to a concentration of MTX that kills over 99% of the cells. The survivors are then pooled and subjected to a slightly higher MTX concentration, and the process is repeated. Over a period of several months, cultures evolve that are resistant to MTX concentrations many orders of magnitude greater than that which would be toxic to wild-type cells. Clones or populations removed from the selective conditions at intermediate concentrations of the drug exhibit resistance only to that intermediate level (Fischer, 1961; Littlefield, 1969; Courtenay and Robins, 1972; Biedler et al., 1978). Selection for mutants resistant to high levels of MTX in a single step is rarely successful. In one study of this type where relatively highly resistant mutants were obtained after a single mutagenic step, DHFR levels were not affected (Orkin and Littlefield, 1971). Resistant mutants generally show a correlation among the concentration of MTX used in the selection, the level of resistance, and amount of DHFR activity per cell. Mutants resistant to the highest concentrations of MTX, generally 0.5–1.0 m*M*, achieve DHFR levels more than 100 times that of wild type; in these cases the proportion of soluble protein synthesis devoted to DHFR can be several percent (Alt et al., 1976; Hangii and Littlefield, 1976; Nunberg et al., 1978; Melera et al., 1982). Apparently the high intracellular levels of DHFR can bind the MTX and leave enough free enzyme available to function in one-carbon metabolism. However, the quantitative relationship between the degree of drug resistance and the level of DHFR catalytic activity in the cell is far from clear. For example, in a series of independently isolated CHL mutants described by Biedler and her colleagues (Biedler et al., 1972) the ratio of relative resistance to relative enzyme level varied over a wide range. Similarly, in hybrids between resistant and sensitive cells, a modest decrease in DHFR-specific activity can be accompanied by a large increase in MTX sensitivity (Littlefield, 1969; Sobel et al., 1971; Flintoff et al., 1976b; Urlaub et al., 1981).

Littlefield and his colleagues, using BHK Syrian hamster cells, were responsible for much of the work demonstrating that the increased DHFR activity in these MTX-resistant mutants is due to an increased amount of DHFR protein (as opposed to the activation of preexisting enzyme molecules) and that this protein is indistinguishable from the enzyme found in wild-type cells (Nakamura and Littlefield, 1972). These workers, as well as Schimke and his colleagues working with mouse S180 cells, then went on to show that that the increased enzyme levels are not due to decreased degradation of the protein (Alt et al., 1976; Hangii and Littlefield, 1976) and are mirrored by increased levels of translatable *dhfr* mRNA in the resistant cells (Chang and Littlefield, 1976; Kellems et al., 1976). Thus by the late 1970s it became clear that high-level MTX resistance can be traced to high levels of *dhfr* mRNA.

Two key studies then connected the high *dhfr* mRNA content to a high *dhfr* gene content. Biedler and Spengler (1976) showed that the development of DHFR overproduction in CHL cells was often accompanied by the appearance of expanded regions on certain chromosomes representing material not present in sensitive cells. This metaphase chromatin did not give rise to the irregularly banded appearance typical of trypsin-Giemsa-stained chromosomes, and so was called a "homogeneously staining region," or HSR. Remarkably, the length of the HSR in a given clone was roughly proportional to the degree of DHFR overproduction, and these authors suggested that the HSRs contained amplified copies of the *dhfr* gene. In 1978, Alt et al. provided direct evidence that *dhfr* genes were amplified in mouse S180 cells. Using an elegant isolation scheme based on mRNA hybridization kinetics, they were able to purify *dhfr* cDNA and use it as a probe to quantitate *dhfr* mRNA and DNA sequences in MTX-sensitive and -resistant cells. Solution hybridization measurements confirmed that *dhfr* mRNA levels were elevated in resistant cells. These authors then went on to show that *dhfr* gene sequences were present in high copy number in the resistant cells but not in the sensitive cells. In fact, there was good quantitative agreement for the level of DHFR activity, the rate of DHFR protein synthesis, the abundance of *dhfr* mRNA, and the degree of *dhfr* gene amplification (Alt et al., 1978). This proportionality has also been documented in several Chinese hamster cell amplified mutants (Nunberg et al., 1978; Melera et al., 1980b; Milbrandt et al., 1981; Flintoff et al., 1982).

B. Cytological Localization of Amplified *dhfr* Genes

As mentioned previously, the first indication that amplified genes were associated with unusual chromosomal structures was the description of HSRs in CHL cells that overproduce DHFR (Bielder and Spengler, 1976; Biedler et al., 1978, 1980). Direct evidence that *dhfr* DNA sequences are physically located in an HSR came from *in situ* hybridization studies on a CHO mutant using the mouse cDNA probe described above (Nunberg et al., 1978). The HSR in this mutant is located on chromosome 2, the site of the wild-type *dhfr* locus (Fig. 16.5). Grains representing hybridized radioactive cDNA were spread evenly throughout the length, which represents 3.5% of the total length of the metaphase chromosomes in this case. Assuming a model in which one *dhfr* allele has been amplified to yield tandem repeats in the HSR, a value of 500 kb can be estimated for the length of one repeated unit. In this calculation it is assumed that one *dhfr* allele has been amplified 300-fold to yield the observed 150-fold increase in *dhfr* DNA and that 160,000 kb (3.5%) of the total cellular DNA is present in the HSR. HSRs containing amplified *dhfr* sequences have also been demonstrated in CHL cells (Lewis et al., 1982a), in additional CHO mutants (Milbrandt et al., 1981), and mouse cells (Dolnick et al., 1979). Moreover, several other amplified genes have now been shown to reside in HSRs [e.g., the CAD gene in Syrian hamster cell mutants (Wahl et al., 1982); rRNA genes in rat hepatoma cells (Tantravahi et al., 1981); oncogenes in neuroblastoma cells (Schwab et al., 1983); for reviews, see Cowell (1982) and Hamlin et al. (1983)].

Amplified *dhfr* genes can also be located on double minute chromosomes. These are supernumerary paired chromosome structures that stain like normal chromosomes but are very much smaller and lack functional centromeres [see Levan and Levan (1978), Cowell (1982), and Hamlin et al. (1983)]. Amplified *dhfr* genes in mouse cells can be located either exclusively in HSRs or exclusively in double minutes, depending on the particular cell line. The number of double minutes per cell can range from one or two to several hundred in any given cell line. As resistance to MTX increases during the evolution of a drug-resistant mutant, the number of amplified *dhfr* genes and the number of double minutes per cell increase in parallel (Brown et al., 1981). *dhfr* genes associated with double minute chromosomes are usually unstable in the absence of continuous selective pressure (Kaufman et al., 1979; Brown et al., 1981), presumably due to segregation aberrancies caused by the lack of

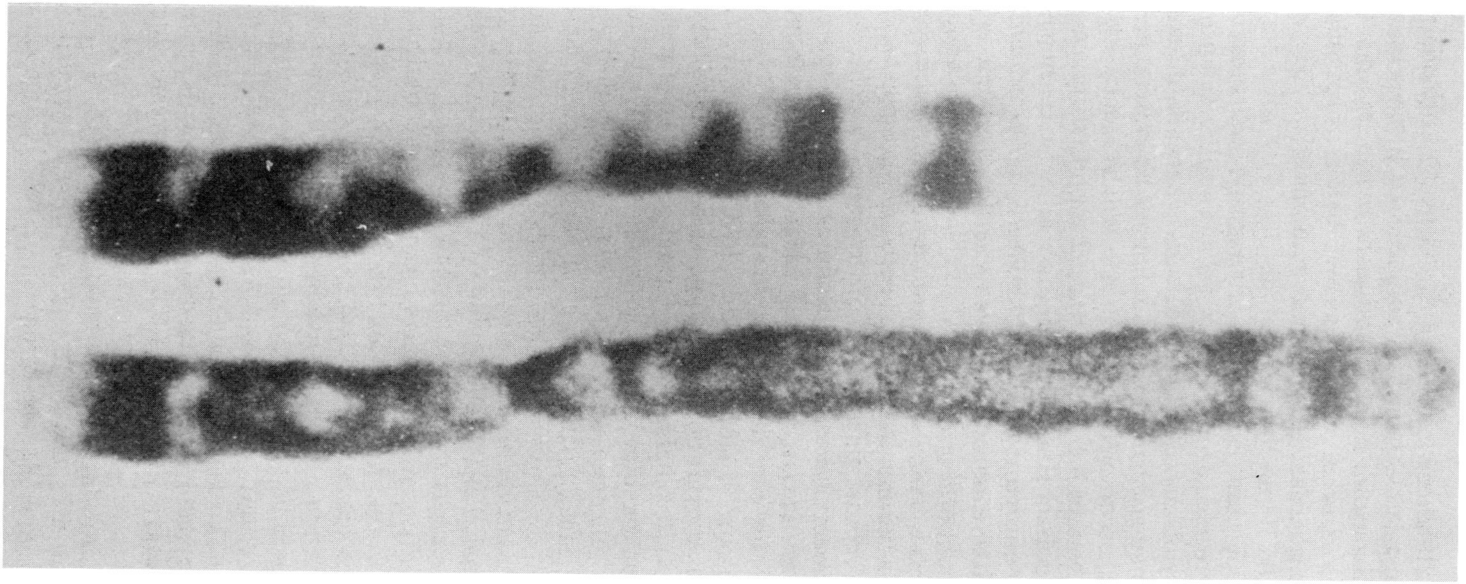

Figure 16.5. Example of an HSR from the amplified CHO mutant MK42 (Nunberg et al., 1978). The normal chromosome 2 is shown at the top. Approximately 300 copies of the *dhfr* gene are estimated to be present in the expanded chromosome. The photograph is from Schimke (1980).

centromeres. In contrast to mouse cells, *dhfr* genes have so far been found only in true chromosomes (in HSRs) in amplified Chinese hamster or human cells (Wolman et al., 1983).

C. More Than One *dhfr* Allele Can Be Amplified in Chinese Hamster Cells

An extensive series of amplified mutants derived by Biedler and her colleagues from CHL cells has been characterized in detail at the enzyme and mRNA level by Melera and co-workers. These mutants fall into two classes: some mutants overproduce a DHFR protein of apparent molecular weight 20,000 whereas others overproduce a 21,000 molecular-weight species (Melera et al., 1980*a*). Three different size *dhfr* mRNA molecules were shown to be present in CHL cells (Lewis et al., 1981, 1982b). In cell-free translation experiments, all three mRNAs from one class of amplified mutant were shown to specify the 20,000-dalton DHFR species, whereas all three mRNAs from the other class specified only the 21,000-dalton molecule (Melera et al., 1982). Thus the differences between these two DHFR species is not due to posttranslational modification nor is it due the activity of one particular size class of mRNA. In more recent experiments, cDNAs have been cloned starting with mRNA from each class of amplified mutant (Melera et al., 1984). Expression of some of these cloned cDNAs has been achieved in *E. coli*; the recombinants express either one or the other form of the DHFR enzyme. Sequence analysis reveals that the two mRNAs code for two proteins that differ in amino acid sequence at two positions; thus the different eletrophoretic mobilities on SDS gels are not due to a difference in molecular weight. Additional nucleotide sequence differences are found outside of the protein coding region.

The two classes of amplified mutants can be further correlated by the restriction fragment pattern of their amplified *dhfr* sequences (Lewis et al., 1982b). Using a probe for the 3′ end of the gene, it was shown that the mutants producing the 20,000-dalton protein contain an 8-kb *Hin*dIII amplified fragment whereas in mutants of the second class this region of the gene yields a 20-kb fragment. The wild-type CHL cells contain both restriction fragments and produce both species of DHFR. The conclusion reached is that CHL cells are polymorphic for the *dhfr* gene, and that this polymorphism is demonstrable at both the protein and the DNA level. DNA derived from Chinese hamster tissue and CHO cells contain only the 21,000-dalton enzyme and yield the corresponding DNA restriction pattern. These results show that CHL cells can acquire MTX resistance by the amplification of either one of (at least) two *dhfr* genes, which probably represent two alleles. Furthermore, once the amplification process starts it remains exclusively with that allele. Interestingly, the two mutant classes also differ in a third characteristic: the relative abundance of the three *dhfr* mRNA species is different in each class. Thus the effects of this polymorphism appear to extend to RNA processing as well.

D. Stability of the Amplified State in Chinese Hamster Cells

Genes amplified in HSR structures are stable relative to those in double minute chromosomes, and, in general, gene amplification in Chinese hamster cells has been associated with HSRs rather than double minutes. Biedler and her co-workers measured stability of individual CHL mutants with respect to MTX-resistance levels, HSR length, and DHFR-specific activity (Biedler et al., 1980). They found that all three parameters decreased together, but at a low rate: the half-lives for DHFR levels ranged from 50 to 250 generations (7–33 weeks). Similarly in CHO cells, highly amplified mutants have been grown in nonselective medium for over 100 generations without any loss of the amplified phenotype (Nunberg et al., 1978). Kaufman and Schimke (1981) did find evidence for initial instability in CHO cells that had been selected for low-level MTX resistance; however, after maintenance in selective medium for 100 generations, these populations became stabilized. In contrast, when similar experiments were performed with mouse 3T6 cells, no such stabilization was found (Kaufman et al., 1981).

A different sort of experiment was used by Urlaub et al. (1981) to detect possible low frequencies of MTX-sensitive cells in a population of amplified MTX-resistant cells. Applying the MTX plus [^{3}H]-dUrd method described in Section II A 4 to a population of mutant CHO cells that was 150-fold amplified, they found no "deamplified" revertants among 10^5 cells screened. Thus after growth for more than a year under nonselective conditions, very few cells had accumulated that had lost most of their amplified *dhfr* genes. However, this selection method would not have detected cells that had lost only a portion of the amplified genes.

E. Agents that Induce *dhfr* Gene Amplification

In most cases, MTX-resistant mutants are selected in a gradual multistep procedure; clones are then derived from populations that have already achieved some degree of resistance to the drug. In several cases, single-step selections for low-level MTX resistance have been carried out. In some of these experiments the effects of various agents on the frequency of resistant mutants has been determined. The results of such selections are often difficult to interpret because of the appearance of wild-type surviving colonies and of some drug-resistant clones that are not due to *dhfr* gene amplification. However, in several cases the results are sufficiently dramatic to indicate a stimulatory effect on gene amplification.

A study of this type using CHO cells was carried out by Flintoff et al. (1976a, 1982) starting with a mutant that was already resistant to a low level of MTX by virtue of a structural change in the DHFR enzyme. Selection for a higher level of resistance yielded mutants that had amplified the altered allele. A frequency of approximately 10^{-6} was found for the amplification step; the degree of amplification for one clone tested was 15-fold.

Experiments with mouse cell lines have identified agents that increase *dhfr* amplification frequencies. Varshavsky (1981) found that the tumor-promoting phorbol ester TPA can increase the frequency of 3T6 cells resistant to MTX by a factor of 100, and that most of these resistant clones contain amplified *dhfr* sequences. In extending this work, Barsoum and Varshavsky (1983) found that several growth-stimulating hormones—insulin, epidermal growth factor, and arginine vasopressin—are also capable of producing this effect. Treatments that disrupt DNA synthesis can have dramatic effects in initiating *dhfr* gene amplification. Both in mouse 3T6 cells and in CHO cells pretreatment with the DNA synthetic inhibitors hydroxyurea, cytosine arabinoside, or methotrexate itself can increase the frequency of MTX-resistant colonies (Tlsty et al., 1982; Brown et al., 1983). Agents that react with DNA are also effective: namely, UV-irradiation and the carcinogen *N*-acetoxy-*N*-acetylaminofluorene. The tumor-promoter TPA acted synergistically with some of the above agents, but had no activity on its own in this study. Many of the resistant clones obtained could not be shown to harbor additional copies of the *dhfr* gene; however, a significant proportion did show an increase, and the frequency of these mutants was stimulated along with the rest.

That pretreatment with MTX itself can induce MTX-resistant mutants raises the possibility that the drug is acting in some direct way to promote *dhfr* gene amplification. Alternatively, MTX may simply be acting as a general inhibitor of DNA synthesis during the pretreatment and only as a selective agent when it is subsequently added. Evidence in favor of the latter comes from the isolation of a CHO mutant that underwent *dhfr* gene amplification in the absence of exposure to MTX. In this case, the starting cell line was a leaky DHFR-deficient mutant that regained the ability to synthesize glycine, purines, and thymidine by amplifying the gene for its partially defective enzyme (Chasin et al., 1982).

F. Structure of the Amplified Region

The minimum size of the unit that is amplified must be larger than the *dhfr* gene itself: 26 kb for the Chinese hamster and 31 kb in the mouse. In most cases a considerable length of flanking sequence is also amplified, so that the size of the repeated unit is many times the size of the gene. In the case of mutants in which an HSR has been shown to contain *dhfr* sequences, it is possible to estimate this size by dividing the amount of DNA in an HSR (estimated from the length of the HSR) by the number of extra copies of the *dhfr* gene it contains (estimated from the fold amplification compared to wild type). Examples of amplified unit sizes calculated in this way are 500 kb for a CHO mutant (Nunberg et al., 1978), 300–700 kb in CHL mutants (Biedler et al., 1980; Lewis et al., 1982a), 800 kb in a mouse L5178Y mutant (Dolnick et al., 1979), and as much as 3000 kb in a mouse PG19 derivative (Bostock and Clark, 1980).

In two cases, MTX-resistant mutants have been isolated in which the degree of amplification has been extensive enough to allow the direct visualization of amplified restriction fragments in agarose gels. The sum of the molecular weights of all fragments then provide an independent method of estimating the size of the amplified unit. This method assumes that most of the amplified units have the same structure, for if flanking regions were not always repeated, they might not be sufficiently amplified to be visible in the ethidium-stained gels (Hamlin et al., 1983). Milbrandt et al. (1981) analyzed a CHO mutant (CHOK-400) estimated to contain 1000 copies of the *dhfr* gene using both methods: the HSR length measurement yielded a value of 270 kb for the repeated unit, whereas the fragment summation method yielded 135 kb. Tyler-Smith and Alderson (1981) applied the summation method to an amplified mouse lymphoma cell line and calculated a value of 500 kb for the repeated unit. The values obtained by both methods may not be significantly different when the probable errors involved in making the necessary measurements are taken into account.

These same two laboratories have offered evidence that the structure of the amplified unit can be the same after independent amplifications. Hamlin and her colleagues cloned 110 kb of the 135 kb seen to be amplified in CHOK-400 and constructed a restriction map of the region (Milbrandt et al., 1983a). They then used these clones to probe the DNA of four independent amplified mutants of CHO and CHL cells isolated in other laboratories. They found that the same sequences had been amplified in all mutants in this 135-kb region (Hamlin et al., 1983). In a related experiment, Bostock and Tyler-Smith (1982) transferred a *dhfr* sequence present in double minute chromosomes from resistant to sensitive cells and subsequently selected for further amplification in the recipient cells. The sequences amplified were the same as those present in high copy number in the original amplified line. Thus a common sequence can be defined that is substantially larger than the *dhfr* gene itself and that is usually amplified along with it.

The amplified sequences flanking the *dhfr* gene in mouse have also been examined. Schilling et al. (1982) have cloned sequences flanking the *dhfr* gene in amplified mouse S-180 cells by "chromosome walking" in both directions, using the previously cloned *dhfr* gene as an initial probe. In all, 200 kb of DNA was cloned and mapped, indicating that the amplified unit in these S-180 cells is at least that large. When these sequences were used as probes against the DNA from independently amplified mouse cell lines, some of the sequences were found not to be amplified. In the most extreme case, it was found that only 95 kb of this DNA had been amplified. Thus in the mouse, independent amplification events can result in repeat units of different structure, although of course all must include the 31-kb *dhfr* gene itself.

Rearrangements within the *dhfr* gene itself can also take place during gene amplification. Crouse et al. (1982) constructed a library of *dhfr* sequences from an amplified mouse S-180 cell mutant. In addition to overlapping fragments that defined the wild-type *dhfr* gene, they also found sequences that corresponded to only the 3′ half of the gene. This truncated gene was also shown to be amplified in the mutant from which the library was made, but was not present in other amplified mutants. The conclusion reached was that the *dhfr* gene had been interrupted by an incomplete amplification event at one point during the evolution of this line, and that this mutated sequence continued to be amplified along with functional *dhfr* genes. A mutated gene also arose during *dhfr* gene amplification in mouse 3T6 cells. This gene carries a missense mutation (Simonsen et al., 1983) that renders the DHFR enzyme more resistant to inhibition by MTX (Haber et al., 1981). In this case the altered gene confers a selective advantage and so its continued presence is easy to understand. However, amplified copies of the wild-type gene are also present, so the mutant gene must have arisen after considerable amplification of the wild-type gene had already occurred. The MTX-resistant enzyme of the A3 line of CHL cells may also have arisen in this way (Albrecht et al., 1972). Other cases of *dhfr* gene rearrangement during amplification have been reported in human breast tumor cells (Cowan et al., 1982) and mouse melanoma cells (Bostock and Tyler-Smith, 1981). These examples are more the exception than the rule, however. In most cases the amplified *dhfr* genes have the same structure as the progenitor, whether in Chinese hamster (Hamlin et al., 1983; Urlaub et al., 1983) or mouse (Crouse et al., 1982) cells.

G. Possible Mechanisms of *dhfr* Gene Amplification

Although various models have been proposed, the initial events leading to gene amplification are not known. The models generally fall into two categories: chromatid exchange and overreplication. Unequal sister chromatid exchange is the simplest version of the first category. Mispairing between homologous sequences that flank the *dhfr* gene and subsequent crossing-over would lead to one cell that has two copies of the *dhfr* gene at its normal location and one cell that has none. Repetition of this scenario could lead to geometrically increasing *dhfr* genes on one chromosome arm (Hamlin et al., 1983). The fact that HSRs bearing *dhfr* genes are often located on the short arm of chromosome 2, at or near the position of the wild-type locus, is in agreement with this model. On the other hand, there are many exceptions to this location of *dhfr* HSRs in Chinese hamster cells (Biedler et al., 1978; Milbrandt et al., 1981). The ease with which the *dhfr* gene is amplified could signify that this locus is in a region where sister chromatid exchange occurs at high frequency. However, the exchange rate within the *dhfr* HSR of CHO cells has been found to be the same as that in the rest of the genome (Chasin et al., 1982). Finally, according to this model, agents that increase the frequency of *dhfr* gene amplification should also yield an increased frequency of *dhfr* deletion mutants (the reciprocal event). As discussed in Section II D 3., DHFR-deficient mutations can be induced by UV-irradiation, a treatment that increases gene amplification in mouse cells (Tlsty et al., 1982). Two out of 15 mutants did lose the gene; these could have arisen from an unequal sister chromatid exchange. Related models invoke recombinations between individual chromosomes rather than sister chromatids. The appearance of ring and dicentric chromosomes during amplification of a mouse *dhfr* sequence transferred to CHO cells prompted Kaufman et al. (1983) to propose a model in which the random breakage of a dicentric structure results in two *dhfr* genes residing on the same rearranged chromosome. Translocations accompanying *dhfr* amplification have been reported by Flintoff et al. (1984).

In the second type of model, the wild-type *dhfr* gene would be replicated more than once during a given S period of the cell cycle (Varshavsky, 1981; Schimke, 1982; Hamlin et al., 1983). Many such replications would lead to an "onion-skin" picture of extra copies of the overreplicated region [e.g., Botchan et al (1979)]. The extra copies would then be joined by recombination to yield a tandem array of amplified genes at the wild-type locus. Evidence that the amplification of simple genes (as opposed to viral genomes) can occur by this sort of mechanism has come from the study of thymidine kinase gene amplification after transfer of a partially disabled version of this gene into mouse L cells (Roberts et al., 1983). Selection in HAT medium necessitates an amplification of the thymidine kinase gene, and an analysis of the resulting DNA organization shows a tandem array of thymidine kinase genes flanked by carrier DNA sequences. Variation in the end points of the repeated unit can be detected within the amplified flanks. The structure can be explained by unscheduled replication of the transferred sequences followed by recombination between homologous flanking regions (pBR322 sequences in this case). This type of model predicts that each amplified unit should contain an origin of DNA replication. Heintz and Hamlin (1982) have in fact been able to test this idea using an amplified CHO mutant carrying an unusually high number of *dhfr* genes. The high degree of amplification in this mutant (CHOK-400) enabled these authors to detect specific amplified restriction fragments that were labeled by [^{3}H]thymidine during replication. The pattern of labeling indicated that each repeated unit contains a unique origin of replication. The existence of a common signal for replication in each of the amplified units is supported by the fact that all of the amplified *dhfr* genes are replicated synchronously early in S phase (Milbrandt et al., 1981). Synchronous early replication of *dhfr* HSRs had previously been shown in CHL mutants as well (Hamlin and Biedler, 1981). While these results are consistent with the overreplication model, they cannot be considered strong support for it: The existence of an origin of replication in a stretch of DNA this large (135 kb) is not unexpected.

Lewis et al. (1982a) recently carried out a cytogenetic study of early events in *dhfr* amplification in CHL cells. Using *in situ* hybridization, they were able to determine the location of the first few copies of the amplified *dhfr* gene. These were usually found in abnormally banded regions (ABRs) on the same arm of chromosome 2 that carries the wild-type *dhfr* locus. Interestingly, the autoradiographic grains were clustered in discrete bands in this region, with clear areas devoid of *dhfr* sequences in between. Since these clear areas are visible at the light microscope level, they obviously represent very long stretches of DNA. This result could be accounted for in the overreplication model if the initial repeated unit is very large, much larger than the final state seen in highly amplified lines. Alternatively, a smaller amplified unit could have reintegrated in the same chromosome at some distance from the original *dhfr* locus. This result could also be explained by the unequal sister chromosome exchange model, however. The course of events early in *dhfr* gene amplification in CHO cells was studied by Kaufman and Schimke (1981). Cell populations that had been subjected to low levels of MTX were analyzed for DHFR content by detecting the amount of bound fluorescent MTX using a fluorescence-activated cell sorter. The earliest differences detected represented increases in DHFR per cell of about fivefold over wild type. These populations were very heterogeneous with respect to enzyme level per cell; in contrast to the stability of more highly amplified CHO lines, these low-level amplified cells rapidly returned to wild-type DHFR levels. Stabilization was routinely established after growth in MTX for about 100 generations. The authors suggest that initial events in gene amplification may involve unstable extrachromosomal genes and that stabilization occurs upon reintegration.

VI. TRANSFER OF GENES SPECIFYING DHFR

A. Chromosome Transfer

Microcell-mediated cell fusion was used to transfer amplified copies of an altered CHO *dhfr* gene to wild-type (MTX-sensitive) cells (Worton et al., 1981). The MTX-resistant phenotype could be transferred by this technique, in which one or a small number of chromosomes are transferred intact. The transferred chromosome was identified as number 2, which also carries the wild-type *dhfr* locus in Chinese hamster cells (Roberts et al., 1983). Isolated metaphase chromosomes have also been used to transfer amplified copies of structurally altered Chinese hamster (Lewis et al., 1980) and mouse (Haber and Schimke, 1982) *dhfr* genes to wild-type recipients. In the latter study, these *dhfr* genes were located on double minute chromosomes; the amplified *dhfr* sequences remained in the form of double minute chromosomes in the recipient clones, and they were rapidly lost if selective conditions were not maintained. Bostock and Tyler-Smith (1982) transferred amplified mouse *dhfr* sequences coding for wild-type DHFR to mouse L cells by the direct uptake of HSR-bearing or double minute chromosomes. After selection for an initial level of MTX resistance, the transferred sequences could be further amplified by exposing the recipients to gradually increasing concentrations of the drug. The amplified sequences in the transformants were present on HSRs or double minutes, respectively, and retained idiosyncratic rearrangements that had taken place during the original amplification process.

B. Gene Transfer Using Total Cellular DNA

Total high-molecular-weight DNA from a CHO mutant carrying amplified, altered *dhfr* genes has been used to transform L cells to low-level MTX resistance (Wigler et al., 1980). By subjecting the transformants to higher levels of MTX, amplification of the tranferred sequences could be effected. Not only were the transferred *dhfr* genes amplified, but carrier prokaryotic DNA that had been cointegrated with the *dhfr* sequences was also amplified. This phenomenon of coamplification was exploited by Christman et al. (1982) who used total DNA from an amplified line together with cloned hepatitis virus DNA to isolate transformants that could be subsequently amplified to produce increased levels of hepatitis B surface antigen.

The *dhfr* gene has also been transferred to intact animals. DNA from a mouse 3T6 cell line carrying amplified copies of a mutant *dhfr* gene was applied to suspensions of mouse peripheral lymphocytes (Cline et al., 1980). The DHFR specified by the mouse mutant exhibits an intrinsic resistance to MTX (Haber et al., 1981). Irradiated mice were injected with the lymphocyte suspension and then periodically treated with MTX. The cells that had been exposed to the DNA, which could be karyotypically distinguished from untreated control cells, were observed to predominate in MTX-treated mice but not in untreated mice. These experiments suggest that it may be feasible to remove cells from a patient for genetic engineering and then reintroduce them for therapeutic purposes.

Milbrandt et al. (1983b) were the first to transfer a functional genomic *dhfr* gene using a cloned sequence. Using a cosmid vector, these authors succeeded in cloning a genomic fragment containing over 90% of the 26 kb CHO *dhfr* gene. The cloned fragment lacks about 1 kb of the most 3′ exon, but most of this exon represents untranslated sequences. The cloned fragment must contain all necessary protein coding information as this DNA is able to transform a DHFR-deficient CHO mutant to a DHFR-positive phenotype. The cloned sequence may lack the first polyadenylation site in the *dhfr* gene, since Northern blot analysis showed that the *dhfr* mRNA in transformants terminated in flanking pBR322 vector DNA. The transferred gene could be amplified in the transformants by exposure to increasing concentrations of MTX. The amplified sequences were present on HSRs, as shown by *in situ* hybridization. Flanking prokaryotic sequences present in the vector were amplified along with *dhfr*, as were additional CHO sequences, presumably from the integration sites.

C. Transfer of Cloned *dhfr* cDNA Sequences

Expression of a cloned *dhfr* cDNA sequence was first achieved in *E. coli* (Chang et al., 1978). The mouse enzyme is less sensitive than its bacterial counterpart to inhibition by trimethoprim, and bacterial transformants can be selected on the basis of their resistance to this drug. Expression in mammalian cells was achieved when Subramani et al. (1981) attached the mouse cDNA to the SV40 early promoter and SV40 splicing and polyadenylation sequences in a pBR322-based plasmid. This DNA was used to transform a DHFR-deficient CHO mutant to prototrophy for glycine, purines, and thymidine. The transferred sequences were shown to be transcribed into mRNA molecules according to the SV40-directed transcription and processing signals.

In an experiment designed to test for the presence of a hormone responsive promoter, the long terminal repeat (LTR) of mouse mammary tumor virus was attached to the mouse *dhfr* cDNA (Lee et al., 1981). Transcription of the proviral genome is known to be inducible by glucocorticoids. The chimeric gene was used to transform DHFR-deficient CHO cells: The transformation frequency was much higher when the glucocorticoid hormone dexamethasone was included in the selection medium. Moreover, the level of DHFR was inducible in individual transformants. Thus the viral promoter is located in the LTR and can function with downstream foreign sequences.

A series of mouse *dhfr* cDNA constructs were assembled by Kaufman and Sharp (1982*a*) using the adenovirus major late promoter, immunoglobulin gene splicing signals, and SV40 polyadenylation sites. The mouse *dhfr* gene has three major polyadenylation sites, and the cDNA used in these experiments contained the first two of these. Transformation of DHFR-deficient CHO cells was very inefficient if the only polyadenylation sites were those present in the mouse cDNA itself. If an SV40 polyadenylation site was inserted at the 3′ end of the cDNA sequence, transformation frequencies increased up to 1000-fold. The transferred modular gene could be amplified by exposure of individual transformant clones to increasing concentrations of MTX (Kaufman and Sharp, 1982b). Amplification of up to 1000-fold could be achieved not only for the *dhfr* gene, but also for other genes included in the plasmid. Moreover, these coamplified genes could be expressed at the same high levels as the *dhfr* gene. For instance, in this manner CHO cells producing several percent of their soluble protein as SV40 small t antigen were isolated. The process of transforming a DHFR-deficient mutant with a modular *dhfr* gene followed by the coamplification of ligated sequences provides a general method for isolating mammalian cells that produce high levels of any cloned gene product. This strategy has been applied by Scahill et al. (1983) to isolate CHO cells producing high levels of human immune interferon.

Cytogenetic studies of the amplified transformants isolated by Kaufman and Sharp showed that HSRs were regularly produced upon gene amplification, as is characteristic of the CHO cells used as recipients (Kaufman et al., 1983). These HSRs contained the amplified *dhfr* sequences and were associated with different chromosomes in different transformants, consistent with the idea of random integration of the transforming DNA. A variety of chromosome abnormalities were seen during the evolution of MTX resistance in these lines, especially ring and dicentric chromosomes. These structures, plus the fact that the HSRs were often found near telomeres, prompted the authors to propose a model for gene amplification in which chromatids fused after breakage of the telomeres. The resulting dicentrics could be resolved only by further breakage, often resulting in unequal assortment of genes.

Kaufman and Sharp (1983) have also analyzed the regulation of the transferred *dhfr* genes in amplified transformant clones. By comparing the levels of *dhfr* mRNA in resting versus growing cells, they were able to show that some transformants exhibited the same type of cell cycle control found in wild-type mouse cells (higher levels of expression in growing cells). When the integrated genes used SV40 polyadenylation sites, it was subject to cell cycle control; when a polyadenylation site in flanking DNA was used, no such control was evident. Surprisingly, none of the transferred genes tested used the internal polyadenylation sites present in the mouse cDNA. Although only a small number of clones were tested, the results suggest that cell cycle control of gene expression may be exerted via 3′ sequences. It is interesting to note that expression during the cell cycle of cloned histone genes in yeast is influenced by 3′ autonomous replication sequences (Osley and Hereford, 1982). It may be that similar 3′ sequences are necessary for the periodic expression of the *dhfr* gene; the different polyadenylation sites could be a secondary consequence of these 3′ sequences.

Modular genes have also been constructed that allow the efficient selection of transformants in wild-type, rather than DHFR-deficient, cells. The LTR from Harvey sarcoma virus has been attached to the mouse *dhfr* cDNA and used to transform mouse 3T3 cells (Murray et al., 1983). Under these conditions enough copies of the provirus are integrated to confer MTX resistance to the wild-type sensitive recipient cells. When these transformants are subsequently amplified, the multiple copies of the *dhfr* gene are found on double minute chromosomes. Simonsen and Levinson (1983) have provided a more general solution: they cloned a cDNA from a mouse mutant that codes for an altered DHFR (Haber et al. 1981). The mutant DHFR contains a single amino acid substitution that renders it resistant to inhibition by MTX. When this cloned cDNA sequence is introduced into wild-type cells, transformants can be efficiently selected in low levels of MTX. The transformation frequency is comparable to that obtained in the absence of MTX with a DHFR-deficient recipient.

D. Construction and Transfer of *dhfr* "Minigenes"

The term "minigene" here is being limited to genes that have been made smaller by the elimination of introns. They represent a category distinct from the modular cDNA genes described above in that their 5′ flanking sequences are those of the original gene, rather than a chimera with foreign controlling elements. Gasser et al. (1982) constructed several different "abbreviated" versions of the mouse *dhfr* gene by ligating restriction fragments from the 5′ end of a cloned genomic sequence to fragments from the 3′ end of a cloned cDNA. Each minigene contained 1 kb of 5′ flank. Multiple polyadenylation sites are present in *dhfr* transcripts, and so the cDNA (derived from one of the large mRNAs) was able to contribute at least one such site in all of the minigenes constructed. The largest of these minigenes contained only the first two introns of the gene; these sequences were capable of transforming CHO DHFR-deficient mutants with high efficiency (1.5×10^{-4} per recipient cell). Removal of these introns reduced the transforming efficiency by a factor of more than 10. However, even the cDNA clone itself was active in transformation at a low frequency (3×10^{-6} per recipient cell). The transformed cells were shown to contain mouse-specific *dhfr* sequences; the mouse sequences were amplifiable upon subsequent MTX selection. Interestingly, more than half of the transformants had apparently lost the pBR322 sequences of the vector. A CHO minigene containing only intron 1 has been similarly constructed; this gene is also active in transformation (Venolia et al., unpublished results).

A novel technique was used by Crouse et al. (1983) to construct similar mouse *dhfr* minigenes. These authors cloned the 5′ genomic sequence and a cDNA sequence into two lambda phage strains that had been developed for recombination studies. The phage contained markers such that only recombinants would form plaques in a mixed infection under the right conditions. The only regions of homology allowing such recombination lay in the *dhfr* exon sequences. Thus recombinant phage were isolated that had undergone a forced crossover between an exon in the genomic sequence and the cDNA sequence. In this way two minigenes were isolated: each contained 1.5 kb of 5′ flank and the first poladenylation site at the 3′ end. The two versions differed in that one contained the small intron 1 (300 bp) whereas the other had no introns. In contrast to the results of Gasser et al. described above, both minigenes transformed DHFR-deficient cells with high efficiency. However, a quantitative analysis of DHFR levels in a number of independent transformants revealed that gene expression was usually less than that expected for a cell carrying one active copy of the gene (often as little as 5% of that amount). Moreover, about half of the transformants surveyed contained multiple copies of the *dhfr* minigene (up to 450); these transformants nevertheless produced DHFR at levels less than or equal to that of wild-type cells. The presence or absence of intron 1 did not greatly influence the degree of gene expression. Increased DHFR levels could be achieved by isolating amplified MTX-resistant derivatives of these transformants. The amplified lines contained 10- to 100-fold higher levels of DHFR, and the *dhfr* minigene copy number also increased in this range. However, these two parameters did not always increase proportionately. In some cases a rearrangement of flanking sequences accompanied gene amplification.

It is clear from all of these transformation studies that the *dhfr* coding sequence can be expressed in many different contexts. In view of the finding that the natural polyadenylation site is not used when a modular cDNA gene is transferred, it will be interesting to see the results of a similar analysis of minigene transformants. Perhaps 5′ or intronic sequence play a role in the determination of polyadenylation sites. On the other hand, the *dhfr* gene can be expressed even when it is devoid of introns. It remains to be seen whether the presence of introns has more subtle effects, such as a role in cell cycle control or in the efficiency of *dhfr* gene expression. Site-directed mutagenesis of these cloned *dhfr* genes, as well as the sequence analysis of *in vivo* mutations at the *dhfr* locus, should define those aspects of gene structure that are important for the expression of this ubiquitous household gene.

ACKNOWLEDGMENTS

The author's work was supported by grant GM22629 from the National Institutes of Health. I am grateful to Gail Urlaub for helpful criticisms of this manuscript, and to Peter Melera and Wayne Flintoff for communicating unpublished results.

REFERENCES

Adair, G. M., Carver, J. H. and Wandres, D. L. (1980). *Mut. Res.* **72**, 187–205.

Albrecht, A. M., Biedler, J. L., and Hutchison, D. J. (1972). *Cancer Res.* **32**, 1539–1546.

Alt, F.W., Kellems, R.E., Bertino, J.R., and Schimke, R.T. (1978). *J. Biol. Chem.* **253**, 1357–1370.

Alt, F.W., Kellems, R.E., and Schimke, R.T. (1976). *J. Biol. Chem.* **251**, 3063–3074.

Ayusawa, D., Koyama, H., Iwata, K., and Seno, T. (1980). *Somat. Cell Genet.* **6**, 261–270.

Barsoum, J. and Varshavsky, A. (1983). *Proc. Natl. Acad. Sci. USA* **80**, 5330–5334.

Bertino, J. (1979). *Cancer Res.* **39**, 293–304.

Biedler, J. L., Albrecht, A. M., Hutchison, D. J., and Spengler, B. A. (1972). *Cancer Res.* **32**, 153–161.

Biedler, J. L., Albrecht, A. M., and Spengler, B. A. (1978). *Eur. J. Cancer* **14**, 41–49.

Biedler, J. L., Melera, P. W., and Spengler, B. A. (1980). *Cancer Genet. Cytogenet.* **2**, 47–60.

Biedler, J. L. and Spengler, B. A. (1976). *Science* **191**, 185–187.

Bostock, C. J. and Clark, E. M. (1980). *Cell* **19**, 709–715.

Bostock, C. J. and Tyler-Smith, C. (1981). *J. Mol. Biol.* **153**, 319–236.

Bostock, C. J. and Tyler-Smith, C. (1982). In *Gene Amplification* (R.T. Schimke, ed.), Cold Spring Harbor Laboratory, New York, pp. 15–21.

Botchan, M., Topp, W., and Sambrook, J. (1979). *Cold Spring Harbor Symp. Quant. Biol.* **43**, 709–719.

Bourgeois, S. and Newby, R. F. (1977). *Cell* **11**, 423–430.

Bradley, W. E. C. and Letovanec, D. (1982). *Somat. Cell Genet.* **8**, 51–66.

Brown, P. C., Beverly, S. M., and Schimke, R. T. (1981). *Molec. Cell. Biol.* **1**, 1077–1083.

Brown, P. C., Tlsty, T. D., and Schimke, R. T. (1983). *Molec. Cell. Biol.* **3**, 1097–1107.

Carothers, A. M., Urlaub, G., Ellis, N., and Chasin, L. A. (1983). *Nucl. Acids Res.* **11**, 1997–2012.

Chang, S. and Littlefield, J.W. (1978). *Cell* **7**, 391–396.

Chang, A. C. Y., Nunberg, J. H., Kaufman, R. J. Erlich, H.A., Schimke, R. T., and Cohen, S. N. (1978). *Nature* **275**, 617–624.

Chasin, L. A. (1973). *J. Cell Physiol.* **82**, 299–308.

Chasin, L. A., Feldman, A., Konstam, M., and Urlaub, G. (1974). *Proc. Natl. Acad. Sci. USA* **71**, 718–722.

Chasin, L. A., Graf, L., Ellis, N., Landzberg, M., and Urlaub, G. (1982). In *Gene Amplification"* (R.T.Schimke, ed.), Cold Spring Harbor Laboratory, New York, pp. 161–165.

Chen, M.-J., Shimada, T., Moulton, A. D., Harrison, M. and Nienhuis, A. W. (1982). *Proc. Natl. Acad. Sci. USA* **79**, 7435–7439.

Christman, J. K., Gerber, M., Price, P. M., Flordellis, C., Edelman, J., and Acs, G. (1982). *Proc. Natl. Acad. Sci. USA* **79**, 1815–1819.

Cline, M. J., Stang, H., Mercola, K., Morse, L., Huprecht, H., Browne, J., and Salser, W. (1980). *Nature* **284**, 422–425.

Collins, M. L., Wu, J.-S. R., Santiago, C. L., Hendrickson, S. L., and Johnson, L. F. (1983). *Molec. Cell. Biol.* **3**, 1792–1802.

Courtenay, V. D. and Robins, A. B. (1972). *J. Natl. Cancer Inst.* **49**, 45–53.

Cowan, K., Goldsmith, M., Levine, R., Aitken, S., Douglass, E., Clendeninn, N., Nienhuis, A., and Lippman, M. E. (1982). *J. Biol Chem.*, 15079–15086.

Cowell, J. K. (1982). *Ann. Rev. Genet.* **16**, 21–59.

Crouse, G. F., McEwan, R. N., and Pearson, M. L. (1983). *Molec. Cell. Biol.* **3**, 257–266.

Crouse, G. F., Simonsen, C. C., McEwan, R. N., and Schimke, R. T. (1982). *J. Biol. Chem.* **257**, 7887–7897.

Deaven, L.L. and Peterson, D.F. (1973). *Chromosoma* **41**, 129–144.

Dolnick, B. J., Berenson, R. J., Bertino, J. R., Kaufman, R. J., Nunberg, J. H., and Schimke, R. T. (1979). *J. Cell Biol.* **83**, 394–402.

Fischer, G. A. (1961). *Biochem. Pharmacol.* **7**, 75–80.

Fischer, G. A. (1962). *Biochem. Pharmacol.* **11**, 1233–1237.

Flintoff, W.F., Davidson, S.V., and Siminovitch. L. (1976a). *Somat. Cell Genet.* **2**, 245–262.

Flintoff, W.F. and Essani, K. (1980). *Biochem.* **19**, 4321–4327.

Flintoff, W. F., Livingston, E., Duff, C. and Worton, R. G. (1984) Molec. Cell. Biol., **4**, 69–76.

Flintoff, W. F., Spindler, S. M., and Siminovitch, L. (1976b). *In Vitro* **12**, 749–757.

Flintoff, W.F. and Weber, M. (1980). *Somat. Cell Genet.* **6**, 517–528.

Flintoff, W.F., Weber, M.K., Nagainis, C.R., Essani, A.K., Robertson, D., and Salser, W. (1982). *Molec. Cell. Biol.* **2**, 275–285.

Frearson, P. M., Kit, S., and Dubbs, D. R. (1966). *Cancer Res.* **26**, 1653–1660.

Gasser, C. S., Simonsen, C. C., Schilling, J. W., and Schimke, R. T. (1982). *Proc. Natl. Acad. Sci. USA* **79**, 6522–6526.

Goldie, J. H., Dedhar, S., and Krystal, G. (1981). *J. Biol. Chem.* **256**, 11629–11635.

Graf, Jr., L.H. and Chasin, L.A. (1982). *Molec. Cell. Biol.* **2**, 93–96.

Gupta, R.S., Flintoff, W.F., and Siminovitch, L. (1977). *Can. J. Biochem.* **55**, 445–452.

Haber, D. A., Beverly, S. M., Kiely, M. L., and Schimke, R. T. (1981). *J. Biol. Chem.* **256**, 9501–9510.

Haber, D. A. and Schimke, R. T. (1982). *Somat. Cell Genet.* **8**, 499–508.

Hakala, N. J. and Taylor, E. (1959). *J. Biol. Chem.* **234**, 126–128.

Hamlin, J. L. and Biedler, J. L. (1981). *J. Cell. Physiol.* **107**, 101–114.

Hamlin, J. L., Milbrandt, J. D., Heintz, N. H., and Azizkhan, J. C. (1983). *Int. Rev. Cytol.* (in press).

Hangii, V. J. and Littlefield, J. W. (1976). *J. Biol. Chem.* **251**, 3075–3080.

Heintz, N. H. and Hamlin, J. L. (1982). *Proc. Natl. Acad. Sci. USA* **79**, 4083–4087.

Hillcoat, B. L., Sweet, V., and Bertino, J. R. (1967). *Proc. Natl. Acad. Sci. USA* **58**, 1632–1636.

Huennekens, F. M., Vitols, K. S., Whitely, J. M., and Neef, V. G. (1976). *Methods Cancer Res.* **13**, 199–225.

Jackson, R. C., Hart, L. I., and Harrap, K. R. (1976). *Cancer Res.* **36**, 1991–1997.

Johnson, L. F., Fuhrman, C. L., and Wiedemann, L. M. (1978). *J. Cell. Physiol.* **97**, 397–406.

Jones, G. E. and Sargent, P. (1974). *Cell* **2**, 43–54.

Kao, F.-T. and Puck, T. T. (1968). *Proc. Natl. Acad. Sci. USA* **60**, 1275–1281.

Kao, F.-T., Chasin, L., and Puck, T. T. (1969). *Proc. Natl. Acad. Sci. USA* **64**, 1284–1291.

Kaufman, R. J., Brown, P. C., and Schimke, R. T. (1979). *Proc. Natl. Acad. Sci. USA* **76**, 5669–5673.

Kaufman, R. J., Brown, P. C., and Schimke, R. T. (1981). *Molec. Cell. Biol.* **1**, 1084–1093.

Kaufman, R. J. and Schimke, R. T. (1981). *Molec. Cell. Biol.* **1**, 1069–1076.

Kaufman, R. J. and Sharp, P. A. (1982a). *J. Mol. Biol.* **159**, 601–621.

Kaufman, R. J. and Sharp, P. A. (1982*b*). *Molec. Cell. Biol.* **2**, 1304–1319.

Kaufman, R. J. and Sharp. P. A. (1983). *Molec. Cell. Genet.* **3**, 1598–1608.
Kaufman, R. J., Sharp, P. A., and Latt, S. A. (1983). *Molec. Cell. Biol.* **3**, 699–711.
Kellems, R. E., Alt., F. W., and Schimke, R. T. (1976). *J. Biol. Chem.* **251**, 6987–6993.
Kellems, R. E., Harper, M. E., and Smith, L. (1982). *J. Cell Biol.* **92**, 531–539.
Lee, F., Mulligan, R., Berg, P., and Ringold, G. (1981). *Nature* **294**, 228–322.
Levan, A. and Levan, G. (1978). *Hereditas* **88**, 91–92.
Lewis, J. A., Biedler, J. L., and Melera, P. W. (1982*a*). *J. Cell Biol.* **94**, 418–424.
Lewis, J. A., Davide, J. P., and Melera, P. W. (1982*b*). *Proc. Natl. Acad. Sci. USA* **79**, 6961–6965.
Lewis, J. A., Kurtz, D. T., and Melera, P. W. (1981). *Nucl. Acids Res.* **9**, 1311–1322.
Lewis, W. H., Srinivasan, P. R. Stokue, N., and Siminovitch, L. (1980). *Somat. Cell Genet.* **6**, 333–347.
Leys, E. J. and Kellems, R. (1981). *Molec. Cell Biol.* **1**, 961–971.
Littlefield, J. W. (1969). *Proc. Natl. Acad. Sci. USA* **62**, 88–95.
Maniatis, T., Fritsch, E.F., Lauer, J., and Lawn, R.M. (1980). *Ann. Rev. Genet.* **14**, 145–178.
Mariani, B. D., Slate, D. L., and Schimke, R. T. (1981). *Proc. Natl. Acad. Sci. USA* **78**, 4985–4989.
Melera, P., Davide, J. P., Hession, C. A., and Scotto, K. W. (1984). *Molec. Cell. Biol.* **4**, 38–48.
Melera, P.W., Hession, C.A., Davide, J.P., Scotto, K.W., Biedler, J.W., Myers, M.B., and Shanske, S. (1982). *J. Biol. Chem.* **257**, 12939–12949.
Melera, P. W., Lewis, J. A., Biedler, J. L., and Hession, C. (1980*a*). *J. Biol. Chem.* **255**, 7024–7028.
Melera, P. W., Wolgemuth, D., Biedler, J. L., and Hession, C. (1980*b*). *J. Biol. Chem.* **255**, 319–322.
Milbrandt, J.D., Azizkhan, J.C., Greisen, K.S., and Hamlin, J.L. (1983*a*). *Molec. Cell. Biol.* **3**, 1266–1273.
Milbrandt, J. D., Azizkhan, J. C., and Hamlin, J. L. (1983*b*). *Molec. Cell. Biol.* **3**, 1274–1282.
Milbrandt, J. D., Heintz, N. H., White, W. C., Rothman, S. M., and Hamlin, J. L. (1981). *Proc. Natl. Acad. Sci. USA* **78**, 6043–6047.
Morandi, C., Masters, J. N., Mottes, M., and Attardi, G. (1982). *J. Mol. Biol.* **156**, 583–607.
Murray, M. J., Kaufman, R. J., Latt, S. A., and Weinberg, R. A. (1983). *Molec. Cell. Biol.* **3**, 32–43.
Nakamura, H. and Littlefield, J.W. (1972). *J. Biol. Chem.* **247**, 179–187.
Naylor, S. L., Busby, L. L., and Klebe, R. J. (1976). *Somat. Cell Genet.* **2**, 93–111.
Nichol, C. A., Lee, C. L., Edelstein, M. P., Chao, J. Y., and Duch, D. S. (1983). *Proc. Natl. Acad. Sci. USA* **80**, 1546–1550.
Nunberg, J.H., Kaufman, R.J., Schimke, R.T., Urlaub, G., and Chasin, L.A. (1978). *Proc. Nat. Acad. Sci. USA* **75**, 5553–5556.
Orkin, S. H. and Littlefield, J. W. (1971). *Exp. Cell Res.* **69**, 174–180.
Osley, M. A. and Hereford, L. (1982). *Proc. Natl. Acad. Sci. USA* **76**, 7689–7693.
Prem veer Reddy, G. and Pardee, A. B. (1980). *Proc. Natl. Acad. Sci. USA* **77**, 3312–3316.
Prem veer Reddy, G. and Pardee, A. B. (1982). *J. Biol. Chem.* **257**, 12526–12531.
Ray, M. and Mohandas, T. (1976). *Cytogenet. Cell Genet.* **16**, 83–91.
Roberts, J. and Axel, R. (1982). *Cell* **29**, 109–119.

Roberts, J. M., Buck, L. B., and Axel, R. (1983). *Cell* **33**, 53–63.

Roberts, M., Huttner, K. M., Schimke, R.T., and Ruddle, F. H. (1980). *J. Cell Biol.* **87**, 288a.

Scahill, S. J., Devos, R., Van der Heyden, J., and Fiers, W. (1983). *Proc. Natl. Acad. Sci. USA* **80**, 4654–4658.

Schilling, J., Beverly, S., Siminsen, C., Crouse, G., Setzer, D., Feagin, J., McGrogan, M., Kohlmiller, N., and Schimke, R. T. (1982). In *Gene Amplification* (R. T. Schimke, ed.), Cold Spring Harbor Laboratory, New York, pp. 149–153.

Schimke, R. T. (1980). *Sci. Amer.* **243** (Nov.), 600–669.

Schimke, R. T. (1982). In *Gene Amplification* (R.T. Schimke, ed.), Cold Spring Harbor Laboratory, New York, pp. 317–333.

Schirch, L. (1982). *Adv. Enzymol.* **53**, 83–112.

Schwab, M., Alitalo, K., Klempnauer, K.-H., Varmus, H.E., Bishop, J.M., Gilbert, F., Brodeur, G., Goldstein, M., and Trent, J. (1983). *Nature* **305**, 245–248.

Setzer, D. R., McGrogan, M., Nunberg, J. H., and Schimke, R. T. (1980). *Cell* **22**, 361–370.

Setzer, D. R., McGrogan, M. and Schimke, R. T. (1982). *J. Biol. Chem.* **257**, 5143–5147.

Simonsen, C. C. and Levinson, A. D. (1983). *Proc. Natl. Acad. Sci. USA* **80**, 2495–2499.

Sobel, J. S., Albrecht, A. M., Riehm, H., and Biedler, J. L. (1971). *Cancer Res.* **31**, 297–307.

Stein, R., Sciaky-Gallili, N., Razin, A., and Cedar, H. (1983). *Proc. Natl. Acad. Sci. USA* **80**, 2422–2426.

Subramani, S., Mulligan, R., and Berg, P. (1981). *Molec. Cell. Biol.* **1**, 854–864.

Tantravahi, U., Guntaka, R. V., Erlanger, B. F., and Miller, O. J. (1981). *Proc. Natl. Acad. Sci. USA* **78**, 489–493.

Tlsty, R., Brown, P. C., Johnston, R., and Schimke, R. T. (1982). In *Gene Amplification* (R.T. Schimke, ed.) Cold Spring Harbor Laboratory, New York, pp. 231–238.

Tyler-Smith, C. and Alderson, T. (1981). *J. Mol. Biol.* **153**, 203–218.

Urlaub, G. and Chasin, L.A. (1980). *Proc. Nat. Acad. Sci. USA* **77**, 4216–4220.

Urlaub, G., Kas, E., Carothers, A. M., and Chasin, L. A. (1983). *Cell* **33**, 405–412.

Urlaub, G., Landzberg, M., and Chasin, L. A. (1981). *Cancer Res.* **41**, 1594–1601.

Varshavsky, A. (1981). *Cell* **25**, 561–572.

Wahl, G. M., Vitto, L., Padgett, R. A., and Stark, G. R. (1982). *Molec. Cell. Biol.* **21**, 308–319.

Weidemann, L. M. and Johnson, L. F. (1979). *Proc. Natl. Acad. Sci. USA* **76**, 2818–2822.

Wigler, M., Perucho, M., Kurtz, D., Dana, S., Pellicer, A., Axel, R., and Silverstein, S. (1980). *Proc. Nat. Acad. Sci. USA* **77**, 3567–3570.

Williams, J. W., Morrison, J. F., and Duggleby, R. G. (1979). *Biochem.* **18**, 2567–2573.

Wolman, S. R., Craven, M. L., Grill, S. P., Domin, B. A., and Cheng, Y.-C. (1983). *Proc. Natl. Acad. Sci. USA* **80**, 807–809.

Worton, R., Duff, C., and Flintoff, W. (1981). *Molec. Cell. Biol.* **1**, 330–335.

Wu, J.-S. R. and Johnson, L. F. (1982). *J. Cell. Physiol.* **110**, 183–189.

Wu, J.-S. R., Weidemann, L. M., and Johnson, L. F. (1982). *Exp. Cell Res.* **141**, 159–169.

Yoder, S. S., Robberson, B. L., Leys, E. J., Hook, A. G., Al-Ubaidi, M., Yeung, C.-Y., and Kellems, R.E. (1983). *Molec. Cell. Biol.* **3**, 819–828.

CHAPTER 17

REGULATION AND AMPLIFICATION OF ASPARAGINE SYNTHETASE

Irene L. Andrulis
Department of Genetics
The Hospital for Sick Children
Toronto, Ontario, Canada

I. INTRODUCTION

The regulation of amino acid biosynthesis has been well characterized in bacteria and simple eukaryotes [for review see Umbarger (1978)]. Over the last 10–15 years developments in the technology and concepts of biochemical genetics in mammalian cells have made an examination of the regulation and expression of many eukaryotic enzyme systems possible. Our own interest has focused on the biosynthetic enzyme for asparagine synthetase (AS).

Asparagine synthetase catalyzes the conversion of glutamine and asparatic acid to form asparagine and glutamate:

$$\text{Aspartic Acid} + \text{Glutamine} + \text{ATP} + H_2O \overset{\text{AS}}{\rightleftharpoons} \text{Asparagine} + \text{Glutamic Acid} + \text{AMP} + \text{PPi}$$

The major purpose of asparagine is to be acylated to $tRNA^{Asn}$ and incorporated into protein by asparaginyl-tRNA synthetase (AsnRS):

$$\text{Asparagine} + tRNA^{Asn} + \text{ATP} \overset{\text{AsnRS}}{\rightleftharpoons} \text{Asparaginyl} - tRNA^{Asn} + \text{AMP} + \text{PPi} \rightarrow \text{Protein}$$

We chose this particular system for study for several reasons. First, animal cells cannot synthesize all of the amino acids at a rate required for growth; therefore, the biosynthesis of only the nonrequired amino acids can be investigated. Since asparagine is a nonessential amino acid in Chinese hamster ovary (CHO) cells, and since asparagine synthetase is solely responsible for its biosynthesis, this enzyme represents one case where regulation of amino acid biosynthesis in mammalian cells can be studied.

Second, we were interested in the involvement of tRNA in regulation of amino acid biosynthesis because of the work on several amino acid biosynthetic operons in bacteria. Amino acid biosynthesis in *Escherichia coli* and *S. typhimurium* has been shown to be controlled in a complex manner. The expression and regulation of the biosynthetic enzymes of several operons depends on feedback inhibition by the end product as well as the extent of aminoacylation of tRNA. For example, in the case of the tryptophan operon it has been well documented by Yanofsky and colleagues that the levels of the tryptophan biosynthetic enzymes are controlled at two levels; at the initiation of transcription, by the amount of the tryptophan-repressor complex (Squires et al., 1975), and by attenuation at the termination of transcription [for review see Kolter and Yanofsky (1982)] by the amount of charging of $tRNA^{Trp}$ (Yanofsky and Soll, 1977).

Studies on the mechanisms that control the amino acid biosynthetic operons in bacteria were facilitated by the isolation of the appropriate mutations: those affecting the genes for the biosynthetic enzymes, the repressor, the tRNA structural sequences, the tRNA modifying enzymes, and the aminoacyl-tRNA synthetases.

Third, the CHO asparagine synthetase system was attractive for study by somatic cell technology because of the availability of amino acid analogs of the substrates for the biosynthetic enzyme and the availability of mutants defective in asparaginyl-tRNA synthetase (Thompson et al., 1975; Adair et al., 1978).

Fourth, asparagine synthetase is of interest because of its altered activity in certain leukemias, mainly acute lymphocytic leukemia. About 30 years ago a factor was discovered in guinea pig serum which could be used to treat some leukemic tumors (Kidd, 1953). This molecule was found to be the enzyme which hydrolyzes asparagine—asparaginase (Broome, 1968). The asparaginase-sensitive tumors tend to have little or no asparagine synthetase activity and obtain asparagine from their surroundings (Horowitz et al., 1968). They are therefore susceptible to asparaginase which depletes the external pools of asparagine. Cells which become resistant to this chemotherapeutic drug often have elevations in asparagine synthetase activity (Horowitz et al., 1968; Patterson and Orr, 1967).

II. AVAILABILITY OF APPROPRIATE MUTANTS TO STUDY ASPARAGINE SYNTHETASE REGULATION IN CHO CELLS

Four genetic systems have been developed in CHO cells for the study of asparagine synthetase regulation. These involve:

1. A temperature-sensitive mutant in asparaginyl-tRNA synthetase.
2. Auxotrophic asparagine synthetase mutants.

3. Mutants resistant to β-aspartyl hydroxamate, an amino acid analog of aspartic acid.
4. Mutants resistant to albizziin, an amino acid analog of glutamine.

All of the cell lines discussed in this review are described in Table 17.1.

In this chapter I propose to describe the general genetic and biochemical characteristics of each of these systems and to indicate, where applicable, how these systems have been useful in examining asparagine synthetase regulation. The review will be more concerned with the two types of amino-acid-analog-resistant mutants since these have been explored in greater detail than the other two systems.

It was of primary importance to determine whether asparagine synthetase was a constitutive or regulated enzyme in wild-type CHO cells. This question was examined by growing CHO cells in medium containing asparagine or depleted of asparagine. Arfin et al. (1977) found that wild-type CHO cells grown in medium containing asparagine exhibited a basal level of activity which was elevated two- to threefold when cells were grown in medium lacking asparagine. To determine whether this was a consequence of lack of free asparagine or a decrease in the amount of charged $tRNA^{Asn}$, we made use of a temperature-sensitive asparaginyl-tRNA synthetase mutant.

A. Temperature-Sensitive (ts) Mutant in Asparaginyl-tRNA Synthetase

The ts asparaginyl-tRNA synthetase mutant (Asn-5) was isolated by the ^3H suicide selection method (Thompson et al., 1975) commonly employed to obtain ts mutations (see Chapter 14). The mutation in Asn-5 involves a temperature-sensitive defect in asparaginyl-tRNA synthetase such that the Asn-5 cells can grow in complete medium (Stanners et al., 1971) at 34.5°C but not at 38.5°C unlike the parental cells which are equally suited to growth at either temperature. The selection of Asn-5 and other ts asparaginyl-tRNA synthetase mutants was of interest genetically since such mutants, as well as ts leucyl-tRNA synthetase mutants, were observed at far higher frequency than other ts aminoacyl-tRNA synthetase isolates even though the selection procedure was not discriminatory in respect to the types of aminoacyl-tRNA synthetases sought. Evidence has been provided that in the case of the asparaginyl-tRNA synthetase, this comparatively higher frequency was probably due to functional hemizygosity at this locus (Adair et al., 1979). As with most of the other ts aminoacyl-tRNA synthetase mutants, the defect in Asn-5 cells could be overcome by adding an excess (10×) of the relevant amino acid (asparagine in this case) to the medium.

An examination of the biochemical properties of the asparaginyl-tRNA synthetase from the Asn-5 cell line showed that the mutant en-

TABLE 17.1
Description and Origins of Cell Lines

Cell Line	Characteristics	Reference
WTT	Wild-type CHO (Toronto strain)	Stanley et al. (1975)
WT1–WT5	Five independent clonal isolates of WTT	
GAT^-	Glycine, adenosine, and thymidine auxotroph of WTT	McBurney and Whitmore (1974)
TsH1	Temperature-sensitive leucyl-tRNA synthetase mutant	Thompson et al. (1973)
Asn-5	Temperature-sensitive asparaginyl-tRNA synthetase mutant derived from GAT^-	Adair et al (1978)
AH2, AH4	β-AHA-resistant mutants isolated from TsH1 after one round of mutagenesis	Andrulis and Siminovitch (1982a)
AH2 300, AH2 800	β-AHA-resistant mutants isolated by growing AH2 in progressively increasing concentrations (300 μ*M*, 800 μ*M*) of β-AHA	Andrulis and Siminovitch (1982a)
Alb^R 1,6,10,27	Albizziin-resistant mutants isolated from GAT^- after one round of mutagenesis	Andrulis et al. (1983)
Alb^R 31,613	Multistep low-level albizziin-resistant cell lines (isolated from the WT3 and WT4 cell lines, respectively)	Andrulis et al. (1983)
Alb^R 24	Multistep albizziin-resistant cell line isolated from the WT2 cell line	Andrulis et al. (1983)
Alb^R 42, Alb^R 43	Multistep albizziin-resistant cell lines isolated from the GAT^- cell line	Andrulis et al. (1983)
Alb^R 52	Multistep albizziin-resistant cell line isolated from the WT5 cell line	Andrulis et al. (1983)
134/7a	Asparagine-synthetase-deficient cell line isolated from the DON line of Chinese hamster lung fibroblasts	Goldfarb et al. (1977)
N3	Asparagine-synthetase-deficient CHO cell line	Waye and Stanners (1979)

zyme differed from that of the parental line in catalytic capacity at 38.5°C and a threefold increase in K_m for asparagine (Andrulis et al., 1978). Because of this defect, the level of charging of $tRNA^{Asn}$ in Asn-5 cells grown in reduced asparagine at 38.5°C was only 12% that of wild-type cells grown under the same conditions (Andrulis et al., 1979).

By comparing the activity of this cell line and that of wild-type CHO cells the effects of no free asparagine versus no charged $tRNA^{Asn}$ could be distinguished. The activity of asparagine synthetase from Asn-5 cells grown at 34.5°C in αMEM containing 330 μM asparagine was compared to levels of the enzyme from cells grown at 38.5°C in the same medium. The concentration of free asparagine was identical but the ability of the enzyme to attach asparagine to $tRNA^{Asn}$ was reduced. Under the restrictive conditions (38.5°C), the levels of asparagine synthetase increased two- to threefold indicating that either Asn-tRNA or asparaginyl-tRNA synthetase was involved in the regulation of asparagine synthetase. Controls for these experiments included the determination of specific activities of other amino acid biosynthetic enzymes that did not change. On the other hand, other aminoacyl-tRNA synthetase mutants, for example, leucyl-tRNA synthetase and histidyl-tRNA synthetase, also exhibited increased levels of asparagine synthetase under restrictive conditions (Andrulis et al., 1979). The reasons for these findings remain unclear but could be similar to "cross-pathway" regulation in *Neurospora crassa* (Carsiotis et al., 1974) or involve a regulatory "alarmone" molecule similar to "magic spot" in bacteria (Stephens et al., 1975).

B. Auxotrophic Asparagine Synthetase Chinese Hamster Cell Lines

Two laboratories (Goldfarb et al., 1977; Waye and Stanners, 1979) have been successful in isolating auxotrophic Chinese hamster cell lines which require asparagine at all temperatures. Goldfarb et al. (1977) selected asparagine-requiring mutants (e.g., 134/7a) from the DON line of Chinese hamster lung fibroblasts. The DON cells were mutagenized with ethyl methane sulfonate (EMS), recovered in complete medium, and selected in medium lacking asparagine by treatment with FUdR and uridine to enrich for cells that were inhibited in DNA synthesis in the absence of asparagine. After three rounds of selection most of the isolates proved to be asparagine synthetase deficient. Asparagine-synthetase-deficient CHO cells (e.g., N3) were also obtained by Waye and Stanners (1979) who modified the above procedure by treatment with [^{3}H]thymidine instead of FUdR to kill the nonauxotrophic cells.

The properties of both of these systems have been examined in a preliminary way. Since indirect methods were used for selection, the mutation frequency of the locus is difficult to estimate, but appeared to be of the order of 2×10^{-6}. The lesion in the mutant lines appears to in-

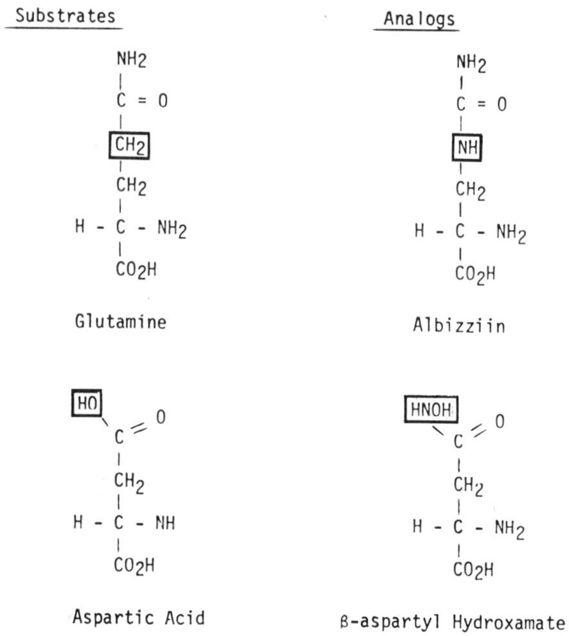

Figure 17.1. Structures of substrates and analogs.

volve the structural gene for asparagine synthetase, since such cells produce little or no enzyme. Because of the auxotrophic nature of the isolates, revertants were obtained relatively easily, and in the case of the CHO system, prototrophs arose at a rate of 10^{-6}/cell/generation. Such revertants contained 64–77% wild-type asparagine synthetase enzyme in the DON cell system and 64–127% in CHO cells. As expected, the mutants behave recessively, and it is of interest that the DON mutation was in the same complementation group as that of the rat Jenson sarcoma asparagine-requiring line.

One of these mutants, N3, has been used to make somatic cell hybrids with human B lymphocytes. By analysis of the human chromosomes retained in hybrids which did not require asparagine, the human gene for asparagine synthetase was mapped to chromosome 7 (Arfin et al., 1983). In addition, the auxotrophic mutants could provide useful recipient cells for DNA-mediated transfer of the asparagine synthetase gene, since transfer and selection would simply involve plating recipient cells in the absence of asparagine.

C. Isolation of Mutant Lines With Altered Levels of Asparagine Synthetase

Since β-AHA and albizziin are analogs of aspartic acid and glutamine (Fig. 17.1), the substrates of asparagine synthetase, they were obvious

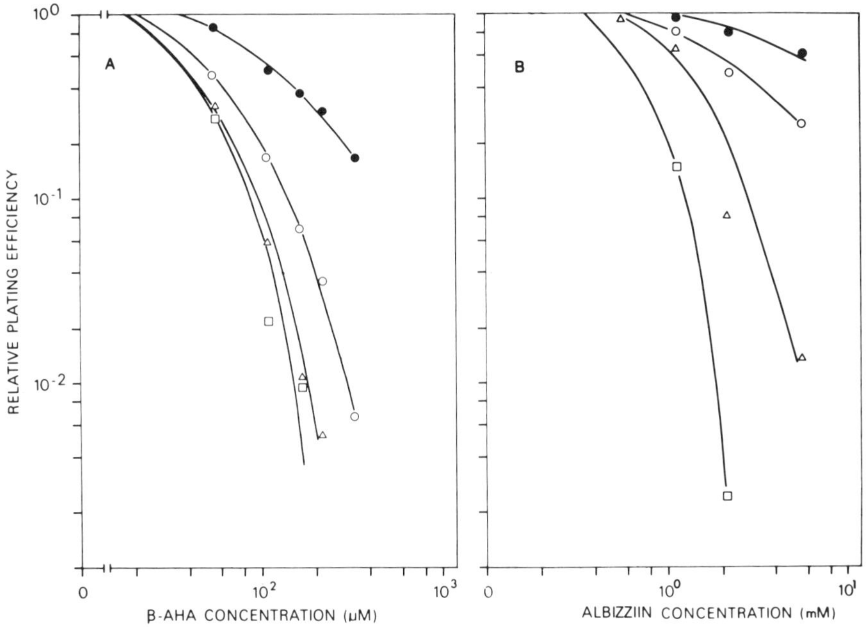

Figure 17.2. Effects of β-AHA and albizziin on the growth of wild-type cells. Cells were plated in medium containing various concentrations of drug. After 24 h (●), 48 h (○), 72 h (Δ), and 1 week (□) the selective media were removed and surviving cells were rescued by growth in nonselective media for 1 week.

choices in attempts to isolate mutants with alterations in the enzyme. Neither drug had been used previously to isolate mutants in cultured cells; therefore, the appropriate selection conditions were determined by plating cells in various concentrations of the drug. As shown in Figure 17.2 neither drug causes rapid cell death as cells could be rescued after 24–48 hr of growth in selective media. After 1 week of growth in 110 μ*M* β-AHA or 2 m*M* albizziin less than 2% and 0.3%, respectively, of the cells were recovered by growth in complete medium. However long-term growth (up to 1 month) of the cells in medium containing 100 μ*M* β-AHA or 800 μ*M* albizziin with medium changes every 4–7 days resulted in a reduction in relative plating efficiency to less than 1×10^{-6}. The complete methods for successful selection of such mutants and the nature of the isolates have been described in detail elsewhere (Andrulis and Siminovitch, 1982*a*; Andrulis et al., 1983), and only highlights and some specific features will be described here.

1. β-AHA-Resistant Lines

a. Selections of β-AHA-resistant lines. Certain culture manipulations were required to obtain β-AHA^r mutants. First, β-AHA does not cause rapid CHO cell death (Fig. 17.2), thus selection requires the con-

TABLE 17.2
Mutation Frequency for β-AHA and Albizziin Resistance

Parental Line	Selection	EMS	Frequency
GAT^-	150 μ*M* β-AHA	+	1×10^{-7}
GAT^-	400 μ*M* β-AHA	+	6×10^{-8}
GAT^-	400 μ*M* β-AHA	−	$<2 \times 10^{-8}$
TsH1	100 μ*M* β-AHA	+	3×10^{-6}
AH2	300 μ*M* β-AHA	+	2×10^{-3}
GAT^-	2 m*M* Albizziin	+	2×10^{-6}
GAT^-	2 m*M* Albizziin	−	5×10^{-7}
WT2	2 m*M* Albizziin	+	3×10^{-6}
WT2	2 m*M* Albizziin	−	4×10^{-7}

tinuous presence of the drug. Second, since β-AHA and asparagine are analogs, it is necessary to carry out the selections in medium lacking asparagine with dialyzed serum. Third, cell density effects were avoided by using (2–5) × 10^5 target cells/10-cm plate and by changing the medium every 4–7 days.

Using such procedures, single-step β-AHA-resistant mutants were obtained at a frequency of 1×10^{-7} to 3×10^{-6} after EMS mutagenesis (Table 17.2) when 100–150 μ*M* β-AHA was used for selection. Higher concentrations of selective drug reduced the frequency but did not result in increased levels of resistance.

As seen in Table 17.3 first-step β-AHA-resistant isolates were only two- to threefold more resistant to the drug as compared to parental cells. Two methods have been used to obtain isolates with increased resistance. One involved a second single-step isolation of resistant mutants derived from the first-step mutants. The second involved passage of the first-step isolates in increasing concentrations of the drug. The latter methodology was of particular interest because in several systems it has resulted in isolates with amplification of the target genes (see later).

β-AHA-resistant lines with increased resistance were obtained using both methods, and the derivation and frequencies of such isolates are shown in Tables 17.1 and 17.2. All of the single-step β-AHA^r mutants were stable in the absence of selection.

b. Genetic and Physiological Properties of β-AHA-Resistant Lines. At least two genetically different isolates could be distinguished by hybrid cell analysis among first-step β-AHA-resistant mutants (Andrulis and Siminovitch, 1982a). One group (80%) behaved recessively in hybrids with parental cells, whereas the others acted codominantly under these conditions. No complementation was observed between independent recessive isolates indicating that they all involved similar genotypes.

TABLE 17.3
Degree of Drug Resistance and Level of Asparagine Synthetase Activity in Parental and Mutant Cell Lines

Cell Line	D_{10} Albizziin (mM)	D_{10} β-AHA (μM)	Asparagine Synthetase Activity (fold increase)
Parental Lines			
GAT⁻	0.4	66	1
TsH1	0.3	70	1
WTT	0.3	ND[a]	1
β-AHA^r Mutants			
AH2	2	180	5
AH4	ND	180	6
AH2 300	ND	460	9
AH2 800	ND	1050	9
Alb^R Mutants			
Alb^R 1	18	330	17
Alb^R 6	21	120	12
Alb^R 10	>30	ND	12
Alb^R 24	25	ND	40
Alb^R 42	21	380	260
Alb^R 52	>50	ND	270

[a]Not determined.

Similar analyses for the second-step mutants indicated that the isolates included mutants which carried combinations of recessive and codominant lesions. The multistep β-AHA-resistant lines were derived from a single-step codominant parent and behaved codominantly in hybrids.

The degree of resistance of all of the β-AHA-derived lines has been assessed by D_{10} (dose of drug required to reduce survival to 10%) values obtained from survival curves. As shown in Figure 17.3 single-step mutants were approximately two to four times more resistant than the parental lines. The most resistant multistep isolates had up to 12-fold increases in D_{10}'s (Fig. 17.3).

2. *Isolation of CHO Mutants Resistant to Albizziin (Alb)*

a. Selection of Alb-Resistant Mutants. The selection procedures used to obtain single-step Alb^R isolates were similar to those employed with β-AHA. Using 2 mM albizziin, such mutants were obtained after continuous exposure to the drug at frequencies of $(1-5) \times 10^{-7}$ without

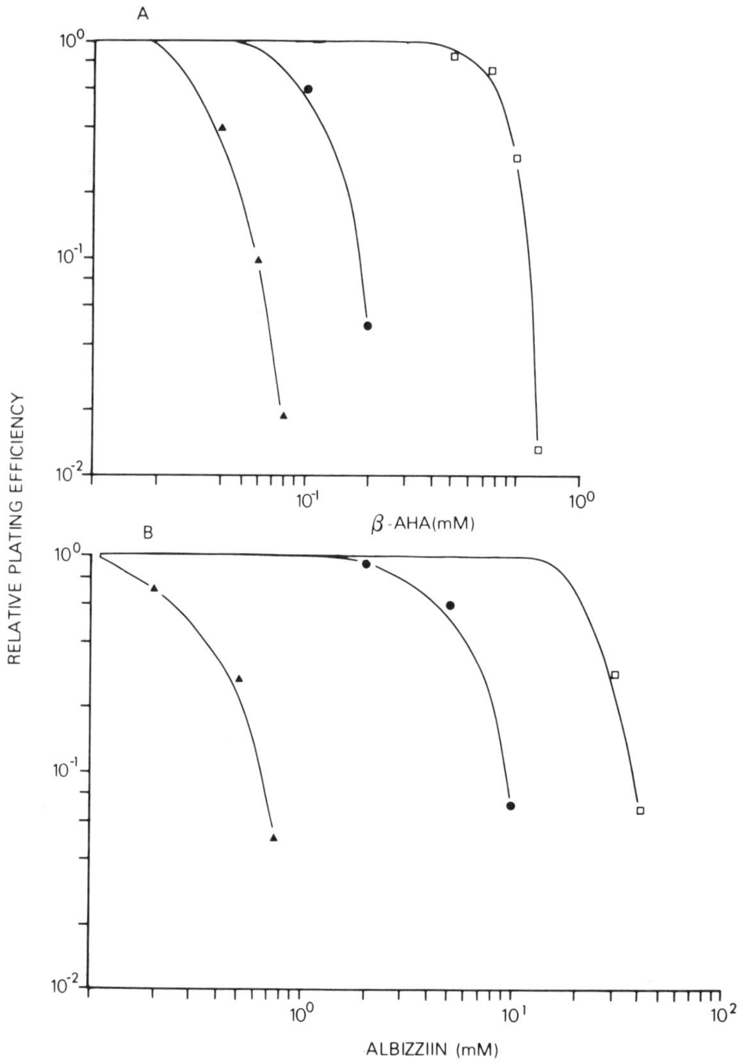

Figure 17.3. Dose response curves of wild-type and mutant lines in β-AHA and albizziin. (A) Plating efficiency of parental (▲), single-step AH2 (●), and multistep AH2 800 (□) in β-AHA. (B) Plating efficiency of parental (▲), single-step Alb (●) and multistep Alb^R 42 lines (□).

mutagenesis, and at about 10-fold higher frequencies when EMS was used (Table 17.2). As with β-AHA, all such mutants were stable in the absence of further exposure to albizziin. Multistep mutants with highly increased resistance to the drug were obtained by continuous subculture of parental cells in increasing concentrations of the drug, without mutagenesis: 2m*M* albizziin was used at the outset and when cell lines grew well in the presence of 2 m*M* albizziin, they were subcultured into 4 m*M* albizziin. Similar subculture at increasing concentrations resulted in cells resistant to 50 m*M* albizziin. At this point, the glutamine concentration of the medium was reduced 10-fold, thus increasing the effective concentration of albizziin (as indicated earlier, albizziin is an analog of glutamine). Further passages under these conditions resulted in highly resistant cell lines (Table 17.3; Fig. 17.3).

b. Genetic and Physiological Properties of Alb^R Lines. In contrast to β-AHA-resistant isolates, both first-step and multistep albizziin-resistant mutants behaved codominantly in somatic cell hybrids. The D_{10}'s of single-step and multistep albizziin mutants were up to, respectively, 10- and 100-fold higher than those of the parental lines.

Since both β-AHA and Alb affect asparagine synthetase, it was of interest to determine whether the mutants would be resistant to both drugs. All of the albizziin-resistant mutants were cross-resistant to β-AHA, but three classes could be distinguished. Most of the mutants were highly resistant to both drugs; one mutant (Alb^R 6), although quite resistant to albizziin, was only slightly resistant to β-AHA; and another group was slightly resistant to both analogs (Table 17.3).

3. Biochemical Analysis of Alb- and β-AHA-Resistant Mutants

a. Overproduction of Asparagine Synthetase Activity. Asparagine synthetase activity was assayed in mutant lines grown in the presence and absence of asparagine to determine whether analog resistance was associated with altered regulation of the enzyme. All of the single-step $\beta\text{-AHA}^r$ lines expressed elevated levels of asparagine synthetase activity up to six-fold, whereas the single-step albizziin-resistant lines had enzyme levels up to 27 times that of the parental lines (Table 17.3). However, regulation of the enzyme by asparagine in all of the mutants differed from that of the parental lines. The parental cells grown in medium containing asparagine expressed a basal level (1×) of enzyme activity which was increased twofold by growth in asparagine-free medium, but the enzyme levels in the mutants were constitutively elevated despite the presence of asparagine in the medium.

Using antibody raised against beef pancreas asparagine synthetase, Gantt and Arfin (1981) showed that the increase in enzyme activity observed in cells grown in the absence of asparagine or under conditions of uncharged tRNA was due to increased synthesis of asparagine synthetase. In addition they found that the constitutive elevation in enzyme activity in several $\beta\text{-AHA}^r$ lines was also correlated with an increased rate of enzyme synthesis and not a change in catalytic efficiency.

Although the selection for multistep $\beta\text{-AHA}^r$ lines produced cell lines that were highly resistant to the drug, there was not a concomitant increase in the level of asparagine synthetase activity. The enzyme activity of cells grown in 300 μ*M* β-AHA increased from five- to ninefold over wild-type levels, but no further elevation was observed for cells grown in 800 μM β-AHA (Table 17.3). Even though the protocol was designed to isolate cell lines with high levels of asparagine synthetase, some mechanism in addition to overproduction of the enzyme was responsible for the high levels of resistance to β-AHA observed in the multistep lines (see below).

In contrast, multistep albizziin-resistant lines did show highly elevated levels of asparagine synthetase activity. In this case increased drug resistance was correlated with elevated enzyme activity (Andrulis and Siminovitch, 1982b; Andrulis et al., 1983) and the multistep Alb^R lines expressed elevations in asparagine synthetase activity of up to 270-fold (Table 17.3).

b. Studies on Asparagine Synthetase in Parental and Mutant Lines. The actions of the analogs have been investigated in order to assess the types of mutants which might be expected. The *in vivo* effects of the addition of albizziin and β-AHA to the medium on enzyme levels and the *in vitro* consequences of the analogs on enzyme activity have been examined. Albizziin and β-AHA affected the normal regulation of asparagine synthetase in wild-type CHO cells. The addition of either β-AHA or albizziin to asparagine-free medium prevents the derepression of asparagine synthetase normally observed in wild-type cells. This suggests that the analogs mimic free asparagine or become attached to tRNA to repress the level of asparagine synthetase activity. Of several Alb^R mutants representing three possible classes, only one mutant (Alb^R 10) was found to respond to the addition of albizziin by a decrease in the level of asparagine synthetase, the other mutants maintained constitutively elevated levels of enzyme activity (Andrulis et al., in preparation).

The effects of β-AHA and albizziin as inhibitors of the enzyme with respect to the substrates have also been examined. Both analogs proved to be competitive inhibitors of asparagine synthetase with respect to glutamine (Andrulis and Siminovitch, 1982a; Andrulis et al., in preparation) and β-AHA is a noncompetitive inhibitor with respect to aspartic acid (Andrulis and Siminovitch, 1982a). The mutant lines were then examined to determine whether any of the lesions were structural gene mutations. None of the single-step β-AHA^r lines had significant changes in kinetic constants for substrates or K_I for β-AHA. However, as shown in Table 17.4, the multistep β-AHA^R lines which had been isolated from AH2 and grown in medium containing up to 800 μM β-AHA exhibited increases in K_m for glutamine and K_I for β-AHA of two- to threefold (Andrulis and Simonovitch, 1982a). Some slight change may have been present in the enzyme from the AH2 cell line as indicated by a small increase in K_m for glutamine and K_I for β-AHA, and change in temperature stability of the enzyme at 47.2°C (Andrulis and Siminovitch, 1981).

TABLE 17.4
Kinetic Constants of Asparagine Synthetase from Parental and Mutant Lines

	K_m (mM)	K_I (mM)	
	Glutamine	β-AHA	Albizziin
Parental (tsH1)	0.6	0.6	
AH2	0.9	0.8	
AH2 800	1.7	2.0	
Parental (GAT$^-$)	0.6		0.9
Alb^R 1	0.8		1.6
Alb^R 6	1.9		4.2
Alb^R 10	0.9		1.7

In addition, one of the single-step albizziin-resistant mutants (Alb^{R}6) appears to have a structural change in asparagine synthetase (Table 17.4). The K_m for glutamine was increased three-fold and K_I for albizziin showed a fivefold increase over the values of the enzyme from the parental line (Andrulis et al., in preparation). This mutant had been classified separately on the basis of cross-resistance to β-AHA and may represent a structural gene mutation. Since these studies were performed on crude cell extracts, this remains to be proven.

c. Studies on Asparaginyl-tRNA Synthetase. If the analogs do indeed mimic asparagine by being acylated to tRNA, other types of mutations that might be expected would be lesions in asparaginyl-tRNA synthetase (AsnRS). It has been shown that both β-AHA (Gantt et al., 1980) and albizziin (Andrulis et al., in preparation) act as competitive inhibitors of AsnRS with respect to asparagine. However, the K_I's for albizziin (6 mM) and β-AHA (27 mM) are 40- and 100-fold greater than the K_m for asparagine. Albizziin acts not only as an inhibitor of AsnRS but also as a substrate for tRNA acylation. Nevertheless, none of the mutants have alterations in AsnRS (Gantt et al., 1980; Andrulis et al., in preparation).

d. Overproduction of Glutamine Synthetase and Glutamine Phosphoribosyl Pyrophosphate (PRPP) Amidotransferase. During the initial selections of Alb^R lines, the culture conditions were designed to enrich for asparagine synthetase mutants by the use of asparagine-free medium which contained 2 mM glutamine. Since albizziin is a glutamine analog and glutamine is a substrate for numerous reactions in addition to asparagine biosynthesis (see Pinkus, 1977), it was important to select against the other targets. As the selective albizziin concentration for the multistep lines approached 100 mM, it was necessary to reduce the concentration of glutamine to 0.2 mM so that the effective albizziin concentration could be lowered simultaneously. After selection under these

TABLE 17.5
Concomitant Increase in Activity of Asn Synthetase and Gln Synthetase or PRPP Amidotransferase in Multistep Alb^R Lines

Cell Line	Asparagine Synthetase (Fold Increase)	Glutamine Synthetase (Fold Increase)	PRPP Amidotransferase (Fold Increase)
Wild type	1	1	1
Alb^R 42	258	8.8	ND[a]
Alb^R 52	287	1	7.5

[a]Not determined.

conditions it was found that some multistep lines had elevations in other enzymes in addition to asparagine synthetase. As shown in Table 17.5 Alb^R 42 cells exhibited a ninefold increase in glutamine synthetase and Alb^R 52 cells express eight-fold elevations in glutamine phosphoribosyl pyrophosphate amidotransferase (Ray et al., unpublished observations). It is obvious that the activities of other enzymes which utilize glutamine as a substrate [e.g., formylglycinamide ribonucleotide amidotransferase (Schroeder et al., 1969)] may also be affected but these have not yet been investigated.

4. *Karyotypic Analysis of β-AHA^r and Alb^R Mutants*

The karyotypes of the mutant lines were examined to determine whether chromosomal changes were associated with resistance to β-AHA and albizziin. All of the single-step and multistep β-AHA^r lines that were studied showed trypsin-Giemsa banding patterns similar to those of the parental lines.

In contrast the single-step Alb^R lines had alterations including breaks and translocations associated with the long arm of chromosome 1 (Fig. 17.4). The number 1 chromosomes are one of several pairs of chromosomes that have remained unchanged in culture from those of the normal Chinese hamster cell (Worton and Duff, 1979) (see Chapter 5). Since they are unmarked by translocations or rearrangements, they cannot be distinguished from one another as some of the members of altered pairs can be. The chromosomal analyses of the multistep Alb^R lines showed one normal chromosome 1 and one aberrant chromosome 1 (Fig. 17.5). There was generally found to be a translocation between one chromosome 1 and other chromosomes. In addition, homogeneously staining regions (HSRs, regions of the chromosome which stain differently from the wild-type, sometimes homogeneously, and which indicate expanded chromosomal material) were observed usually on chromosome 1 at the breakpoint of the translocations. This latter observation was of interest

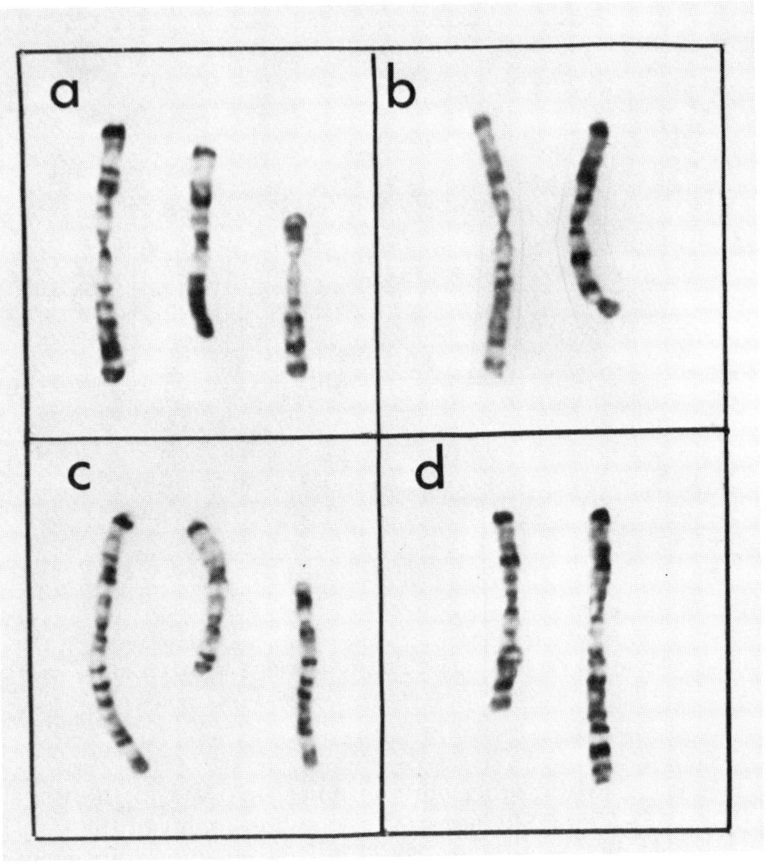

Figure 17.4. Abnormal chromosomes 1 observed in low-level albizziin-resistant lines. Each panel shows the one normal chromosome 1 at the left and the other altered chromosomes in (*a*) Alb^R 31,1q with HSR and t (1q; X); (*b*) Alb^R 613, dicentric 1; (*c*) Alb^R 1, $1q^-$, and t (1q; Z7); (*d*) Alb^R 27, $1q^+$.

since such regions have been shown by *in situ* hybridization to be the sites of the amplified genes for dihydrofolate reductase (Nunberg et al., 1978) and the CAD sequences (Wahl et al., 1981). Since the presence of double minutes (DMs), acentromeric chromosomes which replicate autonomously, have also been associated with amplified genes in some cases, and have consequently facilitated the cloning of certain enriched amplified adrenal tumor cell sequences (George and Powers, 1981), we examined our resistant lines for such entities. However, DMs were only observed in one cell line (Alb^R 52), and, even in this case, there were less than 10/cell and they did not appear in every spread.

The alterations observed in Alb^R lines were always associated with the long arm of chromosome 1, and there did not seem to be any speci-

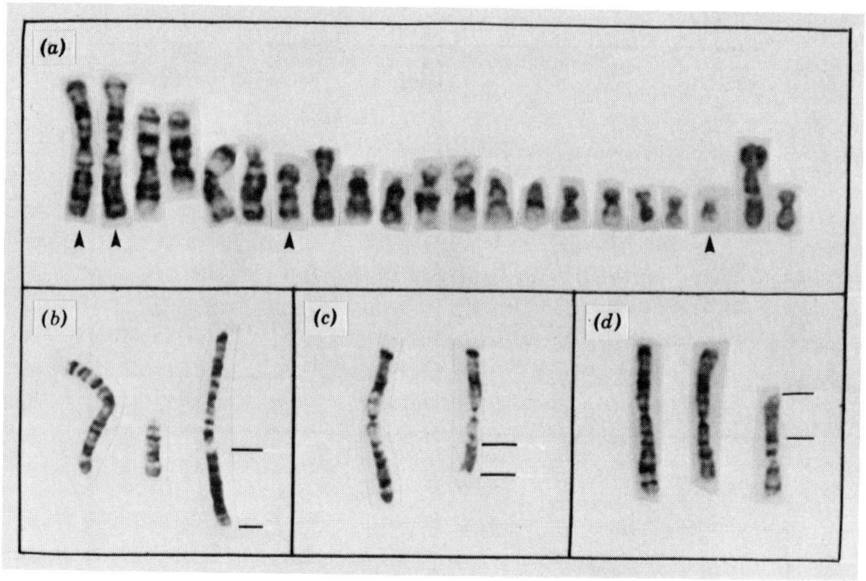

Figure 17.5. Karyotypes of wild-type CHO cells (*a*) and abnormal chromosomes of multistep Alb^R lines. The left chromosome in each case is the normal chromosome 1. The abnormal chromosomes with a bar indicating the HSR are in (*b*) Alb^R 24, t (Z13; 1q) and HSR on $1q^-$; (*c*) Alb^R 43, HSR on $1q^-$; (*d* Alb^R 42, $1q^-$ and HSR on Z7.

ficity in respect to the other chromosomes involved in the translocations (Fig. 17.5). The HSRs occurred at the breakpoint of the translocation; although the specific site could not be unequivocally identified, the rearrangements were generally in the center of the long arm of chromosome 1 between two darkly staining bands. A compilation of the karyotypic data indicate that the single-step Alb^R lines with increases in asparagine synthetase activity of less than 40-fold had breaks and translocations affecting the long arm of chromosome 1, whereas the multistep Alb^R lines with highly elevated levels of enzyme activity (40–300-fold) have HSRs, in many cases on the long arm of chromosome 1.

5. *Stability of the Alb^R Phenotype*

The stability of the highly amplified Alb^R lines in the absence of selection has been studied with the view of examining the association between DMs, HSRs, and the resistant-cell phenotype. In general, DMs have been shown to be associated with unstable methotrexate resistance (Kaufman et al., 1979) and HSRs with a stable phenotype. Consequently, the DM-containing line (Alb^R 52) and an HSR-containing line (Alb^R 24) were compared for stability in the absence of selection. After

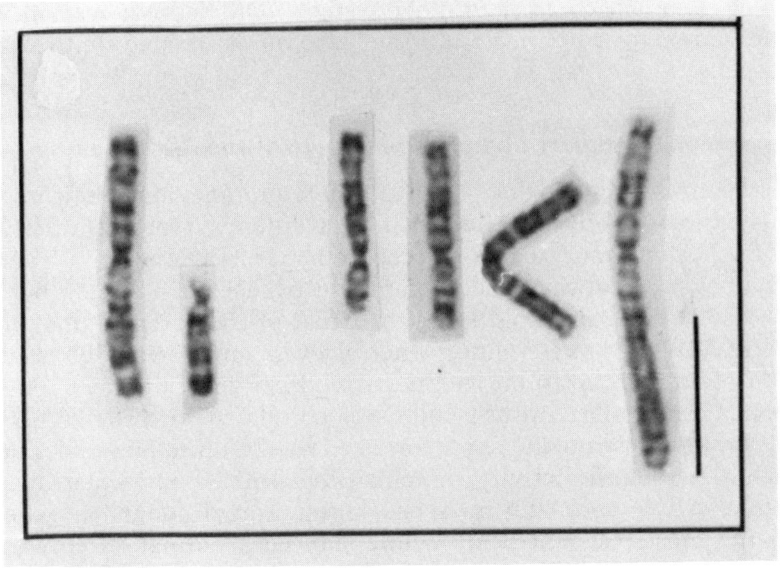

Figure 17.6. Decrease in the size of the HSR in Alb^{R} 24 cells which had reduced levels of asparagine synthetase after growth in the absence of drug. Chromosomes are from left to right the normal chromosomes 1, t Z13; 1q), three chromosomes with reduced HSRs after growth in the absence of selection, and, on the far right, chromosome 1 containing a large HSR from cells maintained in drug.

3 months of growth in the absence of drug, the resistance of the Alb^{R} 24 cell line was surprisingly lost, whereas that of the Alb^{R} 52 line was maintained. This phenotypic behavior of the lines became understandable when the enzyme was assayed and karyotypes were examined. It was found that the Alb^{R} 24 line showed a reduction in the initial level of asparagine synthetase by 10-fold and had lost the HSR from chromosome 1 (Fig. 17.6), whereas the Alb^{R} 52 cell line maintained a high level of asparagine synthetase activity. The cell line did not, however, retain the DMs but instead gained an HSR on chromosome 1. Although these results were unexpected, they were consistent with the hypothesis that the HSR was the site of the amplified genes for asparagine synthetase since high levels of asparagine synthetase activity and resistance to albizziin were associated with the presence of an HSR-containing chromosome in Alb^{R} 52 and a reduction in enzyme activity was correlated with a decrease in the size of the HSR in Alb^{R} 24.

As a further attempt to examine the role of DMs the original Alb^{R} 52 cell population was cloned and grown in continuous albizziin selection in an attempt to isolate cell lines that might be more highly enriched for DMs. After 2 months of selection, instead of detecting cells which contained many DMs, the Alb^{R} 52 cells contained a chromosome with an HSR. Thus, in contrast to other systems, DMs do not seem to be a major vehicle for amplification of the gene for asparagine synthetase in CHO cells.

6. *Summary Comparison of β-AHA^r and Alb^R Lines*

The two amino acid analogs used to isolate mutants affected in asparagine synthetase activity were specific for different types of mutations. The β-AHA^r mutants could be codominant or recessive in hybrids; Alb^R lines were always codominant. Resistance in first-step β-AHA^r lines was two- to threefold more than that of the parental line and at most 12-fold in multistep lines, whereas even single-step Alb^R mutants were 10-fold more resistant and multistep lines approximately 100-fold. The increase in albizziin resistance was correlated with the highly elevated levels of asparagine synthetase activity of up to 300-fold, but asparagine synthetase activity in multi-step β-AHA^r lines plateaued at ninefold. All of the Alb^R lines had chromosomal anomalies whereas even the most resistant β-AHA^r line showed a normal karyotype. Finally, the high levels of resistance and enzyme activity in the multistep Alb^R lines appears to be due to gene amplification while multistep β-AHA^r lines show structural gene changes. These results suggest that the two analogs have very different effects on cellular processes even though they are both competitive inhibitors of asparagine synthetase. It is likely that albizziin as an analog of glutamine, which has a central role in cellular metabolism, inhibits other enzyme activities which causes errors in replication and enhances the possibility of gene amplification.

III. USE OF MUTANT LINES TO ISOLATE THE GENE FOR ASPARAGINE SYNTHETASE

A. Alb^R Cells Have Elevated Levels of Asparagine Synthetase mRNA

It seemed likely that the multistep Alb^R 52 lines had amplified genes for asparagine synthetase because they were isolated by a selection regimen designed to enrich for gene amplification mutants, they expressed highly elevated levels of asparagine synthetase, and they showed HSR containing chromosomes. Further support for the contention that the lines contained amplified genes for asparagine synthetase came from an investigation of the levels of mRNA in wild-type and multistep Alb^R lines. Total RNA was isolated and poly A^+ RNA, purified on an oligo dT cellulose column, was translated in a rabbit reticulocyte lysate system. As shown in Figure 17.7 there is a band at 57,000 MW which is synthesized by the Alb^R 52 message but not by the wild-type mRNA.

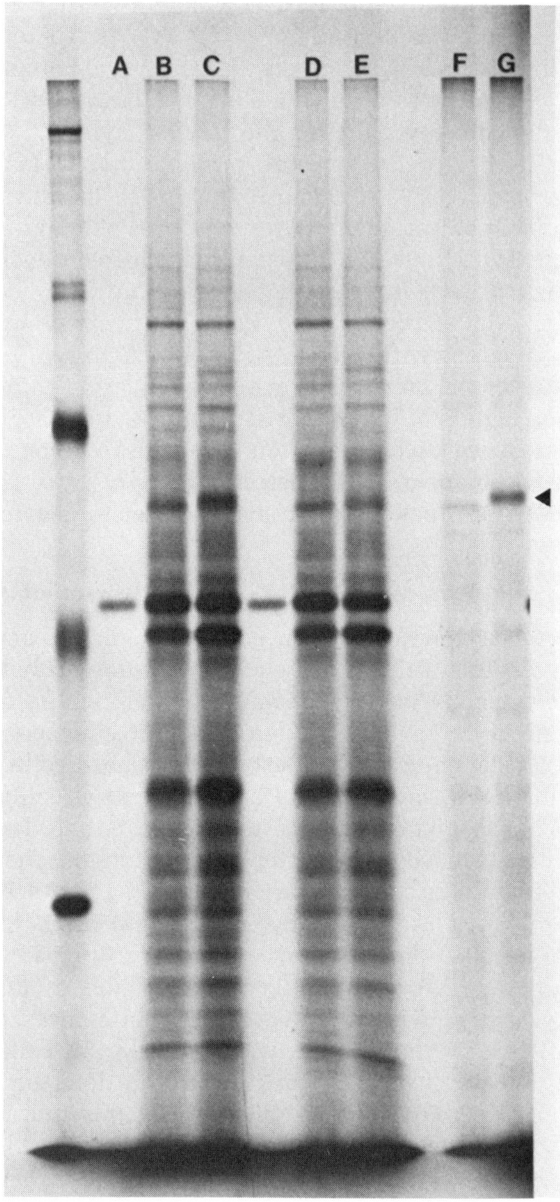

Figure 17.7. Alb^R 52 cells possess more mRNA for asparagine synthetase than wild-type cells. SDS-PAGE of total *in vitro* translation products before immunoprecipitation: (A) no exogenous mRNA, (B) wild-type mRNA, (C) Alb^R 52 mRNA. After immunoprecipitation with antiasparagine synthetase: the supernatants from (D) wild-type and (E) Alb^R 52 and the immunoprecipitates from (F) wild-type and (G) Alb^R 52. The arrowhead indicates the position of asparagine synthetase at 57,000 MW. The molecular weight standards shown on the far left are 200,000; 92,500; 69,000; 46,000; 30,000.

This protein has been shown to have the same migration pattern as purified CHO asparagine synthetase.

Because Alb^R 52 cells have greatly elevated levels of asparagine synthetase, this cell line was used as a source of enzyme for purification. The subunit molecular weight for the CHO enzyme of 57,000 was in agreement with that of the enzyme from rat liver (Hongo and Sato, 1981) and beef pancreas (Gantt and Arfin, 1981). In addition, antibody raised against purified asparagine synthetase was able to immunoprecipitate asparagine synthetase synthesized in a cell-free translation system (Fig. 17.7). These results indicate that in addition to an elevation in enzyme activity, Alb^R 52 cells also have an increase in translatable mRNA for asparagine synthetase most likely due to gene amplification. Further support for the presence of gene amplification comes from work on the isolation of a cDNA probe for asparagine synthetase.

Several approaches have been taken to isolate the gene for asparagine synthetase. Two involve the construction of cDNAs from Alb^R 52 mRNA and the third makes use of gene transfer to isolate genomic sequences.

B. Isolation of cDNAs for Asparagine Synthetase

Poly A^+ RNA was fractionated on sucrose gradients, and those fractions which contained mRNA enriched for asparagine synthetase were used to construct cDNAs. To determine which cDNAs contained the asparagine synthetase sequences, two different approaches were used. The cDNSs for asparagine synthetase could be identified by the standard method of differential screening and hybridization selection or by complementation of asparagine synthetase deficient *E. coli*. For both methods it is desirable to have full length cDNAs, therefore the procedure of Land et al. (1981) was used to prepare full length cDNAs. Colonies containing recombinant plasmids were streaked onto filters for differential hybridization and also grown to mass culture to prepare plasmids for transfection of AS^- *E. coli*. To screen the colonies by differential hybridization, we again took advantage of the differences in asparagine synthetase mRNA levels in wild-type and Alb^R 52 cells. ^{32}P-labeled cDNAs were prepared from both mRNA preparations and individually hybridized to filters containing the recombinant plasmid DNAs. Those colonies showing differential signals to wild-type and Alb^R 52 probes (e.g., Fig. 17.8) were then analyzed by hybridization of the recombinant DNA to poly A^+ mRNA from Alb^R 52 cells and translation of the hybridized mRNA. The translation products were immunoprecipitated and run on SDS-PAGE to determine which recombinant plasmids contained asparagine synthetase sequences. More detailed information about this and other approaches to cDNA cloning and expression can be found in Chapter 10.

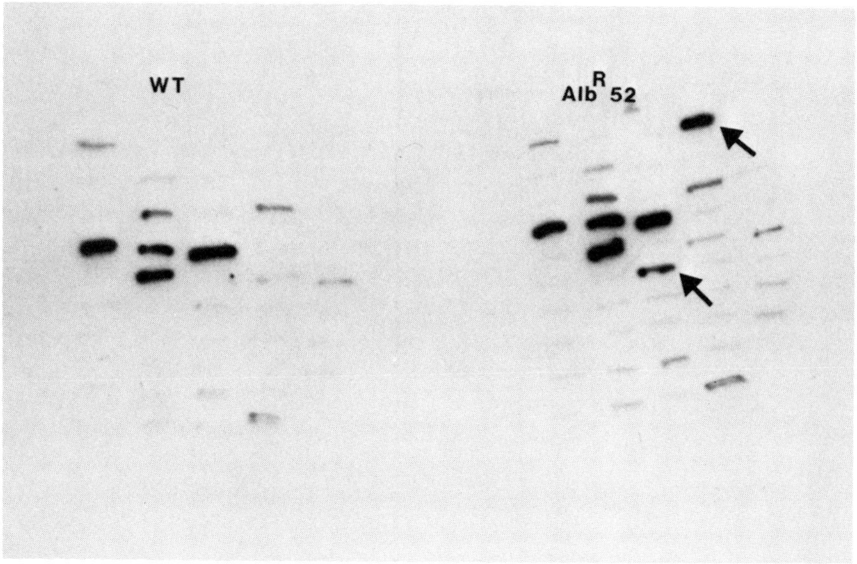

Figure 17.8. Replica filters of colonies containing plasmids with cDNA inserts on which the DNA had been fixed and hybridized. Arrows indicate the colonies containing recombinants that show strong signals using Alb^R 52 cDNA probe and weak signals with GAT^- cDNA probe.

Figure 17.9 shows the identification of two plasmids containing asparagine synthetase cDNAs. The plasmids were used to probe Southern blots of DNA from wild-type, Alb^R 1, and Alb^R 52 cells and demonstrate conclusively that Alb^R 52 cells contain an amplified number of copies of the gene for asparagine synthetase (Ray et al., in preparation). The cDNA probe is being used to isolate genomic sequences for asparagine synthetase in order to study the organization of the wild-type gene and characterize the lesions in the mutant lines at the molecular level.

In addition, all of the initial recombinants were pooled and used to make rapid preparations of plasmid DNA to transfect AS^- *E. coli* cells. In order to allow for expression of the CHO enzyme in bacterial cells, the orientation, reading frame, and sequences of cDNA must be correct. Phenotypic expression of the cDNAs for mouse dihydrofolate reductase (Chang et al., 1978) and human purine-nucleoside phosphorylase (Goddard et al., 1983) has been achieved in bacteria, and this method was used successfully to clone purine-nucleoside phosphorylase. If successful, a full length cDNA probe would be obtained by this method. In this manner a combination of the techniques of somatic cell genetics and bacterial genetics can be used to isolate mammalian genes which can be selected for in a bacterial system.

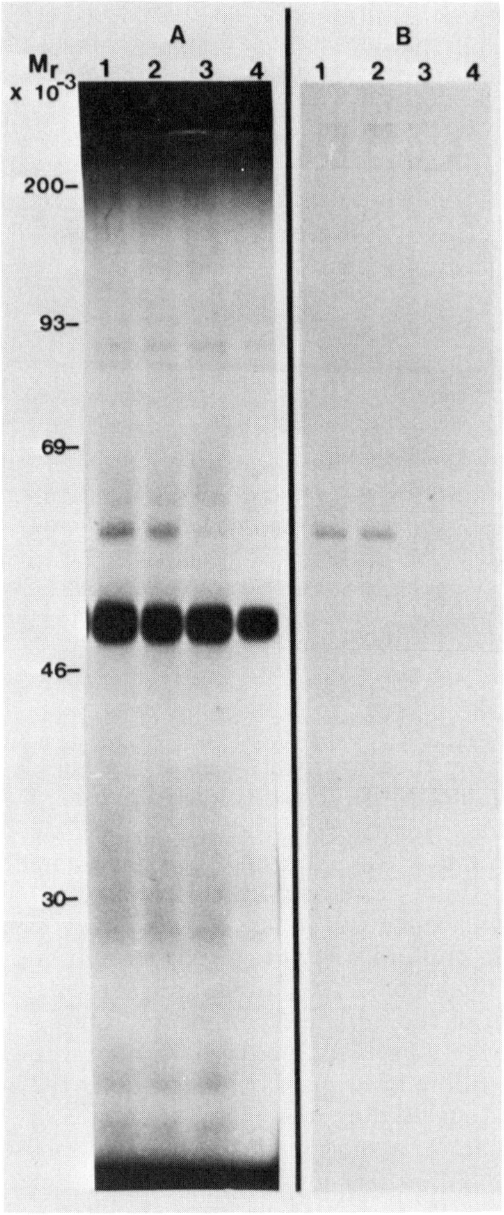

Figure 17.9. SDS-PAGE of translation products of mRNA selected by hybridization to recombinant plasmid cDNAs. Total *in vitro* translation products (A) before immunoprecipitation and (B) after immunoprecipitation of mRNA selected by hybridization to filters containing (1) plasmid 11, (2) plasmid 12, (3) pB, and (4) translation without exogenous message.

C. DNA-Mediated Gene Transfer of Asparagine Synthetase

As an alternative method to isolate the gene for asparagine synthetase, we developed a DNA-mediated gene transfer system for β-AHA resistance (Andrulis and Siminovitch, 1981). As detailed in Chapter 8, DNA transfection has been used to clone the genes for thymidine kinase (Perucho et al., 1980), adenosine phosphoribosyl transferase (Lowy et al., 1980), and several oncogenes (Goldfarb et al., 1982; Shih and Weinberg, 1982; Diamond et al., 1983). In addition, direct transfer of genomic DNA has been used to screen for expression of dominantly acting selectable markers such as methotrexate ^{R} (Wigler et al, 1980).

Since DNA-mediated transfer of genomic sequences occurs at a low frequency (10^{-6}–10^{-5}) even in the most well-characterized systems, it is necessary to choose the donor and recipient cell lines carefully. The usual recipients for DNA transfer are LMTK^{-} cells for dominant markers and NIH 3T3 cells for oncogenes. Unfortunately LMTK^{-} cells could not be used as recipients for transfer of asparagine synthetase since they normally have high levels of the enzyme and are resistant to β-AHA. The ideal recipient would be an asparagine-synthetase-deficient LMTK^{-} cell line, but we were unable to isolate such mutants after several selections. Although there was an asparagine-synthetase-deficient CHO cell line (N3) available, we and others (Waye and Stanners, 1983) have not been able to obtain DNA transferents using this line. This may be due to the high reversion frequency of N3 and/or the stringency of the selective conditions which were employed. Even though CHO cells are, in general, relatively poor recipients for genomic DNA transfer, we were successful in utilizing them for DNA-mediated gene transfer of asparagine synthetase. DNA from multistep β-AHA^{R} lines was transfected into β-AHA^{S} CHO cells and relatively high concentrations of β-AHA were used for selection. Because the donor DNA came from multistep β-AHA^{R} lines, the transferents were expected to be resistant to very high concentrations of drug. Transferants were obtained at a frequency of 2×10^{-7}, greater than 10-fold over the control frequency of less than 2×10^{-8} (Andrulis and Siminovitch, 1981). When multistep Alb^{R} cell lines with higher levels of asparagine synthetase activities and presumed amplification of the gene were used as a source of donor DNA, higher transfer frequencies were observed (Ray et al., 1983). In order to find an enzyme which would not cut the gene for asparagine synthetase and which could be used in further cloning procedures, donor DNA was digested with several restriction enzymes and used to transfect β-AHA^{S} cells. Digestion with Hind III did not destroy the ability of donor DNA to be expressed in β-AHA^{S} cells. Although transer of β-AHA resistance was successful, the frequency proved to be very low. These systems should provide material for cloning by procedures used previously for other genes.

In addition to the transfer of the mammalian gene for asparagine synthetase, Waye and Stanners (1983) found that the bacterial gene could be transferred into mammalian cells. Using a bacteriophage containing cloned bacterial asparagine snythetase they were able to transform asparagine synthetase$^-$ Jenson rat sarcoma cells.

IV. CONCLUDING REMARKS

We are investigating the regulation and expression of asparagine synthetase by a combination of the techniques of somatic cell genetics, biochemistry, and molecular biology. By manipulating the culture conditions for wild-type and ts asparaginyl-tRNA synthetase mutant cells, it was found that the levels of asparagine synthetase activity were controlled not only by the concentration of free asparagine in the medium but also by the extent of charging of tRNA.

To further elucidate the mechanism of expression of asparagine synthetase, mutants resistant to analogs of the substrates were isolated and characterized. The β-AHA^r and Alb^R mutants fell into at least two groups genetically. β-AHA resistance could be either codominant (20%) or recessive in hybrids with wild-type cells, where Alb resistance always acted in a codominant manner. Further subgroups could be described on the basis of differences in cross resistance to the analogs. When the activity of asparagine synthetase was examined in the mutants, it was found that most of the mutants expressed constitutively elevated levels of enzyme activity which were not subject to control by the amount of free asparagine or charaged tRNA. It is possible that the activity of the enzyme was always repressed in these cells due to the elevated levels of asparagine synthetase. Nevertheless, in every line it was far above the basal wild-type level. Some of the mutants exhibited alterations in the inhibition of the enzyme by the analog most likely due to structural gene mutations. Other mutations still remain unidentified but do not appear to be due to lesions in glutamine synthetase, glutamine PRPP amidotransferase, asparaginyl-tRNA synthetase, or transport (unpublished observations).

At this time the most useful type of mutation has been that of amplification of the gene for asparagine synthetase, although classical geneticists might protest against calling this form of genetic alteration a mutation. The amplified lines provided a good source of enzyme for purification, poly (A$^+$) RNA for preparation of double-stranded cDNAs, and single-stranged cDNA probe which was used in combination with wild-type single-stranded cDNA probe to screen the recombinant plasmids by differential hybridization. This facilitated the cloning of cDNA for CHO asparagine synthetase. The asparagine synthetase cDNA will be useful in the isolation of genomic DNA sequences. These are necessary to study the mutations in the single-step mutants which affect the regulation of asparagine synthetase as well as examine the gene structure and regulation in normals and patients with acute lymphocytic leukemia.

What types of genetic alterations might we expect to find in the single-step mutants? Since it has been shown that asparagine synthetase is regulated by the degree of charging of tRNA in CHO cells, similar to that of amino acid biosynthetic operons in bacteria, it is tempting to imagine that attenuation may play a role in asparagine synthetase regulation. In bacteria several amino acid biosynthetic operons are preceded by a regulatory reigion which encodes a short leader peptide enriched in the cognate amino acids of the pathway. The levels of the biosynthetic enzymes are regulated by coupling of transcription and translation of the leader region, the mRNA of which can form various secondary structures. Which of the alternative secondary strauctures is formed is influenced by the extent of charging of the cognate tRNA and ribosome stalling and in concert determines whether transcription is attenuated or proceeds into the structural sequences (Yanofsky, 1981). It has been difficult to conceive of how the coupling of transcription and translation involved in attenuation in bacteria might be feasible in eukaryotic cells since transcription occurs in the nucleus and translation in the cytoplasm. Recently, however, two eukaryotic systems which may utilize attenuation have been described. Andreadis et al. (1982) have sequenced the yeast *leu-2* gene which codes for an enzyme in the leucine biosynthetic pathway, β-isopropylmalate dehydrogenase. They found that the 5′-noncoding region contains an abundance of codons for leucine and speculate that the leader sequence may be translated in addition to the structural sequences from a "long" message. It would code for a leucine-rich peptide which might travel from the cytoplasm to the nucleus where it would affect the secondary structure of the attenuator region to produce transcription termination. In this case when the cytoplasmic leucine concentration is sufficient for the translation of the leucine-rich leader peptide, the peptide would serve to transfer information to the nucleus and attenuate transcription of the *leu-2* gene. Although Andreadis et al. (1982) point out that they have no conclusive evidence which provides a function for the leader peptide, the model offers an intriguing way in which attenuation might occur in eukaryotic cells. Further support comes from the work of Hay et al. (1982) who have described another eukaryotic system which might utilize attenuation. They found that the sequences for the SV40 structural protein VP1 are preceded by a leader region encoding a protein (agnoprotein) which they suggest could transfer information from the cytoplasm to the nucleus.

Since regulatory sequences are expected to precede the structural gene sequences, it will be extremely interesting to determine the 5′-noncoding sequences of the gene for asparagine synthetase. Because the enzyme is regulated by tRNA, and in view of the reports on attenuation in eukaryotic cells, we expect that some of the single-step mutations affecting asparagine synthetase regulation might lie in the 5′-noncoding region. If there is a leader region message capable of forming alternative secondary structures, then mutations in these sequences which prevented the formation of certain configurations could result in constitutive expression of the gene. The mutations would be codominant in hybrids. In addition, mutations in the leader peptide (or a repressor protein) could result in a peptide no longer capable of repression, also causing constitutive expression. In hybrids made with these lines the wild-type cell could provide a functional leader peptide (or repressor) and the mutation would be recessive. This would account for both the codominant and recessive mutations which have been isolated. It is interesting that although β-AHA resistance could be either codominant or recessive, albizziin resistance was always codominant.

How do the effects of albizziin differ from those of β-AHA? We have demonstrated that cell lines resistant to high levels of albizziin have amplifications in the gene for asparagine synthetase (and depending on the selection conditions may overproduce glutamine synthetase or glutamine PRPP amidotransferase as well). Some of the single-step Alb^R mutants (e.g., Alb^R 1) also appear to have amplified copies of the gene. However, none of the β-AHA^r mutations appear to be due solely (if at all) to gene amplification. If the amplification of asparagine synthetase was present in the population before selection, we would expect to obtain the same type of mutation with either drug. Since we do not observe this result, it seems likely that the drugs have different mechanisms of action. It is possible that since, as a glutamine analog, albizziin inhibits other enzymes which use glutamine as a substrate, including those involved in purine biosynthesis, it may consequently affect DNA synthesis. This is supported by the observation that all of the albizziin-resistant mutants have chromosomal alterations, whereas none are observed in β-AHA^r lines, and suggests that albizziin, but not β-AHA, may enhance chromosomal alterations. This obviously remains to be proven.

The amplification protocol using albizziin which has been described for asparagine synthetase should also be useful in obtaining amplifications of genes for other glutamine-utilizing enzymes by manipulating the culture conditions and taking advantage of the tools available in CHO genetics.

ACKNOWLEDGMENT

I would like to thank Louis Siminovitch for critical reading of the manuscript. Much of the work described in this article was carried out in his laboratory with support from the Medical Research Council and National Cancer Institute of Canada.

REFERENCES

Adair, G. M., Thompson, L. H., and Fong, S. (1979). *Somat. Cell Genet.* **5**, 329–344.

Adair, G. M., Thompson, L. H., and Lindl, P. A. (1978). *Somat. Cell Genet.* **4**, 27–44.

Andreadis, A., Hsu, Y-P., Kohlaw, G. B., And Schimmel, P. (1982). *Cell* **31**, 319–325.

Andrulis, I. L., Chiang, C. S., Arfin, S. M., Miner, T. A., and Hatfield, G. W. (1978). *J. Biol. Chem.* **253**, 58–62.

Andrulis, I. L., Duff, C., Evans-Blackler, S., Worton, R., and Siminovitch, L. (1983). *Mol. Cell. Biol.* **3**, 391–398.

Andrulis, I. L., Hatfield, G. W., and Arfin, S. M. (1979). *J. Biol. Chem.* **254**, 10629–10633.

Andrulis, I. L. and Siminovitch, L. (1981). *Proc. Natl. Acad. Sci. USA* **78**, 5724–5728.

Andrulis, I. L. and Siminovitch, L. (1982a). *Somat. Cell Genet.* **8**, 533–545.

Andrulis, I. L. and Siminovitch, L. (1982b). In *Gene Amplification* (R. T. Schimke, ed.), Cold Spring Harbor Laboratory, New York, pp. 75–80.

Arfin, S. M., Cirullo, R. E., Arredondo-Vega, F. X., and Smith, M. (1983). *Somat. Cell Genet.* **9**, 517–531.

Arfin, S. M., Simpson, D. R., Chiang, C. S., Andrulis, I. L., and Hatfield, G. W. (1977). *Proc. Natl. Acad. Sci. USA* **74**, 2367–2369.

Broome, J. D. (1968). *J. Exp. Med.* **127**, 1055–1072.

Carsiotis, M., Jones, R. F., and Wesseling, A. C. (1974). *J. Bacteriol.* **119**, 893–898.

Chang, A. C. Y., Nunberg, J. H., Kaufman, R. J., Erlich, H. A., Schimke, R. T., and Cohen, S. N. (1978). *Nature* **275**, 617–624.

Diamond, A., Cooper, G. M., Ritz, J., and Lane, M. A. (1983). *Nature* **305**, 112–116.

Gantt, J. S. and Arfin, S. M. (1981). *J. Biol. Chem.* **256**, 7311–7315.

Gantt, J. S., Chiang, C-S., Hatfield, G. W., and Arfin, S. M. (1980). *J. Biol. Chem.* **255**, 4808–4813.

George, D. L. and Powers, V. E. (1981). *Cell* **24**, 117–123.

Goddard, J. M., Caput, D., Williams, S. R., and Martin, Jr., D. W. (1983). *Proc. Natl. Acad. Sci. USA* **80**, 4281–4285.

Goldfarb, P. S. G., Garritt, B., Hooper, M. L., and Slack, C. (1977). *Exp. Cell. Res.* **104**, 357–367.

Goldfarb, M., Shimizu, K., Perucho, M., and Wigler, M. (1982). *Nature* **296**, 404–409.

Hay, N., Skolnick-David, H., and Aloni, Y. (1982). *Cell* **29**, 183–193.

Hongo, S. and Sato, T. (1981). *Anal. Biochem.* **114**, 163–166.

Horowitz, B., Madras, B. K., Meister, A., Old, L. J., Boyse, E. A., and Stockert, E. (1968). *Science* **160**, 533–535.

Kaufman, R. J., Brown, P. C., and Schimke, R. T. (1979). *Proc. Natl. Acad. Sci. USA* **76**, 5669–5673.

Kidd, J. G. (1953). *J. Exp. Med.* **98**, 565–582.

Kolter, R. and Yanofsky, C. (1982). In *Annual Review of Genetics* (H. L. Roman, A. Campbell, and L. M. Sandler, eds.), Annual Reviews Inc., Palo Alto, California Vol. 16, pp. 113–134.

Land, H., Grez, M., Hauser, H., Lindenmaier, W., and Schutz, G. (1981). *NAR* **9**, 2251–2266.

Lowy, I., Pellicer, A., Jackson, J. F., Sim, G-K., Silverstein, S., and Axel, R. (1980). *Cell* **22**, 817–823.

McBurney, M. W. and Whitmore, G. F. (1974). *Cell* **2**, 173–182.

Nunberg, J. H., Kaufman, R. J., Schimke, R. T., Urlaub, G., and Chasin, L. (1978). *Proc. Natl. Acad. Sci. USA* **75**, 5553–5556.

Patterson, M. K. and Orr, G. (1967). *Biochem. Biophys. Res. Commun.* **26**, 228–233.

Perucho, M., Hanahan, D., Lipsich, L., and Wigler, M. (1980). *Nature* **285**, 207–210.

Pinkus, L. M. (1977). In *Methods in Enzymology* (W. G. Jakoby and M. Wilchek, eds.), Academic Press, New York, Vol. 46, pp. 414–427.

Ray, P. N., Andrulis, I. L., and Siminovitch, L. (1983). In *Gene Transfer and Cancer* (M. Pearson and N. Sternberg, eds., Raven Press, New York.

Schroeder, D. D., Allison, A. J., and Buchanan, J. M. (1969). *J. Biol. Chem.* **244**, 5856–5865.

Shih, C. and Weinberg, R. (1982). *Cell* **29**, 161–169.

Squires, C. L., Lee, F., and Yanofsky, C. (1975). *J. Mol. Biol.* **92**, 93–111.

Stanley, P., Caillibot, V., and Siminovitch, L. (1975). *Somat. Cell Genet.* **1**, 3–26.

Stanners, C. P., Elicieri, G. L., and Green, H. (1971). *Nature New Biol.* **230**, 52–54.

Stephens, J. C., Artz, S. W., and Ames, B. N. (1975). *Proc. Natl. Acad. Sci. USA* **72**, 4389–4393.

Thompson, L. H., Harkins, J. L., and Stanners, C. P. (1973). *Proc. Natl. Acad. Sci. USA* **70**, 3094–3098.

Thompson, L. H., Stanners, C., and Siminovitch, L. (1975). *Somat. Cell Genet.* **1**, 187–208.

Umbarger, H. E. (1978). In *Annual Review of Biochemistry* (E. E. Snell, P. D. Boyer, A. Meister and C. C. Richardson, eds.), Annual Reviews Inc., Palo Alto, California, Vol. 47, pp. 533–606.

Wahl, G. M., Vitto, L., Padgett, R. A., and Stark, G. R. (1981). *Mol. Cell. Biol.* **2**, 308–319.

Waye, M. M. Y. and Stanners, C. P. (1979). *Somat. Cell Genet.* **5**, 625–639.

Waye, M. M. Y. and Stanners, C. P. (1983). *J. Mol. Appl. Genet.* **2**, 69–82.

Wigler, M., Perucho, M., Kurtz, D., Dana, S., Pellicer, A., Axel, R., and Silverstein, S. (1980). *Proc. Natl. Acad. Sci. USA* **77**, 3567–3570.

Worton, R. G. and Duff, C. (1979). In *Methods in Enzymology* (W. B. Jakoby and I. H. Pastan, eds.), Academic Press, New York Vol. 58, pp. 322–344.

Yanofsky, C. (1981). *Nature* **289**, 751–758.

Yanofsky, C. and Soll, L. (1977). *J. Mol. Biol.* **113**, 663–677.

CHAPTER 18

CHINESE HAMSTER CELL MUTANTS WITH ALTERED LEVELS OF ORNITHINE DECARBOXYLASE: OVERPRODUCERS AND NULL MUTANTS

Carolyn Steglich
Jung Choi
Immo E. Scheffler
Department of Biology
University of California, San Diego
La Jolla, California

I. INTRODUCTION

Polyamines have been found in all living cells from prokaryotes to humans. In cells of higher organisms they are absolutely essential for proliferation and growth, and even prokaryotes grow only very slowly, if at all in their absence (Tabor and Tabor, 1976; Whitney and Morris, 1978; Cohn et al., 1978; Canellakis et al., 1979; Hafner et al., 1979; Cohn et al., 1980; Tabor et al., 1980; Bachrach, 1981; Pegg and McCann, 1982). Under normal conditions intracellular concentrations of specific polyamines reach the millimolar range. In mammalian cells putrescine is typically present at low concentrations, while spermidine and spermine are the abundant polyamines.

A large number of studies have sought to elucidate a specific role for polyamines in a variety of reactions. *In vitro* systems for macromolecular synthesis can often be improved significantly by the addition of polyamines (Abraham and Pihl, 1981), and this is particularly noteworthy in the case of *in vitro* protein synthesis and the charging of tRNAs. In fact, the tRNAs are a specific example where polyamines have been found to be associated stoichiometrically and with a specific orientation in the tertiary structure of the molecule (Sakai and Cohen, 1976). It is believed that the fidelity of translation owes much to the presence of polyamines [e.g., Atkins et al. (1975)]. More recently, a number of examples have been described where polyamines influence the activity of specific protein kinases (Kuehn et al., 1979; Atmar and Kuehn, 1981; Daniels et al., 1981; Criss et al., 1983), but the full biological implications are far from clear.

Thus, polyamines may fulfill multiple functions inside the cell, from

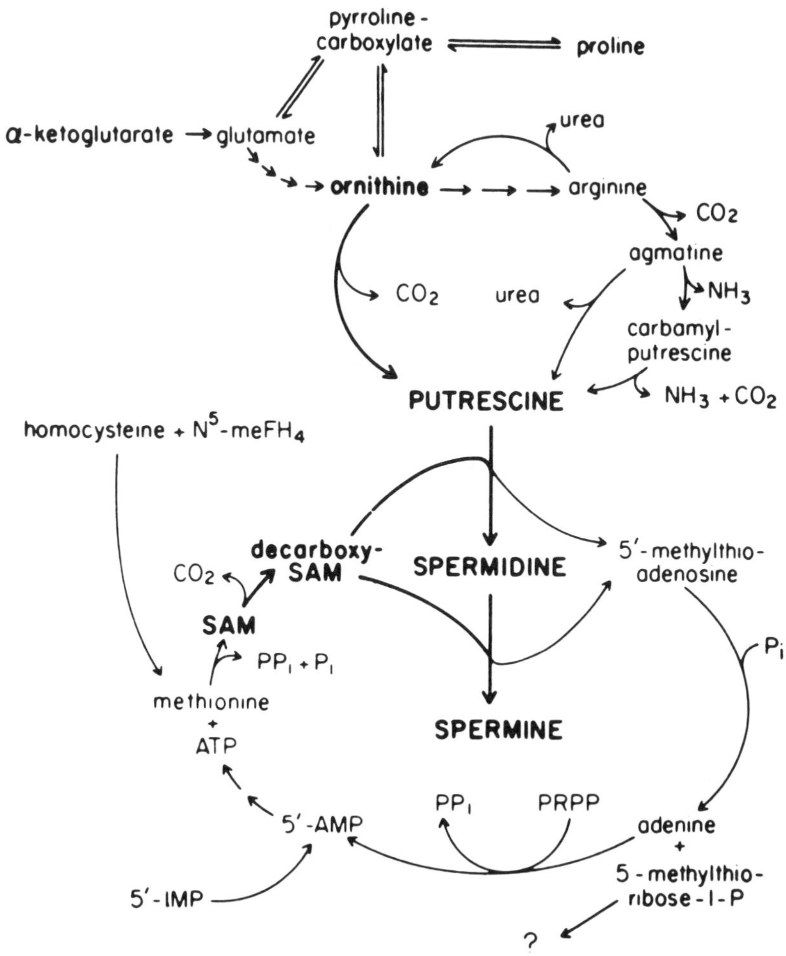

Figure 18.1. Pathways of polyamine biosynthesis.

serving as relatively nonspecific, highly charged counter ions to being specifically involved in the stabilization of macromolecules, in which their structure and charge distribution become relevant.

In animal cells the rate-controlling enzyme in polyamine biosynthesis is ornithine decarboxylase (ODC). Other closely related enzymes are *S*-adenosylmethionine decarboxylase, and the spermidine and spermine synthases. The most significant reactions pertaining to the synthesis of polyamines and their precursors are shown in Figure 18.1.

Ornithine decarboxylase has received considerable attention in the past few years after it was learned that its activity is highly regulated under a variety of conditions and by a number of different mechanisms

(Canellakis et al., 1979; Bachrach, 1981; Pegg and McCann, 1982). Regulation occurs at several levels, including enzyme synthesis, posttranslational modification, and turnover (degradation).

In quiescent, or nondividing, cells, the activity of ODC is extremely low or undetectable both *in vivo* and *in vitro*. In growing tumors, or in regenerating liver following hepatectomy, ODC activity is markedly induced (Kallio et al., 1977; Bachrach, 1981; Scalabrino and Ferioli, 1981). Many other examples of induction *in vivo* have been described, including treatment of animals with thioacetamide or chloroform [induction in liver (Haddox and Russell, 1981)], or with tumor promoters (Weekes et al., 1980; Butler-Gralla and Herschman, 1983). A particularly noteworthy case is the induction of ODC activity in the kidney of male mice treated with testosterone, Pajunen et al., 1982; Isomaa et al., 1983; Seely et al., 1982; Seely and Pegg, 1983). It should be noted, however, that even in the most favorable tissues the amount of enzyme induced represents a very small fraction of the total cellular protein (0.001% or less); hundreds of mice or rats have been sacrificed to obtain less than milligram quantities of purified ODC.

In cells grown in tissue culture, ODC activity declines markedly as cells reach stationary phase or quiescence, but a number of stimuli can induce ODC activity: serum and mitogens are generally used (Canellakis et al., 1979; Russell and Haddox, 1979; Guroff et al., 1980; Bachrach, 1981; Butler-Gralla and Herschman, 1983), and nonphysiological treatments such as exposure to hypotonic buffer are also effective. A most interesting aspect of this induction is that ODC activity increases dramatically to a maximum in several hours after which the activity declines equally rapidly. The peak of induction is reached several hours before the cells enter S-phase, in the case of stimulation with serum or growth factors (Clark, 1974; Landy-Otsuka and Scheffler, 1978; McCormick, 1978). Several experiments have shown that the increase in ODC activity requires mRNA synthesis [e.g., Landy-Otsuka and Scheffler (1978)], but there are also indications that an increased rate of translation plays a role in this ascending part of the curve (McCormick, 1977; Clark and Greenspan, 1979; and see below).

When inhibitors of protein synthesis are added at some point in the curve where ODC activity is still rising, the activity declines rapidly. It has been recognized for some time that ODC activity has one of the shortest half-lives of all mammalian enzymes examined so far (Russell and Snyder, 1969; Clark, 1974; Hogan and Murden, 1974). Estimates for $t_{1/2}$ of 15–90 min are typical. While some earlier measurements with antisera may have been suspect, some very recent studies confirm that the loss of ODC activity is tightly coupled to the loss of immunologically detectable ODC protein (Erwin et al., 1983; Seely and Pegg, 1983; Steglich and Scheffler, 1985). There is speculation but no firm evidence that the inactivation and/or degradation of ODC is preceded by a posttranslational modification. Phosphorylation or transglutamination have been hypothesized to be involved (Scott et al., 1982; Isomaa et al., 1983; McConlogue et al., 1983).

When putrescine, the end product of the ODC-catalyzed reaction, is added together with the serum, the induction of ODC activity is completely suppressed (Heller et al., 1978; Bethell and Pegg, 1979; Canellakis et al., 1979; Weekes et al., 1980; Bachrach, 1981). This effect is seen at concentrations where simple feedback inhibition is not an adequate explanation. In fact, putrescine has been shown to induce an activity, termed an antizyme, which specifically and stoichiometrically binds and inhibits ODC (McCann et al., 1979; Heller and Canellakis, 1981). The very limited availability of this antizyme has made its complete characterization rather difficult, but it is estimated to be a protein of molecular weight 27,000–30,000. It is not known whether antizyme binding is associated with an enzymatic reaction. An argument against such a hypothesis is that it has been possible to separate the antizyme from active ODC by gel filtration at high salt concentration (McCann et al., 1979).

When our studies on ODC in CHO cells were initiated 4 years ago, we asked ourselves how the approaches of somatic cell genetics could contribute to a solution of some of the problems raised by studies of ODC. Several aims could be stated: (1) to provide a more abundant source of the enzyme; (2) to obtain mutants which might be useful in physiological and biochemical studies; and (3) to select mutants which might be useful in the cloning of the ODC gene or the corresponding cDNA. It was clear that overproducers of ODC would be useful in enzyme purification, and at the same time the mechanism leading to overproduction was expected to be of interest. Gene amplification would obviously be helpful in the attempt to clone the gene, and other potential mechanisms would certainly be of interest for the study of the regulation of ODC activity. For example, one study described elevated enzyme levels in hepatoma cells as the result of a decreased turnover rate of the enzyme (Mamont et al., 1978; Pritchard et al., 1982).

A second class of mutants which seemed desirable were mutants with no ODC activity and hence auxotrophs for putrescine. Again, several mechanisms could be envisioned to lead to such a phenotype. Mutations in the structural gene(s) are first to come to mind, but regulatory elements might also be affected by mutations.

II. VARIANTS THAT OVERPRODUCE THE ENZYME

Two specific inhibitors of ODC activity became available a few years ago. Alpha-methylornithine (αMO) is a competitive inhibitor of the enzyme, while alpha-difluoromethylornithine (αDFMO) was shown to be an irreversible "suicide" inhibitor, which becomes covalently attached to the enzyme at the active site (Abdel-Monem et al., 1974; Mamont et al., 1976; Bey et al., 1978; Metcalf et al., 1978; Mamont et al., 1982). With expectations based on the experience with inhibitors of other enzymes [e.g., Padgett et al. (1979) and Schimke et al. (1978)] we initiated a search for CHO variants which could proliferate at elevated concentrations of these analogs.

αDFMO was initially not readily available, and selections were carried out in αMO. Variants/mutants resistant to lethal concentrations were soon found. Single-step selections yielded clones growing at 2 mg/mL αMO, but higher concentrations seemed impractical, because of expected complications with the osmolarity of the medium. The frequency at which clones arose in a population was not markedly elevated by mutagenesis. These first variants, when compared to wild type, were able to form colonies at 10 times higher inhibitor concentrations. Both mutant and wild-type cells were significantly more sensitive to the irreversible inhibitor αDFMO than to αMO, but an order of magnitude difference in sensitivity between wild-type and mutant cells was also observed with this analog. As αDFMO became more readily available through the generosity of Dow/Merrell, the selections could be extended to higher analog concentrations. The first cloned variant was resistant to 10 μg/mL αDFMO (Choi and Scheffler, 1981), and further selections were made at 200 μg/mL, then at 500 μg/mL, and most recently at 1000 μg/mL. This concentration is almost 1000 times higher than the lethal concentration for wild-type CHO cells. The resistant variants are designated DF1, DF2, and DF3, respectively, depending on whether they were selected at 10,100, or 1000 μg/mL. A comparison of plating efficiencies of several of these variants is shown in Figure 18.2 (Choi and Scheffler, 1983).

As far as we have been able to test it, the phenotype of these variants is extremely stable when the cells are carried in the absence of analog in the medium.

Hybrid hamster cells constructed with the help of independent genetic markers between wild-type and DF1 cells were as resistant as the resistant parental cells (Choi and Scheffler, 1981).

It was quickly established that resistance is associated with increased enzyme levels. Two important points need to be made in this connection: (a) When cells have been subcultured in the presence of αDFMO, they have to be maintained in a drug-free medium for at least 1 week before significant ODC activity can be measured in extracts from such cells. We have recently shown by indirect immunofluorescence on sectioned cells that, in the presence of αDFMO, DF3 cells contain significant quantities of inactive enzyme (Dirks and Tokuyasu, unpublished observations). (b) Since ODC activity is so much dependent on the distribution of cells in the phases of the cell cycle, comparisons of specific

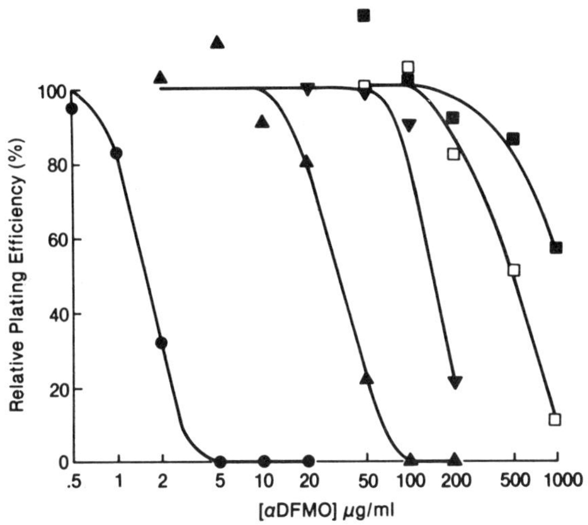

Figure 18.2. Relative plating efficiency in the presence of α-difluoromethylornithine (DFMO). Cells were plated at a low density (100–200 cells/60-mm plate) in the presence of varying concentrations of α-difluoromethylornithine. Ten to eleven days later the plates were stained with 0.1% crystal violet and colonies larger than 50 cells were counted. Results are plotted as the percentage of plating efficiency relative to control plates with no drug added. The points represent the average of duplicate plates. ●, CHO; ▲, DF1; ▼, DF2; □, DF2.5; ■, DF3.

activities should be made in synchronized populations of cells. For many of our experiments we have therefore adopted a protocol which includes a 15–18-hr serum starvation in medium without serum, followed by a refeeding with fresh medium containing 10% fetal calf serum.

ODC activities measured in such cultures of the variants DF1, DF2, DF3, and parental CHO cells are shown in Figure 18.3. In general it is clear that the inducible enzyme activity is higher the more resistant the variant, although not strictly proportional; the most resistant variant has 30–50 times more activity compared to the normal levels of the parents. The peak activities have been somewhat variable, and we believe that one of the determining factors is the cell density at the time of serum stimulation. Two other observations deserve special mention. All cells reached the peak of induction at about the same time, with the activity declining rapidly thereafter. On the other hand, the variants always had a significant basal activity at zero time (see below), while the activity in wild-type cells was frequently not detectable at this time (Choi and Scheffler, 1981, 1983).

A variety of experiments have supported the idea that an unaltered enzyme is produced in these variants. A determination of the K_m for the substrate ornithine is shown in Figure 18.4, and no difference be-

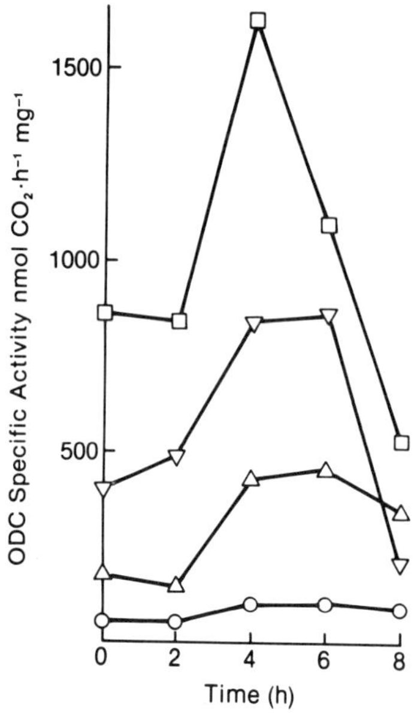

Figure 18.3. ODC activity induced in wild-type and mutant cells treated with α-methylornithine. The wild-type CHO (○) cells and the mutant cells DF1 (Δ), DF2 (∇), and DF3 (□) grown in the absence of drug for 7 days were exposed for 48 hr to 2.0 mg/mL of α-methylornithine, washed, synchronized by serum starvation for 14 hr, and induced by changing to fresh media with 10% serum.

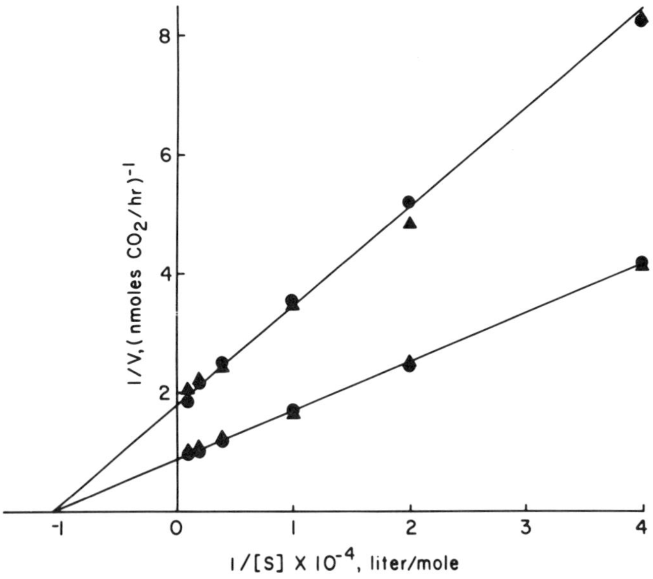

Figure 18.4. Determination of the K_m values with respect to the substrate ornithine for the wild-type and mutant enzymes. Extracts from each cell type were adjusted to give two different maximal rates for each, and then the substrate concentration was varied as indicated; the concentration of pyridoxal phosphate was kept constant at 50 μM. Both curves could be extrapolated to a K_m value of 0.93×10^{-4} M.

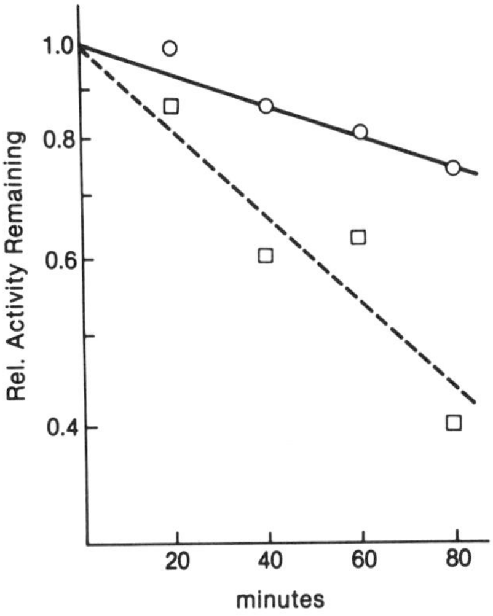

Figure 18.5. Decline of ornithine decarboxylase activity in induced cells treated with cycloheximide. The inhibitor (50 μg/mL) was added to cultures 4.5 hr after release from serum starvation. Activities were normalized with respect to the activity at zero time; wild-type CHO cells (○) had an initial specific activity of 60 units/mg of protein, and the apparent half-life was 190 min; extracts from DF3 cells (□) had an initial specific activity of 2000 units/mg of protein and the measured half-life was 70 min.

tween wild-type and variant enzymes was detectable. Similar determinations for the cofactor pyridoxal phosphate have been difficult for us, but we have some indications from the operation of affinity columns during enzyme purification that the enzyme from DF3 cells may exhibit a slightly altered interaction with the cofactor (Dirks, unpublished observations).

We recently obtained a cDNA clone for mouse ODC (Kontula et al., 1984) and were able to test directly for gene amplification in the overproducing mutants. On DNA dot blots, ODC genes in the DF3 mutant were about 200-fold amplified compared to parental CHO cells (Steglich and Scheffler, 1985).

The half-life of the enzyme was routinely determined by measuring the decay of activity after the addition of emetine or cycloheximide to partially induced cultures. In comparable situations we have not seen any difference in enzyme half-life between the wild-type parents and either DF1 or DF3 cells which could account for the increased enzyme levels in the variants. In the experiments shown (Fig. 18.5), the half-life in the DF3 cells was in fact shorter, but this difference was not reproducibly observed (Choi and Scheffler, 1983).

A direct measurement of the absolute specific activity has become possible with the availability of radioactive αDFMO. Crude extracts of cells can be reacted with [^{3}H]αDFMO, and the derivatized protein can be precipitated and counted. From the known specific activity of the analogue, one can calculate the concentration of enzyme and correlate it with the measured enzyme activity (Fig. 18.6). Again, no differences were observed between normal and drug-resistant cells (Table 18.1; Choi and Scheffler, 1983).

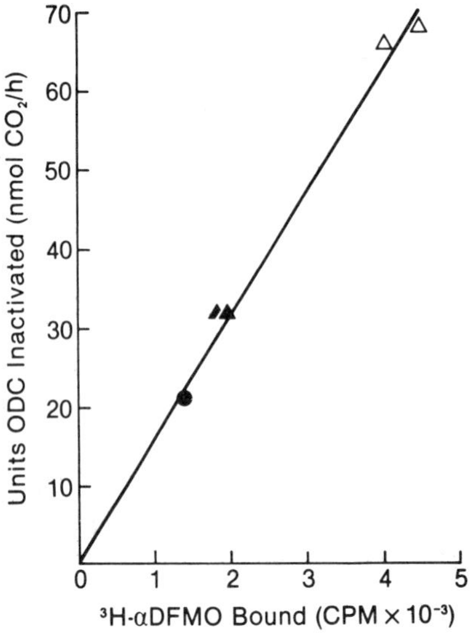

Figure 18.6. Correlation of [^{3}H]DFMO binding with ornithine decarboxylase activity inactivated by the analog: (1) ●, highly purified enzyme from CHO wild-type cells; (2) Δ, crude extract from DF1 cells; (3) ▲, crude extract from DF2 cells.

We were also interested in determining whether the enzyme from the variants showed any altered behavior in its interaction with the antizyme. Purified enzyme as well as crude extracts with ODC activity were reacted with standard crude extracts containing antizyme to obtain a series of inhibition curves (Fig. 18.7). The variant enzymes were inhibited at the same rate or as effectively as wild-type enzyme. An-

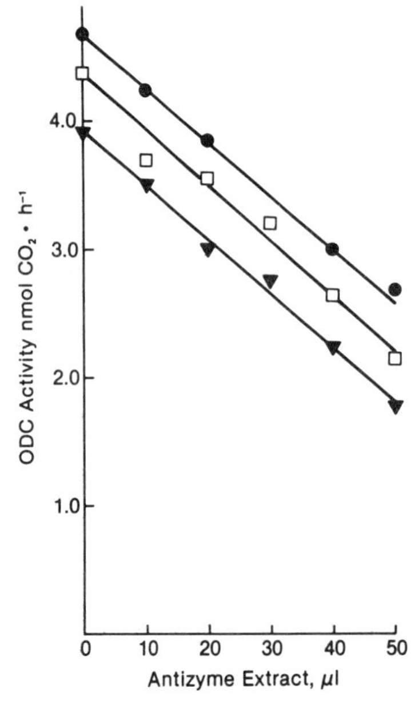

Figure 18.7. Titrations of mutant and wild-type ODC activities by crude extracts of antizyme. Constant amounts of purified ODC from CHO cells or of crude extracts from the overproducing variant cells were mixed with increasing amounts of antizyme with assay buffer plus 0.01% Triton X-100 added as necessary to keep the total volume constant. The mixtures were assayed for the remaining ODC activity. Values represent averages of duplicate assays: ●, purified wild-type enzyme; ▼ extracts from DF2 cells; □, extracts from DF3 cells.

TABLE 18.1
Binding of Tritiated α-Difluoromethylornithine

	Units of Ornithine Decarboxylase Inactivated	$[^3H]$DFMO[a] Precipitated (cpm)	Units, Ornithine Decarboxylase/mmol DFMO $\times 10^{-11}$	Ornithine Decarboxylase, Percentage of Cytosol Protein
CHO	11.9	545	2.27	0.001
DF1	44.3	1780	2.78	0.003
DF3	122	5080	2.50	0.04

[a] DFMO, α-difluoromethylornithine.

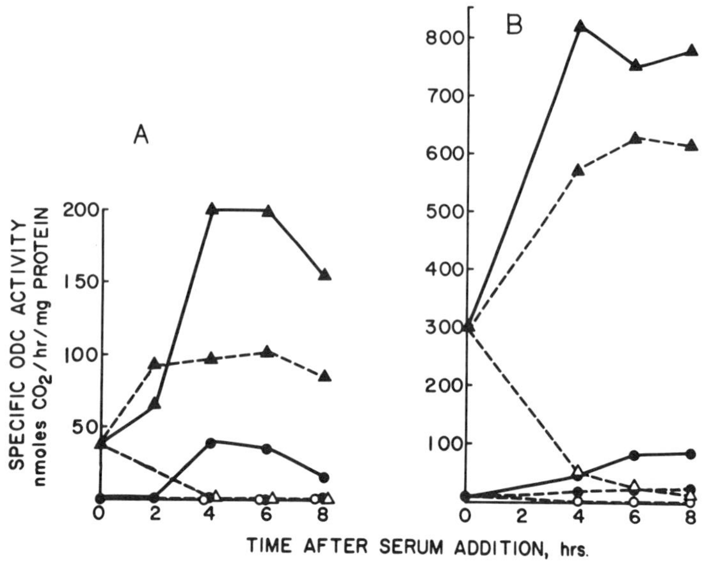

Figure 18.8. Induction of ornithine decarboxylase in wild-type and mutant cells in the presence of various drugs: (A) without αMO treatment, (B) with 48-hr pulse of αMO (2 mg/mL) prior to serum starvation. Cells were grown to near confluence in medium containing 5% fetal calf serum, then switched to medium containing 0.1% serum after washing with TD buffer. After 24 hr of serum starvation, cells were again washed and stimulated to grow by adding growth medium with 5% fetal calf serum, or with the same medium with 5 m*M* DRB, or with 100 μg/mL emetine: ●—●, wild-type control; ● - - - ●, wild type of DRB; ○, wild type with emetine; ▲—▲, C51 control; ▲---▲, C51 with DRB; Δ, C51 with emetine.

other question was whether the amount of antizyme inducible in these variants was altered in any way, but we found no significant difference (Choi and Scheffler, 1983).

It is clear that the overproducing hamster cells are a convenient source of enzyme, yielding crude extracts with specific activities comparable to those of the other good source in use today: kidneys from androgen-treated male mice. Our purifications have given us quantities of enzyme sufficient for the immunization of two rabbits. We have also purified the enzyme by standard procedures in use for this protein (Boucek and Lembach, 1977) which has allowed us to confirm that the molecular weight of the subunit is 54,000 (Seely et al., 1982). Finally, we have been able to localize the protein on two-dimensional polyacrylamide gels, which will help in future studies on the turnover of the protein (Choi and Scheffler, 1983).

Some early studies with the variant DF1 have given some interesting information on the regulation of the ODC activity in serum-stimulated cells (Fig. 18.8). The results obtained with an inhibitor of protein synthesis at the time of addition of serum showed clearly that continued protein synthesis was required to maintain the basal level of ODC activity in the variants. This suggested that there were elevated message levels in these variant cells at $t=0$ hr. When an inhibitor of mRNA synthesis was added at $t=0$ hr, no ODC induction was observed in wild-type cells because there was no message, but in the variant cells ODC induction was readily seen. These results suggested that ODC induction is governed both by the synthesis of new message and by an increased rate of protein synthesis following serum stimulation. Interestingly, the time of peak activity was independent of the amount of enzyme in these cells, as if the timing was not determined by the amount of product (see also Fig. 18.3). From a practical as well as theoretical point of view another observation deserves mention. As stated earlier, cells taken out of αDFMO have no measurable activity. However, when cells, even wild-type cells, were briefly (48 hr) exposed to the reversible inhibitor αMO before the serum starvation, the inducible ODC activity was increased by about a factor of 4 (Choi and Scheffler, 1981, 1983). One possible explanation is that the cells become somewhat depleted in their polyamine pools during this treatment, which leads subsequently to an overcompensation. In conflict with such an interpretation is the observation that ODC activity was superinduced by pretreatment in DF3 cells by concentrations of αMO which would be expected to be relatively ineffective in these cells.

III. ODC-DEFICIENT MUTANTS

The selection of ODC-deficient mutants (which are also auxotrophs for putrescine) was based on a suicide enrichment with [^{3}H]ornithine (Steglich and Scheffler, 1982). Cells were mutagenized with ethylmethane sulfonate. After a 2- or 3-day recovery period the mutagenized cells were allowed to incorporate [^{3}H]ornithine for 2 days followed by 2 weeks storage at -70°C. A wild-type cell that has incorporated large amounts of tritium into its polyamines is killed by radioactive-decay-induced lethal events occurring during the storage period (Fig. 18.9). A mutant cell lacking ODC activity incorporates no significant amount of tritium and is therefore spared from lethal irradiation from [^{3}H]polyamines. Under optimal conditions the survival of wild-type cells was less than 10^{-4} and that of mutants nearly 1. Thus, only a single round of enrichment has been found to be necessary for isolating mutants from mutagenized CHO cells.

CHO cells are especially well suited for this selection since the CHO-*K1* line we used is also a proline auxotroph (Kao and Puck, 1967). In particular, it is lacking the enzyme ornithine transaminase, which is part of the pathway converting ornithine to proline (Smith and Phang,

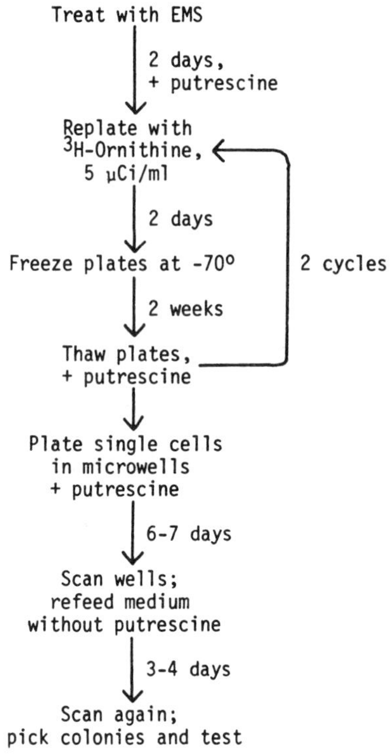

Figure 18.9. Protocol for selecting putrescine auxotrophs.

1979; see Fig. 18.1). Therefore, CHO pro$^-$ cells use ornithine only for polyamine biosynthesis.

Approximately 97% of label from [^{3}H]ornithine is incorporated into acid-soluble components in wild-type CHO cells. This label was shown to be exclusively in the polyamines putrescine, spermidine, and spermine. Virtually no label was found in acid-precipitable material (i.e., proteins, Fig. 18.10). A pro$^+$ hamster cell (V79) incorporated about 15% of total label from ornithine into acid precipitable material (presumably as proline in proteins), significantly reducing the effectiveness of the selection. We have in fact been unable to obtain similar mutants from a selection with this line.

Another significant feature of CHO cells contributing to the success of the selection is that the cells appear to be functionally hemizygous at the ODC locus (see below). The frequency of recessive ODC-deficient mutants in a population of CHO cells mutagenized with EMS was about 5×10^{-4}/survivor (Steglich and Scheffler, 1982), as expected from a mutation induced at a single locus.

All of the mutants which have been characterized to date (a total of seven) have several features in common (Steglich and Scheffler, 1985). All are putrescine auxotrophs, with an absolute requirement for putrescine for proliferation (Fig. 18.11). The dependence of growth on putres-

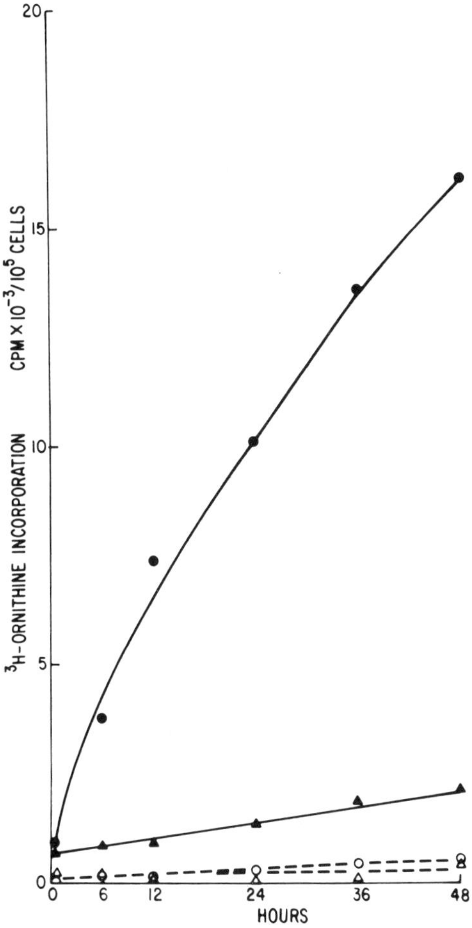

Figure 18.10. Ornithine incorporation. Incorporation of [^{3}H]ornithine by wild-type CHO (●, ○) and mutant C54 (▲, Δ) into acid-precipitable (○, Δ) and acid-soluble (●, ▲) fractions.

cine could be shown both in mass cultures and by plating efficiency tests in varying concentrations of putrescine. All mutants required at least 10^{-5} *M* putrescine to maintain normal growth. They are routinely maintained in medium containing 5×10^{-4} *M* putrescine. Figure 18.12 shows the kinetics of growth arrest from polyamine starvation. Growth arrest was slow, and the cell population increased about 10-fold before proliferation ceased. This reflects the very large amounts of polyamines present in the cells and the relatively slow turnover of the polyamines, especially of spermine.

Table 18.2 illustrates the levels of polyamines found in one of the mutants (C54) after a shift into putrescine-free medium. These levels

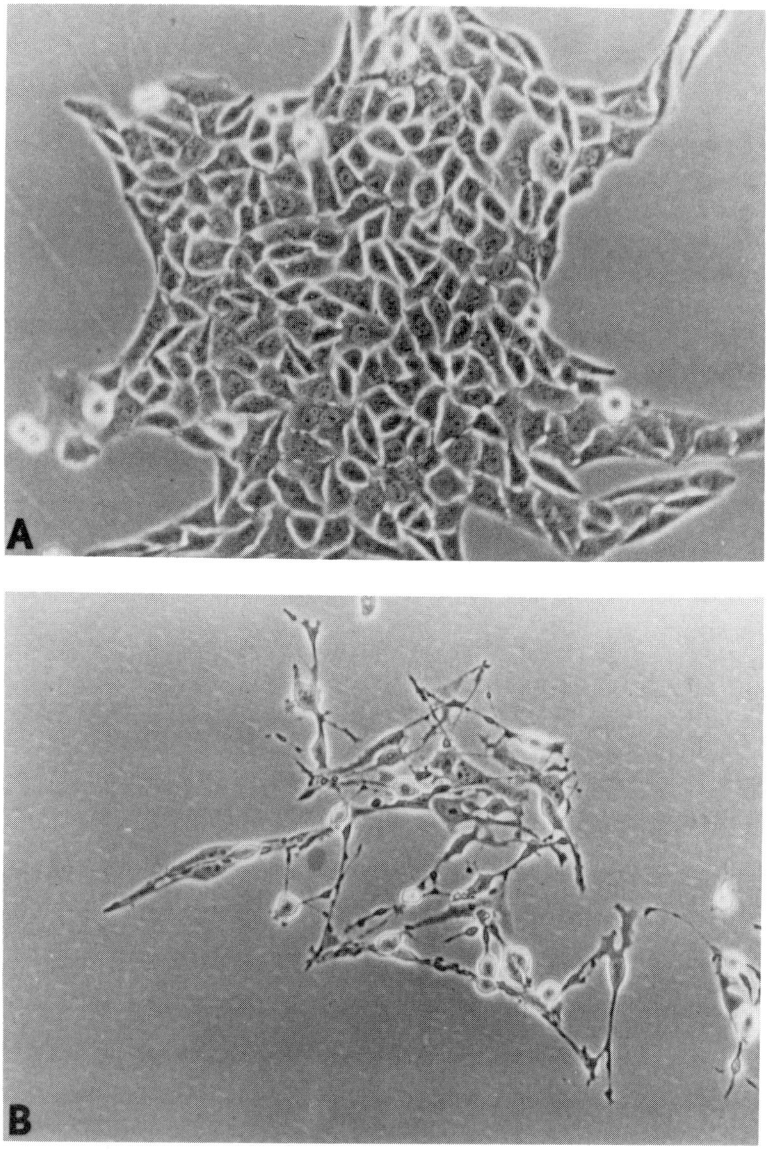

Figure 18.11. Photographs of 6-day-old colonies of the mutant C54 grown (A) with and (B) without putrescine supplementation.

TABLE 18.2
Polyamine Content of Mutant C54 and Wild-Type CHO Cells (Steglich et al., 1983)

Cell Line	Time after Removal of Putrescine (hr)	Putrescine (nmol/10^6 cells)	Spermidine (nmol/10^6 cells)	Spermine (nmol/10^6 cells)
C54	12	1.35	3.30	1.15
	24	0.08	2.50	1.54
	48	0.05	0.18	1.91
	72	0.03	0.09	1.15
	96	0.05	0.10	1.15
	96 + putrescine	1.55	1.65	0.65
CHO parent	96	0.44	1.72	0.92
	96 + putrescine	1.45	1.85	1.00

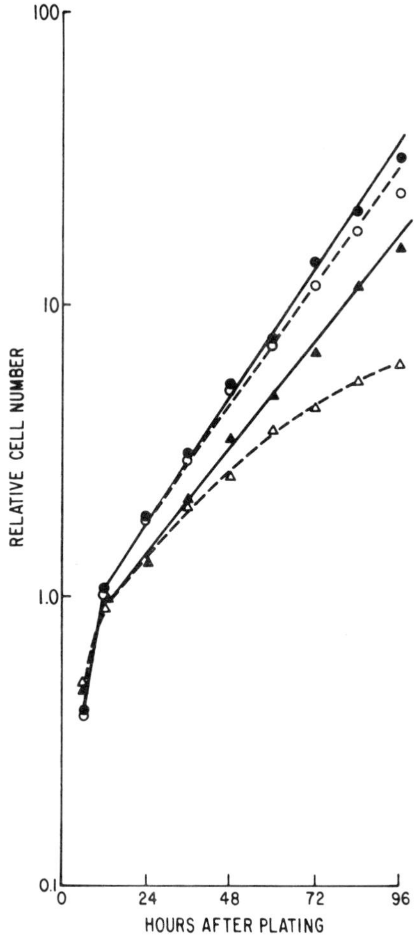

Figure 18.12. Growth rate of wild-type CHO (○, ●) and mutant C54 (Δ, ▲) in the absence (○, Δ) and presence (●, ▲) of 5 x 10^{-4} M putrescine.

were determined for us by Dr. L. Marton and associates in San Francisco. Polyamines in cell extracts were separated on an amino acid analyzer with detection of fluorescent derivatives of polyamines after reaction with *o*-phthalaldehyde (Seidenfeld and Marton, 1980). Putrescine was undetectable by 24 hr, and spermidine was below the limit of detection after 48 hr. Only at this time was the growth rate of the cells affected (Fig. 18.12), indicating that the cells apparently contained sufficiently large pools of polyamines to sustain growth and cell division for at least one or two generations in the absence of new polyamine synthesis. After 96 hr, when cell proliferation had finally ceased, putrescine and spermidine were no longer detectable, but spermine levels were virtually unchanged. Evidently spermine alone was not sufficient.

The mutants differ from one another in other characteristics. Most

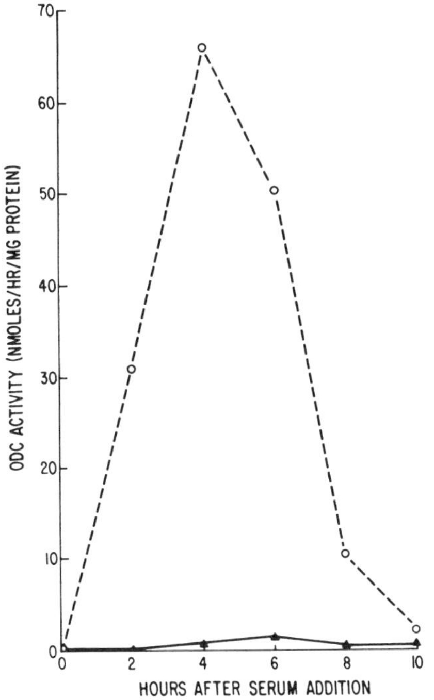

Figure 18.13. Cell-cycle-dependent ornithine decarboxylase activity of (○) wild-type CHO and (▲) mutant C54. Cells that had been growing in medium containing 5×10^{-4} *M* putrescine were replated in medium without putrescine. After 36 hr the cells were washed and fed Dulbecco's modified Eagle's medium without serum for 12 hr. At zero time the cells were refed serum-containing medium. Plates were harvested at 2-hr intervals following serum addition for ODC assays.

notably, they can be divided into two classes based on the amount of residual ODC activity present. Most of the mutants (five of seven) have some residual ODC activity, ranging in amount from about 2% to about 10% of wild-type levels. Enzyme levels in the mutants were determined in cultures grown for 2 days in the absence of putrescine supplementation, and then serum-starved and refed serum-containing medium to synchronize the cells. Enzyme activity at the peak of the induction curve was used to define the amount of residual activity. The kinetics of induction of the residual activity in the mutants are similar to the kinetics of induction in wild-type cells (Fig. 18.13).

These mutants also exhibited a small rate of incorporation of labeled, exogenous [^{3}H]ornithine into polyamines, and they were able to grow slowly if given high concentrations of ornithine in the medium (>1 m*M*). These *in vivo* manifestations of ODC activity correspond roughly to the measurable levels of ODC activity *in vitro* (Fig. 18.13).

To determine if any of these leaky mutants are the result of structural gene mutations, the residual enzyme activity was examined for any evidence for a physical alteration in the protein. We have not yet found any *in vitro* evidence for an altered protein. Several parameters were examined: the K_m for ornithine, heat sensitivity, or inhibition by ornithine analogs.

The other class of mutants (two of seven) had no detectable *in vitro* ODC activity. These mutants also were unable to incorporate any detectable label from [^{3}H]ornithine or to use ornithine for growth.

Two of the mutants with no residual activity and two of the leaky mutants were tested for complementation. One leaky mutant, C54, was further mutagenized to obtain an HPRT-deficient, ouabain-resistant clone which was then fused with polyethylene glycol to each of the other three mutants. Hybrids were selected in HAT-ouabain medium containing putrescine, and then tested for growth in the absence of putrescine. In none of the three crosses were any hybrids produced which were capable of sustained growth in putrescine-free medium. The hybrids from the cross between two leaky mutants had an even slower growth arrest than either parent due to the elevated enzyme in the hybrid, but no proliferating clones were found. We conclude that all the mutants are members of the same complementation group.

The two kinds of mutants also differ radically in their rates of reversion to prototrophy. Revertants are detected by plating 10^5 or fewer mutant cells per 100-mm-diameter plate in medium lacking putrescine. ODC-deficient mutants with no residual activity produce spontaneous revertants at a frequency of less than 10^{-7}. This reversion frequency is in the range usually expected for mutations in mammalian cells in culture. The leaky mutants, on the other hand, produce revertants at a frequency $>10^{-4}$. The rate of reversion in two leaky mutants was measured by a fluctuation test (Luria and Delbruck, 1943), and found to be $(2-7) \times 10^{-5}$ per cell generation, which is an extraordinarily high rate for reversion by standard mutational mechanisms (Steglich and Scheffler, 1985). We speculate that these revertants derived from the leaky mutants may instead be produced by a mechanism similar to that leading to overproduction of ODC in response to the analogs αMO and αDFMO. That is, overproduction of a defective enzyme may lead to a level of enzyme activity which is sufficient for the cells to survive and proliferate in putrescine-free medium. The revertants did contain increased amounts of enzyme activity relative to the parent mutants. DNA dot blots of one of the mutants and its revertant using the mouse cDNA clone as probe indicated that the ODC genes in the revertant were amplified about 2-fold. α-MO resistant variants selected from that revertant were amplified another 10-fold (Steglich and Scheffler, 1985).

The "nonreverting" mutant, C55.7, can be induced to revert if treated with 5-azacytidine (Steglich and Scheffler, 1985). This pyrimidine analog causes hypomethylation of DNA and is thought to reactivate silent genes previously inactivated by an epigenetic mechanism which includes methylation (Jones and Taylor, 1980; Harris, 1982). We have isolated several of the revertants induced by 5-azacytidine and subjected one of them to mutagenesis and reselection of ODC$^-$ mutants. These secondary ODC$^-$ mutants were similar to the first mutants in being able to revert to an ODC$^+$ phenotype spontaneously and after treatment with 5-azacytidine at frequencies comparable to those of the original mutants (Steglich and Scheffler, 1985). Therefore, if an epigenetic mechanism is responsible for these off and on switches of activity, it is readily inducible in both directions.

We have also attempted to find immunologically cross-reacting material (CRM) in the null mutants using a rabbit antiserum raised against mouse kidney ODC by Dr. Lo Persson (Persson, 1982). We have used a radioimmune assay procedure based on the one described by Seely and Pegg (1983), using precipitation of a labeled ODC as a measure of reaction with antibody. ODC reacted with [^{3}H]αDFMO, which becomes covalently bound, was the labeled antigen. Addition of unlabeled ODC in a cell extract resulted in less precipitation of the labeled antigen as the unlabeled antigen competed with it for precipitation by the antibody. In wild type CHO cells there is a good correspondence between ODC activity and ODC protein in synchronized, induced cells. However, the null mutant C55.7, which has no measurable enzyme activity, has little or no measurable CRM either, within the limits of detection of the assay. If any ODC protein is being produced, the amounts are drastically less than in wild type cells (Steglich and Scheffler, 1985).

Very recently, an ODC cDNA clone became available to us from the laboratory of Dr. O. Janne (Kontula et al., 1984). Northern blots of mRNA from wild-type cells and from one of the very stable null mutants as well as from one of the overproducers were probed with the mouse ODC cDNA plasmid, with the following results: the wild-type cells and null mutants had comparable amounts of ODC message at the peak of induction, while the overproducers had substantially more ODC mRNA (Steglich and Scheffler, 1985). Thus, we have strong, direct evidence for transcription of at least one ODC gene—with no detectable enzyme being produced—and we conclude that at least this allele has suffered a structural gene mutation.

One simple model which could account for a CRM$^+$ mutant that is revertible by 5-azacytidine is that CHO cells may be diploid at the locus coding for ODC but only one copy is actively expressed. ODC$^-$ mutants would have one mutated allele and one silent allele. The silent allele can be reactivated to produce the revertants. The secondary mutants derived from the revertants after EMS treatment would contain two mutated alleles and should be stable. However, since these secondary mutants can also be induced to revert with 5-azacytidine, this model may not be correct. Many of these questions may be answered in the near future using molecular probes for the ODC gene to determine directly gene number, organization, and expression in these cells.

IV. CONCLUSIONS

The successful isolation of variants that overproduce normal ODC activity subject to normal control mechanisms concludes the first phase of our studies. We believe that these clones with much elevated ODC levels will represent a convenient source of enzyme, and will constitute a good system for further biochemical studies on the regulation of its activity during the cell cycle. A major challenge for the future is to elucidate the mechanisms which are responsible for the rapid down regulation and turnover of the enzyme, and the significance of the antizyme.

A similar series of studies with mouse S49 lymphoma cells has been reported by McConlogue and Coffino (1983). Overproduction of normal enzyme was demonstrated directly on two-dimensional polyacrylamide gels with extracts from [^{35}S]methionine (pulse) proteins. Since then, a clone was discovered which makes ~15% of its total protein as ODC, and both cDNA and genomic DNA clones have been prepared and are being characterized. Overproduction of ODC is at least in part due to gene amplification in these cells. [P. Coffino, personal communication, and *J. Biol. Chem.* (in press)].

A rat hepatoma cell line selected in αMO and having elevated levels of ODC has already been mentioned (Mamont et al., 1978). In contrast to our mutants, the turnover mechanism of ODC may be disturbed in these cells, because it has been observed that the cells produce an apparently normal ODC enzyme which has a much longer half-life (Pritchard et al., 1982).

The ODC-deficient mutants (or variants) represent the other side of the coin, and the stable, nonleaky mutants could find useful application in physiological studies on polyamine requirements. Another kind of mutant in Chinese hamster cells with similar application has been described by Hölttä and Pohjanpelto (1982). This mutant is deficient in arginase (see Fig. 18.2) and therefore lacks sufficient ornithine in serum-free medium to manufacture polyamines. This mutant can also be useful in studying physiological roles of polyamines.

A major challenge still is to understand the nature of the change giving rise to the ODC$^-$ phenotype and to distinguish between epigenetic and genetic events. The high spontaneous reversion rate of some mutants can be explained in terms of gene amplification and overproduction of a defective but partially active protein. Our results on the selection of variants resistant to the ornithine analogs αMO and αDFMO suggest that clones with elevated enzyme levels arise at comparable frequencies. In this view, the stable ODC$^-$ mutants have structural gene mutations that completely inactivate the ODC activity. On the other hand, the reversion induced by 5-azacytidine is clearly indicative of a mechanism involving gene inactivation and reactivation, that is, alterations that affect gene expression rather than the structure of the gene product.

Since we have been able to reisolate ODC^- variants from 5-azacytidine-induced ODC^+ revertants at a frequency of roughly 10^{-5}, and since these secondary ODC^- variants again produce ODC^+ clones of frequencies $\geq 10^{-5}$, the presumed epigenetic changes observed here must be capable of turning the ODC gene(s) on or off at approximately equal rates. With efficient selections for both phenotypes being available, and with the cloned segments of the ODC gene becoming available, this system promises to be an interesting one in future studies on basic epigenetic mechanisms.

ACKNOWLEGMENTS

The work described in this review was supported by a grant from the United States Public Health Service to I.E.S. C.S. wishes to acknowledge support from a fellowship from USPHS.

REFERENCES

Abdel-Monem, M. M., Newton, N. E., and Weeks, C. E. (1974). *J. Med. Chem.* **17**, 447–451.

Abraham, A. K. and Pihl, A. (1981). *Trends Biochem. Sci.* **6**, 106–107.

Atkins, J. F., Lewis, J. B., Anderson, C. W., and Gesteland, R. F. (1975). *J. Biol. Chem.* **250**, 5688–5695.

Atmar, V. J. and Kuehn, G. D. (1981). *Proc. Natl. Acad. Sci.* **78**, 5518–5522.

Bachrach, U. (1981). *Polyamines in Biomedical Research.* (J. M. Gaugas, ed), Wiley, New York, pp. 81–107.

Bethell, D. R. and Pegg, A. E. (1979). *Biochem. J.* **180**, 87–94.

Bey, P., Danzin, C., Van Dorsselaer, V., Mamont, P., Jung, M., and Tardif, C. (1978). *J. Med. Chem.* **21,** 50–55.

Boucek, Jr., R. J. and Lembach, K. J. (1977). *Arch. Biochem. Biophys.* **184**, 408–415.

Butler-Gralla, E. and Herschman, H. P. (1983). *J. Cell Physiol.* **114**, 317–320.

Canellakis, E. S., Viceps-Madore, D., Kyriakidis, D. A., and Heller, J. S. (1979). *Current Topics in Cellular Regulation.* (B. L. Horecker and E. R. Stadtman, eds.), Academic Press, New York, pp. 155–202.

Choi, J., and Scheffler, I. E. (1981). *Somat. Cell. Genet.* **7**, 219–233.

Choi, J. and Scheffler, I. E. (1983). *J. Biol. Chem.* **258**, 12601–12608.

Clark, J. L. (1974). *Biochem.* **13**, 4668–4674.

Clark, J. L. and Greenspan, S. (1979). *Exp. Cell Res.* **118**, 253–260.

Cohn, M. S., Tabor, C. W., and Tabor, H. (1978). *J. Bacteriol.* **134**, 208–213.

Cohn, M. S., Tabor, C. W. and Tabor, H. (1980). *J. Bacteriol.* **142**, 791–799.

Criss, W. E., Morishita, Y., Watanabe, Q., Akogyeram, C., Sahai, A., Deu, B., and Oka, T. (1983). *Adv. Polyamine Res.* **4**, 647–654.

Daniels, G. R., Atmar, V. J., and Kuehn, G. D. (1981). *Biochem.* **20**, 2525–2532.

Erwin, B. G., Seely, J. E., and Pegg, A. E. (1983). *Biochem.* **22**, 3027–3032.

Guroff, G., Montgomery, P., Tolson, N., Lewis, M. E., and End, D. (1980). *Proc. Natl. Acad. Sci.* **77**, 4607–4609.

Haddox, M. K. and Russell, D. H. (1981). *Biochem.* **20**, 6721–6728.

Hafner, E. W., Tabor, C. W., and Tabor, H. (1979). *J. Biol. Chem.* **254**, 12419–12426.

Harris, M. (1982). *Cell* **29**, 483–492.

Heller, J. S. and Canellakis, E. S. (1981). *J. Cell. Physiol.* **107**, 209–217.

Heller, J. S., Chen, K. Y., Kyriakidis, D. A., Fong, W. F., and Canellakis, E. S. (1978). *J. Cell Physiol.* **96**, 225–234.

Hogan, B. L. M. and Murden, S. (1974). *J. Cell. Physiol.* **83**, 345–352.

Hölttä, E. and Pohjanpelto, P. (1982). *Biochem. Biophys. Acta.* **721**, 321–327.

Isomaa, V. V., Pajunen, A. E. I., Bardin, C. W., and Janne, O. A. (1983). *J. Biol. Chem.* **258**, 6735–6740.

Jones, P. A. and Taylor, S. M. (1980). *Cell* **20**, 85–93.

Kallio, A., Scalabrino, G., and Janne, J. (1977). *FEBS Lett.* **73**, 229.

Kao, F. T. and Puck, T. T. (1967). *Genet.* **55**, 513–524.

Kontula, K. K., Toskkeli, T. K., Bardin, C. W., and Janne, O. A. (1984). *Proc. Natl. Acad. Sci.* **81,** 731–735.

Kuehn, G. D., Affolter, H. U., Atmar, V. J., Seebeck, T., Gubler, U., and Braun, R. (1979). *Proc. Natl. Acad. Sci.* **76**, 2541–2545.

Landy-Otsuka, F. and Scheffler, I. E. (1978). *Proc. Natl. Acad. Sci.* **75**, 5001–5005.

Luria, S. E. and Delbruck, M. (1943). *Genetics* **28**, 491–511.

Mamont, P. S., Bohlen, P., McCann, P. P., Bey, P., Schuber, F. and Tardif, C. (1976). *Proc. Natl. Acad. Sci.* **76**, 1626–1630.

Mamont, P. S., Danzin, C., Wagner, J., Siat, M., Joder-Ohlenbush, A. M., and Claverie, N. (1982). *Eur. J. Biochem.* **123**, 499–520.

Mamont, P. S., Duchesne, M. C., Grove, J., and Tardif, C. (1978). *Exp. Cell Res.* **115**, 387–393.

McCann, P. P., Tardif, C., Hornsperger, J. M., and Davis, J. C. (1979). *J. Cell. Physiol.* **99**, 183–190.

McConlogue, L. and Coffino, P. (1983). *J. Biol. Chem.* **258**, 8384–8388.

McConlogue, L. C., Marton, L. J., and Coffino, P. (1983). *J. Cell. Biol.* **96**, 762–767.

McCormick, F. (1977). *J. Cell. Physiol.* **93**, 285–292.

McCormick, F. (1978). *Biochem. J.* **174**, 427–434.

Metcalf, B. W., Bey, P., Danzin, C., Jung, M. J., Casara, P., and Vevert, J. P. (1978). *J. Am. Chem. Soc.* **100**, 2551–2553.

Padgett, R. A., Wahl, G. M., Coleman, P. F., and Stark, G. R. (1979). *J. Biol. Chem.* **254**, 974–980.

Pajunen, A. E. I., Isomaa, V. V., Janne, O. A., and Bardin, C. W. (1982). *J. Biol. Chem.* **257**, 8190.

Pegg, A. E. and McCann, P. P. (1982). *Am. J. Physiol.* **243**, c212–c221.

Persson, L. (1982). *Acta Chem. Scand.* **36**, 685–688.

Pritchard, M. L., Pegg. A. E. and Jefferson, L. S. (1982). *J. Biol. Chem.* **257**, 5892–5899.

Russell, D. H. and Haddox, M. K. (1979). *Adv. Enzyme Regulation* **17**, 61–87.

Russell, D. H. and Snyder, S. H. (1969). *Molec. Pharmacol.* **5**, 253–262.

Sakai, T. T. and Cohen, S. S. (1976). *Prog. Nuc. Acid Res. Mol. Biol.* **17**, 15–42.

Scalabrino, G. and Ferioli, M. E. (1981). *Adv. Cancer Res.* **35**, 151–268.

Schimke, R. T., Kaufman, R. J., Ah, F. W., and Kellems, R. F. (1978). *Science* **202**, 1051–1055.

Scott, K. F. F., Meyskens, Jr., F. L., and Haddock Russell, D. (1982). *Proc. Natl. Acad. Sci.* **79**, 4093–4097.

Seely, J. E. and Pegg, A. E. (1983). *J. Biol. Chem.* **258**, 2496–2514.

Seely, J. E., Poso, H., and Pegg, A. E. (1982). *Biochem.* **21**, 3394–3399.

Seidenfeld, J. and Marton, L. J. (1980). *Cancer Res.* **40**, 1961–1966.

Smith, R. J. and Phang, J. M. (1979). *J. Cell. Physiol.* **98**, 475–482.

Steglich, C. and Scheffler, I. E. (1982). *J. Biol. Chem.* **257**, 4603–4609.

Steglich, C., Choi, J. H., and Scheffler, I. E. (1983). *Advances in Polyamine Research* (U. Bachrach, A. Kaye and R. Chayen, eds.), Raven Press, New York, pp. 591–602.

Steglich, C. and Scheffler, I. E. (1985). *Somat. Cell Molec. Genetics*, in press.

Tabor, H., Hafner, E. W., and Tabor, C. W. (1980). *J. Bacteriol.* **144**, 952–956.

Tabor, C. W. and Tabor, H. (1976). *Ann. Rev. Biochem.* **45**, 285–306.

Weekes, R. G., Verma, A. K., and Boutwell, R. K. (1980). *Cancer Res.* **40**, 4013–4018.

Whitney, P. A. and Morris, D. R. (1978). *J. Bacteriol.* **134**, 214–220.

CHAPTER 19

MITOCHONDRIAL MUTANTS

Carolyn D. Whitfield
Department of Biochemistry
Howard University
College of Medicine
Washington, D.C.

Both the nuclear and mitochondrial genetic systems contribute to the biogenesis of mitochondria. Several mitochondrial respiratory complexes (cytochrome *c* oxidase; ubiquinol-cytochrome *c* reductase or Complex III; oligomycin-sensitive ATPase) have subunits that are both nuclearly and mitochondrially encoded. In yeast and fungi, mutations in both genetic systems have been invaluable in studying the structure, function, synthesis, and assembly of respiratory complexes [for review see Borst and Grivell (1978) and Tzagoloff et al. (1979)]. Until recently, these studies have been limited by the paucity of mitochondrial mutants in mammalian cells. In this chapter, nuclearly and cytoplasmically inherited mutants, primarily in Chinese hamster cells, affecting the mitochondrial respiratory chain, oxidative phosphorylation, and mitochondrial protein synthesis will be discussed. Before addressing the specifics of mitochondrial mutants, mitochondrial DNA (mtDNA) structure, replication, transcription, and mitochondrial translation will be presented for background information.

I. THE MITOCHONDRIAL GENOME

Mammalian cells contain several thousand copies of a closed circular independently replicating mtDNA with each mitochondrion having 2–10 copies of mtDNA (Clayton, 1982; Oliver and Wallace, 1982). Mitochondria have unique systems for DNA replication, transcription, and translation. The human, beef, and mouse mitochondrial genomes (~16,500 base pairs) have been completely sequenced, and each carries information for 22 tRNAs, 2 rRNAs, and only 13 peptides (Anderson et al., 1981, 1982; Bibb et al., 1981). Thus, the majority of proteins (~95%) in mitochondria are nuclearly encoded, cytoplasmically synthesized, and imported into the mitochondrion.

A. Structure of mtDNA

The mammalian mitochondrial genome exhibits an exceptionally compact organization which contains an almost continuous coding sequence. Only the heavy (H) and light (L) origins of replication do not code for a functional RNA. Transcriptional mapping (Battey and Clayton, 1978; Ojala et al., 1980; Van Etten et al., 1982) and comparison of the DNA sequence to the mRNA and rRNA terminal sequences (Van Etten et al., 1980; Anderson et al., 1981; Montoya et al., 1981; Ojala et al., 1981) have revealed that tRNA genes are interspersed between rRNAs and protein coding sequences with few or no intergenic nucleotides and no introns (Fig. 19.1). The heavy DNA strand is the major coding strand, coding for 12S and 16S rRNAs, and 12 protein coding se-

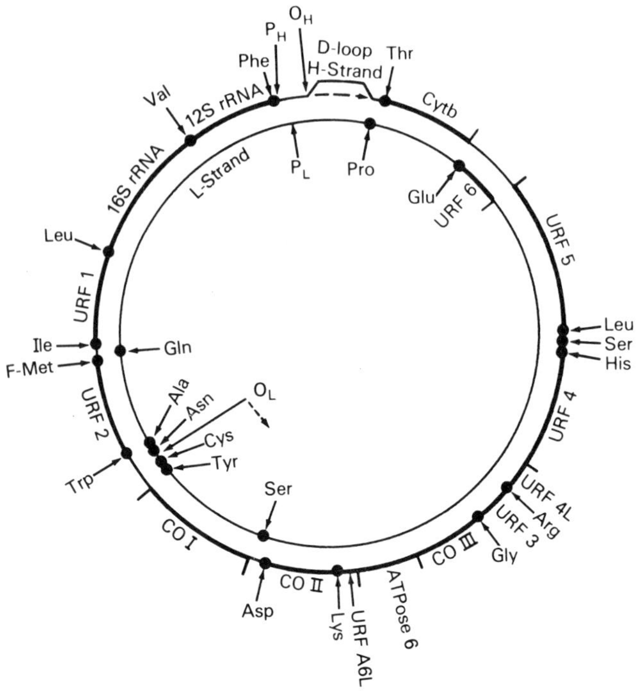

Figure 19.1. Structure of mammalian mitochondrial DNA (Anderson et al., 1981, 1982; Bibb et al., 1981). H-strand, heavy DNA strand; L-strand, light DNA strand; O_H, origin of heavy strand replication; D-loop, displacement loop formed during heavy strand replication. Genes for tRNAs are designated by three-letter abbreviations for amino acids. Coding strand is shown in darker print. Cyt, cytochrome; URF, unidentified reading frame; CO, cytochrome *c* oxidase; O_L, origin of light strand replication; P_L, light strand promoter for transcription (Montoya et al., 1982). There are two heavy strand promoters (P_H) for transcription near the $tRNA^{Phe}$ gene (Montoya et al., 1982, 1983), as discussed in Section I C 1.

quences, while the light DNA strand codes for 8 tRNAs and 1 peptide (Anderson et al., 1981, 1982; Bibb et al., 1981).

Five of the protein coding sequences have been unambiguously identified as the genes for subunits of the respiratory complexes: apocytochrome *b*, subunit 6 of ATPase, and subunits I, II, III of cytochrome oxidase. The products of the eight remaining protein coding sequences have not been functionally identified.

B. mtDNA Replication

mtDNA replication initiates with H-strand synthesis in one direction from a specific origin (0_H) and is catalyzed by mitochondrial DNA polymerase γ [for review see Clayton (1982)]. A triplex displacement (D) loop structure occurs at the H-strand origin due to the synthesis of a short H-strand and the displacement of the parental H-strand as a single-stranded region in the closed circular DNA (Fig. 19-1). In mouse cells, a high portion of mtDNA is in the D-loop configuration where the newly synthesized H-strands, called 7S DNA, have lengths from 520 to 690 nucleotides and are rapidly turned over. A few newly synthesized H-strands in mouse cells continue replication around the circle which exposes the origin of L-strand replication and only then allows synthesis of the L-strand in the opposite direction. Interestingly, the mouse L- and H-origin of replication contain ribonucleotides as does the 5′ end of one species of 7S DNA.

The D-loop region accounts for the differences in sizes of human, mouse, and beef mtDNA and is the least conserved region in mammalian mtDNA in size or sequence (Anderson et al., 1981, 1982; Bibb et al., 1981). Nevertheless, there are several small blocks of sequences which are conserved in the D-loop region and may function in controlling H-strand replication or L- and H-strand transcription. On the other hand, the origin of L-strand replication is highly conserved in human, bovine, and mouse mtDNA, both in sequence and in its hairpin-loop structure which may function in the initiation of replication (Anderson et al., 1981, 1982; Bibb et al., 1981).

C. Transcription of mtDNA

1. tRNA Excision Model of RNA Processing

The order of genes and the absence of noncoding, intergenic sequences between rRNA, tRNA, and protein coding genes have led to the suggestion that both DNA strands are transcribed into a single polycistronic RNA molecule from promoters in the D-loop region. It has been postulated that the sense and antisense tRNAs fold up into their secondary structure as transcription occurs and that a processing RNase P-like enzyme(s) generates mature mRNAs, tRNAs, and rRNAs by endonucleolytic cleavages precisely at the 5′ and 3′ termini of the tRNA genes (Anderson et al., 1981; Bibb et al., 1981; Montoya et al., 1981; Ojala et al., 1981). Maturation of tRNAs posttranscriptionally involves base modification and addition of the characteristic CCA end to their 3′ termini. Posttranscriptional adenylation of both mRNAs (~50 A residues) and rRNAs (1–5 A residues) suggests a role for 3′adenylation in the processing or termination (see below) of mitochondrial RNAs (Dubin et al., 1982; Van Etten et al., 1983).

The following evidence supports the transcription model described above. In HeLa cells, a polycistronic H-strand transcript has been identified corresponding to the $tRNA^{Phe}$ gene, 12S rRNA gene, $tRNA^{Val}$ gene, and 16S rRNA gene (Montoya et al., 1983) and another corresponding to the 12S rRNA gene, $tRNA^{Val}$ gene and 16S rRNA gene (Ojala et al., 1980). Three overlapping, giant polycistronic L-strand transcripts have also been observed in HeLa cells. The 5′ ends of these transcripts are near the tRNA^{Glu} gene and their 3′ termini are near other sense or antisense tRNA genes located one-third to one-half the distance around the genome (Ojala et al., 1980).

The initiation site for human L-strand transcription has been located near the D-loop region, between tRNA^{Phe} and O_H (the origin of H-strand replication); this site corresponds to the 5′ end of a small L-strand transcript called 7S RNA (Montoya et al., 1982). A purified mitochondrial RNA polymerase also initiates L-strand transcription from this site *in vitro* (Walberg and Clayton, 1983; Chang and Clayton, 1984).

Two initiation sites for human H-strand transcription have been located near the D-loop region (Montoya et al., 1982, 1983; Yoza and Bogenhagen, 1984). The initial order of genes on the H-strand is: D loop–tRNA^{Phe}–12S rRNA–tRNA^{Val}-16S rRNA–tRNA$^{Leu}_{UUR}$ –unidentified reading frame (URF) 1 (Anderson et al., 1981, 1982; Bibb et al., 1981). Most rRNA is synthesized from a transcript which begins 16–19 base pairs upstream (toward the D-loop) from the tRNA^{Phe} gene (Montoya et al., 1983; Chang and Clayton, 1984; Yoza and Bogenhagen, 1984). There is controversy over whether the mRNAs and most tRNAs encoded by the H-strand are synthesized from a second overlapping transcript that initiates within the tRNA^{Phe} gene, close to the 5′ end of the 12S rRNA gene (Montoya et al., 1982, 1983) or from the same transcript which begins upstream from the tRNA^{Phe} gene (Chang and Clayton, 1984).

The entire H-strand is transcribed approximately 1–2 times per cell generation, whereas the rRNA genes are transcribed about 50 times per cell generation (Battey and Clayton, 1978; Gelfand and Attardi, 1981). The existence of two distinct initiation sites might explain the differential rate of synthesis of rRNAs and H-strand mRNAs (Montoya et al., 1983); alternatively, transcription termination might explain the difference. The 3′-terminal heterogeneity of the human and mouse 16S rRNA, which extends for seven nucleotides into the mouse tRNA^{Leu} gene, is consistent with a transcription termination site in this region (Dubin et al., 1982; Van Etten et al., 1983). In contrast, the 12S rRNA and mRNAs have precise 3′ termini. Furthermore, the hairpin structure at the 3′ terminus of several mammalian 16S rRNAs resembles the hairpin-oligo(U) signal proposed for bacterial termination attenuation (Attardi et al., 1982; Dubin et al., 1982).

Interestingly, the L-strand initiation site for transcription and both H-strand initiation sites are located within a region having a similar sequence that is rich in repetitive C and repetitive A residues (Bogenhagen et al., 1984; Chang and Clayton, 1984).

2. mRNA Transcripts

Polyadenylated RNA transcripts have been mapped to each protein coding sequence in mtDNA including the eight unidentified reading frames (URF) (Battey and Clayton, 1978; Ojala et al., 1980; Van Etten et al., 1982). There are individual RNA transcripts for each protein coding sequence except that URF A6L and ATPase 6 share a single RNA transcript; likewise, URF 4L and URF 4 are transcribed into a single RNA. Interestingly, URF A6L overlaps out of phase by 46 nucleotides the ATPase 6 reading frame and URF 4L overlaps out of phase by seven nucleotides the URF 4 reading frame (Anderson et al., 1981).

Unlike most eukaryotic mRNAs, mammalian mitochondrial mRNAs do not have a 5′ or 3′ untranslated region. Although polyadenylated, the 3′ termini of mitochondrial mRNAs do not contain a polyadenylation signal. The initiation codon lies within six nucleotides of the 5′ end of the mRNA, and the termination codon is at the 3′ end of the mRNA (Montoya et al., 1981, Ojala et al., 1981). In fact, several mammalian mitochondrial mRNAs have a U or UA at the end of the open reading frame and polyadenylation results in the creation of a UAA termination codon (Anderson et al., 1981, 1982; Bibb et al., 1981).

D. Mitochondrial Translation

1. The Mitochondrial Genetic Code

The mammalian mitochondrial genetic code differs from the "universal" genetic code. In the human, mouse, and bovine mitochondrial code, UGA is used for tryptophan instead of termination and AUA for methionine instead of isoleucine (Anderson et al., 1981, 1982; Bibb et al., 1981). In contrast to the universal genetic code where AGA and AGG code for arginine, these codons are not used in the mouse mitochondrial code and specify termination in human mtDNA (Anderson et al., 1981; Bibb et al., 1981). In beef mitochondria, AGA is termination and AGG is not used (Anderson et al., 1982). In addition to AUG, which is an initiator codon in the "universal" genetic code, AUA and AUU are initiator codons for *N*-formylmethionine in the human mitochondrial code (Anderson et al., 1981). AUG and AUA are initiator codons for *N*-formylmethionine in bovine mtDNA, and AUG, AUU, AUA, and AUC in the mouse mitochondrial code (Anderson et al., 1982; Bibb et al., 1981).

2. Mitochondrial tRNAs

Since only 22 tRNA genes have been identified in the mammalian mitochondrial genome and no tRNAs are imported from the cytoplasm (Aujame and Freeman, 1979), it has been concluded that there are families of two codons (such as UUU and UUC, where the tRNA anticodon is GAA) and families of four codons (such as CUU, CUC, CUA, CUG where the tRNA anticodon is UAG) and each family is read by a single tRNA (Anderson et al., 1981, 1982; Bibb et al., 1981). Two-codon families utilize G:U base pairing in the wobble position; while four-codon families depend on U:N (any nucleotide) wobble or a two-out-of-three base-pairing mechanism. There are two serine tRNA genes and two leucine tRNA genes and single tRNA genes for the other amino acids (Anderson et al., 1981, 1982; Bibb et al., 1981). Since methionyl tRNA and formylmethionyl tRNA are encoded by the same gene, it has been postulated that they differ by base modification. Mammalian mitochondrial tRNAs lack many of the conserved sequences found in other tRNAs; as a result, mitochondrial tRNAs are less stabilized by tertiary interactions (Anderson et al., 1981, 1982; Bibb et al., 1981).

3. *Initiation of Protein Synthesis*

The initiation of protein synthesis is thought to occur by a scanning mechanism whereby the small ribosomal subunit recognizes the 5′ end of the mRNA and migrates to the first initiation codon, at which point the large ribosomal subunit is added and translation proceeds (Montoya et al., 1981). A similar mechanism has been proposed for other eukaryotic mRNAs since translation occurs from the first AUG closest to the 5′ end (Kozak, 1978). RNA sequencing has shown that the initiation codon is within the first six nucleotides at the 5′ end of human mitochondrial mRNAs; thus there is no ribosomal binding site (Montoya et al., 1981). Exceptions are the ATPase 6 and URF 4 genes where translation occurs from an internal initiation codon.

4. *Mitochondrially Encoded Proteins*

Mitochondrially synthesized proteins can be detected by labeling cells with radioactive amino acids in the presence of cycloheximide or emetine, inhibitors of cytoplasmic protein synthesis. In the presence of chloramphenicol, mitochondrial protein synthesis is specifically inhibited. The number (13 or more) of mitochondrial translation products is consistent with the expression of all 13 protein coding sequences (Yatscoff et al., 1978, Burnett and Scheffler, 1981; Ching and Attardi, 1982; Hare and Hodges, 1982). There is direct evidence that proteins specified by URF genes 1, 3, 6, and A6L are expressed in human mitochondria. Using a small C-terminal peptide synthesized from the DNA sequence, antibody was prepared and employed to immunoprecipitate the mitochondrial translation product encoded by URF genes 1, 3, and A6L (Chomyn et al., 1983; Mariottini et al., 1983). Another approach by Oliver et al. (1983) identified the translation products of URF 3 and 6 by comparing the size of peptides observed after proteolytic digestion with those expected from the DNA sequence. URF 3 codes for two polymorphic proteins of human mtDNA (Oliver et al., 1983). The mitochondrial translation products, like the respiratory complexes, are located in the inner mitochondrial membrane (Hare and Hodges, 1982).

The URF A6L, which shares an RNA transcript with ATPase 6, may code for a second subunit of the oligomycin-sensitive ATPase complex. In rat liver and Chinese hamster cells two peptides are immunoprecipitated with antibody to mitochondrial ATPase (Kuzela et al., 1980; Kolarov et al., 1981; Malczewski and Whitfield, 1984). One peptide (molecular weight 18,300) corresponds to subunit 6 of ATPase and the second peptide (molecular weight 4900) corresponds in size to the product of human URF A6L (Mariottini et al., 1983). Furthermore, in yeast (Novitski et al., 1983) and *Aspergillus nidulans* (Grisi et al., 1982) mtDNA, an URF has recently been found to precede the ATPase 6 gene; mutations in this URF affect ATPase assembly or function in yeast (Macreadie et al., 1982).

II. RESPIRATION-DEFICIENT MUTANTS

Respiration-deficient (Res^-) mutants have the following phenotype: a growth requirement for CO_2/bicarbonate and asparagine; inability to grow on galactose, fructose, and other alternate hexoses; and inability to survive on hexose-free media for short periods of time. In order to understand this phenotype and how it was used to select Res^- mutants, energy sources in cultured cells are discussed next.

A. The Respiratory Chain

The mitochondrial respiratory chain, located in the inner mitochondrial membrane, consists of the flavoprotein NADH dehydrogenase; ubiquinone (Q), a small redox active molecule with a long hydrophobic tail containing 9 or 10 isoprenoid units; and cytochromes *b*, c_1, *c*, and aa_3 (Fig. 19.2). As electrons pass down the chain to the terminal acceptor O_2, an electrochemical gradient is generated which results in the synthesis of ATP at three distinct sites. Electrons are derived from succinate and NADH, which are produced in the soluble mitochondrial matrix by the citric acid cycle; from α-glycerol-phosphate, which is generated in the cytoplasm; and from β-oxidation of fatty acids, which occurs in the mitochondrial matrix. In cultured Chinese hamster cells, NADH is the main source of electrons since inhibition of electron transfer from NADH to Q by rotenone or mutation prevents nearly all O_2 consumption by the respiratory chain (DeFrancesco et al., 1976; Breen and Scheffler, 1979; Maiti et al., 1981; Whitfield et al., 1981). Rotenone and antimycin are specific inhibitors of electron transfer, while

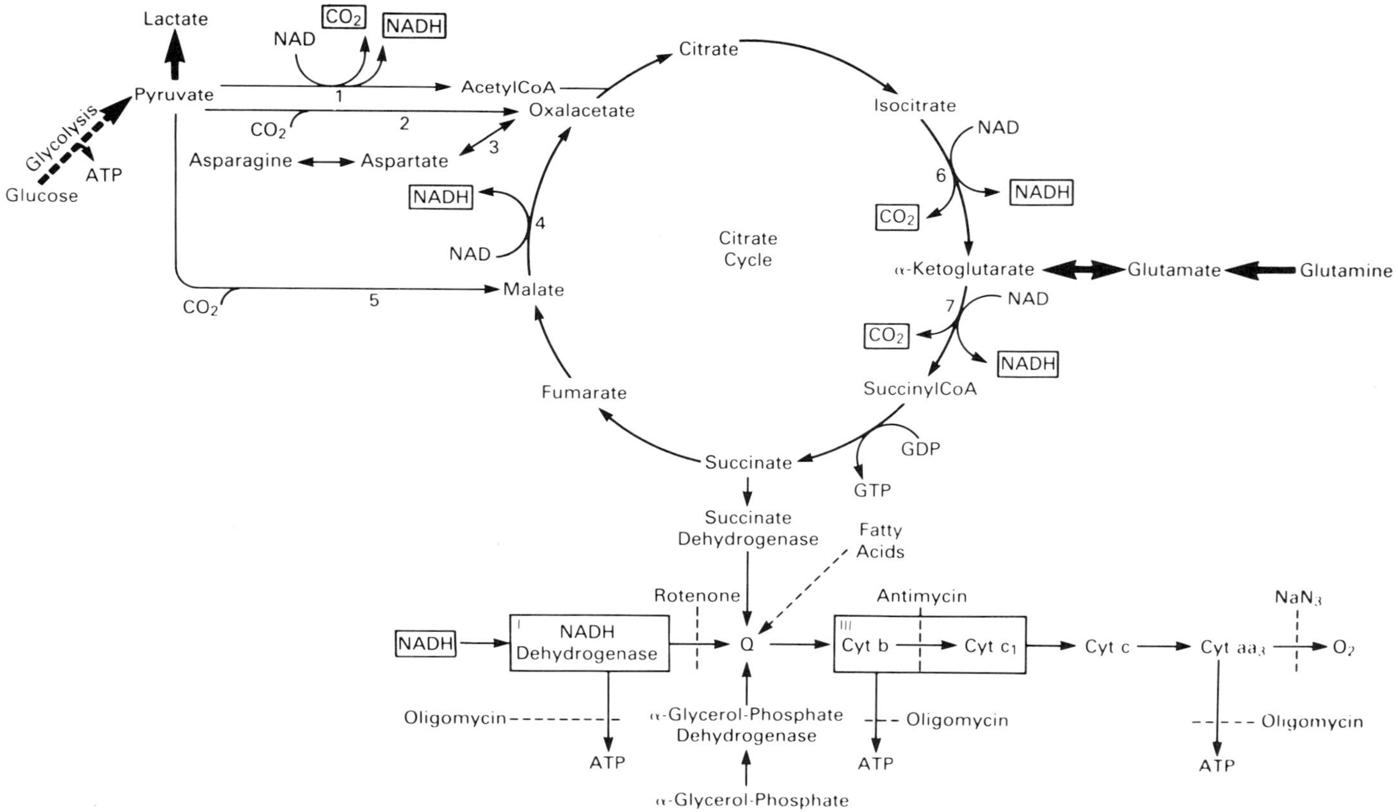

Figure 19.2. Citric acid cycle and respiratory chain. Heavy arrows show major fluxes in cultured mammalian cells (Donnelly and Scheffler, 1976; Reitzer et al., 1979). CO_2 released from pyruvate or by the citric acid cycle is highlighted in boxes; NADH, produced at several points in the cycle, feeds into the respiratory chain. Enzymes involved: 1, pyruvate dehydrogenase; 2, pyruvate carboxylase; 3, aspartate-oxalacetate transaminase; 4, malic dehydrogenase; 5, malic enzyme; 6, isocitrate dehydrogenase; 7, α-ketoglutarate dehydrogenase. Respiratory complexes: I, Complex I (NADH-Q reductase); III, Complex III (ubiquinol-cytochrome *c* reductase); cytochrome *c* oxidase (reduced cytochrome $c \rightarrow$ cytochrome $aa_3 \rightarrow O_2$); succinate-cytochrome *c* reductase (succinate → Q → cytochromes bc_1 → cytochrome *c*); oligomycin-sensitive ATPase, mitochondrial ATP synthase involved in coupling of respiration to phosphorylation of ADP.

oligomycin is a specific inhibitor of the inner-membrane-bound mitochondrial ATPase associated with oxidative phosphorylation.

Energy can be supplied anaerobically by the glycolytic pathway or aerobically by the respiratory chain. Glucose is converted to pyruvate by the glycolytic pathway; pyruvate usually enters the citric acid cycle and is a source of electrons for the respiratory chain (Fig. 19.2). Fifteen times more ATP per molecule of glucose is produced by the aerobic pathway than by the anaerobic pathway. Nevertheless, in cultured cells nearly all of the pyruvate is converted to lactic acid and does not enter the citric acid cycle (Donnelly and Scheffler, 1976; Reitzer et al., 1979). Instead, glutamine, which feeds into the citric acid cycle (Fig. 19.2), is the major source of aerobic energy in human and Chinese hamster cells (Donnelly and Scheffler, 1976; Zielke et al., 1978; Reitzer et al., 1979).

Cultured HeLa cells growing on high levels of glucose (> 1 mM) obtain 35% of their energy from glucose via glycolysis and 65% from oxidative phosphorylation via glutamine, the citric acid cycle, and respiratory chain (Reitzer et al., 1979). On the other hand, nearly all (>98%) energy is from the aerobic glutamine pathway when HeLa cells are grown on galactose or fructose, which are poor glycolytic substrates (Reitzer et al., 1979). The oxidation of glutamine in Chinese hamster cells increases when the glucose concentration is lowered, a situation comparable to growth on galactose or fructose (Donnelly and Scheffler, 1976). In HeLa cells galactose and fructose are primarily metabolized through the hexose monophosphate shunt and serve as precursors for biosynthesis of nucleic acids or glycoproteins (Reitzer et al., 1979). Thus, the hexose requirement of cultured cells is for biosynthetic purposes, not energy. In fact, HeLa cells can grow in the absence of hexoses, provided the media are supplemented with high concentrations of uridine (Wice et al., 1981).

B. Selection of Respiration-Deficient Mutants

1. Auxotrophs for CO_2 and Asparagine

After ethylmethane sulfonate (EMS) mutagenesis, Scheffler (1974) isolated a Res$^-$ mutant from a male Chinese hamster lung (CHL) cell line CCL16 using bromodeoxyuridine (BrdU) and light to kill cells growing at 39°C. A CO_2 auxotroph, CCL16-B2, survived the selection because there was insufficient CO_2 in the media at 39°C.

Wild-type Chinese hamster cells can grow without added CO_2/bicarbonate when media are supplemented with hypoxanthine and uridine, and without added asparagine when glutamine is included (Scheffler, 1974; DeFrancesco et al., 1975). CO_2 is normally produced by the citric acid cycle and asparagine is mainly produced from glutamine by a series of reactions involving the citric acid cycle (Fig. 19.2). In the presence of hypoxanthine and uridine, the mutant, CCL16-B2, required both CO_2/bicarbonate and asparagine for growth because the citric acid cycle was not functioning (DeFrancesco et al., 1975). In the first characterization of a respiration-deficient mutant, DeFrancesco et al. (1976) demonstrated that CCL16-B2 was defective in the first part of the respiratory chain, Complex I (NADH-Q reductase). As a result, the accumulated NADH led to feedback inhibition of citric acid cycle enzymes α-ketoglutarate dehydrogenase and pyruvate dehydrogenase.

Unlike wild-type cells, the respiration-deficient mutant, CCL16-B2, did not grow on galactose (Gal^-) or low concentrations of glucose and did not survive in hexose-free media for short periods of time (Ditta et al., 1976; Donnelly and Scheffler, 1976). This is understandable since the mutant obtains all of its energy from glycolysis and cells growing in galactose or in low concentrations of glucose obtain most of their energy from glutamine via the citric acid cycle and respiratory chain (Donnelly and Scheffler, 1976; Reitzer et al., 1979). The mutant is able to grow in high concentration of glucose by increasing its rate of glycolysis by 40% (Donnelly and Scheffler, 1976). The percentage of high-energy phosphate (energy charge) is the same but the total adenylate pool in CCL16-B2 is 59% of the wild-type level (Soderberg et al., 1980). Considering the rates of glycolysis and total adenylate pool in mutant and wild-type cells, it is likely that wild-type Chinese hamster cells also obtain about 35% of their energy from glycolysis and 65% from oxidative phosphorylation.

Scheffler and colleagues next designed a selection for respiration-deficient mutants based on the phenotype of CCL16-B2. EMS-mutagenized cells were treated with BrdU in media lacking CO_2/bicarbonate and asparagine (Ditta et al., 1976). After irradiation with black light, growing wild-type cells which incorporated BrdU were killed while CO_2/asparagine auxotrophs which did not grow and incorporate BrdU survived. Surviving cells were further tested for their ability to live in hexose-free media for a short period of time (16 hr). With this selection only two new glucose-dependent clones auxotrophic for CO_2/asparagine were isolated from the CCL16 (DON) male Chinese hamster lung cell line, while a large number (up to 65% of the survivors) were isolated from the V79 male Chinese hamster lung cell line. All of the glucose-dependent clones auxotrophic for CO_2/asparagine which were tested were respiration defective. Most of these Res^- mutants reverted to growth on galactose with mutagens or spontaneously at a low frequency (less than 1 in 10^7 cells), except V79-G20 which did not revert (Soderberg et al., 1979).

Recessive respiration-deficient mutants have been isolated from V79 cells at a remarkably high frequency by three different laboratories (Chu et al., 1972; Ditta et al., 1976; Thirion et al., 1976; Maiti et al., 1981; Whitfield et al., 1981). Possible explanations for the high frequency of recessive respiration-deficient mutations in V79 compared to CCL16 are the following. First, one of the best explanations is that of Siminovitch (1976). He has proposed that the high frequency of autosomal, recessive mutations in Chinese hamster cell lines is due to hemizygosity at a number of loci. Chromosomal rearrangements occurring during establishment of these permanent cell lines may have inactivated or deleted one autosomal gene leaving a single active gene. There is evidence for hemizygosity in Chinese hamster ovary cells (Gupta et al., 1978). Hemizygosity at only certain loci might explain why relatively few different types of respiration-deficient mutations have been observed in V79 and CCL16 considering the total ones possible. Second, Day and Scheffler (1982) have shown that three recessive respiration-deficient mutants, representing different complementation groups, are X-linked. This would explain why a large number of mutants are found in one of these complementation units (group I) in V79 cells (Table 19.1); however, it is surprising that no mutants in this complementation group were detected in CCL16 since the same genes are X-linked in mammals (Ohno, 1969). Third, it may be of significance that V79 has a lower rate of oxygen consumption and lower citric acid cycle activity than CCL16 (Ditta et al., 1976).

Soderberg et al. (1979) identified seven complementation groups when pairwise combinations of their 34 respiration-deficient mutants were fused using polyethylene glycol and the resulting tetraploid hybrids were tested for growth on galactose (Table 19.1). All of the mutants behaved recessively, since only certain combinations of mutant hybrids were able to grow on galactose. Surprisingly, of the 34 mutants tested, 21 fell into a single complementation group (I). One mutant, V79-G20, exhibited overlapping complementation because it failed to complement two different complementation groups (I and II) even when hybrid formation was verified using unrelated markers. Later, Day and Scheffler (1982) found that V79-G20 complemented one member of group I, V79-G8, but not another member of group I, V79-G4. Therefore, V79-G20 was not a deletion, as originally proposed, and V79-G4 might represent another overlapping complementation unit connecting V79-G8 and V79-G20. Maiti et al. (1981) discovered that one of their respiration-deficient mutants, 2A13G14, failed to complement V79-G11 (group VI), V79-G14 (group II), and V79-G20; thus, groups II and VI may also be related by overlapping complementation. In conclusion, like Chu's respiration-deficient mutants (Chu, 1974; Whitfield et al., 1981), it appears that several of Scheffler's complementation groups are related by an overlapping complementation pattern.

The molecular basis for the overlapping complementation pattern, which is unusual in somatic cell genetics, is unknown. The situation is complicated by the fact that the respiratory complexes are membrane-bound lipoprotein complexes containing multiple nonidentical sub-

TABLE 19.1
Respiration-Deficient Chinese Hamster Cell Mutants

Complementation Group[a]	Mutants[b]	Notes on Complementation	Biochemical Deficiency	Mapping	Reference[c]
I	V79-G4		Complex I	X-linked, CHL	1,2,3
	V79-G5,G6, G8,G9,G10, G12,G13, G19,G21, G22,G23, G26,G28, G32,G37, G38,G43, G49,G50, G52				
I–II	V79-G20	Does not complement groups I,II except complements G8	Complex I	X-linked,CHL, mouse	1,2,3
II	CCL 16-B2		Complex I		4
	V79-G14		Complex I	X-linked, CHL, mouse	1,2,3
	V79-G24, G31,G42				
II–VI	2A13G14	Does not complement G14, G11,G20	Complex I		5

III	CCL 16-B10 V79-G18,G35		Deficient in O_2 uptake, citrate cycle activity		1,6
IV	CCL16-B9		Succinate dehydrogenase	Human Chromosome 1	1,7,8
	P12GX1				5
V	V79-G7		Mt Protein synthesis	Nuclear	1,9,10
VI	V79-G11		Complex I	Autosomal,CHL	1,3,6
VII	V79-G29		Complex I		1,2
VIII	A13G9 34A13G32		Complex I		5
IX	V61G15		Complex I		5
Gal 50	Gal 50,51	Does not complement Gal 13	Complex I		11,12
Gal 13	Gal 13	Does not complement Gal 50,49; complements groups I–VII	Complex I		1,11,12
Gal 49	Gal 1,9,46, 47,48,49, 58,63,64, 65,66,70	Does not complement Gal 13,17,73	Complex III		11,12

TABLE 19.1 (*continued*)

Complementation Group[a]	Mutants[b]	Notes on Complementation	Biochemical Deficiency	Mapping	Reference[c]
Gal 73	Gal 60,68 71,73	Does not compelement Gal 49	Complex III		11,12
Gal 17	Gal 17	Does not complement Gal 3,49	Ubiquinone		11,12
Gal 3	Gal 3,18	Does not complement Gal 17	Ubiquinone		11,12
Gal 32	Gal 32		Mt protein synthesis		11,13,14
Gal 5	Gal 4,5,6		Similar to Gal 32, Mt protein synthesis?		11,13
Gal 19	Gal 2, 19, 35	Temperature sensitive for growth on galactose	Temperature sensitive for O_2 uptake		15

[a]"Gal" mutants were only tested for complementation with each other except Gal 13; therefore, some Gal mutants may belong to groups I–IX.

[b]All mutants are recessive; mutants were derived from CHL-V79 except those designated CCL16, another male CHL cell line (DON).

[c]References: 1, Soderberg et al. (1979). 2, Breen and Scheffler (1979). 3, Day and Scheffler (1982). 4, DeFrancesco et al. (1976). 5, Maiti et al. (1981). 6, Ditta et al. (1976). 7, Soderberg et al. (1977). 8. Mascarello et al. (1980). 9, Ditta et al. (1977). 10, Burnett and Scheffler (1981). 11, Chu (1974). 12, Whitfield et al. (1981). 13, Malczewski and Whitfield (1982). 14, Malczewski and Whitfield (1984). 15, Whitfield (unpublished).

units. Chu (1974) has suggested that the overlapping complementation pattern may be due to interallelic complementation of missense mutations within a single cistron which codes for a multimeric protein of identical subunits; this is the case in microorganisms where the mutations have been mapped by genetic recombination and the structure of the affected protein has been investigated (Fincham, 1966). Soderberg et al. (1979) have proposed that certain overlapping mutants, such as V79-G20, may represent double mutations since only one mutant was found in each overlapping complementation unit. Nevertheless, this seems unlikely because these mutants were isolated at a relatively high frequency after mutagenesis: approximately 1 in 100 survivors with selection for V79-G20 (Ditta et al., 1976; Soderberg et al., 1979), about 1 in 700 survivors without selection for Gal 13 or Gal 17 (Chu et al., 1972; Chu, 1974). Another explanation is that complementing units may be linked and transcribed into a polycistronic mRNA, and the mutation in the overlapping unit may involve incorrect splicing of the mRNA, thereby affecting the activity of two or more cistrons (Maiti et al., 1981). The possibility that the overlapping mutant is a deletion extending through linked complementation units seems unlikely since V79-G20 does complement V79-G8 and all of Chu's Gal$^-$ mutants revert spontaneously to growth on galactose (Sun et al., 1975). Lastly, in the case of 2A13G14 (Maiti et al., 1981) and Gal 13 and Gal 17 (Chu, 1974), the lack of complementation might be due to failure to form hybrids since hybrid formation was not tested using independent markers.

In mouse–hamster hybrids, the mouse X-chromosome complements the mutation in V79-G14 and V79-G20 (Day and Scheffler, 1982). Using cosegregational analysis of intraspecific tetraploid hamster hybrids, it was shown that the defective gene is X-linked in V79-G14, V79-G20, and V79-G4 and autosomal in V79-G11. Therefore, an explanation of the overlapping complementation involving linkage or interallelic complementation is still possible for V79-G4, V79-G20, and V79-G14 which are X-linked and defective in Complex I activity; however, this is not the case for V79-G11 and V79-G14, which are unlinked even though they fail to complement 2A13G14.

2. *Inability to Grow on Galactose*

With the intention of isolating mutants in the Leloir pathway for galactose metabolism, Chu et al. (1972) isolated 67 CHL V79 mutants which were unable to utilize galactose as a hexose source in place of glucose. The mutants (Gal$^-$) were also unable to grow on other alternate hexoses such as fructose, glucose-6-phosphate, galactose-1-phosphate, and, in some cases, mannose (Sun et al., 1975). Analysis of Gal$^-$ mutants from each complementation group revealed that they are all defective in O_2 consumption by the mitochondrial respiratory chain (Whitfield et

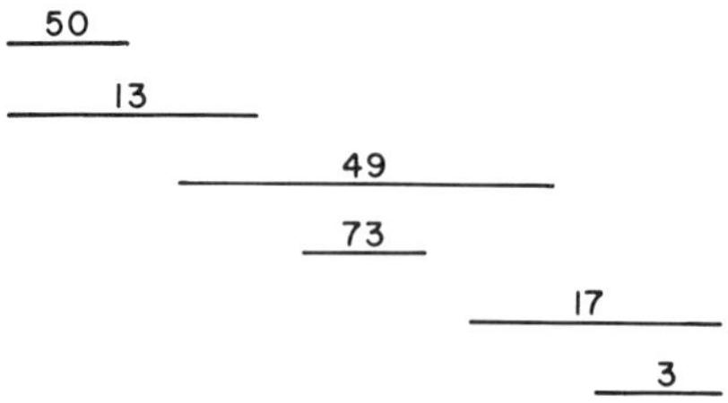

Figure 19.3. Overlapping complementation pattern among 6 Gal$^-$ mutants (Whitfield et al., 1981). Mutants represented by overlapping lines do not complement.

al., 1981; Malczewski and Whitfield, 1982; Whitfield et al., unpublished). This is expected since cultured cells growing in galactose or fructose obtain nearly all of their energy from glutamine via the citric acid cycle, respiratory chain, and oxidative phosphorylation (Reitzer et al., 1979; see Section II A).

Following mutagenesis with 5-bromodeoxyuridine (BrdU) and black light, surviving cells were tested for their ability to grow on glucose but not galactose; a surprisingly high proportion (1–15%) were Gal$^-$ (Chu et al., 1972). The inability of a mutant to grow on alternate hexoses was attributed to a mutation at a single locus, as judged by the spontaneous reversion frequency, one in 10^5–10^7 cells (Sun et al., 1975) and the high frequency at which these mutants were isolated after mutagenesis. The isolation of mutants without selective medium and the recessive behavior in hybrids suggested that the Gal$^-$ mutations are nuclearly encoded. In fact, a single human chromosome, A2, restored growth on galactose in a hybrid formed with a hamster mutant, Gal 2, and human lymphocytes (Sun et al., 1974).

Nine complementation units were identified when pairwise combinations of 29 independently derived Gal$^-$ mutants were fused with inactivated Sendai virus and the intraspecific tetraploid cell hybrids were tested for galactose utilization (Table 19-1) (Chu, 1974). Six of the nine complementation units are related to each other by an overlapping complementation pattern, where complementing units are connected by a series of noncomplementing units (Fig. 19.3). Most of these complementation units contained one to three mutants, whereas one, represented by Gal 49, contained 12 mutants. The ability to grow on galactose exhibited by complementing mutant hybrids and by a hybrid formed from mutant and wild-type cells demonstrated that these mutations are recessive (Chu, 1974; Sun et al., 1975). Interestingly, one complementation unit (represented by Gal 2, 19, 35) was temperature sensitive for growth on galactose. One of the mutants showing overlapping complementation, Gal 13, is in a different complementation group from Scheffler's mutants (Soderberg et al., 1979). Chu's other Gal$^-$ mutants have not been tested for complementation with Scheffler's.

Thirion et al. (1976) have also isolated a large number of Gal$^-$ mutants from CHL V79 cells (V79). After mutagenesis with EMS, cells were plated in galactose and then exposed to BrdU and black light. This procedure enriched for Gal$^-$ cells which stopped growing while wild-type cells incorporated BrdU and were killed. About 5% of the survivors were Gal$^-$. Four of these Gal$^-$ mutants are respiration deficient (Table 19-1) (Maiti et al., 1981). When Thirion's Gal$^-$ mutants were fused to Scheffler's respiration-deficient mutants and those hybrids were tested on galactose, two new Res$^-$ complementation units were defined (Maiti et al., 1981). Furthermore, one of Thirion's mutants (2A13G14) exhibited an overlapping complementation pattern since it failed to complement members of two different complementation groups (G11, G14) as well as the overlapping mutant V79-G20. Thirion's mutants have not been tested in complementation analysis with Chu's.

C. Biochemical Characterization of Res$^-$ Mutants

1. *Mutants Deficient in Complex I*

Respiration-deficient mutants in the first segment of the respiratory chain, Complex I (Fig. 19.2), were dissected as follows. The aerobic energy pathway from glutamine to the citric acid cycle to the respiratory chain to O_2 can be measured in whole cells by determining O_2 uptake from endogenous substrates that are sensitive to specific inhibitors of the respiratory chain, such as rotenone or antimycin. Rotenone-sensitive or antimycin-sensitive O_2 consumption in whole cells was 0–4% of the wild-type level in Complex I mutants (DeFrancesco et al., 1975; Ditta et al, 1976; Maiti et al., 1981; Whitfield et al., 1981). As expected from the selection for CO_2/asparagine auxotrophs, *in vivo* activity of the citric acid cycle measured in whole cells by the formation of $^{14}CO_2$ from 1-[^{14}C]pyruvate or 2-[^{14}C] pyruvate was greatly depressed at less than 5% of the wild-type level (DeFrancesco et al., 1975; Breen and Scheffler, 1979). Citric acid cycle activity was decreased to somewhat lesser extent in Complex I mutants selected as Gal$^-$ (Maiti et al., 1981). $^{14}CO_2$ released by isocitrate dehydrogenase (measured with U-[^{14}C]aspartate) was not substantially affected, but $^{14}CO_2$ released by pyruvate dehydrogenase (measured with 1-[^{14}C]pyruvate) and by α-ketoglutarate dehydrogenase (measured with 1-[^{14}C]glutamate) was decreased. It was concluded that the citric acid cycle was blocked at pyruvate dehyrogenase and α-ketoglutarate dehydrogenase in certain CO_2 asparagine auxotrophs (DeFrancesco, 1975; Breen and Scheffler, 1979). Nevertheless, *in vitro* assays of the citric acid cycle enzymes, α-ketoglutarate dehydrogenase, isocitrate dehydrogenase, malate dehydrogenase, and pyruvate dehydrogenase, revealed no major alterations (DeFrancesco et al., 1976). The citric acid cycle was blocked by the accumulation of NADH, not oxidized by the respiratory chain, which resulted in a feedback inhibition of pyruvate dehydrogenase and α-ketoglutarate dehydrogenase.

In isolated mitochondria O_2 uptake using NADH-linked substrates such as α-ketoglutarate, glutamate, malate, and pyruvate measures the respiratory chain from NADH to O_2 as well as the respective citric acid cycle enzymes (see Fig. 19.2). Activity from NADH-linked substrates was not detected in Complex I mutants (DeFrancesco et al., 1976; Breen and Scheffler, 1979; Maiti et al., 1981). The respiratory chain from NADH to O_2 is measured directly using disrupted mitochondria, NADH, and sensitivity to the specific inhibitors, rotenone or antimycin. Rotenone-sensitive NADH oxidase activity was 9% or less of the wild-type level in CCL16-B2, V79-G4, G14, G29, G20 (DeFrancesco et al, 1976; Breen and Scheffler, 1979), A13G9, 34A13G32, 2A13G14, and V61G15 (Maiti et al., 1981).

On the other hand, these mutants were not altered in O_2 uptake from α-glycerol-phosphate or succinate which measures the respiratory chain between Q and O_2 as well as α-glycerol-phosphate dehydrogenase or succinate dehydrogenase, respectively. Since the respiratory chain activity from NADH to O_2 was decreased, but activity from Q to O_2 was not affected, it was concluded that these mutants were defective in the first segment of the respiratory chain, Complex I (NADH-Q reductase). In the case of Chu's Gal^- mutants, Complex I activity was also assayed directly using Q-1 as an electron acceptor in the presence and absence of rotenone, which interrupts electron flow between NADH dehydrogenase and Q-1 (Singer and Gutman, 1971). Rotenone-sensitive NADH-Q-1 reductase activity was extremely low (5–7% of wild type) in Gal 13 and Gal 50 (Whitfield et al., 1981).

Seven complementation groups and three mutants showing overlapping complementation are primarily deficient in Complex I activity: group I (V79-G4); group II (CCL16-B2, V79-G14); group VII (V79-G29); V79-G20 (which fails to complement groups I and II) (DeFrancesco et al., 1976; Breen and Scheffler, 1979); group VI (V79-G11) (Scheffler, unpublished); group VIII (A13G9, 34A13G32); group IX (V61G15); 2A13G14 (which fails to complement groups II, VI, and V79-G20) (Maiti et al., 1981); Gal 50 complementation group; Gal 13 (which fails to complement Gal 50 and Gal 49) (Whitfield et al., 1981). Since beef heart Complex I contains 26 nonidentical subunits (Heron et al., 1979), it is conceivable that a large number of complementation groups might result; mutants in different subunits might correct one another in the tetraploid hybrid cell. Most (29 out of 34) of the mutants selected as CO_2/asparagine auxotrophs (Ditta et al., 1976) and most (four out of five) of Thirion's Gal^- mutants are deficient in Complex I (Maiti et al., 1981) while only a few (3 out of 29) of Chu's Gal^- mutants are primarily affected in Complex I. Mutants in Complex I are rare in other mammalian cell lines and in yeast; a mutant has been reported in

Escherichia coli (Young and Wallace, 1976). Lichtor et al. (1979) have isolated a cytoplasmic mouse cell mutant which is deficient in Complex I and is rutamycin resistant. Further studies using chromosome mapping, gene transfer, and gene cloning should provide important new information about the mammalian Complex I.

2. Mutants Related by Overlapping Complementation and Deficient in Complex I, Complex III, and Q

The mitochondrial respiratory chain is defective in Gal$^-$ mutants from six complementation units related by an overlapping pattern (Fig. 19.3). Antimycin-sensitive O_2 consumption in whole cells was decreased in all six mutants (Gal 50, 13, 73, 49, 17, 3) as was rotenone-sensitive NADH oxidase activity in disrupted mitochondria (Whitfield et al., 1981). Since cytochrome *c* was lost during purification of mitochondria, it was important to readd cytochrome *c* for maximal activity of succinate oxidase, α-glycerol-phosphate oxidase, and rotenone-sensitive NADH oxidase. Without added cytochrome *c*, the differences between wild type and mutant were obliterated in succinate oxidase and α-glycerol-phosphate oxidase assays. For example, the respiratory chain from Q to O_2 appeared normal in Gal 3 when assayed in the absence of added cytochrome *c*, but it was clearly defective in assays with added cytochrome *c* (Whitfield et al., 1981). Further analysis showed that Gal 3 was deficient in Q.

Interestingly, there is a high degree of correlation between the biochemical data and the complementation map. The component of the respiratory chain most greatly modified varies with the mutant according to its position in the complementation map (Whitfield et al., 1981). For example, Gal 13 and Gal 50, which are located at the left-hand end of the map, are most drastically modified in Complex I activity; rotenone-sensitive NADH-Q-1 reductase activity is 5 and 7%, respectively, of the wild-type value. Gal 49 and Gal 73, which are located in the center of the complementation map, are most greatly affected in the middle of the respiratory chain, Complex III; ubiquinol-cytochrome *c* reductase activity (measured directly with an analog of Q-2 having a decyl side chain) is 9 and 17%, respectively, of the wild-type level. Gal 17 and Gal 3, which are located at the right-hand end of the map, are primarily deficient in the level of ubiquinone. The content of Q is substantially decreased in Gal 3, at 8% of the wild-type value, and in Gal 17, at 33%. Most (12 out of 29) of Chu's Gal$^-$ mutants are in the Gal 49 complementation group; therefore, they are deficient in Complex III. In contrast, the majority of the Res$^-$ mutants isolated by Scheffler and Thirion are defective in Complex I.

Furthermore, most of the Gal$^-$ mutants related by overlapping complementation are simultaneously deficient in some or all of the following: Complex I, Complex III, ubiquinone content, or coupling of respiration to phosphorylation of ADP. For example, Complex I mutant, Gal 13, has only 21% of the wild-type activity for Complex III and a slightly lowered Q content (64%). The Q deficient mutant, Gal 3, also exhibits low Complex III activity (34% of the wild type) and somewhat decreased Complex I activity (64%) even though these activities do not involve endogenous Q as shown with Q-deficient yeast mutants (DeKok and Slater, 1975; Brown and Beattie, 1977). In both Gal 3 and Gal 13 phosphorylation is uncoupled from respiration. The Q deficiency in Gal 3 is probably not in the synthesis of Q, since Q-6 or Q-10 did not stimulate succinate-cytochrome *c* reductase activity as in the yeast Q-deficient mutants (DeKok and Slater, 1975; Brown and Beattie, 1977). The multiple deficiencies in Gal 13 are not due to a defect in phospholipid synthesis, since the different classes of phospholipids are not altered (Whitfield, unpublished).

The pleiotropic alterations in the Gal^- mutants appear to be caused by a single mutation, considering that components diminished in the mutants are simultaneously elevated in their revertants. Therefore, it is unlikely that the failure of Gal 13 to complement Gal 49 is due to a double mutation. The overlapping complementation pattern might be explained by interallelic complementation.

The possibility that all six mutants showing overlapping complementation have different mutations within the same cistron is supported by the observations that complementing mutants at opposite ends of the complementation map (Gal 13 and Gal 3) are simultaneously reduced in Complex I and III activities and in Q content albeit to different extents. This cistron could code for a polypeptide, which is common to Complexes I and III and needed for incorporation of Q, or for a membrane protein needed for proper incorporation of Complexes I, III, and Q. It is conceivable that mutations at different sites in this polypeptide could primarily affect Complex I in one mutant, Complex III in another mutant, or Q content in a third mutant.

3. *Succinate Dehydrogenase Mutant*

One mutant, CCL16-B9, selected for CO_2/asparagine auxotrophy differed from the others in having a high level (72% of wild type) of rotenone-sensitive O_2 consumption in whole cells due to oxidation from NADH-linked substrates (Ditta et al., 1976; Soderberg et al., 1977). Negligible citric acid cycle activity, measured by $^{14}CO_2$ formation, was observed from 1,4-[^{14}C]succinate and from substrates (2-[^{14}C]pyruvate, 5-[^{14}C]glutamate) which require more than one turn of the cycle to release $^{14}CO_2$ (Soderberg et al., 1977).

Since there was no O_2 uptake from succinate in mitochondria but O_2 consumption from α-glycerol-phosphate was normal, it was concluded that CCL16-B9 was defective in succinate dehydrogenase. When succinate dehydrogenase was assayed directly with artifical electron acceptors, activity in the mutant was only 2–4% of the wild-type level (Soderberg et al., 1977).

CCL16-B9 is in a different complementation group (IV) from the other mutants selected as CO_2/asparagine auxotrophs (Soderberg et al., 1979). Maiti et al. (1981) found that one of their Gal$^-$ mutants (P12GX1) did not complement CCL16-B9; however, it did not exhibit low succinate oxidase activity.

A gene for succinate dehydrogenase has been mapped on human chromosome 1 using hamster–human cell hybrids selected for growth on galactose (Mascarello et al., 1980). The appearance of succinate dehydrogenase activity in hybrids correlated with the presence of human chromosome 1, identified by cytological and isozyme analysis. Segregation of chromosome 1 from the hybrids resulted in a loss of succinate dehydrogenase activity. Succinate dehydrogenase consists of two subunits, a peptide containing the flavin with a molecular weight of 70,000, and a peptide with a molecular weight of 27,000 (Davis and Hatefi, 1971). A 70,000-dalton peptide and a 100,000-dalton peptide were immunoprecipitated from solubilized and denatured Chinese hamster wild-type and mutant mitochondria and human mitochondria using antibody to the beef heart 70,000-dalton subunit and cells labeled with [^{35}S]methionine (Mascarello et al., 1980). The 100,000-dalton peptide might be a precursor of the 70,000-dalton subunit or a precursor of both the 70,000- and 27,000-dalton subunits. The presence of cross-reacting material in the mutant deficient for succinate dehydrogenase activity suggests that the mutation has occurred in the structural gene. A regulatory gene mutation or other change in gene expression might be expected to eliminate both 27,000- and 70,000-dalton subunits. Thus, a gene for the 70,000- or 27,000-dalton peptides of succinate dehydrogenase has been mapped on human chromosome 1.

4. Res$^-$ Mutants in Mitochondrial Protein Synthesis

Like other Res$^-$ mutants, V79-G7 and Gal 32 exhibit very low O_2 uptake in whole cells, and G7 has minimal overall citric acid cycle activity (Ditta et al., 1976; Malczewski and Whitfield, 1982). In contrast to other Res$^-$ mutants, cytochrome *c* oxidase activity (<7%) and cytochrome aa_3 content (2%) are negligible in V79-G7 and Gal 32 (Ditta et al., 1977; Malczewski and Whitfield, 1982). Cytochrome *c* oxidase activity measures the terminal portion of the respiratory chain from cytochrome *c* to cytochrome aa_3 to O_2. V79-G7 differs from Gal 32 in its greatly decreased cytochrome *b* content and decreased oligomycin-sensitive ATPase activity, which measures the membrane-bound ATPase associated with oxidative phosphorylation (Ditta et al., 1977;

Malczewski and Whitfield, 1982). Since cytochrome *c* oxidase, oligomycin-sensitive ATPase, and apocytochrome *b* have mitochondrially encoded subunits, mitochondrial protein synthesis was measured.

Cells were labeled with radioactive amino acids in the presence of cycloheximide or emetine, or in the presence of cycloheximide or emetine plus chloramphenicol; mitochondria were isolated and peptides separated by SDS gel electrophoresis. Cycloheximide and emetine inhibit cytoplasmic protein synthesis; chloramphenicol inhibits mitochondrial protein synthesis. All mitochondrially synthesized proteins were drastically reduced in V79-G7; chloramphenicol-sensitive protein synthesis was 13% of the wild-type level (Ditta et al., 1977; Burnett and Scheffler, 1981). *In vitro* protein synthesis in isolated mitochondria was also decreased in this mutant.

A comparable number of mitochondria per cell are observed by electron microscopy in V79-G7 and wild-type cells; however, V79-G7 mitochondria have a unique ultrastructure distinguished by tubular cristae (Burnett and Scheffler, 1981). Tubular cristae and deficient mitochondrial protein synthesis are not observed in Complex I mutant, CCL16-B2 (Ditta et al., 1977; Burnett and Scheffler, 1981). The ultrastructural changes in V79-G7 are not reflected in the overall mitochondrial protein composition. When mitochondrially or nuclearly encoded mitochondrial proteins are separated by two-dimensional electrophoresis, V79-G7 is very similar to the wild-type except that two mitochondrially synthesized subunits of cytochrome oxidase are absent (Burnett and Scheffler, 1981). Since the number of mitochondria and mitochondrial protein composition are similar in the mutant and wild type, Burnett and Scheffler (1981) have proposed that mammalian mitochondria may be required for some function other than respiration, perhaps involving the mitochondrial ATPase. Using immunoprecipitation with antibody to the purified beef heart protein, Burnett and Scheffler (1981) found that at least two of the three largest Chinese hamster cytochrome oxidase subunits are mitochondrially synthesized, whereas the four smallest subunits are not. It is likely that their electrophoretic system did not separate subunits II and III. Both mitochondrially and cytoplasmically synthesized subunits of cytochrome *c* oxidase are drastically reduced in V79-G7. Assuming that the antibody recognizes the cytoplasmically synthesized subunits of cytochrome *c* oxidase, the subunits of cytoplasmic origin are either not synthesized or not incorporated into mitochondria grossly deficient in mitochondrial protein synthesis.

V79-G7 mitochondrial 16S and 12S rRNAs were synthesized normally and incorporated into ribosomal subunits with the same sedimentation as the wild type. Assuming that mitochondrial rRNA transcription is the same as mRNA transcription, Burnett and Scheffler (1981) concluded that the primary defect in V79-G7 is not mitochondrial transcription but mitochondrial translation, possibly a mutation in a ribosomal protein, or in a factor involved in initiation or elongation of mitochondrial protein synthesis, or transport of one of these cytoplasmic factors into the mitochondria. Recent studies of Montoya et al. (1982, 1983) have shown that in human cells ribosomal transcription differs from mRNA transcription in the rate, initiation site, and degree of polyadenylation. The results of hybridization of V79-G7 with an enucleated Chinese hamster cell resistant to chloramphenicol are consistent with a nuclear mutation in V79-G7. Spontaneous revertants which grow on galactose occur at a high frequency, suggesting that V79-G7 has a point mutation.

Mitochondrial protein synthesis is also deficient in Gal 32. Analysis of this mutant has revealed a differential reduction in mitochondrially synthesized peptides (Malczewski and Whitfield, 1982, 1984). Cytoplasmically synthesized mitochondrial proteins separated by two-dimensional electrophoresis are not affected in the mutant. There is a direct correlation between the activity of specific segments of the respiratory chain and the level of their mitochondrially encoded subunits (Malczewski and Whitfield, 1982, 1984). For example, cytochrome *c* oxidase activity and its mitochondrially encoded peptides are drastically reduced in Gal 32, at 1–17% of the wild-type level. In contrast, ATPase activity and its mitochondrially synthesized subunits are present at high levels (53–64% of wild type) in the mutant. Activity from succinate to cytochromes bc_1 to cytochrome *c*; cytochrome *b*, measured spectrophotometrically; and apocytochrome *b* (tentatively identified by comigration with the beef heart protein) are only marginally affected in the mutant.

Two mitochondrially synthesized peptides are immunoprecipitated with antibody to beef heart oligomycin-sensitive ATPase (Malczewski and Whitfield, 1984); the larger one corresponds to ATPase subunit 6 and the smaller corresponds in size to the human peptide encoded by unidentified reading frame A6L (Mariottini et al., 1983). Both peptides, like oligomycin-sensitive ATPase activity, are present at high levels in Gal 32; thus, it is likely that the small peptide is a subunit of the membrane-bound ATPase complex.

Using immunoprecipitation with antibody to the beef heart protein, Malczewski and Whitfield (1984) found that the three largest subunits (I, II, III) of CHL cytochrome *c* oxidase are synthesized within the mitochondria, as in other species (Sebald et al., 1973; Borst and Grivell, 1978; Rascati and Parsons, 1979; Hare et al., 1980; Kolarov et al., 1981). All three mitochondrially synthesized subunits of cytochrome *c* oxidase are substantially decreased in the mutant, but a cytoplasmically synthesized subunit (IV) is present at its normal size and level. Subunit IV was identified by Western immunoblotting, in which mitochondrial proteins are separated by SDS gel electrophoresis, blotted onto nitrocellulose paper, and then reacted with antibody to the beef heart protein. The cytoplasmically synthesized subunit IV of cytochrome *c* oxidase is proteolytically processed from a larger precursor in rat liver (Schmelzer and Heinrich, 1980). If CHL cells are similar to rat liver, the presence of subunit IV in Gal 32 at its proper size indicates that synthesis, proteolytic processing, and association with the mitochondrion are occurring in the mutant in the absence of synthesis of the mitochondrial subunits.

In addition to drastically decreased cytochrome *c* oxidase activity, Gal 32 has negligible Complex I activity; rotenone-sensitive NADH-Q_1 reductase activity is 1% of the wild-type level (Malczewski and Whitfield, 1982). Five subunits of Complex I, identified by Western immunoblotting using antibody to the beef heart protein, are greatly diminished in the mutant; three of these subunits comigrate on SDS gel electrophoresis with mitochondrially synthesized peptides which are diminished in the mutant (Malczewski and Whitfield, 1984). Biochemical and genetic data suggest that a single nuclear mutation is responsible for the dual defect in Complex I and cytochrome *c* oxidase activities in Gal 32, especially since these multiple defects are simultaneously corrected in a revertant cell line that was selected for its ability to grow on galactose. It is clear that the Gal 32 mutation decreases mitochondrially synthesized subunits of cytochrome *c* oxidase but not a cytoplasmically synthesized subunit. In Gal 32, the activity of specific segments of the respiratory chain is related directly to the level of their mitochondrially encoded subunits. Therefore, Malczewski and Whitfield (1984) concluded that Complex I may contain a previously unrecognized mitochondrially synthesized subunit(s) and that the lowered activities of Complex I and cytochrome *c* oxidase in the mutant are due to decreased levels of their mitochondrially encoded subunits. Although indirect, this is the first evidence that mammalian Complex I might contain a mitochondrially synthesized subunit. This conclusion is supported by the cytoplasmic mouse cell mutant which is specifically deficient in Complex I activity and resistant to rutamycin (Lichtor et al., 1979; see Section III.F).

The primary defect responsible for the decreased levels of mitochondrially synthesized proteins in Gal 32 is not known. A defect in a factor involved directly in mitochondrial protein synthesis seems unlikely since certain peptides are present at high levels in the mutant. A possible explanation is a defect in a nuclease involved in processing the polycistronic mitochondrial RNA transcript. Although RNA processing enzymes are known to recognize secondary and tertiary structures, certain sequences and structures are more efficiently processed than others (Guarneros et al., 1982; Gegenheimer and Aprion, 1981). The differential reduction in levels of different mitochondrially synthesized proteins in Gal 32 might be explained by the frequency of a particular amino acid or codon in a gene and the efficiency of cleavage of the corresponding tRNA as well as the efficiency of cleavage of the corresponding mRNA. For example, from the mouse mitochondrial DNA sequence, ATPase subunit 6 (sequence molecular weight 25,100) has many fewer cysteine and tyrosine residues than cytochrome oxidase subunits II and III (sequence molecular weights 26,000 and 29,900 (Bibb et al., 1981). Therefore, if $tRNA^{Cys}$ and $tRNA^{Tyr}$ were cleaved less efficiently, the synthesis of cytochrome oxidase subunits II and III would be significantly more decreased than that of ATPase subunit 6, as observed in Gal 32. Interestingly, the putative genes (ATPase 6 and A6L) for the peptides immunoprecipitated with ATPase antibody are adjacent to each other and give rise to a single mature RNA transcript in human cells (Anderson et al., 1981; Montoya et al., 1981). Perhaps this mRNA transcript is cleaved more efficiently in the mutant resulting in high levels of the ATPase subunit(s).

Another mutant, Gal 5, from a different complementation group is quite similar to Gal 32 in having low cytochrome *c* oxidase activity, decreased cytochrome aa_3 content, and negligible Complex I activity. Like V79-G7, oligomycin-sensitive ATPase activity is also reduced in Gal 5. It is highly likely that Gal 5 is also deficient in mitochondrial protein synthesis.

Res^- mutants deficient in mitochondrial protein synthesis (10% or less of the wild-type level) have been isolated from a human cell line, VA_2-B, after reduction of mtDNA with ethidium bromide and mutagenesis with agents specific for mtDNA (Wiseman and Attardi, 1979). Cells resistant to antimycin or requiring high concentrations of glucose for growth were selected. These mutants are very similar to V79-G7 except that the resistance is cytoplasmically inheritied. Further investigations of these mithochondrial protein synthesis mutants will contribute to our understanding of mitochondrial biogenesis in terms of the assembly of respiratory complexes from cytoplasmically and mitochondrially synthesized subunits, mitochondrial transcription and translation, and the function of peptides encoded by the unidentified reading frames in mammalian mtDNA.

III. MUTANTS RESISTANT TO MITOCHONDRIAL INHIBITORS

Mutants of mitochondrial or nuclear origin have been isolated from mammalian cells that are resistant to inhibitors of mitochondrial protein synthesis (chloramphenicol, erythromycin, carbomycin), the respiratory chain (rotenone, antimycin), and oxidative phosphorylation (oligomycin, rutamycin, venturicidin) (Table 19.2). In general, mutants resistant to mitochondrial inhibitors are the following types: (1) mutation in a protein or rRNA associated with inhibitor binding; as a result, *in vitro* activity is resistant to the inhibitor; cross-resistance to closely

TABLE 19.2
Chinese Hamster Cell Mutants Resistant to Mitochondrial Inhibitors

Cell Line[a]	Pretreatment[b]	Selection[c]	Cross Resistance[d]	Inheritance[e]	Other Properties[f]	Reference[g]
CHL V79						
201,204		CAP^{100},HG		C,cyb		1
CHEF/16						
213-21-3		CAP^{50},Pyr		C,cyb		2
CHEF/18						
294-7	MNNG	CAP^{50},Pyr,HG		C,cyb		2
CHL V79						
5-3		CAP^{100},Pyr		C,cyb,R6G		3,4
CHO BT_3		TEV^{100},1 mo	CAP	C,cyb	$MtPS^R$	5
CHL V79						
CR77A 1-17		CAP^{50},Pyr,HG		C,cyb	Pyr-dep	6
CR77A 23-28		CAP^{50},HG		C,cyb		6
CHL V79 703-3	EMS	$ANT^{0.1}$,HG,3 mo		C,cyb		7
CHO AuxB1[h]						
Ant^r1	EMS	ANT^5,Pyr,2 wk	OLI,RUT	N,rec	O_2^S	8
Olg^r1,30	EMS	OLI^1,Pyr,2 wk	PEL,OSS		Gly ↑	8
Ven^r5	EMS	VEN^5,Pyr,2 wk	EFR,AUR			8
Rut^r10	EMS	RUT^5,Pyr,2 wk	VEN,ANT,CAP			8
CHO Pro^-						
OLI^R8.1	EtBr,EMS	$OLI^{0.001,0.005}$ Pyr,3–4 wk	RUT	C,cyb	$ATPase^R$	9

CHO Pro$^-$ OLI^{R}2.2	EtBr,EMS	OLI$^{0.001,0.005}$ Pyr,3–4 wk	OSS	C,cyb + N,dom	ATPase^R Res^S	10
			EFR,LEU	N,dom		
			VEN,ANT	N,dom		
			CAP,PEL	dom		
			RUT,ROT			
CHL V79 w.t.			OLI,OSS	C,cyb + N,dom	ATPase^R	11
			EFR,LEU			
			PEL			

[a]CHL, Chinese hamster lung; CHEF/18, nontumorigenic Chinese hamster fibroblast cell line; CHEF/16, tumorigenic Chinese hamster fibroblast cell line; CHO, Chinese hamster ovary; Ant^r, antimycin resistant; Olg^r, oligomycin resistant, Ven^r, venturicidin resistant; Rut^r, rutamycin resistant; w. t., wild type.

[b]MNNG, *N*-methyl-*N'*-nitro-*N*-nitrosoguanidine; EMS, ethylmethane sulfonate; EtBr, ethidium bromide.

[c]Superscripts, concentrations in μg/ML; CAP, chloramphenicol; media contained 5.5 m*M* glucose except HG (high glucose) media was 25 m*M*; Pyr, pyruvate; TEV, tevenel; ANT, antimycin A; OLI, oligomycin; VEN, venturicidin; RUT, rutamycin; mo, month; wk, weeks.

[d]Cross resistance in plating experiments. See abbreviations in footnote *c*. PEL, peliomycin; OSS, ossamycin; EFR, efrapeptin; AUR, aurovertin D; LEU, leucinostatin; ROT, rotenone.

[e]C, cytoplasmic; cyb, cybrid (dominant); R6G, rhodamine-6G; N, nuclear; rec, recesive; dom, dominant

[f]Other properties. MtPs^R, resistant to inhibitors of mitochondrial protein synthesis in mitochondria; Pyr-dep, CAP^R is dependent on pyruvate in medium; O_2^S, oxygen consumption in whole cells is sensitive to inhibition by ANT, OLI, RUT; Gly ↑, increased rate of glycolysis; ATPase^R, resistant to inhibitors of mitochondrial ATPase in mitochondria or submitochondria. Res^S, respiration in mitochondria is sensitive to ANT and ROT.

[g]References: 1, Yen and Harris (1978). 2, Howell and Sager (1978). 3, Ziegler and Davidson (1979). 4, Ziegler and Davidson (1981). 5, Yatscoff et al. (1981). 6, Howell (1983). 7, Harris (1978). 8, Lagarde and Siminovitch (1979). 9, Breen and Scheffler (1980). 10, Breen (1982). 11, Simmons and Breen (1983).

[h]CHO Aux Bl Ant^r, Olg^r, Ven^r, Rut^r all have the same cross resistance, inheritance, and other properties. CHO AuxBl is about 3.5-fold more resistant to OLI than CHO Pro$^-$.

related inhibitors is observed; (2) respiratory deficiency; mutant obtains all of its energy from glycolysis and may be cross resistant to a variety of mitochondrial inhibitors; (3) increased glycolysis; mutant can obtain all of its energy from glycolysis even though the respiratory chain is functioning and is cross-resistant to a variety of mitochondrial inhibitors. Examples of these mutants will be presented in this section.

A. Isolation of Cytoplasmically Inherited Mutants

In lower eukaryotes, mutants in the mitochondrial genome have been important in mapping mitochondrial genes and studying their transmission, recombination, and segregation (Birky, 1978; Borst and Grivell, 1978). A number of procedures have been employed to isolate mammalian cytoplasmic mutants; however, it has not been conclusively shown that the frequency of mutants is increased. One difficulty in isolating mammalian cytoplasmic mutants is the large number (over 1000) of mtDNA molecules per cell. Procedures to decrease the mtDNA are treatment with ethidium bromide (Spolsky and Eisenstadt, 1972; Wiseman and Attardi, 1979); the use of minicells, which contain a nucleus and 10% of the cytoplasm and are produced by enucleation (Kuhns and Eisenstadt, 1979); and labeling thymidine kinase deficient (TK^-) cells with BrdU followed by irradiation, which specifically destroys mtDNA since the mitochondrial TK is functioning (Croizat and Attardi, 1975; Lichtor and Getz, 1978).

Another factor to be considered in the isolation of cytoplasmically inherited mutations is specific mutagenesis of mtDNA. Strategies that have been employed involve mutagenizing mtDNA with BrdU in TK^- cells (Wallace and Freeman, 1975; Yatscoff et al., 1981) or with Mn^{2+} (Wiseman and Attardi, 1979), or mutagenizing replicating mtDNA while inhibiting nuclear DNA replication with cycloheximide (Wiseman and Attardi, 1979).

The conditions used for selection are particularly important in the isolation of cytoplasmically inherited mutations. Prolonged exposure (weeks to months) to the selective medium while the cell is dividing is optimal to allow time for the replication and segregation of the mutant mtDNA. In the presence of chloramphenicol (CAP), antimycin, or oligomycin, more cell doublings occur with pyruvate and high glucose (>5 m*M*) than with lower glucose (1 m*M*), galactose, mannose, or high glucose without pyruvate (Howell and Sager, 1979; Ziegler and Davidson, 1979; Harris, 1980). As a result, CAP^R cells have been isolated at a higher frequency (1 in 10^4 CHL-V79 cells, 1 in 10^5 mouse SVT2 or A31 cells) from medium containing pyruvate and high glucose than from medium containing high glucose and no pyruvate (1 in 10^5 CHL-V79 cells, less than 1 in 10^6 mouse SVT2 or A31 cells) (Howell and Sager,

1979; Howell, 1983). No CAP^R CHL cells were isolated from medium containing galactose and CAP (Ziegler and Davidson, 1979).

Tevenel, the sulfamoyl analog of CAP, inhibits mitochondrial protein synthesis to the same extent as CAP, but inhibits Complex I activity directly 10 times less than CAP (Freeman and Haldar, 1968; Freeman, 1970). Perhaps, for this reason Tevenel permits more cell doublings than CAP (Fettes et al., 1972) and results in the isolation of CAP^R mutants from mouse LM(TK^-) cells at a very high frequency (6 in 10^3 cells) (Wallace and Freeman, 1975). In these cases, the frequency of CAP^R cells may reflect the rate of segregation of the mutant mtDNA rather than the mutation frequency.

Considering the energy sources with different sugars and the role of the citric acid cycle in producing nutrients, the growth of cells in the presence of mitochondrial inhibitors and various media is explainable. Mitochondrial inhibitors prevent oxidative phosphorylation, respiratory chain activity, and citric acid cycle activity. In the absence of citric acid cycle activity, aspartate is no longer produced from glutamine but is probably synthesized from pyruvate and CO_2 (involving pyruvate carboxylase or the malic enzyme and oxalacetate-aspartate transaminase; see Fig. 19.2). Therefore, more cell doublings occur in the presence of pyruvate than in its absence. In CHL V79 cells asparagine can substitute for pyruvate (Howell, 1983). Cells growing on low glucose, galactose, or other alternate sugars cannot survive without oxidative phosphorylation (Donnelly and Scheffler, 1976; Reitzer et al., 1979). Thus, it is not surprising that cells are more sensitive to CAP, antimycin, or oligomycin in the presence of low glucose, galactose, or mannose than in high glucose. CAP^R cells do not grow in CAP plus galactose (Howell and Sager, 1979; Ziegler and Davidson, 1979) as might be expected, since CAP inhibits Complex I activity directly and thereby oxidative phosphorylation (Freeman and Haldar, 1968).

B. Evidence for Cytoplasmic Inheritance

Cytoplasmic inheritance is usually demonstrated by transmitting the resistant phenotype when enucleated resistant cells are fused with nucleated sensitive cells (Bunn et al., 1974; Howell and Sager, 1978; Yen and Harris, 1978) or minicells (karyoplasts) (Shay, 1977). Enucleation is performed by disrupting the cell cytoskeleton with cytochalasin B and removing the nuclei by centrifuging cells in a Ficoll density gradient (Wigler and Weinstein, 1975) or centrifuging inverted, attached cells in a tube (Veomett et al., 1976). Using polyethylene glycol (Yatscoff et al., 1981) or inactivated Sendai virus (Bunn et al., 1974), the membrane-bound enucleated cells (cytoplast) are fused to nucleated cells containing a recessive, selectable mutation (such as TK^- or thioguanine-resistant, TG^R) and the cybrids are selected in media that kills both parents and the tetraploid hybrids. For example, fusing CAP^RTG^S enucleated cells with a CAP^STG^R cells and plating in CAP plus TG specifically selects the cybrids (CAP^RTG^R) since CAP resistance is dominant.

One problem with using a recessive marker (TG^R) in the recipient nucleated cell is that the enucleated cell contains a dominant protein that will initially sensitize the cybrid to the selective conditions and decrease the cybrid frequency. Using a dominant marker, resistance to 5,6-dichloro-1-β-D-ribofuranosylbenzimidazole (DRB), in the recipient nucleated Chinese hamster ovary (CHO) cells increased the cybrid frequency 16-fold compared to a recessive marker (3.9×10^{-2} and 2.4×10^{-3}) (Yatscoff et al., 1981). Since the tetraploid hybrid was able to grow in the DRB-CAP selective medium, any remaining DRB^SCAP^R nucleated cells were separated from the smaller enucleated cells by passage through a 5-μm unipore filter.

It is important to characterize the CAP^R cybrid. Abnormal chromosomes, other unselected markers, and isozyme variants have been used to verify the nuclear origin of these intraspecific cybrids (Bunn et al., 1974). Biochemical properties or cross resistance to other drugs might demonstrate that CAP resistance in the cybrid is characteristic of the donor CAP^R cell.

Cytoplasmic transfer of mtDNA between different species is usually not observed, except with closely related mouse species (Ho and Coon, 1979), because the nucleus of one species does not maintain the replication of mtDNA from another species [for review see Wallace (1982)]. Thus, interspecific hybrids generally retain the mtDNA of the parent whose chromosomes are retained (DeFrancesco et al., 1980). In crosses between closely related species (mouse–Chinese hamster or rat–mouse) where both sets of chromosomes are retained, mtDNAs from both parents are also retained; however, there is no evidence for mtDNA recombination (Hayashi et al., 1982; DeFrancesco, 1983; Zuckerman et al., 1984).

Cytoplasmic inheritance has also been confirmed by using rhodamine-6G (R6G) (Ziegler and Davidson, 1981). R6G is a toxic fluorescent dye which preferentially binds to the mitochondrial membrane (Johnson et al., 1980) and inhibits oxidative phosphorylation and adenine nucleotide translocation (Gear, 1974). As a result, R6G specifically inactivates or depletes the mtDNA. Although R6G is toxic to cells, R6G-treated CHL V79 cells are rescued by fusion with untreated cells or untreated cytoplasts (Ziegler and Davidson, 1981). When CAP^R R6G-treated CHL cells are fused with untreated CAP^S Chinese hamster cells or cytoplasts, the resulting hybrids or cybrids are CAP^S compared to hybrids formed from untreated CAP^R cells. Therefore, R6G inactivation of the dominant resistant phenotype in hybrids or cybrids is evidence for cytoplasmic inheritance.

Cytoplasmically inherited mutations often exhibit mitotic segregation from cybrids or hybrids in the absence of selective medium without loss of chromosomes (Bunn et al., 1977; Wallace et al., 1977; Yatscoff et al., 1981). For this reason, the cybrid frequency is highest when selective medium is applied immediately after cybrid formation (Yatscoff et al., 1981). The degree of mitotic segregation is variable and may reflect the initial ratio of mutant mtDNA to normal mtDNA in the cybrid. For example, cybrids formed from less resistant CAP^R mutants exhibit more mitotic segregation than cybrids from highly resistant CAP^R mutants (Wallace et al., 1977; Wallace, 1981).

Restriction-endonuclease site polymorphisms in the mtDNA of mouse subspecies (Ho and Coon, 1979) or human strains (Wallace, 1981) have demonstrated that CAP resistance is associated with the mtDNA. Using two human cell lines with different mtDNA restriction endonuclease cleavage patterns, Wallace (1981) quantitated the levels of mtDNA from each parent in cybrids and hybrids. There was a 1.5- to 10-fold excess of the mtDNA from the CAP^R parent compared to the CAP^S parent in cybrids and hybrids formed from CAP^S and CAP^R cells and selected for growth in CAP. Therefore, the expression of CAP resistance is at the cellular level and depends on the ratio of CAP^R to CAP^S mtDNA. CAP^S mtDNA remained in the hybrids or cybrids after prolonged growth (50 doublings) in CAP and explains why mitotic segregation of CAP resistance can occur in the absence of selection.

C. Chloramphenicol-Resistant Mutants

Mutants resistant to CAP, an inhibitor of mitochondrial protein synthesis, have been isolated from a number of mammalian cell lines; mouse (Bunn et al., 1974), human (Spolsky and Eisenstadt, 1972), and Chinese hamster (Howell and Sager, 1978; Yen and Harris, 1978; Ziegler and Davidson, 1979; Yatscoff et al., 1981; Howell, 1983) (Table 19.2). In most cases, CAP resistance is cytoplasmically inherited due to a mutation in the mtDNA [see Wallace (1982) for a comprehensive review].

CAP inhibits the peptidyl transferase reaction of mitochondrial protein synthesis by binding to a large mitochondrial ribosomal protein and blocking the binding of the charged tRNA (O'Brien et al., 1980). Mitochondrial protein synthesis measured *in vitro* (Spolsky and Eisenstadt, 1972; Yatscoff et al., 1981) or *in vivo* (Siegel et al., 1976) is no longer sensitive to CAP in most CAP^R mutants. In contrast, cytoplasmically inherited human VA_2-B mutants selected as Res^- or antimycin A resistant are deficient in mitochondrial protein synthesis and are also CAP^R (Wiseman and Attardi, 1979). Cross resistance to other inhibitors of mitochondrial ribosomes (carbomycin or mikamycin) is observed in some CAP^R cells (Siegel et al., 1976).

Different single base changes at the 3′ end of the large (16S) mitochondrial rRNA gene are responsible for CAP resistance in three human and two mouse cell lines (Blanc et al., 1981a,b; Kearsey and Craig, 1981). In yeast, mouse, and human cells, base changes conferring CAP^R occur in two regions of the large mitochondrial rRNA gene which are highly conserved from *E. coli* to human mtDNA (Dujon, 1980; Blanc et al., 1981a,b; Kearsey and Craig, 1981). The same T to C transition occurs in three CAP^R mammalian cell lines, one nucleotide removed from the yeast A to C transversion. The two highly conserved rRNA regions can be brought in close proximity by a stem-loop structure. It is likely that the highly conserved regions in the large rRNA are associated with a protein(s) involved in the peptidyl transferase reaction and in binding CAP. Howell (1983) finds that a small (9 base pair) insertion, probably into the 16S rRNA, is associated with CAP resistance in mouse cells.

When CAP^S and CAP^R cells are fused, do their mtDNAs remain in separate mitochondria or do they exist within the same mitochondrion? To answer this question, Oliver and Wallace (1982) examined the expression of two mitochondrially synthesized human variant proteins (MV1 and MV2) in cybrids and hybrids (MV1 CAP^R × MV2 CAP^S and vice versa). Without CAP, the level of variant proteins correlated with the ratio of their parental mtDNAs, indicating that these variant proteins are encoded by the mtDNA. In the presence of CAP, the variant protein encoded by the CAP^S mtDNA was still expressed, even when the CAP^S mtDNA was in great excess. Therefore, CAP^S and CAP^R mtDNAs are mixed, probably by mitochondrial fusion, allowing mRNAs encoded by CAP^S mtDNA to be translated by CAP^R ribosomes. It is not clear if recombination occurs between mammalian mtDNA as is the case with yeast (Lewin et al., 1979). These results also illustrate that CAP^R is codominant over CAP^S.

D. Erythromycin- and Carbomycin-Resistant Mutants

The macrolide antibiotics erythromycin (ERY), carbomycin (CAR), and spiramycin (SPI) inhibit mitochondrial protein synthesis. There is evidence that ERY and CAR bind to the same site in the large ribosomal subunit (Pestka, 1971). ERY and SPI resistance are cytoplasmically inherited in yeast and map in the region of the large rRNA subunit (Borst and Grivell, 1978).

ERY^R mutants have been isolated from HeLa cells (Doersen and Stanbridge, 1979, 1982) and mouse $LMTK^-$ cells (Molloy and Eisenstadt, 1979), and a CAR^R mutant from mouse cells (Bunn and Eisenstadt, 1977) after treatment with ethidium bromide or EMS and selecting in the presence of ERY or CAR. Some mouse cell lines are naturally ERY^R (Molloy and Eisenstadt, 1979). In ERY^R cells, mitochondrial protein synthesis measured *in vivo* [HeLa (Doersen and Stanbridge, 1979, 1982) and mouse (Molloy and Eisenstadt, 1979)] or *in vitro* [HeLa

(Doersen and Stanbridge, 1979, 1982)] is resistant to ERY and cross resistant to CAR and SPI (mouse cells) or CAR (HeLa cells). ERY^R mutants are not cross resistant to CAP.

When tested by cybrid formation, ERY resistance was dominant and cytoplasmically inherited in one HeLa cell mutant (ERY^R 2301) (Doersen and Stanbridge, 1979). ERY resistance in other HeLa cell mutants (Doersen and Stanbridge, 1982) and in mouse cells (Molloy and Eisenstadt, 1979) and CAR resistence in the mouse cell mutant (Bunn and Eisenstadt, 1977) were not cytoplasmically inherited in cybrids: instead, they appear to be nuclear.

E. Mutants Resistant to Antimycin, HQNO, Myxothiazol

Antimycin A (ANT), 2-*N*-heptyl-4-hydroxy-quinoline-*N*-oxide (HQNO), myxothiazol (MYX), and funiculosin all inhibit respiratory-chain O_2 uptake by binding to cytochrome *b* (Rieske, 1980; Thierbach and Reichenbach, 1981). Resistance of these cytochrome-*b*-binding compounds in mammalian cells has been associated with (1) an increased rate of glycolysis, or (2) respiratory deficiency, or (3) an alteration in cytochrome *b*. LaGarde and Siminovitch (1979) isolated ANT^R CHO cell mutants that are cross resistant to inhibitors of oxidative phosphorylation in plating experiments; however, O_2 consumption in intact mutant cells is not greatly altered and is still sensitive to these inhibitors (Table 19.2). The resistance might be explained by an increased rate of glycolysis. In contrast, the ANT^R mutants selected by Wiseman and Attardi (1979) in human VA_2-B cells are cytoplasmic, cross resistant to CAP, and deficient in respiration and mitochondrial protein synthesis.

ANT^R mutants have been isolated from EMS-treated CHL V79 cells (Harris, 1978) and from mouse LA9 cells (Howell et al., 1983a) by prolonged selection (2–3 months) with ANT in pyruvate-free medium. In parental cells, growth is barely inhibited by ANT in the presence of pyruvate (Harris, 1980; Howell et al., 1983a). ANT resistance in the CHL (Harris, 1978) and mouse (Howell et al., 1983a) cell mutants is codominant and cytoplasmically inherited when cybrids were formed.

In contrast to other ANT^R mammalian cell mutants, the ANT^R mouse cell mutant appears to be specifically altered in cytochrome *b* (Howell et al., 1983a). O_2 uptake in whole cells of the mouse ANT^R mutant is the same level as in the ANT^S parent, but resistance to inhibition by ANT is dramatically increased in the mutant. Likewise, in sonicated mitochondria, succinate-cytochrome *c* reductase activity, which measures electron transfer from succinate to Q to cytochromes bc_1 to cytochrome *c*, is not altered in the mutant; however, 100 times more ANT was required for 50% inhibition of activity in the ANT^R mutant compared to the ANT^S parent. Succinate-cytochrome *c* reductase activity is also cross resistant to inhibition by funiculosin in the ANT^R mutant (Howell et al., 1983*a*).

Howell et al. (1983*b*) have also selected mouse cell mutants resistant to HQNO and MYX. MYX and HQNO resistances are cytoplasmically transferred in cybrids. Succinate-cytochrome *c* reductase activity in sonicated mitochondria is greatly resistant to inhibition by MYX in the MYX^R mutant or to HQNO, ANT, and funiculosin in the $HQNO^R$ mutants. Cross resistance to other compounds which bind to cytochrome *b* suggests an alteration in this cytochrome. In the case of the $HQNO^R$ mutants and their cybrids, restriction digestion of the mtDNA revealed a new fragment corresponding to a small (10 base pair) insertion or duplication into the middle of the apocytochrome *b* gene (Howell et al., 1983b). The ratio of the mutant mtDNA restriction fragment to the normal fragment was higher in a $HQNO^R$ mutant with increased resistance.

F. Mutants Resistant to Inhibitors of Oxidative Phosphorylation

The mitochondrial membrane-bound ATPase complex functions in the synthesis of ATP which is coupled to respiration. In this complex, the soluble F_1 subunits, which catalyze ATP synthesis, protrude into the mitochondrial matrix and are connected to the hydrophobic membrane-bound F_0 subunits, which function in proton translocation (Senior, 1973). Efrapeptin and aurovertin D inhibit ATPase activity by binding to F_1, while leucinostatin, oligomycin (OLI), and venturicidin inhibit by binding to F_0, probably at different sites. Dicyclohexyl carbodiimide (DCCD), oligomycin and its analogs (rutamycin and peliomycin), and possibly ossamycin bind to the same proteolipid component, designated ATPase subunit 9 in yeast or DCCD-binding protein (Lardy et al., 1975; Linnett and Beechey, 1979; Sebald et al., 1979).

Most ATPase subunits are nuclearly encoded except ATPase subunit 6 which is mitochondrially encoded in yeast and mammals (Borst and Grivell, 1978; Anderson et al., 1981, 1982; Bibb et al., 1981). The DCCD-binding subunit is mitochondrially encoded in yeast (Borst and Grivell, 1978), nuclearly encoded in *Neurospora crassa* (Sebald et al., 1977), and probably nuclearly encoded in mammals. There is no sequence corresponding to ATPase subunit 9 in mammalian mtDNA (Anderson et al., 1981, 1982). Another membrane-bound ATPase subunit might be mitochondrially encoded in mammalian cells (Kuzela et al., 1980; Kolarov et al., 1981; Malczewski and Whitfield, 1984). In yeast, OLI resistance has been associated with mitochondrial mutations in ATPase subunit 6 (Macino and Tzagoloff, 1980) and ATPase subunit 9 (Sebald et al., 1979), as well as a nuclear mutation (Saunders et al., 1979).

Both nuclear and cytoplasmic mammalian cell mutants selected for resistance to oligomycin, rutamycin, or venturicidin have been isolated. Resistance has been associated with (1) a mitochondrial ATPase with increased resistance to inhibitors, or (2) an increased rate of glycolysis, or (3) respiratory deficiency, or (4) an alteration in mtDNA which may also be responsible for CAP resistance.

Breen and Scheffler (1980) isolated oligomycin-resistant mutants from CHO cells treated with ethidium bromide and EMS. One of these mutants, OLI^R 8.1, is 100-fold more resistant to OLI than the wild type and cross resistant to rutamycin but not to venturicidin or DCCD in plating experiments. ATPase activity is 32-fold more resistant to inhibition by OLI in the mutant mitochondria or submitochondria than in the wild type. Respiratory chain activity measured in mitochondria is normal in OLI^R 8.1. OLI resistance in this mutant is stable in the absence of selection and dominant in hybrid crosses with sensitive cells. Cytoplasmic inheritance of OLI resistance is observed; cybrids formed from OLI^R 8.1 cytoplasts and sensitive cells are as resistant to OLI as the mutant.

Another CHO mutant, OLI^R 2.2 (Breen, 1982) and naturally occurring Chinese hamster cell lines, lung V79 and bone marrow M3-1 (Simmons and Breen, 1983), exhibit extremely high OLI resistance which is inherited both cytoplasmically and nuclearly. OLI^R 2.2 is 50,000-fold and V79 and M3-1 are 10,000-fold more resistant to OLI than the CHO wild type in plating experiments. In sonicated mitochondria, resistance of the ATPase to inhibition by oligomycin is increased only 25-fold in OLI^R 2.2 or 100-fold in V79 compared to the CHO wild type with little differences in total ATPase activity. The V79 ATPase is more sensitive to heat inactivation than the CHO complex, indicating a difference in their structure. OLI resistance is inherited codominantly; hybrids formed from V79 (or OLI^R 2.2) and OLI^S CHO cells are resistant but not as resistant as the parental cells. V79 × OLI^S CHO cybrids or OLI^R 2.2 × OLI^S CHO cybrids are more resistant to OLI in plating experiments than OLI^S CHO cells but less resistant than the corresponding hybrids and much less resistant than the OLI^R parent. Similarly, in V79 × OLI^S CHO cybrids, OLI inhibition of ATPase activity and ATPase heat inactivation are intermediate between that for OLI^S CHO cells and the corresponding hybrid. Therefore, it was concluded that both nuclear and cytoplasmic determinants are responsible for oligomycin resistance of growing V79 and OLI^R 2.2 cells, and for the V79 mitochondrial ATPase.

The pattern of cross resistance in V79 is different from that in OLI^R 2.2. V79 is cross resistant to inhibitors of the mitochondrial ATPase (efrapeptin, ossamycin, peliomycin, and leucinostatin) but not to venturicidin, and other mitochondrial inhibitors (antimycin A, rotenone, and CAP). Cross resistance to leucinostatin, efrapeptin, and ossamycin is also codominant and due to both a nuclear and cytoplasmic gene. On the other hand, OLI^R 2.2 is cross resistant in plating experiments to inhibitors of oxidative phosphorylation (rutamycin, ossamycin, peliomycin, venturicidin, leucinostatin, and efrapeptin) as well as to other mitochondrial inhibitors (CAP, rotenone, and antimycin A). Respiratory activity is normal in OLI^R 2.2 mitochondria and still sensitive to rotenone and antimycin. In OLI^R 2.2, resistance to ossamycin, like OLI, is due to both codominant nuclear and cytoplasmic mutations; however, cross resistance to efrapeptin, leucinostatin, venturicidin, and antimycin resulted from a codominant nuclear mutation. Although not resistant to colchicine, the nuclear mutation in OLI^R 2.2 might affect the permeability of the plasma membrane.

After treatment with EMS (but not ethidium bromide), Lagarde and Siminovitch (1979) selected CHO mutants resistant to OLI, rutamycin, venturicidin, and antimycin. All of these mutants are cross resistant in plating experiments to inhibitors of oxidative phosphorylation, (OLI, rutamycin, peliomycin, ossamycin, venturicidin, efrapeptin, and aurovertin) and cross resistant to other mitochondrial inhibitors (antimycin A and CAP) but not to colchicine. Nevertheless, O_2 uptake in resistant whole cells is not affected and remains sensitive to these inhibitors. The resistant phenotype might result from an increased rate of glycolysis; the specific activities for certain glycolytic enzymes are increased when assayed *in vitro*. The recessive behavior in hybrids and the short period for selection suggested that the resistant mutation(s) are nuclearly encoded.

OLI^R or rutamycin-resistant mutants of nuclear or cytoplasmic origin have also been isolated from mouse cells. Kuhns and Eisenstadt (1979, 1981) have selected an OLI^R mouse ($LMTK^-$) cell line which is the result of a codominant nuclear mutation. The mutant mitochondrial ATPase exhibits decreased activity but increased resistance to OLI, DCCD, and venturicidin.

A rutamycin-resistant mouse cell mutant having a mitochondrial ATPase with increased resistance to rutamycin and leucinostatin and decreased total activity was isolated by Lichtor and Getz (1978). In addition, the mutant is respiratory deficient and its growth is resistant to mitochondrial respiratory-chain inhibitors, antimycin and rotenone, as might be expected since no energy is provided by oxidative phosphorylation (Lichtor et al., 1979). Rotenone-sensitive NADH-cytochrome *c* reductase activity is decreased in sonicated mutant mitochondria, while succinate-cytochrome *c* reductase and cytochrome *c* oxidase activities are not significantly affected, suggesting a specific defect in Complex I (Lichtor et al., 1979). Both rutamycin-resistance and Complex I deficiency are codominant and cytoplasmically inherited in cybrids. If Complex I contains a mitochondrially encoded subunit(s), as suggested by Malczewski and Whitfield (1984), then a mutation in this Complex I subunit and a mutation in a mitochondrially encoded subunit of the ATPase complex might explain the phenotype.

OLI^R mouse cell mutants were isolated in a multistep procedure: first, CAP^R cells dependent on pyruvate (PYR^{DEP}) to grow in CAP were selected; second, CAP^R cells not dependent on pyruvate (PYR^{IND}); third, $OLI^R CAP^R PYR^{IND}$ cells (Howell and Sager, 1979; Howell, 1983). Restriction endonuclease digestion of the $OLI^R CAP^R PYR^{IND}$ mtDNA revealed a new fragment corresponding to a small insertion (9 base pairs) or duplication probably in the 16S rRNA (Howell, 1983). The ratio of the mutant mtDNA to normal mtDNA increased at each step in going from CAP^S cells to $OLI^R CAP^R PYR^{IND}$ cells. In this case, the CAP^R, PYR^{IND}, OLI^R phenotypes appear to correspond to increasing proportions of the same mutant mtDNA; therefore, Howell (1983) has proposed a "threshold" model where mitochondrial phenotypes depend on the ratio of mutant mtDNA to normal mtDNA. The PYR^{IND} phenotype may reflect a higher level of mitochondrial protein synthesis and citric acid cycle activity compared to PYR^{DEP} (see Section III A); however, it is not clear how a mutation in the large rRNA confers OLI resistance.

Webster et al. (1982) have isolated an OLI^R human cell mutant that is codominant and nuclear. ATPase activity in disrupted mitochondria is partially resistant to OLI. In hybrids formed from OLI^R human cells and OLI^S mouse cells, OLI resistance is associated with two copies of human chromosome 10 and no detectable human mtDNA.

Freeman et al. (1983) have isolated a CHO cell line which is resistant to uncouplers of respiration from oxidative phosphorylation, but not cross-resistant to mitochondrial inhibitors. Uncoupler resistance, which is recessive in this pseudotetraploid cell line, is observed with cell growth, whole cell respiration, and respiration of isolated mitochondria.

IV. CONCLUSIONS

In summary, nuclear and cytoplasmic mutants affecting Complex I, succinate dehydrogenase, Q, Complex III, cytochrome *b*, mitochondrial ATPase, mitochondrial protein synthesis, and 16S rRNA have been isolated from mammalian cells. Mutants affecting the structural genes for cytochrome *c* oxidase, cytochromes c, c_1, and most citric acid cycle enzymes have not been reported in mammalian cells. There are still relatively few mammalian mitochondrial mutants. Many questions concerning mitochondrial biogenesis remain unanswered, such as the transmission, segregation, and recombination of mtDNA. Recombination between mtDNAs, and nuclear–mitochondrial interactions might be investigated with interspecific hybrids (DeFrancesco, 1983; Zuckerman et al., 1984). Mitochondrial transcription and translation, and the processing and import of cytoplasmically synthesized mitochondrial proteins are not fully understood. The molecular basis for the overlapping complementation pattern among respiration-deficient mutants is unknown and could be investigated by gene mapping and cloning. Further biochemical and genetic investigations of these mammalian mitochondrial mutants, including the identification of the mutation at the molecular level, will contribute to our understanding of the structure and function of the respiratory complexes, the coupling of respiration to ATP synthesis, the synthesis and assembly of respiratory complexes, and the function of proteins encoded by the unidentified reading frames in mtDNA.

REFERENCES

Anderson, S., Bankier, A. T., Barrell, B. G., de Bruijn, M. H. L., Coulson, A. R., Drouin, J., Eperon, I. C., Nierlich, D. P., Roe, B. A., Sanger, F., Schreier, P. H., Smith, A. J. H., Staden, R., and Young, I. G. (1981). *Nature* **290**, 457–465.

Anderson, S., de Bruijn, M. H. L., Coulson, A. R., Eperon, I. C., Sanger, F., and Young, I. G. (1982). *J. Mol. Biol.* **156**, 683–717.

Attardi, G., Cantatore, P., Chomyn, A., Crewes, S., Gelfand, R., Merkel, C., Montoya, J., and Ojala, D. (1982). In *Mitochondrial Genes* (P. Sloninski, P. Borst, and G. Attardi, eds.), Cold Spring Harbor Laboratory, Cold Spring Harbor, New York, pp. 51–71.

Aujame, L. and Freeman, K. B. (1979). *Nuc. Acid Res.* **6**, 455–469.

Battey, J. and Clayton, D. A. (1978). *Cell* **14**, 143–156.

Bibb, M. J., Van Etten, R. A., Wright, C. T., Walberg, M. W., and Clayton, D. A. (1981). *Cell* **26**, 167–180.

Birky, Jr., C. W. (1978). *Ann. Rev. Genet.* **12**, 471–512.

Blanc, H., Adams, C. A., and Wallace, D. C. (1981a). *Nucl. Acids Res.* **9**, 5785–5795.

Blanc, H., Wright, C. T., Bibb, M. J., Wallace, D. C., and Clayton, D. A. (1981b). *Proc. Natl. Acad. Sci. USA* **78**, 3789–3793.

Bogenhagen, D. F., Applegate, E. F., and Yoza, B. K. (1984). *Cell* **36**, 1105–1113.

Borst, P. and Grivell, L. A. (1978). *Cell* **15**, 705–723.

Breen, G. A. M. (1982). *Mol. Cell. Biol.* **2**, 772–781.

Breen, G. A. M. and Scheffler, I. E. (1979). *Somat. Cell Genet.* **5**, 441–451.

Breen, G. A. M. and Scheffler, I. E. (1980). *J. Cell Biol.* **86**, 723–729.

Brown, G. G. and Beattie, D. S. (1977). *Biochem.* **16**, 4449–4454.

Bunn, C. L. and Eisenstadt, J. M. (1977). *Somat. Cell Genet.* **3**, 611–627.

Bunn, C. L., Wallace, D. C., and Eisenstadt, J. M. (1974). *Proc. Natl. Acad. Sci. USA* **71**, 1681–1685.

Bunn, C. L., Wallace, D. C., and Eisenstadt, J. M. (1977). *Somat. Cell Genet.* **3**, 71–92.

Burnett, K. G. and Scheffler, I. E. (1981). *J. Cell Biol.* **90**, 108–115.

Chang, D. D. and Clayton, D. A. (1984). *Cell* **36**, 635–643.

Ching, E. and Attardi, G. (1982). *Biochem.* **21**, 3188–3195.

Chomyn, A., Mariottini, P., Gonzalez-Cadavid, N., Attardi, G., Strong, D. D., Trovato, D., Riley, M., and Doolittle, R. F. (1983). *Proc. Natl. Acad. Sci. USA* **80**, 5535–5539.

Chu, E. H. Y. (1974). *Genetics* **78**, 115–132.

Chu, E. H. Y., Sun, N. C., and Chang, C. C. (1972). *Proc. Natl. Acad. Sci. USA* **69**, 3459–3463.

Clayton, D. A. (1982). *Cell* **28**, 693–705.

Croizat, B. and Attardi, G. (1975). *J. Cell Sci.* **19**, 69–84.

Davis, K. A. and Hatefi, Y. (1971). *Biochem.* **10**, 2509–2516.

Day, C. E. and Scheffler, I. E. (1982). *Som. Cell Genet.* **8**, 691–707.

DeFrancesco, L., (1983). *Somat. Cell Genet.* **9**, 133–139.

DeFrancesco, L., Attardi, G., and Croce, C. M. (1980). *Proc. Natl. Acad. Sci. USA* **77**, 4079–4083.

DeFrancesco, L., Scheffler, I. E., and Bissell, M. J. (1976). *J. Biol. Chem.* **251**, 4588–4595.

DeFrancesco, L., Werntz, D., and Scheffler, I. E. (1975). *J. Cell Physiol.* **85**, 293–306.

DeKok, J. and Slater, E. C. (1975). *Biochim. Biophys. Acta* **376**, 27–41.

Ditta, G., Soderberg, K., Landy, F., and Scheffler I. E. (1976). *Somat. Cell Genet.* **2**, 331–344.

Ditta, G., Soderberg, K., and Scheffler, I. E. (1977). *Nature* **268**, 64–66.

Doersen, C.-J. and Stanbridge, E. J. (1979). *Proc. Natl. Acad. Sci. USA* **76**, 4549–4553.

Doersen, C.-J. and Stanbridge, E. J. (1982). In *Techniques in Somatic Cell Genetics* (J. W. Shay, ed.), Plenum Press, New York, pp. 139–157.

Donnelly, M. and Scheffler, I. E. (1976). *J. Cell. Physiol.* **89**, 39–52.

Dubin, D. T., Montoya, J., Timko, K. D., and Attardi, G. (1982). *J. Mol. Biol.* **157**, 1–19.

Dujon, B. (1980). *Cell* **20**, 185–197.

Fettes I. M., Haldar, D., and Freeman, K. B. (1972). *Can. J. Biochem.* **50**, 200–209.

Fincham, J. R. S. (1966). *Genetic Complementation*, Benjamin, New York.

Freeman, K. B. (1970). *Can. J. Biochem.* **48**, 469–478.

Freeman, K. B. and Haldar, D. (1968). *Can. J. Biochem.* **46**, 1003–1008.

Freeman, K. B., Yatscoff, R. W., Mason, J. R., Patel, H. V., and Buckle, M. (1983). *Eur. J. Biochem.* **134**, 215–222.

Gear, A. R. L. (1974). *J. Biol. Chem.* **249**, 3628–3637.

Gegenheimer, P. and Aprion, D. (1981). *Microbiol. Rev.* **45**, 502–541.

Gelfand, R. and Attardi, G. (1981). *Mol. Cell. Biol.* **1**, 497–511.

Grisi, E., Brown, T. A., Waring, R. B., Scazzocchio, C., and Davies, R. W. (1982). *Nucleic Acid Res.* **10**, 3531–3539.

Guarneros, G., Montanez, C., Hernandez, T., and Court, D. (1982). *Proc. Natl. Acad. Sci. USA* **79**, 238–242.

Gupta, R. S., Chan, D. T. H., and Siminovitch, L. (1978). *Cell* **14**, 1007–1013.

Hare, J. F., Ching, E., and Attardi, G. (1980). *Biochem.* **19**, 2023–2030.

Hare, J. F. and Hodges, R. (1982). *J. Biol. Chem.* **257**, 3575–3580.

Harris, M. (1978). *Proc. Natl. Acad. Sci. USA* **75**, 5604–5608.

Harris, M. (1980). *Somat. Cell Genet.* **6**, 699–708.

Hayashi, J. I., Gotoh, O., Tagashira, Y., Tosu, M., Sekiguchi, T., and Yoshida, M. C. (1982). *Somat. Cell Genet.* **8**, 67–81.

Heron, C., Smith, S., and Ragan, I. C. (1979). *Biochem. J.* **181**, 435–443.

Ho, C. and Coon, H. G. (1979). In *Extrachromosomal DNA, ICN-UCLA Symposium on Molecular and Cellular Biology* (D. J. Cummings, P. Borst, I. B. Dawid, S. M. Weissman, and C. F. Fox, eds), Academic Press, New York, Vol. XV, pp. 501–514.

Howell, N. (1983). *Somat. Cell Genet.* **9**, 1–24.

Howell, N., Bantel, A., and Huang, P. (1983*b*). *Somat. Cell Genet.* **9**, 721–743.

Howell, N., Huang, P., Kelliher, K., and Ryan, M. L. (1983*a*). *Somat. Cell Genet.* **9**, 143–163.

Howell, N. and Sager, R. (1978). *Proc. Natl. Acad. Sci. USA* **75**, 2358–2362.

Howell, N. and Sager, R. (1979). *Somat. Cell Genet.* **5**, 833–845.

Johnson, L. V., Walsh, M. L., and Chen, L. B. (1980). *Proc. Natl. Acad. Sci. USA* **77**, 990–994.

Kearsey, S. E. and Craig, I. W. (1981). *Nature* **290**, 607–608.

Kolarov, J., Kuzela, S., Wielburski, A., and Nelson, B. D. (1981). *FEBS Lett.* **126**, 61–65.

Kozak, M. (1978). *Cell* **15**, 1109–1123.

Kuhns, M. C. and Eisenstadt, J. M. (1979). *Somat. Cell Genet.* **5**, 821–832.

Kuhns, M. C. and Eisenstadt, J. M. (1981). *Somat. Cell Genet.* **7**, 737–750.

Kuzela, S., Luciakova, K., and Lakota, J. (1980). *FEBS Lett.* **114**, 197–201.

Lagarde, A. E. and Siminovitch, L. (1979). *Somat. Cell Genet.* **5**, 847–871.

Lardy, H., Reed, P., and Lin, C. H. C. (1975). *Fed. Proc.* **34**, 1707–1710.

Lewin, A. S., Morimoto, R., and Rabinowitz, M. (1979). *Plasmid* **2**, 155–181.

Lichtor, T. and Getz, G. S. (1978). *Proc. Natl. Acad. Sci. USA* **75**, 323–328.

Lichtor, T., Tung, B., and Getz, G. S. (1979). *Biochem.* **18**, 2582–2590.

Linnett, P. E. and Beechey, R. B. (1979). *Methods Enzymol.* **55**, 472–519.

Macino, G. and Tzagoloff, A. (1980). *Cell* **20**, 507–517.

Macreadie, I. G., Choo, W. M., Novitski, C. E., Marzuki, S., Nagley, P., Linnane, A. W., and Lukins H. B. (1982). *Biochem. Int.* **5**, 129–136.

Maiti, I. B., Comlan de Souza, A., and Thirion, J.-P. (1981). *Somat. Cell Genet.* **7**, 567–582.

Malczewski, R. M. and Whitfield, C. D. (1982). *J. Biol. Chem.* **257**, 8137–8142.

Malczewski, R. M. and Whitfield, C. D. (1984). *J. Biol. Chem.* **259**, 11103–11113.

Mariottini, P., Chomyn, A., Attardi, G., Trovato, D., Strong, D. D., and Doolittle, R. F. (1983). *Cell* **32**, 1269–1277.

Mascarello, J. T., Soderberg, K., and Scheffler, I. E. (1980). *Cytogenet. Cell Genet.* **28**, 121–135.

Molloy, P. L. and Eisenstadt, J. M. (1979). *Somat. Cell Genet.* **5**, 585–595.

Montoya, J., Ojala, D., and Attardi, G. (1981). *Nature* **290**, 465–470.

Montoya, J., Christianson, T., Levens, D., Rabinowitz, M., and Attardi, G. (1982). *Proc. Natl. Acad. Sci. USA* **79**, 7195–7199.

Montoya, J., Gaines, G. L., and Attardi, G. (1983). *Cell* **34**, 151–159.

Novitski, C. E., Macreadie, I. G.., Maxwell, R. J., Lukins, H. B., Linnane, A. W., and Nagley, P. (1983). In *Manipulation and Expression of Genes in Eukaryotes* (P. Nagley, A. W. Linnane, W. J. Peacock, and J. A. Pateman, eds.), Academic, Sydney, Australia, pp. 257–268.

O'Brien, T. W., Denslow, N. D., Harville, T. O., Hessler, R. A., and Matthews, D. E. (1980). In *The Organization and Expression of the Mitochondrial Genome* (A. M. Kroon and C. Saccone, eds.), Elsevier/North Holland Biomedical Press, Amsterdam, pp. 301–305.

Ohno, S. (1969). *Ann. Rev. Genet.* **3**, 495–524.

Ojala, D., Merkel, C., Gelfand, R., and Attardi, G. (1980). *Cell* **22**, 393–403.

Ojala, D., Montoya, J., and Attardi, G. (1981). *Nature* **290**, 470–474.

Oliver, N. A., Greenberg, B. D., and Wallace, D. C. (1983). *J. Biol. Chem.* **258**, 5834–5839.

Oliver, N. A. and Wallace, D. C. (1982). *Mol. Cell. Biol.* **2**, 30–41.

Pestka, S. (1971). *Ann. Rev. Microbiol.* **25**, 487–562.

Rascati, R. J. and Parsons, P. (1979). *J. Biol. Chem.* **254**, 1594–1599.

Reitzer, L. J., Wice, B. M., and Kennell, D. (1979). *J. Biol. Chem.* **254**, 2669–2676.

Rieski, J. S. (1980). *Pharmacol. Ther.* **11**, 415–450.

Saunders, G. W., Rank, G. H., Kustermann-Kuhn, B., and Hollenberg, C. P. (1979). *Mol. Gen. Genet.* **175**, 45–52.

Scheffler, I. E. (1974). *J. Cell. Physiol.* **23**, 219–230.

Schmelzer, E. and Heinrich, P. C. (1980). *J. Biol. Chem.* **255**, 7503–7506.

Sebald, W., Machleidt, W., and Otto, J. (1973). *Eur. J. Biochem.* **38**, 311–324.

Sebald, W., Sebald-Althanus, M., and Wachter, E. (1977). In *Mitochondria 1977: Genetics and Biogenesis of Mitochondria* (W. Bandlow, R. J. Scheweyen, K. Wolf, and F. Kandewitz, eds.), W. deGruyter and Co., Berlin, Vol. 2, pp. 433–440.

Sebald, W., Wachter, E., and Tzagoloff, A. (1979). *Eur. J. Biochem* **100**, 559–607.

Senior, A. E. (1973). *Biochim. Biophys. Acta* **301**, 249–277.

Shay, J. W. (1977). *Proc. Natl. Acad. Sci. USA* **74**, 2461–2464.

Siegel, R. L., Jeffreys, A. J., Sly, W., and Craig, I. W. (1976). *Exp. Cell Res.* **102**, 298–310.

Siminovitch, L. (1976). *Cell* **7**, 1–11.

Simmons, W. A. and Breen, G. A. M. (1983). *Somat. Cell Genet.* **9**, 549–566.

Singer, T. P. and Gutman, M. (1971). *Adv. Enzymol. Relat. Areas Mol. Biol.* **34**, 79–153.

Soderberg, K. L., Ditta, G. S., and Scheffler, I. E. (1977). *Cell* **10**, 697–702.

Soderberg, K., Mascarello, J. T., Breen, G. A. M., and Scheffler, I. E. (1979). *Somat. Cell Genet.* **5**, 225–240.

Soderberg, K., Nissinen, E., Bakay, B., and Scheffler, I. E. (1980). *J. Cell. Physiol.* **103**, 169–172.

Spolsky, C. M. and Eisenstadt, J. M. (1972). *FEBS Lett.* **25**, 319–324.

Sun, N. C., Chang, C. C., and Chu, E. H. Y. (1974). *Proc. Natl. Acad. Sci. USA* **71**, 404–407.

Sun, N. C., Chang, C. C., and Chu, E. H. Y. (1975). *Proc. Natl. Acad. Sci. USA* **72**, 469–473.

Thierbach, G. and Reichenbach, H. (1981). *Biochim. Biophys. Acta* **638**, 282–289.

Thirion, J.-P., Labrecque, R., and Vu, T. P. (1976). *J. Cell. Physiol.* **87**, 135–139.

Tzagoloff, A., Macino, G., and Sebald, W. (1979). *Ann. Rev. Biochem.* **48**, 419–441.

Van Etten, R. A., Bird, J. W., and Clayton, D. A. (1983). *J. Biol. Chem.* **258**, 10104–10110.

Van Etten, R. A., Michael, N. L., Bibb, M. J., Brennicke, A., and Clayton, D. A. (1982). In *Mitochondrial Genes* (P. Slonimski, P. Borst, and G. Attardi, eds.), Cold Spring Harbor, New York, pp. 73–88.

Van Etten, R. A., Walberg, M. W., and Clayton, D. A. (1980). *Cell* **22**, 157–170.

Veomett, G., Shay, J., Hough, P. V. C. and Prescott, D. M. (1976). In *Methods in Cell Biology* (D. M. Prescott, ed.), Academic Press, New York, Vol. XIII, pp. 1–6.

Walberg, M. W. and Clayton, D. A. (1983). *J. Biol. Chem.* **258**, 1268–1275.

Wallace, D. C. (1981). *Mol. Cell. Biol.* **1**, 697–710.

Wallace, D. C. (1982). In *Techniques in Somatic Cell Genetics* (J. W. Shay, ed.), Plenum Press, New York, pp. 159–187.

Wallace, D. C., Bunn, C. L., and Eisenstadt, J. M. (1977). *Somat. Cell Genet.* **3**, 93–119.

Wallace, R. B. and Freeman, K. B. (1975). *J. Cell Biol.* **65**, 492–498.

Webster, K. A., Oliver, N. A., and Wallace, D. C. (1982). *Somat. Cell Genet.* **8**, 223–244.

Whitfield, C. D., Bostedor, R., Goodrum, D., Haak, M., and Chu, E. H. Y. (1981). *J. Biol. Chem.* **256**, 6651–6656.

Wice, B. M., Reitzer, L. J., and Kennell, D. (1981). *J. Biol. Chem.* **256**, 7812–7819.

Wigler, M. H. and Weinstein, I. B. (1975). *Biochem. Biophys. Res. Commun.* **63**, 669–674.

Wiseman, A. and Attardi, G. (1979). *Som. Cell Genet.* **5**, 241–262.

Yatscoff, R. W., Aujame, L., Freeman, K. B., and Goldstein, L. (1978). *Can. J. Biochem.* **56**, 939–942.

Yatscoff, R. W., Mason, J. R., Patel, H. V., and Freeman, K. B. (1981). *Somat. Cell Genet.* **7**, 1–9.

Yen, R. C. K. and Harris, M. (1978). *Cell Struct. Funct.* **3**, 79–88.

Young, I. G. and Wallace, B. J. (1976). *Biochim. Biophys. Acta* **449**, 376–385.

Yoza, B. K. and Bogenhagen, D. F. (1984). *J. Biol. Chem.* **259**, 3909–3915.

Ziegler, M. L. and Davidson, R. L. (1979). *J. Cell Physiol.* **98**, 627–635.

Ziegler, M. L. and Davidson, R. L. (1981). *Somat. Cell Genet.* **7**, 73–88.

Zielke, H. R., Ozand, P. T., Tildon, J. T., Sevdalian, D. A., and Cornblath, M. (1978). *J. Cell. Physiol.* **95**, 41–48.

Zuckerman, S. H., Solus, J. F., Gillespie, F. P., and Eisenstadt, J. M. (1984). *Somat. Cell Mol. Genet.* **10**, 85–92.

B. Cell Structure and Behavior

CHAPTER 20

CELL CYCLE MUTANTS

Menashe Marcus
Department of Genetics
The Hebrew University of Jerusalem
Jerusalem, Israel

I. INTRODUCTION

The rule of thumb for a geneticist says that mutants must be acquired because they allow the identification of normal traits and the analysis of complex biological systems. In fact, there are good reasons for the existence of the above unwritten law. The fantastic progress in molecular biology and genetics is due to a very large extent to the elegant isolation and usage of mutants of *Drosophila*, bacteria, bacteriophages, and many other organisms.

Isolation of mutants from mammalian cells in cell culture had for many years been a difficult and futile task. In the last 15 years or so this situation has changed dramatically. The number of isolated mutants has increased considerably (Siminovitch, 1976). In addition, new procedures for selection and screening of mutants have been introduced (Hochstadt et al., 1981). Several widely used cell lines including many Chinese hamster cell lines show functional hemizygosity for many genes located on different chromosomes in addition to the X chromosome (Gupta, 1980). This hemizygosity has been shown in many cases to be due to gene inactivation (see Chapter 28; Chasin and Urlaub, 1975; Siminovitch, 1976,1979; Simon et al.,1982). As a result, recessive mutations are expressed and can be selected. Siminovitch presents a thorough discussion of this subject in Chapter 29. Let us then agree that isolation and usage of mutants is feasible and desirable in the analysis of the mammalian cell cycle. What kinds of mutants should we seek? To answer this question we have to try and decide in what ways the cell cycle interests us.

II. CELL CYCLE IN EUKARYOTES

The cell cycle constitutes a contiguous and complex array of processes whose controls are not yet understood. For convenience, investigators of cell cycle divide the cell cycle into four phases: Mitosis (M), in which each cell divides into two daughter cells; Gap 1 (G1), which starts at the end of Mitosis and ends with the beginning of DNA synthesis; DNA Synthesis (S), during which nuclear DNA is replicated; Gap 2 (G2), which starts at the end of phase S and ends at the beginning of the mitotic prophase. Many excellent books and reviews which concentrate on the cell cycle have been written (Mitchison, 1971; Baserga, 1976; Prescott, 1976a,b; Pardee et al., 1978; Hochhauser et al., 1981; Yanishevsky and Stein, 1981). I shall therefore limit myself to a very short review of the main problems which face the student of the cell cycle who wishes to study this intriguing subject with the use of mutants.

A. Phase G1

In animal cells and in many other eukaryotes, this is the phase in which the cell cycle is controlled; namely, the decision is made in G1 to go through another cycle or to stop (Hochhauser et al., 1981; Yanishevsky and Stein, 1981). The notion that G1 is the phase in which cell proliferation is controlled is based on several observations. First, the length of this phase is highly variable, while the length of the other phases is quite constant. In addition, the length of G1 alone is affected dramatically by the environment. Thus, under starvation conditions the cells arrest at G1 and stay viable for some time. Most differentiated cells in the organism which do not cycle are also in G1 judging from their content of DNA per cell. Aging cells also arrest at G1 (Rao and Palumbo, 1978). On the other hand, under these conditions, tumor cells continue to cycle at a slower and slower rate and eventually arrest in all cell cycle phases and die rapidly (Pardee, 1974).

Many studies of G1 have been performed. Some investigators tend to look at all cells which contain half the amount of DNA in mitotic cells as cells in G1, whether they progress from Mitosis to the S phase or stop growing due to starvation or differentiation (Dell'Orco et al., 1975; Rubin and Steiner, 1975). Others suggest that cells arrested in G1 leave the cycle and are in a phase named G0 or quiescent stage (Baserga, 1976; Prescott, 1976a,b). Cells in G0 reenter G1 after a suitable induction and then they progress to phase S. Some investigators suggest that all cells which complete mitosis and enter G1 reach a point at which the decision to progress toward S phase is made (Smith and Martin, 1973; Shilo et al., 1976). At this point each cell has a chance to progress in G1 which may be high under optimal environmental conditions or low under starvation conditions. This stage, which has been called the "restriction point" by Padree (1974), is defective in tumor cells, and, therefore, under starvation conditions they continue to grow, although at a slower rate until they arrest along the whole cell cycle and die. Aging cells, on the other hand, stop growing and enter a quiescent stage at a low density under conditions that would support active growth of young cells (Hayflick, 1965).

Many kinetic studies have been performed to analyze the regulatory determinants that control cell proliferation (Yen et al., 1978; Cherington et al., 1979; Yen and Pardee, 1979; Hochhauser et al., 1981). However, our knowledge of these mechanisms is still very limited.

Two interesting parameters with respect to the cell cycle are worth mentioning here. Studies conducted by Darzynkiewicz et al. (1982) have shown that quiescent cells, including Chinese hamster cells, contain a low concentration of RNA in comparison to cycling cells. When such cells are induced to proliferate, the concentration of RNA increases and reaches a specific threshold before the cells resume their progress in G1 toward S phase. Hittelman and Rao (1978) have shown that the chromosomes decondense as the cells advance in G1 and that decondensation reaches its peak at the G1/S boundary. In mouse 3T3 cells that have been arrested in G0 by starvation, the chromosomes remain condensed, the level of condensation being characteristic of early G1 phase. This condensed state does not change for about 8 hr after the cells are fed with complete medium. Chromosome decondensation then starts to occur and the cells begin to advance toward S phase at a normal rate (Marcus and Meiss, 1981). It is possible that during this 8-hr lag the cells accumulate RNA to the level necessary before progress in G1 is allowed. While these two interesting observations tell us about changes which occur when quiescent cells are induced to grow, they do not reveal the mechanism that controls cell cycle.

B. Phase S

Initiation of DNA replication signals the end of G1 and the beginning of phase S. Histone proteins which start to be synthesized in G1 (Groppi and Coffino, 1980), and continue to be synthesized in phase S, make up the chromatin core. Several studies have been performed to determine the dependence of histone synthesis on DNA replication (Sheinin and Lewis, 1980). These investigators observed that the arrest of two mouse ts cell cycle mutants in phase S was followed by a cessation of histone synthesis, and suggested a dependence of histone synthesis on DNA synthesis. In another ts cell cycle mutant, ts2, a DNA negative mutant from mouse origin, this dependence of histone synthesis on DNA synthesis does not hold. Sheinin and Lewis (1980) suggest that such a dependence occurs only in mutants arrested in phase S and they suggest that ts2 is arrested in G1.

Many studies have dealt with the fate of the histone core in the single chromatid chromosome after the DNA is replicated. The question was whether the old histones are redistributed in the two chromatid chromosomes in a specific manner. Several studies (Seidman et al., 1979) suggest that the old histone core is conserved in one chromatid and the other chromatid contains the newly made histone core. However, these results are controversial (DePamphilis and Wassarman, 1980).

The S phase represents a highly ordered and controlled stage, however, knowledge of it is fragmentary and limited. Owing to thorough studies of DNA synthesis in prokaryotes and the existence of technology developed to study DNA synthesis, it seems easier to tackle directly specific problems in analyzing phase S in comparison to G1 or G2.

The mechanism that controls and triggers initiation of DNA synthesis and the nature of the first chromosomal regions to be replicated are not yet well understood. Such regions may share a common base sequence or a specific chromatin structure that is recognized by the DNA synthesizing machinery. In yeast, specific DNA sequences which seem to act as origins of replication have been found and isolated. These DNA sequences are called "autonomously replicating sequences" (ARS) (Stinchcomb et al., 1979). In mammalian cells, Heintz and Hamlin (1982) have identified in an amplified chromosomal sequence, sites in which DNA synthesis is initiated when the amplified sequence is replicated indicating the existence of sequence-specific origins of replication. Origins of replication are well known in viruses like SV40 (Nathans, 1979) and others (Dhar et al., 1978). On the other hand, studies performed by Harland and Laskey (1980) have shown that any DNA fragment introduced into *Xenopus* eggs which are induced to proliferate will replicate every S phase, indicating that no specific DNA sequence is necessary for initiating DNA replication. Taken together, these results indicate that we are still far from understanding at the molecular level the mechanism that triggers DNA synthesis and the nature of the origins of replication. After initiation, DNA synthesis continues bidirectionally and simultaneously in many replicons distributed in many chromosomes. Replication of replicons in each replicon cluster is simultaneous (Hand, 1978).

DNA synthesis in different clusters of replicons shows a specific order (Hand, 1978). Thus, some clusters of replicons replicate early in S and others replicate late. Isogenic chromosomes and to a lesser extent homologous chromosomes replicate synchronously (Marcus, unpublished results). However, the inactive X chromosome in female cells replicates later than the active X chromosome while maintaining in it a pattern of replication similar to that of the active X chromosome (Wahrman et al., 1983). Human cells that originated from different tissues exhibit different patterns of late replication indicating that specific pattern of replication reflects specific state of differentiation (Farber and Davidson, 1977). Lin and Davidson (1975), who studied the pattern of replication of human chromosomes in human mouse cell hybrids, have found that the few human chromosomes in each of the hybrid strains continued to replicate at the order characteristic of their replication in human cells, indicating that the order of replication is determined autonomously by every chromosome. These human chromosomes originated from human lymphocytes. However, in human mouse cell hybrids, in which the human chromosome originated from fibroblast cells, the pattern of human chromosome replication changed and resembled the pattern of replication in human lymphocyte cells, as if this is a basic pattern (Farber and Davidson, 1978). The techniques of analyzing the pattern of DNA synthesis are not very accurate and they are very tedious. Many more studies are required to reveal the nature of the mechanism that controls the time of replication of each chromosome. However, all of these results show that an intricate mechanism that regulates the pattern of replication of clusters of replicons in each chromosome and the order of replication of the chromosomes themselves certainly exists.

Recent studies indicate that the S phase can be subdivided into three stages (Klevecz et al., 1975). In the first stage, a small amount of DNA is synthesized, while in the second and third stages the bulk of the DNA is synthesized. Holmquist et al. (1982) have shown that in Chinese hamster cells in the first two stages only the "light Giemsa bands" are synthesized while the "dark Giemsa bands" are synthesized only in the third stage. They also showed that DNA synthesis is suppressed between the substages, indicating that the DNA synthesizing machinery finishes synthesizing one group of replicon clusters, then reorients itself and starts to synthesize the other group of replicon clusters. Holmquist et al. (1982) suggest that a replicon cluster comprises a Giemsa band. These studies indicate the existence of control mechanisms which regulate completion of one subdivision in S and the initiation of the following subdivision.

Extensive studies of DNA synthesis in prokaryotes and eukaryotes revealed many enzymes that participate in DNA synthesis (Kornberg, 1980, 1982). Recent studies (Reddy and Pardee, 1980,1983) indicate that at least some of these enzymes organize as a complex in phase S to synthesize DNA. In this complex, the "replitase," the different enzymes may affect each other by allosteric changes. Several groups of investigators have shown that DNA synthesis occurs on a proteinaceous cytoskeleton structure in the nucleus termed the "nuclear matrix" (Pardoll et al., 1980). It is possible that the replitase, being attached to the nuclear matrix, synthesizes DNA, and in between the different substages in DNA replication it dissociates from the nuclear matrix and must reassociate again before DNA synthesis can be resumed (Holmquist et al., 1982).

Usually DNA is replicated only once in a cell cycle. This fact seems trivial, but it is hardly so. A single replication per cycle requires that every replicated replicon become unavailable for a second round of replication until the next cell cycle. We learn that this process is controlled, from the rare examples where this rule has been violated. Thus, it has been found that at a low frequency cells in a culture go through two or sometimes even three successive rounds of DNA replication and reach mitosis with four or eight chromatid chromosomes, respectively (Schwarzacher and Schnedl, 1965). This process, termed "endoreduplication," can be induced by many agents which interfere with the normal progress of the cell cycle (Hirschberg et al., 1980; Hirschberg, 1981); however, it also occurs normally in some tissues during embryogenesis (Siegel and Kalf, 1982). Partial endoreduplication in which a part of a chromosome is endoreduplicated occurs too. In such cases, endoreduplication ends at the centromere (Lejeune et al., 1968; Noël et al., 1977). Cloning of a centromere into a plasmid in yeast has been found to reduce the number of plasmid copies per haploid genome to about one (Tschumper and Carbon, 1983), indicating that in yeast and probably in mammalian cells too, the centromere plays a role in the control of chromosome replication.

C. Phase G2

Phase G2 is usually a short phase in which the cell prepares for mitosis. RNA and protein synthesis are required for normal progress through this phase to mitosis (Tobey et al., 1966). In addition this phase is marked by an extensive chromosome condensation (Sperling and Rao, 1974). Several studies (Marcus et al., 1979; Marcus and Sperling, 1979) indicate that the condensation of the heterochromatic chromosomal regions which occurs in G2 has a specific order. Other studies indicate that in rat kangaroo cells, the pattern of condensation of the heterochromatic regions is positively correlated with the pattern of replication of these regions in phase S (Goitein et al., 1984). It remains to be seen whether such a pattern exists for euchromatic regions too.

It has been observed that cells whose DNA has been damaged by irradiation or chemical agents tend to be arrested in G2 (Dewey and Highfield, 1976). It seems plausible to assume that the delay or arrest in G2 is required to allow repair mechanisms to function. However, the nature of this phenomenon is not yet understood. It is also interesting to note that plant cells in quiescence state are usually in G2 (Kudirka and Van't Hof, 1980).

D. Mitosis

The complex mitotic process has been analyzed extensively by cytologists in many lower and higher eukaryotes (Pickett-Heaps et al., 1982). In mammalian cells during the first stage of mitosis, namely the prophase, the chromosomes can be seen under the light microscope as they go through a considerable condensation. During this stage the nucleolus disappears and, at the end, the nuclear membrane disappears, too. The centrioles migrate to the poles, and the spindle is organized and reaches from one pole to the other. As the second stage, metaphase, begins after the nuclear membrane disappears, the spindle fibers interact with the kinetochores in each chromosome and function in aligning the chromosomes along the cell equator. The role of spindle fibers in this process is detailed in Chapter 22. Each chromosome is made of two sister chromatids which are held together by the kinetochores. As the kinetochores separate, the third stage, anaphase, begins. The chromosomes move toward the poles by virtue of the interaction between the kinetochores and the spindle fibers. At the poles the two groups of single chromatid chromosomes aggregate, and this aggregation signals the last stage of mitosis, namely, the telophase. During this stage the nuclear membrane is reconstructed around the chromosomes. In addition, cytokinesis occurs, dividing the mitotic cell into two daughter cells.

The many studies of the ultrastructure of the spindle and the analysis of it at the biochemical level have not yet resolved many of the mysteries of mitosis. I therefore advise the reader to turn to the recent reviews (Chapter 22; Inoué, 1981; Pickett-Heaps et al., 1982) for a more thorough description of mitosis. However, the main open questions that may be tackled with suitable mutants are clear. They include the controlled movements of the chromosomes toward the equator during prometaphase; separation of the kinetochores and the controlled movement of sister chromatids to opposite poles; disaggregation of the nuclear membrane at the end of the prophase and its reaggregation around the two groups of chromosomes at the end of the telophase; cytokinesis of the cell to two similar daughter cells; the mode of action of both the centrioles and the kinetochores with respect to the construction of the spindle and its function; the structure of the spindle and the different elements which comprise it.

III. CELL CYCLE MUTANTS

A. Concept of a Cell Cycle Mutation

Mutations that perturb the cell's capability to divide and proliferate are lethal. Most of the cell cycle mutations fall into this group and can therefore be acquired only as conditional mutations. The most common group of conditional mutants in eukaryotes is the group of temperature-sensitive mutants. The majority of them are heat sensitive and a few are cold sensitive. Cell cycle mutants isolated from mammalian cells have recently been reviewed (Simchen, 1978; Hochstadt et al., 1981; Wissinger and Wang, 1983).

Theoretically, any mutation that disturbs the growth of cells can be regarded as a cell cycle mutation; however, such a definition makes any attempt to analyze cell cycle anachronistic. Hartwell (1974) suggested an operative definition by which a mutant is considered to be a cell cycle mutant only if the mutation affects a single specific stage of the cell cycle. This empirical definition is still too broad. Thus, a mutation that affects the transport of different substrates including serum growth factors, which are essential for the movement of cells from G1 to S, will be considered as a cell cycle mutation, since the mutant cells will be arrested in G1 when incubated at the restrictive temperature. However, our knowledge of the cell cycle is very limited and for the moment there are no better definitions. The virtue of this definition is that it allows a rapid distinction between the large group of temperature-sensitive mutants for growth and the much smaller group of temperature-sensitive cell cycle mutants, since the growth of the former group of mutants would be arrested under restrictive temperature at all stages of the cell cycle, while the growth of the mutants in the latter group would be arrested under restrictive temperature at a specific cell cycle phase.

In the future, when the different systems of cell cycle genes will be better understood, it will probably be possible to devise better selection and screening techniques to obtain cell cycle mutants.

B. Execution Point

When randomly growing cells of a temperature-sensitive cell cycle mutant are transferred to the restrictive temperature for growth, the thermosensitive gene product is inactivated. Cells in which this gene product has already completed its function can complete the cell cycle and divide. The growth of all the rest of the cells in the culture is arrested at a specific stage along the ongoing cycle. The point in the cell cycle beyond which the transfer of cells to the restrictive temperature cannot stop the cells from completing the cycle is defined as the execution point (Hartwell, 1974). Determination of execution points in cell cycle mutants constitutes an important parameter in the study of such mutants. It helps in the analysis of the temporal sequence of events in the cell cycle. It is important to remember the limitations of using the execution point of a specific mutant for further analysis of the cell cycle. Execution points are allele specific and not gene specific. Various mutations in the same gene may exhibit different execution points owing to differences in temperature sensitivity of each of the mutated gene products in the mutants. Thus, a mutant in which the synthesis of the temperature-sensitive gene product is thermolabile but the final product is temperature resistant may have an execution point that will differ from that exhibited by a mutant in which the same gene is mutated so that the final gene product is thermolabile. In the former mutant the execution point describes the point at which the gene product is being synthesized and accumulated, while in the latter mutant the execution point describes the point at which the final product fulfills its function. Differences between different mutated alleles with respect to the kinetics of inactivation of their respective temperature-sensitive gene products add to the diversity of execution points exhibited by cell cycle mutants in any specific gene.

As a consequence, efficient usage of execution points in the study of the nature of cell cycle mutants may be very hard. It remains easier for cell cycle mutants in which the execution points are close to their respective points of arrest. In such cases it is reasonable to assume that the mutated gene product is required for completion of the stage at which the progress of the mutant cell has been arrested. A thorough discussion of the information which can be obtained by studying execution points is presented elsewhere (Byers and Goetsch, 1976a,b; Nurse et al., 1976; Orr and Rosenberger, 1976; Bisson and Thorner, 1977; Hartwell, 1978). Some cell cycle mutants arrest at the first cell cycle after the shift to the restrictive temperature for growth, while others arrest only at the second cell cycle or even later. It is very difficult to perform kinetic studies of the latter group of mutants. However, the situation is different when several cell cycle mutations are known to affect the same gene and the different mutants fall into the above two groups. In these cases it can be assumed that the difference between the alleles is due to the fact that the allele which results in an arrest in the first cell cycle determines the synthesis of a thermolabile product, while the allele which causes an arrest in the second or a later cell cycle determines the synthesis of a thermostable final product whose synthesis is temperature sensitive. Such data are informative since they suggest that the gene product is present throughout the cell cycle and therefore its function cannot be controlled by synthesis at a specific cell cycle phase.

C. Point of Arrest and Terminal Phenotype

As mentioned above, when temperature-sensitive (ts) cell cycle mutants are transferred to the restrictive temperature, their growth is arrested at a specific phase of the cell cycle and usually at a specific point along this phase entitled "the point of arrest." The point of arrest is a landmark of every mutant. The cells in an asynchronous cell culture of any specific ts cell cycle mutant transferred to the restrictive temperature will eventually accumulate at the point of arrest. There the cells acquire a characteristic morphological, physiological, and biochemical phenotype which forms the "terminal phenotype" of the mutant cells (Hartwell, 1974). Analysis of the terminal phenotype of a mutant is yet another aspect in the search for the nature and function of cell cycle genes. Here, again, there are many pitfalls that must be realized. Many of the cell cycle mutants enter an unbalanced growth and lose viability at the point of arrest (Hirschberg and Marcus, 1982). In fact, some mutants are committed to die even before that, once they pass their execution point at the restrictive temperature (Hirschberg and Marcus, 1982). Hence, at the restrictive temperature, randomly growing mutant cells, which accumulate during a generation time at the point of arrest, form a synchronized cell culture only with respect to the localization of the cells along the cell cycle. Otherwise these cells constitute a heterogeneous population of cells because each one of them reaches the point of arrest at a different time. There they may enter an unbalanced growth causing secondary effects for various periods of time in different cells before the whole culture can be analyzed as a "synchronized" cell culture at the point of arrest. It is therefore advisable to try and synchronize cell cycle mutants at the permissive temperature, before they are transferred to the restrictive temperature. Under these conditions the cells will reach their point of arrest synchronously and there the sequence of abnormal events which occur can be better studied. Nonetheless, in many cases, even without a previous synchronization of the mutant cells at the permissive temperature, an informative specific abnormal behavior can be recognized at the point of arrest (Hartwell, 1974; Hochstadt et al., 1981; Hirschberg and Marcus, 1982; Wissinger and Wang, 1983).

D. Interrelationship Among Various Cell Cycle Processes

As reviewed very briefly, proliferation of cells represents a very complex multiprocess phenomenon which requires a stringent order, and most probably depends on the coordinated function of many hundreds of genes (Hartwell, 1978; Hirschberg and Marcus, 1982). With the use of cell cycle mutants it is possible to study the interactions among different cell cycle genes and construct independent, interdependent, and dependent cell cycle pathways (Hartwell, 1978). The most comprehensive study of dependent pathway of landmarks was performed in the study of the cell cycle of yeast (Hartwell, 1974; Nurse et al., 1976). Both independent and dependent pathways have been found. Some pathways have already been studied to some extent in mammalian cells. Thus, nuclear division can occur without cytokinesis to produce multinucleate cells. (Wang and Yin, 1976) and successive rounds of DNA replication in one S period can occur producing endoreduplicated chromosomes (Schwarzacher and Schnedl, 1965; Siegel and Kalf, 1982). However, cytokinesis does not occur without previous nuclear division, which in itself depends on a preceding DNA synthesis phase. For a comprehensive study of this nature it is essential to isolate a large number of cell cycle mutants. Unfortunately, only a small number of ts cell cycle mutants have already been isolated and even these mutants originated from many cell lines of different species and not from one species. Analysis of many cell cycle mutants will be required for a full understanding of the mechanisms that regulate cell cycle, and it is advisable to isolate them from Chinese hamster cells which easily yield many different mutants (Hirschberg and Marcus, 1982).

IV. ISOLATION OF TEMPERATURE-SENSITIVE CELL CYCLE MUTANTS

A. Mutagenesis

Investigators have successfully used different mutagenic agents which include alkylating agents like ethyl methane sulfonate (EMS) and *N*-methyl-*N*′-nitro-*N*-nitrosoguanidine (NG)(Basilico and Meiss, 1974), UV irradiation (Chang et al., 1978), X rays (Chang et al., 1978), [^{3}H]thymidine (Cleaver, 1978), and others.

The frequency of mutations per gene in many different genes after a mutagen treatment reaches a rate of 1×10^{-6}–1×10^{-4}. Yet, it is important to remember that the efficiency of induction of mutations in different genes by specific mutagens was found to vary drastically (Cleaver, 1978). Thus, to obtain mutations in many different genes it may be advisable to use more than one mutagen.

Investigators usually treat cells with mutagens under conditions in which only 20–50% of the mutagenized cells survive (Hochstadt et al., 1981). These conditions were adapted both from bacterial mutagenesis techniques and from experiments in which the correlation between the rates of killing and induction of specific mutations in mammalian cells were determined. This method of optimizing the treatment of cells with a mutagen is far from being accurate. The differences in the rate of mutant induction between different experiments can be on the order of 100-fold (Orkin and Littlefield, 1971).

A better method for determining the efficiency of a specific mutagenic treatment is an empirical one in which the rate of mutation for an easily selectable dominant allele is measured before and after mutagenesis. Resistance to ouabain (Chang et al., 1978) can play the role of an indicator for the efficiency of mutagenesis.

In experiments in which the investigator has to screen the mutants from the mutagenized cell culture because no selection methods are available, it is essential to know whether the mutagenic treatment was successful, as screening requires a lot of work and it is better to repeat mutagenesis rather than screen mutants from a culture in which the rate of induction of mutants was low. In this case, determination of the rate of induction of ouabain resistance as an indicator sign for a successful mutagenic treatment has a big disadvantage, since the investigator has to wait for resistant colonies to appear. During this period the mutagenized cells either have to be grown continuously or frozen. Both treatments may lower the fraction of mutants in the culture as the surviving wild-type cells may grow faster and survive freezing and thawing better. Recently, a new and fast technique to determine the rate of mutation in a mutagenized culture has been developed by Ronen et al. (1984). These investigators study mutagenesis and use for their studies the mutation to diphtheria toxin resistance (Gupta and Siminovitch, 1978). Appearance of this mutation can be determined in single cells, namely, immediately after the fixation of mutations in the mutagenized culture. Thus, the investigator can directly decide whether to use the culture for screening mutants or repeat the mutagenic treatment. Determination of resistance to diphtheria toxin is based on incubation of a sample of cells seeded on a slide in the presence of [^{3}H]leucine and the toxin which inhibits protein synthesis. Resistant mutant cells incorporate the radioactive leucine and can be easily located under the microscope by autoradiography, using a dark field and a low-magnification objective. Even a few mutant cells can easily be screened on a slide containing 5×10^5 cells by this technique since they shine on the dark background. The exposure time required for autoradiography is very short (8–16 hr). Hence, in one day the investigator obtains a good evaluation of the rate of induction of mutations by the mutagenic treatment and can decide whether to repeat the experiment or use the mutageized culture for screening of mutants.

B. Fixation of Mutations

Usually, the mutagenized cells are incubated for about three generations to allow for "fixation" of the mutations before any selection pressure is applied. During this time interval the change which occurred in one DNA strand is copied in the other strand, the treated cells recover from various deleterious effects caused by the mutagenic treatment, and the normal gene product is diluted or destroyed while the mutated gene product appears (Hochstadt et al., 1981).

C. Negative Selection of Mutants Whose Growth Is Heat Sensitive

As mentioned before, most of the cell cycle mutants are conditional lethal mutants whose growth is arrested at a high restrictive temperature, which is permissive for wild-type cells. Investigators of the cell cycle took advantage of this general phenotype of cell cycle mutants and developed mutant selection techniques to enrich mutagenized cell cultures with cells whose growth is arrested at the restrictive temperature. These techniques are all based on killing of growing cells. They have been discussed thoroughly by Basilico (1978), and I shall therefore limit myself to a brief discussion of the principle behind all of them. The mutagenized cells are incubated for about one to two generations at the supposedly restrictive temperature for growth of the potential temperature-sensitive mutants to arrest their growth. Then, agents that kill growing cells are added and incubation of the cells at the restrictive temperature continues for about one to two additional generations. These agents include inhibitors of DNA synthesis like 5-fluorodeoxyuridine (5-FUdR) and cytosine arabinoside, and "DNA poisons" such as [^{3}H]thymidine and 5-bromodeoxyuridine (5-BrdU). Disintegration of the radioactive thymidine incorporated into DNA of cells causes their death. Similarly, irradiation with flurorescent light of cells that incorporated 5-BrdU into DNA induces breaks in the labeled DNA which causes their death. Similar approaches have been used to isolate auxotrophic mutants (Chapter 1) and revertants of malignantly transformed Chinese hamster cells (Chapter 26).

All the above mentioned agents are supposed to kill cells which synthesize DNA; namely, proliferating cells. The killing efficiency of the different agents is usually not high enough to allow an easy screening of mutants after one cycle of selective killing of wild-type cells. Therefore, it is customary to perform several cycles of selective killing of proliferating cells before the final screening of mutants is performed (Basilico and Meiss, 1974). The screening is based on growing clones from surviving single cells and checking each clone for temperature sensitivity for growth before further studies are made.

These techniques have been fruitful, giving rise to the isolation of a number of temperature-sensitive cell cycle mutants. Most of the mutants isolated have been found to be arrested in phase G1 (Basilico, 1978). The big disadvantage of the techniques based on a mutant-enrichment step results from the long exposure of the cells to the restrictive temperature of growth of the potential mutants. Many studies have shown that most of the temperature-sensitive cell cycle mutants lose viability when incubated at the restrictive temperature (Basilico, 1978; Simchen, 1978; Hirschberg and Marcus, 1982). The more tolerant temperature-sensitive cell cycle mutants have been found to be G1 mutants (Simchen, 1978; Hirschberg and Marcus, 1982), namely, mutants which arrest at G1 under restrictive conditions. This observation explains why most cell cycle mutants isolated by the above techniques have been G1 mutants.

Meiss et al. (1978) have modified the enrichment step in order to try and isolate S phase mutants from the Syrian hamster BHK cells. Their modification is based on synchronizing the cells in G1 at the permissive temperature by starving them for either serum or isoleucine. The arrest in G1 is reversed by the addition of the missing nutrients and simultaneously transferring the cells to the restrictive temperature of growth of the potential mutants in the presence of 5-FUdR or 5-BrdU throughout the S phase. The 5-BrdU-treated cells are then irradiated with "black light." The rate of killing in these experiments is very high. Among the survivors, they isolated several S phase mutants. Nishimoto and Basilico (1978) studying BHK cells, have added another modification to the above procedure in order to obtain tight mutants, as the problem of leakiness of many of the isolated mutants has greatly interfered with many of the studies with existing cell cycle mutants. Their modification was based on performing the mutant enrichment step at 37.5°C instead of the usual 39°C or 40°C used by most other investigators. Their notion has been that temperature-sensitive cell cycle mutants whose growth would be arrested at 37.5°C well enough to allow them to survive the step of selective killing of proliferating cells would no doubt arrest very tightly at higher temperatures like 39°C or 40°C, which would still allow wild-type cells to grow. Indeed, they obtained cell cycle mutants which have been very tight at 39.5°C. However, most of the mutants they have isolated have been G1 mutants.

Even under the modifications specified the exposure to the restrictive temperature of growth of potential mutants during the mutant-enrichment step is very long, so that mutants which lose viability immediately or a short time after they pass through the execution point at the restrictive temperature are selected against by the above procedures.

Roscoe et al. (1973) have tried to isolate temperature-sensitive cell cycle mutants from Chinese hamster cells without using the selective killing step. Their technique was based on the well-known observation that mitotic cells round up and detach from the solid surface on which the cells grow, and attach again only after completion of mitosis and entry to G1. They transferred the mutagenized cells to the restrictive temperature of growth for potential cell cycle mutants and then arrested the proliferating cells with hydroxyurea in phase S. Under these conditions, supposedly only cell cycle mutants that have been arrested in mitosis should remain round and detached. However, the cell cycle mutant they isolated has been found to be a G1 mutant.

D. Isolation of Temperature-Sensitive Mutants for Growth by a Replica Plating Technique

Screening and isolating of mutants by a replica plating technique is a well-known procedure introduced more than 30 years ago (Lederberg and Lederberg, 1952) for bacterial genetic studies. Several investigators of mammalian cells in culture have suggested and used various replica plating techniques to isolate mutants (Goldsby and Mandell, 1973; Stamato and Hohmann, 1975; Esko and Raetz, 1978). Yet, cell cycle investigators did not tend to follow these procedures, because the number of mutants obtained by the negative selection techniques discussed above, indicated that the frequency of cell cycle mutants was too low for screening by replica plating techniques, unless automatic facilities are introduced (Basilico, 1978).

We tried to screen temperature-sensitive mutants for growth by a replica plating technique based on a simple multisyringe injector we built which fits the 96-well microtest plates commercially available (Hirschberg and Marcus, 1982). We used the Chinese hamster established cell line E36 (Gillin et al., 1972), which is a hypoxanthine guanosine phosphoribosyl transferase negative ($HPRT^-$) mutant derived from the Chinese hamster lung cell line V79. (See Chapter 3 for a detailed description of this cell line.) The cells were treated with the mutagen EMS to the normal survival rate of 20–50% and were incubated at 34°C for 3 days for fixation of the mutations and recovery of the mutagenized cells as described above. These cells were cloned with the multisyringe injector in microtest plates with 96 wells at 34°C and the resultant colonies were replicated with the same injector to sister microtest plates. One of each pair of plates was incubated at 34°C and the other at 40°C. The plates were screened 48–72 hr later. Every colony of cells in the microtest plates incubated at 40°C which either died or did not grow normally was suspected of being a mutant colony and was further studied. Among 6500 colonies screened, 84 were found to be temperature-sensitive mutants for growth; namely, more than 1% of the survivors of the mutagenic treatment were found to be temperature-sensitive mutants for growth. This frequency seems high at a first glance; however, it becomes more reasonable if we take into consideration that after a good mutagenic treatment, the frequency of a mutation in a specific gene may be in the order of 1×10^{-5}, and that a mutation in any of at least 1000 different genes can cause a defect that will result in temperature sensitivity for growth. These rough calculations are based on the notion that many genes in E36, like in CHO cells, are in a functionally hemizygous state owing to loss or inactivation of their homologous loci (Gupta, 1980; Siminovitch, 1976, 1979), thus enabling recessive mutations to be expressed. Indeed most of the mutations we studied so far proved to be recessive (Hirschberg, 1981). Out of the 84 temperature-sensitive mutants for growth, six were found to be cell cycle mutants that will be discussed later. The successful isolation of many temperature-sensitive mutants for growth with the use of a replica plating technique, namely, under conditions in which all the potential mutants are recovered, suggests the feasibility and advisability of this technique in comparison to the techniques based on selection of mutants after a long exposure of the cells to the restrictive temperature.

E. Screening of Temperature-Sensitive Cell Cycle Mutants from Temperature-Sensitive Mutants for Growth

As stated above, the definition of a cell cycle mutant is operative and it includes all mutants for growth whose gene product is required only once along the cell cycle at a specific phase. Thus, under restrictive conditions all the mutant cells arrest at a specific phase of the cycle and in many cases the arrest occurs at a specific point in this phase. There are several techniques which allow the localization of cells along the cell cycle. Each of these techniques has its own advantages and disadvantages, which will be discussed below.

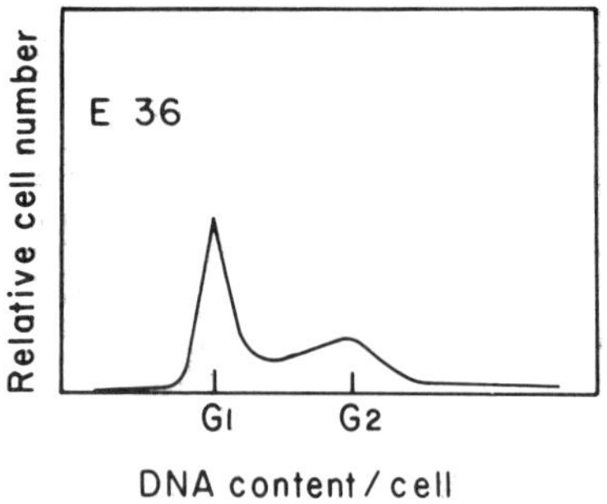

Figure 20.1. Flow microfluorimetry (FMF) measurement of the DNA content distribution of a random population of wild-type E36 cells. The cells were processed for FMF as described by Crissman et al. (1975).

1. The Flow Microfluorimetric (FMF) Technique

The cytofluorograph and the more-sophisticated cell sorter (Crissman et al., 1975) allows a very rapid and accurate measurement of the amount of DNA per cell in a very large population of cells. Measurement is based on specifically staining the DNA of a fixed suspension of cells with propidium iodide or other fluorescent dyes. The stained cells are then passed at a rate of up to 5000 cells/sec through a laser beam, and the amount of fluorescence emitted by each cell, proportional to the amount of DNA in it, is determined. The distribution of cells with different amounts of DNA appears on a screen. A normal distribution of cells of the Chinese hamster cell line E36 is illustrated in Figure 20.1. Such an asynchronous cell culture includes cells in G1 in which the amount of DNA per cell is 2C, cells in G2 and mitosis in which the amount of DNA per cell is 4C, and cells in phase S in which the amount of DNA per cell depends on its location along S and runs between 2C and 4C. A population of cells of a temperature-sensitive cell cycle mutant incubated at the restrictive temperature for growth and arrested at a specific phase of the cell cycle can thus be easily distinguished from a normal cell culture or a temperature-sensitive mutant for growth which is not a cell cycle mutant and thus arrests along the whole cell cycle.

2. The Quinacrine Dihydrochloride (QDH) Technique

The quinacrine dihydrochloride (QDH) technique, which has been developed by Moser and Meiss (1977), is based on quantitative and qualitative differences in the fluorescence of nuclei in cells positioned at different points along the cell cycle. These differences result from the continuous changes in the level of condensation and organization of chromatin along the cell cycle. The use of this technique allows quite accurately and easily the localization of cells at different points along phase G1. The nuclei in cells in early G1 are small and fluoresce very brightly. Fluorescence of nuclei in cells at later stages in G1 is duller and is very faint in cells at the

G1/S boundary. Thereafter, fluorescence begins to increase and is very high again in cells in G2. In this phase, the nuclei exhibit a characteristic granular organization. Additional differentiation of cells in phase S from cells in G1 can be performed by pulse labeling of a studied cell culture with [^{3}H]thymidine. The labeled cells are first stained with QDH and photographed and then processed for autoradiography and rephotographed.

3. The Premature Chromosome Condensation Technique

The premature chromosome condensation technique and its use in the analysis of cell cycle has been recently described in detail (Marcus and Hirschberg, 1982; Sperling, 1982). It is based on fusing mitotic cells with interphase cells, usually by UV-inactivated Sendai virus. As yet undefined mitotic factors induce an immediate mitosis in the fused interphase cells. This process, which can be observed as early as 30 min after fusion, includes the disappearance of the nuclear membrane and the marked condensation of the interphase chromosomes. The prematurely condensed chromosomes (PCC) can be visualized under the light microscope and their organization can be studied. It is very easy to differentiate with this technique between cells in G1 which have one chromatid chromosomes, cells in G2 which have two chromatid chromosomes, and cells in S which have chromosomes that appear pulverized. This technique reveals many more details concerning the position of cells along the cell cycle and the organization of chromatin in comparison to the other two techniques. This fine analysis is based on the ability of the investigator to visualize the interphase chromosomes under the microscope. The position of cells in G1 and to some extent in G2 can be determined by the level of chromosome condensation in the PCC. The position of cells in phase S is determined by the number of replicated chromosomal segments and their lengths. The appearance of PCC throughout the cell cycle in the Chinese hamster cell line E36 is presented in Figure 20.2. As can be seen, the G1 PCC includes six substages which depict the progress of cells from very early G1 to the G1/S boundary. These substages were first described by Hittelman and Rao (1978). They show that the final level of chromosome condensation in the PCC decreases in accordance with the position of the cells along phase G1. Thus, very early in G1 the PCC are very short. They appear longer in cells progressing toward S and very long in cells at the G1/S boundary. In S phase PCC, the unreplicated chromosomal segments appear as long lightly stained threads; the replicated chromosomal segments appear as highly condensed, darkly stained, and double chromatid regions, and the replicating regions, which cannot condense and are therefore hardly seen, give these PCC the characteristic pulverized appearance. In early S phase PCC, only very small replicated chromosomal

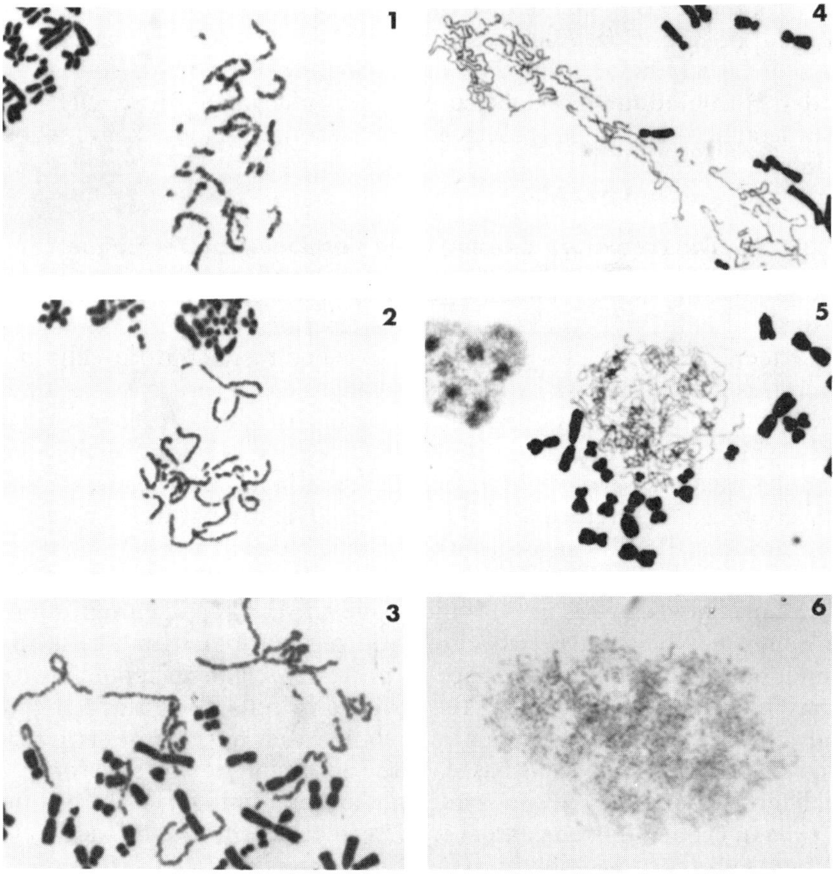

Figure 20.2. Illustration in E36 PCC of different identifiable stages along the cell cycle. 1–6, G1 PCC showing the varying degrees of chromosome condensation.

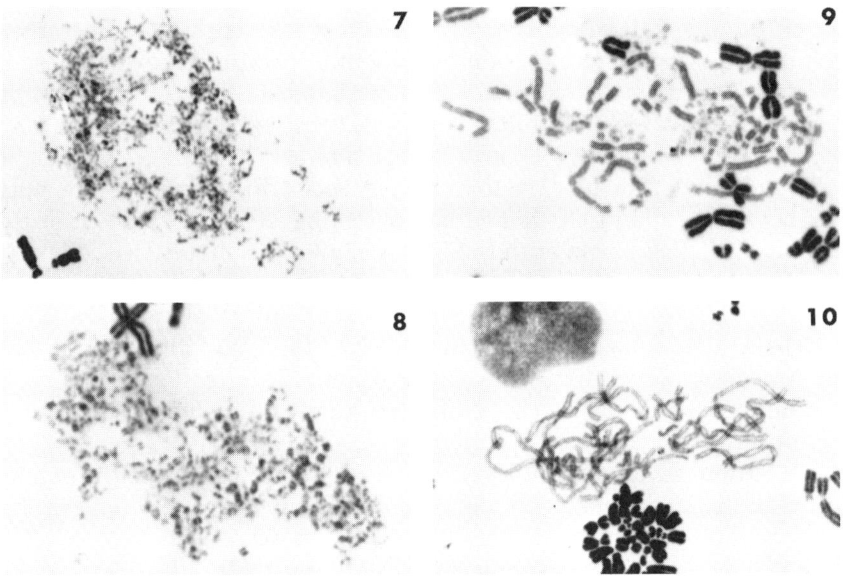

Figure 20.2. (continued) Illustration in E36 PCC of different identifiable stages along the cell cycle. 7–9, S PCC exhibiting different numbers and amounts of replicated highly condensed and darkly stained chromosomal segments. 10, G2 PCC. Mitotic CHO cells were fused to randomly growing E36 cells and induced them to enter premature chromosome condensation. The mitotic CHO chromosomes appear in most of the photographs as short and darkly stained chromosomes.

segments are seen. In middle S, the replicated regions become longer, and, in late S, the number of unreplicated or replicating regions becomes small.

Even in PCC it is hard to determine the exact phase of the cell cycle near the G1/S boundary and the S/G2 boundary. It is possible to overcome this difficulty by pulse labeling the interphase cells with [^{3}H]thymidine just before fusion and analyzing the PCC by autoradiography (Hirschberg et al., 1980).

F. Advantages and Disadvantages of the Different Techniques Used in the Analysis of the Distribution of Cells Along the Cell Cycle

The FMF technique allows a very fast screening of very large populations of cells. It should be the method of choice for a preliminary analysis of all mutants whose growth is temperature sensitive, in order to choose from them those mutants which accumulate at a specific cell cycle phase when incubated at the restrictive temperature. It may also be advisable to choose for further study ts mutants for growth that exhibit an abnormal distribution of cells along the cell cycle at the restrictive temperature, even if no distinct accumulation of cells in a specific cell cycle phase is observed. We have isolated such a mutant (E36 ts41) which turned out to be a very interesting and unique cell cycle mutant (Hirschberg and Marcus, 1982). This mutant will be described later.

Darzynkiewicz et al. (1982) have succeeded in differentiating with FMF between cells in G0 and cells progressing in G1 toward S. Yet the analysis of the cell cycle with the FMF technique is mainly based on the quantitative determination of the amount of DNA per cell and, therefore, it cannot differentiate among many different points along G1 and G2 (Marcus and Hirschberg, 1982). On the other hand, and for the same reason, it allows a much better and more accurate localization of cell cycle mutants arrested in phase S in comparison to the QDH and the premature chromosome condensation techniques.

The FMF technique is not informative at all with respect to the organization of chromosomes in the nucleus. It also does not differentiate between tetraploid cells in G1 and diploid cells in G2.

The QDH technique is simple and easy to master. It allows a good differentiation between cell cycle mutants arrested in different positions along G1. With the use of [^{3}H]thymidine labeling and autoradiography the difference between cells in different positions along S and in G1 can be obtained. Temperature-sensitive cell cycle mutants, which have been incubated at the restrictive temperature and accumulated at their specific point of arrest, sometimes reveal phenotypic changes in the organization of chromatin that affect the QDH staining and make futile any attempt to use the QDH technique to determine their point of arrest.

The PCC technique is the most tedious technique of the three. It requires the preparation of a cell culture in mitosis and a good batch of Sendai virus as a fusogen, although some investigators have reported good results with the use of polyethylene glycol as a fusogen (Lau and Arrighi, 1981). Screening of the PCC under the microscope is also a tedious task; however, this is the only method that provides much information concerning the organization of chromatin in the mutant cells. As shall be described later, this technique added important information to our studies of cell cycle mutants, which could not have been obtained by any of the other techniques with respect to the analysis of an S phase cell cycle mutant. A possible inherent disadvantage of this technique is due to the fact that the cells that are studied are only those cells which have fused and whose chromosomes are induced to prematurely condense. Studies performed by Rao et al. (1977) showed that under specified conditions cells throughout the cell cycle have a similar chance to fuse and be induced to go through premature chromosome condensation. This assumption may not be true for some cell cycle mutants grown at the restrictive temperature for growth. Many of these mutants lose viability at the point of arrest

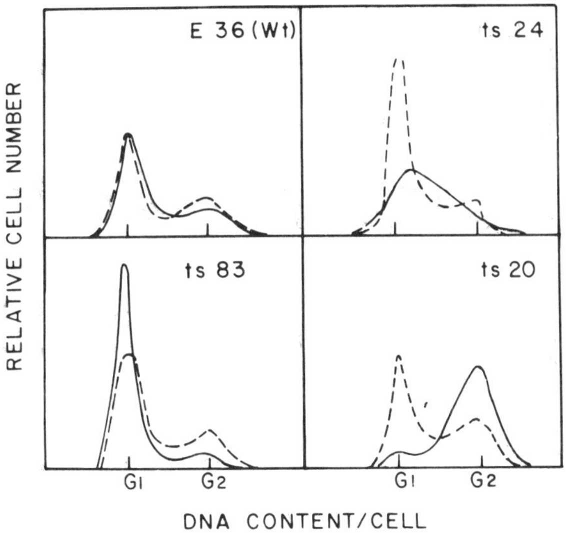

Figure 20.3. FMF measurement of the DNA content distribution in E36 (wild type) cells and three ts cell cycle mutants—ts83, ts24, and ts20—which were derived from it. Cell cultures were grown for 24 hr at 34°C (- - -) or 40.3°C(——) and processed for FMF as described by Crissman et al. (1975). Note, that at the restrictive temperature ts83 cells arrest in G1, ts24 cells arrest early in S, and ts24 cells arrest at G2.

and this loss may be accompanied by, or even be preceded by, a loss in the ability to fuse or to be induced to go through premature chromosome condensation. Under such conditions, the only cells that will show PCC will be those cells which have not yet reached the point of arrest or have reached it just before fusion. As a result, the distribution of the mutant cells along the cell cycle based on PCC will be artifactual.

In conclusion, it may be advisable to use more than one technique to analyze the point of arrest of a cell cycle mutant, and one of these techniques should be premature chromosome condensation.

With the use of the FMF technique we have studied the 84 temperature-sensitive mutants for growth that we have isolated (Hirschberg and Marcus, 1982; Fainsod et al., 1984a,b). Five of them exhibited accumulation of cells at a specific stage of the cell cycle. Two mutants were arrested in G1, two arrested in S, and one in G2. An illustration of the distribution along the cell cycle of three ts cell cycle mutants and their parental wild-type strain based on the amount of DNA per cell is presented in Figure 20.3.

As can be seen, cells of ts83 arrest with 2C DNA per cell when grown at the restrictive temperature suggesting an arrest in G1. Most ts24 cells arrest with about 2.3C DNA per cell under similar growth conditions

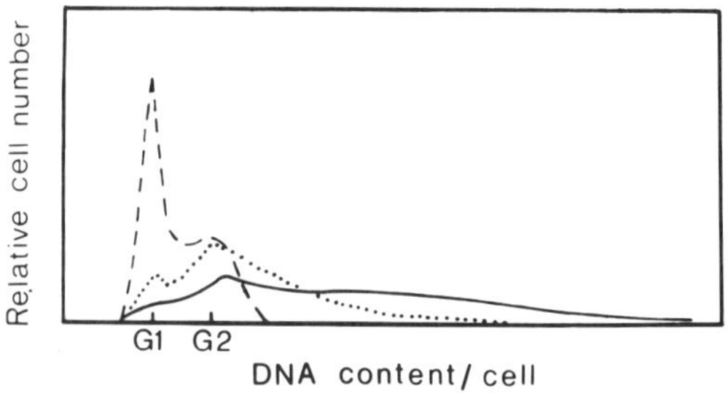

Figure 20.4. FMF measurement of the DNA content distribution of ts41 cells grown at 34°C (- - -), 40°C (····) for 24 hr, and 40°C (——) for 48 hr. The cell cultures were harvested and processed for FMF as described by Crissman et al. (1975). Note that at 40°C cells with an abnormally high content of DNA accumulate.

suggesting an arrest early in S, and ts20 arrests with 4C DNA per cell under these growth conditions, suggesting an arrest in G2.

Premature chromosome condensation analysis of these mutants revealed that ts83 mutant cells grown at the restrictive temperature arrest very early in G1 (Fig. 20.2, stage 2); ts24 cells under similar growth conditions arrest at early S (Fig. 20.2, stage 7); and ts20 cells arrest at G2 (Fig. 20.2, stage 10). Thus, the results obtained by both techniques support each other.

Another temperature-sensitive mutant for growth, ts41, represents a new type of cell cycle mutant. At the restrictive temperature for growth cells of this mutant accumulate in phase S, however, DNA synthesis in the mutant cells continues for a long time—at least for 48 hr—at the restrictive temperature and during this time interval the nuclei enlarge considerably and the amount of DNA per cell in some cells reaches values of 8C–16C. The results obtained by FMF are presented in Figure 20.4. Premature chromosome condensation analysis of the mutant cells incubated at the restrictive temperature for 24–48 hr revealed that the mutant cells continued to synthesize DNA in an ordered manner which resembled endoreduplication producing first four and then eight chromatid chromosomes. A typical endoreduplicated PCC in late S phase is presented in Figure 20.5. These data could not be obtained by either the FMF or the QDH techniques. In addition, all the techniques to isolate cell cycle mutants which are based on a mutant enrichment step and in which DNA synthesizing cells are preferentially killed would have eliminated ts41.

V. ANALYSIS OF TEMPERATURE-SENSITIVE CELL CYCLE MUTANTS

A. Determination of the Execution Point

After the point of arrest of a ts cell cycle mutant is known, its execution point should be determined. As stated above, execution points are allele

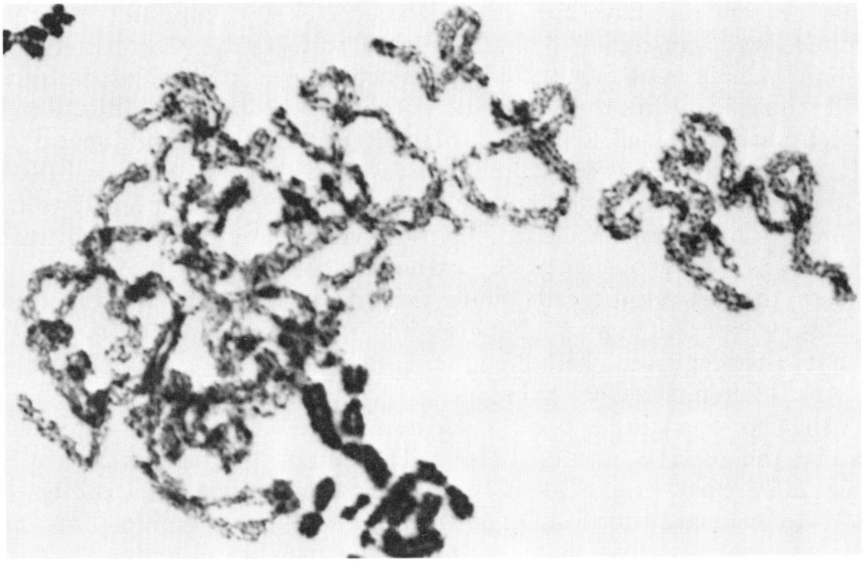

Figure 20.5. PCC of ts41 cell grown for 24 hr at 40°C. This cell has gone through two continuous and almost complete rounds of replication resulting in the appearance of four chromatid chromosomes. Note the undercondensed appearance of the replicated PCC in comparison to a normal late S PCC (Fig. 20.2, stage 9).

specific and not gene specific. A thorough discussion of the information which can be gathered by determining execution points is beyond the scope of this chapter. Those who are interested in it should refer to the excellent reviews by Hartwell (1978), Pringle (1978), and Pringle and Hartwell (1981).

The execution point of a ts cell cycle mutant is determined by kinetic studies. First, the length of the different cell cycle phases of the wild-type cells grown at the restrictive temperature for the ts mutant cells is determined by conventional techniques (Quastler and Sherman, 1959). Next, at this temperature, the mutant cells are analyzed for the distance in time units of their execution point from mitosis and completion of cell division. This can be achieved as follows: A randomly growing mutant cell culture is transferred to the restrictive temperature. The time at which the constant mitotic index drops and cell proliferation ceases marks the distance of the execution point from mitosis and cell division. Mapping of the execution point along the cell cycle is then achieved by subtracting the execution point parameter from the total cell cycle length (Fainsod et al., 1984a).

There is another way to determine the execution point in ts cell cycle mutants. According to this technique, the mutant cells are transferred to their restrictive temperature for growth and simultaneously 5-BrdU is added to the medium. The last cells which reach mitosis at this temperature are cells that have been just beyond the execution point when the culture was transferred to the restrictive temperature. The pattern of 5-BrdU replication bands (Latt,1973) is determined in these last mitotic cells and it identifies the localization of the execution point. Thus, if the execution point is in G2, no 5-BrdU-labeled mitotic cells should appear. If the execution point is in G1, then the last mitotic cells to appear before the mitotic index drops should be labeled throughout their length with 5-BrdU. A much more accurate localization of the execution point by this technique is possible when the execution point occurs in S. In such a case, the last mitotic cells would exhibit a specific pattern of labeled bands, which differentiate between the unlabeled chromosomal segments that replicated before the execution point and the labeled segments that replicated beyond it.

The Chinese hamster ts cell cycle mutant ts24 may serve as an example for the analysis discussed above. The mutant cells arrest at early S (Fig. 20.2, stage 7). At 40.3°C, the parental wild-type cells exhibit a G2 = 2 hr and an S = 11 hr (G2 + S = 13 hr). Proliferation of the mutant cells stops about 10–11 hr after the shift in temperature to the restrictive temperature and, at the same time, the mitotic index drops (Fainsod et al., 1984a). These kinetic studies indicate that the execution point is localized 2–3 hr after the beginning of S. Analysis of the last mutant cells which reached mitosis at the restrictive temperature in an experiment in which 5-BrdU was added at the time the temperature was shifted up revealed a pattern of replication bands in which most chromosomal regions were labeled with 5-BrdU, indicating again that the execution point is localized at early S. As previously described, the point of arrest of this mutant is also at early S, indicating that in this mutant the execution point is very close to the point of arrest, suggesting that the final gene product in this mutant is thermolabile.

Accurate mapping of execution points in cell cycle mutants can be accomplished by the reciprocal shift experiments (Hartwell 1978). These experiments are based on the transfer of cell cultures from one set of conditions which arrest growth at a specific point along the cell cycle to another set of conditions which arrest cells at another point of the cycle. Under these conditions, one determines the ability of the transferred cells to progress in the cell cycle, complete it, and divide, Transfer of such cells from the conditions which have to be executed late in the cycle to those which have to be executed early in the cycle allows them to complete the cycle and divide. On the other hand, in the reciprocal experiment, the cells will just move from the first point of arrest to the second one. An example may clarify this strategy. Cells of a ts cell cycle mutant grown at the permissive temperature are starved for serum and arrest at early G1 (Hittelman and Rao, 1978). The arrested cells are transferred to the restrictive temperature and simultaneously serum is added. The ability of the cells to complete the ongoing cell cycle and divide is determined. If the execution point of the ts cell cycle mutation occurs before the point at which serum-starved cells arrest, then the ts cell cycle function has been completed in the starved cells and, therefore, these cells will be able to complete a cycle even at the restrictive temperature of growth, once serum is added. On the other hand, if serum starvation arrests the mutant cells at a point which precedes the ts execution point, then the addition of serum followed by transfer of the cells to the restrictive temperature will allow the cells to progress along the cycle only up to the point of arrest they exhibit at the restrictive temperature. Thus, such a strategy can help in localizing execution points. Other landmarks, in addition to serum starvation, that have been used are isoleucine starvation, which arrests cells at mid G1, and hydroxyurea treatment, which arrest cells near the G1/S boundary (Hittelman and Rao, 1978). For such a study, of course, the arrest of cells has to be reversible.

We have mapped by the above strategy the execution points in two ts G1 mutants, ts82 and ts83. Premature chromosome condensation of these two mutants showed that both of them arrest early in G1 (Fig. 20.2, stage 2). Serum-starved cells arrest near or at this point (unpublished results; Hittelman and Rao, 1978). The reciprocal shift experiment clearly showed that arrest of cells by serum starvation at the permissive temperature allowed the arrested cells to complete a cycle when they were transferred to the restrictive temperature in the presence of serum (Hirschberg, 1981). On the other hand, in the reciprocal experiment, the cells did not complete the cycle. These results suggested that the execution points in both ts mutants precede the point at which serum starvation results in the arrest of cells.

B. Suppression and Complementation of Cell Cycle Mutations

Temperature-sensitive mutations are usually recessive and can therefore be complemented by the wild-type alleles in hybrid cells. Hence, they are amenable to mapping on chromosomes by synteny (Ruddle, 1973; Chapter 5) in unstable cell hybrids. In such hybrids, the intact mutant genome is maintained, and superimposed on it is a partial karyotype of a wild-type cell genome which includes the chromosome containing the normal allele of the mutated gene. Mapping of a cell cycle gene on a chromosome may be a first stage in its complete characterization at the molecular level by techniques which I shall describe below. Using this technique, Ming et al. (1976) mapped in human chromosome 3 the normal allele mutated in the ts cell cycle mutant tsAF8, which originated from the Syrian hamster cell line BHK. In addition, fusion of pairs of ts cell cycle mutants and selection of temperature-resistant hybrid clones allow the construction of complementation groups (Nishimoto and Basilico, 1978).

A promising avenue has been taken by Jonak and Baserga (1979) studying ts cell cycle mutants that originated from the Syrian hamster cell line

BHK. These investigators used the two ts G1 cell cycle mutants, tsAF8 and ts13, which arrest at the restrictive temperature beyond the point of arrest induced by serum starvation, and complement each other in cell hybrids. They wished to determine whether the nucleus was required for the expression of the ts cell cycle functions during the transition from quiescence (G0) to the proliferation state, and progress of cells toward S. To do so, they fused cytoplasts of G0 tsAF8 cells with G0 ts13 whole cells, and cytoplasts of G0 ts13 cells with G0 tsAF8 whole cells. The fused cells were stimulated by the addition of serum at the restrictive temperature of growth for the mutant cells, and entry to S was monitored by [^{3}H]thymidine incorporation. In both fusion experiments the fused cells reached S and synthesized DNA, demonstrating that both ts functions were already present in the cytoplasm of the G0 cells. These experiments indicate that execution of the two cell cycle functions discussed above is controlled at the translational level or the level of activity (Jonak and Baserga, 1979).

C. Virus—Host Interactions in Cell Cycle Mutants

The complexity of chromatin in eukaryotes makes the study of its ordered and regulated duplication very difficult. One way of overcoming this obstacle is the use of small viruses as probes to analyze this complex process. Thus, several investigators (Cherington et al., 1979; Dubrow et al., 1979) have studied the ability of small RNA and DNA viruses to induce quiescent cells under different starvation conditions to enter the proliferative state. Their results indicate that cells transformed with the small DNA viruses like Polyoma or SV-40 bypass the mechanisms which regulate in G1 the proliferation of cells. On the other hand, cells transformed chemically, "spontaneously," or by RNA viruses, still show a dependency on growth factors for their traverse through G1 and a loss of requirement for a specific growth factor.

Several investigators determined the ability of DNA viruses to bypass ts mutations in G1 cell cycle mutants and induce the arrested cells to advance through G1 to S and synthesize DNA at the restrictive temperature. Burstin and Basilico (1975) used this technique in their study of the tsAF8 G1 cell cycle mutant. They infected the mutant cells with Polyoma virus and isolated transformed cells. The transformed cells, Py-AF8, continued to synthesize DNA at the restrictive temperature for tsAF8. However, they did not reach mitosis. Rather, the cells which entered S at the restrictive temperature and synthesized DNA died. Rossini et al. (1979) found that infection of tsAF8 cells at 40°C with adenovirus 2 but not with adenovirus 12 induced cellular DNA synthesis indicating that the infection of mutant cells with adenovirus 2 resulted in a bypass of the ts mutation which causes an arrest in G1. Similarly, Floros et al. (1981) studying the G1 ts cell cycle mutant ts13 which originated from

BHK have shown that microinjection of SV-40 DNA into cells arrested in G1 induced cellular DNA synthesis. SV-40 T-antigen was implicated in this induction. In recent studies, Soprano et al. (1983) microinjected SV-40 T-antigen deletion mutants and mapped the regions in this antigen which are essential to induce cellular DNA synthesis.

In other studies, several investigators infected ts cell cycle mutants arrested either in G1 or S with different viruses, and determined the ability of the viruses to replicate their DNA under these conditions. Thus, DNA replication of Polyoma virus has been found to be temperature sensitive in two ts S phase cell cycle mutants of mouse origin (Slater and Ozer, 1976; Sheinin et al., 1978) and temperature insensitive in another ts S phase cell cycle mutant (Sheinin, 1976). In other studies (Yanagi et al., 1978), even the replication of the large DNA virus Herpes simplex type 1 (HSV-1) has been found to be temperature sensitive in ts cell cycle mutants of Syrian hamster origin. The progress in our knowledge of the life cycle of different viruses, and especially of the small and simple DNA viruses, allows us to predict the nature of the additional cellular genes which may be required for the replication of these viruses. As a result, a temperature-sensitive or temperature-resistant viral DNA replication in a ts cell cycle mutant may point to the nature of the ts cell cycle mutation and lead to the final characterization of the mutated gene product.

D. Additional Studies of Some Temperature-Sensitive Cell Cycle Mutants

Population density dependence, leakiness of the mutant phenotype, and a high reversion rate may make the efficient usage of a specific cell cycle mutant an impossible task. Thus, in any preliminary cell cycle analysis these parameters should be determined and the decision whether to continue studies with any specific mutant should be reached only later. It may be possible to study to some extent mutants which exhibit their mutant phenotype only at low density of cells (Smith and Wigglesworth, 1972). However, in many cases, biochemical analysis may require masses of cells and make the use of population-dependent cells very hard. Leakiness of temperature-sensitive mutations is a very frequent phenomenon. In many cases it is possible to overcome it by raising the temperature of growth (Nishimoto et al., 1980). In addition, as discussed in the preceding sections, Nishimoto and Basilico (1978) found that very tight ts cell cycle mutants could be obtained if the restrictive temperature used during the mutant-isolation procedure had been 37.5°C and the mutants were tested at 39°C. Very high reversion rates (0.2%) have been observed in the analysis of some mutants (Wang, 1976). High reversion rates make almost impossible the use of such mutants as recipients in gene transfer experiments where selection of transformations requires that the mutation be stable.

Several routine kinetic studies of cell cycle mutants are required in the preliminary analysis, in addition to the determination of the points of arrest and the execution points. These studies include determination of the rates of synthesis of DNA, RNA, and protein at both permissive and restrictive temperature (Srinivarsan et al., 1980; Hirschberg, 1981), and the kinetics of loss of viability at the restrictive temperature (Srinivasan et al., 1980; Hirschberg and Marcus, 1982). All of these results add valuable information which may suggest to the investigator the most possible direct appraoches to take, in order to reveal at the molecular level the function which is thermosensitive in the studied mutant, and isolate the specific gene product for a final analysis.

I shall describe in the following section several experiments performed in the analysis of specific ts cell cycle mutants, and concentrate on studies performed with Chinese hamster cells whenever possible. Mutants defective in mitosis are also discussed in Chapter 22.

1. Mutants Exhibiting a Defective Mitosis

Wang (1976) isolated from the Syrian hamster cell line HM-1 a temperature-sensitive cell cycle mutant, ts655, which exhibits an aberrant prophase. At the restrictive temperature, the growth of the mutant cells diminishes, and after 3 days all of them round up and detach. Cytological examinations of the mutant cells under these conditions revealed that the chromatin condensed considerably and the nuclear membrane disappeared. However, no discrete chromosomes could be observed. Instead, clumps of condensed chromatin appeared, and they coalesced to form aggregates which were concentrated in the cells, and occupied a region not larger than the nucleus. The nuclear membrane failed to re-form. Cells which could grow at the restrictive temperature appeared with a very high rate (~0.2%).

Wang (1974), and later Wang and Yin (1976), studied ts546, which is a ts cell cycle mutant defective in metaphase. This mutant has originated from the cell line HM-1. At the restrictive temperature the mutant cells arrest at metaphase and exhibit scattered chromosomes. After several hours, the chromosomes, which are very condensed, form aggregates, and nuclear membranes form around the aggregates resulting in the formation of one to many aberrant nuclei per cell. Similar results were observed in cells treated with colcemid. Shiomi and Sato (1976) have reported a similar ts cell cycle mutant of mouse origin. This mutant, ts2, originated from the cell line L5178Y. At the restrictive temperature, the mutant cells arrest temporarily in metaphase, and as a result the mitotic index rises to 24%. The metaphases are abnormal, showing scattered chromosomes. Eventually, mitosis seems to end and interphase like cells with many nuclei appear.

Wissinger and Wang (1978) have isolated a ts cell cycle mutant, ts687, defective in anaphase chromosome movement from the Syrian hamster cell line HM-1. At the restrictive temperature, the anaphase becomes aberrant. Many lagging chromosomes are seen, and as a consequence cytokinesis is aberrant and not complete, resulting in the appearance of multinucleate cells.

Wang et al. (1983) have isolated from HM-1 cells a ts cell cycle mutant, ts745, which exhibits abnormal centriole separation and chromosome movement. The mutant cells arrest temporarily in mitosis when grown at the restrictive temperature giving rise to a high mitotic index (~35%). The chromosomes are distributed in the periphery with four centrioles in the center of the cells. The chromatids do not separate. Microtubules can be seen radiating from the centrioles to the chromosomes. After a few hours, the nuclear membrane forms around the chromosomes and interphaselike cells appear. These cells can cycle several times, giving rise to giant cells with nuclei that may contain several hundreds of chromosomes.

Several ts mutants defective in cytokinesis have been isolated. Hatzfeld and Buttin (1975) have isolated such a mutant, ts111, from the Chinese hamster cell line GM7S which originated from CCL39. They observed that at the restrictive temperature, owing to the aberrant cytokinesis and the ability of such cells to continue to cycle, giant cells with one giant nucleus or many small nuclei appear. In addition, many anucleated cells appear. Cytochalasin B treatment affects mitotic cells similarly. The possible relations between this ts mutation and the target of cytochalasin B was therefore studied by Hatzfeld and Buttin (1975) who found that ts111 is indeed more sensitive to cytochalasin B than its parental strain. Thompson and Lindl (1976) have isolated from the Chinese hamster cell line CHO, a ts mutant, MS1-1, which exhibits defects in cytokinesis. In cells grown at the restrictive temperature, mitosis almost ends and the furrow between the cells appears. However, cytokinesis is not complete and as a consequence cells with two or more nuclei appear. Even at the permissive temperature about 34% of the cells are binucleated. Partial manifestation of the ts mutations at the permissive temperature occurs also in some of the other mitotic mutants. The sensitivity of MS1-1 mutant cells to cytochalasin B and to colcemid was determined and found to be similar to that exhibited by the wild-type parental cells. Hence, most probably this mutant affects a function which is different from that defective in ts111.

A completely different approach has been taken by Abraham et al. (1983). These investigators isolated CHO mutant cells resistant to different mitogenic inhibitors due to mutations in either α tubulin or β tubulin (Cabral et al., 1981, 1982). Some of these mutants have been found to be temperature sensitive for growth. At the restrictive temperature, these mutants exhibit aberrant and long mitosis. The spindles are abberrant and the chromosomes are scattered. Cytokinesis is also aberrant and in many cases not complete. As a consequence, multinucleated cells appear and eventually the cells die. The α tubulin mutants and the β tubulin mutants behave similarly. The progress of these cells along the interphase seems normal, indicating that the spindle formation is the limiting function in these mutant cells. Cabral (1983) has also isolated taxol-dependent mutants with a similar defect in spindle formation in the absence of taxol (see Chapter 22).

The questions asked with respect to mitosis remain unanswered for the moment, yet the advance made with the battery of available mutants supports the notion that this is the right way to study this phenomenon.

2. *Mutants Arrested in G1*

The majority of ts cell cycle mutants isolated arrest at G1, however, none of the mutants studied revealed the nature of the mechanism that controls cell proliferation. Many studies of mutants arrested in G1 concentrated on the kinetics of growth arrest and resumption of growth after a shift to the permissive temperature. Using such techniques and synchronized cultures, Melero (1979) mapped several G1 ts cell cycle mutants isolated from the Chinese hamster established cell line WglA, which originated from DON cells. Roufa et al. (1979) isolated a Chinese hamster temperature sensitive mutant, ts154, from the established cell line V-79 (clone HT-1). This cell cycle mutant arrests at the restrictive temperature in G1 only after it completes two rounds of cell division. Other studies of G1 ts cell cycle mutants, which have been presented in the previous sections, include the induction of DNA synthesis in mutant cells infected with viruses.

An interesting study concerning the construction of complementation groups of ts cell cycle mutations causing an arrest in G1 has been performed by Meiss et al. (1978). As described in another section, these investigators designed a method of negative selection to isolate ts DNA synthesis mutants. They isolated G1 mutants which were recessive but did not complement each other, as if all these mutants belong to one complementation group even though they arose independently. Additional studies (Schwartz et al., 1979) revealed that the mutations in the five noncomplementing mutants are assigned to human and hamster chromosome X, and thus may be allelic.

Melero and Fincham (1978) have studied the Chinese hamster G1 ts cell cycle mutant K12 (Roscoe et al., 1973), whose execution point is mapped 4 hr before its entry into S, and found that the synthesis of three polypeptides is markedly induced at the restrictive temperature. The synthesis of one of these polypeptides which is specifically induced in K12 cells, occurs during the time at which the mutated function is expressed, and like the cell arrest is irreversible. The investigators suggest that the ts mutated gene product is responsible for the commitment of cells in G1 to enter S, and the regulation of synthesis of the above three polypeptides. The enhanced and specific synthesis of one of them can serve as a marker in further studies of the nature of the mutation in K12.

Tenner et al. (1976) and Tenner and Scheffler (1979) have thoroughly studied the Chinese hamster G1 ts cell cycle mutant, tsK/34C, which originated from the established cell line WglA. At the restrictive temperature, the mutant cells arrest at G1 and show a marked reduction in the synthesis of glycoproteins, which is caused by a reduction in the rate of transfer of the oligosaccharide core from the lipid–oligosaccharide intermediates to the nascent polypeptide chain. Revertants of this mutant grow at the restrictive temperature and synthesize glycoproteins normally, indicating that both the ts cell cycle phenotype and the ts glycoprotein synthesis result from the same mutation. Landy-Otsuka and Scheffler (1980) performed additional studies with the same mutant, in order to determine if and how a ts cell cycle function affects cell-cycle-specific mRNA synthesis, and influences coordinately the synthesis of cell-cycle-specific enzymes. The enzymes studied were ornithine decarboxylase (ODC), *S*-adenosyl-methionine decarboxylase (SAMDC), and thymidine kinase (TK). The results show that SAMDC activity is not affected at all by the incubation of the mutant cells at the restrictive temperature, while ODC activity is reduced and TK synthesis is blocked. Further study is required to reveal the nature of this ts cell cycle function and its effects on the synthesis of cell cycle specific enzymes.

A well-characterized G1 ts cell cycle mutant is tsAF8, which originated from BHK cells (Meiss and Basilico, 1972). This ts cell cycle mutant arrests at middle G1. A mutation in RNA polymerase II seemed to have occurred in this mutant (Shales et al., 1980). Recently, Ingles and Shales (1982) have shown with the use of DNA-mediated gene transfer that the gene defective in tsAF8 is that of RNA polymerase II. Recognition of revertants from transformants in these experiments was based on the use of donor and recipient strains which differed with respect to the sensitivity of their RNA polymerases II to α amanitin.

3. *Mutants Arrested in Phase S*

Hirschberg (1981) and Hirschberg and Marcus (1982) isolated and studied three S phase ts cell cycle mutants from the Chinese hamster lung cell line E36. One mutant, ts41, seems to be defective in the traverse from S to G2 when incubated at the restrictive temperature for growth. At this temperature, the mutant cells reach the end of S and initiate immediately a second round of DNA replication followed by a third and a fourth. As a result, the nuclei enlarge, and with the use of the technique of premature chromosome condensation, cells with four and eight chromatid chromosomes are observed. Cells with very large nuclei could not be induced to go through premature chromosome condensation. DNA synthesis in ts41 incubated at the restrictive temperature continues for about 48 hr and ceases. Eventually the mutant giant cells disintegrate. The duplicated chromosomal segments in ts41 cells, grown for 24 hr at the restrictive temperature and observed by the premature chromosome condensation technique, seem to be much less condensed in comparison to similar PCC in wild-type cells. The significance of this undercondensed appearance of the PCC to the nature of the ts function in these cells is not yet understood. The ordered synthesis of DNA in each round of replication exhibited by these cells at the restrictive temperature indicates that at least two mechanisms control a single replication of DNA in a cycle. One operates at the replicon level, and inhibits a second replication of each replicated chromosomal segment, while the other operates at the level of the whole genome and inhibits a second round of replication after the first round has ended. In ts41 cells, the mechanism which operates at the level of the whole genome is defective and allows additional rounds of replication at the restrictive temperature. However, in each round, DNA replication is ordered, indicating that another mechanism which controls a single replication at the level of each replicon functions normally. Endoreduplication (Schwarzacher and Schnedl, 1965) is a manifestation of perturbations of the mechanism which controls DNA replication at the genome level, while gene amplification (Schimke, 1982) manifests perturbations in DNA replication at the replicon level. The second mutant, ts24, is being studied in detail by Fainsod et al. (1984a,b). This ts mutant arrests in early S at the restrictive temperature. Its execution point is also in early S, close to the point of arrest. Up to this point and beyond it DNA syntyhsis occurs even at the restrictive temperature. At the point of arrest DNA synthesis continues almost at the normal rate exhibited by the parental wild-type cells. However, practically no net DNA synthesis is observed and additional studies show that the newly synthesized DNA is degraded. Ligase activity seems to be normal. Thus, this mutant seems to be defective in a function which is not a direct part of the DNA synthesizing machinery but is essential for the progress of cells from early S to middle S. As mentioned before, S phase seems to include three stages (Klevecz et al., 1975; Holmquist et al., 1982). ts24 may be defective in a function which is required for the normal progress from the first stage in which only about 10–15% of the DNA is replicated to the second stage. Cloning of the normal gene defective in ts24 has been successful and the results will be described in the last section.

McCracken (1982) isolated an S phase ts cell cycle mutant, tsC8, from the CHO established cell line. The rate of incorporation of radioactive thymidine into DNA in tsC8 cells transferred to the restrictive temperature for growth was unaffected for about 2 hr and then decreased considerably, indicating a ts defect in DNA synthesis. The execution point of tsC8 was mapped to the interval between G1/S and middle S. The mutant cells incubated at the restrictive temperature lost viability rapidly. Complementation studies in cell hybrids show that the mutation in tsC8 is recessive.

Srinivasan et al. (1980) isolated from CHO cells two ts mutants for growth, ts13A and ts15C, in which DNA synthesis was shut off rapidly when the mutant cells were transferred to the restrictive temperature. The mutations in these mutants seem to be allelic and recessive. The mutants are more sensitive to alkylating agents like EMS as compared to their parental cells but show normal sensitivity to UV irradiation. The high toxicity of alkylating agents is not correlated with a change in the rate of induction of mutations by these agents. In temperature-insensitive revertants, the high sensitivity to the alkylating agents is changed to normal, indicating that both lesions resulted from a single mutation. Biochemical analysis of these two mutants may shed light on genes which participate both in DNA repair and DNA synthesis.

Nishimoto et al. (1978) studied an interesting S phase ts cell cycle mutant, tsBN-2, which originated from the BHK cell line. They showed that the rate of DNA synthesis declines rapidly in synchronized cultures of tsBN-2 cells transferred during early S to the restrictive temperature. The cells arrest at about middle S, lose the nuclear membrane, and exhibit PCC, which is followed later by the appearance of cells with micronuclei. Eventually these cells die. Cells transferred to the restrictive temperature in G1 arrest there, implicating the temperature-sensitive function in the initiation of DNA synthesis. Indeed, Eilen et al. (1980) have shown with the use of fiber autoradiography that the rate of fork movement is normal in tsBN-2 cells incubated at the restrictive temperature, while the frequency of initiation events is reduced. An additional result, indicating that this ts function is required for DNA replication, was based on the temperature sensitivity of Herpes simplex DNA replication in these cells (Yanagi et al., 1978).

However, recent studies indicate that the primary defect in tsNB2 cells concerns the regulation of chromosome condensation. Studies performed by Nishimoto et al. (1981), showed that the chromosomes in tsBN2 cells were induced to prematurely condense in cells arrested at the permissive temperature by DNA synthesis inhibitors, and shifted up to the restrictive temperature. tsBn2 cells incubated in G2 phase at the restrictive temperature also showed PCC. These results clearly showed that DNA synthesis was not required to induce the abnormal PCC phenotype. Suppression of PCC by cycloheximide, on the other hand, indicated that protein synthesis was required to induce PCC. Additional studies (Ajiro et al., 1983) of tsBN2 PCC were performed in synchronized cells in S phase which were shifted up to the restrictive temperature. They showed that histone H1 and H3 were phosphorylated extensively, in a manner similar to that which occurs in normal cells in late G2 phase and mitosis, when dramatic chromosome condensation takes place. Furthermore (Yamashita et al., 1984), an acidic protein associated with chromosome condensation has been identified in tsBn2 PCC. All of these results indicate that a condensing protein is prematurely synthesized in tsBN2 cells incubated at the restrictive temperature (Nishimoto et al., 1981). Preliminary studies indicate that the ts mutation in tsBN2 cells can be complemented by DNA mediated gene transfer (Kai et al., 1983).

Three S phase ts cell cycle mutants of mouse origin have been studied in detail. They include ts2 isolated by Slater and Ozer (1976) from the established cell line Balb/C-3T3, and tsC1 and tsA1S9 isolated by Thompson et al. (1970) from L cells. Randomly growing cells of ts2 transferred to the restrictive temperature arrest in S and at the G1/S boundary (Slater and Ozer, 1976). The ts defect does not appear to result from a deficiency in precursors for DNA synthesis. Polyoma virus DNA synthesis in the mutant cells is temperature sensitive, while the replication of mouse adenovirus is unaffected by the mutation. Jha et al. (1980) performed complementation studies with ts2 cells and assigned this gene to human chromosome X. Hybrids between ts2 and several DNA negative ts mutants, including tsA1S9 and tsC1, did not show complementation, even though ts2, tsA1S9, and tsC1 arose independently and show differences in phenotype, indicative of mutations in different genes. These somewhat puzzling results concerning the noncomplementing mutations, add to the other noncomplementing group of mutations described by Meiss et al. (1978), which have also been assigned to human chromosome X. Further studies are required to determine the nature of noncomplementation exhibited by these two groups of mutants.

Sheinin (1976) studied the S phase ts cell cycle mutant tsA1S9 and showed that at the restrictive temperature the cells could not convert the newly synthesized DNA (molecular weight 5 x 10^6 daltons) to high-molecular-weight chromosomal DNA. The organization of the chromatin in the mutant cells arrested in S was changed (Setterfield et al., 1978) and condensed chromatin bodies seemed to disaggregate. The mutant cells resumed DNA synthesis upon their transfer to the permissive temperature and, concomitantly, the normal organization of chromatin with condensed bodies reappeared. It is assumed that the normal disaggregation and reaggregation of chromatin during S is upset in the mutant-arrested cells, so that the disaggregation necessary for replication occurs, but the reaggregation is prevented by the lack of DNA synthesis. Polyoma virus is replicated in tsA1S9 at the restrictive temperature for growth, indicating that the functions required for the virus DNA replication are not affected by the ts mutation. Sparkuhl and Sheinin (1980) observed a marked increase in cytoplasmic protein degradation in tsA1S9 cells transferred to the restrictive temperature. Enhancement of protein degradation was observed as early as 2–4 hr after the temperature upshift, before any reduction in DNA synthesis occurred. The investigators proposed that the inactivation of the ts cell cycle gene product results in the activation of a lysosome-mediated mechanism for rapid protein degradation.

Colwill and Sheinin (1982, 1983) who continued to study the mutation in tsA1S9 cells, concluded from the terminal phenotype of the cells arrested in S phase, that the defect in these mutant cells result in modification of chromatin structure and supercoiling of nuclear DNA. They have therefore studied the topoisomerase function in tsA1S9 cells and found that these mutant cells exhibit an enhanced sensitivity to novobiocin which inhibits topoisomerase II (Miller et al., 1981). Furthermore, the topoisomerase II activity in nuclear novobiocin binding proteins purified from tsA1S9 cells grown at permissive temperature showed high sensitivity to novobiocin, and lacked a 30,000-dalton polypeptide component present in the wild type cells and a temperature resistant revertant strain. At nonpermissive temperature, purified novobiocin binding proteins from tsA1S9 cells showed no topoisomerase II activity. On the basis of all of these results, Colwill and Sheinin (1983) suggested that tsA1S9 cells encode a temperature sensitive polypeptide that is required for normal DNA-topoisomeras II activity.

Guttman and Sheinin (1979) studied the tsC1 S phase mutant and found by kinetic studies that this mutant is defective in a function which is required during S, probably during the late part of S. In tsC1-arrested cells, the chromatin shows changes in aggregation similar to those observed in tsA1S9. The mutation is reversible and, upon the transfer of cells to the permissive temperature, DNA synthesis is resumed.

Hyodo and Suzuki (1982) isolated from mouse FM3A cells an S phase ts cell cycle mutant, Ts131B, which arrests throughout S when incubated at the restrictive temperature. At the restrictive temperature DNA synthesis is shut off. DNA fiber autoradiography was performed in order to determine whether the decrease in the rate of DNA synthesis observed in Ts131B cells, incubated at the restrictive temperature, results from a decrease in the rate of DNA chain elongation or the rate of initiation of DNA replication. The results show a twofold decrease in the rate of elongation in the mutant cells grown for 6 hr at the restrictive temperature. The rate of incorporation of thymidine into the DNA of the mutant cells incubated at the restrictive temperature for 6 hr is reduced to 25% of the rate exhibited by them at the permissive temperature. Thus, the investigators concluded that the mutation resulted in a reduction both in the rate of elongation and in the rate of initiation of DNA replication. In another ts S phase cell cycle mutant, T244, which originated from the mouse cell line FM3A, Tsai et al. (1979) observed that a rapid decrease in the rate of DNA synthesis which occurred in the mutant cells grown at the restrictive temperature, was accompanied by a loss of DNA polymerase α activity in the nuclei of the mutant cells.

4. *Mutants Arrested in G2*

Previously, all efforts to isolate mammalian ts cell cycle mutants which arrest in G2 have been unsuccessful. Basilico (1978) suggested that such mutants could have been very sensitive to incubation at the restrictive temperature of growth, and thus were lost during the mutant-enrichment step which required exposure of the cells to the restrictive temperature. Recently, Hirschberg (1981) and Hirschberg and Marcus (1982) isolated a G2 ts cell cycle mutant, ts20, from the Chinese hamster lung established cell line E36. Their studies show that ts20 cells transferred to the restrictive temperature begin immediately to lose viability, indicating that passage of cells through the execution point at the restrictive temperature commits them irreversibly to death. Most of the ts20 cells detach and float in the medium within 24 hr of incubation at the restrictive temperature, and many of them already begin to disintegrate.

Mita et al. (1980) and Matsumoto et al. (1980) isolated a G2 ts cell cycle mutant, ts85, of mouse origin from the established cell line FM3A. At the restrictive temperature, most mutant cells arrest at G2 and some in late S phase. Electron micrographs showed that the chromatin in the arrested cells is under-condensed, and biochemical analysis showed a marked deficiency in histone H1 phosphorylation. However, *in vitro* studies clearly showed that the activity of nuclear and cytoplasmic protein kinase with histone H1 as a substrate behaved as wild type, indicating that the reduction in histone H1 phosphorylation at the restrictive temperature for growth is secondary to some other primary defects (Yasuda et al., 1981). In addition, another modification of histones, the ubiquitination of histone H_2A, has also been found to be drastically reduced in ts85 cells grown at the restrictive temperature (Marunouchi et al., 1982; Matsumoto et al., 1983). A temperature resistant revertant clone, ts85R-MN3, isolated from ts85 mutant cells, exhibited wild type levels of both modifications of the histones suggesting that a single cell cycle mutation was responsible for the complex mutant phenotype.

Recently, Finley et al. (1984) and Ciechanover et al. (1984) showed that the primary leison in ts85 cells is in the ubiquitin-activating-enzyme (E1) which is a component of the ubiquitin-protein ligase system. At the restrictive temperature, due to the thermolability of E1, H_2A histone is not ubiquitinated. Ciechanover et al. (1984) further showed that in ts85 cells at the restrictive temperature, short-lived proteins were not degraded. Their studies suggest that degradation of the bulk of short-lived intracellular proteins in these mouse cells proceeds through a ubiquitin dependent pathway.

The nature of the late S and G2 arrest of these mutant cells is not yet clear. Ciechanover et al. (1984) suggest that the cells arrest at the restrictive temperature because termination of DNA replication fails, or because G2 regulatory proteins which are normally labile fail to be degraded in ts85 cells.

5. *Mutants Arrested in Both G1 and G2*

Ohlsson-Wilhelm et al. (1980) reported on the isolation of cold-sensitive (cs) cell cycle mutants, II29 and II32, from CHO cells which arrest both at

G1 and G2 when incubated at the restrictive temperature. Shifting the arrested cells back to the permissive temperature was found to allow the cells arrested in G1 to resume growth; however, the arrest in G2 was found to be irreversible. Chen and Wang (1982) have also isolated a ts cell cycle mutant which arrests at both G1 and G2 when incubated at the restrictive temperature. Their ts mutant, ts550, has originated from the Syrian hamster established cell line HM1-1. Several studies reviewed by Mitchison (1971) and Pardee et al. (1978) suggested that controlling events may be present in both G1 and G2. Cell cycle mutants defective in the execution of such events should be arrested in both G1 and G2. Isolation of mutants which arrest at both of these two phases seem to support this notion. However, the specific irreversibilty of the arrest in G2, exhibited by the ts mutants II29 and II32, in contrast to the reversibility of the G1 arrest of the same mutants, is puzzling in this respect if the arrest in both phases is caused by a single mutation. Further study of these mutants is required to answer these questions.

A summary of cell cycle mutants isolated in Chinese hamster and other cell lines is given in Table 20.1.

E. The G1-Less G2-Less Cell Cycle

Two chinese hamster cell lines, DON and V79-8, which lack G1, yield interesting and intriguing data concerning the general idea implicating the G1 period or a part of it in the control of cell proliferation. V79-8 has been studied considerably by several investigators (Robbins and Scharff, 1967; Liskay, 1978; Liskay and Prescott, 1978; Rao et al., 1978; Liskay et al., 1979; Liskay et al., 1980; Stancel et al., 1981). It is now well established that S phase cells contain a cytoplasmic factor(s) that induces DNA synthesis in G1 phase nuclei upon fusion between the cells in S and those in G1 (Rao and Johnson, 1970). This factor(s) is absent in G2 and mitosis in cells which exhibit both G2 and G1, and accumulates after mitosis during G1 (Rao et al., 1978). In V79-8 this factor(s) may be made during mitosis (Rao and Sunkara, 1980). As a result, when HeLa cells in G1 are fused with mitotic V79-8 cells, they are induced to go through premature chromosome condensation, on the one hand, and concomitantly are induced to initiate DNA synthesis. Fusion of CHO G1$^+$ cells with the V79-8 G1$^-$ cells results in the formation of G1$^-$ hybrid cells indicating that the G1$^-$ phenotype behaves as a dominant allele in intraspecific hybrids (Liskay et al., 1979). This result indicates that V79-8 cells supply a factor(s) necessary for the initiation of DNA synthesis which is limiting in the G1$^+$ CHO cells.

G1$^+$ mutants have been isolated from V79-8 cells. The different mutants complemented each other upon fusion (Liskay and Prescott, 1978). Similarly, fusion of cells from several known Chinese hamster cell lines which grow slower than V79-8 and are G1$^+$ resulted in the appearance of G1$^-$ hybrid cells (Liskay et al., 1979). In addition, it has

TABLE 20.1

Cell Line	Mutant Designation	Point of Arrest or Pause	Function Affected	References
A List of Chinese Hamster Temperature-Sensitive Cell Cycle Mutants				
(V79)E36	ts82	Early G1		Hirschberg (1981)
	ts83	Early G1		Hirschberg and Marcus (1982)
	ts79	S		
	ts20	G2		
	ts41	S	Control of a single S/cycle	
	ts24	early S	Progress from early S to middle S	Fainsod et al. (1984a,b)
HT-1	ts154	G1		Roufa et al. (1979)
GM-7	ts111	Mitosis	Cytokinesis	Hatzfeld & Buttin (1975)
CHO	MS1-1	Mitosis	Cytokinesis	Thompson and Lindl (1976)
	Grs-2	Mitosis	β-tubulin	Abraham et al. (1983)
	Cmd-4		β-tubulin	
	Tax-1		α tubulin	
	Tax-18		Spindle formation	Cabral et al. (1983)
	tsC8	S	DNA synthesis	McCracken (1982)
	ts13A	S	DNA synthesis and	Srinivasan et al. (1980)
	ts15C		sensitivity to alkylating agents	
	II29 (cold sensitive)	G1 and G2		Ohlsson-Wilhelm et al. (1980)
	II32 (cold sensitive)			
(DON)Wg1A	K12	G1		Roscoe et al. (1973)
	K18			
	K27			
	K33			
	4/3	G1		Melero (1979)
	4/2			
	3/1			
	5/2			
	tSK/34C	G1	Membrane glycoprotein synthesis	Tenner et al. (1976) Tenner and Scheffler (1979)

Organism	Cell Line	Mutant Designation	Point of Arrest or Pause	Function Affected	References
A List of Temperature-Sensitive Cell Cycle Mutants Studied in Detail which Are Not of Chinese Hamster Origin					
Syrian hamster	BHK21	tsAF8	G1	RNA polymerase II	Meiss and Basilico (1972)
					Ingles and Shales (1982)
		ts13	G1		Jonak and Baserga (1979)
		tsBN-2	G1,S	Initiation of DNA synthesis	Nishimoto et al. (1978) Eilen et al. (1980)
	HM-1	ts655	Mitosis	Prophase progression	Wang (1976)
		ts546		Metaphase	Wang and Yin (1976)
		ts687		Anaphase chromosome movement	Wissinger and Wang (1978)
		ts745		Centriole separation chromosome movement	Wang et al. (1983)
		ts550	G1 and G2		Chen and Wang (1982)
Murine	L5178Y	ts2	Mitosis	Metaphase	Shiomy and Sato (1976)
	Balb/C-3&3	ts2	S	DNA synthesis	Slater and Ozer (1976)
	L	tsCl	S	DNA synthesis	Guttman and Sheinin (1979)
		tsA1S9	S	DNA synthesis	Sheinin (1976)
	FM3A	Ts131b	S	DNA synthesis	Hyodo and Suzuki (1982)
		T244	S	DNA polymerase α	Tsai et al. (1979)
		ts85	S and G2	Histone H1 phosphorylation	Matsumoto et al. (1980)

been found that reduction in the rate of protein synthesis in V79-8 cells resulted in the appearance of a G1 period (Liskay et al., 1980). All these results seem to indicate that any mutation or change in the environment which slows down growth rate will result in the appearance of a G1 period. In interspecific hybrids between G1$^-$ V79-8 cells, and mouse or human G1$^+$ cells, the G1$^-$ behaved as a recessive gene (Liskay, 1978), indicating that the factor(s) required for the synthesis of DNA are species specific. This is a somewhat puzzling result in view of the ability of many genes including cell cycle genes to function normally in cell hybrids.

Naha (1979) has isolated G1$^-$ mutants at a high frequency from a ts cell cycle mutant arrested in G1 at the restrictive temperature. One such mutant, R18, exhibits a G1$^+$ phenotype at 33°C and a G1$^-$ phenotype at 38°C. The cell lines used in this study are of mouse Balb/C-3T3 origin; the parental G1 ts cell cycle mutant was AB-3. Naha's results support the results obtained with the Chinese hamster V79-8 cells.

The absence of a measurable G2 phase in V79-8 cells indicates that in this cell line chromosome condensation must reach the level exhibited by cells in prophase already at the end of the S phase (Liskay, 1977). As the DNA has to be fully stretched during its replication, it most probably begins to condense immediately after it has replicated. Such a condensation may act as the mechanism which controls a single replication of every replicon in a single S phase, making the DNA unavailable to the DNA synthesizing machinery. Chromatin condensation occurs along S phase also in cells which exhibit G2 (Moser and Meiss, 1977); however, in such cells it continues along G2 to the level exhibited in prophase cells (Marcus et al., 1979).

F. Cloning of Cell Cycle Genes

In spite of the many investigations and the extensive effort invested in the analysis of cell cycle, only a few cell cycle genes have been completely characterized (Hartwell, 1978). The advances made in genetic engineering techniques make feasible the cloning of cell cycle genes for a full analysis. In yeast, this approach has already been successfully used (Nasmyth and Reed, 1980; Beach et al., 1982). Having cloned cell cycle genes at hand opens several new ways of studying cell cycle including their *in vitro* transcription and translation, to identify and understand their functions at the molecular level. Cloning can be achieved by DNA-mediated gene transfer (Wigler et al., 1977; Wigler et al., 1978; Lowy et al., 1980; Jolly et al., 1982; Lund et al., 1982; Chapter 8). Successful cloning by this technique requires that the mutant recipient cells be good recipients. This has been proven true for some cell lines, mouse L cells being the best; however, Chinese hamster cells have been found to be poor recipients (Cirullo et al., 1983). Recently, a modification in DNA transformation has improved the efficiency of DNA transformation in Chinese hamster cells (Abraham et al., 1982). Nevertheless, the DNA transformation experiments in which total donor DNA is used, the efficiency of transformation may still be too low to use efficiently if unique genes are to be selected. Cirullo et al. (1983) have circumvented this problem and transferred human DNA to the Chinese hamster mutant cells by an elegant modification of the classical technique of interspecific cell hybridization (Harris and Watkins, 1965). Their procedure included three cycles of hybridization experiments. In the first cycle, human cells were fused with the Chinese hamster mutant cells and hybrid cells were selected. In such hybrids, usually only a few human chromosomes are maintained and can be identified. These hybrids are irradiated with a high dosage of gamma irradiation, in order to fragment their chromosomes. They are then fused to new mutant cells and secondary hybrids are selected. Owing to preferential loss of the fragmented chromosomes in these hybrids, only small translocated human chromosomal fragments remain, including the fragment that contains the normal human allele of the mutated Chinese hamster gene. This procedure has been repeated again. In the resulting hybrid, which still expressed the selected gene, less than 0.1% of the DNA was of human origin. A genomic DNA library can be prepared from this hybrid and human DNA fragments can be screened and isolated by the methods described by Gusella et al. (1980). With this method, cloning of unique genes which complement mutations in Chinese hamster cells may be achieved.

Another method which may be used to study cell cycle mutants and eventually to clone the normal alleles of the cell cycle genes is based on the isolation and purification of mRNA of these genes. Gene expression has been successfully studied in mammalian cells microinjected with mRNA (Graessmann and Graessmann, 1976; Stacy and Allfrey, 1976). Specific mRNA has been found to be efficiently translated *in vivo* and transiently complemented a mutant phenotype (Liu et al., 1979). In addition, Lin et al. (1982) succeeded in fractionating mRNA according to its molecular weight and obtained biologically active fractions enriched for specific mRNAs. We have used this technique to study the Chinese hamster ts cell cycle mutant ts24. This mutant, which I described above, is arrested in early S phase when incubated at the restrictive temperature. We microinjected the mutant cells arrested in S with total poly-A mRNA isolated from the wild-type parental cell line E36 and found that the microinjected mutant cells resumed net DNA synthesis at the restrictive temperature (Fainsod et al., 1984b). Next, we franctionated the mRNA and succeeded in finding a fraction which contained the mRNA capable upon microinjection of transiently complementing the ts cell cycle mutation in ts24 cells. This finding has already advanced our studies of this mutant considerably. We now know that this cell cycle gene is transcribed into a poly-A mRNA which can be obtained from the cytoplasm and is made of about 940 bases. If the untranscribed leader sequence and the poly-A tail added together include 240 bases, then the coding potential of this mRNA is a protein of about 230 amino acids. In addition, we obtained a fraction of mRNA which is enriched by about 100-fold for the specific mRNA which complements transiently the mutation in the ts cell cycle mutant ts24. Our assay for transient complementation was based on the resumption of net DNA synthesis measured by [^{3}H]thymidine incorporation and determined by autoradiography. This may be the method of choice in determining in such experiments transient complementation of ts cell cycle mutations in mutants arrested in any one of the cell cycle phases.

The mRNA fraction enriched for the specific cell cycle gene can be used to clone the gene in an expression vector for a final characterization of its product (Okayama and Berg, 1983; see Chapter 10). This approach should be an extremely useful one for extending the analysis of gene products involved in regulating the cell cycle.

Recently (Fainsod et al., 1984c), we have isolated from a human genomic library in λ charon 4a, a phage containing an insert (14.2 kb) which complements the ts mutation in mutant ts24. The procedure we used to screen this library is described below. It can be used to screen any gene whose function is required for growth.

The human library was first screened with labeled cDNA synthesized from the mRNA fraction enriched for the cell cycle message that transiently complemented the mutation in ts24 cells. The 845 plaques we picked were screened in DNA transformation experiments for transient complementation manifested by the ability of ts24 cells arrested at the restrictive temperature to resume net DNA synthesis and incorporate [^{3}H]-thymidine. Screening was first performed on groups of 85 different phages, and one active group was found. This group was subdivided and after two additional cycles of screening, one phage was isolated. Its human DNA insert was found to complement stably the ts mutation in ts24 cells. The homology of the human gene in this insert to the defective Chinese hamster gene is under study.

ACKNOWLEDGMENTS

I thank Ruth Goitein, Abraham Fainsod, and Gill Diamond for their helpful comments and excellent assistance in preparing this manuscript.

This work was supported by the United States–Israel Binational Science Foundation No. 2895-82.

REFERENCES

Abraham, I., Marcus, M., Cabral, F., and Gottesman, M. M. (1983). J. *Cell Biol.* **97**, 1055–1061.

Abraham, I., Tyagi, J. S., and Gottesman, M. M., (1982). *Somat. Cell Genet.* **8**, 23–39.

Ajiro, K., Nishimoto, T., and Takahashi, T. (1983). *J. Biol. Chem.* **258**, 4534–4583.

Baserga, R. (1976). *Multiplication and Division in Mammalian Cells,* Marcel Dekker, New York, p. 239.

Basilico, C. (1978). *J. Cell. Physiol.* **95**, 367–376.

Basilico, C. and Meiss, H. K. (1974). *Met. Cell Biol.* (D. M. Prescott, ed.), Academic Press, New York, Vol. 8, pp. 1–22.

Beach, D., Durkacz, B., and Nurse, P. (1982). *Nature* **300**, 706–709.

Bisson, L. and Thorner, J. (1977). *J. Bacteriol.* **132**, 44–50.

Burstin, S. J. and Basilico, C. (1975). *Proc. Natl. Acad. Sci. USA* **72**, 2540–2544.

Byers, B. and Goetsch, L. (1976a). *J. Cell Biol.* **69**, 717–721.

Byers, B. and Goetsch, L. (1976b). *J. Cell Biol.* **70**, 35a.

Cabral, F. (1983). *J. Cell Biol.* **97**, 22–29.

Cabral, F., Abraham, I., and Gottesman, M. M. (1981). *Proc Natl. Acad. Sci. USA* **78**, 4388–4391

Cabral, F., Abraham, I., and Gottesman, M. M. (1982). *Mol. Cell. Biol.* **2**, 720–729.

Cabral, F., Wible, L., Brenners, S., and Brinkley, B. R. (1983). *J. Cell Biol.* **97**, 30–39.

Chang, C. C., Trosko, T. E., and Akera, T. (1978). *Mutat. Res.* **51**, 85–98.

Chasin, L. A. and Urlaub, G. (1975). *Science* **187**, 1091–1093.

Chen, D. J-C. and Wang, R. J. (1982). *Somat Cell Genet.* **8**, 653–666.

Cherington, P. V., Smith, B. L., and Pardee, A. B. (1979). *Proc. Natl. Acad. Sci. USA* **76**, 3937–3941.

Ciechanover, A., Finley, D., and Varshavsky, A. (1984). *Cell* **37**, 57–66.

Cirullo, R. E., Dana, S., and Wasmuth, J. J. (1983). *Mol. Cell. Biol.* **3**, 892–902.

Cleaver, G. E. (1978). *Genetics* **87**, 129–138.

Colwill, R. W. and Sheinin, R. (1982). *Can. J. Biochem.* **60**, 195–203.

Colwill, R. W. and Sheinin, R. (1983). *Proc. Natl. Acad. Sci. USA* **80**, 4644–4648.

Crissman, H. A., Mullany, P. F., and Steinkamp, J. A. (1975). *Mel. Cell. Biol.* **9**, 179–246.

Darzynkiewicz, Z., Crissman, H., Traganos, F., and Steinkamp, J. (1982). *J. Cell. Physiol.* **113**, 465–474.

Dell'Orco, R. T., Crissman, H. A., Steinkamp J. A., and Kraemer, P. M. (1975). *Exp. Cell Res.* **92**, 271–274.

DePamphilis, M. L. and Wassarman, P. M. (1980). *Ann. Rev. Biochem.* **49**, 627–666.

Dewey, W. C. and Highfield, D. P. (1976). *Radiat. Res.* **65**, 511–528.

Dhar, R., Lai, C. J., and Khoury, G. (1978). *Cell* **13**, 345–358.

Dubrow, R., Riddle, V. G. H., and Pardee, A. B. (1979). *Cancer Res.* **39**, 2718–2726.

Eilen, E., Hand, R., and Basilico, C. (1980). *J. Cell. Physiol.* **105**, 259–266.

Esko, J. D. and Raetz, C. R. H. (1978). *Proc. Natl. Acad. Sci. USA* **75**, 1190–1193.

Fainsod, A, Goitein, R., and Marcus, M. (1984a). *Exp. Cell Res.* 152, 77–90.

Fainsod, A., Marcus, M., Lin, P. F., and Ruddle, F. H. (1984b). *Proc. Natl. Acad. Sci. USA* **81**, 2393–2395.

Fainsod, A., Marcus, M., and Ruddle, F. H. (1984c). (submitted).

Farber, R. A. and Davidson, R. L. (1977). *Cytogenet. Cell Genet.* **18**, 349–363.

Farber, R. A. and Davidson, R. L. (1978). *Proc. Natl. Acad. Sci. USA* **75**, 1470–1474.

Finley, D., Ciechanover, A., and Varshavsky, A. (1984). *Cell* **37**, 43–55.

Floros, J., Jonak, G., Galati, N., and Baserga, R. (1981). *Exp. Cell Res.* **132**, 215–223.

Gillin, F. D., Roufa, D. J., Beaudet, A. L., and Caskey, C. T. (1972). *Genetics* **72**, 239–252.

Goitein, R., Marcus, M., and Sperling, K. (1984). *Cytogenet. Cell Genet.* **38**, 116–121.

Goldsby, R. A. and Mandell, N. (1973). *Methods Cell Biol.* **7**, 261–268.

Graessmann, M. and Graessman, A. (1976). *Proc. Natl. Acad. Sci. USA* **73**, 366–370.

Groppi, Jr., V. E. and Coffino, P. (1980). *Cell* **21**, 195–204.

Gupta, R. S. (1980). *Somatic Cell Genet.* **5**, 115–125.

Gupta, R. S. and Siminovitch, L. (1978). *Somat. Cell Genet.* **4**, 553–572.

Gusella, J. F., Keys, C., Varsanyi-Breiner, A., Kao, F.-T., Jones, C., Puck, T., and Housman, D. (1980). *Proc. Natl. Acad. Sci. USA* **77**, 2829–2833.

Guttman, S. A. and Sheinin, R. (1979). *Exp. Cell Res.* **123**, 191–205.

Hand, R. (1978). *Cell* **15**, 317–325.

Harland, R. M. and Laskey, R. A. (1980), *Cell* **21**, 761–771.

Harris, H. and Watkins, J. S. (1965). *Nature* **205**, 640.

Hartwell, L. H. (1974). *Bact. Rev.* **38**, 164–198.

Hartwell, L. H. (1978). *J. Cell. Biol.* **77**, 627–637.

Hatzfeld, J. and Buttin, G. (1975). *Cell* **5**, 123–129.

Hayflick, L. (1965). *Exp. Cell Res.* **37**, 614–636.

Heintz, N. H. and Hamlin, J. L. (1982). *Proc. Natl. Acad. Sci. USA* **79**, 4083–4087.

Hirschberg, J. (1981). Ph.D. Thesis, The Hebrew University, pp. 1–103.

Hirschberg, J., Lavi, U., Goitein, R., and Marcus, M. (1980). *Exp. Cell Res.* **130**, 63–72.

Hirschberg, J. and Marcus, M. (1982). *J. Cell. Physiol.* **113**, 159–166.

Hittelman, W. N. and Rao, P. N. (1978). *J. Cell. Physiol.* **95**, 333–342.

Hochhauser, S. J., Stein, J. L., and Stein, G. S. (1981). *Inter. Rev. Cytol.* **71**, 95–243.

Hochstadt, J., Ozer, H. L., and Shopsis, C. (1981). *Current Topics in Microbiology and Immunology* (W. Henle et al., eds.), Springer-Verlag, New York, Vol. 94/95, pp. 243–308.

Holmquist, G., Gray, M., Porter, T., and Jordan, J. (1982). *Cell* **31**, 121–129.

Hyodo, M. and Suzuki, K. (1982). *Exp. Cell Res.* **137**, 31–38.

Ingles, C. J. and Shales, M. (1982). *Mol. Cell. Biol.* **2**, 666–673.

Inoué, S. (1981). *J. Cell Biol.* **91**, 131s–147s.

Jha, K. K., Siniscalco, M., and Ozer, H. L. (1980). *Somat. Cell Genet.* **6**, 603–614.

Jolly, D. J., Esty, A. C., Bernard, H. U., and Friedman, T. (1982). *Proc. Natl. Acad. Sci. USA* **79**, 5038–5041.

Jonak, G. J. and Baserga, R. (1979). *Cell* **18**, 117–123.

Kai, R., Sekisuchi, T., Yamashita, K., Sekisuchi, M., and Nishimoto, T. (1983). *Somatic Cell Genet.* **9**, 673–680.

Kao, F.-T. and Puck, T. T. (1974). *Met. Cell Biol.* (D. M. Prescott, ed.), Academic Press, New York, Vol. 8, pp. 23–39.

Klevecz, R. R., Keniston, B., and Deaven, L. (1975). *Cell* **5**, 195–203.

Kornberg, A. (1980). *DNA Replication* W. H. Freeman, San Francisco, pp. 724.

Kornberg, A. (1982). *DNA Replication Supplement*, W. H. Freeman, San Francisco, pp. 273.

Kudirka, D. T. and Van't Hof, J. (1980). *Exp. Cell Res.* **130**, 443–450.

Lajtha, L. G. (1963) *J. Cell. Comp. Physiol.* **62**, 143–149.

Landy-Otsuka, F. and Scheffler, I. E. (1980). *J. Cell. Physiol.* **105**, 209–220.

Latt, S. A. (1973). *Proc. Natl. Acad. Sci. USA* **70**, 3395–3399.

Lau, Y.-F. and Arrighi, F. E. (1981). *Chromosoma* **83**, 721–741.

Lederberg, J. and Lederberg, E. M. (1952). *J. Bacteriol.* **63**, 399–406.

Lejeune, J., Dutrillaux, B., Lafourcade, J., Berger, R., Abonyi, D., and Rethoré, M. D. (1968). *Compt. Rend. Acad. Sci.* **266**, 24–26.

Lin, M. S. and Davidson, R. L. (1975). *Somat. Cell Genet.* **1**, 111–122.

Lin, P. F., Yamaizumi, M., Murphy, P. D., Egg, A., and Ruddle, F. H. (1982). *Proc. Natl. Acad. Sci. USA* **79**, 4290–4294.

Liskay, R. M. (1977). *Proc. Natl. Acad. Sci. USA* **74**, 1622–1625.

Liskay, R. M. (1978). *Exp. Cell Res.* **114**, 69–77.

Liskay, R. M., Fullerton, P., and Kornfeld, B. (1980). *Exp. Cell Res.* **128**, 191–197.

Liskay, R. M., Leonard, K. E., and Prescott, D. M. (1979). *Somat. Cell Genet.* **5**, 615–623.

Liskay, R. M. and Prescott, D. M. (1978). *Proc. Natl. Acad. Sci. USA* **75**, 2873–2877.

Liu, C. P., Slate, D. L., Gravel, R., and Ruddle, F. H. (1979). *Proc. Natl. Acad. Sci. USA* **76**, 4503–4506.

Lowy, I., Pellicer, A., Jackson, J. F., Sim, G.-K. Silverstein, S., and Axel, R. (1980). *Cell* **22**, 817–823.

Lund, T., Grosveld, F. G., and Flavell, R. A. (1982). *Proc. Natl. Acad. Sci. USA* **79**, 520–524.

Marcus, M. and Hirschberg, J. (1982). In *Premature Chromosome Condensation,*(P. N. Rao, and R. T. Johnson, and K. Sperling, eds.) Academic Press, New York, pp. 173–194.

Marcus, M. and Meiss, H. K. (1981). *J. Cell Biol.* **91**, 7a.

Marcus, M., Nielsen, K., Goitein, R., and Gropp, A. (1979). *Exp. Cell Res.* **122**, 191–201.

Marcus, M. and Sperling, K. (1979). *Exp. Cell Res.* **123**, 406–411.

Marunouchi, T., Mita, S., Matsumoto, Y., and Yasuda, H. (1982). In *Premature Chromosome Condensation* (P. N. Rao, R. T. Johnson, and K. Sperling, eds.), Academic Press, New York, pp. 195–206.

Matsumoto, Y., Yasuda, H., Mita, S., Marunouchi, T., and Yamada, M. (1980). *Nature* **284**, 181–183.

Matsumoto, Y., Yasuda, H., Marunouchi, T., and Yamada, M. (1983). *FEBS Lett.* **151**, 139–142.

McCracken, A. A. (1982). *Somat. Cell Genet.* **8**, 179–195.

Meiss, H. K. and Basilico, C. (1972). *Nature New Biol.* **239**, 66–68.

Meiss, H. K., Talavera, A., and Nishimoto, T. (1978). *Somat. Cell Genet.* **4**, 125–130.

Melero, J. A. (1979). *J. Cell. Physiol.* **78**, 17–30.

Melero, J. A. and Fincham, V. (1978). *J. Cell. Physiol.* **95**, 295–306.

Miller, K. G., Liu, L. F., and Englund, P. T. (1981). *J. Biol. Chem.* **256**, 9334–9339.

Ming, P. L., Chang, N. L., and Baserga, R. (1976). *Proc. Natl. Acad. Sci. USA* **73**, 2052–2055.

Mita, S., Yasuda, H., Marunouchi, T., Ishiko, S., and Yamada, M. (1980). *Exp. Cell Res.* **126**, 407–417.

Mitchison, J. M. (1971). *The Biology of the Cell Cycle,* The University Press, Cambridge, p. 313.

Moser, G. C. and Meiss, H. K. (1977). *Somat. Cell Genet.* **3**, 449–456.

Naha, P. M. (1979). *J. Cell Sci.* **40**, 33–42.

Nasmyth, K. A. and Reed, S. I. (1980). *Proc. Natl. Acad. Sci. USA* **77**, 2119–2123.

Nathans, D. (1979). *Science* **206**, 903–909.

Nishimoto, T. and Basilico, C. (1978). *Somat. Cell Genet.* **4**, 323–340.

Nishimoto, T., Eilen, E., and Basilico, C. (1978). *Cell* **15**, 475–483.

Nishimoto, T., Takahashi, T., and Basilico, C. (1980). *Somat. Cell Genet.* **6**, 465–476.

Nishiomoto, T., Ishida, R., Ajiro, K., Yamamoto, S., and Takahashi, T. (1981). *J. Cell. Physiol.* **109**, 299–308.

Noel, B., Quack, B., Mottet, J., Nantois, Y., and Dutrillaux, B. (1977). *Exp. Cell Res.* **104**, 423–426.

Nurse, P., Thuriaux, P., and Nasmyth, K. (1976). *Mol. Gen. Genet.* **146**, 167–178.

Ohlsson-Wilhelm, B. M., Leary, J. F., Pacilio, M., and Martin, T. (1980). *Somat. Cell Genet.* **6**, 349–359.

Okayama, H. and Berg, P. (1983). *Mol. Cell. Biol.* **3**, 280–289.

Orkin, S. H. and Littlefield, J. W. (1971). *Exp. Cell Res.* **66**, 69–74.

Orr, E. and Rosenberger, R. F. (1976). *J. Bacteriol.* **126**, 903–906.

Pardee, A. B. (1974). *Proc. Natl. Acad. Sci. USA* **71**, 1286–1290.

Pardee, A. B., Dubrow, R., Hamlin, J. L., and Kletzien, R. F. (1978). *Ann. Rev. Biochem.* **47**, 715–750.

Pardoll, D. M., Vogelstein, B., and Coffey, D. S. (1980). *Cell* **19**, 527–536.

Pickett-Heaps, J. D., Tippit, D. H., and Porter, K. R. (1982). *Cell* **29**, 729–744.

Prescott, D. M. (1976a). *Adv. Genet.* **18**, 99–177.

Prescott, D. M. (1976b). *Reproduction of Eukaryotic Cells*, Academic Press, New York, p. 177.

Pringle, J. R. (1978). *J. Cell. Physiol.* **95**, 393–406.

Pringle, J. R. and Hartwell, L. H. (1981). In *The Molecular Biology of the Yeast Saccharomyces* (E. Strathern W. Jones and J. J. Broach eds.), Cold Spring Harbor, New York, pp. 97–142.

Quastler, H. and Sherman, F. G. (1959). *Exp. Cell Res.* **17**, 420–438.

Rao, P. N. and Johnson, R. T. (1970). *Nature* **225**, 159–164.

Rao, P. N. and Palumbo, M. E. (1978). *J. Cell Biol.* **79**, 2a.

Rao, P. N. and Sunkara, P. S. (1980). *Exp. Cell Res.* **125**, 507–511.

Rao, P. N., Wilson, B., and Puck, T. T. (1977). *J. Cell. Physiol.* **91**, 131–142.

Rao, P. N., Wilson, B. A., and Sunkara, P. S. (1978). *Proc. Natl. Acad. Sci. USA* **75**, 5043–5047.

Reddy, G. P. V. and Pardee, A. B. (1980). *Proc. Natl. Acad. Sci. USA* **77**, 3312–3316.

Reddy, G. P. V. and Pardee, A. B. (1983). *Nature* **304**, 86–88.

Robbins, E. and Scharff, M. D. (1967). *J. Cell. Biol.* **34**, 684–686.

Ronen, A., Gingerich, J. D., Duncan, A. M. V., and Heddle, J. A. (1984). *Proc. Natl. Acad. Sci. USA.* **81**, 6124–6128.

Roscoe, D. H., Read, M., and Robinson, H. (1973). *J. Cell. Physiol.* **82**, 325–332.

Rossini, M., Floros, J., Baserga, R., and Weinmann, R. (1979). In *Hormones and Cell Culture* (G. H. Sato and R. Ross, eds.), Cold Spring Harbor Conferences on Cell Proliferation 6, Cold Spring Harbor Lab, New York, pp. 393–402.

Roufa, D. J., McGill, S. M., and Mollenkamp, J. W. (1979). *Somat. Cell Genet.* **5**, 97–115.

Rubin, H. and Steiner, R. (1975). *J. Cell. Physiol.* **85**, 261–270.

Ruddle, F. H. (1973). *Nature* **242**, 165–169.

Schwartz, H. E., Moser, G. C., Holmes, S., and Meiss, H. K. (1979). *Somat. Cell Genet.* **5**, 217–224.

Schwarzacher, H. G. and Schnedl, W. (1965). *Cytogenetics* **4**, 1–18.

Schimke, R. T. (ed.) (1982). *Gene Amplification,* Cold Spring Harbor Laboratory, Cold Spring Harbor, New York, p. 337.

Seidman, M. M., Levine, A. J., and Weintroub, H. (1979). *Cell* **18**, 439–449.

Setterfield, G., Sheinin, R., Dardick, I., Kiss, G., and Dubsky, M. (1978). *J. Cell Biol.* **77**, 246–262.

Shales, M., Bergsagel, J., and Ingles, C. J. (1980). *J. Cell. Physiol.* **105**, 527–532.

Sheinin, R. (1976). *Cell* **7**, 49–57.

Sheinin, R., Humbert, J., and Pearlman, R. E. (1978). *Ann. Rev. Biochem.* **47**, 277–316.

Sheinin, R. and Lewis, P. N. (1980). *Somat. Cell. Genet.* **6**, 225–239.

Shilo, B., Shilo, V., and Simchen, G. (1976). *Nature* **264**, 767–770.

Shiomi, T. and Sato, K. (1976). *Exp. Cell Res.* **100**, 297–302.

Siegel, R. L. and Kalf, G. F. (1982). *J. Biol. Chem.* **257**, 1785–1790.

Simchen, G. (1978). *Ann. Rev. Genet.* **12**, 161–191.

Siminovitch, L. (1976). *Cell* **7**, 1–11.

Siminovitch, L. (1979). *ICN-UCLA Symposium on Molecular and Cellular Biology,* Alan R. Liss, New York, pp. 433–443.

Simon, A. E., Taylor, M. W., Bradley, W. E. C., and Thompson, L. H. (1982). *Mol. Cell. Biol.* **2**, 1126–1133.

Slater, M. L., and Ozer, H. (1976). *Cell* **7**, 289–295.

Smith, B. J. and Wigglesworth, N. M. (1972). *J. Cell. Physiol.* **80**, 253–259.

Smith, J. A. and Martin L. (1973). *Proc. Natl. Acad. Sci. USA* **70**, 1263–1267.

Soprano, K. J., Galanti, N., Jonak, G. J., McKercher, S., Pipas, J. M., Peden, K. W. C., and Baserga, R. (1983). *Mol. Cell. Biol.* **3**, 214–219.

Sparkuhl, J. and Sheinin, R. (1980). *J. Cell. Physiol.* **105**, 247–258.

Sperling, K. (1982). In *Premature Chromosome Condensation* (P.N. Rao, R. T. Johnson, and K. Sperling, eds.), Academic Press, New York, pp. 43–78.

Sperling, K. and Rao, P. N. (1974). *Chromosoma* **45**, 121–131.

Srinivasan, P. R., Gupta, R. S., and Siminovitch, L. (1980). *Somat. Cell Genet.* **6**, 567–582.

Stacey, D. W. and Allfrey, V. G. (1976). *Cell* **9**, 725–732.

Stamato, T. D. and Hohmann, L. K. (1975). *Cytogenet. Cell Genet.* **15**, 372–379.

Stancel, G. M., Prescott, D. M., and Liskay, R. M. (1981). *Proc. Natl. Acad. Sci. USA* **78**, 6295–6298.

Stinchcomb, D. T., Struhl, K., and Davis, R. W. (1979). *Nature* **282**, 39–43.

Tenner, A. J. and Scheffler, I. E. (1979). *J. Cell. Physiol.* **98**, 251–266.

Tenner, A., Zieg, J., and Scheffler, I. E. (1976). *J. Cell. Physiol.* **90**, 145–160.

Thompson, L. H., Mankowitz, R., Baker, R. M., Till, J. E., Siminovitch, L., and Whitmore, G. F. (1970). *Proc. Natl. Acad. Sci. U.S.A.* **23**, 377–384.

Thompson, L. H. and Lindl, P. A. (1976). *Somat. Cell Genet.* **2**, 387–400.

Tobey, R. A., Petersen, D. F., and Anderson, E. C. (1966). *Biophys. J.* **6**, 567–581.

Tsai, Y-J., Hanaoka, F., Nakano, M. M., and Yamada, M-A. (1979). *Biochem. Biophys. Res. Comm.* **91**, 1190–1195.

Tschumper, G. and Carbon, J. (1983). *Gene* **23**, 221–232.

Wahrman, J., Richler, C., Neufeld, E., and Friedmann, A. (1983). *Cytogenet. Cell Genet.* **35**, 161–180.

Wang, R. J. (1974). *Nature* **248**, 76–78.

Wang, R. J. (1976). *Cell* **8**, 257–261.

Wang, R. J. and Yin, L. (1976). *Exp. Cell Res.* **101**, 331–336.

Wang, R. J., Wissinger, W., King, E. J., and Wang, G. (1983). *J. Cell Biol.* **96**, 301–306.

Wigler, M., Silverstein, S., Lee, L.-S., Pellicer, A., Cheng, Y.-C., and Axel, R. (1977). *Cell* **11**, 223–232.

Wigler, M., Pellicer, A., Silverstein, S., and Axel, R. (1978). *Cell* **14**, 725–731.

Wissinger, W. and Wang, R. J. (1978). *Exp. Cell Res.* **112**, 89–94.

Wissinger, W. L. and Wang, R. J. (1983). *Inter. Rev. Cytol. Suppl.* **15**, 91–113.

Yamashita, K., Nishimoto, T., and Sekisuchi, M. (1984). *J. Biol. Chem.* **259**, 4667–4671.

Yanagi, K., Talavera, A., Nishimoto, T., and Rush, M. G. (1978). *J. Virol.* **25**, 42–50.

Yanishevsky, R. M. and Stein, G. H. (1981). *Inter. Rev. Cytol.* **69**, 223–259.

Yasuda, H., Matsumoto, Y., Mita, S., Marunouchi, T., and Yamada, M. (1981). *Biochemistry* **20**, 4414–4419.

Yen, A. and Pardee, A. B. (1979). *Science* **204**, 1315–1317.

Yen, A., Warrington, C. W., and Pardee, A. B. (1978). *Exp. Cell Res.* **114**, 458–462.

CHAPTER 21

DNA REPAIR MUTANTS

Lawrence H. Thompson
Lawrence Livermore National Laboratory
Biomedical Sciences Division
University of California
Livermore, California

I. INTRODUCTION

During evolution virtually all living cells acquired sophisticated biochemical machinery for preserving the integrity of their DNA molecules in the face of radiation damage or chemical alteration. Enzyme systems exist for recognizing a wide variety of structural changes in DNA and restoring the molecules to their original configuration. Some repair enzymes are highly specific for the lesions they recognize (e.g., certain pyrimidine dimer endonucleases and *N*-glycosylases), while other repair systems detect generalized distortions in structure. To develop some understanding of how these processes work, it has been necessary historically to rely heavily on a combined genetic/biochemical approach in which mutations affecting individual repair genes play a key role. Much is now known about the molecular detail of repair in the bacterium *Escherichia coli*, in yeast, and in other microbial systems (Friedberg and Bridges, 1983). This progress has been possible largely because of the relative ease with which these systems can be genetically manipulated. The prospect of similar progress in the study of animal cells depends on both the availability of mutant strains and suitable genetic techniques.

The first mutations shown to affect repair in human cells were those resulting in the syndrome known as xeroderma pigmentosum (XP). Cleaver (1968, 1969) found that the biochemical defect in XP causes defective excision repair in response to ultraviolet (UV) radiation. XP individuals are extremely prone to skin cancer resulting from exposure to sunlight (Cleaver, 1978). The finding that XP cells in culture are hypermutable by UV and chemical mutagens (Maher et al., 1979; Yang et al., 1980) has provided strong evidence to support the hypothesis that cancer is initiated by mutation resulting from unrepaired damage. Other cancer-prone genetic disorders such as Fanconi's anemia, Bloom's syndrome, and ataxia telangiectasia also involve defects in the repair or metabolism of DNA (Setlow, 1978; Friedberg et al., 1979). Therefore, an analysis of the molecular nature of these defects is of obvious importance. The biochemical bases of these syndromes is undoubtedly complex. In those cases that have been examined by genetic complementation tests, there appear to be multiple genes involved in a given syndrome, namely, XP (Keijzer et al., 1979; Arase et al., 1979), AT (Murnane & Painter, 1982), and Cockayne syndrome (Tanaka et al., 1981; Lehmann, 1982).

Although the study of these human mutant cells in culture is proving valuable in understanding mutagenesis, there are certain limitations in this approach. For instance, some genes involved in repair may not be represented among the human genetic diseases if the gene products are essential during development of the embryo. Also, many of the human strains are not transformed, making them difficult to manipulate genetically. Therefore, producing repair mutants in established cell lines may help offset such shortcomings. During the last 4 years considerable progress has been made in isolating and characterizing such mutants in Chinese hamster cells (strains CHO and V79) and mouse lymphoma cells. The hamster cells are particularly advantageous because they are highly versatile in handling and growth properties, they have an attractive karyotype for cytogenetic studies, and they have a proven history of yielding

"CHO" — The culture obtained from Dr. W.C. Dewey at Ft. Collins, CO, in 1971. He obtained his cells from R. Tobey in Los Alamos, NM.

↓

CHO-S — A culture derived by L. Thompson, without subcloning, by adapting the original stock to rapid, single-cell growth in suspension culture. This was the stock used by L. Siminovitch's laboratory in Toronto.

↓

SC1 — A subclone selected on the basis of its colony morphology on soft agar by Konrad et al., 1977.

↓

AA8-3 (abbr. AA8) — A subclone isolated on the basis of resistance to 8 μg/ml 8-azaadenine (Thompson, et al., 1980). This clone was the stock for isolating DNA repair mutants (Thompson et al., 1980a; Busch, 1980; Busch et al., 1980; Thompson et al., 1981).

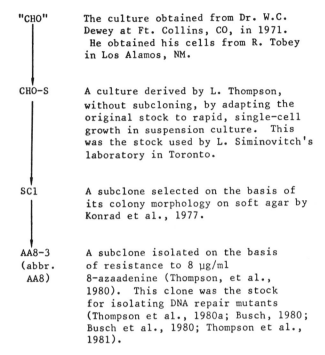

Figure 21.1. Pedigree and derivation of the CHO strain used by L. Thompson, D. Busch, and co-workers for isolating DNA repair mutants.

a wide variety of mutant phenotypes. For these reasons we sought to obtain DNA repair mutants using CHO cells, which we had previously used for obtaining temperature-sensitive mutants. The method of isolation and properties of these mutants are described here and compared with the mutations from human repair disorders, and also with mutant cells from other rodent systems. For our studies we have used a strain of CHO cells that was adapted for rapid, single-cell growth in suspension culture while retaining its ability to grow as a monolayer. The origin of this strain is given in Figure 21.1. It is referred to as AA8 and is functionally heterozygous at the *aprt* locus (Thompson et al., 1980b), a useful property for assaying induced mutations (see Chapter 12).

II. METHODS OF ISOLATING DNA REPAIR MUTANTS

A. Direct Screening Based on Hypersensitivity to a Mutagen

Experience with microbial systems has shown that mutations conferring reduced DNA repair capacity will often have the property of increased

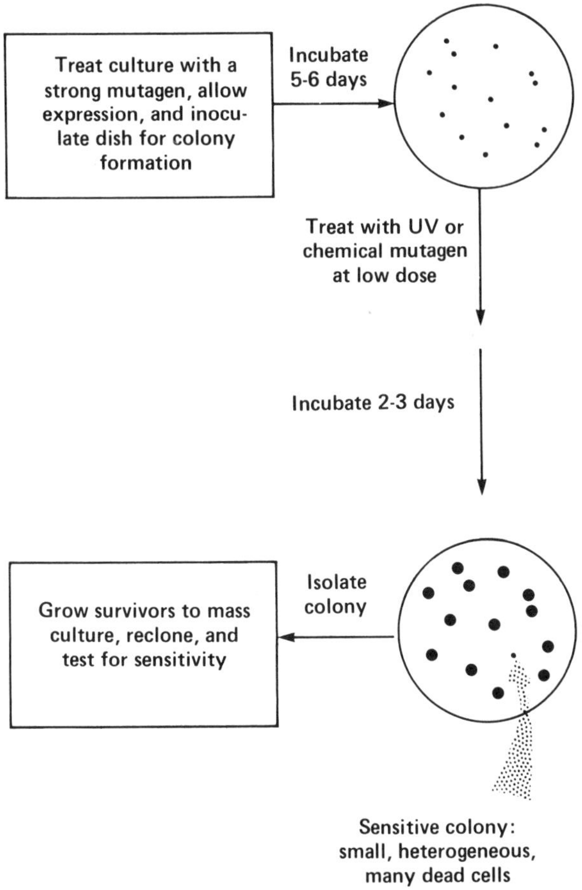

Figure 21.2. Direct screening procedure for mutagen-sensitive mutants.

sensitivity to killing by a mutagenic agent. Although the property of hypersensitivity provides a basis for isolating such mutants, in practice there are substantial difficulties in recovering the rare cells that are preferentially killed by mutagen treatment. Historically, replica plating has been the widely used approach in microbial systems, but many of the DNA repair mutants of CHO cells were obtained using other means because replica plating of mammalian cells is cumbersome. The first method to be discussed can be referred to as a direct screening procedure. It involves simply screening colonies for candidate mutants, based on morphological criteria, after the appropriate treatment, and then recovering live cells directly from the affected colonies. This method is illustrated in Figure 21.2.

Direct screening for mutants has been performed using both manual and automated procedures. Our laboratory devised a procedure that did not require any special equipment (Thompson et al., 1980a). In this study clones were obtained that had the property of being hypersensitive to killing by either UV radiation, ethyl methanesulfonate (EMS), or mitomycin C. Cultures were first exposed to a chemical mutagen (usually EMS) to produce mutations in the population and allowed a period of growth for expression. The cells were then plated into plastic Petri dishes so that there would be several hundred colonies per dish. After about 5 days growth the colonies, which were still small, were exposed to a low dose of UV radiation (3–4 J/m^2). This dose had no discernible effect on most colonies. After 2 days further incubation to allow for expression of damage in hypersensitive clones, the dishes were screened visually under a dissection microscope to identify abnormal, potentially mutant colonies. Candidate mutant colonies were chosen by three criteria: (1) unusually small size, (2) heterogeneous appearance among the cells, and (3) the presence of dead cells as indicated by the dye erythrosin B. (Before starting searches for mutant colonies the properties of such colonies were simulated and studied by exposing the normal cells to much higher UV exposures.) Whenever it was possible to obtain viable cells from such a colony, they were expanded and tested for UV sensitivity. Using this procedure six highly UV-sensitive clones were recovered from 54,000 colonies screened, and they fell into two genetic complementation groups (Thompson et al., 1981). For a more complete description of how complementation testing may be done in CHO cells, see Chapter 11. The utility of the procedure for isolating UV mutants was confirmed by Adair and Clarkson (1980) using a different strain of CHO cells that is heterozygous at the *aprt* and *tk* loci.

Clones that were hypersensitive to EMS or to mitomycin C (MMC) were obtained similarly except that an interval of 3 days was allowed for expression of the damage before screening. (Cells die and disintegrate more slowly after treatment with the chemical mutagens than with UV radiation.) The EMS-sensitive clones proved to be 10-fold more sensitive than the parental cells and had a novel phenotype that differed from that of the cancer-prone human genetic diseases. An attempt to obtain mutants sensitive to another alkylating agent, ethyl nitrosourea, which produces a different spectrum of modified bases than EMS, was unsuccessful because no sensitive colonies were detected. Several limitations in using this screening procedure should be noted: (1) Because of the tendency of CHO cells to detach from the plastic surface, it is difficult to avoid cross contamination of mutant clones with stray cells from normal colonies. (2) It is not feasible to screen enough colonies to detect the rarer classes of mutants. (3) The technique may not be equally successful with other cell types that differ in their morphological characteristics, although Shiomi et al. (1982a) have used it to isolate UV-sensitive clones of mouse L5178Y cells.

Busch et al. (1980) devised a semiautomated system for recovering UV-sensitive clones of CHO cells, which was based on principles similar to those used in the manual screening. In Busch's study the colonies were grown on the surface of soft agar, irradiation was performed after removing the lids of the Petri dishes, and photographs were taken before, and several days after, irradiation to identify putative mutant colonies. Colonies that did not grow in the interval between the two photographs were isolated and tested for UV sensitivity. The screening of several million colonies after various mutagenesis treatments yielded about 150 clones having enough sensitivity to allow complementation testing (Busch, 1980). These mutants were divided into five complementation groups in our laboratory (Thompson et al., 1981; Thompson and Carrano, 1983). Groups 1 and 2 contained 94% of all mutants tested, including the clones derived from the manual method. Groups 2 and 3 were each represented by two or more clones, but for group 5 there was only one isolate. The prototype mutants from each complementation group that have been studied (see Table 21.1) all show similar UV sensitivity and quantitatively similar defects in nucleotide excision repair after UV exposure (Thompson et al., 1981, 1982a).

B. Replica Plating Procedures

Several laboratories have used different versions of replica plating to obtain UV-sensitive and X-ray-sensitive mutants of CHO or other Chinese hamster cells. Stamato and Waldren (1977) reported the isolation of a UV-sensitive CHO clone (UV-1) using nylon cloth to replicate colonies. After removing the medium from a dish, the cloth was placed in contact with the colonies for a few hours and then transferred to another dish with medium. After further incubation, cells that attached to the cloth provided a replica of the original colonies (Stamato and Hohmann, 1975). The original dishes were UV irradiated with 6 J/m^2, seeded with a large excess of unirradiated cells, and incubated for 2 additional days. UV-sensitive colonies were evident as plaques (areas of detached cells) after staining the dishes. One clone identified this way, UV-1, proved to be defective in postreplication recovery or maturation of newly synthesized DNA after UV exposure. It also shows hypomutability after treatment with UV or EMS when induced mutation is plotted versus survival (Stamato et al., 1981). Excision repair as measured by unscheduled DNA synthesis is normal in UV-1. Thus, the properties of UV-1 are completely different from those of the UV-sensitive mutants described above that were obtained using the direct screening procedures (cf. Table 21.1).

Wood and Burki (1982) used a filter-paper replica plating technique combined with the plaque method. Two clones having high UV sensitivity were obtained, and one of these, which belonged to complementation group 2 of the UV-sensitive clones analyzed by Thompson et al. (1981), was characterized in terms of its inability to recover DNA synthesis after UV treatment (Wood et al., 1982).

Two laboratories have used replica plating to screen for mutants that are hypersensitive to ionizing radiation. Stamato et al. (1983) used a modified version of the nylon-cloth procedure in which photographs were taken before irradiation with ^{137}Cs gamma rays and again after 6 days incubation. By superimposing the negative of the first photograph with a print of the second, Stamato identified colonies that failed to grow after irradiation. One mutant (XR-1) studied in detail showed an interesting cell cycle dependence with cells in the G1 period being extremely sensitive to irradiation (e.g., 10-fold).

Jeggo et al. (1982) used a tooth-pick transfer method to make replicas of CHO colonies growing under an overlay of agar. One set of colonies was X-irradiated and then scanned 6 days later with an inverted microscope to identify colonies whose growth was severely inhibited. Clones having up to 10-fold increased X-ray sensitivity were obtained at a frequency of about 10^{-3} from cultures initially treated with EMS. Later studies on seven X-ray-sensitive clones (Jeggo and Kemp, 1984) showed that the mutants were heterogeneous with respect to their cross sensitivities to various chemical mutagens and UV, suggesting that multiple gene loci are responsible for the phenotypes.

C. Enrichment Procedures: Radioactivity Suicide and BrdUrd–Light Photolysis

In principle, enrichment procedures have the advantage of reducing greatly the number of clones needed to be tested. In practice, however, this approach is difficult with repair mutants because it requires using a dose of damaging agent that will produce enough repair synthesis in the normal cells without killing many of the sensitive mutant cells. Two enrichment schemes have been reported. Using CHO cells, Stefanini et al. (1982) preferentially killed the repair-competent cells with a BrdUrd–light treatment. After a challenge dose of UV, the culture, composed predominantly of G1 cells, was allowed to incorporate BrdUrd for 2 hr, was treated with the sensitizing dye Hoechst 33258 for 1 hr, and then was exposed to black light. Surviving colonies were replica plated, and UV-sensitive colonies were detected by either their lack of unscheduled DNA synthesis or failure to grow. Survival curves of several of the mutants showed that they were 2.5- to 3-fold more sensitive than the parental cells to killing by UV. These mutants are not as hypersensitive to UV killing as other CHO mutants mentioned above (Busch et al., 1980; Thompson et al., 1982a; Wood and Burki, 1982), and they showed 10–15% residual repair synthesis. Thus, it may be difficult to obtain fully deficient mutants using the BrdUrd–light procedure because cells having no repair capacity would likely be killed by the challenge dose of UV.

Schultz et al. (1981) used a tritiated-thymidine suicide procedure to selectively kill V79 cells performing repair synthesis after UV exposure. One mutant clone showed slightly increased sensitivity and reduced repair synthesis. Other slightly UV-sensitive clones showed cross sensitivity to other agents (X-rays, *N*-acetoxy-2-acetylaminofluorene, and nitrosoguanidine). Because these latter clones did not have reduced levels of repair synthesis after UV exposure, it is unclear how their UV sensitivity relates to their having survived the selection procedure.

D. Resistance to Drugs That Inhibit Repair

Since mutations that confer resistance to drugs are readily selectable, drugs that inhibit enzymes involved in repair can provide a way of obtaining genetic alterations that perturb repair processes. An example of this approach involves the compound aphidicolin, which effectively inhibits polymerase α, the enzyme considered responsible for normal DNA replication (Huberman, 1981). However, evidence also points to a role of polymerase α in repair synthesis (Downes et al., 1983). Chang et al. (1981) isolated mutants of V79 cells that were resistant to aphidicolin. One of these clones, aph^r-4, had an interesting pleiotropic phenotype that included thymidine auxotrophy, UV sensitivity and hypermutability, and increased chromosomal aberrations in the presence of BrdUrd. In addition to its hypersensitivity to UV mutagenesis (Liu et al., 1982b), this mutant exhibits a clear-cut mutator phenotype at the *hprt*, ouabain resistance, and diphtheria-toxin-resistance loci (Liu et al., 1982a, 1983). These properties of aph^r-4 (summarized in Table 21.1) appear to be caused by an alteration in the structure of the α polymerase, which was shown to be resistant to aphidicolin inhibition and to have a reduced K_m for dCTP when purified from cell extracts (Liu et al., 1983). These studies provide evidence that polymerase α is involved in the repair of UV-induced damage and that an altered enzyme can cause erroneous repair synthesis.

III. PROPERTIES OF SEVERAL CLASSES OF MUTANTS

Table 21.1 lists the properties of the best-characterized repair mutants, which are discussed in this section. The phenotypes of the CHO mutants are compared with the cells from human genetic disorders that have similar properties. A summary of DNA repair mutants reported for mouse cells is also given for comparison.

A. Mutants That Are Defective in Nucleotide Excision Repair (Hypersensitivity to Killing and Mutation Induction by UV)

It is of much interest to compare the properties of UV-sensitive CHO mutants with the human mutants and other mammalian-cell mutants that

TABLE 21.1
Properties of DNA Repair Mutants of CHO and V79 Cells

Strain	Hypersensitivity to	Biochemical Defect	Notable Features
UV5 (1–6)[a]	UV, monofunctional large-adduct chemicals	Incision step of nucleotide excision repair	Resembles XP; group 1[b]
UV20 (1–7)	UV, bifunctional and some monofunctional large-adduct chemicals	Incision step after UV and some aspect of cross-link repair	Resembles both XP and FA; group 2
UV24 (2,3,5)	UV, monofunctional large-adduct chemicals	Incision step of nucelotide excision repair	Resembles XP; group 3
UV41 (2,3,5)	UV, bifunctional and some monofunctional large-adduct chemicals	Incision step after UV and some aspect of cross-link repair	Resembles both XP and FA; group 4
UV135 (3,5,8)	UV, monofunctional large-adduct chemicals	Incision step of nucleotide excision repair	Resembles XP; group 5
EM9 (9–11)	MMS, EMS, CldUrd, X-rays	DNA strand-break rejoining (DNA ligase appears normal)	Very high SCE like BS cells
UV-1 (12–14)	UV, MNNG, EMS, MMC, *N*-OAc-2-AAF	Maturation of newly replicated DNA after UV	Normal excision repair with UV
XR-1 (15)	Gamma rays, bleomycin	Repair of double-strand breaks?	Cell cycle dependence
XRS-7 (16)	X-rays, bleomycin, EMS, MMS, MNNG, and UV	Unknown	Broad spectrum of sensitivity
Aph^r-4 (17–19)	UV, BrdUrd	Altered α polymerase (Resistant to aphidicolin)	Mutator phenotype

[a]Numbers in parentheses indicate literature references as follows: (1) Thompson et al., 1980a; (2) Thompson et al., 1981; (3) Thompson et al., 1982a; (4) Thompson et al., 1983b; (5) Thompson et al., 1984a; (6) Hoy et al., 1983; (7) Meyn et al., 1982; (8) Thompson and Carrano, 1983; (9) Thompson et al., 1982b; (10) Dillehay et al., 1983; (11) Dillehay et al., 1984; (12) Stamato and Waldren, 1977; (13) Stamato et al., 1981; (14) Waldren et al., 1983; (15) Stamato et al., 1983; (16) Jeggo and Kemp, 1984; (17) Chang et al., 1981; (18) Liu et al., 1982b; (19) Liu et al., 1983.

[b]The group number refers to the genetic complementation group for UV excision repair mutants of CHO cells as defined in Thompson et al. (1981) and Thompson and Carrano (1983).

have similar properties. As mentioned in Section I, sensitivity to UV radiation in human cells has been historically associated with the syndrome xeroderma pigmentosum. XP individuals have been shown to be of two types, classical XP and a variant form (Bootsma, 1978; Cleaver, 1978). Cells of the classical type tend to be highly UV sensitive and are characteristically reduced in their capacity for repair synthesis after UV exposure. Different XP strains vary considerably in the degree of repair deficiency, with some complementation groups being consistently more deficient than others (Takebe et al., 1977). All classical XP mutations (seven complementation groups) seem to affect the incision step of the nucleotide excision repair process (Bootsma, 1978; Zelle and Lohman, 1979). XP variant cells have normal excision repair but are defective in the recovery of DNA synthesis after UV (Lehmann et al., 1975). Another human disorder involving UV sensitivity is Cockayne syndrome (Schmickel et al., 1977), in which excision repair also is normal (Mayne et al., 1982). When classical XP cells are compared with the excision-repair CHO mutants, there is a close parallel between their phenotypes in response to UV radiation. The CHO mutants fall into five genetic complementation groups (Thompson et al., 1981; Thompson and Carrano, 1983). Representative mutants from each group were shown to be at least 10-fold hypermutable by UV (Thompson et al., 1982a), and each was grossly defective in performing the incision step (Thompson et al., 1982a), such that little or no repair synthesis occurs after UV exposure (Thompson et al., 1980a). Figure 21.3 presents data showing the defective incision (or nicking) in these mutants. The mutants from each complementation group are similar in their response to UV, which may be explained partly because only clones having strong sensitivity to killing were initially chosen for further analysis.

In response to chemical mutagens, the CHO excision-repair mutants and XP mutants show both important similarities and differences. XP cells seem to be invariably hypersensitive to killing and mutagenesis by monofunctional chemicals that form bulky covalent DNA adducts (Maher et al., 1975, 1977; San et al., 1978), and the CHO mutants share this property (Thompson et al., 1982c, 1983b; Hoy et al., 1983). For the large-adduct mutagen 7-bromomethylbenz(*a*)anthracene, our laboratory showed that mutants in each of the five UV complementation groups were unable to remove significant amounts of adducts during the first 24 hr after exposure (Fig. 21.4). These results showing lack of *excision* of damage are fully consistent with the data indicating defective *incision* described above.

In contrast to the results with monofunctional bulky mutagens, the behavior of both the CHO and XP mutants differs with bifunctional chemicals. For example, with the DNA cross-linking compound mitomycin C (MMC), several XP strains have been reported to have a normal killing response and normal removal of the DNA interstrand cross-links

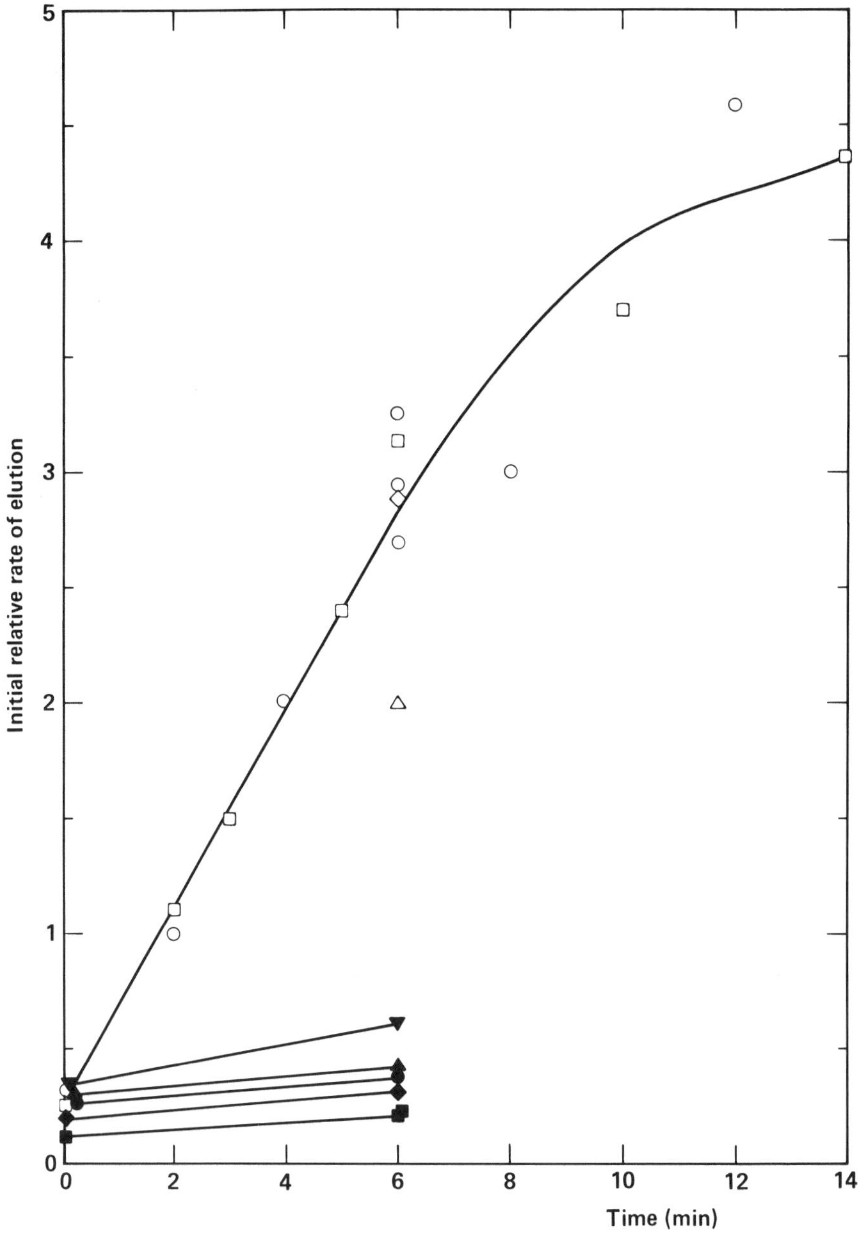

Figure 21.3. Defective incision in UV-sensitive mutants. The amount of incision after treating cells with 6 J/m^2 was estimated from the slopes of alkaline elution profiles as described by Thompson et al. (1982*a*). Open sysmbols, normal CHO cells (strain AA8); closed symbols, mutants representing five genetic complementation groups (UV5, ●; UV20, ▼; UV24, ▲; UV41, ♦; UV135, ■).

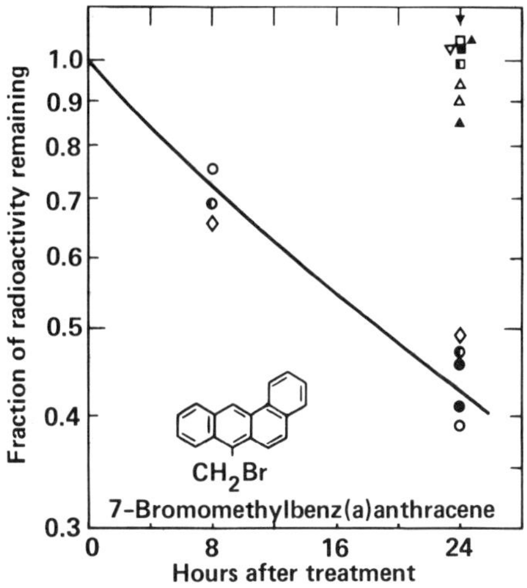

Figure 21.4. Defective excision in UV-sensitive mutants. The amount of 7-bromomethylbenz(*a*)anthracene bound to nuclear DNA was measured using tritium-labeled compound as described by Thompson et al. (1984a). Circles, normal CHO cells, AA8; diamonds, strain EM9 (see text); other symbols, mutants representing five genetic complementation groups (UV5, □, □■, ■; UV20, Δ; UV24, ▲; UV41, ∇; UV135, ▼).

caused by this chemical (Fujiwara et al., 1977; Fujiwara, 1982). To our knowledge no XP strains have been reported to be extremely hypersensitive (e.g., more than 10-fold) to killing by DNA cross-linking agents. However, cells derived from Fanconi's anemia (FA) patients are characteristically highly sensitive to killing by a wide variety of bifunctional chemicals (Sasaki, 1978; Ishida and Buchwald, 1982). FA cells have also been shown to be defective in the repair of DNA cross-links produced by MMC (Fujiwara et al., 1977; Fujiwara, 1982). The UV sensitivity of FA cells is in the normal range (Sasaki et al., 1978), and FA cells show normal repair synthesis in response to UV (Poon et al., 1974).

The UV-sensitive CHO mutants from complementation groups 2 and 4 are unusual in that they exhibit extreme sensitivity (i.e., more than 10-fold) to all cross-linking agents tested (Hoy et al., 1983; Hoy and Thompson, unpublished data). This hypersensitivity correlates with a deficiency in repair of the DNA cross-link lesions for mitomycin C, *cis*-diamminedichloroplatinum (Meyn et al., 1982), and diepoxybutane (Hoy and Thompson, unpublished data). These CHO mutants thus combine properties of both XP and FA cells, that is, the UV sensitivity of XP cells and the sensitivity to cross-linking agents seen in FA cells. These hamster mutants may well represent new classes of mutations not recognized or present in humans. Because of the complexity of repair, however, the value of phenotype comparisons is limited, and complementation studies

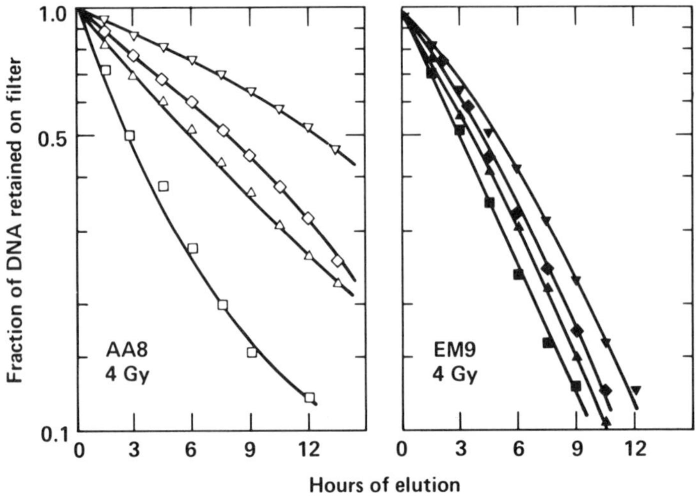

Figure 21.5. Defective repair of DNA strand breaks in mutant EM9. The method of alkaline elution of DNA from filters was used to measure the rate at which X-irradiated DNA was restored to high-molecular-weight, slowly eluting material as a function of incubation time after treatment (Thompson et al., 1982b). Cells were exposed to 4 Gy of 40 KVp X-rays on ice, and then time for repair at 25°C was allowed for 0 min (□, ■), 4 min (Δ, ▲), 8 min (◇, ◆), or 12 min (∇, ▼). Left, normal CHO cells; right, mutant EM9.

are needed to determine whether these hamster mutations have occurred in different genes from those affected in XP and FA.

B. Mutants That Have a Defect in Rejoining DNA Strand Breaks (Hypersensitivity to Certain Simple Alkylating Agents)

A particularly interesting mutation that confers a pleiotropic phenotype was isolated in CHO cells using the direct screening method (Thompson et al., 1980a). Two clones were originally obtained by their 10-fold hypersensitivity to the alkylating agent EMS. Because these clones did not complement in genetic tests based on cell fusion, only one, EM9, was chosen for further study. EM9 was found to have two other important properties (Thompson et al., 1982b): (1) a defect in rejoining strand breaks in the DNA and (2) an extremely high rate of sister-chromatid exchange (SCE) in untreated cultures.

The defect in strand-break repair was seen with agents that produce different types of DNA lesions, namely, EMS, MMS, and X-rays, suggesting that the defect lies in some critical step that is common to the repair of the lesions caused by these agents. This defect is illustrated in Figure 21.5, which shows the results of an experiment in which the alkaline elution technique was used to assess the size of the DNA molecules. The data indicate that EM9 cells are unable to restore the DNA to high molecular weight at a normal rate when incubated postirradiation. EM9 appears to be the only known mammalian cell mutant that is defective in repairing radiation-induced strand breaks. A human mutant fibroblast strain, BR46, shows defective strand-break repair after dimethyl sulfate but is normal in its repair of X-ray-induced breaks (Teo et al., 1983). Although the repair defect in EM9 is pronounced after exposure to X-rays, the increased sensitivity to killing by X-rays or gamma rays is only about twofold (which suggests that single-strand breaks are not the only cause of lethality with these agents). Recently, EM9 was found to be about 30-fold more sensitive than the parental cells to killing by CldUrd incorporated into the DNA, apparently because of numerous strand breaks that are associated with such incorporation (Dillehay et al., 1984). This latter discovery is important because it provides the basis of an efficient method for selectively killing the EM9 cells.

The primary biochemical defect in EM9 has not been determined. Because many of the properties of this mutant are like the phenotype of normal CHO cells treated with benzamides, which inhibit poly(ADP-ribose) polymerase [see Shall (1984)], the activity of this enzyme and the substrate profiles of poly(ADP-ribose) were examined in EM9. However, by all criteria studied the poly(ADP-ribose) system, including the NAD pool, appears normal in EM9 (Ikejimma et al., 1984). DNA ligases were also examined in EM9 for their level of activity (in both normal and mutagen-treated cells), their heat lability, and salt sensitivity. EM9 did not show any differences from the parental AA8 cells for any of these endpoints (Chan et al., 1984). Further studies on endonucleases and other enzymes are clearly needed, since knowledge of the primary defect in this mutant could be crucial in understanding how SCEs arise.

The baseline frequency of SCE in EM9 cells is about 12-fold elevated compared with the normal CHO, that is, 110 vs 9 SCE/cell, when measured under the standard conditions of 10 μM BrdUrd in the medium for two cycles of incorporation (Thompson et al., 1982b). These SCEs in EM9 appear to occur predominately in the second cycle when DNA is replicating on a template that contains BrdUrd (Dillehay et al., 1983). Under these conditions the size maturation of newly synthesized DNA is significantly delayed in EM9 compared with normal cells (Dillehay et al., 1983), suggesting this delay may be one factor that underlies the occurrence of the SCEs.

The high SCE frequency in EM9 resembles that seen in certain cultures derived from patients with Bloom's syndrome (BS) (Chaganti et al., 1974; Shiraishi et al., 1983a). In BS cells, as in EM9, the SCEs occur predominantly in the second cycle when BrdUrd is in the parental strand (Shiraishi et al., 1983b). However, the biochemical defect(s) may differ in BS cells and EM9 because BS cells do not show a defect in DNA strand-break repair after EMS treatment (Thompson and Carrano, 1983) or after ionizing radiation (Vincent et al., 1978). BS cells also show unique quadraradial configurations in metaphase figures (German and Schonberg, 1980) that are not seen at an elevated level in EM9 (Carrano, personal communication). Genetic complementation tests are needed to determine whether the mutation in EM9 is in the same gene(s) as BS cells.

C. Mutants That Are Hypersensitive to Ionizing Radiation

As described in a preceding section, two laboratories have succeeded in obtaining mutant phenotypes of CHO cells involving hypersensitivity to ionizing radiation by plating replicas and screening for this property. Stamato et al. (1983) isolated an interesting gamma-ray-sensitive mutant, which appears to have different properties from any of the human mutagen-sensitivity syndromes. This mutant (strain XR-1) was also sensitive to bleomycin, an agent that is known to produce double-strand DNA breaks. The response of XR-1 to EMS, which produces abundant single-strand breaks, was normal. These results raise the intriguing possibility that XR-1 has a defect in the repair of double-strand breaks. Recent experiments have provided data in support of this interpretation (Stamato, personal communication).

Jeggo and co-workers (1982, 1984) reported isolating several CHO clones of varying X-ray sensitivity, which had various patterns of cross sensitivity to chemical mutagens. One mutant (clone xrs-7) showed noticeable hypersensitivity to all agents tested: bleomycin, EMS, MMS, MNNG, and UV radiation. No information is yet available about whether multiple complementation groups are involved or about the nature of the biochemical defects in the strains.

D. DNA Repair Mutants Isolated in Mouse Cells

It is important to point out that interesting mutants have been isolated in non-Chinese-hamster cells. These also should prove useful in studying the molecular genetics of DNA repair processes. In particular, Sato and co-workers at the National Institute of Radiological Sciences, Chiba, Japan, have obtained a wide variety of mutant phenotypes in mouse lymphoma cells (L5178Y) by using replica plating procedures. They isolated four complementation groups of UV-sensitive mutants, and these showed cross sensitivity to 4-nitroquinoline-1-oxide (4-NQO) and MMC, but not to X-rays or MNNG (Sato and Hieda, 1979a; Shiomi et al., 1982a). One of these mutants (Q31) was shown by an assay for pyrimidine dimer endonuclease-sensitive sites to be defective in dimer removal and unable to do the incision step (Sato and Setlow, 1981). Therefore, these mutants closely resemble the UV excision-repair mutants of CHO cells, which are cross sensitive to large-adduct chemicals and defective in the incision step. Two L5178Y mutants belonging to different complementation groups have an MMS-sensitive phenotype (Shiomi et al., 1982b,c), but only one is also X-ray sensitive (Sato and Hieda, 1979b; Shiomi et al., 1982d). Mutants were also isolated by hypersensitivity (5- to 10-fold) to MMC and shown to fall into two complementation groups (Hama-Inaba et al., 1983). Although these MMC mutants were not cross sensitive to UV, X-rays, 4-NQO, or MMS, they were highly sensitive to the monofunctional derivative of MMC, decarbamoyl MMC, which should not form DNA cross-links. It will be important to determine whether these mutants are genetically equivalent to cells of FA syndrome.

Two other noteworthy features of the L5178Y mutants pertain to their responses to the mutagen 4-NQO, which is known to produce several DNA adducts that are repaired by the nucleotide excision-repair pathway (Ikenaga et al., 1977). Strain M10 is hypersensitive to MMS, X-rays, and 4-NQO, while strain Q31 is hypersensitive to UV and 4-NQO (Shiomi et al., 1982b). With respect to 4-NQO sensitivity both strains show a distinct codominant phenotype in tetraploid hybrid cells even though the phenotypes are recessive with respect to MMS or UV sensitivity. Double mutants were also constructed by starting with strain M10 and mutating it to a UV-sensitive phenotype that belonged to the same complementation group as Q31 (Shiomi et al., 1983). These double mutants were extremely hypersensitive to killing by 4-NQO (30-fold), indicating synergism between the two defects. These results point toward considerable complexity in the repair of chemical adducts but may also lead to better models of repair for testing.

IV. APPLICATIONS OF REPAIR MUTANTS OF CHO CELLS

As indicated in the preceding section, repair mutants are obviously needed for analyzing the biochemistry and enzymology of DNA repair pathways. This section illustrates other, and related, applications of the mutants.

A. Complementation Tests for Genetic Relationships Between CHO Mutants and Human Cells

Complementation tests are needed to determine whether mutations isolated in CHO cells or other mammalian lines are genetically equivalent to any of the human mutagen-sensitivity syndromes such as XP, FA, ataxia telangiectasia, or Bloom's syndrome. As a prerequisite to such tests, it is important first to show that DNA repair functions are not species specific, that is, that rodent and human gene products can substitute biochemically for each other. We have approached this problem using procedures for making hybrid cells by fusion with polyethylene glycol and then selecting directly for the cells in which complementation of the repair function occurs. Mutant CHO cells have been fused with diploid human fibroblasts from either normal or XP individuals. For example, with the UV-sensitive mutant strain UV20, which is 70-fold hypersensitive to killing by MMC, the selection conditions are MMC at a low concentration (10 n*M*) that does not affect normal CHO cells and ouabain at a concentration (10 μ*M*) that specifically kills the human cells. Under these conditions hybrid cells that have acquired resistance to MMC (and UV) can be isolated, but the frequency is low (i.e., $\sim 10^{-6}$). Cytological examination shows that these hybrids contain the hamster chromosomes and a variable number of human chromosomes. In several hybrids examined in detail and compared with the karyotypes of the parental cells, either a small amount of human chromosomal material was visible, or no human material could be seen (Carrano and Thompson, unpublished results). A fragment of a human chromosome was sometimes translocated onto a hamster chromosome. Whether this breakage and rearrangement of the human chromosomes is a consequence of growing the cells in MMC, which may produce a low level of unrepaired damage, is not known. In several cases putative hybrids were grown in nonselective medium for several weeks to allow for the loss of complementing human genetic material. Subclones were then isolated and shown to be either resistant to MMC like the parent hybrid or sensitive like the parent CHO mutant. This result provides evidence that bona fide complementation does occur between the hamster and human cells and that segregation of the responsible locus can occur under nonselective conditions. The results obtained with XP fibroblasts from various complementation groups suggest that CHO mutants UV20 and UV41 (groups 2 and 4) can be complemented by most of the XP strains (Thompson, unpublished data). Because of the typically low frequency of hybrids and the occurrence of occasional revertants or colonies having pseudoresistance, it has not been possible to interpret those crosses in which no colonies arose.

For the UV-sensitive CHO mutants from complementation groups 1, 3, and 5, it has been more difficult to devise an efficient selection for complementation events. Because the agent used should ideally act over a period of several days to ensure killing of virtually all sensitive cells, UV itself is awkward to use, and none of the chemical mutagens tested have shown as clear-cut differential toxicity as MMC for normal versus mutant cells. However, for the mutant EM9 (see Table 21.1) the use of CldUrd in the medium has proved highly effective in killing the cells without producing noticeable toxicity in the normal CHO cells (Dillehay et al., 1984). Preliminary results (Thompson and Carrano, unpublished data) show that it is possible to obtain complementing hybrid cells from fusions between EM9 and normal human diploid fibroblasts, based on the criteria of CldUrd resistance and normalization of SCE frequency. From one hybrid clone, segregant subclones that had regained the mutant phenotype were obtained after growing the cells in nonselective medium.

As an alternative to using whole cells and fusion/hybridization to test for genetic complementation, mutant cells can be treated with isolated DNA obtained from the cells of interest and analyzed for incorporation and expression of functional repair genes. MacInnes et al. (1983, 1984) found that the CHO mutant UV135 (group 5) could be complemented by DNA obtained from normal CHO cells. Rubin et al. (1983) used DNA extracted from human HeLa cells to transfect the CHO mutant UV20 and obtained clones that were resistant to MMC and UV. These transformed lines were shown to contain human DNA sequences, providing strong evidence that the UV-resistance phenotype was not caused by a reversion event or some other anomaly. As discussed below, this approach has important applications in attempts to isolate human repair genes. In studies with the CHO mutant UV-1, Waldren et al. (1983) have obtained evidence that the defect here also can be complemented with DNA from human cells.

From the above results one can conclude that many, if not all, of the mutations in the hamster cells will be amenable to complementation by human repair genes.

B. Chromosomal Mapping of Human Repair Genes

Mutant cells of rodent origin that are defective in DNA repair, combined with the appropriate selective systems, provide a useful means of determining which human chromosomes carry the corresponding repair genes. It is well known that hybrid cells formed by crossing mouse or hamster cells with human cells tend to lose human chromosomes preferentially (see Chapter 5). By growing such hybrids under conditions that select for a particular phenotype, one can test for a positive correlation between the phenotype and the retention of a specific human chromosome. This approach can be used with the UV excision-repair mutants of CHO cells and other DNA repair mutants of rodent origin. It should be possible to use the selective systems that involve MMC or CldUrd, described in the preceding section, to isolate reduced hybrid cells in which specific human chromosomes complement the repair deficiencies. For example, we performed an experiment in which mutant UV20 was fused with human peripheral blood lymphocytes and repair-proficient hybrids selected using MMC in the medium. The hybrids obtained appeared to lose human chromosomes less rapidly than those from crosses using human fibroblasts. For many of the human chromosomes it is possible to establish their presence or absence in the hybrids by analyzing electrophoretic markers. By a combination of isozyme and cytogenetic analysis the gene that complements the defect in mutant UV20 was mapped to human chromosome 19 (Thompson et al., 1984b). Surprisingly, chromosome 19 also corrected the deficiency in mutant EM9 (Siciliano, Thompson, and Carrano, unpublished results). The UV-sensitive mouse lymphoma mutant Q31 was used to show that human chromosome 13 carries a gene that corrects the defect in nucleotide excision repair (Hori et al., 1983).

C. Isolating Human Repair Genes

The hamster mutants hold great promise as a tool for isolating human repair genes. Once a hamster mutation has been shown to be complemented by a human gene through cell fusion or DNA transfection, one should be able to apply recombinant DNA technology to clone the human gene. Detailed descriptions of this approach can be found in Chapter 8. In the human genome the Alu family of repetitive interspersed sequences is so common that such sequences reside within a few kilobases of almost any gene (Schmid and Jelinek, 1982). Moreover, the hamster Alu family sequences differ sufficiently from those of the human that they can be distinguished by differential efficiency of molecular hybridization. Thus, the Alu-family sequences constitute a built-in physical genetic marker that should allow one to detect a human repair gene in a transformed hamster cell, by using the appropriate radiolabeled probe, and to retrieve the DNA segment containing the repair gene by recombinant methods. The hybridization probe can be a cloned Alu-family sequence (Shih et al., 1981) or simply total nick-translated human DNA (Murray et al., 1981). (This approach has been used to isolate human oncogenes in transfected mouse cells [e.g., Pulciani et al. (1982)]). Rubin et al. (1983) are using this approach to isolate the human gene(s) that corrects the repair defect in the CHO mutant UV20. UV-resistant transformants were obtained by transfecting with HeLa-cell DNA, and secondary transformants were made by using the DNA from the primary clones to transfect UV20 a second time. When restriction digests of DNAs on Southern blots were probed with nick-translated HeLa DNA, human-specific bands were seen in both primary and secondary clones, with some bands common to both. These results provide convincing evidence that a functional human repair gene, closely linked to human Alu-type sequences, is present in the transformants.

Recently a human gene complementing a CHO mutant in the same complementation group as UV20 was isolated from a cosmid library (Westerveld et al., 1984). The gene was identified by its linkage to the *Ecogpt* gene, which was ligated to partially cleaved HeLa cell DNA before the primary transfection.

D. Mutagenesis Testing

The general property of hypersensitivity to mutagens confers advantages to the repair mutants in studies designed to detect mutagens and evaluate their potency. Because repair-deficient cell lines were not available when a major effort was being directed at validating reliable systems for quantifying mutation induction in mammalian cells (Hsie et al., 1979), mutant strains have not been widely used in testing. However, our laboratory has evaluated several aspects of using the repair mutants for genetic toxicology.

Several of the CHO mutants were examined in our laboratory and found to be hypersensitive to mutation induction. Strain UV5 (group 1) was compared with the parental strain AA8 and found to be more sensitive to mutagenesis at the *hprt* and *aprt* loci in response to a variety of monofunctional chemicals that react with DNA to form large adducts: benzo(*a*)pyrene, 7,12-dimethylbenz(*a*)anthracene, 3-methylcholanthrene, 2-acetylaminofluorene, aflatoxin B1, antidiolepoxide of benzo(*a*)pyrene, and 7-bromomethylbenz(*a*)anthracene (Thompson et al., 1982c, 1984a). For the one chemical tested, strain UV20 (complementation group 2) behaved similarly to UV5 (Thompson et al., 1982c). This enhanced sensitivity is useful in situations where a limited amount of sample is available for testing or where certain other complications exist. For example, in evaluating the genotoxicity of two compounds known to be present after cooking certain types of protein foods, we found that the results obtained with UV5 helped clarify the genotoxicity of these compounds (Thompson et al., 1983a). The compound Trp-P-2 (3-amino-1-methyl-5*H*-pyrido-[4,3-*b*]indole a potent bacterial mutagen, was highly toxic and mutagenic in both normal and UV5 CHO cells, with UV5 being two-fold more sensitive. In contrast the compound IQ (2-amino-3-methylimidazo[4,5-*f*]quinoline), an equally effective bacterial mutagen, was weakly toxic in both CHO cell types and nonmutagenic in the normal CHO cells. However, in UV5, IQ gave a clear-cut dose response for mutation induction. Thus, IQ was classed as a weak mutagen in CHO cells, detectable only in repair-deficient cells. Analysis of chromosomal aberrations suggested that the toxicity produced by IQ did not occur through DNA damage, leading us to conclude that the compound has nongenetic toxicity superimposed on its weak genetic effects.

In another study we evaluated the endpoint of differential cytotoxicity between normal and repair-deficient cells as a simple, rapid, and efficient indicator of genetic damage (Hoy and Thompson, unpublished data). The assay (illustrated in Fig. 21.6) involves inoculating each row of a 24-well tray with either the normal CHO or one of the mutants UV5, UV4, and EM9. (The properties of these mutants are given in Table 21.1; UV4 belongs to complementation group 2, like UV20.) After exposure to the chemical for either 4 hr or overnight, the tray is incubated for 3 days and then stained. The staining intensity provides a measure of the amount of cell growth; lack of staining indicates cell killing. In the example shown in Figure 21.6, cells were treated with thio-TEPA [Tris(1-aziridinyl)phosphine sulfide], a DNA cross-linking agent, at the concentrations shown. Strains AA8, EM9, and UV5 had identical responses with complete killing at 25 μg/mL. UV4, which was treated with a lower dose

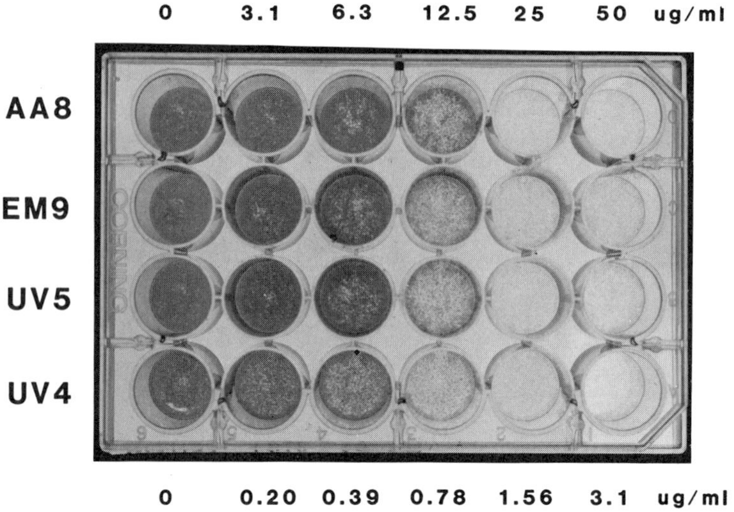

Figure 21.6. Differential cytotoxicity assay for detecting genotoxic chemicals. About 4 $\times$ 10^4 cells were inoculated into each well and allowed several hours to attach. The compound thio-TEPA was added and removed after 4-hr exposure. The tray was then incubated for 72 hr, fixed with ethanol, and stained with crystal violet. The intensity of staining provides a measure of cell growth.

range, showed complete killing at 1.6 μg/mL. These results clearly illustrate both the specificity of the UV4 strain in responding to cross-link damage and its high degree of sensitivity (16-fold in this instance). This assay for differential killing has proved reproducible and sensitive. Although some classes of mutagens (intercalating agents and many simple alkylating agents) are not efficiently detected with the mutants shown in Figure 21.6, adding other types of mutants could broaden the assay's sensitivity and practical value.

E. Studying Cell Cycle Effects of Mutagens

It is well known that the sensitivity of cells to killing by various DNA-damaging agents varies through the cell cycle. An improved understanding of this phenomenon is especially important in cancer chemotherapy, since many of the cytotoxic agents used are cycle dependent and act by first binding to DNA. Although DNA repair processes are thought to play a major role in determining cell cycle responses, little information on this matter has been available. The use of repair-deficient mutant lines affords an opportunity to determine what components of a cell cycle response are dependent on repair. One can compare the normal versus mutant strains in their response to an agent using synchronized cultures. Two studies of this type have been performed using the UV excision repair mutants of

CHO cells. Wood and Burki (1982) analyzed the survival after UV treatment of normal CHO cells and a mutant belonging to complementation group 2 (clone 43-3B). The normal cells showed a characteristic cell cycle response to UV with maximum sensitivity in early S and maximum resistance in late S. Repair-deficient cells, given a lower dose that resulted in similar asynchronous survival, had an almost constant survival throughout the cycle. These results suggested that repair is the major factor governing the variations in survival that occur in the normal cells.

Our laboratory did a similar type of study using the mutagen 7-bromomethylbenz(*a*)anthracene, which reacts with DNA to give repairable adducts on the extracyclic amino groups of guanine and adenine (Dipple and Roberts, 1977). We compared the response of the wild-type cells with that of the mutant UV5 (Table 21.1) in cultures that were synchronized by centrifugal elutriation (Thompson et al., 1984a). In these highly synchronous cultures, the normal cells were most sensitive in mid S, most resistant in late S/G2, and intermediate in sensitivity in G1. Maximal sensitivity occurred well beyond the G1/S boundary. The repair-deficient UV5 cells were most sensitive in early G1, and most resistant in mid S, but the magnitude of the overall changes was much less (about threefold) than that seen in the normal cells (about 20-fold) for doses of mutagen that produced about 20% survival in the asynchronous cultures of each line. The response pattern of the normal cells could be interpreted if unrepaired adducts interfere with the DNA replication machinery. The fact that the survival was not constant in UV5 suggested that interference with transcription by adducts may be an important pathway leading to cell killing. In both cell lines the binding efficiency of the mutagen through the cycle did not change significantly. Although further studies of this type are needed, our results show that repair is a major, but not the only, determinant of the cell cycle variations in survival for this agent.

V. CONCLUSIONS

The studies discussed here show that DNA repair mutants of CHO and other rodent lines are proving valuable for a variety of applications, especially as one views these mutants as an adjunct to the study of repair and metabolism of DNA in human cells. Several of the hamster mutants described differ phenotypically from the cells obtained from the human repair syndromes. However, so far it is not known whether any of the mutants isolated in cultured rodent cells involves the same locus as one of the human genetic disorders such as xeroderma pigmentosum. Extensive complementation tests are needed to address this issue, which is experimentally difficult because of the many genetic complementation groups that have already been identified for XP and other syndromes. As the genetic relationships between the repair genes of hamster and human cells become better understood, the hamster mutants can be used as a powerful tool for isolating human repair genes. Several of the hamster and mouse mutant lines have clear-cut phenotypes that can be effectively selected against in complementation assays. These features allow one to introduce a human gene into a rodent genetic background by DNA transfection procedures. The use of established recombinant DNA techniques should then allow the retrieval and molecular cloning of the repair gene. At present none of the mutants of CHO cells (Table 21.1) has been identified in terms of a specific gene product, and the same situation holds with the mutations in the human repair syndromes. Because one does not have direct assays for the functions that are defective in the various mutants, one must rely on indirect tests for the presence of a functional gene, that is, repair activity or cellular resistance to a mutagen. However, as human genes become available in appropriate expression vectors, these systems may then provide better ways of identifying and isolating the repair proteins that have been so illusive.

ACKNOWLEDGMENT

Work performed under the auspices of the U.S. Department of Energy by the Lawrence Livermore National Laboratory under contract number W-7405-ENG-48.

REFERENCES

Adair, G. M. and Clarkson, J. M. (1980). *Environ. Mutagen.* **2**, 267 (abstract).

Arase, S., Kozuka, T., Tanaka, K., Ikenaga, M., and Takebe, H. (1979). *Mutat. Res.* **59**, 143–146.

Bootsma, D. (1978). In *DNA Repair Mechanisms* (P. C. Hanawalt, E. C. Friedberg, and C. F. Fox, eds.), Academic Press, New York, pp. 589–601.

Busch, D. B. (1980). Ph.D. Dissertation, University of California, Berkeley, California.

Busch, D. B., Cleaver, J. E., and Glaser, D. A. (1980). *Somat. Cell Genet.* **6**, 407–418.

Chaganti, R. S. K., Schonberg, S., and German, J. (1974). *Proc. Natl. Acad. Sci. USA* **71**, 4508–4512.

Chan, J. Y. H., Thompson, L. H., and Becker, F. F. (1984). *Mutat. Res.* **131**, 209–214.

Chang, C.-C., Boezi, J. A., Warren, S. T., Sabourin, C. L. K., Liu, P. K., Glatzer, L., and Trosko, J. E. (1981). *Somat. Cell Genet.* **7**, 235–253.

Cleaver, J. E. (1968). *Nature* **218**, 652–656.

Cleaver, J. E. (1969). *Proc. Natl. Acad. Sci. USA* **63**, 428–435.

Cleaver, J. E. (1978). In *Metabolic Basis of Inherited Disease*, 4th ed. (J. B. Stanbury, J. B. Wyngaarden, and D. S. Frederickson, eds.), McGraw-Hill, New York, pp. 1072–1095.

Dillehay, L. E., Thompson, L. H., Minkler, J. L., and Carrano, A. V. (1983). *Mutat. Res.* **109**, 283–296.

Dillehay, L. E., Thompson, L. H., and Carrano, A. V. (1984). *Mutat. Res.* **131**, 129–136.

Dipple, A. and Roberts, J. J. (1977). *Biochem.* **16**, 1499–1503.

Downes, C. S., Collins, A. R. S., and Johnson, R. T. (1983). *Mutat. Res.* **112**, 75–83.

Friedberg, E. C. and Bridges, B. R. (eds.) (1983). *Cellular Responses to DNA Damage, UCLA Symp. on Molec. and Cell. Biol., New Series*, Alan R. Liss, New York, Vol. 11.

Friedberg, E. C., Ehmann, U. K., and Williams, J. J. (1979). *Adv. Rad. Biol.* **8**, 85–174.

Fujiwara, Y. (1982). *Biochim. Biophys. Acta* **699**, 217–225.

Fujiwara, Y., Tatsumi, M., and Sasaki, M. S. (1977). *J. Mol. Biol.* **113**, 635–649.

German, J. and Schonberg, S. (1980). In *Genetic and Environmental Factors in Experimental and Human Cancer* (H. V. Gelboin et al., eds.), Japan Scientific Society Press, Tokyo, pp. 175–186.

Hama-Inaba, H., Hieda-Shiomi, N., Shiomi, T., and Sato, K. (1983). *Mutat. Res.* **108**, 405–416.

Hori, T., Shiomi, T., and Sato, K. (1983). *Proc. Natl. Acad. Sci. USA* **80**, 5655–5659.

Hoy, C., Thompson, L. H., and Salazar, E. (1983). *J. Cell. Biochem. Suppl.* **7B**, 184.

Hsie, A. W., O'Neill, J. P., and McElheny, V. K. (eds.) (1979). *Banbury Report 2, Mammalian Cell Mutagenesis: The Maturation of Test Systems*, Cold Spring Harbor Laboratory, Cold Spring Harbor, New York.

Huberman, J. A. (1981). *Cell* **23**, 647–648.

Ikejima, M., Bohannon, D., Gill, D. M., and Thompson, L. H. (1984). *Mutat. Res.* (in press).

Ikenaga, M., Takebe, H., and Ishii, Y. (1977). *Mutat. Res.* **43**, 415–427.

Ishida, R. and Buchwald, M. (1982). *Cancer Res.* **42**, 4000–4006.

Jeggo, P. A. and Kemp, L. M. (1984). *Mutat. Res.* **112**, 313–327.

Jeggo, P. A., Kemp, L. M., and Holiday, R. (1982). *Biochimie* **64**, 713–715.

Keijzer, W., Jaspers, N. G. J., Abrahams, P. J., Taylor, A. M. R., Arlett, C. F., Zelle, B., Takebe, H., Kinmont, P. D. S, and Bootsma, D. (1979). *Mutat. Res.* **62**, 183–190.

Konrad, M. W., Storrie, B., Glaser, D. A., and Thompson, L. H. (1977). *Cell* **10**, 305–312.

Lehmann, A. R. (1982). *Mutat. Res.* **106**, 347–356.

Lehmann, A. R., Kirk-Bell, S., Arlett, C. F., Patterson, M. C., Lohman, P. H. M., de Weerd-Kastelein, E. A., and Bootsma, D. (1975). *Proc. Natl. Acad. Sci. USA* **72**, 219–223.

Liu, P. K., Chang, C.-C., and Trosko, J. E. (1982a). *Mutat. Res.* **106**, 317–332.

Liu, P. K., Chang, C.-C., Trosko, J. E., Dube, D. K., Martin, G. M., and Loeb, L. A. (1983). *Proc. Natl. Acad. Sci. USA* **80**, 797–801.

Liu, P. K., Trosko, J. E., and Chang, C.-C. (1982b). *Mutat. Res.* **106**, 333–345.

MacInnes, M. A., Bingham, J. M., Strniste, G. F., and Thompson, L. H. (1983). In *Cellular Responses to DNA Damage, UCLA Symposium on Molecular and Cellular Biology, New Series* (E. C. Friedberg and B. A. Bridges, eds.), Alan R. Liss, New York, Vol. 11, pp. 593–602.

MacInnes, M. A., Bingham, J. M., Thompson, L. H., and Strinste, G. F. (1984). *Molec. Cell. Biol.* **4**, 1152–1158.

Maher, V. M., Birch, H., Otto, J. R., and McCormick, J. J. (1975). *J. Natl. Cancer Inst.* **54**, 1287–1294.

Maher, V. M., Dorney, D. J., Mendrala, A. L., Konze-Thomas, B., and McCormick, J. J. (1979). *Mutat. Res.* **62**, 311–323.

Maher, V. M., McCormick, J. J., Grover, P. L., and Sims, P. (1977). *Mutat. Res.* **43**, 117–138.

Mayne, L. V., Lehmann, A. R., and Waters, R. (1982). *Mutat. Res.* **106**, 179–189.

Meyn, R. E., Jenkins, S. F., and Thompson, L. H. (1982). *Cancer Res.* **42**, 3106–3110.

Murnane, J. P. and Painter, R. B. (1982). *Proc. Natl. Acad. Sci. USA* **79**, 1960–1963.

Murray, M. J., Shilo, B.-Z., Shih, C., Cowing, D., Hsu, H. W., and Weinberg, R. (1981). *Cell* **25**, 355–361.

Poon, P. K., O'Brien, R. L., and Parker, J. W. (1974). *Nature* **250**, 223–225.

Pulciani, S., Santos, E., Lauver, A. V., Long, L. K., Robbins, K. C., and Barbacid, M. (1982). *Proc. Natl. Acad. Sci. USA* **79**, 2845–2849.

Rubin, J. S., Joyner, A. L., Bernstein, A., and Whitmore, G. F. (1983). *Nature* **306**, 206–208.

San, R. H. C., Stich, W., and Stich, H. F. (1978). *Int. J. Cancer* **20**, 181–187.

Sasaki, M. S. (1978). In *DNA Repair Mechanisms* (P. C. Hanawalt, E. C. Friedberg, and C. F. Fox, eds.), Academic Press, New York, pp. 675–684.

Sato, K. and Hieda, N. (1979*a*). *Int. J. Radiat. Biol.* **35**, 83–87.

Sato, K. and Hieda, N. (1979*b*). *Radiat. Res.* **78**, 167–171.

Sato, K. and Setlow, R. B. (1981). *Mutat. Res.* **84**, 443–455.

Schmickel, R. E., Chu, E. H. Y., Trosko, J. E., and Chang, C. C. (1977). *Pediatrics* **60**, 135–139.

Schmid, C. W. and Jelinek, W. R. (1982). *Science* **216**, 1065–1070.

Schultz, R. A., Trosko, J. E., and Chang, C.-C. (1981). *Environ. Mutagen.* **3**, 53–64.

Setlow, R. B. (1978). *Nature* **271**, 713–717.

Shall, S. (1984). *Adv. Radiat. Biol.* **11** (in press).

Shih, C., Padhy, L. C., Murray, M., and Weinberg, R. A. (1981). *Nature* **290**, 261–264.

Shiomi, T., Hieda-Shiomi, N., and Sato, K. (1982a). *Somat. Cell Genet.* **8**, 329–345.

Shiomi, T., Hieda-Shiomi, N., and Sato, K. (1982b). *Mutat. Res.* **95**, 313–325.

Shiomi, T., Hieda-Shiomi, N., and Sato, K. (1982c). *Mutat. Res.* **95**, 327–337.

Shiomi, T., Hieda-Shiomi, N., and Sato, K. (1982d). *Mutat. Res.* **103**, 61–69.

Shiomi, T., Hieda-Shiomi, N., and Sato, K. (1983). *Mutat. Res.* **108**, 395–404.

Shiraishi, Y., Yashimoto, S., Miyoshi, I., Kondo, N., Orii, T., and Sandberg, A. A. (1983a). *Cancer Res.* **43**, 3836–3840.

Shiraishi, Y., Yosida, T. H., and Sandberg, A. A. (1983b). *Proc. Natl. Acad. Sci. USA* **80**, 4369–4373.

Stamato, T. D. and Hohmann, L. K. (1975). *Cytogenet. Cell Genet.* **15**, 372–379.

Stamato, T. D., Hinkle, L., Collins, A. R. S., and Waldren, C. A. (1981). *Somat. Cell Genet.* **7**, 307–320.

Stamato, T. D. and Waldren, C. A. (1977). *Somat. Cell Genet.* **3**, 431–440.

Stamato, T. D., Weinstein, R., Giaccia, A., and Mackenzie, L. (1983). *Somat. Cell Genet.* **9**, 165–173.

Stefanini, M., Reuser, A., and Bootsma, D. (1982). *Somat. Cell Genet.* **8**, 635–642.

Takebe, H., Miki, Y., Kozuka, T., Furuyama, J., Tanaka, K., Sasaki, M., Fujiwara, Y., and Akiba, H. (1977). *Cancer Res.* **37**, 490–495.

Tanaka, K., Kawai, K., Kumahara, Y., Ikenaga, M., and Okada, Y. (1981). *Somat. Cell Genet.* **7**, 445–455.

Teo, I. A., Broughton, B. C., Day, R. S., James, M. R., Karran, P., Mayne, L. V., and Lehmann, A. R. (1983). *Carcinogen.* **4**, 559–564.

Thompson, L. H., Brookman, K. W., Carrano, A. V., and Dillehay, L. E. (1982c) *Proc. Natl. Acad. Sci. USA* **79**, 534–538.

Thompson, L. H., Brookman, K. W., Dillehay, L. E., Carrano, A. V., Mazarimas, J. A., Mooney, C. L., and Minkler, J. L. (1982b). *Mutat. Res.* **95**, 427–440.

Thompson, L. H., Brookman, K. W., Dillehay, L. E., Mooney, C. L., and Carrano, A. V. (1982a). *Somat. Cell Genet.* **8**, 759–773.

Thompson, L. H., Brookman, K. W., and Mooney, C. L. (1984a). *Somat. Cell Molec. Genet.* **10**, 183–194.

Thompson, L. H., Busch, D. B., Brookman, K., Mooney, C. L., and Glaser, D. A. (1981). *Proc. Natl. Acad. Sci. USA* **78**, 3734–3737.

Thompson, L. H. and Carrano, A. V. (1983). In *Cellular Responses to DNA Damage, UCLA Symposium on Molecular and Cellular Biology, New Series* (E. C. Friedberg and B. R. Bridges, eds.), Alan R. Liss, New York, Vol. 11, pp. 125–143.

Thompson, L. H., Carrano, A. V., Salazar, E., Felton, J. S., and Hatch, F. T. (1983a). *Mutat. Res.* **117**, 243–257.

Thompson, L. H., Fong, S., and Brookman, K. (1980b). *Mutat. Res.* **74**, 21–36.

Thompson, L. H., Mooney, C. L., Burkhart-Schultz, K., Carrano, A. V., and Siciliano, M. J. (1984b). *Somat. Cell Molec. Genet.* (in press).

Thompson, L. H., Rubin, J. S., Cleaver, J. E., Whitmore, G. F., and Brookman, K. (1980a). *Somat. Cell Genet.* **6**, 391–405.

Thompson, L. H., Salazar, E. P., Brookman, K. W., and Hoy, C. A. (1983b). *Mutat. Res.* **112**, 329–344.

Vincent, Jr., R. A., Hays, M. D., and Johnson, R. C. (1978). In *DNA Repair Mechanisms* (P. C. Hanawalt, E. C. Friedberg, and C. F. Fox, eds.), Academic Press, New York, pp. 663–666.

Waldren, C., Snead, D., and Stamato, T. (1983). In *Cellular Responses to DNA Damage, UCLA Symposium on Molecular and Cellular Biology, New Series* (E. C. Friedberg and B. R. Bridges, eds.), Alan R. Liss, New York, Vol. 11, pp. 637–646.

Wood, R. D. and Burki, H. J. (1982). *Mutat. Res.* **95**, 505–514.

Wood, R. D., de Veciana, M., and Presson-Tincknell, B. (1982). *Photochem. Photobiol.* **36**, 169–174.

Westerveld, A., Hoeijmakers, J. H. J., van Duin, M., de Wit, J., Odijk, H., Pastink, A., Wood, R. D., and Bootsma, D. (1984). *Nature* **310**, 425–429.

Yang, L. L., Maher, V. M., and McCormick, J. J. (1980). *Proc. Natl. Acad. Sci. USA* **77**, 5933–5937.

Zelle, B. and Lohman, P. H. M. (1979). *Mutat. Res.* **62**, 363–368.

CHAPTER 22

MICROTUBULE MUTANTS

Matthew J. Schibler
Fernando Cabral
Division of Endocrinology
University of Texas Medical School
Houston, Texas

I. INTRODUCTION

A. General Features of Microtubule Structure, Function, and Distribution

Microtubules are long cylindrical filaments found in virtually all eukaryotic cells. As seen by electron microscopy, they measure approximately 25 nm in diameter with walls about 5 nm thick. Their lengths are variable, ranging from a fraction of a micrometer up to several micrometers and in a few instances, for example, brain axons, they may be up to several millimeters long (Porter, 1966). Microtubules comprise the major fibrous component of the mitotic spindle and play a vital role in cell division (Inoue, 1981). They also are one of the three major fibrous systems which are collectively named the cytoskeleton.

Microtubules participate in several types of cell motility. For example, they are believed to play a role in the movements of feeding arms of protozoa, protozoan axostyles, and flagellar axonemes, and movements within neuronal axons and mitotic spindles (Dustin, 1978). They have also been implicated in the movement of vesicles and organelles in the cytoplasm of cells by a process called saltatory motion (Rebhun, 1972).

Microtubules (MTs) are polymers of a globular protein, tubulin, which is present in the cell as a 6S heterodimer of two similar but nonidentical 55,000-dalton subunits, α and β (Luduena, 1979). The cell maintains a "dynamic" equilibrium between free and polymerized tubulin. Polymerization does not occur at random in the cells, but begins at specific foci called microtubule organizing centers (MTOC) (Pickett-Heaps, 1969). In the electron microscope, these are visualized as either specific structures such as the centrosomes of interphase cells and the poles of the spindle apparatus or in some organisms as amorphous dark-staining densities with little or no structural detail. Virtually all microtubules in a cell have one of their ends complexed to an MTOC.

In interphase mammalian cells, microtubules are present throughout the cytoplasm in the form of a cytoplasmic microtubule complex (CMTC) (Fuller et al., 1975; Weber et al., 1975). They originate in or near the centrosome, a perinuclear organelle which consists of a pair of centrioles and associated pericentriolar material, and radiate out to the cell periphery. During S phase, the pair of centrioles is duplicated in preparation for their role as spindle poles in mitosis (Robbins et al., 1968). As the cell enters mitosis, the microtubules undergo a dramatic rearrangement. The centrosomes migrate to opposite sides of the nucleus and the CMTC depolymerizes, thus increasing the pool of free tubulin dimers. The centrosomes now act as spindle poles and nucleate the assembly of spindle microtubules utilizing the tubulin dimers from the depolymerized CMTC (Fulton and Simpson, 1979; Bibring and Baxandall, 1977). Microtubules continue to grow out in every direction from each centrosome. Those situated in the region between the centrosomes will eventually overlap and form the mitotic spindle.

As the cell enters prometaphase, the nuclear envelope breaks down, and microtubules are soon found attached to the condensed chromosomes at specialized structures called kinetochores. The chromosomes are eventually aligned along an equatorial plane, equidistant from either pole. At this time, two classes of spindle microtubules can be distinguished: the kinetochore microtubules (those attached to the chromosomes) and the nonkinetochore or interpolar microtubules (those which are not attached to the chromosomes). The nonkinetochore MTs usually extend from one of the two poles to the equator. MTs from one pole often interdigitate with those from the other so that the spindle is actually two half spindles which overlap one another. When anaphase begins and the chromosomes start their poleward migration, the kinetochore microtubules begin to shorten. Later, as the poles begin to separate, several overlapping nonkinetochore microtubules begin to lengthen. The number of interpolar microtubules decreases at this stage.

In telophase, the chromosomes have reached the poles and have begun to decondense, and as the cells progress into G_1, the nuclear envelope reforms around them creating the daughter cell nuclei. Simultaneously, the spindle microtubules begin to depolymerize so that in late telophase and early G_1 only a small remnant of the original spindle microtubules remain as a bridge between the two daughter cells. The CMTC is then reformed as the daughter cells revert to the interphase state.

B. Approaches Used to Study Microtubules

A variety of approaches are used in the study of microtubules. These include morphological and biochemical techniques, including *in vitro* polymerization, as well as the use of microtubule specific drugs.

1. Morphological Techniques

Microtubules were first seen but were not recognized as such by the early light microscopists in the late 1800s. With the introduction of elec-

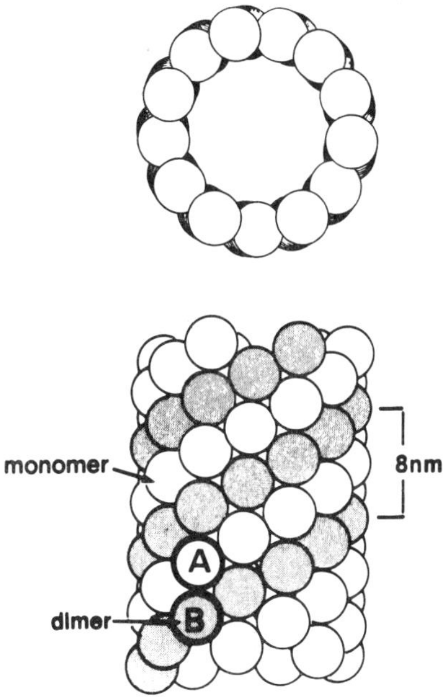

Figure 22.1. Schematic diagram of a microtubule. Top, cross-sectional view showing the 13 protofilament substructure. Bottom, lateral view showing the end-to-end and lateral associations of tubulin heterodimers to form the tubular structure. Reprinted with permission of W. B. Saunders Co.

tron microscopy, they were observed to be tubular structures. They were only occasionally seen in ultrastructural specimens and not believed to be common structures until better fixatives, such as glutaraldehyde, came into common use in the early 1960s (Sabatini et al., 1963). Since then microtubules have been found in virtually every eukaryotic cell examined. Electron microscopy has also revealed much about the substructure of microtubules. A schematic representation of the microtubule structure is shown in Figure 22.1. In favorable cross sections of embedded material, 13 subunits are discernible. These subunits run longitudinally and are known as protofilaments. When tannic acid is added to fixatives, these protofilaments show up well as a negative image. This method was used to demonstrate that the 13 protofilament substructure is common to microtubules from most sources (Tilney et al., 1973). Negatively stained preparations of microtubules show the protofilament structure in longitudinal view quite well, allowing the helical structure of the microtubule to be seen. The reader is referred to Amos (1979) for a more detailed discussion of these techniques.

Light microscopic techniques used in microtubule study include polarization and dark-field microscopy as well as epifluorescence microscopy with fluorescently tagged antibodies. The growth, development, and breakdown of the microtubule patterns in the mitotic spindle have been extensively studied with the polarizing microscope because microtubules in parallel array exhibit strong birefringence in polarized light allowing them to be easily seen in living cells. With dark-field optics, measurements have actually been made of microtubule growth from sequential photographs (Summers and Kirschner, 1979), although one cannot determine whether the image seen represents more than one microtubule.

Light microscopic fluorescence techniques with antibodies raised against tubulin have revealed the distribution of microtubules in the cell at the various stages of the cell cycle (see earlier section), although it is not yet possible to use this technique for prolonged observation of living cells. Such studies are most often done using indirect techniques where cells are first fixed and permeabilized and then treated in succession first with antitubulin antibodies raised in one animal and second with a fluorescently tagged preparation from a second animal raised against the antibodies of the first. The most difficult step in this procedure is to obtain the antibodies against tubulin, because tubulin is a poor antigen. A number of investigators, however, have succeeded in obtaining specific antibodies to tubulin, and the fluorescence images they obtained are spectacular and very clearly show the organization of microtubules in the cell. The reader is referred to Brinkley et al. (1980) and Fujiwara and Pollard (1980) for a more detailed treatment of this subject.

2. *Microtubule Biochemistry*

Tubulin can be isolated easily from brain tissue, where it makes up 10–15% of the soluble protein. The most commonly used technique is to prepare a cold high-speed supernatant of a brain tissue homogenate. Since MTs are largely cold-labile, this supernatant will contain a major portion of the cell's tubulin. By adding GTP and warming the supernatant to 37° in the presence of calcium chelators, the formation of MTs is induced. These MTs can then be separated from contaminating proteins by low-speed centrifugation. Further purification can be achieved by resuspending the microtubule pellet in fresh buffer and keeping the solution on ice to induce MT disassembly (Borisy and Olmsted, 1972; Weisenberg, 1972). High-speed centrifugation will then remove proteins which cosediment with the MTs at low speed. Addition of GTP and warming the solution to 37°C will then again induce the formation of MTs. After several cycles of such assembly/disassembly, reasonably pure preparations are obtained consisting of 80–85% tubulin and 15–20% microtubule-associated proteins (MAPs). These latter proteins have been reported to enhance the polymerization of microtubules *in vitro*. The *in vivo* significance of these proteins is, however, still an open question (see below). Other methods for the purification of tubulin take advantage of the ability of certain drugs to precipitate tubulin or enhance its polymerization into MTs. Vinblastine at micromolar concentrations causes tubulin to aggregate into paracrystalline arrays, and procedures for the isolation of these paracrystals have been published (Bryan, 1971, 1972). More recently, the drug taxol, which promotes tubulin assembly into MTs (Schiff et al., 1979; Schiff and Horwitz, 1980), has been used to purify tubulin. Addition of taxol to cell or brain tissue extracts at 37°C induces the formation of MTs, and these may be readily purified by centrifugation. MTs obtained this way are as pure as three-times-cycled tubulin (Vallee, 1982).

Guanine nucleotides have been shown to be associated with tubulin dimers. Each dimer has two binding sites for the nucleotide, one of which exchanges easily with GTP (the E site) and the other which exchanges slowly or not at all (Jacobs et al., 1974). When microtubules are purified by cycles of assembly/disassembly, it has been found that in order to obtain maximal rates of polymerization, the concentration of GTP must be equimolar with tubulin.

As mentioned earlier, two different subunits, α and β, make up the 6S tubulin dimer. The β subunit has the greater mobility on SDS gels and is the more acidic subunit by isoelectric focusing. In our laboratory, we have taken advantage of these properties using two-dimensional gels to separate the subunits and to identify mutant species of both α- and β-tubulin (see below).

3. Amino Acid Sequence and Tubulin Gene Isolation

Complete amino acid sequences have been worked out for α- and β-tubulins from porcine brain (Krauhs et al., 1980; Ponstingl et al., 1981), a rat α-tubulin (Lemischka et al., 1981; Lemischka and Sharp, 1982), a chicken α-tubulin (Valenzuela et al., 1981), a yeast β-tubulin (Neff et al., 1983), and a human β-tubulin (Lee et al., 1983). Although most of these sequences were deduced from cloned tubulin genes, the porcine brain sequences were deduced from protease digests and Edman degradation. The technology of gene cloning combined with the newer methods of nucleic acid sequencing now available have made amino acid sequence analysis a considerably easier task than it has been in the past. Once the DNA base sequence is known, the amino acid sequence can often be directly deduced.

A major problem with this method is that isolated natural tubulin DNAs often contain intervening sequences which do not encode the tubulin mRNA, and the possibility exists that the gene under study may represent a nonexpressed pseudogene. These problems can be avoided by using cloned cDNAs, but these seldom contain the sequence of the complete gene. In spite of these problems, DNA sequencing has produced much information about tubulin amino acid sequences. Use of cloned cDNA sequences has also increased our knowledge of the numbers and organization of tubulin genes in different species, for example, in chicken and *Drosophila* DNA, four unique genes have been found which encode α- and β-tubulin (Lopata et al., 1983).

So far, cloned tubulin DNAs have been isolated from a number of organisms, in the forms of cDNAs and natural genes. Cloned cDNAs have been obtained for a rat α-tubulin (Lemischka et al., 1981), chicken α- and β-tubulins (Cleveland et al., 1980; Valenzuela et al., 1981), sea urchin α- and β-tubulins (Alexandraki and Ruderman, 1981), and α- and β-tubulins of *Chlamydomonas* (Silflow and Rosenbaum, 1981). Genomic DNAs for chicken α- and β-tubulins (Lopata et al., 1983), human α- and β-tubulins (Wilde et al., 1982a, b; Lee et al., 1983), rat α-tubulin (Lemischka and Sharp, 1982), yeast β-tubulin (Neff et al., 1983), and *Drosophila* α-tubulin (Kalfayan and Wensink, 1981; Mischke and Pardue, 1982) have also been obtained.

4. Drugs That Affect Microtubule Polymerization

Several drugs are known to directly affect the polymerization of MTs (Luduena, 1979) and are useful tools for studying MT regulation in cells; among them are colcemid, colchicine, nocodazole, vinblastine, vincristine, podophyllotoxin, griseofulvin, maytansine, and taxol. In this chapter, these drugs will be referred to as MT-active drugs. Many of these are used in the chemotherapy of cancer (Suffness and Douros, 1979) and the best studied one, colchicine, has been used in the treatment of gout for centuries. Most of these drugs inhibit the polymerization of MTs or cause them to depolymerize; but they do so in a variety of ways. Colchicine and colcemid bind to the same site on the tubulin dimers (Shelanski and Taylor, 1967; Weisenberg et al., 1968; Wilson et al., 1974). It is thought that MTs depolymerize when these poisoned dimers bind to the assembling end, thereby inhibiting further elongation. The other end disassembles and the MT shortens (Margolis and Wilson, 1977). Nocodozole acts in a similar manner, but its binding is more quickly reversible than that of colcemid or colchicine (Hoebeke et al., 1976). As mentioned earlier, vinblastine and vincristine cause depolymerization of MTs by forming paracrystalline arrays or aggregates of tubulin (Bryan, 1971, 1972; Grisham et at., 1973). Griseofulvin causes MT disorganization and has been suggested to act on MAPs rather than on the tubulin itself (Roobol et al., 1977). Podophyllotoxin apparently overlaps the colchicine binding site and is quite similar to the latter in its action on MTs (Cortese et al., 1977). Maytansine also causes MT disruption and is extremely potent *in vivo* (Remillard et al., 1975). It is believed to interact with the vinblastine site on the protein. Taxol is a drug of great current interest for the study of MTs because unlike other known MT-active drugs, it stabilizes MTs both *in vitro* and *in vivo* to treatment with cold, colchicine, and high Ca^{2+}, conditions which normally lead to MT depolymerization (Schiff et al., 1979; Schiff and Horwitz, 1980). Taxol also inhibits mitosis by a mechanism that is not yet well understood, although spindle assembly is clearly affected. The ability of these drugs to interact with MTs and their resultant toxicity to the cell have made them useful agents for studying MT function and regulation [reviewed by Dustin (1978)] and for the isolation of mutants with altered MTs.

C. Problems Remaining in Microtubule Research

1. *Function*

As mentioned above, MTs have been implicated in a wide variety of physiological processes. The majority of the evidence, however, comes from studies employing MT-active drugs. While much of the data accumulated in these studies are no doubt correct, caution should be employed in their interpretation. In many cases, doses of the inhibitors far in excess of that required to poison MT assembly have been used. Thus, it is difficult to decide if the observed effects are the direct result of microtubule inhibition, or whether they result from side effects of the drug or even from a generalized cytotoxic state of the cells. Indeed, some of the functions attributed to MTs on the basis of inhibitor sensitivity have recently been questioned (Trifaro et al., 1972; Beebe et al., 1979; Brady et al., 1980; Skoda et al., 1983). Even in those cases (e.g., mitosis) in which a role for MTs has been clearly established, the mechanism of MT involvement remains to be elucidated. A genetic approach has the advantage of implicating MTs in a given process without the use of inhibitors, and is further capable of providing an avenue for obtaining detailed molecular information on the mechanism of involvement. Examples of this will be discussed below.

2. *Are MTs Essential to Eukaryotic Cell Life?*

The known involvement of MTs in mitosis and the necessity of mitosis for cell propagation make it very likely that MTs are essential to the life of the cell. Still, this question has been directly addressed only recently. Evidence that long-term treatment with agents that disrupt MTs results in cell death is not convincing, since these agents may have side effects distinct from their action on MTs. Again, the best way to approach this question is through genetics. As discussed below, recent data show that MTs are indeed essential components in the cell and further indicate why these structures are vital.

3. *Regulation of MT Assembly*

Development of an *in vitro* system for MT assembly has led to the identification of a number of proteins which copurify in a constant stoichiometric ratio to tubulin through several rounds of assembly/disassembly and which lower the critical concentration of tubulin necessary for assembly when mixed with phosphocellulose-purified tubulin (Scheele and

Borisy, 1979). These proteins are called microtubule-associated proteins (MAPs). While the existence and probable function of these proteins is well worked out *in vitro* in a few systems such as brain (Kirschner, 1978) and HeLa cells (Weatherbee et al., 1978; Bulinski and Borisy, 1979), their actual *in vivo* significance is still questionable. Most of the evidence for an *in vivo* role for these proteins derives from the observation that antibodies prepared against these proteins decorate microtubules (Sherline and Schiavone, 1977; Conolly et al., 1978) in fixed cells by immunofluorescence and that projections with a periodic distribution on MTs are seen by electron microscopy and are apparently composed of MAPs (Dentler et al., 1975; Murphy and Borisy, 1975). Still, it cannot be excluded that these localizations are fixation artifacts or that the association of MAPs with MTs is serendipitous.

Ever since Weisenberg (1972) worked out the conditions for *in vitro* MT assembly, it has been recognized that calcium is a potent inhibitor of MT assembly. The mechanism by which calcium inhibits MT assembly, however, is still not understood. A few years ago, Welsh et al. (1978) showed that calmodulin, a ubiquitous calcium-binding protein, is a component of the mitotic spindle apparatus; more recently, it has also been shown to localize to centrosomes in interphase cells (Willingham et al., 1983). These results suggest that calcium effects on MT assembly may be mediated by calmodulin. Direct demonstration of this possibility *in vitro* using purified tubulin and calmodulin, however, has been inconclusive (Kumagai and Nishida, 1979). One possible explanation is that the components in the reaction are "too pure," that is, other proteins necessary for the action of Ca^{2+}/calmodulin on tubulin are missing. These other proteins may be termed calmodulin-acceptor proteins (CAPs) (Pardue et al., 1981). It will be crucial to our understanding of MT regulation to determine if these proteins exist and, if so, to purify and characterize them.

Finally, we wish to discuss the possibility that distinct microtubule-associated proteins exist for spindle as opposed to cytoplasmic MTs. We propose to call such proteins spindle-associated proteins (SAPs). Evidence for the existence of such proteins has in fact already appeared. Izant et al. (1982) prepared monoclonal antibodies against purified HeLa MAPs and have found that some of these monoclonals localize to the mitotic spindle apparatus but not to cytoplasmic MTs. One should not necessarily conclude from these results, however, that distinct proteins associate with spindle MTs. Recent work by Pardue et al. (1983) has shown that monoclonal antibodies produced against calmodulin can selectively localize to calmodulin bound to distinct subcellular structures; that is, some monoclonal antibodies react only with spindle calmodulin, others with mitochondrial calmodulin, etc.; even though there is no biochemical evidence that these calmodulins are discrete molecules. Biochemical studies supporting the existence of SAPs have been reported by Zieve and Solomon (1982), who isolated the mitotic spindle apparatus, ran two-dimensional gels of the preparations, and determined which proteins were released by colchicine and by calcium, two agents known to depolymerize MTs. In this way a SAP of approximately 150,000 daltons has been identified.

Thus far genetics has contributed little to resolving the question of whether spindle-specific microtubule-associated proteins exist, and this is an important area for future experiments. Preliminary experiments from our laboratory and others (Gottesman, personal communication), however, suggest that tubulin subunits altered by mutation assemble into both cytoplasmic and spindle MTs. This argues against the involvement of unique tubulin subunits in either of these two structures.

4. Do Multiple Tubulin Genes Code for Tubulins Performing Specific Functions?

This question is an expression of the multitubulin hypothesis first proposed by Fulton and Simpson (1976). Evidence for multiple tubulins has in fact appeared. For example, genetic studies in *Aspergillus* suggest the existence of multiple genes for α-tubulin (Morris et al., 1979); similar work in *Drosophila* demonstrates the existence of multiple β-tubulin genes in that organism (Kemphues et al., 1979; Raff et al., 1982); and our own work suggests that there are at least three distinct alleles or loci coding for both α- and β-tubulin in CHO cells (Cabral et al., 1980, 1981). Experiments using cloned cDNA sequences have demonstrated that multiple sequences complementary to α- and β-tubulin mRNA are dispersed through the genome in chick, *Drosophila, Chylamydomonas,* rat, sea urchin, and human cells (Cleveland et al., 1980; Sanchez et al., 1980; Alexandraki and Ruderman, 1981, Cowan et al., 1981; Kalfayan and Wensink, 1981; Silflow and Rosenbaum, 1981; Mischke and Pardue, 1982; Wilde et al., 1982b, c; Lopata et al., 1983). While these studies demonstrate that multiple tubulin genes exist in many eukaryotic cells, little evidence for diverse function of these gene products has appeared. In fact, a testis-specific β-tubulin described by Raff and her colleagues (Kemphues et al., 1979, 1980) has recently been shown to assemble into sperm axonemes, cytoplasmic MTs, and the meiotic but not mitotic spindle apparatus (Kemphues et al., 1982). Thus, a single developmentally regulated, tissue-specific tubulin has been shown to assemble into and function in diverse MT structures, thus arguing against the concept of unique tubulins for specific functions. Further genetic studies should help resolve the question of whether this is a general phenomenon or whether examples of function-specific tubulins exists.

5. Regulation of Tubulin Expression

One of the factors that controls MT assembly is the concentration of tubulin in the cell. Thus, it should not be surprising that the level of tubulin is tightly regulated. A few years ago, Ben-Ze'ev et al. (1979) reported that new tubulin synthesis is sensitive to the existing level of free but not polymerized tubulin. Cells treated with colchicine and other drugs that lead to depolymerization of MTs and an increase in the pool of 6S tubulin dimers show depressed rates of tubulin synthesis while cells treated with vinblastine, which precipitates tubulin into paracrystalline inclusions and reduces the pool of 6S tubulin dimers, show somewhat enhanced rates of tubulin synthesis. This apparent ability of free tubulin to inhibit tubulin expression was confirmed by Cleveland et al. (1981), who further showed that the regulation occurs at a pretranslational level, and does not appear to result from transcriptional control (Cleveland and Havercroft, 1983).

A number of questions concerning this mechanism of regulation remain. Does the regulation operate at the level of mRNA processing, nuclear export, or stability? This question can be approached using the well-developed techniques of biochemistry and molecular biology. A second question, "Is regulation of tubulin expression gene-specific or coordinate?," is less easily answered with conventional techniques owing to the multiplicity of tubulin genes and the frequent comigration of these gene products on one- and two-dimensional gels. Genetics can help by providing mutant tubulins with altered electrophoretic mobilities so that effects on specific gene products can be monitored. It is also hoped that mutants with altered rates of tubulin expression in the absence of any exogenously added drugs might be obtained. Such mutants would be very useful in studying *in vivo* mechanisms of tubulin gene expression.

D. Rationale for Using Somatic Cell Genetics

The contributions which somatic cell genetics can make in answering the questions posed above have already been described. In all cases, the rationale is to look for cells with alterations in tubulin or microtubules and then study a given phenomenon in both wild-type and mutant cells. If the phenomenon is affected in mutant cells, then microtubules presumably play a role in the process being studied. The use of revertants can establish the role of MTs in the process with little or no ambiguity. Examples of how this may be done will be given below.

II. HOW MUTANTS MAY BE OBTAINED

A. Functional Selections

In an effort to probe the process of mitosis, many early workers selected cells with an impaired ability to progress through this stage of the cell cycle. The mutants were selected either by their ability to grow (divide) at a permissive but not at a nonpermissive temperature, or by using the observation that mitotic cells in many cell lines round up and become less adhesive. A complete description of techniques for the isolation and analysis of cell cycle mutants is given in Chapter 20. In an early example of this approach Smith and Wigglesworth (1972) selected temperature-sensitive (ts) BHK-21 cells by killing the cells able to grow at the nonpermissive temperature with cytosine arabinoside. Cells which were not cycling at the nonpermissive temperature (and which, therefore, were not killed by the drug) could then be recovered by washing out the drug and shifting the cells to the permissive temperature at which they could grow and form colonies. In this way, a number of mutants with a defect in cytokinesis were isolated. These cells grow poorly and become binucleated at the nonpermissive temperature, but protein and DNA synthesis is not affected. Similar ts mutants in a Chinese hamster fibroblast line were found by Hatzfeld and Buttin (1975) using a BrdU suicide selection. Again, cytokinesis is defective at the nonpermissive temperature and giant multinucleated cells are produced. Interestingly, cytokinesis resumes in 25% of the cells after shifting back to the permissive temperature even after a prolonged (96 hr) arrest in the restrictive state. The authors suggest that the cells which are able to recover represent karyoplasts derived from the giant cells. Somewhat later, Shiomi and Sato (1976) isolated a similar mutant, ts2, in a murine leukemic cell line which again is defective in mitosis and cytokinesis but in which the biochemical lesion was also not demonstrated.

Using similar methods, Wang (1974) isolated 50 ts mutants in hamster HM-1 cells and of these, one mutant, ts-546, blocked mitosis at the nonpermissive temperature. The plating efficiency of this cell line at the restrictive temperature was 5×10^{-7}; this number represents the reversion frequency of the strain. The mitotic index of the mutant cell population increases with time at the restrictive temperature accumulating C-metaphaselike figures. Anaphase and telophase figures are absent at the nonpermissive temperature. Like the mutants mentioned above, cytokinesis appears to be defective in this mutant and the cells which reenter interphase have multiple nuclei (Wang and Yin, 1976). The author and his colleagues have also described cell lines defective in prophase progression (Wang, 1976) and, more recently, in anaphase chromosome movement (Wissinger and Wang, 1978) and centriole separation (Wang et al., 1983). In no case, however, has the specific biochemical defect in these mutants been defined.

Using a related but somewhat different approach, Thompson and Lindl (1976) exploited the observation that many mammalian cells become less adhesive during mitosis to obtain CHO cells with a temperature-sensitive defect in cytokinesis. A cell population, mutagenized and maintained at 34°C, was enriched for cells with mitotic defects by shifting the cells to 38.5°C. The unattached cells were collected at 30-min intervals and replated at 34°C. Cells collected in this way over a 2-hr period were pooled. This procedure was repeated six times, and the cells were cloned into Linbro trays. Of 100 clones isolated in this manner, one, ts MS1–1, was temperature sensitive and showed a defect in cytokinesis. Similar to the mutants described above, MS1–1 grew poorly and became multinucleated at the nonpermissive temperature (38.5°C); however, a high frequency of polyploid cells were also observed at 34°C. Although the biochemical lesion in this mutant strain was not found, the ts mutation behaved recessively in hybrids with the wild-type strain, suggesting an altered or absent product of the MS1–1 gene. Morphological examination of this mutant shows that a cleavage furrow forms but does not proceed to completion. Microtubules, microfilaments, and other cellular structures appear to be normal.

This short summary demonstrates that ts selections can yield interesting mutants blocked in various stages of mitosis, but it also points out the limited utility of these mutants (thus far) because of a lack of biochemical information regarding the nature of the lesions. This latter situation has led Stanners (1978) to suggest that selections designed to obtain mutants with defects in specific proteins might be a more appropriate way to study control of cell division. This is the approach which is described in the next section.

B. Drug-Resistance Selections

1. *Mechanism of Drug Action*

The more information available about a drug, the more rationally a selection can be devised to obtain the kinds of mutants desired. For example, an agent whose primary action is on a modifying enzyme which in turn affects a variety of cellular systems including microtubules is not a good agent to use for selections of mutants with structural changes in tubulin. To obtain such mutants, drugs which bind directly to tubulin and inhibit its polymerization into microtubules would be more desirable. Fortunately, many such drugs have been described for microtubules (Dustin, 1978). In general, these drugs bind to various sites on free tubulin heterodimers and inhibit their assembly, thereby blocking microtubule function. One notable exception to this mechanism is taxol, which binds to polymerized microtubules and promotes their assembly but also blocks their function. It would seem from this observation concerning the mechanism of taxol toxicity that the state of tubulin polymerization in the cell is carefully controlled and any agent which interferes with this balance in either direction is toxic to the cell. Indeed, this interpretation is supported by a number of genetic studies described below.

Aside from the observation that microtubule-active drugs bind to free or polymerized tubulin and affect assembly, their mechanism of action is not well understood. It has been known for a number of years that drugs like colchicine inhibit microtubule assembly when present in amounts substoichiometric to tubulin. This has led Margolis and Wilson (1977) to propose a mechanism in which colchicine binds only to free tubulin dimers. When these poisoned dimers assemble onto the end of a growing microtubule, they block further elongation. Since MTs are believed to have higher rates of disassembly than of assembly at the opposite end of the MT, the net effect of poisoning the assembling end is to promote the dissolution of the MTs in the cell. This is a very appealing mechanism which has gained wide popularity since its proposal; however, a number of recent kinetic experiments cannot be explained with this model nor can this model explain the recent observation that colchicine binds all along the length of a MT and not just at its ends (Sternlicht and Ringel, 1979; Farrell and Wilson, 1980; Sternlicht et al., 1980; Saltarelli and Pantaloni, 1983).

Perhaps a simpler way to think about the action of these various drugs is to consider that the equilibrium between free and polymerized tubulin is carefully maintained in the cell. Agents which bind to the tubulin dimers could lower the affinity of these dimers for the MT and thereby shift the equilibrium to the dimer form. Agents which bind to the MTs would diminish the ability of polymerized tubulin to disassemble and thereby shift the equilibrium to the polymerized form. Agents of the first type include the drugs described earlier, colchicine, colcemid, griseofulvin, nocodazole, podophyllotoxin, vinblastine, maytansine, and perhaps calcium or a calcium-regulated protein. Agents of the second type include taxol and endogenous MAPs. Thus, agents such as colchicine might be thought of as "anti-MAPs" and taxol might be thought of as "MAPlike." Again, the genetic studies described below support this way of looking at the effect of drugs on MT assembly.

This explanation for the mechanism of drug action is still rather simpleminded. It is hoped that a close analysis of drug-resistant mutants will provide more details as to how these drugs act. For the purpose of mutant selection, however, it is only essential to know that these drugs bind to tubulin with high affinity and that they are cytotoxic to cells. The cytotoxicity of these agents probably results from the known ability of colchicine and the other drugs to block cells in mitosis. How this leads to cell death will be discussed in more detail below.

2. *How Drugs May Be Used for Selection*

a. Drug Cytotoxicity. The simplest method for obtaining mutants resistant to a microtubule-active drug is to look for cells which survive at a concentration of the drug that kills the vast majority of the cells. We generally use the lowest drug concentration that will reduce the plating efficiency of our cells to below 10^{-5}–10^{-4}. A less accurate but simpler method for determining the optimal selecting drug concentration is to set up a 24-well dish with increasing concentrations of the drug from 0 to 10 μg/mL in 2 ml of medium per well. To each well is then added a constant number of wild-type cells (100–1000 cells/well) and the dish is placed at 37°C for 5–7 days. At the end of this time, each well is examined microscopically for cell growth and normal or aberrant morphology. The lowest concentration of drug that inhibits the growth and alters the morphology of the cells is the concentration used for the selection. It should be realized, however, that single-step selections have yielded mutants which are only two- to threefold resistant to the selecting drug, and thus it is often necessary to use very small (two-fold) incremental steps in looking for the appropriate selecting drug concentration.

Once the appropriate selecting drug concentration is known, the cells are mutagenized with EMS or UV light, to a survival of approximately 20%, allowed to recover for 2–5 days, and then plated at a density of 5×10^5 cells in a 100-mm tissue culture dish in 20 ml of α-MEM containing 10% fetal bovine serum and the concentration of the selecting drug that was determined to be optimal. More detailed methods for mutant isolation have been previously published (Thompson, 1979; Cabral et al., 1980). After 7–10 days at 37°C in a 5% humidified atmosphere, individual colonies can be observed growing over a background of dead cells. These drug-resistant mutant colonies generally arise at a frequency of approximately 10^{-5} in a mutagenized cell population and at a frequency below 10^{-7}–10^{-6} in nonmutagenized cells.

Colonies are picked from these dishes by mechanical scraping with a sterile Pasteur pipette and transferred to individual wells of a 24-well dish. They are then allowed to grow in selective medium until the wells are nearly confluent. The cells from each well may then be trypsinized, recloned in soft agarose or agar (0.3%), and grown to mass culture for biochemical and genetic analysis. It is wise to freeze the cells as soon as possible after isolating the mutant clones to insure against loss by contamination, incubator malfunction, reversion, etc.

b. Resistance to Loss of Function. The isolation scheme described above depends on the cytotoxic properties of the microtubule-active drugs. As a result, cells surviving this selection might be expected to have alterations in processes vital to the growth of the cells. MTs, however, are believed to participate in a large number of cellular processes such as mobility of cell surface proteins, cell morphology, cell locomotion, and saltatory motion, which may not be critical to cell survival in a cell culture environment. In fact, as will be discussed below, the conditional mutants isolated in our laboratory thus far exhibit defects only in mitosis at the nonpermissive condition. Thus it is possible that for CHO cells growing in culture, mitosis is the only process in which MT involvement is crucial for cell survival.

To obtain mutants in other processes in which MTs are involved, it will probably be necessary to devise new selections. One potential selection would be the use of MT-active drugs coupled with a selection based on the putative MT function. For example, if MT-active drugs inhibited cell adhesion, then a selection could be devised based on the ability of mutant cells to adhere even in the presence of the drug. In general, these selections would have to be short term, since long-term selections would give predominantly mitotic mutants. While this type of selection is conceivable, it has not been carried out in practice. More effort needs to be directed toward the development of such selections.

c. Single Versus Multiple Round Selections. As mentioned previously, single-step selections generally yield cells that are only two- to three-fold resistant to the selecting drugs compared to the wild-type parental cells. Higher levels of resistance may be achieved by subjecting cells to multiple rounds of selection at increasing concentrations of the drug; but the resulting mutant cells have phenotypes that are difficult to interpret because of the likelihood of multiple genetic lesions in the cells. To our knowledge, no mutant with a high level of resistance to a microtubule-active drug has been clearly shown to have achieved that level of resistance solely through a mutation affecting tubulin. One possible way of selecting such mutants is to perform stepwise selections to increasing drug concentrations but to clone and characterize the mutants at each step before proceeding to the next step. Multiple-step selections have been useful, however, for obtaining mutants with karyotypic alterations suggestive of gene amplification (Meyers and Biedler, 1981; Kuo et al., 1982). The amplified gene products in these mutants have not yet been clearly identified, however, nor have the cells been shown to have alterations in their microtubule structure.

The reasons for the low level of resistance in single-step mutants have not yet been determined. It may be rationalized, however, that the reasons result from the delicate balance between free and polymerized tubulin, which we and others believe exists in mammalian cells. Mutations leading to resistance to microtubule-active drugs that act to destabilize MTs probably result in MTs which are inherently more stable. This situation has been found in *Aspergillus* (Oakley and Morris, 1981) and CHO cells (Gupta et al., 1982; Warr et al., 1982). In many of our own mutants, cells resistant to these drugs are supersensitive to taxol, a MT-stabilizing drug (Schiff et al., 1979; Schiff and Horwitz, 1980). The converse, that mutations leading to resistance to MT-active drugs which act to stabilize MTs probably result in MTs that are inherently less stable, has also been demonstrated (Cabral, 1983). Thus, we argue that any mutation which changes the fold resistance too greatly through an alteration in the MTs, would upset the carefully controlled balance of free and polymerized tubulin to such an extent that the cell would not be able to survive. Thus, only mutants with low levels of resistance are found.

d. Screening Methods. Once a drug-resistant mutant is isolated, it must be identified as a tubulin or MT mutant. This is not a trivial task, since most drug-resistant mutants in CHO cells are permeability mutants (Ling and Thompson, 1974; Cabral et al., 1980). These permeability mutants have been shown to exhibit cross resistance to a wide variety of unrelated hydrophobic drugs (Bech-Hansen et al., 1976) and this cross resistance provides a convenient method for screening out permeability mutants. A drug-resistant clone is simply tested for resistance to a drug such as puromycin whose site of action is distinct from MT-active drugs. If the clone is resistant to puromycin, it is probably a permeability mutant; if it is puromycin sensitive, then it has a much greater chance of being a tubulin mutant.

The puromycin-sensitive clones must still be shown to be tubulin mutants, however, and this can be accomplished in a number of ways. The method we prefer in our laboratory is to screen the clones by two-dimensional gel electrophoresis (Cabral and Schatz, 1979). Examples of such gels are shown in Figure 22.2 and will be discussed more fully below. The advantage of this technique is that it is simple, highly reproducible, requires no sample purification, and can unambiguously show an alteration in a particular tubulin subunit. Thus, if a particular mutant exhibits an altered tubulin on two-dimensional gels, one can be reasonably certain that the mutant has an alteration in the structural gene for tubulin, provided the proper controls are performed (discussed below). The major disadvantage is that not all alterations in primary amino acid sequence will result in an electrophoretic shift on two-dimensional gels and thus many potential mutants will be lost. Still, this approach has the decided advantage that when an electrophoretic shift is observed, one can be reasonably sure of the nature of the mutation.

Another approach to screen for MT mutants among the drug-resistant, puromycin-sensitive cells is to ascertain which of these are ts for growth. If MTs are necessary for cell viability, it might be expected that mutations affecting MTs would cause the MTs to become defective at some temperature other than the selecting temperature and therefore lead to problems in cell growth. We have, in fact, observed this for most of the tubulin mutants isolated in our laboratory. To make use of this observation, one would simply look at MT structure at the nonpermissive temperature using a technique such as indirect immunofluorescence with antibodies to tubulin. This approach has the advantage of identifying mutants with alterations in tubulin which do not cause electrophoretic shifts and mutants which have alterations in MT proteins other than tubulin. It has the disadvantage of not identifying the altered protein and of missing mutants with alterations in MT proteins including tubulin which do not result in temperature sensitivity.

A final screening method is to look for changes in drug binding affinity using crude homogenates from the mutant cells. These determinations can be performed using a simple filter binding assay (Wilson, 1970). Such a screening method has been employed by Ling et al. (1979) to identify several colcemid-resistant mutants with altered colcemid-binding affinity. This procedure is useful for identifying subclasses of tubulin mutants and providing an explanation for the mechanism of drug resistance. Its major disadvantages are the lack of simplicity and reproducibility of the assay and its inability to recognize MT mutants which have no alteration in drug binding. Indeed, none of the mutants isolated in our laboratory to date has shown an alteration in drug binding and we suspect that such mutants may be a minor subclass of all tubulin mutants obtainable in a given selection.

III. CHARACTERIZATION OF MUTANTS

A. Biochemical Methods

As mentioned, after preliminary screening of drug-resistant mutants for puromycin sensitivity, we make positive identification of tubulin mutants using two-dimensional gel electrophoresis. This procedure is made practical by the fact that the tubulins comprise approximately 3–5% of the total protein in CHO cells and therefore appear as prominent spots on two-dimensional gels of whole cell lysates. Examples of such gels for wild-type (A), a colcemid-resistant β-tubulin mutant (B), and a taxol-resistant α-tubulin mutant (C) are shown in Figure 22.2. In the wild type, both α- and β-tubulin migrate as single spots on the gel. In the colcemid-resistant β-tubulin mutant, as well as the taxol-resistant α-tubulin mutant, however, there are "extra spots" (arrows) which do not appear in the wild-type gels and presumably reflect the fact that CHO cells have multiple loci encoding both tubulins. It is relatively simple to confirm that these "extra spots" are actually electrophoretically variant α- or β-tubulins, by excising [^{35}S]methionine-labeled mutant polypeptides from the gels and subjecting them to one- and two-dimensional peptide mapping (Cabral et al., 1980, 1981).

We can estimate the number of loci by measuring the stoichiometry of the wild-type and mutant polypeptides in the mutant cells. This is most easily accomplished by labeling the cells with [^{35}S]methionine, running a whole cell lysate on a two-dimensional gel, cutting out the wild-type and mutant spots, and measuring the radioactivity in each spot by liquid scintillation counting (Cabral and Gottesman, 1978; Cabral et al., 1980). Analysis of several α- and β-tubulin mutants by this technique leads to a

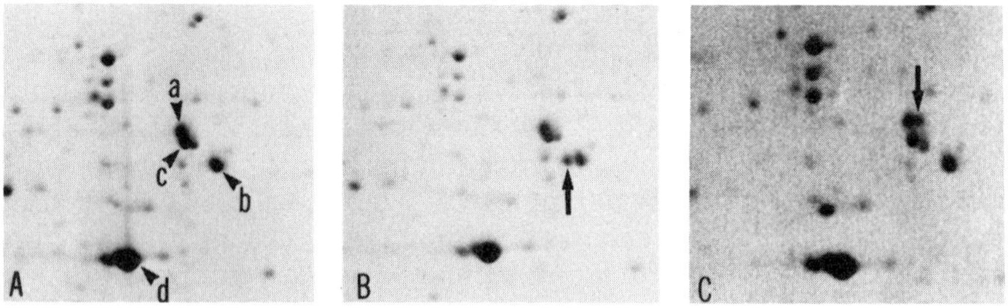

Figure 22.2. Two-dimensional gel autoradiograms of wild-type and mutant CHO cells. Cells were metabolically labeled for 30 min with [^{35}S]methionine, lysed in SDS, and run on two-dimensional gels. Autoradiograms (A,B) or the Coomassie blue stained gel (*C*) are shown. (A) wild-type cells: *a*, α-tubulin; *b*, β-tubulin; *c*, vimentin; *d*, actin. (B) Mutant Cmd-4. Note the presence of an "extra" spot (arrow) to the basic side of β-tubulin. (C) Mutant Tax-1. Note the presence of an "extra" spot (arrow) to the acidic side of α-tubulin. Only a portion of each gel is shown.

stoichiometry of 2:1, wild-type: mutant polypeptides. These results suggest strongly that there is more than one gene coding for α- and β-tubulin in CHO cells, and is most consistent with the presence of three distinct loci or alleles for both α- and β-tubulin. This conclusion agrees with *in situ* hybridization experiments which predict four genes for both α- and β-tubulin in chicken and *Drosophila* using cloned cDNA probes for the tubulin genes (Cleveland et al., 1980; Sanchez et al., 1980). The possibility that the unequal stoichiometry between wild-type and mutant tubulins might be due to enhanced degradation of the mutant polypeptides was ruled out by pulse-labeling experiments and by *in vitro* translation of isolated RNA from these cells. The latter technique in combination with somatic cell hybridization experiments also negates the possibility that the mutant tubulins might have arisen by posttranslational modification of the wild-type subunits. Direct confirmation for the number of genetic loci coding for tubulin in CHO cells is still lacking, but development of the appropriate probes is well under way in several laboratories and this information should be available soon.

Demonstration of an altered tubulin species in a mutant is good evidence that an MT mutant has been isolated, but this is not sufficient to reveal the mechanism of resistance to the drug. The simplest mechanism for drug resistance is an alteration in the binding site for the drug. This possibility is most easily tested by measuring the capacity of crude wild-type and mutant cell homogenates to bind the radiolabeled drug using a filter binding assay (Wilson, 1970). Such experiments have proven negative for our mutant strains, but a multiple-step colcemid-resistant mutant isolated by Ling appears to exhibit reduced binding affinity for [^{3}H]colcemid. Still it is our experience that such mutants are rare and most MT mutants are drug resistant for other reasons.

In vitro assembly experiments such as those pioneered by Weisenberg (1972) and Borisy and Olmsted (1972) for brain tubulin are potentially very informative, but have thus far not been greatly used in mutant analysis. A major problem is the difficulty in adapting the technique to cultured cells where tubulin is in much lower abundance and where factors which inhibit *in vitro* polymerization may exist (Nagle et al., 1977). Still, some laboratories have had at least limited success in assembling MTs from cell extracts (Nagle et al., 1977; Keates et al., 1981) and more effort is needed in this direction. Once conditions for *in vitro* assembly are worked out, experiments to determine the effects of mutant tubulin on the polymerization characteristics of wild-type tubulin can be explored. Furthermore, dosage experiments to determine how much mutant tubulin is required to counteract a given drug dose can be carried out far more easily than *in vivo*. Other methods for *in vitro* assembly of MTs utilizing MT stabilizers such as taxol (Vallee, 1982) are much more adaptable to cultured cells but are objectionable because of the presence of interfering MT binding agents. Still, these methods are useful for large-scale purification of tubulin from cultured cells for immunological or chemical studies.

Analysis of a number of our mutants has recently shown that defective MTs primarily affect mitosis and spindle assembly in particular (Abraham et al., 1983; Cabral, 1983; Cabral et al., 1983). We therefore feel that it is increasingly important to focus our attention on the spindle apparatus. This is especially crucial for the analysis of mutants which may not have alterations in tubulin itself, but rather in MAPs. Spindle isolation and characterization will also be important for analysis of mitotic mutants in which MTs are not affected but other SAPs are altered.

Spindle-isolation procedures have been published by a number of laboratories (Chu and Sisken, 1977; Kuriyama, 1982; Mullins and McIntosh, 1982), and the isolated spindles have been well-characterized morphologically. A more careful examination of the biochemical composition of the spindle, however, is still required. In particular, a determination of identities of the structural proteins of the spindle is needed; that is, which of these are MT associated, which are components of the spindle poles, which are chromatin associated, which comprise the kinetochore region of the chromosome, which are involved in contractile ring formation, and which enzymes affect the assembly or function of the spindle apparatus. A number of laboratories are now addressing these questions by refining isolation procedures to obtain spindles of the highest biochemical purity, by examining the composition of isolated spindles using two-dimensional gel electrophoresis in combination with drug and calcium treatments to determine which proteins are MT associated (Zieve and Solomon, 1982) and by preparing monoclonal antibodies against isolated spindles. These latter reagents will be useful for sorting out the various SAPs and for analyzing mutants to determine if certain proteins fail to be synthesized or to assemble. Already, immunological approaches have demonstrated the existence of MAPs which associate with spindle but not with cytoplasmic MTs (Lydersen and Pettijohn, 1980; McCarty et al., 1981; Izant et al., 1982). It is expected that other such proteins will be discovered in the future.

B. Genetic Methods

Once a MT mutant has been identified, one must determine if the drug resistance behaves in a dominant or recessive manner. This can be accomplished by fusing the mutant cell line with wild-type cells carrying ouabain-resistance and 8-azaguanine-resistance (HAT-sensitive) markers and selecting hybrid cells in HAT medium containing ouabain (Jha and Ozer, 1976). The drug resistance of these hybrid cells is then compared to wild-type/wild-type hybrids and to mutant/mutant hybrids to determine if they behave more like wild type or mutant. As may be seen in Table 22.1, most of the mutants isolated behave in a codominant manner, that is, the mutant phenotype is expressed in mutant/wild-type hybrid cells. This result is not surprising since, as was discussed, CHO cells appear to have multiple loci coding for tubulin. Unless specific gene products serve unique functions (even though they all comigrate on two-dimensional gels) or gene dosage has profound effects on MT assembly, recessive mutants with alterations in tubulin would not be expected. One mutant, Tax-18, does appear to be recessive and this will be discussed in some detail below.

Complementation analysis of MT mutants in CHO cells is not yet feasible because of the paucity of recessive mutants that have been isolated. The existence of Tax-18, however, suggests that other recessive mutants will be found and complementation studies will be performed in the future. Such studies will help to determine the number of genetic loci involved in MT assembly in the cells.

C. Reversion Analysis

Demonstration of an altered protein in a mutant does not prove that the altered protein is responsible for the mutant phenotype. It can be argued that the mutant cells have more than a single altered protein and that the one under observation may not be the one responsible for the phenotype being studied. This argument is even more compelling in mutagenized cells where more than a single genetic lesion almost certainly exists. The simplest and most direct method to show that an altered protein is responsible for a given phenotype is to obtain revertants of the phenotype and show further alterations in the protein at high frequency. If the protein is unrelated to the phenotype, then there is a very low probability that it will again be altered in revertants. If, however, the protein is in

TABLE 22.1
Microtubule Mutants in CHO Cells

Strain	Phenotype	Biochemical Alteration	Dominance	Reference
Cmd-4	Colcemid^R	β-Tubulin	D	Cabral et al., 1980
Col-2	Colchicine^R	β-Tubulin	D	Cabral et al., 1980
Grs-1	Griseofulvin^R	β-Tubulin	D	Cabral et al., 1980
Grs-2	Griseofulvin^R	β-Tubulin	D	Cabral et al., 1980
Tax-1	Taxol^R	α-Tubulin	ND	Cabral et al., 1981
Tax-2	Taxol^R	β-Tubulin	ND	Unpublished
Tax-18	Taxol^R	ND	R	Cabral, 1983; Cabral et al., 1983
CM^{R}795	Colcemid^R	α-Tubulin	D	Ling et al., 1979; Keates et al., 1981
CM^{R}761	Colcemid^R	ND	D	Ling et al., 1979
CM^{R}117	Colcemid^R	ND	D	Ling et al., 1979
Tax101[a]	Taxol^R	ND	ND	Warr et al., 1982
Ben 1	Benzimidazole carbamate^R	ND	ND	Warr et al., 1982
Pod^{RII}	Podophyllotoxin^R	66,000 dalton Protein	D	Gupta, 1981; Gupta et al., 1982
Tax^R-1	Taxol^R	ND	D	Gupta, 1983

[a]The strain designation in the original publication is Tax_1 but has been changed to Tax101 to avoid confusion with other Tax 1 mutants (J. R. Warr, personal communication). D, dominant or codominant; R, recessive; ND, not determined.

some way responsible for the phenotype, then the probability that it will again show alterations in the revertants is very high.

The isolation of revertants of drug-resistant cells is not straightforward. Ideally, one would like to select directly for drug-sensitive revertants. This cannot be done, however, because those are the cells which are dying in the selection. One way around this problem is to screen for drug-sensitive cells using replica plating techniques, but this method is tedious. Fortunately, we have found that a number of our tubulin mutants are also temperature sensitive (ts) for growth. Thus, we argued, as Morris et al. (1979) argued earlier with their *Aspergillus* mutants, that if the altered tubulin is responsible for both the drug resistance and temperature sensitivity of the mutant, then among revertants selected for loss of the ts phenotype, we should find cells which have lost their drug resistance and their altered tubulin.

This prediction has proven correct in the two mutants, Cmd-4 and Tax-1, in which we have carried out such an analysis. In the better studied example, 24 "temperature-resistant" revertants were isolated from seven independent cultures of Cmd-4 and these were analyzed for their drug resistance and for the presence or absence of the altered β-tubulin (β*) using two-dimensional gel electrophoresis (Cabral et al., 1982). The 24 revertants could be grouped into three classes. The most abundant class (19/24) contained cells that were still drug resistant and still had β*. We cannot yet fully explain the existence of this class of revertants, but one possibility is that these cells are intragenic revertants that give rise to a second electrophoretically "silent" alteration which restores the stability of the MTs at elevated temperature but does not remove the ability of β* to confer drug resistance. An even more interesting possibility is that this class contains extragenic suppressors, that is, the cells have lesions in other genes coding for proteins which interact with MTs. An alteration in such a protein could compensate for the alteration in β*, making the MTs more stable at the elevated temperature but again not affecting the ability of β* to confer drug resistance.

A second class of revertants (4/24) were no longer drug resistant and no longer expressed β*. This is the revertant class that was predicted if drug resistance, ts, and expression of β* are the result of a single gene mutation and proves that β* is responsible for both phenotypes. Again, the reasons for loss of expression of β* in these revertants is unknown, but some possibilities include intragenic reversion to give a polypeptide with wild-type electrophoretic mobility, gene deletion, or gene inactivation. Chromosome loss cannot explain the loss of β* because the frequency of the event is too low (10^{-5}) and because karyotype analysis reveals the normal number of chromosomes in these cells.

The third class of revertants (1/24) has a single member but is both the most easily explained and the most interesting revertant obtained to date. It has lost its drug resistance and has a further alteration in β* to produce

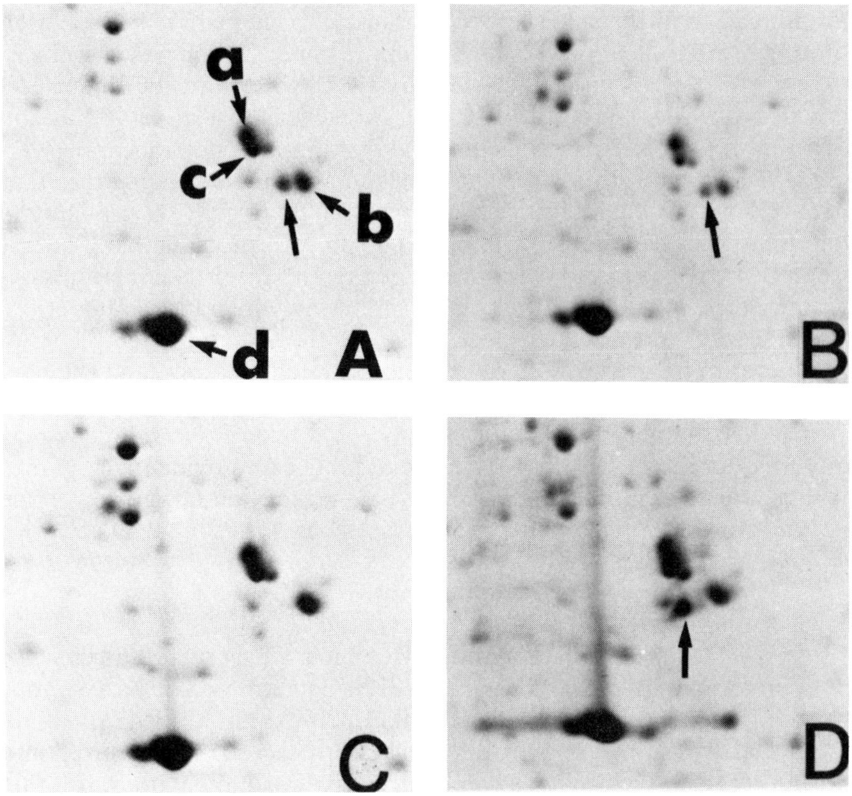

Figure 22.3. Two-dimensional gel autoradiograms of mutant Cmd-4 and three of its revertants. Cells were labeled and run on two-dimensional gels as described in Figure 22.2. (A) Mutant Cmd-4: a,b,c, and d are the same proteins described in Figure 22.2. The arrow denotes the presence of an altered β-tubulin polypeptide. (B) A Class I revertant. Note the continued presence of altered β-tubulin. (C) A Class II revertant. The mutant β-tubulin spot is gone. (D) A Class III revertant. The mutant β-tubulin migrates in a new position (arrow). Only a portion of each gel is shown.

β^{**} with an altered electrophoretic mobility on two-dimensional gels. Examples of two-dimensional gels for the three classes of revertants are shown in Figure 22.3. It is clear from this figure that the position of β^{**}, whose peptide map establishes that it is indeed a tubulin, differs markedly from that of β^{*}. This cell line, then, is a clear example of an intragenic revertant. Determination of the mechanism of reversion, however, required a further experiment. Crude preparations of MTs from CHO cells may be easily obtained by lysing cells with Triton X-100 in an MT stabilizing buffer (Pipeleers et al., 1977; Cabral et al., 1980). After centrifugation, the pellet contains a number of Triton-insoluble proteins, but also contains the cellular MTs which had been polymerized *in vivo*

(Cabral et al., 1980). The soluble proteins as well as the unpolymerized tubulin remain in the supernatant. If crude MTs are prepared in this way from the mutant cells, α-, β-, and β*-tubulin appear in the pellet, indicating that all three species assemble into MTs. A similar experiment performed with the Class III revertant, however, reveals only α- and β-tubulin in the pellet fraction. The β** tubulin does not appear in polymerized MTs. Thus, this revertant is presumably no longer drug resistant or ts for growth because a further alteration in β* to give β** renders the protein unable to assemble into MTs. This in turn suggests that β* must assemble into MTs in order to confer drug resistance. It is not simply sequestering the drug or modifying it in some way which inactivates it.

This section was intended to demonstrate the need for reversion studies in order to prove the linkage between an altered protein and the phenotype under study. At the same time, however, we hope that the discussion has also emphasized that revertants can provide a rich source of new "mutants" in their own right. Thus, revertants can provide insights into the mechanism of mutation in the original mutant line as well as provide new information about the function of MTs and MAPs in CHO cells.

D. Morphology

In many cases, drug-resistant mutants which are puromycin-sensitive and therefore putative MT mutants will be encountered, but these may not show biochemical alterations in tubulin or exhibit altered drug binding. In such cases, the mutant may still be characterized as an MT mutant if alterations in the MTs can be demonstrated. Such alterations are most likely to be seen under nonpermissive conditions in conditional mutants. For example, the α- and β-tubulin mutants isolated in our laboratory have normal MT morphology at normal growth temperature (37°C), but have clear defects in spindle morphology at the restrictive temperature (40.5°C) (Abraham et al., 1983). Similarly, Tax-18, a taxol-dependent mutant, is normal in the presence of the selecting drug, taxol, but has defective spindle assembly if taxol is omitted from the growth medium (Cabral et al., 1983). This mutant will be discussed in more detail below.

The simplest method for looking at MT morphology is indirect immunofluorescence using antibodies to tubulin (Brinkley et al., 1980). This technique is simple and rapid, and has the advantage of allowing one to examine the overall distribution of MTs in a large number of cells very easily. It is important when using this method to examine both interphase and mitotic cells, since it has been our experience that mutations can affect MTs in mitosis while leaving the cytoplasmic MTs seemingly unperturbed. The major disadvantage of immunofluorescence is that it is incapable of giving a detailed view of MT structure. If a closer look at MT structure is desired, then electron microscopy is the method of choice.

This technique has a number of disadvantages, however, including tediousness, expensive equipment, and slow pace, and the fact that only small sections of cells can be examined at one time. Still, the information obtained is very detailed and may not be easily obtained by any other method.

E. Gene Structure

None of the mutants isolated thus far has been extensively examined at the genetic level. One reason that they have not been examined is, as mentioned before, that the appropriate probes are not yet available. Although a full length chicken cDNA for β-tubulin (Cleveland et al., 1980) and a full length rat cDNA for α-tubulin (Lemischka et al., 1981) have been reported, similar probes for CHO tubulin have not yet been isolated.

The development of such probes will be useful for determining the complete amino acid sequences of CHO tubulins and for subsequent mapping of the lesions in mutant genes and determination of the nature of the genetic changes. Such probes can also be used in *in situ* hybridization studies to determine the chromosomal arrangement of tubulin genes and should allow exploration of the mechanism of tubulin regulation in mutants with altered tubulin expression.

IV. SURVEY OF MUTANTS ISOLATED

Temperature-sensitive mutants blocked in mitosis have been described in an earlier section. To date, the biochemical lesions in these cells have not been found, but it is likely that at least some of them have alterations which affect MTs. We will not discuss these mutants any further in this chapter. Survivors of drug-resistance selections are mostly permeability mutants. This class of mutants has been extensively studied by Ling and his co-workers and will be discussed in Chapter 25. Among the survivors of drug-resistance selections that are not permeability mutants, a number of MT mutants have been isolated in CHO cells and these are summarized in Table 22.1. As may be seen, the number of such mutants is relatively small and not all have been sufficiently characterized to be certain of the biochemical lesion. Nonetheless, this list illustrates that the selection protocol for obtaining such mutants works, although MT mutants appear at low frequency. A procedure to increase the frequency of MT mutants among the drug-resistant cells would be a welcome improvement.

The majority of MT mutants isolated in our laboratory have alterations in β- rather than α-tubulin. An example of a two-dimensional gel for each of these two kinds of mutant are shown in Figure 22.2. Too few mutants have been obtained to be able to speculate whether certain drugs yield mutants with alterations in a particular subunit. This is almost certainly not generally true, however, since we have isolated two taxol-resistant mutants, one of which has an alteration in α-tubulin (Cabral et al., 1981) and the other in β-tubulin (unpublished work). Also, we have isolated a β-tubulin colcemid-resistant mutant (Cabral et al., 1980) while Keates et al. (1981) have reported the isolation of a colcemid-resistant mutant with an altered α-tubulin. Many more mutants with defined alterations in tubulin will be needed to determine whether certain drugs yield mutants with alterations in particular tubulin subunits. Such information coupled with studies to determine the mechanism of drug resistance in the mutants would greatly aid in understanding how the drugs inhibit (or promote) MT assembly.

It might be anticipated that a large number of the MT mutants would have alterations in drug binding. We have not found this to be the case, however. None of the mutants we have tested have shown any alterations in drug binding and other evidence also argues against alterations in drug binding accounting for the drug resistance. For example, as will be discussed below, many of our mutants are cross resistant to other MT-active drugs even though those drugs are believed to bind to different sites on tubulin. It may be more appropriate to think of these mutants as having conformational alterations in tubulin which affect the ability of tubulin to assembly even in the presence of diverse MT-active drugs.

Only a single mutant with an alteration in tubulin has been shown to have altered drug binding. This mutant, $CM^{R}795$, was isolated by Ling et al. (1979) using a multiple-step colcemid-resistance selection. Crude extracts from this cell line exhibit a reduced binding affinity for radiolabeled colcemid compared to wild-type cell extracts. Furthermore, this reduced binding affinity was retained when tubulin was purified from the mutant cell line by two cycles of polymerization and depolymerization, although the degree of reduced binding was not as great in the purified sample. Later experiments (Keates et al., 1981) designed to examine the changes at each step of the multistep selection demonstrated the existence of an altered α-tubulin after the first step of the selection. This step did not significantly alter the colcemid binding affinity, however, making it unlikely that the alteration in α-tubulin is responsible for the altered colcemid-binding parameters. Subsequent selection steps did alter the colcemid binding but no further changes in α-tubulin or β-tubulin were seen.

One of the mutants listed in Table 22.1, Pod^{RII6}, is reported to have an altered 66,000-dalton protein and the authors suggest that this protein may be a MAP. As discussed previously, however, demonstration of an altered protein in a mutant is not sufficient to prove that the altered protein is responsible for the drug-resistance phenotype. It would thus be premature to assume that this mutant represents a MAP mutant. It seems likely, however, that among the mutants and revertants which have already been isolated, MAP mutants exist. We believe that Tax-18 may represent such a mutant, and this will be discussed in the next section. The difficulty in obtaining MAP mutants at present does not lie in the selections being used but rather in our inability to identify MAPs in CHO cells. Since MAPs are minor proteins in cells, it will be difficult to identify mutants in these proteins until methods are developed for their purification. We are approaching this problem by purifying MTs from cell extracts and the spindle apparatus from mitotic cells and determining the protein composition of these preparations on two-dimensional gels. This information should allow the identification of MAP mutants in the future and provide insights into the mechanism of MAP regulation of MT assembly *in vivo.*

V. WHAT MICROTUBULE MUTANTS HAVE DEMONSTRATED

A. Microtubule Function

Microtubule mutants in CHO cells and in other organisms have already proven useful for identifying processes in which MTs are involved. Furthermore, in some cases, insights into the mechanism of MT involvement have been obtained as a result of mutant characterization. For example, elegant genetic studies to determine the role of various proteins and structures in flagellar motion in *Chlamydomonas* have been performed (Witman et al., 1978; Huang et al., 1981, 1982). Tubulin has been shown to be necessary for cell viability in yeast (Neff et al., 1983). In *Drosophila* a testis-specific β-tubulin has been described (Kemphues et al., 1979) and shown to participate in assembly of sperm axonemes and in meiotic but not mitotic spindle assembly (Kemphues et al., 1982). This is the first convincing demonstration that a single tubulin species can participate in the assembly of multiple structures.

Another way in which mutants have been used to argue for MT involvement in a given process is to make use of the drug-resistance phenotype. If the mutant cell is more resistant to a drug because of a MT defect, then processes which requires more drug for inhibition in the mutant than in wild-type cells must have MT involvement. This paradigm has been used by Oakley and Morris (1980) to show that nuclear migration in *Aspergillus* is MT dependent and by Aubin et al. (1980) using mutant CM^R795, to show that capping of cell surface antigens in CHO cells involves MTs.

A final method for using mutants to study MT function is to isolate conditional mutants, show that the conditionality is due to an MT defect, and then examine various cellular processes under the restrictive conditions. For example, the α- and β-tubulin mutants isolated in our laboratory have proven to be ts for growth, and this phenotype has been shown to result from an alteration in β-tubulin for mutant Cmd-4 (Cabral et al., 1982) and from an alteration in α-tubulin for mutant Tax-1 (Cabral et al., 1981). Examination of these cells at the nonpermissive temperature has shown that spindle assembly is defective (Abraham et al., 1983). These results indicate that α- and β-tubulin are necessary for cell viability and that it is their role in spindle assembly during mitosis which constitutes their most crucial function.

Spindle assembly is also defective in Tax-18, and the relative ease with which permissive and nonpermissive conditions can be manipulated has allowed us to pinpoint the defective stage in spindle assembly more precisely. Tax-18 was selected for taxol resistance but was also found to require taxol for normal cell division. In the absence of the drug the cells fail to divide and become larger and multinucleated (Cabral, 1983). In the presence of taxol, the cells have normal growth and morphology, implying that they cannot be permeability mutants. As confirmation of this assumption, we find that the cells do not exhibit cross resistance to any other drug we have tested.

Since taxol is known to promote MT assembly (Schiff et al., 1979; Schiff and Horwitz, 1980), we argued that Tax-18 might require the drug because it is unable to assemble MTs in its absence. To test this idea, cells grown with or without taxol were examined by indirect immunofluorescence with antibodies to tubulin (Cabral et al., 1983). The results shown in Figure 22.4 demonstrated that Tax-18 can assemble cytoplasmic MTs with or without taxol. When mitotic cells were examined, however, it was discovered that mutant cells grown in taxol display all stages of mitotic spindle assembly including prophase, metaphase, anaphase, and telophase; but mutant cells cultured without taxol display only a "prophaselike" spindle (see Figure 22.5). Thus, Tax-18 appears to block in the prophase to metaphase transition when taxol is omitted from the medium. This block is reminiscent of a colchicine-blocked cell but differs in that prominent MTs persist at the spindle poles and the spindle poles are separated.

Electron microscopic examination of these cells demonstrates that the microtubule-organizing centers (centrioles, kinetochores) are structurally intact and that kinetochore-to-pole MTs persist. Evidence for the presence of interpolar MTs, however, could not be found. Thus, we tentatively suggest that the defect in Tax-18 may affect the formation of interpolar MTs and that this in turn prevents the cell from achieving metaphase, anaphase, and telophase configurations.

The biochemical lesion which gives rise to these morphologically observable defects is presently unknown. Two-dimensional gel analysis of several mutants with the Tax-18 phenotype has thus far failed to reveal any alterations in α- or β-tubulin. Also, the Tax-18 phenotype behaves recessively (is not expressed) when mutant cells are fused with the wild-type cells. Since most of the tubulin mutants selected to date behave in a

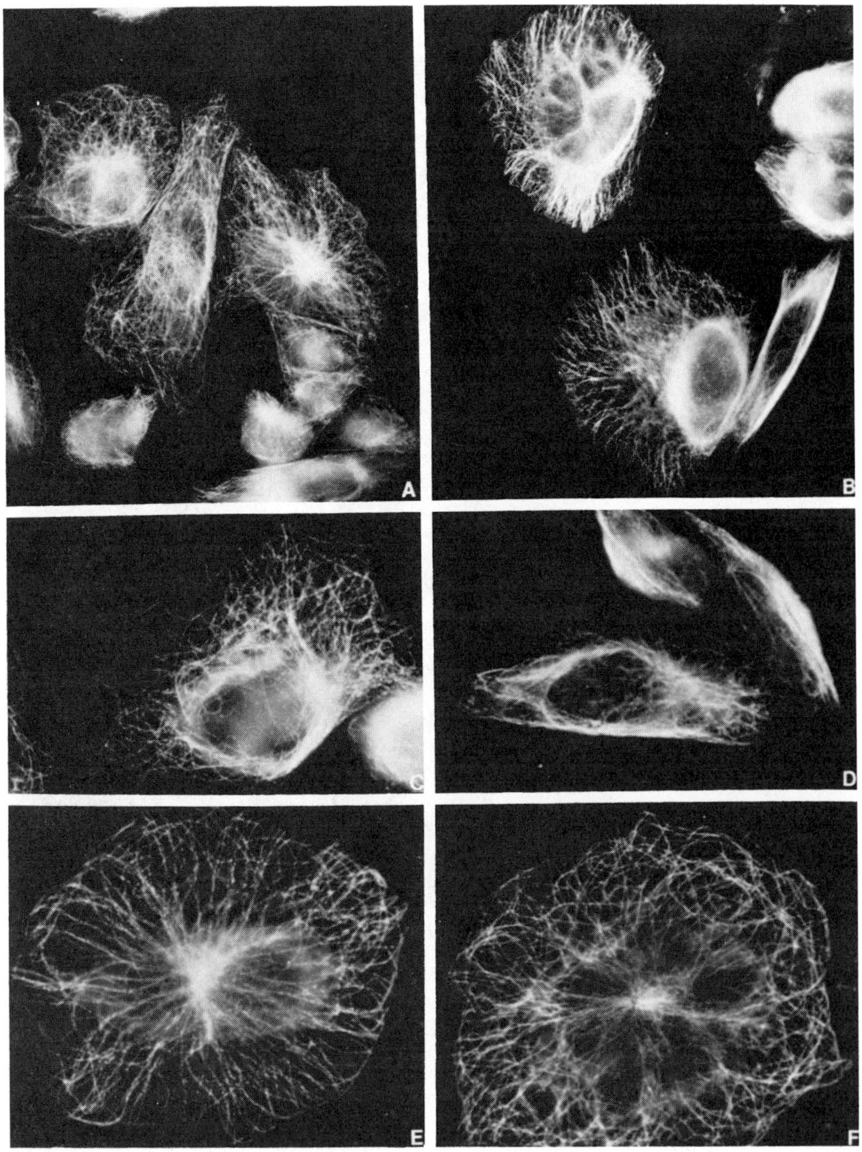

Figure 22.4. Cytoplasmic microtubule complex in wild-type and Tax-18 interphase cells. Wild-type cells were grown without taxol (A) or in the presence of 0.2 μg/mL taxol for 48 hr (B). They were then permeabilized with Triton X-100, fixed, stained with antitubulin antibodies, and viewed by fluorescence microscopy [see Cabral et al. (1983) for details]. (C,D) Tax-18 grown continuously in 0.2 μg/mL taxol. (E,F) Tax-18 deprived of taxol for 2 days. Reprinted with permission of the Rockefeller University Press.

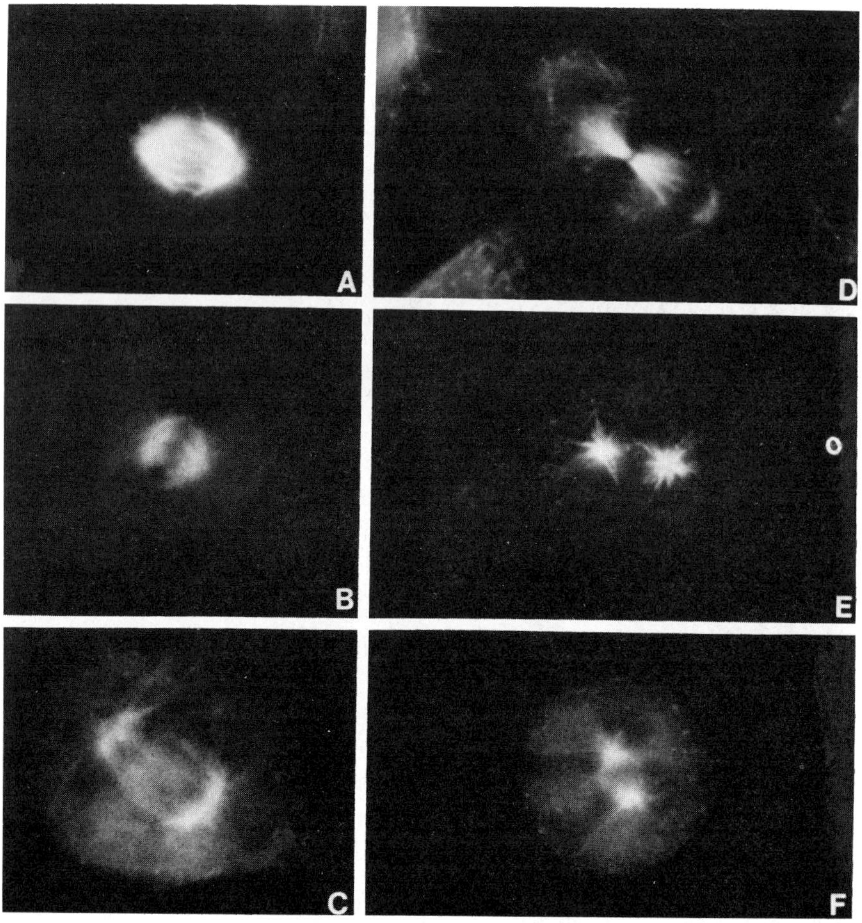

Figure 22.5. Spindle microtubules in wild-type and Tax-18 mitotic cells. Conditions were similar to those employed in Figure 22.4. (A) Wild-type cell at metaphase. (B) Tax-18 at metaphase. (C) Tax-18 at anaphase. (D) Tax-18 at telophase. (E,F) Tax-18 blocked in a prometaphaselike state. (A,E,F) No taxol present. (B,C,D) Taxol present in the growth medium at 0.2 μg/mL. Reprinted with permission of the Rockefeller University Press.

codominant manner (see Table 22.1 and preceding discussion of multiple loci for tubulin in CHO and other cells), we feel that Tax-18 may not have a defect in tubulin itself, but rather in a MAP which may or may not be specific for interpolar spindle MTs. We are currently searching for such a defective MAP by isolating spindles from wild-type and taxol-supplemented mutant cells. Although the biochemical lesion in Tax-18 has not been identified, we have little doubt that spindle MT assembly is the primary defect in the cell, since taxol is known to bind to MTs with a high affinity and to promote MT assembly *in vitro* (Manfredi et al., 1981). Fur-

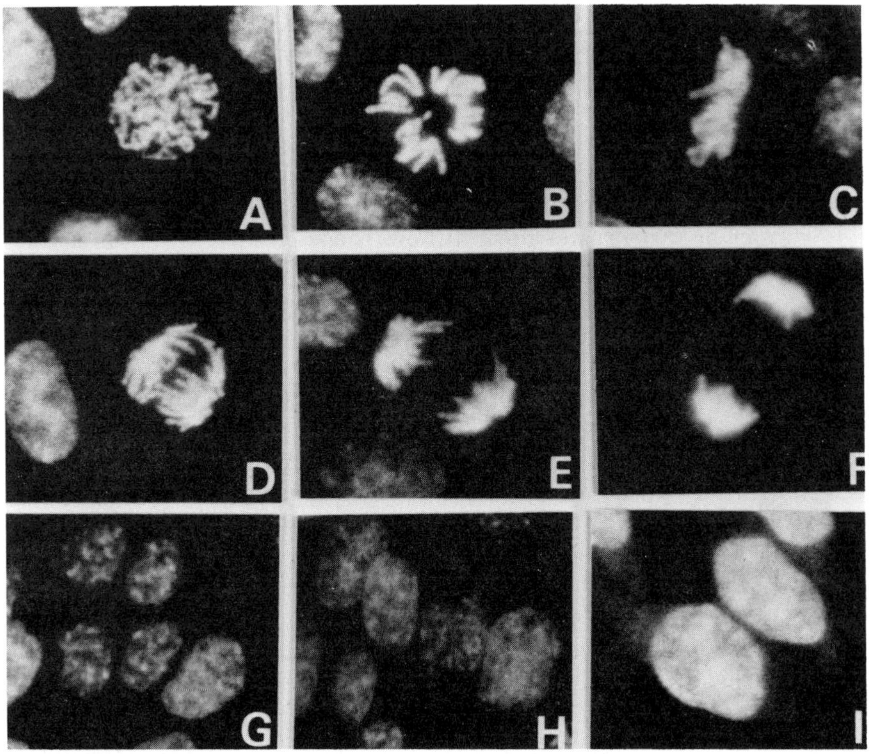

Figure 22.6. Chromosome organization in Tax-18 grown in the continuous presence of 0.2 μg/mL taxol. The cells were seeded onto coverslips and after 2 days of growth were fixed and stained with Hoescht dye 33258. When viewed by fluorescence microscopy, interphase nuclei and mitotic chromosomes were clearly visible. Various fields were photographed and the composite was put together as follows: (A) chromosomes in prophase; (B) prometaphase; (C) metaphase; (D) anaphase; (E) late anaphase; (F) telophase; (G,H,I) interphase nuclei. Note the well-organized chromosomes during mitosis and the presence of a single nucleus per cell in interphase.

thermore, the earliest morphological alterations we can find in these cells involve spindle MTs (Cabral et al., 1983), and these cells have been found to be supersensitive to other MT active drugs (Cabral, 1983).

The consequences of the defective spindle assembly are profound. First, chromosome organization during mitosis is disrupted, and this leads to the eventual formation of micronuclei. Examples of chromosome organization during mitosis for Tax-18 grown in taxol is shown in Figure 22.6 and for Tax-18 deprived of taxol for 24 hr is shown in Figure 22.7. The mutant cells in taxol (Fig. 22.6) have well-organized chromosomes characteristic of prophase (A), prometaphase (B), metaphase (C), anaphase (D,E), and telophase (F). In addition, interphase cells have a single normal appearing nucleus (G–I). When mutant cells are deprived of taxol for 1 day (Fig. 22.7), prophase chromosomes (A) already exhibit disorganization which is carried over into prometaphase (B). Metaphase and anaphase figures are not seen, but "telophaselike" aggregates of chromosomes are common (C). These do not represent true telophase figures since the chromsomes are much thicker (chromatid separation has presumably not occurred) and the chromosome arms are not oriented toward an opposite spindle pole as would be the case in a normal telophase situation. Restitution nuclear membranes form around these clumps of chromosomes leading to interphase cells with multinucleated or strangely shaped nuclei (D–F).

A second consequence of defective spindle assembly is the absence of cytokinesis. This is not as easily explained as the loss of chromosome organization. It has long been observed that the cleavage furrow in dividing cells always occurs along a plane perpendicular to and bisecting the spindle axis, suggesting that the location of the spindle or the spindle poles determines the site of cleavage (Conrad and Rappaport, 1981). This idea is supported by studies in which mechanical displacement of the spindle during mitosis causes a repositioning of the contractile ring or even induces the formation of a second contractile ring if a first has already formed (Conrad and Rappaport, 1981). Other experiments, however, suggest that a cleavage furrow will form and cytokinesis will proceed even if the spindle is mechanically removed from the cell or destroyed provided the cell has already reached metaphase or anaphase (Hiramoto, 1971; Rappaport, 1971). In Tax-18 deprived of taxol, cells never reach metaphase, yet time lapse observation reveals the presence of a cleavage furrow. This cleavage furrow constricts the cytoplasm, and cytokinesis appears to be proceeding albeit with greater than normal membrane blebbing. After a prolonged period, however, membrane movement and cytoplasmic flow diminish, the contractile ring relaxes, and cell division fails. We thus believe, although further experiments are needed to be certain of the details, that in Tax-18 a contractile ring can form and function in the absence of a spindle but the final stages of cytokinesis are defective. We speculate that for cell division to occur a nonreversible interaction between the contractile ring and the spindle must take place. Interpolar MTs (which are apparently missing in Tax-18) may be an integral component in this interaction.

In spite of the inability of these cells to divide, they are still able to progress through the cell cycle. This may be seen by the increase in size of these cells, by time lapse observation, by measuring the continued synthesis of protein and DNA, and by monitoring the continued replication of centrioles (Cabral, 1983; Cabral et al., 1983). The simplest method, however, is to simply look at the number of chromosomes per cell as the duration of taxol deprivation increases. Cells with octaploid and greater chromosome numbers become common by 2 days, and as taxol starvation proceeds, chromosomal abnormalities become more apparent (unpub-

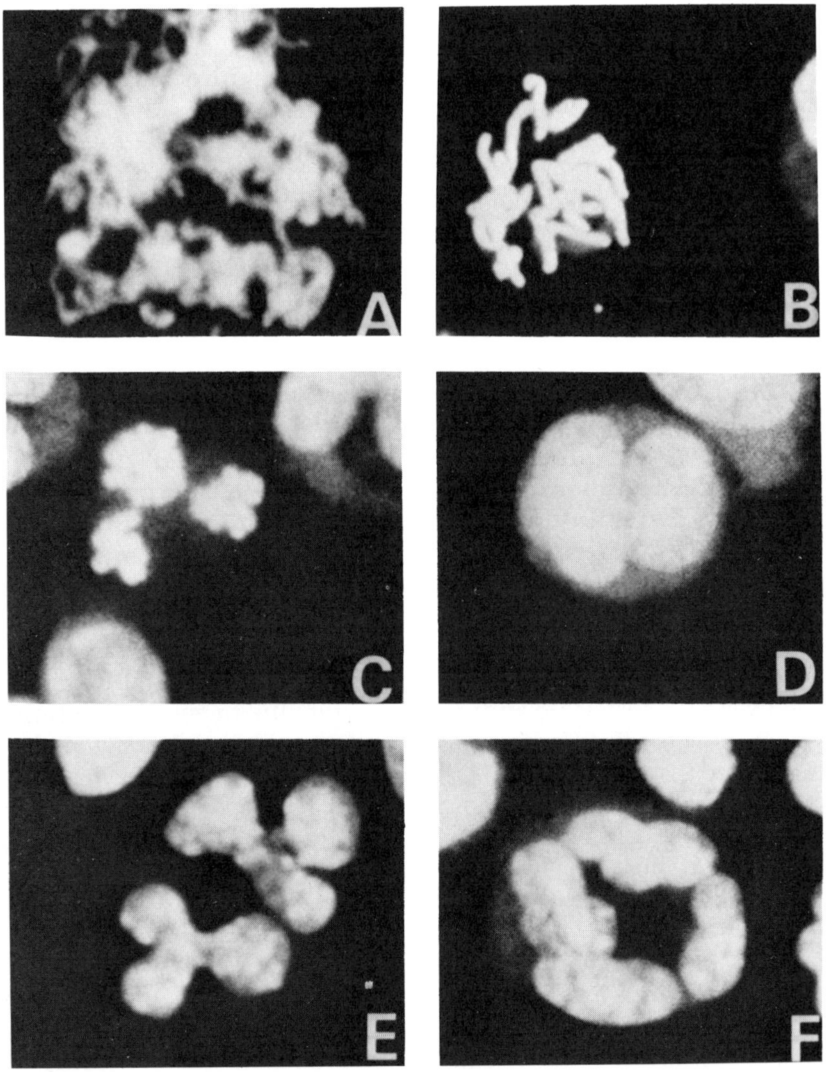

Figure 22.7. Chromosome organization in Tax-18 cultured without taxol for 24 hr. The cells were treated as described in Figure 22.6 except that they were deprived of taxol 24 hr before fixation: (A) prophase; (B) prometaphase; (C) "telophaselike" cluster; (D,E,F) interphase nuclei. Note the aberrant chromosome organization in prophase and prometaphase, the absence of metaphase and anaphase figures, and the presence of clusters of chromosomes reminiscent of, but distinct from, chromosomes in telophase. Reformation of nuclear membranes around randomly distributed clumps of chromosomes in the absence of cytokinesis, leads to binuclear (D), trilobed (E), circular (F), and other oddly shaped nuclei in interphase.

lished observations). We feel that the inability of the cells to handle the vastly increased amount of DNA probably leads to their eventual death.

It may be seen that a study of this single mutant has taught us much about the role of spindle microtubules in mitotic events. Spindle microtubules are not essential for short-term survival of cells; however, their role in chromosome segregation and cytokinesis makes them indispensable for long-term viability. Without these MTs, a proliferative cell loses its ability to carry out normal replicative processes and cannot maintain the gene dosage necessary for cell survival. It is likely that microtubule-active drugs which inhibit cells in a prometaphaselike state kill cells by a very similar mechanism.

B. Mechanism of Drug Action

One of the great potential uses of drug-resistant mutants is the elucidation of the mechanism of drug action. An explanation of why a mutant is resistant to a particular drug should provide insight as to how that drug normally works. For example, the demonstration that many resistant mutants have alterations in tubulin provides genetic proof that these drugs affect cell viability through their action on microtubules. In one report an alteration in drug binding has been demonstrated in a mutant (Ling et al., 1979). In CHO cells, however, an alteration in drug binding has not yet been correlated with an altered tubulin, therefore, genetic evidence for drug binding sites on specific tubulin subunits has not yet been demonstrated.

As discussed above, most of our mutants appear to be resistant to microtubule-active drugs as a result of increased stability or lability of MTs in these cells. Thus, cells resistant to colchicine, a MT-destabilizing drug, have hyperstable MTs, making them more sensitive to taxol, a MT-stabilizing drug. Conversely, cells resistant to taxol have hyperlabile MTs, making them more sensitive to colchicine. While we cannot say that all our mutants are of this type, many of the mutants we have examined exhibit this property. Mutants of this type have previously been reported in *Aspergillus* (Oakley and Morris, 1981) and in mutants CM^R797, Tax_1, and BEN_1 (Gupta et al., 1982; Warr et al., 1982). These changes in MT stability could result from alterations in tubulin which affect the stability of α/β interactions or the affinity of the MTs for MAPs; or they could result from changes in the MAPs which increase or decrease their affinity for MTs. An increased or decreased accumulation of MAPs could also affect MT stability by altering the extent of their interaction with the MTs.

The pattern of drug cross resistance in the mutants which have been studied is still complex and not easily explained. For example, Cmd-4 is cross resistant to colchicine and vinblastine but poorly cross resistant to griseofulvin. Mutant Grs-2, on the other hand, is cross resistant to colcemid and vinblastine but poorly cross resistant to colchicine (Cabral et al., 1980). These observations cannot be explained by alterations in the drug-binding sites and suggest that these drugs may have distinct mechanisms of action. Indeed, it has been reported that the binding site for griseofulvin resides on MAPs rather than on tubulin itself (Roobol et al., 1977). As more information on drug-binding patterns is accumulated, it may be possible to formulate a hypothesis to explain how these diverse drugs affect MT assembly and stability. The isolation of MAPs from these mutant cells and an examination of their binding to MTs *in vitro* should greatly increase our understanding of the mechanism of resistance in the various mutants.

C. Microtubule Assembly

The use of mutants to study microtubule assembly is still in its infancy. We have described the isolation of a revertant, Cmd4-A5, which has an altered β-tubulin incapable of assembling (Cabral et al., 1982) and the isolation of this revertant is discussed above. The site of the alteration in this peptide is not yet known, but its identification should prove useful in determining the domain on the molecule important in assembly. More specifically, we believe the defect in the protein prevents the interaction of this altered β-tubulin with α-tubulin so that α/β heterodimers cannot form. We are currently developing the tools to test this hypothesis. If this hypothesis is true, we would be in a position to chemically define one of the first steps necessary for MT assembly.

Revertant Cmd4-A5 is also a good starting point for the isolation of second-site suppressor mutations. The rationale proceeds as follows. The mutant parent, Cmd-4, has an altered β-tubulin (β^*) which can confer colcemid resistance upon the cell. Revertant Cmd4-A5 is no longer colcemid resistant because β^* has acquired a second alteration (to give β^{**}) which prevents it from assembling into MTs and accumulating in the cell. If β^{**} can be induced to again assemble into MTs, then the cell should regain its colcemid resistance. One way to bring this about is to alter a second protein which normally interacts with β-tubulin in MT assembly in such a way that it is able to interact with β^{**} and allow it to assemble. Thus, a selection for second site suppressor mutations simply involves selecting for colcemid-resistant mutants of Cmd4-A5. We have performed such a selection, and among the colcemid-resistant cells, we have found a cell line with an altered α-tubulin (α^*) (unpublished work). We are now attempting to show directly that α^* is able to interact with β^{**} and cause it to reassemble into MTs. If this scheme is correct, then complementary sites for α/β interactions have been genetically defined. The isolation of other such mutants and revertants should provide us with a wealth of information about molecular domains involved in MT assembly as well as with clues about how the entire process is regulated.

VI. FUTURE TRENDS

A good deal of progress in the development of methods for obtaining MT mutants has been made in the last few years. Although only a handful of mutants have thus far been isolated, many of them have given us a glimpse of the kinds of detailed information they can provide about MT structure, assembly, function, and regulation; and about the mechanism of drug action in mammalian cells. The mutants already isolated must be further characterized to understand the biochemical lesion and how it relates to MT function or drug action and a search for new mutants should continue. The unexpected discovery of taxol-requiring mutants demonstrates how repeated selections can yield mutant classes that could not have been predicted. Furthermore, new ideas about how to select mutants are needed in order to enrich the collection of mutant phenotypes available.

Finally, examination of the mutants already has made it clear that a good deal of effort needs to be expended in order to develop the tools necessary to fully characterize mutant phenotypes. For example, although many tubulin antibodies are available, few are effective for immunoprecipitation. Furthermore, specific antibody reagents able to discriminate between α- and β-tubulin would be very desirable. Another example is the lack of information concerning the identity of MAPs and SAPs in CHO cells. Although a number of laboratories are exploring spindle and MT structure by preparing monoclonal antibodies against these structures, these antibodies frequently do not cross react with CHO cell antigens. The development of similar antibodies for CHO cells is desperately needed in order to advance our understanding of the mutant phenotypes. Once the necessary tools are available, genetic analysis of the MT system should progress rapidly, providing us with new insights into how the synthesis, assembly, and function of these intricate organelles are regulated.

ACKNOWLEDGMENTS

We wish to thank Diane Sutcliffe for excellent technical assistance. Portions of this work were funded by grants GM29955 from the NIH and CD154 from the American Cancer Society. F.C. is the recipient of a Jr. Faculty Research Award from the ACS.

REFERENCES

Abraham, I., Marcus, M., Cabral, F., and Gottesman, M. M. (1983). *J. Cell Biol.* **97**, 1055–1061.

Alexandraki, D. and Ruderman, J. V. (1981). *Mol. Cell. Biol.* **1**, 1125–1137.

Amos, L. A. (1979). In *Microtubules* (K. Roberts and J. S. Hyams, eds.), Academic Press, New York, pp. 1–64.

Aubin, J. E., Tolson, N., and Ling, V. (1980). *Exp. Cell Res.* **126**, 75–85.

Bech-Hansen, N. T., Till, J. E., and Ling, V. (1976). *J. Cell. Physiol.* **88**, 23–32.

Beebe, D. C., Feagans, D. E., Blanchette-Mackie, E. J., and Nau, M. E. (1979). *Science* **206**, 836–837.

Ben-Ze'ev, A., Farmer, S. R., and Penman, S. (1979). *Cell* **17**, 319–325.

Bibring, T. and Baxandall, J. (1977). *Dev. Biol.* **55**, 191–195.

Borisy, G. G. and Olmsted, J. B. (1972). *Science* **177**, 1196-1197.

Brady, S. T., Crothers, S. D., Nosal, C., and McClure, W. O. (1980). *Proc. Natl. Acad. Sci. USA* **77**, 5909–5913.

Brinkley, B. R., Fistel, S. H., Marcum, J. M., and Pardue, R. L. (1980). *Int. Rev. Cytol.* **63**, 59–95.

Bryan, J. (1971). *Exp. Cell. Res.* **66**, 129–136.

Bryan, J. (1972). *Biochem.* **11**, 2611–2615.

Bulinski, J. C. and Borisy, G. G. (1979). *Proc. Natl. Acad. Sci. USA* **76**, 293–297.

Cabral, F. (1983). *J. Cell Biol.* **97**, 22–29.

Cabral, F. and Gottesman, M. M. (1978). *Anal. Biochem.* **91**, 548–556.

Cabral, F. and Schatz, G. (1979). *Met. Enzymol.* **56**G, 602–613.

Cabral, F., Sobel, M., and Gottesman, M. M. (1980). *Cell* **20**, 29–36.

Cabral, F., Abraham, I., and Gottesman, M. M. (1981). *Proc. Natl. Acad. Sci. USA* **78**, 4388–4391.

Cabral, F., Abraham, I., and Gottesman, M. M. (1982). *Mol. Cell. Biol.* **2**, 720–729.

Cabral, F., Wible, L., Brenner, S., and Brinkley, B. R. (1983). *J. Cell Biol.* **97**, 30–39.

Chu, L. K. and Sisken, J. E. (1977). *Exp. Cell Res.* **107**, 71–77.

Cleveland, D. W. and Havercroft, J. C. (1983). *J. Cell Biol.* **97**, 919–924.

Cleveland, D. W., Lopata, M. A., MacDonald, R. J., Cowan, N. J., Rutler, W. J., and Kirschner, M. W. (1980). *Cell* **20**, 95–105.

Cleveland, D. W., Lopata, M. A., Sherline, P., and Kirschner, M. W. (1981). *Cell* **25**, 537–546.

Conolly, J. A., Kalnins, V. I., Cleveland, D. W., and Kirschner, M. W. (1978). *J. Cell Biol.* **76**, 781–786.

Conrad, G. W. and Rappaport, R. (1981). In *Mitosis/Cytokinesis* (A. Zimmerman and A. Forer, eds.), Academic Press, New York, pp. 365–396.

Cortese, F., Bhattacharyya, B., and Wolff, J. (1977). *J. Biol. Chem.* **252**, 1134–1140.

Cowan, N. J., Wilde, C. D., Chow, L. T., and Wefald, F. C. (1981). *Proc. Natl. Acad. Sci. USA* **78**, 4877–4881.

Dentler, W. L., Granett, S., and Rosenbaum, J. L. (1975). *J. Cell Biol.* **65**, 237–241.

Dustin, P. (1978). *Microtubules*, Springer-Verlag, Berlin.

Farrell, K. W. and Wilson, L. (1980). *Biochem.* **19**, 3048–3054.

Fujiwara, K. and Pollard, T. D. (1980). *Curr. Top. Devel. Biol.* **14**, 271–296.

Fuller, G. M., Brinkley, B. R., and Boughter, J. M. (1975). *Science* **187**, 948–950.

Fulton, C. and Simpson, P. A. (1976). In *Cell Motility* (R. Goldman, T. Pollard, and J. Rosenbaum, eds.), Cold Spring Harbor Laboratory Press, New York, pp. 987–1005.

Fulton, C. and Simpson, P. A. (1979). In *Microtubules* (K. Roberts and J. S. Hyams, eds.), Academic Press, New York, pp. 117–174.

Grisham, L. M., Wilson, L., and Bensch, K. G. (1973). *Nature* **244**, 294–296.

Gupta, R. S. (1981). *Somat. Cell Genet.* **7**, 59–71.

Gupta, R. S. (1983). *J. Cell. Physiol.* **114**, 137–144.

Gupta, R. S., Ho, T. K. W., Moffat, M. R. K., and Gupta, R. (1982). *J. Biol. Chem.* **257**, 1071–1078.

Hatzfeld, J. and Buttin, G. (1975). *Cell* **5**, 123–129.

Hiramoto, Y. (1971). *Exp. Cell. Res.* **68**, 291–298.

Hoebeke, J., Van Nijen, A., and DeBrabander, M. (1976). *Biochem. Biophys. Res. Commun.* **69**, 319–324.

Huang, B., Piperno, G., Ramanis, Z., and Luck, D. J. L. (1981). *J. Cell Biol.* **88**, 80–88.

Huang, B., Ramanis, Z., and Luck, D. J. L. (1982). *Cell* **28**, 115–124.

Inoue, S. (1981) *J. Cell Biol.* **91**, 131S–147S.

Izant, J. G., Weatherbee, J. A., and McIntosh, J. R. (1982). *Nature* **295**, 248–250.

Jacobs, M., Smith, H., and Taylor, E. W. (1974). *J. Mol. Biol.* **89**, 455–468.

Jha, K. K. and Ozer, H. L. (1976). *Somat. Cell Genet.* **2**, 215–223.

Kalfagan, L. and Wensink, P. C. (1981). *Cell* **24**, 97–106.

Keates, R. A. B., Sarangi, F., and Ling, V. (1981). *Proc. Natl. Acad. Sci. USA* **78**, 5638–5642.

Kemphues, K. J., Raff, R. A., Kaufman, T. C., and Raff, E. C. (1979). *Proc. Natl. Acad. Sci. USA* **76**, 3991–3995.

Kemphues, K. J., Raff, E. C., Raff, R. A., and Kaufman, T. C. (1980). *Cell* **21**, 445–451.

Kemphues, K. J., Kaufman, T. C., Raff, R. A., and Raff, E. C. (1982). *Cell* **31**, 655–670.

Kirschner, M. W. (1978). *Int. Rev. Cytol.* **54**, 1–71.

Krauhs, E., Little, M., Kempf, T., Hofer-Warbinek, R., Ade, W., and Ponstingl, H. (1981). *Proc. Natl. Acad. Sci. USA* **78**, 4156–4160.

Kumagai, H. and Nishida, E. (1979). *J. Biochem. (Japan)* **85**, 1267–1274.

Kuo, T., Pathak, S., Ramagli, L., Rodriguez, L., and Hsu, T. C. (1982). In *Gene Amplification* (R. T. Schimke, ed.), Cold Spring Harbor Laboratory Press, New York, pp. 53–57.

Kuriyama, R. (1982). *Cell Struct. Funct.* **7**, 307–315.

Lee, M. A.-S., Lewis, S. A., Wilde, C. D., and Cowan, N. J. (1983). *Cell* **33**, 477–487.

Lemischka, I. R., Farmer, S., Racaniello, V. R., and Sharp, P. A. (1981). *J. Mol. Biol.* **151**, 101–120.

Lemischka, I. and Sharp, P. A. (1982). *Nature* **300**, 330–335.

Ling, V., Aubin, J. E., Chase, A., and Sarangi, F. (1979). *Cell* **18**, 423–430.

Ling, V. and Thompson, L. H. (1974). *J. Cell Physiol.* **83**, 103–116.

Lopata, M. A., Havercroft, J. C., Chow, L. T., and Cleveland, D. W. (1983). *Cell* **32**, 713–724.

Luduena, R. F. (1979). In *Microtubules* (K. Roberts and J. S. Hyams, eds.), Academic Press, New York, pp. 65–116.

Lydersen, B. K. and Pettijohn, D. E. (1980). *Cell* **22**, 489–499.

Manfredi, J. J., Parness, J., and Horwitz, S. B. (1981). *J. Cell Biol.* **94**, 688–696.

Margolis, R. L. and Wilson, L. (1977). *Proc. Natl. Acad. Sci. USA* **74**, 3466–3470.

McCarty, G. A., Valencia, D. W., Fritzler, M. J., and Barada, F. A. (1981). *N. Eng. J. Med.* **305**, 703.

Meyers, M. B. and Biedler, J. L. (1981). *Biochem. Biophys. Res. Commun.* **99**, 228–235.

Mischke, D. and Pardue, M. L. (1982). *J. Mol. Biol.* **156**, 449–466.

Morris, N. R., Lai, M. H., and Oakley, C. E. (1979). *Cell* **16**, 437–442.

Mullins, J. M. and McIntosh, J. R. (1982). *J. Cell Biol.* **94**, 654–661.

Murphy, D. B. and Borisy, G. G. (1975). *Proc. Natl. Acad. Sci. USA* **72**, 2696–2700.

Nagle, B. W., Doenges, K. H., and Bryan, J. (1977). *Cell* **12**, 573–586.

Neff, N. F., Thomas, J. H., Grisafi, P., and Botstein, D. (1983). *Cell* **33**, 211–219.

Oakley, B. R. and Morris, N. R. (1980). *Cell* **19**, 255–262.

Oakley, B. R. and Morris, N. R. (1981). *Cell* **24**, 837–845.

Pardue, R. L., Kaetzel, M. A., Hahn, S. H., Brinkley, B. R., and Dedman, J. R. (1981). *Cell* **23**, 533–542.

Pardue, R. L., Brady, R. C., Perry, G. W., and Dedman, J. R. (1983). *J. Cell Biol.* **96**, 1149–1154.

Pickett-Heaps, J. D. (1969). *Cytobiol.* **1**, 257–280.

Pipeleers, D. G., Pipeleers-Marichal, M. A., Sherline, P., and Kipnis, D. M. (1977). *J. Cell Biol.* **74**, 341–350.

Ponstingl, H., Krauhs, E., Little, M., and Kempf, T. (1981). *Proc. Natl. Acad. Sci. USA* **78**, 2757–2761.

Porter, K. R. (1966). In *Ciba Foundation Symposium on Principles of Biomolecular Organization*, Churchill Press, London, pp. 308–345.

Raff, E. C., Fuller, M. T., Kaufman, T. C., Kemphues, K. J., Rudolph, J. E., and Raff, R. A. (1982). *Cell* **28**, 33–40.

Rappaport, R. (1971). *Int. Rev. Cytol.* **31**, 169–213.

Rebhun, L. I. (1972). *Intern. Rev. Cytol.* **32**, 93–139.

Remillard, S., Rebhun, L. I., Howie, G. A., and Kupchan, S. M. (1975), *Science* **189**, 1002–1005.

Robbins, E., Jenksch, A., and Micall, A. (1968). *J. Cell Biol.* **36**, 329–339.

Roobol, A., Gull, K., and Pogson, C. I. (1977). *FEBS Lett.* **75**, 149–153.

Sabatini, D. D., Bensch, K., and Barrnett, R. J. (1963), *J. Cell Biol.* **17**, 19–58.

Saltarelli, D. and Pantaloni, D. (1983). *Biochem.* **22**, 4607–4614.

Sanchez, F., Natzle, J. E., Cleveland, D. W., Kirschner, M. W., and McCarthy, B. J. (1980). *Cell* **22**, 845–854.

Scheele, R. and Borisy G. (1979). In *Microtubules* (K. Roberts and J. S. Hyams, eds.), Academic Press, New York, pp. 175–254.

Schiff, P. B., Fant, J. and Horwitz, S. B. (1979). *Nature* **277**, 665–667.

Schiff, P. B. and Horwitz, S. B. (1980). *Proc. Natl. Acad. Sci. USA* **77**, 1561–1565.

Shelanski, M. L. and Taylor, E. W. (1967), *J. Cell Biol.* **34**, 549–554.

Sherline, P. and Schiavone, K. (1977). *Science* **198**, 1038–1040.

Shiomi, T. and Sato, K. (1976). *Exp. Cell Res.* **100**, 297–302.

Silflow, C. D. and Rosenbaum, J. L. (1981). *Cell* **24**, 81–88.

Skoda, R. C., Jaussi R., and Christen, P. (1983). *Biochem. Biophys. Res. Commun.* **115**, 144–152.

Smith, B. J. and Wigglesworth, N. M. (1972). *J. Cell Physiol.* **80**, 253–260.

Stanners, C. P. (1978). *J. Cell Physiol.* **95**, 407–409.

Sternlicht, H. and Ringel, I. (1979). *J. Biol. Chem.* **254**, 10540–10550.

Sternlicht, H., Ringel, I., and Szasz, J. (1980). *J. Biol. Chem*, **255**, 9138–9148.

Suffness, M. and Douros, J. (1979). *Met. Cancer Res.* **16**, 73–126.

Summers, K. and Kirschner, M. W. (1979). *J. Cell Biol.* **83**, 205–217.

Thompson, L. (1979). *Met. Enzymol.* **58**, 308–322.

Thompson, L. H. and Lindl, P. A. (1976). *Somat. Cell Genet.* **2**, 387–400.

Tilney, L. A., Bryan, J., Bush, D. J., Fujiwara, K., Mooseker, M. S., Murphy, D. B., and Snyder, D. H. (1973). *J. Cell Biol.* **59**, 267–275.

Trifaro, J. M., Collier, B., Lastowecka, A., and Stern, D. (1972). *Molec. Pharmacol.* **8**, 264–267.

Valenzuela, P., Quiroga, M., Zaldivar, J., Rutler, W. J., Kirschner, M. W., and Cleveland, D. W. (1981). *Nature* **289**, 650–655.

Vallee, R. B. (1982). *J. Cell Biol.* **92**, 435–442.

Wang, R. J. (1974). *Nature* **248**, 76–78.

Wang, R. J. (1976). *Cell* **8**, 257–261.

Wang, R. J., Wissinger, W., King, E. J., and Wang, G. (1983). *J. Cell Biol.* **96**, 301–306.

Wang, R. J. and Yin, L. (1976). *Exp. Cell Res.* **101**, 331–336.

Warr, J. R., Flanagan, D. J., and Anderson, M. (1982). *Cell Biol. Internatl. Rep.* **6**, 455–460.

Weatherbee, J. A., Luftig, R. B., and Weihing, R. R. (1978). *J. Cell Biol.* **78**, 47–57.

Weber, K., Bibring, T. H., and Osborn, M. (1975), *Exp. Cell Res.* **95**, 111–120.

Weisenberg, R. C. (1972). *Science* **177**, 1104–1105.

Weisenberg, R. C., Borisy, G. G., and Taylor, E. W. (1968), *Biochem.* **9**, 4466–4479.

Welsh, M. J., Dedman, J. R., Brinkley, B. R., and Means, A. R. (1978), *Proc. Natl. Acad. Sci. USA* **75**, 1867–1871.

Wilde, C. D., Chow, L. T., Wefald, F. C., and Cowan, N. J. (1982a). *Proc. Natl. Acad. Sci. USA* **79**, 96–100.

Wilde, C. D., Crowther, C. E., and Cowan, N. J. (1982b). *J. Mol. Biol.* **155**, 533–538.

Wilde, C. D., Crowther, C. E., and Cowan, N. J. (1982c), *Science* **217**, 549–552.

Willingham, M. C., Wehland, J., Klee, C. B., Richert, N. D., Rutherford, A. V., and Pastan, I. H. (1983). *J. Histochem. Cytochem.* **31**, 445–461.

Wilson, L. (1970). *Biochem.* **9**, 4999–5007.

Wilson, L., Bamburg, J. R., Migel, S. B., Grisham, L. M., and Creswell, K. M. (1974). *Fed. Proc.* **33**, 158–166.

Wissinger, W. and Wang, R. J. (1978). *Exp. Cell Res.* **112**, 89–94.

Witman, G. B., Plummer, J., and Sander, G. (1978). *J. Cell Biol.* **76**, 729–747.

Zieve, G. and Solomon, F. (1982). *Cell* **28**, 233–242.

CHAPTER 23

GENETICS OF CYCLIC-AMP-DEPENDENT PROTEIN KINASES

Michael M. Gottesman
Laboratory of Molecular Biology
National Cancer Institute
National Institutes of Health
Bethesda, Maryland

I. INTRODUCTION

A. The Second Messenger Hypothesis and Protein Kinase

The second messenger hypothesis proposed by Sutherland [reviewed in Robison et al. (1971)] states that cyclic 3′, 5′-adenosine monophosphate (cAMP) is an intracellular mediator of the action of many peptide hormones. An enumeration of the many hormones which raise intracellular cAMP levels can be found in Sutherland (1971). Recently, bacterial toxins, such as cholera toxin (Holmgren, 1981) and pertussis toxin (Katada and Ui, 1982), and the diterpene, forskolin (Seamon et al., 1981), have also been shown to raise cAMP levels in most mammalian cells. Specificity in tissue response to a specific hormone or toxin depends both on (1) the presence on the cell surface of specific receptors and (2) the differential response of diverse cell types to increased intracellular cAMP levels. In recent years, considerable information has accumulated indicating that the binding of a hormone to its receptor results in stimulation of the membrane-associated enzyme adenylate cyclase, which converts ATP to cAMP with release of pyrophosphate. Cyclic AMP is released within the cell and activates the enzyme cAMP-dependent protein kinase (cADepPK). There are two forms of this tetrameric enzyme within most cells. Each form is composed of two catalytic subunits and two regulatory subunits (shown schematically in Fig. 23.1). The regulatory subunits, known as Type I cADepPK regulatory subunit (RI) and Type II cADepPK regulatory subunit (RII) are different for the two forms of cADepPK, while the catalytic subunits (C) are thought to be the same [reviewed in Beavo et al. (1975)].

Once cAMP activates cADepPK by binding to RI or RII and releasing the free, active C subunit, the C subunits phosphorylate specific protein substrates within the cells responsible for mediating the effects of cyclic AMP. Recent evidence has accumulated that each regulatory subunit has two distinct binding sites for cAMP which differ in affinity depending on the analog of cAMP used (Corbin et al., 1981).

There are many ways in which specificity of the response to cAMP in any individual cell type may be determined. Differentiated cells may contain certain substrates which are missing in other cell types. For example, in liver, glycogen synthase is an abundant substrate for cADepPK. It is phosphorylated after cADepPK is activated by increased cAMP levels resulting from glucagon or epinephrine stimulation of adenylate cyclase [reviewed in Cohen (1980)]. The phosphorylated form of glycogen synthase is less active than the unphosphorylated form, so that glycogen accumulation slows down or stops. Cyclic AMP also stimulates breakdown of glycogen by activating phosphorylase kinase which stimulates Ca^{2+}-dependent phosphorylase-mediated hydrolysis of glycogen. These well-studied reactions occur principally in muscle, whereas in other tis-

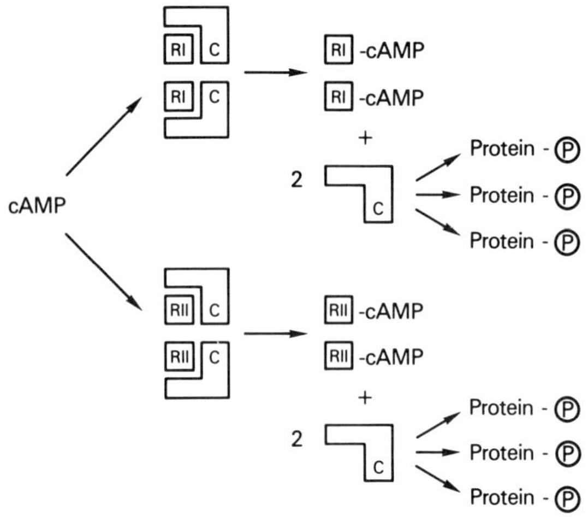

Figure 23.1. Schematic representation of the activation and effects of cADepPK in CHO cells.

sues, or in CHO cells, the enzymes of glycogen metabolism are present in very small amounts and treatments affecting cAMP levels have little or no effect on glycogen metabolism (LeCam et al., 1981). There are many other examples of cAMP-dependent phosphorylations affecting enzymes of intermediary metabolism [reviewed in Cohen, (1980); Cohen (1982)].

Another way in which specificity of response to cAMP may be determined is by the presence or absence of Type I or Type II cADepPK. Some tissues, such as brain and adipose tissue, have predominantly Type II cADepPK, whereas others, such as liver, testis, and skeletal muscle, have predominantly Type I cADepPK (Corbin et al., 1975). Species differences in the distribution of Type I and Type II cADepPK have been demonstrated for cardiac tissue (Corbin and Keely, 1977). Furthermore, there are at least two forms of Type II cADepPK, since brain and heart forms of RII differ immunologically (Erlichman et al., 1980). It has been assumed that these various forms of cADepPK subserve different functions in different tissues, but this has not been rigorously proved in a system in which the presence or absence of each of these forms of cADepPK can be manipulated independently and the effect on cAMP responsiveness determined. This kind of question is especially susceptible to genetic analysis in CHO cells (see below).

A third way in which cAMP may exert specific effects on cells presumes that the Type I and Type II cADepPKs have different subcellular localizations in different cell types so that the C subunit is released in its active form in the vicinity of different substrates in these various cell types. For example, recent evidence indicates that RII may be specifically associated with microtubule-associated protein substrates for Type II cADepPK within brain (Theurkauf and Vallee, 1982). There is also some immunofluorescence evidence that C localization may shift from the cytoplasm to the nucleus after activation of cADepPK by cAMP (Byus and Fletcher, 1982), but no mutants have yet been isolated which are defective in this putative relocalization of C.

B. The Contribution of Genetics to Understanding cAMP Effects

What contribution has genetic analysis made to our understanding of how cAMP works within cells? Most of our knowledge comes from the study of cells whose growth is inhibited by cAMP. Since the pioneering work of Tomkins and Coffino who isolated cAMP-resistant mutants of mouse S49 lymphoma cells [reviewed in Insel et al. (1975)], a number of laboratories have selected mutants of different cell lines which are resistant to cAMP [reviewed in Gottesman (1980) and summarized in Table 23.1]. A few general principles have resulted from these studies. First, it is possible to isolate mutants which are close to totally defective in cADepPK activity, indicating that this enzyme is probably not needed in normal amounts for the survival of individual cells. This conclusion in no way rules out a critical role for cADepPK in modulating the interactions among cells and tissues in a whole organism, nor does it rule out an essential role for small amounts of cADepPK within cells. Second, all of the effects of cAMP which are known in mammalian cells are blocked in mutants with defective cADepPKs. This finding suggests, but does not prove, that cADepPK may be the only intracellular mediator of cAMP action. Finally, mutations affecting both R and C have been isolated, indicating the importance of an intact cADepPK system for the mediation of the complete response to cAMP.

C. The Effects of cAMP on CHO Cells

CHO cells have many advantages for the genetic analysis of cAMP effects in cultured cells. Because of the genetic properties of CHO cells, it was expected that cAMP-resistant mutants could be isolated and that, when isolated, these mutants would have discrete, meaningful phenotypes. Even more significantly, CHO cells respond to cAMP or agents which raise intraceullular cAMP levels in an interesting variety of ways.

The agents which affect intracellular cAMP levels include analogs of cAMP, such as 8-Br-cAMP and $N^6,O^{2'}$-dibutyryl cAMP (dbcAMP). The cell membrane is relatively permeable to both of these analogs, and both are resistant to the effects of cyclic nucleotide phosphodiesterase, the enzyme which degrades cAMP within the cell. Cyclic AMP itself is never used in these studies because it is subject to rapid degradation by

TABLE 23.1
Mammalian Cell Mutants With Altered cAMP-Dependent Protein Kinases

Affected Subunit	Cell Type	Mutant Designation	Reference
Regulatory subunit	CHO	10248	Gottesman et al. (1981)
			Gottesman (unpublished)
	Mouse S49 lymphoma	*kin*A	Steinberg et al. (1977)
	Mouse Y-1 adrenal	kin8	Doherty et al. (1982)
	Mouse J774.2 macrophage		Rosen et al. (1979)
Catalytic subunit	CHO	10215	Evain et al. (1979)
	CHO	10260, 10265	Singh et al. (1981)
			Murtaugh et al. (1982)
	Mouse S49 lymphoma	*kin*C	Steinberg et al. (1978)
Affected subunit not known	CHO	10226	Gottesman et al. (1980)
	CHO	10987	Gottesman et al. (1983)
	CHO	11348	Gottesman (1983)
	CHO	10223	Singh et al. (1981)
	Mouse neuroblastoma		Simantov and Sachs (1975)
	Mouse Cloudman melanoma		Pawelek (1979)

phosphodiesterase with the formation of 5′-AMP, which is highly toxic to most mammalian cells. Another way to increase cAMP levels involves the use of hormones or toxins which stimulate adenylate cyclase. We routinely use cholera toxin because it produces high intracellular cAMP levels for long periods of time in CHO cells (Li et al., 1977). The growth inhibition and other cellular changes which result from cholera toxin treatment are mediated by cADepPK as shown by genetic studies (see below) in which cADepPK mutants are completely resistant to cholera toxin effects. CHO cells have very few receptors for other hormones which affect adenylate cyclase such as epinephrine, or have these receptors uncoupled from the cyclase, but they do respond somewhat to follicle-stimulating hormone (FSH) and luteinizing hormone (LH) (Evain and Anderson, 1979) and prostaglandin E1 (Johnson and Gottesman, unpublished data). The effects of FSH and LH are small, and prostaglandin E1 elevation of cAMP levels is brief (Johnson and Mukku, 1979), presumably because of turn-off of the cyclase and activity of phosphodiesterase. For these reasons, these three hormones are generally not used to elevate cAMP levels in CHO cells.

Another class of agents which raise cAMP levels within CHO cells are phosphodiesterase inhibitors such as methylisobutylxanthine and theophylline. The inhibitory effects of these drugs are not entirely specific, and at concentrations greater than 1 m*M* they appear to have undefined toxic effects, as well as inhibiting phosphodiesterase. In CHO cells at 0.5 m*M* or less, most of these effects are specific since they are blocked in cADepPK mutants, and hence, are mediated by cADepPK. When used in combination with cholera toxin, extremely high and prolonged levels of cAMP can be achieved for long periods of time (several days). Addition of cholera toxin and methylisobutylxanthine creates the most stringent conditions for selection of cAMP-resistant mutants. This selection has yielded our most resistant mutants.

As summarized in Table 23.2, addition of cAMP analogs to CHO cells results in inhibition of cell growth within 12 hr after treatment. This effect forms the basis of the selection of cAMP-resistant mutants and allows the isolation of mutants which continue to grow normally despite the presence of elevated intracellular levels of cAMP. Cyclic AMP also causes a profound change in the morphology of most fibroblastic cells, including CHO cells, which go from their normal triangular shape, to elongated, aligned cells within 1–2 hr (Hsie and Puck, 1971; Johnson et al., 1971; Gottesman et al., 1980). This process, which has been termed "reverse transformation" by Puck and his co-workers (see Chapter 2), implies that cAMP promotes a profound reorganization of the cytoskeleton of the CHO cell. Associated with this shape change is the almost complete elimination of cell surface ruffling and blebbing activity found in the untreated CHO cell (Gottesman et al., 1983). A recent report describes the increase of a cell-surface glycoprotein after cAMP treatment,

TABLE 23.2
Effects of Cyclic AMP on CHO Cells

Effect Observed	Time to Onset	Blocked in Mutants?	Reference
Decreased blebbing	Seconds	Yes	Gottesman et al. (1983)
Cell elongation and alignment	1–2 hr	Yes	Gottesman et al. (1980)
Stimulation of ornithine decarboxylase activity	2–6 hr	Yes	Lichti and Gottesman (1982)
Stimulation of transglutaminase activity	2–6 hr	Yes	Milhaud et al. (1980)
Stimulation of phosphodiesterase activity	2–6 hr	Yes	Klee and Gottesman (unpublished data)
Growth inhibition	6–12 hr	Yes	Gottesman et al. (1980)
Decreased V_{max} for glucose transport	10–24 hr	Yes	LeCam et al. (1980)
Decreased V_{max} for amino acid transport (system A)	24 hr	Yes	LeCam et al. (1980)
Formation of gap junctions	Not determined	Yes	Weiner and Loewenstein (1983)

but its role in affecting cell shape or adhesion is not known (Imada and Imada, 1982). An understanding of the molecular mechanism of the growth and morphological effects of cAMP which might result from a complete genetic analysis of cAMP response in CHO cells would be expected to yield important information about the regulation of cell shape and cell growth and the interaction of these phenomena.

Another intriguing physiological effect of cAMP on CHO cells is the stimulation of electrically coupled junctions by cAMP. Since transformed cells such as CHO cells tend to have reduced intercellular communication, this stimulation of junction formation by cAMP can be considered a component of the "reverse transformation" response. Genetic analysis of cADepPK mutants has indicated a role for this enzyme in stimulation of the formation of electrically coupled junctions (Wiener and Loewenstein, 1983).

Cyclic AMP has other significant effects on CHO cells. Twelve hours of cAMP treatment results in a reduction by 50% of the maximum velocity (V_{max}) of transport of both glucose and amino acids (A system), without an effect on the affinity of the transport systems (K_m) for these important metabolites. In the case of amino acid transport, this reduction in V_{max} affects only the A system which is the transport system for α-aminoisobutyric acid whose activity is absolutely dependent on sodium ions. The relationship between reduced nutrient transport and reduced growth rate is not yet clear. Since the time of appearance of both phenomena is delayed for at least 10 hr after initiation of cAMP treatment, it is assumed that both of these effects are the result of a complex series of metabolic changes induced by cAMP, rather than direct, immediate effects of single phosphorylation events. Whether decreased nutrient transport reduces growth rate or vice versa is also not known, but the magnitude of the reduction in transport does not seem sufficient to account for the more profound reduction in growth rate. The growth inhibition also seems to precede the peak effects on nutrient transport, suggesting that transport effects may be secondary to growth inhibition or even unrelated to growth inhibition. Since both growth and nutrient transport effects are delayed in onset after cAMP treatment, it is reasonable to question whether cAMP is producing these effects by activation of protein kinase or by another mechanism. As shown below, a genetic analysis of cAMP resistance in CHO cells indicates that cADepPK does mediate both the growth and transport effects of cAMP.

The effects of cAMP in CHO cells which are most immediate and easiest to measure involve the stimulation of activity of various enzymes. To date, three enzymes of intermediary metabolism have been studied in CHO cells: transglutaminase, ornithine decarboxylase, and cyclic nucleotide phosphodiesterase. Transglutaminase is an enzyme which covalently cross-links amino groups of small molecules and proteins to the lysine ε-amino groups of other proteins. It has been postulated that this enzyme is involved in clustering of cell surface ligands after binding to receptors (Milhaud et al., 1980a). Cyclic AMP stimulates activity of this enzyme by a factor of 2-fold (Milhaud et al., 1980a). Ornithine decarboxylase (ODC) is the first enzyme in the metabolic pathway leading to the formation of polyamines, which are thought to play a key role in DNA metabolism in the cell (see Chapter 18). Cyclic AMP stimulates ODC activity by a factor of 10 to several-hundred-fold, depending on the conditions of treatment (Costa, 1978; Meloni et al., 1980). ODC activity has been observed to vary throughout the cell cycle, and it has been postulated that similar variations in cAMP levels might trigger changes in ODC levels, which regulate cell growth throughout the cell cycle (Russell and Haddox, 1979). Genetic analysis of both ODC (Chapter 18) and the cADepPKs has indicated that this hypothesis is not likely to be correct (see below).

Cyclic nucleotide phosphodiesterase (PDE) activity is responsible for the degradation of cAMP and cGMP within the cell. It was of interest to discover that cAMP treatment of cells results in stimulation of PDE activity, resulting in more rapid breakdown of the cAMP itself [d'Armiento et al. (1972); reviewed in Wells and Hardman (1977)]. This provides an autoregulatory mechanism which prevents the accumulation of high levels of cAMP and subsequent continuous activation of cADepPK within cells. Since PDE is a cAMP-binding protein itself, it seemed possible that its activation by cAMP might not require the involvement of cADepPK. Genetic analysis of CHO (Klee and Gottesman, unpublished data) and S49 (Bourne et al., 1973) cADepPK mutants, however, indicate an important role for cADepPK in this PDE stimulation.

Another mechanism which achieves a similar effect is the turn-off of adenylate cyclase that occurs shortly after hormonal stimulation so as to prevent continuous synthesis of cAMP after a single stimulus. It is of interest that toxins such as cholera toxin and pertussis toxin, which result in pathological stimulation of high levels of cAMP, achieve this effect by subversion of the normal mechanism of adenylate cyclase turn-off. In the case of cholera toxin, the toxin has an ADP-ribosylating activity which inactivates the GTPase activity associated with the GTP-binding subunit of cyclase and leaves this GTP-activated enzyme in a constant state of stimulation (Moss et al., 1976; Cassel and Selinger, 1977); for pertussis toxin the ADP-ribosylating activity inhibits an inhibitory regulatory subunit of the cyclase, resulting in activation (Katada and Ui, 1982).

Many other enzyme activities have been shown to be stimulated or inhibited by cAMP in cell types other than CHO cells [reviewed in Cohen (1980)]. In some cases these effects of cAMP can be shown to be a direct effect of phosphorylation of the enzyme. In many cases, such as for ODC and PDE, the stimulatory effect seems to be due to induction of new synthesis of the enzyme. One of the most striking stimulatory effects of cAMP is observed with phosphoenolpyruvate carboxykinase (PEPCK) in liver cells where many-fold more enzyme is synthesized after treatment with hormones that raise cAMP levels (Wicks, 1969). In this case, cAMP treatment results in a sequential increase in nuclear and cytoplasmic mRNA levels for PEPCK (Beale et al., 1982; Chrapkiewicz et al., 1982), indicating that cAMP can stimulate enzyme synthesis in liver by a transcriptional mechanism, as it does in bacterial cells (Pastan and Adhya, 1976). Whether cADepPK is involved in this case of PEPCK induction is not known, and whether true increased enzyme synthesis resulting from elevated mRNA levels occurs for ODC and PDE, is also not yet known. Induction of mRNAs after cAMP treatment has been shown for other enzymes, such as tyrosine aminotransferase (Ernest and Feigelson, 1978), alkaline phosphatase (Firestone and Heath, 1981), and lactic dehydrogenase-5 (Derda et al., 1980). The generality and mechanism of cAMP induction of increased mRNA levels will be matters of active investigation by molecular geneticists interested in the mechanisms of cAMP action in the near future. One approach will be to study the expression of chimeric genes with cAMP-responsive promoters transfected into tissue culture cells (Wynshaw-Boris et al., 1984).

A variety of other effects of dbcAMP in CHO cells have also been described including the stimulation of collagen synthesis (Hsie et al., 1971) and increased synthesis of fibronectin (Nielson and Puck, 1980). Both of these responses have been attributed to the "reverse transformation" effects of cAMP; that is, decreased fibronectin and collagen synthesis result from malignant transformation of cells and cAMP acts to make cells behave in a more "normal" manner. There are, however, several examples of cell types in which collagen synthesis does not respond to cAMP treatment. Even in CHO cells, the effects of cAMP treatment on collagen and fibronectin synthesis are small and variable, and the fibronectin effects can be mimicked by sodium butyrate (Milhaud et al., 1980b). The significance of these effects is not yet understood.

II. SELECTION OF CHO CELLS RESISTANT TO cAMP

A. Rationale for Selection

As discussed above, one of the most striking effects of cAMP on CHO cells is the inhibition of cell growth which occurs within 12 hr after treatment. Unlike S49 lymphoma cells, which have been used extensively to isolate cAMP-resistant mutants, CHO cells are not killed by cAMP. This means that if CHO cells are simply plated in medium containing a cAMP analog such as 8-Br-cAMP or an agent which stimulates adenylate cyclase, such as cholera toxin, the cells will grow more slowly, but mutants will be hard to see against a dense background of living, cAMP-sensitive cells. This background can be substantially reduced by plating the cells in suspension in semisolid medium such as methylcellulose, agarose, or agar. The requirement for growth in suspension appears to act synergistically with the cAMP to inhibit the growth of cAMP-sensitive cells, and the resistant colonies stand out clearly in the semisolid medium and can be easily picked and repeatedly recloned to eliminate any contaminating sensitive cells. Detailed protocols for the isolation of cAMP-resistant CHO mutants have recently been published (Gottesman, 1983).

Spontaneous appearance of cAMP-resistant CHO mutants is rare, with a frequency of less than 10^{-6}. Cell populations which have been mutagenized with ethylmethanesulfonate or ultraviolet light give rise to cAMP-resistant mutants at a frequency of approximately 10^{-5} (Gottesman et al., 1980). To date, we have characterized the cADepPK activity in extracts from 30 independent mutants, and the great majority of them, perhaps all, have alterations in their cADepPK activity.

B. Methods for Characterizing Protein Kinase Activity in cAMP-Resistant CHO Mutants

A variety of methods are available for characterizing the cADepPK activity of wild-type and mutant CHO cells. Initially, we screen our potential mutants for activity by preparing Dounce homogenates and assaying kinase activity in 30,000*g* supernatants using $[^{32}P]\gamma$-ATP and calf thymus histone as substrates in the presence or absence of cAMP as described by Corbin and Reimann (1976). The sensitivity of this screening procedure can be increased greatly by generating a dose response curve by measuring kinase activity in the presence of increasing amounts of cAMP. This approach detects mutants that have cADepPK activity, but in which it is necessary to add more cAMP than for wild-type cells in order to see an equivalent level of activity. The great majority of our mutants fall into this class (approximately 25/30). The remaining mutants either have lower levels of activity even after maximal cAMP stimulation (3/30) or have apparently normal kinase activity in crude extracts (2/30).

The next step in analysis of mutant CHO cells involves the fractionation of CHO extracts on DEAE cellulose columns to determine the relative levels of Type I and Type II cADepPKs (Evain et al., 1979). Type I kinase elutes at a lower salt concentration than Type II kinase and can be readily separated from it by a linear salt gradient. cADepPK activity eluting from the column can be easily detected using the assay described above, and wild-type cells are seen to have two approximately equal peaks of cADepPK activity. The eight mutants we have examined by this technique to date have either lost Type I cADepPK activity or Type II cADepPK activity, or have reduced levels of both activities [reviewed in Gottesman et al. (1981) and Gottesman (1983)]. This relatively simple analysis has allowed us to conclude that both Type I and Type II cADepPKs are needed for cAMP inhibition of growth of CHO cells, and both enzymes are also probably needed for the morphological effects of cAMP and the effects on nutrient transport. This analysis supports the preliminary conclusion reached above on the basis of the relatively delayed appearance of these phenomena that they are likely to be complex and are probably the result of at least two, and perhaps many more, independent phosphorylation events.

Once the individual types of cADepPK activity have been isolated, it is desirable to learn as much as possible about the regulatory (R) and catalytic (C) subunits in the mutants. The R subunit can be studied in a variety of ways which depend upon its ability to bind to cAMP and its analogs. It is possible to measure the binding of [^{3}H]cAMP directly to partially purified cADepPKs, using a filter binding assay developed by Gilman (1970). Recently, the development of the photoaffinity label 8-azido[^{32}P]cAMP has allowed both the direct quantitation of R subunits in purified or crude extracts containing cADepPK activity, as well as the visualization of these subunits on polyacrylamide gels (Walter et al., 1977). Using this technique, we have found that Type I cADepPK from CHO cells has an R subunit with a M_r of 49,000 daltons, while RII consists of two species with M_rs of 54,000 and 56,000 daltons (Gottesman et al., 1981). Similar species of RII have been found in bovine cardiac muscle, where the higher-molecular-weight form of RII is thought to be a phosphorylated form of the lower-molecular-weight species (Rangel-Aldao et al., 1979).

When these photoaffinity-labeled species of RI and RII are analyzed on two-dimensional gels, the RI band is seen to consist of four major species with slightly different isoelectric points (Gottesman, et al., 1981). Preliminary experiments based on analysis of cAMP-binding proteins synthesized from mRNA in a cell-free system suggest that two of these species may represent different forms of RI synthesized in CHO cells, whereas the other two species may be phosphorylated derivatives of these. Phosphorylation of RI has previously been described in S49 lymphoma cells (Steinberg et al., 1977; Steinberg and Agard, 1981a). In addition to the RI and RII species associated with their respective cADepPK activities on the DEAE column, there is a third peak of 8-azido[^{32}P]cAMP binding activity which elutes between the two peaks of enzyme activity. This peak consists entirely of RI not associated with any C subunit and has been referred to as "free" RI. In CHO cells there is as much "free" RI or more than there is RI associated with C. In CHO cells, as well as in neuroblastoma cells (Prashad et al., 1979; Walter et al., 1979) and S49 cells (Steinberg and Agard, 1981b), this form of RI is increased dramatically (three- to fivefold) after treatment of cells with dbcAMP. This result, which appears to be due primarily to stabilization of RI by dbcAMP binding (Gottesman, unpublished data; Steinberg and Agard, 1981*b*) suggests that "free" RI may have a function independent of its role in inhibition of the activity of C. However, no cAMP-resistant mutant in any cell system so far examined has had an alteration in "free" RI unaccompanied by a change in cADepPK activity, so there is no genetic evidence to prove a critical role for "free" RI in determining response to cAMP.

The avidity with which cAMP binds to RI and RII has facilitated the development of affinity chromatographic procedures that allow the direct isolation of pure preparations of these regulatory subunits from DEAE fractions, or even crude extracts. In our laboratory, we use affinity columns of N^6-(2-ethyl)-cAMP to purify RI from DEAE fractions containing Type I cADepPK activity. Coffino and Steinberg and their co-workers have used a similar approach to examine the mutant RI subunits in many of their S49 cAMP-resistant mutants labeled with [^{35}S]methionine (Steinberg et al., 1977). Relatively pure preparations of catalytic subunit can be prepared by dissociating R and C with cAMP and rechromatographing on DEAE. The C subunit is relatively basic and does not bind to the DEAE column in the absence of R, whereas other proteins in relatively crude preparations of Type I cADepPK will not be affected by treatment with cAMP and will bind to DEAE.

Availability of relatively pure preparations of C and R makes it possible to do reconstitution experiments in which putative mutant and wild-type subunits can be mixed in order to determine which subunits are defective. For example, if an R subunit from a mutant extract is mixed with a wild-type C, formation of a mutant enzyme (e.g., an enzyme with a shifted dose response to cAMP) would indicate that the R subunit was mutated. An important control in such an experiment would include mixing C from the mutant with wild-type R and confirming formation of a wild-type enzyme. This approach has been used successfully to demonstrate the presence of a mutant R subunit in an S49 cAMP-resistant line (Hochman et al., 1975). Unfortunately, the C subunits isolated from CHO cells appear to be very unstable and complete reconstitution experiments in this system have so far proved impossible.

C. Dominance and Complementation Analysis of cAMP-Resistant Mutants

Once having established that most of our mutants had alterations in their cADepPK system and that some of these alterations affected Type I and Type II cADepPKs, we sought to classify them further with respect to their behavior in somatic cell hybrids. The principles of hybrid formation between two cell types have been outlined in Chapter 7. To reiterate, the basic idea is to fuse cells with polyethylene glycol or Sendai virus and to select for the formation of true hybrid cells in medium which would kill either cell alone, but in which genetic combinations of the two cells are able to survive. We selected hybrid cells between cAMP-resistant mutants and wild-type cells in two ways (Gottesman et al., 1980). First, we fused our resistant mutants with a cAMP-sensitive strain which was ouabain resistant [a dominant selectable marker, see Baker et al. (1974)] and an 8-azaguanine-resistant cell line which would not grow in HAT medium (see Chapter 13). Selection for growth in medium containing 2 m*M* ouabain and HAT (Jha and Ozer, 1976) resulted in hybrid cells that had close to a tetraploid chromosome number (42 for CHO cells). Second, we hybridized cAMP-resistant cells with a cAMP-sensitive cell line which was auxotrophic for glycine, adenosine, and thymidine (GAT^-, see Chapter 24) but which was no longer auxotrophic for proline as is the parent cell line from which our mutants were derived (see Chapter 2). In both cases, we examined several independent hybrids and asked whether they were still resistant to cAMP. The great majority of hybrids were cAMP resistant, indicating that resistance to cAMP is a dominant phenotype. All of the mutants which had a shifted dose response curve for cAMP and lacked Type II cADepPK were found to be dominant in somatic cell hybrids. The three mutants which had reduced levels of cADepPK activity were found to be recessive, confirming that these mutants belonged to a separate class (see below for details).

Because we had three recessive mutants, it was possible to determine whether or not these three fell into the same complementation group. To do this experiment, it was necessary to introduce ouabain resistance and HAT sensitivity into some of the cAMP-resistant strains. This was done relatively easily by sequential selection for growth in 2 m*M* ouabain followed by selection in 8-azaguanine (see Chapter 13). Hybrids formed between two of our recessive mutants failed to complement, that is, hybrid cells were as resistant to cAMP as were the parents (Singh et al., 1981). Hybrids formed between one of these recessive mutants (10260) and a third mutant (10223) did complement; that is, these hybrids were sensitive to cAMP. This result indicates that there are two complementation groups of recessive mutants.

In order to confirm that these three mutants were really recessive we performed a segregation analysis. As indicated in Chapter 28, hybrid cell lines frequently lose markers at a high rate, indicative of a variety of segregation events, including chromosome loss. When our hybrids between wild-type cells and the recessive cAMP-resistant mutants were selected for cAMP resistance, we were able to recover cAMP-resistant cells at a frequency of approximately 1%. This frequency is much higher than the expected mutation frequency to cAMP resistance and is consistent with loss of a dominant wild-type allele in these cells. This analysis indicates that the original mutant allele was indeed still present in the hybrid cells and that the mutations were truly recessive.

The three recessive mutants (10260 and 10265 in one complementation group and 10223 in the other) were all mutants that had showed reduced total cADepPK activity when this activity was assayed in crude extracts. The small amount of activity they did have was associated only with the Type II cADepPK peak from a DEAE column. Thus, the genetic analysis confirmed our biochemical analysis that these three mutants were different from the majority of dominant mutants which we had isolated. At this point, we had firm evidence that we had at least three distinct classes of mutants resistant to cAMP in CHO cells consisting of one large class of dominant mutants with shifted cAMP dose response curves for activation and little or no Type II cADepPK and two complementing groups of recessive mutants with greatly reduced cADepPK activity in crude extracts and no detectable Type I kinase activity.

D. Mutations Affecting Catalytic and Regulatory Subunits of cAMP-Dependent Protein Kinase

After establishing that cAMP-resistant CHO cells contained mutations that affected the activity of the cADepPKs, we began to examine the mutants in more detail to learn more about the subunits affected in each of the mutants. For the purpose of this review, three mutants will be discussed in greater detail than the others: two of these appear to have alterations affecting the catalytic subunit (10260 and 10215), while one appears to have an altered RI subunit. A summary of all of the CHO mutants discussed in this chapter is given in Table 23.1. These mutants will be compared to well-characterized cADepPK mutants in other cultured cell systems.

1. Regulatory Subunit Mutants

To date we have found only one CHO cAMP-resistant mutant that appears to have a bona fide alteration in the Type I regulatory subunit. This finding is in contrast to experience in other cultured cell systems in which the great majority of mutants appear to have altered regulatory subunits [see Gottesman (1980), Table 23.1, and below]. Possible reasons for this difference in relative frequency of mutants will be discussed below.

Mutant 10248 carries a dominant mutation conferring resistance to cAMP and was isolated after ethylmethanesulfonate mutagenesis and selection in the presence of 1 μg/mL cholera toxin and 1 m*M* theophylline in monolayer culture (Gottesman et al., 1980). As indicated in Section I C, this selection medium is quite stringent and allows the growth of only the most cAMP-resistant mutants. In addition, as opposed to selections that use cAMP analogs, the growth inhibition is due to cAMP itself and would be expected to proceed by a physiological pathway. As originally characterized, mutant 10248 was thought to have a relatively normal cADepPK system (Gottesman et al., 1980). However, on more detailed analysis (Singh and Gottesman, unpublished data) it became apparent that extracts from mutant 10248 show a subtle shift in dose response to cAMP and a dramatic loss of Type II cADepPK when extracts were analyzed on DEAE columns. Type I cADepPK is present, but requires at least 10-fold more cAMP for activation. When RI is examined by photoaffinity labeling using 8-azido[^{32}P]cAMP the binding affinity of the Type I holoenzyme is much reduced compared to the binding affinity of the wild-type I cADepPK. In contrast to this finding, the "free" RI in the mutant is normal in amount and binding affinity. When RI is dissociated from the mutant holoenzyme, it too has an altered binding affinity for 8-azido[^{32}P]cAMP. This finding suggests that there are two forms of RI in this mutant CHO cell, only one of which is altered.

Two-dimensional gel analysis of 8-azido[^{32}P]cAMP-labeled RI from mutant 10248 shows no major alteration in the isoelectric point of the major RI species in this cell line, and two-dimensional tryptic peptide maps of the labeled mutant RI are not distinguishable from the wild type. Preliminary attempts to reconstitute mutant RI with bovine C have produced a mutant holoenzyme with reduced response to cAMP (Verna and Gottesman, unpublished data), but control experiments using C from the mutant cell line or wild-type CHO cells have not been possible because of the instability of C in extracts from CHO cells. Type II regulatory subunit is present in this cell line and its molecular weight as determined by sucrose density gradient centrifugation is the same as wild-type RII dissociated from Type II cADepPK; that is, it exists as a dimer of M_r 110,000 daltons. These data taken together argue that there are two species of RI in mutant 10248, only one of which is altered by mutation, and that this altered RI is tightly bound to C to form the predominant form of Type I cADepPK within the mutant cells. We propose that the "free" RI in this cell line consists of a normal RI subunit which does not associate with C because C is tightly bound to the mutant subunit. Similarly, RII cannot compete favorably with mutant RI for C and also exists in this mutant in the "free" form. This model accounts for the presence of two species of RI in wild-type cells (see Section II B) which might represent different alleles of RI or perhaps two different genes. (see Section IV A for evidence that there is only one gene for RI.) In addition, the model explains why the mutation in mutant 10248 is dominant in somatic cell hybrids since the presence of a mutant RI which cannot be easily dissociated from C even in the presence of excess amounts of wild-type RI or RII would result in a cAMP-resistant phenotype for reasons given above. This model fits the data well, but is speculative and does not account for complexities such as the possibility of heterodimers of RI.

Regulatory subunit mutants have also been isolated from mouse S49 lymphoma cells, mouse Y-1 adrenal cells, mouse Cloudman melanoma cells, mouse J774.2 macrophages, and mouse neuroblastoma cells (Pawelek et al., 1975; Simantov and Sachs, 1975; Steinberg et al., 1977; Rosen et al., 1979; Doherty et al., 1982). In the case of the S49 mutants, dbcAMP was used as the selecting agent, and a large number of mutants have been extensively characterized and shown to have altered RI subunits characterized by altered isoelectric points on two-dimensional gels (Steinberg et al., 1977). Recent studies have used partial proteolysis to pinpoint the mutations to the cAMP-binding domains of RI (Steinberg, 1983). These elegant studies show unequivocally that mutations in RI can produce cAMP-resistant phenotypes in mammalian cells.

Why are RI mutants common in many cultured cell systems, but relatively rare in CHO cells? The answer to this question is complex. First, all of the other systems studied utilize mouse cells. It is quite possible that the genes encoding elements of the cADepPK system are differently organized or different in copy number in these aneuploid cell lines than in CHO cells. However, S49 cells also appear to have two alleles encoding RI (as indicated by the coexistence of mutant and wild-type RI spots on two-dimensional gels and the reversion analysis discussed in Section II E). Second, dbcAMP is a different selective agent than 8-Br-cAMP, which we have used to isolate most of our mutants. It is striking that the one mutant which does have an RI alteration was isolated in the presence of cholera toxin and theophylline and so was thus actually selected for resistance to endogenous cAMP. Ample evidence exists that there are two cyclic-AMP-binding sites on each RI subunit (Builder et al., 1980; Corbin et al., 1981), and that these differ in specificity for dbcAMP and 8-Br-cAMP, whereas cAMP binds to both sites. In addition, S49 mutants selected for resistance to dbcAMP are apparently somewhat sensitive to 8-Br-cAMP (R. Steinberg, personal communication), strongly suggesting that the physiological effect of these agents may be different. Dibutyryl cAMP has not been used as a selective agent in CHO cells because of the sensitivity of these cells to the butyrate released by hydrolysis of this analog (Gottesman et al., 1980). By this model, of the two sites on RI available for binding to cyclic AMP only one is subject to a high frequency of mutation (the one which also binds dbcAMP), whereas the other might mutate at a much lower frequency.

This model would appear to be disproved by the finding of Doherty et al. (1982) that RI mutants can be isolated in mouse adrenal Y-1 cells after selection in 8-Br-cAMP. However, this selection also yields adenylate cyclase mutants in this cell line, suggesting that 8-Br-cAMP is in some way able to stimulate adenylate cyclase (Martin and Kowalchyk, 1981) and that growth inhibition may be due to elevations of intracellular cAMP levels as a result of the cyclase stimulation. Also according to this hypothesis, mutant 10248 should be sensitive to 8-Br-cAMP, but it is not, presumably because it lacks Type II cADepPK.

2. Catalytic Subunit Mutants

Two classes of cAMP-resistant CHO mutants appear to have alterations affecting the C subunit: a class consisting of dominant mutants with a shifted dose response to cAMP, represented in this discussion by mutant

10215 (Evain et al., 1979), and a complementing group of two recessive mutants, represented by mutant 10260 (Singh et al., 1981). Since the existence of these two mutants proves that it is the catalytic subunit of the kinase itself which is essential for cAMP responsiveness, the finding of these two mutant classes in the CHO system has been of great significance. In contrast, these mutant classes are not found among other cAMP-resistant mammalian cells. In the S49 system it has been possible to isolate, at low frequency, very interesting mutants which have a dominant phenotype and appear to affect regulation of expression of the C subunit in a transdominant manner (*kin*C) (Steinberg et al., 1978). Each of these classes of mutants affecting the C subunit will be discussed in turn.

CHO mutant 10215 carries a dominant mutation and was isolated after ethylmethylsulfonate mutagenesis in suspension in the presence of 1 μg/mL cholera toxin. The biochemical alteration in this mutant is in many ways similar to the defect found in mutant 10248: Type II cADepPK is missing, but RII is present and Type I kinase is poorly dissociated by high concentrations of cAMP. However, when Type I cADepPK is examined in detail, there is no evidence for a defect in RI (Evain et al., 1979); instead, the C subunit released from RI in the presence of high concentrations of cAMP is found to have an altered substrate specificity with a greater preference for basic protein substrates, such as histone, than for acid protein substrates, such as phosvitin, than is seen for wild-type C. In addition, the K_m for ATP of the isolated catalytic subunit is increased. Although this C subunit cannot be isolated in pure form owing to its extreme instability, evidence points to a primary defect in this subunit characterized by altered substrate specificity, altered binding of ATP, and increased affinity for RI. The failure to see Type II cADepPK holoenzyme in this mutant may be due to decreased affinity of C for RII, or, as appears more likely in the light of our analysis of mutant 10248, the failure of RII to compete with RI for a C subunit which binds more avidly to RI. Again, the dominance of this mutation may relate to the increased tightness of binding of mutant C to normal RI subunits which excludes wild-type C from Type I cADepPK, followed by rapid degradation of wild-type C.

One of our recessive mutants, 10260, has very low levels of cADepPK activity reflected in the total absence of Type I cADepPK and the presence of a very small amount of Type II cADepPK. This mutant has very low levels (< 10% of wild-type) of catalytic subunit as detected on nitrocellulose blots of cell extracts using an affinity purified anti-C subunit antibody (Murtaugh et al., 1982). Binding of [^{3}H]cAMP or 8-azido[^{32}P]cAMP by mutant extracts is also somewhat reduced (Gottesman et al., 1980, 1981) to approximately 50% of wild-type levels, but this phenotype would appear to be secondary to the reduction in levels of C which is far more profound. The simplest explanation for the defect in this mutant is that it has a *cis*-dominant mutation in a genetic element controlling expression of the C subunit (i.e., promoter or operator), but other mechanisms, such as a structural alteration in C which makes it highly unstable, are possible. These data also suggest that levels of cADepPK R subunits are controlled to some extent by activity of C.

Coffino and his co-workers have reported on a class of S49 cAMP-resistant mutants (*kin*C) with a dominant phenotype which have undetectable levels of cADepPK activity and somewhat reduced levels of R (Steinberg et al., 1978). These appear to be regulatory mutations which turn off expression of C, and are analagous to superrepressor mutations in bacterial systems. Their existence suggests that C subunit expression is under complex, perhaps negative, control and that the further characterization of this genetic system may shed some light on genetic regulation of an important enzyme in a mammalian system.

3. *Mutants with Undefined Defects*

Many of the cADepPK mutants we have isolated do not fall readily into the categories of C or R subunit mutants. Several dominant mutants, such as 10226 (Gottesman, et al., 1980), 10987 (Gottesman, et al., 1983), and 11348 (Gottesman, 1983) lack Type II cADepPK activity, but a defect in the R or C subunits of the remaining Type I cADepPK cannot be demonstrated. Similarly, the recessive mutant 10223, which belongs to a different complementation group then C subunit mutant 10260, lacks Type I cADepPK but has substantial levels of RI, RII, and C. The existence of these mutants suggests that Type I or Type II cADepPK alone is not sufficient to catalyze the growth-inhibitory effects of cAMP. The behavior of some of these mutants suggests that there may be independent factors that control the association of RI and RII with C in the cell. The isolation and further characterization of these genetic elements may be possible using techniques of DNA-mediated gene transfer (see Section IV B).

Other groups have described mutants of cultured mammalian cells resistant to cAMP which cannot as yet be classified as R or C cADepPK mutants. Puck and his co-workers recently isolated a series of CHO-K1 mutants which do not respond to cAMP, and some of these have decreased ability to phosphorylate specific substrates (Gabrielson et al., 1982). Sisskin and Weinstein (1980) have described a CHO line with a very flat morphology and elevated levels of cAMP. In the Y-1 adrenal cell system (Rae et al., 1979) and in S49 cells (Daniel et al., 1973; Bourne et al., 1975), mutants with altered adenylate cyclase activity have been isolated. Because S49 cells are killed by cAMP and it is relatively easy to isolate cAMP-resistant S49 cells, a variety of other mutant types have been isolated in this system. Some of these mutants have elevated phosphodiesterase activity (Brothers et al., 1982), some excrete dbcAMP, the selecting agent (Steinberg et al., 1979), and others (deathless) cannot be killed by cAMP but have normal cADepPK activity (Lemaire and Coffino, 1977). These S49 mutants have proved extremely valuable for the analysis of other steps in cAMP response.

Pawelek and his co-workers have described a variety of mutants in the mouse Cloudman melanoma line which are cAMP resistant and others which are cAMP dependent for growth. Apparently, the Cloudman melanoma requires small amounts of cAMP to grow in tissue culture, whereas growth is inhibited at higher concentrations of cAMP (Pawelek et al., 1975). One mutant has been described in which an alteration in protein kinase results in the complete dependence of the cell line on cAMP for growth (Pawelek, 1979). The precise defect in this cADepPK mutant is not yet known.

E. Revertants of cAMP-Resistant Mutants

The isolation of revertants of mutant mammalian cells offers a powerful way to prove that the mutations are responsible for a particular biochemical phenotype, and in the case of revertants that involve outside suppressor mutations, to demonstrate interaction among the products of different genetic loci within the cell. In this volume, an example of this latter approach can be found in the work of Cabral et al. (1982) (Chapter 22) who isolated revertants of tubulin mutants which appear to affect microtubule-associated proteins in the cell.

The isolation of revertants of cADepPK mutants involves development of a selection scheme that enriches for revertant cells which fail to grow in the presence of cAMP, whereas cAMP-resistant cells are killed if they do grow. "Suicide selections" such as this, using BrdU to kill growing cells, described in detail in Chapter 2 and Chapter 11 for the selection of auxotrophic cells, Chapter 20 for the selection of cell cycle mutants, and Chapter 26 for the selection of nontransformed cell lines, have been successfully employed to isolate revertants of S49 cAMP-resistant cells (van Daalen Wetters and Coffino, 1982). This is a clean selection since S49 cells, which are sensitive to cAMP, are growth arrested quickly by dbcAMP and hence can survive a suicide selection with high efficiency. In the S49 system, revertants of the cADepPK mutants with altered RI (*kin*A mutants) have been used to demonstrate that only one of two loci had been mutated in the first step and to produce cell lines with only one wild-type locus (van Daalen Wetters and Coffino, 1983). These haploid cells have in turn been used to generate cAMP-resistant mutants with only a single mutant RI gene. In the case of the *kin*C mutants discussed above (Section II D 2), revertants that are temperature sensitive for suppression of the *kin*C mutation have been isolated, suggesting strongly that the product of the *kin*C locus is a protein (van Daalen Wetters et al., 1983).

In CHO cells, unfortunately, it has not been possible to select revertant cells by use of a suicide selection since cAMP does not produce a rapid or complete inhibition of growth. Efforts to isolate revertants have been directed toward specific phenotypes of cADepPK mutants other than resistance to growth inhibition. For example, it has recently been found that two of our cAMP-resistant mutants (10215 and 10248) are more sensitive than wild-type cells to the microtubule-depolymerizing drugs colcemid and maytansine (Abraham and Gottesman, unpublished data). This increased sensitivity is presumably due to a reduced stability of microtubules in the absence of a completely active cADepPK system. This phenomenon has allowed us to select cADepPK mutants which are as colcemid resistant as the parental cell line. We have found two classes of such revertants; one class is cAMP resistant and may be composed of mutants with altered microtubules, and a second class is cAMP sensitive. This latter class of mutants appears to have acquired normal levels of Type I and Type II cADepPK activity, but the precise mechanism of this reversion has not been determined. This reversion analysis provides convincing genetic evidence that cADepPK interacts with elements of the microtubule system in CHO cells.

F. Resistance of Rous-Sarcoma-Virus-Transformed CHO Cells to cAMP

Although wild-type CHO cells are quite sensitive to the growth-inhibitory effects of cAMP, there are many other cells types such as epidermal cells (Green, 1978), melanocytes (Eisinger and Marko, 1982), mammary cells (Taylor-Papadimitrou et al., 1980), and Swiss 3T3 cells (Pruss and Herschman, 1979; Rozengurt et al., 1981), whose growth or DNA replication is stimulated by cAMP under some conditions. In the course of studies on the physiology of CHO cells transformed by Rous sarcoma virus (RSV) (see Chapter 26), we found that these cells, and other mammalian cells transformed by RSV, are relatively resistant to the growth-inhibitory effects of cholera toxin and 8-Br-cAMP (Roth et al., 1982). It should be emphasized that this resistance is relative rather than absolute, and other authors have noted that growth inhibition and morphological changes can be demonstrated after treatment with high levels of dbcAMP (Puck et al., 1981; Meek, 1982).

The molecular basis of this resistance of RSV-transformed cells has been analyzed, and there does not appear to be an alteration in either the quantity or behavior of the cADepPKs in RSV-transformed cells (Roth et al., 1982). We next directed our attention to the possibility that the relative resistance of RSV-transformed cells is related to the activity of $pp60^{src}$, the product of the oncogene encoded by RSV [reviewed in Bishop and Varmus (1982)]. It has been possible to show that $pp60^{src}$ is a substrate for cADepPK in cell extracts (Collett et al., 1979) and, recently, increased phosphorylation of $pp60^{src}$ was demonstrated in intact cells (Roth et al., 1983). This increased phosphorylation can be correlated with stimulation of the tyrosine protein kinase activity of $pp60^{src}$ measured as increased transfer of [^{32}P] from γ[^{32}P]ATP to the heavy chain of a $pp60^{src}$-specific immunoglobulin in immunoprecipitates from RSV–CHO cell extracts treated with 8-Br-cAMP (Roth et al., 1983).

CHO–RSV cells form tumors poorly in nude mice. However, when these cells are treated with cholera toxin prior to inoculation into the animals, the yield of tumors is quite high (Gottesman, Roth, Vlahakis, and Pastan, 1984). These results argue that RSV–CHO cells can have their growth stimulated by cAMP under some conditions. It seems plausible that this growth stimulation is related to the activation of $pp60^{src}$ by a cAMP-dependent phosphorylation event. It is interesting to speculate that other cell types whose growth is stimulated by cAMP may also have their growth regulated by cellular analogs of oncogenes or growth factor receptors whose activity can be increased by cAMP-dependent phosphorylation.

G. Mutations Affecting cAMP Metabolism in Nonmammalian Cells

Dictyostelium discoideum (slime mold), *Sacchromyces cerevesiae* (Baker's yeast), and *Drosophila melanogaster* (fruit fly) have been used as genetic systems to analyze cAMP metabolism. In the case of *Dictyostelium*, cAMP is used as an intercellular signal which mediates cell aggregation and is essential for differentiation [reviewed in Rossier et al. (1980)]. A variety of mutants affected in cAMP signaling have been isolated, but specific molecular defects have not yet been assigned to these mutants. In *Drosophila*, it is possible to reduce levels of cyclic nucleotide phosphodiesterase activity and increase cAMP levels by increasing the dosage of chromomere 3D4 (Davis and Kiger, 1978). Alterations in dosage of chromomere 3D4 have profound effects on the development of many organ systems in *Drosophila*, suggesting the exciting possibility that further genetic analysis of this organism may shed some light on the involvement of cAMP in the embryological development of higher organisms.

Yeast has proved a fertile ground for the isolation of mutants affecting cAMP metabolism, since cAMP appears to be essential for normal cell cycling in yeast [reviewed in Thorner (1982)]. Using a set of yeast mutants permeable to cAMP and a suicide selection, Matsumoto et al. (1982) were able to isolate yeast that required cAMP for growth because of a defect in adenylate cyclase. Revertants of these mutants which were cAMP independent had reduced levels of R and high levels of cAMP-independent protein kinase (Matsumoto et al., 1982). One class of dominant cAMP-requiring yeast consisting of two mutants, one of which has been shown to have an altered R subunit with a reduced affinity for cAMP (Uno et al., 1982) in direct analogy with the mutants in CHO, S49 and Y-1 cells discussed earlier (Section II D 1). Another mutant has been described in this system with decreased phosphodiesterase activity which suppresses this R subunit defect (Uno et al., 1983). Yeast mutants carrying temperature-sensitive mutations affecting adenylate cyclase and cADepPK are defective in regulation of initiation of meiosis (Matsumoto et al., 1983).

III. USE OF MUTANTS TO STUDY CYCLIC-AMP-MEDIATED PROCESSES

Having characterized a variety of mutants affecting cADepPK in CHO cells, it became feasible to examine the effects of these mutations on cell behavior in the presence or absence of cAMP. It has also been possible to use the cADepPK mutants to begin to define phosphorylation events which are mediated by cAMP in CHO cells.

As noted in Section I B, cADepPK mutants in all systems so far analyzed including CHO cells, appear to grow normally under usual culture conditions. This result argues that the cADepPKs are not likely to play a critical role in the timing events in individual cell cycles. Tumor growth in nude mice of cADepPK mutants appears to be at least as good as wild-type cells. S49 cADepPK mutants actually form tumors with lower inoculums and grow faster than wild-type S49 cells (J. Hochman, personal communication). When cAMP levels are elevated in cells, as occurs after hormone stimulation, activation of cADepPK clearly plays a major role in inhibiting cell growth both in tissue culture and in tumors. There are, in addition, some stressful conditions under which cADepPK mutants survive less well than wild-type cells, for example, Section II E discusses the sensitivity of cADepPK mutants to microtubule-disrupting agents.

A. Failure of Mutants to Respond to cAMP

The availability of numerous cADepPK mutants in CHO cells, and in the other cell systems discussed above, has allowed the analysis of cADepPK involvement in the wide variety of responses cells have to cAMP. In the CHO system, these responses are listed in Table 23.2 and discussed extensively in Section I. In every case so far examined, mutants selected on the basis of resistance to growth inhibition and found to contain alterations in cADepPK have been found to be resistant to all of the other effects of cAMP. Similar results have been found for other cell systems analyzed in a variety of laboratories. These results argue strongly that cADepPK mediates all known effects of cAMP, but does not prove that cADepPK is the only mediator of such effects (i.e., cADepPK is necessary, but may or may not be sufficient).

B. Phosphorylation in cAMP-Resistant Mutants

One obvious way to use mutants with altered cADepPKs is to determine which phosphorylation events in cells are cAMP mediated. The two simplest approaches to this problem involve study of cAMP-dependent phosphorylation events in crude cell extracts in the presence of $\gamma[^{32}P]ATP$ and study of cAMP-dependent phosphorylation events in intact cells metabolically labeled with [^{32}P]orthophosphate. Both approaches have been taken using wild-type and mutant CHO cells (LeCam et al., 1981). When analyzed by polyacrylamide gel electrophoresis, a large number of proteins are seen to be phosphorylated in the presence of γ[^{32}P]ATP in Dounce homogenates of CHO cells. Addition of cAMP to these extracts results in the appearance of four new phosphoproteins, of molecular weight 300,000, 34,000, 26,000, and 17,000 daltons, and there is frequently a general increase in phosphorylation of all of the other phosphoproteins. The function of none of these phosphoproteins has been determined, although the highest M_r protein (approximately 300,000 daltons) is likely to be a microtubule-associated protein or another high-molecular-weight cytoskeletal protein. This result supports the conclusion reached by genetic means (Section II E) and other biochemical analyses (Theurkauf and Vallee, 1982) that a component or components of the microtubular system is affected by cADepPK. All of the cAMP-dependent phosphorylation events seen in crude extracts of CHO cells are reduced dramatically in extracts prepared from cAMP-resistant mutants (LeCam et al., 1981).

Genetic evidence has been obtained indicating that Type I and Type II cADepPKs in CHO cells may have different functions (see Section II B). Since both are found in CHO cells as soluble enzymes in crude extracts such as Dounce homogenates, it is reasonable to assume that within intact cells Type I and Type II cADepPKs may be compartmentalized and that phosphorylation events observed in crude extracts may not be truly representative of events occurring within the cells. For this reason, intact cells were either exposed or not exposed to cAMP and labeled with [^{32}P]orthophosphate. In these studies, one additional protein of approximate M_r 52,000 daltons was phosphorylated after cAMP treatment (LeCam et al., 1981). Once again, there were a large number of proteins phosphorylated in the absence of cAMP whose general level of phosphorylation seemed to be increased by cAMP treatment. By two-dimensional electrophoresis, this protein was shown to consist of a family of phosphoproteins (or a single protein phosphorylated on multiple sites) with pI 5.0–5.3. The function of these proteins has not been determined, but they appear to be quantitatively a very minor cell population as determined by Coomassie Blue staining or [^{35}S]methionine labeling of total cell protein, and they do not comigrate with vimentin, tubulin, or either RI or RII (Gottesman et al., 1981). This cAMP-dependent phosphorylation of the 52,000-dalton protein is missing in CHO mutants lacking Type II cADepPK (10215, 10248), but is found in mutants lacking Type I cADepPK which have some Type II cADepPK (10223, 10260, 10265), suggesting that this phosphorylation event is mediated by Type II cADepPK. Bloom and Lockwood (1980) and Gabrielson et al. (1982) have described a similar cAMP-dependent phosphorylation event in intact CHO cells.

Why is there apparently only one protein whose phosphorylation is absolutely dependent on addition of exogenous cAMP? It seems likely that many other proteins are substrates for cADepPK, but many of these proteins are already phosphorylated by cAMP-independent mechanisms and hence would not be readily detectable by analysis on polyacrylamide gels. To examine this hypothesis, we studied vimentin, a phosphoprotein in CHO cells (Cabral and Gottesman, 1979) which is the major component of intermediate filaments in these cells (Cabral et al., 1981). This protein, when labeled with ^{32}P and analyzed by two-dimensional gels, shows a single phosphorylated spot from cells or cell extracts treated or not treated with cAMP. If this spot is excised from the gel and subjected to two-dimensional tryptic peptide mapping, only one major phosphopeptide is found in the protein in the absence of cAMP treatment, whereas two phosphopeptides are seen in cells treated with cAMP, but are not found in one cADepPK mutant (10215) tested (Gottesman et al., 1981). This result indicates that vimentin, another cytoskeletal protein, is a substrate for cADepPK, and argues that many more cellular phosphoproteins may be substrates for cADepPK but that these can only be detected by peptide analysis of individual phosphoproteins.

Another approach to the examination of phosphorylation events in intact cells was taken by Steinberg and Coffino (1979) with S49 cells. They reasoned that phosphorylation of [^{35}S]methionine-labeled cell proteins should result in generation of slightly more acidic derivatives of cellular proteins which could be resolved on two-dimensional gels. A large number of proteins with the predicted charge shift after cAMP treatment were observed in S49 cells and are seen as closely spaced doublets consisting of phosphorylated and unphosphorylated proteins on two-dimensional gels. They also found that cAMP treatment altered synthesis of some proteins consistent with induction or repression by cAMP. Some of the putative phosphorylation events were confirmed by phosphate labeling and/or were shown not to occur in S49 cADepPK mutants (Steinberg and Coffino, 1979). Identification of these phosphoproteins is in progress, but at least one appears to be vimentin, one is a nascent form of actin (Steinberg, 1980), and one is a subunit of F_1-ATPase (Steinberg, 1984).

Although early hope of using cADepPK mutants for the rapid identification of physiologically significant phosphorylation events in cultured cells seem naive now, the mutants have certainly proved helpful in confirming the cAMP dependence of phosphorylation of some specific proteins.

C. Use of Mutants to Determine Mechanism of Drug Action

A great many drugs have been shown to elevate cAMP levels within cultured cells, but is has not been clear whether their pharmacologic effects are secondary to these elevations of cAMP with subsequent activation of cADepPK or whether the elevations in cAMP are secondary to the pharmacologic or toxic effects of the drugs. This question can be easily answered using cADepPK mutants, since such mutants should not respond to the drug in question if cADepPK is mediating the effect of the agent.

A variety of drugs have been analyzed by this approach in CHO cells. Sodium butyrate, which raises cAMP levels in CHO cells (Storrie et al., 1978), reduces cell growth and causes increased cellular adhesion to substrate, has the same effects on cADepPK mutants (Gottesman et al., 1980), and hence these effects cannot be mediated by cADepPK. Because treatment with sodium butyrate mimics effects of cAMP on CHO cells, we have avoided using dbcAMP as a selecting agent. 8-Br-AMP, an agent which may stimulate adenylate cyclase activity in some cell types (Martin and Kowalchyk, 1981), also inhibits cell growth and changes the morphology of CHO cells, but these effects also occur in the cADepPK mutants (Gottesman, unpublished data).

Two more physiologically interesting materials, interferon and the tumor promoter TPA, have also been studied in CHO cells. In the case of interferon, a survey of rodent and human interferons indicated that CHO cells were most sensitive to human β-interferon which suppressed cell growth, altered cell shape, and inhibited proliferation of encephalomyocarditis virus (EMCV), and, to a lesser extent, vesicular stomatitis virus (Banerjee et al., 1983). These effects also occurred in the cADepPK mutants which were studied, although to a somewhat reduced extent. Similar results have been reported for S49 mutants (Atwater and Samuel, 1982), but in mutant J774.2 macrophages, there appears to be significantly reduced growth inhibition by interferon (Schnek et al., 1982). One interesting sidelight of these studies was the finding that EMCV grows poorly in some of the cADepPK mutants, suggesting that a cAMP-dependent phosphorylation step is essential for normal growth and/or maturation of the virus.

The tumor promoter TPA stimulates activity of the enzyme ornithine decarboxylase in CHO cells, as does cAMP (see Section I C). This effect of TPA persists in cADepPK mutants, although to a somewhat reduced extent, whereas the cAMP-mediated effect does not (Lichti and Gottesman, 1982). This result argues that there are two independent pathways for stimulation of ornithine decarboxylase activity in CHO cells. The recent finding that TPA stimulates activity of the cAMP-independent protein kinase C, which is also the phorbol ester receptor (Castagna et al., 1982; Niedel et al., 1983), provides an intriguing biochemical mechanism for a second pathway of ornithine decarboxylase activation. As noted, in the case of both interferon and TPA, effects are blunted somewhat in cADepPK mutants, indicating that cAMP elevations may have a facilitating, but not essential role in promoting response to these agents.

IV. ISOLATION OF GENES AFFECTING cAMP-DEPENDENT PROTEIN KINASE ACTIVITY

A. Cloning of Regulatory and Catalytic Subunit Genes

A combination of exploitation of the genetic systems described in this chapter and more classical techniques of cDNA cloning (see Chapter 9) should allow the isolation of all of the genes involved in regulation of expression of cADepPK. The isolation of a cDNA probe encoding RI from bovine testis has been recently reported (Lee et al., 1983). Although not yet achieved, the same technology (cDNA cloning from a tissue with high levels of mRNA encoding cADepPK subunits) should allow the isolation of cDNAs encoding the C subunit. An alternate approach, involving use of gene transfer from mutants with dominant mutations affecting C, will be described below (Section IV B).

The RI subunit probe isolated by Lee et al. (1983) can be used to demonstrate that there is probably only one gene in CHO cells which encodes RI. Using a variety of restriction endonucleases to digest genomic DNA, and then probing Southern blots of agarose gels of this digested DNA, only one or two bands which hybridize with the RI probe are seen (Abraham, McKnight, and Gottesman, unpublished data). These data do not rule out the possibility of two closely linked genes encoding RI or a second gene which hybridizes poorly to the bovine probe. However, they suggest that only one RI gene is present in CHO cells, presumably with two active alleles, since mutant 10248 seems to have one normal RI subunit and one mutant subunit (see Section II D 1). The existence of mutants expressing both wild-type and altered RI in both S49 cells (Steinberg et al., 1977) and Y-1 cells (Doherty et al., 1982) is also strong evidence for at least two transcriptionally active alleles in these cell types as well. The presence of two active alleles for RI in CHO and S49 cells would explain why all the RI mutations isolated to date in these systems are dominant, since they must be expressed in the presence of a wild-type allele.

B. Gene Transfer of cAMP Resistance

As was pointed out in Section II C, the great majority of CHO cAMP-resistant mutants are dominant. This property has allowed us to transfer the phenotype of cAMP resistance by the technique of DNA-mediated gene transfer (see Chapter 8). To date, we have transferred resistance from one RI subunit mutant (10248), one C subunit mutant (10215), and several mutants with no demonstrable defect in either R or C (10226, 10987, and 11348) (Abraham and Gottesman, unpublished data).

The approach we have employed for gene transfer of cAMP resistance is outlined in Chapter 8, Section IV E, and involves two steps: (1) cotransfer of a cloned gene encoding resistance to the antibiotic G418 (pSV$_2$*neo*) with DNA encoding cAMP resistance, followed by selection for G418 resistance; and (2) selection of cAMP-resistant recipients among the G418-resistant transformants. This two-step approach has several advantages. First, it eliminates the background of cAMP-sensitive cells which survive the selective conditions by increasing the frequency of cAMP resistance in the second step to approximately 10^{-4}–10^{-3} as opposed to 10^{-7}–10^{-6} or less in a one-step transfer. Second, the cotransfer of cloned pSV$_2$*neo* sequences provides a molecular marker of successful gene transfer. Finally, these cotransferred sequences when present in high copy number, should be linked to the transferred genomic sequences, thereby providing one means for rescuing these sequences from DNA libraries made from the transferents.

V. FUTURE PROSPECTS

The use of genetic analysis of cultured somatic cells has already contributed significant information to our understanding of the role of cADepPK in the pleiotropic response to cAMP. Future genetic studies will be aimed at isolating mutants defective in specific responses to cAMP, with the goal of determining more of the physiologically significant substrates for this enzyme. Continued reversion analysis should generate useful information about systems which interact with cADepPK in intact cells. The cloning of cAMP responsive genes such as PEPCK should allow determination of the factors needed for cAMP induction of specific genes (i.e., the role of cADepPK in this induction) as well as the development of *in vitro* transcription systems to analyze which transcription factors are phosphorylated by cADepPK.

Within a relatively short period of time, R and C subunit genes will be isolated. The structure of these genes, whose expression appears to be closely regulated in mammalian cells, is of considerable interest. Genes which affect cADepPK activity, such as the *kin*C gene in S49 cells, and genes affected in several of our CHO mutants with apparently normal R and C subunits, will be isolated using techniques of gene transfer. These cloned, dominant genes can be introduced into a variety of differentiated cells to study the effect of inactivation of cADepPK on differentiated cell function. Finally, the introduction of these genes into mouse oocytes should result in cADepPK-deficient embryos, and perhaps an intact animal model for cAMP resistance.

ACKNOWLEDGMENTS

I would like to thank the many colleagues who have contributed to the studies on cAMP resistance in CHO cells, especially Irene Abraham,

Wayne Anderson, Dipak Banerjee, Fernando Cabral, Daniele Evain, Alphonse LeCam, Ulrike Lichti, Pierre Milhaud, Charles Roth, and Toolsee Singh. Special thanks go to Ira Pastan for advice, encouragement, and support.

REFERENCES

Atwater, J. A. and Samuel, C. E. (1982). *Virology* **123**, 206–211.

Baker, R. M., Brunette, D. M., Mankovitz, C. H., Thompson, L. H., Whitmore, G. F., Siminovitch, L., and Till, J. E. (1974). *Cell* **1**, 9–21.

Banerjee, D. K., Baksi, K., and Gottesman, M. M. (1983). *Virology* **129**, 230–238.

Beale, E. G., Hartley, J. L., and Granner, D. K. (1982). *J. Biol. Chem.* **257**, 2022–2028.

Beavo, J. A., Bechtel, P. J., and Krebs, E. G. (1975). *Adv. Cyclic Nucleotide Res.* **5**, 241–251.

Bishop, J. M. and Varmus, H. (1982). In *RNA Tumor Viruses* (R. Weiss, N. Teich, H. Varmus, and J. Coffin, eds.), Cold Spring Harbor Laboratory, Cold Spring Harbor, New York, pp. 999–1108.

Bloom, G. S. and Lockwood, A. H. (1980). *J. Supramol. Struct.* **14**, 241–250.

Bourne, H. R., Tomkins, G. M., and Dion, S. (1973). *Science* **181**, 952–954.

Bourne, H. R., Coffino, P., and Tomkins, G. M. (1975). *Science* **187**, 750–752.

Brothers, V. M., Walker, N., and Bourne, H. R. (1982). *J. Biol. Chem.* **257**, 9349–9355.

Builder, S. E., Beavo, J. A., and Krebs, E. G. (1980). *J. Biol. Chem.* **255**, 2350–2354.

Byus, C. V. and Fletcher, W. H. (1982). *J. Cell Biol.* **93**, 727–734.

Cabral, F. and Gottesman, M. M. (1979). *J. Biol. Chem.* **254**, 6203–6206.

Cabral, F., Abraham, I., and Gottesman, M. M. (1982). *Mol. Cell. Biol.* **2**, 720–729.

Cassel, D. and Selinger, Z. (1977). *Proc. Natl. Acad. Sci. USA* **74**, 3307–3311.

Castagna, M., Takai, Y., Kaibuchi, K., Sano, K., Kikkawa, U., and Nishizuka, Y. (1982). *J. Biol. Chem.* **257**, 7847–7851.

Chrapkiewicz, N. B., Beale, E. G., and Granner, D. K. (1982). *J. Biol. Chem.* **257**, 14428–14432.

Cohen, P. (1980). In *Recently Discovered Systems of Enzyme Regulation by Reversible Phosphorylation* (P. Cohen, ed.), Elsevier/North Holland Biomedical Press, Amsterdam, Vol. 1.

Cohen, P. (1982). *Nature* **296**, 613–620.

Collett, M. S., Erikson, E., and Erikson, R. L. (1979). *J. Virol.* **29**, 770–781.

Corbin, J. D. and Keely, S. L. (1977). *J. Biol. Chem.* **252**, 910–918.

Corbin, J. D., Keely, S. L., and Park, C. R. (1975). *J. Biol. Chem.* **250**, 216–225.

Corbin, J. D., Rannels, S. R., Flockhart, D. A., Robinson, A. M., and Atkins, P. D. (1981). In *Protein Phosphorylation* (O. M. Rosen and E. G. Krebs, eds.), Cold Spring Harbor Laboratory, Cold Spring Harbor, New York, Vol. 1, pp. 45–53.

Corbin, J. D. and Reimann, E. M. (1976). In *Methods in Enzymology* (J. G. Hardman and B. W. O'Malley, eds.), Academic Press, New York, Vol. 38, pp. 287–290.

Costa, M. (1978). *Biochem. Biophys. Res. Commun.* **81**, 832–840.

Daniel, V., Litwack, G., and Tomkins, G. (1973). *Proc. Natl. Acad. Sci. USA* **70**, 76–79.

D'Armiento, M., Johnson, G. S., and Pastan, I. (1972). *Proc. Natl. Acad. Sci. USA* **69**, 459–462.

Davis, R. L. and Kiger, J. A. (1978). *Biochem. Biophys. Res. Commun.* **81**, 1180–1186.

Derda, D., Miles, M., Schweppe, J., and Jungmann, R. (1980). *J. Biol. Chem.* **255**, 11112–11121.

Doherty, P. J., Tsao, J., Schimmer, B. P., Mumby, M. C., and Beavo, J. A. (1982). *J. Biol. Chem.* **257**, 5877–5883.

Eisinger, M. and Marko, O. (1982). *Proc. Natl. Acad. Sci. USA* **79**, 2018–2022.

Erlichman, J., Sarkar, D., Fleischer, N., and Rubin, C. S. (1980). *J. Biol. Chem.* **255**, 8179–8184.

Ernest, M. J. and Feigelson, P. (1978). *J. Biol. Chem.* **253**, 319–322.

Evain, D. and Anderson, W. B. (1979). *J. Cell. Physiol.* **99**, 153–158.

Evain, D., Gottesman, M., Pastan, I., and Anderson, W. B. (1979). *J. Biol. Chem.* **254**, 6931–6937.

Firestone, G. and Heath, E. (1981). *J. Biol. Chem.* **256**, 1396–1403.

Gabrielson, E. G., Scoggin, C. H., and Puck, T. T. (1982). *Exp. Cell Res.* **142**, 63–68.

Gilman, A. G. (1970). *Proc. Natl. Acad. Sci. USA* **67**, 305–312.

Gottesman, M. M. (1983). In *Methods in Enzymology* (J. D. Corbin and J. G. Hardman, eds.), Academic Press, New York, Vol. 99, pp. 197–206.

Gottesman, M. M., LeCam, A., Bukowski, M., and Pastan, I. (1980). *Somat. Cell Genet.* **6**, 45–61.

Gottesman, M. M., Roth, C., Leitschuh, M., Richert, N., and Pastan, I. (1983). In *Viruses and Differentiation* (E. Scolnick and A. Levine, eds.) Alan R. Liss, New York, pp. 365–380.

Gottesman, M. M., Singh, T., LeCam, A., Roth, C., Nicolas, J.-C., Cabral, F., and Pastan, I. (1981). In *Protein Phosphorylation* (O. M. Rosen and E. G. Krebs, eds.), Cold Spring Harbor Laboratory, Cold Spring Harbor, New York, Vol. 1, pp. 195–209.

Gottesman, M. M., Roth, C., Vlahakis, G., and Pastan, I. (1984). *Mol. Cell. Biol.* **4**, 2639–2642.

Green, H. (1978). *Cell* **15**, 801–811.

Hochman, J., Insel, P. A., Bourne, H. R., Coffino, P., and Tomkins, G. M. (1975). *Proc. Natl. Acad. Sci. USA* **72**, 5051–5055.

Holmgren, J. (1981). *Nature* **292**, 413–417.

Hsie, A. W., Jones, J., and Puck, T. T. (1971). *Proc. Natl. Acad. Sci. USA* **68**, 1648–1652.

Hsie, A. W. and Puck, T. T. (1971). *Proc. Natl. Acad. Sci. USA* **68**, 358–361.

Imada, S. and Imada, M. (1982). *J. Biol. Chem.* **257**, 9108–9113.

Insel, P. A., Bourne, H. R., Coffino, P., and Tomkins, G. M. (1975). *Science* **190**, 896–898.

Jha, K. K. and Ozer, H. L. (1976). *Somat. Cell Genet.* **2**, 215–223.

Johnson, G. S., Friedman, R. M., and Pastan, I. (1971). *Proc. Natl. Acad. Sci. USA* **68**, 425–429.

Johnson, G. S. and Mukku, V. R. (1979). *J. Biol. Chem.* **254**, 95–100.

Katada, T. and Ui, M. (1982). *Proc. Natl. Acad. Sci. USA* **79**, 3129–3133.

LeCam, A., Gottesman, M. M., and Pastan, I. (1980). *J. Biol. Chem.* **255**, 8103–8108.

LeCam, A., Gottesman, M. M., and Pastan, I. (1981). *Biochim. Biophys. Acta* **675**, 94–100.

Lee, D. C., Carmichael, D. F., Krebs, E. G., and McKnight, G. S. (1983). *Proc. Natl. Acad. Sci. USA* **80**, 3608–3612.

Lemaire, I. and Coffino, P. (1977). *Cell* **11**, 149–155.

Li, A. P., O'Neill, J. P., Kawashima, K., and Hsie, A. W. (1977). *Arch. Biochem. Biophys.* **182**, 181–187.

Lichti, U. and Gottesman, M. M. (1982). *J. Cell. Physiol.* **113**, 433–439.

Martin, T. and Kowalchyk, J. A. (1981). *Science* **213**, 1120–1122.

Matsumoto, K., Uno. I., and Ishikawa, T. (1983). *Cell* **32**, 417–423.

Matsumoto, K., Uno, I., Oshima, Y., and Ishikawa, T. (1982). *Proc. Natl. Acad. Sci. USA* **79**, 2355–2359.

Meek, W. D. (1982). *Mol. Cell. Biol.* **2**, 863–874.

Meloni, M., Perra, M., and Costa, M. (1980). *Exp. Cell Res.* **126**, 465–469.

Milhaud, P. G., Davies, P. J. A., Pastan, I., and Gottesman, M. M. (1980a). *Biochim. Biophys. Acta* **630**, 476–484.

Milhaud, P., Yamada, K. M., and Gottesman, M. M. (1980b). *J. Cell. Physiol.* **104**, 163–170.

Moss, J., Manganiello, V. C., and Vaughn, M. (1976). *Proc. Natl. Acad. Sci. USA* **73**, 4424–4427.

Murtaugh, M. P., Steiner, A. L., and Davies, P. J. A. (1982). *J. Cell Biol.* **95**, 64–72.

Niedel, J. E., Kuhn, L. J., and Vandenbark, G. R. (1983). *Proc. Natl. Acad. Sci. USA* **80**, 36–40.

Nielson, S. E. and Puck, T. T. (1980). *Proc. Natl. Acad. Sci. USA* **77**, 985–989.

Pastan, I. and Adhya, S. (1976). *Bacteriol. Rev.* **40**, 527–551.

Pawelek, J. M. (1979). *J. Cell. Physiol.* **98**, 619–626.

Pawelek, J., Halaban, R., and Christie, G. (1975). *Nature* **258**, 539–540.

Prashad, N., Rosenberg, R. N., Wischmeyer, B., Ulrich, C., and Sparkman, D. (1979). *Biochem.* **18**, 2717–2725.

Pruss, R. M. and Herschman, H. R. (1979). *J. Cell. Physiol.* **98**, 469–473.

Puck, T. T., Erickson, R. L., Meek, W. D., and Nielson, S. E. (1981). *J. Cell. Physiol.* **107**, 399–412.

Rae, P. A., Gutman, N. S., Tsao, J., and Schimmer, B. P. (1979). *Proc. Natl. Acad. Sci. USA* **76**, 1896–1900.

Rangel-Aldao, R., Kupiec, J. W., and Rosen, O. M. (1979). *J. Biol. Chem.* **254**, 2499–2508.

Robison, G. A., Butcher, R. W., and Sutherland, E. W. (1971). *Cyclic AMP*, Academic Press, New York.

Rosen, N., Piscitello, J., Schneck, J., Muschel, R. J., Bloom, B. R., and Rosen, O. M. (1979). *J. Cell. Physiol.* **98**, 125–136.

Rossier, C., Eitle, E., Van Driel, R., and Gerisch, G. (1980). In *The Eukaryotic Microbial Cell* (G. W. Gooday, D. Lloyd, and A. P. J. Trinci, eds.), Cambridge University Press, Cambridge, England, pp. 405–424.

Roth, C., Pastan, I., and Gottesman, M. M. (1982). *J. Cell. Physiol.* **111**, 42–48.

Roth, C. W., Richert, N. D., Pastan, I., and Gottesman, M. M. (1983). *J. Biol. Chem.* **258**, 10768–10773.

Rozengurt, E., Legg, A., Strang, G., and Courtenay-Luck, N. (1981). *Proc. Natl. Acad. Sci. USA* **78**, 4392–4396.

Russell, D. H. and Haddox, M. K. (1979). In *Advances in Enzyme Regulation* (G. Weber, ed.), Oxford Press, England, Vol. 17, pp. 61–87.

Schnek, J., Rager-Zisman, B., Rosen, O. M., and Bloom, B. R. (1982). *Proc. Natl. Acad. Sci. USA* **79**, 1879–1883.

Seamon, K. B., Padgett, W., and Daly, J. W. (1981). *Proc. Natl. Acad. Sci. USA* **78**, 3363–3367.

Simantov, R. and Sachs, L. (1975). *J. Biol. Chem.* **250**, 3236–3242.

Singh, T. J., Roth, C., Gottesman, M. M., and Pastan, I. (1981). *J. Biol. Chem.* **256**, 926–932.

Sisskin, E. E. and Weinstein, I. B. (1980). *J. Cell. Physiol.* **102**, 141–153.

Steinberg, R. A. (1980). *Proc. Natl. Acad. Sci. USA* **77**, 910–914.

Steinberg, R. A. (1983). *J. Cell Biol.* **97**, 1072–1080.

Steinberg, R. A. (1984). *J. Cell Biol.* **98**, 2174–2178.

Steinberg, R. A. and Agard, D. A. (1981a). *J. Biol. Chem.* **256**, 11356–11364.

Steinberg, R. A. and Agard, D. A. (1981b). *J. Biol. Chem.* **256**, 10731–10734.

Steinberg, R. and Coffino, P. (1979). *Cell* **18**, 719–733.

Steinberg, R. A., O'Farrell, P. H., Friedrich, U., and Coffino, P. (1977). *Cell* **10**, 381–391.

Steinberg, R. A., Steinberg, M. G., and van Daalen Wetters, T. (1979). *J. Cell. Physiol.* **100**, 579–588.

Steinberg, R. A., van Daalen Wetters, T., and Coffino, P. (1978). *Cell* **15**, 1351–1361.

Storrie, B., Puck, T. T., and Wenger, L. (1978). *J. Cell. Physiol.* **94**, 69–76.

Sutherland, E. W. (1971). In *Cyclic AMP,* (G. A. Robison, R. W. Butcher, and E. W. Sutherland, eds.), Academic Press, New York, p. 23.

Taylor-Papadimitrou, J., Purkis, P., and Fentiman, I. S. (1980). *J. Cell. Physiol.* **102**, 317–321.

Theurkauf, W. and Vallee, R. B. (1982). *J. Biol. Chem.* **257**, 3284–3290.

Thorner, J. (1982). *Cell* **30**, 5–6.

Uno, I., Matsumoto, K., and Ishikawa, T. (1982). *J. Biol. Chem.* **257**, 14110–14115.

Uno, I., Matsumoto, K., and Ishikawa, T. (1983). *J. Biol. Chem.* **258**, 3539–3542.

Van Daalen Wetters, T. and Coffino, P. (1982). *Mol. Cell. Biol.* **2**, 1229–1237.

Van Daalen Wetters, T. and Coffino, P. (1983). *Mol. Cell. Biol.* **3**, 250–256.

Van Daalen Wetters, T., Murtaugh, M. and Coffino, P. (1983). *Cell* **35**, 311–320.

Walter, U., Uno, I. Liu, A. Y.-C., and Greengard, P. (1977). *J. Biol. Chem.* **252**, 6494–6500.

Walter, U., Costa, M. R. C., Breakefield, X. O., and Greengard, P. (1979). *Proc. Natl. Acad. Sci. USA* **76**, 3251–3255.

Weiner, E. C. and Loewenstein, W. R. (1983). *Nature* **305**, 433–435.

Wells, J. N. and Hardman, J. G. (1977). *Adv. Cyclic Nucleotide Res.* **8**, 119–143.

Wicks, W. (1969). *J. Biol. Chem.* **244**, 3941–3950.

Wynshaw-Boris, A., Lugo, T. G., Short, J. M., Fournier, R. E. K., and Hanson, R. W. (1984). *J. Biol. Chem.* **259**, 12161–12169.

CHAPTER 24

LECTIN-RESISTANT GLYCOSYLATION MUTANTS

Pamela Stanley
Department of Cell Biology
Albert Einstein College of Medicine
Bronx, New York

I. INTRODUCTION

Lectins are proteins or glycoproteins of nonimmune origin that bind to specific carbohydrate configurations. Although lectins have been isolated from animal tissues, the best characterized molecules with lectin activity have been purified from the seeds of plants (Goldstein and Hayes, 1978). They bind to the surface of animal cells through their interactions with the carbohydrate moieties of membrane glycoproteins and glycolipids. Many of the lectins subsequently exert a cytotoxicity which makes them suitable as selective agents for isolating lectin-resistant (Lec^{R}) mutants. Such cell lines were initially sought from transformed cell populations in the hope of selecting revertants of transformation (Ozanne and Sambrook, 1971; Culp and Black, 1972). Subsequently, lectins were applied to CHO cells to select for stable "membrane" mutants of somatic cells (Wright, 1973). Over the last 10 years, Lec^{R} mutants have been isolated from a variety of cultured cell lines, although the largest range of different

Lec^R types has been obtained from CHO cells [reviewed by Baker and Ling (1978); Stanley (1980*a*); Wright et al. (1980); Briles (1982); Kerbel et al. (1982)].

The mechanism by which lectins exert their cytotoxicity is known only for ricin, abrin, and related lectins (Olsnes and Pihl, 1982*a*). These toxins consist of two subunits (A and B) joined by disulfide bonds. The B chain is the carbohydrate-binding subunit, while the A chain inhibits protein synthesis following an enzymic interaction with the 60S subunit of ribosomes. It seems likely that the majority of the other cytotoxic lectins cause cell death by disrupting membrane function in a nonspecific fashion as has been reported for the agglutinins from wheat germ (Greene et al., 1976). Whatever the molecular bases of lectin toxicity, it is apparent that binding to carbohydrate moieties at the cell surface is a required first step. Most lectin-resistant cells are "lectin-receptor" mutants. They express altered carbohydrates at the cell surface and, as a result, exhibit reduced binding of the selective lectin (Stanley and Carver, 1977).

At least 20 distinct phenotypes with lesions in glycosylation pathways have now been identified among Lec^R CHO cells (discussed in Section III). Only two Lec^R CHO mutant types have been reported to arise from alternative lesions: the ricin-resistant mutants described by Ray and Wu (1982), which are defective in ricin internalization, are thought to possess a lesion in endocytosis; and the mutant AR100-9 of Ono et al. (1982) which, in addition to a loss of ricin receptors, possesses an altered 60S ribosomal subunit accounting for its very high degree of ricin resistance. For the most part, therefore, the lectins are highly specific and varied reagents for selecting glycosylation mutants of animal cells. These mutants provide an incisive approach to defining glycosylation pathways as well as to investigating the diverse functional roles of carbohydrates. This chapter will describe the use of lectins in the selection and initial characterization of Lec^R CHO cells, the molecular bases of many of the mutations that lead to lectin resistance, and the functional consequences of expressing altered carbohydrates at the cell surface. The directions of future research involving Lec^R CHO cells are also discussed.

A. Carbohydrate-Binding Properties of Lectins

Although the binding of lectins to cells or glycosylated molecules can be inhibited by simple sugars (Goldstein and Hayes, 1978), it is now clear that the carbohydrate moieties with highest affinity for a particular lectin have strict structural requirements. For example, L-PHA, the leukoagglutinin from *Phaseolus vulgaris,* exhibits highest binding affinity for the galactose residue(s) in a carbohydrate which contains galactose bound to a β1,6-linked *N*-acetylglucosamine residue (Cummings and Kornfeld, 1982; Hammarstrom et al., 1982; Fig. 24.1). Galactose residue(s) in alternative configurations exhibit a markedly reduced affinity for L-PHA.

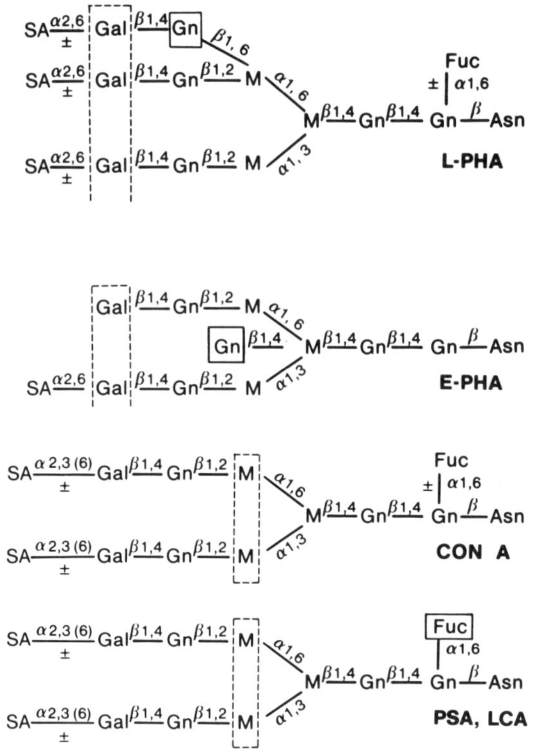

Figure 24.1. Carbohydrate recognition by lectins. The Asn-linked carbohydrate structures to which the indicated lectins bind with high affinity. Dotted-line boxes, sugar(s) with which the lectin interacts directly; solid-line boxes, sugars required for correct "presentation" of the sugars to which the lectin binds. Abbreviations: Asn = asparagine; Gn = *N*-acetylglucosamine; Fuc = fucose; M = mannose; Gal = galactose; SA = sialic acid; ± = may be present or absent. See text for lectin abbreviations.

The second structure shown in Figure 24.1 binds specifically to E-PHA—the erythroagglutinin purified from the same *Phaseolus vulgaris* seeds (Irimura et al., 1981; Cummings and Kornfeld, 1982). The galactose residue(s) are again key to specific recognition, but, in this case, they must be presented on a carbohydrate which contains the "bisecting" *N*-acetylglucosamine residue (the *N*-acetylglucosamine in β1,4 linkage to the β-linked mannose residue).

The third carbohydrate in Figure 24.1 binds with high affinity to CON A—the lectin from *Canavalia ensiformis.* The α-linked mannose residues are important since substitution of these residues at a position other than C2 abolishes CON A binding activity (Ogata et al., 1975). Carbohydrates containing the "bisecting" *N*-acetylglucosamine also exhibit a markedly reduced affinity for CON A (Baenziger and Fiete, 1979; Narasimhan, 1982). However, the absence of the α1,6-linked fucose in the core region does not appear to affect binding to CON A. This is in striking contrast to the binding specificity of the closely related lectins PSA (from *Pisum sativum*) and LCA (from *Lens culinaris*), which exhibit an absolute requirement for the presence of the α1,6-linked fucose residue in complex biantennary structures (Kornfeld et al., 1981). In addition, both PSA and LCA are able to bind at least one "branched" carbohydrate structure—the fucosylated, triantennary structure in Figure 24.1 which is recognized by L-PHA (Cummings and Kornfeld, 1982; Stanley et al., 1984). This branched moiety is not able to bind to CON A (Cummings and Kornfeld, 1982; Stanley et al., 1984).

The carbohydrate recognition properties of the lectins summarized in Figure 24.1 lead to an important conclusion: the overall *conformation* of a carbohydrate structure is often crucial to the ability of any one of its constituent sugars to be recognized by a lectin. Evidence that carbohydrates of the type shown in Figure 24.1 assume unique conformations in solution has been obtained by ^{1}H-NMR spectroscopy from Nuclear Overhauser Enhancement measurements (Brisson and Carver, 1983). It is apparent that the conformation of a carbohydrate moiety may be dramatically changed by the addition or deletion of a single sugar. Therefore, the ability of a cell to bind certain lectins may depend on its ability to synthesize particular carbohydrate structures. In practice, extremely complex lectin binding curves are obtained for CHO cells [see Stanley et al., (1980)] indicating the existence of numerous classes of binding sites presumably composed of related carbohydrate structures possessing a range of lectin-binding affinities.

B. Carbohydrates at the CHO Cell Surface

The carbohydrates of CHO cells have been partially characterized, although few carbohydrate structures have actually been determined (Li and Kornfeld, 1978, 1979, Li et al., 1980*a*). It is clear, however, that CHO cells synthesize all the major classes of carbohydrates which have been found associated with glycoproteins, glycolipids, and proteoglycans.

Examples of the variety of carbohydrates found *N*-glycosidically linked to asparagine (Asn) residues of glycoproteins are shown in Figure 24.2. The first step in the synthesis of these carbohydrates is

$$\text{Dol-P} + \text{UDP-GlcNAc} \rightleftharpoons \text{Dol-P-P-GlcNAc} + \text{UMP}$$

The transferase that catalyzes this reaction is inhibited by the antibiotic tunicamycin which thereby prohibits the synthesis of all Asn-linked structures (Tkacz and Lampen, 1975). Sequential sugar additions to Dol-P-P-GlcNAc to form the mature lipid-linked oligosaccharide (Glc_3Man_9-Gn_2-P-P-Dol) involve UDP-GlcNAc, GDP-Man, Dol-P-Man, and Dol-P-

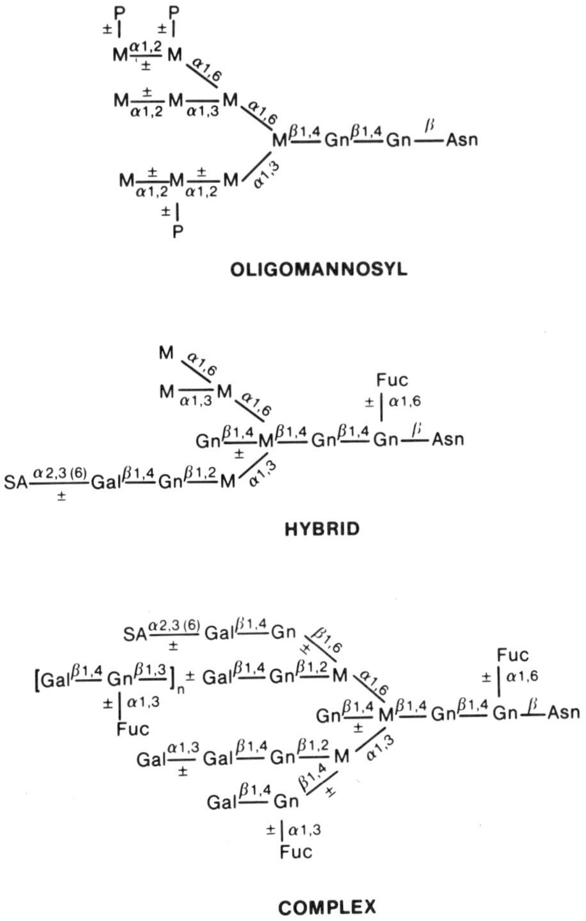

Figure 24.2. Carbohydrates at the cell surface. Structures representative of the range of Asn-linked carbohydrates reported in the literature showing the potential structural variation in the Asn-linked carbohydrates at the CHO cell surface. The ± signs indicate linkages at which heterogeneity has been observed. Only a few of these possibilities have actually been documented for CHO cells (Li and Kornfeld, 1978, 1979; Li et al., 1980). The overall complexity of carbohydrate structures at the cell surface is actually much greater since Ser/Thr-linked carbohydrates of glycoproteins, as well as glycolipids and proteoglycans, are present. P = phosphate.

Glc intermediates (Spiro and Spiro, 1982). It is possible that a unique transferase catalyzes the addition of each sugar. After transfer of the completed oligosaccharide to protein, a family of glycosidases removes (or "processes") the glucose residues and the four α1,2-linked outer mannose residues. The intermediate Man_5Gn_2-Asn is an acceptor for GlcNAc which is added from UDP-GlcNAc by the enzyme termed

GlcNAc-TI (Narasimhan et al., 1977). The addition of this GlcNAc residue provides a substrate for α-mannosidase II which removes the terminal Man residues (Tabas and Kornfeld, 1978). The subsequent action of a variety of specific glycosyltransferases results in the transfer (from nucleotide sugars) of GlcNAc, Gal, Fuc, and sialic acid to form mature complex structures (Schachter, 1981; see Fig. 24.2). Failure of α-mannosidase II to act is thought to give rise to hybrid structures, whereas incomplete processing is thought to give rise to oligomannosyl structures (Fig. 24.2). The heterogeneity observed among the Asn-linked carbohydrates of glycoproteins reflects the presence (or availability) of glycosyltransferase and "processing" enzymes as well as the actual position of each glycosylation site in the glycoprotein. Although individual conformational constraints are presumably important, a general correlation between the proximity of a glycosylation site to the *N*-terminus and the extent of processing has been noted (Pollack and Atkinson, 1983).

Structures of the oligomannosyl type containing between five and nine mannose residues (Fig. 24.2) have been found at the CHO cell surface (Li and Kornfeld, 1979). It is likely that these moieties contain phosphate residues attached to certain mannoses (Varki and Kornfeld, 1983). In addition, CHO cells synthesize complex moieties which vary in their branching patterns and their terminal sugars (Li and Kornfeld, 1978, 1979; Hunt, 1982; Stanley, 1982). A significant proportion of the complex moieties in CHO cells carry polylactosamine sequences [(Galβ1,4GlcNAcβ1,3)$_n$; Li et al. (1980a); Carver et al., unpublished observations]. The existence of hybrid structures in CHO cells has not so far been demonstrated.

Because of the number of different sugars and linkages involved, many different carbohydrate configurations are feasible. To what extent this variety is realized at the CHO cell surface is not known. Recently, novel sugar residues, which are not expressed by wild-type CHO cells, have been found associated with the glycoproteins synthesized by three dominant Lec^R CHO mutants (Campbell and Stanley, 1983; and unpublished observations). These mutants appear to have sustained regulatory mutations which have induced the expression of glycosyltransferases not expressed in parental CHO cells (see Section III). Therefore, the complete range of structures which CHO cells are *capable* of synthesizing cannot be deduced solely from a knowledge of the carbohydrates synthesized by wild-type cells.

In addition to Asn-linked carbohydrates, cellular glycoproteins contain sugars linked *O*-glycosidically to serine or threonine (Ser/Thr) residues. Although these carbohydrates tend to be more simple, linear sugar sequences, they may also occur as branched moieties with a high degree of structural complexity (Kornfeld and Kornfeld, 1980). The same is true for the carbohydrates attached to ceramide to form glycolipids (Schachter and Roseman, 1980). In CHO cells, the major glycolipid is the simple ganglioside GM_3 [$SA^{\alpha2,3}Gal^{\beta1,4}$Glc-ceramide (Yogeeswaran et al., 1974; Briles et al., 1977; Stanley et al., 1980)]. However more complex carbohydrates might be present among the minor glycolipids.

The third class of sugar-containing molecules is the proteoglycans. CHO cells express chondroitin sulfate, heparan sulfate, and hyaluronic acid at the cell surface (Kraemer, 1971; Atherly, et al., 1977; Kraemer and Barnhart, 1978; Barnhart et al., 1979). Like Ser/Thr-linked carbohydrates, glycolipids and the terminal regions of Asn-linked carbohydrates, most of the proteoglycan carbohydrates are synthesized by direct sequential addition of sugars catalyzed by specific glycosyltransferases. One exception is keratan sulfate which contains a population of carbohydrates linked to Asn that are synthesized on dolichol intermediates in the same manner as the complex moieties shown in Figure 24.2 (Roden, 1980).

II. SELECTION AND PHENOTYPIC CHARACTERIZATION OF Lec^R CHO CELLS

As mentioned previously, many lectins are cytotoxic and may be used directly as selective agents. In addition, lectins which bind to the cell surface but do not kill cells at a reasonable concentration ($\leq$100 μg/mL) may be rendered toxic by conjugation to a molecule such as the A chain of ricin (Olsnes and Pihl, 1982b). Most lectins may, therefore, be used as direct cytotoxins. Detailed methods for the selection of Lec^R mutants have been described by Baker et al. (1982) and Stanley (1983a).

Lec^R mutants arise spontaneously in CHO populations at frequencies of approximately 10^{-7}–10^{-5} and often at much higher frequencies when the cells are pretreated with a mutagen (Stanley et al., 1975a,b; Stanley, 1981). For the selection of Lec^R CHO cells, lectin-containing medium and cells are incubated together in culture dishes in a CO_2 incubator until resistant colonies appear (about 8–10 days). The minimum dose of lectin used in selection should be twice the concentration that reduces survival of parental cells to 10% (termed the D_{10} value). The number of cells exposed per dish should be high enough to include rare mutants but low enough so that chance survival due to cell–cell cooperation (cell-density effects) is minimized. The colonies which occur following exposure to a single lectin are a heterogeneous mixture of different phenotypes resulting from mutations in different genes. Survivors include mutants resistant to various levels of the selective lectin (i.e., any concentration equal to or higher than that used in selection). For example, at least nine distinct CHO phenotypes resistant to different concentrations of WGA (the agglutinins from *Triticum vulgaris*) have been identified (Stanley, 1983a,b).

The number of different mutants in a lectin selection is greatly influenced by the frequency at which particular mutations occur spontaneously or may be induced by a given mutagen. In a random selection of survivors, only the most common mutations will be obtained. In order to identify rare mutants or to reisolate known mutant phenotypes at will, more specific selection protocols are required. Such protocols rely on the unique property of most Lec^R cells to exhibit a pleiotropic lectin-resistance phenotype. This characteristic is also key to the rapid and accurate phenotypic classification of Lec^R isolates.

The pleiotropic nature of the Lec^R phenotype is manifested by an increased resistance and/or hypersensitivity to lectins not present during the selection (Table 24.1). This property is readily understood when the carbohydrate structural changes expressed by different Lec^R mutants are known (see Section III). In practical terms, it is a very useful characteristic which may be exploited in designing specific selection protocols for obtaining particular Lec^R phenotypes or revertants, in defining the phenotype of new isolates and in determining the completeness of dominance in the dominant Lec^R mutants.

All the mutant types in Table 24.1 are resistant to WGA, yet each one exhibits a distinct Lec^R phenotype on a qualitative and/or a quantitative basis. Since the reproducibility of D_{10} values is ±30% (Stanley and Carver, 1977), it is clear that a comparison of sensitivities to a few key lectins allows rapid phenotypic assignment of new WGA-resistant isolates. In addition, by choosing appropriate concentrations of selective lectins, it is possible to select for particular phenotypes (e.g., new mutants) and against unwanted phenotypes. For example, a selection using 20 μg/mL WGA which is replaced after 6 days by 7.5 μg/mL CON A is highly specific for the Lec8 phenotype (Stanley, 1981). The high concentration of WGA selects against Lec2 and Lec3 phenotypes, while CON A selects against the Lec1 phenotype. Both LEC11 and LEC12 mutants are rare (Stanley, 1983b) and would be expected to represent at most a small fraction of the survivors.

A. Nomenclature and Genetic Characterization of Lec^R CHO Cells

The phenotypic complexity of different Lec^R mutants has led to the evolution of a complicated nomenclature. Many mutants were originally isolated from different selections and therefore given different names, although they were subsequently found to express identical phenotypes and to fall into the same complementation groups. To overcome this problem among CHO isolates, we have developed a nomenclature in which CHO mutants with altered lectin-resistance or lectin-binding abilities are designated Lec (recessive) or LEC (dominant), and each distinct phenotype is given a number (Stanley, 1983a). Lec^R lines which are not glycosylation defective (e.g., endocytosis mutants) would be included in this definition.

The different genes represented by the recessive phenotypes are deter-

TABLE 24.1
Lectin-Resistance Properties of Lec^R CHO Cells[a]

Mutant Type	Lec^R Phenotype				
	L-PHA (3)	WGA (2)	CON A (18)	RIC (0.005)	LCA (18)
Lec1	R >1000	R 30	S 6	R 100	R >200
Lec2	(S)	R 11	—	S 100	S 2
Lec3	(S)	R 5	—	S 10	S 2
Lec8	R 10	R 100	(S)	(R)	S 10
LEC11	R 4	R 8	—	S 25	R 3
LEC12	R 3	R 50	—	S 4	R 2

[a]The lectin-resistance properties of six WGA-resistant CHO mutants [see Stanley (1983a,b)]. The D_{10} values (μg/mL) for parental CHO cells for each lectin are shown in brackets. The values for the mutants are given as fold-resistance (R) or fold-sensitivity (S) compared with parental cells. Differences less than two-fold are denoted by parentheses. No difference from wild type is denoted by —.

mined by complementation analysis. This is accomplished by fusing two Lec^R mutants which are resistant to the same lectin but which carry the complementary auxotrophic markers Pro$^-$ (requires proline for growth) or Gat$^-$ (requires glycine, adenosine, and thymidine for growth). Viable hybrids are selected in deficient medium which will not support the growth of either of the original fusion partners. The hybrids are subsequently tested for their lectin resistance. If they are resistant, the mutants used in their formation are assigned to the same complementation group. If, on the other hand, the lectin resistance of the hybrids is similar to wild type, the mutants are assigned to different complementation groups. In this manner, 11 complementation groups have been defined for the recessive Lec^R CHO mutants (Stanley and Siminovitch, 1977; Stanley 1981, 1983a, and unpublished observations; Ripka and Stanley, unpublished observations) and 5 dominant mutants have been identified (Stanley, 1983a,b; Ripka and Stanley, unpublished observations). Based on numerous phenotypic differences among the latter, it seems likely that each of the dominant mutant types represents a mutation in a different gene. Four other mutants (Lec1A, Lec2A, Lec2B, and

Lec13A), which fall into established recessive complementation groups by genetic analysis but which, nevertheless, exhibit unique Lec^R or lectin-binding properties, have been given alphabetical designations within their respective complementation groups. Such phenotypes might reflect the existence of different mutations in the same gene affected in other members of the complementation group.

Although the genetic characterization of Lec^R CHO mutants cannot be carried far without extensive mapping studies, it is clearly important to clone isolates rigorously (because of the problem of different mutants coexisting on selection plates), to determine their chromosome content (because of the risk of isolating pseudotetraploids), and to select each new phenotype idependently from different parental populations (so that phenotypes and genotypes may be precisely correlated). Following the establishment of a heritable Lec^R phenotype which is stable in the absence of selection, the dominance characteristics of the phenotype should be determined in somatic cell hybrids. If recessive, the new mutant should be subjected to complementation analysis with all previously identified, recessive mutants of related phenotype.

B. Alternative Selections Which Give Rise to Glycosylation Mutants

Since carbohydrate biosynthesis requires sugars, nucleotides, proteins, lipids, and cofactors, perturbation of almost any major metabolic pathway might, in theory, disrupt glycosylation patterns. For example, the temperature-sensitive mutant of Chinese hamster lung cells, which was isolated from a nonspecific cell-cycle suicide protocol, was subsequently found to exhibit a defect in carbohydrate biosynthesis (Tenner et al., 1977; Tenner and Scheffler, 1979). Another selection not originally intended for the isolation of glycosylation mutants was for Thy 1^- mouse lymphoma cell lines using cytotoxic antibody to Thy 1 and complement (Hyman, 1973). The molecular basis of the Thy 1 deficiency in class E mutants was traced to a defect in glycosylation (Chapman et al., 1979, 1980). The altered carbohydrates associated with the Thy 1 molecule in this class of mutants increase its susceptibility to intracellular proteolysis (Trowbridge et al., 1978b). Thus few molecules reach the cell surface, resulting in the Thy 1^- phenotype. Thy 1^- mutants of this type were subsequently found to be resistant to and selectable by CON A (Trowbridge et al., 1978*a*). However, not all Lec^R lymphoma cell lines exhibit the Thy 1^- phenotype even though they possess specific glycosylation defects (Trowbridge et al., 1978a). Therefore, protocols designed to select against any antigenic determinant, enzyme, or other glycosylated molecule which relies on specific carbohydrate structures for correct intracellular compartmentation or for ultimate function may give rise to certain types of glycosylation mutants.

In addition to the chance isolation of glycosylation mutants, alternative selections aimed at obtaining different classes of glycosylation mutants have been applied to CHO populations. A number of groups have selected cell lines resistant to tunicamycin (Sudo and Onodera, 1979; Kuwano et al., 1981; Criscuolo and Krag, 1982). In one case the mechanism of resistance appears to be increased synthesis ("overproduction") of the enzyme UDP-*N*-acetylglucosamine:dolichol phosphate *N*-acetylglucosamine-1-phosphate transferase (Criscuolo and Krag, 1982), while, in another case, a structural mutation in the transferase is indicated (Kuwano et al., 1981). Suicide selections using radiolabeled sugars have also been successful in obtaining glycosylation mutants (Gottesman, 1979; Hirschberg et al., 1981), as have replica plating screens for isolating colonies with altered carbohydrate metabolism (Hirschberg et al., 1982). In applying both the latter approaches for the selection of CHO cells defective in fucose incorporation, Hirschberg et al. (1981, 1982) consistently obtained the Lec1 phenotype. Considering the metabolic ramifications of the GlcNAc-TI enzyme defect expressed by Lec1 cells (see Section III), it is not surprising that this mutation has been selected by a variety of different protocols. The other mutant type identified by Hirschberg et al. (1981) as Fuc^-, expressed a Lec^R phenotype very similar to that of Lec1A CHO mutants (Stanley, 1983b).

These findings reveal the importance of determining the lectin-resistance properties of any mutant which might conceivably arise from a glycosylation defect. An altered Lec^R phenotype is diagnostic of all the glycosylation mutants examined to date and can be extremely useful in classification. Although different cell lines may exhibit some variation in lectin-resistance properties because of their particular spectrum of glycosyltransferase enzymes, qualitative phenotypic relationships are usually consistent among mutants that express the same glycosyltransferase lesion. For example, the BHK mutant Ric^R14 and Lec1 CHO cells have both lost the *N*-acetylglucosaminyltransferase GlcNAc-TI and exhibit qualitatively similar Lec^R phenotypes (Stanley et al., 1975c; Meager et al., 1976). Lec4 CHO cells and the mouse lymphoma mutant $Pha^R2.1$ are highly resistant to L-PHA and both mutants appear to have lost the ability to synthesize the carbohydrate moiety recognized by L-PHA (see Fig. 24.2; Cummings et al., 1982; Stanley et al., 1984). Increased synthesis of specific $\alpha 1,3$ fucosyltransferase activities correlates with resistance to WGA in both the CHO mutants LEC11 and LEC12 (Campbell and Stanley, 1983) and the mutant Wa-4 isolated from B16 melanoma cells (Finne et al., 1980). However, these correlations do not always exist. The mouse lymphoma mutant Thy 1E is resistant to CON A (Trowbridge et al., 1978*a*), whereas the CHO mutant B4-2-1 (or Lec15) which exhibits an apparently identical loss of Dol-P-Man synthetase (Stoll et al., 1982) is not resistant to CON A but is three- to five-fold resistant to ricin (Robbins et al., 1981; Stanley, unpublished observations).

III. BIOCHEMICAL BASES OF LECTIN RESISTANCE

Except for the mutant defective in internalization of ricin (Ray and Wu, 1982) and the double mutant which possesses a ricin-resistant 60S ribosomal subunit (Ono et al., 1982), the biochemical bases of lectin-resistance in CHO cells have been correlated with specific alterations in glycosylation reactions (see Table 24.2). The definition of structural carbohydrate changes and the enzymic lesions that give rise to them therefore require a detailed analysis of carbohydrate biosynthesis in mutant compared with parental cells.

A. Determination of Structural Lesions

As mentioned in Section I B, the glycopeptides derived from CHO cells include a broad range of carbohydrate structures. The identification of a particular alteration might therefore pose a significant problem in purification. This difficulty may be alleviated to a large extent by investigating the simplified subset of carbohydrates associated with a viral glycoprotein. Viral infection inhibits the synthesis of cellular macromolecules so that radiolabeled precursors (added a few hours postinfection) are incorporated almost exclusively into proteins specified by the viral genome. Most animal viruses do not encode information for glycosylation enzymes. In the small enveloped viruses, the number of viral glycoproteins is one or two, so that the carbohydrates expressed on the virus are few compared with the array found at the cell surface. Provided a glycosylation defect is reflected in viral carbohydrates, the use of the virus to enrich for altered structures is an enormous advantage in the initial localization of a structural change and in subsequent purification for direct structural studies.

The virus most commonly used for studying glycosylation mutations is vesicular stomatitis virus (VSV), which encodes a single glycoprotein termed G. The amino acid sequence of G is known (Rose and Gallione, 1981). The Asn-linked glycosylation sites at residues 178 and 335 carry complex carbohydrates, although the range of structures observed and their relative proportions depend on the host cell (Etchison and Holland, 1974: Etchison et al., 1981). The major carbohydrate from VSV grown in Pro^-5 CHO cells has been shown by ^{1}H-NMR spectroscopy to be a biantennary, complex structure with heterogeneity in terminal α2,3-linked sialic acid and α1,6-linked fucose residues (Stanley et al., 1984). About one-fifth of the total carbohydrates are branched structures. One of the latter is a triantennary moiety very similar to the structure at the top of Figure 24.1. This triantennary carbohydrate is not synthesized by Lec4 CHO mutants as shown from studies of G glycopeptides

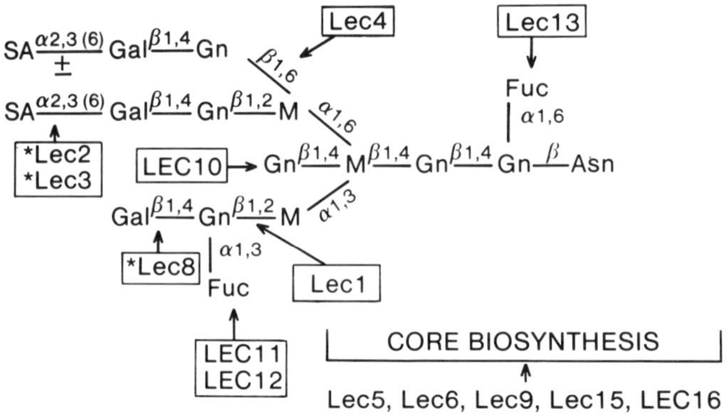

Figure 24.3. CHO glycosylation mutants. The site on Asn-linked carbohydrates of the structural lesions expressed by each of the Lec^R glycosylation mutants described in Table 24.2. The asterisks indicate mutants in which glycoproteins and glycolipids are affected.

of VSV grown in parental CHO compared with Lec4 mutants (Stanley, 1982; Stanley et al., 1984).

The structural defects expressed by many of the CHO Lec^R mutant types have been successfully identified by this approach (Schlesinger et al., 1976; Robertson et al., 1978; Tabas and Kornfeld, 1978; Hunt, 1980a,b; Stanley and Sudo, 1981; Stanley, 1982; Campbell and Stanley, 1983; Davidson and Hunt, 1983). However, altered structures associated with VSV cannot be quantitatively extrapolated to the cell surface and, in certain cases, may conceivably be qualitatively different. In addition, the lesion expressed at the cell surface might affect more than one class of glycosylated molecule. For example, the CHO mutants Lec2 and Lec3 are sialylation mutants which synthesize glycoproteins and glycolipids with reduced sialic acid (Stanley et al., 1980). It is likely that their sialylation defects extend to Ser/Thr-linked moieties. To determine the full effects of a glycosylation defect, therefore, cell surface carbohydrates must be examined directly. In any event, the conclusions deduced from viral glycopeptides should be corroborated wherever possible by showing that the same altered structure exists among cellular carbohydrates.

B. Molecular Bases of Glycosylation Mutations

Most CHO Lec^R mutants have been partially characterized at the biochemical level and, in many cases, an enzymic lesion which appears to provide the basis of the glycosylation defect has been identified (Table 24.2 and Fig. 24.3). Independently isolated clones with phenotypic identity have been described. However, no mutations have been characterized at a molecular level. Therefore, a direct link between a mutated gene and any particular Lec^R phenotype has not been established at this time.

TABLE 24.2
Glycosylation Defects of Lec^R CHO Cells[a]

Mutant Type	Surface Molecules Affected: GPs	GLs	Carbohydrate Synthetic Lesion	Enzymic Basis of Carbohydrate Alteration
Lec1	+	–	Blocked in "processing" Asn-linked carbohydrates at Man_5Gn_2Asn intermediate	G1cNAc-T1 (lost)
Lec2	+	+	Deficient in addition of sialic acid. Cannot translocate CMP-SA to Golgi.	?
Lec3	+	+	Deficient in addition of sialic acid. Similar to Lec2.	?
Lec4	+	–	Blocked in synthesis Asn-linked moieties containing a β1,6-linked Gn "branching" residue	? β1,6 G1cNAc-T (lost)
Lec5	+	(–)[b]	Reduced glucosylation of Man_9-P-P-Dol precursor	?
Lec1.Lec6	+	(–)[b]	Synthesizes shortened oligosaccharide-dolichol Man_7Gn_2P-P-Dol instead of Man_9Gn_2-P-P-Dol. Complex carbohydrates are blocked at Man_4Gn_2Asn	G1cNac-T1 (lost) + ?
Lec8	+	+	Deficient in addition sialic acid and galactose. Numerous changes in glycosyltransferase and glycosidase activities	?
Lec9	+	(–)[b]	Defective in synthesis of $G1c_3Man_9Gn_2$-P-P-Dol	?

TABLE 24.2 (*continued*)

Mutant Type	Surface Molecules Affected		Carbohydrate Synthetic Lesion	Enzymic Basis of Carbohydrate Alteration
	GPs	GLs		
LEC10	+	–	Synthesizes carbohydrates containing the "bisecting" GlcNAc	GlcNAc-TIII (acquired)
LEC11	+	?	Synthesizes carbohydrates containing Fuc linked α1,3- to β1,2-linked GlcNAc residues. Generates the mouse embryonic antigen SSEA-1	α1,3 fucosyltransferase I (acquired)
LEC12	+	?	As above but properties of α1,3 fucosyltransferase differ markedly from that expressed by LEC11 cells	α1,3 fucosyltransferase II (acquired)
Lec13	+	$(-)^b$	Synthesizes carbohydrates which lack α1,6-linked fucose residues. Cannot convert GDP-Man to GDP-Fuc in an *in vitro* assay	?
Lec15	+	$(-)^b$	Synthesizes $Glc_3Man_5Gn_2$-P-P-Dol instead of the $Glc_3 = Man_9Gn_2$P-P-Dol intermediate	Dol-P-Man Synthetase (lost)
LEC16	+	$(-)^b$	Affected in synthesis of $Glc_3Man_9Gn_2$P-P-Dol	?

[a]The glycosylation defects of the Lec^R CHO mutants are summarized. Biochemical phenotypes associated with Lec7, LEC14, Lec1A, Lec2A, Lec13A, or Lec2B are not known at this time. A discussion of each mutant type with appropriate references is given in the text.

[b]Although not yet tested, the nature of the defects in these mutants suggests that they should not exhibit altered synthesis of glycolipids.

Each recessive mutant (Lec) has been shown to complement all other mutants with related Lec^R phenotypes and the dominant mutants (LEC) have been shown to exhibit distinct phenotypes (Stanley and Siminovitch, 1977; Stanley, 1981, 1983a,b; Ripka and Stanley, unpublished observations). Most of the Lec^R CHO mutants isolated independently by other laboratories fall into the groups defined in Table 24.2 as determined directly by complementation analyses (Stanley, unpublished observations) or indirectly by phenotypic comparisons from published reports.

Many laboratories have isolated the Lec1 phenotype from CHO populations (Gottlieb et al.; 1974, Stanley et al., 1975a,b; Stanley, 1981; Hirshberg et al., 1981, 1982; Ray and Wu, 1982; Ono et al., 1982, C. A. Jones, unpublished observations). This is presumably because it is resistant to a variety of different lectins (Table 24.1) and also because it is a frequently occurring mutation among CHO cells (Stanley, 1981). Although this phenotype has been isolated from BHK 21 cells (Meager et al., 1975, 1976), it was not observed among the BW5147 Lec^R mutants described by Trowbridge et al. (1978a). This may reflect the variable hemizygosity of this locus or its relative susceptibility to mutation in different cultured cell lines. Lec1 mutants lack GlcNAc-Tl transferase activity (Gottlieb et al., 1975; Stanley et al., 1975c). The synthesis of Asn-linked carbohydrates is consequently blocked at the oligomannosyl intermediate Man_5Gn_2Asn (Robertson et al., 1978; Tabas and Kornfeld 1978). This carbohydrate change profoundly affects the charge, size, and composition of the Asn-linked carbohydrates associated with the membrane glycoproteins of Lec1 cells (Juliano and Stanley, 1975; Li and Kornfeld, 1978; Stanley et al., 1980).

Mutants with the Lec2 phenotype have also been isolated independently (Stanley et al., 1975b; Briles, et al., 1977, Stanley, 1981). Complementation analyses have shown that clone 1021 of Briles et al. (1977) belongs to complementation group 2 (Deutscher et al., 1984). Lec2 mutants exhibit a broad sialylation defect which arises from defective Golgi compartmentation of CMP-sialic acid required for the addition of sialic acid to glycoproteins and glycolipids (Briles et al., 1977; Stanley et al., 1980).

Mutants belonging to complementation group 3 also exhibit a sialylation defect affecting glycolipids and glycoproteins (Stanley et al., 1980). This mutation is comparatively rare compared with Lec1 and Lec2 mutations (Stanley, 1981). However, Lec3 mutants have been isolated independently from Pro$^-$5 and Gat$^-$2 CHO cells (Stanley, 1981) and from CHO-Kl cells (Stanley and Jones, unpublished observations). The molecular basis of the sialylation defect in Lec3 mutants is not known at this time.

Lec4 mutants are rare among CHO cells, requiring prior mutagenesis and elimination of Lec1 mutants for their deliberate isolation (Stanley and Sudo, 1981). These mutants do not make triantennary Asn-linked moieties containing a β1,6-linked GlcNAc "branch" [i.e., the structure shown in Fig. 24.1 to which L-PHA binds (Stanley, 1982; Stanley et al., 1984)]. The biochemical basis of this defect is probably a deficiency in the β1,6-GlcNAc-transferase enzyme since the mutant $Pha^R2.1$ from BW5147 cells exhibits a similar carbohydrate structural lesion and lacks this particular glycosyltransferase activity (Cummings et al., 1982).

Mutants with the phenotype of Lec5 cells were originally isolated by R. M. Baker (Cifone and Baker, 1976). They appear to have been independently isolated by Wright (1975) who describes mutants with a similar phenotype which probably fall into the same complementation group. Lec5 mutants are temperature sensitive for growth, and the properties of many of the "temperature" revertants (selected for the ability to grow at 39.5°C) suggest that the temperature sensitivity may be directly associated with the Lec^R phenotype (Cifone et al., 1979). The biochemical phenotype expressed by Lec5 mutants is complicated, but seems to stem from a defect in glucosylation of the oligomannosyl dolichol intermediate Man_9Gn_2-P-P-Dol (Krag, 1979; Wright et al., 1979; Krag and Robbins, 1982).

The Lec6 mutation has not so far been identified as a single mutation on a wild-type background. It was isolated as a second mutation in a Lec1 mutant (Stanley et al., 1975b). Lec1.Lec6 cells synthesize a truncated dolichol precursor (Man_7-Gn_2-P-P-Dol), which is found with and without associated glucose and appears to be the major oligosaccharide moiety transferred to protein in these cells (Hunt, 1980a,b). The mature carbohydrate synthesized by Lec1.Lec6 mutants is Man_4Gn_2Asn (Robertson et al., 1978; Hunt, 1982). This moiety lacks the outer α1,6-linked mannose residue of the Man_5Gn_2Asn synthesized by Lec1 CHO cells (Etchison and Summers, 1979) and is not bound by CON A Sepharose (Hunt, 1982). As in the case of Lec1 cells, many glycoproteins of Lec1.Lec6 mutants are significantly altered in size, charge, and composition (Stanley and Sudo, unpublished observations).

CHO cells with the Lec8 phenotype have been described independently by Briles et al. (1977; clone 13), Li et al. (1980b; Abr^R D4.1b), and Stanley (1980b). Complementation analyses have shown that clone 13 cells fall into complementation group 8 (Stanley, unpublished observations). The biochemical basis of the Lec8 mutation is not known, but, like Lec2 and Lec3 mutations, it exerts a wide-ranging effect reducing the addition of Gal residues and therefore sialic acid residues to both glycoproteins and glycolipids (Briles et al., 1977; Stanley, 1980b). Since galactosyltransferase activity is present in Lec8 cells, a compartmentation defect might also explain this phenotype.

Lec9 CHO cells are recent isolates (Stanley, 1983b) which have not yet been well characterized at the biochemical level. However, preliminary studies have shown that while they have normal levels of Dol-P-Man synthetase activity, they synthesize reduced amounts of labeled

$Glc_3Man_9Gn_2$-P-P-Dol (Rosenwald, Krag, and Stanley, unpublished observations). The mutation in Lec9 cells may reside in a mannosyltransferase involved in the biosynthesis of the large oligomannogyl-dolichol precursor. The Lec9 mutation results in a temperature-sensitive phenotype (Stanley, 1983b). Temperature-resistant revertants exhibit almost wild-type sensitivity to RIC and also synthesize the usual oligosaccharide dolichol intermediates.

The LEC10 mutation in CHO cells behaves dominantly in somatic cell hybrids (Stanley and Siminovitch, 1977). It appears to be the result of a regulatory mutation which induces the activity of a GlcNAc transferase termed GlcNAc-TIII (Campbell and Stanley, 1984a). This enzyme adds the "bisecting" GlcNAc to the β1,4-linked core mannose residue of Asn-linked carbohydrates (Narasimhan, 1982). The enzyme activity is not detectable in parental cell extracts nor is the product of the enzyme found among the glycopeptides synthesized by parental cells (Campbell and Stanley, 1984a).

The LEC11 cell line is another example of a dominant mutant of CHO cells (Stanley, 1983b). The biochemical basis of the LEC11 phenotype is the increased activity of a specific fucosyltransferase which adds fucose in α1,3 linkage to GlcNAc residues in the outer branches of Asn-linked complex carbohydrates (Campbell and Stanley, 1983, 1984b). Parental cell extracts exhibit no detectable α1,3 fucosyltransferase activity (a minimum 400-fold difference between LEC11 and parental cells has been demonstrated). In addition, parental cells do not express α1,3 fucose residues at the cell surface as measured by a sensitive radioimmunoassay using monoclonal antibody to SSEA-1 [the stage-specific embryonic antigen: Galβ1,4(Fuc α1,3)GlcNAcβ]. The evidence suggests that the mutation expressed by LEC11 cells is a regulatory mutation which induces the expression of a fucosyltransferase termed α1,3 fucosyltransferase I (Campbell and Stanley, 1984b).

LEC12 mutants are also the result of a regulatory mutation which induces an α1,3 fucosyltransferase activity. LEC12 cells bind high levels of the radioiodinated SSEA-1 monoclonal antibody and exhibit a minimum 500-fold difference compared with parental cell extracts in α1,3 fucosyltransferase activity (Campbell and Stanley, 1983). However the α1,3 fucosyltransferase expressed by LEC12 cells is different from that found in LEC11 cells and is therefore termed α1,3 fucosyltransferase II (Campbell and Stanley, 1984b). In addition to qualitative differences in their substrate specificities, the LEC11 and LEC12 enzymes are differentially inhibited by *N*-ethylmaleimide and exhibit different abilities to bind to GDP-hexanolamine Sepharose. Therefore, it would appear that at least two α1,3 fucosyltransferase genes exist in the CHO genome and that both have been induced to give rise to the LEC11 and LEC12 phenotypes, respectively. A similar mutant type has been described in a WGA-resistant B16 melanoma cell line (Finne et al., 1980, 1982). However, the combined properties of the α1,3 fucosyltransferase enzyme in this mutant (Prieels et al., 1983) suggest it differs from the α1,3 fucosyltransferases I and II expressed by LEC11 and LEC12 CHO cells.

Lec13 CHO cells are recent isolates (Ripka and Stanley, unpublished observations) which exhibit a phenotype similar to the pea lectin-resistant BW5147 mutant termed $PL^R1.3$ (Reitman et al., 1980). $PL^R1.3$ cells are defective in the first enzyme (GDP-mannose-4,6-reductase) involved in converting GDP-mannose to GDP-fucose. The Lec13 CHO mutants exhibit markedly reduced incorporation of α1,6-linked fucose into Asn-linked complex moieties and cell extracts from Lec13 mutants cannot convert GDP-mannose to GDP-fucose *in vitro* (Ripka and Stanley, unpublished observations). It is likely, therefore, that Lec 13 CHO cells are defective in one of the enzymes in the pathway for the synthesis of GDP-fucose from GDP-mannose.

The CHO mutants designated Lec15 were isolated by Robbins et al. (1981) in a search for a mutant deficient in Man-6-phosphate (Man-6-P) receptor activity. One of the receptor deficient lines (B4-2-1) was subsequently shown to lack Dol-P-Man synthetase activity, resulting in the synthesis of a truncated oligosaccharide dolichol containing only five mannose residues (Stoll et al., 1982). The B4-2-1 mutant which is three- to five-fold resistant to RIC (Robbins et al., 1981) is recessive and complements all other Ric^R CHO mutants and therefore has been assigned to a new complementation group (Stanley, unpublished observations). The properties of Lec15 CHO cells appear identical to those of Thy1E BW5147 mutants (Chapman et al., 1979, 1980).

LEC16 CHO cells represent a recent isolate which displays altered ratios of oligosaccharide-dolichol precursors (Krag and Stanley, unpublished observations). The phenotype confers a three- to five-fold resistance to RIC, which is similar to that of Lec15 CHO cells. However, unlike the Lec15, mutation, the LEC16 mutation behaves dominantly in somatic cell hybrids (Stanley, unpublished observations).

The glycosylation lesions summarized in Table 24.2 are presented diagrammatically in Figure 24.3. Although other CHO Lec^R mutants have been reported in the literature (Stanley 1981, 1983b; Hirschberg et al., 1982; Baker et al., 1982), they are not included in these summaries because their biochemical phenotypes remain to be characterized. The conditional-lethal mutant of DON Chinese hamster cells which exhibits a defect in glycosylation (Tenner et al., 1977; Tenner and Sheffler, 1979) might represent another complementation group. Likewise, the hyaluronic-acid-deficient or -elevated mutants described by Atherley et al. (1977), Kraemer and Barnhart, (1979), and Barnhart et al. (1979) would be expected to represent distinct complementation groups. However, as with most glycosylation mutants selected by chance, the lectin-resistance properties of these mutants were not determined so that tentative assignment to a complementation group is not possible. At this time, therefore, the number of Lec^R CHO mutants which represent mutations in different genes (including double mutants and those which are not glycosylation defective) is at least 20 .

IV. FUNCTIONAL CONSEQUENCES OF GLYCOSYLATION MUTATIONS

Despite the marked changes in cell surface carbohydrates expressed by Lec^R cells, most mutants continue to grow well in suspension and monolayer culture. Many phenotypes exhibit altered adhesive abilities or slightly reduced growth rates but, by and large, these effects are comparatively mild (Gottlieb et al., 1974; Stanley et al., 1975a; Cifone et al., 1979; Stanley and Sudo, 1981). This result is to be expected perhaps, if one of the key roles of surface carbohydrate is in specific recognition of carbohydrate-binding proteins during development and differentiation. The lectin-binding properties of Lec^R cells provide a striking example of the potential for carbohydrate involvement in specific recognition phenomena. The fact that membrane molecules carrying altered carbohydrates are functionally competent in Lec^R cells argues teleologically in favor of a model in which profound changes in the array of carbohydrates presented for a recognition event would be possible without seriously affecting membrane function.

However, some carbohydrate changes clearly are not compatible with cell viability. Both Lec5 and Lec9 CHO mutants exhibit a conditional—lethal phenotype being temperature sensitive (ts) for growth at 39.5°C. Revertants selected for the ability to grow at high temperature are reverted for lectin-resistance as well as for their biochemical phenotypes, suggesting that their glycosylation defects confer their respective ts phenotypes. Further studies of the biochemical bases and metabolic consequences of these mutations should identify glycosylation reactions that are essential for cell viability.

Other carbohydrate changes, although not reducing cell viability, do affect the function of particular membrane molecules. For example, Thy 1 molecules carrying altered carbohydrates due to the lack of Dol-P-Man synthetase actitivity in Thy 1E mutants exhibit a decreased intracellular half-life (Trowbridge et al., 1978b). This presumably accounts for the functional lack of Thy 1 molecules at the lymphoma cell surface. Similarly, a defect in the same enzyme in Lec15 CHO cells results in functional impairment of the Man-6-P receptor at the cell surface (Robbins et al., 1981; Stoll et al., 1982). It is not known whether this is due to an inability of the Man-6-P receptor to reach the plasma membrane. However, altered intracellular compartmentalization does seem to be a consequence of the glycosylation lesion expressed by Lec5 CHO mutants. In these cells certain lysosomal enzymes are not primarily localized in lysosomes (Krag and Robbins, 1982); they fail to acquire the Man-6-P recognition marker and possess carbohydrates which are fully processed to complex moieties. By contrast, the lysosomal enzymes of Lec15 mutants which synthesize a truncated oligosaccharide-dolichol precursor appear to acquire the Man-6-P recognition marker (Stoll et al., 1982; Gabel and Kornfeld, 1982). However, they are not compartmentalized correctly in the Lec15 CHO cells due to an altered Man-6-P receptor (Robbins and Myerowitz, 1981).

One widely observed consequence of altered glycosylation is a loss of tumorigenicity and/or metastatic ability that often accompanies the acquisition of lectin resistance (Tao and Burger, 1977, 1982; Bramwell and Harris, 1978; Kerbel, 1979; Reading et al., 1980; Kerbel et al., 1982; Lin et al., 1983). However, this is not always the case and the combined data suggest that prior mutagenesis may be a significant factor in generating "antigen-loss" variants that are deficient in tumor-forming or metastatic abilities (see Kerbel et al., 1982). Clearly the analysis of revertants or segregants generated *in vitro* and *in vivo* is required to establish a causal relationship between cell surface glycosylation and tumorigenic potential. Revertants obtained from rare metastases of a Lec^R MDAY-D2 line that exhibits reduced tumorigenicity were indeed lectin sensitive, but apparently due to fusion at the site of inoculation with a host cell of bone marrow origin. Although B16 melanoma revertants for lectin resistance have been described by Finne et al. (1980), their tumor-forming and metastasizing properties have not been reported. Studies of the ability of the Lec^R CHO mutants to form tumors in the nude mouse initially identified Lec1A CHO cells as unable to form tumors (Ripka, Shin, and Stanley, unpublished observations). However, an independent Lec1A isolate obtained without prior mutagenesis was subsequently found to be as tumorigenic as parental CHO cells. Clearly there are numerous events which might lead to loss of a complex phenotype such as tumorigenicity or metastasis. To determine whether cell surface carbohydrates play a role in any one of the multitude of steps required either to establish a tumor or to metastasize to a specific organ requires a rigorous somatic cell genetic approach.

Finally there are a number of Lec^R mutations that affect the membrane fusion ability of the mutants. Lec2 CHO cells do not fuse to form somatic cell hybrids with inactivated Sendai virus, whereas hybrids can be obtained using polyethylene glycol (Stanley and Siminovitch, 1977). In addition, infectious Sendai virus is produced in very low amounts from Lec2 cells and the block to infection has been localized to the absorption/penetration stage (Markwell, unpublished observations). Polos and Gallaher (1979) have shown that 15B (Lec1) CHO cells are refractory to fusion by another paramyxovirus, Newcastle disease virus. In myoblast cell lines, which will fuse to form mytotubes under certain culture conditions, the acquisition of resistance to concanavalin-A has been correlated with an inability to become myotubes (Parfett et al., 1981). Thus different cell surface lesions lead, in a number of instances, to a reduction in fusion potential. Investigation of the bases of phenotypes affecting virus–cell or cell–cell fusion events should provide a valuable approach to the delineation of the molecules involved in membrane fusion phenomena.

V. FUTURE DIRECTIONS

The Lec^R CHO mutants, as well as similar mutants from other cell lines, provide novel biological material with which to investigate the role of cellular carbohydrates in structure/function relationships of glycoproteins and glycolipids. The mutants studied so far have shown that certain carbohydrate structural changes have profound effects on the intracellular localization or function of particular molecules. In addition, it is apparent that changes in the complement of cellular glycosylation enzymes leads to alterations in the array of carbohydrates expressed at the cell surface resulting in dramatic changes in the ability of the cells to interact with carbohydrate binding proteins. These properties of Lec^R cells reveal the enormous potential for directing structural changes in carbohydrates at the cell surface and suggest that the developmental regulation of such changes during embryogenesis and differentiation might be due to the controlled expression of glycosylation genes.

Clearly Lec^R mutants also provide an incisive approach to investigating the role of carbohydrates in the biosynthesis and function of molecules which are not normally synthesized by CHO cells. Transfection of Lec^R CHO cells with cloned genes should allow the construction of glycoproteins with "tailor-made" carbohydrates for structure/function studies. Equally important is the exploitation of Lec^R CHO mutants in the identification and characterization of the glycosylation genes of animal cells. Since glycosyltransferases are ill-characterized membrane-associated enzymes present in only catalytic amounts, cloning by coventional approaches would be tedious. Cloning by transfection into Lec^R CHO mutants is therefore an attractive prospect. Selections for both mutant and parental Lec^R phenotypes are available in most cases due to the lectin hypersensitivities conferred by many of the Lec^R glycosylation mutations (Table 24.1). The recessive mutants should be useful as recipients for wild-type genes, while the dominant mutants should act as donors for the isolation of glycosylation genes which exert a dominant affect. It is hoped that the dominant mutants will also provide an approach to defining regulatory signals which affect the expression of glycosylation genes.

Obtaining probes for genes involved in glycosylation should allow the study of their expression during embryogenesis and differentiation. Although numerous reports provide evidence for temporal changes in carbohydrate structure during development (Yoshima et al., 1982; Hakomori and Kannagi, 1983; Kannagi et al., 1983), causal relationships have not been established. The ability to dissect and manipulate glycosylation genes should enable the direct investigation of the biological significance of developmentally regulated carbohydrate structural changes.

In addition to cloning glycosylation-related genes, it is clearly important to pursue the somatic cell genetic approach. The generation of new mutants, families of revertants, and cell lines carrying multiple glycosylation mutations is critical to an eventual understanding of the biological roles of carbohydrates. The plant lectins remain excellent selective agents with enormous potential for the isolation of new mutants: new cytotoxic lectins and those which require conjugation to a toxin such as ricin A to improve their cytotoxicity are still to be exploited; lectins of different binding specificities may be combined to eliminate unwanted phenotypes (a significant advantage since reisolating known mutations is a serious problem in searching for new glycosylation mutants); and the availability of approximately 20 different CHO Lec^R mutants with distinct Lec^R phenotypes allows new selection protocols to be monitored in reconstruction experiments so that the efficiency of isolating new phenotypes may be maximized. Finally, it seems likely that many more mutants remain to be identified in CHO cells. This prophesy is based not only on a knowledge of the many reactions involved in carbohydrate biosynthesis and metabolism but also on the existence of glycosylation mutants in other cell lines which have not yet been isolated from CHO populations. For example, Ric^R21 BHK cells which exhibit reduced GlcNAc-TII transferase activity (Hughes et al., 1983), a BW5147 mutant ($Pha^R2.7$) which lacks α-glucosidase II processing activity (Reitman et al., 1982), and an L cell mutant (CL6) which appears to lack α1,2 mannosidase activity (Tabas and Kornfeld, 1978) have been described. The selection of each of these mutations from CHO cells should be possible and, in fact, represents an important goal since the CHO genome is the best characterized in structural and genetic terms of all animal cell lines.

ACKNOWLEDGMENTS

During the writing of this chapter, the author's laboratory was supported by grants from the National Institutes of Health (R01-CA30645), the American Cancer Society (BC-332A), and the National Science Foundation (PCM 80-23672). Partial support was also provided by Core Cancer Grant (1P01-CA13330). The author is the recipient of a faculty award from the American Cancer Society. Thanks are extended to Chris Campbell, Sharon Krag, William Chaney, and James Ripka for helpful comments on the manuscript.

REFERENCES

Atherly, A. G., Barnhart, B. J., and Kraemer, P. M. (1977). *J. Cell Physiol.* **89**, 375–386.

Baenziger, J. U. and Fiete, D. (1979). *J. Biol. Chem.* **254**, 2400–2407.

Baker, R. M., Hirschberg, C. B., O'Brien, W. A., Awerbuch, T. E., and Watson D. (1982). *Met. Enzymology* **83**, 444–458.

Baker, R. M., and Ling, V. (1978). *Met. Membr. Biol.* **9**, 337–384.

Barnhart, B. J., Cox, S. H., and Kraemer, P. M. (1979). *Exp. Cell Res.* **119**, 327–332.

Bramwell, M. E. and Harris, H. (1978). *Proc. R. Soc. Lond. B.* **201**, 87–106.

Briles, E. B. (1982). *Int. Rev. Cytol.* **75**, 101–165.

Briles, E. B., Li, E., and Kornfeld, S. (1977). *J. Biol. Chem.* **252**, 1107–1116.

Brisson, J-R., and Carver, J. P. (1983). *Can. J. Biochem. Cell Biol.* **61**, 1067-1078.

Campbell, C. and Stanley, P. (1983). *Cell* **35**, 303–309.

Campbell, C. and Stanley, P. (1984a). *J. Biol. Chem.* **261**, 13370–13378.

Campbell, C. and Stanley, P. (1984b). *J. Biol. Chem.* **259**, 11208–11214.

Chapman, A., Fujimoto, K., and Kornfeld, S. (1980). *J. Biol. Chem.* **255**, 4441–4446.

Chapman, A., Trowbridge, I. S., Hyman, R., and Kornfeld, S. (1979). *Cell* **17**, 509–515.

Cifone, M. A. and Baker, R. M. (1976). *J. Cell Biol.* **70**, 77a.

Cifone, M. A., Hynes, R. O., and Baker, R. M. (1979). *J. Cell Physiol.* **100**, 39–54.

Criscuolo, B. A. and Krag, S. S. (1982). *J. Cell Biol.* **94**, 586-591.

Culp, L. A. and Black, P. H. (1972). *J. Virol.* **9**, 611–620.

Cummings, R. D. and Kornfeld, S. (1982). *J. Biol. Chem.* **257**, 11230–11234.

Cummings, R. D., Trowbridge, I. S., and Kornfeld, S. (1982). *J. Biol. Chem.* **257**, 13421–13427.

Davidson, S. K. and Hunt, L. A. (1983). *J. Gen. Virol.* **64**, 613–625.

Deutscher, S. L., Nuwayid, N., Stanley, P., Briles, E. B., and Hirschberg, C. B. (1984). *Cell* (in press).

Etchison, J. R. and Holland, J. J. (1974). *Proc. Natl. Acad. Sci. U.S.A.* **71**, 4011–4014.

Etchison, J. R., Robertson, J. S., and Summers, D. F. (1981). *J. Gen. Virol.* **57**, 43–52.

Etchison, J. R. and Summers, D. F. (1979). *J. Supramol. Struct.* **10**, 205.

Finne, J., Burger, M. M., and Prieels, J-P. (1982). *J. Cell Biol.* **92**, 277–282.

Finne, J., Tao, T-W., and Burger, M. M. (1980). *Cancer Res.* **40**, 2580–2587.

Gabel, C. A. and Kornfeld, S. (1982). *J. Biol. Chem.* **257**, 10605–10612.

Goldstein, I. J. and Hayes, C. E. (1978). *Adv. Carb. Chem and Biochem.* **35**, 127–340.

Gottesman, M. M. (1979). *Somat. Cell Genet.* **5**, 665–671.

Gottlieb, C., Baenziger, J., and Kornfeld, S. (1975). *J. Biol. Chem.* **250**, 3303–3309.

Gottlieb, C., Skinner, A. M., and Kornfeld, S. (1974). *Proc. Natl. Acad. Sci. U.S.A.* **71**, 1078–1082.

Greene, W. C., Parker, C. M., and Parker, C. W. (1976). *J. Biol. Chem.* **251**, 4017–4025.

Hakomori, S.-I. and Kannagi, R. (1983). *J. Natl. Cancer Inst.* **71**, 231–251.

Hammarstrom, S., Hammarstrom, M-L. (nee Dillner), Sundbland, G., Arnarp, J., and Lonngren, J. (1982). *Proc. Natl. Acad. Sci.* U.S.A. **79**, 1611–1615.

Hirschberg, C. B., Baker, R. M., Perez, M., Spencer, L. A., and Watson, D. (1981). *Mol. Cell. Biol.* **1**, 902–909.

Hirschberg, C. B., Perez, M., Snider, M., Hanneman, W. L., Esko, J., and Raetz, C. R. H. (1982). *J. Cell. Physiol.* **111**, 255–263.

Hughes, R. C., Mills, G., and Stojanovic. D. (1983). *Carbohydrate Research* **120**, 215–234.

Hunt, L. A. (1980a). *Cell* **21**, 407–415.

Hunt, L. A. (1980b). *J. Virol.* **35**, 362–370.

Hunt, L. A. (1982). *Biochem. J.* **205**, 623–630.

Hyman, R. (1973). *J. Natl. Canc. Inst.* **50**, 415–422.

Irimura, T., Tsuji, T., Tagami, S., Yamamoto, K., and Osawa, T. (1981). *Biochem.* **20**, 560–566.

Juliano, R. L. and Stanley, P. (1975). *Biochim. Biophys. Acta* **389**, 401–406.

Kannagi, R., Levery, S. B., and Hakomori, S-I. (1983). *Proc. Natl. Acad. Sci. U.S.A.* **80**, 2844–2848.

Kerbel, R. S. (1979). *Am. J. Pathol.* **97**, 609–622.

Kerbel, R. S., Dennis, J. W., Lagarde, A. E., and Frost, P. (1982). *Cancer Metastasis Reviews* **1**, 99–140.

Kornfeld, R. and Kornfeld, S. (1980). In *The Biochemistry of Glycoproteins and Proteoglycans* (W. J. Lennarz, ed.), Plenum, New York, pp. 1–34.

Kornfeld, K., Reitman, M. L., and Kornfeld, R. (1981). *J. Biol. Chem.* **256**, 6633–6640.

Kraemer, P. M. (1971). *Biochem.* **10**, 1437–1445.

Kraemer, P. M. and Barnhart, B. J. (1978). *Expt. Cell Res.* **114**, 153–157.

Krag, S. S. (1979). *J. Biol. Chem.* **254**, 9167–9177.

Krag, S. S. and Robbins, A. R. (1982). *J. Biol. Chem.* **257**, 8424–8431.

Kuwano, M., Tabuki, T., Akiyama, S., Mifune, K., Takatsuki, A., Tamura, G., and Ikehara, Y. (1981). *Somat. Cell Genet.* **7**, 507–521.

Li, E., Gibson, R., and Kornfeld, S. (1980a). *Arch. Biochem. Biophys.* **199**, 393–399.

Li, E. and Kornfeld, S. (1978). *J. Biol. Chem.* **253**, 6426–6431.

Li, E. and Kornfeld, S. (1979). *J. Biol. Chem.* **254**, 1600–1605.

Li, I-C., Blake, D. A., Goldstein, I. J., and Chu, E. H. Y. (1980b). *Exp. Cell Res.* **129**, 351–360.

Lin, L-H., Stern, J. L., and Davidson, E. A. (1983). *Carbohydrate Research* **111**, 257–271.

Meager, A., Ungkitchanukit, A., and Hughes, R. C. (1976). *Biochem. J.* **154**, 113–124.

Meager, A., Ungkitchanukit, A., Nairn, R., and Hughes, R. C. (1975). *Nature* **257**, 137–139.

Narasimhan, S. (1982). *J. Biol. Chem.* **257**, 10235–10242.

Narasimhan, S., Stanley, P., and Schachter, H. (1977). *J. Biol. Chem.* **252**, 3926–3933.

Ogata, S., Muramatsu, T., and Kobata, A. (1975). *J. Biochem. (Tokyo)* **78**, 687–696.

Olsnes, S., and Pihl, A. (1982a). In *The Molecular Actions of Toxins and Viruses* (P. L. Cohen, and S. Van Heyningen, eds.), Elsevier/North Holland, Amsterdam, pp. 51–105.

Olsnes, S. and Pihl, A. (1982*b*). *Pharmac. Ther.* **15**, 355–381.

Ono, M., Kuwano, M., Watanabe, K-I., and Funatsu G. (1982). *Mol. Cell Biol.* **2**, 599–606.

Ozanne, B. and Sambrook, J. (1971). In *The Biology of Oncogenic Viruses* (L. G. Silvestri, ed), North-Holland Publishing, Amsterdam, pp. 248–257.

Parfett, C. L. J., Jamieson, J. C., and Wright, J. A. (1981). *Exp. Cell Res.* **136**, 1–14.

Pollack, L. and Atkinson, P. H. (1983). *J. Cell Biol.* **97**, 293–300.

Polos, P. G. and Gallaher, W. R. (1979). *J. Virol.* **30**, 69–75.

Prieels, J-P., Monnom, D., Perraudin, J-P., Finne, J., and Burger, M. (1983). *Eur. J. Biochem.* **130**, 347–351.

Ray, B. and Wu, H. C. (1982). *Mol. Cell. Biol.* **2**, 535–544.

Reading, C. L., Belloni, P. N., and Nicolson, G. L. (1980). *J. Nat. Cancer Inst.* **64**, 1241–1249.

Reitman, M. L., Trowbridge, I. S., and Kornfeld, S. (1980). *J. Biol. Chem.* **255**, 9900–9906.

Reitman, M. L., Trowbridge, I. S., and Kornfeld, S. (1982). *J. Biol. Chem.* **257**, 10357–10363.

Robbins, A. R. and Myerowitz, R. (1981). *J. Biol. Chem.* **256**, 10623–10627.

Robbins, A. R., Myerowitz, R., Youle, R. J., Murray, G. J., and Neville, Jr., D. M. (1981). *J. Biol. Chem.* **256**, 10618–10622.

Robertson, M. A., Etchison, J. R., Robertson, J. S., Summers, D. F., and Stanley, P. (1978). *Cell* **13**, 515–526.

Roden, L. (1980). In *The Biochemistry of Glycoproteins and Proteoglycans* (W. J. Lennarz, ed.), Plenum, New York, pp. 267–372.

Rose, J. K. and Gallione, C. J. (1981). *J. Virol.* **39**, 519–528.

Schachter, H. (1981). In *Lysosomes and Lysosomal Storage Diseases* (J. W. Callahan and J. A. Lowden, eds.) Raven Press, New York, pp. 73–93.

Schachter, H. and Roseman, S. (1980). In *The Biochemistry of Glycoproteins and Proteoglycans* (W. J. Lennarz, ed.), Plenum, New York, pp. 85–160.

Schlesinger, S., Gottlieb, C.., Feil, P., Gelb, N., and Kornfeld, S. (1976). *J. Virol.* **17**, 239–246.

Spiro, R. G. and Spiro, M. J. (1982). *Philos. Trans. R. Soc. London Series B* **300**, 117–127.

Stanley, P. (1980a). In *Biochemistry of Glycoproteins and Proteoglycans* (W. J. Lennarz, ed.), Plenum, New York, pp. 161–189.

Stanley, P. (1980b). *ACS Symposium Series* **128**, 213–221.

Stanley, P. (1981). *Mol. Cell. Biol.* **1**, 687–696.

Stanley, P. (1982). *Arch. Biochem. Biophys.* **219**, 128–139.

Stanley, P. (1983a). In *Methods in Enzymology* (S. Fleischer and B. Fleischer, eds.), Academic Press, New York, Vol. 96, pp. 157–184.

Stanley, P. (1983b). *Somat. Cell Genet.* **9**, 593–608.

Stanley, P., Caillibot, V., and Siminovitch, L. (1975a). *Somat. Cell Genet.* **1**, 3–26.

Stanley, P., Caillibot, V., and Siminovitch, L. (1975b). *Cell* **6**, 121–128.

Stanley, P. and Carver, J. P. (1977). *Adv. Expt. Med. Biol.* **84**, 265–282.

Stanley, P., Narasimhan, S., Siminovitch, L., and Schachter, H. (1975c). *P.N.A.S.* **72**, 3323–3327.

Stanley, P. and Siminovitch, L. (1977). *Somat. Cell Genet.* **3**, 391–405.

Stanley, P. and Sudo, T. (1981). *Cell* **23**, 763–769.

Stanley, P., Sudo, T., and Carver, J. P. (1980). *J. Cell Biol.* **85**, 60–69.

Stanley, P., Vivona, G., and Atkinson, P. H. (1984). *Arch. Biochem. Biophys.* **230**, 363–374.

Stoll, J., Robbins, A. R., and Krag, S. S. (1982). *Proc. Natl. Acad. Sci. U.S.A.* **79**, 2296–2300.

Sudo, T., and Onodera, K. (1979). *J. Cell. Physiol.* **101**, 149–156.

Tabas, I. and Kornfeld, S. (1978). *J. Biol. Chem.* **253**, 7779–7786.

Tao, T-W. and Burger, M. M. (1977). *Nature* **270**, 437–438.

Tao, T.-W.. and Burger, M. M. (1982). *Int. J. Cancer* **29**, 425–430.

Tenner, A. J. and Scheffler, I. E. (1979). *J. Cell. Physiol.* **98** 251–266.

Tenner, A. J., Zieg, J., and Scheffler, I. E. (1977). *J. Cell. Physiol.* **90**, 145–160.

Tkacz, J. S. and Lampen, J. O. (1975). *Biochem. Biophys. Res. Commun.* **65**, 248–257.

Trowbridge, I. S., Hyman, R., Ferson, T., and Mazauskas, C. (1978a). *Eur. J. Immunol.* **8**, 716–723.

Trowbridge, I. S., Hyman, R., and Mazauskas, C. (1978b). *Cell* **14**, 21–32.

Varki, A. and Kornfeld, S. (1983). *J. Biol. Chem.* **258**, 2808–2818.

Wright, J. A. (1973). *J. Cell. Biol.* **56**, 666–675.

Wright, J. A. (1975). *Can. J. Microbiol.* **21**, 1650–1654.

Wright, J. A., Jamieson, J. C., and Ceri, H. (1979). *Exp. Cell Res.* **121**, 1–8.

Wright, J. A., Lewis, W. H., and Parfett, C. L. (1980). *Can. J. Genet. Cytol.* **22**, 443–496.

Yogeeswaran, G., Murray, R. K., and Wright, J. A. (1974). *Biochem. Biophys. Res. Commun.* **56**, 1010–1016.

Yoshima, H., Shiraishi, N., Matsumoto, A., Maeda, S., Sugiyama, T., and Kobata A. (1982). *J. Biochem.* [*Tokyo*] **91**, 233–246.

CHAPTER 25

MULTIDRUG-RESISTANT MUTANTS

Victor Ling
Department of Medical Biophysics
University of Toronto and the
Ontario Cancer Institute
Toronto, Ontario, Canada

I. INTRODUCTION

Selection for drug resistance in mammalian cell lines often yields variants which express a wide-ranging resistance to a variety of unrelated compounds (Ling et al., 1983). This is called the "multidrug-resistance" phenotype. In this chapter, a class of membrane-altered colchicine-resistant (CH^R) mutants isolated in Chinese hamster ovary (CHO) cells is described. These mutants were selected for resistance to the antimitotic drug colchicine and found to display a concomitant resistance to structurally and functionally unrelated drugs such as adriamycin, daunorubicin, actinomycin D, gramicidin D, emetine, puromycin, vinblastine, and others (Ling and Thompson, 1974; Bech-Hansen et al., 1976). These mutants are therefore classified as multidrug-resistant mutants, and they are distinct from the microtubule-altered colchicine-resistant mutants, which do not display this extensive phenotype (Ling et al., 1979; see also Chapter 22). In addition, increased sensitivity to certain compounds and temperature sensitivity for growth are also observed in some isolates (Bech-Hansen et al., 1976; Ling, 1977).

Multidrug-resistant mutants are thus interesting because they express complex, unanticipated multiple phenotypes. Genetic analysis was undertaken in order to determine the basis of these phenotypes. One question at issue was whether the multiple phenotypes are consequences of multiple genetic changes, or whether they result from single pleiotropic mutations. In order to discriminate between these two possibilities, aspects of methodology, for example, mutant isolation, and genetic characterization, are particularly important, and they are emphasized in this chapter.

Establishing that a complex variant phenotype stems from a pleiotropic mutation is significant since expression of pleiotropy often reflects alterations in functions requiring complex multimolecular interactions, and characterization of pleiotropic mutants could reveal cooperativity between cellular components not previously suspected (Baker and Ling, 1978). Understanding the origin and nature of the multidrug-resistance phenotype also has clinical relevance. It is now apparent that drug-resistant variants do arise in malignant cells, and they could play a major role contributing to a disease resistant to chemotherapy (Ling, 1982). The occurrence of multidrug-resistance mutations in cancer cells could be especially insidious, since, in theory, they could render such cells resistant even to combination chemotherapy.

II. MUTANT ISOLATION

The guiding principle for selecting mutants in which expression of multiple phenotypes is anticipated is to employ procedures that will most likely result in isolates with a single mutational change. This will allow for a more confident interpretation of pleiotropy. Some fundamental points are addressed here but for more comprehensive dicussions, see reviews by Thompson and Baker (1973) and Baker and Ling (1978) where details of choice of cell lines, selecting conditions, cell hybrid analysis, use of selective markers, etc., are outlined.

As a general rule, selections are best initiated with purified clonal populations so that variants isolated can be compared to a homogeneous parental population. Adequate frozen stocks of the parental clonal line should be maintained so that this reference population can be renewed in the event of phenotypic drift. Purity of the selecting agent should also be ensured. Selections using a single-step procedure and isolating independent clones for analysis are preferred over multiple or continuous selections. Mutagen treatment should initially be avoided, if possible, in order

AUXB1 $\xrightarrow[\text{EMS}]{0.1}$ CH^RA3 $\xrightarrow{3.0}$ CH^RB3 $\xrightarrow[\text{EMS}]{10}$ CH^RC5

AUXB1 $\xrightarrow{0.1}$ CH^R2H $\xrightarrow[\text{EMS}]{0.5}$ CH^R2HA $\xrightarrow{5}$ CH^RC4

Figure 25.1. Stepwise selection of colchicine-resistant clones. Independent series of related clones selected for increased colchicine resistance were isolated in CHO cells as described (Ling and Thompson, 1974). The selecting drug concentrations (μg/mL) were as indicated. Mutagen treatment with ethyl methanesulfonate (EMS) was employed in some selection steps.

to minimize potential complications arising from multiple changes. Mutagen treatment, however, could increase the frequency of some mutations that are relatively rare. Such a finding would of course support a genetic basis for the phenotype under consideration.

The isolation of methotrexate-resistant mutants in CHO cells serves as a good example of these principles (Flintoff et al., 1976). By using clonal selections in single-step drug treatments, Flintoff and co-workers have been able to isolate and characterize stable mutants possessing three different mechanisms of resistance. One mechanism, such as reduced methotrexate binding to the dihydrofolate reductase (DHFR) enzyme, resulted in a relatively low level of resistance of about 10-fold. Such an alteration might have been masked if a continuous selection protocol had been employed, since a more dominant mutation such as increased level of DHFR, which results in much higher levels of resistance, could also occur in the same cells.

Multiple-step selections are best exploited when they are carried out in the form of a sequence of discrete single steps so that clonal populations selected at intermediate stages may be sampled in order to monitor progressive phenotypic and molecular alterations. Such an approach was used for selection of the colchicine-resistant multidrug-resistance lines in CHO cells (Ling and Thompson, 1974), as illustrated in Figure 25.1. Analysis of independent clones isolated in this manner indicates that increased resistance to colchicine correlates with increased resistance to other unrelated drugs involved in the multidrug-resistance phenotype. This concordant expression of colchicine resistance and multidrug resistance in independent clones strongly indicates that the resistance to unrelated drugs is a pleiotropic response.

As previously mentioned, continuous selections involving long-term exposure of a population of cells to increasing drug concentrations, and waiting for outgrowth of resistant populations, should usually be avoided. However, in some instances, it appears that variants with unstable phenotypes are more easily isolated by this approach. In such instances, rigorous characterization is required to substantiate interpretation of pleiotropy.

III. MULTIDRUG-RESISTANCE PHENOTYPE

A. Cross Resistance

The drug sensitivity of a series of related CHO cell clones selected for increased resistance (Fig. 25.1) to colchicine is shown in Table 25.1. These data show that the cross resistance involves compounds that are structurally and functionally unrelated. Also, it demonstrates that an increase in resistance in colchicine is accompanied by an increased cross resistance. This indicates that increased resistance at each selection step is mediated by basically the same mechanism, and that increased resistance is likely achieved via a gene dosage effect (see below). From analysis of scores of clones selected under different conditions, it appears that colchicine-resistant mutations resulting in multidrug resistance is a common mechanism (Ling and Thompson, 1974). A similar conclusion can be drawn for clones selected for resistance to other drugs such as daunorubicin, taxol, VM-26, VP16-213, vinblastine, auromomycin, etc. (Kartner et al., 1983b, Gupta, 1983a,b; Rauscher III et al., 1984).

Although expression of cross resistance to unrelated drugs is a common feature of colchicine-resistant isolates, the pattern and degree of resistance among clones derived even from a single selection are not identical. For multidrug-resistant mutants isolated with different drugs, differences in the cross-resistance phenotype are even more marked (compare, for example, the cross-resistance pattern of lines CH^RC5, and CH^RC4 selected with colchicine, with those of lines DNR^R51, and DNR^R159 selected with daunorubicin described in Kartner at al. (1983b).

TABLE 25.1
Relative Cross Resistance[a]

Drugs	Cell Lines				
	AUXB1	CH^RA3	CH^RB3	CH^RC5	CH^RC4
Colchicine	1	2	6	184	74
Colcemid	1	NT	NT	16	11
Puromycin	1	3	13	105	29
Emetine	1	1	2	29	15
Daunomycin	1	2	8	76	32
Adriamycin	1	NT	NT	7	25
Gramicidin D	1	NT	NT	144	91
Cytochalasin B	1	1	2	11	6

[a]Relative resistance was determined by the concentration of drug required to inhibit growth in the CH^R drug-resistant line divided by that required for the parental AUXB1 line (Bech-Hansen et al., 1976). The derivation of these lines is as indicated in Figure 25.1. AUXB1, CH^RA3, CH^RB3, and CH^RC5 are one related series of clones. NT = not tested. Data for this table are compiled from Bech-Hansen et al. (1976).

TABLE 25.2
Relative Collateral Sensitivity and Response to Nonionic Triton X Detergents[a]

Drugs	Cell Lines		
	AUXB1	CH^RC4	CH^RC5
Procaine	1	0.3	0.4
Tetracaine	1	0.3	0.2
Xylocaine	1	0.3	0.1
Deoxycorticosterone	1	0.7	0.1
1-Dehydrotestosterone	1	0.8	0.1
Acryonycine	1	NT	≤0.06
Triton X 45 (EOU = 5)	1	0.9	0.1
Triton X 114 (EOU = 7–8)	1	0.8	0.2
Triton X 102 (EOU = 12–13)	1	1	0.4
Triton X 165 (EOU = 16)	1	2	2
Triton X 305 (EOU = 30)	1	3	2

[a]Relative resistance was determined as in Table 25.1 (Bech-Hansen et al., 1976). Collateral sensitivity is denoted by a value less than 1 and cross resistance by a value greater than 1. Derivation of the lines is as indicated in Figure 25.1. NT = not tested. The average number of ethylene oxide units (EOU) for each Triton X detergent is indicated (Rohm and Haas). Data for this table are compiled from Bech-Hansen et al. (1976) and Ling (1982).

Quite often, the mutant line displays the highest degree of resistance to the selecting compound relative to a number of drugs tested. This however is not always the case. In CH^RC4 cells, selected for resistance to colchicine, a significant higher resistance to gramicidin D is observed (Table 25.1). Along the same vein, in a human cell line selected for vinblastine resistance, much higher resistance to vincristine is observed (Ling et al., 1984).

B. Collateral Sensitivity

The complexity of the response of multidrug-resistant lines to drugs is manifested in another way. This is the observation that increased sensitivity (collateral sensitivity) to certain compounds is expressed, especially in lines which were selected for a high degree of resistance. This feature is not quite as common as cross resistance to unrelated drugs, but it is just as enigmatic. Some compounds to which colchicine-resistant lines display collaterally sensitivity are listed in Table 25.2. One aspect which all these compounds have in common is that they are hydrophobic and may interact with the cellular membrane. These properties are not unique to the collaterally sensitive compounds, however, since many of the cross-resistant compounds have similar properties.

The ability of multidrug-resistant cells to discriminate between relatively subtle changes in the structure of cytotoxic compounds is illustrated by the response of colchicine-resistant lines to a series of related non- ionic detergents of the Triton X family (Table 25.2). In CH^RC5 cells, degree of cross resistance or collateral sensitivity are observed depending on the number of ethylene oxide residues in the side chain of the detergent molecule involved. Thus cross resistance is observed to Triton X-165 with an average of 16 ethylene oxide residues and increasing collateral sensitivity is observed to the detergents with reduced numbers of ethylene oxide residues. A similar degree of collateral sensitivity to these detergents is not observed in the colchicine-resistant line CH^RC4.

Taken together, all the above findings indicate that at the present state of our understanding of the multidrug-resistance phenotype, it is not possible to predict *a priori* the drug sensitivity profile of a given multidrug-resistant line.

C. Cold Sensitivity for Growth

About 3% of the colchicine-resistant mutants isolated in a single step display cold sensitivity for growth (Ling, 1977). One line, CH^RE5, was characterized in detail. It expresses cross resistance to unrelated drugs such as puromycin, actinomycin D, and vinblastine. At 38.5°C CH^RE5 cells grow with a doubling time of about 22 hr while at 34°C they do not double in number. Employing synchronous cultures, the cold-sensitive block in CH^RE5 has been determined to be located prior to S in the G1 phase of the cell cycle. Revertants of CH^RE5 able to grow at 34°C have been isolated and found to display increased colchicine sensitivity.

D. Altered Drug Uptake

The basis of the resistant phenotype in the colchicine-resistant CHO lines has been examined by uptake and binding studies using labeled colchicine (Ling and Thompson, 1974; Ling et al., 1979). The mutant cells accumulated decreased amounts of drug proportionate to the degree of resistance, while the drug-binding ability of cytoplasmic extracts made from the same cells was not altered. Thus it was concluded that the basis of colchicine resistance resulted from reduced colchicine uptake, and that the mutants were altered at the plasma membrane level. Uptake of drugs such as puromycin, actinomycin D, and daunorubicin was also decreased in the colchicine-resistant lines (See et al., 1974; Kartner et al., 1983b). This indicates that the basis for the cross-resistance phenotype also results from a decreased accumulation of the drugs involved. The basis of the collateral-sensitivity phenotype is not known at this time, but it may involve increased drug permeability (O'Leary and Ling, unpublished observation).

The kinetics of colchicine uptake was examined in greater detail as an approach toward investigating the nature of the presumptive membrane alteration in the mutant lines. Kinetics consistent with unmediated diffusion was observed (Carlsen et al., 1976); moreover, no competition was observed with related compounds such as colcemid. "Membrane-active" agents such as nonionic detergents, local anesthetics, and a variety of drugs stimulated the uptake of colchicine in both the mutant and wild-type cell lines with the mutant cells requiring proportionately higher concentrations of these agents for maximum stimulation (Carlsen et al., 1976). The difference in drug accumulation and in cellular resistance between mutant and parental line is greatly reduced under these conditions (Carlsen et al., 1976; Ling et al., 1979). In a similar manner, compounds that interfere with calcium uptake are also able to circumvent the drug resistance phenotype in some lines (Tsuruo, 1983). It is not yet clear whether calcium is specifically involved since many calcium antagonists have membrane "fluidizing" properties and may thus stimulate drug permeability (Ramu et al., 1983).

Unexpectedly, it was also discovered that metabolic inhibitors such as cyanide or dinitrophenol, in the absence of metabolizable sugars, also stimulated the uptake of colchicine into CHO cells such that the difference between mutant and wild-type cells was eliminated (See et al., 1974). The uptake of other drugs such as puromycin and actinomycin D was similarly affected by metabolic inhibitors. This stimulation of drug uptake may be dependent on the ATP level inside the cell (Carlsen et al., 1977). The activation energy of colchicine uptake in the presence or absence of metabolic inhibitors was identical however, indicating that the mechanism of drug uptake was the same under both conditions (Ling et al., 1977).

These findings indicate that the net uptake of drugs into CHO cells is mediated by a complex process and may involve an energy-dependent mechanism not previously suspected. The possibility that a drug efflux mechanism may be involved has been proposed by some investigators (Danø, 1973; Skovsgaard, 1978; Inaba et al., 1981). Further studies of the multidrug-resistant mutants may provide insights into this process.

E. Cell Surface Alterations

Analysis of cell surface components of colchicine-resistant mutant lines indicated that an increased amount of a glycoprotein of approximately 170,000 daltons (*P*-glycoprotein) was found associated with the degree of drug resistance. This surface component was identified by labeling cell surface carbohydrate by incorporation of metabolic precursors of labeled amino acids and sugars, and by plasma membrane isolation and subsequent biochemical purification (Juliano et al., 1976; Juliano and Ling, 1976; Riordan and Ling, 1979). Moreover, antisera raised against mem-

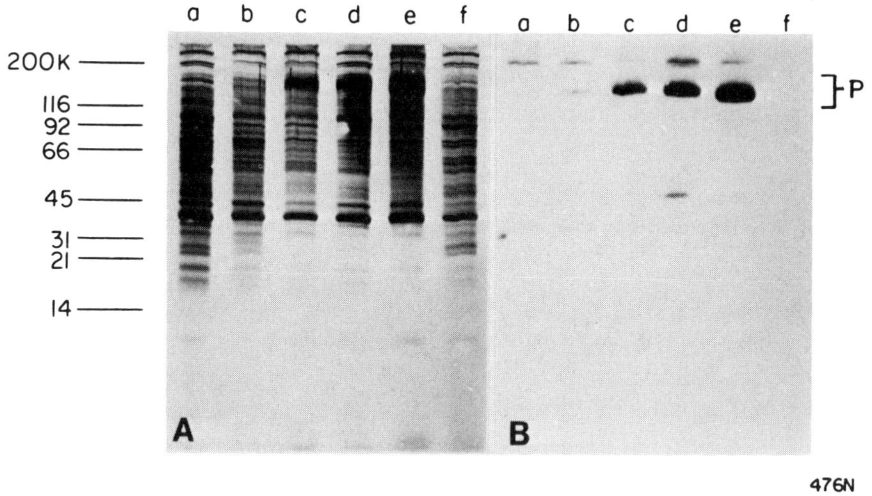

Figure 25.2. Analysis of membrane components. Cell membranes were prepared and fractionated by gel electrophoresis as described (Kartner et al., 1983a). In Panel A, membrane components were stained by the silver-staining methods of Switzer et al. (1979). In Panel B, membrane components were separated as in Panel A and electrophoretically transferred onto nitrocellulose paper. Western blot (Towbin et al., 1979) analysis was then performed using an absorbed rabbit antiserum specific for *P*-glycoprotein (Kartner et al., 1983b). Approximate molecular weights in kilodaltons are indicated. The position of the gel at 170,000 daltons where *P*-glycoprotein migrates is shown. Sources of membranes are in lane (a) AUXB1, (b) CH^RA3, (c) CH^RB3, (d) CH^RC5, (e) daunomycin-resistant line DNR^R51, and (f) a revertant of CH^RC5. See Kartner et al. (1983a) for details.

branes of resistant cells stained the *P*-glycoprotein component in an immunoblot (Western blot) procedure (Fig. 25.2). The amount of staining correlated with the degree of colchicine resistance. The use of such antisera to probe membrane components of multidrug-resistant lines from hamster, mouse, and human cells revealed the presence of "*P*-glycoprotein" in all these lines; moreover, the *P*-glycoprotein from each species fractionated at an identical apparent molecular weight of 170,000 daltons (Kartner et al., 1973a). These findings suggest that *P*-glycoprotein may be a highly conserved cell surface component and that its overexpression is closely associated with the multidrug-resistance phenotype. How this is mediated, or whether *P*-glycoprotein alone is able to mediate all aspects of the complex multidrug-resistance phenotype is not known at present.

Analysis of other membrane components such as the lipids and fatty acids in the colchicine-resistant mutants failed to reveal a consistent alteration associated with the resistance phenotype (Ling et al., 1979). Examination of the surface carbohydrates was undertaken by labeling as previously mentioned; however, analysis of possible contribution of carbohydrates to the multidrug-resistance phenotype was undertaken by further selecting for resistance to phytohemagglutinin, a mutation affecting the carbohydrate moiety in *P*-glycoprotein and a number of other carbohydrate-containing components (Ling et al., 1983). No obvious alteration in the multidrug-resistance phenotype was observed, suggesting that the bulk of the cell surface carbohydrates do not play a major role in the multidrug-resistance phenotype. Beck and Cirtain (1982) arrived at a similar conclusion in their study in which a multidrug-resistant human line was treated with tunicamycin or with pronase.

A variety of cell surface changes have been reported in different multidrug-resistant lines (Bosmann 1971; Juliano and Ling, 1976; Beck et al., 1979; Biedler and Peterson, 1981; Garman and Center, 1982; Wheeler et al., 1982; Biedler et al., 1983; Ramu et al., 1983). Many of the changes described are consistent with an overexpression of a *P*-glycoproteinlike component. Whether or not other changes identified in the lipid domain are related to *P*-glycoprotein expression is not known.

F. Other Molecular Alterations

An increased expression of a cytoplasmic component of about 20,000 daltons has been reported in some multidrug-resistance lines (Meyers and Biedler, 1981; Kuo et al., 1982; Beck, 1983). How such a component participates in the multidrug-resistance phenotype or whether or not it is related to the cell surface *P*-glycoprotein is not known. At present, it appears that overexpression of the surface *P*-glycoprotein is the most consistent molecular indicator of the multidrug-resistance phenotype.

IV. GENETIC STUDIES

As can be appreciated from the preceding review, the multidrug-resistance phenotype is complex. Genetic analyses are particularly useful for delineating alterations that are necessary and sufficient for a particular phenotype. As mentioned above, independent single-step clonal isolates express the multidrug-resistance phenotype. This provides strong evidence that the multiple expressions of drug resistance and collateral sensitivity are the consequence of the same genetic change. The fact that high resistance (100-fold or more) could be obtained only from multiple selection steps (Ling and Thompson, 1974) implicates the involvement of a gene dosage mechanism in generating cells of high multidrug resistance.

A. Revertants

Revertants were isolated from the colchicine-resistant lines by taking advantage of the multidrug-resistance phenotype. They were obtained directly by selecting for resistance to collaterally sensitive compounds such as 1-dehydrotestosterone (Baker and Ling, 1978). An indirect approach, selecting for sensitivity to inhibition of labeled amino acid incorporation by puromycin, was also used. Cells resistant to puromycin were able to incorporate enough labeled amino acid within a few hours in the presence of puromycin to be killed by radiation. Under the same conditions, revertant cells sensitive to puromycin incorporated little labeled amino acids and survive (Ling, 1975). In cases where colchicine resistance was accompanied by cold sensitivity for growth, revertants were obtained by selection for the wild-type ability to grow at high temperatures (Ling, 1977). In each case, reversion of the multidrug-resistance phenotype was observed such that cross resistance and collateral sensitivity to different drugs reverted toward the wild-type phenotype. In a few instances, revertant lines were examined for expression of *P*-glycoprotein. Invariably, the amount of *P*-glycoprotein was reduced (Ling, 1975; Riordan and Ling, 1979; Kartner et al., 1983a). Such revertant cells were selected in the absence of prior mutagen treatment using clonal isolation techniques. Thus these results are completely consistent with the multidrug-resistance phenotype being a pleiotropic phenotype.

B. Cell Hybrids

Cell–cell hybrid studies were undertaken to examine the dominant or recessive nature of the multidrug-resistance phenotype in the colchicine-resistant lines (Ling and Baker, 1978). Multidrug resistance was expressed in a dominant, but incomplete manner. *P*-Glycoprotein was also expressed in the hybrid cells. The dominant nature of the multidrug-resistance phenotype precludes the analysis of this phenotype by complementation but does allow for the DNA-mediated transfer and selection of this phenotype.

C. DNA Transfection

DNA-mediated transfer using genomic DNA from colchicine-resistant cells CH^RC5 or from a more highly selected variant line (B30) was performed using the calcium phosphate precipitation technique and mouse L cells as recipient (Debenham et al. 1982; Robertson et al., 1984). Colchicine-resistant transfectants were obtained at frequencies of about 10^{-6} when DNAs from resistant cells were used, but with DNAs from sensitive cells, no transfectants were observed, resulting in a frequency calculated to be less than 3×10^{-8}. Although the transfection frequency for colchicine resistance was low under the stringent selective conditions used, there was a clear difference between DNAs from mutant and wild-type cells. Markers unlinked to colchicine resistance such as methotrexate resistance, ouabain resistance, and the thymidine kinase locus, were not coexpressed in the colchicine-resistant transfectants. Thus the fact that expression of *P*-glycoprotein and cross resistance to unrelated drugs are found in the transfectants is significant, and provides strong evidence that the expression of these phenotypes is intimately associated. Moreover, the transfectants were not phenotypically stable for colchicine resistance when cultured in the absence of drug. Both colchicine resistance, resistance to unrelated drugs, and *P*-glycoprotein overexpression were lost concordantly (Robertson et al., 1984). We conclude from the above series of experiments that the multidrug-resistance phenotype is most likely coded by a single pleiotropic multidrug-resistance gene.

It could not be concluded, however, that this presumptive multidrug resistance gene is the *P*-glycoprotein gene, since it was not possible to determine whether the *P*-glycoprotein expressed in the transfectants was of mouse or hamster origin due to the conserved nature of *P*-glycoprotein expressed (Debenham et al., 1982). For example, if *P*-glycoprotein expressed is of mouse origin, then the transfected DNA probably codes for another gene(s) which mediates the expression of mouse *P*-glycoprotein and the multidrug-resistance phenotype in mouse cells. Cloning of the hamster *P*-glycoprotein gene and subsequent transfection of the cloned gene to test for function should provide a more definitive answer to this question.

D. Karyotypic Analysis

Selection for increased colchicine resistance in a series of transfectants by continuous exposure to increasing drug concentrations resulted in lines able to grow in drugs 20-fold higher than the original selecting dose. Increased cross resistance to unrelated drugs and increased *P*-glycoprotein expression were observed along with increased double minute chromosomes (Robertson et al., 1984). Double minutes (DMs) along with homogeneously staining regions (HSRs) are thought to be karyotypic expression of gene amplification in a number of well-studied systems (Cowell, 1982). The observation that the number of DMs in the transfectants quantitatively correlates with multidrug resistance and *P*-glycoprotein expression allows the speculation that the amplification of *P*-glycoprotein gene leads to *P*-glycoprotein overexpression and results in multidrug resistance (Robertson et al., 1984). Such karyotypic markers have been observed in different drug-resistant lines expressing the multidrug-resistance phenotype (Biedler and Peterson, 1981; Baskin et al., 1981; Kopnin, 1981; Kuo et al., 1982).

E. DNA Analysis

Recently, Roninson (1983) has developed an elegant *in situ* renaturation technique to detect fractionated amplified DNA sequences in mammalian cells. In examining two independently selected highly multidrug-resistant hamster lines (one of them is CH^RC5 of Fig. 25.1) with this approach, common amplified sequences could be found associated with these mutants (Roninson et al., 1984).

V. CONCLUDING REMARKS

Multidrug resistance is clearly a complex but important phenotype. Observation of multidrug resistance has been made in a variety of cell lines and in transplantable tumors (Skipper et al., 1972; Ling, 1982; Ling et al., 1983; Schabel Jr. et al., 1983). As reviewed in this chapter, progress made in elucidating aspects of this phenotype was greatly facilitated by genetic analysis of the colchicine-resistant CHO cell lines. Findings from these and other studies point to several areas which deserve to be investigated further. First, genetic analyses indicate that the variety of phenotypes expressed in such mutants likely result from a *pleiotropic* mutation rather than from different multiple genetic changes. Gene amplification appears to be the genetic mechanism involved. In this context, it is recognized that the amplified unit can contain many genes; however, the cell surface *P*-glycoprotein appears to be overexpressed, and is the most consistent molecular marker associated with multidrug resistance. Whether or not overexpression of *P*-glycoprotein alone is sufficient to effect the repertoire of phenotypes found in multidrug-resistant mutants is not known and remains to be demonstrated. Isolation of the *P*-glycoprotein gene by cloning along with gene transfer into a variety of cells to test for function will be required to answer this question in a rigorous manner.

Second, characterization of the pleiotropic phenotypes displayed by multidrug-resistant mutants has revealed the ability of mammalian cells to modulate permeability to a wide variety of apparently passively diffusing compounds in a manner that had not been appreciated previously. It apparently involves an energy-dependent mechanism. Further study to delineate the components of this mechanism and what role *P*-glycoprotein plays will add significantly to our understanding of membrane function in mammalian cells.

Third, the demonstration that multidrug-resistance mutations are relatively common in mammalian cell lines and in transplantable tumors suggests that such mutations could occur during progression of human neoplasms (Goldie and Coldman, 1982; Ling, 1982; Curt et al., 1983; De Vita Jr., 1983). This possibility has wide clinical implications and is currently being investigated (Ling, 1982; Ling et al., 1984). Multidrug-resistance mutations could render a malignant stem cell resistant to combination chemotherapy. In advanced malignancies, development of resistant disease to a combination of drugs is frequently encountered, thus it seems possible that multidrug-resistance mutations play a role in this aspect of cancer progression. In this context, the finding that overexpression of the highly conserved *P*-glycoprotein is intimately associated with the multidrug-resistance phenotype provides a convenient and sensitive molecular marker to investigate this question with human biopsy samples. Continued studies of well-defined model systems of multidrug resistance will be required to provide new tools and insights to investigate the more difficult problem of drug resistance in patients.

ACKNOWLEDGMENTS

I thank my colleagues for contributing to the work described here. This work was supported by the Medical Research Council of Canada, the National Cancer Institute of Canada, and the Ontario Cancer Treatment and Research Foundation.

REFERENCES

Baker, R. M. and Ling, V. (1978). In *Methods in Membrane Biology* (E. Korn, ed.), Plenum, New York, London, Vol. 9, pp. 337–384.

Baskin, F., Rosenberg, R. N., and Dev, V. (1981). *Proc. Natl. Acad. Sci. USA* **78**, 3654–3658.

Bech-Hansen, N. T., Till, J. E., and Ling, V. (1976). *J. Cell Physiol.* **88**, 23–31.

Beck, W. T. (1983). *Cancer Treat. Rep.* **67**, 875–882.

Beck, W. T. and Cirtain, M. C. (1982). *Cancer Res.* **42**, 184–189.

Beck, W. T., Mueller, T. J., and Tanzer, L. R. (1979). *Cancer Res.* **39**, 2070–2076.

Biedler, J. L., Chang, T., Meyers, M. B., Peterson, R. H. F., and Spengler, B. A. (1983). *Cancer Treat. Rep.* **67**, 859–867.

Biedler, J. L. and Peterson, R. H. F. (1981). In *Molecular Actions and Targets for Cancer Chemotherapeutic Agents* (A. C. Sartorelli, J. S. Lazo, and J. R. Bertino, eds.), Academic Press, New York, pp. 453–482.

Bosmann, H. B. (1971). *Nature* **233**, 566–569.

Carlsen, S. A., Till, J. E., and Ling, V. (1976). *Biochim. Biophys. Acta.* **455**, 900–912.

Carlsen, S. A., Till, J. E., and Ling, V. (1977). *Biochim. Biophys. Acta.* **467**, 238–250.

Cowell, J. K. (1982). *Ann. Rev. Genet.* **16**, 21–59.

Curt, G. A., Clendeninn, N. J., and Chabner, B. A. (1983). *Cancer Treat. Rep.* **68**, 87–99.

Danø, K. (1973). *Biochim. Biophys. Acta.* **323**, 466–483.

Debenham, P. G., Kartner, N., Ling, V., Siminovitch, L., and Riordan, J. R. (1982). *Mol. Cell. Biol.* **2**, 881–889.

De Vita, Jr., V. (1983). *Cancer* **51**, 1209–1220.

Flintoff, W. F., Davidson, S. V., and Siminovitch, L. (1976). *Somatic Cell Genet.* **2**, 245–261.

Garman, D. and Center, M. S. (1982). *Biochem. Biophys. Res. Commun.* **105**, 157–163.

Goldie, J. H. and Coldman, A. J. (1982). In *Drug and Hormone Resistance in Neoplasia* (N. Bruchovsky and J. H. Goldie, eds.), CRC Press, Miami, Vol. II, pp. 111–127.

Gupta, R. S. (1983a). *J. Cell. Physiol.* **114**, 137–144.

Gupta, R. S. (1983b). *Cancer Res.* **43**, 1568–1574.

Inaba, M., Fujikura, R. and Sakurai, Y. (1981). *Biochem. Pharmacol.* **30**, 1863–1865.

Juliano, R. L. and Ling, V. (1976). *Biochim. Biophys. Acta.* **455**, 152–162.

Juliano, R., Ling, V., and Graves, J. (1976). *J. Supramolec. Struct.* **4**, 521–526.

Kartner, N., Riordan, J. R., and Ling, V. (1983a). *Science* **221**, 1285–1288.

Kartner, N., Shales, M., Riordan, J. R., and Ling, V. (1983b). *Cancer Res.* **43**, 4413–4419.

Kopnin, B. P. (1981). *Cytogenet. Cell Gent.* **30**, 11–14.

Kuo, T., Pathak, S., Ramagli, L., Rodiguez, L., and Hsu, T. C. (1982). In *Gene Amplification* (R. T. Schimke, ed.), Cold Spring Harbor Laboratory, New York, pp. 53–57.

Ling, V. (1975). *Canad. J. Genet. Cytology* **17**, 503–515.

Ling, V. (1977). *J. Cell. Physiol.* **91**, 209–224.

Ling, V. (1982). In *Drug and Hormone Resistance in Neoplasia* (N. Bruchovsky and J. H. Goldie, eds.), CRC Press, Miami, Vol. 1, pp. 1–19.

Ling, V., Aubin, J. E., Chase, A., and Sarangi, F. (1979). *Cell* **18**, 423–430.

Ling, V. and Baker, R. M. (1978). *Somat. Cell Genet.* **4**, 193–200.

Ling, V., Carlsen, S. A., and See, Y. P. (1977). In *Advances in Experimental Medicine and Biology* (M. W. Miller and R. Shamoo, eds.), Plenum, New York, Vol. 84, pp. 247–264.

Ling, V., Gerlach, J., and Kartner, N. (1984). *Breast Cancer Res. Treat.* **4**, 89–94.

Ling, V., Kartner, N., Sudo, T., Siminovitch, L., and Riordan, J. R. (1983). *Cancer Treat. Rep.* **67**, 869–874.

Ling, V. and Thompson, L. H. (1974). *J. Cell. Physiol.* **83**, 103–116.

Meyers, M. B. and Biedler, J. L. (1981). *Biochem. Biophys. Res. Commun.* **99**, 228–235.

Ramu, A., Shan, T., and Glaubiger, D. (1983). *Cancer Treat. Rep.* **67**, 895–899.

Rauscher III, F. J., Beerman, A., and Baker, R. M. (1984). (Abstract) *Proceedings of the AACR* (in press).

Riordan, J. R. and Ling, V. (1979). *J. Biol. Chem.* **254**, 12701–12705.

Robertson, S. M., Ling, V., and Stanners, C. P. (1984). *Molec. Cell. Biol.* **4**, 500–506.

Roninson, I. B. (1983). *Nucleic Acids Res.* **11**, 5413–5431.

Roninson, I., Abelson, H. T., Hausman, D. E., Howell, N., and Varshavsky, A. (1984). *Nature* **309**, 626–628.

Schabel, Jr., F. M., Skipper, H. E., Trader, M. W., Laster, Jr., W. R., Griswold, D. P., and Corbett, T. H. (1983). *Cancer Treat. Rep.* **67**, 905–922.

See, Y. P., Carlsen, S. A., Till, J. E., and Ling, V. (1974). *Biochim. Biophys. Acta.* **373**, 242–252.

Skipper, H. E., Hutchison, D. J., Schabel, F. M., Schmidt, L. H., Goldin, A., Brockman, R. W., Venditti, J. M., and Wodinsky, I. (1972). *Cancer Chemo. Report part 1* **56**, 493–498.

Skovsgaard, T. (1978). *Cancer Res.* **38**, 4722–4727.

Switzer, R. C., Merril, C. R., and Shifrin, S. (1979). *Anal. Biochem.* **98**, 231–237.

Thompson, L. H. and Baker, R. M. (1973). In *Methods in Cell Biology* (D. M. Prescott, ed.), Academic Press, New York, Vol. 6, pp. 209–281.

Tsuruo, T. (1983). *Cancer Treat. Rep.* **67**, 889–894.

Wheeler, C., Rader, R., and Kessel, D. (1982). *Biochem. Pharmacol.* **31**, 2691–2693.

CHAPTER 26

USING CHINESE HAMSTER CELLS TO STUDY MALIGNANT TRANSFORMATION

Charles W. Roth
Unité d'Immunoparasitologie
Institut Pasteur
Paris, France

I. INTRODUCTION

Recent studies have provided illuminating information on genes that can transform normal differentiated cells into cells that have lost their differentiated characteristics and growth control. These studies have shown that several of the RNA tumor viruses contain "transforming genes" or "oncogenes" that were derived from normal cellular genes. Other tumor viruses may initiate tumors by "activating" cellular homologs of those viral oncogenes. Another exciting observation has been the finding that the protein product encoded by some of the oncogenes has an unusual protein kinase activity that phosphorylates proteins on a tyrosine residue. These results have provided new insight into the questions of what oncogenes are and how some tumor viruses have developed. In contrast, our knowledge of how viral or cellular oncogenes bring about transformation is much less advanced.

This chapter will discuss the use of Chinese hamster cells to examine the question of how mammalian cells become transformed. The first section will present a short review of genetic studies that use somatic cell hybrids to examine spontaneous transformation in mammalian cell lines. In the second section, studies to examine how a specific viral oncogene, the *src* gene of Rous sarcoma virus, interacts with Chinese hamster ovary cells to bring about a transformed phenotype will be presented.

II. THE USE OF SOMATIC CELL HYBRIDS TO STUDY TRANSFORMATION

A. Background

Numerous studies have shown that transformed cells have many different properties from their counterpart nontransformed cells. In this chapter, we will define a transformed cell by its ability to form tumors in any suitable animal system, that is, we are equating transformation with tumorigenicity. In general, there are certain characteristics of cultured cells which are good, but not perfect, predictors of their ability to form tumors in animals. The tissue culture phenotype of transformation includes (1) the ability of the cells to grow in the presence of low concentrations of serum or growth factors, (2) the ability of the cells to grow in suspension or in semisolid medium, (3) the failure of the cells to stop growing at confluence which results in overlapping of nuclei as the cells grow on top of each other, and (4) an altered cellular morphology that is usually characterized by the loss of adhesion to the substrate and a reorganization of the cytoskeleton resulting in a fusiform or round shape.

One of the basic questions in cell biology has been to determine what type of genetic change in mammalian cells can bring about this large assortment of new phenotypes or properties seen in transformed cells. In a simple case one could imagine that transformed cells had "lost" a repressor that normally kept the genes coding for the transformed phenotype turned off. Alternatively, transformed cells might contain a "new" gene that coded for a product that "turned on" the transformation phenotype. The techniques of somatic cell hybridization and gene transfer have been used to distinguish between such alternatives and provide a better understanding of the types of genetic changes that take place in tumor cells.

As has been discussed in Chapter 7, somatic cells can be fused to form hybrid cells that contain both the nuclei and cytoplasm of the parent cells. By starting with cell lines containing the appropriate selectable

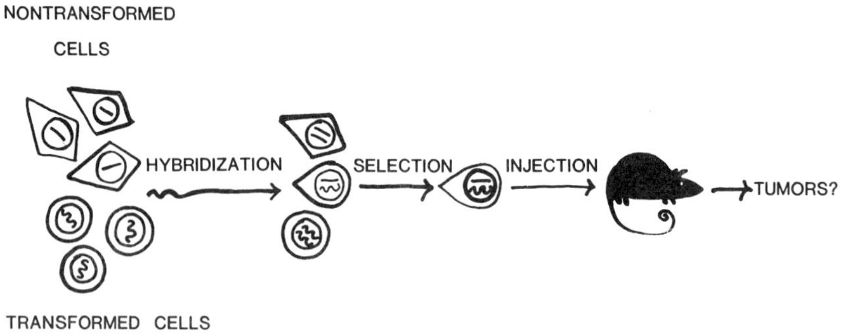

Figure 26.1. Schematic representation of how somatic cell hybrids are used to test for dominance or recessiveness of the transformation phenotype.

markers, the fusion product of cells from each of the two lines that one wants to examine can be isolated. Thus cell lines having one copy of the genes from a transformed cell and another set of genes from a nontransformed cell can be isolated. This scheme is shown diagramatically in Figure 26.1. If transformation is a dominant phenotype such as one would expect in our example where a new gene turned on transformation, then the hybrid cells should have a transformed phenotype. If transformation results from a lost gene product that can be supplied by the nontransformed partner, then the transformed phenotype should be recessive and the hybrid cells should be nontumorigenic.

Although the theory is simple, there are numerous technical difficulties in carrying out an analysis with hybrid tetraploid cells. The largest problem is the instability of the tetraploid karyotype of the cells. Although hybrid cells might be expected to have a karyotype equal to the sum of the karyotypes of the two parent cells, this expectation is usually not met. Even without deliberate selection, there is chromosomal assortment and loss during culture. Additionally, it is very difficult to predict or control the selection process that cells experience when they are injected into animals. Large numbers of cells are usually injected into animals to test for tumorigenicity, and there is often a question of whether the tumor resulted from the growth of a random portion of the injected cells or from the selection of a specific subset of cells in the injected population. Bearing in mind the problems that may be encountered in this type of analysis, informative and interesting experiments can be conducted.

Early experiments to examine the question of the dominance of the transformed phenotype were carried out by Barski's laboratory and by Ephrussi and his collaborators. Since hybrid cells formed from tumorigenic and nontumorigenic mouse cells formed tumors when they were injected into mice, these workers concluded that "transformation" is a dominant genetic trait (Barski et al., 1961; Defendi et al., 1967). In a large series of similar experiments, Harris and his co-workers also observed tumors when hybrid mouse cells were injected into animals. When these latter workers reisolated the cells from the tumors, they observed that the cells from the tumors consistently had fewer chromosomes than the hybrid cells with which they had inoculated the animals (Harris et al., 1969; Weiner et al., 1971).This group proposed that the initial hybrid cells were not tumorigenic but that a subset of cells that had lost chromosomes were tumorigenic and grew into tumors. In this analysis, transformation behaves like a recessive genetic trait. More recent studies by Stanbridge and his co-workers (Stanbridge et al., 1982) using a human carcinoma cell line (HeLa) and normal human fibroblastic cell lines (IMR-90 and WI-38) with careful attention being paid to the karyotype of the hybrid cell lines suggest that, at least in this case, the tumorigenicity of the HeLa cell line is a recessive genetic trait. It is important to point out that it is not proven that tumorigenicity is a recessive trait in all transformed cells. There is strong evidence that tumorigenicity behaves like a dominant genetic trait in some virally transformed cell lines. For a more detailed survey of the literature on this topic the reader is referred to the review of Ozer and Jha (1977).

B. The Use of Chinese Hamster Cell Hybrids

1. Transformation and Tumorigenesis

Because they recognized the problems associated with the extreme instability of the karyotype of the mouse cells which they had been studying, Barski's laboratory started using Chinese hamster ovary (CHO) cells, which have a more stable karyotype. Berebbi and Barski (1971) observed that a hybrid cell line formed from a tumorigenic CHO derivative (DC-3F) and a nontumorigenic cell line derived from DC-3F formed tumors at high frequency and maintained a near tetraploid chromosome number. A single hybrid cell line termed HyC was isolated and used to examine tumorigenicity. When 2×10^6 Hy-C cells were injected into the cheek pouch of Syrian hamsters treated with cortisone, 22 of 25 animals developed tumors. When the cells from several of the tumors were examined, they were shown to have modal chromosome counts of 42–44 chromosomes, whereas the hybrid cells used to inject the animal had a modal chromosome number of 39–40. The hybrid, Hy-C, when grown in tissue culture, gave off subclones that were nontumorigenic. Blanchard et al. (1973) examined six random subclones of Hy-C and found that one subclone gave no tumors in 10 animals, while another was highly tumorigenic (14 of 16 animals developed tumors). The other clones gave intermediate results. The subclones had lost chromosomes, but no correlation could be made between the loss of a specific chromosome and the loss of tumorigenicity. From these results the authors concluded that tumorigenicity is a dominant trait in CHO cells.

The one obvious weakness in the study of Barski and his co-workers is that they worked with such a limited sample size, that is, one hybrid cell line. One can argue that their result is based on an unusual hybrid that has undergone some alterations that simply are not detected by the karyotype analysis they used. As is discussed in Chapter 27, Sager's laboratory has studied a large number of Chinese hamster cell hybrids to examine the question of genetic control of tumorigenesis, and they have concluded that in general the ability to form tumors behaves as a recessive genetic trait. However, in one unusual hybrid this trait appears to be dominant. Sager and Kovac (1978,1979) studied numerous properties of this unusual tumor-derived cell line termed H401-7t clone C or simply clone tC. They observed that this tumor-derived cell line was resistant to BrdU (5-bromo-2-deoxyuridine) and sensitive to TG (6-thioguanine), properties that are the same as those of one of the original parents in the initial hybrid and the opposite of those selected for when the hybrid cell lines were formed. Primarily based on this change in the pattern of drug resistance, Sager and Kovac (1979) suggested that the tumor arose from a tetraploid cell that was TG-resistant and BrdU sensitive and that this hybrid cell lost a large number of chromosomes and became TG sensitive and BrdU resistant. This nearly diploid cell then underwent endoreduplication to become tetraploid and grew out to form the tumor. This interesting speculation has the problem of explaining what happened to the intermediate nearly diploid cells from which the tetraploid tumor cells arose, since there is no obvious selection against the diploid cells.

The result of Sager's studies with Chinese hamster fibroblast cell lines indicating that transformation generally behaves like a recessive trait during hybrid cell analysis is in accordance with many other studies with other cell lines. These results give rise to the exciting prediction that genes that can suppress the transformation phenotype can be isolated and used to convert tumor cells to a nontransformed phenotype. The identification of genes that can suppress the transformed phenotype would have profound effects on our understanding of how oncogenes function to create the transformation phenotype.

At the same time one must be cautious not to lump all transformed cells in one category. There are numerous ways in which cells can become transformed and some pathways to transformation may behave as a recessive trait while others may act as a dominant trait. For example, it is not easy to understand how some viruses, such as Rous sarcoma virus, can transform appropriate host cells with a high efficiency if these cells contain general repressors of transformation. Additionally, in gene transfer experiments such as will be discussed in the next section, the genes governing some types of transformation appear to act in a dominant manner. Attention needs to be given to considerations of how a particular oncogene functions and whether the concentration of the oncogene product is important for converting a normal cell into a transformed cell. It is possible that hybrid cells suppress the transformed phenotype simply by diluting the concentration of the transforming gene product below its effective concentration. Until the molecular mechanism by which the cells being studied in the hybrid analyses is understood, great care should be taken before extrapolating conclusions from tetraploid cells to the conditions in normal diploid cells.

C. Gene Transfer Studies

Despite the weight of evidence in the literature suggesting that, in most cases, spontaneously transformed cells carry suppressible genes encoding the transformation phenotype, a number of laboratories undertook the isolation of transforming genes using the technique of gene transfer. This technique is described in detail in Chapter 8. The logic behind the attempts to isolate transforming genes was that in an appropriate genetic background, an oncogene could be expressed and would turn a "normal" cell into a transformed cell, that is, act in a dominant manner. Dominant is used here in the sense of a gene which is expressed and produces a recognizable phenotype irrespective of the presence of a normal second allele of the gene. In these experiments DNA isolated from transformed cells of various origins (chemically transformed, spontaneously transformed, or virally transformed) is applied, generally in the form of a calcium phosphate precipitate, to "normal" cells. The recipient cells in most of the experiments have been NIH 3T3 cells which have "normal" growth properties in tissue culture, that is, they show normal density-dependent growth control, require high concentrations of serum or growth factors, and do not form tumors when injected into immune-deficient animals. In addition, these cells take up exogenous DNA and become "transfected" for a number of genetic markers including transformation at a relatively high frequency. Using this cell line, a number of laboratories including those of Cooper, Weinberg, and Wigler have been able to show that many tumorigenic human and nonhuman cultured cells contain genes able to transform NIH 3T3 cells. These studies, which have recently been reviewed by Weinberg (1983), have greatly expanded our understanding of oncogenes and their activation in oncogenesis. Several genes have been isolated with a potential role in the development of human cancer.

It is worth noting that not all tumorigenic cell types yield genes that transform indicator cells such as NIH 3T3 cells. For example, DNA from HeLa cells does not transform the NIH cell line. One interpretation is that the HeLa cells contains a recessive gene(s) that is suppressed in the "normal" NIH cells as was concluded from the somatic cell hybridization studies. Another way to look at the result is to conclude that some types of oncogenes only function in certain cell types, thus some cells are resistant to or suppress the action of certain oncogenes. One curious aspect of the gene isolation studies is that most of the isolated oncogenes fall into a family of genes termed *ras,* which were originally identified as the oncogenes in several rodent RNA tumor viruses. Several recent studies suggest that there are at least two oncogenes involved in tumor formation, one of the genes is assayable in the NIH 3T3 system and one which is not detected by that system (Diamond et al., 1983; Land et al., 1983).

Sager's laboratory has shown that a clone of the *ras* gene isolated from a human bladder carcinoma cell line can transform the "normal" Chinese hamster embryo fibroblast cell line (CHEF-18) but not a line of diploid fibroblasts derived from human foreskin (Sager et al., 1983). The authors conclude that since the cloned *ras* gene does not transform the human line while other oncogenes can, the *ras* genes isolated from tumors may have nothing to do with initiating human tumors. While their point that one should not rush to conclusions on the basis of correlations is valid, it is also important to realize that a negative result in one heterologous experimental system does not truly indicate how a gene might function in another cell type under different hormonal stimuli, etc.

III. THE USE OF KNOWN ONCOGENES TO STUDY TRANSFORMATION AND TUMORIGENESIS

A. The Rous Sarcoma Virus *src* Gene in CHO Cells

1. Background

A second approach toward studying the transformation process in mammalian cells is to transform cells by introducing a known oncogene into nontransformed cells and to examine how that gene functions. The oncogene might be introduced either as part of a transforming virus or by another method such as DNA gene transfer. The important concept is that the cells are transformed by a known gene(s) so that specific interactions between the gene product and cellular components can be studied. Several approaches can be used to study the interactions between the oncogene and the cell. These approaches are, broadly speaking, a genetic approach and a biochemical approach. The genetic approach is to isolate cellular mutants that do not respond to the transforming effect of the oncogene and to determine what cellular components have been altered in the mutant cells. The biochemical approach is to study transformed and nontransformed cells to determine what biochemical changes have taken place in the transformed cells. This section will discuss the use of these approaches to study the transformation of CHO cells by Rous sarcoma virus.

Rous sarcoma virus (RSV) has numerous advantages for the study of transformation in mammalian cells. These advantages include the following: RSV is a small RNA tumor virus that has been extensively studied. The virus can transform both mammalian and avian cells but grows only in avian cells. The virus has a single oncogene which is known to code for a protein with an unusual protein kinase activity that phosphorylates tyrosine residues. The protein encoded by the oncogene has been purified and antibodies to the protein are available. The transforming protein is itself a phosphoprotein with a molecular weight of approximately 60,000 daltons and is called $pp60^{v\text{-}src}$ to distinguish it from a cellular homolog with similar properties (Collett et al., 1978).

This section will discuss the use of the genetic approach to study the interaction between RSV and CHO cells. The goals of the genetic approach are to determine which cellular components are in the "pathway" through which the transforming protein *src* acts and to determine whether there is a single pathway leading to the transformed phenotypes. The procedure is to transform a CHO cell line with RSV and to then select mutants that are no longer transformed from the transformed cells. Among the "revertant" cells isolated should be some that have alterations that make the cell unable to respond to the *src* protein. The altered gene products in the revertant cells can then be identified to establish a pathway through which the *src* protein works. If there is a single pathway of *src* action, then all the revertant cells should have lost all the mutant phenotypes simultaneously. If there are multiple pathways, then at least some of the isolated revertants should have lost some of their transformed properties while retaining others that are in the unaffected pathway(s). This genetic approach toward studying the mechanism of cellular transformation by RSV was first used by Deng et al. (1974) and Boettiger (1974) with baby hamster kidney cells. They isolated revertants with reduced expression of the RSV genome. More recently, Faras' laboratory has studied revertants of RSV transformed field vole cells (Lau et al., 1979) and the laboratories of Wyke (Wyke et al., 1980) and Varmus (Varmus et al., 1981) have isolated revertants of RSV-transformed rat cells. Lau et al. (1979) studied a revertant of RSV-transformed vole cells that was morphologically like nontransformed cells but which still formed tumors when injected into immune-deficient nude mice. Wyke and Varmus have isolated a large number of revertant rat cells which primarily have alterations in the tumor virus. As can be seen from this short summary, there are numerous ways in which an RSV-transformed cell can revert to a nontransformed phenotype.

The primary decision to be made before commencing this type of study is to choose the cell line to be used. Since one wants to be able to compare the transformants and revertants with the parent cell line, an established cell line that has many of the characteristics of a primary nontransformed cell culture is valuable even though it is "partially" transformed, that is, it can be maintained indefinitely in cell culture. The cell line (LR-73) chosen for the work to be described subsequently is one isolated by Pollard and Stanners (1979) as a revertant of the Toronto line of CHO cells. This cell line morphologically resembles nontransformed cells, does not grow in suspension, and does not form tumors in nude mice. Cells of a subclone of cell line LR-73, termed 10900, have been infected by RSV subgroup D, and transformed cells were isolated either based on their ability to form foci on the surface of culture dishes or on their ability to form colonies in suspension. The cells have been judged to be transformed by RSV based on the fact that the cells contain RSV DNA, produce $pp60^{v\text{-}src}$, and lose the transformed phenotype when they lose the RSV genome. Most of the studies have been done with a single clone of RSV-transformed cells, clone 11001 (Roth et al., unpublished data).

Starting with a transformed cell population, the number of revertant cells in the population is generally very small even if the cells have been mutagenized, so a method to select or enrich for revertant cells is necessary. The most commonly employed selection method takes advantage of the fact that transformed cells have different growth control mechanisms than nontransformed cells. Thus RSV-transformed CHO cells can grow in suspension while the revertant cells should not be able to grow under those conditions. We have found that when RSV-transformed CHO cells growing on a solid surface are transferred to a suspension culture at 10^5 cells/mL in growth medium containing BrdU and incubated with shaking for 48 hr, 99% of the transformed cells are killed upon exposure to fluorescent light. Using this selection method the surviving cells are plated out and the colonies are screened by eye to detect colonies containing cells with a nontransformed morphology, that is, "flat" cells. These colonies can be easily detected because the transformed cells form very round colonies that do not attach tightly to the culture dish, while the desired colonies form large colonies with the cells in a monolayer. After confirming the cell morphology microscopically, the "flat" colonies are picked and purified.

An alternative procedure is to select colonies that stick tightly to the surface of culture flasks. In this procedure, the transformed cell population is plated in plastic culture flasks, and when the cells are at high density, the medium is removed and replaced with buffered saline. The flasks are then hit sharply against the palm of the hand to knock the loosely attached transformed cells into suspension. The suspended cells are removed, the flask is washed several times with buffered saline, and fresh growth medium is added. The cells are again allowed to grow to confluence and the procedure is repeated. After several rounds of enrichment, flat colonies containing cells with a nontransformed morphology are detected and isolated.

As mentioned above, transformed cells can revert to a nontransformed phenotype for a number of reasons. The revertant clones must be studied

TABLE 26.1
Classes of Revertants Isolated from RSV-Transformed Chinese Hamster Ovary Cells

Class	Alteration	Distinguishing Characteristics	Number
Viral changes	1. Lost Virus	No viral DNA in cells No viral rescue	16
	2. Viral Mutation	Viral DNA in cells Altered pp60 kinase *Cannot* rescue transforming virus Cells can be retransformed	3
Cellular alterations	1. Host Inactivation	Viral DNA in cells; Can rescue transforming virus Low pp60 kinase activity	2
	2. Mutation in Transformation Pathway	Cells *cannot* grow in suspension Cells *cannot* be retransformed Can rescue transforming virus High pp60 kinase activity Viral DNA in cells	"3"[a]

[a]The number in quotation marks indicates that the placement of the revertants in that class is tentative.

to determine the mechanism of their reversion in order to select those that reverted due to changes in cellular components rather than due to viral alterations. The mechanisms by which revertant cells might be generated along with features that can be used to distinguish among the various classes of mutants are listed in Table 26.1. One potential class of revertants includes cells that are not listed because they are not of interest for these studies since they contain cells that have become tetraploid. These clones appear much flatter than the diploid transformed cells but maintain a transformed phenotype.

Another group that is not relevant to the present study, although it is interesting for other reasons, is the class with viral alterations. This class is the largest class of revertants isolated in this study. Most of the revertants in this class have lost the RSV genome as judged by DNA hybridization experiments, although the mechanism of the deletion is not known. Another group in this class has retained viral DNA, but the cells do not have $pp60^{v\text{-}src}$ kinase activity, and the virus particles recovered when the revertant cells were fused with chicken embryo fibroblasts where the virus can replicate do not transform chicken cells. Presumably the virus in these cells either has a mutation in the *src* gene or in another viral gene that alters the expression of the *src* gene. Another group of mutants appears to have a cellular alteration that reduces the expression of the RSV genome in the cells. This conclusion is based on the observation that the expression of pp60$^{v\text{-}src}$ kinase activity is reduced about twofold in the revertant cells, but normal transforming viruses can be recovered when these cells are fused with chicken embryo fibroblasts. Small differences in level of expression of pp60$^{v\text{-}src}$ can also affect transformation of rat cells as shown in studies in which the *src* gene was placed under control of a steroid-inducible MMTV promoter (Jakobovits et al. 1984).

Revertants have also been recovered which appear to have alterations in host cell components, although this assumption is difficult to prove. The vital test, to demonstrate that the revertant cells are resistant to retransformation by RSV, cannot be carried out because the phenotype of the mutants is genetically very unstable and the rate of reversion to the transformed phenotype is greater than the rate of transformation of the CHO cells by RSV. In the absence of formal proof that the mutants are altered in a gene specifically related to transformation, the mutants have been tentatively placed into this category based on the following observations: (1) the cells contain high levels of pp60$^{v\text{-}src}$ kinase activity; (2) the cells have reverted based on several criteria, including morphology and ability to grow in suspension; and (3) subclones that retain the revertant morphology have variable levels of the pp60$^{v\text{-}src}$ kinase activity ranging from none to very high. The nature of the host cell alterations responsible for this revertant phenotype has not been determined.

These studies are only the first step toward understanding how pp60$^{v\text{-}src}$ interacts with cells to bring about the transformed phenotype. Many more mutants will have to be isolated and analyzed, and the biochemical alterations identified, before the goals of this work can be accomplished. The important conclusion of the work presented is that CHO cells can be transformed by RSV and that numerous classes of revertants can be isolated from the transformed cells. This conclusion indicates that RSV-transformed CHO cells are a workable system for accomplishing the difficult goals that were discussed in the beginning of this section.

The revertants that have been analyzed show that the favored method of reversion from the transformed phenotype is different in different cell lines. For example, in the Rat-1 cell line most of the revertants studied had mutations in the RSV genome (Varmus et al., 1981) and in the baby hamster kidney cells most of the revertants had reduced expression of the RSV genome (Boettiger, 1974), while in the CHO cells discussed here the main mode of reversion was loss of the virus. It must be kept in mind, however, that different clones of the same cell line transformed with the same virus may show a different pattern of reversion. This result has been observed with RSV-transformed Rat-1 cells (Chiswell et al., 1982) and in studies with RSV-transformed CHO cells (Roth, unpublished data).

With the information gained from these initial studies it should be possible to engineer CHO cell lines that yield fewer viral alterations and more cellular alterations. An example might be a cell containing multiple active copies of the RSV genome so that the probability of isolating a viral mutation is reduced.

Another very promising genetic approach for studying how the *src* gene product acts in cells is to alter the gene by site-specific mutagenesis. Several laboratories (Bryant and Parsons, 1983; Cross and Hanafusa, 1983; Snyder et al., 1983) have made specific alterations in the sites of the *src* gene thought to be important for the kinase activity and the transforming ability of $pp60^{v\text{-}src}$. In these studies alterations of the *src* protein such that the C-terminal phosphotyrosine was removed did not prevent the *src* genes from transforming chick (Bryant and Parsons, 1983; Cross and Hanafusa, 1983) or mouse cells (Snyder et al., 1983). Likewise, an alteration removing the N-terminal phosphoserine did not prevent the altered *src* gene from transforming chick cells (Cross and Hanafusa, 1983). This genetic technique is providing a powerful tool for studying which part of viral genes interact with cellular components and bring about transformation.

2. Biochemical Approach: Cyclic AMP

a. Cell Growth. Numerous studies such as those mentioned earlier concerning the suppression of the transformed phenotype in hybrid cells formed between transformed and nontransformed cells suggest that the physiological state of the cell has an effect on the transformed phenotype. Since cyclic AMP (cAMP) is known to have a wide variety of regulatory functions in mammalian cells, the effect of cAMP on RSV-transformed CHO cells was examined. When normal CHO cell lines, which are spontaneously transformed cells, are treated with cAMP, cell growth is inhibited and the cells undergo a rapid change in shape (see Chapter 23). When RSV-transformed CHO cells were treated with 1 m*M* 8-bromo-cyclic AMP (8Br-cAMP), an analog of cAMP that is taken up by mammalian cells and mimics the action of cAMP, they were not growth inhibited (Roth et al., 1982). In addition, none of the transformation characteristics such as cell shape or the ability to grow in suspension were altered by cAMP. The RSV-transformed CHO cells were also resistant to the growth-inhibitory effects of cholera toxin, which functions by activating adenylate cyclase and increasing the intracellular concentration of cAMP. This result indicates that the observed growth effects with 8Br-cAMP were not simply related to using a cAMP analog in the experiments. The nontransformed parent cell line of the RSV-transformed cells, spontaneously transformed cells from the same parent cell line, and revertants of the RSV-transformed cell line that had lost the virus all were sensitive to the morphological and growth-inhibitory effects of cAMP. Thus the resistance to growth inhibition and the other effects of cAMP by

RSV-transformed CHO cells are specifically related to transformation by RSV and not general properties of the parent cell line or of transformation.

Cyclic AMP acts in mammalian cells by activating cAMP-dependent protein kinases which then phosphorylate specific proteins. One reason why RSV-transformed CHO cells might be resistant to the effects of cAMP is that the cAMP-dependent protein kinases have been altered. This possibility was examined in the RSV transformed CHO cells by examining the two cAMP-dependent protein kinases present in CHO cells. No alterations in the activation of cAMP-dependent protein kinase activity, the chromotographic pattern of the enzymes on DEAE columns, or the ability of cAMP to stimulate the phosphorylation of a specific substrate *in vivo* could be detected. Based on these data, it was concluded that transformation of CHO cells by RSV was not making them resistant to the effects of cAMP through an easily detected alteration in cAMP-dependent protein kinase activity in the cells (Roth et al., 1982).

b. pp60$^{v\text{-}src}$ Kinase. Another approach to studying the effect of cAMP on RSV-transformed CHO cells was to examine whether cAMP had any effect on the transforming protein pp60$^{v\text{-}src}$. This protein has two sites of phosphorylation, one a serine residue near the N-terminus (Collett et al., 1979) and the other a tyrosine residue near the C-terminus (Hunter and Sefton, 1980). Since cAMP-dependent protein kinase normally phosphorylates proteins on serine or threonine residues, the phosphorylation of the N-terminus site might be altered in cells treated with cAMP. This possibility was examined by growing control and cAMP-treated cells in medium containing $^{32}P_i$ and then immunoprecipitating pp60$^{v\text{-}src}$ and measuring the relative incorporation of ^{32}P into each of the two sights. When determined either by separating the phosphorylated tryptic peptides or the two ends of the protein after digestion with protease V-8, phosphorylation of the serine site of phosphorylation was increased three- to fourfold in cAMP-treated cells relative to that of the tyrosine site (Roth et al., 1983). This increase in phosphorylation takes place in a time-dependent manner, increasing over the first 3–4 hr of cAMP treatment and then the stimulation decreases with longer periods of treatment. Although the reason for the decrease in the rate of phosphorylation is not clear, the same result was observed when the kinetics of another protein phosphorylated by cAMP-dependent protein kinase was examined in CHO cells (LeCam et al., 1981). This result suggests that cAMP acting through cAMP-dependent protein kinase can regulate the phosphorylation and possibly the activity of pp60$^{v\text{-}src}$.

To examine the suggestion that cAMP treatment of CHO cells might regulate the activity of pp60$^{v\text{-}src}$, the tyrosine-specific protein kinase activity of pp60 immunoprecipitated from control and cAMP-treated cells was measured. The immunoprecipitated protein kinase activity was stimulated several-fold in cAMP-treated cells, and like the increase in phosphorylation, the increase in kinase activity was time dependent. Treatment of the cells with cAMP increased the activity of both the soluble and the membrane-bound pp60$^{v\text{-}src}$ and did not have a significant effect on the distribution of the enzyme between the two intracellular locations. In addition, treatment of the cells with cAMP had no effect on the rate of synthesis of pp60$^{v\text{-}src}$, so the change in specific activity is most probably related to alterations of the enzyme.

c. Tumorigenesis. The stimulation of the phosphotyrosine transferase activity of pp60$^{v\text{-}src}$ in RSV-transformed cells treated with cAMP suggests that the transforming activity of pp60$^{v\text{-}src}$ might be increased in cAMP-treated cells. This suggestion is difficult to test directly because of the limited tests for transformation in tissue culture where conditions can be closely regulated and because it is impossible to regulate the level of cAMP in animals. The approach that was used was to treat cells with cholera toxin which binds to the cells and stimulates adenylate cyclase. This stimulation causes the cells to produce large amounts of cAMP for an extended period of time (Li et al., 1977). When the ability of the RSV-transformed cells to form tumors in immune-deficient nude mice was measured, pretreatment of the cells with cholera toxin dramatically increased the tumorigenicity of the cells.

The untreated RSV-transformed cells are only slightly tumorigenic in nude mice. Only 2 of 24 mice injected with 10^7 cells developed measurable tumors within 21 days. In contrast when the cells were treated with cholera toxin (100 ng/mL) for 30 min and then washed to remove unbound toxin before injection, 9 of the 25 animals injected developed tumors (Gottesman et al., 1984). This fourfold stimulation of tumor formation was specific for RSV-transformed cells. No stimulation of tumor formation was observed when the parent nontransformed cells, spontaneously transformed cells of the parent cell line, or revertants of the RSV-transformed cells that had lost the virus were tested in the same way. In fact, the tumorigenicity of the spontaneously transformed cells was reduced when they were treated with cholera toxin before being injected. This result correlates with the inhibition of the growth of spontaneously transformed CHO cell in tissue culture by cholera toxin and suggests that cholera toxin is having a direct effect on the treated cells.

When cells were isolated from several cholera-toxin-stimulated tumors, they were found to have the same karyotype as the injected RSV-transformed cells. In addition, the cells isolated from the tumors demonstrated the same stimulation of tumorigenicity by cholera toxin as the cell line from which they were derived. The conclusion from these experiments was that cholera toxin specifically stimulated the tumorigenicity of RSV-transformed CHO cells and that cholera toxin was acting directly on the physiology of the treated cells and not acting by altering the susceptibility of the host to tumors or by selecting tumorigenic mutants in the injected population.

d. Discussion. The experiments discussed above provide a set of correlations: (a) cAMP treatment of RSV-transformed CHO cells stimulates the phosphorylation of pp60$^{v\text{-}src}$ at the major serine site of the protein; (b) the phosphotyrosine kinase activity of pp60$^{v\text{-}src}$ is increased in cAMP-treated cells; (c) the tumorigenicity of RSV-transformed cells is increased by treatment with cholera toxin.

The hypothesis that has been drawn from these data is the following: When RSV-transformed CHO cells are treated with cAMP, the cAMP stimulates cAMP-dependent protein kinase which phosphorylates pp60$^{v\text{-}src}$. The phosphorylation of pp60$^{v\text{-}src}$ at the serine site increases the phosphotyrosine kinase activity of pp60$^{v\text{-}src}$ which phosphorylates a larger amount of its substrates in the cell and the cell becomes more tumorigenic. This is a simple, straightforward interpretation of the results but other hypotheses can be imagined that could account for the data.

The RSV-transformed CHO cells may provide a unique system for examining the role of phosphorylation on the mode of action of pp60$^{v\text{-}src}$. In most of the cell lines that are transformed by RSV, pp60$^{v\text{-}src}$ is highly phosphorylated. In chick cells (Sefton et al., 1980), vole cells (Lau et al., 1979) and rat cells (Gottesman et al., 1983), the N-terminal serine site has about 2.5 times more phosphate incorporated than does the C-terminal site. In chicken cells, Sefton et al. (1980) estimate that as much as 60% of the pp60$^{v\text{-}src}$ molecules are phosphorylated at the serine site. In the case of the RSV-transformed CHO cells, the two sites are normally phosphorylated to nearly the same extent and the N-terminal serine site only becomes much more phosphorylated than the C-terminal site after treatment of the cells with cAMP. Thus the effects of modulation of the phosphorylation are more dramatic in the CHO cells and may be more easily observed. In addition, because the RSV-transformed CHO cells are only weakly tumorigenic, it is easier to observe changes in their tumorigenicity. In contrast, chicken cells infected with RSV contain multiple copies if the virus in their genome (Saban et al., 1979) and probably have several times the amount of pp60$^{v\text{-}src}$ activity needed to transform the cells. In this situation only alterations in pp60$^{v\text{-}src}$ that drastically change its transforming ability will be observed.

The ability to observe changes in the tumorigenicity of RSV-transformed CHO cells has demonstrated the role that the cellular environment can make on the ability of cells to become tumors. Although the role of the physiological environment stimulating tumor selection or growth is not understood, the results with the RSV-transformed CHO cells indicate that such factors can play an important role. The observed effects of cAMP on tumorigenicity suggest that substances such as caffeine which can influence the level of cAMP in mammalian cells by reducing its turnover through inhibition of cyclic nucleotide phosphodiesterase may play a role in allowing some tumors to become established.

The great genetic stability of Chinese hamster cells is another factor that makes them an extremely valuable system for studying transformation. In studies such as those presented here, the ability to establish and compare nontransformed parental cells, transformed cells, and revertant cells without major chromosome rearrangements is very important. With the advances being made in gene transfer techniques, it should soon be possible to introduce well-characterized mutations from other CHO cell lines into the cells transformed with RSV or other defined oncogenes. This ability will allow studies on the effect of specific cellular alterations on the transformed phenotype of the recipient cells and should provide new insight into the transformation process.

B. Other Genetic Systems for Studying Transformation

As was discussed in the previous section, several cell lines other than Chinese hamster cell lines have been successfully used to study the interaction between Rous sarcoma virus and the host transformed cell. In addition, the interaction between other tumor viruses and their host mammalian cells has been studied. Recently, there has been a paper by Noda et al., (1983) reporting the isolation of phenotypic revertants of Kirsten murine sarcoma virus (KiMuSV) transformed NIH-3T3 cells. The two revertants studied produced approximately normal levels of the protein p21, the product of the KiMuSV oncogene *v-Ki-ras,* and normal transforming virus can be recovered from the cells. These results indicate that the oncogene is functional in the revertant cells. These cells are also resistant to retransformation by exogenously added KiMuSV particles. Thus, the revertant cells meet the criteria for reversion due to a cellular alteration rather than a viral mutation.

The behavior of the revertant phenotype was studied in somatic cell hybrids between the revertant cells and NIH-3T3 cells transformed by a number of agents. Although control crosses between the KiMuSV-transformed NIH-3T3 cells and nontransformed NIH-3T3 cells indicated that the transformed phenotype acted like a dominant genetic trait, the cell hybrids formed with the revertants and KiMuSV-transformed cells had a nontransformed phenotype. The revertant phenotype also appeared to be dominant in cell hybrids between the revertant cells and NIH-3T3 cells transformed with Harvey MuSV and Balb MuSV, which are closely related to KiMuSV. More surprisingly, the revertant phenotype was generally dominant in cell hybrids of the revertant cells and NIH-3T3 cells transformed with a number of other tumor viruses unrelated to KiMuSV. The notable exception was the hybrids formed with RSV-transformed NIH-3T3 cells and the revertant cells.

Although care must be taken in the interpretation of these results, since, as was mentioned in Section II, the nontransformed phenotype is dominant in many crosses between transformed and nontransformed cells, the results seem to indicate that a single cellular alteration can make a cell resistant to transformation by a number of different RNA tumor viruses. It also must be noted that transformation in the hybrid cells was measured by the tissue culture criterion of growth in suspension. Although the biochemical basis of the alterations in the revertant cells is unknown, it is very encouraging that revertants that meet the criteria for cellular mutations can be isolated and that they make the cells resistant to transformation by a number of different agents. Such mutants should provide important tools for understanding the biochemical process of transformation.

The more typical result has been the isolation of revertant cells that appear to lose their transformed phenotype because expression of the virus is reduced. One example of this type of reversion is demonstrated by the revertants of Snyder–Theilen feline sarcoma virus (ST-FeSV) transformed mink lung cells isolated in Aaronson's laboratory (Porzig et al., 1979). They found that some clones of ST-FeSV transformed cells reverted at a very high frequency. For example, in one clone (25a3) approximately 10% of the subclones had a revertant phenotype, and these revertant cells retransformed spontaneously with a frequency varying between 10^{-8} and 10^{-3} depending on how long the revertants were in culture. These revertants had lost the transformed phenotype based both on tissue culture criteria and on the loss of tumorigenicity. Normal ST-FeSV could be recovered from the revertant cells, but the cells could not be retransformed by infection with either ST-FeSV or KiMuSV. These results suggest that the revertant cells are suppressing the expression of the original transforming virus and measurements of viral RNA synthesis supported this suggestion.

Since nontransforming helper virus could stimulate the production of ST-FeSV particles, the effect of helper virus infection on the rate of retransformation by the revertant cells was examined and a large increase in the number of foci formed was found. Dilution experiments where the revertant cells were diluted after infection with the helper virus demonstrated that high cell densities were necessary in order to see this stimulation of retransformation. These experiments were interpreted to mean that the revertant cells reduced the expression of the original transforming virus in a way that could be overcome by the helper virus for the formation of new virus particles but that retransformation required the infection of a neighboring cell with the newly synthesized ST-FeSV.

Although the numerous revertants with phenotypes like that described above for the ST-FeSV transformed mink cells do not appear to be mutants altered in the mechanism used by oncogenes to bring about the transformed phenotype, they should be of interest for studies on the mechanisms cells use to turn off genes. One possible mechanism, methylation of DNA regions, has been implicated in the regulation of both developmental and endogenous viral genes. Such mechanisms of general gene control are interesting, but they probably will not provide insight into the mechanisms by which *oncogenes* function to transform cells.

The more informative mutants should be those with alterations that make them resistant to retransformation with the same or other closely related oncogenes. Understanding the biochemical basis for such mutants would dramatically increase our understanding of how oncogenes function in cells to bring about transformation. Unfortunately, unraveling the biochemical basis of such mutations will probably be quite difficult. The oncogene *ras* is very interesting because of its association with a number of human tumors and it has recently been shown that it has a GTPase activity (Sweet et al., 1984). With oncogenes such as *src*, where a kinase activity is known for their protein products, there is also hope for identifying the alteration in the mutant cells. For example, in the case of *src* the protein has a protein kinase activity which is thought to be important for its function in bringing about transformation. In this case it is hoped that alterations in the phosphorylation of one of the relatively small number of cellular proteins containing phosphotyrosine residues will be seen in the mutants. Such an observation would provide a target protein that could be identified and whose cellular function could be identified. By identifying a number of such proteins associated with the function of an oncogene, one should be able to understand the "pathway" of that oncogene's action in cells.

The work that has been described in this section indicates that the first step toward understanding how oncogenes function, the isolation of transformation-resistant mutants, can be accomplished in a number of cell systems. One of those systems which is particularly appealing because of its relatively stable genetic background is the Chinese hamster cell transformed by specific oncogenes. These cells should become increasingly popular for studies on the mechanism of cell transformation and it is hoped will help unravel the mystery of how oncogenes function to bring about cellular transformation.

ACKNOWLEDGMENTS

The author's work with RSV-transformed CHO cells was performed in the Laboratory of Molecular Biology, National Cancer Institute, The National Institutes of Health, Bethesda, Maryland. I would like to thank the following people who contributed to these studies: B. Lovelace, N. Richert, G. Merlino, J. Sivaswami, G. Vlahakis, I. Pastan, and M. M. Gottesman.

REFERENCES

Barski, G., Sorieul, S., and Cornefert, F. (1961). *J. Natl. Cancer Inst.* **51**, 781–785.

Berebbi, M. and Barski, G. (1971). *C. R. Hebd. Seances Acad. Sci.* **272**, 351–356.

Blanchard, M.-G., Barski, G., Leon, B., and Hemon, D. (1973). *Int. J. Can.* **11**, 178–181.

Boettiger, D. (1974). *Virology* **62**, 522–529.

Bryant, D. and Parsons J. T. (1983). *J. Virol.* **44**, 683–685.

Chiswell, D. J., Enrietto, P. J., Evans, S., Quade, K., and Wyke, J. (1982). *Virology* **116**, 428.

Collett, M. S., Brugge, J. S., and Erikson, R. L. (1978). *Cell* **15**, 1363–1369.

Collett, M. S., Erikson, E., and, Erikson, R. L. (1979). *J. Virol.* **29**, 770–781.

Cross, R. and Hanafusa, H. (1983). *Cell* **34**, 597–607.

Defendi, V., Ephrussi, B., Koprowski, H., and Yoshida, M. C. (1967) *Proc. Natl. Acad. Sci. USA* **57**, 299–303.

Deng, C.-T., Boettiger, D., MacPhearson, I., and Varmus, H. (1974). *Virology* **62**, 512–521.

Diamond, A., Cooper, G. M., Ritz, J., and Lane, M.-A. (1983). *Nature* **305**, 112–116.

Gottesman, M. M., Roth, C., Leitshuh, M., Richert, N., and Pastan, I. (1983). in *Tumor Viruses and Differentiation* (E. Scolnick and A. Levine, eds.), Alan R. Liss, New York, pp. 365–380.

Gottesman, M. M., Roth, C., Vlahakis, G., and Pastan, I. (1984). *Mol. Cell. Biol.* **4**, 2639–2642.

Harris, H., Miller, O., Klein, G., Worst, P., and Tachihana, T. (1969). *Nature* **233**, 363–364.

Hunter, T. and Sefton, B. W. (1980). *Proc. Natl. Acad. Sci. USA* **77**, 1311–1315.

Jakobovits, E. B., Majors, J. E., and Varmus, H. E. (1984). *Cell* **38**, 757–765.

Land, H., Parada, L. F., and Weinberg, R. A. (1983). *Nature* **304**, 596–602.

Lau, A. F., Krzyzek, R., Brugge, J., Erikson, R., Schollmeyer, J., and Faras, A., (1979). *Proc. Natl. Acad. Sci. USA* **76**, 3904–3908.

LeCam, A., Nicholas, J.-C., Singh, T.S., Cabral, F., Pastan, I., and Gottesman, M. M. (1981). *J. Biol. Chem.* **256**, 933–941.

Li, A. P., O'Neill, P., Kawashima, K., and Hsie, A. W. (1977). *Arch. Biochem. Biophys.* **182**, 181–189.

Noda, M., Selinger, Z., Scolnick, E. M., and Bassin, R. H. (1983). *Proc. Natl. Acad. Sci. USA* **80**, 5602–5606.

Ozer, H. L. and Jha, K. K. (1977) *Advances in Cancer Research* (G. Klein and S. Weinhouse, eds.), Academic Press, New York, vol. 25, pp. 53–93.

Pollard, J. W. and Stanners, C. P. (1979). *J. Cell. Physiol.* **98**, 571–586.

Porzig, K. J., Robbins, K. C., and Aaronson, S. A. (1979). *Cell* **16**, 875–884.

Roth, C. W., Richert, N., Pastan, I., and Gottesman, M. M. (1983). *J. Biol. Chem.* **258**, 10768–10773.

Roth, C., Singh, T., Pastan, I., and Gottesman, M. M. (1982). *J. Cell. Physiol.* **111**, 42–48.

Sabran, J. L., Hsu, T. W., Yeater, C., Kaji, A., Mason, W. S., and Taylor, J. M. (1979). *J. Virol.* **29**, 170–179.

Sager, R. and Kovac, P. E. (1978). *Somat. Cell Genet.* **4**, 375–392.

Sager, R. and Kovac, P. E. (1979). *Somat. Cell Genet.* **5**, 491–502.

Sager, R., Tanka, K., Ebina, Y., and Anisowicz, A. (1983). *Proc. Natl. Acad. Sci. USA* **80**, 7601–7605.

Sefton, B. M., Hunter, T., Beemon, K., and Eckhart, W. (1980). *Cell* **20**, 807–816.

Snyder, M. A., Bishop, J. M., Colby, W. W., and Levinson, A. D. (1983). *Cell* **32**, 891–901.

Stanbridge, E. J., Der, C. J., Doersen, C.-J., Nishimi, R. Y., Peehl, D. M., Weissman, B. E. and Wilkinson, J. E. (1982). *Science* **215**, 252–259.

Sweet, R. W., Yokoyama, S., Kamata, T., Feramisco, J. R., Rosenberg, M., and Gross, M. (1984). *Nature* **311**, 273–275.

Varmus, H. E., Quintrell, N., and Wyke, J. (1981). *Virology* **108**, 28–46.

Weinberg, R. A. (1983). *J. Cell Biol.* **97**, 1661–1662.

Weiner, F., Klein, G., and Harris, H. (1971). *J. Cell Sci.* **8**, 681–687.

Wyke, J. A., Beamand, J. A., and Varmus, H. E. (1980). *Cold Spring Harbor Symp. Quant. Biol.* **44**, 1065–1073.

CHAPTER 27

GENETIC STUDIES OF TRANSFORMATION AND TUMORIGENESIS IN CHINESE HAMSTER EMBRYO FIBROBLASTS

Ruth Sager
Dana–Farber Cancer Institute
Harvard Medical School
Boston, Massachusetts

INTRODUCTION

A. Cancer as a Genetic Disease of Somatic Cells

Cancer is a genetic disease, resulting from genomic changes that occur in rare somatic cells. All subsequent manifestations of the disease are brought about by multiplying clones of such cells, which undergo further genomic changes during progression of the disease. A major goal of cancer research is to identify the genetic changes that underlie the origin and progression of cancer and to determine the consequent changes in gene expression.

Some forms of cancer show a strong hereditary predisposition that can be followed in family pedigrees as single gene mutations. Two classes have been distinguished genetically.

1. The chromosome instability syndromes are inherited as recessive genes, with little or no expression in heterozygotes. These diseases, principally Bloom's syndrome, Fanconi's anemia, ataxia telangiectasia, and xeroderma pigmentosum, recently reviewed and discussed in detail (German, 1982), have in common a high frequency of chromosome breakage, rearrangements, and, in Bloom's syndrome and in xeroderma pigmentosum, sister chromatid exchanges. The recessive pattern of inheritance suggests that the mutants have lost an enzyme involved in normal DNA repair, but as yet no specific enzymes have been identified in association with any of the diseases. Their relevance to cancer lies in the high probability of malignancy occurring in individuals with any one of these syndromes, thus presenting a strong correlation between DNA damage and the onset of cancer.
2. A series of mutations or rearrangements that lead to cancer have been classified as dominant genetically in the sense that in each syndrome one copy of the mutation is sufficient to ensure expression of the disease with high probability (Knudson, 1977, 1983). However, at the cellular level, these diseases are recessive since only a rare cell expresses the cancer phenotype (Benedict et al., 1983). In detailed studies with one of these diseases, retinoblastoma, it has become clear that genetic changes occurring at the same locus on the other chromosome (Cavenee et al., 1983), and possibly elsewhere in the genome as well during growth of the individual (Balaban et al., 1982), are required for tumor expression. The fact that the mutation is carried in every cell of the body, but expressed in only one or a few cells, indicates that further event(s) are required for neoplastic transformation.

Thus, the genetic changes that ultimately lead to the growth of clones of cancer cells occur in somatic cells. The experimental approaches to genetic analysis of tumorigenesis are appropriately carried out with cells in culture using the methods of (i) somatic cell genetics, employing cell fusion and subsequent analysis of hybrids and (ii) molecular genetics in which fragments of DNA are transferred from cancer cells into normal cells. Some of the principles of this type of analysis are discussed in Chapter 26.

B. Use of Somatic Cell Genetics to Analyze Tumorigenesis

Historically, the development of somatic cell genetics began with experiments designed to examine tumorigenicity of cell hybrids from fusions of different tumor cells or of normal cells with tumor cells [reviewed in Ephrussi (1972)]. Initially, it appeared that tumorigenicity was a dominant trait, since hybrid clones from tumor × normal (T × N) cell fusions, when grown up and injected into animals, gave rise to tumors. Subsequently, with the use of inactivated Sendai virus (Harris and Watkins, 1965) and later PEG (Chapter 7) to facilitate cell fusion, and with rapid testing of hybrids, it has become evident that most hybrids from T × N fusions are initially nontumor forming, and that as the hybrid clones are grown in culture, they became increasingly tumorigenic. The simplest interpretation of these findings is that tumorigenicity is recessive and unexpressed until the normal homologous gene which suppresses tumorigenicity is lost in the process of chromosome elimination that occurs during growth of hybrid cells (Harris, 1971; Jonasson et al., 1977).

The generality of this interpretation is limited by experiments in which cell hybrids are tumorigenic despite little or no apparent chromosome loss (e.g., Giacomoni, 1979; Aviles et al., 1980). Also, the failure of proponents of the chromosome loss hypothesis to identify the suppressing chromosome has greatly weakened the argument. Recently, in a heroic series of difficult chromosome studies, Harris and colleagues (Evans et al., 1982) have been able to identify mouse chromosome 4 as carrier of a suppressing gene or genes active in hybrids from fusions of normal with highly malignant tumor cells. A serious limitation in the use of cell hybrids for genetic analysis is the occurrence of chromosome elimination, which can drastically alter results.

These difficulties are more severe with mouse cells which have 40 hard-to-distinguish chromosomes than with Chinese hamster cells which contain 22 easily distinguishable chromosomes. In addition, most mouse cell lines are much more rearranged and aneuploid than are Chinese hamster cell lines. CHEF cells are diploid and thus possess a considerably advantage over other cell lines in the use of cell hybrids for genetic analysis, as will be presented below.

A cellular assay for tumorigenicity would be a great advance over animal testing. Numerous changes in cellular properties were examined by many investigators looking for correlation with tumorigenicity, using various cell lines over the years beginning in the late 1960s. Most of them are no longer in use because of unreliability or lack of correlation in more than a few cell lines (Sato et al., 1982). The single parameter that has fared best in extensive testing is anchorage independence, the loss of the attachment requirement for growth (Shin et al., 1975; Kahn and Shin, 1979). In practice this means that populations of cells can be seeded into soft agar, agarose, or methylcellulose and only the anchorage-independent cells will produce clones that can be counted and selected for further study. Widespread use has been made of this test, and many investigators now accept anchorage independence as an adequate and reliable indicator of tumorigenicity.

As discussed below, however, our studies with CHEF cells have demonstrated that anchorage independence is not an adequate indicator of tumor-forming ability since anchorage-independent one-step mutants are nontumorigenic (Smith and Sager, 1982). *Indeed, at this time no cellular parameters are known that provide a definitive identification of tumorigenic cells.* Furthermore, when the chromosomes of injected cells to be assayed in the animal are compared with the chromosomes of cells from the resulting tumors, one routinely finds chromosome aberrations that occurred during tumor growth, or were present in a very small fraction of injected cells. For this reason, one must suspect that additional genetic changes are required for tumor formation beyond those present in the vast majority of injected cells. This problem plagues the interpretation of both cellular and molecular studies. In DNA transfer experiments, focus formation is taken as the indicator of tumorigenicity, but is it more reliable than anchorage independence? No detailed analysis has been reported.

Our genetic studies of transformation and tumorigenicity with CHEF cells have taken the following form. (i) Cell fusion studies between nontumorigenic CHEF/18 cells and tumorigenic CHEF/16 cells were designed to look for expression of tumor-forming ability in hybrids, and for linkage between tumorigenicity and transformation traits such as anchorage and serum factor requirements for growth (Sager and Kovac, 1978; Marshall and Sager, 1981; Smith and Sager, 1985). (ii) Cybrids were engineered by fusion of cytoplasts from CHEF/18 cells with tumorigenic CHEF/16 cells to look for suppression of tumor formation determined by cytoplasmic genes, for example, mitochondrial genes from normal cells (Howell and Sager, 1978). (iii) Mutation studies were designed to select transformed mutants of CHEF/18 cells, in order to find out whether anchorage or serum mutants were also tumorigenic, that is, another test of linkage between cell culture traits and tumor-forming ability (Smith and Sager, 1982). (iv) Extensive chromosome studies were undertaken to identify specific chromosomes associated with either transformation or tumorigenicity (Kitchin and Sager, 1980; Kitchin et al., 1982; Gadi et al., 1984). (v) Transfection of the mutant c-Ha-ras gene was undertaken to determine whether CHEF cells could be made tumorigenic by this gene alone. The results of these studies will now be summarized, and their implications will be discussed.

II. SUPPRESSION OF TUMOR FORMATION IN CHEF CELL HYBRIDS

The first experiments with CHEF cells designed to examine the genetics of transformation and tumorigenesis were carried out with cell hybrids formed by fusions between the two mutant cell lines 205-30 (thioguanine-resistant mutant of CHEF/18) and 204-Bu50 (BrdU-resistant mutant of CHEF/16). Cells were mixed 1:1 and hybrid yields after HAT selection in several experiments were in the range of 10^{-4}–10^{-3} of the fusion mixture. Hybrid colonies were transferred from the fusion dishes and grown for one passage in selective medium, then for one or two passages with added HT (hypoxanthine plus thymidine) to ensure recovery from the effects of aminopterin, and then grown to high titer for storage. Further studies were initiated from frozen ampoules.

Twenty hybrids were characterized; their properties are summarized in Table 27.1. Each of them resembled the CHEF/18 parent in morphology, high serum requirement, and anchorage dependence, indicating that the normal phenotypes are dominant over the transformed. However, when injected into nude mice at 4×10^6 cells/site to test for tumor-forming ability, each hybrid eventually gave rise to tumors, usually after a lag of 1–2 months. These results suggested that the tumors formed from some small subpopulation of cells and that the majority, that is, those seen in cell culture tests, were nontumorigenic.

This possibility was supported by coinjection experiments carried out with three of the hybrids (Sager and Kovac, 1979). Each was coinjected at 10^2, 10^4, and 10^6 cells mixed with 10^7 X-irradiated CHEF/18 cells. None produced tumors at 10^2, two of the three produced tumors with a long lag when injected at 10^4, and more rapidly at 10^6, and the third produced tumors only when injected 4×10^6. In the same experiment, other subclones of these hybrids, with reduced numbers of chromosomes derived by back-selection in BrdU or growth in methylcellulose, or tumor derived, were also examined. The tumor-derived cells produced tumors rapidly at 10^2, whereas the other subclones required larger inocula for tumor formation. However, all produced tumors at 10^6. Thus, the hybrids were mixtures of tumorigenic and nontumorigenic cells.

In a correlated chromosome study, it was shown that the HAT-selected hybrids were tetraploid, with modal chromosome numbers in the range of 37–44, whereas the chromosome counts of tumor-derived cells were in the diploid range, 22–29, with one tetraploid exception. Chromosome studies subsequently showed that the tetraploid arose by endoreduplication from a reduced hybrid in the diploid range. Thus, extensive chromosome reduction occurred in the hybrid cells that produced tumors, consistent with the evidence that the initial hybrids are nontumorigenic, and that tumor-forming ability was reexpressed coincident with chromosome loss. These results provide evidence for the segregation of chromosomes carrying genes that suppress tumor forming ability, but do not identify any specific chromosomes.

III. SUPPRESSION OF TUMOR FORMATION IN CHEF CYBRIDS

Cybrids are fusion products between cell *A* and an enucleated cell *B*, and thus contain the nucleus of cell *A* in the combined cytoplasms from *A* and *B*. If the cell lines *A* and *B* differ by some discrete selectable phenotypic difference, one can prepare reciprocal cybrids to determine whether the gene that determines the difference is nuclear or cytoplasmic. This methodology has been successfully applied to the identification of mitochondrial genes in mouse cell lines (Bunn et al., 1974; Wallace et al., 1975; Howell and Sager, 1978; 1979; Shay, 1982), by fusing cytoplasts from mitochondrial drug-resistant mutants (e.g., chloramphenicol) with cells sensitive to that drug but resistant to a nuclear-coded drug resistance (e.g., BrdU or TG resistance). Cybrids are then selected, for example, in BrdU or TG plus chloramphenicol.

Howell and Sager (1978) fused CHEF/18 nontumorigenic CAP-R cytoplasts of CHEF/16 tumorigenic TK^- (BrdU-resistant) CAP-S cells, and recovered chloramphenicol-resistant cybrid colonies at a frequency of about 1×10^{-4} per cell in fusion mixture plated. On subsequent testing in nude mice, both tumorigenic and nontumorigenic TK^- cybrid clones were identified. The experiment was repeated using reconstructed cells (recons) in which the enucleated cells of one parent were fused with karyoplasts (nuclei surrounded by a thin layer of cytoplasm within a plasma membrane envelope) of the other parent in an effort to decrease the contribution of cytoplasm from the tumorigenic parent (Sager and Spudich, unpublished). As in the cybrids, both tumor-forming and suppressed hybrid clones were identified in the nude mouse assay (Sager and Spudich, unpublished), in recon clones derived from fusion of a T karyoplast with an N cytoplast but not in the reciprocal recons. After further passage in cell culture, however, the suppressed cybrids and recons became tumorigenic (Sager, unpublished). The suppressed phenotype was maintained for at least 20–30 doublings after the initial cybrid fusion.

In similar studies with mouse cell lines, cybrid suppression of tumorigenicity has been found by some investigators but not by others. Since different cell lines were used in each investigation, the reported inconsistency in results may be understandable when the molecular basis of cybrid suppression becomes known. If the basis lies in mitochondrial DNA, then the difficulty in maintaining stable suppression may result from competing growth rates or different functional efficiencies of mitochondria from the two parental lines.

The cybrid suppression results suggest that mitochondrial DNA or other cytoplasmic DNAs may represent contributing elements in the genetic control of tumorigenicity.

IV. MUTATIONAL STUDIES OF TRANSFORMATION AND TUMORIGENICITY

The primary object of these studies was to determine whether CHEF/18 cells could be made tumorigenic in a single mutational step, and if not, to find out how many steps were required and if possible to identify the steps (Smith and Sager, 1982). Three mutagens were used: EMS, MNNG, and 4-NQO. After 9–15 doublings in mass culture following mutagen treatment, cells were injected into nude mice at 10^7 cells/site for tumor testing, and plated in selective media at appropriate cell densities to select anchorage-independent or low-serum mutants.

Anchorage-indepenent mutants that formed colonies in methylcellulose were selected after EMS mutagenesis and compared with spontaneous anchorage mutants arising in untreated CHEF/18 populations. Most mutants, both spontaneous and induced, retained the parental requirement for serum while losing the anchorage requirement. A few double mutants with reduced anchorage and serum requirements were recovered, but they were no more tumorigenic than the others. Of the 21 anchorage mutants examined, only four made any tumors at all, and those came up very late and at only a fraction of sites (2/29, 1/8, 1/7, and 2/8 tumors/total sites tested). Similarly, of 10 low-serum mutants selected at 1% of 3% serum, only two were tumorigenic. In subsequent chromosome studies discussed below, it was shown that tumors arose from cells carrying chromosome rearrangements.

Tumor-forming ability of mutagenized surviving cells was tested by subcutaneous injection of unselected populations into nude mice. Nineteen sites were injected after EMS mutagenesis and only two tumors arose, each after 5 months. Six sites were injected after MNNG mutagenesis and one tumor arose after 6.5 months. Eighteen sites were injected after 4-NQO mutagenesis and no tumors arose. In total only three tumors arose in tests at 43 sites. In untreated controls, 34 sites were injected at various times and no tumors arose. The very slow appearance of tumors after a lag of several months indicates that multiple genetic changes and selection preceded appearance of a palpable tumor.

These results demonstrate unequivocally that tumors do not arise in a single step following mutagenesis of CHEF/18 cells, and that single-step mutations to anchorage independence or to loss of a stringent serum requirement for growth do not produce tumor-forming cells. Thus malignant transformation is a multistep process with CHEF/18 cells, and the genetic changes underlying the transformation phenotypes of anchorage independence and partial loss of the serum requirement do not lead *per se* to tumorigenicity.

V. GENETIC ANALYSIS OF ANCHORAGE INDEPENDENCE

Anchorage-independent mutants of CHEF cells were derived in two ways: (1) CHEF/18 mutants (Smith and Sager, 1982) after mutagenesis; and (2) CHEF/16 cells present in the original *F4224A* cell line. To find out whether the same or different mutations were responsible for anchorage independence in these cell lines, crosses were made to look for complementation (Marshall and Sager, 1981; Marshall et al., 1982). Examination of many different crosses was facilitated by a rapid screening procedure, in which the fusion mixture was plated directly into selective medium containing methylcellulose, as well as into selective medium on plastic. The ratio of hybrid colonies formed on methylcellulose to those formed on plastic provides a direct assay of the anchorage independence of the hybrids.

The results are summarized in Table 27.1. The interpretation of suppression is based on the fact that crosses of anchorage-dependent CHEF/18 × anchorage-independent CHEF/16 cells gives anchorage-dependent hybrids showing that dependence is dominant and the transformed phenotype is suppressed. Crosses between different anchorage-independent CHEF/18 and CHEF/16 mutants gives anchorage dependence, demonstrating complementation between different recessive mutants. Interpretating the results in terms of dominance and recessiveness is justified by the fact that the anchorage-independent CHEF/18's were recovered as one-step mutants after mutagenesis. These experiments demonstrate the presence of at least two different complementing genes regulating the anchorage requirement for growth.

In a further study of suppression and complementation (Smith and Sager, 1985), we found similar interactions between the low-serum mutants derived from CHEF/18 and the low-serum trait present in CHEF/16. Here, too, hybrids between CHEF/16 and CHEF/18 low-serum mutants showed complementation, that is, the hybrids required high serum, as shown by comparative plating of fusion mixtures on selective media with 10% or 1% serum. This result demonstrates not only that at least two loci are involved, but also that each of the transformed traits is recessive.

The localization of one of these genes, the one from CHEF/16 cells, was established in the following way (Marshall et al., 1982). We have already seen that CHEF/18 × CHEF/16 hybrids are initially anchorage-dependent, and that during growth of hybrid populations anchorage-independent hybrid subclones arise and can be recovered by cloning in methylcellulose. In this experiment, five tetraploid hybrids were plated into methylcellulose and many anchorage-independent subclones were selected. Under these conditions, minimal chromosome reduction occurs; subclones retaining most chromosomes from both parents were chosen and found to remain stably in the tetraploid range. Karyotype analysis of suppressed hybrids and their anchorage-independent subclones showed that segregation of the anchorage requirement was correlated with the loss of one copy of chromosome 1.

Additionally, in one of the hybrids, breaks and rearrangements affecting chromosome 1 had occurred, making it possible systematically to localize the region of chromosome 1 involved in control of anchorage requirement to the terminal portion of the short arm (lp10 to lpter).

TABLE 27.1
Anchorage Assay of Fusion Mixtures by Colony Formation on Plastic and in Methylcellulose (MC)[a]

Fusion	Number of Experiments	Mean Number of Hybrid Colonies per 10^4 Plated[b]		Anchorage Phenotype of Hybrids	Interpretation[c]
		Plastic	MC		
Anchorage dependent × anchorage independent					
205-30 × 204-Bu50	6	7.0 ± 1.86	0.1 ± 0.001	Dependent	Suppression
18-1TK$^-$ × 294-7MC3/10	2	10.0	$<10^{-6}$	Dependent	Suppression
18-1TK$^-$ × 16dm	1	5.5	$<10^{-6}$	Dependent	Suppression
Anchorage independent × anchorage independent					
204-Bu50 × 16dm	7	24.5 ± 3.27	15.4 ± 1.9	Independent	NC
204-Bu50 × 294-7MC3/10	5	22.8 ± 4.3	23.7 ± 5.4	Independent	NC
1T2 MC6 × 16dm	2	4.7	1.65	Independent	NC
T4A MC6 × 16dm	6	17.1 ± 3.8	0.3 ± 0.02	Dependent	C
20-5 × 16dm	2	7.9	$<10^{-6}$	Dependent	C
21-2 × 16dm	6	15.1 ± 4.5	0.05 ± 0.02	Dependent	C
18m3 × 16dm	1	4.3	0.02	Dependent	C
18m4 × 16dm	1	12.3	0.7	Dependent	C

[a]From Marshall and Sager (1981).

[b]Every experiment included control plates that were not treated with PEG; no colonies were recovered from any control plates. Where four or more experiments were performed, standard errors are shown.

[c]NC, no complementation; C, complementation.

VI. CHROMOSOME STUDIES

Because the Chinese hamster chromosome complement is relatively small and distinctive (Chapter 5) with 10 pairs of distinguishable autosomes plus the sex chromosomes, XX or XY, it is possible to analyze changes occurring in mutants, in hybrids, and in tumor-derived cells by Giemsa banding and C banding techniques. An example is the localization of an anchorage gene to chromosome lp as discussed above.

Chromosome changes arising in transformed and tumor-derived cells have been reported with CHEF/16 cells derived from tumors (Kitchin and Sager, 1980), with anchorage and low-serum mutants (Kitchin et al., 1982), and with azacytidine- and insulin-derived preadipocytes of CHEF/18 origin (Harrison et al., 1983; Gadi et al., 1984). Each of these studies will be summarized here.

A. Tumor-Derived CHEF/16 Cells

CHEF/16 cells are unique in being diploid, while every cell is potentially tumorigenic as shown by coinjection experiments (Kitchin and Sager, 1980). Are any specific chromosome changes characteristic of either growth in methylcellulose or tumor formation? To answer this question, chromosomes from 11 colonies recovered from methylcellulose and 10 tumor-derived cell populations were examined by Giemsa banding. In brief, the clones from methylcellulose remained diploid or close to diploid. The commonest change seen was loss of the heterochromatic long arm of the X chromosome.

The tumor-derived cells, however, provide a dramatic contrast with the anchorage clones. No diploid cells were found in any of the 10 populations, although the modal chromosome counts for six of the samples were in the diploid range: 22–25. The others were hypotetraploid. All of the cells had undergone chromosome rearrangements, regardless of chromosome number. Each tumor contained a relatively homogeneous population of cells judged by their chromosome rearrangement patterns, clearly indicating that each of the tumors was clonal in origin. The tumors came from a coinjection experiment in which either 10^2, 10^4, or 10^6 CHEF/16 cells were coinjected with 10^7 X-irradiated CHEF/18. Nonetheless, whether 100 or 10^6 tumorigenic CHEF/16 cells were injected, the tumors arose from one or (in two instances) two stemlines identified by chromosome pattern. This result is consistent with the clonal origin of neoplasia seen in animal studies and in human cancer.

The detailed patterns of chromosome rearrangements were complex. However, one chromosome was involved in recognizable changes in 9 of the 10 tumors: chromosome 3. The commonest rearrangement was the presence of an extra copy of the long arm of chromosome 3 and translocations were also found. In the tumor with no detectable changes in chromosome 3, rearrangements of chromosome 1 and 9 were present.

These results are particularly interesting in the light of subsequent studies discussed below, in which similar changes were seen in CHEF/18-derived tumor cells. Also, in a parallel study of CHEF/16-derived tumors, Deaven et al. (1981) found similar changes in chromosome 3 in four out of five tumor-derived cell populations examined.

In summary, these studies reveal the occurrence of chromosome changes especially involving chromosome 3 in tumorigenic cells during the process of tumor growth.

B. Chromosome Changes in CHEF/18 Mutants

The chromosome constitution was examined of a series of 19 anchorage mutants, 10 low-serum mutants, 3 anchorage–low-serum double mutants, and in tumor-derived cells from some of the mutants (Kitchin et al., 1982). Most of the anchorage mutants were nontumorigenic, as discussed above, and 7 of the 19 mutants remained diploid after mutagenesis and selection. The rest of them (all but two) were pseudodiploid with few rearrangements. The commonest rearrangement in the anchorage mutants involved chromosome 1, consistent with the result described above localizing an anchorage gene to chromosome lp. These rearrangements did not induce tumorigenicity *per se*, although two of the mutants with a t(1;7) translocation were slightly tumorigenic. In contrast, among the tumorigenic anchorage mutants two of the four examined had an extra copy of chromosome 3q, and in the tumor-derived cells from these anchorage mutants three of the four had an extra copy of 3q. Similarly, among the eight tumors analyzed from low-serum mutants, all but two contained rearrangements involving chromosome 3.

C. Chromosome Changes in Preadipocytes from CHEF/18

As previously described (Chapter 4) CHEF/18 cells can be induced to form preadipocytes by either 5-azacytidine or insulin (Harrison et al., 1983; Gadi et al., 1984). AzaC had previously been reported to have low carcinogenic action (Benedict et al., 1977) as well as being used as an anticancer drug in treatment of certain leukemias (Saiki et al., 1978). More recently, azaC has been shown to inhibit DNA methylation and to do so following its incorporation into DNA in place of cytosine, by binding the methyl transferase enzyme very tightly (Santi et al., 1983).

We found that CHEF cells treated with azaC for one or two rounds of the cell cycle became tumorigenic with a low frequency, such that populations of CHEF/18 cells grown up after pretreatment with azaC gave rise to tumors at a fraction of sites injected subcutaneously into nude mice (Harrison et al., 1983). A series of 41 preadipocyte clones and subclones derived from azaC or insulin induction were then examined for tumorigenicity, for chromosome changes, and for methylation changes using six different cDNA gene probes. We found an unexpectedly clearcut correlation linking decreased methylation, a unique chromosome change, and tumorigenesis. All the azaC-induced clones and subclones with one exception were highly tumorigenic, hypomethylated, and contained an extra copy of the long arm of chromosome 3. The exception was a subclone that was nontumorigenic and diploid. Among the insulin-induced lines, four of the five original preadipocytes were nontumorigenic, diploid, and normally methylated as in controls. The fifth line was tumorigenic at a fraction of sites, and both tumorigenic and nontumorigenic subclones were recovered from it. Their chromosomes and methylation patterns were similar to the azaC pattern described above.

These results are remarkably clearcut in demonstrating certain important features of tumorigenesis. First, a simple and unambiguous identification of trisomy for chromosome 3 associated with tumor-forming ability is consistent with all of our other studies. Judging from chromosome rearrangements, the relevant region lies between 3q4 and the centromere. It seems likely that an oncogene is located in this region.

Another striking difference between the tumorigenic and nontumorigenic preadipocytes is in the extent of DNA methylation. Differences were shown by examining the frequency of methylation at HpaII sites ($C^{m}CGG$) at a set of gene loci randomly selected. This method was chosen as an arbitrary index of widespread methylation changes. Hypomethylation has been correlated with tumor formation by other investigators (Feinberg and Vogelstein, 1983). The possibility that methylation changes as well as trisomy of chromosome 3q contribute to the tumorigenicity seen in this study must be considered.

Furthermore, the much increased tumorigenicity of the azaC preadipocytes compared with azaC-treated unselected CHEF/18 populations merits consideration. Has the differentiation commitment of the cells increased their tumorigenic potential? An analogy can be found in the increased susceptibility of certain differentiating cell types to carcinogens.

D. Summary

In human cancer, particular chromosome changes including translocations, deletions, and trisomies have been associated with particular tissue-specific forms of the disease (Yunis, 1983). Within the past few years, oncogenes have been identified at or near the site of breakage in several of the translocations, and it has been widely assumed that oncogene activation resulting from chromosome rearrangement is the carcinogenic agent, driving cell growth and selection of increasingly malignant cells.

Our results with CHEF cells, that identify a region on chromosome lp with anchorage independence and a region on chromosome 3q with tumor formation, are in line with an oncogene interpretation. However, the resistance of CHEF cells to one-step and in some instances two-step induction of tumor-forming ability by mutagenesis, suggests that DNA changes in addition to trisomy for chromosome 3q may be required. Perhaps hypomethylation of particular sites in DNA is an example of a requisite change at the DNA level.

A direct approach to evaluating the role of oncogenes vis-à-vis chromosome rearrangements is provided by DNA transfer experiments, discussed in the next section.

VII. TRANSFECTION OF CHEF/18 CELLS WITH ONCOGENIC DNA

A direct answer to the question of whether CHEF/18 cells can be made tumorigenic in a single step could presumably come from transfection with a known oncogene. Experiments of this kind have been carried out with genomic DNA from the human bladder carcinoma cell line EJ (Hastings and Franks, 1981) as well as with plasmid constructions carrying the mutant EJ gene (Sager et al., 1983) or the normal human allele of EJ, c-Ha-ras, in the plasmid J132 (Lau and Sager, unpublished).

Using genomic DNA from the EJ cell line, transfectants of CHEF/18 were obtained (Smith et al., 1982) and shown to contain the transfected EJ gene by blot hybridization. However, the integrated gene was rearranged in various ways in tumor-derived cell lines from different foci, including one line with amplified sequences detected initially by blot hybridization and subsequently by *in situ* hybridization. A unique amplification was identified as an HSR region on chromosome 2q, that hybridized both with an *Alu* probe and with the EJ probe (Gadi, Anisowicz, and Sager, unpublished).

The occurrence of rearrangements and amplifiction of EJ DNA in the transfected cell lines left unclear the answer to the single-step question with which the project began. Were the rearrangements essential or unrelated to tumorigenesis? For clarification, we turned to the use of plasmid rather than genomic DNA. The pEJ plasmid (Shih and Weinberg, 1982) was used, as well as the EJ gene inserted into pSV2gpt (Sager et al., 1983) and the J132 plasmid containing a normal c-Ha-ras gene linked to the Moloney LTR (Chang et al., 1982). Both pEJ and pSV2gpt-EJ gave high yields of foci in the range of 10^3/pmol transforming DNA. The transfecting DNA was identified in the foci sampled, cells from these foci were tumorigenic in nude mice, and the tumor-derived cells contained EJ DNA. Thus, it appeared that the EJ gene was fully capable of making CHEF/18 cells tumorigenic.

Two kinds of evidence raise doubt about this simple conclusion. (i) Chromosomes from the tumor-derived cells show an array of rearrangements that may or may not be relevant to tumorigenesis. Further work is required to evaluate the significance of these aberrations. (ii) In transfections using either J132 to look at LTR-driven transcription of a normal c-Ha-ras gene, or pSVgpt without EJ as a control, foci were recovered on the experimental dishes at a frequency of 1–2% that of the EJ-transfections. Control dishes treated with salmon sperm DNA contained no foci or few small foci distinct from those induced in the experiment. None of the foci from the experimental dishes, treated with either J132 or pSVgpt, contained donor DNA, but did give rise to tumors, which also did not contain donor DNA. These results are a clear example of "hit and run" tumorigenesis, comparable to that previously reported with DNAs derived from adenovirus (Paraskeva et al., 1982) and from Herpes virus (Galloway and McDougall, 1983).

These results suggest that transient transformation with either J132 or pSV2gpt is sufficient to initiate a process leading to stable tumorigenesis by endogenous oncogenes in the absence of the inducing DNA. In these circumstances, one is forced to consider whether the same process may also be at work in the pEJ and pSV2gpt-EJ transfectants in which the inducing DNA is still present. Thus, DNA transfection experiments may be deceptive in their biology and simplistic in their interpretation when taken as evidence that the integrated donor DNA alone determined tumorigenesis.

Nonetheless, DNA transfer experiments provide the most direct approach to the identification of oncogenes. As yet, only the *ras* (Shih and Weinberg, 1982) and *Blym* (Diamond et al., 1984) genes have been initially identified in this way. All other putative oncogenes so far identified have been found by hybridization with retroviral-transforming genes. Still, the methodology of recognizing and cloning genes by means of DNA transfer into suitable recipient cells is by far the most powerful now available, and should provide access to as yet unidentified oncogenes. The CHEF cells provide excellent material for this search, since they form foci in response to either genomic or plasmid DNAs. With the use of genomic DNAs from tumor-derived cells, it should be possible to identify new genes with oncogenic potential.

VIII. SUMMARY

The CHEF cell lines—CHEF/18 and CHEF/16—and their derivatives have been used in a series of investigations, still in progress, designed to identify specific genes and/or processes (e.g., rearrangement, methylation) involved in tumorigenesis. The experimental system is simple: the conversion of diploid, nontumorigenic, nontransformed cells into a population of cells each capable of forming a tumor in the nude mouse assay.

A. Transformation

The transformation phenotypes of anchorage independence and growth in low serum were found to be under genetically independent control, and each was shown to be independent of tumorigenicity. All three traits: anchorage independence, growth in low serum, and tumor formation were suppressed in cell hybrids formed by fusions between normal and transformed cells. On this basis, the transformed phenotypes and tumorigenicity are considered recessive traits.

At least two different genes that regulate the anchorage requirement were identified by complementation; and parallel results were obtained with the low-serum requirement. Each hybrid showing complementation expressed the nontransformed phenotype: for example, fusions of different anchorage-independent lines gave anchorage-dependent hybrids; and fusions of different low-serum lines gave hybrids with a high-serum requirement. These results carry the implication that the dominant allele of each gene can suppress its recessive allele but not the recessive allele of the other gene, that is, that the genes are epistatic. In these studies, one parent was of CHEF/16 origin, and the other was a mutant of CHEF/18 origin. Fusions between different CHEF/18 mutants have not yet been feasible. Among the low-serum mutants, some have lost their EGF requirement and others have a diminished insulin requirement, suggesting that these at least will complement with each other.

The available data are consistent with the hypothesis that suppression of the transformed phenotype is an expression of dominance: that the normal alleles whether acting via coding or cis-acting regulatory sequences, control the expression of these phenotypes, and that the mutant or recessive alleles, either by altered coding or by loss of regulation, determine the transformed phenotypes.

B. Tumorigenicity

The transformed cells recovered in mutagenesis experiments are almost without exception nontumorigenic, whether anchorage independent, low-serum requiring, or doubly transformed. Most tumor-derived cells are transformed, thus suggesting that acquisition of tumor-forming abiity requires additional genetic changes. However, some tumor-derived cells are exceptional, in retaining anchorage or serum growth requirements. In these instances, specific suppressors of transformed traits may be active.

Tumor-forming ability itself can be suppressed. Cell hybrids from fusions between CHEF/18 and CHEF/16 are initially nontumorigenic, and tumorigenic subclones arise during growth as shown by nude mouse co-injection assays that reveal hybrid clones as mixtures of tumorigenic and nontumorigenic cells. A prevalent interpretation is that suppressors of tumorigenicity, which are present in normal cells, are lost by chromosome elimination during growth of hybrid clones, permitting expression of the recessive phenotype, namely, tumor-forming ability. These genes cannot be simply epistatic, since tumorigenesis does not arise in a single step. That is, the loss of one suppressor is not sufficient to alter the phenotype.

If transformed cells require additional genetic changes to become tumorigenic, then tumor-derived cells should provide genomic DNA that will induce tumorigenicity in transformed but nontumorigenic cells. The strongest clue we have as to the identity of a tumor-inducing oncogene comes from chromosome studies that strongly implicate a region on the long arm of chromosome 3. Rearrangements of chromosome 3 were seen with a high frequency in all of our CHEF studies of tumor-derived cells. We need now to clone out this potential oncogene, hopefully from FACS sorted copies of chromosome 3q.

In summary, CHEF cells provide unique material for cell and molecular genetic analysis of tumorigenesis. The cells are stably nontumorigenic, they are diploid with $2n = 22$ readily identifiable chromosomes, they grow well in culture including serum-free medium, and they respond well to transfection with genomic or plasmid DNA. Our goal, the identification of tumor-inducing genes and their suppressors, distinct from genes that regulate the transformed phenotype, should be readily approachable with CHEF cells.

REFERENCES

Aviles, D., Ritz, E. and Jami, J. (1980). *Somat. Cell Genet.* **6**, 171–186.

Balaban, G., Gilbert, F., Nichols, W., Meadows, A.T., and Shields, J. (1982). *Cancer Genet. Cytogenet.* **6**, 213–221.

Benedict, W. F., Banerjee, A., Gardner, A., and Jones, P. A. (1977). *Cancer Res.* **37**, 2202–2208.

Benedict, W. F., Murphree, A. L., Banerjee, A., Spina, C. A., Sparkes, M. C., and Sparkes, R. S. (1983). *Science* **219**, 973–975.

Bunn, C. L., Wallace, D. C., and Eisenstadt, J. M. (1974). *Proc. Natl. Acad. Sci. USA* **71**, 1681–1685.

Cavenee, W. K., Dryja, T. P., Phillips, R. A., Benedict, W. F., Godbout, R., Gallie, B. L., Murphree, A. L., Strong, L. C., and White, R. L. (1983). *Nature* **305**, 779–784.

Chang, E. H., Furth, M. E., Scolnick, E. M., and Lowy, D. R. (1982). *Nature* **2987**, 479–483.

Deaven, L. L., Cram, L. S., Wells, R. S., and Kraemer, P. M. (1981). In *Genes, Chromosomes and Neoplasia* (F. E. Arrighi, P. N. Rao, and E. Stubblefield, eds.), Raven Press, New York, pp. 419–449.

Diamond, A., Cooper, G. M., Ritz, J., and Lane, M.-A. (1984). *Nature* **305**, 112–116.

Ephrussi, B. (1972). *Hybridization of Somatic Cells*, Princeton University Press, Princeton, New Jersey.

Evans, E. P., Burtenshaw, M. D., Brown, B. B., Hennion, R., and Harris, H. (1982). *J. Cell Sci.* **56**, 113–130.

Feinberg, A. P. and Vogelstein, B. (1983). *Nature* **301**, 89–92.

Gadi, I. K., Harrison, J. J., and Sager, R. (1984). *Somat. Cell Mol. Genet.* **10**, 521–529.

Galloway, D. A. and McDougall, J. K. (1983). *Nature* **302**, 21–24.

German, J. (1982). *Chromosome Breakage and Neoplasia*, Alan R. Liss, New York.

Giacomoni, D. (1979). *Canc. Res.* **39**, 4481–4484.

Harris, H. (1971). *Proc. Roy. Soc. Lond.* **179**, 1–20.

Harris, H. and Watkins, J. F. (1965). *Nature* **205**, 640–646.

Harrison, J. J., Anisowicz, A., Gadi, I. K., Raffeld, M. and Sager, R. (1983). *Proc. Natl. Acad. Sci. USA* **80**, 6606–6610.

Hastings, R. J. and Franks, L. M. (1981). *Int. J. Cancer* **27**, 15–21.

Howell, N. and Sager, R. (1978). *Proc. Natl. Acad. Sci. USA* **75**, 2358–2362.

Howell, N. and Sager, R. (1979). *Somat. Cell Genet.* **5**, 833–845.

Jonasson, J., Povey, S., and Harris, H. (1977). *J. Cell Sci.* **24**, 217–254.

Kahn, P. and Shin, S.-I. (1979). *J. Cell Biol.* **82**, 1–16.

Kitchin, R. and Sager, R. (1980). *Somat. Cell Genet.* **6**, 615–629.

Kitchin, R., Gadi, I. K., Smith, B. L., and Sager, R. (1982). *Somat. Cell Genet.* **8**, 677–689.

Knudson, A. G. (1977). In *Advances in Human Genetics*, Vol. 8, (H. Harris and K. Kirschborn, eds.), Plenum, New York, Vol. 8, pp. 1–66.

Knudson, A. G. (1983). *Prog. Nucl. Acids Res. Mol. Biol.* **29**, 17–25.

Marshall, C. J., Kitchin, R., and Sager, R. (1982). *Somat. Cell Genet.* **8**, 709–722.

Marshall, C. J. and Sager, R. (1981). *Somat. Cell Genet.* **7** 713-723.

Paraskeva, C., Brown, K., Dunn, A., and Gallimore, P. (1982). *J. Virol.* **44**, 759–764.

Sager, R. and Kovac, P. (1978). *Somat. Cell Genet.* **4**, 375–392.

Sager, R. and Kovac, P. (1979). *Somat. Cell Genet.* **5**, 491–502.

Sager, R., Tanaka, K., Lau, C. C., Ebina, Y., and Anisowicz, A. (1983). *Proc. Natl. Acad. Sci. USA* **80**, 7601–7605.

Saiki, J. J., McCready, K. B., Veitti, T. J., Hewlett, J. S., Morrison, F. S., Costanzi, J. J., Stuckey, W. J., Whitecar, J., and Hoogstraten, B. (1978). *Cancer* **42**, 2111–2114.

Santi, D. V., Garrett, C. E., and Barr, P. J. (1983). *Cell* **33**, 9–10.

Sato, G., Pardee, A. B., and Sirbasku, D. (eds.) (1982). *Growth of Cells in Hormonally Defined Media*, Cold Spring Habor Laboratory, Cold Spring Harbor, New York.

Shay, J. W., (ed.) (1982). *Techniques in Somatic Cell Genetics*, Plenum Press, New York.

Shih, C. and Weinberg, R. A. (1982). *Cell* **29**, 161–169.

Shin, S., Freedman, V. H., Risser, R., and Pollack, R. (1975). *Proc. Natl. Acad. Sci. USA* **72**, 4435–4439.

Smith, B. L. and Sager, R. (1982). *Canc. Res.* **42**, 389–396.

Smith, B. L., Anisowicz, A., Chodosh, L. A., and Sager, R. (1982). *Proc. Natl. Acad. Sci. USA* **79**, 1964–1968.

Smith, B. L. and Sager, R. (1985) *Somat. Cell and Molec. Genet* ., **11** (in press).

Wallace, D. C., Bunn, C. L., and Eisenstadt, J. M. (1975). *J. Cell Biol.* **67**, 174–188.

Yunis, J. J. (1983). *Science* **221**, 227–236.

C. Mechanism of Genetic Variation

CHAPTER 28

SEGREGATION-LIKE EVENTS IN CHINESE HAMSTER CELLS

Ronald G. Worton
Stephen G. Grant
Genetics Department and Research Institute
The Hospital for Sick Children
and Departments of Medical Genetics
and Medical Biophysics
University of Toronto
Toronto, Ontario, Canada

I. THE NATURE OF SEGREGATION IN SOMATIC CELLS

The tenets of classical Mendelian genetics are based on the processes involved in meiotic segregation. The separation or segregation of distinct genetic units which occurs between generations is determined by two major mechanisms. The great majority of genetic loci segregate at meiotic metaphase I when the chromosomes they are located upon undergo random assortment. To a lesser degree, and dependent on their relative location, genes on the same chromosome segregate at meiosis I by the process of crossing-over, a recombinational event between homologous chromatids. These genetic mechanisms are very well characterized, and have been widely exploited to identify and map genetic loci; indeed, they were the basis on which the genetic unit, or gene, was defined.

The term segregation has also been used in relation to genetic events in somatic cells, but here its definition is somewhat less precise. Segregation-like events were first observed as the spontaneous loss of whole chromosomes from somatic cell hybrids (Ephrussi et al., 1964). An effective system of gene segregation in hybrids was later described by Weiss and Green (1967), who demonstrated that in human/mouse hybrids the majority of human chromosomes were rapidly lost. This unidirectional loss of human chromosomes has been exploited by correlating the loss of human enzymes or other detectable gene products with the loss of human chromosomes to map human genes to specific chromosomes or chromosomal regions (McKusick and Ruddle, 1977; Ruddle and Creagan, 1978). Many Chinese hamster genes have also been mapped by this type of analysis in Chinese hamster/mouse hybrids segregating hamster chromosomes (Chapter 5); and conversely, Chinese hamster/human hybrids segregating human chromosomes have been valuable for mapping onto human chromosomes those human genes that complement specific mutant alleles of the Chinese hamster.

Intraspecies hybrids have also played a major role in somatic cell genetics providing tests for the dominant or recessive nature of a mutant phenotype and for complementation between independent isolates with a particular mutant phenotype (Thompson and Baker, 1973). These intraspecies hybrids also undergo segregational events such as the loss of intact chromosomes; however, in such hybrids chromosome loss is not confined to the chromosomes of one parent, nor is it progressive as it is in interspecies hybrids (Worton et al., 1977). In addition, because the genomes that are combined in intraspecies hybrids are homologous, there may be opportunities for regulatory and/or recombinational events, that will, at the level of a single gene locus, mimic gene segregation by chromosome loss. These segregation-like events require that we adopt an operational definition of segregation in such hybrids.

We, as well as others, have therefore adopted an *operational definition of segregation* in intraspecies hybrids as *any event which leads to the reappearance of the recessive phenotype* in a hybrid line formed by fusing a cell line carrying a recessive marker with one that is wild type (or otherwise dominant) at the same locus. Thus, consider a mutant line that is resistant to a particular drug (genotype r) fused to a wild-type line (genotype +) to yield a hybrid of genotype +/r (Fig. 28.1). Ignore, for the moment, the fact that one expects each parental line to have two copies of autosomal genes, so the hybrid may have four copies in total. Assuming the drug resistance is recessive, this hypothetical +/r hybrid has a wild-type phenotype (drug sensitivity). Such hybrids have been shown to generate drug resistant cells at a high rate, orders of magnitude higher than the mutation rate for the drug-resistant marker. Since chromosomal segregation could lead to resistance by loss of the chromosome carrying the + allele, one usually thinks of such cells as drug-resistant segregants. Our knowledge of other systems, however, intimates that other possible mechanisms may exist for generating drug-resistant cells, and by our operational definition all would be considered as "segregants." Some of these other segregation-like events might be detectable if not overshadowed by a very high rate of chromosome loss.

Six possible segregation-like events are shown schematically in Figure 28.1. They are divided into two groups: those which can occur at any time during the cell cycle and those which must (or most likely) occur during mitosis. Among the nonmitotic events we have included mutation, which, although expected to be rare, could on occasion result in a hybrid cell displaying the resistant phenotype; inactivation of the wild-type allele due to an epigenetic event which could also give rise to a drug-resistant segregant; and, since chromosomal rearrangement seems to take place more frequently in near-tetraploid hybrid cells than it does in near-diploid Chinese hamster cells (Worton et al., 1977), deletion of the

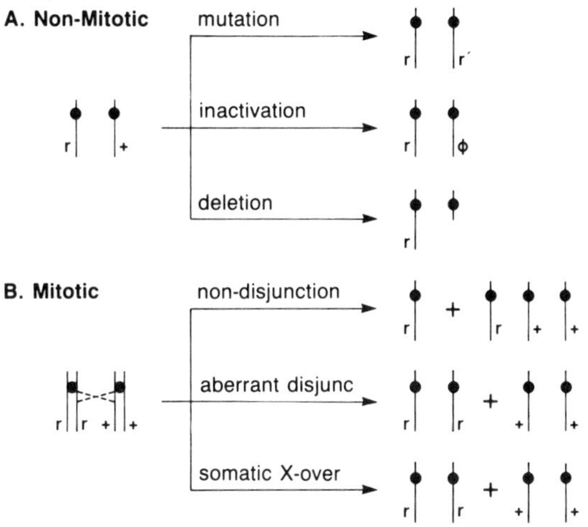

Figure 28.1. Schematic diagrams of six mechanisms which may convert a drug-sensitive cell, heterozygous for a recessive drug-resistant marker (r), into a drug-resistant cell. The drug-sensitive heterozygote (r/+) is depicted on the left, the drug-resistant derivatives on the right. For simplicity the first three mechanisms are depicted as occurring in the G1 phase of the cell cycle, and the last three in G2 or at mitosis. Mutation converts a wild-type allele (+) to a drug-resistant allele (r'). Gene inactivation converts the + allele to a null (ϕ) allele allowing expression of the recessive r allele. Deletion of the + allele also allows expression of the r allele. Nondisjunction produces two cell lines, one trisomic and one monosomic, the latter selectable as a drug-resistant segregant. In an aberrant disjunction both chromatids of one chromosome go to one pole, and both chromatids of the homologous chromosome go to the other. The homozygous (r/r) cell line will be drug resistant. Somatic crossing-over, as indicated between the second and third chromatids (broken lines), will yield the daughter cells shown 50% of the time, whenever one recombinant and one nonrecombinant chromosome segregate to the same pole. Such daughter cells then become homozygous (r/r) or (+/+) for all markers distal to the point of exchange, the r/r line being selectable as a drug-resistant segregant.

wild-type allele has also been considered as a possible segregational mechanism.

Among the mitotic events, the simplest is loss of a whole chromosome by nondisjunction, allowing the recessive allele on the remaining homologue to be expressed. We have also considered an aberrant segregation at mitosis in which homologs instead of sister chromatids separate and segregate to opposite poles, having the effect of deleting one chromosome and duplicating the homolog. Such an event, known to occur in fungi (Pontecorvo and Kafer, 1958), would result in a line homozygous for the drug-resistant allele. Finally, mitotic recombination (somatic crossing-over at mitosis), an event known to occur at appreciable frequency in fungi, yeast, and insects (Stern, 1936; Catcheside, 1977), might also result in homozygosis of any drug-resistance allele distal to the point of crossing-over.

A possible segregational mechanism not depicted in Figure 28.1 is the existence of a parasexual cycle whereby tetraploid cells might segregate diploids. Martin and Sprague (1969) have observed such segregation in human fibroblast cultures carrying heterozygous cytogenetic markers, and Rizzoni et al. (1974) subsequently observed multipolar mitoses which might account for this segregation. Perticone et al. (1976) have also described this type of segregation in cultured cells derived from kidney of hybrid mice in whom the parental and maternal chromosomes were clearly distinguishable.

It is the purpose of this chapter to review briefly the uses of chromosomal segregation for gene mapping in both interspecies and intraspecies cell hybrids involving Chinese hamster cell lines. A major section will then discuss the evidence for or against the "segregation-like" events described in Figure 28.1. Finally, we will discuss the possible role of segregation-like events in the expression of recessive mutations in diploid cells (a topic also discussed by Siminovitch in Chapter 29) and in the expression of recessive cancer genes in the cells of man.

II. SEGREGATION AS A TEST FOR SYNTENIC MARKERS

A. Early Studies in Intraspecies Hybrids

Markers located on the same chromosome are said to be *syntenic* (Renwick, 1969). In somatic cell hybrids segregation of genetic markers often takes place by chromosome loss so that pairs of markers on the same chromosome segregate together while markers on different chromosomes segregate independently, thus providing a definitive test for synteny.

The first demonstration of marker segregation in intraspecies hybrids was by Littlefield (1964) in his now-classic paper on the use of drug-resistance markers for the selection of somatic cell hybrids. Littlefield used two clonal derivatives of mouse L cells, one lacking the enzyme hypoxanthine phosphoribosyltransferase ($HPRT^-$) and the other lacking the enzyme thymidine kinase (TK^-). Hybrids were selected in HAT medium (HAT selections for such hybrids are described in by Fenwick in Chapter 13) and were shown to have both enzymes and about double the normal number of chromosomes. Selection of hybrid cells in azaguanine gave $HPRT^-$ variants at a frequency of about 7×10^{-4}. Since this frequency was much higher than that expected for new mutation, Littlefield argued that the resistant hybrids must be due to segregation, with chromosome loss the implied mechanism.

In early studies with Chinese hamster cells, Kao et al. (1969) generated hybrids between CHO cell mutants with different nutritional growth requirements (auxotrophs). Hybrids formed between a glycine-requiring (GLY^-) and a hypoxanthine-requiring (HYP^-) mutant were found to be phenotypically GLY^+ and HYP^+, indicating that the mutant phenotypes were recessive. Other hybrids formed between two independent GLY^- strains were found to be GLY^+, thereby demonstrating complementation between the two mutant alleles responsible for glycine auxotrophy. This early test for dominance versus recessiveness, and for complementation between markers, set the stage for later work wherein such tests became an important part of the characterization of every new mutant isolated (Thompson and Baker, 1973; Siminovitch, 1976). Although Kao et al. (1969) did not select segregants, they did count chromosomes and reported that the parental lines had 20–21 chromosomes, whereas hybrid lines had approximately double that number soon after fusion but lost 5–10% of their chromosomes stabilizing at a modal number of about 37–38 after 100 generations in culture.

B. Mapping by Segregation in Intraspecies Hybrids

The first attempt to establish synteny in Chinese hamster cells came from Chasin (1972) who constructed hybrids between thioguanine-resistant ($HPRT^-$) CHO cells and diaminopurine-resistant ($APRT^-$; lacking adenosine phosphoribosyltransferase) CHO cells. The parental lines were also glycine auxotrophs, but with complementing mutations affecting different steps in the glycine biosynthetic pathway. Hybrid lines carrying the markers $HPRT^-$, $GlyA^-$, $GlyB^+$, $APRT^+$/ $HPRT^+$, $GlyA^+$, $GlyB^-$, $APRT^-$ were selected in glycine-deficient medium. Segregants that had lost HPRT or APRT activity were selected in medium containing thioguanine or diaminopurine and appeared at frequencies of approximately 10^{-3} and 10^{-4}, respectively. Since drug-resistant mutants arose at a frequency of only 10^{-7} in the parental lines, the explanation for the high frequency of recovery of drug-resistant colonies from hybrid lines appeared to be chromosomal segregation, although the loss of these markers was not associated with extensive loss of chromosomes, and the specific loss of particular chromosomes could not be detected. Segregants of both the $HPRT^-$ and $APRT^-$ classes were found to be glycine independent, suggesting they had retained both the $glyA^+$ and the $glyB^+$ gene. If segregation was due to chromosome loss, this result implied no synteny between the *hprt* and *glyA* loci and between the *aprt* and *glyB* loci. In a further hybrid constructed by fusion of an $HPRT^-$ $APRT^-$ line with an $HPRT^+$ $APRT^+$ line, segregants that had lost the $hprt^+$ marker retained the $aprt^+$ marker and vice versa, suggesting that these genes were also not syntenic.

The analysis described above became the prototype for several subsequent studies involving tests for synteny between pairs of genetic loci in Chinese hamster cells. In independent studies using gene dosage estimates for G6PD (glucose-6-phosphate dehydrogenase) and PGK (3-phosphoglycerate kinase) in Chinese hamster diploid cells, hybrids, and segregants, synteny was reported between *g6pd* and *hprt* (Westerveld et al., 1971) and between *pgk* and *hprt* (Westerveld et al., 1972). The *g6pd-hprt* synteny was later established more rigorously by Rosenstrauss and Chasin (1975) by segregation in hybrids constructed with double ($HPRT^-$ $G6PD^-$) mutants. Chasin and Urlaub (1976) showed that segregation (defined by reappearance of the recessive $HPRT^-$ phenotype) is not due to a new mutation but rather to the reappearance of a characteristic preexisting mutation at the *hprt* locus. Farrell and Worton (1977) applied banded chromosome analysis to the study of Chinese hamster hybrids of genotype $hprt^+/hprt^-$ and thioguanine-resistant segregants derived from these hybrids, to show concordance between segregation of the $HPRT^-$ phenotype and loss of an X chromosome. This provided the first direct evidence that chromosome loss is indeed a mechanism of segregation in intraspecies hybrids and mapped the *hprt* gene (and by inference the syntenic *g6pd* and *pgk* loci) to the Chinese hamster X chromosome. A few segregants that had deleted a portion of the X chromosome short arm allowed assignment of the *hprt* gene to the distal end of the short arm. This assignment was consistent with earlier studies mapping the human (Bengtsson et al., 1975) and mouse (Hashmi and Miller, 1976) *hprt* locus to the X chromosome in those species.

Similar studies with autosomal markers have provided chromosomal assignments for several loci. Worton et al. (1980) constructed hybrids between emetine-resistant (Emt^r) and wild-type (Emt^+) lines of Chinese hamster cells and found that the reappearance of the recessive emetine-resistant phenotype correlated in certain hybrids with loss of a chromosome 2, thus mapping the gene coding for emetine sensitivity to chromosome 2. A few deletion-type segregants mapped the *emt* gene to the proximal half of the long arm. Campbell and Worton (1980) constructed double mutants resistant to both emetine and chromate (presumed genotype emt^r chr^r), fused these to wild-type (emt^+ chr^+) lines, and found concurrent reappearance of both resistant phenotypes upon selection in medium containing either emetine or chromate alone, thus demonstrating synteny between these two loci. Similarly, Wasmuth and Chu (1980) constructed double mutants that were emetine resistant and temperature sensitive, the latter due to a heat-labile leucyl-tRNA synthetase (see Wasmuth, Chapter 14) and constructed hybrids of the type emt^r $leuS^{ts}/emt^+$ $leuS^+$. A high degree of cosegregation of the recessive temperature-sensitive characteristic in cells selected for emetine resistance alone demonstrated synteny of these two markers. Since the emetine-resistance marker in this study was of the same complementation class as that in studies from our own laboratory (the *emtB* locus—see Wasmuth, Chapter 14), the three loci *emtB*, *chr* and *leuS* were all assigned to chromosome 2 in the Chinese hamster.

C. Mapping by Segregation in Interspecies Hybrids

As stated in Section I, interspecies hybrids often show a rapid and progressive loss of the chromosomes contributed by one of the parental cell lines (Weiss and Green, 1967; Scaletta et al., 1967; Kao and Puck, 1970). This loss is less pronounced when the parental lines are closely related in evolutionary terms (Davidson et al., 1966; Yerganian and Nell, 1966). However, in hybrids between two lines, one of which is transformed, it is generally the chromosomes of the nontransformed parent that are lost, regardless of species (Laserre et al., 1980). While agents such as X or γ radiation (Pontecorvo, 1971), UV (Graves, 1980), and drug selection have been found not to influence the direction or rate of segregation, recent studies have shown the direction of loss may be determined by the presence of a specific chromosome from one of the parental lines (Pravtcheva and Ruddle, 1983).

Thus, in cell hybrids formed between established mouse lines and Chinese hamster primary fibroblast lines, it is the Chinese hamster chromosomes that are lost. Using such hybrids formed between mouse cells with a recessive resistance to diphtheria toxin and toxin-sensitive Chinese hamster fibroblasts, Roberts and Ruddle (1980) demonstrated that hybrid clones that had lost diphtheria-toxin sensitivity had also lost Chinese hamster chromosome 2 as well as three hamster isozymes—PGM2, 6PGD, and Enol—thus mapping all four loci to chromosome 2. Similar studies from Siciliano's laboratory (Stallings et al., 1982) confirmed these four loci on chromosome 2 and in further work from these two laboratories (Stallings and Siciliano, 1982; Adair et al., 1983a; Roberts et al., 1983; Siciliano et al., 1983; Stallings et al., 1984), a total of 39 markers have been mapped onto 9 of the 11 Chinese hamster chromosomes (for a review see Siciliano, Chapter 5).

Another mapping strategy that is essentially the reverse of the above utilizes hybrid lines formed by fusion of an established Chinese hamster line with a human primary cell type (lymphocyte, cultured fibroblast, etc.). Such hybrids lose human chromosomes, and again, genes may be mapped by correlating the retention or loss of a particular phenotype with the presence or absence of specific human chromosomes. The phenotype under consideration may simply be an electrophoretically distinguishable human enzyme activity (see the series of reports of the seven Human Gene Mapping Conferences: *Human Gene Mapping* 1–7). Alternatively, by choosing as the Chinese hamster parent a line carrying a recessive mutant allele for temperature sensitivity, auxotrophy, or drug resistance, one can determine if there is a human gene that will correct the Chinese hamster defect and then map this complementing allele.

Mapped human genes complementing auxotrophic Chinese hamster markers include, for example, one which corrects proline auxotrophy (*pro*$^-$) on human chromosome 10 (Jones, 1975), serine hydroxymethyltransferase (*glyA*$^-$, glycine auxotrophy) on chromosome 12 (Jones et al., 1972), and folylpolyglutamate synthetase (*gat*$^-$, auxotrophic for glycine, adenosine and thymidine) on chromosome 9 (Jones et al., 1980). Also mapped are several genes encoding the enzymes of the purine biosynthetic pathway, which correct the defect in adenosine-requiring mutants. These include GAR synthetase (*adeC*) and AIR synthetase (*adeG*) on chromosome 21, FGAR transferase (*adeB*) and GAR transferase (*adeE*) on chromosome 14, and PRPP amidotransferase (*adeA*) on chromosome 4 [for a review see Jones et al. (1981) and Patterson, Chapter 11]. Among temperature-sensitive mutants whose human complementing alleles have been mapped by this strategy are the transfer RNA synthetase genes *leuS* on chromosome 5 (Giles et al., 1980) and *asnS* on chromosome 18 (Cirullo et al., 1983).

Of particular interest for future studies of segregation in human cells is the establishment of several selectable markers on human chromosome 5. Thus, having previously determined that the *leuS* gene is syntenic with the *emtB* and *chr* loci on Chinese hamster chromosome 2 (Wasmuth and Chu, 1980), Dana and Wasmuth (1982a) subsequently found cosegregation of all three phenotypes with human chromosome 5 in hybrids formed between human leukocytes and CHO cells carrying mutations in all three genes. This confirmed linkage of the three complementing wild-type alleles on human chromosome 5. In further studies selecting simultaneously for retention of the human *leuS* allele and against retention of the *emt* and *chr* genes, they found deletion-type segregants that allowed regional mapping of these selectable markers (Dana and Wasmuth, 1982b).

A variety of other markers defined through mutations in Chinese hamster cells have been mapped onto human chromosomes (see *Human Gene Mapping* 1–7) and together with those described above provide dramatic evidence of how studies in Chinese hamster cells contribute to knowledge in other organisms, including humans.

D. Mapping by Microcell-Mediated Chromosome Transfer

As described above, segregation in inter- and intraspecies hybrids has been widely exploited for gene mapping. A third type of hybrid is also useful in this regard and may be thought of as segregation *preceding* hybrid formation instead of following it, since the microcell technique allows direct selection of hybrids with only one or a few transferred chromosomes (see also Chapter 7).

Microcells are generally constructed by inhibiting mitotic spindle formation (Johnson et al., 1975), ideally with a high level of colcemid (Ege and Ringertz, 1974), resulting in small groups of chromosomes, even individual chromosomes, becoming encased by a nuclear membrane to form micronuclei. After treatment of the cell membrane with cytochalasin B, the micronuclei are removed from the cell by centrifugation, taking with them a small portion of cytoplasm and cell membrane (Schor et al., 1975; Wigler and Weinstein, 1975). This preparation of microcells is then separated from whole cells by size fractionation (Fournier and Ruddle, 1977a; McNeill and Brown, 1980), microcells are fused to intact recipient cells and hybrids are then plated in selective medium to allow growth of only those hybrids that have received a particular dominant selectable marker (Fournier, 1981). As an alternative to size fractionation, a double selection procedure can be used selecting for the presence of one donor marker and against another, thus favoring the growth of those hybrids which have received only a small number of donor chromosomes (Worton and Duff, 1981).

Using the above procedure several mammalian genes, and their associated chromosomes, have been transferred. These include the human *hprt* gene on the X chromosome (Tourian et al., 1978), the mouse ouabain resistance (*Oua-1*) gene on chromosome 3 (Kozak et al., 1979), the mouse *aprt* and *GR-1* (glutathione reductase) genes on chromosome 8 and the *GLO-1* (glyoxylase-1) gene on chromosome 17 (Fournier and Frelinger, 1982), the *FPGS* (complementary to the Chinese hamster gat^- mutation) and *Ak-1* (adenosine kinase) genes on mouse chromosome 2 (Fournier and Moran, 1983), the X-linked Chinese hamster *hprt* gene (Worton and Duff, 1981), and the linked *mtx*, *emt*, and *chr* loci on Chinese hamster chromosome 2 (Worton et al., 1981). Microcell hybrids have also been utilized to determine the site(s) of transgenome integration (Fournier and Ruddle, 1977b; Fournier et al., 1979).

We have found microcell hybrids to be very unstable, losing chromosomes rapidly in an attempt to reestablish a balanced karyotype. While this lack of stability may seriously limit the utility of microcell transfer for gene mapping, in some cases it can be turned to advantage. For example, by maintaining selection for the transferred chromosome the homologue of the transferred chromosome may be the one to be lost (unpublished observations), thus effectively providing replacement of the recipient chromosome by a donor chromosome. Such chromosome replacement may have an important place in the future of somatic cell genetics.

III. THE USE OF SYNTENIC MARKERS TO ASSESS SEGREGATIONAL MECHANISMS

A. Segregation of X-Linked Markers in Chinese Hamster Cell Hybrids

1. Evidence for Chromosome Loss and Deletion

As depicted in Figure 28.1, a variety of mechanisms might lead to the reappearance of the recessive phenotype in hybrid lines that are heterozygous for the recessive marker. A few studies have attempted to utilize selectable markers in Chinese hamster cells as tools for assessing the relative importance of the various mechanisms shown in Figure 28.1. As has already been outlined above, segregation of the X-linked *hprt* gene locus seems to involve primarily chromosome loss, with some segregants attributable to chromosome deletion (Farrell and Worton, 1977). Two different intraspecies Chinese hamster cell hybrids were examined in these experiments, and it was found that chromosome loss accounted for 60–80% of all segregation, with the remaining 20–40% attributable to deletion. Neither set of segregants gave any evidence for a mechanism other than these two.

The mammalian X chromosome is unique in that it is functionally hemizygous in all somatic cells. The second X chromosome is absent in males and inactivated in females to compensate for the excess gene dosage (Lyon, 1961). Thus, variation in the number of X chromosome may be tolerated more easily by cultured cells than variation in the number of autosomes. *In vivo*, the phenotypic manifestations of X chromosome monosomy (resulting in Turner's syndrome) are far less severe than that of autosomal monosomies, which are usually lethal events (Hamerton, 1971).

The hybrids described above were specifically constructed from lines lacking an inactive X chromosome. Segregation of X-linked markers in hybrids containing an inactive X will be discussed later in this section.

2. Tests for Mitotic Recombination on the X Chromosome

Several investigators have made direct attempts to detect mitotic recombination in somatic cells. An early study by Tarrant and Holliday (1977) utilized hybrids constructed by fusing different $HPRT^-$ hamster mutants and selection for restoration of the $HPRT^+$ phenotype. The rationale for this experiment was that recombination between *hprt* genes with mutations at different sites within the structural gene could result in a functional gene. Although no spontaneous recombinants were observed, a small number (three) of putative recombinants were found among 10 hybrid cell lines following treatment with the recombinogenic antibiotic, mitomycin C. Mutation was not ruled out as a source of these rare $HPRT^+$ segregants. In a similar study Rosenstraus and Chasin (1978) were also unable to detect intragenic recombination at the *hprt* locus.

In the same study, Rosenstraus and Chasin (1978) also looked for intergenic recombination between the *hprt* and *g6pd* loci on the Chinese hamster X chromosome, utilizing hybrids heterozygous at both loci. In CHO cell hybrids with genotype $hprt^+ \ g6pd^+/hprt^B \ g6pd^-$ (one parent was a double mutant with deficient G6PD activity and an electrophoretically altered weak HPRT activity), the two markers were found to cosegregate over 90% of the time. A few rare segregants of phenotype

$HPRT^B$ $G6PD^+$ were identified and were considered to have arisen either by deletion to give the genotype *g6pd*$^+$/*hprt*^{B} *g6pd*$^-$, or by recombination to give new chromatids *hprt*$^+$ *g6pd*$^-$/*hprt*^{B} *g6pd*$^+$ followed by segregation at mitosis of one recombinant and one nonrecombinant chromatid into the same daughter cell to yield a cell of genotype *hprt*B *g6pd*$^-$/*hprt*^{B} *g6pd*$^+$. In segregants generated by deletion, the *hprt*^{B} allele would be linked to *g6pd*$^-$, whereas in segregants generated by recombination it would be linked to *g6pd*$^-$ on one X chromosome and to *g6pd*$^+$ on the other. Demonstrating linkage of *hprt*^{B} to *g6pd*$^+$ in such rare segregants, therefore, would essentially prove a recombination mechanism. To test this they fused several such rare segregants to another CHO line of genotype *hprt*^{A} *g6pd*$^-$ where the *hprt*^{A} allele is a deficient gene that complements the B allele and is electrophoretically distinguishable from it. These near-hexaploid hybrids, $HPRT^+$ by virtue of intragenic complementation (Chasin and Urlaub, 1976), were selected for loss of the $HPRT^+$ phenotype by plating in thioguanine and then tested for HPRT and G6PD phenotype. If the original segregation event was a deletion, the hexaploid hybrid would have genotype *g6pd*$^+$/*hprt*^{B} *g6pd*$^-$/*hprt*^{A} *g6pd*$^-$ and thioguanine-resistant segregants should arise primarily by loss of the second or third chromosome and result in secondary segregants of either phenotype $HPRT^A$ $G6PD^+$ or $HPRT^B$ $G6PD^+$ with approximately equal frequency. If, on the other hand, the original segregation event was mitotic recombination, the hexaploid hybrid would have the expected genotype *hprt*^{B} *g6pd*$^-$/*hprt*^{B} *g6pd*$^+$/*hprt*^{A} *g6pd*$^-$ and thioguanine-resistant segregants should arise primarily by loss of the third chromosome to give the phenotype $HPRT^B$ $G6PD^+$ and rarely by the loss of both the first and second chromosomes to give the phenotype $HPRT^A$ $G6PD^-$. Furthermore, in this latter model, all or most segregants of phenotype $HPRT^A$ should also be $G6PD^-$. The results of such segregation analysis showed that all of the hexaploid hybrids yielded $HPRT^A$ segregants with a frequency equal to or greater than those of phenotype $HPRT^B$ and, furthermore, most of the $HPRT^A$ segregants were also $G6PD^+$. This result favored deletion (or perhaps mutation or inactivation) of the *hprt* locus as the mechanism for the apparent linkage disruption in the initial segregants, and ruled out mitotic recombination as a common mechanism.

Other sets of hybrids with different marker combinations gave similar results, and allowed the authors to conclude that "if mitotic recombination takes place at all in this system, the rate must be less than about 10^{-6} per call per generation."

3. *Evidence for Epigenetic Events*

Other studies involving segregation of the *hprt* gene have given clear evidence for events other than chromosome loss and deletion. For example, Harris (1979) found adaptive shifts to thioguanine resistance in a series of

Chinese hamster hybrids heterozygous at the *hprt* locus following removal of HAT selective medium. Clonal analysis suggested that this was a true population shift in response to the drug, and not the accumulation of HPRT$^-$ segregants. Despite long-term propagation in selective medium (either HAT or medium containing thioguanine), this population of cells never attained a truly stable phenotype. He suggested that in his hybrid cells the *hprt*$^+$ gene may be partially repressed, or, alternatively, the *hprt*$^-$ gene may be partially derepressed, the degree of repression or derepression depending on the culture conditions. Such an interpretation appeared reasonable in light of clear evidence from interspecies hybrids showing that HPRT activity can often be derived from the enzyme-deficient parent (Watson et al., 1972; Bakay et al., 1973; Croce et al., 1973; Shin et al., 1973; Bakay et al., 1975). In three studies, two involving mouse/human hybrids (Kahan and DeMars, 1975, 1980; Hellkuhl and Grzeschik, 1978) and one a Chinese hamster/human hybrid (Hors-Cayla et al., 1983) the reestablishment of HAT resistance in segregants was dependent on the presence of the human inactive X chromosome.

B. Segregation of Autosomal Markers in Chinese Hamster Cell Hybrids

1. Evidence for Epigenetic Events

As shown above for the X chromosome, segregation may involve a number of different processes. In some autosomal systems, epigenetic mechanisms seem to predominate. Thus, Harris (1975) reported that Chinese hamster hybrids of presumed genotype *tk*$^+$/*tk*$^-$ underwent a gradual evolution of phenotype during long term propagation in bromodeoxyuridine (BrdU) containing medium. This change toward increased plating efficiency in BrdU, and decreased plating efficiency in HAT medium, could be reversed by transferring the mass culture to HAT medium. Such changes are not consistent with conventional genetic mechanisms and led Harris to invoke the possibility of epigenetic mechanisms.

A similar study by Harris and Whitmore (1977) with segregation at the locus *aza*^{r}*ts* (temperature-sensitive azaguanine resistance) in CHO cell hybrids demonstrated a similar intermediate phenotype in segregants selected at low doses of azaguanine but a clearcut and nonreversible resistant phenotype in segregants selected at high doses of the drug. Thus, it appears that the type of segregant selected may depend on the selection protocol. More will be said about epigenetic events in a later section dealing with the role of segregation-like events in the expression of recessive mutation in near diploid cells.

2. Segregation of the emt Locus on Chromosome 2—Evidence for Chromosome Loss and Deletion

Studies of segregation mechanisms from this laboratory have concentrated on using the several selective markers mapped to Chinese hamster chromosome 2. Segregation studies with the *emetine* locus were first carried out by Gupta et al. (1978) in an attempt to estimate the number of *emt* alleles in CHO cells. Emetine-resistant mutants can be generated in CHO cells at a high frequency, yet the mutation behaves recessively in somatic cell hybrids (Gupta and Siminovitch, 1977). Thus, if CHO, being near-diploid in its chromosome complement (Deaven and Petersen, 1973), has two copies of the *emt* gene, both would have to be mutated for the recessive drug-resistance phenotype to be expressed. It is unlikely that two independent mutation events would give mutants at a measurable frequency, hence CHO was considered to perhaps be hemizygous at the *emt* locus. The experiments of Gupta et al. (1978) were designed to measure segregation rate in hybrids between an Emt^r CHO line and a set of four wild-type Chinese hamster lines including CHO and V-79. Segregants resistant to emetine were obtained from CHO × CHO hybrids at a frequency of about 10^{-3}, whereas segregants from CHO × V-79 (and the other two) occurred at frequencies of 10^{-6}–10^{-4}. Their interpretation was that the CHO line is hemizygous at the *emt* locus and, therefore, the CHO × CHO hybrids needed to lose only one *emt* allele whereas the other hamster lines, presumed to be dizygous, had to segregate both of their *emt* alleles. In parallel studies Campbell and Worton (1979) measured Emt^r mutation frequencies for CHO cells as well as the other Chinese hamster cell lines, and finding CHO to be uniquely high in its mutation rate, reached similar conclusions regarding hemizygosity at the *emt* locus in CHO.

In subsequent studies from this laboratory attempting to map the *emt* locus by karyotype analysis of emetine-resistant segregants, it was expected that CHO (emt^r) × (emt^+) segregants should lose one single chromosome, whereas segregants of hybrids such as CHO (emt^r) × V-79 (emt^+/emt^+) were expected to lose a pair of chromosomes. This was not the result obtained. Instead, it was found that the CHO × CHO hybrids generated segregants 80% of which had no apparent chromosome loss and 20% of which had lost a chromosome 2. The CHO × V-79 type of hybrids generated segregants at a much lower frequency, most of which had lost a single chromosome 2, or suffered a deletion of the proximal portion of the long arm of chromosome 2. It was concluded that the *emt* locus is located on the proximal half of the long arm of chromosome 2, and that simple chromosome loss and deletion could not account for all segregation (Worton et al., 1980). This suggested the possibility that mutation, inactivation, mitotic recombination, or some other mechanism played a role in the generation of many of these segregants.

In demonstrating cosegregation of the *emt* and *chr* loci in hybrids of the type CHO (emt^r chr^r) × CHO (emt^+ chr^+), Campbell and Worton (1980) reported that while all segregants selected for emetine resistance were also chromate resistant, only 75% of segregants selected for chromate resistance were also emetine resistant. It was suggested that this polarity of linkage disruption might be accounted for by a mitotic recombination mechanism with *chr* distal to *emt* on the 2q arm, or, alternatively, by gene inactivation from a spreading center of inactivation nearer to the *chr* locus. This polarity effect has not been observed, however, in similar hybrids constructed from different strains (unpublished observations).

The mapping of *emt* and *chr* on 2q was significant because CHO cells were known to have a single intact chromosome 2, the other homologue (called the Z2) having a major deletion of the proximal two-thirds of its long arm (Deaven and Petersen, 1973; Worton et al., 1977), as depicted in Figure 28.2. This confirmed the hemizygosity of the *emt* locus in CHO and explained the high mutation rate at this locus.

3. Segregation of Multiple Markers on Chromosome 2—Evidence for Chromosome Loss Plus Duplication

In order to adequately examine mechanisms of segregation with a level of sophistication approaching that available in lower organisms, more markers were needed on chromosome 2; at least one on the short arm, and one distal to the *emt* or *chr* locus on the long arm. Two cell lines were identified with cytogenetic alterations of chromosome 2, both of which provided useful cytological markers in studies of segregation. The first was a cell line called Mtx^{RIII}, a methotrexate-resistant derivative of CHO with a moderate (10-fold) amplification of sequences encoding the target enzyme, dihydrofolate reductase (Flintoff et al., 1976, 1984; see also Chasin, Chapter 16). This line was found to have a translocation between 2p (short arm) and 5q (long arm) generating two marker chromosomes, $2p^-$ and $5q^+$ (Worton et al., 1981). The *mtx* locus (amplified *dhfr* gene) was mapped by microcell-mediated gene transfer (see above and Chapter 7) to the remaining portion of the short arm of the $2p^-$ chromosome (Worton et al., 1981), thus providing a dominant selectable marker (methotrexate resistance) and a cytogenetic marker on 2p.

The other marked chromosome was found in a line called LR73, a phenotypic revertant of the temperature-sensitive tsH1 line (Pollard and Stanners, 1979), and consisted of a $2q^+$ marker with an apparent tandem duplication or insertion in the long arm of chromosome 2. The $2q^+$ and $2p^-$ markers are pictured in Figure 28.2.

Using these marker chromosomes in somatic cell hybrids, Campbell and Worton (1981) were able to distinguish among most potential segregation mechanisms. One set of hybrids, for example, was constructed by fusion of a CHO line carrying both the dominant methotrexate resistance gene (mtx^{RIII}) and the recessive emetine resistance gene (emt^r) on the $2p^-$ chromosome, with a second line carrying the corresponding wild-type alleles on the $2q^+$ marker chromosome. The chromosomes 2 of this hybrid are depicted in Figure 28.3. In the hybrid, the presence of two

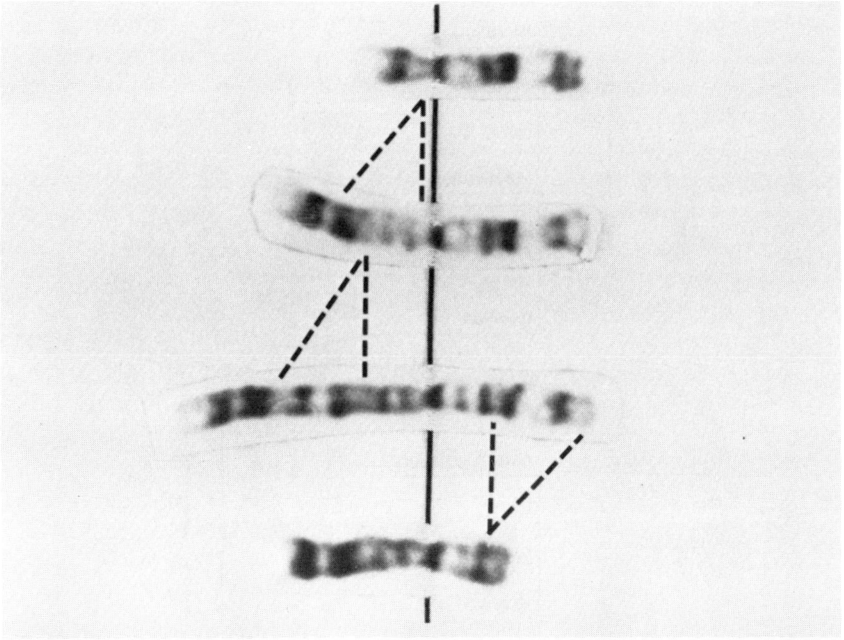

Figure 28.2. Cytogenetic markers of Chinese hamster chromosome 2. Left to right: Z2; 2; $2q^+$; $2p^-$. The dotted lines indicate the long arm interstitial deletion of the Z2, the long arm insertion or tandem duplication of the $2q^+$ marker, and the short arm deletion of the $2q^-$ marker, the latter resulting from a balanced reciprocal translocation with a chromosome 5. The *emt*, *chr*, and *leuS* loci are on 2q above the upper edge of the insertion in the $2q^+$ marker, within the region deleted from the Z2. The *mtx* locus is on 2p below the break point in the $2p^-$ marker.

Z2 chromosomes (one from each parent) could be essentially ignored since the Z2 carries no *emt* locus and since its mtx^+ gene is recessive. Emetine-resistant segregants were selected and examined by karyotype analysis. Figure 28.3 depicts the predicted fate of the $2p^-$ and the $2q^+$ chromosomes in segregants arising by various mechanisms. As can be seen, mutation and gene inactivation yield emetine-resistant segregants with no change in karyotype, deletion yields segregants that have lost a portion of the long arm of the $2q^+$ marker, and nondisjunction yields segregants that have lost the entire $2q^+$ marker. The type of aberrant disjunction discussed in Figure 28.1 yields segregants with two copies of the $2p^-$ and none of the $2q^+$, and somatic crossing-over generates segregants that contain one parental chromosome, the $2p^-$, and a new recombinant chromosome that combines the short arm of the $2q^+$ and the long arm of the $2p^-$ and thereby resembles a normal chromosome 2. In examining segregants from this particular hybrid, plus others in which the *chr* marker was used to select chromate-resistant

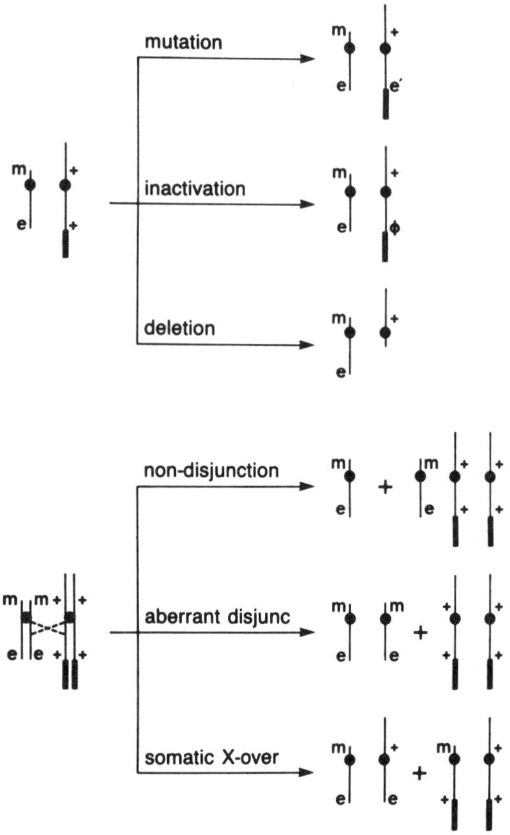

Figure 28.3 Schematic diagram of six segregation mechanisms, depicting expected genotype and karyotype for each mechanism. The hybrid cell line depicted on the left is heterozygous for two markers, methotrexate resistance (m) and emetine resistance (e). The e and m alleles are on a chromosome 2 with a deletion of the short arm ($2p^-$), while the wild-type alleles (+) are on a chromosome 2 with a normal short arm but an elongated long arm ($2q^+$) (heavy bar). The hybrid line also contains a marker 2 (Z2) from each parent, but these are left off the diagram since they have no emetine-resistance locus (see text). Segregants selectable in medium containing emetine are depicted on the right. The rationale is the same as for Figure 28.1. Although mutation and gene inactivation do not alter the karyotype, the other four mechanisms result in karyotypic change, the change being different in each case. Note that segregants derived by somatic crossing-over should contain a normal chromosome 2 (normal short arm from 2q+, normal long arm from $2p^-$) that did not exist before in the hybrid.

segregants, Campbell and Worton (1981) reached the following conclusions. Approximately 75% of the segregants were generated by a mechanism that involved loss of the chromosome bearing the emt^+ (or chr^+) gene and duplication of the chromosome bearing the resistant gene; a further 10% were generated by simple loss of the homolog bearing the wild-type gene; 10% involved deletion of the emt^+ (or chr^+)

gene; and the remaining 5% resulted from gene inactivation, new mutation, or a cytologically undetectable small deletion. Mitotic recombination was not detected, and, therefore, must occur at a rate of less than 10^{-6} in this system.

The mechanism for generating segregants with loss of one chromosome and duplication of the homologous chromosome was investigated by segregation of hybrids with three copies of the *emt* and *chr* loci. Drug-resistant segregants from hybrids of the type CHO (emt^r or chr^r) × V-79 (emt^+ chr^+/emt^+ chr^+) arose primarily by loss of both V-79 chromosomes and duplication of the CHO chromosome. Segregants from hybrids of the type CHO (emt^+ chr^+) × V-79 (emt^r/emt^r or chr^r/chr^r) arose primarily by loss of the single CHO chromosome and duplication of a V-79 chromosome. If loss plus duplication occurred as two separate events with the duplication event providing a selective advantage due to restoration of chromosome balance, one would have expected the chromosome bearing the resistant allele to be duplicated in the first set of segregants but not in the second. The fact that loss and duplication predominated in both sets of segregants suggested that the duplication event was coupled to one of the loss events, and thus led to the model of abberant disjunction shown in Figures 28.1 and 28.3 (Worton et al., 1980; Campbell and Worton, 1981). Regardless of mechanism, this study firmly established loss and duplication of homologous chromosomes as a major mechanism of genetic variation in somatic cells.

4. Segregation of Multiple Markers on Chromosome 2 —Evidence for Gene Inactivation

One problem in using cytogenetically marked chromosomes for studies such as those described above is that the deletions and duplications used as markers may inhibit the very processes (e.g., somatic crossing-over) that the experiments are designed to detect. This is especially true since chromosomal rearrangement is known to block mitotic recombination in both fungi (Kafer, 1977) and *Drosophila* (Ronen, 1964; Merriam and Garcia-Bellido, 1972). More recent studies from this laboratory have therefore focused on repeating studies of the type described above, but with more structurally normal chromosomes.

Thus, Worton et al. (1984) utilized hybrids of the type $2p^-(mtx^{RIII}$ $emt^+)/2(mtx^+$ $emt^r)$ with no q^+ marker on the long arm. The problem in this type of hybrid is that while one can easily distinguish aberrant disjunction (gives 2/2) from somatic crossing-over (gives $2p^-/2$), one cannot distinguish the latter from gene inactivation (or mutation or small deletion) since the chromosomes are of the parental type cytogenetically. In emetine-resistant segregants from this type of hybrid, karyotype analysis revealed that about 75% were due to chromosome loss, loss plus duplication, or deletion; the remaining 25% were due to either gene inactivation, mitotic recombination, new mutation, or small deletion. Thus, elimination of the q^+ marker on one chromosome 2 did seem to enhance a particular form of segregation. To distinguish segregants arising by gene inactivation (or small deletion) with presumed genotype $2p^-(mtx^{RIII}\ emt^{\varphi})/2(mtx^+\ emt^r)$ from those arising by new mutation or mitotic recombination with presumed genotype $2p^-\ (mtx^{RIII}\ emt^r)/2(mtx^+\ emt^r)$ three tests were applied (Worton et al., 1984; Grant et al., manuscript in preparation). First, the $2p^-$ chromosome from such a segregant was examined by transferring it into a new host using the microcell-mediated chromosome transfer technique selecting for the dominant Mtx^{RIII} phenotype. The new host was a second-level emetine-resistant mutant (Emt^{rII}), such resistance being recessive to both rI level resistance and wild-type sensitivity (Gupta and Siminovitch, 1978*a*). The $2p^-$ chromosome from a gene inactivation segregant carries a null (emt^{φ}) allele and should not change the phenotype of the host after transfer, whereas the $2p^-$ chromosome from a mitotic recombination segregant carries an rI (emt^{rI}) allele and should convert the host cell from an rII to an rI phenotype. Two segregants were examined in this manner and in both cases the host phenotype was not affected by the transferred chromosome, suggesting that these two segregants arose by gene inactivation. The second test was to measure the induced mutation frequency for the conversion of the rI level segregants to the rII phenotype. Gene inactivation segregants should mutate at high frequency since only one allele must be mutated, whereas mitotic recombination segregants should yield rII mutants at low frequency since two rI genes must be mutated (or one mutated and one segregated). Based on comparison with control cell lines carrying one or two copies of the rI allele, the mutation frequency in 10 segregants examined was found to be consistent with one allele and not with two, suggesting that all 10 segregants examined arose by gene inactivation. While these results rule out mutation and recombination as the mechanism in those segregants tested, they fail to distinguish gene inactivation from a small, cytologically undetectable deletion. The ultimate proof for any gene inactivation event must rely on the successful reactivation of the putative null allele.

5-Azacytidine (5azaCR) is a cytidine analog that acts as a strong inhibitor of DNA methylation in mammalian cells (Creusot et al., 1982). Treatment with 5azaCR has resulted in reactivation of genes on the inactive human X chromosome (Mohandas et al., 1981), genes which define products of differentiated tissues (Taylor and Jones, 1979), and genes not expressed on endogenous viruses (Groudine et al., 1981). In order to determine whether this drug might reactivate the putative null *emt* allele in our segregants, several segregants were treated overnight with 5azaCR. Control segregants arising via chromosome loss or loss plus duplication failed to respond to this treatment, whereas two Emt^r and two Chr^r segregants, which were karyotypically consistent with either gene inactivation or mitotic recombination, showed reversion to the wild-type phenotype in 15–20% of cells. This strongly suggests that the majority of segregants in this class were generated by a gene inactivation mechanism, and further that the inactivation may have been caused by methylation at or near the *emt* locus. Further studies with appropriate cloned DNA segments will be necessary to prove this hypothesis.

Gene inactivation has previously been shown to act as a segregational mechanism in ciliates, where it was called "allelic repression" (Nanney, 1968). In mammals inactivation of the second X chromosome in female somatic cells occurs at the level of the DNA, as evidenced by the lack of expression of inactive X DNA in transformation experiments (Liskay and Evans, 1980; Chapman et al., 1982; Venolia and Gartler, 1983). Gene inactivation may also be an important mechanism in differentiation (Taylor and Jones, 1979) and in control of endogenous (and infectious exogenous) viruses (Niwa and Sugahara, 1981). The molecular mechanism of gene inactivation has recently been the subject of much experimental work; inactivation has been linked to DNA methylation (Waalwijk and Flavell, 1978) and chromatin structure (Groudine et al., 1981), and to the presence of left-handed Z DNA (Nordheim et al., 1981).

5. Segregation of Multiple Markers on Chromosome 2 —Evidence for Mitotic Recombination

Despite genetic evidence against mitotic recombination on the X chromosome there is considerable circumstantial evidence for the occurrence of somatic recombination in mammalian cells. Enzymes necessary for breakage and recombination are available in Chinese hamster cells, as evidenced by a high frequency of sister chromatid exchange (Rommelaere et al., 1973), and detection of associated hybrid DNA (Rommelaere and Miller-Faures, 1975; Moore and Holiday, 1976). Quadriradial figures, thought to be recombination intermediates, have been observed in many cell types, especially those from patients with Bloom's syndrome (German, 1964; Therman and Kuhn, 1976). In some cell types the frequency of quadriradial figures was found to be increased following treatment of the cells with a known recombinogen such as mitomycin C (Huttner and Ruddle, 1976). Furthermore, animal viruses such as SV40 undergo homologous recombination *in vivo* (Brockman, 1977), and, recently, plasmids containing virus DNA and selectable markers have been shown to recombine upon superinfection (Kucherlapati et al., 1984) or after integration into the genome (Liskay and Stachelek, 1983).

It is not surprising therefore that Wasmuth and Hall (1984) have now reported that mitotic recombination occurs as a mechanism of segregation in Chinese hamster hybrid cells. Utilizing the *emt* and *leuS* genes mapped to the deleted portion of Chinese hamster chromosome 2 (see Fig. 28.2), as well as an mtx^{R} allele with an associated cytogenetic marker (a homogeneously staining region or HSR) on the short arm of the 2, hybrids were constructed of genotype 2(*mtx*$^+$ *leuS*^{ts} *emt*^{r})/2p^{HSR}(*mtx*^{R} *leuS*$^+$ *emt*$^+$). Twenty segregants were then selected in medium containing emetine and three distinct types were recovered. Fifteen had lost all three dominant alleles and the 2p^{HSR} marker chromosome and were found to carry two copies of the normal chromosome 2 (chromosome loss and duplication). Three of the segregants had lost their emetine sensitivity, but retained both methotrexate resistance and the wild-type level of temperature sensitivity and were consistent with the mechanisms of gene inactivation, new mutation, or small deletion. The remaining two segregants had retained their marker chromosome 2 and methotrexate resistance, but were emetine resistant and temperature sensitive. These segregants could have occurred via an inactivation event which encompassed both the *emt* gene and the *leuS* gene, or by a mitotic recombination event occurring proximal to these two loci on the long arm of chromosome 2.

To distinguish between these two possibilities the segregants were each treated with 5azaCR and plated at 39°C to test for reactivation of the putative *leuS*$^{\varphi}$ gene. The frequency of phenotypic reversion was not affected by 5azaCR, an indication that the *leuS*$^+$ gene had not been inactivated. The segregants were then fused to a line of genotype 2(*mtx*$^+$ *leuS*$^+$ *emt*^{rII}) and secondary segregants from these triparental hybrids were selected in a high concentration (rII level) of emetine. If the original segregants had arisen by gene inactivation, only the *emt*^{rI} gene on the normal chromosome 2 would have to be segregated out to allow expression of the *emt*^{rII} gene. If the original segregants had arisen by mitotic recombination, both the normal 2 and the 2p^{HSR} chromosome would carry an *emt*^{rI} gene, and both of these chromosomes would have to be segregated from the triparental hybrid to attain an Emt^{rII} phenotype. Not only did secondary segregants arise at a very low frequency consistent with a double segregation process, but the segregants that did occur had lost both the chromosome 2 and the marker 2p^{HSR} supplied by the original hybrid. This appears to be good evidence for a mitotic recombination event in the two original segregants, and constitutes the first genetic evidence for the occurrence of mitotic recombination in mammalian cells.

IV. THE ROLE OF SEGREGATION-LIKE EVENTS IN THE EXPRESSION OF RECESSIVE MUTATIONS IN CULTURED CELLS

Many of the mutations generated at high frequency in cultured Chinese hamster cells, particularly CHO cells, have phenotypes that behave recessively in somatic cell hybrids. This has raised some doubt about the diploid nature of the CHO line and led Siminovitch (1976) to propose that CHO might be characterized by extensive regions of the genome that are functionally hemizygous so that for many markers only the one remaining functional allele must be mutated (see Siminovitch, Chapter 29). The *emt, chr,* and *leuS* map to the structurally hemizygous region of chromosome 2 (Wasmuth and Chu, 1980; Worton et al., 1981) so that the isolation of recessive mutations at these three loci is readily explained. These may be exceptional, however, since the other CHO chromosomes do not show extensive deletions (Deaven and Petersen, 1973; Worton et al., 1977) and since many of the recessive markers in CHO are not linked to the *emt* locus on chromosome 2, nor to the *hprt* locus on the X chromosome [see, for example, Gupta (1980)]. Isozyme analysis has revealed that most regions of the CHO genome are not functionally hemizygous (Siciliano et al., 1978), so that some other explanation is required for the recessive mutations generated at high frequency on other chromosomes. One possibility is that two events occur, one a mutation at one of the two alleles, and the second a segregation-like event such as nondisjunction, aberrant disjunction, inactivation of the second allele, or somatic crossing-over between the homologs carrying the mutant and wild-type alleles. Several mutation studies support this concept in Chinese hamster cells, although many of them were done in tetraploid cells.

A. Mutation Studies in Diploid versus Tetraploid Cells

Several mutation studies have been carried out to test the hypothesis that if a near-diploid Chinese hamster line yields a particular recessive mutant phenotype with a mutagen-induced frequency μ, then, if mutation is the only event taking place, a near-tetraploid line should yield the same mutant phenotype at a frequency μ^2, since there are twice as many alleles to be mutated. If μ is in the range of 10^{-7}–10^{-5}, then μ^2 will be 10^{-14}–10^{-10}, a frequency too low to detect in typical platings of 10^7–10^8 mutagenized cells. Early studies in frog (Mezger-Freed, 1971, 1972) and rat (Prickett et al., 1974) cells found mutation to occur at similar frequencies in cells of different ploidy. Initial studies in Chinese hamster cells by Harris (1971, 1973, 1974), Van Zeeland and Simons (1975), and Harris and Whitmore (1977) also failed to demonstrate this μ and μ^2 relationship, suggesting a role for mechanisms other than mutation. McBurney and Whitmore (1974), on the other hand, attempted to select temperature-sensitive mutants from near-diploid and near-tetraploid CHO cells, but failed to detect any ts mutants in the latter, a finding consistent with the expected μ and μ^2 relationship.

Perhaps the clearest results came from Chasin (1973) who found that thioguanine-resistant mutants (with an $HPRT^-$ phenotype) arose following mutagenesis with ethyl methanesulfonate (EMS) with a 25-fold reduced frequency in near-tetraploid CHO cells compared to the normal (near-diploid) CHO line. This result suggested the possibility of a low-frequency mutational event coupled with a high-frequency event such as segregation acting together to give thioguanine resistance in the tetraploid cells. This was later verified by Chasin and Urlaub (1975) who showed that in a near-tetraploid CHO cell hybrid with two X chromosomes and genotype *hprt*$^+$ *g6pd*$^-$/*hprt*$^+$ *g6pd*$^+$, approximately 50% of the HPRT$^-$ mutants selected were phenotypically G6PD$^-$. This strongly suggested mutation of one *hprt* allele in cells which had previously or subsequently lost the other allele by chromosomal segregation. Similar results were obtained in tetraploid mouse cells by Raskind and Gartler (1978) who demonstrated the loss of an X chromosome from thioguanine-resistant variants, and by Farber (1973) who found electrophoretic patterns consistent with chromosome loss and loss plus duplication in cells carrying three and four copies of the chromosomes carrying the associated enzyme loci, respectively. These results suggest that a similar mechanism might exist to generate recessive mutant phenotypes for autosomal loci in diploid cells.

B. Role of Segregation-like Events in the Generation of Recessive Mutants in Diploid Cells

If certain recessive mutants in diploid cells arise by a two-step process of mutation plus segregation, then any of the segregation mechanisms depicted in Figure 28.1 could be considered. Indeed, evidence exists to support gene inactivation, deletion, recombination, chromosome loss, and loss plus duplication as mechanisms occurring in diploid cells, which when coupled with mutation lead to the expression of recessive phenotypes.

1. Evidence for Gene Inactivation

Gene inactivation has been implicated as a mechanism of high-frequency mutation at the *aprt*, *tk*, *glk*, *pro*, *emt*, and *hprt* loci in Chinese hamster cells. In early studies at the *aprt* locus Jones and Sargent (1974) reported APRT$^-$ mutants, fully resistant to diaminopurine, which reverted to APRT$^+$ (presumed genotype *aprt*$^+$/*aprt*$^-$) at high frequency. More recently, Turker et al., 1984) described *aprt* heterozygotes directly selected from mouse teratocarcinoma cells which could undergo facile conversion to an APRT$^-$ phenotype and back again. While these two examples suggest a gene inactivation/reactivation event at this locus, most of the definitive evidence for such events come from other genetic loci.

In studies at the *tk* locus Bradley (1979) looked for segregation-like events in CHO by selecting TK$^-$ mutants in BrdU, then back-selecting a TK$^+$ revertant of presumed genotype *tk*$^+$/*tk*$^-$ in HAT medium. He then found that TK$^-$ "segregants" appeared at high frequency when the heterozygote was again plated in BrdU. These segregants easily reacquired the TK$^+$ phenotype in nonselective medium, suggesting an inactivation and reactivation at this locus. By constructing lines that were also heterozygous at the closely linked galactokinase (*glk*) locus, he showed coordinate loss of both phenotypes. He termed the phenomenon allelic silence.

Siminovitch (1979) then reported a subline of CHO in which both TK^- and GLK^- mutants arose at high frequency. Similarly, in TK^- mutants selected at low frequency from V-79 cells, GLK^- mutants arose at high frequency, suggesting that mutation at the *tk* locus occurs by a mechanism which renders the linked *glk* locus functionally hemizygous.

Subsequently *tk* heterozygotes were directly selected in low levels of BrdU (Harris and Collier, 1980; Bradley et al., 1982) and these also yielded fully resistant mutants in a manner (high frequency, initially unstable) consistent with gene inactivation. In later studies, Bradley (1983) generated glk^+/glk^- heterozygotes either directly (by selection in low doses of 2-deoxygalactose) or by sib-selection (without exposure to the drug). Directly selected heterozygotes arose at high frequency and were functionally hemizygous at the linked *tk* locus, again suggesting coordinate inactivation of these two genes. These heterozygotes underwent mutation to the GLK^- phenotype at low frequency. The sib-selected heterozygotes, on the other hand, arose at low frequency, remained dizygous for *tk*, and gave rise to GLK^- variants at high frequency, half of which had lost the expression of one *tk* allele. Thus, two different events occur to produce the GLK^- phenotype, and the high-frequency event seems to be a gene inactivation which often results in the simultaneous silencing of the neighboring *tk* allele.

Harris (1982) took a different approach to the study of the *tk* gene. He exposed TK^- Chinese hamster cell lines from various sources to 5azaCR and was able to recover TK^+ "revertants" at very high frequency. As the revertants had approximately 50% wild-type activity he concluded that the original mutants had most likely arisen via a two-step procedure involving both a mutation and a gene inactivation event..Bradley (1983) repreated this type of experiment with his tk^-/tk^- homozygous lines (described above) and found that they also could revert at high frequency when treated with 5azaCR.

There is similar evidence for gene inactivation at the *pro* locus. Whereas diploid primary Chinese hamster cells can make their own proline from glutamate or ornithine, the CHO line cannot and is therefore a PRO^- auxotroph (Kao and Puck, 1967). Morgan Harris (personal communication) has recently found that overnight treatment of CHO cells with 5azaCR restores the proline synthetic pathway in 10% of cells, a dramatic demonstration that the auxotrophic condition was due, at least in part, to a reversible inactivation of one or more genes coding for enzymes in the biosynthetic pathway. In our laboratory we have confirmed the 5azaCR-induced reversion at the *pro* locus but failed to reverse the auxotrophy for a number of mutants in the glycine and adenosine biosynthetic pathways, suggesting that the results with proline cannot be generalized to other loci.

Harris and co-workers have recently demonstrated 5azaCR-induced reactivation of the gene for glutamine prototrophy in Chinese hamster V-79 cells (Harris, 1984), and for asparagine prototrophy in Jensen rat sarcoma cells (Sugiyama et al., 1983). It has also been shown that classical mutagens, such as EMS can induce gene inactivation by causing DNA hypermethylation (Ivarie and Morris, 1982; Nakamura and Okada, 1983), leading Harris to suggest the widespread presence of so-called "methylation mutants" among cultured cell lines.

We have observed that mutation in V-79 cells resistant to low doses of emetine (putative homozygotes emt^{rI}/emt^{rI}) to the rII level of resistance can occur at a frequency consistent with a single gene mutation. This implies that the original EMT^{rI} mutation can occur by a mechanism that renders the locus hemizygous. Siminovitch (1981) reported a strain of wild type, emetine sensitive V-79 cells which gave rise to EMT^{rI} mutants at high frequency. Mutation at the *emt* gene in V-79 therefore can occur in two steps, mutation plus segregation (not necessarily in that order), the segregation leading to functional hemizygosity. As all of these lines retain a normal V-79 karyotype, with two chromosome 2s, this segregation event cannot be chromosome loss or gross deletion, so is most likely gene inactivation.

Although the CHO line is of female origin, X-linked genes such as *hprt* are hemizygous, since these cells carry only one intact X chromosome (Deaven and Petersen, 1973; Worton et al., 1977). Thioguanine-resistant mutants may be selected from CHO in a single step, and these mutants fall into two classes. The first, which occurs at relatively low frequency, shows a stable phenotype and probably consists of true mutants. The second, more frequent class produces unstable variants consistent with gene inactivation (Bradley, 1980). It has been shown that 5azaCR can reactivate the *hprt* gene on the inactive human X chromosome in somatic cell hybrids (Mohandas et al., 1981; Lester et al., 1982; Hors-Cayla et al., 1983) and on the inactive mouse X chromosome in tetraploid cells (Graves, 1982). We have shown a similar reactivation of the inactive X chromosome in diploid female cells of Chinese hamster, in both primary cells and transformed derivatives (Grant and Worton, manuscript in preparation). Thus, it appears that for at least five or six gene loci a gene inactivation event can suppress the wild-type gene activity, allowing expression of the recessive mutant allele on the homologous chromosome, and in some cases the inactivation is reversible by treatment of the cells with 5azaCR. Clearly, much work remains to be done on the molecular mechanism of these inactivation and reactivation processes.

One possible explanation for some of the extensive hemizygosity, apparently due in part to gene inactivation, of the CHO line is suggested by recent work mapping enzyme loci in normal Chinese hamster cells and in CHO. The gene locus for isocitrate dehydrogenase-2 (*IDH2*) has been assigned to Chinese hamster chromosome 3 (Adair et al., 1983a). In CHO cells this chromosome exists only in a rearranged state (Deaven and Petersen, 1973), with material from chromosome 3 identified cytogenetically on the Z3, Z4, and Z7 marker chromosomes peculiar to this line (Worton et al., 1977). The *IDH2* locus is hemizygous in CHO, and maps to the Z3 chromosome (Adair et al., 1984). The homologous gene locus should map to a region present on the Z4 but very close to the inversion break point. It is possible that this second locus has been inactivated by virtue of its proximity to the break site. Such a "position effect" has been documented in *Drosophila* (Baker, 1968) and was shown recently to occur in human cells with an X-autosome translocation (Mohandas et al., 1982).

2. Evidence for Deletion and Structural Mutation

In studies at the *aprt* locus in CHO cells $aprt^{+}/aprt^{-}$ heterozygotes have been generated by both direct selection and by sib-selection (Bradley and Letovanec, 1982; Simon et al., 1982). Sib-selected heterozygotes arose at low frequency, and in one case an altered APRT protein was demonstrated by two-dimensional electrophoresis. Second-step mutants from these lines arose at high frequency, had lost the wild-type spot on two-dimensional gels, and were completely nonrevertible. Quantitative Southern blot analysis has determined that only a single *aprt* gene remains in these mutants, and, furthermore, by demonstration of a unique restriction fragment, the remaining gene has been identified as a mutant allele (Simon and Taylor, 1983). In this system, therefore, two events have taken place: the first a structural gene mutation, and the second a deletion of the remaining wild-type allele.

The Chinese hamster *aprt* gene, like the *IDH2* locus described above, maps to chromosome 3 (Adair et al., 1983a) and therefore to the Z7 (far from any chromosomal break points) and Z4 (close to the point of inversion) marker chromosomes in CHO (Adair et al., 1984). In certain CHO strains the *aprt* gene has been shown to be functionally hemizygous (Taylor et al., 1977), and in one such line deletion of the allele located on the Z4 has been demonstrated karyologically (Adair et al., 1983b).

Directly selected $aprt^{+}/aprt^{-}$ heterozygotes occur at high frequency in CHO. Half of these heterozygotes have lost an *aprt* gene. Mutants fully resistant to diaminopurine have been derived from these cells only at low frequency, and have been shown to carry a single mutant allele. Thus, in three separate systems segregation at the *aprt* locus in CHO has been shown to occur by an allele-specific structural gene deletion (Simon et al., 1983).

Other directly selected heterozygotes retain two copies of the *aprt* gene and so are consistent with gene inactivation. One such heterozygote, however, has been shown to produce an altered APRT protein, and thus arose by point mutation (Simon et al., 1983). This result raises the possibility that a high-frequency mutation event may occur at this locus.

The existance of high-frequency mutational events has been suggested by the fact that certain mutant phenotypes arise at very high frequency in cultured cells. These include thymidine-uptake mutants of Chinese hamster DON cells (Breslow and Goldsby, 1969) and tubercidin-resistant (AK^-) mutants of CHO (Gupta and Siminovitch, 1978b; Rabin and Gottesman, 1979). In only one system, however, the *aprt* gene in mouse L cells, has strong evidence (the presence of an enzymatically inactive, immunologically cross-reacting protein) been presented for high-frequency mutation actually involving a structural mutation (Tischfield et al., 1982).

3. Evidence for Chromosome Loss, Loss Plus Duplication, and Structural Rearrangement

Evidence for chromosomal events playing a role in mutation of diploid cells comes, not from Chinese hamster, but from mouse. Thus, Eves and Farber (1981) isolated mutants of CAK mouse cells deficient in adenosine kinase (AK^-) and examined them for phenotype at the esterase 10 (*ES-10*) locus, which is syntenic on mouse chromosome 14. The near-diploid CAK line is heterozygous for a pair of electrophoretic variants of ES-10, and approximately half of all AK^- mutants isolated were missing one of the two electrophoretic forms of ES-10. Karyotype analysis revealed that of these, about half had lost a chromosome 14, and a number of others had undergone rearrangement of chromosome 14, demonstrating that the expression of an autosomal recessive mutation in near-diploid mouse cells is frequently associated with the segregation of a physically linked marker and part or all of the chromosome carrying the two markers.

In these experiments the mutants with a normal pair of chromosomes 14 but with only one ES-10 phenotype may have undergone a coordinate gene inactivation or they might have lost one chromosome and duplicated the homologue. To determine if such a loss and duplication event could indeed occur chromosomal markers were required. For this reason, Eves and Farber (1983) turned to the *aprt* locus on mouse chromosome 8, selecting mutants from CAK cells with a morphologically marked chromosome 8. Of about 20 independent $APRT^-$ mutants, 20% had only a single chromosome 8, another 20% had lost one chromosome 8 and duplicated the other, and an additional 25% had chromosomal rearrangements affecting a number 8 chromosome, thus demonstrating that chromosome loss, loss plus duplication, and structural rearrangement all play a role in the generation of recessive mutation at this locus in mouse cells.

V. THE ROLE OF SEGREGATION-LIKE EVENTS IN TUMORIGENESIS

A. Segregation of Genes Associated with the Transformed Phenotype

Segregation has also been used to study tumorigenesis and the nature of transformation in cultured cells. Certain traits associated with the transformed phenotype have been found to be suppressed in hybrids formed between tumorigenic and nontumorigenic Chinese hamster cells. These include colony morphology, reduced serum dependence, anchorage independence, and tumorigenicity itself. Hybrids that have segregated the tumorigenic phenotype have been selected by passage through nude mice (Sager and Kovac, 1978) while those segregating anchorage independence have been selected by plating in methylcellulose (Sager and Kovac, 1979; Marshall and Sager, 1981). Using this approach, Marshall et al. (1982) have mapped the gene responsible for anchorage dependence to chromosome 1. The complexity of the transformed phenotype is indicated by the lack of cosegregation of any of the traits mentioned above (see Chapters 26 and 27 for a more complete description of this work).

B. Segregation-like Events in the Development of Retinoblastoma

The segregation-like events described above for Chinese hamster and other cells have been observed primarily through segregation analysis of hybrid cells or through mutation analysis in near-diploid cells. It is important to ask whether the observed processes occur in true diploid cells and whether they have any important medical consequences.

That these events do have important medical consequences is demonstrated dramatically by recent studies on retinoblastoma, a tumor of the retina occurring in early childhood. In the hereditary form of this disease it is thought that a germ-line mutation is present at the *Rb-1* locus on chromosome 13 in all cells, acting as a recessive allele (Sparkes et al., 1983). The initiation of the tumor requires a second event taking place in the somatic cells of the developing retina. In one study cells from retinoblastoma tumors were examined karyologically and one line was found to have lost a chromosome 13, presumably allowing a mutant allele at the *Rb-1* locus on the remaining homolog to be expressed (Benedict et al., 1983). In a separate study tumors were examined from patients heterozygous at the esterase D (*ESD*) locus, which has been found to be closely linked to *Rb-1* (Sparkes et al., 1983). Whereas both ESD variants were expressed in normal tissue, tumor cells from these patients often expressed a single ESD allele. This result was presented as evidence for an inactivation event affecting the *RB-1* locus and often encompassing the *ESD* gene as well (Godbout et al., 1983). The most definitive study utilized seven cloned DNA segments from chromosome 13, each of which revealed a restriction fragment length polymorphism (RFLP) in the human population (Cavenee et al., 1984). In several retinoblastoma patients heterozygous at RFLP markers widely separated on chromosome 13, all markers were rendered homozygous in tumor cells. Since it is unlikely that gene inactivation would extend over the whole chromosome, it was postulated that these tumors arose by chromosome loss and duplication. In other patients, markers heterozygous in normal tissue remained heterozygous in the tumor. These tumors may have arisen by gene inactivation, new mutation, or small deletion at the *Rb-1* locus. In one tumor three markers distal to *Rb-1* were rendered homozygous while one located between *Rb-1* and the centromere remained heterozygous. This suggested that a recombinational event between this marker and the retinoblastoma locus had given rise to the tumor (Cavenee et al., 1983, 1984).

Thus, the expression of the recessive *Rb-1* mutation requires a segregation event to unmask the recessive phenotype and, here again, the mechanisms seem to include chromosome loss, gene inactivation, loss and duplication of homologous chromosomes, and mitotic recombination.

VI. CONCLUDING REMARKS

Somatic cell geneticists are only now developing the tools and systems required to investigate mechanisms of genetic variation in detail. Segregation has been and will continue to be useful for gene mapping, in hamster as in other animals, and with the assignment of more genetic loci, especially selectable markers, the ability to manipulate animal cell systems will increase dramatically. The level of sophistication available for genetic manipulation should soon approach that available in lower organisms. New applications for segregation techniques are inevitable.

In this review it was necessary to adopt an operational definition of segregation as any event leading to the reexpression of a recessive phenotype in a cell line heterozygous at the locus responsible for the phenotype. This led us to consider a number of genetic and epigenetic mechanisms that are not usually thought of as "segregation" mechanisms. Thus, while deletion is usually regarded as a type of mutation, and gene inactivation as an epigenetic pheonmenon, both can and do act to uncover recessive phenotypes in somatic cells, both in hybrid cells and in diploid cells in culture or *in vivo*.

Perhaps the most common mechanism involved in expressing recessive alleles is loss of the chromosome carrying the dominant allele. Surprisingly, this is very often accompanied by a duplication of the homolog bearing the recessive allele. Although the mechanism of this process is poorly understood, it does appear to occur as a single event rather than as separate loss and duplication events. It occurs not only in near-tetraploid hybrid cells, but also in near-diploid cells and perhaps also in diploid cells *in vivo*.

Mitotic recombination has long been known to occur in lower eukaryotes including yeast and *Drosophila*. It has now been demonstrated in two mammalian cell types—on Chinese hamster chromosome 2 in somatic cell hybrids and on human chromosome 13 in retinoblastoma tumors. Presumably these represent homologous recombination due to pairing of whole chromosomes or chromosomal parts at some point during the mitotic cell cycle.

Gene inactivation also serves to uncover recessive phenotypes. Inactivation of a dominant allele can expose the activity of the homologous recessive allele, and in more than one system the inactive allele has been restored by treatment with 5azaCR, suggesting hypermethylation as being responsible for the primary inactivation event. This is the case for inactivation at the *emt* and *chr* locus on chromosome 2 and for the *tk* locus on chromosome 7 in Chinese hamster cells. It also appears to be responsible for the proline auxotrophy in CHO cells and is related to the maintenance of the inactive X chromosome in female somatic cells. Just as inactivation uncovers recessive alleles in hybrid cells, a similar event appears to act in near-diploid cells to allow the expression of new mutant phenotypes, and in diploid cells *in vivo* to reveal recessive cancer genes. Whether these *in vivo* inactivation events also involve a mechanism that can be reversed by 5azaCR remains to be determined. In some cases inactivation might involve random events, while in others it might be the result of coordinated gene inactivation as part of a normal differentiation pathway.

The study of gene segregation in somatic cells has come full circle. Initially, many genetic markers were mapped to specific chromosomes utilizing chromosomal segregation from interspecies, and to a lesser extent intraspecies, hybrids. Subsequently, as selectable markers were mapped on Chinese hamster chromosomes, these became invaluable for studying the segregational mechanism leading to reexpression of recessive markers in heterozygous cells. Now that we have several mapped markers and a general knowledge of the various segregational mechanisms, we can begin to ask more direct questions concerning the role of such mechanisms in animal cell biology and disease. We have seen that the segregational mechanisms occurring on chromosome 2 in Chinese hamster cell hybrids also occur in Chinese hamster and mouse cells with a near-diploid karyotype and play a role in the reexpression of selectable mutations in these cells. The same mechanisms also appear to operate *in vivo* to unmask recessive cancer genes leading to retinoblastoma. This latter observation raises a whole new challange for cancer biologists; to determine the extent to which segregational mechanisms are responsible for initiation of malignancy, and whether such mechanisms may operate to modify the growth or metastatic properties of a tumor once it is formed. It seems to us that the use of mapped selectable markers coupled with sets of RFLP markers along each chromosome will allow detailed study of segregational mechanisms in a variety of systems, leading to a better understanding of such "chromosome gymnastics" in natural biological processes.

ACKNOWLEDGMENTS

We are grateful to our colleagues Dr. Christine Campbell, Dr. L. Siminovitch, and Ms. Catherine Duff whose continued efforts have contributed to the experiments and ideas expressed in this review.

We are pleased to acknowledge financial support from the Medical Research Council of Canada.

REFERENCES

Adair, G. M., Stallings, R. L., Friend, K. K., and Siciliano, M. J. (1983a). *Somat. Cell Genet.* **9**, 477–487.

Adair, G. M., Stallings, R. L., Nairn, R. S., and Siciliano, M. J. (1983b) *Proc. Natl. Acad. Sci. USA* **80**, 5961–5964.

Adair, G. M., Stallings, R. L., and Siciliano, M. J. (1984). *Somat. Cell and Molec. Genet.* **10**, 283–295.

Bakay, B., Croce, C. M., Koprowski, H., and Nyhan, W. L. (1973). *Proc. Natl. Acad. Sci. USA* **70**, 1998–2002.

Bakay, B., Nyhan, W. L., Croce, C. M., and Koprowski, H. (1975). *J. Cell Sci.* **17**, 567–578.

Baker, W. K. (1968). *Adv. Genet.* **14**, 133–169.

Benedict, W. F., Murphree, A. L., Banerjee, A., Spina, C. A., Sparkes, M. C., and Sparkes, R. S. (1983). *Science* **219**, 973–975.

Bengtsson, B. O., Nabholz, M., Kennett, R., Bodmer, W. F., Povey, S., and Swallow, D. (1975). *Somat. Cell Genet.* **1**, 41–64.

Bradley, W. E. C. (1979). *J. Cell. Physiol.* **101**, 325–340.

Bradley, W. E. C. (1980). *Exp. Cell Res.* **129**, 251–263.

Bradley, W. E. C. (1983). *Mol. Cell. Biol.* **3**, 1172–1181.

Bradley, W. E. C., Dinelle, C., Charron, J., and Langlier, Y. (1982). *Somat. Cell Genet.* **8**, 207–222.

Bradley, W. E. C. and Letovanec, D. (1982). *Somat. Cell Genet.* **8**, 51–66.

Breslow, R. E. and Goldsby, R. A. (1969). *Exp. Cell Res.* **55**, 339–346.

Brockman, W. W. (1977). *Prog. Med. Virol.* **23**, 69–95.

Campbell, C. E. and Worton, R. G. (1979). *Somat. Cell Genet.* **5**, 51–65.

Campbell, C. E. and Worton, R. G. (1980). *Somat. Cell Genet.* **6**, 215–224.

Campbell, C. E. and Worton, R. G. (1981). *Mol. Cell. Biol.* **1**, 336–346.

Catcheside, D. G. (1977). *The Genetics of Recombination*, University Park Press, Baltimore.

Cavenee, W. K., Dryja, T. P., Phillips, R. A., Benedict, W. F., Godbout, R., Gallie, B. L., Murphree, A. L., Strong, L. C., and White, R. L. (1983). Nature **305**, 779–784.

Cavenee, W., Leach, R., Mohandas, R., Pearson, P., and White, R. (1984). *Am. J. Hum. Genet.* **36**, 10–24.

Chapman, V. M., Kratzer, P. G., Siracusa, L. D., Quarantillo, B. A., Evans, R., and Liskay, R. M. (1982). *Proc. Natl. Acad. Sci. USA* **79**, 5357–5361.

Chasin, L. A. (1972). *Nature New Biol.* **240**, 50–52.

Chasin, L. A. (1973). *J. Cell. Physiol.* **82**, 299–308.

Chasin, L. A. and Urlaub, G. (1975). *Science* **187**, 1091–1093.

Chasin, L. A. and Urlaub, G. (1976). *Somat. Cell Genet.* **2**, 453–467.

Cirullo, R. E., Arredondo-Vega, F. X., Smith, M., and Wasmuth, J. J. (1983). *Somat. Cell Genet.* **9**, 215–233.

Creusot, F., Acs, G., and Christman, J. K. (1982). *J. Biol. Chem* **257**, 2041–2048.

Croce, C. M., Bakay, B., Nyhan, W. L., and Koprowski, H. (1973). *Proc. Natl. Acad. Sci. USA* **70**, 2590–2594.

Dana, S. and Wasmuth, J. J. (1982a). *Somat. Cell Genet.* **8**, 245–264.

Dana, S. and Wasmuth, J. J. (1982b). *Mol. Cell. Biol.* **2**, 1220–1228.

Davidson, R. L., Ephrussi, B., and Yamamoto, K. (1966). *Proc. Natl. Acad. Sci. USA* **56**, 1437–1440.

Deaven, L. L. and Petersen, D. F. (1973). *Chromosoma* **41**, 129–144.

Ege, T. and Ringertz, N. R. (1974). *Exp. Cell Res.* **87**, 378–382.

Ephrussi, B., Scaletta, L. J., Stenchever, M. A., and Yoshida, M. (1964). *Symp. Intern. Soc. Cell Biol.* **3**, 13–25.

Eves, E. M. and Farber, R. A. (1981). *Proc. Natl. Acad. Sci. USA* **78**, 1768–1772.

Eves, E. M. and Farber, R. A. (1983). *Somat. Cell Genet.* **9**, 771–778.

Farber, R. A. (1973). *Genetics* **74**, 521–531.

Farrell, S. A. and Worton, R. G. (1977). *Somat. Cell Genet.* **3**, 539–551.

Flintoff, W. F., Davidson, S. V., and Siminovitch, L. (1976). *Somat. Cell Genet.* **2**, 245–261.

Flintoff, W. F., Livingston, E., Duff, C., and Worton, R. G. (1984). *Mol. Cell. Biol.* **4**, 69–76.

Fournier, R. E. K. (1981). *Proc. Natl. Acad. Sci. USA* **78**, 6349–6353.

Fournier, R. E. K., and Frelinger, J. A. (1982). *Mol. Cell. Biol.* **2**, 526–534.

Fournier, R. E. K., Juricek, D. K., and Ruddle, F. H. (1979). *Somat. Cell Genet.* **5**, 1061–1077.

Fournier, R. E. K. and Moran, R. G. (1983). *Somat. Cell Genet.* **9**, 69–84.

Fournier, R. E. K. and Ruddle, F. H. (1977a). *Proc. Natl. Acad. Sci. USA* **74**, 319–323.

Fournier, R. E. K. and Ruddle, F. H. (1977b). *Proc. Natl. Acad. Sci. USA* **74**, 3937–3941.

German, J. (1964). *Science* **144**, 298–301.

Giles, R. E., Shimizu, N., and Ruddle, F. H. (1980). *Somat. Cell Genet.* **6**, 667–687.

Godbout, R., Dryja, T. P., Squire, J., Gallie, B. L., and Phillips, R. A. (1983). *Nature* **304**, 451–453.

Graves, J. A. M. (1980). *Exp. Cell Res.* **125**, 483–486.

Graves, J. A. M. (1982). *Exp. Cell Res.* **141**, 99–105.

Groudine, M., Eisenman, R., and Weintraub, H. (1981). *Nature* **292**, 311–317.

Gupta, R. S. (1980). *Somat. Cell Genet.* **6**, 115–125.

Gupta, R. S., Chan, D. Y. H., and Siminovitch, L. (1978). *Cell* **14**, 1007–1013.

Gupta, R. S. and Siminovitch, L. (1977). *Cell* **10**, 61–66.

Gupta, R. S. and Siminovitch, L. (1978a). *Somat. Cell Genet.* **4**, 77–93.

Gupta, R. S. and Siminovitch, L. (1978b). *Somat. Cell Genet.* **4**, 715–735.

Hamerton, J. (1971). *Human Cytogenetics: General Cytogenetics 1*, Academic Press, New York.

Harris, J. F. and Whitmore, G. F. (1977). *Somat. Cell Genet.* **3**, 173–193.

Harris, M. (1971). *J. Cell. Physiol.* **78**, 177–184.

Harris, M. (1973). *Genetics Suppl.* **73**, 181–185.

Harris, M. (1974). *J. Natl. Cancer Inst.* **52**, 1811–1816.

Harris, M. (1975). *J. Cell. Physiol.* **86**, 413–430.

Harris, M. (1979). *Somat. Cell Genet.* **5**, 793–808.

Harris, M. (1982). *Cell* **29**, 483–492.

Harris, M. (1984). *Somat. Cell Mol. Genet.* (in press).

Harris, M. and Collier, K. (1980). *Proc. Natl. Acad. Sci USA* **77**, 4206–4210.

Hashmi, S. and Miller, O. J. (1976). *Cytogenet. Cell Genet.* **17**, 35–41.

Hellkuhl, B. and Grzeschik, K.-H. (1978). *Cytogenet. Cell Genet.* **22**, 527–530.

Hors-Cayla, M. C., Heuertz, S., and Frezal, J. (1983). *Somat. Cell Genet.* **9**, 645–657.

Human Gene Mapping (1974). Ruddle, F. H., Bootsma, D., McKusick, V. A., and Klinger, H. P. (eds.) *Birth Defects: Original Article Series*, Vol. X No. 3.

Human Gene Mapping 2 (1975). Bootsma, D., Ruddle, F. H., McKusick, V. A., and Klinger, H. P. (eds.) *Cytogenet. Cell Genet.* **14**, No. 3-6.

Human Gene Mapping 3 (1976). McKusick, V. A., Klinger, H. P., Bootsma, D., and Ruddle, F. H. (eds.) *Cytogenet. Cell Genet.* **16**, No. 1-5.

Human Gene Mapping 4 (1978). Hamerton, J. L., Klinger, H. P., McKusick, V. A., and Evans, H. J. (eds.) *Cytogenet. Cell Genet.* **22**, No. 1-6.

Human Gene Mapping 5 (1979). Evans, H. J., Hamerton, J. L., Klinger, H. P., and McKusick, V. A. (eds.) *Cytogenet. Cell Genet.* **25**, No. 1-4.

Human Gene Mapping 6 (1982). Berg, K., Evans, H. J., Hamerton, J. L., and Klinger, H. P. (eds.) *Cytogenet. Cell Genet.* **32**, No. 1-4.

Human Gene Mapping 7 (1984). Sparkes, R. S., Berg, K., Evans, H. J., and Klinger, H. P. (eds.) *Cytogenet. Cell Genet.* **37**, No. 1-4.

Huttner, K. M. and Ruddle, F. H. (1976). *Chromosoma* **56**, 1–13.

Ivarie, R. D. and Morris, J. A. (1982). *Proc. Natl. Acad. Sci. USA* **79**, 2967–2970.

Johnson, R. T., Mullinger, A. M., and Skaer, R. J. (1975). *Proc. R. Soc. Lond.* **189**, 591–602.

Jones, C. (1975). *Somat. Cell Genet.* **1**, 345–354.

Jones, C., Kao, F.-T., and Taylor, R. T. (1980). *Cytogenet. Cell Genet.* **28**, 181–194.

Jones, C., Patterson, D., and Kao, F.-T. (1981). *Somat. Cell Genet.* **7**, 399–409.

Jones, C., Wuthier, P., Kao, F.-T., and Puck, T. T. (1972). *J. Cell. Physiol.* **80**, 291–298.

Jones, G. E. and Sargent, P. A. (1974). *Cell* **2**, 43–54.

Kahan, B. and DeMars, R. (1975). *Proc. Natl. Acad. Sci. USA* **72**, 1510–1514.

Kahan, B. and DeMars, R. (1980). *Somat. Cell Genet.* **6**, 309–323.

Kafer, E. (1977). *Adv. Genet.* **19**, 33–131.

Kao, F.-T., Johnson, R. T., and Puck, T. T. (1969). *Science* **164**, 312–314.

Kao, F.-T. and Puck, T. T. (1967). *Genetics* **55**, 513–524.

Kao, F.-T. and Puck, T. T. (1970). *Nature* **228**, 329–332.

Kozak, C. A., Fournier, R. E. K., Leinwand, L. A., and Ruddle, F. H. (1979). *Biochem. Cenet.* **17**, 23–34.

Kucherlapati, R. S., Eves, E. M., Song, K.-Y., Morse, B. S., and Smithies, O. (1984). *Proc. Natl. Acad. Sci. USA* (in press).

Lasserre, C., Jami., and Wiles, D. (1980). *J. Cell. Physiol.* **104**, 403–413.

Lester, S. C., Korn, N. J., and DeMars, R. (1982). *Somat. Cell Genet.* **8**, 265–284.

Liskay, R. M., and Evans, R. J. (1980). *Proc. Natl. Acad. Sci. USA* **77**, 4895–4898.

Liskay, R. M., and Stachelek, J. L. (1983). *Cell* **35**, 157–165.

Littlefield, J. W. (1964). *Science* **145**, 709–710.

Lyon, M. F. (1961). *Nature* **190**, 372–373

Marshall, C. J., Kitchin, R. M., and Sager, R. (1982). *Somat. Cell Genet.* **8**, 709–721.

Marshall, C. J. and Sager, R. (1981). *Somat Cell Genet.* **7**, 713–723.

Martin, G. M. and Sprague, C. A. (1969), *Science* **166**, 761–763.

McBurney, M. W. and Whitmore, G. F. (1974). *J. Cell. Physiol.* **83**, 69–74.

McNeill, C. A. and Brown, R. L. (1980). *Proc. Natl. Acad. Sci. USA* **77**, 5394–5398.

McKusick, V. A. and Ruddle, F. H. (1977). *Science* **196**, 390–405.

Merriam, J. R. and Garcia-Bellido, A. (1972). *Mol. Gen. Genet.* **115**, 302–313.

Mezger-Freed, L. (1971). *J. Cell Biol.* **51**, 742–751.

Mezger-Freed, L. (1972). *Nature New Biol.* **235**, 245–246.

Mohandas, T., Sparkes, R. S., and Shapiro, L. J. (1981). *Science* **211**, 393–396.

Mohandas, T., Sparkes, R. S., and Shapiro, L. J. (1982). *Am. J. Hum. Genet.* **34**, 811–817.

Moore, P. D., and Holliday, R. (1976). *Cell* **8**, 573–579.

Nakamura, N. and Okada, S. (1983). *Mutat. Res.* **111**, 353–364.

Nanney, D. L. (1968). *Ann. Rev. Genet.* **2**, 121–140.

Niwa, O. and Sugahara, T. (1981). *Proc. Natl. Acad. Sci. USA* **78**, 6290–6294.

Nordheim, A., Pardue, M. L., Lafer, E. M., Moller, A., Stollar, B. D., and Rich, A. (1981). *Nature* **294**, 417–422.

Perticone, P., Tanzarella, C., Palitti, F., Ricordy, R., DiChiara, P., DiPietro, G., Spirito, F., Diana, G., De Salivia, R., and Rizzoni, M. (1976). *Chromosoma* **56**, 243–248.

Pollard, J. W. and Stanners, C. P. (1979). *J. Cell. Physiol.* **98**, 571–585.

Pontecorvo, G. (1971). *Nature* **230**, 367–369.

Pontecorvo, G. and Kafer, E. (1958). *Adv. Genet.* **9**, 71–104.

Pravtcheva, D. D., and Ruddle, F. H. (1983). *Exp. Cell Res.* **146**, 401–416.

Prickett, M. S., Coultrip, L., Patterson, M. K., and Morrow, J. (1974). *J. Cell. Physiol.* **85**, 621–626.

Rabin, M. S. and Gottesman, M. M. (1979). *Somat. Cell Genet.* **5**, 571–583.

Raskind, W. H. and Gartler, S. M. (1978). *Somat. Cell Genet.* **4**, 491–506.

Renwick, J. H. (1969). *Br. Med. Bull.* **25**, 65–73.

Rizzoni, M., Palitti, F., and Perticone, P. (1974). *Chromosoma* **45**, 151–162.

Roberts, M. and Ruddle, F. H. (1980). *Exp. Cell Res.* **127**, 47–54.

Roberts, M., Scangos, G. A., Hart, J. T., and Ruddle, F. H. (1983). *Somat. Cell Genet.* **9**, 235–248.

Rommelaere, J. and Miller-Faures, A. (1975). *J. Mol. Biol.* **98**, 195–218.

Rommelaers, J., Susskind, M., and Errera, M. (1973). *Chromosoma* **41**, 243–257.

Ronen, A. (1964). *Genetics* **50**, 649–658.

Rosenstraus, M. and Chasin, L. A. (1975). *Proc. Natl. Acad. Sci. USA* **72**, 493–497.

Rosenstraus, M. J. and Chasin, L. A. (1978). *Genetics* **90**, 735–760.

Ruddle, F. H., and Creagan, R. P. (1978). *Ann. Rev. Genet.* **9**, 407–486.

Sager, R. and Kovac, P. E. (1978). *Somat. Cell Genet.* **4**, 375–392.

Sager, R. and Kovac, P. E. (1979). *Somat. Cell. Genet.* **5**, 491–502.

Scaletta, L. J., Rushforth, N. B., and Ephrussi, B. (1967). *Genetics* **37**., 107–124.

Schor, S. L., Johnson, R. T., and Mullinger, A. M. (1975). *J. Cell Sci.* **19**, 281–303.

Shin, S., Caneva, R., Schildkraut, C. L., Klinger, H. P., and Siniscalco, M. (1973). *Nature New Biol.* **241**, 194–196.

Siciliano, M. J., Siciliano, J., and Humphrey, R. M. (1978). *Proc. Natl. Acad. Sci. USA* **75**, 1919–1923.

Siciliano, M. J., Stallings, R. L., Adair, G. M., Humphrey, R. M., and Siciliano, J. (1983). *Cytogenet. Cell Genet.* **35**, 15–20.

Siminovitch, L. (1976). *Cell* **7**, 1–11.

Siminovitch, L. (1979). In *Eucaryotic Gene Regulation* ICN-UCLA Symposia on Molecular and Cellular Biology, (R. Axell, T. Maniatis, and C. F. Fox, eds.), Academic Press, New York, pp. 433–443.

Siminovitch, L. (1981). In *Genes, Chromosomes and Neoplasia* (F. E. Arrighi, P. N. Rao, and E. Stubblefield, eds.), Raven Press, New York, pp. 157–174.

Simon, A. E. and Taylor, M. W. (1983). *Proc. Natl. Acad. Sci. USA* **80**, 810–814.

Simon, A. E., Taylor, M. W., and Bradley, W. E. C. (1983). *Mol. Cell. Biol.* **3**, 1703–1710.

Simon, A. E., Taylor, M. W., Bradley, W. E. C., and Thompson, L. H. (1982). *Mol. Cell. Biol.* **2**, 1126–1133.

Sparkes, R. S., Murphree, A. L., Lingua, R. W., Sparkes, M. C., Field, L. L., Funderburk, S. J., and Benedict, W. F. (1983). *Science* **219**, 971–973.

Stallings, R. L., Adair, G. M., and Siciliano, M. J. (1984). *Somat. Cell Mol. Genet.* **10**, 109–110.

Stallings, R. L. and Siciliano, M. J. (1982). *H. Hered* **73**, 399–404.

Stallings, R. L., Siciliano, M. J., Adair, G. M., and Humphrey, R. M. (1982). *Somat. Cell Genet.* **8**, 413–422.

Stern, C. (1936). *Genetics* **21**, 625–730.

Sugiyama, R. H., Arfin, S. M., and Harris, M. (1983). *Mol. Cell. Biol.* **3**, 1937–1942.

Tarrant, G. M., and Holliday, R. (1977). *Mol. Gen. Genet.* **156**, 273–279.

Taylor, M. W., Pipkorn, J. H., Tokito, M. K., and Pozzatti, R. O. (1977). *Somat. Cell Genet.* **3**, 195–206.

Taylor, S. M. and Jones, P. A. (1979). *Cell* **17**, 771–779.

Therman, E. and Kuhn, E. M. (1976). *Cytogenet. Cell Genet.* **17**, 254–267.

Thompson, L. H. and Baker, R. M. (1973). *Methods Cell Biol.* **6**, 209–281.

Tischfield, J. A., Trill, J. J., Lee, Y. I., Coy, K., and Taylor, M. W. (1982). *Mol. Cell. Biol.* **2**, 250–257.

Tourian, A., Johnson, R. T., Burg, K., Nicolson, S. W., and Sperling, K. (1978). *J. Cell Sci.* **30**, 193–209.

Turker, M. S., Smith, A. C., and Martin, G. M. (1984). *Somat. Cell Mol. Genet.* **10**, 55–69.

Van Zeeland, A. A. and Simons, J. W. I. M. (1975). *Mutat. Res.* **28**, 239–250.

Venolia, L. and Gartler, S. M. (1983). *Nature* **302**, 82–83.

Waalwijk, C. and Flavell, R. A. (1978). *Nucleic Acids Res.* **5**, 4631–4641.

Wasmuth, J. J. and Chu, L.-Y. (1980). *J. Cell Biol.* **87**, 697–702.

Wasmuth, J. J. and Hall, L. V. (1984). *Cell* **36**, 697–707.

Watson, B., Gormley, I. P., Gardiner, S. E., Evans, H. J., and Harris, H. (1972). *Exp. Cell Res.* **75**, 401–409.

Weiss, M. C. and Green, H. (1967). *Proc. Natl. Acad. Sci. USA* **58**, 1104–1111.

Westerveld, A., Visser, R. P. L. S., and Freeke, M. A. (1971). *Biochem. Genet.* **5**, 591–599.

Westerveld, A., Visser, R. P. L. S., Freeke, M. A., and Bootsma, D. (1972). *Biochem. Genet.* **7**, 33–40.

Wigler, M. H. and Weinstein, I. B. (1975). *Biochem. Biophys. Res. Commun.* **63**, 669–674.

Worton, R., Duff, C., and Flintoff, W. (1981). *Mol. Cell. Biol.* **1**, 330–335.

Worton, R. G. and Duff, C. (1981). *Cytogenet. Cell Genet.* **29**, 184–188.

Worton, R. G., Duff, C., and Campbell, C. E. (1980). *Somat. Cell Genet.* **6**, 199–213.

Worton, R. G., Grant, S. G., and Duff, C. (1984). In *Gene Transfer and Cancer* (N. L. Sternberg and M. L. Pearson, ed.), Raven Press, New York.

Worton, R. G., Ho, C. C., and Duff, C. (1977). *Somat. Cell Genet.* **3**, 27–45.

Yerganian, G. and Nell, M. B. (1966). *Proc. Natl. Acad. Sci. USA* **55**, 1066–1073.

CHAPTER 29

MECHANISMS OF GENETIC VARIATION IN CHINESE HAMSTER OVARY CELLS

Louis Siminovitch
Department of Genetics
and Research Institute
Hospital for Sick Children
and Department of Medical Genetics
University of Toronto
Toronto, Ontario, Canada

INTRODUCTION

Rapid advances have been made over the last two decades in comprehension of the mechanisms whereby mammalian cells can alter their genotype. This information has been generated by research using a variety of cell lines, but as the many papers in this volume indicate, the primary and most useful experimental vehicle has been the Chinese hamster ovary (CHO) line, first developed by Ted Puck (Puck and Kao, 1967). The reasons for this are both historical and practical. The early successes of Puck and his collaborators with CHO cells showing the feasibility of conducting quantitative somatic cell genetics in this system encouraged other investigators to adopt this line for such studies. The practical advantages offered by CHO cells are well known, but include ease of manipulation in both suspension and solid-phase culture, ease of quantitation, and a relatively easily analyzed karyotype.

It was natural, therefore, that in an earlier review published in 1976 (Siminovitch, 1976), where I set out some thoughts on genetic variation in mammalian cells, I relied heavily on data generated with CHO cells. In this chapter, I plan to review the current state of knowledge on this topic. Although I shall again profit extensively from the large body of knowledge obtained with the CHO system, in the interests of comprehensiveness, any discussion of genetic variation in mammalian cells must include insights obtained with other systems.

II. HISTORY

In the early years of the field of somatic cell genetics, it was natural to believe that recessive variants isolated in diploid or quasidiploid cells could not be due to classical mechanisms of mutation. Most publications appearing at that time supported this contention, and it was suggested that variants arose by some type of epigenetic mechanism. However, as further information appeared, it became clear that an increasing number of variants were indeed due to classical structural genetic mechanisms, as indicated first by inferential criteria, but later by molecular analysis. Although this was already evident in 1976 when I wrote my earlier review (Siminovitch, 1976), the emphasis in most somatic cell genetic work at that time was to validate the "structural gene" thesis, and there was a tendency to place less emphasis on aspects of variation which did not fit this model. In fact, until very recently, most of the information generated since 1976 has solidified the conclusion that much of the genetic variation seen in somatic cells involves a change in DNA sequence at the nucleotide level. It is now essentially not a subject of serious debate and many of the reviews in other chapters of this volume provide striking examples of structural genetic variation. However, as I'll indicate later, the subject has come full circle. Whereas in 1976 there was actually no solid evidence for epigenetic variation, and it was not even clear what this term meant, recent studies have shown that such variation can indeed lead to relatively stable new phenotypes. In addition, the nature of the structural changes that may give rise to different genotypes in somatic cells has become much clearer.

My purpose in this chapter is to simply provide an outline of the many different mechanisms that have been well documented to give rise to genetic variation in somatic cells. Several of the chapters in this volume provide detailed examples of these systems.

III. MISSENSE, NONSENSE, AND DELETION MUTATIONS

There are now several examples where the evidence for missense, nonsense, or deletion mutations is overwhelming. They include the *ts* systems, where a temperature-sensitive protein has been identified, the auxotrophs, particularly those in which reversion gives rise to a temperature-sensitive protein (McBurney and Whitmore, 1974, Chasin et al., 1974, Waye and Stanners, 1979), the *aprt* (Chapter 12), and *hprt* (Chapter 13) systems which have been analyzed at the molecular level, and several of the drug-resistant systems such as those involving RNA polymerase II (Chan et al., 1972, Ingles et al., 1976, Ingles, 1978; Chapter 15), dihydrofolate reductase (Flintoff et al., 1976, Gupta et al., 1977, Urlaub and Chasin, 1980; Chapter 16), asparagine synthetase (Andrulis and Siminovitch, 1982, Andrulis et al., 1983; Chapter 17), Na^{+}/K^{+} ATPase (Baker et al., 1974) and so on. However, it has to be emphasized that, as of now, most of the relevant genes have not been cloned, and there is consequently little detailed analysis at the nucleotide or amino acid level in most systems. Thus the evidence is often inferential. For example, it was assumed that the pro^{-} auxotroph originally described by Puck's group was due to a structural gene mutation (Puck and Kao, 1967). However, it has now been demonstrated that this phenotype can be reversed at high efficiency by 5-azacytidine treatment, presumably due to gene inactivation by hypermethylation (Harris, personal communication). As another example of uncertainty, several reports have appeared in which the mutation frequencies are much higher (10^{-3}–10^{-2}) than would be expected for well-characterized structural mutations (10^{-7}–10^{-5}). These include, for example, the mutations in CHO cells which give rise to isolates lacking adenosine kinase activity (Gupta and Siminovitch, 1978; Rabin and Gottesman, 1979), or lacking UDP-*N*-acetylglucosamine-glycoprotein *N*-acetyl-glucosaminyltransferase (Stanley et al., 1975). Little or no residual activity is observed in these mutants, and no reversions have been observed. It seems quite possible, therefore, that the mutations are not due to a structural gene change, but to some other mechanisms, such as segregation or gene inactivation (see Chapter 28 and below). As with all somatic cell genetic systems, the definitive resolution of the mechanism of variation in these systems will require the cloning of the relevant genes.

IV. AMPLIFICATION MUTANTS

Perhaps the most dramatic advances in the somatic cell genetics field in recent years have involved the isolation and characterization of amplification mutants. It is now clear that somatic cells can often acquire resistance to a drug by amplification of the relevant gene. In many such systems, the assumption that amplification has occurred is based on the presence of increased levels of the target protein in the mutant cells. However, in some systems, the fact that amplification of the gene has occurred has been solidly documented by molecular analysis. These include the genes for dihydrofolate reductase (Alt et al., 1978; Kaufman et al., 1979), asparagine synthetase (Andrulis et al., 1983; Ray et al., 1984), the trienzyme complex, carbamyl phosphate synthetase, aspartate transcarbamylase, dihydroorotase (Kempe et al., 1976; Wahl et al., 1979), metallothionein I (Beach and Palmiter 1981; Gick and McCarty, 1982), and several others (cf. Schimke, 1982; Stark and Wahl, 1984).

Several aspects of gene amplification are of particular interest. First, in all of the above systems, the amplification is associated with either the presence of a chromosomal homogeneously staining region (HSR), or small so-called double minute (DM) chromosomes (Biedler and Spengler, 1976; Nunberg et al., 1978; Kaufman et al., 1979). The HSRs obviously span a region far greater than the simple amplification of the gene, indicating that a much larger amount of genetic material is amplified at the same time (cf. Brison et al., 1982; Caizzi and Bostock, 1982; Cowell, 1982; Crouse et al., 1982; Padgett et al., 1982). This has been confirmed molecularly.

Second, the very existence of the amplification, and the often associated increased amounts of specific mRNA, have facilitated cloning of the genes (Nunberg et al., 1980; Padgett et al., 1982; Schilling et al., 1982; Ray et al., 1984). It is obvious that amplification of a gene offers immediate advantages in respect to gene cloning, and this in turn has allowed for detailed molecular analyses of the structure of the amplified regions. Usually, amplification systems have been developed by continuous selection in increasing concentrations of a drug and nearly always have involved a dominant marker. But only a limited number of genes can be amplified in this way. Recently, an alternative selection strategy has been described in which amplification is observed after reversion (1) of *hprt* mutants with *ts* (Fuscoe et al., 1983) or kinetically modified enzymes (Brennand et al., 1982), or (2) of a mutant with altered *dhfr* expressed at low levels in murine cells (Haber et al., 1981), or (3) of integrated SV40 genomes containing a temperature-sensitive T antigen (Hiscott et al., 1980). Such procedures may enhance considerably the spectrum of genes which can be amplified, and which can then be cloned.

Third, amplification has taken on added potential significance because of recent findings which indicate that many tumors contain amplified oncogene sequences (George and Powers, 1983; Alitalo et al., 1983; Schwab et al., 1983). It is not clear whether these amplifications represent early or primary events in oncogenesis, whether they are consequent to chromosomal rearrangements which are observed in the tumors, or whether the mechanisms of amplification are the same as in those in which selection procedures are used. It is of course possible that the amplified oncogene sequences facilitate escape from growth control, and this also involves a selection of sorts. In any event HSRs and DMs are observed in the tumors.

Fourth, amplification is often observed after gene transfer, and a great deal of study has been made of the processes involved in such amplification. So far such studies have usually involved previously cloned genes, and, consequently, have not been exploited for the cloning of "new" genes (cf. Schimke 1982; Stark and Wahl, 1984).

A great deal of work has been done on the molecular mechanisms of gene amplification but such a discussion is beyond the scope of this chapter. It is covered in other articles in this volume as well as in a recent review by Stark and Wahl (1984).

V. SEGREGATION AND GENE INACTIVATION

In the last few years it has been clear that some of the variation seen in somatic cells results from segregation and gene inactivation-like events. This topic is covered in greater detail by Worton in Chapter 28. Here I propose to discuss only that aspect that relates to the role these events play in the origin of mutants in quasi-diploid cells.

In 1976, I proposed that one could explain the high frequency with which recessive mutations were observed in CHO cells by assuming that such cells were functionally hemizygous for the loci involved (Siminovitch, 1976). A necessary correlate to this hypothesis was that the other allele had been "inactivated" during the evolution of the CHO cell due to processes one could not define.

Two types of studies have provided information on the possible mechanisms that lead to functional loss of one allele and consequent hemizygosity. In the case of emetine and chromate resistance, and the *ts* leucyl-tRNA synthetase mutation, all of which are recessive markers, one of the chromosome 2 homologs in CHO cells (called Z2), was shown to contain a deletion in the region of the three genes (Deaven and Peterson, 1973, Worton et al., 1977). Deletions of this kind, even if not discernible by chromosome banding, could of course account for functional hemizygosity for other loci.

Several studies have provided evidence for another mechanism that may lead to functional hemizygosity. The frequency of Emt^r in V79 cells is about 1/100th of that found in CHO cells. Analysis of hundreds of unselected subclones of V79 cells in our laboratory showed that in about 1 in a 100 clones the frequency of EMt^r mutants was of the same order as that found in CHO cells. Thus, in V79 cells, there must be two steps which result in emetine resistance, one occurring at a frequency of 10^{-2} leading to functional hemizygosity, and the other at 10^{-5} (after mutagenesis) leading to the eventual Emt^r phenotype (Siminovitch, 1979, 1981). Thompson and his collaborators reached the same type of conclusion for two *ts* mutations, using a different type of analysis (Adair et al., 1979). The nature of the high-frequency event is not understood, but processes such as segregation and gene inactivation come immediately to mind.

Evidence exists for both of these latter events. In terms of segregation, Eves and Farber showed that recessive adenosine-kinase-deficient mutants in near-diploid cells had lost one of their chromosome 14's, the chromosome on which this gene is located (Eves and Farber, 1981). These workers have therefore clearly documented the loss of a chromosome, and a mutation at the other allele, in a two event process.

There is also good evidence for gene inactivation. Using both immunological and electrophoretic methods, Simon et al. (1982) demonstrated that completely deficient $aprt^-$ recessive mutants also arose in two steps, involving a high-frequency event leading to allelic inactivation and a low-frequency event involving a structural gene change. The two events could occur in either order. Bradley, Taylor, and others have also documented gene inactivation at the *aprt* locus (see Chapter 12).

It is important to recognize that in two of the above systems the mutants were generated in single-step selections. Thus these latter procedures do not preclude multiple events during the selection process, and it is unclear at the moment how many of the existing recessive mutants arose either from one or from two steps during selection.

Although the relatively high success in isolation of recessive mutants in CHO cells suggested extensive functional hemizygosity in CHO cells, it is probably not as common as one would have predicted. In fact, Siciliano et al. (1978) have demonstrated that many loci in CHO cells are present as two functional alleles.

The demonstration of a two-step process in the evolution of mutant phenotypes has some rather interesting implications. Although the process has been demonstrated in two cell types, it is clear that the frequency of isolation of a particular recessive mutant in a cell containing two functional alleles will depend on the frequency of the segregation or inactivation-like event for that marker or that cell, rather than just mutation rate *per se*. At the moment we know little about the determining factors for the former processes. On the one hand, the relative success with CHO or V79 cells may be attributable to their propensity for such events, and, on the other hand, some of the mutant isolations that have failed may be due to low frequencies of segregation or inactivation for that gene.

It is of interest as well that in spite of the high frequency at which segregation or inactivation-like processes take place, the occurrence of two such events giving rise to a mutant phenotype seems to be rare. This is understandable in terms of chromosome segregation, since the loss of both homologs will obviously be lethal, but the same reasoning may apply to gene inactivation, since such inactivation may be regional rather than local, and the loss of both alleles for other genes in that region would also likely be lethal.

The adenosine kinase locus is of interest in this regard because the frequency of mutation in this gene is about 10^{-3} in CHO cells but can range down to a 10^{-6} in a variety of cell lines (Gupta and Siminovitch, 1978). This system may therefore involve all the possible parameters, including one or two steps, and different frequencies of gene inactivation or segregation in various cell lines.

The molecular nature of the gene inactivation events described above is not as yet understood. However, information is available in some specific cases. As indicated earlier, Harris has shown that the recessive pro^- phenotype in CHO cells can be reversed at high efficiency by treatment with 5-azacytidine, likely by hypomethylation of a gene that had previously undergone gene inactivation. Presumably in this case, the second allele had also been inactivated by mutation or otherwise at some stage. The interpretation of this result is complicated because the pro^- phenotype was never selected as such but was a property of the CHO line when it was first established (Puck and Kao, 1967). It may not be possible to generalize the observation in terms of auxotrophic mutants, since Grant and Worton have been unsuccessful in reversing a number of other auxotrophic isolates (Worton, Chapter 28). Nevertheless, Harris (1982) was also able to revert Tk^- cells, isolated by selection, to a Tk^+ phenotype by 5-azacytidine treatment. Since the revertants showed about 50% levels of thymidine kinase, he concluded that the conversion of Tk^+/Tk^+ cells to Tk^-/Tk^- occurred in two steps, one of which involved gene inactivation, and the other mutation, reminiscent of the two-step systems described earlier. Evidence for gene inactivation has also been obtained for the Emt^r and Chr^r loci in specific CHO cell hybrids by Grant and Worton (Worton, this volume). All of the above experiments on gene inactivation have involved "reactivation" of the loci with 5-azacytidine, and rest on the hypothesis that hypomethylation has occurred at the DNA level. Validation of this hypothesis will of course involve analysis with the cloned genes.

In summary, the frequencies of mutation for recessive and dominant mutations for many structural genes are often of the same order of magnitude. It seems likely, therefore, that the recessive alteration in these cases is due to a single step, and that the parental cell has become hemizygous for that locus due to gene inactivation, segregation, or deletion of that locus, during its culture history. However, for other loci, both alleles seem to be functional in the parental cell, the changes in genotype here occur as a two-step process, usually involving a mutation of the structural gene, and again a segregation or inactivation event. In at least one example (Simon et al., 1982), it has been shown that the order of the latter steps is reversible. Although the mutationlike events are manipulable by mutagens, etc., the conditions that control or affect the segregation or inactivation steps are not understood, either in terms of the specific genes or of the cell types. Such information is of course of importance in terms of the potential for selection of recessive mutants in quasidiploid cells carrying two active alleles.

VI. CONCLUSION

As I have indicated before, and as this volume attests, there has been considerable progress in the field of somatic cell genetics in the last decade or so. A plethora of genetic systems involving a large number of important enzymes or organelles have been identified and characterized at the biochemical level. We are also much more knowledgeable about the mechanisms of variation in somatic cells, at least at the decriptive level, as I have tried to point out in this chapter.

However, it has been fairly obvious over the last few years that the future evolution of this field will essentially depend on the intensive application of molecular genetic technology. There has been some progress in this direction, particularly with genes which have been amplified. In these latter systems, the amplification has resulted in highly increased mRNA levels for the relevant locus, and has thus greatly facilitated the cloning of the gene. However, progress for unamplified genes has been rather limited and success has only been achieved in such cases after considerable effort (Lowy et al., 1980; Perucho et al., 1980; Goldfarb et al., 1982; Shih and Weinberg, 1982; Diamond et al., 1983; Goddard et al., 1983; Lewis et al., 1983). It is clear, therefore, that if one is to progress in terms of application of molecular technology and the detailed analysis of "housekeeping" genes, it will be important either to develop methods of amplifying the relevant gene (see amplification section earlier), or to develop more efficient ways of cloning single copy genes. Although there is still much to be learned using the standard genetic methods, the most efficient and direct approach to examining the problems in somatic cell genetics to which I have alluded in this review will depend on cloning of the genetic material. This is clearly the challenge in this field in the next few years.

ACKNOWLEDGMENTS

I wish to thank Dr. P. Ray, M. Breitman, and I. Andrulis for help in preparation of this manuscript. My research has been funded by grants from the Medical Research Council and the National Cancer Institute of Canada.

REFERENCES

Adair, G. M., Thompson, L. H., and Font, S. (1979). *Somat. Cell Genet.* **5**, 3291

Alitalo, K., Schwab, M., Lin, C. C. Varmus, H. E., and Bishop, J. M. (1983). *Proc. Natl. Acad. Sci. USA* **80**, 1707.

Alt, F., Kellems, R., Bertino, J. R., and Schimke; R. T. (1978). *J. Biol. Chem.* **253**, 1357.

Andrulis, I. L. and Siminovitch, L. (1982). *Somat. Cell Genet.* **8**, 533.

Andrulis, I. L., Duff, C., Evans-Blackler, S., Worton, R., and Siminovitch, L. (1983). *Mol. Cell. Biol.* **3**, 391.

Baker, R. M., Brunette, D. M., Mankovitz, R., Thompson, L. H., Whitmore, G. F., Till, J. E., and Siminovitch, L. (1974). *Cell* **1**, 9.

Beach, L. R. and Palmiter, R. D. (1981). Proc. Natl. Acad. Sci. USA **78**, 211.

Biedler, J. L. and Spengler, B. A. (1976). *Science* **191**, 185.

Brennard, J., Chinault, A. C., Konechi, D. S., Melton, D. W., and Caskey, C. T. (1982). *Proc. Natl. Acad. Sci. USA* **79**, 1950.

Brison, O., Andeshin, F., and Stark, G. R. (1982). *Mol. Cell. Biol.* **2**, 578.

Chan, V., Whitmore, F. G., and Siminovitch, L. (1972). *Proc. Natl. Acad. Sci. USA* **69**, 119.

Chasin, L. A., Feldman, A., Konstam, M. and Urlaub, G. (1974). *Proc. Natl. Sci. USA* **71**, 718.

Caizzi, R. and Bostock, C. J. (1982). *Nucl. Acid Res.* **10**, 6597.

Cowell, J. K. (1982). *Ann. Rev. Genet.* **16**, 12.

Crouse, G. F., Simonsen, C. C., McEwan, R. N., and Schimke, R. T. (1982). *J. Biol. Chem.* **257**, 7887.

Deaven, L. L. and Peterson, D. F. (1973). *Chromosoma* **41**, 129.

Diamond, A., Cooper, G. M., Ritz, J., and Lane, M. A. (1983). *Nature* **305**, 112.

Eves, E. M. and Farber, R. A. (1981). *Proc. Natl. Acad. Sci. USA* **78**, 1768.

Flintoff, W. F., Davidson, S. V., and Siminovitch, L. (1976). *Somat. Cell Genet.* **2**, 245.

Fuscoe, J. C., Fenwick, Jr., R. G., Ledbetter, D. H., and Caskey, C. T. (1983). *Mol. Cell. Biol.* **3**, 1086.

George, D. L. and Powers, V. E. (1983). *Cell* **24**, 117.

Gick, G. G. and McCarty, K. S. (1982). *J. Biol. Chem.* **257**, 9049.

Goddard, J. M., Caput, D., Williams, S. R., and Martin, D. W. (1983). *Proc. Natl. Acad. Sci. USA* **80**, 4281.

Goldfarb, M., Shimizu, K., Perucho, M., and Wigler, M. (1982). *Nature* **296**, 404.

Gupta, R. S., Flintoff, W. F., and Siminovitch, L. (1977). *Canad. J. Biochem.* **55**, 445.

Gupta, R. S. and Siminovitch, L. (1978). *Somat. Cell Genet.* **4**, 715.

Haber, D. A., Beverly, S. M., Kelley, M. L., and Schimke, R. T. (1981). *J. Biol. Chem.* **256**, 9501.

Harris, M. (1982). *Cell* **29**, 483.

Hiscott, J., Murphy, D., and Defendi, V. (1980). *Cell* **22**, 535.

Ingles, C. J. (1978). *Proc. Natl. Acad. Sci. USA* **75**, 405.

Ingles, C. J., Pearson, M. L., Buchwald, M., Beatty, B. C., Crerar, M. M., Guialis, A., Lobban, P. E., Siminovitch, L., and Somers, D. C. (1976). In *RNA Polymerase* (M. Chamberlin and R. Hosick, eds.) Cold Spring Harbor Laboratory, New York, p. 835.

Kaufman, R. J., Brown, P. C., and Schimke, R. T. (1979). *Proc. Natl. Acad. Sci. USA* **76**, 5669.

Kempe, T. D., Swyryd, E. A., Bruist, M., and Stark, G. R. (1976). *Cell* **9**, 541.

Lewis, J. A., Shimizu, K., and Zipser, D. (1983). *Mol. Cell. Biol.* **3**, 1815.

Lowy, I., Pellicer, A., Jackson, J. F. Sim, G-K., Silverstein, S., and Axel, R. (1980). *Cell* **22**, 817.

McBurney, M. and Whitmore, G. F. (1974). *Cell* **2**, 183.

Nunberg, J. H., Kaufman, R. J., Schimke, R. T., Urlaub, G., and Chasin, L. A. (1978). *Proc. Natl. Acad. Sci. USA* **75**, 5553.

Nunberg, J. H., Kaufman, R. J., Chong, A. C. Y., Cohen, S. N., and Schimke, R. T. (1980). *Cell* **19**, 355.

Padgett, R. A., Wahl, G. M., and Stark, G. R. (1982). *Mol. Cell. Biol.* **2**, 293.

Perucho, M. Hanahan, D., Lipsich, L., and Wigler, M. (1980). *Nature* **385**, 207.

Puck, T. T. and Kao, F. T. (1967). *Proc. Natl. Acad. Sci. USA* **58**, 1227.

Rabin, M. S. and Gottesman, M. M. (1979). *Somat. Cell Genet.* **5**, 571.

Ray, P., Siminovitch, L., and Andrulis, I. L. (1984). *Gene* **30**, 1–9.

Schilling, J., Beverley, S., Simonsen, C., Crouse, G., Setzer, D., Fegin, J., McGrogan, M., Kohlmiller, N., and Schimke, R. T. (1982). In *Gene Amplification* (R. T. Schimke, ed.), Cold Spring Harbor Laboratory, New York, p. 149.

Schimke, R. T. (1982). *Gene Amplification*, Cold Spring Harbor Laboratory, New York, p. 317.

Schwab, M., Alitalo, K., Varmus, H. E., Bishop, J. M., and George, D. (1983). *Nature* **303**, 497.

Shih, C. and Weinberg, R. (1982). *Cell* **29**, 161.

Siciliano, M. J., Siciliano, J., and Humphrey, R. M. (1978). *Proc. Natl. Acad. Sci. USA* **4**, 1919.

Siminovitch, L. (1976). *Cell* **7**, 1.

Siminovitch, L. (1979). In *Eucaryotic Gene Regulation*, ICN-UCLA Symposia on Molecular and Cellular Biology, p. 433.

Siminovitch, L. (1981). In *Chromosomes and Neoplasia* (F. Arrighi, P. Rao, and E. Stubblefield, eds.) Raven Press, New York, p.157.

Simon, A. E., Taylor, M. W., Bradley, W. E. C., and Thompson, L. H. (1982). *Mol. Cell Biol.* **2**, 1126.

Stanley, P., Narasimhan, S., Siminovitch, L., and Schachter, H. (1975). *Proc. Natl. Acad. Sci. USA* **72**, 3323.

Stark, G. R. and Wahl, G. M. (1984). *Ann. Rev. Biochem.* (in press).

Urlaub, G. and Chasin, L. A. (1980). *Proc. Natl. Acad. Sci. USA* **77**, 4216.

Wahl, G. M., Padgett, R. A., and Stark, G. R. (1979). *J. Biol. Chem.* **254**, 8679.

Waye, M. M. Y. and Stanners, C. P. (1979). *Somat. Cell Genetics* **5**, 625.

Worton, R. G. Ho, C. C., and Duff, C. (1977). *Somat. Cell Genet.* **3**, 27.

APPENDIXES

APPENDIX I

LINEAGES OF CHINESE HAMSTER CELL LINES

Michael M. Gottesman
Laboratory of Molecular Biology
National Cancer Institute
National Institutes of Health
Bethesda Maryland

Outbred Chinese Hamsters
Imported from China, December 1948 (Hu and Watson, Chapter 1)

Date	Chinese Hamster Ovary Lines		Date	Chinese Hamster Lung and Other Lines
1957	CHO	(developed by Puck[a] from outbred hamster of Yerganian's, Chap. 2, ATCC CCL 61)	1958	CHL, V79 (Ford and Yerganian[f], Elkind) (Chaps. 3, 19, 20)
			1962	CHL, Dede (female, CCL 39, T. C. Hsu)
			1964	CHL, Don (male, CCL 16, Hsu and Zenzes[g])
1967		CHO-KI (Denver, Kao) (Chaps. 2, 5, 11, 12, 13, 16, 18)		
	CHO	(Los Alamos, Tobey) (Chap. 7)		
1970	CHO	(Ft. Collins, Colorado)		
1971	CHO-S	(adapted to suspension, Thompson, Toronto) (Chap. 14)	1970s	Various cell lines derived from inbred Chinese hamsters (Yerganian, Chap. 1)
1975		CHO GAT$^-$ (McBurney and Whitmore,[b] Toronto) (Chaps. 14, 24)		

1975 WTT (WT-1 to 5)
Pro$^-$5
(Siminovitch,[d] Toronto)
(Chaps 15, 17, 24, 25, 28, 29)

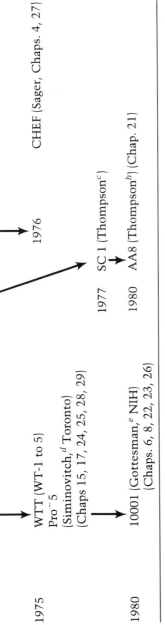

1976 CHEF (Sager, Chaps. 4, 27)

1977 SC 1 (Thompson[c])

1980 10001 (Gottesman,[e] NIH)
(Chaps. 6, 8, 22, 23, 26)

1980 AA8 (Thompson[h]) (Chap. 21)

---

[a]Puck, T. T., Ciecuira, S. J., and Robinson, A. (1958). *J. Exp. Med.* **108**, 945–955.

[b]McBurney, M.W. and Whitmore, G.F. (1974). *Cell* **2**, 173–182.

[c]Konrad, M. W., Storrie, B., Glaser, D.A., and Thompson, L.H. (1977). *Cell* **10**, 305–312.

[d]Stanley, P., Caillibot, V. and Siminovitch, L. (1975). *Somat. Cell Genet.* **1**, 3–26.

[e]Gottesman, M. M., LeCam, A., Bukowski, M., and Pastan, I. (1980). *Somat. Cell Genet.* **6**, 45–61.

[f]Ford, D. K. and Yerganian, G. (1958). *J. Natl. Canc. Inst.* **21**, 393–425.

[g]Hsu, T. C. and Zenzes, M. T. (1964). *J. Natl. Canc. Inst.* **32**, 857–869.

[h]Thompson, L. H., Fong, S., and Brookman, K. (1980). *Mutation Res.* **74**, 21–36.

APPENDIX II

CHINESE HAMSTER CELL MUTANTS

Michael M. Gottesman
Laboratory of Molecular Biology
National Cancer Institute
National Institutes of Health
Bethesda Maryland

This table lists mutants isolated in cultured Chinese hamster cells only. In most cases, the tables referred to in individual Chapters contain information about similar mutants in other cultured mammalian cells. The table is loosely based on an earlier tabulation of somatic cell mutants by Ray and Siminovitch (1982).

Function Affected[a]	Cell Line[b]	Selection[c]	Lesion[d]	Reference[e]
Amino Acid, Purine, Pyrimidine and Polyamine Biosynthesis				
Proline$^-$	CHO	Spontaneous		Chap. 2, Table 2.1
Glycine$^-$	CHO-K1	BUdR–visible-light suicide	Four complementation groups; serine hydroxymethyltransferase	Chap. 2, Table 2.1
Adenine$^-$	CHO-K1	BUdR–visible light suicide	Nine complementation groups	Chap. 11, Table 11.3
Glycine, adenine, and thymidine$^-$	CHO	BUdR–visible-light suicide	Folylpolyglutamate synthetase	Chap. 11, Table 11.3
Adenine and thymidine$^-$	CHO-K1	BUdR–visible-light suicide	Folate metabolism	Chap. 11, Table 11.3
Glycine, hypoxanthine, thymidine$^-$	CHO-K1	[^{3}H]uridine suicide	Dihydrofolate reductase	Chap. 16, Table 16.1
		Methotrexate resistant	Amplified dihydrofolate reductase	Chap. 16
			Altered dihydrofolate reductase	Chap. 16
Thymidine$^-$	CHO	1-β-D-Arabinofuranosyl cytosine resistant	Ribonucleotide diphosphate reductase	Meuth et al. (1979)

Uridine⁻	CHO-K1	BUdR–visible light	Three complementing groups	Chap. 11
		N-(phosphonacetyl)-L-aspartate resistant	Amplification of carbamyl phosphate synthetase, aspartate transcarbamylase, dihydroorotase	Chap. 11
		5-Fluoroorotic acid, 5-fluorouracil resistant	Orotate phosphoribosyl transferase, Orotidylate decarboxylase increase	Chap. 11
Serine⁻	CHO-K1	BUdR–visible-light suicide	Unknown	Chap. 2, Table 2.1
Glutamate⁻	CHO-K1	BUdR–visible-light suicide	Unknown	Chap. 2, Table 2.1
Glutamine	CHO-K1	Methionine sulfoximine resistant	Glutamine synthetase overproduced	Sanders and Wilson (1984)
Alanine⁻	CHO-K1	BUdR–visible-light suicide	Alanyl tRNA synthetase	Chap. 2, Table 2.1
Asparagine	CHO	β-Aspartyl-hydroxymate, albizziin resistant	Asparagine synthetase overproduction	Chap. 17
Valine, leucine, isoleucine⁻	CHO-K1	BUdR–visible-light suicide	Branched chain amino acid transaminase	Chap. 2, Table 2.1
Arginine⁻	CHO-K1	Spontaneous	Argininosuccinate synthetase or argininosuccinase	Chap. 2, Table 2.1

(continued)

Function Affected[a]	Cell Line[b]	Selection[c]	Lesion[d]	Reference[e]
Cysteine$^-$	CHO-K1	Spontaneous	Cystathionase	Chap. 2, Table 2.1
Polyamine	CHO	[^{3}H]ornithine suicide	Ornithine decarboxylase	Chap. 18
		α-Diflouro-methylornithine resistant	Ornithine decarboxylase overproduced	Chap. 18
Methionine$^-$	CHO	Cycloleucine resistant	Methionine adenosyl transferase overproduced	Caboche and Mulsant (1978)
		Macromolecular Sythesis		
Amino-acyl tRNA synthesis	CHO, V79	[^{3}H]amino acid suicide, ts; thialysine selection, ts	Nine complementing groups; asp$^-$, leu$^-$, met$^-$, arg$^-$, his$^-$, gluNH$_2$$^-$, lys$^-$, ala$^-$, and trp-tRNA synthetases	Chap. 14, Section III A
	CHO-K1	[^{3}H]TdR + ara-C, ts	Three complementing groups; val$^-$, his$^-$, leu$^-$	Ashman (1978)
		Borrelidin resistant	thr-tRNA synthetase	Gantt et al. (1981)
Ribosome function	CHO, V79, CH peritoneal cells	Emetin resistant	Three complementing groups; 40S ribosomal subunit	Chap. 14, Section III B
	CHO	Trichodermin resistant	60S ribosomal subunit	Gupta and Siminovitch (1978*a*)
	CHO	Ricin resistant	60S ribosomal subunit	Ono et al. (1982)

Translation	CHO, CHO-K1	Diphtheria-toxin resistant	several groups; EF-2 and EF-2 modifying enzymes	Moehring, et al. (1980); Draper et al. (1979); Gupta and Siminovitch (1980a)
	CHO-K1	*Pseudomonas*-exotoxin resistant	Translation	Moehring and Moehring (1977)
Synthesis of DNA and DNA precursors	CCL 39	1-β-D-Arabinofuranosyl cytosine resistant	Deoxycytidine kinase	Robert de Saint Vincent and Buttin (1979)
			CTP-synthetase	Buttin et al. (1982)
	CHO	1-β-D-Arabinofuranosyl cytosine resistant	Ribonucleotide diphosphate reductase	Meuth et al. (1979)
	CHO	Aphidicolin resistant	Ribonucleotide reductase	Sabourin et al. (1981)
	CHO	Hydroxyurea resistant	Ribonucleotide reductase	Lewis and Wright (1979)
	CHO	6(ρ-hydroxyphenylazo)ura (HPUra) resistant	Ribonucleotide reductase	Arpaia et al. (1983)
	CCL39	Spontaneous	Deoxycytidine deaminase	Robert de Saint Vincent et al (1980)
	CCL39	Coformycin resistant	Adenylate deaminase overproduction	Buttin et al. (1982)
	V79	Aphidicolin resistant	DNA polymerase	Liu et al. (1983)
	V79	DNA-ts	Unknown	Roufa et al. (1979)
	V79, CHO	DNA-ts	Unknown	Chap. 20, Table 20.1

(*continued*)

Function Affected[a]	Cell Line[b]	Selection[c]	Lesion[d]	Reference[e]
DNA repair	CHO,V79	Sensitivity to UV, MMS, X-rays, bleomycin	Defective in incision step, DNA ligation, replication after UV damage	Chap. 21, Table 21.1
Transcription	CHO	α-Amanitin resistant	RNA polymerase II	Chap. 15, Section III A
	CHL	ara-C, suicide, ts	mRNA synthesis	Wong and Scheffler (1982)
	CHO-K1	5,6-Dichloro-1-β-D-ribofuranosyl benzimidazole (DRB) resistant	RNA polymerase II complex	Funanage (1982)
		Nucleotide Scavenger Pathways		
Hypoxanthine	CHO	8-Azaguanine, 6-thioguanine	Hypoxanthine-guanine phosphoribosyl transferase	Chap. 13
Adenine	CHO	2,6-Diaminopurine, 8-azaadenine, 8-fluoroadenine resistant	Adenine phosphoribosyl transferase	Chap. 12
Adenine, hypoxanthine, guanine	V79	[^{3}H]adenine suicide	Phosphoribosyl pyrophosphate synthetase	Crawford and Bryant (1982)

Adenosine	CHO	Tubercidin, toyocamycin, 2-fluoroadenosine resistant	Adenosine kinase	McBurney and Whitmore (1975); Gupta and Siminovitch (1978*b*); Rabin and Gottesman (1979); Plagemann and Wohlhueter (1983)
	CHO	Formycin A; formycin B resistant	Unknown	Mehta and Gupta (1983)
Thymidine	CHO	5-Bromodeoxyuridine; trifluorothymidine resistant	Thymidine kinase	Adair and Carver (1979); Bradley et al. (1982)
	CHO	HPUra resistant	Resistant to thymidine in presence of BUdR; mechanism unknown	Arpaia et al. (1983)
		Cell Cycle and Cell Division		
Early G1 block; DNA synthesis; G2 block; Cytokinesis	V79	Resistant to inhibitors of DNA synthesis or DNA poisons at nonpermissive temperature	Exact lesions unknown	Chap. 20, Table 20.1
Mitotic spindle	CHO	Colcemid, colchicine, griseofulvin, taxol, podophyllotoxin resistant	α- and β-tubulin; 66,000-dalton protein	Chap. 22, Table 22.1

(*continued*)

Function Affected[a]	Cell Line[b]	Selection[c]	Lesion[d]	Reference[e]
cAMP growth control	CHO	8-Br-cAMP, cholera toxin, methylisobutylxanthine resistant	Regulatory and catalytic subunits of cAMP dependent protein kinase (Type I and Type II)	Chap. 23, Table 23.1
		Sugar and Glycoprotein Metablism		
Glycoprotein synthesis	CHO	L-PHA, WGA, Con A, Ric, and LCA resistant	Diverse defects associated with synthesis of carbohydrate moiety of glycoproteins	Chap. 24, Table 24.2
	CHO	Tunicamycin resistant	Synthesis of *N*-acetylglucosamine lipid intermediate	Sudo and Onodera (1979); Kuwano et al. (1981)
Glucose metabolism	CHO-K1	Screen for enzyme activity	Glucose-6-phosphate dehydrogenase	Rosenstraus and Chasin (1975); Stamato et al. (1982)
	CHO-K1	Replica plating	Lactic dehydrogenase A	Stamato and Jones (1977)
	CCL 39	2-Deoxy[^{3}H]glucose suicide	Phosphoglucose isomerase	Pouysségur et al. (1980); Whitfield et al. (1978)

Galactose metabolism	CHO	2-Deoxyglucose resistant	Galactokinase	Thirion et al. (1976)
	V79, Don, CHO	BUdR–visible-light suicide	See mitochondrial mutants	Chap. 19, Table 19.1
Glucosamine$^-$	CHL	BUdR–visible-light suicide	L-glutamine: D-fructose-6-phosphate transaminase	Onoda et al. (1975)
		Membrane Function: Transport and Endocytosis		
Glucose transport	CHO-K1	3-*O*-methyl-D-glucose resistant	Increased hexose carrier	Whitfield et al. (1982)
	CCL 39	2-Deoxy[^{3}H]glucose suicide	Reduced hexose carrier	Pouysségur et al. (1980)
pH regulation	CCL 39	Proton suicide	Na^+-proton antiport	Pouysségur et al. (1984)
Sulfate transport	CHO	Chromate resistant	Sulfate carrier(?)	Campbell et al. (1981)
Polyamine transport	CHO	Methylglyocal bis-guanyl hydrazone resistant	Polyamine carrier (?)	Mandel and Flintoff (1978)
Amino acid transport	CHO-K1	Alanine resistant	A, ASC, P transport systems	Curriden and Englesberg (1981)
	CHO-K1	MeAIB resistant	ASC transport increased	Ertsey and Englesberg (1984)
	CHO	[^{3}H]proline suicide	A system transport	Moffett et al. (1983); Dantzig et al. (1982); Moffett and Englesberg (1984)

(*continued*)

Function Affected[a]	Cell Line[b]	Selection[c]	Lesion[d]	Reference[e]
Amino acid transport	CHO	tr revertants of leu-tRNA synthetase mutants	Increased L system activity	Shotwell et al. (1983)
	CHO	L-phenylalanine mustard resistant	A, L, Ly^+ system transport	Dantzig et al. (1984)
Methotrexate transport	CHO	Methotrexate resistant	Methotrexate carrier(?)	Flintoff et al. (1976)
Hydrophobic drugs	CHO, V79	Colchicine, vincristine, adriamycin resistant	Unknown; overproduce p170	Chap. 25
Ouabain action	CHO	Ouabain resistant	Na^+-K^+ ATPase	Baker et al. (1974)
Virus resistant	CHO	Sindbis virus resistant	Viral mRNA translation	Mento and Siminovitch (1981)
		Diphtheria-toxin resistant	Endocytosis	Robbins et al. (1983)
Lysosomal enzymes	CHO	Replica plating; screen for loss of enzyme activity	α-Mannosidase	Robbins (1979)
	CHO	Con A resistant	α-L-iduronidase; β-hexosaminidase	Krag and Robbins (1982)
	CHO	Man 6P-ricin resistant	Man 6-P receptor; synthesis of mannosylphosphoryldolichol	Robbins et al. (1981); Stoll et al. 1982)

Lysosomal enzymes	V79	Chloroquine resistant	Cathepsin B; lysosomes altered	Ono et al. (1983)
Adhesion	CHO	Inability to adhere to culture dishes or collagen	Unknown	Harper and Juliano (1980); Cabral and Gottesman (1977); Klebe et al. (1977)
Morphology	CHO	Pactomycin resistant	Unknown	Gupta and Siminovitch (1980b)
Receptor mediated endocytosis	CHO-K1	Diphtheria-toxin resistant	Acidification of endosomes (receptosomes)	Didsbury et al. (1983); Merion et al. (1983); Robbins et al. (1983); Klausner et al. (1984)
	CHO	Ricin- and *Pseudomonas*-toxin resistant	Internalization	Ray and Wu (1982)
	CHO-K1	Diphtheria-toxin and modeccin resistant; ts for growth	Receptosome function(?)	Marnell et al. (1984)
Receptors	CHO-K1	Toxic or fluorescent LDL derivatives	LDL receptor; at least 2 classes	Krieger et al. (1981); Krieger (1983)
	CHO-K1	Compactin and amphotericin B resistant in mevalonic acid/LDL medium	LDL receptor; endocytosis (?)	Krieger et al. (1983)

(continued)

Function Affected[a]	Cell Line[b]	Selection[c]	Lesion[d]	Reference[e]
		Lipid Metabolism		
Myoinositol$^-$	CHO-K1	Replica plating	Myoinositol synthesis	Esko and Raetz (1978) Chap. 2, Table 2.1
Inositol$^-$	CHO-K1	BUdR–visible-light suicide	Inositol synthesis	Chap. 2, Table 2.1
Phosphatidylcholine synthesis	CHO-K1	Replica plate and assay enzyme	Phosphatidyl-choline synthesis	Esko and Raetz (1980)
Cholesterol synthesis	CHO-K1, CCL 39, CHO	25-OH cholesterol resistant	Hydroxymethyl-glutaryl CoA (HMGCoA) reductase overproduced	Sinensky (1977); Cavenee and Baker (1982)
	V79	Compactin resistant	HMG CoA reductase overproduced	Masuda et al. (1982)
Cholesterol$^-$	CHO-K1	BUdR–visible-light suicide	HMG CoA reductase not induced	Limanek et al. (1978); Saito et al. (1977)
	CHO-K1	BUdR/bisbenzimide suicide	HMG CoA reductase	Mosley et al. (1983)
	CHO-K1	BUdR–visible-light suicide	Demethylation of lanosterol	Chang et al. (1977)
Mevalonic acid$^-$	CHO-K1	BUdR–visible-light suicide	HMG CoA synthase	Schnitzer-Polokoff et al. (1982)

Unsaturated fatty acids	CHO-K1	BUdR–visible-light suicide	Microsomal stearoyl-CoA desaturase	Chang and Vagelos (1976)
		Mitochondria and Oxidative Phosphorylation		
Respiratory chain	V79, Don	BudR–visible-light suicide in media lacking CO_2/HCO_3^- and asparagine; gal auxotrophs	Eighteen or more overlapping complementation groups; Complex I, succinate dehydrogenase, mitochondrial protein synthesis, Complex III, ubiquinone, O_2 uptake	Chap. 19, Table 19.1
	CHO, V79	Antimycin	Cytochrome *b*	Chap. 19, Table 19.2
	CHO	Oligomycin venturicidin, rutamycin	Increased mitochondrial ATPase; other changes in respiration	Chap. 19, Table 19.2
Mitochondrial protein synthesis	V79, CHEF	Chloramphenicol resistant	Mitochondrial 16S rRNA	Chap. 19, Table 19.1

(*continued*)

Function Affected[a]	Cell Line[b]	Selection[c]	Lesion[d]	Reference[e]
		Miscellaneous		
Heavy-metal metabolism	CHO	Cd^{2+}-resistant	Increased induction of metallothionein	Hildebrand et al. (1979)
	CHO-K1	Cd^{2+}, Zn^{2+} resistant	Metallothionein overproduced due to gene amplification	Gick and McCarty (1982)

[a]The presence of a superscript (−) indicates a nutritional auxotrophy.

[b]CHO, Toronto subline; CHO-K1, Denver subline; V79, CHL, CCL 39, and Don are Chinese hamster lung lines. See Appendix I for more detailed pedigrees.

[c]The general approach to selection is given here. The term "spontaneous" means that the Chinese hamster line as originally isolated was defective in these functions without selection.

[d]A defective enzyme, structure, or process is listed here. If the protein is overproduced, as opposed to being decreased in amount or qualitatively altered, this is indicated.

[e]Wherever possible, reference is given to the chapter, section, or table in this book where additional information about the mutant can be obtained. In cases where there is no extensive reference to the mutant, reference is given to the literature.

REFERENCES

Adair, G. M. and Carver, J. H. (1979). *Mut. Res.* **60**, 207–213.

Arpaia, E., Ray, P. N., and Siminovitch, L. (1983). *Somat. Cell Genet.* **9**, 287–297.

Ashman, C. R. (1978). *Somat. Cell Genet.* **4**, 299–312.

Baker, R. M., Brunette, D. M., Mankovitz, R., Thompson, L. H., Whitmore, G. F., Siminovitch, L., and Till, J. E. (1974). *Cell* **1**, 9–21.

Bradley, W. E. C., Dinelle, C., Charron, J., and Langelier, Y. (1982). *Somat. Cell Genet.* **8**, 207–222.

Buttin, G., Debatisse, M. and Robert de Saint Vincent, B. (1982). In *Somatic Cell Genetics.* (C. T. Caskey and D. C. Robbins, eds.), Plenum, New York, pp. 1–18.

Caboche, M. and Mulsant, P. (1978). *Somat. Cell Genet.* **4**, 407–421.

Cabral, F. and Gottesman, M. M. (1977). *J. Cell Biol.* **75**, 391a.

Campbell, C. E., Gravel, R. A., and Worton, R. G. (1981). *Somat. Cell Genet.* **7**, 535–546.

Cavenee, W. K. and Baker, R. M. (1982). *Somat. Cell Genet.* **8**, 557–574.

Chang, T.-Y., Telakowski, C., Vanden Heuvel, W., Alberts, A. W., and Vagelos, P. R. (1977). *Proc. Natl. Acad. Sci. USA* **74**, 832–836.

Chang, T.-Y. and Vagelos, R. (1976). *Proc. Natl. Acad. Sci. USA* **73**, 24–28.

Crawford, C. R. and Bryant, R. E. (1982). *Mol. Cell. Biol.* **2**, 1619–1627.

Curriden, S. A. and Englesberg, E. (1981). *J. Cell. Physiol.* **106**, 245–252.

Dantzig, A. H., Slayman, C. W., and Adelberg, E. A. (1982). *Somat. Cell Genet.* **8**, 509–520.

Dantzig, A. H., Fairgrieve, M., Slayman, C. W. and Adelberg, E. A. (1984). *Somat. Cell and Molec. Genet.* **10**, 113–121.

Didsbury, J. R., Moehring, J. M., and Moehring, T. J. (1983). *Mol. Cell. Biol.* **3**, 1283–1294.

Draper, R. K., Chin, D., Eurey-Owens, D., Scheffler, I. E., and Simon, M. I. (1979). *J. Cell Biol.* **83**, 116–125.

Ertsey, R. and Englesberg, E. (1984). *Somat. Cell and Molec. Genet.* **10**, 171–182.

Esko, J. D. and Raetz, C. R. H. (1978). *Proc. Natl. Acad. Sci. USA* **75**, 1190–1193.

Esko, J. D. and Raetz, C. R. H. (1980). *Proc. Natl. Acad. Sci. USA* **77**, 5192–5196.

Flintoff, W. F., Davidson, S. V., and Siminovitch, L. (1976). *Somat. Cell Genet.* **2**, 245–261.

Funanage, V. L. (1982). *Mol. Cell. Biol.* **2**, 467–477.

Gantt, J. S., Bennett, C. A., and Arfin, S. M. (1981). *Proc. Natl. Acad. Sci. USA* **78**, 5367–5370.

Gick, G. G. and McCarty, K. S. (1982). *J. Biol. Chem.* **257**, 9049–9053.

Gupta, R. S. and Siminovitch, L. (1978*a*). *Somat. Cell Genet.* **4**, 355–375.

Gupta, R. S. and Siminovitch, L. (1978*b*). *Somat. Cell Genet.* **4**, 715–736.

Gupta, R. S. and Siminovitch, L. (1980a). *Somat. Cell Genet.* **6**, 361–379.

Gupta, R. S. and Siminovitch, L. (1980*b*). *J. Cell. Physiol.* **102**, 305–316.

Harper, P. A. and Juliano, R. L. (1980). *J. Cell Biol.* **87**, 755–763.

Hildebrand, C. E., Tobey, R. A., Campbell, E. W., and Enger, M. D. (1979). *Exp. Cell Res.* **124**, 237–246.

Klausner, R. D., Renswoude, J., Kempf, C., Rao, K., Bateman, J. L., and Robbins, A. R. (1984). *J. Cell Biol.* **98**, 1098–1101.

Klebe, R. J., Rosenberger, P. G., Naylor, S. L. Burns, R. L., Novak, R., and Kleinman, H. (1977). *Exp. Cell Res.* **104**, 119–125.

Krag, S. S. and Robbins, A. R. (1982). *J. Biol. Chem.* **257**, 8424–8431.

Krieger, M., Brown, M. S. and Goldstein, J. L. (1981). *J. Mol. Biol.* **150**, 167–184.

Krieger, M. (1983). *Cell* **33**, 413–422.

Krieger, M., Martin, J., Segal, M., and Kingsley, D. (1983). *Proc. Natl. Acad. Sci. USA* **80**, 5607–5611.

Kuwano, M., Tabuki, T., Akiyama, S., Mifune, K., Takatsuki, A., Tamura, G., and Kiehara, Y. (1981). *Somat. Cell Genet.* **7**, 507–521.

Lewis, W. H. and Wright, J. A. (1979). *Somat. Cell Genet.* **5**, 83–96.

Limanek, J. S., Chin, J., and Chang, T. Y. (1978). *Proc. Natl. Acad. Sci. USA* **75**, 5452–5456.

Liu, P. K. Chang, C.-C., Trosko, J. E., Dube, D. K., Martin, G. M., and Loeb, L. (1983). *Proc. Natl. Acad. Sci. USA* **80**, 797–801.

Mandel, J. and Flintoff, W. I. (1978). *J. Cell Physiol.* **97**, 335–344.

Marnell, M. H., Mathis, L. S., Stookey, M., Shia, S.-P., Stone, D. K., and Draper, R. K. (1984). *J. Cell Biol.* **97**, 1907–1916.

Masuda, A., Akiyama, S., and Kuwano, M. (1982). *Mol. Cell. Biol.* **2**, 1354–1362.

McBurney, M. W. and Whitmore, G. F. (1975). *J. Cell Physiol.* **85**, 87–100.

Mehta, K. D. and Gupta, R. S. (1983). *Mol. Cell. Biol.* **3**, 1468–1477.

Mento, S. J. and Siminovitch, L. (1981). *Virology* **111**, 320–330.

Merion, M., Schlesinger, P., Brooks, R. M., Moehring, J. M., Moehring, T. J., and Sly, W. S. (1983). *Proc. Natl. Acad. Sci. USA* **80**, 5315–5319.

Meuth, M., L'Heureux-Huard, N., and Trudel, M. (1979). *Proc. Natl. Acad. Sci. USA* **76**, 6505–6509.

Moehring, T.J. and Moehring, J.M. (1977). *Cell* **11**, 447–454.

Moehring, J. M., Moehring, T. J., and Danley, D. E. (1980). *Proc. Natl. Acad. Sci. USA* **77**, 1010–1014.

Moffett, J., Curriden, S., Ertsey, R., Mendiaz, E., and Englesberg, E. (1983). *Somat. Cell Genet.* **9**, 189–213.

Moffett, J. and Englesberg, E. (1984). *Molec. Cell. Biol.* **4**, 799–808.

Mosley, S. T., Brown, M. S., Anderson, R. G., and Goldstein, J. L. (1983). *J. Biol. Chem.* **258**, 13875–13881.

Ono, M., Ando, M., Shimada, T., Furuno, K., Kato, K., and Kuwano, M. (1983) *J. Biochem. (Tokyo)* **94**, 1493–1503.

Ono, M., Kuwano, M., Watanabe, K.-I., and Funatsu, G. (1982). *Mol. Cell. Biol.* **2**, 599–606.

Onoda, T., Kajiwara, N., and Matsuo, Y. (1975). *Hiroshima Journal of Medical Sciences* **24**, 97–107.

Plagemann, P. G. W. and Wohlhueter, R. M. (1983). *J. Cell. Physiol.* **116**, 236–246.

Pouysségur, J., Franchi, A., Salomon, J.-C., and Silvestre, P. (1980). *Proc. Natl. Acad. Sci. USA* **77**, 2698–2701.

Pouysségur, J., Sardet, C., Franchi, A., L'Allemain, G., and Paris, S. (1984). *Proc. Natl. Acad. Sci. USA* **81**, 4833–4837.

Rabin, M. S. and Gottesman, M. M. (1979). *Somat. Cell Genet.* **5**, 571–583.

Ray, B. and Wu, H. C. (1982). *Mol. Cell. Biol.* **2**, 535–544.

Ray, P. N. and Siminovitch, L. (1982). In *Somatic Cell Genetics* (C. T. Caskey and D. C. Robbins, eds.), Plenum, New York, pp. 127–167.

Robbins, A. R. (1979). *Proc. Natl. Acad. Sci. USA* **76**, 1911–1915.

Robbins, A. R., Myerowitz, R., Youle, R. J., Murray, G. J., and Neville, D. M. (1981). *J. Biol. Chem.* **256**, 10618–10622.

Robbins, A. R., Peng, S. S., and Marshall, J. L. (1983). *J. Cell Biol.* **96**, 1064–1071.

Robert de Saint Vincent, B. and Buttin, G. (1979). *Somat. Cell Genet.* **5**, 67–82.

Robert de Saint Vincent, B., Dechamps, M., and Buttin, G. (1980). *J. Biol. Chem.* **255**, 162–167.

Rosenstraus, M. and Chasin, L. A. (1975). *Proc. Natl. Acad. Sci. USA* **72**, 493–497.

Roufa, D. J., McGill, S. M., and Mollenkamp, D. (1979). *Somat. Cell Genet.* **5**, 97–115.

Sabourin, C. L. K., Bates, P. F., Glatzer, L., Chang, C.-C., Trosko, J. E., and Boezi, J. A. (1981). *Somat. Cell Genet.* **7**, 255–268.

Saito, Y., Chou, S. M., and Silbert, D. F. (1977). *Proc. Natl. Acad. Sci. USA* **74**, 3730–3734.

Sanders, P. G. and Wilson, R. H. (1984). *The EMBO Journal* **3**, 65–71.

Schnitzer-Polokoff, R., von Gunten, C., Logel, J., Torget, R., and Sinenski, M. (1982). *J. Biol. Chem* **257**, 472–476.

Shotwell, M. A., Collarini, E. J., Mansukhani, A., Hampel, A. E., and Oxender, D. L. (1983). *J. Biol. Chem.* **258**, 8183–8187.

Sinensky, M. (1977). *Biochem. Biophys. Res. Commun.* **78**, 863–867.

Stamato, T. D. and Jones, C. (1977). *Somat. Cell Genet.* **3**, 639–647.

Stamato, T. D., MacKenzie, L., Pagani, J. M., and Weinstein, R. (1982) *Somat. Cell Genet.* **8**, 643–651.

Stoll, J., Robbins, A. R., and Krag, S. S. (1982). *Proc. Natl. Acad. Sci. USA* **79**, 2296–2300.

Sudo, T. and Onodera, K. (1979). *J. Cell Physiol.* **101**, 149–156.

Thirion, J. P., Banville, D., and Noel, H. (1976). *Genetics* **83**, 137–147.

Whitfield, C. D. Buchsbaum, B., Bostedor, R., and Chu, E. H. Y. (1978). *Somat. Cell Genet.* **4**, 699–713.

Whitfield, C. D., Hupe, L. M., Nugent, C., Urbani, K. E., and Whitfield, H. J. (1982). *J. Biol. Chem.* **257**, 4902–4906.

Wong, E. A. and Scheffler, I. E. (1982). *Mol. Cell. Biol.* **2**, 1558–1573.

INDEX